MW00681139

OMC | Stern Drive
1986-98 REPAIR MANUAL
GASOLINE ENGINES & DRIVE SYSTEMS

SELOC ®

Managing Partners	Dean F. Morgantini, S.A.E.
	Barry L. Beck
Executive Editor	Kevin M. G. Maher, A.S.E.
Production Managers	Melinda Possinger
	Ronald Webb

Manufactured in USA
© 2004 Seloc Publishing
104 Willowbrook Lane
West Chester, PA 19382
ISBN 13: 978-0-89330-056-2
ISBN 10: 0-89330-056-X
7890123456 8765432109

www.selocmarine.com
1-866-SELOC55

CONTENTS

CONTENTS

FUEL SYSTEMS FUEL INJECTION

9

IGNITION AND ELECTRICAL SYSTEMS

10

COOLING SYSTEM

11

STERN DRIVE SP/SX/SX COBRA, KING COBRA AND DP/DP DUOPROP

12

TRIM AND TILT

13

STEERING SYSTEMS

14

REMOTE CONTROLS

15

SAFETY NOTICE

Proper service and repair procedures are vital to the safe, reliable operation of all marine engines, as well as the personal safety of those performing repairs. This manual outlines procedures for servicing and repairing engines and drive systems using safe, effective methods. The procedures contain many NOTES, CAUTIONS and WARNINGS which should be followed, along with standard procedures, to minimize the possibility of personal injury or improper service which could damage the vehicle or compromise its safety.

It is important to note that repair procedures and techniques, tools and parts for servicing these engines, as well as the skill and experience of the individual performing the work, vary widely. It is not possible to anticipate all of the conceivable ways or conditions under which the engine may be serviced, or to provide cautions as to all possible hazards that may result. Standard and accepted safety precautions and equipment should be used during cutting, grinding, chiseling, prying, or any other process that can cause material removal or projectiles.

Some procedures require the use of tools specially designed for a specific task. Before substituting another tool or procedure, you must be completely satisfied that neither your personal safety, nor the performance of the vessel, will be endangered. All procedures covered in this manual requiring the use of special tools will be noted at the beginning of the procedure by means of an **OEM symbol**

Additionally, any procedure requiring the use of an electronic tester or scan tool will be noted at the beginning of the procedure by means of a **DVOM symbol**

Although information in this manual is based on industry sources and is complete as possible at the time of publication, the possibility exists that some manufacturers made later changes which could not be included here. While striving for total accuracy, Seloc Publishing cannot assume responsibility for any errors, changes or omissions that may occur in the compilation of this data. We must therefore warn you to follow instructions carefully, using common sense. If you are uncertain of a procedure, seek help by inquiring with someone in your area who is familiar with these motors before proceeding.

PART NUMBERS

Part numbers listed in this reference are not recommendations by Seloc Publishing for any particular product brand name, simply iterations of the manufacturer's suggestions. They are also references that can be used with interchange manuals and aftermarket supplier catalogs to locate each brand supplier's discrete part number.

SPECIAL TOOLS

Special tools are recommended by the manufacturers to perform a specific job. Use has been kept to a minimum, but, where absolutely necessary, they are referred to in the text by the part number of the manufacturer if at all possible; and also noted at the beginning of each procedure with one of the following symbols: **OEM** or **DVOM.**

The **OEM** symbol usually denotes the need for a unique tool purposely designed to accomplish a specific task, it will also be used, less frequently, to notify the reader of the need for a tool that is not commonly found in the average tool box.

The **DVOM** symbol is used to denote the need for an electronic test tool like an ohmmeter, multi-meter or, on certain later engines, a scan tool.

These tools can be purchased, under the appropriate part number, from your local dealer or regional distributor, or an equivalent tool can be purchased locally from a tool supplier or parts outlet. Before substituting any tool for the one recommended, read the SAFETY NOTICE at the top of this page.

Providing the correct mix of service and repair procedures is an endless battle for any publisher of "How-To" information. Users range from first time do-it yourselfers to professionally trained marine technicians, and information important to one is frequently irrelevant to the other. The editors at Seloc Publishing strive to provide accurate and articulate information on all facets of marine engine repair, from the simplest procedure to the most complex. In doing this, we understand that certain procedures may be outside the capabilities of the average DIYer. Conversely we are aware that many procedures are unnecessary for a trained technician.

SKILL LEVELS

In order to provide all of our users, particularly the DIYers, with a feeling for the scope of a given procedure or task before tackling it we have included a rating system denoting the suggested skill level needed when performing a particular procedure. One of the following icons will be included at the beginning of most procedures:

① EASY

EASY. These procedures are aimed primarily at the DIYer and can be classified, for the most part, as basic maintenance procedures; battery, fluids, filters, plugs, etc. Although certainly valuable to any experience level, they will generally be of little importance to a technician.

② MODERATE

MODERATE. These procedures are suited for a DIYer with experience and a working knowledge of mechanical procedures. Even an advanced DIYer or professional technician will occasionally refer to these procedures. They will generally consist of component repair and service procedures, adjustments and minor rebuilds.

③ DIFFICULT

DIFFICULT. These procedures are aimed at the advanced DIYer and professional technician. They will deal with diagnostics, rebuilds and internal engine/drive components and will frequently require special tools.

④ SKILLED

SKILLED. These procedures are aimed at highly skilled technicians and should not be attempted without previous experience. They will usually consist of machine work, internal engine work and gear case rebuilds.

Please remember one thing when considering the above ratings—they are a guide for judging the complexity of a given procedure and are subjective in nature. Only you will know what your experience level is, and only you will know when a procedure may be outside the realm of your capability. First time DIYer, or life-long marine technician, we all approach repair and service differently so an easy procedure for one person may be a difficult procedure for another, regardless of experience level. All skill level ratings are meant to be used as a guide only! Use them to help make a judgement before undertaking a particular procedure, but by all means read through the procedure first and make your own decision—after all, our mission at Seloc is to make boat maintenance and repair easier for everyone whether you are changing the oil or rebuilding an engine. Enjoy boating!

ACKNOWLEDGMENTS

Seloc Publishing expresses appreciation to the following companies who supported the production of this book:

- Belks Marine—Holmes, PA
- Marine Mechanics Institute—Orlando, FL
- Stern Drive, LTD—Everett, WA

Thanks to John Hartung and Judy Belk of Belk's Marine for for their assistance, guidance, patience and access to some of the motors photographed for this manual.

Thanks to Wayne at Sterndrive.com for his assistance and patience during the creation of this manual.

Seloc Publishing would like to express thanks to the fine companies who participate in the production of all our books:

- Hand tools supplied by Craftsman are used during all phases of our vehicle teardown and photography.
- Many of the fine specialty tools used in our procedures were provided courtesy of Lisle Corporation.
- Much of our shop's electronic testing equipment was supplied by Universal Enterprises Inc. (UEI).

1

GENERAL INFORMATION, SAFETY AND TOOLS

HOW TO USE THIS MANUAL

This manual is designed to be a handy reference guide to maintaining and repairing your OMC engine and drive system. We strongly believe that regardless of how many or how few year's experience you may have, there is something new waiting here for you.

This manual covers the topics that a factory service manual (designed for factory trained mechanics) and a manufacturer owner's manual (designed more by lawyers than boat owners these days) covers. It will take you through the basics of maintaining and repairing your engine/drive system, step-by-step, to help you understand what the factory trained mechanics already know by heart. By using the information in this manual, any boat owner should be able to make better informed decisions about what they need to do to maintain and enjoy their OMC.

Even if you never plan on touching a wrench (and if so, we hope that we can change your mind), this manual will still help you understand what a mechanic needs to do in order to maintain your engine.

Can You Do It?

If you are not the type who is prone to taking a wrench to something, NEVER FEAR. The procedures provided here cover topics at a level virtually anyone will be able to handle. And just the fact that you purchased this manual shows your interest in better understanding your OMC.

You may find that maintaining your engine/drive system yourself is preferable in most cases. From a monetary standpoint, it could also be beneficial. The money spent on hauling your boat to a marina and paying a tech to service the engine could buy you fuel for a whole weekend's boating. If you are unsure of your own mechanical abilities, at the very least you should fully understand what a marine mechanic does to your boat. You may decide that anything other than maintenance and adjustments should be performed by a mechanic (and that's your call), but know that every time you board your boat, you are placing faith in the mechanic's work and trusting him or her with your well-being, and maybe your life.

It should also be noted that in most areas a factory-trained mechanic will command a hefty hourly rate for off site service. This hourly rate is often charged from the time they leave their shop to the time that they return home. The cost savings in doing the job yourself might be readily apparent at this point.

Of course, if even you're already a seasoned Do-It-Yourselfer or a Professional Technician, you'll find the procedures, specifications, special tips as well as the schematics and illustrations helpful when tackling a new job on an engine.

■ **To help you decide if a task is within your skill level, procedures will often be rated using a wrench symbol in the text. When present, the number of wrenches designates how difficult we feel the procedure to be on a 1-4 scale. For more details on the wrench icon rating system, please refer to the information under Skill Levels at the beginning of this manual.**

Where to Begin

Before spending any money on parts, and before removing any nuts or bolts, read through the entire procedure or topic. This will give you the overall view of what tools and supplies will be required to perform the procedure or what questions need to be answered before purchasing parts. So read ahead and plan ahead. Each operation should be approached logically and all procedures thoroughly understood before attempting any work.

Avoiding Trouble

Some procedures in this manual may require you to "label and disconnect . . . " a group of lines, hoses or wires. Don't be lulled into thinking you can remember where everything goes — you won't. If you reconnect or install a part incorrectly, the engine may operate poorly, if at all. If you hook up electrical wiring incorrectly, you may instantly learn a very expensive lesson.

A piece of masking tape, for example, placed on a hose and another on its fitting will allow you to assign your own label such as the letter "**A**", or a short name. As long as you remember your own code, you can reconnect the lines by matching letters or names. Do remember that tape will dissolve when saturated in some fluids (especially cleaning solvents).

If a component is to be washed or cleaned, use another method of identification. A permanent felt-tipped marker can be very handy for marking metal parts; but remember that some solvents will remove permanent marker. A scribe can be used to carefully etch a small mark in some metal parts, but be sure NOT to do that on a gasket-making surface.

SAFETY is the most important thing to remember when performing maintenance or repairs. Be sure to read the information on safety in this manual.

Maintenance or Repair?

Proper maintenance is the key to long and trouble-free engine life, and the work can yield its own rewards. A properly maintained engine performs better than one that is neglected. As a conscientious boat owner, set aside a Saturday morning, at least once a month, to perform a thorough check of items that could cause problems. Keep your own personal log to jot down which services you performed, how much the parts cost you, the date, and the amount of hours on the engine at the time. Keep all receipts for parts purchased, so that they may be referred to in case of related problems or to determine operating expenses. As a do-it-yourselfer, these receipts are the only proof you have that the required maintenance was performed. In the event of a warranty problem (on new engines), these receipts can be invaluable.

It's necessary to mention the difference between maintenance and repair. Maintenance includes routine inspections, adjustments, and replacement of parts that show signs of normal wear. Maintenance compensates for wear or deterioration. Repair implies that something has broken or is not working. A need for repair is often caused by lack of maintenance.

For example: draining and refilling the gearcase oil is maintenance recommended by all manufacturers at specific intervals. Failure to do this can allow internal corrosion or damage and impair the operation of the engine, requiring expensive repairs. While no maintenance program can prevent items from breaking or wearing out, a general rule can be stated: MAINTENANCE IS CHEAPER THAN REPAIR.

Directions and Locations

◆ **See Figure 1**

Two basic rules should be mentioned here. First, whenever the Port side of the engine (or boat) is referred to, it is meant to specify the left side of the engine when you are sitting at the helm. Conversely, the Starboard means your right side. The Bow is the front of the boat and the Stern or Aft is the rear.

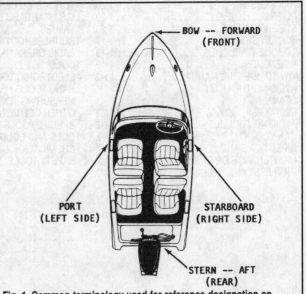

Fig. 1 Common terminology used for reference designation on boats of all size. These terms are used though out the manual

Most screws and bolts are removed by turning counterclockwise, and tightened by turning clockwise. An easy way to remember this is: righty-tighty; lefty-loosey. Corny, but effective. And if you are really dense (and we have all been so at one time or another), buy a ratchet that is marked ON and OFF (like Snap-on® ratchets), or mark your own. This can be especially helpful when you are bent over backwards, upside down or otherwise turned around when working on a boat-mounted component.

Professional Help

Occasionally, there are some things when working on an engine or drive system that are beyond the capabilities or tools of the average Do-It-Yourselfer (DIYer). This shouldn't include most of the topics of this manual, but you will have to be the judge. Some engines require special tools or a selection of special parts, even for some basic maintenance tasks.

Talk to other boaters who use the same model of engine and speak with a trusted marina to find if there is a particular system or component on your engine that is difficult to maintain.

You will have to decide for yourself where basic maintenance ends and where professional service should begin. Take your time and do your research first (starting with the information in this manual) and then make your own decision. If you really don't feel comfortable with attempting a procedure, DON'T DO IT. If you've gotten into something that may be over your head, don't panic. Tuck your tail between your legs and call a marine mechanic. Marinas and independent shops will be able to finish a job for you. Your ego may be damaged, but your boat will be properly restored to its full running order. So, as long as you approach jobs slowly and carefully, you really have nothing to lose and everything to gain by doing it yourself.

On the other hand, even the most complicated repair is within the ability of a person who takes their time and follows the steps of a procedure. A rock climber doesn't run up the side of a cliff, he/she takes it one step at a time and in the end, what looked difficult or impossible was conquerable. Worry about one step at a time.

Purchasing Parts

◆ See Figures 2 and 3

When purchasing parts there are two things to consider. The first is quality and the second is to be sure to get the correct part for your engine. To get quality parts, always deal directly with a reputable retailer. To get the proper parts always refer to the information tag on your engine prior to calling the parts counter. An incorrect part can adversely affect your engine performance and fuel economy, and will cost you more money and aggravation in the end.

Just remember, a tow back to shore will cost plenty. That charge is per hour from the time the towboat leaves their home port, to the time they return to their home port. Get the picture...$$$?

Fig. 2 By far the most important asset in purchasing parts is a knowledgeable and enthusiastic parts person

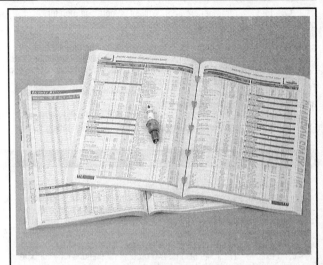

Fig. 3 Parts catalogs, giving application and part number information, are provided by manufacturers for most replacement parts

So whom should you call for parts? Well, there are many sources for the parts you will need. Where you shop for parts will be determined by what kind of parts you need, how much you want to pay, and the types of stores in your neighborhood.

Your marina can supply you with many of the common parts you require. Using a marina as your parts supplier may be handy because of location (just walk right down the dock) or because the marina specializes in your particular brand of engine. In addition, it is always a good idea to get to know the marina staff (especially the marine mechanic).

The marine parts jobber, who is usually listed in the yellow pages or whose name can be obtained from the marina, is another excellent source for parts. In addition to supplying local marinas, they also do a sizeable business in over-the-counter parts sales for the do-it-yourselfer.

Almost every boating community has one or more convenient marine chain stores. These stores often offer the best retail prices and the convenience of one-stop shopping for all your needs. Since they cater to the do-it-yourselfer, these stores are almost always open weeknights, Saturdays, and Sundays, when the jobbers are usually closed.

The lowest prices for parts are most often found in discount stores or the auto department of mass merchandisers. Parts sold here are name and private brand parts bought in huge quantities, so they can offer a competitive price. Private brand parts are made by major manufacturers and sold to large chains under a store label. And, of course, more and more large automotive parts retailers are stocking basic marine supplies.

Avoiding the Most Common Mistakes

There are 3 common mistakes in mechanical work:

1. Following the incorrect order of assembly, disassembly or adjustment. When taking something apart or putting it together, performing steps in the wrong order usually just costs you extra time; however, it CAN break something. Read the entire procedure before beginning disassembly. Perform everything in the order in which the instructions say you should, even if you can't immediately see a reason for it. When you're taking apart something that is very intricate, you might want to draw a picture of how it looks when assembled at one point in order to make sure you get everything back in its proper position. When making adjustments, perform them in the proper order; often, one adjustment affects another, and you cannot expect satisfactory results unless each adjustment is made only when it cannot be changed by subsequent adjustments.

■ Digital cameras are handy. If you've got access to one, take pictures of intricate assemblies during the disassembly process and refer to them during assembly for tips on part orientation.

2. Overtorquing (or undertorquing). While it is more common for overtorquing to cause damage, undertorquing may allow a fastener to vibrate loose causing serious damage. Especially when dealing with plastic and aluminum parts, pay attention to torque specifications and utilize a torque wrench in assembly. If a torque figure is not available, remember that if you are using the right tool to perform the job, you will probably not have to strain yourself to get a fastener tight enough. The pitch of most threads is so slight that the tension you put on the wrench will be multiplied many times in actual force on what you are tightening.

3. Cross-threading. This occurs when a part such as a bolt is screwed into a nut or casting at the wrong angle and forced. Cross-threading is more likely to occur if access is difficult. It helps to clean and lubricate fasteners, then to start threading with the part to be installed positioned straight inward. Always start a fastener, etc. with your fingers. If you encounter resistance, unscrew the part and start over again at a different angle until it can be inserted and turned several times without much effort. Keep in mind that some parts may have tapered threads, so that gentle turning will automatically bring the part you're threading to the proper angle, but only if you don't force it or resist a change in angle. Don't put a wrench on the part until it has been tightened a couple of turns by hand. If you suddenly encounter resistance, and the part has not seated fully, don't force it. Pull it back out to make sure it's clean and threading properly.

BOATING SAFETY

In 1971 Congress ordered the U.S. Coast Guard to improve recreational boating safety. In response, the Coast Guard drew up a set of regulations.

Aside from these federal regulations, there are state and local laws you must follow. These sometimes exceed the Coast Guard requirements. This section discusses only the federal laws. State and local laws are available from your local Coast Guard. As with other laws, "Ignorance of the boating laws is no excuse." The rules fall into two groups: regulations for your boat and required safety equipment on your boat.

Regulations For Your Boat

Most boats on waters within Federal jurisdiction must be registered or documented. These waters are those that provide a means of transportation between two or more states or to the sea. They also include the territorial waters of the United States.

DOCUMENTING OF VESSELS

A vessel of five or more net tons may be documented as a yacht. In this process, papers are issued by the U.S. Coast Guard as they are for large ships. Documentation is a form of national registration. The boat must be used solely for pleasure. Its owner must be a citizen of the U.S., a partnership of U.S. citizens, or a corporation controlled by U.S. citizens. The captain and other officers must also be U.S. citizens. The crew need not be.

If you document your yacht, you have the legal authority to fly the yacht ensign. You also may record bills of sale, mortgages, and other papers of title with federal authorities. Doing so gives legal notice that such instruments exist. Documentation also permits preferred status for mortgages. This gives you additional security, and it aids in financing and transfer of title. You must carry the original documentation papers aboard your vessel. Copies will not suffice.

REGISTRATION OF BOATS

If your boat is not documented, registration in the state of its principal use is probably required. If you use it mainly on an ocean, a gulf, or other similar water, register it in the state where you moor it.

If you use your boat solely for racing, it may be exempt from the requirement in your state. Some states may also exclude dinghies, while others require registration of documented vessels and non-power driven boats.

All states, except Alaska, register boats. In Alaska, the U.S. Coast Guard issues the registration numbers. If you move your vessel to a new state of principal use, a valid registration certificate is good for 60 days. You must have the registration certificate (certificate of number) aboard your vessel when it is in use. A copy will not suffice. You may be cited if you do not have the original on board.

NUMBERING OF VESSELS

A registration number is on your registration certificate. You must paint or permanently attach this number to both sides of the forward half of your boat. Do not display any other number there.

The registration number must be clearly visible. It must not be placed on the obscured underside of a flared bow. If you can't place the number on the bow, place it on the forward half of the hull. If that doesn't work, put it on the superstructure. Put the number for an inflatable boat on a bracket or fixture. Then, firmly attach it to the forward half of the boat. The letters and numbers must be plain block characters and must read from left to right. Use a space or a hyphen to separate the prefix and suffix letters from the numerals.

The color of the characters must contrast with that of the background, and they must be at least three inches high.

In some states your registration is good for only one year. In others, it is good for as long as three years. Renew your registration before it expires. At that time you will receive a new decal or decals. Place them as required by state law. You should remove old decals before putting on the new ones. Some states require that you show only the current decal or decals. If your vessel is moored, it must have a current decal even if it is not in use.

If your vessel is lost, destroyed, abandoned, stolen, or transferred, you must inform the issuing authority. If you lose your certificate of number or your address changes, notify the issuing authority as soon as possible.

SALES AND TRANSFERS

Your registration number is not transferable to another boat. The number stays with the boat unless its state of principal use is changed.

HULL IDENTIFICATION NUMBER

A Hull Identification Number (HIN) is like the Vehicle Identification Number (VIN) on your car. Boats built between November 1, 1972 and July 31, 1984 have old format HINs. Since August 1, 1984 a new format has been used.

Your boat's HIN must appear in two places. If it has a transom, the primary number is on its starboard side within two inches of its top. If it does not have a transom or if it was not practical to use the transom, the number is on the starboard side. In this case, it must be within one foot of the stern and within two inches of the top of the hull side. On pontoon boats, it is on the aft crossbeam within one foot of the starboard hull attachment. Your boat also has a duplicate number in an unexposed location. This is on the boat's interior or under a fitting or item of hardware.

LENGTH OF BOATS

For some purposes, boats are classed by length. Required equipment, for example, differs with boat size. Manufacturers may measure a boat's length in several ways. Officially, though, your boat is measured along a straight line from its bow to its stern. This line is parallel to its keel.

The length does not include bowsprits, boomkins, or pulpits. Nor does it include rudders, brackets, outboard motors, outdrives, diving platforms, or other attachments.

CAPACITY INFORMATION

◆ See Figure 4

Manufacturers must put capacity plates on most recreational boats less than 20 feet long. Sailboats, canoes, kayaks, and inflatable boats are usually exempt. Outboard boats must display the maximum permitted horsepower of their engines. The plates must also show the allowable maximum weights of the people on board. And they must show the allowable maximum combined weights of people, engine(s), and gear. Inboards and stern drives need not show the weight of their engines on their capacity plates. The capacity plate must appear where it is clearly visible to the operator when underway. This information serves to remind you of the capacity of your boat under normal circumstances. You should ask yourself, "Is my boat loaded above its recommended capacity" and, "Is my boat overloaded for the present sea and wind conditions?" If you are stopped by a legal authority, you may be cited if you are overloaded.

U.S. COAST GUARD CAPACITY INFORMATION

MAXIMUM PERSON CAPACITY

MAXIMUM WEIGHT CAPACITY (PERSONS AND GEAR)   POUNDS

THIS BOAT COMPLIES WITH U. S. COAST GUARD SAFETY STANDARDS IN EFFECT ON THE DATE OF CERTIFICATION

BAHAMA INDUSTRIES, INC.
5612 LA PALMA ST. ANAHEIM, CALIF. 92807

Fig. 4 A U.S. Coast Guard certification plate indicates the amount of occupants and gear appropriate for safe operation of the vessel

CERTIFICATE OF COMPLIANCE

◆ See Figure 4

Manufacturers are required to put compliance plates on motorboats greater than 20 feet in length. The plates must say, "This boat," or "This equipment complies with the U. S. Coast Guard Safety Standards in effect on the date of certification." Letters and numbers can be no less than one-eighth of an inch high. At the manufacturer's option, the capacity and compliance plates may be combined.

VENTILATION

A cup of gasoline spilled in the bilge has the potential explosive power of 15 sticks of dynamite. This statement, commonly quoted over 20 years ago, may be an exaggeration; however, it illustrates a fact. Gasoline fumes in the bilge of a boat are highly explosive and a serious danger. They are heavier than air and will stay in the bilge until they are vented out.

Because of this danger, Coast Guard regulations require ventilation on many powerboats. There are several ways to supply fresh air to engine and gasoline tank compartments and to remove dangerous vapors. Whatever the choice, it must meet Coast Guard standards.

■ The following is not intended to be a complete discussion of the regulations. It is limited to the majority of recreational vessels. Contact your local Coast Guard office for further information.

General Precautions

Ventilation systems will not remove raw gasoline that leaks from tanks or fuel lines. If you smell gasoline fumes, you need immediate repairs. The best device for sensing gasoline fumes is your nose. Use it! If you smell gasoline in an engine compartment or elsewhere, don't start your engine. The smaller the compartment, the less gasoline it takes to make an explosive mixture.

Ventilation for Open Boats

In open boats, gasoline vapors are dispersed by the air that moves through them. So they are exempt from ventilation requirements.

To be "open," a boat must meet certain conditions. Engine and fuel tank compartments and long narrow compartments that join them must be open to the atmosphere." This means they must have at least 15 square inches of open area for each cubic foot of net compartment volume. The open area must be in direct contact with the atmosphere. There must also be no long, unventilated spaces open to engine and fuel tank compartments into which flames could extend.

Ventilation for All Other Boats

Powered and natural ventilation are required in an enclosed compartment with a permanently installed gasoline engine that has a starter motor. A compartment is exempt if its engine is open to the atmosphere. Diesel powered boats are also exempt.

VENTILATION SYSTEMS

There are two types of ventilation systems. One is "natural ventilation." In it, air circulates through closed spaces due to the boat's motion. The other type is "powered ventilation." In it, air is circulated by a motor-driven fan or fans.

Natural Ventilation System Requirements

A natural ventilation system has an air supply from outside the boat. The air supply may also be from a ventilated compartment or a compartment open to the atmosphere. Intake openings are required. In addition, intake ducts may be required to direct the air to appropriate compartments.

The system must also have an exhaust duct that starts in the lower third of the compartment. The exhaust opening must be into another ventilated compartment or into the atmosphere. Each supply opening and supply duct, if there is one, must be above the usual level of water in the bilge. Exhaust openings and ducts must also be above the bilge water. Openings and ducts must be at least three square inches in area or two inches in diameter. Openings should be placed so exhaust gasses do not enter the fresh air intake. Exhaust fumes must not enter cabins or other enclosed, non-ventilated spaces. The carbon monoxide gas in them is deadly.

Intake and exhaust openings must be covered by cowls or similar devices. These registers keep out rain water and water from breaking seas. Most often, intake registers face forward and exhaust openings aft. This aids the flow of air when the boat is moving or at anchor since most boats face into the wind when properly anchored.

Power Ventilation System Requirements

◆ See Figure 5

Powered ventilation systems must meet the standards of a natural system. They must also have one or more exhaust blowers. The blower duct can serve as the exhaust duct for natural ventilation if fan blades do not obstruct the air flow when not powered. Openings in engine compartment, for carburetion are in addition to ventilation system requirements.

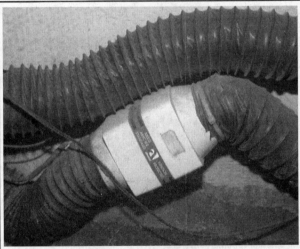

Fig. 5 Typical blower and duct system to vent fumes from the engine compartment

Required Safety Equipment

Coast Guard regulations require that your boat have certain equipment aboard. These requirements are minimums. Exceed them whenever you can.

TYPES OF FIRES

There are four common classes of fires:
• Class A—fires are of ordinary combustible materials such as paper or wood.
• Class B—fires involve gasoline, oil and grease.
• Class C—fires are electrical.
• Class D—fires involve ferrous metals

One of the greatest risks to boaters is fire. This is why it is so important to carry the correct number and type of extinguishers onboard.

The best fire extinguisher for most boats is a Class B extinguisher. Never use water on Class B or Class C fires, as water spreads these types of fires. Additionally, you should never use water on a Class C fire as it may cause you to be electrocuted.

FIRE EXTINGUISHERS

◆ **See Figure 6**

If your boat meets one or more of the following conditions, you must have at least one fire extinguisher aboard. The conditions are:
- Inboard or stern drive engines
- Closed compartments under seats where portable fuel tanks can be stored
- Double bottoms not sealed together or not completely filled with flotation materials
- Closed living spaces
- Closed stowage compartments in which combustible or flammable materials are stored
- Permanently installed fuel tanks
- Boat is 26 feet or more in length.

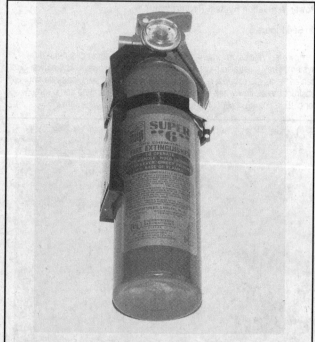

Fig. 6 An approved fire extinguisher should be mounted close to the operator for emergency use

Contents of Extinguishers

Fire extinguishers use a variety of materials. Those used on boats usually contain dry chemicals, Halon, or Carbon Dioxide (CO2). Dry chemical extinguishers contain chemical powders such as Sodium Bicarbonate—baking soda.

Carbon dioxide is a colorless and odorless gas when released from an extinguisher. It is not poisonous but caution must be used in entering compartments filled with it. It will not support life and keeps oxygen from reaching your lungs. A fire-killing concentration of Carbon Dioxide can be lethal. If you are in a compartment with a high concentration of CO2, you will have no difficulty breathing. But the air does not contain enough oxygen to support life. Unconsciousness or death can result.

Halon Extinguishers

Some fire extinguishers and 'built-in' or 'fixed' automatic fire extinguishing systems contain a gas called Halon. Like carbon dioxide it is colorless and odorless and will not support life. Some Halons may be toxic if inhaled.

To be accepted by the Coast Guard, a fixed Halon system must have an indicator light at the vessel's helm. A green light shows the system is ready. Red means it is being discharged or has been discharged. Warning horns are available to let you know the system has been activated. If your fixed Halon system discharges, ventilate the space thoroughly before you enter it. There are no residues from Halon but it will not support life.

Although Halon has excellent fire fighting properties; it is thought to deplete the earth's ozone layer and has not been manufactured since January 1, 1994. Halon extinguishers can be refilled from existing stocks of the gas until they are used up, but high federal excise taxes are being charged for the service. If you discontinue using your Halon extinguisher, take it to a recovery station rather than releasing the gas into the atmosphere. Compounds such as FE 241, designed to replace Halon, are now available.

Fire Extinguisher Approval

Fire extinguishers must be Coast Guard approved. Look for the approval number on the nameplate. Approved extinguishers have the following on their labels: "Marine Type USCG Approved, Size..., Type..., 162.208/," etc. In addition, to be acceptable by the Coast Guard, an extinguisher must be in serviceable condition and mounted in its bracket. An extinguisher not properly mounted in its bracket will not be considered serviceable during a Coast Guard inspection.

Care and Treatment

Make certain your extinguishers are in their stowage brackets and are not damaged. Replace cracked or broken hoses. Nozzles should be free of obstructions. Sometimes, wasps and other insects nest inside nozzles and make them inoperable. Check your extinguishers frequently. If they have pressure gauges, is the pressure within acceptable limits? Do the locking pins and sealing wires show they have not been used since recharging?

Don't try an extinguisher to test it. Its valves will not reseat properly and the remaining gas will leak out. When this happens, the extinguisher is useless.

Weigh and tag carbon dioxide and Halon extinguishers twice a year. If their weight loss exceeds 10 percent of the weight of the charge, recharge them. Check to see that they have not been used. They should have been inspected by a qualified person within the past six months, and they should have tags showing all inspection and service dates. The problem is that they can be partially discharged while appearing to be fully charged.

Some Halon extinguishers have pressure gauges the same as dry chemical extinguishers. Don't rely too heavily on the gauge. The extinguisher can be partially discharged and still show a good gauge reading. Weighing a Halon extinguisher is the only accurate way to assess its contents.

If your dry chemical extinguisher has a pressure indicator, check it frequently. Check the nozzle to see if there is powder in it. If there is, recharge it. Occasionally invert your dry chemical extinguisher and hit the base with the palm of your hand. The chemical in these extinguishers packs and cakes due to the boat's vibration and pounding. There is a difference of opinion about whether hitting the base helps, but it can't hurt. It is known that caking of the chemical powder is a major cause of failure of dry chemical extinguishers. Carry spares in excess of the minimum requirement. If you have guests aboard, make certain they know where the extinguishers are and how to use them.

Using a Fire Extinguisher

A fire extinguisher usually has a device to keep it from being discharged accidentally. This is a metal or plastic pin or loop. If you need to use your extinguisher, take it from its bracket. Remove the pin or the loop and point the nozzle at the base of the flames. Now, squeeze the handle, and discharge the extinguisher's contents while sweeping from side to side. Recharge a used extinguisher as soon as possible.

If you are using a Halon or carbon dioxide extinguisher, keep your hands away from the discharge. The rapidly expanding gas will freeze them. If your fire extinguisher has a horn, hold it by its handle.

Legal Requirements for Extinguishers

You must carry fire extinguishers as defined by Coast Guard regulations. They must be firmly mounted in their brackets and immediately accessible.

A motorboat less than 26 feet long must have at least one approved hand-portable, Type B-1 extinguisher. If the boat has an approved fixed fire extinguishing system, you are not required to have the Type B-1 extinguisher.

Also, if your boat is less than 26 feet long, is propelled by an outboard motor, or motors, and does not have any of the first six conditions described at the beginning of this section, it is not required to have an extinguisher. Even so, it's a good idea to have one, especially if a nearby boat catches fire, or if a fire occurs at a fuel dock.

A motorboat 26 feet to under 40 feet long, must have at least two Type B-1 approved hand-portable extinguishers. It can, instead, have at least one Coast Guard approved Type B-2. If you have an approved fire extinguishing system, only one Type B-1 is required.

A motorboat 40 to 65 feet long must have at least three Type B-1 approved portable extinguishers. It may have, instead, at least one Type B-1 plus a Type B-2. If there is an approved fixed fire extinguishing system, two Type B-1 or one Type B-2 is required.

WARNING SYSTEM

Various devices are available to alert you to danger. These include fire, smoke, gasoline fumes, and carbon monoxide detectors. If your boat has a galley, it should have a smoke detector. Where possible, use wired detectors. Household batteries often corrode rapidly on a boat.

There are many ways in which carbon monoxide (a by-product of the combustion that occurs in an engine) can enter your boat. You can't see, smell, or taste carbon monoxide gas, but it is lethal. As little as one part in 10,000 parts of air can bring on a headache. The symptoms of carbon monoxide poisoning—headaches, dizziness, and nausea—are like seasickness. By the time you realize what is happening to you, it may be too late to take action. If you have enclosed living spaces on your boat, protect yourself with a detector.

PERSONAL FLOTATION DEVICES

Personal Flotation Devices (PFDs) are commonly called life preservers or life jackets. You can get them in a variety of types and sizes. They vary with their intended uses. To be acceptable, PFDs must be Coast Guard approved.

Type I PFDs

A Type I life jacket is also called an offshore life jacket. Type I life jackets will turn most unconscious people from facedown to a vertical or slightly backward position. The adult size gives a minimum of 22 pounds of buoyancy. The child size has at least 11 pounds. Type I jackets provide more protection to their wearers than any other type of life jacket. Type I life jackets are bulkier and less comfortable than other types. Furthermore, there are only two sizes, one for children and one for adults.

Type I life jackets will keep their wearers afloat for extended periods in rough water. They are recommended for offshore cruising where a delayed rescue is probable.

Type II PFDs

◆ See Figure 7

A Type II life jacket is also called a near-shore buoyant vest. It is an approved, wearable device. Type II life jackets will turn some unconscious people from facedown to vertical or slightly backward positions. The adult size gives at least 15.5 pounds of buoyancy. The medium child size has a minimum of 11 pounds. And the small child and infant sizes give seven pounds. A Type II life jacket is more comfortable than a Type I but it does not have as much buoyancy. It is not recommended for long hours in rough water. Because of this, Type IIs are recommended for inshore and inland cruising on calm water. Use them only where there is a good chance of fast rescue.

Type III PFDs

Type III life jackets or marine buoyant devices are also known as flotation aids. Like Type IIs, they are designed for calm inland or close offshore water where there is a good chance of fast rescue. Their minimum buoyancy is 15.5 pounds. They will **NOT** turn their wearers face up.

Type III devices are usually worn where freedom of movement is necessary. Thus, they are used for water skiing, small boat sailing, and fishing among other activities. They are available as vests and flotation coats. Flotation coats are useful in cold weather. Type IIIs come in many sizes from small child through large adult.

Life jackets come in a variety of colors and patterns—red, blue, green, camouflage, and cartoon characters. From purely a safety standpoint, the best color is bright orange. It is easier to see in the water, especially if the water is rough.

Type IV PFDs

◆ See Figures 8 and 9

Type IV ring life buoys, buoyant cushions and horseshoe buoys are Coast Guard approved devices called throwables. They are made to be thrown to people in the water, and should not be worn. Type IV cushions are often used as seat cushions. But, keep in mind that cushions are hard to hold onto in the water, thus, they do not afford as much protection as wearable life jackets.

The straps on buoyant cushions are for you to hold onto either in the water or when throwing them, they are **NOT** for your arms. A cushion should never be worn on your back, as it will turn you face down in the water.

Type IV throwables are not designed as personal flotation devices for unconscious people, non-swimmers, or children. Use them only in emergencies. They should not be used for, long periods in rough water.

Ring life buoys come in 18, 20, 24, and 30 inch diameter sizes. They usually have grab lines, but you will need to attach about 60 feet of polypropylene line to the grab rope to aid in retrieving someone in the water. If you throw a ring, be careful not to hit the person. Ring buoys can knock people unconscious

Type V PFDs

Type V PFDs are of two kinds, special use devices and hybrids. Special use devices include boardsailing vests, deck suits, work vests, and others. They are approved only for the special uses or conditions indicated on their labels. Each is designed and intended for the particular application shown on its label. They do not meet legal requirements for general use aboard recreational boats.

Hybrid life jackets are inflatable devices with some built-in buoyancy provided by plastic foam or kapok. They can be inflated orally or by cylinders of compressed gas to give additional buoyancy. In some hybrids the gas is released manually. In others it is released automatically when the life jacket is immersed in water.

The inherent buoyancy of a hybrid may be insufficient to float a person unless it is inflated. The only way to find this out is for the user to try it in the water. Because of its limited buoyancy when deflated, a hybrid is recommended for use by a non-swimmer only if it is worn with enough inflation to float the wearer.

If they are to count against the legal requirement for the number of life jackets you must carry, hybrids manufactured before February 8, 1995 must be worn whenever a boat is underway and the wearer must not go below decks or in an enclosed space. To find out if your Type V hybrid must be worn to satisfy the legal requirement, read its label. If its use is restricted it will say, "REQUIRED TO BE WORN" in capital letters.

Hybrids cost more than other life jackets, but this factor must be weighed against the fact that they are more comfortable than Types I, II or III life jackets. Because of their greater comfort, their owners are more likely to wear them than are the owners of Type I, II or III life jackets.

The Coast Guard has determined that improved, less costly hybrids can save lives since they will be bought and used more frequently. For these reasons, a new federal regulation was adopted effective February 8, 1995. The regulation increases both the deflated and inflated buoyancys of hybrids, makes them available in a greater variety of sizes and types, and reduces their costs by reducing production costs.

Even though it may not be required, the wearing of a hybrid or a life jacket is encouraged whenever a vessel is underway. Like life jackets, hybrids are now available in three types. To meet legal requirements, a Type I hybrid can be substituted for a Type I life jacket. Similarly Type II and III hybrids can be substituted for Type II and Type III life jackets. A Type I hybrid, when inflated, will turn most unconscious people from facedown to vertical or slightly backward positions just like a Type I life jacket. Type II and III hybrids function like Type II and III life jackets. If you purchase a new hybrid, it should have an owner's manual attached that describes its life jacket type and its deflated and inflated buoyancys. It warns you that it may have to be inflated to float you. The manual also tells you how to don the life jacket and how to inflate it. It also tells you how to change its inflation mechanism, recommended testing exercises, and inspection or maintenance procedures.

Fig. 7 Type II PFDs are recommended for inshore and inland cruising on calm water (where there is a good chance of fast rescue)

Fig. 8 Type IV buoyant cushions are thrown to people in the water. If you can squeeze air out of the cushion, it should be replaced

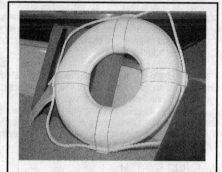

Fig. 9 Type IV throwables, such as this ring life buoy, are not designed for unconscious people, non-swimmers, or children

The manual also tells you why you need a life jacket and why you should wear it. A new hybrid must be packaged with at least three gas cartridges. One of these may already be loaded into the inflation mechanism. Likewise, if it has an automatic inflation mechanism, it must be packaged with at least three of these water sensitive elements. One of these elements may be installed.

Legal Requirements

A Coast Guard approved life jacket must show the manufacturer's name and approval number. Most are marked as Type I, II, III, IV or V. All of the newer hybrids are marked for type.

You are required to carry at least one wearable life jacket or hybrid for each person on board your recreational vessel. If your vessel is 16 feet or more in length and is not a canoe or a kayak, you must also have at least one Type IV on board. These requirements apply to all recreational vessels that are propelled or controlled by machinery, sails, oars, paddles, poles, or another vessel. Sailboards are not required to carry life jackets.

You can substitute an older Type V hybrid for any required Type I, II or III life jacket provided:

1. Its approval label shows it is approved for the activity the vessel is engaged in

2. It's approved as a substitute for a life jacket of the type required on the vessel

3. It's used as required on the labels
and

4. It's used in accordance with any requirements in its owner's manual (if the approval label makes reference to such a manual.)

A water skier being towed is considered to be on board the vessel when judging compliance with legal requirements.

You are required to keep your Type I, II or III life jackets or equivalent hybrids readily accessible, which means you must be able to reach out and get them when needed. All life jackets must be in good, serviceable condition.

General Considerations

The proper use of a life jacket requires the wearer to know how it will perform. You can gain this knowledge only through experience. Each person on your boat should be assigned a life jacket. Next, it should be fitted to the person who will wear it. Only then can you be sure that it will be ready for use in an emergency. This advice is good even if the water is calm, and you intend to boat near shore.

Boats can sink fast. There may be no time to look around for a life jacket. Fitting one on you in the water is almost impossible. Most drownings occur in inland waters within a few feet of safety. Most victims had life jackets, but they weren't wearing them.

Keeping life jackets in the plastic covers they came wrapped in, and in a cabin, assures that they will stay clean and unfaded. But this is no way to keep them when you are on the water. When you need a life jacket it must be readily accessible and adjusted to fit you. You can't spend time hunting for it or learning how to fit it.

There is no substitute for the experience of entering the water while wearing a life jacket. Children, especially, need practice. If possible, give your guests this experience. Tell them they should keep their arms to their sides when jumping in to keep the life jacket from riding up. Let them jump in and see how the life jacket responds. Is it adjusted so it does not ride up? Is it the proper size? Are all straps snug? Are children's life jackets the right sizes for them? Are they adjusted properly? If a child's life jacket fits correctly, you can lift the child by the jacket's shoulder straps and the child's chin and ears will not slip through. Non-swimmers, children, handicapped persons, elderly persons and even pets should always wear life jackets when they are aboard. Many states require that everyone aboard wear them in hazardous waters.

Inspect your lifesaving equipment from time to time. Leave any questionable or unsatisfactory equipment on shore. An emergency is no time for you to conduct an inspection.

Indelibly mark your life jackets with your vessel's name, number, and calling port. This can be important in a search and rescue effort. It could help concentrate effort where it will do the most good.

Care of Life Jackets

Given reasonable care, life jackets last many years. Thoroughly dry them before putting them away. Stow them in dry, well-ventilated places. Avoid the bottoms of lockers and deck storage boxes where moisture may collect. Air and dry them frequently.

Life jackets should not be tossed about or used as fenders or cushions. Many contain kapok or fibrous glass material enclosed in plastic bags. The bags can rupture and are then unserviceable. Squeeze your life jacket gently. Does air leak out? If so, water can leak in and it will no longer be safe to use. Cut it up so no one will use it, and throw it away. The covers of some life jackets are made of nylon or polyester. These materials are plastics. Like many plastics, they break down after extended exposure to the ultraviolet light in sunlight. This process may be more rapid when the materials are dyed with bright dyes such as "neon" shades.

Ripped and badly faded fabrics are clues that the covering of your life jacket is deteriorating. A simple test is to pinch the fabric between your thumbs and forefingers. Now try to tear the fabric. If it can be torn, it should definitely be destroyed and discarded. Compare the colors in protected places to those exposed to the sun. If the colors have faded, the materials have been weakened. A life jacket covered in fabric should ordinarily last several boating seasons with normal use. A life jacket used every day in direct sunlight should probably be replaced more often.

SOUND PRODUCING DEVICES

All boats are required to carry some means of making an efficient sound signal. Devices for making the whistle or horn noises required by the Navigation Rules must be capable of a four-second blast. The blast should be audible for at least one-half mile. Athletic whistles are not acceptable on boats 12 meters or longer. Use caution with athletic whistles. When wet, some of them come apart and loose their "pea." When this happens, they are useless.

If your vessel is 12 meters long and less than 20 meters, you must have a power whistle (or power horn) and a bell on board. The bell must be in operating condition and have a minimum diameter of at least 200mm (7.9 in.) at its mouth.

VISUAL DISTRESS SIGNALS

◆ See Figure 10

Visual Distress Signals (VDS) attract attention to your vessel if you need help. They also help to guide searchers in search and rescue situations. Be sure you have the right types, and learn how to use them properly.

It is illegal to fire flares improperly. In addition, they cost the Coast Guard and its Auxiliary many wasted hours in fruitless searches. If you signal a distress with flares and then someone helps you, please let the Coast Guard or the appropriate Search And Rescue (SAR) Agency know so the distress report will be canceled.

Recreational boats less than 16 feet long must carry visual distress signals on coastal waters at night. Coastal waters are:
• The ocean (territorial sea)
• The Great Lakes
• Bays or sounds that empty into oceans
• Rivers over two miles across at their mouths upstream to where they narrow to two miles.

Recreational boats 16 feet or longer must carry VDS at all times on coastal waters. The same requirement applies to boats carrying six or fewer passengers for hire. Open sailboats less than 26 feet long without engines are exempt in the daytime as are manually propelled boats. Also exempt are boats in organized races, regattas, parades, etc. Boats owned in the United States and operating on the high seas must be equipped with VDS.

A wide variety of signaling devices meet Coast Guard regulations. For pyrotechnic devices, a minimum of three must be carried. Any combination can be carried as long as it adds up to at least three signals for day use and at least three signals for night use. Three day/night signals meet both requirements. If possible, carry more than the legal requirement.

■ The American flag flying upside down is a commonly recognized distress signal. It is not recognized in the Coast Guard regulations, though. In an emergency, your efforts would probably be better used in more effective signaling methods.

Types of VDS

VDS are divided into two groups; daytime and nighttime use. Each of these groups is subdivided into pyrotechnic and non-pyrotechnic devices.

Daytime Non-Pyrotechnic Signals

A bright orange flag with a black square over a black circle is the simplest VDS. It is usable, of course, only in daylight. It has the advantage of being a continuous signal. A mirror can be used to good advantage on sunny days. It can attract the attention of other boaters and of aircraft from great distances.

Mirrors are available with holes in their centers to aid in "aiming." In the absence of a mirror, any shiny object can be used. When another boat is in sight, an effective VDS is to extend your arms from your sides and move them up and down. Do it slowly. If you do it too fast the other people may think you are just being friendly. This simple gesture is seldom misunderstood, and requires no equipment.

Daytime Pyrotechnic Devices

Orange smoke is a useful daytime signal. Hand-held or floating smoke flares are very effective in attracting attention from aircraft. Smoke flares don't last long, and are not very effective in high wind or poor visibility. As with other pyrotechnic devices, use them only when you know there is a possibility that someone will see the display.

To be usable, smoke flares must be kept dry. Keep them in airtight containers and store them in dry places. If the "striker" is damp, dry it out before trying to ignite the device. Some pyrotechnic devices require a forceful "strike" to ignite them.

All hand-held pyrotechnic devices may produce hot ashes or slag when burning. Hold them over the side of your boat in such a way that they do not burn your hand or drip into your boat.

Nighttime Non-Pyrotechnic Signals

An electric distress light is available. This light automatically flashes the international morse code SOS distress signal (••• --- •••). Flashed four to six times a minute, it is an unmistakable distress signal. It must show that it is approved by the Coast Guard. Be sure the batteries are fresh. Dated batteries give assurance that they are current.

Under the Inland Navigation Rules, a high intensity white light flashing 50–70 times per minute is a distress signal. Therefore, use strobe lights on inland waters only for distress signals.

Nighttime Pyrotechnic Devices

◆ See Figure 11

Aerial and hand-held flares can be used at night or in the daytime. Obviously, they are more effective at night.

Currently, the serviceable life of a pyrotechnic device is rated at 42 months from its date of manufacture. Pyrotechnic devices are expensive. Look at their dates before you buy them. Buy them with as much time remaining as possible.

Like smoke flares, aerial and hand-held flares may fail to work if they have been damaged or abused. They will not function if they are or have been wet. Store them in dry, airtight containers in dry places. But store them where they are readily accessible.

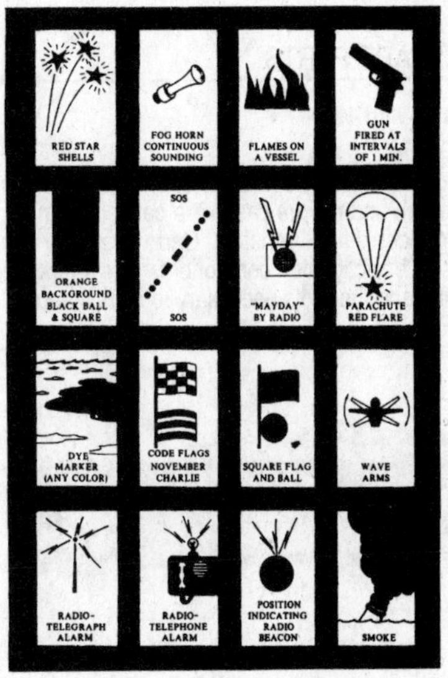

Fig. 10 Internationally accepted distress signals

Fig. 11 Moisture protected flares should be carried onboard any vessel for use as a distress signal

Fig. 12 Always carry an adequately stocked first aid kit on board for the safety of the crew and guests

Fig. 13 Choose an anchor of sufficient weight to secure the boat without dragging

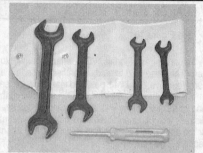

Fig 14 A few wrenches, a screwdriver and maybe a pair of pliers can be very helpful to make emergency repairs

Aerial VDSs, depending on their type and the conditions they are used in, may not go very high. Again, use them only when there is a good chance they will be seen.

A serious disadvantage of aerial flares is that they burn for only a short time; most burn for less than 10 seconds. Most parachute flares burn for less than 45 seconds. If you use a VDS in an emergency, do so carefully. Hold hand-held flares over the side of the boat when in use. Never use a road hazard flare on a boat, it can easily start a fire. Marine type flares are specifically designed to lessen risk, but they still must be used carefully.

Aerial flares should be given the same respect as firearms since they are firearms! Never point them at another person. Don't allow children to play with them or around them. When you fire one, face away from the wind. Aim it downwind and upward at an angle of about 60 degrees to the horizon. If there is a strong wind, aim it somewhat more vertically. Never fire it straight up. Before you discharge a flare pistol, check for overhead obstructions that might be damaged by the flare. An obstruction might deflect the flare to where it will cause injury or damage.

Disposal of VDS

Keep outdated flares when you get new ones. They do not meet legal requirements, but you might need them sometime, and they may work. It is illegal to fire a VDS on federal navigable waters unless an emergency exists. Many states have similar laws.

Emergency Position Indicating Radio Beacon (EPIRB)

There is no requirement for recreational boats to have EPIRBs. Some commercial and fishing vessels, though, must have them if they operate beyond the three-mile limit. Vessels carrying six or fewer passengers for hire must have EPIRBs under some circumstances when operating beyond the three-mile limit. If you boat in a remote area or offshore, you should have an EPIRB. An EPIRB is a small (about 6 to 20 inches high), battery-powered, radio transmitting buoy-like device. It is a radio transmitter and requires a license or an endorsement on your radio station license by the Federal Communications Commission (FCC). EPIRBs are either automatically activated by being immersed in water or manually by a switch.

Equipment Not Required But Recommended

Although not required by law, there are other pieces of equipment that are good to have onboard.

SECOND MEANS OF PROPULSION

All boats less than 16 feet long should carry a second means of propulsion. A paddle or oar can come in handy at times. For most small boats, a spare trolling or outboard motor is an excellent idea. If you carry a spare motor, it should have its own fuel tank and starting power. If you use an electric trolling motor, it should have its own battery.

BAILING DEVICES

All boats should carry at least one effective manual bailing device in addition to any installed electric bilge pump. This can be a bucket, can, scoop, hand-operated pump, etc. If your battery "goes dead" it will not operate your electric pump.

FIRST AID KIT

◆ See Figure 12

All boats should carry a first aid kit. It should contain adhesive bandages, gauze, adhesive tape, antiseptic, aspirin, etc. Check your first aid kit from time to time. Replace anything that is outdated. It is to your advantage to know how to use your first aid kit. Another good idea would be to take a Red Cross first aid course.

ANCHORS

◆ See Figure 13

All boats should have anchors. Choose one of suitable size for your boat. Better still, have two anchors of different sizes. Use the smaller one in calm water or when anchoring for a short time to fish or eat. Use the larger one when the water is rougher or for overnight anchoring.

Carry enough anchor line, of suitable size, for your boat and the waters in which you will operate. If your engine fails you, the first thing you usually should do is lower your anchor. This is good advice in shallow water where you may be driven aground by the wind or water. It is also good advice in windy weather or rough water, as the anchor, when properly affixed, will usually hold your bow into the waves.

VHF-FM RADIO

Your best means of summoning help in an emergency or in case of a breakdown is a VHF-FM radio. You can use it to get advice or assistance from the Coast Guard. In the event of a serious illness or injury aboard your boat, the Coast Guard can have emergency medical equipment meet you ashore.

TOOLS AND SPARE PARTS

◆ See Figures 14 and 15

Carry a few tools and some spare parts, and learn how to make minor repairs. Many search and rescue cases are caused by minor breakdowns that boat operators could have repaired. Carry spare parts such as propellers, fuses or basic ignition components (like spark plugs, wires or even ignition coils) and the tools necessary to install them.

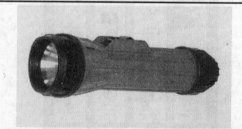

Fig. 15 A flashlight with a fresh set of batteries is handy when repairs are needed at night. It can also double as a signaling device

Courtesy Marine Examinations

One of the roles of the Coast Guard Auxiliary is to promote recreational boating safety. This is why they conduct thousands of Courtesy Marine Examinations each year. The auxiliarists who do these examinations are well-trained and knowledgeable in the field.

These examinations are free and done only at the consent of boat owners. To pass the examination, a vessel must satisfy federal equipment requirements and certain additional requirements of the coast guard auxiliary. If your vessel does not pass the Courtesy Marine Examination, no report of the failure is made. Instead, you will be told what you need to correct the deficiencies. The examiner will return at your convenience to redo the examination.

If your vessel qualifies, you will be awarded a safety decal. The decal does not carry any special privileges, it simply attests to your interest in safe boating.

SAFETY IN SERVICE

It is virtually impossible to anticipate all of the hazards involved with maintenance and service, but care and common sense will prevent most accidents.

The rules of safety for mechanics range from "don't smoke around gasoline," to "use the proper tool(s) for the job." The trick to avoiding injuries is to develop safe work habits and to take every possible precaution. Whenever you are working on your boat, pay attention to what you are doing. The more you pay attention to details and what is going on around you, the less likely you will be to hurt yourself or damage your boat.

Do's

- Do keep a fire extinguisher and first aid kit handy.
- Do wear safety glasses or goggles when cutting, drilling, grinding or prying, even if you have 20–20 vision. If you wear glasses for the sake of vision, wear safety goggles over your regular glasses.
- Do shield your eyes whenever you work around the battery. Batteries contain sulfuric acid. In case of contact with the eyes or skin, flush the area with water or a mixture of water and baking soda; then seek immediate medical attention.
- Do use adequate ventilation when working with any chemicals or hazardous materials.
- Do disconnect the negative battery cable when working on the electrical system. The secondary ignition system contains EXTREMELY HIGH VOLTAGE. In some cases it can even exceed 50,000 volts. Furthermore, an accidental attempt to start the engine could cause the propeller or other components to rotate suddenly causing a potentially dangerous situation.
- Do follow manufacturer's directions whenever working with potentially hazardous materials. Most chemicals and fluids are poisonous if taken internally.
- Do properly maintain your tools. Loose hammerheads, mushroomed punches and chisels, frayed or poorly grounded electrical cords, excessively worn screwdrivers, spread wrenches (open end), cracked sockets, or slipping ratchets can cause accidents.
- Likewise, keep your tools clean; a greasy wrench can slip off a bolt head, ruining the bolt and often harming your knuckles in the process.
- Do use the proper size and type of tool for the job at hand. Do select a wrench or socket that fits the nut or bolt. The wrench or socket should sit straight, not cocked.
- Do, when possible, pull on a wrench handle rather than push on it, and adjust your stance to prevent a fall.
- Do be sure that adjustable wrenches are tightly closed on the nut or bolt and pulled so that the force is on the side of the fixed jaw. Better yet, avoid the use of an adjustable if you have a fixed wrench that will fit.
- Do strike squarely with a hammer; avoid glancing blows.
- Do use common sense whenever you work on your boat or motor. If a situation arises that doesn't seem right, sit back and have a second look. It may save an embarrassing moment or potential damage to your beloved boat.

Don'ts

- Don't run the engine in an enclosed area or anywhere else without proper ventilation—EVER! Carbon monoxide is poisonous; it takes a long time to leave the human body and you can build up a deadly supply of it in your system by simply breathing in a little every day. You may not realize you are slowly poisoning yourself.
- Don't work around moving parts while wearing loose clothing. Short sleeves are much safer than long, loose sleeves. Hard-toed shoes with neoprene soles protect your toes and give a better grip on slippery surfaces. Jewelry, watches, large belt buckles, or body adornment of any kind is not safe working around any craft or vehicle. Long hair should be tied back under a hat.
- Don't use pockets for toolboxes. A fall or bump can drive a screwdriver deep into your body. Even a rag hanging from your back pocket can wrap around a spinning shaft.
- Don't smoke when working around gasoline, cleaning solvent or other flammable material.
- Don't smoke when working around the battery. When the battery is being charged, it gives off explosive hydrogen gas. Actually, you shouldn't smoke anyway. Save the cigarette money and put it into your boat!
- Don't use gasoline to wash your hands; there are excellent soaps available. Gasoline contains dangerous additives that can enter the body through a cut or through your pores. Gasoline also removes all the natural oils from the skin so that bone dry hands will suck up oil and grease.
- Don't use screwdrivers for anything other than driving screws! A screwdriver used as an prying tool can snap when you least expect it, causing injuries. At the very least, you'll ruin a good screwdriver.

TROUBLESHOOTING

Troubleshooting can be defined as a methodical process during which one discovers what is causing a problem with engine operation. Although it is often a feared process to the uninitiated, there is no reason to believe that you cannot figure out what is wrong with an engine, as long as you follow a few basic rules.

To begin with, troubleshooting must be systematic. Haphazardly testing one component, then another, **MIGHT** uncover the problem, but it will more likely waste a lot of time. True troubleshooting starts by defining the problem and performing systematic tests to eliminate the largest and most likely causes first.

Start all troubleshooting by eliminating the most basic possible causes; begin with a visual inspection of the boat, engine and drive. If the engine won't crank, make sure that the kill switch or safety lanyard is in the proper position. Make sure there is fuel in the tank and the fuel system is primed before condemning the carburetor or fuel injection system. Make sure there are no blown fuses, the battery is fully charged, and the cable connections (at both ends) are clean and tight before suspecting a bad starter, solenoid or switch.

The majority of problems that occur suddenly can be fixed by simply identifying the one small item that brought them on. A loose wire, a clogged passage or a broken component can cause a lot of trouble and are often the cause of a sudden performance problem.

The next most basic step in troubleshooting is to test systems before components. For example, if the engine doesn't crank on an electric start motor, determine if the battery is in good condition (fully charged and properly connected) before testing the starting system. If the engine cranks, but doesn't start, you know already know the starting system and battery (if it cranks fast enough) are in good condition, now it is time to look at the ignition or fuel systems. Once you've isolated the problem to a particular system, follow the troubleshooting/testing procedures in the section for that system to test either sub-systems (if applicable, for example: the starter circuit) or components (starter solenoid).

Basic Operating Principles

◆ **See Figures 16 and 17**

Before attempting to troubleshoot a problem with your engine, it is important that you understand how it operates. Once normal engine or system operation is understood, it will be easier to determine what might be causing the trouble or irregular operation in the first place. System descriptions are found throughout this manual, but the basic mechanical operating principles for both 2-stroke engines and 4-stroke engines (like yours) are given here. A basic understanding of both types of engines is useful not only in understanding and troubleshooting your OMC, but also for dealing with other motors in your life.

All engines covered by this manual (and probably MOST of the motors you own) operate according to the Otto cycle principle of engine operation. This means that all engines follow the stages of intake, compression, power

and exhaust. But, the difference between a 2- and 4-stroke engine is in how many times the piston moves up and down within the cylinder to accomplish this. On 2-stroke motors (as the name suggests) the four cycles take place in 2 movements (one up and one down) of the piston. Again, as the name suggests, the cycles take place in 4 movements of the piston for 4-stroke engines.

2-STROKE MOTORS

The 2-stroke engine differs in several ways from a conventional four-stroke (automobile or marine) engine.
1. The intake/exhaust method by which the fuel-air mixture is delivered to the combustion chamber.
2. The complete lubrication system.
3. The frequency of the power stroke.
Let's discuss these differences briefly (and compare 2-stroke engine operation with 4-stroke engine operation.)

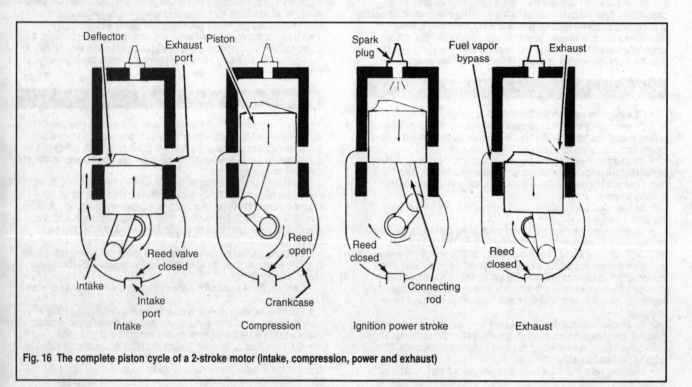

Fig. 16 The complete piston cycle of a 2-stroke motor (intake, compression, power and exhaust)

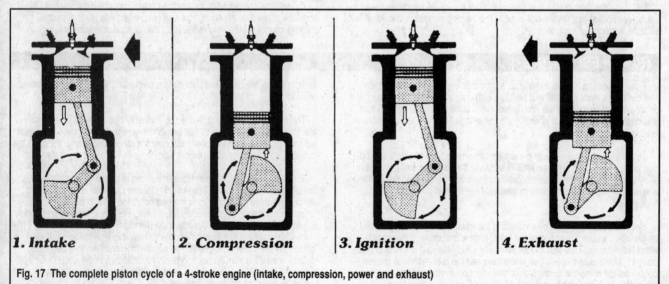

Fig. 17 The complete piston cycle of a 4-stroke engine (intake, compression, power and exhaust)

Intake/Exhaust

◆ **See Figure 18**

Two-stroke engines utilize an arrangement of port openings to admit fuel to the combustion chamber and to purge the exhaust gases after burning has been completed. The ports are located in a precise pattern in order for them to be open and closed off at an exact moment by the piston as it moves up and down in the cylinder. The exhaust port is located slightly higher than the fuel intake port. This arrangement opens the exhaust port first as the piston starts downward and therefore, the exhaust phase begins a fraction of a second before the intake phase.

Actually, the intake and exhaust ports are spaced so closely together that both open almost simultaneously. For this reason, many 2-stroke engines utilize deflector-type pistons. This design of the piston top serves two purposes very effectively.

First, it creates turbulence when the incoming charge of fuel enters the combustion chamber. This turbulence results in a more complete burning of the fuel than if the piston top were flat. The second effect of the deflector-type piston crown is to force the exhaust gases from the cylinder more rapidly.

These systems of intake and exhaust are in marked contrast to individual intake and exhaust valve arrangement employed on four-stroke engines (and the mechanical methods of opening and closing these valves).

■ **It should be noted here that there are some 2-stroke engines that utilize a mechanical valve train, though it is very different from the valve train employed by most 4-stroke engines. Rotary 2-stroke engines use a circular valve or rotating disc that contains a port opening around part of one edge of the disc. As the engine (and disc) turns, the opening aligns with the intake port at and for a predetermined amount of time, closing off the port again as the opening passes by and the solid portion of the disc covers the port.**

Lubrication

A 2-stroke engine is lubricated by mixing oil with the fuel. Therefore, various parts are lubricated as the fuel mixture passes through the crankcase and the cylinder. In contrast, four-stroke engines have a crankcase containing oil. This oil is pumped through a circulating system and returned to the crankcase to begin the routing again.

Power Stroke

The combustion cycle of a 2-stroke engine has four distinct phases.
1. Intake
2. Compression
3. Power
4. Exhaust

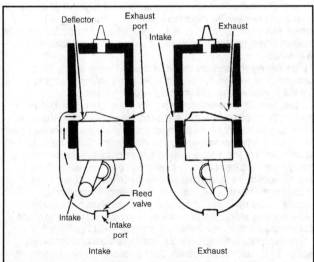

Fig. 18 The intake and exhaust cycles of a two-stroke engine— Cross flow (CV) design shown

The four phases of the cycle are accomplished with each up and down stroke of the piston, and the power stroke occurs with each complete revolution of the crankshaft. Compare this system with a four-stroke engine. A separate stroke of the piston is required to accomplish each phase of the cycle and the power stroke occurs only every other revolution of the crankshaft. Stated another way, two revolutions of the four-stroke engine crankshaft are required to complete one full cycle, the four phases.

Physical Laws

◆ **See Figure 19**

The 2-stroke engine is able to function because of two very simple physical laws.

One: Gases will flow from an area of high pressure to an area of lower pressure. A tire blowout is an example of this principle. The high-pressure air escapes rapidly if the tube is punctured.

Two: If a gas is compressed into a smaller area, the pressure increases, and if a gas expands into a larger area, the pressure is decreased.

If these two laws are kept in mind, the operation of the 2-stroke engine will be easier understood.

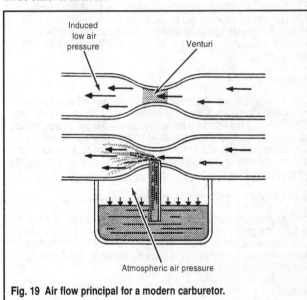

Fig. 19 Air flow principal for a modern carburetor.

Actual Operation

◆ **See Figure 16**

■ **The engine described here is of a carbureted type.**

Beginning with the piston approaching top dead center on the compression stroke: the intake and exhaust ports are physically closed (blocked) by the piston. During this stroke, the reed valve is open (because as the piston moves upward, the crankcase volume increases, which reduces the crankcase pressure to less than the outside atmosphere (creates a vacuum under the piston). The spark plug fires; the compressed fuel-air mixture is ignited; and the power stroke begins.

As the piston moves downward on the power stroke, the combustion chamber is filled with burning gases. As the exhaust port is uncovered, the gases, which are under great pressure, escape rapidly through the exhaust ports. The piston continues its downward movement. Pressure within the crankcase (again, under the piston) increases, closing the reed valves against their seats. The crankcase then becomes a sealed chamber so the air-fuel mixture becomes compressed (pressurized) and ready for delivery to the combustion chamber. As the piston continues to move downward, the intake port is uncovered. The fresh fuel mixture rushes through the intake port into the combustion chamber striking the top of the piston where it is deflected along the cylinder wall. The reed valve remains closed until the piston moves upward again.

When the piston begins to move upward on the compression stroke, the reed valve opens because the crankcase volume has been increased, reducing crankcase pressure to less than the outside atmosphere. The intake and exhaust ports are closed and the fresh fuel charge is compressed inside the combustion chamber.

Pressure in the crankcase (beneath the piston) decreases as the piston moves upward and a fresh charge of air flows through the carburetor picking up fuel. As the piston approaches top dead center, the spark plug ignites the air-fuel mixture, the power stroke begins and one complete Otto cycle has been completed.

4-STROKE ENGINES

The 4-stroke engine may be easier to understand for some people, either because of its prevalence in automobile and street motorcycle engines today, or perhaps because each of the four strokes corresponds to one distinct phase of the Otto cycle. Essentially, a 4-stroke engine completes one Otto cycle of intake, compression, ignition/power and exhaust using two full revolutions of the crankshaft and four distinct movements of the piston (down, up, down and up).

Intake

The intake stroke begins with the piston near the top of its travel. As crankshaft rotation begins to pull the piston downward, the exhaust valve closes and the intake opens. As volume of the combustion chamber increases, a vacuum is created that draws in the air/fuel mixture from the intake manifold.

Compression

Once the piston reaches the bottom of its travel, crankshaft rotation will begin to force it upward. At this point the intake valve closes. As the piston rises in the bore, the volume of the sealed (both intake and exhaust valves are closed) combustion chamber decreases and the air/fuel mixture is compressed. This raises the temperature and pressure of the mixture and increases the amount of force generated by the expanding gases during the Ignition/Power stroke.

Ignition/Power

As the piston approaches top dead center (the highest point of travel in the bore), the spark plug will fire, igniting the air/fuel mixture. The resulting combustion of the air/fuel mixture forces the piston downward, rotating the crankshaft (causing other pistons to move in other phases/strokes of the Otto cycle on multi-cylinder engines).

Exhaust

As the piston approaches the bottom of the Ignition/Power stroke, the exhaust valve opens. When the piston begins its upward path of travel once again, any remaining unburned gasses are forced out through the exhaust valve. This completes one Otto cycle, which begins again as the piston passes top dead center, the intake valve opens and the Intake stroke starts.

COMBUSTION

Whether we are talking about a 2- or 4-stroke engine, all Otto cycle, internal combustion engines require three basic conditions to operate properly,
1. Compression
2. Ignition (Spark)
3. Fuel
A lack of any one of these conditions will prevent the engine from operating. A problem with any one of these will manifest itself in hard-starting or poor performance.

Compression

An engine that has insufficient compression will not draw an adequate supply of air/fuel mixture into the combustion chamber and, subsequently, will not make sufficient power on the power stroke. A lack of compression in just one cylinder of a multi-cylinder engine will cause the engine to stumble or run irregularly.

But, keep in mind that a sudden change in compression is unlikely in 2-stroke motors (unless something major breaks inside the crankcase, but that would usually be accompanied by other symptoms such as a loud noise

when it occurred or noises during operation). On 4-stroke engines, a sudden change in compression is also unlikely, but could occur if the timing belt or chain was to suddenly break. Remember that the timing belt/chain is used to synchronize the valve train with the crankshaft. If the valve train suddenly ceases to turn, some intake and some exhaust valves will remain open, relieving compression in that cylinder.

Ignition (Spark)

Traditionally, the ignition system is the weakest link in the chain of conditions necessary for engine operation. Spark plugs may become worn or fouled, wires will deteriorate allowing arcing or misfiring, and poor connections can place an undue load on coils leading to weak spark or even a failed coil. The most common question asked by a technician under a no-start condition is: "do I have spark and fuel" (as they've already determined that they have compression).

A quick visual inspection of the spark plug(s) will answer the question as to whether or not the plug(s) is/art worn or fouled. While the engine is shut **OFF** a physical check of the connections could show a loose primary or secondary ignition circuit wire. An obviously physically damaged wire may also be an indication of system problems and certainly encourages one to inspect the related system more closely.

If nothing is turned up by the visual inspection, perform the Spark Test provided in the Ignition System section to determine if the problem is a lack of or a weak spark. If the problem is not compression or spark, it's time to look at the fuel system.

Fuel

If compression and spark is present (and within spec), but the engine won't start or won't run properly, the only remaining condition to fulfill is fuel. As usual, start with the basics. Is the fuel tank full? Is the fuel stale? If the engine has not been run in some time (a matter of months, not weeks), there is a good chance that the fuel is stale and should be properly disposed of and replaced.

■ Depending on how stale or contaminated (with moisture) the fuel is, it may be burned in an automobile or in yard equipment, though it would be wise to mix it well with a much larger supply of fresh gasoline to prevent moving your driveability problems to that engine. But it is better to get the lawn tractor stuck on stale gasoline than it would be to have your boat engine quit in the middle of the bay or lake.

For hard-starting engines, is the primer system operating properly. Remember that the prime should only be used for **cold** starts. A true cold start is really only the first start of the day, but it may be applicable to subsequent starts on cooler days, if the engine sat for more than a few hours and completely cooled off since the last use. Applying the primer to the engine for a hot start may flood the engine, preventing it from starting properly. One method to clear a flood is to crank the engine while at wide-open throttle (allowing the maximum amount of air into the engine to compensate for the excess fuel). But, keep in mind that the throttle should be returned to idle immediately upon engine start-up to prevent damage from over-revving.

Fuel delivery and pressure should be checked before delving into the carburetor(s) or fuel injection system. Make sure there are no clogs in the fuel line or vacuum leaks that would starve the engine of fuel.

Make sure that all other possible problems have been eliminated before touching the carburetor. It is rare that a carburetor will suddenly require an adjustment in order for the engine to run properly. It is much more likely that an improperly stored engine (one stored with untreated fuel in the carburetor) would suffer from one or more clogged carburetor passages sometime after shortly returning to service. Fuel will evaporate over time, leaving behind gummy deposits. If untreated fuel is left in the carburetor for some time (again typically months more than weeks), the varnish left behind by evaporating fuel will likely clog the small passages of the carburetor and cause problems with engine performance. If you suspect this, remove and disassemble the carburetor following procedures under Fuel System.

The electronics of the fuel injection system used on some models will monitor the condition of the circuitry. Don't suspect a fuel injection problem unless the CHECK ENGINE indicator of gauge remains illuminated during engine operation. If so, refer to the information on the Fuel Injection system regarding Trouble Codes and fuel injection diagnostics.

TOOLS AND EQUIPMENT

Safety Tools

WORK GLOVES

◆ See Figure 20

Unless you think scars on your hands are cool, enjoy pain and like wearing bandages, get a good pair of work gloves. Canvas or leather gloves are the best. And yes, we realize that there are some jobs involving small parts that can't be done while wearing work gloves. These jobs are not the ones usually associated with hand injuries.

A good pair of rubber gloves (such as those usually associated with dish washing) or vinyl gloves is also a great idea. There are some liquids such as solvents and penetrants that don't belong on your skin. Avoid burns and rashes. Wear these gloves.

And lastly, an option. If you're tired of being greasy and dirty all the time, go to the drug store and buy a box of disposable latex gloves like medical professionals wear. You can handle greasy parts, perform small tasks, wash parts, etc. all without getting dirty! These gloves take a surprising amount of abuse without tearing and aren't expensive. Note however, that it has been reported that some people are allergic to the latex or the powder used inside some gloves, so pay attention to what you buy.

EYE AND EAR PROTECTION

◆ See Figures 21 and 22

Don't begin any job without a good pair of work goggles or impact resistant glasses! When doing any kind of work, it's all too easy to avoid eye injury through this simple precaution. And don't just buy eye protection and leave it on the shelf. Wear it all the time! Things have a habit of breaking, chipping, splashing, spraying, splintering and flying around. And, for some reason, your eye is always in the way!

If you wear vision-correcting glasses as a matter of routine, get a pair made with polycarbonate lenses. These lenses are impact resistant and are available at any optometrist.

Often overlooked is hearing protection. Engines and power tools are noisy! Loud noises damage your ears. It's as simple as that! The simplest and cheapest form of ear protection is a pair of noise-reducing ear plugs. Cheap insurance for your ears! And, they may even come with their own, cute little carrying case.

More substantial, more protection and more money is a good pair of noise reducing earmuffs. They protect from all but the loudest sounds. Hopefully those are sounds that you'll never encounter since they're usually associated with disasters.

WORK CLOTHES

Everyone has "work clothes." Usually these consist of old jeans and a shirt that has seen better days. That's fine. In addition, a denim work apron is a nice accessory. It's rugged, can hold some spare bolts, and you don't feel bad wiping your hands or tools on it. That's what it's for.

When working in cold weather, a one-piece, thermal work outfit is invaluable. Most are rated to below freezing temperatures and are ruggedly constructed. Just look at what local marine mechanics are wearing and that should give you a clue as to what type of clothing is good.

Chemicals

There is a whole range of chemicals that you'll find handy for maintenance and repair work. The most common types are: lubricants, penetrants and sealers. Keep these handy. There are also many chemicals that are used for detailing or cleaning.

When a particular chemical is not being used, keep it capped, upright and in a safe place. These substances may be flammable, may be irritants or might even be caustic and should always be stored properly, used properly and handled with care. Always read and follow all label directions and be sure to wear hand and eye protection!

LUBRICANTS & PENETRANTS

◆ See Figure 23

Anti-seize is used to coat certain fasteners prior to installation. This can be especially helpful when two dissimilar metals are in contact (to help prevent corrosion that might lock the fastener in place). This is a good practice on a lot of different fasteners, BUT, NOT on any fastener that might vibrate loose causing a problem. If anti-seize is used on a fastener, it should be checked periodically for proper tightness.

Lithium grease, chassis lube, silicone grease or synthetic brake caliper grease can all be used pretty much interchangeably. All can be used for coating rust-prone fasteners and for facilitating the assembly of parts that are a tight fit. Silicone and synthetic greases are the most versatile.

■ Silicone dielectric grease is a non-conductor that is often used to coat the terminals of wiring connectors before fastening them. It may sound odd to coat metal portions of a terminal with something that won't conduct electricity, but here is it how it works. When the connector is fastened the metal-to-metal contact between the terminals will displace the grease (allowing the circuit to be completed). The grease that is displaced will then coat the non-contacted surface and the cavity around the terminals, SEALING them from atmospheric moisture that could cause corrosion.

Silicone spray is a good lubricant for hard-to-reach places and parts that shouldn't be gooped up with grease.

Penetrating oil may turn out to be one of your best friends when taking something apart that has corroded fasteners. Not only can they make a job easier, they can really help to avoid broken and stripped fasteners. The most familiar penetrating oils are Liquid Wrench® and WD-40®. A newer penetrant, PB Blaster® works very well (and has become a mainstay in our shops). These products have hundreds of uses. For your purposes, they are vital!

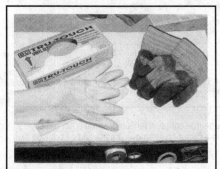
Fig. 20 Three different types of work gloves. The box contains latex gloves

Fig. 21 Don't begin major repairs without a pair of goggles for your eyes and earmuffs to protect your hearing

Fig. 22 Things have a habit of, splashing, spraying, splintering and flying around during repairs

Fig. 23 Anti-seize, penetrating oil, lithium grease, electronic cleaner and silicone spray should be a part of your chemical collection

Fig. 24 Sealants are essential for preventing leaks

Fig. 25 On some engines, RTV is used instead of gasket material to seal components

Before disassembling any part, check the fasteners. If any appear rusted, soak them thoroughly with the penetrant and let them stand while you do something else (for particularly rusted or frozen parts you may need to soak them a few days in advance). This simple act can save you hours of tedious work trying to extract a broken bolt or stud.

SEALANTS

◆ See Figures 24 and 25

Sealants are an indispensable part for certain tasks, especially if you are trying to avoid leaks. The purpose of sealants is to establish a leak-proof bond between or around assembled parts. Most sealers are used in conjunction with gaskets, but some are used instead of conventional gasket material.

The most common sealers are the non-hardening types such as Permatex® No. 2 or its equivalents. These sealers are applied to the mating surfaces of each part to be joined, then a gasket is put in place and the parts are assembled.

■ A sometimes overlooked use for sealants like RTV is on the threads of vibration prone fasteners.

One very helpful type of non-hardening sealer is the "high tack" type. This type is a very sticky material that holds the gasket in place while the parts are being assembled. This stuff is really a good idea when you don't have enough hands or fingers to keep everything where it should be.

The stand-alone sealers are the Room Temperature Vulcanizing (RTV) silicone gasket makers. On some engines, this material is used instead of a gasket. In those instances, a gasket may not be available or, because of the shape of the mating surfaces, a gasket shouldn't be used. This stuff, when used in conjunction with a conventional gasket, produces the surest bonds.

RTV does have its limitations though. When using this material, you will have a time limit. It starts to set-up within 15 minutes or so, so you have to assemble the parts without delay. In addition, when squeezing the material out of the tube, don't drop any glops into the engine. The stuff will form and set and travel around a cooling passage, possibly blocking it. Also, most types are not fuel-proof. Check the tube for all cautions.

CLEANERS

◆ See Figures 26 and 27

There are two basic types of cleaners on the market today: parts cleaners and hand cleaners. The parts cleaners are for the parts; the hand cleaners are for you. They are NOT interchangeable—please resist that urge to use parts cleaner on your hands when they are really bad!!

There are many good, non-flammable, biodegradable parts cleaners on the market. These cleaning agents are safe for you, the parts and the environment. Therefore, there is no reason to use flammable, caustic or toxic substances to clean your parts or tools.

As far as hand cleaners go; the waterless types are the best. They have always been efficient at cleaning, but they used to all leave a pretty smelly odor. Recently though, most of them have eliminated the odor and added stuff that actually smells good.

Fig. 26 Citrus hand cleaners not only work well, but they smell pretty good too. Choose one with pumice for added cleaning power

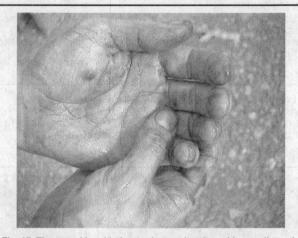

Fig. 27 The use of hand lotion seals your hands and keeps dirt and grease from sticking to your skin

Make sure that you pick one that contains lanolin or some other moisture-replenishing additive. Cleaners not only remove grease and oil but also skin oil.

■ Most women already know to use a hand lotion when you're all cleaned up. It's okay. Real men DO use hand lotion too! Believe it or not, using hand lotion BEFORE your hands are dirty will actually make them easier to clean when you're finished with a dirty job. Lotion seals your hands, and keeps dirt and grease from sticking to your skin.

TOOLS

◆ **See Figure 28**

Tools; this subject could fill a completely separate manual. The first thing you will need to ask yourself, is just how involved do you plan to get. If you are serious about maintenance and repair you will want to gather a quality set of tools to make the job easier, and more enjoyable. BESIDES, TOOLS ARE FUN!!!

Almost every do-it-yourselfer loves to accumulate tools. Though most find a way to perform jobs with only a few common tools, they tend to buy more over time, as money allows. So gathering the tools necessary for maintenance or repair does not have to be an expensive, overnight proposition.

When buying tools, the saying "You get what you pay for ..." is absolutely true! Don't go cheap! Any hand tool that you buy should be drop forged and/or chrome vanadium. These two qualities tell you that the tool is strong enough for the job. With any tool, go with a name that you've heard of before, or, that is recommended buy your local professional retailer. Let's go over a list of tools that you'll need.

Most of the world uses the metric system. However, many American-built engines and aftermarket accessories use standard fasteners. So, accumulate your tools accordingly. Any good DIYer should have a decent set of both U.S. and metric measure tools.

■ **Don't be confused by terminology. Most advertising refers to "SAE and metric", or "standard and metric." Both are misnomers. The Society of Automotive Engineers (SAE) did not invent the English system of measurement; the English did. The SAE likes metrics just fine. Both English (U.S.) and metric measurements are SAE approved. Also, the current "standard" measurement IS metric. So, if it's not metric, it's U.S. measurement.**

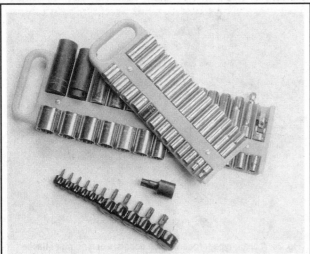

Fig. 28 Socket holders, especially the magnetic type, are handy items to keep tools in order

Hand Tools

SOCKET SETS

◆ **See Figures 29, 30, 31, 32, 33, 34 and 35**

Socket sets are the most basic hand tools necessary for repair and maintenance work. For our purposes, socket sets come in three drive sizes: 1/4 in., 3/8 in. and 1/2 in. Drive size refers to the size of the drive lug on the ratchet, breaker bar or speed handle.

A 3/8 in. set is probably the most versatile set in any mechanic's toolbox. It allows you to get into tight places that the larger drive ratchets can't and gives you a range of larger sockets that are still strong enough for heavy-duty work. The socket set that you'll need should range in sizes from 1/4 in. through 1 in. for standard fasteners, and a 6mm through 19mm for metric fasteners.

You'll need a good 1/2 in. set since this size drive lug assures that you won't break a ratchet or socket on large or heavy fasteners. Also, torque wrenches with a torque scale high enough for larger fasteners are usually 1/2 in. drive.

Plus, 1/4 in. drive sets can be very handy in tight places. Though they usually duplicate functions of the 3/8 in. set, 1/4 in. drive sets are easier to use for smaller bolts and nuts.

As for the sockets themselves, they come in shallow (standard) and deep lengths as well as 6 or 12 point. The 6 and 12 points designation refers to how many sides are in the socket itself. Each has advantages. The 6 point socket is stronger and less prone to slipping which would strip a bolt head or nut. 12 point sockets are more common, usually less expensive and can operate better in tight places where the ratchet handle can't swing far.

Standard length sockets are good for just about all jobs, however, some stud-head bolts, hard-to-reach bolts, nuts on long studs, etc., require the deep sockets.

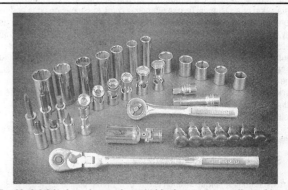

Fig. 29 A 3/8 inch socket set is probably the most versatile tool in any mechanic's tool box

Fig. 30 A swivel (U-joint) adapter (left), a 1/4 in.-to-3/8 in. adapter (center) and a 3/8 in.-to-1/4 in. adapter (right)

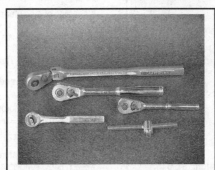

Fig. 31 Ratchets come in all sizes and configurations from rigid to swivel-headed

Fig. 32 Shallow sockets (top) are good for most jobs. But, some bolts require deep sockets (bottom)

See Fig. 33 Hex-head fasteners require a socket with a hex shaped driver

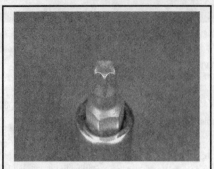

See Fig. 34 Torx® drivers . . .

See Fig. 35 . . . and tamper resistant drivers are required to remove special fasteners

Most marine manufacturers use recessed hex-head fasteners to retain many of the engine parts. These fasteners require a socket with a hex shaped driver or a large sturdy hex key. To help prevent torn knuckles, we would recommend that you stick to the sockets on any tight fastener and leave the hex keys for lighter applications. Hex driver sockets are available individually or in sets just like conventional sockets.

More and more, manufacturers are using Torx® head fasteners, which were once known as tamper resistant fasteners (because many people did not have tools with the necessary odd driver shape). Since Torx® fasteners have become commonplace in many DIYer tool boxes, manufacturers designed newer tamper resistant fasteners that are essentially Torx® head bolts that contain a small protrusion in the center (requiring the driver to contain a small hole to slide over the protrusion. Tamper resistant fasteners are often used where the manufacturer would prefer only knowledgeable mechanics or advanced Do-It-Yourselfers (DIYers) work.

Torque Wrenches

◆ See Figure 36

In most applications, a torque wrench can be used to ensure proper installation of a fastener. Torque wrenches come in various designs and most stores will carry a variety to suit your needs. A torque wrench should be used any time you have a specific torque value for a fastener. Keep in mind that because there is no worldwide standardization of fasteners, so charts or figure found in each repair section refer to the manufacturer's fasteners. Any general guideline charts that you might come across based on fastener size (they are sometimes included in a repair manual or with torque wrench packaging) should be used with caution. Just keep in mind that if you are using the right tool for the job, you should not have to strain to tighten a fastener.

Beam Type

◆ See Figures 37 and 38

The beam type torque wrench is one of the most popular styles in use. If used properly, it can be the most accurate also.

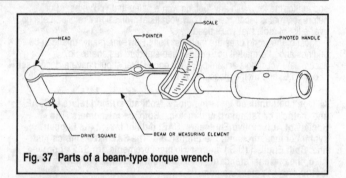

Fig. 37 Parts of a beam-type torque wrench

Fig. 38 A beam-type torque wrench consists of a pointer attached to the head that runs the length of the flexible beam (shaft) to a scale located near the handle

It consists of a pointer attached to the head that runs the length of the flexible beam (shaft) to a scale located near the handle. As the wrench is pulled, the beam bends and the pointer indicates the torque using the scale.

Click (Breakaway) Type

◆ See Figures 39 and 40

Another popular torque wrench design is the click-type. The clicking mechanism makes achieving the proper torque easy and most use a ratcheting head for ease of bolt installation. To use the click-type wrench you pre-adjust it to a torque setting. Once the torque is reached, the wrench has a reflex signaling feature that causes a momentary breakaway of the torque wrench body, sending an impulse to the operator's hand. But be careful, as continuing the turn the wrench after the momentary release will increase torque on the fastener beyond the specified setting.

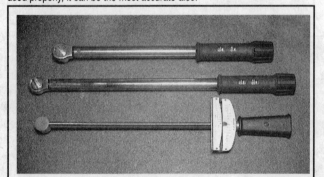

Fig. 36 Three types of torque wrenches. Top to bottom: a 3/8 in. drive beam-type that reads in inch lbs., a 1/2 in. drive clicker-type and a 1/2 in. drive beam-type

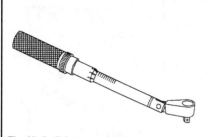

Fig. 39 A click-type or breakaway torque wrench—note this one has a pivoting head

Fig. 40 Setting the torque on a click-type wrench involves turning the handle until the specification appears on the dial

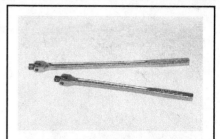

Fig. 41 Breaker bars are great for loosening large or stuck fasteners

Breaker Bars

◆ See Figure 41

Breaker bars are long handles with a drive lug. Their main purpose is to provide extra turning force when breaking loose tight bolts or nuts. They come in all drive sizes and lengths. Always take extra precautions and use the proper technique when using a breaker bar (pull on the bar, don't push, to prevent skinned knuckles).

WRENCHES

◆ See Figures 42, 43, 44, 45 and 46

Basically, there are 3 kinds of fixed wrenches: open end, box end, and combination.

Open-end wrenches have 2-jawed openings at each end of the wrench. These wrenches are able to fit onto just about any nut or bolt. They are extremely versatile but have one major drawback. They can slip on a worn or rounded bolt head or nut, causing bleeding knuckles and a useless fastener.

■ Line wrenches are a special type of open-end wrench designed to fit onto more of the fastener than standard open-end wrenches, thus reducing the chance of rounding the corners of the fastener.

Box-end wrenches have a 360° circular jaw at each end of the wrench. They come in both 6 and 12 point versions just like sockets and each type has some of the same advantages and disadvantages as sockets.

Combination wrenches have the best of both. They have a 2-jawed open end and a box end. These wrenches are probably the most versatile.

As for sizes, you'll probably need a range similar to that of the sockets, about 1/4 in. through 1 in. for standard fasteners, or 6mm through 19mm for metric fasteners. As for numbers, you'll need 2 of each size, since, in many instances, one wrench holds the nut while the other turns the bolt. On most fasteners, the nut and bolt are the same size so having two wrenches of the same size comes in handy.

■ Although you will typically just need the sizes we specified, there are some exceptions. Occasionally you will find a nut that is larger. For these, you will need to buy ONE expensive wrench or a very large adjustable. Or you can always just convince the spouse that we are talking about SAFETY here and buy a whole (read expensive) large wrench set.

One extremely valuable type of wrench is the adjustable wrench. An adjustable wrench has a fixed upper jaw and a moveable lower jaw. The lower jaw is moved by turning a threaded drum. The advantage of an adjustable wrench is its ability to be adjusted to just about any size fastener.

The main drawback of an adjustable wrench is the lower jaw's tendency to move slightly under heavy pressure. This can cause the wrench to slip if it is not facing the right way.

INCHES	DECIMAL		DECIMAL	MILLIMETERS
1/8″	.125		.118	3mm
3/16″	.187		.157	4mm
1/4″	.250		.236	6mm
5/16″	.312		.354	9mm
3/8″	.375		.394	10mm
7/16″	.437		.472	12mm
1/2″	.500		.512	13mm
9/16″	.562		.590	15mm
5/8″	.625		.630	16mm
11/16″	.687		.709	18mm
3/4″	.750		.748	19mm
13/16″	.812		.787	20mm
7/8″	.875		.866	22mm
15/16″	.937		.945	24mm
1″	1.00		.984	25mm

Fig. 42 Comparison of U.S. measure and metric wrench sizes

Fig. 43 Always use a back-up wrench to prevent rounding flare nut fittings

Fig. 44 Note how the flare wrench sides are extended to grip the fitting tighter and prevent rounding

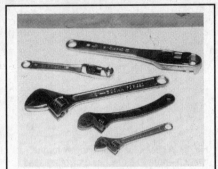

Fig. 45 Several types and sizes of adjustable wrenches

Pulling on an adjustable wrench in the proper direction will cause the jaws to lock in place. Adjustable wrenches come in a large range of sizes, measured by the wrench length.

PLIERS

◆ See Figure 47

Pliers are simply mechanical fingers. They are, more than anything, an extension of your hand. At least 3 pairs of pliers are an absolute necessity—standard, needle nose and slip-joint.

In addition to standard pliers there are the slip-joint, multi-position pliers such as ChannelLock® pliers and locking pliers, such as Vise Grips®.

Slip-joint pliers are extremely valuable in grasping oddly sized parts and fasteners. Just make sure that you don't use them instead of a wrench too often since they can easily round off a bolt head or nut.

Locking pliers are usually used for gripping bolts or studs that can't be removed conventionally. You can get locking pliers in square jawed, needle-nosed and pipe-jawed. Locking pliers can rank right up behind duct tape as the handy-man's best friend.

SCREWDRIVERS

You can't have too many screwdrivers. They come in 2 basic flavors, either standard or Phillips. Standard blades come in various sizes and thickness for all types of slotted fasteners. Phillips screwdrivers come in sizes with number designations from 1 on up, with the lower number designating the smaller size. Screwdrivers can be purchased separately or in sets.

HAMMERS

◆ See Figure 48

You need a hammer for just about any kind of work. You need a ball-peen hammer for most metal work when using drivers and other like tools. A plastic hammer comes in handy for hitting things safely. A soft-faced dead-blow hammer is used for hitting things safely and hard. Hammers are also VERY useful with non air-powered impact drivers.

Other Common Tools

There are a lot of other tools that every DIYer will eventually need (though not all for basic maintenance). They include:

- Funnels
- Chisels
- Punches
- Files
- Hacksaw
- Portable Bench Vise
- Tap and Die Set
- Flashlight
- Magnetic Bolt Retriever
- Gasket scraper
- Putty Knife
- Screw/Bolt Extractors
- Prybars

Hacksaws have just one use—cutting things off. You may wonder why you'd need one for something as simple as maintenance or repair, but you never know. Among other things, guide studs to ease parts installation can be made from old bolts with their heads cut off.

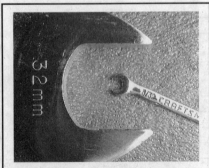

Fig. 46 You may find a nut that requires a particularly large or small wrench (it is usually available at your local tool store

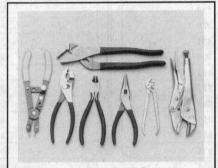

Fig. 47 Pliers and cutters come in many shapes and sizes. You should have an assortment on hand

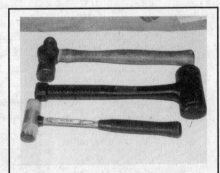

Fig. 48 Three types of hammers. Top to bottom: ball peen, rubber dead-blow, and plastic

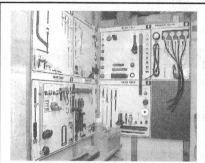

Fig. 49 Almost every marine engine around today requires at least one special tool to perform a certain task

Fig. 50 The Battery Tender® is more than just a battery charger; when left connected it keeps your battery fully charged

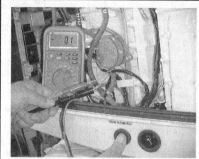

Fig. 51 Multimeters, such as this one from UEI, are an extremely useful tool for troubleshooting electrical problems

A tap and die set might be something you've never needed, but you will eventually. It's a good rule, when everything is apart, to clean-up all threads, on bolts, screws or threaded holes. Also, you'll likely run across a situation in which you will encounter stripped threads. The tap and die set will handle that for you.

Gasket scrapers are just what you'd think, tools made for scraping old gasket material off of parts. You don't absolutely need one. Old gasket material can be removed with a putty knife or single edge razor blade. However, putty knives may not be sharp enough for some really stubborn gaskets and razor blades have a knack of breaking just when you don't want them to, inevitably slicing the nearest body part! As the old saying goes, "always use the proper tool for the job". If you're going to use a razor to scrape a gasket, be sure to always use a blade holder.

Putty knives really do have a use in a repair shop. Just because you remove all the bolts from a component sealed with a gasket doesn't mean it's going to come off. Most of the time, the gasket and sealer will hold it tightly. Lightly driving a putty knife at various points between the two parts will break the seal without damage to the parts.

A small — 8–10 inches (20–25 centimeters) long — prybar is extremely useful for removing stuck parts.

■ Never use a screwdriver as a prybar! Screwdrivers are not meant for prying. Screwdrivers, used for prying, can break, sending the broken shaft flying!

Screw/bolt extractors are used for removing broken bolts or studs that have broke off flush with the surface of the part.

Special Tools

◆ See Figure 49

Almost every marine engine around today requires at least one special tool to perform a certain task. In most cases, these tools are specially designed to overcome some unique problem or to fit on some oddly sized component.

When manufacturers go through the trouble of making a special tool, it is usually necessary to use it to ensure that the job will be done right. A special tool might be designed to make a job easier, or it might be used to keep you from damaging or breaking a part.

Don't worry, MOST maintenance procedures can either be performed without any special tools OR, because the tools must be used for such basic things, they are commonly available for a reasonable price. It is usually just the low production, highly specialized tools (like a super thin 7-point star-shaped socket capable of 150 ft. lbs. (203 Nm) of torque that is used only on the crankshaft nut of the limited production what-dya-callit engine) that tend to be outrageously expensive and hard to find. Hopefully, you will probably never need such a tool.

Special tools can be as inexpensive and simple as an adjustable strap wrench or as complicated as an ignition tester. A few common specialty tools are listed here, but check with your dealer or with other boaters for help in determining if there are any special tools for YOUR particular engine. There is an added advantage in seeking advice from others, chances are they may have already found the special tool you will need, and know how to get it cheaper (or even let you borrow it).

Electronic Tools

BATTERY TESTERS

The best way to test a non-sealed battery is using a hydrometer to check the specific gravity of the acid. Luckily, these are usually inexpensive and are available at most parts stores. Just be careful because the larger testers are usually designed for larger batteries and may require more acid than you will be able to draw from the battery cell. Smaller testers (usually a short, squeeze bulb type) will require less acid and should work on most batteries.

Electronic testers are available and are often necessary to tell if a sealed battery is usable. Luckily, many parts stores have them on hand and are willing to test your battery for you.

BATTERY CHARGERS

◆ See Figure 50

If you are a weekend boater and take your boat out every week, then you will most likely want to buy a battery charger to keep your battery fresh. There are many types available, from low amperage trickle chargers to electronically controlled battery maintenance tools that monitor the battery voltage to prevent over or undercharging. This last type is especially useful if you store your boat for any length of time (such as during the severe winter months found in many Northern climates).

Even if you use your boat on a regular basis, you will eventually need a battery charger. The charger should be used anytime the boat is going to be in storage for more than a few weeks or so. Never leave the dock or loading ramp without a battery that is fully charged.

Also, some batteries are shipped dry and in a partial charged state. Before placing a new battery of this type into service it must be filled and properly charged. Failure to properly charge a battery (which was shipped dry) before it is put into service will prevent it from ever reaching a fully charged state.

MULTIMETERS (DVOMS)

◆ See Figure 51

Multimeters or Digital Volt Ohmmeter (DVOMs) are an extremely useful tool for troubleshooting electrical problems. They can be purchased in analog or digital form and have a price range to suit any budget. A multimeter is a voltmeter, ammeter and ohmmeter (along with other features) combined into one instrument. It is often used when testing solid state circuits because of its high input impedance (usually 10 megaohms or more). A brief description of the multimeter main test functions follows:

• Voltmeter—the voltmeter is used to measure voltage at any point in a circuit, or to measure the voltage drop across any part of a circuit. Voltmeters usually have various scales and a selector switch to allow the reading of different voltage ranges. The voltmeter has a positive and a negative lead. To avoid the possibility of damage to the meter, whenever possible, connect the negative lead to the negative (-) side of the circuit (to ground or nearest the ground side of the circuit) and connect the positive lead to the positive (+) side of the circuit (to the power source or the nearest power source). Luckily, most quality DVOMs can adjust their own polarity internally and will indicate (without damage) if the leads are reversed.

Note that the negative voltmeter lead will always be black and that the positive voltmeter will always be some color other than black (usually red).

• Ohmmeter—the ohmmeter is designed to read resistance (measured in ohms) in a circuit or component. Most ohmmeters will have a selector switch which permits the measurement of different ranges of resistance (usually the selector switch allows the multiplication of the meter reading by 10, 100, 1,000 and 10,000). Some ohmmeters are "auto-ranging" which means the meter itself will determine which scale to use. Since the meters are powered by an internal battery, the ohmmeter can be used like a self-powered test light. When the ohmmeter is connected, current from the ohmmeter flows through the circuit or component being tested. Since the ohmmeter's internal resistance and voltage are known values, the amount of current flow through the meter depends on the resistance of the circuit or component being tested. The ohmmeter can also be used to perform a continuity test for suspected open circuits. In using the meter for making continuity checks, do not be concerned with the actual resistance readings. Zero resistance, or any ohm reading, indicates continuity in the circuit. Infinite resistance indicates an opening in the circuit. A high resistance reading where there should be little or none indicates a problem in the circuit. Checks for short circuits are made in the same manner as checks for open circuits, except that the circuit must be isolated from both power and normal ground. Infinite resistance indicates no continuity, while zero resistance indicates a dead short.

✳ WARNING

Never use an ohmmeter to check the resistance of a component or wire while there is voltage applied to the circuit.

• Ammeter—an ammeter measures the amount of current flowing through a circuit in units called amperes or amps. At normal operating voltage, most circuits have a characteristic amount of amperes, called "current draw" which can be measured using an ammeter. By referring to a specified current draw rating, then measuring the amperes and comparing the two values, one can determine what is happening within the circuit to aid in diagnosis. An open circuit, for example, will not allow any current to flow, so the ammeter reading will be zero. A damaged component or circuit will have an increased current draw, so the reading will be high. The ammeter is always connected in series with the circuit being tested. All of the current that normally flows through the circuit must also flow through the ammeter; if there is any other path for the current to follow, the ammeter reading will not be accurate. The ammeter itself has very little resistance to current flow and, therefore, will not affect the circuit, but, it will measure current draw only when the circuit is closed and electricity is flowing. Excessive current draw can blow fuses and drain the battery, while a reduced current draw can cause engines to run slowly, lights to dim and other components to not operate properly.

GAUGES

Compression Gauge

◆ See Figure 52

An important element in checking the overall condition of your engine is to check compression. This becomes increasingly more important on engines with high hours. Compression gauges are available as screw-in types and hold-in types. The screw-in type is slower to use, but eliminates the possibility of a faulty reading due to pressure escaping by the seal. A compression reading will uncover many problems that can cause rough running. Normally, these are not the sort of problems that can be cured by a tune-up.

Vacuum Gauge

◆ See Figures 53 and 54

Measuring Tools

Eventually, you are going to have to measure something. To do this, you will need at least a few precision tools.

MICROMETERS & CALIPERS

Micrometers and calipers are devices used to make extremely precise measurements. The simple truth is that you really won't have the need for many of these items just for routine maintenance. But, measuring tools, such as an outside caliper can be handy during repairs. And, if you decide to tackle a major overhaul, a micrometer will absolutely be necessary.

Should you decide on becoming more involved in boat engine mechanics, such as repair or rebuilding, then these tools will become very important. The success of any rebuild is dependent, to a great extent on the ability to check the size and fit of components as specified by the manufacturer. These measurements are often made in thousandths and ten-thousandths of an inch.

Micrometers

◆ See Figure 55

A micrometer is an instrument made up of a precisely machined spindle that is rotated in a fixed nut, opening and closing the distance between the end of the spindle and a fixed anvil. When measuring using a micrometer, don't overtighten the tool on the part as either the component or tool may be damaged, and either way, an incorrect reading will result. Most micrometers are equipped with some form of thumbwheel on the spindle that is designed to freewheel over a certain light touch (automatically adjusting the spindle and preventing it from overtightening).

Outside micrometers can be used to check the thickness of parts such shims or the outside diameter of components like the crankshaft journals. They are also used during many rebuild and repair procedures to measure the diameter of components such as the pistons. The most common type of micrometer reads in 1/1000 of an inch. Micrometers that use a vernier scale can estimate to 1/10 of an inch.

Inside micrometers are used to measure the distance between two parallel surfaces. For example, in engine rebuilding work, the "inside mike" measures cylinder bore wear and taper. Inside mikes are graduated the same way as outside mikes and are read the same way as well.

Remember that an inside mike must be absolutely perpendicular to the work being measured. When you measure with an inside mike, rock the mike gently from side to side and tip it back and forth slightly so that you span the widest part of the bore. Just to be on the safe side, take several readings. It takes a certain amount of experience to work any mike with confidence.

Metric micrometers are read in the same way as inch micrometers, except that the measurements are in millimeters. Each line on the main scale equals 1mm. Each fifth line is stamped 5, 10, 15 and so on. Each line on the thimble scale equals 0.01 mm. It will take a little practice, but if you can read an inch mike, you can read a metric mike.

Fig. 52 Cylinder compression test results are extremely valuable indicators of internal engine condition

Fig. 53 Vacuum gauges are useful for troubleshooting including testing some fuel pumps

Fig. 54 You can also use the vacuum gauge on a hand-operated vacuum pump for tests

Fig. 55 Outside micrometers measure the thickness of parts like shims or the diameter of a shaft

Fig. 56 Calipers are the fast and easy way to make precise measurements

Fig. 57 Calipers can also be used to measure depth . . .

Fig. 58 . . . and inside diameter measurements, usually to 0.001 inch accuracy

Fig. 59 This dial indicator is measuring the end-play of a crankshaft during an engine rebuild

Fig. 60 Telescoping gauges are used during engine rebuilding procedures to measure the inside diameter of bores

Calipers

◆ **See Figures 56, 57 and 58**

Inside and outside calipers are useful devices to have if you need to measure something quickly and absolute precise measurement is not necessary. Simply take the reading and then hold the calipers on an accurate steel rule. Calipers, like micrometers, will often contain a thumbwheel to help ensure accurate measurement.

DIAL INDICATORS

◆ **See Figure 59**

A dial indicator is a gauge that utilizes a dial face and a needle to register measurements. There is a movable contact arm on the dial indicator. When the arms moves, the needle rotates on the dial. Dial indicators are calibrated to show readings in thousandths of an inch and typically, are used to measure end-play and runout on various shafts and other components.

Dial indicators are quite easy to use, although they are relatively expensive. A variety of mounting devices are available so that the indicator can be used in a number of situations. Make certain that the contact arm is always parallel to the movement of the work being measured.

TELESCOPING GAUGES

◆ **See Figure 60**

A telescope gauge is really only used during rebuilding procedures (NOT during basic maintenance or routine repairs) to measure the inside of bores. It can take the place of an inside mike for some of these jobs. Simply insert the gauge in the hole to be measured and lock the plungers after they have contacted the walls. Remove the tool and measure across the plungers with an outside micrometer.

DEPTH GAUGES

◆ **See Figure 61**

A depth gauge can be inserted into a bore or other small hole to determine exactly how deep it is. One common use for a depth gauge is measuring the distance the piston sits below the deck of the block at top dead center. Some outside calipers contain a built-in depth gauge so you can save money and buy just one tool.

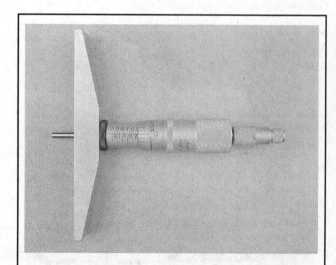

Fig. 61 Depth gauges are used to measure the depth of bore or other small holes

FASTENERS, MEASUREMENTS AND CONVERSIONS

Bolts, Nuts and Other Threaded Retainers

◆ See Figures 62 and 63

Although there are a great variety of fasteners found in the modern boat engine, the most commonly used retainer is the threaded fastener (nuts, bolts, screws, studs, etc). Most threaded retainers may be reused, provided that they are not damaged in use or during the repair.

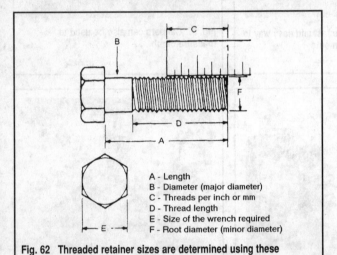

A - Length
B - Diameter (major diameter)
C - Threads per inch or mm
D - Thread length
E - Size of the wrench required
F - Root diameter (minor diameter)

Fig. 62 Threaded retainer sizes are determined using these measurements

■ Some retainers (such as stretch bolts or torque prevailing nuts) are designed to deform when tightened or in use and should not be reused.

Whenever possible, we will note any special retainers which should be replaced during a procedure. But you should always inspect the condition of a retainer when it is removed and you should replace any that show signs of damage. Check all threads for rust or corrosion that can increase the torque necessary to achieve the desired clamp load for which that fastener was originally selected. Additionally, be sure that the driver surface itself (on the fastener) is not compromised from rounding or other damage. In some cases a driver surface may become only partially rounded, allowing the driver to catch in only one direction. In many of these occurrences, a fastener may be installed and tightened, but the driver would not be able to grip and loosen the fastener again. (This could lead to frustration down the line should that component ever need to be disassembled again).

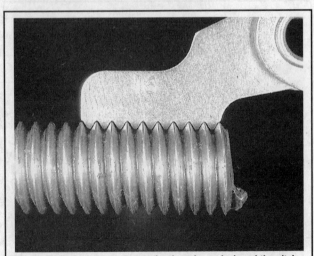

Fig. 63 Thread gauges measure the threads-per-inch and the pitch of a bolt or stud's threads

If you must replace a fastener, whether due to design or damage, you must always be sure to use the proper replacement. In all cases, a retainer of the same design, material and strength should be used. Markings on the heads of most bolts will help determine the proper strength of the fastener. The same material, thread and pitch must be selected to assure proper installation and safe operation of the engine afterwards.

Thread gauges are available to help measure a bolt or stud's thread. Most part or hardware stores keep gauges available to help you select the proper size. In a pinch, you can use another nut or bolt for a thread gauge. If the bolt you are replacing is not too badly damaged, you can select a match by finding another bolt that will thread in its place. If you find a nut that will thread properly onto the damaged bolt, then use that nut as a gauge to help select the replacement bolt. If however, the bolt you are replacing is so badly damaged (broken or drilled out) that its threads cannot be used as a gauge, you might start by looking for another bolt (from the same assembly or a similar location) which will thread into the damaged bolt's mounting. If so, the other bolt can be used to select a nut; the nut can then be used to select the replacement bolt.

In all cases, be absolutely sure you have selected the proper replacement. Don't be shy, you can always ask the store clerk for help.

✳✳ WARNING

Be aware that when you find a bolt with damaged threads, you may also find the nut or tapped bore into which it was threaded has also been damaged. If this is the case, you may have to drill and tap the hole, replace the nut or otherwise repair the threads. Never try to force a replacement bolt to fit into the damaged threads.

Torque

Torque is defined as the measurement of resistance to turning or rotating. It tends to twist a body about an axis of rotation. A common example of this would be tightening a threaded retainer such as a nut, bolt or screw. Measuring torque is one of the most common ways to help assure that a threaded retainer has been properly fastened.

When tightening a threaded fastener, torque is applied in three distinct areas, the head, the bearing surface and the clamp load. About 50 percent of the measured torque is used in overcoming bearing friction. This is the friction between the bearing surface of the bolt head, screw head or nut face and the base material or washer (the surface on which the fastener is rotating). Approximately 40 percent of the applied torque is used in overcoming thread friction. This leaves only about 10 percent of the applied torque to develop a useful clamp load (the force that holds a joint together). This means that friction can account for as much as 90 percent of the applied torque on a fastener.

Standard and Metric Measurements

Specifications are often used to help you determine the condition of various components, or to assist you in their installation. Some of the most common measurements include length (in. or cm/mm), torque (ft. lbs., inch lbs. or Nm) and pressure (psi, in. Hg, kPa or mm Hg).

In some cases, that value may not be conveniently measured with what is available in your toolbox. Luckily, many of the measuring devices that are available today will have two scales so U.S. or Metric measurements may easily be taken. If any of the various measuring tools that are available to you do not contain the same scale as listed in your specifications, use the accompanying conversion factors to determine the proper value.

The conversion factor chart is used by taking the given specification and multiplying it by the necessary conversion factor. For instance, looking at the first line, if you have a measurement in inches such as "free-play should be 2 in." but your ruler reads only in millimeters, multiply 2 in. by the conversion factor of 25.4 to get the metric equivalent of 50.8mm. Likewise, if a specification was given only in a Metric measurement, for example in Newton Meters (Nm), then look at the center column first. If the measurement is 100 Nm, multiply it by the conversion factor of 0.738 to get 73.8 ft. lbs.

CONVERSION FACTORS

LENGTH–DISTANCE

Inches (in.)	x 25.4	= Millimeters (mm)	x .0394	= Inches
Feet (ft.)	x .305	= Meters (m)	x 3.281	= Feet
Miles	x 1.609	= Kilometers (km)	x .0621	= Miles

VOLUME

Cubic Inches (in3)	x 16.387	= Cubic Centimeters	x .061	= in3
IMP Pints (IMP pt.)	x .568	= Liters (L)	x 1.76	= IMP pt.
IMP Quarts (IMP qt.)	x 1.137	= Liters (L)	x .88	= IMP qt.
IMP Gallons (IMP gal.)	x 4.546	= Liters (L)	x .22	= IMP gal.
IMP Quarts (IMP qt.)	x 1.201	= US Quarts (US qt.)	x .833	= IMP qt.
IMP Gallons (IMP gal.)	x 1.201	= US Gallons (US gal.)	x .833	= IMP gal.
Fl. Ounces	x 29.573	= Milliliters	x .034	= Ounces
US Pints (US pt.)	x .473	= Liters (L)	x 2.113	= Pints
US Quarts (US qt.)	x .946	= Liters (L)	x 1.057	= Quarts
US Gallons (US gal.)	x 3.785	= Liters (L)	x .264	= Gallons

MASS–WEIGHT

Ounces (oz.)	x 28.35	= Grams (g)	x .035	= Ounces
Pounds (lb.)	x .454	= Kilograms (kg)	x 2.205	= Pounds

PRESSURE

Pounds Per Sq. In. (psi)	x 6.895	= Kilopascals (kPa)	x .145	= psi
Inches of Mercury (Hg)	x .4912	= psi	x 2.036	= Hg
Inches of Mercury (Hg)	x 3.377	= Kilopascals (kPa)	x .2961	= Hg
Inches of Water (H_2O)	x .07355	= Inches of Mercury	x 13.783	= H_2O
Inches of Water (H_2O)	x .03613	= psi	x 27.684	= H_2O
Inches of Water (H_2O)	x .248	= Kilopascals (kPa)	x 4.026	= H_2O

TORQUE

Pounds–Force Inches (in–lb)	x .113	= Newton Meters (N·m)	x 8.85	= in–lb
Pounds–Force Feet (ft–lb)	x 1.356	= Newton Meters (N·m)	x .738	= ft–lb

VELOCITY

Miles Per Hour (MPH)	x 1.609	= Kilometers Per Hour (KPH)	x .621	= MPH

POWER

Horsepower (Hp)	x .745	= Kilowatts	x 1.34	= Horsepower

FUEL CONSUMPTION*

Miles Per Gallon IMP (MPG)	x .354	= Kilometers Per Liter (Km/L)
Kilometers Per Liter (Km/L)	x 2.352	= IMP MPG
Miles Per Gallon US (MPG)	x .425	= Kilometers Per Liter (Km/L)
Kilometers Per Liter (Km/L)	x 2.352	= US MPG

*It is common to covert from miles per gallon (mpg) to liters/100 kilometers (1/100 km), where mpg (IMP) x 1/100 km = 282 and mpg (US) x 1/100 km = 235.

TEMPERATURE

Degree Fahrenheit (°F)	= (°C x 1.8) + 32
Degree Celsius (°C)	= (°F – 32) x .56

Metric Bolts

Relative Strength Marking	4.6, 4.8			8.8		
Bolt Markings						
Usage	Frequent			Infrequent		
Bolt Size	Maximum Torque			Maximum Torque		
Thread Size x Pitch (mm)	Ft-Lb	Kgm	Nm	Ft-Lb	Kgm	Nm
6 x 1.0	2–3	.2–.4	3–4	3–6	.4–.8	5–8
8 x 1.25	6–8	.8–1	8–12	9–14	1.2–1.9	13–19
10 x 1.25	12–17	1.5–2.3	16–23	20–29	2.7–4.0	27–39
12 x 1.25	21–32	2.9–4.4	29–43	35–53	4.8–7.3	47–72
14 x 1.5	35–52	4.8–7.1	48–70	57–85	7.8–11.7	77–110
16 x 1.5	51–77	7.0–10.6	67–100	90–120	12.4–16.5	130–160
18 x 1.5	74–110	10.2–15.1	100–150	130–170	17.9–23.4	180–230
20 x 1.5	110–140	15.1–19.3	150–190	190–240	26.2–46.9	160–320
22 x 1.5	150–190	22.0–26.2	200–260	250–320	34.5–44.1	340–430
24 x 1.5	190–240	26.2–46.9	260–320	310–410	42.7–56.5	420–550

SAE Bolts

SAE Grade Number	1 or 2			5			6 or 7		
Bolt Markings Manufacturers' marks may vary—number of lines always two less than the grade number.									
Usage	Frequent			Frequent			Infrequent		
Bolt Size (inches)—(Thread)	Maximum Torque			Maximum Torque			Maximum Torque		
	Ft-Lb	kgm	Nm	Ft-Lb	kgm	Nm	Ft-Lb	kgm	Nm
1/4—20	5	0.7	6.8	8	1.1	10.8	10	1.4	13.5
—28	6	0.8	8.1	10	1.4	13.6			
5/16—18	11	1.5	14.9	17	2.3	23.0	19	2.6	25.8
—24	13	1.8	17.6	19	2.6	25.7			
3/8—16	18	2.5	24.4	31	4.3	42.0	34	4.7	46.0
—24	20	2.75	27.1	35	4.8	47.5			
7/16—14	28	3.8	37.0	49	6.8	66.4	55	7.6	74.5
—20	30	4.2	40.7	55	7.6	74.5			
1/2—13	39	5.4	52.8	75	10.4	101.7	85	11.75	115.2
—20	41	5.7	55.6	85	11.7	115.2			
9/16—12	51	7.0	69.2	110	15.2	149.1	120	16.6	162.7
—18	55	7.6	74.5	120	16.6	162.7			
5/8—11	83	11.5	112.5	150	20.7	203.3	167	23.0	226.5
—18	95	13.1	128.8	170	23.5	230.5			
3/4—10	105	14.5	142.3	270	37.3	366.0	280	38.7	379.6
—16	115	15.9	155.9	295	40.8	400.0			
7/8— 9	160	22.1	216.9	395	54.6	535.5	440	60.9	596.5
—14	175	24.2	237.2	435	60.1	589.7			
1— 8	236	32.5	318.6	590	81.6	799.9	660	91.3	894.8
—14	250	34.6	338.9	660	91.3	849.8			

2

ENGINE AND DRIVE MAINTENANCE

ENGINE AND DRIVE MAINTENANCE

Overview

At Seloc, we estimate that 75% of engine repair work can be directly or indirectly attributed to lack of proper care for the engine. This is especially true of care during the off-season period. There is no way on this green earth for a mechanical engine to be left sitting idle for an extended period of time, say for six months, and then be ready for instant satisfactory service.

Imagine, if you will, leaving your car or truck for six months, and then expecting to turn the key, having it roar to life, and being able to drive off in the same manner as a daily occurrence.

Therefore it is critical for an engine to either be run (at least once a month), preferably, in the water and properly maintained between uses or for it to be specifically prepared for storage and serviced again immediately before the start of the season.

Only through a regular maintenance program can the owner expect to receive long life and satisfactory performance at minimum cost.

Many times, if an engine is not performing properly, the owner will "nurse" it through the season with good intentions of working on the unit once it is no longer being used. As with many New Year's resolutions, the good intentions are not completed and the engine may lie for many months before the work is begun or the unit is taken to the marine shop for repair.

Imagine, if you will, the cause of the problem being a blown head gasket. And let us assume water has found its way into a cylinder. This water, allowed to remain over a long period of time, will do considerably more damage than it would have if the unit had been disassembled and the repair work performed immediately. Therefore, if an engine is not functioning properly, do not stow it away with promises to get at it when you get time, because the work and expense will only get worse, the longer corrective action is postponed. In the example of the blown head gasket, a relatively simple and inexpensive repair job could very well develop into major overhaul and rebuild work.

OK, perhaps no one thing that we do as boaters will protect us from risks involved with enjoying the wind and the water on a powerboat. But, each time we perform maintenance on our boat or motor, we increase the likelihood that we will find a potential hazard before it becomes a problem. Each time we inspect our boat and motor, we decrease the possibility that it could leave us stranded on the water.

In this way, performing boat and engine service is one of the most important ways that we, as boaters, can help protect ourselves, our boats, and the friends and family that we bring aboard.

Serial Number Identification

An engine specifications decal can generally be found on top of the flame arrestor, on the side of the thermostat housing, or on the top/side of the rocker arm cover, usually near the breather/PCV line (most models); all pertinent serial number information can be found here—engine and drive designations, serial numbers and model numbers. Unfortunately this decal is not always legible on older engines and it can be quite difficult to find, so please refer to the following procedures for each individual unit's serial number location.

■ Serial numbers tags are frequently difficult to see when the engine is installed in the boat; a mirror can be a handy way to read all the numbers.

ENGINE

◆ See Figures, 1, 2, 3, 4, 5, 6 and 7

The engine serial numbers are the manufacturer's key to engine changes. These alpha-numeric codes identify the year of manufacture, the horsepower rating and various model/option differences. If any correspondence or parts are required, the engine serial number must be used for proper identification.

Remember that the serial number establishes the year in which the engine was produced, which is often not the year of first installation.

The engine specifications decal contains information such as the model number or code, the serial number (a unique sequential identifier given ONLY to that one engine) as well as other useful information.

An engine specifications decal can generally be found on top of the flame arrestor, on the side of the thermostat housing (early V6/V8 engines), or on the inner side of the rocker arm cover, usually near the breather/PCV line (port side on most models) - all pertinent serial number information can be found here—engine and drive designations, serial numbers and model

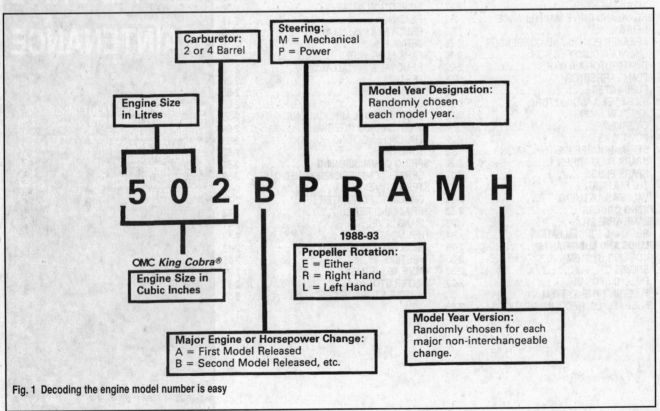

Fig. 1 Decoding the engine model number is easy

numbers. Unfortunately this decal is not always legible on older boats and it's also quite difficult to find, so please refer to the following procedures for each individuals unit's serial number location.

■ **Serial numbers tags are frequently difficult to see when the engine is installed in the boat; a mirror can be a handy way to read all the numbers.**

The engine serial/model number is sometimes also stamped on the port rear side of the engine where it attaches to the bell housing; although on most later models it may instead be a metal plate attached in the same location. If your engine has a stamped number it will simply be the serial number; if you have a plate (and you should), it will always show a Model number and then the actual Serial number. Additionally, most models will also have this plate or sticker on the transom bracket.

• The first two characters identify the engine size in liters (L); **30** represents the 3.0L, **43** represents the 4.3L, **50** represents the 5.0L and so forth. Please note though, that there were a few King Cobra engines that listed the engine size in cubic inches; the 1989 460 for example used the first three characters and showed **460** rather than the **75** that it would normally show.

• The third character identifies the fuel delivery system; **2** designates a 2 bbl carburetor, **4** is a 4 bbl carburetor, and **F** is a fuel injected engine. Remember those King Cobras we just discussed? If the engine designation is in cubic inches rather than liters, the third character will not be used to designate the type of fuel delivery.

• The fourth character designates a major engine or horsepower change—it doesn't let you know what the change was, just that there was some sort of change. **A** means it is the first model released, **B** would be the second, and so forth

• The fifth character designates what type of steering system was used; **M** would be manual steering and **P** would be power steering.

• Now here's where it gets interesting; on 1986-87 engines and 1994-98 engines, the sixth, seventh and eighth characters designate the model year. The sixth and seventh actually show the model year, while the eighth is a random model year version code. **KWB** and **WXS** represent 1986; and **ARJ**, **ARF**, **FTC**, **SRC** or **SRY** show 1987. **MDA** is 1994, **HUB** is 1995, **NCA** is 1996, **LKD** is 1997 and **BYC** is 1998.

• On 1988-93 engines, the sixth character designates the direction of propeller rotation. **R** is right hand, **L** is left hand and **E** is either.

• Also on 1988-93 engines, the seventh, eighth and ninth characters designate the model year. The seventh and eighth actually show the model year, while the nineth is a random model year version code. **GDE** or **GDP** is 1988, **MED** or **MEF** is 1989, **PWC**, **PWR** or **PWS** is 1990, **RGD** or **RGF** is 1991, **AMH** or **AMK** is 1992 and **JVB** or **JVN** is 1993.

Any remaining characters are proprietary. So in example, a Model number on the ID plate that reads **574AMFTC** would designate a 1987 5.7L engine with a 4 bbl carburetor and manual steering, first model released. A number reading **58FAPRJVB** would designate a 1993 5.8L engine with fuel injection, power steering and a right hand propeller, first release; get the picture?

STERN DRIVE

◆ **See Figures 8 and 9**

The stern drive serial/model number plate can be found on the center port side of the unit, just under and beneath the trim/tilt cylinder. Both the serial number and the gear ratio should be on the tag. Make sure you don't confuse the two! You may also find the model number included on the transom assembly tag.

TRANSOM ASSEMBLY

◆ **See Figure 9**

The transom assembly serial number decal/plate can be found on the upper starboard end of the unit.

Flame Arrestor

In a marine engine compartment, the minimal amount of dust and dirt in the air mean that a marine air filter requires less maintenance than its counterpart in the automotive world. However, the maintenance of a marine air filter is equally important.

Fig. 2 Engine serial number sticker—3.0L engines (2.3L and 2.5L engine similar)

Fig. 3 Engine serial number sticker—4.3L engines

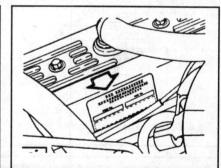

Fig. 4 Engine serial number sticker—V6 and V8 engines

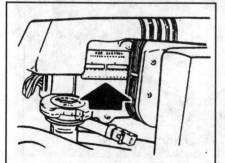

Fig. 5 The engine serial number sticker on injected Ford motors can be found on the Thick Film module bracket engines

Fig. 6 On many early engines, the sticker is found on the thermostat housing

Fig. 7 You should also be able to find a plate on the transom bracket

The marine filter prevents dirt from entering the engine. This lessens oil/fuel consumption and extends the engine's life. The air filter on some engines is also used as an intake silencer to quiet the intake air sound as it rushes into the cylinder head from the intake ports.

Over time, the air filter element will become clogged with dirt and oil, decreasing the amount of air entering the engine and lowering engine output. If an excessive amount of oil is clogging the filter, this could be an indication of worn cylinders or piston ring failure causing high pressure in the crankcase.

The maintenance interval for flame arrestor cleaning is at the end of the first boating season, and then every 100 hours of engine operation or once a year, whichever comes first on 1986-90 engines. On 1991-98 engines, the interval is decreased to every 50 hours of engine operation or once a season, whichever comes first.

REMOVAL & INSTALLATION

All Models W/Carburetor Or TBI

◆ **See Figures 10, 11, 12, 13 and 14**

1. Remove or open the engine compartment cover.
2. Many models utilize a plastic cover over the flame arrestor—if so equipped, remove the retaining nut and lift off the cover.
3. Tag and disconnect the crankcase ventilation hose and bracket from the arrestor and the rocker arm cover. TBI models and later carbureted models will have two hoses and brackets.
2. Remove the nut and washer securing the flame arrestor cover to the carburetor.
3. Lift off the flame arrestor. Unscrew the stud if necessary.

■ **Early 2.5L and 3.0L engine utilize a gasket between the arrestor and the carburetor—remove this gasket and discard it.**

4. On TBI models, lift the injector baffle off of the stud.
To install:
5. Clean the arrestor in solvent and dry with compressed air if possible; otherwise make sure that it dries completely by air. Clean the hose(s) and then inspect them for cracks or deterioration. Replace if necessary.
6. Install the injector baffle if removed.
7. If you remove the stud, thread it into the carburetor (or throttle body) and tighten it to 65-80 inch lbs. (7.3-9 Nm).
8. Install a new gasket onto the carburetor on 2.5L/3.0L engines if equipped.
9. Position the arrestor over the stud and reconnect the ventilation hose(s) and bracket(s).
10. Install the washer and nut, tightening it to 30-40 inch lbs. (3.4-4.5 Nm) on carbureted models; or 25-35 inch lbs. (2.8-4 Nm) on TBI models.
11. Install the flame arrestor cover (if equipped) and tighten the nut 30-35 inch lbs. (3.4-4 Nm) on carbureted models; or 25-35 inch lbs. (2.8-4 Nm) on TBI models.
12. Close the engine compartment.

Ford Engines With MFI

◆ **See Figures 15, 16 and 17**

1. Remove or open the engine compartment cover.
2. Remove the four screws securing the plastic flame arrestor cover to the engine and lift off the cover. Set it aside where you won't step on it— they break easy and they're expensive!
3. Remove the two bolts (1) and washers securing the flame arrestor to the throttle body.

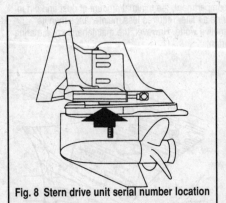

Fig. 8 Stern drive unit serial number location

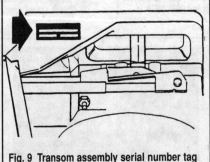

Fig. 9 Transom assembly serial number tag

Fig. 10 Many models will have a decorative cover over the arrestor cover...

Fig. 11 ...make sure to remove the nut(s) before lifting it off

Fig. 12 Once the plastic cover is removed, loosen the arrestor retaining nut and remove the breather hose(s). Notice that not all models have a bracket for the hose— this late model 5.7L has the connection built-in...

Fig. 13 ...and then lift off the arrestor

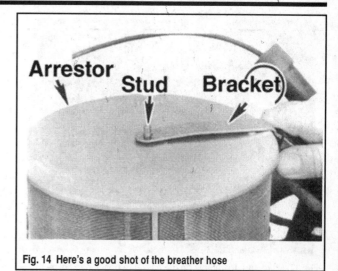

Fig. 14 Here's a good shot of the breather hose

■ Make sure you secure the lower bracket while removing the bolts as it will fall free when the bolts are removed.

4. Cut the plastic tie strap securing the breather hose to the oil filler fitting and wiggle the hose off the nipple.
5. Lift off the flame arrestor.
To install:
6. Clean the arrestor in solvent and dry with compressed air if possible; otherwise make sure that it dries completely by air.
7. Check the rubber gasket around the throttle body throat for any cracks, tears or other obvious signs of deterioration. Replace if necessary and make sure that it seats properly. The little nipple must be facing downward. We would replace it whether it needs it or not—cheap insurance; but you make the call.
8. Position the arrestor over the throttle body throat and move the lower bracket into position so the nipple in the gasket (1) fits into the hole on the bracket. Install the bolts finger-tight.

✳✳ WARNING

When tightening the flame arrestor, you must have pressure on it while tightening the mounting bolts in order to ensure against any air leaks.

9. Have an assistant press the arrestor in, and against, the throttle body while you tighten the two mounting bolts to 24-48 inch lbs. (2.75-4.8 Nm).
10. Wiggle the breather hose back into position on the filler nipple and secure it with a new plastic tie.
11. Install the plastic cover and tighten the four screws securely. Make sure you don't tighten them too much or you'll crack the cover.
12. Close the engine compartment.

GM Engines With MFI Except 1998 7.4Gi Engines

◆ See Figures 18, 19 and 20

1. Remove or open the engine compartment cover.
2. Carefully wiggle the crankcase ventilation hoses off of their necks on the sides of the arrestor.
3. Remove the nut securing the flame arrestor to the throttle body.
4. Lift off the flame arrestor.

■ Take note of the positioning of the arrestor prior to removal. Many models use a baffle to ensure correct air distribution and must be installed in the same position that they were in when removed.

5. Remove the rubber gasket from the lip on the arrestor.
To install:
6. Clean the arrestor in solvent and dry with compressed air if possible; otherwise make sure that it dries completely by air.
7. Check the rubber gasket for any cracks, tears or other obvious signs of deterioration. Replace if necessary and make sure that it seats properly. We would replace it whether it needs it or not—its cheap insurance; but you make the call.
8. Position the arrestor, install the retaining nut and washer and tighten securely but not so tightly as to dimple the cover.
9. Check the breather hose ends for cracks or deterioration and then slide then back into position.
10. Close the engine compartment.

Fig. 15 Remove the 4 screws and lift off the plastic cover

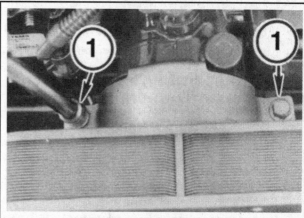

Fig. 16 Remove the 2 bolts and lift off the flame arrestor

Fig. 17 Make sure the nipple on the gasket in positioned correctly

Fig. 18 Remove the gasket before cleaning the arrestor...

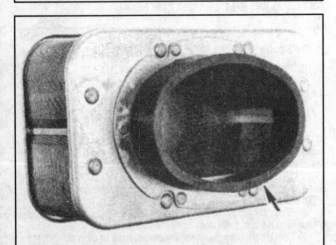

Fig. 19 ...and then make sure it is positioned correctly before reinstalling

Fig. 20 Tighten the mounting bolt before installing the hoses

1998 7.4Gi Engines With MFI

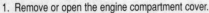

◆ See Figures 21 and 22

1. Remove or open the engine compartment cover.
2. Remove the retaining nuts (one or two, depending on application) from the plastic cover and lift off the cover. Set it aside where you won't step on it— they break easy and they're expensive!
3. Tag and disconnect the MAP sensor harness connector on the 8.1L and move it aside. Press inward on the grey plastic lip while carefully pulling down on the lower half of the connector.
4. Back off the adjusting screw on the arrestor clamp until the clamp is loose.
5. Lift off the flame arrestor.

■ Take note of the positioning of the arrestor prior to removal. Many models use a baffle to ensure correct air distribution and must be installed in the same position that they were in when removed.

To install:

6. Clean the arrestor in solvent and dry with compressed air if possible; otherwise make sure that it dries completely by air.
7. Position the arrestor, install the retaining clamp and tighten the adjusting screw securely.
8. Reconnect the MAP sensor on the 8.1L.
9. Install the plastic cover and tighten the nut(s) securely
10. Close the engine compartment.

Fuel Filter

A fuel filter is designed to keep particles of dirt and debris from entering the carburetors or the fuel injection system and clogging the tiny internal passages of either. A small speck of dirt or sand can drastically affect the ability of the fuel system to deliver the proper amount of air and fuel/oil to the engine. If a filter becomes clogged, the flow of gasoline will be impeded. This could cause lean fuel mixtures, hesitation and stumbling and idle problems in carburetors. Although a clogged fuel passage in a fuel injected engine could also cause lean symptoms and idle problems, dirt can also prevent a fuel injector from closing properly. A fuel injector that is stuck partially open by debris would likely cause the engine to run rich due to the unregulated fuel constantly spraying from the pressurized injector.

Regular cleaning or replacement of the fuel filter (depending on the type or types used) will decrease the risk of blocking the flow of fuel to the engine, which could leave you stranded on the water. It will also decrease the risk of damage to the small passages of a carburetor or fuel injector that could require more extensive and expensive replacement. Keep in mind that fuel filters are usually pretty inexpensive (at lease when compared to a tow) and replacement is a simple task. Service your fuel filter on a regular basis to avoid fuel delivery problems. All filters should be replaced no less than once a season or every 50 hours of operation, although halving this interval is cheap insurance!

The type of fuel filter used on your engine will vary with the year and model. Because of the number of possible variations it is impossible to accurately give instructions based on model. Instead, we will provide instructions for the different types of filters the manufacturer used on various families of motors or systems with which they are equipped. To determine what filter(s) are utilized by your engine, trace the fuel line from the tank to the fuel pump and then from the pump to the carburetors or throttle body. As a general rule of thumb, the majority of engines covered here utilize a canister-type in-line water separating filter. Most 2.5L and 3.0L engines have the filter incorporated into the fuel pump; but may also have an in-line water separating filter. Additionally, most carbureted engines will further utilize a small filter/screen in the carburetor.

As mentioned previously, most new engines have a factory-installed water separating fuel filter. This type filter is also available as an accessory for all other engines and should be installed at the earliest possible convenience. Such a kit is not expensive, and contains instructions for correct installation.

A water separating filter, as its name suggests, removes water and other fuel system contaminants before they reach the carburetor and helps minimize potential problems. The presence of water in the fuel will alter the proportion of air/fuel mixture to the "lean" side, resulting in a higher operating temperature and possible damage to pistons, if not corrected.

The filter consists of a mounting plate and disposable canister filter (much like an oil filter). The filter is installed between the fuel tank and the fuel pump.

REMOVAL & INSTALLATION

✳✳ CAUTION

Observe all applicable safety precautions when working around fuel. Whenever servicing the fuel system, always work in a well-ventilated area. Do not allow fuel spray or vapors to come in contact with a spark or open flame. Do not smoke while working around gasoline. Keep a dry chemical fire extinguisher near the work area. Always keep fuel in a container specifically designed for fuel storage; also, always properly seal fuel containers to avoid the possibility of fire or explosion.

Fuel Pump Filter—2.5L And 3.0L Engines Only

◆ See Figures 23 and 24   ② ⟋ MODERATE

■ All 2.5L engines and MOST 3.0L engines are equipped with this style filter, but not all—1986-90 engines should have the filter canister in the top half of the fuel pump, 1991-94 engines should have no filter in the fuel pump and use a standard inline canister filter, while 1995-98 engines should have the filter in the lower half of the fuel pump. Further, any of these engine may also use a standard in-line canister filter, and they all should have a small in-line filter at the carburetor.

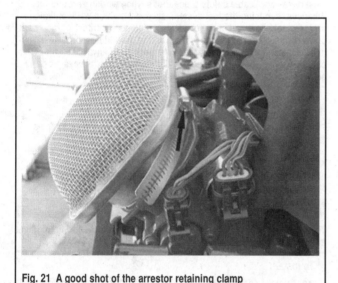

Fig. 21 A good shot of the arrestor retaining clamp

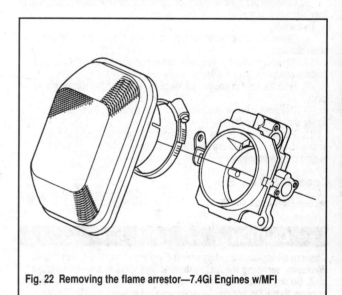

Fig. 22 Removing the flame arrestor—7.4Gi Engines w/MFI

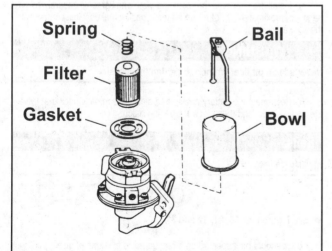

Fig .23 Some engines have the filter in the top half of the fuel pump...

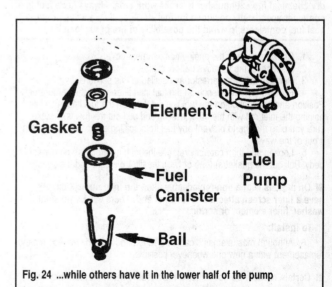

Fig. 24 ...while others have it in the lower half of the pump

✳✳ CAUTION

Observe all applicable safety precautions when working around fuel. Whenever servicing the fuel system, always work in a well-ventilated area. Do not allow fuel spray or vapors to come in contact with a spark or open flame. Do not smoke while working around gasoline. Keep a dry chemical fire extinguisher near the work area. Always keep fuel in a container specifically designed for fuel storage; also, always properly seal fuel containers to avoid the possibility of fire or explosion.

1. Remove or open the engine compartment hatch.
2. Disconnect the negative battery cable and then remove the flame arrestor.
3. Remove the safety wire from the screw at the bottom (or top) of the pump.
4. Loosen the screw and release the filter bowl bail from the housing. It's not necessary to remove the bail completely, you can usually swing the bail to the side far enough to allow removal of the bowl/canister.

✳✳ CAUTION

There will be fuel in the bowl, so have plenty of rags available.

5. Carefully pry the bowl from the pump housing and remove the spring, filter element and gasket.
To install:
6. Clean all parts carefully and check for any cracks, deterioration or other damage.
7. Position the spring, new filter and gasket into the bowl. Make sure that the open end faces the pump.
8. Hold the bowl in position on the pump and snap the retaining bail into place.
9. Tighten the screw securely, but not too tight, and then install the safety wire (if equipped).
10. Reconnect the battery cable, install the flame arrestor and start the engine. Check that there are no fuel leaks and then install or close the engine compartment hatch.

Carburetor Fuel Inlet Filter

◆ See Figures 25 and 26

 MODERATE

✳✳ CAUTION

Observe all applicable safety precautions when working around fuel. Whenever servicing the fuel system, always work in a well-ventilated area. Do not allow fuel spray or vapors to come in contact with a spark or open flame. Do not smoke while working around gasoline. Keep a dry chemical fire extinguisher near the work area. Always keep fuel in a container specifically designed for fuel storage; also, always properly seal fuel containers to avoid the possibility of fire or explosion.

1. Remove or open the engine compartment cover.
2. Disconnect the negative battery cable.
3. Remove the flame arrestor as detailed in this section.
4. Position an flare wrench on the fuel inlet filter nut at the carburetor. Position another wrench on the fuel line nut. Loosen the fuel line nut while holding the inlet nut with the other wrench and pull out the fuel line. Make sure you plug the line to prevent any fuel from spilling and carefully position it out of the way.
5. Loosen and then carefully remove the inlet nut from the carburetor body. Pull out the gasket(s) (one or two), the filter element and the spring.

■ **On models with a non-replaceable filter, the inlet nut will simply have a filter screen attached to its inner end. There will be no small washer, filter element or spring.**

To install:
6. Although most elements can be cleaned and reused, we recommend replacement with a new one whenever possible.

■ **Certain models do not utilize a replaceable filter, instead they will have a filter screen attached to the inner end of the inlet nut which can only be cleaned and reused.**

7. On models with a replaceable filter:
  a. Insert the spring into the carburetor and then slide in the filter element. Be sure that the open end of the filter faces out (toward the inlet nut).
  b. Position the large gasket over the inlet nut threads and the small one inside the nut. Screw the nut into the carburetor and tighten it securely; but not too tight.
8. On models with a non-replaceable filter:
  a. Slide the washer over the inlet nut threads.
  b. Screw the nut into the carburetor and tighten it securely; but not too tight.
9. Clean the threads and position the fuel line nut into the inlet, seat it a few turns with your fingers and then tighten it to 11-13 ft. lbs. (14.9-17.6 Nm).
10. Install the flame arrestor and connect the battery cable. Start the engine and check for fuel leaks.
11. Install or close the engine compartment.

Water Separating Canister Filter

◆ See Figures 27 and 28

 MODERATE

✳✳ CAUTION

Observe all applicable safety precautions when working around fuel. Whenever servicing the fuel system, always work in a well-ventilated area. Do not allow fuel spray or vapors to come in contact with a spark or open flame. Do not smoke while working around gasoline. Keep a dry chemical fire extinguisher near the work area. Always keep fuel in a container specifically designed for fuel storage; also, always properly seal fuel containers to avoid the possibility of fire or explosion.

1. Remove or open the engine compartment hatch.
2. Disconnect the negative battery cables.
3. Remove the canister filter from the mounting plate by rotating the canister counterclockwise—as viewed from the bottom end of the filter canister. An oil filter strap wrench may be necessary to break the filter free. Keep the filter upright to avoid spilling fuel. Properly dispose of the fuel and fuel saturated canister. Make sure you have plenty of rags handy just in case!

✳✳ WARNING

The old canister filter cannot be cleaned and used a second time. Never attempt to reuse the filter!

To install:
4. Coat the sealing ring(s) of a NEW canister filter with clean engine oil (there may be two rings so make sure that the old one comes out and the new one goes in!).
5. Install the filter onto the mounting plate and tighten it securely by hand—approximately 1/2 of a turn after the gasket makes initial contact with the flange.

✳✳ WARNING

Never use an oil filter wrench to tighten the canister.

6. Reconnect the battery cables and start the engine. Check that there are no fuel leaks and then close the engine hatch.

Belts

INSPECTION

OEM ①  EASY

◆ See Figures 29, 30, 31, 32 and 33

V-belts should be inspected on a regular basis for signs of glazing or cracking. A glazed belt will be perfectly smooth from slippage, while a good belt will have a slight texture of fabric visible. Cracks will usually start at the inner edge of the belt and run outward. All worn or damaged drive belts

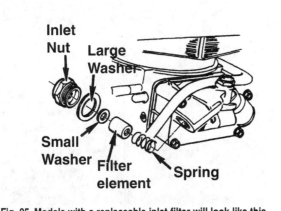

Fig. 25 Models with a replaceable inlet filter will look like this...

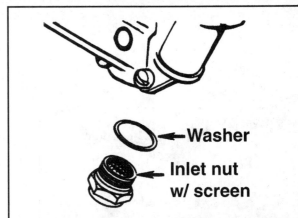

Fig. 26 ...while models with a non-replaceable inlet filter will look like this

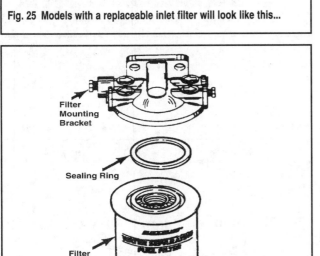

Fig. 27 Typical water separating fuel filter

Fig. 28 The filter canister is usually mounted at the front of the engine, on either side depending upon the application

should be replaced immediately. It is best to replace all drive belts at one time, as a preventive maintenance measure, during this service operation.

Inspect the alternator, power steering, sea-water supply pump and water pump V-belts every 50 hours or twice a season (whichever comes first) for evidence of wear such as cracking, fraying, and incorrect tension. New belts should be checked after the first 10 hours of operation.

Determine the V-belt tension at a point halfway between the pulleys by pressing on the belt with moderate thumb pressure. The belt should deflect 1/4-1/2 in. (6-13mm) on all engines. If the defection is found to be too much or too little, make adjustments as necessary. Always test the tension between the component pulley and the engine circulating pump (water pump) pulley

An alternate method (and actually much more accurate) is to use an OMC tension gauge (#1159660-8) or an equivalent tension gauge and check that the tension is 44-55 lbs. (20-25 kg).

■ When replacing belts, we recommend cleaning the inside of the belt pulleys to extend the service life of the belts. Never use automotive belts, marine belts used on your engine are heavy duty and not interchangeable.

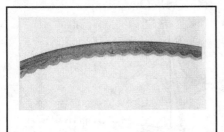

Fig. 29 An example of a healthy drive belt

Fig 30 Deep cracks in this belt will cause flex, building up heat that will eventually lead to belt failure

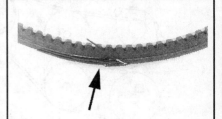

Fig. 31 The cover of this belt is worn, exposing the critical reinforcing cords to excessive wear

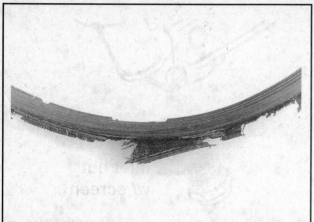

Fig. 32 Installing too wide a belt can result in serious belt wear and/or breakage

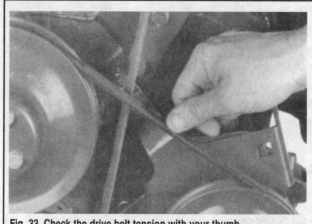

Fig. 33 Check the drive belt tension with your thumb

ROUTING DIAGRAMS

◆ See Figures 34 thru 40

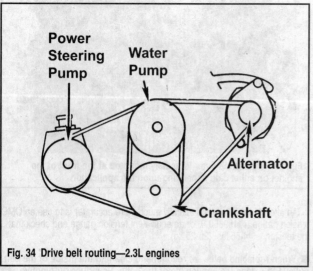

Fig. 34 Drive belt routing—2.3L engines

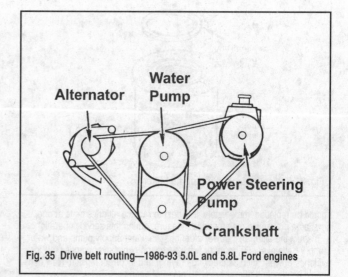

Fig. 35 Drive belt routing—1986-93 5.0L and 5.8L Ford engines

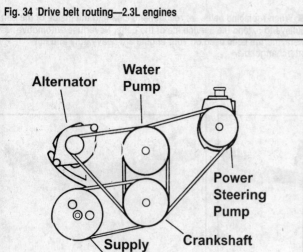

Fig. 36 Drive belt routing—1994-96 5.0L and 5.8L Ford engines

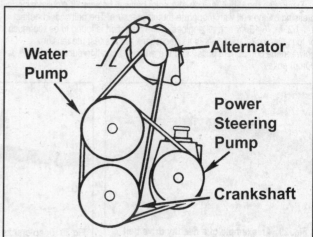

Fig. 37 Drive belt routing—1987-93 2.5L, 3.0L, 4.3L and 5.7L GM engines and 1990-95 7.4L GM engines

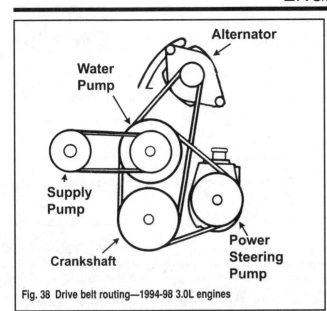

Fig. 38 Drive belt routing—1994-98 3.0L engines

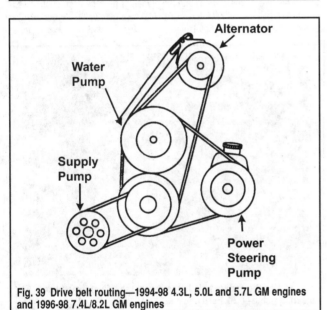

Fig. 39 Drive belt routing—1994-98 4.3L, 5.0L and 5.7L GM engines and 1996-98 7.4L/8.2L GM engines

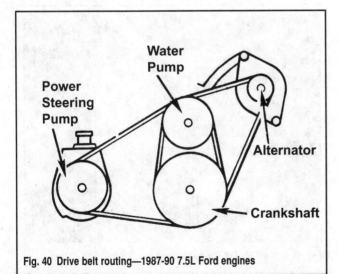

Fig. 40 Drive belt routing—1987-90 7.5L Ford engines

ADJUSTMENT

Alternator

◆ See Figure 41

Although most alternators can be found on the upper port side of the engine, certain Ford engines will have it located on the lower port side.
1. Remove or open the engine hatch and disconnect the battery cables.
2. Loosen the nut on the lower pivot bolt at the bottom of the alternator.
3. Loosen the bolt on the adjustment bracket behind the assembly.
4. Carefully insert a 1/2 in. breaker bar into the space between the alternator and a sturdy spot on the engine block, and then pivot the alternator away from the engine.
5. Check that the tension is now within specifications and tighten the bolt and nut while keeping steady pressure on the breaker bar.
6. Recheck the belt tension and connect the battery cable.
7. Close the engine compartment.

Fig. 41 A good shot of the adjustment bracket and bolt (upper) and the pivot bolt (lower)—4.3L shown

Power Steering Pump

All Engines Except 7.4L/7.5L/8.2L

◆ See Figures 42 and 43

Although on most engines the power steering pump can be found on the lower port side of the engine; 2.3L engines have it located on the lower starboard side, while 5.0L/5.8L Ford engines will have it on the upper port side.
1. Remove or open the engine hatch and disconnect the battery cables.
2. Loosen the power steering pump mounting bolts and/or nuts—usually two, but the 2.3L engine use four (3 in the front and 1 in the back).
3. Insert a 1/2 in. breaker bar into the square hole on the mounting bracket and then pivot the pump assembly away from the engine.

✳✳ CAUTION

Never pry against the reservoir or filler neck.

4. Check that the tension is now within specifications and tighten the mounting bolts or nuts while keeping steady pressure on the breaker bar.
5. Recheck the belt tension and connect the battery cable.
6. Close the engine compartment.

7.4L, 7.5L And 8.2L Engines

◆ See Figure 44

Although on the two GM engines the power steering pump can be found on the lower port side of the engine; 7.5L engines will have it located on the lower starboard side.

7. Remove or open the engine hatch and disconnect the battery cables.

8. Loosen the pump mounting bracket bolts or nuts.

9. Insert a long screw driver-like breaker bar into the space between the pump and the engine and then pry against the corner of the timing chain cover and the tab on the pump mounting bracket.

✳✳ CAUTION

Never pry against the reservoir or filler neck.

10. Check that the tension is now within specifications and tighten the bolt closest to the tab. Tighten the remaining bolts while keeping steady pressure on the breaker bar.

11. Recheck the belt tension and connect the battery cable.

12. Close the engine compartment.

Fig. 42 Use the square hole in the bracket to adjust the power steering pump—most GM engine similar

Fig. 43 Use the square hole in the bracket to adjust the power steering pump—5.0L/5.8L Ford engines

Raw Water Supply Pump

◆ See Figure 45

The supply pump is always located on the starboard side of the engine on all applications.

1. Remove or open the engine hatch and disconnect the battery cables.

2. Loosen the mounting bracket bolts.

3. Insert a 1/2 in. breaker bar (we actually prefer a wooden broom handle) between the pump housing and a sturdy point on the engine and then pivot the pump assembly away from the engine.

4. Check that the tension is now within specifications and tighten the bolts while keeping steady pressure on the breaker bar.

5. Recheck the belt tension and connect the battery cable.

6. Close the engine compartment.

REMOVAL & INSTALLATION

The replacement of the inner belt on multi-belted engines may require the removal of the outer belts.

To replace a drive belt, loosen the pivot and mounting bolts of the component which the belt is driving, then, using a wooden lever or equivalent, pry the component inward to relieve the tension on the drive belt; always be careful where you locate the prybar, or damage to components may result. Slip the belt off the component pulley, and match the new belt with the old belt for length and width.

Fig. 44 Use the square hole in the bracket to adjust the power steering pump—7.5L Ford engines

Fig. 45 A good shot of the adjustment bracket and bolt

These measurements must be equal. It is normal for an old belt to be slightly longer than a new one. After a new belt is installed correctly, properly adjust the tension.

■ **When removing more than one belt, be sure to mark them for identification. This will help avoid confusion when replacing the belts.**

Thermostat

The thermostat is a simple temperature sensitive valve that opens and closes to control cooling water flow through the engine. In operation, the thermostat hovers somewhere between open and closed. As engine load and temperature increase, the thermostat opens to allow more cooling water into the engine. As temperature and load decrease, the thermostat closes.

A sticking thermostat will either allow the temperature to rise well above the normal operating temperature before it opens, or, if stuck in the open position, will never allow the engine to reach operating temperature.

All thermostats are rated based on the temperature at which they open and this rating should always be stamped somewhere on the thermostat, usually on the flange area. All engines covered here utilize a 160°F thermostat except for the 7.4L/8.2L engines which use a thermostat rated at 140°F.; or models with a closed cooling system which generally use a 170°F thermostat.

■ **On all engines covered here, the thermostat housing can be found on the front, top of the engine—easily identifiable by the large hoses attached to it.**

✳✳ CAUTION

Serious damage may result from operating your engine without a thermostat! Don't even consider this!

✳✳ CAUTION

NEVER use an automotive thermostat in a marine engine. No matter how tempting this may seem, forget it!!

REMOVAL & INSTALLATION

2.3L, 2.5L And 3.0L Engines—1986-96

◆ See Figures 46 and 47

■ **On all engines covered here, the thermostat housing can be found on the front, top of the engine—easily identifiable by the large hoses attached to it.**

✳✳ CAUTION

Serious damage may result from operating your engine without a thermostat! Don't even consider this!

✳✳ CAUTION

NEVER use an automotive thermostat in a marine engine.

1. Open or remove the engine hatch cover and disconnect the negative battery cables.
2. Drain all water from the cylinder block and exhaust manifold(s) as detailed in the Cooling System section.
3. Locate the thermostat housing at the front of the engine, loosen the hose clamps and then wiggle all coolant hoses off of the thermostat housing. In many cases you may have to use a small prybar to persuade them off of the fitting—be careful that you don't damage the hose end in the process.

■ **Although it is not absolutely necessary to remove the coolant hoses, we feel that it makes the job easier.**

4. Loosen and remove the alternator bracket mounting bolt on 2.5L/3.0L engines.
5. Remove the 4 mounting bolts (2 on the 2.3L) with their lock washers and then remove the upper thermostat housing cover. Some models may have a lifting eye incorporated in the housing—take note of its positioning.
6. Remove the gasket and be sure to scrape of any remaining material from the two mating surfaces.
7. Lift out the thermostat and discard it. If you are not sure that it is inoperable, perform the testing procedures outlined in this section. Don't forget to remove the cork seal if it doesn't come out with the thermostat.

To install:
8. Once again, make sure that any remaining gasket material has been removed from the thermostat housing and cover.
9. Position a **NEW** cork seal in the housing.
10. Insert a new thermostat (160°F) into the housing. The element must be pointing into the housing so that the pointy end of the thermostat is facing upward on the 2.5L/3.0L or outward on the 2.3L. Make sure that a new cork seal is positioned over the base of the thermostat.
11. Coat both sides of a new 2.5L/3.0L housing gasket with OMC Gasket Sealing compound (or similar) and position it onto the housing so that the holes line up.

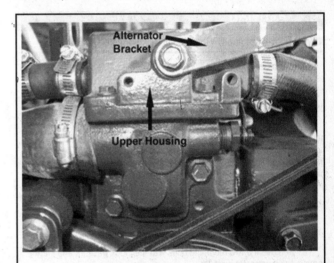

Fig. 46 A good shot of the thermostat housing—3.0L engines

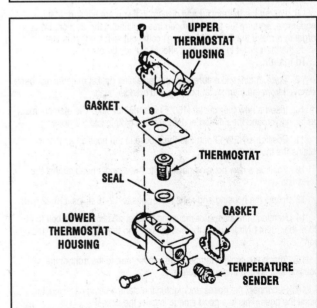

Fig. 47 Exploded view of the thermostat and housing—3.0L engines

12. Install the upper housing cover and mounting bolts/washers and tighten to 5-7 ft. lbs. (7-9 Nm) on the 2.5L/3.0L. On the 2.3L engine, install the housing and bolts, long bolt in the lower hole, and then tighten to 20-25 ft. lbs. (27-34 Nm).

13. Reconnect the hoses and tighten the clamps being careful not to pinch the hose. This is a good time to inspect the hoses!

14. Reattach the alternator mounting bracket, recheck the belt tension adjustment and then tighten the bolt securely.

15. Connect the batteries and then start the engine and check for leaks.

3.0L Engines—1997-98

■ On all engines covered here, the thermostat housing can be found on the front, top of the engine—easily identifiable by the large hoses attached to it.

✲✲ CAUTION

Serious damage may result from operating your engine without a thermostat! Don't even consider this!

✲✲ CAUTION

NEVER use an automotive thermostat in a marine engine.

1. Open or remove the engine hatch cover and disconnect the negative battery cables.

2. Drain all water from the cylinder block and exhaust manifold(s) as detailed in the Cooling System section.

3. Locate the thermostat housing at the front of the engine, loosen the hose clamps and then wiggle the four coolant hoses off of the thermostat housing. In many cases you may have to use a small prybar to persuade them off of the fitting—be careful that you don't damage the hose end in the process.

4. Tag and disconnect (unscrew the nut) the coolant temperature sender lead from the front of the housing and position it out of the way.

5. Loosen the alternator adjusting bolt. Loosen and remove the alternator bracket mounting bolt and pivot it out of the way.

6. Remove the two mounting bolts with their lock washers and then remove the thermostat housing.

7. Remove the gasket and be sure to scrape of any remaining material from the two mating surfaces.

8. Lift out the thermostat and discard it. If you are not sure that it is inoperable, perform the testing procedures outlined in this section. Don't forget to remove the O-ring if it doesn't come out with the thermostat—it holds the thermostat in place, so it should already be out.

To install:

9. Once again, make sure that any remaining gasket material has been removed from the thermostat housing and cylinder head.

10. Insert a new thermostat (160°F) into the housing. The element must be facing you and the pointed end should be pointing into the housing.

11. Position a **NEW** O-ring into the groove in the housing so that it retains the thermostat.

12. Position a new housing gasket onto the cylinder head so that the holes line up.

13. Install the housing and tighten the bolts to 12-16 ft. lbs. (16-22 Nm).

14. Reattach the alternator mounting bracket and tighten the bolt to 26-30 ft. lbs. (35-41 Nm). Adjust the belt tension and then tighten the adjusting bolt.

15. Attach the coolant temperature sender lead to the front of the housing and tighten the nut securely.

16. Reconnect the hoses and tighten the clamps being careful not to pinch the hose. This is a good time to inspect the hoses!

17. Connect the batteries and then start the engine and check for leaks.

V6 and V8 Engines

◆ See Figures 48, 49, 50, 51 and 52

■ On all engines covered here, the thermostat housing can be found on the front, top of the engine—easily identifiable by the large hoses attached to it.

✲✲ CAUTION

Serious damage may result from operating your engine without a thermostat! Don't even consider this!

✲✲ CAUTION

NEVER use an automotive thermostat in a marine engine.

1. Open or remove the engine hatch cover and disconnect the negative battery cables.

2. Drain all water from the cylinder block and exhaust manifold(s) as detailed in the Cooling System section.

3. Locate the thermostat housing at the front of the engine, loosen the hose clamps and then wiggle the four coolant hoses off of the thermostat housing. In many cases you may have to use a small prybar to persuade them off of the fitting—be careful that you don't damage the hose end in the process.

■ EFI models will have five hoses that need to be removed.

1. Remove the two mounting bolts with their lock washers and then remove the thermostat housing. Some models may have a lifting eye attached to the housing—take note of its positioning before removing it. On certain later models you will also need to unscrew the temperature sender and move it out of the way.

2. Remove the gasket and be sure to scrape off any remaining material from the two mating surfaces.

3. Pry out the O-ring that holds the thermostat in place and then lift out the thermostat and discard it. If you are not sure that it is inoperable, perform the testing procedures outlined in this section.

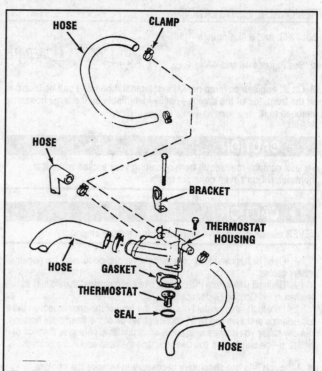

Fig. 48 Exploded view of the thermostat and housing—typical, but most models similar

Fig. 49 Thermostat housing—4.3L engines (Gi shown)

Fig. 50 Thermostat housing—5.7L/7.4L/8.2L engines

Fig. 51 Thermostat housing—5.0L/5.8L Ford engines

Fig. 52 Thermostat housing—7.5L engines

To install:

4. Once again, make sure that any remaining gasket material has been removed from the thermostat housing and cylinder head.

5. Insert a new thermostat (160°F except 7.4L/8.2L which use a 140°F) into the housing. The element must be facing you when installed in the housing so that it goes into the engine when installed.

6. Position a **NEW** O-ring into the groove in the housing so that it retains the thermostat. Make sure the O-ring is fully seated in the groove!

7. Coat both sides of a new housing gasket with OMC Gasket Sealing compound (or similar) and position it onto the manifold so that the holes line up.

8. Install the housing and lifting eye, then tighten the bolts to 20-25 ft. lbs. (27-34 Nm).

9. Reconnect the hoses and tighten the clamps being careful not to pinch the hose. This is a good time to inspect the hoses!

✸✸ CAUTION

Always ensure that the water supply hose is attached to the correct nipple. On the side with three hoses (or four) it's the lower one in the middle.

10. Connect the batteries and then start the engine and check for leaks.

TESTING

◆ **See Figures 53, 54, 55 and 56**

1. Inspect the thermostat at room temperature. If the thermostat is fully open, it is defective and must be replaced. Hold the thermostat up to the light and check it for leaks. A light leak around the perimeter indicates the thermostat is not closing, and therefore, it must be replaced.

2. Attach a length of thread to the thermostat. Now, suspend the thermostat and a thermometer inside a container filled with water (do not use distilled water or ethylene glycol!). Take care to be sure neither the thermostat or the thermometer touches the container. If either one does touch the container, the test will be unreliable

3. Heat the water until the thermostat just begins to open - when this happens confirm that the temperature is the same as the thermostat rating. The thermometer reading must agree with the rating stamped on the thermostat; 157-163°F for a 160°F thermostat, or 138-142°F for the 140°F unit. If the unit fails the test, it must be replaced.

4. Continue to heat the water until a temperature of 182°F is reached; on 7.4L/8.2L engines, it should reach 162°F. At this time the thermostat should be completely open to 5/32 in. (3.96mm); if not, replace it.

5. Turn the heat off and allow the water to cool to a temperature 10° below the rating. The thermostat should now be completely closed; if not, replace it.

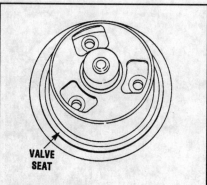

Fig. 53 Examine the perimeter of the thermostat for any visible light

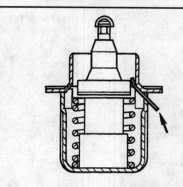

Fig. 54 Check the opening gap here on inline engine thermostats

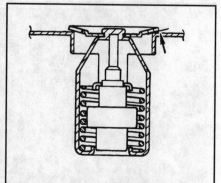

Fig. 55 Check the opening gap here on V6 and V8 engine thermostats

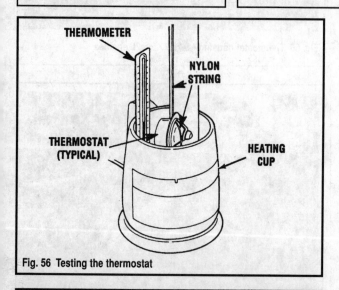

Fig. 56 Testing the thermostat

Fig. 57 When checking the compression, always use a quality gauge

Cylinder Compression

Cylinder compression test results are extremely valuable indicators of internal engine condition. The best marine mechanics automatically check an engine's compression as the first step in a comprehensive tune-up. A compression test will uncover many mechanical problems that can cause rough running or poor performance.

CHECKING COMPRESSION

◆ See Figures 57 and 58

1. Make sure that the proper amount and viscosity of engine oil is in the crankcase, then ensure the battery is fully charged.
2. Warm-up the engine to normal operating temperature, then shut the engine OFF. If the boat is out of the water, make sure to install a flush test kit.
3. Remove the flame arrestor and open the choke or throttle fully.
4. Disable the ignition system by doing the following:

1986-89 engines
• Disconnect the high tension lead running between the ignition coil and the distributor at the distributor. Ground the lead to the engine to prevent any sparking.

1990 engines
• All Cobra engines except 4.3L HO: Disconnect the high tension lead running between the ignition coil and the distributor at the distributor. Ground the lead to the engine to prevent any sparking.
• 4.3L HO: Tag and disconnect the 14-pin connector at the ignition module. Position it out of the way.
• All King Cobra engines: Remove the mounting bolt and then disconnect the lead at the ignition module. Position it out of the way.

1991-93 engines
• On all 3.0L engines: tag and disconnect the upper 2-wire connector at the ignition coil and position it out of the way.
• V6 and V8 carbureted Cobra engines except the 1991 4.3L HO: remove both distributor primary wires at the coil. Tape the terminals to avoid grounding.
• 1991 4.3L HO: tag and disconnect the 14-pin connector at the ignition module. Position it out of the way.
• V8 carbureted King Cobra engines except the 5.7 LE: tag and disconnect the 14-pin connector at the ignition module. Position it out of the way.
• 5.7 LE: remove both distributor primary wires at the coil. Tape the terminals to avoid grounding.
• EFI engines: disconnect the 2-way connector at the ignition coil.

1994-95 engines
• 3.0L and 4.3L engines and 7.4L EFI engines: tag and disconnect the upper 2-wire connector at the ignition coil and position it out of the way..
• 5.0FL, 5.7GL and 5.8FL engines: remove both distributor primary wires at the coil. Tape the terminals to avoid grounding
• 7.4GL engines: tag and disconnect the 14-wire connector at the ignition module.
• EFI engines except the 7.4 EFI: tag and disconnect the 2-way connector at the ignition coil

1996 engines
• 3.0L, 4.3L, 5.7Gi and 7.4Gi engines: tag and disconnect the upper (grey) 2-wire connector at the ignition coil and position it out of the way..
• 5.0FL and 5.8FL engines: remove both distributor primary wires at the coil. Tape the terminals to avoid grounding

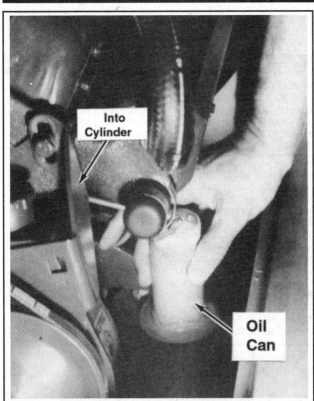

Fig. 58 If a cylinder exhibits unusually low compression, try squirting about a tablespoon of oil into it

• 5.0Fi, 5.8Fi and 5.8FSi engines: tag and disconnect the 2-way connector at the ignition coil
• 7.4GL engines: tag and disconnect the 14-wire connector at the ignition module.

1997-98 engines
• All except 5.7GL/GS and 7.4GL engines: disconnect the grey 2-wire connector at the ignition coil, it has purple and grey wires
• 5.7GL/GS engines: remove both distributor primary wires at the coil. Tape the terminals to avoid grounding
• 7.4GL: Disconnect the 14-wire connector at the ignition module.

1. Tag and disconnect all spark plug wires and then remove the plugs themselves.

2. Install a screw-in type compression gauge into the No. 1 cylinder spark plug hole until the fitting is snug. Please refer to the firing order illustrations for location of the No. 1 cylinder. When fitting the compression gauge adapter to the cylinder head, make sure the bleeder of the gauge (if equipped) is closed.

3. According to the tool manufacturer's instructions, connect a remote starting switch to the starting circuit.

4. With the ignition switch in the **OFF** position, use the remote starting switch to crank the engine through at least five compression strokes (approximately 5 seconds of cranking) and record the highest reading on the gauge.

5. Repeat the test on each cylinder, cranking the engine approximately the same number of compression strokes and/or times as the first.

6. Compare the highest readings from each cylinder to that of the others. The indicated compression pressures are considered within specifications if the lowest reading cylinder is within 75 percent of the pressure recorded for the highest reading cylinder. For example, if your highest reading cylinder pressure was 150 psi (1034 kPa), then 75 percent of that would be 113 psi (779 kPa). So the lowest reading cylinder should be no less than 113 psi (779 kPa).

7. Compression readings that are generally low indicate worn, broken, or sticking piston rings, scored pistons or worn cylinders.

8. If a cylinder exhibits an unusually low compression reading, squirt a tablespoon of clean engine oil into the cylinder through the plug or injector

hole and repeat the compression test. If the compression rises after adding oil, it means that the cylinder's piston rings and/or cylinder bore are damaged or worn. If the pressure remains low, the valves may not be seating properly (a valve job is needed), or the head gasket may be blown near that cylinder.

9. If compression in any two adjacent cylinders is low (with normal compression in the other cylinders), and if the addition of oil doesn't help raise compression, there is leakage past the head gasket. Oil and coolant in the combustion chamber, combined with blue or constant white smoke from the tailpipe, are symptoms of this problem. However, don't be alarmed by the normal white smoke emitted from the tailpipe during engine warm-up during cold weather. There may be evidence of water droplets on the engine oil dipstick and/or oil droplets in the cooling system if a head gasket is blown.

Spark Plugs

The spark plug performs four main functions:
• It fills a hole in the cylinder head.
• It acts as a dielectric insulator for the ignition system.
• It provides spark for the combustion process to occur.
• It removes heat from the combustion chamber.

It is important to remember that spark plugs do not create heat, they help remove it. Anything that prevents a spark plug from removing the proper amount of heat can lead to pre-ignition, detonation, premature spark plug failure and even internal engine damage.

In the simplest of terms, the spark plug acts as the thermometer of the engine. Much like a doctor examining a patient, this "thermometer" can be used to effectively diagnose the amount of heat present in each combustion chamber.

Spark plugs are valuable tuning tools, when interpreted correctly. They will show symptoms of other problems and can reveal a great deal about the engine's overall condition. By evaluating the appearance of the spark plug's firing tip, visual cues can be seen to accurately determine the engine's overall operating condition, get a feel for air/fuel ratios and even diagnose driveability problems.

As spark plugs grow older, they lose their sharp edges and material from the center and ground electrodes is slowly eroded away. As the gap between these two points grows, the voltage required to bridge this gap increases proportionally. The ignition system must work harder to compensate for this higher voltage requirement and hence there is a greater rate of misfires or incomplete combustion cycles. Each misfire means lost horsepower, reduced fuel economy and higher emissions. Replacing worn out spark plugs with new ones (with sharp new edges) effectively restores the ignition system's efficiency and reduces the percentage of misfires, restoring power, economy and reducing emissions.

How long spark plugs last will depend on a variety of factors, including engine compression, fuel used, gap, center/ground electrode material and the conditions in which the engine is operated.

SPARK PLUG HEAT RANGE

◆ **See Figure 59**

Spark plug heat range is the ability of the plug to dissipate heat from the combustion chamber. The longer the insulator (or the farther it extends into the engine), the hotter the plug will operate; the shorter the insulator (the closer the electrode is to the block's cooling passages) the cooler it will operate.

Selecting a spark plug with the proper heat range will ensure that the tip will maintain a temperature high enough to prevent fouling, yet be cool enough to prevent pre-ignition. A plug that absorbs little heat and remains too cool will quickly accumulate deposits of oil and carbon since it is not hot enough to burn them off. This leads to plug fouling and consequently to misfiring. A plug that absorbs too much heat will have no deposits but, due to the excessive heat, the electrodes will burn away quickly and might possibly lead to pre-ignition or ignition problems.

Pre-ignition takes place when plug tips get so hot that they glow sufficiently to ignite the air/fuel mixture before the actual spark occurs. This early ignition will usually cause a pinging during heavy loads and if not corrected will result in severe engine damage. While there are many other things that can cause pre-ignition, selecting the proper heat range spark plug will ensure that the spark plug itself is not a hot-spot source.

SPARK PLUG SERVICE

■ **New technologies in spark plug and ignition system design have pushed the recommended replacement interval higher and higher. However, this depends on usage and conditions.**

Spark plugs should only require replacement once a season. The electrode on a new spark plug has a sharp edge but with use, this edge becomes rounded by wear, causing the plug gap to increase. As the gap increases, the plug's voltage requirement also increases. It requires a greater voltage to jump the wider gap and about two to three times as much voltage to fire a plug at high speeds than at idle.

Tools needed for spark plug replacement include: a ratchet, short extension, spark plug socket (there are two types; either 13/16 inch or 5/8 inch, depending upon the type of plug), a combination spark plug gauge and gapping tool and a can of anti-seize type compound.

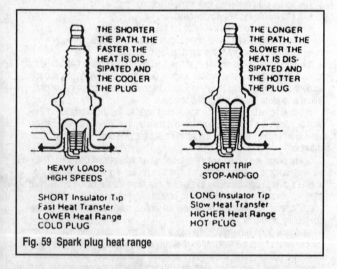

THE SHORTER THE PATH, THE FASTER THE HEAT IS DISSIPATED AND THE COOLER THE PLUG

THE LONGER THE PATH, THE SLOWER THE HEAT IS DISSIPATED AND THE HOTTER THE PLUG

HEAVY LOADS, HIGH SPEEDS

SHORT TRIP STOP-AND-GO

SHORT Insulator Tip
Fast Heat Transfer
LOWER Heat Range
COLD PLUG

LONG Insulator Tip
Slow Heat Transfer
HIGHER Heat Range
HOT PLUG

Fig. 59 Spark plug heat range

REMOVAL & INSTALLATION

◆ **See Figures 60 and 61**

 EASY

1. When removing spark plugs, work on one at a time. Don't start by removing the plug wires all at once, because unless you number them, they may become mixed up. Take a minute before you begin and number the wires with tape.

2. Disconnect the negative battery cable or turn the battery switch **OFF**.

3. If the engine has been run recently, allow the engine to thoroughly cool. Attempting to remove plugs from a hot cylinder head could cause the plugs to seize and damage the threads in the cylinder head, especially on aluminum heads!

4. Carefully twist the spark plug wire boot to loosen it, then pull the boot using a twisting motion to remove it from the plug. Be sure to pull on the boot and not on the wire, otherwise the connector located inside the boot may become separated from the high-tension wire.

■ **A spark plug wire removal tool is recommended as it will make removal easier and help prevent damage to the boot and wire assembly.**

5. Using compressed air (and safety glasses), blow debris from the spark plug area to assure that no harmful contaminants are allowed to enter the combustion chamber when the spark plug is removed. If compressed air is not available, use a rag or a brush to clean the area. Compressed air is available from both an air compressor or from compressed air in cans available at photography stores.

■ **Remove the spark plugs when the engine is cold, if possible, to prevent damage to the threads. If plug removal is difficult, apply a few drops of penetrating oil to the area around the base of the plug and allow it a few minutes to work.**

6. Using a 5/8 in. spark plug socket that is equipped with a rubber insert to properly hold the plug, turn the spark plug counterclockwise to loosen and remove the spark plug from the bore.

✳✳ WARNING

Avoid the use of a flexible extension on the socket. Use of a flexible extension may allow a shear force to be applied to the plug. A shear force could break the plug off in the cylinder head, leading to costly and frustrating repairs. In addition, be sure to support the ratchet with your other hand—this will also help prevent the socket from damaging the plug.

7. Evaluate each cylinder's performance by comparing the spark plug condition. Check each spark plug to be sure they are all of the same manufacturer and have the same heat range rating. Inspect the threads in the spark plug opening of the block and clean the threads before installing the plug.

8. When purchasing new spark plugs, always ask the dealer if there has been a spark plug change for the engine being serviced. Many times manufacturers will update the type of spark plug used in an engine to offer better efficiency or performance.

9. Crank the engine through several revolutions to blow out any material that might have become dislodged during cleaning. Always use a new gasket (if applicable), but never use gaskets on taper seat plugs. The gasket must be fully compressed on clean seats to complete the heat transfer process and to provide a gas tight seal in the cylinder.

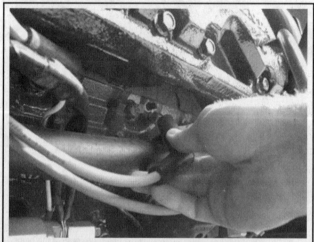

Fig. 60 Grab the plug wire boot and twist while removing it

Fig. 61 Removing the spark plug

10. Inspect the spark plug boot for tears or damage. If a damaged boot is found, the spark plug boot and possible the entire wire will need replacement.

11. Check the spark plug gap prior to installing the plug. Most spark plugs do not come gapped to the proper specification.

12. Apply a thin coating of anti-seize on the thread of the plug. This is extremely important on aluminum head engines.

13. Carefully thread the plug into the bore by hand. If resistance is felt before the plug completely bottomed, back the plug out and begin threading again.

✳✳ WARNING

Do not use the spark plug socket to thread the plugs. Always carefully thread the plug by hand or using an old plug wire to prevent the possibility of cross-threading and damaging the cylinder head bore.

14. Carefully tighten the spark plug. If the plug you are installing is equipped with a crush washer, tighten the plug until the washer seats, then turn it 1/4 turn to crush the washer. Whenever possible, spark plugs should be tightened to the factory torque specification:

1986-88 engines
- All engines—15 ft. lbs. (20 Nm)

1989 engines
- 2.3L—5-10 ft. lbs. (7-13 Nm)
- 3.0L, 4.3L, 5.7L, 262, 350—22 ft. lbs. (30 Nm)
- Ford 5.0L/5.8L—15-20 ft. lbs. (20-27 Nm)
- 460 King Cobra—5-10 ft. lbs. (7-13 Nm)

1990 engines
- 2.3L—5-10 ft. lbs. (7-13 Nm)
- GM Cobras—22 ft. lbs. (30 Nm)
- GM King Cobras—20 ft. lbs. (27 Nm)
- Ford V8—15-20 ft. lbs. (20-27 Nm)

1991-98 engines
- GM—20 ft. lbs. (27 Nm)
- Ford—5-10 ft. lbs. (7-13 Nm)

15. Apply a small amount of silicone dielectric grease to the end of the spark plug lead or inside the spark plug boot to prevent sticking, then install the boot to the spark plug and push until it clicks into place. The click may be felt or heard. Gently pull back on the boot to assure proper contact.

16. Connect the negative battery cable or turn the battery switch **ON**.

17. Start the engine and insure proper operation.

READING SPARK PLUGS

◆ **See Figures 62 thru 68**

Reading spark plugs can be a valuable tuning aid. By examining the insulator firing nose color, you can determine much about the engine's overall operating condition.

In general, a light tan/gray color tells you that the spark plug is at the optimum temperature and that the engine is in good operating condition.

Dark coloring, such as heavy black wet or dry deposits usually indicate a fouling problem. Heavy, dry deposits can indicate an overly rich condition, too cold a heat range spark plug, possible vacuum leak, low compression, overly retarded timing or too large a plug gap.

If the deposits are wet, it can be an indication of a breached head gasket, oil control from ring problems or an extremely rich condition, depending on what liquid is present at the firing tip.

Look for signs of detonation, such as silver specs, black specs or melting or breakage at the firing tip.

Compare your plugs to the illustrations shown to identify the most common plug conditions.

Fouled Spark Plugs

A spark plug is fouled when the insulator nose at the firing tip becomes coated with a foreign substance, such as fuel, oil or carbon. This coating makes it easier for the voltage to follow along the insulator nose and leach back down into the metal shell, grounding out, rather than bridging the gap normally.

Fuel, oil and carbon fouling can all be caused by different things but in any case, once a spark plug is fouled, it will not provide voltage to the firing tip and that cylinder will not fire properly. In many cases, the spark plug cannot be cleaned sufficiently to restore normal operation. It is therefore recommended that fouled plugs be replaced.

Signs of fouling or excessive heat must be traced quickly to prevent further deterioration of performance and to prevent possible engine damage.

Overheated Spark Plugs

When a spark plug tip shows signs of melting or is broken, it usually means that excessive heat and/or detonation was present in that particular combustion chamber or that the spark plug was suffering from thermal shock.

Fig. 62 A normally worn spark plug should have light tan or gray deposits on the firing tip (electrode)

Fig. 63 A carbon-fouled plug, identified by soft, sooty black deposits, may indicate an improperly tuned engine

Fig. 64 A physically damaged spark plug may be evidence of severe detonation in that cylinder. Watch the cylinder carefully between services, as a continued detonation will not only damage the plug but will most likely damage the engine

Fig. 65 An oil-fouled spark plug indicates an engine with worn piston rings

Fig. 66 This spark plug has been left in the engine too long, as evidenced by the extreme gap. Plugs with such an extreme gap can cause misfiring and stumbling accompanied by a noticeable lack of power

Fig. 67 A bridged or almost bridged spark plug, identified by the build-up between the electrodes caused by excessive carbon or oil build-up on the plug

Since spark plugs do not create heat by themselves, one must use this visual clue to track down the root cause of the problem. In any case, damaged firing tips most often indicate that cylinder pressures or temperatures were too high. Left unresolved, this condition usually results in more serious engine damage.

Detonation refers to a type of abnormal combustion that is usually preceded by pre-ignition. It is most often caused by a hot spot formed in the combustion chamber.

As air and fuel is drawn into the combustion chamber during the intake stroke, this hot spot will "pre-ignite" the air fuel mixture without any spark from the spark plugs.

Detonation

Detonation exerts a great deal of downward force on the pistons as they are being forced upward by the mechanical action of the connecting rods. When this occurs, the resulting concussion, shock waves and heat can be severe. Spark plug tips can be broken or melted and other internal engine components such as the pistons or connecting rods themselves can be damaged.

Left unresolved, engine damage is almost certain to occur, with the spark plug usually suffering the first signs of damage.

■ When signs of detonation or pre-ignition are observed, they are symptom of another problem. You must determine and correct the situation that caused the hot spot to form in the first place.

INSPECTION & GAPPING

◆ See Figures 69 and 70

A particular spark plug might fit hundreds of engines and although the factory will typically set the gap to a pre-selected setting, this gap may not be the right one for your particular engine.

Insufficient spark plug gap can cause pre-ignition, detonation, even engine damage. Too much gap can result in a higher rate of misfires, noticeable loss of power, plug fouling and poor economy.

Check the spark plug gap before installation. The ground electrode (the L-shaped one connected to the body of the plug) must be parallel to the center electrode and the specified size wire gauge must pass between the electrodes with a slight drag.

Do not use a flat feeler gauge when measuring the gap on a used plug,

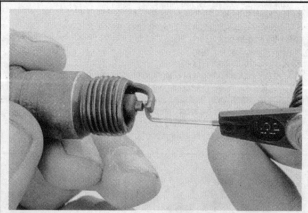

Fig. 69 Using a wire-type spark plug gapping tool to check the distance between center and ground electrodes

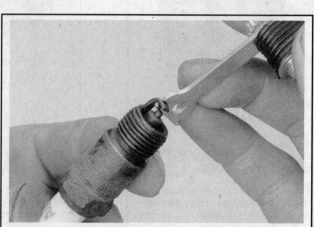

Fig. 70 Most spark plug gapping tools have an adjusting tool used to bend the ground electrode. USE IT! This tool greatly reduces the chance of breaking off the electrode and is much more accurate

Tracking Arc
High voltage arcs between a fouling deposit on the insulator tip and spark plug shell. This ignites the fuel/air mixture at some point along the insulator tip, retarding the ignition timing which causes a power and fuel loss.

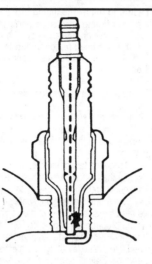

Wide Gap
Spark plug electrodes are worn so that the high voltage charge cannot arc across the electrodes. Improper gapping of electrodes on new or "cleaned" spark plugs could cause a similar condition. Fuel remains unburned and a power loss results.

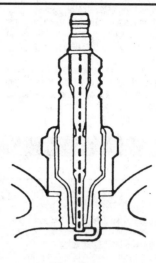

Flashover
A damaged spark plug boot, along with dirt and moisture, could permit the high voltage charge to short over the insulator to the spark plug shell or the engine. AC's buttress insulator design helps prevent high voltage flashover.

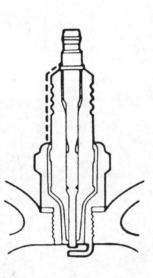

Fouled Spark Plug
Deposits that have formed on the insulator tip may become conductive and provide a "shunt" path to the shell. This prevents the high voltage from arcing between the electrodes. A power and fuel loss is the result.

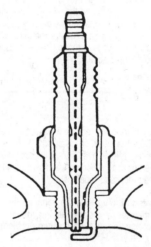

Bridged Electrodes
Fouling deposits between the electrodes "ground out" the high voltage needed to fire the spark plug. The arc between the electrodes does not occur and the fuel air mixture is not ignited. This causes a power loss and exhausting of raw fuel.

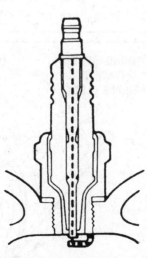

Cracked Insulator
A crack in the spark plug insulator could cause the high voltage charge to "ground out." Here, the spark does not jump the electrode gap and the fuel air mixture is not ignited. This causes a power loss and raw fuel is exhausted.

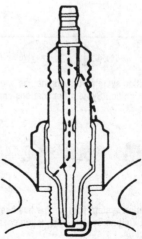

Fig. 68 Typical spark plug problems showing damage that may indicate engine problems

because the reading may be inaccurate. A round wire-type gapping tool is the best way to check the gap. The correct gauge should pass through the electrode gap with a slight drag. If you're in doubt, try a wire that is one size smaller or larger. The smaller gauge should go through easily, while the larger one shouldn't go through at all.

Wire gapping tools usually have a bending tool attached. Use this tool to adjust the side electrode until the proper distance is obtained. Never attempt to bend the center electrode. Also, be careful not to bend the side electrode too far or too often as it may weaken and break off within the engine, requiring removal of the cylinder head to retrieve it.

Spark Plug Wires

TESTING DVOM

Each time you remove the engine cover, visually inspect the spark plug wires for burns, cuts or breaks in the insulation. Check the boots on the coil and at the spark plug end. Replace any wire that is damaged.

Once a year, usually when you change your spark plugs, check the resistance of the spark plug wires with an ohmmeter. Wires with excessive resistance will cause misfiring and may make the engine difficult to start. In addition worn wires will allow arcing and misfiring in humid conditions.

Remove the spark plug wire from the engine. Test the wires by connecting one lead of an ohmmeter to the coil end of the wire and the other lead to the spark plug end of the wire. Resistance should measure approximately 3000-7000 ohms per foot of wire. If a spark plug wire is found to have excessive (high) resistance, the entire set should be replaced.

REMOVAL & INSTALLATION

When installing a new set of spark plug wires, replace the wires one at a time so there will be no confusion. Coat the inside of the boots with dielectric grease to prevent sticking. Install the boot firmly over the spark plug until it clicks into place. The click may be felt or heard. Gently pull back on the boot to assure proper contact. Repeat the process for each wire.

■ **It is important to route the new spark plug wire the same as the original and install it in a similar manner on the engine. Improper routing of spark plug wires may cause engine performance problems.**

Ignition Timing

As the engine must be running while performing this operation we recommend that it is undertaken with the boat in the water. If not, make certain that an engine flushing kit has been installed.

ADJUSTMENT

■ **If you are unsure of which ignition system your engine has been equipped with, please refer to the Ignition System Applications chart.**

Breaker Point Ignition Systems

◆ See Figures 71, 72 and 73

■ **If you are unsure of which ignition system your engine has been equipped with, please refer to the Ignition System Application chart found in the Ignition Systems section.**

■ **Failure to follow the timing procedure instructions exactly will result in improper timing and cause performance problems at the least and possibly severe engine damage. If the timing is completely off, or if the distributor has been removed and the engine rotated, please refer to the initial timing procedures detailed in the Distributor Installation section.**

1. Confirm that the dwell angle is within specifications, adjust if necessary.
2. Connect a suitable timing light to the No. 1 spark plug lead (see Firing Order illustrations for location of the No. 1 cylinder). Connect the power supply lead to the battery as detailed in the light manufacturer's instructions.
3. Connect a tachometer to the engine as detailed by the manufacturer. Do not use the tachometer on the instrument panel as it will not provide the necessary accuracy.
4. Locate the timing mark scale on the engine's timing chain cover (just above the crankshaft pulley/harmonic balancer) and place a bit of white paint where the proper mark should be (TDC is usually marked with a **0**, while each mark or gradation should equal 2°). Also, paint a dab on the mark stamped into the pulley (4 cyl.) or harmonic balancer (V6/V8). On Ford engines, the scale is actually on the harmonic balancer and there is a small pointer attached to the front cover.

■ **Please refer to the Tune-Up Specifications chart for the correct timing figure. You may also be able to check the ignition timing specification on the engine tune-up sticker affixed to the engine. If this figure differs from that which is listed here, ALWAYS go with the figure on the sticker.**

Fig. 71 Typical timing mark tab—GM engine

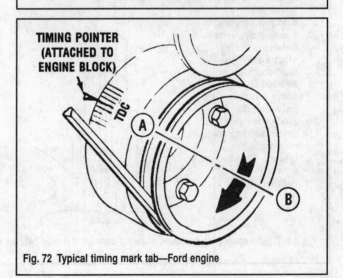

Fig. 72 Typical timing mark tab—Ford engine

5. Start the engine and allow it to reach normal operating temperature at idle—the choke valve should be wide open.

6. Check that the idle speed is to specification (as per the Tune-Up Specifications chart) with the engine in gear and then shift it to Neutral.

7. While still idling, point the light at the timing marks. The strobe will make it appear that the mark on the tab and the mark on the pulley stand still in alignment.

8. If the timing requires adjustment, loosen the clamp bolt at the base of the distributor and then carefully rotate the distributor or sensor until the correct marks line up.

9. Tighten the clamp bolt to 20 ft. lbs. (27 Nm) and check the timing one last time.

10. Turn off the engine.

■ Please refer to the Tune-Up Specifications chart for the correct timing figure. You may also be able to check the ignition timing specification on the engine tune-up sticker affixed to the engine. If this figure differs from that which is listed here, ALWAYS go with the figure on the sticker.

11. Turn off the engine. Disconnect the timing light and the tachometer.

BID And HEI Ignition Systems

◆ See Figures 71, 72 and 73

■ If you are unsure of which ignition system your engine has been equipped with, please refer to the Ignition System Application chart in the Ignition Systems section.

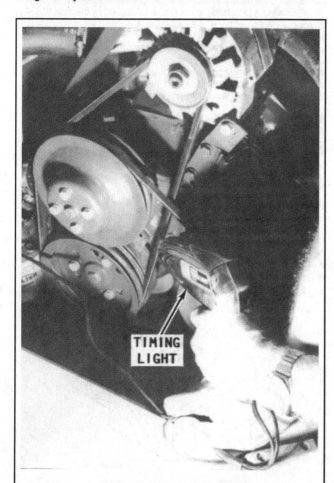

Fig. 73 Always use a timing light to adjust the ignition timing

■ Failure to follow the timing procedure instructions exactly will result in improper timing and cause performance problems at the least and possibly severe engine damage.

1. Connect a suitable timing light to the No. 1 spark plug lead (see firing order illustrations for location of the No. 1 cylinder). Connect the power supply lead to the battery as detailed in the light manufacturer's instructions.

2. Connect a tachometer to the engine as detailed by the manufacturer. Do not use the tachometer on the instrument panel as it will not provide the necessary accuracy.

3. Locate the timing mark scale on the engine's timing chain cover (just above the crankshaft pulley/harmonic balancer) and place a bit of white paint where the proper mark should be (TDC is usually marked with a **0**, while each mark or gradation should equal 2°). Also, paint a dab on the mark stamped into the pulley (4 cyl.) or harmonic balancer (V6/V8). On Ford engines, the scale is actually on the harmonic balancer and there is a small pointer attached to the front cover.

■ Please refer to the Tune-Up Specifications chart for the correct timing figure. You may also be able to check the ignition timing specification on the engine tune-up sticker affixed to the engine. If this figure differs from that which is listed here, ALWAYS go with the figure on the sticker.

4. Start the engine and allow it to reach normal operating temperature at idle—the choke valve should be wide open.

5. Check that the idle speed is to specification (as per the Tune-Up Specifications chart) with the engine in gear and then shift it to Neutral.

■ Always ensure the air gap between the sensor and the trigger wheel is correct before checking ignition timing.

6. While still idling, point the light at the timing marks. The strobe will make it appear that the mark on the tab and the mark on the pulley stand still in alignment.

7. If the timing requires adjustment, loosen the clamp bolt at the base of the distributor and then carefully rotate the distributor or sensor until the correct marks line up.

8. Tighten the clamp bolt to 20 ft. lbs. (27 Nm) and check the timing one last time.

9. Restart the engine and recheck the idle speed and mixture. Adjust as necessary.

■ Please refer to the Tune-Up Specifications chart for the correct specifications. You may also be able to check the ignition timing specification on the engine tune-up sticker affixed to the engine. If this figure differs from that which is listed here, ALWAYS go with the figure on the sticker.

EST Ignition Systems

■ If you are unsure of which ignition system your engine has been equipped with, please refer to the Ignition System Application chart found in the Ignition Systems section.

Carbureted Engines

◆ See Figures 71, 72, 73 and 74

■ Failure to follow the timing procedure instructions exactly will result in improper timing and cause performance problems at the least and possibly severe engine damage.

1. Connect a suitable timing light to the No. 1 spark plug lead (see firing order illustrations for location of the No. 1 cylinder). Connect the power supply lead to the battery as detailed in the light manufacturer's instructions.

2. Connect a tachometer to the engine as detailed by the manufacturer. Do not use the tachometer on the instrument panel as it will not provide the necessary accuracy.

3. Locate the timing mark scale on the engine's timing chain cover (just above the crankshaft pulley/harmonic balancer) and place a bit of white paint where the proper mark should be (TDC is usually marked with a **0**, while each mark or gradation should equal 2°). Also, paint a dab on the mark stamped into the pulley (4 cyl.) or harmonic balancer (V6/V8). On Ford engines, the scale is actually on the harmonic balancer and there is a small pointer attached to the front cover.

■ **Please refer to the Tune-Up Specifications chart for the correct timing figure. You may also be able to check the ignition timing specification on the engine tune-up sticker affixed to the engine. If this figure differs from that which is listed here, ALWAYS go with the figure on the sticker.**

4. Before moving forward, you must bypass the electronic spark advance function of the EST system. Disconnect the 4-wire connector on the side of the distributor and install OMC Adaptor #986662 across the two white leads. Connect the bare wire to a 12-volt engine power source.

■ **Not all models will have a harness connected to the 4-wire connector on the distributor. If there is nothing attached, simply remove the connector cover.**

5. Start the engine and allow it to reach normal operating temperature at idle—the choke valve should be wide open.
6. Check that the idle speed is to specification (as per the Tune-Up Specifications chart) with the engine in gear and then shift it to Neutral.
7. While still idling, point the light at the timing marks. The strobe will make it appear that the mark on the tab and the mark on the pulley stand still in alignment.
8. If the timing requires adjustment, loosen the clamp bolt at the base of the distributor and then carefully rotate the distributor or sensor until the correct marks line up.

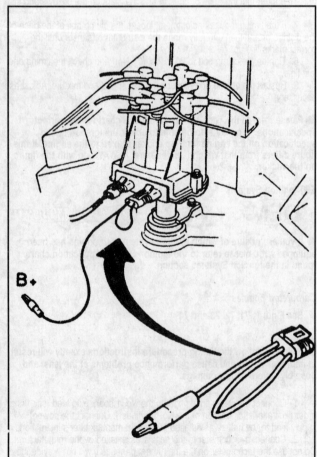

Fig. 74 Before checking the ignition timing on EST systems, you must install the special connector harness

9. Tighten the clamp bolt securely and check the timing one last time.
10. Turn off the engine, remove the adapter plug from the distributor and reconnect the 4-wire connector (or install the cap if there was no connector).
11. Restart the engine. With the timing light still connected and the engine running at 1000 rpm, check that the timing advanced to 15° BTDC on all 1991-94 engines. Re-adjust as necessary by rotating the distributor.

■ **Please refer to the Tune-Up Specifications chart for the correct timing figure. You may also be able to check the ignition timing specification on the engine tune-up sticker affixed to the engine. If this figure differs from that which is listed here, ALWAYS go with the figure on the sticker.**

12. Turn off the engine. Disconnect the timing light and the tachometer.

EFI Engines

◆ **See Figures 71, 72, 73 and 75**

Although ignition timing is adjustable on these models, it is generally controlled by the EFI electronic control module. In order to adjust the timing, the ECM must be forced to enter into its service mode by using a scan tool. This done, the ECM will stabilize the base timing to allow for adjustment by conventional means of rotating the distributor.

■ **The idle speed must be correctly adjusted and within specifications before performing this procedure.**

1. Open or remove the engine compartment hatch.
2. Connect a suitable timing light to the No. 1 spark plug lead (see firing order illustrations for location of the No. 1 cylinder). Connect the power supply lead to the battery as detailed in the light manufacturer's instructions.
3. Connect a tachometer to the engine as detailed by the manufacturer. Do not use the tachometer on the instrument panel, as it will not provide the necessary accuracy.
4. Locate the timing mark scale on the engine's timing chain cover (just above the crankshaft pulley/harmonic balancer) and place a bit of white paint where the proper mark should be (TDC is usually marked with a **0**, while each mark or gradation should equal 2°). Also, paint a dab on the mark stamped into the pulley (4 cyl.) or harmonic balancer (V6/V8).

■ **Please refer to the Tune-Up Specifications chart for the correct timing figure. You may also be able to check the ignition timing specification on the engine tune-up sticker affixed to the engine. If this figure differs from that which is listed here, ALWAYS go with the figure on the sticker.**

5. Start the engine and allow it to reach normal operating temperature at idle. Set the idle speed to 1000 rpm.
6. Turn off the engine, locate the data link connector (DLC) on the EFI/MPI main harness (usually at the front of the engine on the upper starboard side) and plug in OMC's Marine Diagnostic Trouble Code (MDTC) tool—please refer to the Fuel System section for further details on this. If an MDTC is not available, use a jumper wire and connect it between the white/black and black wire terminals on the DLC.

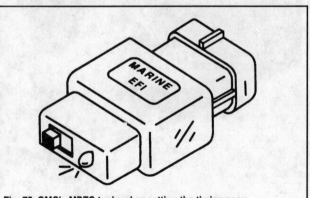

Fig. 75 OMC's MDTC tool makes setting the timing easy

7. Restart the engine and allow the idle to stabilize.

8. Set the MDTC to the **ON** position.

9. While still idling, point the light at the timing marks. The strobe will make it appear that the mark on the tab and the mark on the damper/pulley stand still in alignment.

10. If the timing requires adjustment, loosen the clamp bolt at the base of the distributor and then carefully rotate the distributor until the correct marks line up.

11. Tighten the clamp bolt securely and then recheck the timing.

12. Disconnect the tool or the jumper wire.

13. Check the timing one last time. If still correct, disconnect the light and tachometer.

Spitfire EEM Ignition Systems

These engine utilize a timing ring mounted on the crankshaft between the flywheel and the engine coupler which directs the EEM module when to fire a particular cylinder. No ignition timing adjustment is necessary or required. Please refer to the Ignition Systems section for further information on the Spitfire system.

Thick Film Integrated (TFI-IV) Ignition Systems

◆ **See Figures 72 and 73**

■ **If you are unsure of which ignition system your engine has been equipped with, please refer to the Ignition System Application chart found in the Ignition Systems section.**

■ **Failure to follow the timing procedure instructions exactly will result in improper timing and cause performance problems at the least and possibly severe engine damage.**

1. Position the shift lever in Neutral.

2. Connect a suitable variable advance, inductive timing light to the No. 1 spark plug lead (see Firing Order illustrations for location of the No. 1 cylinder). Connect the power supply lead to the battery as detailed in the light manufacturer's instructions.

3. Connect a tachometer to the engine as detailed by the manufacturer. Do not use the tachometer on the instrument panel as it will not provide the necessary accuracy.

4. Locate the timing mark scale on the harmonic balancer and place a bit of white paint where the proper mark should be (TDC is usually marked with a **0**, while each mark or gradation should equal 2°).

■ **Please refer to the Tune-Up Specifications chart for the correct timing figure. You may also be able to check the ignition timing specification on the engine tune-up sticker affixed to the engine. If this figure differs from that which is listed here, ALWAYS go with the figure on the sticker.**

5. Start the engine and allow it to reach normal operating temperature at idle.

6. Locate the two-wire SPOUT connector (near the front of the engine) and carefully pull out the shorting bar.

7. While still idling, point the light at the timing marks. The strobe will make it appear that the mark on the tab and the mark on the pulley stand still in alignment.

8. If the timing requires adjustment, loosen the clamp bolt at the base of the distributor and then carefully rotate the distributor until the correct marks line up.

9. Apply Electrical Terminal grease to the SPOUT connector terminals and then reinsert the shorting bar.

10. Check the timing advance to ensure it is advancing beyond the initial setting.

11. Disconnect the timing light and tachometer.

12. Tighten the distributor clamp bolt to 17-25 ft. lbs. (23-34 Nm).

13. Restart the engine and recheck the idle speed and mixture. Adjust as necessary.

■ **Please refer to the Tune-Up Specifications chart for the correct specifications. You may also be able to check the ignition timing specification on the engine tune-up sticker affixed to the engine. If this figure differs from that which is listed here, ALWAYS go with the figure on the sticker.**

Breaker Points and Condenser

All 1986-89 models and many 1990 models were equipped with a standard breaker points ignition system. There are two ways to check breaker point gap: with a feeler gauge or with a dwell meter. Either way you choose, you are adjusting the amount of time (in degrees of distributor rotation) that the points will remain open. If you adjust the points with a feeler gauge, you are setting the maximum amount the points will open when the rubbing block on the points is on one of the high points of the distributor cam. When you adjust the points with a dwell meter, you are measuring the number of degrees (of distributor cam rotation) that the points will remain closed before they start to open as a high point of the distributor cam approaches the rubbing block of the points.

Although using a feeler gauge is reasonably accurate when setting new point sets, this method can be unreliable when checking used points due to the rough surface caused by pitting associated with wear and tear. Adjusting the dwell should always be considered the more accurate method.

There are two rules that should always be followed when adjusting or replacing points:

• Points and condenser are a matched set; NEVER replace one without replacing the other.

• When you change the point gap or dwell of the engine, you also change the ignition timing. Always adjust the timing after a point or dwell adjustment.

■ **Marine distributors have a corrosion-resistant coating applied to the return spring on top of the breaker plate and on the two small springs under the plate - NEVER use automotive parts as a replacement!**

REMOVAL AND INSTALLATION

◆ **See Figures 76, 77 and 78**

1. Remove or open the engine compartment hatch cover. Disconnect the negative battery cable or turn the battery switch to OFF.

2. Loosen the distributor cap retaining screws (two) and carefully lift off the cap. Although it is not necessary remove the spark plug wires from the cap, we recommend that you first tag all the wires just to be safe.

3. Note the position of the rotor and pull the rotor straight up and remove it. Check the rotor carefully for a burned or corroded center contact, cracks or carbon tracks.

4. Loosen the primary terminal nut and then disconnect the lead wire. Do the same for the condenser lead wire.

5. Loosen the condenser/breaker point mounting screws and then lift them up and off the breaker plate. Clean any dirt or oil left on the plate.

6. Coat the distributor cam with a small amount of distributor Cam Lubricant (NEVER use grease or oil), wipe the new point set clean and position it on the breaker plate. Tighten the mounting screws, leaving the lock screw slightly loose..

7. Reconnect the lead wires for the condenser and primary.

8. Check that the points are in alignment. If not, carefully bend the stationary arm until they align properly. If you are still not satisfied, get a new set of points. Never adjust alignment on used points.

9. Adjust the point gap as detailed following. Install the rotor in the same position is was removed. Install the cap, connect the battery cables and check for proper operation.

10. Adjust the ignition timing

Fig. 76 Pull the rotor straight up to remove it

Fig. 77 The condenser is held in place by a screw and clamp

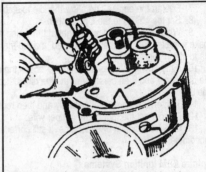

Fig. 78 Install the point set on the breaker plate, then attach the wires

ADJUSTMENT—FEELER GAUGE

◆ See Figures 79, 80 and 81

1. Perform the first three steps of the removal procedure above.
2. Connect a remote starter switch as detailed in the manufacturers' instructions. Have a friend bump the engine over until you see that the breaker point rubbing block is resting on the high point of the distributor cam—the points should open to their fullest extent.
3. Insert an feeler gauge between the points—please refer to the Tune-Up Specifications chart for correct point gap. The gauge should be snug but not tight. If adjustment is required, loosen the lock screw and insert a screwdriver in to the adjustment slot on the breaker plate; move the point set until a slight drag can be felt on the feeler gauge and then tighten the lock screw. Always check the gap a final time after tightening the screw as the points sometimes move when tightening the screw.
4. Install the rotor and distributor cap. Connect the battery cables and adjust the ignition timing.

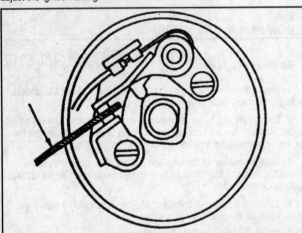

Fig. 79 The arrow points to the feeler gauge used to measure point gap

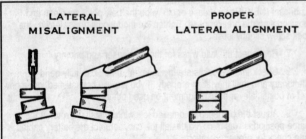

Fig. 80 New contact points must be aligned by bending the fixture contact support. NEVER bend the breaker lever!

ADJUSTMENT—DWELL METER

◆ See Figure 82

1. Perform the first three steps of the removal procedure above.
2. Connect a dwell meter as per the manufacturer's instructions - usually the positive lead of the meter to the negative side of the coil and the negative lead of the meter to ground.
3. Connect a remote starter switch as detailed in the manufacturer's instructions and then crank the engine. Observe the dwell reading on the meter, it should be as listed in the Tune-Up Specifications chart. If not in range, loosen the lock screw slightly and then adjust the point opening by means of the adjustment slot. Increasing the point gap lowers the dwell reading, while decreasing the gap raises it. When the reading is within specifications, tighten the lock screw and then recheck the dwell one final time.
4. Install the rotor and distributor cap. Connect the battery cables and adjust the ignition timing.

✶✶ CAUTION

Dwell should be checked between idle and 1750 rpm. Any dwell variations of more than 3° from idle to 1750 rpm indicate possible wear in the distributor.

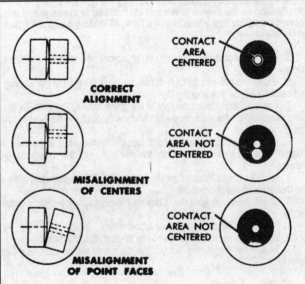

Fig. 81 Check the points for proper alignment after installation

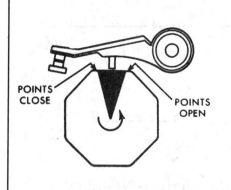

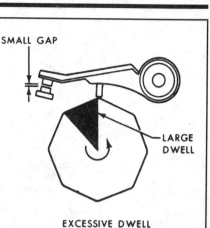

Fig. 82 Three views of differing point dwell conditions

Distributor Air Gap

All models equipped with BID ignition systems require an air gap adjustment between the sensor wheel and the trigger wheel; particularly after the distributor has been removed. It is also a good idea to ensure that the air gap is within specification prior to performing any ignition timing procedures

ADJUSTMENT—BID DISTRIBUTORS ONLY

◆ See Figure 83

1. Open or remove the engine hatches.
2. Remove the distributor cap and pull off the rotor as detailed in the Ignition System section.
3. Connect a remote starter switch as detailed in the manufacturer's instructions.
4. Check the sensor coil, trigger wheel and teeth for signs of wear or other obvious damage. Using the remote starter, crank the engine around until a tooth on the wheel aligns with the contact on the coil.
5. Insert a non-metallic flat-bladed feeler gauge between the sensor and trigger wheel tooth. The gap should be 0.008 in. (0.20mm).
6. If not within specification, loosen the small screw on the trigger coil slightly and move the coil carefully until the gap is within spec. Tighten the screw securely—but not too tight— and recheck that the gap has not changed.
7. Disconnect the remote starter.
8. Install the rotor and distributor cap.

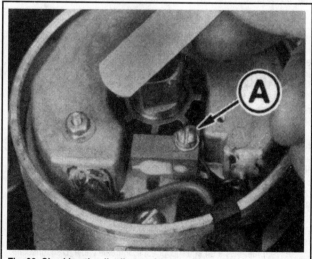

Fig. 83 Checking the distributor air gap

Idle Speed And Mixture

ADJUSTMENT

All Engines W/2bbl Carburetor

◆ See Figures 84 and 85

1. Locate the idle speed screw on your carburetor and turn it in or out until it is just resting on the idle cam, but is not moving it. The idle screw is threaded into the throttle linkage on the side of the carburetor and has a spring between the screw head and the linkage.
2. Locate the idle mixture screw on the base of the carburetor and turn it in (clockwise) until it just lightly seats itself and then back it out:
- 1986-89 2.5L/3.0L engines—2 turns
- 1986-89 2.3L, 4.3L and 5.0L engines—2 1/2 turns
- 1988-90 Rochester—2 turns
- 1988-90 2.3L engines—1/2 turn
- 1988-93 3.0L engines—3/4 turn
- 1988-93 3.0L HO engines—1 turn on starboard screw; 1/2 turn on port screw
- 1994 3.0GL engines—3/4 turn
- 1994-98 3.0GS engines—1 turn on starboard screw; 1/2 turn on port screw
- 1988-94 4.3L engines—5/8 turn
- 1995-98 4.3GL engines—5/8 turn
- 1996-97 4.3GS engines—3/4 turn
- 1988-90 5.0L engines—1 1/2 turns
- 1991-93 5.0L engines—1 1/8 turns
- 1990-93 5.0L HO engines—1 turn
- 1994-96 5.0FL engines—1 turn
- 1994 5.7GL engines—3/4-1 turn
- 1997-98 5.7GL/GS engines—3/4-1 turn

✳✳ CAUTION

Be careful not to turn the idle mixture screw in past the seated position or you risk damaging the seat or needle. Both the screw tip and the casting seat can be affected. Replace any screws suspected of damage.

3. Ensure that the flame arrestor is in place and is free of debris and obstructions. A clogged filter will greatly impact this adjustment.
4. Connect a tachometer as per the manufacturer's instructions—do not use the tach on your boat!
5. Start the engine and allow it to run at idle until it reaches normal operating temperature.

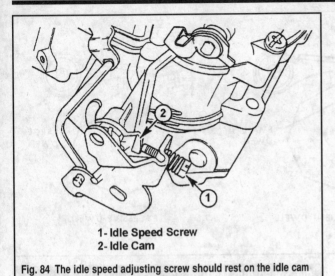

1- Idle Speed Screw
2- Idle Cam

Fig. 84 The idle speed adjusting screw should rest on the idle cam

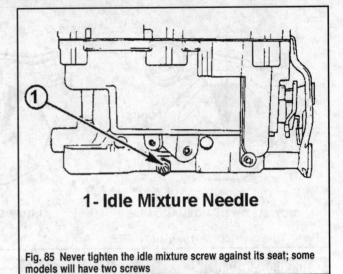

1- Idle Mixture Needle

Fig. 85 Never tighten the idle mixture screw against its seat; some models will have two screws

✳✳ CAUTION

Water must circulate through the lower unit and to the engine anytime the engine is being run. Severe damage could result otherwise. If the boat is not in the water, make sure a flushing attachment has been installed properly.

6. You will need to disconnect the throttle cable from the throttle lever before performing the next step because the boat will need to be in gear.

7. With the engine still running and the tach hooked up, move the shift lever into **F** and turn the mixture screw(s) in (clockwise), carefully and slowly, until the engine speed JUST begins to drop due to a LEAN condition. Jot down the number of turns. If your carburetor has two screws, and it most likely will, there are two schools of thought here—you can adjust them individually or alternately. We prefer the alternating method; that is, turning each screw in a little at a time, the same amount, until you get the LEAN condition. If you go this route, we suggest turning them alternately no more than 1/8th turn at a time.

8. Now, back the screw(s) out (counterclockwise), evenly and alternately, until you achieve the correct idle speed as detailed in the Tune-Up Specifications chart. If the idle begins to drop before you get to the correct speed (due to a RICH condition this time) then back them down slightly until it runs smooth again.

9. Recheck the idle speed and compare it to the recommended speed in the chart. If adjustment is required, now you can use the idle speed stop screw— turn it, gradually and evenly, until the correct idle is achieved.

10. Turn the engine off and reconnect the throttle cable. Disconnect the tachometer.

All Engines W/4BBL Carburetor

 MODERATE

1. Locate the idle speed screw on your carburetor and turn it in or out until it is just resting on the idle cam, but is not moving it. The idle screw is threaded into the throttle linkage on the side of the carburetor and has a spring between the screw head and the linkage.

2. Locate the idle mixture screw on the base of the carburetor and turn it in (clockwise) until it just lightly seats itself and then back it out:
- 1986 engines—3 3/4 turns
- 1987-89 engines w/Rochester—3 3/4 turns
- 1987-89 engines w/Holley—1 1/2 turns
- 1990-91 4.3L HO engines—3/4 turn
- 1996-97 4.3GS engines—3/4 turn
- 1998 4.3GS engines—1 1/8 turns
- 1990-92 5.7L engines—1/4 turn
- 1990-91 5.7L LE engines—1 turn
- 1991-92 5.7L engines—1/2 turn
- 1997 5.7GS engines—1/2 turn

- 1990 350 engines—1 turn
- 1990 5.8L engines—1 1/2 turns
- 1991-96 5.8L engines—7/8 turn
- 1990 454 engines—1 turn
- 1991-97 7.4L/8.2L engines—1/4 turn
- 1990 460 engines—1 1/2 turns

✳✳ CAUTION

Be careful not to turn the idle mixture screw in past the seated position or you risk damaging the seat or needle. Both the screw tip and the casting seat can be affected. Replace any screws suspected of damage.

3. Ensure that the flame arrestor is in place and is free of debris and obstructions. A clogged filter will greatly impact this adjustment.

4. Connect a tachometer as per the manufacturer's instructions—do not use the tach on your boat!

5. Start the engine and allow it to run at idle until it reaches normal operating temperature.

✳✳ CAUTION

Water must circulate through the lower unit and to the engine anytime the engine is being run. Severe damage could result otherwise. If the boat is not in the water, make sure a flushing attachment has been installed properly.

6. You will need to disconnect the throttle cable from the throttle lever before performing the next step because the boat will need to be in gear.

7. With the engine still running and the tach hooked up, move the shift lever into **F** and turn the mixture screw(s) in (clockwise), carefully and slowly, until the engine speed JUST begins to drop due to a LEAN condition. Jot down the number of turns. If your carburetor has two screws, and it most likely will, there are two schools of thought here—you can adjust them individually or alternately. We prefer the alternating method; that is, turning each screw in a little at a time, the same amount, until you get the LEAN condition. If you go this route, we suggest turning them alternately no more than 1/8th turn at a time.

8. Now, back the screw(s) out (counterclockwise), evenly and alternately, until you achieve the correct idle speed as detailed in the Tune-Up Specifications chart. If the idle begins to drop before you get to the correct speed (due to a RICH condition this time) then back them down slightly until it runs smooth again.

9. Recheck the idle speed and compare it to the recommended speed in the chart. If adjustment is required, now you can use the idle speed stop screw— turn it, gradually and evenly, until the correct idle is achieved.

10. Turn the engine off and reconnect the throttle cable. Disconnect the tachometer.

Fuel Injected Engines

Idle speed is constantly monitored by the electronic control module (ECM) and controlled by the idle air control valve (IAC). Idle speed and mixture are not adjustable. Please refer to the Fuel System section for further information on the fuel injection system.

PCV Valve

Many engines are equipped with a positive crankcase ventilation (PCV) circuit that utilizes a PCV valve in the rocker cover in order to ventilate unburned crankcase gases back into the engine via the intake manifold in order that they can be re-burned. The PCV valve should be replaced every boating season or 100 hours of operation.

A PCV system that is malfunctioning can cause rough running or idle, and also increased fuel consumption. Do not attempt to disconnect or bypass the system.

REMOVAL & INSTALLATION

◆ See Figure 86

1. Locate the PCV valve in the cylinder head cover—usually the Port but it could be either.
2. Carefully wiggle it back and forth while pulling upward on the valve itself until it pops out of the cover.
3. Loosen the clamp (if equipped) and disconnect the breather hose from the valve.
4. Reconnect the hose and press the valve back into the cylinder head.

■ PCV valves are not serviceable. If your valve is clogged or otherwise not working properly, it must be replaced with a new one.

INSPECTION

◆ See Figure 87

Start the engine and allow it to reach normal operating temperature. Pop out the PCV valve as detailed above and cover the opening with your thumb. You should be able to feel significant vacuum; if not replace the valve. With the valve still in your fingers, shake it back and forth a few times; you should be able to hear the inside components moving around—a significant clicking sound.

If the valve passes these two tests, it is functioning properly. If it fails either of the tests and there are no leaks in any of the hoses or connections, it will require replacement.

■ PCV valves are not serviceable. If your valve is clogged or otherwise not working properly, it must be replaced with a new one.

FLUIDS AND LUBRICANTS

Fluid Disposal

Used fluids such as engine oil, gear oil, antifreeze and power steering fluid are hazardous waste and must be disposed of properly and responsibly. Before draining any fluids it is always a good idea to check with your local authorities; in many areas there are recycling programs available for easy disposal. Service stations, Parts stores and Marinas also often will accept waste fluids for recycling.

Be sure of your local recyclers' policies before draining any fluids, as many will not accept fluids that have been mixed together.

Fuel And Oil Recommendations

FUEL

All engine covered in this manual are designed to run on unleaded fuel. Never use leaded fuel in your boat's engine. The minimum octane rating of fuel being used for your engine must be at least 86 AKI (outside the US, 90

Fig. 86 Typical PCV valve

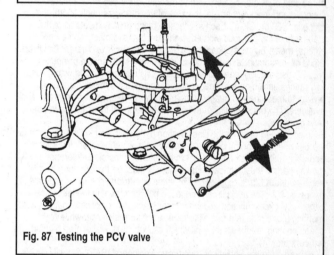
Fig. 87 Testing the PCV valve

Valve Adjustment

All engines covered in this manual are equipped with hydraulic valve lifters and do not require periodic valve adjustment. Adjustment to zero lash is maintained automatically by the hydraulic pressure in the lifters. For initial adjustment procedures after cylinder head work or if a problem is expected, please refer to the appropriate Engine Mechanical section.

RON), on early models and 89 AKI (93 RON) on all others, but some engines may require higher octane ratings. OMC actually recommends the use of 89 AKI (93 RON) fuel as the ideal—in fact, anything less than this on many 4.3L, 5.0GL and 5.7L engines will require a change to the ignition timing. Fuel should be selected for the brand and octane that performs best with your engine. Check your owner's manual if in doubt.

The use of a fuel too low in octane (a measure of anti-knock quality) will result in spark knock. Newer systems have the capability to adjust the engine's ignition timing to compensate to some extent, but if persistent knocking occurs, it may be necessary to switch to a higher grade of fuel. Continuous or heavy knocking may result in engine damage.

ENGINE OIL

Nothing affects the performance and durability of an engine more than the engine oil. If inferior oil is used, or if your engine oil is not changed regularly, the risk of piston seizure, piston ring sticking, accelerated wear of the cylinder walls or liners, bearings and other moving components increases significantly.

Maintaining the correct engine oil level is one of the most basic (and essential) form of engine maintenance. Get into the habit of checking your oil on a regular basis; all engines naturally consume small amounts of oil, and if left neglected, can consume enough oil to damage the internal components of the engine. Assuming the oil level is correct because you "checked it the last time" can be a costly mistake.

If your engine has not been operated for more than 6 months it should be primed prior to starting.

When shutting the engine down, always let the engine idle a few minutes to bring engine temperature down to a normal level. Since the engine is, at least in part, cooled by engine oil, it is necessary to allow the engine oil temperature to stabilize prior to shutdown. Not allowing the temperature to stabilize can damage vital engine components.

Every container of engine oil for sale in the U.S. should have a label describing what standards it meets. Engine oil service classifications are designated by the American Petroleum Institute (API), based on the chemical composition of a given type of oil and testing of samples. The ratings include "S" (normal gasoline engine use) and "C" (commercial, fleet and diesel) applications. Over the years, the "S" rating has been supplemented with various letters, each one representing the latest and greatest rating available at the time of its introduction. During recent years these ratings have changed and most recently (at the time of this manual's publication), the rating is "SG" or "CH-4". Each successive rating usually meets all of the standards of the previous alpha designation, but also meets some new criteria, meets higher standards and/or contains newer or different additives. Since oil is so important to the life of your engine, you should obviously NEVER use an oil of questionable quality. Oils that are labeled with modern API ratings, including the "energy conserving" donut symbol, have been proven to meet the API quality standards. Always use the highest grade of oil available. The better quality of the oil, the better it will lubricate the internals of your engine.

In addition to meeting the classification of the API, your oil should be of a viscosity suitable for the outside temperature in which your engine will be operating. Oil must be thin enough to get between the close-tolerance moving parts it must lubricate. Once there, it must be thick enough to separate them with a slippery oil film. If the oil is too thin, it won't separate the parts; if it's too thick, it can't squeeze between them in the first place—either way, excess friction and wear takes place. To complicate matters, cold-morning starts require a thin oil to reduce engine resistance, while high speed boating requires a thick oil which can lubricate vital engine parts at temperatures.

According to the Society of Automotive Engineers' (SAE) viscosity classification system, an oil with a high viscosity number (such as SAE 40 or SAE 50) will be thicker than one with a lower number (SAE 10W). The "W" in 10W indicates that the oil is desirable for use in winter operation, and does not stand for "weight". Through the use of special additives, multiple-viscosity oils are available to combine easy starting at cold temperatures with engine protection at high speeds. For example, a 10W40 oil is said to have the viscosity of a 10W oil when the engine is cold and that of a 40 oil when the engine is warm. The use of such an oil will decrease engine resistance and improve efficiency.

All OMC engines require the use of SAE 30 oil at temperatures above 32°F (0°C). Engines operating in 0°F (-18°C) to 32°F (0°C) temperatures ranges should use SAE 20W-20 oil; while anyone operating under 0° F (-18°C) should use SAE 10W oil.

■ **Although acceptable in a pinch and also in certain weather conditions, multi-viscosity oils such as 10W-30 and 10W-40 are not recommended. OMC suggests any recommendation that you may find on your engine regarding a multi-viscosity oil should be ignored. These references are for automotive use only and not intended for marine applications.**

Priming

■ **Anytime your boat's engine has not been run for more than 6 months it is a good idea to prime the engine prior to starting it.**

1. Check that the proper amount of oil is in the crankcase (see the Capacities chart).
2. Remove the spark plugs.
3. With the ignition key in the **OFF** position, connect a remote starter according to the manufacturer's instructions.

4. Crank the engine for 15 seconds and then allow the starter to cool for 1 minute. Repeat this sequence 2 more times until a total cranking time of 45 seconds has been reached.
5. Remove the remote starter.
6. Install the spark plugs and start the engine.

OIL LEVEL CHECK

◆ **See Figures 88 thru 94**

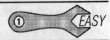

✳✳ CAUTION

The EPA warns that prolonged contact with used engine oil may cause a number of skin disorders, including cancer! You should make every effort to minimize your exposure to used engine oil. Protective gloves should be worn when changing the oil. Wash your hands and any other exposed skin areas as soon as possible after exposure to used engine oil. Soap and water, or waterless hand cleaner should be used.

When checking the oil level, it is best that the boat be level and the oil be at operating temperature. Checking the level immediately after stopping the engine will give a false reading; always wait about 5 minutes before checking.

It is normal for an engine to naturally consume oil during the course of operation, particularly during break-in on a new engine. You should not be alarmed if the oil level in your engine drops slightly between inspections. In fact, certain of OMC's high performance engines may use up to a quart of oil every 5 hours when operated at full throttle.

1. Also the color of the oil is usually a pitch black color. Smelling the oil is a better indicator of oil condition than the color. If the oil smells burned, it should be replaced immediately.

Over-filled crankcases can cause a fluctuation or drop in oil pressure, and, particularly on OMC engines, clattering from the rocker arms. Take great care in checking and filling your engine with oil. Always maintain it between the **ADD** and **FULL** or **OP RANGE** markings on the dipstick. Beware of false readings by checking the level too soon after adding oil.

■ **It takes a little while for fresh oil poured into the engine to reach the crankcase. Wait for about 3 minutes and then check the oil level again.**

Oil level should be checked each day the engine is operated and it is best to check it while in the water.

2. Run the engine until it reaches normal operating temperature. Shut it off and allow the oil to settle for at least 5 minutes.
3. Locate the engine oil dipstick—almost always on either side and toward the front of the engine.
4. Clean the area around the dipstick to prevent dirt from entering the engine.
5. Remove the dipstick and note the color of the oil. Wipe the dipstick clean with a rag.

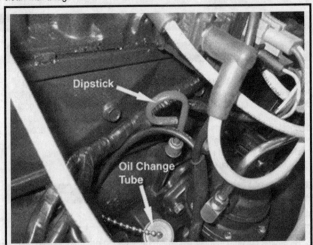

Fig. 88 The oil dipstick on the 3.0L is on the starboard side. Note the top of the oil withdrawal tube

Fig. 89 The oil dipstick on the 4.3L is on the starboard side, near the power steering pump

Fig. 91 The oil dipstick on the 5.0L GM is on the front port side of the engine

Fig. 90 The oil dipstick on the 5.0L Ford is on the front starboard side of the engine

Fig. 92 The oil dipstick on the 5.7L is on the front of the engine, near the alternator

Fig. 93 The oil dipstick on the 7.4L is on the port side of the engine, near the front (8.2L similar)

6. Insert the dipstick fully into the tube and remove it again. Hold the dipstick horizontal and read the level on the dipstick. The level should always be at the upper limit. If the oil level is below the upper limit, sufficient oil should be added to restore the proper level of oil in the crankcase. Most dipsticks are marked with **ADD** and **FULL** or **OP RANGE** gradations.

7. See Engine Oil Recommendations for the proper viscosity and type of oil.

8. Oil is added through the filler port cap in the top of the cylinder head cover. Add oil slowly and check the level frequently to prevent overfilling the engine.

✱✱ WARNING

Do not overfill the engine. If the engine is overfilled, the crankshaft will whip the engine oil into a foam causing loss of lubrication and severe engine damage.

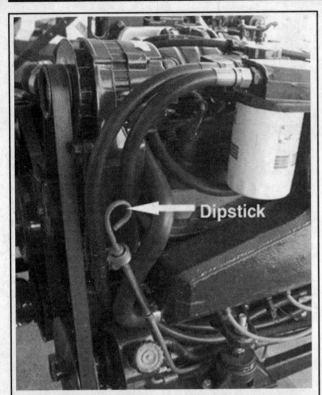

Fig. 94 The oil dipstick on the 7.5L is on the port side of the engine, near the front (8.2L similar)

OIL & FILTER CHANGE

◆ See Figures 95 thru 102

✳✳ CAUTION

The EPA warns that prolonged contact with used engine oil may cause a number of skin disorders, including cancer! You should make every effort to minimize your exposure to used engine oil. Protective gloves should be worn when changing the oil. Wash your hands and any other exposed skin areas as soon as possible after exposure to used engine oil. Soap and water, or waterless hand cleaner should be used.

A few precautions can make the messy job of oil and filter maintenance much easier. By placing oil absorbent pads, available at industrial supply stores, into the area below the engine, you can prevent oil spillage from reaching the bilge.

It is a good idea to warm the engine oil first so it will flow better; and the contaminates in the bottom of the pan willbe suspended in the oil. This is accomplished by starting the engine and allowing it to reach normal operating temperature.

Changing engine oil is sometimes complicated by the location of the drain plug. Most boats equipped with inboard engines use an evacuation pump to remove the used engine oil through the dipstick tube. If you don't have a permanently mounted oil suction pump in your engine compartment, you may want to consider installing one. This pump sucks waste oil out through either the dipstick tube or a connection on the oil drain plug and is available from your OMC dealer or from any number of aftermarket sources. They come in a variety of configurations: motorized, hand-pumped or attachments for an ordinary household drill.

The maintenance interval for oil and filter change is every 100 hours of engine operation or annually, whichever comes first. This interval should be strictly kept; in fact we recommend cutting the interval in half!

■ Since you will be hanging into the engine compartment, gather all the tools and spare parts necessary for the job. Don't forget plenty of rags to clean up any spills, and most importantly, remember a container or plastic milk jug to drain the oil into.

1. Start the engine and run it until it reaches normal operating temperature. Turn the engine off and remove the dipstick and/or the withdrawal tube cap.

2. Connect the evacuation pump hose to the dipstick tube or insert it into the withdrawal tube. Position the other hose, or the outlet at the bottom of the pump in your container. Keep in mind that the fast flowing oil, which will spill out of the pump hose, will flow with enough force that it could miss the container and end up all over the deck or in the bilge. Position the container accordingly and be ready to move it if necessary. Some models are equipped with a quick drain oil hose attachment; be sure to pull the tether through the bilge drain before removing the drain plug from the hose.

✳✳ CAUTION

Use caution around the hot oil; when at operating temperature, it is hot enough to cause a severe burn.

3. Allow the oil to drain until nothing but a few drops come out of the pump. It should be noted that depending on the angle of the engine, some oil may be left in the crankcase. This is normal and should not cause the engine harm.

4. Remove the evacuation pump and reinsert the dipstick or withdrawal tube cap.

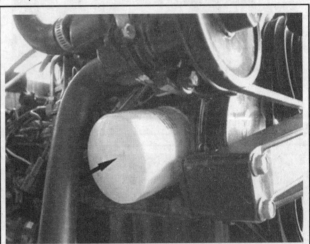

Fig. 95 A good shot of the filter—3.0L shown

Fig. 96 Although most engines use the dipstick tube to remove the old engine oil, certain engines (2.5/3.0L) may use a separate tube

5. Position a drain pan, or a cut-down milk jug under the oil filter. Some filters are mounted horizontally and some vertically. In either case, there is usually oil left in the filter. When the filter is removed, oil will flow out of the engine and the filter. If you are not prepared, you will have a mess on your hands. It may also be necessary to use a funnel or fashion some type of drain shield to guide the oil into the drain pan. This can be as simple as a recycled oil bottle with the bottom cut off or an elaborate creation made of tin. In any case, its purpose is to prevent oil from spilling all over the engine and bilge.

6. To remove the filter, you will probably need an oil filter strap wrench (available at almost any place that sells marine or automotive parts and fluids). Heat from the engine tends to tighten even a properly installed filter and makes it difficult to remove. Place the wrench on the filter as close to the engine as possible, while still leaving room to work. This will put the wrench at the strongest part of the filter (near the threaded end) and prevent crushing the filter. Loosen the filter with the wrench using a counterclockwise turning motion.

■ Some later models may utilize a remote oil filter.

7. Once loosened, wrap a rag around the filter and unscrew it from the boss on the engine. Make sure that the drain pan and shield are positioned properly before you start unscrewing the filter. Should some of the hot oil happen to get on your hands and burn you, dump the filter into the drain pan.

Fig. 97 On the newer V6/V8 engines with fuel injection (5.0L shown), the remote filter is on the port side near the top

Fig. 98 On the older V6/V8 engines and most V6/V8 engine with a carburetor (5.7L shown), look on the side of the engine block

8. Wipe the base of the mounting boss with a clean, dry lint-free rag. If the filter is installed vertically, you may want to fill the filter about half way with oil prior to installation. This will prevent oil starvation when you fire the engine up again. Pre-filling the oil filter is usually not possible with horizontally mounted filters. Smear a little bit of fresh oil on the filter gasket to help it seat properly on the engine.

9. Install the filter and tighten it approximately a quarter-turn after it contacts the mounting boss (always follow the filter manufacturer's instructions). This usually equals "hand tight." Using a wrench to tighten the filter is not required, nor should it be considered.

✳✳ WARNING

Never operate the engine without engine oil. Severe and costly engine damage will result in a matter of seconds without proper lubrication.

10. If any oil has gotten into the bilge, remove it using an oil absorbent pad. These pads are specially formulated to only absorb oil and will not soak up any water in the bilge. Perform a visual inspection and make sure all connections are tight.

11. Carefully remove the drain pan from under the oil filter and transfer the oil into a suitable container for recycling.

12. Refill the engine with the proper quantity and quality of oil immediately (see the Capacities chart). You may laugh at the severity of this warning, but if you wait to refill the engine and someone unknowingly tries to start it, severe and costly engine damage will result.

13. Refill the engine crankcase slowly through the filler cap on the cylinder head cover. Use a funnel as necessary to prevent spilling oil. Check the level often. You may notice that it usually takes less than the amount of oil listed in the Capacities chart to refill the crankcase. This is only until the engine is started and the oil filter is filled with oil.

14. To make sure the proper level is obtained, run the engine to normal operating temperature, turn the engine OFF, allow the oil to drain back into the oil pan and recheck the level after about 5 minutes. Top off the oil to the correct mark on the dipstick.

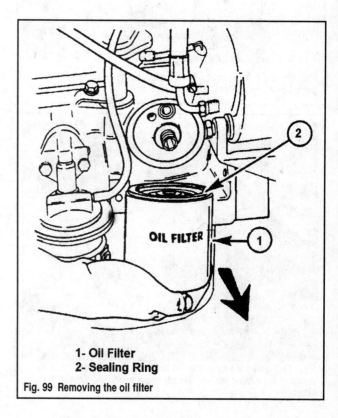

1- Oil Filter
2- Sealing Ring
Fig. 99 Removing the oil filter

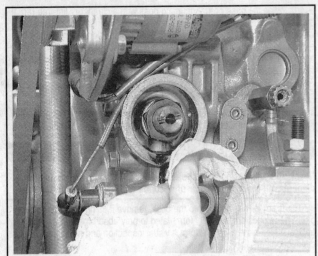

Fig. 100 Before installing the new filter, wipe the gasket mating surface clean

Fig. 101 Dip a finger into a fresh bottle of oil, and lubricate the gasket of the new filter

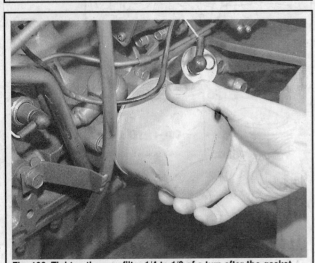

Fig. 102 Tighten the new filter 1/4 to 1/2 of a turn after the gasket contacts the mating surface. Do not use a wrench to tighten the filter!

Power Steering

FLUID LEVEL

◆ See Figures 103 and 104

■ Power steering fluid level should always be checked with the engine hot if at all possible.

1. Start the engine and run it until it reaches normal operating temperature.

2. Rotate the steering wheel lock-to-lock about 10 times to ensure that the fluid is warm.

3. Turn the engine OFF and place the drive unit so that it is straight back.

4. Locate the dipstick cap on top of the power steering fluid reservoir. Unscrew the cap and remove the dipstick. Wipe down the dipstick with a clean rag and reinsert it into the reservoir making sure that the cap is screwed all the way down.

5. Remove the dipstick again and check that the fluid level is up to the **FULL HOT** mark (**FULL COLD** if the level is being checked cold). If the level is between the **FULL** mark and the indent on the stick, it's OK. Below the indent or hash mark and you must add fluid. Use OMC Power Trim/Tilt & Steering Fluid, GM Power Steering Fluid, or Dexron® III ATF fluid. Do not overfill!

6. It is always a good idea to bleed the system after adding fluid.

BLEEDING THE SYSTEM

■ It is important that all air be removed from the system after filling or component removal. If air is left in the system, the fluid in the pump may foam during operation causing discharge or spongy steering.

1. With the engine stopped and the drive unit positioned straight back, check that the fluid level is at the **FULL COLD** mark on the dipstick.

2. Turn the steering wheel all the way over to Port and then check the fluid level again. Add fluid if necessary.

3. Install the cap/dipstick onto the reservoir, start the engine and run it at idle for a minute and then shut it OFF again. Remove the reservoir cap allowing any foam in the pump to escape.

4. Reinsert the dipstick and check the fluid again, making sure that it is at the **FULL HOT** mark this time. Add fluid as necessary. Repeat this step as necessary until the system no longer requires fluid.

5. Start the engine again and move the steering wheel **slowly** from lock to lock several times. Remember the reservoir cap should still be off, so while you're moving the wheel, observe the fluid level in the reservoir—it should be above the pump body, if not, carefully add fluid so that the level stays above the body.

6. Fluid containing air will be foamy and light red or tan in color. This is normal, but you must make sure that the fluid level is high enough so that the foam DOES NOT enter the pump inlet. Continue moving the wheel and adding fluid until there is no foam left, indicating the absence of air bubbles.

■ If you see excessive foaming during the previous step, shut the engine down and let everything sit for 30 minutes before trying it again.

7. Once the foam has disappeared, position the steering wheel at the center position and shut the engine Off. Screw in the cap/dipstick and recheck the fluid level. If not at the **FULL HOT** level, add fluid slowly until it is.

Fig. 103 Typical power steering pump

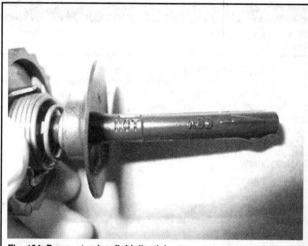

Fig. 104 Power steering fluid dipstick

Cooling System

All engines covered in this manual utilize one of two cooling systems, or variations thereof. Simply, there is a Seawater Cooling system, sometimes called Raw Water system and a Closed Cooling system.

Seawater systems are just that...they utilize the water that the boat is operating in to cool and lubricate the drive and engine. Water is drawn in through the stern drive unit and circulated to, and through, the engine and its components. The heated water is then returned overboard through the exhaust system.

Closed Cooling systems are actually two systems working together; a seawater system and a closed system consisting of antifreeze—very similar to an automotive system, without the radiator. The two systems work together in a variety of ways.

Complete information on individual system operation, and component repair procedures is detailed in the Cooling Systems section. In this section we will only deal with checking fluid levels, draining and filling, and flushing the systems.

LEVEL CHECK

Closed Cooling System Only

◆ See Figure 105

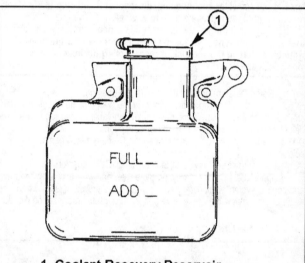

1- **Coolant Recovery Reservoir**

Fig. 105 Check that the coolant level is to the FULL line on the coolant recovery tank when the engine is hot (early model shown, late models utilize a rectangular tank

■ Always allow the engine to cool down before checking the coolant level. If possible, check cold. Opening a pressure cap with the engine hot can cause a violent discharge of steam and water, resulting in severe injury. If the engine is anything but cold, always remove the cap a quarter turn at a time in order to allow any residual pressure to escape slowly.

1. Remove the pressure cap at the top of the heat exchanger. Coolant should be visible at the bottom of the filler neck, or within 1 in. (25mm) of the bottom of the neck.

2. Reinstall the cap on the heat exchanger making sure that it seats on it's stops on the filler neck.

3. Start the engine and run it until it reaches normal operating temperature. Turn off the engine and check that the coolant level is to the **FULL** line on the side of the coolant recovery tank.

4. Fill the system with a 50/50 mixture of distilled water and ethylene glycol antifreeze.

FLUSHING THE SYSTEM

■ If your boat is operated in saltwater, heavy mineral fresh water or severely polluted water, flush the cooling system regularly—after each usage if at all possible! Always flush the system before draining and/or winter storage.

Seawater Systems W/O Fresh Water Flush

Boat Out Of Water
◆ See Figure 106

✳✳ CAUTION

NEVER run the engine when the boat is out of the water without water being supplied to the stern drive.

1. Connect a flushing attachment over the water intake openings in the stern drive gear housing. These devices are available at your local OMC dealer or through a variety of aftermarket suppliers.

2. Connect a garden hose between the flushing attachment and a water spigot.

3. Open the spigot(s) slowly, no more than half way, and allow the drive and cooling system to fill completely. You'll know the system is full when water begins to flow out of the drive unit, or through the propeller on closed systems.

✳✳ CAUTION

Never allow the water to reach the flushing device at anything more than low pressure.

4. With the drive in Neutral, start the engine and let it idle for 10 minutes or until the water being discharged is clear and then turn the engine off. Never run the engine above 1500 rpm with the flushing device attached and always keep an eye on the temperature gauge incase the engine begins to overheat.

5. Turn off the water from the spigot(s), disconnect the hose(s) and remove the flushing attachment.

Boat In Water

◆ See Figure 106

✳✳ CAUTION

If your boat has a seacock (a water inlet valve), it must remain closed during this procedure in order to prevent water from flowing back into the boat. If your boat does not have a seacock, locate the water inlet hose at the seawater supply pump, disconnect it and plug it. We highly recommend that you leave a note to yourself in the vicinity of the ignition key reminding of the fact that this procedure has been done so that you, or someone else doesn't start the engine after flushing without reopening the seacock or reconnecting the inlet hose. Seem silly? Do it anyway!

1. Raise the stern drive unit to the full UP position.
2. Connect a flushing attachment over the water intake openings in the stern drive gear housing. These devices are available at your local OMC dealer or through a variety of aftermarket suppliers.
3. Connect a garden hose between the flushing attachment and a water spigot.
4. Lower the stern drive unit to the full IN/DOWN position.
5. Open the spigot(s) slowly, no more than half way, and allow the drive and cooling system to fill completely. You'll know the system is full when water begins to flow out of the drive unit, or through the propeller on closed systems.

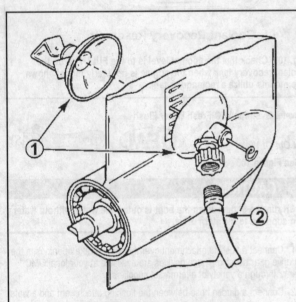

1- Flushing Attachment
2- Water Hose

Fig. 106 Attach a flushing device to the drive unit when flushing the cooling system

✳✳ CAUTION

Never allow the water to reach the flushing device at anything more than low pressure.

6. With the drive in Neutral, start the engine and let it idle for 10 minutes or until the water being discharged is clear and then turn the engine off. Never run the engine above 1500 rpm with the flushing device attached and always keep an eye on the temperature gauge incase the engine begins to overheat.

7. Turn off the water from the spigot(s).
8. Raise the drive unit to the UP position again, disconnect the hose and remove the flushing attachment.
9. Lower the unit and make sure you open the seacock and reconnect the inlet hose (don't forget to unplug it first!).

Seawater Systems With Fresh Water Flush

◆ See Figure 107

✳✳ WARNING

Do not, under any circumstance, start the engine while performing this procedure!

1. Locate the flush valve connected to a black hose on the side of the engine. There should be flush socket attachment with a cap connected to it.
2. Remove the socket from the cap, connect a garden hose and then plug into the flush valve.
3. Turn the hose spigot to its highest pressure and allow the engine to flush for 5 minutes.
4. Turn off the water and remove the hose/socket from the valve.
5. Unscrew the socket from the hose and insert it back into the cap.
6. You're done! That was easy—now you know why people spend the time and money to retro-fit these onto their engines..

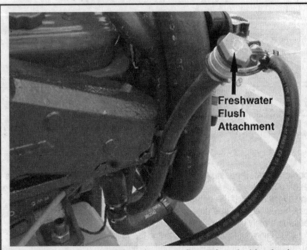

Fig. 107 Certain late model engines may be equipped with a fresh water flushing attachment (or have been retro-fitted), making it much easier to regularly flush your cooling system

Closed Systems

The very name of this systems implies that it is not necessary to flush it. If for some reason you absolutely must flush the closed portion of this system, simply follow the drain and fill procedures. Remember though that the closed system uses a conventional seawater cooling system in conjunction with its closed portion—this system MUST be flushed on a regular basis and all procedures are detailed previously under Seawater Systems.

PRESSURE CAP TESTING

Closed System Only

◆ **See Figures 108, 109 and 110**

The pressure cap on the coolant reservoir is designed to maintain closed system pressure at approximately 14 psi when the engine is at normal operating temperature. Clean, inspect and pressure test the cap at the end of your first season and then every 100 hours (or once a year) thereafter.

✳✳ WARNING

Make sure the engine is cool before attempting to remove the cap. To be safe, always turn the cap 1/4 turn and allow any residual pressure to escape before removing it completely. Use a heavy rag or wear gloves.

1. Remove the cap and wash it thoroughly to remove any debris from the sealing surfaces.
2. Inspect the gasket and rubber seal for tears, cuts or cracks. If you notice anything, we recommend replacing the entire cap although you can replace the gasket on some older engines if you like.
3. Make sure that the locking tabs on the cap are not bent or damaged in any way. If so, replace the cap.
4. Use a cooling system pressure tester and, following the manufacturer's instructions, install the cap. Check that the cap relieves pressure at 14 psi and holds pressure for 30 seconds without falling below 11 psi. Replace the cap if it fails.
5. Check the inside of the filler neck for debris; it should be completely smooth. Check that the lock flanges on the filler neck are not bent or damaged and then install the cap.

PRESSURE TEST

Closed System Only

✳✳ WARN·ING

Make sure the engine is cool before attempting to remove the cap. To be safe, always turn the cap 1/4 turn and allow any residual pressure to escape before removing it completely. Use a heavy rag or wear gloves.

1. Remove the pressure cap from the reservoir or heat exchanger.
2. Perform the pressure cap Testing procedure. If the cap is bad, this may be your problem. Replace the cap and ensure that the problem does not go away; if it persists, move to the next step.
3. Add coolant so that the level is within one inch of the bottom of the filler neck. Attach a cooling system pressure tester to the filler neck and pressurize the system to 17 psi.
4. Watch the gauge for about 2 minutes. If the pressure remains steady, you're OK. If the pressure drops, move to the next step.
5. Maintain the specified pressure and check the entire closed system for any leaks—hoses, plugs, petcocks, pump seals, etc.). Also, listen very closely for any hissing or bubbling.
6. If you've still not found any leaks, test the heat exchanger as detailed in the Cooling section.

7. If the exchanger is OK then you most likely have an problem with loose head bolts, a bad head gasket or a warped cylinder head, not the cooling system.

DRAINING & FILLING

The cooling system should be drained, cleaned, and refilled each season, although OMC's recommendations are for every two years on normal anti-freeze systems. We think its cheap insurance to do it every season, but you certainly can't go wrong by following the factory's suggestion. The bow of the boat must be higher than the stern to properly drain the cooling system. If the bow is not higher than the stern, water will remain in the cylinder block and in the exhaust manifold. Insert a piece of wire into the drain holes, but not in the petcock, to ensure sand, silt, or other foreign material is not blocking the drain opening.

If the engine is not completely drained for winter storage, trapped water can freeze and cause severe damage. The water in the oil cooler—if so equipped—must also be drained.

Seawater System—4 Cylinder Engines

◆ **See Figure 111**

1. If the boat is in the water, close the seacock to prevent water from entering the cooling system. If the boat is not equipped with a seacock, disconnect the seawater inlet line and plug it. Make sure you leave yourself a note by the ignition switch reminding yourself to turn the valve back on or connect the line, especially if the boat is going to sit for a period of time. Sound silly? Do it!
2. Position suitable containers under all drain plugs and hoses (if space permits) to catch any water being drained or else it will collect in the bilge.
3. Loosen the clamp and disconnect the long supply hose at the thermostat housing. Slowly lower the hose and allow the water to drain.
4. Loosen the clamp and disconnect the large house running from the thermostat housing to the engine water pump, at the pump. Allow all water to drain.
5. Remove the drain plug on the side of the cylinder block (port on 2.5L/3.0L; starboard on 2.3L) and allow all water to drain.
6. Remove the drain plug on the exhaust manifold or elbow. If your engine is not equipped with a drain, simply disconnect the hose from the lower fitting.
7. On models equipped with power steering, disconnect the lower seawater hose at the cooler—some models may actually have a drain plug on the bottom side of the cooler. If the cooler is mounted horizontally, loosen the mounting bolt and tilt the cooler to aid in the draining.
8. Loosen the clamps and disconnect both hoses at the seawater pump (if equipped). Allow all water to drain and then reconnect the hoses.
9. On engines with closed cooling, remove the drain plug from the bottom of the heat exchanger. If there is no plug, disconnect the seawater inlet hose at the bottom of the exchanger.
10. Disconnect and remove the lower end of any other hoses and hold them as low as possible while draining. Reconnect when all water has stopped draining. On models with closed cooling, refer to the flow diagrams in the Cooling section so that you do not disconnect any hoses used for the closed portion of the system.

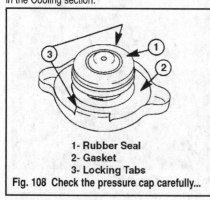

1- Rubber Seal
2- Gasket
3- Locking Tabs
Fig. 108 Check the pressure cap carefully...

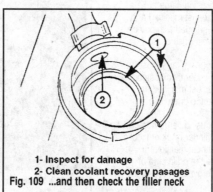

1- Inspect for damage
2- Clean coolant recovery pasages
Fig. 109 ...and then check the filler neck

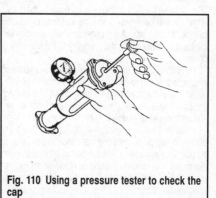

Fig. 110 Using a pressure tester to check the cap

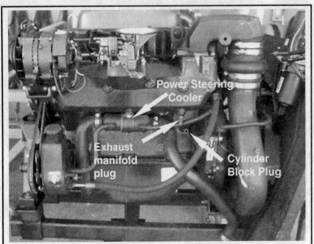

Fig. 111 Both drain plugs are on the port side—3.0L shown

11. Using a stiff piece of wire, clean out all drain holes. Don't forget any drain holes in the drive unit!

12. Tilt the drive unit to the full UP position and then remove the water drain plug from the port side of the pivot housing; this can be found toward the front of the upper housing where it attached to the bell housing.

13. Turn the ignition switch **ON** and crank the engine over a few times to force out any residual water in the block or seawater pump. DO NOT START THE ENGINE!

14. Coat the threads of all drain plugs with sealer and then reinstall them. Tighten the plugs and petcock securely.

15. If you are winterizing the boat and its going to sit for a few months, it's a good idea to remove the thermostat housing and fill the cylinder block and head with a mixture of water and anti-freeze. Remove the inlet line at the exhaust manifold and do the same. Make sure you drain all coolant into a suitable container prior to refilling the system with seawater next season.

16. If you closed the seacock or disconnected the water inlet line, make sure you open or reconnect it PRIOR to restarting the engine.

Seawater System—V6 And V8 Engines

◆ See Figures 112 thru 120   *MODERATE*

Specific drain locations and disconnections vary from engine to engine. Please review the following items for the actual description for your engine before beginning the procedure:

• 1987-90 2.3L engines— long hose at thermostat housing, large hose at engine water pump running from thermostat housing, starboard side exhaust manifold petcock and port side cylinder block petcock. If equipped with power steering or other oil cooler, remove the drain plug or lowest hose at the cooler.

• 2.5L/3.0L engines—long hose at thermostat housing, large hose at engine water pump running from thermostat housing, port side exhaust manifold petcock and port side cylinder block petcock. If equipped with power steering or other oil cooler, remove the drain plug or lowest hose at the cooler.

• 1986-87 4.3L engines—long hose at thermostat housing, large hose at engine water pump running from thermostat housing, exhaust manifold and cylinder block petcocks on port and starboard sides. If equipped with power steering or other oil cooler, remove the drain plug or lowest hose at the cooler.

• 1988-95 4.3L engines—long hose at thermostat housing, large hose at engine water pump running from thermostat housing, cylinder block petcocks on port and starboard sides, hose at aft end of each exhaust manifold. If equipped with power steering or other oil cooler, remove the drain plug or lowest hose at the cooler.

• 1996-98 4.3L engines—long hose at thermostat housing, large hose at engine water pump running from thermostat housing, cylinder block petcocks on port and starboard sides, hose at aft end of each exhaust manifold. If equipped with power steering or other oil cooler, remove the drain plug or lowest hose at the cooler. On 4.3Gi engines, remove the bypass hose at the check valve also.

• 1986-88 5.0L and 5.7L engines—long hose at thermostat housing, large hose at engine water pump running from thermostat housing, cylinder block petcocks on port and starboard sides, two rubber caps at fore and aft sides of each exhaust manifold. If equipped with power steering or other oil cooler, remove the drain plug or lowest hose at the cooler.

• 1998 5.0L engines—large hose at engine water pump running from thermostat housing, cylinder block petcocks on port and starboard sides, rubber hose at aft end of each exhaust manifold. If equipped with power steering or other oil cooler, remove the drain plug or lowest hose at the cooler.

• 1989-96 5.7L engines—long hose at thermostat housing, large hose at engine water pump running from thermostat housing, cylinder block petcocks on port and starboard sides, rubber cap/hose at aft end of each exhaust manifold. If equipped with power steering or other oil cooler, remove the drain plug or lowest hose at the cooler. On 5.7Gi engines, remove the bypass hose at the check valve also.

• 1997-98 5.7L engines—large hose at engine water pump running from thermostat housing, cylinder block petcocks on port and starboard sides, rubber hose at aft end of each exhaust manifold. If equipped with power steering or other oil cooler, remove the drain plug or lowest hose at the cooler. On Gi and GSi engines, remove the small bypass hose at the top of the thermostat housing also.

• 1989-90 5.0L and 5.8L engines—long hose at thermostat housing, large hose at engine water pump running from thermostat housing, cylinder block petcocks on port and starboard sides, exhaust manifold petcocks on port and starboard sides, rubber cap at aft end of each exhaust manifold. If equipped with power steering or other oil cooler, remove the drain plug or lowest hose at the cooler.

Fig. 112 Port drain plug—4.3L engines

Fig. 113 Starboard drain plug—4.3L engines

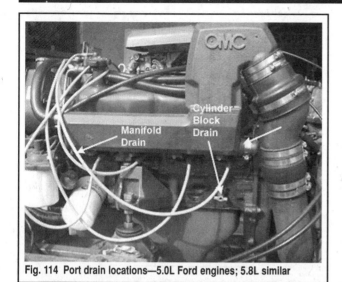
Fig. 114 Port drain locations—5.0L Ford engines; 5.8L similar

Fig. 115 Starboard drain locations—5.0L Ford engines; 5.8L similar

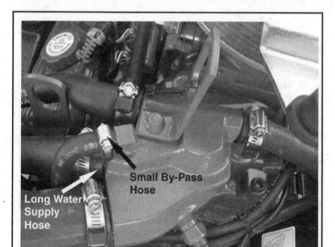
Fig. 116 Disconnect the long water supply hose—5.0L/5.7L GM engines; and the small by-pass hose—EFI engines, others similar

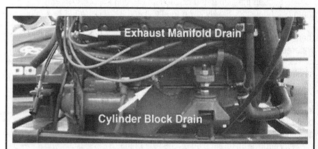
Fig. 117 Loosen the starboard drain plugs—5.0L/5.7L GM engines, others similar

Fig. 118 Loosen the port drain plugs—5.0L/5.7LGM engines, others similar

• 1991-96 5.0L and 5.8L engines—long hose at thermostat housing, large hose at engine water pump running from thermostat housing, cylinder block petcocks on port and starboard sides, rubber cap/hose at aft end of each exhaust manifold. If equipped with power steering or other oil cooler, remove the drain plug or lowest hose at the cooler. On Fi and FSi engines, remove the bypass hose at the check valve also.

• 1988-98 7.4L/8.2L engines—long hose at thermostat housing, large hose at engine water pump running from thermostat housing, cylinder block petcocks on port and starboard sides, rubber cap/hose at aft end of each exhaust manifold. If equipped with power steering or other oil cooler, remove the drain plug or lowest hose at the cooler. On Gi and GSi engines, remove the bypass hose at the check valve also.

• 1987-90 7.5L engines—long hose at thermostat housing, large hose at engine water pump running from thermostat housing, cylinder block petcocks on port and starboard sides, rubber cap/hose at aft end of each exhaust manifold. If equipped with power steering or other oil cooler, remove the drain plug or lowest hose at the cooler.

1. If the boat is in the water, close the seacock to prevent water from entering the cooling system. If the boat is not equipped with a seacock, disconnect the seawater inlet line and plug it. Make sure you leave yourself a note by the ignition switch reminding yourself to open the valve or reconnect the line, especially if the boat is going to sit for a period of time. Sound silly? Do it!

Fig. 119 A nice close-up of a cylinder block drain plug

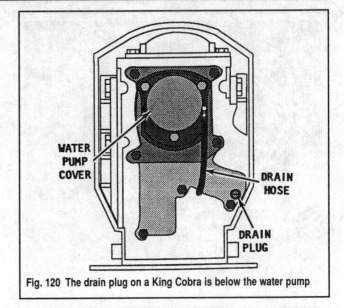

Fig. 120 The drain plug on a King Cobra is below the water pump

2. Position suitable containers under all drain plugs and hoses (if space permits) to catch any water being drained or else it will collect in the bilge.

3. Loosen the clamp and disconnect the long supply hose at the thermostat housing. Slowly lower the hose and allow the water to drain.

4. Loosen the clamp and disconnect the large hose running from the thermostat housing to the engine water pump, at the water pump. Allow all water to drain.

5. On most fuel injected engines, loosen the clamp and disconnect the small water by-pass hose on top of the thermostat housing, bend it down and allow the water to drain.

6. On 1998 4.3L engines, remove the drain plug at the front of the intake manifold and allow the manifold to drain. You may need to lower the bow slightly in order to get all of the water out. Reinstall the plug when draining is complete.

7. Remove the drain plug on the side of the cylinder block and allow all water to drain. Repeat this with the plug on the other side if you have a V-engine.

8. Remove the drain plug on the exhaust manifolds or elbow. If your engine is not equipped with a drain, simply disconnect the hose from the lower fitting. Some engines may have these covered with a rubber cap.

■ On certain late model 7.4Gi engines, there is no port side cylinder block drain plug; simply remove the knock sensor.

9. On models equipped with power steering, disconnect the hoses at the cooler—some models may actually have a drain plug on the bottom side of the cooler. If the cooler is mounted horizontally, loosen the mounting bolt and tilt the cooler to aid in the draining.

10. Loosen the clamps and disconnect both hoses at the seawater pump (if equipped). Allow all water to drain and then reconnect the hoses.

11. On engines with closed cooling, remove the drain plug from the bottom of the heat exchanger. If there is no plug, disconnect the seawater inlet hose at the bottom of the exchanger.

12. Disconnect and remove the lower end of all hoses and hold them as low as possible while draining. Reconnect when all water has stopped draining. On models with closed cooling, refer to the flow diagrams at the rear of this section so that you do not disconnect any hoses used for the closed portion of the system.

13. Using a stiff piece of wire, clean out all drain holes. Don't forget any drain holes in the drive unit!

14. On models with SP drives, tilt the drive unit to the full UP position and then remove the water drain plug from the port side of the pivot housing; this can be found toward the front of the upper housing where it attached to the bell housing. Many models will also have a vent and drain plug on the starboard side of pivot housing as well.

15. On King Cobra models, remove the 3 bolts and pull off the water pump access cover. Remove the water passage drain plug and move the drive to the full Down position.

16. Turn the ignition switch ON and crank the engine over a few times to force out any residual water in the block or seawater pump. DO NOT START THE ENGINE!

17. Install the drain plug and access cover on King Cobra models.

18. Coat the threads of all drain plugs with sealer and then reinstall them. Tighten the plugs and petcock securely.

19. If you are winterizing the boat and its going to sit for a few months, it's a good idea to remove the thermostat housing and fill the cylinder block and head with a mixture of water and anti-freeze. Remove the inlet line at the exhaust manifold and do the same. Make sure you drain all coolant into a suitable container prior to refilling the system with seawater next season.

20. If you closed the seacock or disconnected the water inlet line, make sure you open or reconnect it PRIOR to restarting the engine.

Closed System

This system should be kept filled with the appropriate water and anti-freeze solution year-round; even if laying the boat up for the winter. Coolant should be changed at least every other year.

■ Don't forget to drain the seawater system if winterizing the boat.

1. Carefully remove the pressure cap from the system. Follow the precautions detailed in the Pressure Cap procedures in this section if the engine is hot.

2. Loosen and remove the petcock(s) or drain plug(s) on the cylinder block making sure you have a suitable container under the drains or else the coolant will flow into the bilge,.

3. Disconnect the coolant hose (the other is a seawater hose—refer to the flow diagrams) from the engine circulating pump at the heat exchanger. Loosen and remove the drain plug at the bottom of the exchanger.

4. Clean all drain holes with a stiff piece of wire.

5. If equipped with a coolant recovery or overflow tank, drain it also.

6. Coat the threads of all plugs with sealer and install them securely. Reconnect the coolant hose at the heat exchanger.

7. Pour the appropriate mix of water and anti-freeze into the reservoir filler neck until the level is within 1 in. of the bottom of the neck.

8. With the cap still off, start the engine and run it at 1500-1800 rpm. Add additional coolant until the level is within 1 in. of the bottom of the filler neck. If the boat is out of the water, don't forget to attach a flushing device to ensure proper flow of seawater.

9. After the engine has reached normal operating temperature, check that the level in the filler neck is correct; add coolant if necessary.

10. Install the pressure cap, check the coolant recovery reservoir and fill it to the **FULL** line.

11. If any leakage or overheating is noticed, shut the engine down immediately and look for the cause of the problem.

Stern Drive Unit

FLUID LEVEL

◆ See Figures 121, 122, 123 and 124

Check the fluid level after the first 20 hours of operation and then at least every 50 hours of operation or once a season, whichever comes first. It's never a bad idea to check your unit more frequently, particularly if you use your boat in severe service conditions. We recommend checking the fluid level weekly as air pockets during a fluid fill or change may cause an initially false reading; which is to say that the initial level may drop on its own over the first 10 hours of operation or so. OMC recommends using OMC Hi-Vis Gear Case Lubricant in 1986-93 drives, OMC Ultra HPF Gear Case lube in 1994-95 drives, and Mobilube 1 SHC Fully Synthetic SAE 75W-90 gear lubricant in 1996-98 units. Never substitute regular automotive grease.

1. Position the drive unit in the full IN/DOWN position so that the anti-ventilation plate is level.

■ The oil level should always be checked with the drive unit cold.

2. Remove the oil level dipstick found on top of the upper gear unit wipe it down with a clean lint-free rag and reinsert it. Remove it once again and check that the lubricant is up to the line on the stick. If the level is satisfactory, check the condition of the washer and then install it and the dipstick. It's a good idea to replace the washer on a regular basis regardless of its visible condition.

3. If the level is low, add oil lubricant, very slowly, through the dipstick opening and then reinstall the dipstick and washer.

4. Let it sit for a moment and then recheck it.

✳✳ CAUTION

The unit should never require more than 2 oz. of lubricant. If more is required, you've got an oil leak and the unit should not be operated until it is found and fixed.

5. Clean any excess oil from the housing. Recheck the oil level on the dipstick again a final time after the next use.

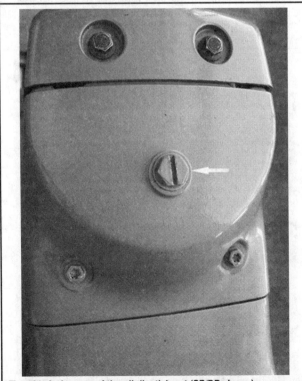
Fig. 121 A close up of the oil dipstick nut (SP/DP shown)

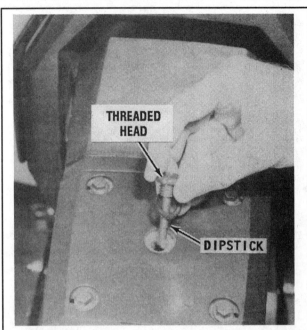
Fig. 122 Use the dipstick to check the fluid level—Cobra and King Cobra models

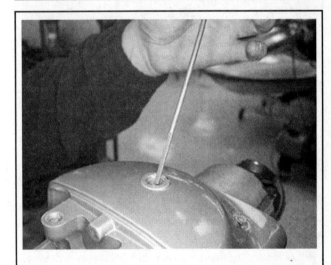
Fig. 123 Use the dipstick to check the fluid level—SP/DP models

CHECKING FOR WATER

It is a good idea to check the unit periodically for water contamination. Water in the drive oil usually indicates a bad seal somewhere on the unit and should be corrected immediately.

1. With the engine off and the stern drive unit cold, trim the drive to the full UP position.

2. Remove the oil fill/drain plug and take a small sample of the lubricant—a teaspoon worth is more than enough.

3. If the oil sample looks milky brown then there is likely a leaking seal in the drive unit and it should be located and replaced before operating the boat again.

4. Most units covered in the manual have magnetic fill/drain plugs, so be sure to check that the end of the plug is free of any metal filings. If you find any metallic particles on the plug it can be an indication of greater problems within the drive unit.

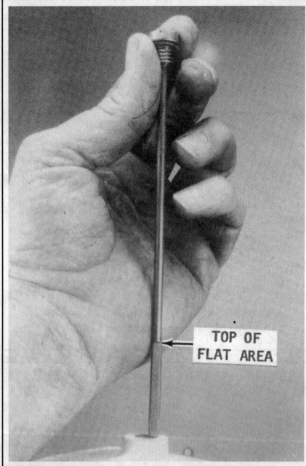

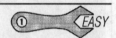

Fig. 124 The top of the flat area indicates the case is full

DRAIN AND REFILL

The oil in the stern drive should be changed at least every 50 hours of operation or twice a season on Cobra and King Cobra drives,100 hours of operation or once a season on SP/DP drives; whichever comes first. It's never a bad idea to change the fluid in your unit more frequently, particularly if you use your boat in severe service conditions. OMC recommends using only OMC Hi-Vis Gear Case Lubricant in 1986-93 drives, OMC Ultra HPF Gear Case lube in 1994-95 drives, and Mobilube 1 SHC Fully Synthetic SAE 75W-90 gear lubricant in 1996-98 units. Never substitute regular automotive grease.

1986-93 Cobra Units

◆ See Figure 125

✳✳ CAUTION

Stern drive oil should always be changed with the unit cold.

■ Take a look at the oil as it is being drained, if the oil is milky brown in color there is a good chance there is a leak in the unit. The stern drive unit should not be operated until the leak is found and corrected.

1. Trim the stern drive to the full IN/DOWN position.
2. Position an oil drain pan or an old plastic milk jug under the drain hole on the bottom of the drive unit and then remove the drain plug (starboard side of lower gearcase). Removing the oil dipstick at this time will help the oil drain quicker.

3. When the old oil has completely drained, install the drain plug and tighten it securely.
4. Remove the oil fill plug on the starboard side of the drive, just above the anti-cavitation plate. Insert the dipstick into the hole so that it is just resting on the upper surface of the bore; do not screw it in.
5. Pump the proper lubricant into the drive through the fill hole until it reaches upper mark on the dipstick. This will give a false reading (slightly high) in order to account for a fluid level drop once the air bubbles are purged when you start the engine.

■ Please refer to the Capacities chart for lubricant capacities on a specific drive unit.

✳✳ WARNING

Fill the drive unit slowly! Adding lubricant too quickly will cause too many air bubbles and could lead to an inaccurate level reading or even more severe damage.

6. With the lube pump still attached, screw in the oil level dipstick securely. Quickly remove the pump and install the fill plug and washer.
7. After allowing the oil to settle for a few minutes, check the level using the dipstick. Add lubricant slowly through the dipstick hole as required.
8. Clean any excess oil on the housing. Check the oil level a final time and then recheck it again after the first use.

1989-92 King Cobra Units

◆ See Figure 126

✳✳ CAUTION

Stern drive oil should always be changed with the unit cold.

■ Take a look at the oil as it is being drained, if the oil is milky brown in color there is a good chance there is a leak in the unit. The stern drive unit should not be operated until the leak is found and corrected.

1. Trim the stern drive to the full IN/DOWN position.
2. Position an oil drain pan or an old plastic milk jug under the drain hole on the bottom of the drive unit and then remove the drain plug (port side of lower gearcase). Removing the oil dipstick at this time will help the oil drain quicker.

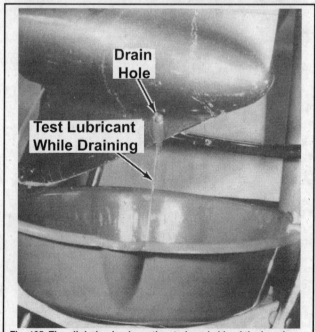

Fig. 125 The oil drain plug is on the starboard side of the housing

3. When the old oil has completely drained, install the drain plug and tighten it securely.

4. Pump the proper lubricant into the drive through the oil dipstick hole until it reaches the upper mark on the dipstick with the dipstick just sitting in the bore—do not screw it into the hole. This will give a false reading (slightly high) in order to account for a fluid level drop once the air bubbles are purged when you start the engine.

■ **Please refer to the Capacities chart for lubricant capacities on a specific drive unit.**

✳✳ WARNING

Fill the drive unit slowly! Adding lubricant too quickly will cause too many air bubbles and could lead to an inaccurate level reading or even more severe damage.

5. After allowing the oil to settle for a few minutes, check the level using the dipstick. Add lubricant slowly through the dipstick hole as required.

6. Clean any excess oil on the housing. Check the oil level a final time and then recheck it again after the first use.

1993-95 King Cobra Units

◆ See Figure 126

✳✳ CAUTION

Stern drive oil should always be changed with the unit cold.

■ **Take a look at the oil as it is being drained, if the oil is milky brown in color there is a good chance there is a leak in the unit. The stern drive unit should not be operated until the leak is found and corrected.**

1. Trim the stern drive to the full IN/DOWN position.

2. Position an oil drain pan or an old plastic milk jug under the drain hole on the bottom of the drive unit and then remove the fill/drain plug (port side of lower gearcase). Removing the oil dipstick at this time will help the oil drain quicker.

3. When the old oil has completely drained, install an appropriate pump into the drain plug and tighten it securely.

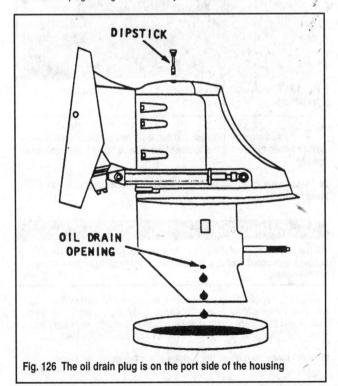

Fig. 126 The oil drain plug is on the port side of the housing

4. Pump the proper lubricant into the drive until it reaches the upper mark on the dipstick with the dipstick just sitting in the bore—do not screw it into the hole. This will give a false reading (slightly high) in order to account for a fluid level drop once the air bubbles are purged when you start the engine.

■ **Please refer to the Capacities chart for lubricant capacities on a specific drive unit.**

✳✳ WARNING

Fill the drive unit slowly! Adding lubricant too quickly will cause too many air bubbles and could lead to an inaccurate level reading or even more severe damage.

5. With the lube pump still attached, thread in the dipstick and tighten it securely. Quickly remove the pump and install the fill/drain plug and washer; tighten it to 60-84 inch lbs. (6.8-9.5 Nm). Reinstall the oil level dipstick with a new seal and tighten it to 48-72 inch lbs. (5.4-8.1 Nm). Use sealing compound on the threads of the oil fill/drain and level plugs.

6. After allowing the oil to settle for a few minutes, check the level using the dipstick. Add lubricant slowly through the dipstick hole as required.

7. Clean any excess oil on the housing. Check the oil level a final time and then recheck it again after the first use.

1994-95 Cobra Units And 1996-98 SP Units

◆ See Figures 127, 128 and 129

✳✳ CAUTION

Stern drive oil should always be changed with the unit cold.

■ **Take a look at the oil as it is being drained, if the oil is milky brown in color there is a good chance there is a leak in the unit. The stern drive unit should not be operated until the leak is found and corrected.**

1. Trim the stern drive to the full IN/Down position.

2. Position an oil drain pan or an old plastic milk jug under the drain hole on the bottom of the drive unit and then remove the fill/drain plug (port side of lower gearcase). Remove the oil dipstick at this time which will help the oil drain quicker.

3. While waiting for the oil to drain, remove the three shift lever cover retaining screws (two on the upper gear housing, one on the intermediate housing). Remove the cover and then remove the oil level plug to starboard of the shift lever.

4. When the old oil has completely drained, position and install a suitable lubricant pump into the drain hole.

5. Pump the proper lubricant into the drive through the fill/drain hole until it reaches the level at the bottom of the oil level hole on the upper gear housing.

■ **Please refer to the Capacities chart for lubricant capacities on a specific drive unit.**

✳✳ WARNING

Fill the drive unit slowly! Adding lubricant too quickly will cause air bubbles that could lead to an inaccurate level reading or even more severe damage.

6. With the lube pump still attached, reinstall the oil level plug with a new seal, tighten it to 48-72 inch lbs. (5.4-8.1 Nm). Do the same with the oil dipstick and then quickly remove the pump and install the fill/drain plug and washer; tighten it to 60-84 inch lbs. (6.8-9.5 Nm). Use sealing compound on the threads of the oil fill/drain and level plugs.

7. After allowing the oil to settle for a few minutes, check the level using the dipstick. Add lubricant slowly through the dipstick hole as required.

8. Install the shift cover and tighten the three bolts to 108-132 inch lbs. (12.2-14.9 Nm).

9. Clean any excess oil on the housing. Check the oil level a final time and then recheck it again after the first use.

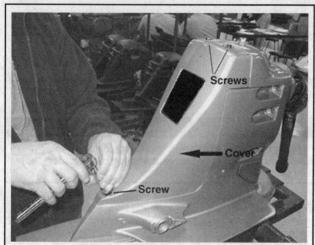
Fig. 127 After removing the shift cover...

Fig. 129 The oil drain plug is on the port side of the housing

Fig. 128 ...remove the oil level plug

Fig. 130 The oil drain plug is inside the housing, behind the propellers...

1996-98 DP Units

◆ See Figures 127 thru 131

✳✳ CAUTION

Stern drive oil should always be changed with the unit cold.

■ Take a look at the oil as it is being drained, if the oil is milky brown in color there is a good chance there is a leak in the unit. The stern drive unit should not be operated until the leak is found and corrected.

1. Remove the propeller(s) as detailed in the Drive Systems section.
2. Trim the stern drive to the full IN/Down position.
3. Position an oil drain pan or an old plastic milk jug under the bottom of the drive unit and then remove the fill/drain plug just inside the lower gear case housing, under the propeller shaft. Remove the oil dipstick at this time which will help the oil drain quicker.
4. While waiting for the oil to drain, remove the shift lever cover retaining screws (one, two or three depending on the model, but usually three). Remove the cover and then remove the oil level plug to starboard of the shift lever.
5. When the old oil has completely drained, position and install a suitable lubricant pump into the drain hole. Almost all models will require first screwing in a Fill Adaptor (# 3855932-4)

6. Pump the proper lubricant into the drive through the fill/drain hole until it reaches the level at the bottom of the oil level hole on the upper gear housing.

■ Please refer to the Capacities chart for lubricant capacities on a specific drive unit.

✳✳ WARNING

Fill the drive unit slowly! Adding lubricant too quickly will cause air bubbles that could lead to an inaccurate level reading or even more severe damage.

7. With the lube pump still attached, reinstall the oil level plug with a new seal, tighten it to 48-72 inch lbs. (5.4-8.1 Nm). Do the same with the oil dipstick and then quickly remove the pump and fill adaptor and install the fill/drain plug and washer; tighten it to 60-84 inch lbs. (6.8-9.5 Nm).

■ Don't forget to replace the O-rings when replacing the plugs!

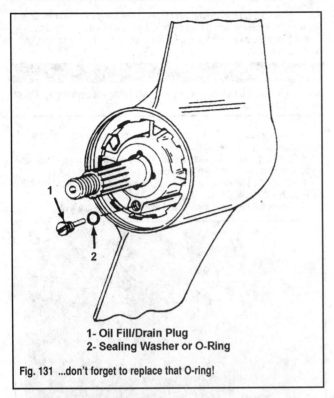

1- Oil Fill/Drain Plug
2- Sealing Washer or O-Ring

Fig. 131 ...don't forget to replace that O-ring!

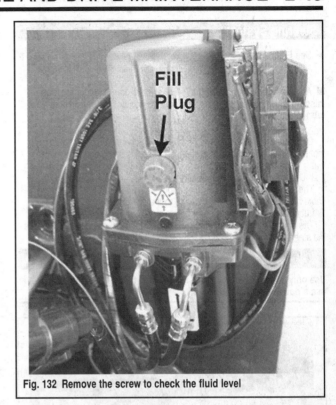

Fig. 132 Remove the screw to check the fluid level

8. Install the propellers.

9. After allowing the oil to settle for a few minutes, check the level using the dipstick. Add lubricant slowly through the dipstick hole as required.

10. Install the shift cover and tighten the retaining bolts to 108-132 inch lbs. (12.2-14.9 Nm).

11. Clean any excess oil on the housing. Check the oil level a final time and then recheck it again after the first use.

Power Trim/Tilt Pump

FLUID LEVEL

◆ See Figure 132

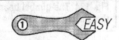

Fluid levels in the power trim/tilt pump should be checked every 100 hours of operation or once a season, whichever comes first. We recommend that you use OMC Power Trim/Tilt & Power Steering Fluid, although OMC suggests that in a pinch you can use Dexron® II ATF.

LUBRICATION POINTS

Propeller Shaft

Although OMC does not publish intervals for propeller shaft lubrication, we recommend lubricating the shaft every 100 hours or 6 months, whichever comes first when the vessel is being operated in fresh water. If the vessel is in saltwater, every 50 hours or 2 months. Please refer to the Drive Systems section for details on propeller removal and installation.

Throttle Cable

Lubricate the cable pivot points every 100 hours or 6 months, whichever comes first when the vessel is being operated in fresh water. In salt water, lubricate every 50 hours or 2 months whichever comes first. Always use OMC Grease.

1. Position the stern drive in the full UP/OUT position in order to relieve all hydraulic pressure.

✷✷ CAUTION

Failure to position the drive unit in the full UP/OUT position before removing the plug could result in hydraulic fluid spraying all over upon plug removal.

2. Loosen and remove the fill screw from the side of the pump housing.

3. Check that the lubricant is up to the bottom of the threads in the hole. If necessary, add more fluid slowly through the hole u0ntil the level is correct.

4. With the fill screw still out, move the drive through its full range 6-10 times in order to purge any air that may have entered the system. Lower the drive fully and recheck the fluid level again; repeat the process if necessary.

5. Install the fill screw and tighten it securely.

Shift Cable And Transmission Linkage Pivot Points

Lubricate the cable pivot points every 100 hours or 6 months, whichever comes first when the vessel is being operated in fresh water. In salt water, lubricate every 50 hours or 2 months whichever comes first. Always use OMC Grease.

Steering System

◆ See Figure 133

■ Always check the steering system and its components for loose, missing or damaged components and fittings before performing your lubrication procedures.

Lubricate the various steering system components every 50 hours or 2 months, whichever comes first when the vessel is being operated in fresh water. In salt water, lubricate every 25 hours or 30 days whichever comes first. Use only OMC Triple-GuardGrease

The transom end of the steering cable must be fully retracted into the cable housing before applying grease to the fitting.

Lubricate all pivot points on the system with SAE 20 or 30 motor oil.

The control valve grease fitting takes OMC Triple-Guard Grease as does the steering ram.

✳✳ CAUTION

Use only a hand operated grease gun when filling the zerk fitting. Never use a power operated gun!

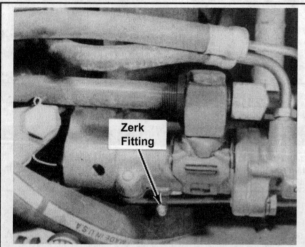

Fig. 133 Always use a hand operated grease gun when filling the control valve

Gimbal Bearing And U-Joints

◆ See Figures 134 and 135

The gimbal bearing and the U-Joints must be lubricated every 100 hours of operation, or once a season; whichever comes first. OMC EP Wheel Bearing Grease should be used in both instances. Certain older units may not be equipped with grease fittings; if yours is not, the U-joints will require disassembly and greasing. Because OMC recommends removing the drive unit prior to performing this service, the perfect time to consider performing it is during your winterization.

✳✳ CAUTION

Failure to perform this service on a regular basis will result in severe damage to the transom assembly and drive unit.

1. Remove the drive unit as detailed in the Drive Unit section. Carefully support it in a holding assembly.

2. Connect a hand operated grease gun to the fitting for the bearing and slowly pump in grease until the **all** of the old grease has been forced out. Take note of the old grease as it is expelled—if any evidence of water is observed, inspect the bearing for damage and replace as necessary.

3. Now connect the grease gun to one of the fittings for the U-joints and slowly pump in grease until the **all** of the old grease has been forced out. Repeat this procedure for the second U-joint, being careful to force ALL of the old grease out.

✳✳ CAUTION

Use only a hand operated grease gun when filling the zerk fitting. Never use a power operated gun!

4. Wipe all old grease off of the driveshaft splines and then coat the splines thoroughly with new grease.

5. Carefully wipe any grease, oil or dirt from the two O-rings at the top of the shaft and then coat them thoroughly with light oil.

6. Check the bellows for any signs of damage or deterioration, replacing it if in doubt (as detailed in the Drive Unit section)

7. Check the engine alignment and then reinstall the drive unit.

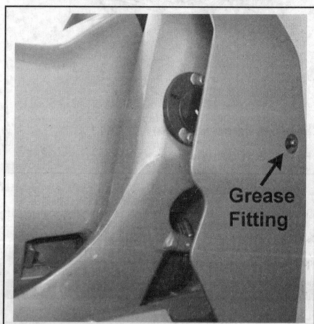

Fig. 134 The gimbal bearing is serviced via a grease (zerk) fitting on the starboard side of the housing

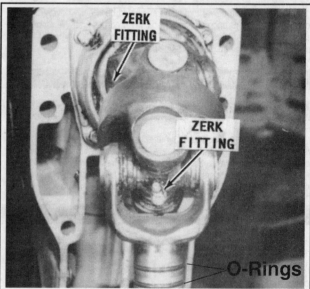

Fig. 135 Each of the U-Joints has a grease fitting. Don't forget to lube the two O-rings at the end of the shaft

BOAT MAINTENANCE

Inside The Boat

Probably the biggest surprise for boat owners is the extent to which mold and mildew develop in a boats interior. Preventing this growth is a two-fold process. First, the boat's interior should be thoroughly cleaned. Second, ventilation should be provided to allow adequate air circulation.

Properly cleaning the boat includes removing as much as possible from it. The less there is on a boat, the less there is to attract mildew. Clothing, foul weather gear, shoes, books, charts, paper goods, leather, bedding, curtains, food stuffs, first aid supplies, odds and ends, etc., should be taken home. They're all fertile soil for mildew, as is dirt, grease, soap scum, etc.

The next step is to vacuum, scrub and polish every surface, especially galley and head surfaces. Use a polish that leaves a protective coating on which it's hard for mold to get a foothold. Mildew-preventive sprays are available for carpets and furniture. The cleaner and more highly polished the insides of lockers, cabinets, refrigerators, icemakers, drawers and shower stalls are when you leave them, the less likely it is you'll find mold. Marine stores sell cleaners and polishes specifically for boats and for the marine environment. Many people find that regular household products are satisfactory. It's the elbow grease that counts.

If dirt and grease are the growing medium, stagnant air is the fertilizer for mold and mildew (mildew is a form of mold). The more freely air can circulate in the interior of your boat the less conducive conditions will be for the growth of mold. Openings or vents at each end near the top can be fashioned to let air in. Flaps or stovepipe elbows can be used to keep rain out. Hatches and windows can then be left partially open. Visiting the boat on dry, sunny days and opening it up as much as possible is a great way to let fresh air in. Do this whenever possible.

The best way to protect navigation and communications equipment from deteriorating is to remove the units from the boat. Wrap them in towels, not plastic, and take them home. Television sets, VCRs and stereos should also be removed to keep them safe and to keep cold and dampness from affecting them adversely. Before storing electronics equipment, clean off the terminals and the connecting plugs and spray them with a moisture-displacing lubricant that specifically says it's intended for use on electronics. Antennas should also be stored in a safe, dry place to protect them from both the elements and accidental damage. All terminal blocks, junction blocks, fuse holders and the back of electrical panels should be cleaned and sprayed with an appropriate protectant. Remove any corrosion that has developed. Anything that could be easily stolen, such as anchors, flare guns, binoculars, etc., is best taken home.

The Boat's Exterior

◆ See Figure 136

Fiberglass reinforced plastic hulls are tough, durable, and highly resistant to impact. However, like any other material they can be damaged. One of the advantages of this type of construction is the relative ease with which it may be repaired. Because of its break characteristics, and the simple techniques used in restoration, these hulls have gained popularity throughout the world. From the most congested urban marina, to isolated lakes in wilderness areas, to the severe cold of northern seas, and in sunny tropic remote rivers of primitive islands or continents, fiberglass boats can be found performing their daily task with a minimum of maintenance.

A fiberglass hull has almost no internal stresses. Therefore, when the hull is broken or stove-in, it retains its true form. It will not dent to take an out-of-shape set. When the hull sustains a severe blow, the impact will either be absorbed by deflection of the laminated panel or the blow will result in a definite, localized break. In addition to hull damage, bulkheads, stringers, and other stiffening structures attached to the hull may also be affected and therefore, should be checked. Repairs are usually confined to the general area of the rupture.

The best way to care for a fiberglass hull is to first wash it thoroughly. Immediately after hauling the boat, while the bottom is still wet, is best, if possible. Remove any growth that has developed on the bottom. Use a pressure cleaner or a stiff brush to remove barnacles, grass, and slime. Pay particular attention to the waterline area. A scraper of some sort may be needed to attack tenacious barnacles. Pot scrubbers work well. Attend to any blisters; don't wait!

Fig. 136 The best way to care for a fiberglass hull is to wash it thoroughly to remove any growth that has developed on the bottom

Remove cushions and all weather curtains and enclosures and take them home. Make sure they are clean before storing them, and don't store them tightly rolled. After washing the topsides, remove any stains that have developed with one of the fiberglass stain removers sold in marine stores. For stubborn stains, wet-sanding with 600-grit paper may be necessary. Remove oxidation and stains from metal parts. Apply a coat of wax to everything.

BELOW WATERLINE

◆ See Figure 137, 138, 139 and 140

A foul bottom can seriously affect boat performance. This is one reason why racers, large and small, both powerboat and sail, are constantly giving attention to the condition of the hull below the waterline.

In areas where marine growth is prevalent, a coating of vinyl, anti-fouling bottom paint should be applied. If growth has developed on the bottom, it can be removed with a solution of Muriatic acid applied with a brush or swab and then rinsed with clear water. Always use rubber gloves when working with Muriatic acid and take extra care to keep it away from your face and hands. The fumes are toxic. Therefore, work in a well-ventilated area, or if outside, keep your face on the windward side of the work.

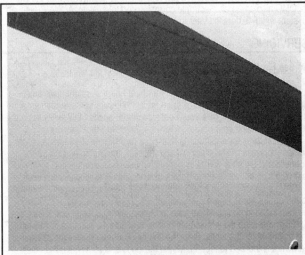

Fig. 137 This beautiful new fiberglass hull will not stay this good looking for long if it is not protected with anti-foul paint

Fig. 138 In areas where marine growth is prevalent, a coating of vinyl, anti-fouling bottom paint should be applied

Fig. 139 This anti-foul paint has seen better days and must be replaced

Fig. 140 This hull is in even worse condition and should be sand blasted and repainted

Barnacles have a nasty habit of making their home on the bottom of boats that have not been treated with anti-fouling paint. Actually they will not harm the fiberglass hull, but can develop into a major nuisance.

If barnacles or other crustaceans have attached themselves to the hull, extra work will be required to bring the bottom back to a satisfactory condition. First, if practical, put the boat into a body of fresh water and allow it to remain for a few days. A large percentage of the growth can be removed in this manner. If this remedy is not possible, wash the bottom thoroughly with a high-pressure fresh water source and use a scraper. Small particles of hard shell may still hold fast. These can be removed with sandpaper.

Anodes (Zincs)

The idea behind anodes is simple: When dissimilar metals are dunked in water and a small current is leaked between or amongst them, the less-noble metal (galvanically speaking) is sacrificed.

The zinc alloy that the anodes are made of is designed to be less noble than the aluminum alloy your drive unit is made from. If there's any electrolysis and there almost always is, the inexpensive zinc anodes are consumed in lieu of the expensive drive.

These zincs need a little attention in order to do their job. Make sure they're there, solidly attached to a clean mounting site and not covered with any kind of paint or wax.

Periodically inspect them to make sure they haven't eroded too much. At a certain point in the erosion process, the mounting holes start to enlarge, which is when the zinc might fall off. Obviously, once that happens, your drive no longer has any protection.

Many OMCs offer an option of a zinc or magnesium anode, magnesium actually works better in fresh water - never use magnesium in saltwater! If you tend to use your vessel in fresh and saltwater, always go with the zincs.

OMC also offers an electronic corrosion control system, Active Protection System which is detailed later in this manual.

SERVICING

◆ See Figure 141, 142 and 143

Depending on what kind of drive your boat has, you might have any number of zincs. Regardless of the number, there are some fundamental rules to follow that will give your boat's sacrificial anodes the ability to do the best job protecting your boat's underwater hardware that they can.

The first thing to remember is that zincs (or magnesiums) are electrical components and like all electrical components, they require good clean connections. So after you've undone the mounting hardware and removed last year's zincs, you want to get the zinc mounting sites clean and shiny.

Get a piece of coarse emery cloth or some 80-grit sandpaper. Thoroughly rough up the areas where the zincs attach (there's often a bit of corrosion residue in these spots). Make sure to remove every trace of corrosion.

Zincs are attached with stainless steel machine screws that thread into the mounting for the zincs. Over the course of a season, this mounting hardware is inclined to loosen. Mount the zincs and tighten the mounting hardware securely. Tap the zincs with a hammer hitting the mounting screws squarely. This process tightens the zincs and allows the mounting hardware to become a bit loose in the process. Now, do the final tightening. This will

insure your zincs stay put for the entire season.

All units utilize anodes in at least two locations, many have three. The following list should point you in the right direction for your particular model:

a. Cobra—At the bottom of the transom shield/assembly; on the leading edge of the lower gear case, where it mates with the upper/intermediate unit; and attached to the bearing carrier inside the lower unit.

b. SP and King Cobra—At the bottom of the transom shield/assembly; on the leading edge of the lower gear case, where it mates with the upper/intermediate unit; and, on the aft edge of the lower gear case, just forward of the propeller.

c. DP—At the bottom of the transom shield/assembly; and, on the leading edge of the lower gear case, where it mates with the upper/intermediate unit.

INSPECTION

If you use your boat in salt water, and your zincs never wear, inspect them carefully. Paint or wax on zincs prevents them from working properly. They must be left bare. If the zincs are installed properly and not painted or waxed, inspect around them for signs of corrosion. If corrosion is found, strip it off immediately and repaint with a rust inhibiting paint. If in doubt, replace the zincs.

On the other hand, if your zinc seems to erode in no time at all, this may be a symptom of the zinc itself. Each manufacturer uses a specific blend of metals in their zincs. If you are using zincs with the wrong blend of metals, they may erode more quickly or leave you with diminished protection.

Batteries

Difficulty in starting accounts for almost half of the service required on boats each year. A survey by a major engine parts company indicated that roughly one third of all boat owners experienced a "won't start" condition in a given year. When an engine won't start, most people blame the battery when, in fact, it may be that the battery has run down in a futile attempt to start an engine with other problems.

Maintaining your battery in peak condition may be thought of as either tune-up or maintenance material. Most wise boaters will consider it to be both. A complete check up of the electrical system in your boat at the beginning of the boating season is a wise move. Continued regular maintenance of the battery will ensure trouble free starting on the water.

Complete battery service procedures are included in this section. The following are a list of basic electrical system service checks that should be performed.

• Check the battery for solid cable connections
• Check the battery and cables for signs of corrosion damage
• Check the battery case for damage or electrolyte leakage
• Check the electrolyte level in each cell
• Check to be sure the battery is fastened securely in position
• Check the battery's state of charge and charge as necessary
• Check battery voltage while cranking the starter. Voltage should remain above 9.5 volts
• Clean the battery, terminals and cables

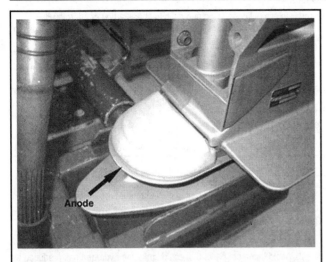

Fig. 141 Almost all drive units covered here utilize an anode on the leading edge of the unit

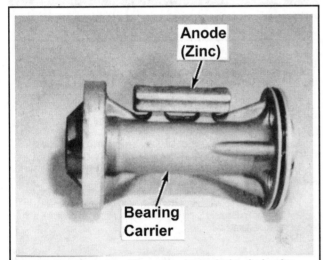

Fig. 142 Certain models utilize an anode attached to the bearing carrier...

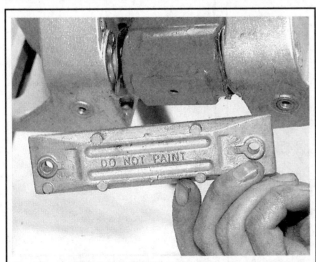

Fig. 143 . . . while most models also use one like this, mounted on the bottom of the transom assembly

- Coat the battery terminals with dielectric grease or terminal protector
- Check the tension on the alternator belt.

Batteries that are not maintained on a regular basis can fall victim to parasitic loads (small current drains which are constantly drawing current from the battery, like clocks, small lights, etc.). Normal parasitic loads may drain a battery on boat that is in storage and not used frequently. Boats that have additional accessories with increased parasitic load may discharge a battery sooner. Storing a boat with the negative battery cable disconnected or battery switch turned **OFF** will minimize discharge due to parasitic loads.

CLEANING

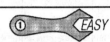

Keep the battery clean; as a film of dirt can help discharge a battery that is not used for long periods. A solution of baking soda and water mixed into a paste may be used for cleaning, but be careful to flush this off with clear water.

■ **Do not let any of the solution into the filler holes on non-sealed batteries. Baking soda neutralizes battery acid and will de-activate a battery cell.**

CHECKING SPECIFIC GRAVITY

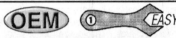

The electrolyte fluid (sulfuric acid solution) contained in the battery cells will tell you many things about the condition of the battery. Because the cell plates must be kept submerged below the fluid level in order to operate, maintaining the fluid level is extremely important. In addition, because the specific gravity of the acid is an indication of electrical charge, testing the fluid can be an aid in determining if the battery must be replaced. A battery in a boat with a properly operating charging system should require little maintenance, but careful, periodic inspection should reveal problems before they leave you stranded.

✳✳ CAUTION

Battery electrolyte contains sulfuric acid. If you should splash any on your skin or in your eyes, flush the affected area with plenty of clear water. If it lands in your eyes, get medical help immediately.

As stated earlier, the specific gravity of a battery's electrolyte level can be used as an indication of battery charge. At least once a year, check the specific gravity of the battery. It should be between 1.20 and 1.26 on the gravity scale. Most parts stores carry a variety of inexpensive battery testing hydrometers. These can be used on any non-sealed battery to test the specific gravity in each cell.

Conventional Battery

◆ See Figure 144

A hydrometer is required to check the specific gravity on all batteries that are not maintenance-free. The hydrometer has a squeeze bulb at one end and a nozzle at the other. Battery electrolyte is sucked into the hydrometer until the float or pointer is lifted from its seat. The specific gravity is then read by noting the position of the float/pointer. If gravity is low in one or more cells, the battery should be slowly charged and checked again to see if the gravity has come up. Generally, if after charging, the specific gravity of any two cells varies more than 50 points (0.50), the battery should be replaced, as it can no longer produce sufficient voltage to guarantee proper operation.

Check the battery electrolyte level at least once a month, or more often in hot weather or during periods of extended operation. Electrolyte level can be checked either through the case on translucent batteries or by removing the cell caps on opaque-case types. The electrolyte level in each cell should be kept filled to the split ring inside each cell, or the line marked on the outside of the case.

■ Never use mineral water or water obtained from a well. The iron content in these types of water is too high and will shorten the life of or damage the battery.

If the level is low, add only distilled water through the opening until the level is correct. Each cell is separate from the others, so each must be checked and filled individually. Distilled water should be used, because the chemicals and minerals found in most drinking water are harmful to the battery and could significantly shorten its life.

If water is added in freezing weather, the battery should be warmed to allow the water to mix with the electrolyte. Otherwise, the battery could freeze.

Maintenance-Free Batteries

Although some maintenance-free batteries have removable cell caps for access to the electrolyte, the electrolyte condition and level is usually checked using the built-in hydrometer "eye." The exact type of eye varies between battery manufacturers, but most apply a sticker to the battery itself explaining the possible readings. When in doubt, refer to the battery manufacturer's instructions to interpret battery condition using the built-in hydrometer.

The readings from built-in hydrometers may vary, however a green eye usually indicates a properly charged battery with sufficient fluid level. A dark eye is normally an indicator of a battery with sufficient fluid, but one that may be low in charge. In addition, a light or yellow eye is usually an indication that electrolyte supply has dropped below the necessary level for battery (and hydrometer) operation. In this last case, sealed batteries with an insufficient electrolyte level must usually be discarded.

BATTERY TERMINALS

◆ See Figures 145 and 146

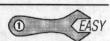

At least once a season, the battery terminals and cable clamps should be cleaned. Loosen the clamps and remove the cables, negative cable first. On batteries with top mounted posts, the use of a puller specially made for this purpose is recommended. These are inexpensive and available in most parts stores.

Clean the cable clamps and the battery terminal with a wire brush, until all corrosion, grease, etc., is removed and the metal is shiny. It is especially important to clean the inside of the clamp thoroughly (a wire brush is useful here), since a small deposit of foreign material or oxidation there will prevent

a sound electrical connection and inhibit either starting or charging. It is also a good idea to apply some dielectric grease to the terminal, as this will aid in the prevention of corrosion.

After the clamps and terminals are clean, reinstall the cables, negative cable last; do not hammer the clamps onto battery posts. Tighten the clamps securely, but do not distort them. To retard corrosion, give the clamps and terminals a thin external coating of clear polyurethane paint after installation.

Check the cables at the same time that the terminals are cleaned. If the insulation is cracked or broken, or if its end is frayed, the cable should be replaced with a new one of the same length and gauge.

BATTERY & CHARGING SAFETY PRECAUTIONS

Always follow these safety precautions when charging or handling a battery.

1. Wear eye protection when working around batteries. Batteries contain corrosive acid and produce explosive gas, a byproduct of their operation. Acid on the skin should be neutralized with a solution of baking soda and water made into a paste. In case acid contacts the eyes, flush with clear water and seek medical attention immediately.

2. Avoid flame or sparks that could ignite the hydrogen gas produced by the battery and cause an explosion. Connection and disconnection of cables to battery terminals is one of the most common causes of sparks.

✳✳ CAUTION

Always make sure that your stored battery is no where near any ignition source!

3. Always turn a battery charger **OFF**, before connecting or disconnecting the leads. When connecting the leads, connect the positive lead first, then the negative lead, to avoid sparks.

4. When lifting a battery, use a battery carrier or lift at opposite corners of the base.

5. Ensure there is good ventilation in a room where the battery is being charged.

6. Do not attempt to charge or load-test a maintenance-free battery when the charge indicator dot is indicating insufficient electrolyte.

7. Disconnect the negative battery cable if the battery is to remain in the boat during the charging process.

8. Be sure the ignition switch is **OFF** before connecting or turning the charger **ON**. Sudden power surges can destroy electronic components.

9. Use proper adapters to connect the charger leads to batteries with non-conventional terminals.

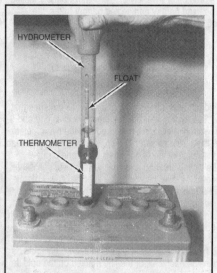

Fig. 144 The best way to determine the condition of a battery is to test the electrolyte with a battery hydrometer

Fig. 145 A battery post cleaner is used to clean the battery posts...

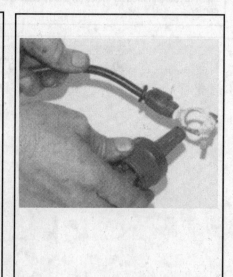

Fig. 146 ...and the battery terminals

BATTERY CHARGERS

Before using any battery charger, consult the manufacturer's instructions for its use. Battery chargers are electrical devices that change Alternating Current (AC) to a lower voltage of Direct Current (DC) that can be used to charge a marine battery. There are two types of battery chargers—manual and automatic.

A manual battery charger must be physically disconnected when the battery has come to a full charge. If not, the battery can be overcharged, and possibly fail. Excess charging current at the end of the charging cycle will heat the electrolyte, resulting in loss of water and active material, substantially reducing battery life.

■ **As a rule, on manual chargers, when the ammeter on the charger registers half the rated amperage of the charger, the battery is fully charged. This can vary, and it is recommended to use a hydrometer to accurately measure state of charge.**

Automatic battery chargers have an important advantage—they can be left connected (for instance, overnight) without the possibility of overcharging the battery. Automatic chargers are equipped with a sensing device to allow the battery charge to taper off to near zero as the battery becomes fully charged. When charging a low or completely discharged battery, the meter will read close to full rated output. If only partially discharged, the initial reading may be less than full rated output, as the charger responds to the condition of the battery. As the battery continues to charge, the sensing device monitors the state of charge and reduces the charging rate. As the rate of charge tapers to zero amps, the charger will continue to supply a few milliamps of current—just enough to maintain a charged condition.

REPLACING BATTERY CABLES

Battery cables don't go bad very often, but like anything else, they can wear out. If the cables on your boat are cracked, frayed or broken, they should be replaced.

When working on any electrical component, it is always a good idea to disconnect the negative (-) battery cable. This will prevent potential damage to many sensitive electrical components

Always replace the battery cables with one of the same length, or you will increase resistance and possibly cause hard starting. Smear the battery posts with a light film of dielectric grease, or a battery terminal protectant spray once you've installed the new cables. If you replace the cables one at a time, you won't mix them up.

■ **Any time you disconnect the battery cables, it is recommended that you disconnect the negative (-) battery cable first. This will prevent you from accidentally grounding the positive (+) terminal when disconnecting it, thereby preventing damage to the electrical system.**

Before you disconnect the cable(s), first turn the ignition to the **OFF** position. This will prevent a draw on the battery which could cause arcing. When the battery cable(s) are reconnected (negative cable last), be sure to check all electrical accessories are all working correctly.

STORAGE

If the boat is to be laid up for the winter or for more than a few weeks, special attention must be given to the battery to prevent complete discharge or possible damage to the terminals and wiring. Before putting the boat in storage, disconnect and remove the batteries. Clean them thoroughly of any dirt or corrosion and then charge them to full specific gravity reading. After they are fully charged, store them in a clean cool dry place where they will not be damaged or knocked over, preferably on a couple blocks of wood. Storing the battery up off the deck, will permit air to circulate freely around and under the battery and will help to prevent condensation.

Never store the battery with anything on top of it or cover the battery in such a manner as to prevent air from circulating around the filler caps. All batteries, both new and old, will discharge during periods of storage, more so if they are hot than if they remain cool. Therefore, the electrolyte level and the specific gravity should be checked at regular intervals. A drop in the specific gravity reading is cause to charge them back to a full reading.

In cold climates, care should be exercised in selecting the battery storage area. A fully-charged battery will freeze at about 60° below zero. A discharged battery, almost dead, will have ice forming at about 19° above zero.

WINTER STORAGE

Taking extra time to store the boat properly at the end of each season will increase the chances of satisfactory service at the next season. Remember, storage is the greatest enemy of a marine engine. In a perfect world the unit should be run on a monthly basis. The steering and shifting mechanism should also be worked through complete cycles several times each month. But who lives in a perfect world!

For most of us, if a small amount of time is spent in winterizing our beloved boats, the reward will be satisfactory performance, increased longevity and greatly reduced maintenance expenses.

Winter Storage Checklist

Proper winterizing involves adequate protection of the unit from physical damage, rust, corrosion, and dirt. The following steps provide guide to winterizing your marine engine at the end of a season.

1. Always keep a note or a checklist of just what maintenance needs to be done prior to starting the engine in the spring.

2. Replace the oil and oil filter. Normal combustion produces corrosive acids that are absorbed by the oil. Leaving dirty oil in the engine for an extended time allows these acids to attack and damage bearing surfaces.

3. Change the engine oil and filter. Used oil contains harmful acids and contaminants that should not be allowed to go to work on the engine all winter long. The drive/transmission oil should also be changed.

4. Replace the fuel filter elements—draining any water from the filter bowls.

5. Keeping the fuel tank full over the winter will help cut down on condensation. Using fuel additives, like OMC 2+4 Marine Fuel Conditioner/Stabilizer, that control bacterial growth or prevent gelling in cold climates are also popular with many boaters and are highly recommended. Be sure to follow the manufacturer's instructions on the side of the can.

■ **On models with a 4bbl carburetor, run the engine under sufficient load to circulate the conditioner through the secondary fuel system.**

6. Bleed the fuel system of air and replace the water separating fuel filter.
    a. Flush the seawater circuit to remove corrosive salts.
    b. Drain the seawater system, taking special care to empty all the low spots.
    c. On closed cooling systems, replace the coolant with a clean, fresh, 50/50 mixture of water and antifreeze. Always mix the antifreeze and water mixture prior to pouring it into the engine.

7. Inspect all hoses for signs of softening, cracking or bulging, especially those routinely exposed to high heat. Check hose clamps for tightness and corrosion.

8. Close the fuel shut-off valve, if equipped, to stop all flow of fuel from the tanks to the engine.

9. On carbureted engines, remove the flame arrestor and start the engine. Allow it to reach normal operating temperature. Run the engine at fast idle and squirt about 2/3 of an 8 oz. (0.24L) of can of Storage Fogging Oil or 12 oz. (355 ml) spray into the carburetor or throttle body. Just as the engine is starting to sputter, squirt the remaining 1/3 of the container into the carb/throttle body and allow the engine to die. Install the flame arrestor.

10. On EFI engines, purchase a 6 gallon external fuel tank for an outboard. Add 5 gallons of fuel, 4 pints (64 oz.) of OMC Storage Fogging Oil and 1/3 cup (2.5 oz.) of 2+4 Marine Fuel Conditioner—mix thoroughly, and carefully. Carefully disconnect and plug the main fuel line from the boats fuel tank at the fuel pump. Hook a line from the 6 gallon fuel tank to the pump. Start the engine (using the fogging mixture as the fuel source) and run it at 1500 rpm for 5 minutes to ensure that all fuel system and internal engine components are completely coated and protected. Turn off the engine and reinstall the main fuel line.

11. It's also a good idea to remove the spark plugs and pour about an ounce of Fogging Oil into each cylinder. You can also use SAE 20 motor oil. Crank the engine with the starter for a few seconds to make sure that the oil coats the cylinder bores completely.

12. On engines so equipped, remove the water separating fuel filter and pour in about 2 oz. of 2 cycle outboard oil. Reinstall the filter, start the engine and run it until it dies. Install a new filter.

13. Perform all the lubrication procedures detailed in this section.

14. Loosen the adjusting bolts or tensioners on all drive or serpentine belts until the tension is relieved.

15. Fully charge the batteries. Disconnect all leads. Unattended, a battery naturally discharges over a period of several weeks. The electrolyte on a discharged battery can freeze at 20°F (-7°C), so keep the batteries fully charged or, better still, remove them to a warmer storage area. Small automatic trickle chargers work well.

16. Treat battery and cable terminals with petroleum jelly, silicone grease, or a heavy-duty corrosion inhibitor.

17. Protect external surfaces with a heavy-duty corrosion inhibitor.

18. Grease all greaseable points on the drivetrain.

19. Lightly coat the alternator and starter with a light lubricant to disperse the water.

20. Ensure that all drain holes on the stern drive are open by inserting a piece of wire into them.

21. Make sure the stern drive is in the full IN/DOWN position.

22. Cover the engine with a waterproof sheet in case there are any leaks from above. Some boaters will take the time to seal all openings to the engine (air inlet, breathers, exhaust), however this traps moisture and may do more harm than good.

23. Place a checklist in a handy spot to remind you of just what maintenance needs to be done prior to starting the engine in the spring. If this sounds silly, DO IT ANYWAY!!

24. If you visit the boat during the winter, additional protection can be achieved using the starter to turn the engine over and circulate oil to the bearings and cylinder walls.

25. If the boat is to be stored out of the water, always remove the drain plug(s) if equipped. We suggest you attach the plug to the ignition key as added insurance that you remember to reinstall it in the spring - sound silly? Why take the chance!

■ Special inhibiting oils are available that provide greater protection. Use these to replace the standard engine oil; run the engine briefly to coat all surfaces, and then drain the oil. Protection remains good, provided the engine is not turned.

26. If the engine cannot be fully winterized, replace oil, coolant, and all filters and run the engine up to operating temperatures monthly.

SPRING COMMISSIONING

Satisfactory performance and maximum enjoyment can be realized if a little time is spent in commissioning your boat in the spring. Assuming you have followed the steps we recommended to winterize your vessel (in addition to any the manufacturer specifies) and the unit has been properly stored, a minimum amount of work should be required to prepare it for use.

After performing the spring commissioning and testing the boat on the water, it is a good idea to perform a full tune-up. Remember, you are relying on your engine to get you where you want to go. Treat it good now and it will treat you good later.

Spring Commissioning Checklist

The following steps outline a logical sequence of tasks to be performed before starting your engine for the first time in a new season.

1. Pick up the checklist you made to remind yourself of just what maintenance needs to be done prior to starting the engine in the spring. You did remember to write yourself a checklist....right?

2. INSTALL THE DRAIN PLUG IF REMOVED!

3. Remove the cover placed over the engine last winter. Unseal any engine openings (air inlet, breathers, exhaust) previously sealed.

4. Replace all zincs.

5. If you took our advice, you removed the battery for the winter. While it was in storage, you should have kept it fully charged. It should be ready to go. So install the battery and connect the battery cables. Treat battery and cable terminals with petroleum jelly, silicone grease, or a heavy-duty corrosion inhibitor. Capacity test the batteries.

6. Tighten the alternator and other belts.

7. As you did in the winter, inspect all hoses for signs of softening, cracking or bulging, especially those routinely exposed to high heat. Check hose clamps for tightness and corrosion.

8. Ensure the exhaust manifolds are tight.

9. On closed system engines, check the condition of the coolant mixture with a coolant tester and adjust the mixture by adding antifreeze or water.

10. Bleed the fuel system of air.

11. Connect the spark plug wires and any other connections.

12. Start the engine and allow it to reach operating temperature.

13. Once running, check the oil pressure, the raw water discharge and the engine for oil and water leaks.

14. Drain water from the filter bowls. If an excessive amount of water is noted in the fuel system, you may want to consider a fuel conditioner or replacing the old fuel with fresh fuel.

15. After testing the boat on the water, it is a good idea to change the engine oil and filter. The drive/transmission oil should also be changed. This eliminates the adverse effects of moisture in the oil.

FIRING ORDERS

◆ See Figures 147 thru 156

■ To avoid confusion, ALWAYS label the spark plug wires with a piece of tape before disconnecting them from the plug or distributor cap.

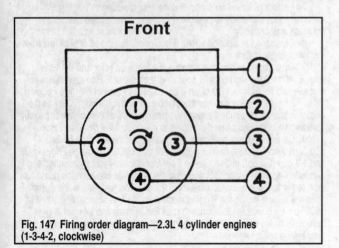

Fig. 147 Firing order diagram—2.3L 4 cylinder engines (1-3-4-2, clockwise)

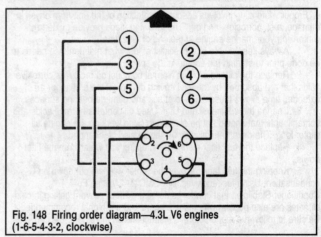

Fig. 148 Firing order diagram—4.3L V6 engines (1-6-5-4-3-2, clockwise)

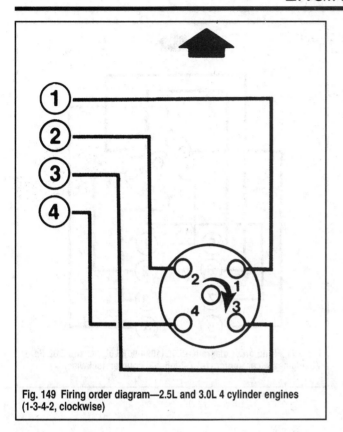

Fig. 149 Firing order diagram—2.5L and 3.0L 4 cylinder engines (1-3-4-2, clockwise)

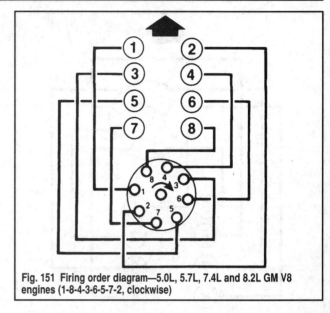

Fig. 151 Firing order diagram—5.0L, 5.7L, 7.4L and 8.2L GM V8 engines (1-8-4-3-6-5-7-2, clockwise)

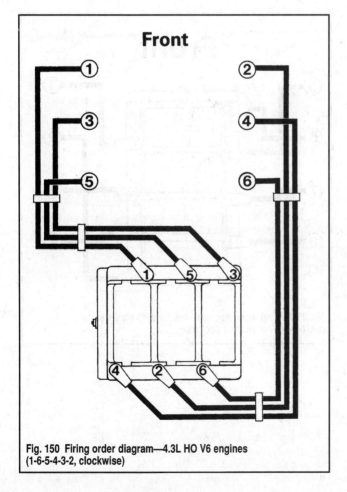

Fig. 150 Firing order diagram—4.3L HO V6 engines (1-6-5-4-3-2, clockwise)

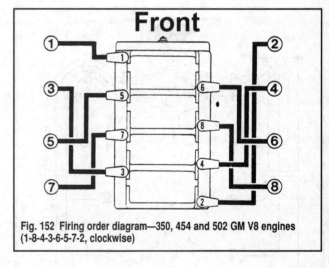

Fig. 152 Firing order diagram—350, 454 and 502 GM V8 engines (1-8-4-3-6-5-7-2, clockwise)

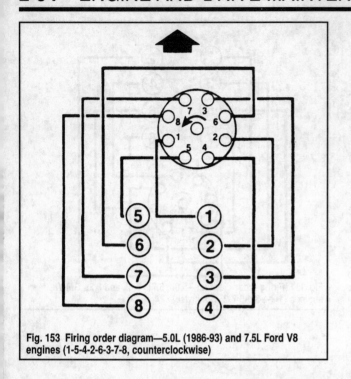

Fig. 153 Firing order diagram—5.0L (1986-93) and 7.5L Ford V8 engines (1-5-4-2-6-3-7-8, counterclockwise)

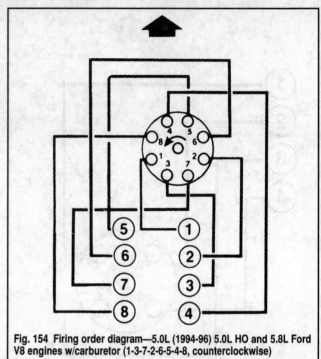

Fig. 154 Firing order diagram—5.0L (1994-96) 5.0L HO and 5.8L Ford V8 engines w/carburetor (1-3-7-2-6-5-4-8, counterclockwise)

Fig. 155 Firing order diagram—5.0L, 5.8L and 351 Ford V8 engines w/EFI (1-3-7-2-6-5-4-8, counterclockwise)

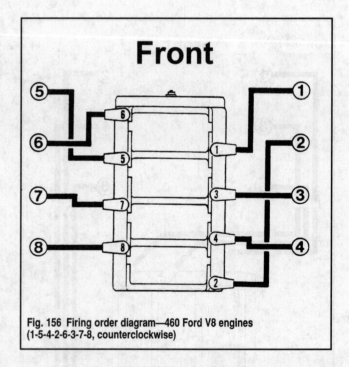

Fig. 156 Firing order diagram—460 Ford V8 engines (1-5-4-2-6-3-7-8, counterclockwise)

Engine Model Applications Applications—1990-92

ENGINE MODEL APPLICATIONS

Year	Model	Model Number	Displacement L/Cu. In.	Engine Type	HorsePower (At Prop.)	Fuel Delivery	Stern Drive
1990	2.3	232BMRPWS, 232BPRPWS	2.3/140	L4	NA	2 bbl	Cobra
	3.0	302AMRPWC, 302BMRPWS, 302AMRPWR, 302APRPWC, 302BPRPWS, 302BPRPWR	3.0/181	L4	120	2 bbl	Cobra
	3.0 HO	302CPRPWS, 302CPRPWR, 302CMLPWS, 302CMLPWR, 302CMRPWR	3.0/181	L4	NA	2 bbl	Cobra
	4.3	432APRPWS, 432AMLPWS	4.3/262	V6	160	2 bbl	Cobra
	4.3 HO	434APRPWS, 434AMLPWS	4.3/262	V6	NA	4 bbl	Cobra
	5.0	502BPRPWS	5.0/302	V8	185	2 bbl	Cobra
	5.0 HO	502APRPWS, 502AMLPWS	5.0/302	V8	200	2 bbl	Cobra
	5.7	574APRPWS, 574AMLPWS	5.7/350	V8	230	4 bbl	Cobra
	5.7 LE	574APPWS	5.7/350	V8	235	4 bbl	King Cobra
	5.8	584APRPWS, 584AMLPWS	5.8/351	V8	235	4 bbl	Cobra
	350	574BPPWS	5.7/350	V8	NA	4 bbl	King Cobra
	454	744APPWS	7.4/454	V8	NA	4 bbl	King Cobra
	460	754APPWS	7.5/460	V8	NA	4 bbl	King Cobra
1991	3.0	302BMRRGD	3.0/181	L4	120	2 bbl	Cobra
	3.0 HO	302CMRRGD, 302CMRGD	3.0/181	L4	NA	2 bbl	Cobra
	4.3	432APRRGD, 432APLRGD	4.3/262	V6	160	2 bbl	Cobra
	4.3 HO	434APRRGD, 434APLRGD	4.3/262	V6	NA	4 bbl	Cobra
	5.0	502BPRRGD	5.0/302	V8	185	2 bbl	Cobra
	5.0 HO	502APRRGD, 502APLRGD	5.0/302	V8	200	2 bbl	Cobra
	5.7	574APRRGD, 574APLRGD	5.7/350	V8	230	4 bbl	Cobra
	5.7 LE, 350	574APERGD, 574BPERGD	5.7/350	V8	NA	4 bbl	King Cobra
	5.8	584APRRGD, 584APLRGD	5.8/351	V8	235	4 bbl	Cobra
	7.4GL, 454, 454 HO	744APERGD, 744BPERGF, 744BPERGD, 744APERG	7.4/454	V8	300	4 bbl	King Cobra
1992	3.0	302BMRAMH	3.0/181	L4	120	2 bbl	Cobra
	3.0 HO	302CPRAMH, 302CMRAMH	3.0/181	L4	NA	2 bbl	Cobra
	4.3	432APRAMH, 432APLAMH	4.3/262	V6	160	2 bbl	Cobra
	5.0	502BPRAMH	5.0/302	V8	185	2 bbl	Cobra
	5.0 HO	502APRAMH, 502APLAMH	5.0/302	V8	200	2 bbl	Cobra
	5.7	574APRAMH, 574APLAMH	5.7/350	V8	230	4 bbl	Cobra
	5.7 LE	574APEAMH, 574BPEAMH	5.7/350	V8	NA	4 bbl	King Cobra
	5.8	584APRAMH, 584APLAMH	5.8/351	V8	235	4 bbl	Cobra
	5.8 LE, 351	584DPEAMH	5.8/351	V8	235	4 bbl	Cobra
	7.4GL, 454, 454 HO	744CPEAMH, 744DPEAMH, 744DPPAMH	7.4/454	V8	300	4 bbl	King Cobra
	8.2GL, 502	824BPEAMH	8.2/502	V8	390	4 bbl	King Cobra

NA Not Available

Engine Model Applications Applications—1986-89

ENGINE MODEL APPLICATIONS

Year	Model	Model Number	Displacement L/Cu. In.	Engine Type	HorsePower (At Prop.)	Fuel Delivery	Stern Drive
1986	2.5	252AMKVB, 252APKVB	2.5/153	L4	NA	2 bbl	Cobra
	3.0	302AMKVB, 302APKVB	3.0/181	L4	NA	2 bbl	Cobra
	4.3	432AMKVB, 432AMVXS, 432APKVB, 432APVXS	4.3/262	V6	NA	4 bbl	Cobra
		434AMKVB, 434AMVXS, 434APKVB, 434APVXS	4.3/262	V6	NA	4 bbl	Cobra
	5.0	504AMKVB, 504AMVXS, 504APKVB, 504APVXS	5.0/305	V8	185	4 bbl	Cobra
	5.7	574AMKVB, 574AMVXS, 574APKVB, 574APVXS,	5.7/350	V8	230	4 bbl	Cobra
1987	2.3	232AMFTC, 232APFTC, 232APSRC	2.3/140	L4	NA	2 bbl	Cobra
	3.0	302AMARJ, 302AMFTC, 302APFTC, 302APSRC, 302APSRY	3.0/181	L4	NA	2 bbl	Cobra
	4.3	432AMFTC, 432AMARY, 432APFTC, 432APARY, 432APSRC, 432APSRY	4.3/262	V6	NA	2 bbl	Cobra
		434AMFTC, 434AMARY, 434APFTC, 434APSRY	4.3/262	V6	NA	4 bbl	Cobra
	5.0	504AMFTC, 504AMARY, 504APFTC, 504APARY	5.0/305	V8	185	4 bbl	Cobra
	5.7	574AMFTC, 574AMARY, 574APFTC, 574APARY, 574APSRC	5.7/350	V8	230	4 bbl	Cobra
	7.5	754APSRC, 754BPSRC	7.5/460	V8	335	4 bbl	Cobra
		754BPFTC, 754BPSRC	7.5/460	V8	440	4 bbl	Cobra
1988	2.3	232AMRGDE, 232AMRGDP, 232APRGDE, 232APRGDP	2.3/140	L4	NA	2 bbl	Cobra
	3.0	302AMRGDP, 302APRGDP	3.0/181	L4	NA	2 bbl	Cobra
	4.3	432AMLGDP, 432APRGDE, 432APRGDP	4.3/262	V6	NA	4 bbl	Cobra
		434AMLGDP, 434APRGDP	4.3/262	V6	NA	4 bbl	Cobra
	5.0	502AMLGDP, 502APRGDP, 504AMLGDP, 504APRGDP	5.0/305	V8	185	4 bbl	Cobra
	5.7	574AMLGDP, 574APRGDP, 574BMLGDP, 574BPLGDP	5.7/350	V8	230	4 bbl	Cobra
	7.4 454	744AMRGDP, 744APRGDP	7.4/454	V8	NA	4 bbl	King Cobra
	7.5 460	754BMLGDP, 754BPRGDP	7.5/460	V8	NA	4 bbl	King Cobra
1989	2.3	232BMRMED, 232BMRMEF, 232BPRMED, 232BPRMEF	2.3/140	L4	NA	2 bbl	Cobra
	3.0	302AMRMED, 302APRMED	3.0/181	L4	120	2 bbl	Cobra
	4.3	432AMLMED, 432APRMED	4.3/262	V6	160	2 bbl	Cobra
	5.0	502AMLMED, 502APRMED	5.0/302	V8	200	2 bbl	Cobra
		502BPRMED	5.0/302	V8	185	4 bbl	Cobra
	5.7	574AMLMED, 574APRMED	5.7/350	V8	230	4 bbl	Cobra
	5.8	584AMLMED, 584APRMED	5.8/351	V8	235	4 bbl	Cobra
	262	262AMLMED, 262APRMED	4.3/262	V6	NA	4 bbl	King Cobra
	350	350APRMED, 350AMLMED	5.7/350	V8	NA	4 bbl	King Cobra
	460	460AMLMED, 460APRMED	7.5/460	V8	NA	4 bbl	King Cobra

NA Not Available

Engine Model Applications Applications—1997-98

ENGINE MODEL APPLICATIONS

Year	Model	Model Number	Displacement L/Cu. In.	Engine Type	HorsePower (At Prop.)	Fuel Delivery	Stern Drive
1997	3.0GS	302CMLKD, 302CPLKD	3.0/181	L4	135	2 bbl	SP
	4.3Gi	43FBPLKD	4.3/262	V6	205	TBI	SP/DP
	4.3GL	432BPLKD	4.3/262	V6	190	2 bbl	SP/DP
	4.3GS	434BPLKD	4.3/262	V6	205	4 bbl	SP/DP
	5.7Gi	57FAPLKD	5.7/350	V8	250	TBI	SP/DP
	5.7GL	572BPLKD, 572BPLKR	5.7/350	V8	215	2 bbl	SP/DP
	5.7GS	574BPLKD	5.7/350	V8	250	4 bbl	SP/DP
	5.7GSi	57FCPLKD	5.7/350	V8	280	TBI	SP/DP
	7.4Gi	74FGPLKD	7.4/454	V8	330	MPI	SP/DP
	7.4GL	744FPLKD	7.4/454	V8	300	4 bbl	SP/DP
1998	3.0GS	302CMBYC, 302CPBYC	3.0/181	L4	135	2 bbl	SP
	4.3Gi	43FBPBYC	4.3/262	V6	205	TBI	SP/DP
	4.3GL	432BPBYC	4.3/262	V6	190	2 bbl	SP/DP
	4.3GS	434BPBYC	4.3/262	V6	205	4 bbl	SP/DP
	5.0GS	502BPBYC	5.0/305	V8	220	2 bbl	SP/DP
	5.7GS	5728BPBYC	5.7/350	V8	250	2 bbl	SP/DP
	5.7GSi	57FCPBYC	5.7/350	V8	280	TBI	SP/DP
	7.4Gi	74FJPBYC	7.4/454	V8	310	MPI	SP/DP
	7.4GSi	74FHPBYC	7.4/454	V8	385	MPI	DP
	8.2GSi	82FCPBYC	8.2/502	V8	415	MPI	DP

EFI Electronic Fuel Injection
MPI Multi-Port Injection
TBI Throttle Body Injection
SP   Single Prop
DP   DuoProp
NA   Not Available

Engine Model Applications Applications—1993-96

ENGINE MODEL APPLICATIONS

Year	Model	Model Number	Displacement L/Cu. In.	Engine Type	HorsePower (At Prop.)	Fuel Delivery	Stern Drive
1993	3.0	302BMRJVB	3.0/181	L4	120	2 bbl	Cobra
	3.0 HO	302CPRJVB, 302CMRJVB	3.0/181	L4	130	2 bbl	Cobra
	4.3	432APRJVB, 432APLJVB, 432APRJVN, 432APLJVN	4.3/262	V6	160	2 bbl	Cobra
	4.3 HO	434APRJVB, 434APLJVB, 434APRJVN, 434APLJVN	4.3/262	V6	NA	2 bbl	Cobra
	5.0	502BPRJVB	5.0/302	V8	185	2 bbl	Cobra
	5.0 HO	502APRJVB, 502APLJVB	5.0/302	V8	200	2 bbl	Cobra
	5.0 EFI	50FAPRJVB, 50FAPLJVB, 50FAPRJVN, 50FAPLJVN	5.0/302	V8	215	EFI	Cobra
	5.8	584APRJVB, 584APLJVB	5.8/351	V8	235	4 bbl	Cobra
	5.8 EFI	58FAPRJVB, 58FAPLJVB, 58FAPRJVN, 58FAPLJVN	5.8/351	V8	240	EFI	Cobra
	5.8 LE, 351	584DPEJVB, 584DPPJVB, 58FBPEJVB, 58FAPPJVB	5.8/351	V8	NA	4 bbl	King Cobra
	7.4GL, 454, 454 HO	744CPEJVB, 744DPEJVB, 744DPPJVB	7.4/454	V8	300	4 bbl	King Cobra
	8.2GL, 502	824BPEJVB	8.2/502	V8	390	4 bbl	King Cobra
1994	3.0GL	302DCMMDA, 302DGMMDA	3.0/181	L4	120	2 bbl	Cobra
	3.0GS	302CCPMDA, 302CGPMDA	3.0/181	L4	135	2 bbl	Cobra
	4.3GL	432ACMMDA, 432ACPMDA, 432AGMMDA, 432AGPMDA	4.3/262	V6	160	2 bbl	Cobra
	5.0FL	502ACPMDA, 502AGPMDA	5.0/302	V8	190	2 bbl	Cobra
	5.0Fi	50FACPMDA, 50FAGPMDA	5.0/302	V8	215	EFI	Cobra
	5.7GL	572ACPMDA, 572AGPMDA	5.7/350	V8	225	2 bbl	Cobra
	5.8FL	584ACPMDA, 584AGPMDA	5.8/351	V8	230	4 bbl	Cobra
	5.8Fi	58FACPMDA, 58FAGPMDA	5.8/351	V8	255	EFI	Cobra
	351 EFI	58FBDPMDA	5.8/351	V8	265	EFI	King Cobra
	7.4GL	744DCPMDA, 744DPPMDA, 744DGPMDA, 744DJPMDA	7.4/454	V8	300	4 bbl	King Cobra
1995	3.0GS	302CMHUB, 302CPHUB	3.0/181	L4	130	2 bbl	Cobra
	4.3GL	432APHUB	4.3/262	V6	160	2 bbl	Cobra
	5.0FL	502APHUB	5.0/302	V8	190	2 bbl	Cobra
	5.0 EFI	50FAPHUB	5.0/302	V8	220	4 bbl	Cobra
	5.8FL	584APHUB	5.8/351	V8	235	2 bbl	Cobra
	5.8 EFI	58FAPHUB	5.8/351	V8	255	EFI	Cobra
	5.8 HO EFI	58FBPHUB	5.8/351	V8	265	EFI	Cobra
	7.4GL	744DPHUB	7.4/454	V8	300	4 bbl	King Cobra
	7.4 EFI	74FEPHUB	7.4/454	V8	330	MPI	King Cobra
1996	3.0GS	302CMNCA, 302CPNCA	3.0/181	L4	130	2 bbl	SP
	4.3GL	432APNCA	4.3/262	V6	160	2 bbl	SP/DP
	4.3Gi	43FAPNCA	4.3/262	V6	180	TBI	SP/DP
	4.3GS	434APNCA	4.3/262	V6	185	4 bbl	SP/DP
	5.0FL	502APNCA	5.0/302	V8	190	2 bbl	SP/DP
	5.0Fi	50FAPNCA	5.0/302	V8	220	EFI	SP/DP
	5.7Gi	57FAPNCA	5.7/350	V8	250	TBI	SP/DP
	5.8FL	584APNCA	5.8/351	V8	235	4 bbl	SP/DP
	5.8Fi	58FCPNCA	5.8/351	V8	255	EFI	SP/DP
	5.8FSi	58FBPNCA	5.8/351	V8	265	EFI	SP/DP
	7.4GL	744EPNCA	7.4/454	V8	300	4 bbl	SP/DP
	7.4Gi	74FFPNCA	7.4/454	V8	330	MPI	SP/DP

EFI Electronic Fuel Injection   SP Single Prop
MPI Multi-Port Injection   DP DuoProp
TBI Throttle Body Injection   NA Not Available

TUNE-UP SPECIFICATIONS
1986-88

Year	Model	Spark Plug Champion	AC	Gap In. (mm)	Ignition Timing (Deg. BTDC @ rpm)	Idle Speed Rpm (In Gear)	Max. WOT (rpm)	Idle Mixture (Turns Out)	Point Gap In. (mm)	Dwell (Deg.)	Fuel Pressure @ Idle (psi)	Compression Pressure (psi)
1986	2.5	RV8-C	MR43T	0.035 (0.9)	①	500-600	4200-4600	2	0.019 (0.48)	32-34	3.5-6	130
	3.0	RV8-C	MR43T	0.035 (0.9)	①	500-600	4200-4600	2	0.019 (0.48)	32-34	3.5-6	130
	4.3 (2bbl)	RV8-C	MR43T	0.035 (0.9)	②	500-600	4200-4600	2.5	0.019 (0.48)	37-41	5.75-7	160-175
	4.3 (4bbl)	RV8-C	MR43T	0.035 (0.9)	②	500-600	4400-4800	3 3/4	0.019 (0.48)	37-41	5.75-7	160-175
	5.0	RV8-C	MR44T	0.035 (0.9)	③	500-600	4200-4600	3 3/4	0.018 (0.45)	28-32	5.75-7	150-160
	5.7	RV8-C	MR43T	0.035 (0.9)	④	500-600	4200-4600	3 3/4	0.018 (0.45)	28-32	5.75-7	150-160
1987	2.3	RS12YC	AWSF-32 ⑤	0.035 (0.9)	10	600-650	5200-5600	2 1/2	0.018 (0.45)	34-38	3.5-6	165
	3.0	RV8-C	MR43T	0.035 (0.9)	①	500-600	4200-4600	2	0.019 (0.48)	31-34	3.5-6	130
	4.3 (2bbl)	RV8-C	MR43T	0.035 (0.9)	②	500-600	4200-4600	2.5	0.019 (0.48)	37-41	5.75-7	160-175
	4.3 (4bbl)	RV8-C	MR44T	0.035 (0.9)	⑥	500-600	4400-4800	3 3/4	0.019 (0.48)	37-41	5.75-7	160-175
	5.0	RV8-C	MR44T	0.035 (0.9)	⑥	500-600	4200-4600	3 3/4	0.018 (0.45)	28-32	5.75-7	150-160
	5.7	RV8-C	MR43T	0.035 (0.9)	⑦	500-600	4200-4600	3 3/4	0.018 (0.45)	28-32	5.75-7	150-160
	7.5	RS12YC	AWSF-32 ⑤	0.035 (0.9)	⑧	600-650	4400-4800	3 3/4	0.017 (0.43)	29-33	5.75-7	165
1988	2.3	RS12YC	AWSF-32 ⑤	0.035 (0.9)	10	600-650	5200-5600	1/2	0.018 (0.45)	34-38	3.5-6	165
	3.0	RV8-C	MR43T	0.035 (0.9)	①	500-600	4200-4600	2	0.019 (0.48)	31-34	3.5-6	130
	4.3 (2bbl)	RV8-C	MR43T	0.035 (0.9)	②	500-600	4200-4600	5/8	0.019 (0.48)	37-41	5.75-7	160-175
	4.3 (4bbl)	RV8-C	MR44T	0.035 (0.9)	②	500-600	4400-4800	1 1/2	0.019 (0.48)	37-41	5.75-7	160-175
	5.0	RV8-C	MR43T	0.035 (0.9)	④	500-600	4200-4600	⑩	0.017 (0.43)	27-31	5.75-7	150-160
	5.7	RV8-C	MR43T	0.035 (0.9)	⑨	500-600	4200-4600	3 3/4	0.017 (0.43)	27-31	5.75-7	150-160
	7.4 454	RV8-C	MR43T	0.035 (0.9)	⑨	500-600	4200-4600	3 3/4	0.017 (0.43)	27-31	5.75-7	150-160
	7.5 460	RV12YC	ARF-32 ⑤	0.035 (0.9)	⑧	600-650	4400-4800	1 1/2	0.017 (0.43)	29-33	5-6	165

NA   Not adjustable

① 1 ATDC w/86 AKI octane; 4 BTDC w/89 AKI octane
② 1 BTDC w/86 AKI octane; 6 BTDC w/89 AKI octane
③ 8 BTDC w/86 AKI octane; 13 BTDC w/89 AKI octane
④ 5 BTDC w/86 AKI octane; 10 BTDC w/89 AKI octane
⑤ Motorcraft

⑥ 504AMFTC & 504APFTC: 8 BTDC w/86 AKI octane; 13 BTDC w/89 AKI octane
   504AMARY & 504APARY: 5 BTDC w/86 AKI octane; 10 BTDC w/89 AKI octane
⑦ 574AMFTC & 574APFTC: 5 BTDC w/86 AKI octane; 10 BTDC w/89 AKI octane
   574AMARY & 574APARY: 3 BTDC w/86 AKI octane; 8 BTDC w/89 AKI octane
⑧ 10 BTDC w/86 AKI octane; 15 BTDC w/89 AKI octane
⑨ 3 BTDC w/86 AKI octane; 8 BTDC w/89 AKI octane
⑩ : 2 1/2, 4 bbl: 3 3/4

TUNE-UP SPECIFICATIONS 1989-90

Year	Model	Spark Plug Champion	Spark Plug AC	Gap In. (mm)	Ignition Timing (Deg. BTDC @ rpm)	Idle Speed Rpm (In Gear)	Max. WOT (rpm)	Idle Mixture (Turns Out)	Point Gap In. (mm)	Dwell (Deg.)	Fuel Pressure @ Idle (psi)	Compression Pressure (psi)
1989	2.3	RS12YC	AWSF-32 ③	0.035 (0.9)	10	600-650	5200-5600	1/2	0.019 (0.48)	34-38	5.75-7	155-165
	3.0	RV8-C	MR43T	0.035 (0.9)	①	550-600	4200-4600	2	0.019 (0.48)	31-34	5.75-7	130
	4.3 (2bbl)	RV8-C	MR43T	0.035 (0.9)	②	550-600	4200-4600	5/8	0.019 (0.48)	37-41	5.75-7	160-175
	4.3 (4bbl)	RV8-C	MR43T	0.035 (0.9)	②	550-600	4400-4800	1 1/2	0.019 (0.48)	37-41	5.75-7	160-175
	5.0	RV8-C	MR44T	0.035 (0.9)	10	550-600	4000-4400	1 1/2	0.018 (0.45)	27-31	5.75-7	145-155
	5.7	RV8-C	MR43T	0.035 (0.9)	⑨	550-600	4200-4600	3 3/4	0.018 (0.45)	27-31	5.75-7	150-160
	5.8	RV8-C	MR43T	0.035 (0.9)	10	550-600	4000-4400	1 1/2	0.018 (0.45)	27-31	5.75-7	145-155
	262	RV8-C	MR43T	0.035 (0.9)	②	550-600	4400-4800	1 1/2	0.019 (0.48)	37-41	5.75-7	160-175
	350	RV8-C	MR43T	0.035 (0.9)	⑨	550-600	4200-4600	3 3/4	0.018 (0.45)	27-31	5.75-7	150-160
	460	RV12YC	ASF-32M ③	0.035 (0.9)	10	600-650	4400-4800	1 1/2	0.018 (0.45)	27-31	5-6	155-165
1990	2.3	RS12YC	AWSF-32 ③	0.035 (0.9)	10	600-650	5200-5600	1/2	0.019 (0.48)	34-38	5.75-7	75% of Highest
	3.0	⑦	⑦	0.035 (0.9)	0 ⑧	550-600	4200-4600	3/4	NA	NA	5.75-7	75% of Highest
	3.0 HO	RS12YC	MR43LTS	0.035 (0.9)	0 ⑧	550-600	4200-4600	1 (Starboard) 1/2 (Port)	NA	NA	5.75-7	75% of Highest
	4.3	RV8-C	MR43T	0.035 (0.9)	②	550-600	4200-4600	5/8	0.019 (0.48)	37-41	5.75-7	75% of Highest
	4.3 HO	RV8-C	MR43T	0.035 (0.9)	NA	550-600	4400-4800	3/4	NA	NA	5.75-7	75% of Highest
	5.0	RV8-C	MR43T	0.030 (0.76)	10	550-600	4200-4600	1 1/2	0.018 (0.45)	27-31	5.75-7	75% of Highest
	5.0 HO	RV8-C	MR43T	0.030 (0.76)	10	550-600	4200-4600	1	0.018 (0.45)	27-31	5.75-7	75% of Highest
	5.7	RV8-C	MR43T	0.035 (0.9)	⑨	550-600	4200-4600	1/4	0.018 (0.45)	27-31	5.75-7	75% of Highest
	5.7 LE	RV8-C	MR43T	0.035 (0.9)	⑨	500-600	4200-4600	1	0.018 (0.45)	27-31	5.75-7	150-160
	5.8	RV8-C	MR43T	0.030 (0.76)	10	550-600	4200-4600	1 1/2	0.018 (0.45)	27-31	5.75-7	75% of Highest
	350	RV8-C	MR43T	0.035 (0.9)	NA	500-600	4200-4600	1	NA	NA	5.75-7	150-160
	454	RV8-C	MR43T	0.035 (0.9)	NA	500-600	4600-5000	1	NA	NA	5.75-7	150-160
	460	RS12YC	MR43TS	0.035 (0.9)	NA	600-650	4400-4800	1 1/2	NA	NA	5-6	155-165

NA Not adjustable
① 1 ATDC w/86-88 AKI octane; 4 BTDC w/89 AKI octane
② 1 BTDC w/86-88 AKI octane; 6 BTDC w/89 AKI octane
③ Motorcraft
④ 504AMFTC & 504APFTC: 8 BTDC w/86 AKI oct; 13 BTDC w/89 AKI oct.
504AMARY & 504APARY: 5 BTDC w/86 AKI oct.; 10 BTDC w/89 AKI oct.
⑤ 574AMFTC & 574APFTC: 5 BTDC w/86 AKI oct; 10 BTDC w/89 AKI oct.
574AMARY & 574APARY: 3 BTDC w/86 AKI oct; 8 BTDC w/89 AKI oct.
⑥ 10 BTDC w/86 AKI octane; 15 BTDC w/89 AKI octane
⑦ PWS: AC-R43TS, Champ.-RV12YC. PWC: AC-MR43T, Champ.-RV8C
⑧ Shunt required
⑨ 3 BTDC w/86-88 AKI octane; 8 BTDC w/89 AKI octane

TUNE-UP SPECIFICATIONS
1991-92

Year	Model	Spark Plug Champion	Spark Plug AC	Gap In. (mm)	Ignition Timing (Deg. BTDC @ rpm)	Idle Speed Rpm (In Gear)	Max. WOT (rpm)	Distributor Sensor Air Gap In. (mm)	Idle Mixture (Turns Out)	Oil Pressure @ 2000 rpm (psi)	Fuel Pressure @ Idle (psi)	Compression Pressure (psi)
1991	3.0	RV12YC	R43TS	0.045 (1.14)	0 ①	650-750	4200-4600	NA	3/4	40-60	5.75-7	75% of Highest
	3.0 HO	RS12YC	R43LTS	0.045 (1.14)	0 ①	650-750	4200-4600	NA	1 (Starboard) 1/2 (Port)	40-60	5.75-7	75% of Highest
	4.3	RV8-C	MR43T	0.035 (0.9)	②	550-650	4200-4600	0.008 (0.203)	5/8	40-60	5.75-7	75% of Highest
	4.3 HO	RV8-C	MR43T	0.045 (1.14)	NA	550-650	4400-4800	NA	3/4	40-60	5.75-7	75% of Highest
	5.0	RV8-C	MR43T	0.035 (0.9)	10	550-650	4000-4400	0.008 (0.203)	1 1/8	40-60	5.75-7	75% of Highest
	5.0 HO	RV8-C	MR43T	0.035 (0.9)	10	550-650	4000-4400	0.008 (0.203)	1	40-60	5.75-7	75% of Highest
	5.7	RV8-C	MR43T	0.035 (0.9)	③	550-650	4200-4600	0.008 (0.203)	1/4	40-60	5.75-7	75% of Highest
	5.7 LE	RV8-C	MR43T	0.035 (0.9)	③	550-650	4200-4600	0.008 (0.203)	1	40-60	5.75-7	75% of Highest
	5.8	RV8-C	MR43T	0.035 (0.9)	10	550-650	4000-4400	0.008 (0.203)	7/8	40-60	5.75-7	75% of Highest
	350	RV8-C	MR43T	0.045 (1.14)	NA	550-650	4200-4600	NA	1	40-60	5.75-7	75% of Highest
	7.4GL, 454	RV8-C	MR43T	0.045 (1.14)	NA	550-650	4000-4400	NA	1/4	40-60	5.75-7	75% of Highest
	454 HO	RV8-C	MR43T	0.045 (1.14)	NA	550-650	4600-5000	NA	1/4	40-60	5.75-7	75% of Highest
1992	3.0	RV12YC	R43TS	0.045 (1.14)	0 ①	650-750	4200-4600	NA	3/4	40-60	5.75-7	75% of Highest
	3.0 HO	RS12YC	R43LTS	0.045 (1.14)	0 ①	650-750	4200-4600	NA	1 (Starboard) 1/2 (Port)	40-60	5.75-7	75% of Highest
	4.3	RV8-C	MR43T	0.035 (0.9)	②	550-650	4200-4600	0.008 (0.203)	5/8	40-60	5.75-7	75% of Highest
	5.0	RV8-C	MR43T	0.035 (0.9)	10	550-650	4000-4400	0.008 (0.203)	1 1/8	40-60	5.75-7	75% of Highest
	5.0 HO	RV8-C	MR43T	0.035 (0.9)	10	550-650	4000-4400	0.008 (0.203)	1	40-60	5.75-7	75% of Highest
	5.7	RV8-C	MR43T	0.035 (0.9)	③	550-650	4200-4600	0.008 (0.203)	1/4	40-60	5.75-7	75% of Highest
	5.7 LE	RV8-C	MR43T	0.035 (0.9)	③	550-650	4200-4600	0.008 (0.203)	1	40-60	5.75-7	75% of Highest
	5.8	RV8-C	MR43T	0.035 (0.9)	10	550-650	4000-4400	0.008 (0.203)	7/8	40-60	5.75-7	75% of Highest
	5.8 LE, 351	RS9-YC	RX2CLTS	0.045 (1.14)	NA	550-650	4600-5000	NA	7/8	40-60	5.75-7	75% of Highest
	7.4GL, 454	RV8-C	MR43T	0.045 (1.14)	NA	550-650	4200-4600	NA	1/4	40-60	5.75-7	75% of Highest
	454 HO	RV8-C	MR43T	0.045 (1.14)	NA	550-650	4600-5000	NA	1/4	40-60	5.75-7	75% of Highest
	8.2GL	RV8-C	MR43T	0.045 (1.14)	NA	550-650	4800-5200	NA	1/4	40-60	5.75-7	75% of Highest

NA Not adjustable
① Shunt required
② 1 BTDC w/86-88 AKI octane; 6 BTDC w/89 AKI octane
③ 3 BTDC w/86-88 AKI octane; 8 BTDC w/89 AKI octane

TUNE-UP SPECIFICATIONS
1993-94

Year	Model	Spark Plug Volvo	Champion	Gap In. (mm)	Ignition Timing (Deg. BTDC @ rpm)	Idle Speed Rpm (In Gear)	Max. WOT (rpm)	Distributor Sensor Air Gap In. (mm)	Idle Mixture (Turns Out)	Oil Pressure @ 2000 rpm (psi)	Fuel Pressure @ Idle (psi)	Compression Pressure (psi)
1993	3.0	RV12YC	R43TS	0.045 (1.14)	0 ①	650-750	4200-4600	NA	3/4	40-60	5.75-7	75% of Highest
	3.0 HO	RS12YC	R43LTS	0.045 (1.14)	0 ①	650-750	4200-4600	NA	1 (Starboard) 1/2 (Port)	40-60	5.75-7	75% of Highest
	4.3	RV8-C	MR43T	0.035 (0.9)	②	550-650	4200-4600	0.008 (0.203)	5/8	40-60	5.75-7	75% of Highest
	4.3 HO	RV8-C	MR43T	0.035 (0.9)	②	550-650	4200-4600	0.008 (0.203)	5/8	40-60	5.75-7	75% of Highest
	5.0	RV8-C	MR43T	0.035 (0.9)	10	550-650	4000-4400	0.008 (0.203)	NA	40-60		75% of Highest
	5.0 HO	RV8-C	MR43T	0.035 (0.9)	10	550-650	4000-4400	0.008 (0.203)	1 1/8	40-60	5.75-7	75% of Highest
	5.0 EFI	RV8-C	MR43T	0.045 (1.14)	5	600 ④	4200-4600	NA	1 1/8	40-60	5.75-7	75% of Highest
	5.8	RV8-C	MR43T	0.035 (0.9)	10	550-650	4000-4400	0.008 (0.203)	7/8	40-60	5.75-7	75% of Highest
	5.8 EFI	RS9YC	RX2CLTS	0.045 (1.14)	5	600 ④	4200-4600	NA	NA	40-60	5.75-7	75% of Highest
	5.8 LE, 351	RS9YC	RX2CLTS	0.045 (1.14)	10 ⑤	550-650	4600-5000	NA	7/8	40-60	5.75-7	75% of Highest
	7.4GL, 454	RV8-C	MR43T	0.045 (1.14)	10 ⑤	550-650	4200-4600	NA	1/4	40-60	5.75-7	75% of Highest
	454 HO	RV8-C	MR43T	0.045 (1.14)	10 ⑤	550-650	4600-5000	NA	1/4	40-60	5.75-7	75% of Highest
	8.2GL, 502	RV8-C	MR43T	0.045 (1.14)	10 ⑤	550-650	4800-5200	NA	1/4	40-60	5.75-7	75% of Highest
1994	3.0GL	RS12YC	R43LTS	0.045 (1.14)	0 ①	650-750	4200-4600	NA	3/4	40-60	5.75-7	75% of Highest
	3.0GS	RS12YC	R43LTS	0.045 (1.14)	0 ①	650-750	4200-4600	NA	1 (Starboard) 1/2 (Port)	40-60	5.75-7	75% of Highest
	4.3GL	RV15YC4	MR43T	0.045 (1.14)	0 ①	550-650	4200-4600	NA	5/8	40-60	5.75-7	75% of Highest
	5.0Fi	RS12YC	R44LTS	0.045 (1.14)	5	600 ④	4200-4600	NA	NA	40-60	31-39	75% of Highest
	5.0FL	RV15YC4	MR43T	0.035 (0.9)	10	550-650	4200-4600	0.008 (0.203)	1	40-60	5.75-7	75% of Highest
	5.7GL	RV15YC4	MR43T	0.035 (0.9)	③	550-650	4200-4600	0.008 (0.203)	3/4-1	40-60	5.75-7	75% of Highest
	5.8Fi	RS9YC	R42CLTS	0.045 (1.14)	5	600 ④	4200-4600	NA	NA	40-60	31-39	75% of Highest
	5.8FL	RV15YC4	MR43T	0.035 (0.9)	10	550-650	4000-4400	0.008 (0.203)	7/8	40-60	5.75-7	75% of Highest
	351 EFI	RS9YC	R42CLTS	0.045 (1.14)	5	600 ④	4600-5000	NA	NA	40-60	31-39	75% of Highest
	7.4GL	RV15YC4	MR43T	0.045 (1.14)	10 ⑤	550-650	4200-4600	NA	1/4	40-60	5.75-7	75% of Highest

NA Not adjustable
① Shunt required
② 1 BTDC w/86-88 AKI octane; 6 BTDC w/89 AKI octane
③ 3 BTDC w/86-88 AKI octane; 8 BTDC w/89 AKI octane
④ Fixed
⑤ Not adjustable

TUNE-UP SPECIFICATIONS
1995-96

Year	Model	Spark Plug Volvo	Champion	Gap In. (mm)	Ignition Timing (Deg. BTDC @ rpm)	Idle Speed Rpm (In Gear)	Max. WOT (rpm)	Distributor Sensor Air Gap In. (mm)	Idle Mixture (Turns Out)	Oil Pressure @ 2000 rpm (psi)	Fuel Pressure @ Idle (psi)	Compression Pressure (psi)
1995	3.0GS	RS12YC	R43LTS	0.045 (1.14)	0 ①	650-750	4200-4600	NA	1 (Starboard) 1/2 (Port)	40-60	5.75-7	75% of Highest
	4.3GL	RV15YC4	MR43T	0.045 (1.14)	②	550-650	4200-4600	NA	5/8	40-60	5.75-7	75% of Highest
	5.0 EFI	RS9YC	R42CLTS	0.045 (1.14)	5	600 ③	4200-4600	NA	NA	40-60	31-39	75% of Highest
	5.0FL	RV15YC4	MR43T	0.035 (0.9)	10	550-650	4400-4800	0.008 (0.203)	1	40-60	5.75-7	75% of Highest
	5.8 EFI	RS9YC	R42CLTS	0.045 (1.14)	5	600 ③	4200-4600	NA	NA	40-60	31-39	75% of Highest
	5.8FL	RV15YC4	MR43T	0.035 (0.9)	10	550-650	4000-4400	0.008 (0.203)	7/8	40-60	5.75-7	75% of Highest
	5.8 HO EFI	RS9YC	R42CLTS	0.045 (1.14)	5	600 ③	4600-5000	NA	NA	40-60	31-39	75% of Highest
	7.4GL	RV15YC4	MR43T	0.045 (1.14)	10 ③	550-650	4200-4600	NA	1/4	40-60	5.75-7	75% of Highest
	7.4 EFI	RV15YC4	MR43T	0.045 (1.14)	10 ④	600 ③	4800-5200	NA	NA	40-60	31-39	75% of Highest
1996	3.0GS	RS12YC	MR43LTS	0.045 (1.14)	0 ①	650-750	4200-4600	NA	1 (Starboard) 1/2 (Port)	40-60	5.75-7	75% of Highest
	4.3Gi	RV15YC4	R43TS	0.045 (1.14)	8 ④	600 ③	4400-4800	NA	NA	40-60	31-39	75% of Highest
	4.3GL	RV15YC4	MR43T	0.045 (1.14)	8 ④	550-650	4200-4600	NA	5/8	40-60	5.75-7	75% of Highest
	4.3GS	RV15YC4	MR43T	0.045 (1.14)	8 ④	550-650	4400-4800	NA	3/4	40-60	5.75-7	75% of Highest
	5.0Fi	RS12YC	MR43LTS	0.045 (1.14)	5	600 ③	4200-4600	NA	NA	40-60	31-39	75% of Highest
	5.0FL	RV15YC4	MR43T	0.035 (0.9)	10	550-650	4200-4600	0.008 (0.203)	1	40-60	5.75-7	75% of Highest
	5.7Gi	RV15YC4	R43TS	0.045 (1.14)	8 ④	600 ③	4200-4600	NA	NA	40-60	31-39	75% of Highest
	5.8Fi	RS9YC	R42CLTS	0.045 (1.14)	5	600 ③	4200-4600	NA	NA	40-60	31-39	75% of Highest
	5.8FL	RV15YC4	MR43T	0.035 (0.9)	10	550-650	4000-4400	0.008 (0.203)	7/8	40-60	5.75-7	75% of Highest
	5.8FSi	RS9YC	R42CLTS	0.045 (1.14)	5	600 ③	4200-4600	NA	NA	40-60	31-39	75% of Highest
	7.4GL	RV15YC4	MR43T	0.045 (1.14)	10 ③	550-650	4200-4600	NA	1/4	40-60	5.75-7	75% of Highest
	7.4Gi	RV15YC4	R43TS	0.045 (1.14)	10 ④	600 ③	4200-4600	NA	NA	40-60	31-39	75% of Highest

NA Not adjustable
① Shunt required
② 5 BTDC w/86-88 AKI octane; 0 TDC w/89 AKI octane; shunt req'd
③ Not adjustable
④ Diagnostic leads shorted

TUNE-UP SPECIFICATIONS
1997-98

Year	Model	Spark Plug Volvo	Spark Plug Champion	Gap In. (mm)	Ignition Timing (Deg. BTDC @ rpm)	Idle Speed Rpm (In Gear)	Max. WOT (rpm)	Distributor Sensor Air Gap In. (mm)	Idle Mixture (Turns Out)	Oil Pressure @ 2000 rpm (psi)	Fuel Pressure @ Idle (psi)	Compression Pressure (psi)
1997	3.0GS	RS12YC	MR43LTS	0.045 (1.14)	0 ①	650-750	4200-4600	NA	1 (Starboard) 1/2 (Port)	40-60	5-7	75% of Highest
	4.3Gi	RS12YC	MR43LTS	0.045 (1.14)	8 ②	600 ③	4400-4800	NA	NA	40-60	31-39	75% of Highest
	4.3GL	RS12YC	MR43LTS	0.045 (1.14)	④	550-650	4200-4600	NA	5/8	40-60	5-7	75% of Highest
	4.3GS	RS12YC	MR43LTS	0.045 (1.14)	④	550-650	4400-4800	NA	3/4	40-60	5-7	75% of Highest
	5.7Gi	RV15YC4	R43TS	0.045 (1.14)	8 ②	600 ③	4200-4600	NA	NA	40-60	31-39	75% of Highest
	5.7GL	RV15YC4	MR43T	0.035 (0.9)	⑤	550-650	4200-4600	0.008 (0.203)	3/4-1	40-60	5-7	75% of Highest
	5.7GS	RV15YC4	MR43T	0.035 (0.9)	⑤	550-650	4200-4600	0.008 (0.203)	3/4-1	40-60	5-7	75% of Highest
	7.4GSi	RS12YC	MR43LTS	0.045 (1.14)	8 ②	600 ③	4600-5000	NA	NA	40-60	31-39	75% of Highest
	7.4Gi	RV15YC4	R43TS	0.045 (1.14)	10 ②	600 ③	4200-4600	NA	NA	40-60	31-39	75% of Highest
	7.4GL	RV15YC4	MR43T	0.045 (1.14)	10 ⑤	550-650	4200-4600	NA	1/4	40-60	5-7	75% of Highest
1998	3.0GS	RS12YC	MR43LTS	0.045 (1.14)	0 ①	650-750	4200-4600	NA	1 (Starboard) 1/2 (Port)	40-60	5-7	75% of Highest
	4.3Gi	RS12YC	MR43LTS	0.045 (1.14)	8 ②	600 ③	4400-4800	NA	NA	40-60	31-39	75% of Highest
	4.3GL	RS12YC	MR43LTS	0.045 (1.14)	④	550-650	4200-4600	NA	5/8	40-60	5-7	75% of Highest
	4.3GS	RS12YC	MR43LTS	0.045 (1.14)	④	550-650	4400-4800	NA	1 1/8	40-60	5-7	75% of Highest
	5.0GL	RS12YC	MR43LTS	0.035 (0.89)	⑥	550-650	4400-4800	0.008 (0.203)	3/4-1	40-60	5-7	75% of Highest
	5.7GS	RS12YC	MR43LTS	0.035 (0.89)	⑦	550-650	4400-4800	0.008 (0.203)	3/4-1	40-60	5-7	75% of Highest
	5.7GSi	RS12YC	MR43LTS	0.045 (1.14)	8 ②	600 ③	4600-5000	NA	NA	40-60	31-39	75% of Highest
	7.4Gi	RS12YC	MR43LTS	0.045 (1.14)	10 ②	600 ③	4200-4600	NA	NA	40-60	31-39	75% of Highest
	7.4GSi	RV15YC4	MR43T	0.045 (1.14)	10 ②	600 ③	4800-5200	NA	NA	40-60	31-39	75% of Highest
	8.2GSi	RV15YC4	MR43T	0.045 (1.14)	10 ②	600 ③	4600-5000	NA	NA	40-60	31-39	75% of Highest

NA Not adjustable
① Shunt required
② Diagnostic leads shorted
③ Not adjustable

④ 1 BTDC w/86-88 AKI octane; 6 BTDC w/89 AKI octane; shunt req'd
⑤ 3 BTDC w/86-88 AKI octane; 8 BTDC w/89 AKI octane
⑥ 5 BTDC w/86-88 AKI octane; 10 BTDC w/89 AKI octane
⑦ 7 BTDC w/86-88 AKI octane; 12 BTDC w/89 AKI octane

CAPACITIES

Capacities—1986-91

Year	Model	Crankcase (With Filter) Qts. (L)	Cobra	King Cobra	SP	DP	Power Trim Reservoir Oz. (ml)
1986	2.5, 3.0	4.0 (3.8)	60 (1774)	X	X	X	54 (1597)
	4.3	4.5 (4.3)	64 (1892)	X	X	X	54 (1597)
	5.0	6.0 (5.7)	64 (1892)	X	X	X	54 (1597)
	5.7	6.0 (5.7)	64 (1892)	X	X	X	54 (1597)
1987	2.3	5.0 (4.7)	60 (1774)	X	X	X	54 (1597)
	3.0	4.0 (3.8)	60 (1774)	X	X	X	54 (1597)
	4.3	4.5 (4.3)	64 (1892)	X	X	X	54 (1597)
	5.0	6.0 (5.7)	64 (1892)	X	X	X	54 (1597)
	5.7	6.0 (5.7)	64 (1892)	X	X	X	54 (1597)
	7.5	6.8 (6.4)	108 (3194)	X	X	X	54 (1597)
1988	2.3	5.0 (4.7)	60 (1774)	X	X	X	54 (1597)
	3.0	4.0 (3.8)	60 (1774)	X	X	X	54 (1597)
	4.3	4.5 (4.3)	64 (1892)	X	X	X	54 (1597)
	5.0	6.0 (5.7)	64 (1892)	X	X	X	54 (1597)
	5.7	6.0 (5.7)	64 (1892)	64 (1892) ①	X	X	54 (1597)
	7.4 454	7.0 (6.6)	X	108 (3194)	X	X	54 (1597)
	7.5 460	6.8 (6.4)	X	108 (3194)	X	X	54 (1597)
1989	2.3	5.0 (4.7)	60 (1774)	X	X	X	54 (1597)
	3.0	4.0 (3.8)	64 (1892)	X	X	X	54 (1597)
	4.3	4.5 (4.3)	64 (1892)	X	X	X	54 (1597)
	5.0, 5.8	6.0 (5.7)	64 (1892)	X	X	X	54 (1597)
	5.7	6.0 (5.7)	64 (1892)	X	X	X	54 (1597)
	262	4.5 (4.3)	X	64 (1892) ①	X	X	54 (1597)
	350	6.0 (5.7)	X	64 (1892) ①	X	X	54 (1597)
	460	6.0 (5.7)	X	108 (3194)	X	X	54 (1597)
1990	2.3	5.0 (4.7)	60 (1774)	X	X	X	54 (1597)
	3.0, 3.0 HO	4.0 (3.8)	64 (1892)	X	X	X	54 (1597)
	4.3, 4.3 HO	4.5 (4.3)	64 (1892)	X	X	X	54 (1597)
	5.0, 5.0 HO	6.0 (5.7)	64 (1892)	X	X	X	54 (1597)
	5.7	6.0 (5.7)	64 (1892)	X	X	X	54 (1597)
	5.7 LE	6.0 (5.7)	X	103 (3050)	X	X	54 (1597)
	5.8	6.0 (5.7)	64 (1892)	X	X	X	54 (1597)
	350	6.0 (5.7)	X	103 (3050)	X	X	54 (1597)
	454	7.0 (6.6)	X	103 (3050)	X	X	54 (1597)
	460	6.0 (5.7)	X	103 (3050)	X	X	54 (1597)
1991	3.0, 3.0 HO	4.0 (3.8)	64 (1892)	X	X	X	54 (1597)
	4.3	4.5 (4.3)	64 (1892)	X	X	X	54 (1597)
	4.3 HO	5.5 (5.2)	64 (1892)	X	X	X	54 (1597)
	5.0, 5.0 HO	6.0 (5.7)	64 (1892)	X	X	X	54 (1597)
	5.7	6.0 (5.7)	X	103 (3050)	X	X	54 (1597)
	5.7 LE, 350	6.0 (5.7)	64 (1892)	X	X	X	54 (1597)
	5.8	6.0 (5.7)	X	103 (3050)	X	X	54 (1597)
	7.4GL, 454, 454 HO	7.0 (6.6)	X	103 (3050)	X	X	54 (1597)

① Although badged as King Cobra, the 262 and 350 are actually Cobras

Capacities—1992-98

Year	Model	Crankcase (With Filter) Qts. (L)	Cobra	King Cobra	SP	DP	Power Trim Reservoir Oz. (ml)
1992	3.0, 3.0 HO	4.0 (3.8)	64 (1892)	X	X	X	54 (1597)
	4.3	4.5 (4.3)	64 (1892)	X	X	X	54 (1597)
	5.0, 5.0 HO	6.0 (5.7)	64 (1892)	X	X	X	54 (1597)
	5.7, 5.8	6.0 (5.7)	64 (1892)	X	X	X	54 (1597)
	5.7 LE, 351	6.0 (5.7)	X	103 (3050)	X	X	54 (1597)
	454, 454 HO, 502	7.0 (6.6)	X	103 (3050)	X	X	54 (1597)
1993	3.0, 3.0 HO	4.0 (3.8)	64 (1892)	X	X	X	54 (1597)
	4.3, 4.3 HO	4.5 (4.3)	64 (1892)	X	X	X	54 (1597)
	5.0, 5.0 HO, 5.0 EFI	6.0 (5.7)	64 (1892)	X	X	X	54 (1597)
	5.8, 5.8 EFI	5.0 (4.7)	64 (1892)	X	X	X	54 (1597)
	5.8 LE, 351	5.0 (4.7)	X	103 (3050)	X	X	54 (1597)
	454, 454 HO, 502	7.0 (6.6)	X	103 (3050)	X	X	54 (1597)
1994	3.0GL, 3.0GS	4.0 (3.8)	71 (2100)	X	X	X	54 (1597)
	4.3GL	4.5 (4.3)	71 (2100)	X	X	X	54 (1597)
	5.0FL, 5.0Fi	6.0 (5.7)	71 (2100)	X	X	X	54 (1597)
	5.7GL	5.0 (4.7)	71 (2100)	X	X	X	54 (1597)
	5.8FL, 5.8Fi	6.0 (5.7)	71 (2100)	X	X	X	54 (1597)
	351 EFI	5.0 (4.7)	X	103 (3050)	X	X	54 (1597)
	7.4GL	7.0 (6.6)	X	103 (3050)	X	X	54 (1597)
1995	3.0GS	4.0 (3.8)	71 (2100)	X	X	X	54 (1597)
	4.3GL	4.5 (4.3)	71 (2100)	X	X	X	54 (1597)
	5.0FL, 5.0 EFI	6.0 (5.7)	71 (2100)	X	X	X	54 (1597)
	5.8FL	6.0 (5.7)	71 (2100)	X	X	X	54 (1597)
	5.8 EFI, 5.8 HO EFI	5.0 (4.7)	71 (2100)	X	X	X	54 (1597)
	7.4GL, 7.4 EFI	7.0 (6.6)	X	103 (3050)	X	X	54 (1597)
1996	3.0GS	4.0 (3.8)	X	X	71 (2100)	X	54 (1597)
	4.3Gi, 4.3GL, 4.3GS	4.5 (4.3)	X	X	71 (2100)	81 (2400)	54 (1597)
	5.0Fi, 5.0FL	6.0 (5.7)	X	X	71 (2100)	81 (2400)	54 (1597)
	5.7Gi	6.0 (5.7)	X	X	71 (2100)	81 (2400)	54 (1597)
	5.8Fi, 5.8FSi	5.0 (4.7)	X	X	71 (2100)	81 (2400)	54 (1597)
	7.4Gi	9.0 (8.5)	X	X	71 (2100)	81 (2400)	54 (1597)
	7.4GL	7.0 (6.6)	X	X	71 (2100)	81 (2400)	54 (1597)
1997	3.0GS	4.0 (3.8)	X	X	71 (2100)	X	54 (1597)
	4.3Gi, 4.3GL, 4.3GS	4.5 (4.3)	X	X	71 (2100)	81 (2400)	54 (1597)
	5.7Gi, 5.7 GS	6.0 (5.7)	X	X	71 (2100)	81 (2400)	54 (1597)
	5.7Gi, 5.7GSi	6.0 (5.7)	X	X	71 (2100)	81 (2400)	54 (1597)
	7.4Gi, 7.4GL	9.0 (8.5)	X	X	71 (2100)	81 (2400)	54 (1597)
1998	3.0GS	4.0 (3.8)	X	X	71 (2100)	X	54 (1597)
	4.3Gi, 4.3GL, 4.3GS	4.5 (4.3)	X	X	71 (2100)	81 (2400)	54 (1597)
	5.0GL	6.0 (5.7)	X	X	71 (2100)	81 (2400)	54 (1597)
	5.7GS, 5.7 GSi	6.0 (5.7)	X	X	71 (2100)	81 (2400)	54 (1597)
	7.4Gi	9.0 (8.5)	X	X	71 (2100)	81 (2400)	54 (1597)
	7.4GSi, 8.2GSi	9.0 (8.5)	X	X	71 (2100)	81 (2400)	54 (1597)

All capacities are approximate. Always use the dipstick or fill/drain hole to determine the exact quantity of lubricant required.

MAINTENANCE INTERVALS

Component	Procedure	Daily/ Before Use	Weekly	25 Hrs.	50 Hrs./ 60 Days	50 Hrs./ 120 Days	100 Hrs./ 120 Days	50 Hrs./ Season	100 Hrs./ Season	Once A Season
Anodes	Inspect		X							
Battery	Check Fluid Level		X							
	Inspect		X							
Bilge	Inspect	X								
Carburetor	Inspect									X
Cooling System	Check Level Closed System	X								
	Check Pressure Cap								X	
	Flush Sea Water	X		X						
	Inspect Hoses & Clamps: Freshwater						X			
	Saltwater				X					
	Replace Coolant									X
Drive Belts	Inspect					X				
Electrical System	Check Wiring: Freshwater						X			
	Saltwater				X					
Engine Alignment	Check									X
Flame Arrestor	Clean/Inspect 1986-90								X	
	1991-98							X		
Fuel Filter	Replace							X		
Fuel Pump - Mechanical	Check Vent Tube			X						

MAINTENANCE INTERVALS

Component	Procedure	Daily/ Before Use	Weekly	25 Hrs.	50 Hrs./ 60 Days	50 Hrs./ 120 Days	100 Hrs./ 120 Days	50 Hrs./ Season	100 Hrs./ Season	Once A Season
Fuel System	Inspect/Adjust System									X
	Replace Filter									X
Fuel Tank	Check For Water				x					
Gimbal Bearing	Lubricate								X	
Ignition System	Clean/Inspect								X	
	Adjust Timing									X
Oil	Check Level:									
	Crankcase	X								
	Power Steering		X							
	Stern Drive							X		
	Trim Pump								X	
	Change:									
	Crankcase								X	
	Power Steering								X	
	Stern Drive								X	
	Trim Pump									X
Oil Filter	Change								X	
PCV Valve	Inspect/Replace								X	
Shift/Throttle Cable	Inspect/Adjust/ Lubricate:									
	Freshwater						X			
	Saltwater				X					
Spark Plugs	Check				X					
	Gap/Replace									X
Spark Plug Wires	Inspect									X
Steering System	Check Fluid Level		X							
	Inspect/Adjust/ Lubricate:				X					

MAINTENANCE INTERVALS

Component	Procedure	Daily/ Before Use	Weekly	25 Hrs.	50 Hrs./ 60 Days	50 Hrs./ 120 Days	100 Hrs./ 120 Days	50 Hrs./ Season	100 Hrs./ Season	Once A Season
Stern Drive	Check Water Pick-up		X							
	Oil									
	Check Level							X		
	Replace								X	
Transom Shield/Bracket	Inspect								X	
Trim/Tilt Pump	Inspect									X
	Check Fluid Level								X	
U-Joints/Shafts	Grease								X	
	Lube Splines								X	

TROUBLESHOOTING

SYMPTOM/CONDITION	SYSTEM/QUESTIONS	CHECK
Engine - Cranks, But Won't Start	Engine	Compression
	Fuel System	Fuel: quantity/condition
		Anti-syphon valve
		Fuel tank vent: unrestricted
		Fuel tank pick-up screen: clean/unrestricted
		Fuel lines
		Fuel shut-off valve
		Fuel tank valves
		Fuel pump: component/relay/circuit breaker
		Carburetor: accelerator pump
	Ignition System	Wiring from ignition switch to coil/module
		Wiring from ignition coil to spark plug
		Spark plugs: gap, fouling
		EEC power lead
		Circuit breaker
		Battery voltage
Engine - Dies	Engine seizure	Drive unit: internal damage
		Oil pressure gauge
		Crankcase: oil level
		Temperature gauge
		Cooling system
		Internal components
	Fuel System	Fuel gauge
		Fuel level
		Fuel: water/dirt contamination
		Fuel pick-up tube/screen: dirt/obstruction
		Fuel tank vent: unrestricted
		Fuel filters
		Fuel system: air leak on suction side
		Fuel system: fuel leak on pressure side
		Anti-syphon valve
		Fuel lines
		Fuel pump: pressure
		Carburetor
	Ignition System	Wiring from ignition switch to coil/module
		Wiring from ignition coil to spark plug
		Ignition switch: shorts or opens
		Circuit breakers
		Wiring: between engine and dash
		Main harness

TROUBLESHOOTING

SYMPTOM/CONDITION	SYSTEM/QUESTIONS	CHECK
Engine - Lubrication System	Engine	Oil filter: clogged or incorrect type
		Oil filter: bad bypass grommet
		Oil pump: worn gears, cover or shaft
		Oil pump: relief valve spring
		Oil pump relief valve: plunger loose
		Oil pump pick-up: clogged screen or broken tube
		Oil galleys: blocked
		Hydraulic lifters: dirty/defective
		Push rods: clogged oil passages
		Oil: low/poor quality/wrong viscosity
		Remote filters: wrong hose routing
		Crankcase: water contamination
	Oil Pressure Warning System	Gauge: operation/wiring
		Warning horn: operation/wiring
		Engine: temperature
		Senders: operation/wiring
Engine - Noise/Vibration	Alternator	Loose pulley
		Worn bearings
		Belt too tight
		Loose mounting bolts
	Cooling system	Raw water pump
		Belts/pulleys: loose/adjustment
	Crankshaft Balancer/Flywheel	Loose bolts
	Drive Unit	U-joints
		Gimbal bearing
		Damaged internal components
		Propeller: hub/blades
		Engine coupler: loose/worn/damaged
	Engine mountings	Engine mounts: Loose/broken/worn
		Engine stringer: loose lag screws
	Ignition system - ping or knock	Tune-up
		Spark plug wires: incorrect routing
		Fuel: octane rating too low
	Valves/Lifters	Valve tap on start only: wrong/dirty oil, lifter varnish
		Intermittent tap: lifter check ball leaking
		Tap at idle: leak down rate, bad check ball seat
		Tapping in general: too much oil, lifter plunger stuck
		Loud noise at normal temp.: scored lifter plunger, fast leak down rate, oil too light for temperature
Engine - Overheats	Overheating	Engine temperature: known good thermometer
		Temperature gauge
		Sending unit: operation/circuit

TROUBLESHOOTING

SYMPTOM/CONDITION	SYSTEM/QUESTIONS	CHECK
Engine - Overheats (cont'd)	Overheating	Water pumps: raw water/engine circulating
		Water intake screen: dirt/obstructions
		Thermostat
		Water supply hoses: leaks/cracks/obstructions
		Ignition timing
		Raw water pump: leaks on pressure side
		Raw water pump: air leaks on suction side
		Engine: compression
Engine - Runs Rough	At slow speed?	Idle speed/mixture
		Ignition timing
		Spark plugs: gap, fouling
		Fuel pump: pressure
		Fuel: water/dirt contamination
		Carburetor: vacuum leak
		Carburetor: internal fuel leak
		Intake manifold: vacuum leak
	At high speed?	Fuel system: air leak on suction side
		Fuel: too low an octane rating
		Wiring from ignition coil to spark plug
		Ignition timing
		Carburetor: model, jets, vacuum diaphragm
		Fuel filters
		Fuel pump: pressure
		Engine: compression
		Fuel: water/dirt contamination
Engine - Will Not Crank	Starter Circuit	Battery condition: charge
		Battery cables: loose/dirty connections
		Ignition switch: shorts or opens
		Starter motor: shorts, grounds, opens
		Starter solenoid/relay
		Circuit breaker
		Wiring: battery to ignition switch
Engine - Won't Reach Temperature	Runs too cool	Fuel: type/octane
		Propeller: pitch/hub/blades
		Crankcase: oil level
		Hull: marine growth
		Drive unit: marine growth
		Drive unit: gear ratio
		Altitude: too high for fuel system
		Carburetor: air intake restricted
		Exhaust outlets: restrictions in engine/drive/transom
		Engine: compression
		Carburetor

TROUBLESHOOTING

SYMPTOM/CONDITION	SYSTEM/QUESTIONS	CHECK
Engine - Won't Reach Temperature (cont'd)	Runs too cool	Fuel pump: pressure/vacuum
		Boat: overloaded
		Engine: overheating
		Ignition timing
		Ignition system
		Cables/linkage: connections/adjustment
Hard Start - Engine Cold	Always been this way?	Choke: operation/adjustment
		Fuel lines
		Fuel tank: dirt in tank
		Fuel filters
	Not used for a long time?	Carburetor: empty float bowl
		Fuel: water/dirt contamination
	Just begun to happen?	Choke: operation/adjustment
		Carburetor: accelerator pump
		Fuel system: leaks, dirt, obstructions
		Ignition system
		Ignition timing
Hard Start - Engine Hot	Always been this way?	Choke: operation/adjustment
		Fuel: condition, brand, type, octane
		Spark plugs: gap, fouling
		Fuel: water/dirt contamination
		Battery: charge/condition
		Starter
	Just begun to happen?	
	Won't start after having been run?	Wiring from ignition switch to coil/module
		Ignition coil/module
		Ignition timing
		Choke: operation/adjustment
Low Battery After Short Storage	Engine/Boat	Electrical accessories: all
		Ignition circuit cutoff
		Battery: disconnect negative cable
		Battery cables: current draw
		Main harness connector

3

ENGINE MECHANICAL FORD 4 CYLINDER ENGINES

ENGINE MECHANICAL

General Information

The OMC 2.3L 140 cubic inch displacement engine is manufactured by Ford and has been a favorite combination when mated to the OMC Cobra stern drives.

✳✳ CAUTION

This in-line four-cylinder overhead camshaft power plant uses a full pressure lubrication system with a disposable flow-thru oil filter cartridge. Oil pressure is furnished by a gear-type oil pump, driven by the auxiliary shaft, which is driven, via the timing belt, by the crankshaft. A regulator on the oil pump controls the amount of oil pressure output. The oil pump scavenges oil from the bottom of the oil pan and feeds it through the oil filter and then to the main oil gallery in the block. Drilled passages in the block and crankshaft distribute oil to the camshaft and crankshaft to lubricate the rod, main and camshaft bearings.

The camshaft runs in 4 bearings and is driven by the crankshaft via the timing belt. The auxiliary shaft, also driven by the crankshaft via the timing belt, drives the fuel pump and distributor in addition to the oil pump. The crankshaft runs in 5 main bearings with the middle bearing (#3) providing thrust.

Cylinder numbering and firing orders are identified in the illustrations at the end of the Maintenance section.

✳✳ CAUTION

NEVER, NEVER attempt to use standard automotive parts when replacing anything on your engine. Due to the uniqueness of the environment in which they are operated in, and the levels at which they are operated at, marine engines require different versions of the same part; even if they look the same. Stock and most aftermarket automotive parts will not hold up for prolonged periods of time under such conditions. Automotive parts may appear identical to marine parts, but be assured, OMC marine parts are specially manufactured to meet OMC marine specifications. Most marine items are super heavy-duty units or are made from special metal alloy to combat against a corrosive salt water atmosphere.

OMC marine electrical and ignition parts are extremely critical. In the United States, all electrical and ignition parts manufactured for marine application must conform to stringent U.S. Coast Guard requirements for spark or flame suppression. A spark from a non-marine cranking motor solenoid could ignite an explosive atmosphere of gasoline vapors in an enclosed engine compartment.

Engine Identification

ENGINE

◆ See Figure 1

The engine serial numbers are the manufacturer's key to engine changes. These alpha-numeric codes identify the year of manufacture, the horsepower rating and various model/option differences. If any correspondence or parts are required, the engine serial number must be used for proper identification.

Remember that the serial number establishes the year in which the engine was produced, which is often not the year of first installation.

The engine specifications decal contains information such as the model number or code, the serial number (a unique sequential identifier given ONLY to that one engine) as well as other useful information.

An engine specifications decal can generally be found on top of the flame arrestor or on the side of the rocker arm cover - all pertinent serial number information can be found here—engine and drive designations, serial numbers and model numbers. Unfortunately this decal is not always legible on older boats and it's also quite difficult to find, so please refer to the following procedures for each individuals unit's serial number location.

■ Serial numbers tags are frequently difficult to see when the engine is installed in the boat; a mirror can be a handy way to read all the numbers.

The engine serial/model number is sometimes also stamped on the port rear side of the engine where it attaches to the bell housing; although on most later models it may instead be a metal plate attached in the same location. If your engine has a stamped number it will simply be the serial number; if you have a plate (and you should), it will always show a Model number and then the actual Serial number. Additionally, most models will also have this plate or sticker on the transom bracket.

- The first two characters identify the engine size in liters (L); **23** represents the 2.3L, **30** represents the 3.0L and so forth.

- The third character identifies the fuel delivery system; **2** designates a 2 bbl carburetor, **4** is a 4 bbl carburetor, and **F** is a fuel injected engine.

- The fourth character designates a major engine or horsepower change—it doesn't let you know what the change was, just that there was some sort of change. **A** means it is the first model released, **B** would be the second, and so forth

- The fifth character designates what type of steering system was used; **M** would be manual steering and **P** would be power steering.

- Now here's where it gets interesting; on 1987 engines, the sixth, seventh and eighth characters designate the model year. The sixth and seventh actually show the model year, while the eighth is a random model year version code. **ARJ, ARF, FTC, SRC** or **SRY** show 1987.

- On 1988-90 engines, the sixth character designates the direction of propeller rotation. **R** is right hand, **L** is left hand and **E** is either.

- Also on 1988-90 engines, the seventh, eighth and ninth characters designate the model year. The seventh and eighth actually show the model year, while the ninth is a random model year version code. **GDE** or **GDP** is 1988, **MED** or **MEF** is 1989, **PWC, PWR** or **PWS** is 1990.

Any remaining characters are proprietary. So in example, a Model number on the ID plate that reads **232AMFTC** would designate a 1987 2.3L engine with a 2 bbl carburetor and manual steering, first model released.

Engine Model Designations

All engines covered here utilize unique identifiers assigned by OMC; surnames if you will—2.3, 3.0GL, 3.0GS, etc. Obviously the first two characters designate the engine size in litres (L). The second letter, a **G** or and **F** designate the engine manufacturer; General Motors (G) or Ford (F). The third through fifth letters can be found in different combinations, but the individual letter designates the same thing regardless of position. **L** designates limited output. **S** and **X** designate superior output—a 3.0GL will always have a lesser horsepower rating than a 3.0GS in a given model year; a 5.7GL will be less than a 5.7GS. **i** designates that the engine is fuel injected, if there is no **i** then you know the engine uses a carburetor.

Engine

REMOVAL & INSTALLATION

◆ See Figures 2, 3 and 4

1. Check the clearance between the front of the engine and the inside edge of the engine compartment bulkhead. If clearance is less than 6 in. (15.2mm), you will need to remove the stern drive unit because there won't be enough room to disengage the driveshaft from the engine coupler. More than 6 in. will provide enough working room to get the engine out without removing the drive, BUT, we recommend removing the drive anyway. If you intend on doing anything to the mounts or stringers, you will need to re-align the engine as detailed in the Engine Alignment section—which requires removing the drive, so remove the drive!

2. Remove the stern drive unit as detailed in the Drive Systems section.

3. Open or remove the engine hatch cover.

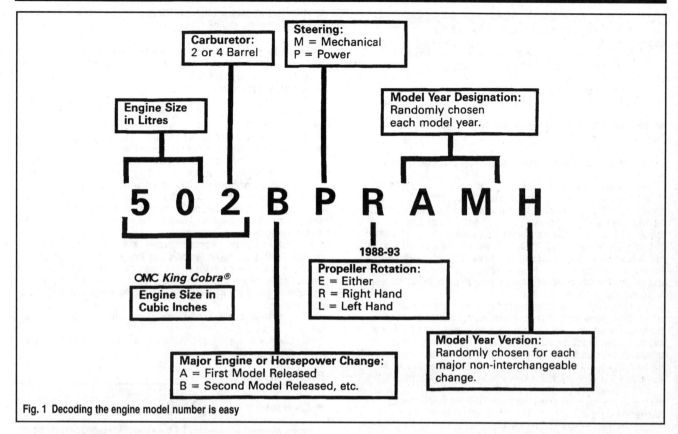

Carburetor:
2 or 4 Barrel

Steering:
M = Mechanical
P = Power

Model Year Designation:
Randomly chosen
each model year.

Engine Size
in Litres

5 0 2 B P R A M H

1988-93

OMC *King Cobra®*

Engine Size in
Cubic Inches

Propeller Rotation:
E = Either
R = Right Hand
L = Left Hand

Model Year Version:
Randomly chosen for each
major non-interchangeable
change.

Major Engine or Horsepower Change:
A = First Model Released
B = Second Model Released, etc.

Fig. 1 Decoding the engine model number is easy

4. Disconnect the battery cables (negative first) at the battery and then disconnect them from the engine block and starter.

✳✳ CAUTION

Make sure that all switches and systems are OFF before disconnecting the battery cables.

5. Disconnect the two power steering hydraulic lines at the steering cylinder (models w/power steering). Carefully plug them and then tie them off somewhere on the engine, making sure that they are higher then the pump to minimize any leakage.

6. Disconnect the fuel inlet line at the fuel pump or filter (whichever comes first on your particular engine) and quickly plug it and the inlet— a clean golf tee and some tape works well in this situation. Make sure you have rags handy, as there will be some spillage.

7. Tag and disconnect the two-wire trim/tilt connector.

8. Pop the two-wire trim/tilt sender connector out of the retainer and then disconnect it. You may have to cut the plastic tie securing the cable in order to move it out of the way.

9. Locate the large rubber coated instrument cable connector (should be on the starboard side), loosen the hose clamp and then disconnect it from the bracket. Move it away from the engine and secure it. On most models, you will also need to unplug the three-wire trim/tilt cable connector just above it.

■ Take note of your throttle arm attachment stud— is it a "push-to-close" or a "pull-to-close"?

10. Disconnect the remote control shift cable and the transom bracket shift cable from the engine shift bracket. Remove the cables from the anchor pockets and the shift lever. Tape the trunnions to the cables so that the adjustment does not get knocked out of kilter.

11. Remove the cotter pin and washer from the throttle arm. Loosen the anchor block retaining nut and then spin the retainer away from the cable trunnion. Remove the throttle cable from the arm and anchor bracket. Be sure to mark the position of the holes that the anchor block was attached to.

12. Loosen the hose clamps (4) and disconnect the exhaust elbow bellows from elbow. You may want to spray some WD-40 around the lip of the hose where it connects to the elbow, grasp it with both hands and wiggle it back and forth while pulling down on it.

13. Drain the cooling system as detailed in the Maintenance section.

Fig. 2 Disconnect the power steering lines at the cylinder

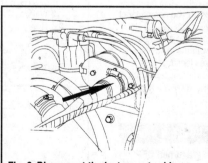

Fig. 3 Disconnect the instrument cable connector

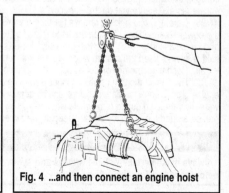

Fig. 4 ...and then connect an engine hoist

14. Loosen the hose clamp on the water supply hose at the transom bracket and carefully slide it off the water tube. Attach the hose to the engine.

15. Disconnect the shift cables and position them out of the way.

16. Tag and disconnect any remaining lines, wires or hoses at the engine.

17. Attach a suitable engine hoist to the lifting eyes (usually front starboard side, rear port side) and take up any line slack until it is just taught.

■ **The engine hoist should have a capacity of at least 1500 lbs. (680 kg).**

18. Locate the rear engine mounts and remove both lock nuts and flat washers.

19. Locate the front engine mounts and remove the two lag bolts.

20. If you listened to us at the beginning of the procedure, the drive unit should be removed. If so, slowly and carefully, lift out the engine. Try not to hit the power steering control valve, or any other accessories, while removing it from the engine compartment. If you didn't listen to us, and you had sufficient clearance in the engine compartment, the drive unit is probably still installed. Raise the hoist slightly until the weight is removed from the mounts and then carefully pull the engine forward until the driveshaft disengages from the coupler, now raise the engine out of the compartment.

To install:

21. Apply Engine Coupler grease to the splines of the coupler.

22. Slowly lower the engine into the compartment. If the drive unit was not removed, AND the crankshaft has not been rotated, insert the driveshaft into the coupler as you push the engine backwards until they engage completely and then lower the engine into position over the rear mounts until the front mounts just touch the stringers. If the shaft and coupler will not align completely, turn the crankshaft or driveshaft slightly until they mate correctly.

If the drive was removed, or the mounts were disturbed in any way, lower the engine into position over the rear mounts until the front mount just touches the stringer.

23. Install the two flat washers into the recess in the engine bracket side of the rear mounts and then install the two lock nuts. Tighten them to 28-30 ft. lbs. (38-40 Nm).

✳✳ CAUTION

Never use an impact wrench or power driver to tighten the locknuts.

24. Install the lag bolts into their holes on the front mounts and tighten each bolt securely.

25. If the drive was removed, the mounts were disturbed or the driveshaft/coupler didn't mate correctly, perform the engine alignment procedure detailed in this section. We think it's a good idea to do this regardless! After you've performed the alignment procedure, make sure that you hold the top adjusting nut and tighten the lower nut on the mount to 100-120 ft. lbs. (136-163 Nm).

26. Reconnect the exhaust bellows by sliding it up and over the elbow, position the clamps between the ribs in the hose and then tighten the clamps securely. Make sure you don't position the clamps into the expanding area.

27. Reconnect the water inlet hose. Lubricate the inside of the hose and wiggle it onto the inlet tube. Slide the clamp over the ridge and tighten it securely. This sounds like an easy step, but it is very important—if the hose, particularly the underside, is not installed correctly the hose itself may collapse or come off. Either scenario will cause severe damage to your engine, so make sure you do this correctly!

28. Carefully, and quickly, remove the tape and plugs so you can connect the power steering lines. Tighten the large fitting to 15-17 ft. lbs. (20-23 Nm) and the small fitting to 10-12 ft. lbs. (14-16 Nm). Don't forget to check the fluid level and bleed the system when you are finished with the installation.

29. Reconnect the trim/tilt connector so the two halves lock together.

30. Reconnect the trim position sender leads, the instrument cable, the engine ground wire, the battery cables and all other wires, lines of hoses that were disconnected during removal. Make sure you swab a light coat of grease around the fitting for the large engine/instrument cable plug.

✳✳ CAUTION

Always make certain that all switches and systems are turned OFF before reconnecting the battery cables.

■ **Make sure all cables, wires and hoses are routed correctly before initially starting the engine.**

31. Unplug the fuel line and pump/filter fitting and reconnect them. Remember to check for leaks as soon as you start the engine.

32. Install and adjust the throttle cable. For complete details, please refer to the Fuel System section.

a. Remember we asked you to determine if you had a "push-to-close" or "pull-to-open" throttle cable (the throttle arm stud)? Position the remote control handle in Neutral—the propeller should rotate freely.

b. Turn the propeller shaft and the shifter into the forward gear detent position and then move the shifter back toward the Neutral position halfway.

c. Position the trunnion over the groove in the throttle cable so the internal bosses align and then snap it into the groove until it is fully seated.

d. Install the trunnion/cable into the anchor block so the open side of the trunnion is against the block. Position the assembly onto the bracket over the original holes (they should be the lower two of the four holes) and then install the retaining bolt and nut. When the nut is securely against the back of the bracket, tighten the bolt securely.

e. Install the connector onto the throttle cable and then pull the connector until all end play is removed from the cable. Turn it sideways until the hole is in alignment with the correct stud on the throttle arm. Slide it over the stud and install the washer and a new cotter pin. Make sure the cable is on the same stud that it was removed from. Tighten the jam nut against the connector

33. Install and adjust the shift cables. Please refer to the Drive Systems section for further details.

34. Check and refill all fluids. Start the engine and check for any fuel or coolant leaks. Go have fun!

ENGINE ALIGNMENT

◆ See Figure 5

Engine alignment is imperative for correct engine installation and also for continued engine and drive operation. It is a good idea to ensure proper alignment every time that the drive or engine have been removed. Engine alignment is checked by using OMC alignment tool #912273 and handle #311880 (Volvo #3851083-0 and #3850609-31). Engine alignment is adjusted by raising or lowering the front engine mount(s).

1. With the drive unit off the vessel, slide the alignment tool through the gimbal bearing and into the engine coupler. It should slide easily, with no binding or force. If not, check the gimbal bearing alignment as detailed in the Drive Systems section. If bearing alignment is correct, move to the next step.

2. If your engine utilizes a jam nut on the bottom of the mount bolt, loosen it and back it off at least 1/2 in.

3. Loosen the lock nut and back it off.

4. Now, determine if the engine requires raising or lowering to facilitate alignment—remember, the alignment tool should still be in position. Tighten or loosen the adjusting nut until the new engine height allows the alignment tool to slide freely.

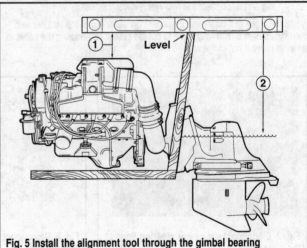

Fig. 5 Install the alignment tool through the gimbal bearing

5. Hold the adjusting nut with a wrench and then tighten the lock nut to 100-120 ft. lbs. (136-163 Nm). If your engine uses a jam nut, cinch it up against the lock nut.

6. Remove the alignment tool and handle.

Front Engine Mounts

REMOVAL & INSTALLATION

◆ See Figures 6 and 7  DIFFICULT

1. Position an engine hoist over the engine and hook it up to the two engine lifting.

2. Remove the two lag screws/bolts on each side of the mount where it rests on the stringer.

3. Raise the engine just enough to allow working room for removing the mount.

4. Remove the 3 mount-to-engine mounting bolts with their lock washers and lift out the mount.

5. Measure the distance between the top of the large washer on the mount and the flat on the lower side of the mounting bracket. Record it.

6. Position a wrench over the bottom nut on the adjusting bolt, just underneath the mount to, to hold the shaft and then remove the top nut. Lift off the bracket.

7. Remove the two lock nuts from the bolt and slide out the adjusting bolt. Remember which washer goes where. The rubber mount and lower foot are serviced as an assembly

To install:

8. Slide the adjusting bolt up through the mount and then position the small and large washers over the bolt—large over small.

9. Spin on the first (lower) lock nut and tighten it to 60-75 ft. lbs. (81-102 Nm).

10. Screw on the upper lock nut and then position the mount bracket over the bolt. Install the washer and adjusting nut and check the measurement taken in Step 5. Move the upper lock nut up or down until the correct specification is achieved and then tighten the adjusting nut to 100-120 ft. lbs. (136-163 Nm).

11. Spray the 3 mounting bolts with Loctite Primer N and allow them to air dry. Once dry, coat the bolts with Loctite or OMC Thread Sealing Agent and attach the mount to the engine. Tighten the bolts to 32-40 ft. lbs. (43-54 Nm).

12. Position the mount over the lag screw holes and then slowly lower the engine until all weight is off the hoist. Install and tighten the lag screws securely.

13. If you're confident that your measurements and subsequent adjustment place the engine exactly where it was prior to removal, then you are through. If you're like us though, you may want to check the engine alignment before you fire up the engine.

Rear Engine Mounts

REMOVAL & INSTALLATION

◆ See Figure 8  DIFFICULT

1. Remove the engine as previously detailed.

2. Loosen the two bolts and remove the mount from the transom plate.

3. Hold the square nut with a wrench and remove the shaft bolt. Be sure to take note of the style and positioning of the two mount washers as you are removing the bolt. Mark them, lay them out, or write it down, but don't forget their orientation!!

To install:

4. Slide the lower of the two washers onto the mount bolt, exactly as it came off (this should be the thin one).

5. Slide the bolt into the flat side of the rubber mount, install the remaining washer (as it came off!) and then spin on the square nut. Do not tighten it yet.

■ **Incorrect washer installation will cause excessive vibration during engine operation.**

6. Turn the assembly upside down and clamp the nut in a vise. Spin the assembly until the holes in the mounting plate are directly opposite any two of the flat sides on the nut. This is important, otherwise the slot on the engine pad will not engage the mount correctly. Secure the mount in this position and tighten the bolt to 18-20 ft. lbs. (24-27 Nm) on 1987-88 engines, or 44-52 ft. lbs. (60-71 Nm) on 1989-90 engines.

7. Remove the mount from the vise and position it on the transom plate. Install the bolts and washers and tighten each to 20-25 ft. lbs. (27-34 Nm).

8. Install the engine, making sure that the slot in the engine pad engages the square nut correctly. Install the two washers and locknut and tighten it to 28-30 ft. lbs. (38-41 Nm).

Fig. 6 A good view of the front mount...

Fig. 7 ...and a closer view

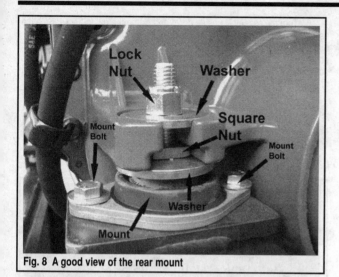

Fig. 8 A good view of the rear mount

Cylinder Head (Camshaft) Cover

REMOVAL & INSTALLATION

 MODERATE

1. Open or remove the engine compartment hatch. Disconnect the negative battery cable.
2. Loosen the clamp and remove the crankcase ventilation hose at the cover (if equipped). Carefully move it out of the way.
3. If your engine has a spark plug wire retainer attached to the cover, unclip the wires or remove the retainer.
4. Loosen the cover mounting bolts (8) and lift off the cylinder head cover. Take note of any harness or hose retainers and clips that might be attached to certain of the mounting bolts; you need to make sure they go back in the same place.

To install:

5. Clean the cylinder head and cover mounting surfaces of any residual gasket material with a scraper or putty knife.
6. Coat the cylinder head mounting surface with high temperature, oil resistant sealer.
7. Position a new gasket on the cylinder head and then position the cover. Tighten the mounting bolts to 48-84 inch lbs. (5.4-9.5 Nm). Make sure any retainers or clips that were removed are back in their original positions.
8. Reconnect the spark plug wires.
9. Connect the crankcase ventilation hose. Check that there were no other wires or hoses you may have repositioned in order to gain access to the cover.
10. Connect the battery cables.

Rocker Arms and Hydraulic Adjusters

REMOVAL & INSTALLATION

OEM ② MODERATE

◆ See Figures 9 and 10

1. Open or remove the engine hatch cover and disconnect the negative battery cable. Remove the cylinder head cover as detailed previously.
2. Rotate the camshaft until the tip of the camshaft lobe over the rocker is facing upward—the heel is resting on the rocker arm.
3. Install the valve tool (4T74P-6565-A) over the valve spring and depress the spring until the rocker arm is released and you can lift it out.
4. Remove the hydraulic valve adjuster.
5. Repeat this procedure for each remaining rocker arm and adjuster.

To install:

6. Clean and inspect the rocker assemblies for nicks, scratches or signs of undue wear. If the pad at the valve end of the arm shows a groove, replace the arm.
7. Coat all bearing surfaces of the rocker assembly and adjuster with engine oil.

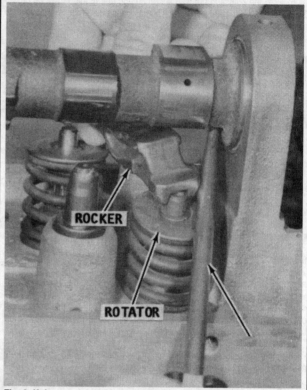

Fig. 9 If the special tool is not available, you can usually compress the spring with a small prybar

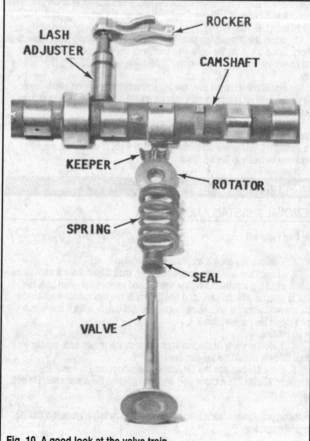

Fig. 10 A good look at the valve train

8. Rotate the camshaft until the lobe is pointing up on the cylinder that you are working on.

9. Depress the valve spring with the special tool again and insert the rocker arm into position.

10. Repeat this procedure for each remaining rocker arm and adjuster, making sure that the spring has been released and the tool removed before moving on to the next arm.

11. Adjust the valves and install the cylinder head cover.

12. Connect the battery cable and check the idle speed.

VALVE ADJUSTMENT

These engines utilize hydraulic valve lifters, although there is no need for periodic valve adjustment, it is necessary to perform a preliminary adjustment after any work on the valve train/rocker assembly. All adjustment should be undertaken while the lifter is on the base circle of the camshaft lobe for that particular cylinder. This means the opposite side of the pointy part of each lobe.

1. Rotate the camshaft by hand until the base circle (heel) of the camshaft lobe is resting on the cam follower at the valve you intend to adjust; the lobe will be pointing straight up.

2. Install the valve tool (#T74P-6565-A) into position and slowly apply pressure to the handle until the until the cam follower has collapsed the hydraulic lash adjuster completely.

3. While using the tool to hold the adjuster in the collapsed position, insert a feeler gauge between the base of the camshaft lobe and the follower. Clearance should be 0.040-0.050 in. (1-1.3mm), although 0.035-0.055 in. (0.89-1.4mm) is acceptable.

4. If clearance is out of range, remove the cam follower again and inspect it for any signs of damage. If the follower appears to be OK and not showing signs of excessive wear, measure the valve spring height between the pad and the retainer—if not within 1 17/32-1 19/32 in., replace the spring.

5. If the valve spring installed height was within specifications, check that the camshaft is within specifications. If the camshaft checks out OK, install a new adjuster and recheck the lash a final time.

Exhaust Manifold

REMOVAL & INSTALLATION

1. Open or remove the engine compartment hatch. Disconnect the negative battery cable.

2. Drain all water from the engine, manifold and exhaust elbow as detailed in the Maintenance section.

3. Loosen the 4 clamps holding the exhaust hose bellows to the high-rise elbow and wiggle the bellows off of the elbow.

4. Disconnect the 2 shift cables at the shift bracket on the manifold. Remove the 2 locknuts and the bolt attaching the bracket to the manifold and lift off the bracket, positioning it on the engine out of the way.

5. Remove the 4 mounting bolts and their lock washers and then lift off the high-rise elbow—you may have to persuade it with a few taps from a rubber mallet.

6. On models with power steering, loosen the drive belt and then remove the pump and bracket from the thermostat housing on the leading edge of the manifold.

7. Disconnect the small water line coming off the starboard side of the thermostat housing. Have a rag handy, because there will probably be some water spillage even though you've already drained the system.

8. Remove the 2 mounting bolts and lock washers securing the thermostat housing to the manifold. Tap the housing a few times lightly with a rubber mallet until it separates from the manifold and then lift it off and position it out of the way with the remaining hoses still attached. Remove the gasket and thermostat.

9. On models equipped with power steering, remove the oil cooler mounting bolts and pull it away from the bottom of the manifold. You should not have to disconnect the hydraulic lines or remove the cooler completely.

10. Remove the 3 remaining locknuts and the upper bolt and then lift off the manifold. Remove the gasket. Once again, you may have to give the manifold a few friendly taps with a rubber mallet.

To install:

11. Clean all gasket and mating surfaces thoroughly (elbow, manifold and thermostat housing) and inspect the components for cracks, undue wear or other signs of damage.

12. Position a new manifold gasket over the studs so the connecting bar is on the bottom and push it up against the block. No sealer or adhesive is necessary.

13. Position the manifold over the studs and into position. Thread on the lock nuts (use new washers) and tighten them to 30-35 ft. lbs. (41-47 Nm). Install the upper bolt and tighten it to 20-25 ft. lbs. (27-34 Nm).

14. Install the elbow with a new gasket and tighten the 4 bolts securely.

15. Reposition the oil cooler if you removed it and tighten the mounting bolt securely.

16. Slide a new cork gasket over the element end of the thermostat and install them into the housing. Coat both sides of a new housing-to-manifold gasket with OMC Sealing Compound, position it on the manifold and then move the housing into place. Slide in the mounting bolts (the longer one goes on the inside) and tighten them to 20-25 ft. lbs. (27-34 Nm). Reconnect the small water line and tighten the clamp securely, being careful not to pinch the hose.

17. Position the shift bracket and tighten the 2 nuts (use new washers if possible) and tighten to 30-35 ft. lbs. (41-47 Nm). Thread in the bolt to the elbow and tighten it to 12-14 ft. lbs. (16-19 Nm). Connect the shift cables.

18. Install the power steering pump, if removed, and adjust the belt tension.

19. Coat the elbow outlet with a bit of soapy water and then wiggle the bellows up and into position. Tighten the hose clamps securely.

20. Reconnect the battery cables and run the engine. Check for any water leaks.

High-Rise Exhaust Elbow

REMOVAL & INSTALLATION

1. Drain the cooling system.

2. Loosen the hose clamps and slide off the exhaust hose (bellows). Lubricating the elbow-to-bellows connection with a bit of soapy water will help break the hose loose.

3. Loosen the 2 shift bracket lock nuts and unscrew them as far as you can without removing the nut from the stud.

4. Remove the 4 elbow mounting bolts on the elbow mating flange and swing the shift cable anchor bracket out of the way.

5. Lift it off the elbow. A little friendly persuasion with a soft rubber mallet may be necessary! Be careful though, no need to take out all your aggressions on the poor thing.

6. Remove the gasket(s) and restrictor plate (if equipped). You can throw away the gasket(s), but keep the plate if your engine uses it.

To install:

7. Clean the mating surfaces of the manifold and elbow thoroughly, coat both sides of a new gasket with Gasket Sealing compound and position it onto the manifold flange.

8. If your engine was so equipped, position the restrictor plate and/or second gasket onto the manifold. Don't forget to coat the gasket with sealing compound.

9. Position the elbow on the manifold, position the anchor bracket and screw the bolts in finger tight. Tighten all four to 12-14 ft. lbs. (16-19 Nm).

10. Tighten the 2 lower nuts on the shift bracket to 30-35 ft. lbs. (41-47 Nm).

11. Slide the exhaust hose, while wiggling it, all the way onto the elbow and tighten the clamp screws securely. Make sure that the 2 upper clamps are riding in the channels on the upper side of the bellows.

Exhaust Hose (Bellows)

REMOVAL & INSTALLATION

◆ **See Figures 11 and 12**

1. Loosen all four hose clamps, two on top of the hose and two on the bottom.
2. Drizzle a soapy water solution over the top of hose where it mates with the exhaust elbow and let it sit for a minute.
3. Grasp the hose with both hands and wiggle it side-to-side while pulling down on it until it separates from the elbow.
4. Now wiggle it while pulling upwards until it pops off the exhaust pipe. Make sure that you secure the exhaust pipe while pulling off the hose.
5. Check the hose for wear, cracks and deterioration.
6. Coat the inside of the lower end of the hose with soapy water and wiggle it into position on the pipe. Remember to install the two clamps before sliding it over the end of the pipe.

■ **There are usually two ribs on one end of the hose—this is the side that attaches to the exhaust elbow. Do not position this end onto the exhaust pipe.**

7. Slide two clamps over the upper end (the side with two ribs!), lubricate the inside with soapy water and wiggle the hose over the elbow until it is in position.
8. Tighten all four clamps securely.

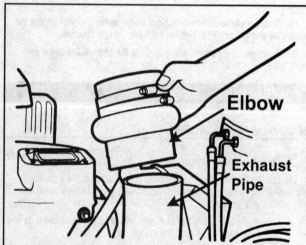

Fig. 11 Loosen the lower clamp screws and then pull the bellows off of the exhaust pipe

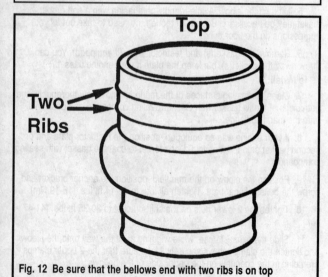

Fig. 12 Be sure that the bellows end with two ribs is on top

Exhaust Pipe

REMOVAL & INSTALLATION

1. It is unlikely you will be able to get the pipe off without removing the engine, so remove the engine as previously detailed.
2. Loosen the four retaining bolts at the transom shield and then remove the exhaust pipe. Carefully scrape any remnants of the seal from the pipe and transom mounting surfaces.

■ **The pipe mounting holes in the transom shield utilize Heli-Coil<reg> locking inserts. NEVER clean the holes or threads with a tapping tool or you risk damaging the locking feature of the threads.**

3. Coat a new seal with OMC Adhesive M or 3M Scotch Grip Rubber Adhesive and position it into the groove on the transom shield mating surface.
4. Coat the mounting bolts with Gasket Sealing Adhesive. Position the exhaust pipe, insert the bolts and tighten them to 10-12 ft. lbs. (14-16 Nm) on 1987-89 engines; or 20-25 ft. lbs. (27-34 Nm) on 1990 engines.

■ **Make sure that the trim/tilt lines are routed correctly (above and behind) before tightening the pipe to the transom.**

5. Install the engine.

EXHAUST VALVE (FLAPPER) REPLACEMENT

1. Remove the exhaust hose from the elbow and exhaust pipe. The flapper is located in the upper end of the pipe.
2. The valve is held in place by means of a pin running through two bushings in the sides of the pipe. Position a small punch over one end of the pin and carefully press the pin out of the pipe.

■ **Make sure you secure the valve while removing the retaining pin so it doesn't fall down into the exhaust pipe.**

3. Press out the two bushings and discard them. Coat two new bushings with Scotch Grip Rubber Adhesive and press them back into the sides of the pipe.
4. Position the new valve into the pipe with the long side DOWN. When looking at the valve, the molded retaining rings are off-center—the side with the rings should face the top of the pipe. When the valve is in place, coat the pin lightly with engine oil and slowly slide it through one of the bushings, through the two retaining holes on the valve and then through the opposite bushing. Make sure the pin ends are flush with the sides of the pipe on both sides.
5. Install the exhaust hose.

Intake Manifold

REMOVAL & INSTALLATION

◆ **See Figure 13**

1. Open the engine covers/hatch and disconnect the battery cables. Drain all water from the engine as detailed in the Maintenance section.
2. Disconnect the throttle cable at the bellcrank and anchor block. Disconnect the swivel linkage at the throttle arm on the carburetor.
3. Carefully disconnect the fuel lines at the carburetor and fuel filter. Plug the lines and be careful to wipe up any spilled fuel.
4. Remove the fuel line clamp bracket at the intake manifold and then disconnect the fuel lines at the fuel pump. Remove the lines. Pull the fuel pump vent tube from the nipple on the carburetor and position it out of the way.
5. Loosen the hose clamp and then wiggle the water line off of the nipple on the end of the manifold. Tie the line back and out of the way with the open end facing upward.

6. Remove the fuel filter mounting bolts and then pull the filter off of the manifold and sit it down somewhere so that it is supported in the upright position. It is not necessary to remove the clamp from the oil dipstick.

7. Remove the plastic cover and then the flame arrestor.

8. Tag and disconnect the electrical lead to the choke at the housing.

9. Remove the 4 mounting nuts and lift off the carburetor and gasket.

10. Disconnect the distributor lead at the ignition coil and then remove the distributor cap, with the leads still attached, and position it out of the way. Pull out the rotor and then cover the top of the distributor with a clean rag.

11. Pop the PCV valve out, with the hose still connected, and position it out of the way.

12. Remove the 2 upper rear manifold bolts and move the throttle cable bracket out of the way.

13. Remove the 2 upper front manifold bolts and move the lifting bracket out of the way.

14. Remove the 4 lower manifold bolts, keeping them separate from the longer upper bolts. Don't forget the clamps on the 2 end bolts when you are removing them.

15. Tap the manifold lightly a few time with a rubber mallet until it breaks loose from the engine and then remove the manifold and gasket.

To install:

16. Clean all mating surfaces thoroughly so that there is no residual gasket material. Inspect the components carefully for cracks, burrs and other signs of wear or damage.

17. Insert a bolt (one of the longer ones) through the upper hole on each end of the manifold and then position the gasket over the bolts and onto the manifold. Position the manifold onto the block and tighten the 2 bolts finger-tight.

18. Thread in the 4 lower bolts (the short ones) until they are finger-tight—don't forget the clamps on the 2 outside bolts.

19. Position the lifting and throttle cable brackets and then install the 2 remaining upper bolts through the brackets and manifold.

20. Tighten the manifold bolts, in the sequence illustrated, to 60-84 inch lbs. (6.8-9.5 Nm). Follow the same sequence and then tighten them once more to 14-18 ft. lbs. (19-24 Nm).

21. Push the PCV valve back in place. Install the rotor and distributor cap. Connect the lead to the ignition coil.

22. Position a new carburetor gasket onto the manifold so that the 2 rounded nubs are facing the engine and then install the carburetor. Tighten the nuts to 12-14 ft. lbs. (16-19 Nm).

23. Connect the electrical lead at the choke housing and the snap the swivel linkage onto the ball stud at the throttle arm.

24. Install the flame arrestor and plastic cover. Install the vent hose bracket.

25. Position the fuel filter and thread in the bolts so the dipstick tube clamp is under the inner bolt. Tighten both bolts securely.

26. Push the water line over the manifold nipple and tighten the clamp securely without pinching the hose.

27. Remove the plugs and connect the fuel lines to the carburetor and filter. Attach the fuel line bracket and reconnect the vent hose at the carburetor.

28. Connect both fuel lines to the fuel pump and tighten the nuts securely.

29. Connect the throttle cable at the bellcrank; position the flat washer and then insert a new cotter pin. Pivot the retainer down until you can get the trunnion into the anchor block and then tighten the nut securely.

30. Connect the battery cables, start the engine and check for any fuel or water leaks.

Oil Pan

REMOVAL & INSTALLATION

◆ See Figures 14 and 15    ⟨**DIFFICULT**

■ **More times than not this procedure will require the removal of the engine. Your boat and its unique engine installation will determine this, but the procedure is almost always easier with the engine removed from the boat.**

1. Remove the engine as previously detailed in this section.

2. If you haven't already drained the engine oil, do it now. Make sure you have a container and lots of rags available.

3. Remove the starter motor as detailed in the Electrical section. Although not absolutely necessary, this step will make the job much easier.

4. Remove the oil withdrawal tube from the oil drain plug fitting.

5. Loosen and remove the oil pan retaining bolts and (22 on 1987-89 engines and 20 on 1990 engines), starting with the center bolts and working out toward the pan ends. Lightly tap the pan with a rubber mallet to break the seal and then lift it off the cylinder block. If your engine stand will allow for rotating the engine, you'll find that this will be easier with the pan facing up.

6. Lift the reinforcement strips off of the pan flange on 1990 engines.

To install:

Clean the pan mating surfaces of any residual gasket material with a scraper or putty knife. Make sure that no old gasket material has been pressed into the retaining bolt holes in the pan, block or front cover. Clean the pan itself thoroughly with solvent.

1987-89 Engines:

7. Apply a 0.125 in. (3.18mm) bead of RTV sealant to the joint where the front cover and block meet. Install new front and rear seals into the front cover and over rear bearing cap— move quickly here as the sealant sets up very quickly. Make sure that you press the tabs on each seal completely into the cylinder block.

8. Apply gasket adhesive to the oil pan flange and also to 2 new side gaskets. Allow the adhesive to set up until it is tacky and then install the gaskets onto the flange on each side of the pan.

9. Insert a guide pin into a hole on each side of the cylinder block mounting surface and then position the oil pan onto the block. Install 4 M8 bolts (two front and two rear) and tighten them to 96-120 inch lbs. (10.8-13.6 Nm).

10. Install the remaining 18 M6 bolts and tighten them to 72-96 inch lbs. (8.1-10.8 Nm), working clockwise around the pan from.

11. Coat a new gasket with RTV sealant and then position the gasket onto the pan being very careful to line up all the holes.

1990 Engines:

12. Apply a 0.125 in. (3.18mm) bead of RTV sealant to the two joints (A) where the front cover meets the pan and the four joints where the oil pan meets the block (B).

13. Apply gasket adhesive to a new oil pan gasket and allow it to set up until tacky. Position the gasket onto the oil pan flange making sure that all the holes line up correctly.

14. Insert a guide pin into a hole on each side of the cylinder block mounting surface and then position the oil pan onto the block. Position the reinforcing strips on each side of the pan and then install 18 M6 bolts finger-tight.

15. Remove the 2 guide pins and thread in the remaining two M6 bolts. Tighten all bolts to 72-96 inch lbs. (8.1-10.8 Nm) in a clockwise pattern working around the pan.

16. Install the oil drain fitting and finger tighten the bolt. Rotate the fitting and then attach the withdrawal tube. Tighten the fitting bolt and flare nut to 15-18 ft. lbs. (20-24 Nm).

17. Install the starter motor and then install the engine (if removed). Run the engine up to normal operating temperature, shut it off and check the pan for any leaks.

Fig. 13 Intake manifold tightening sequence

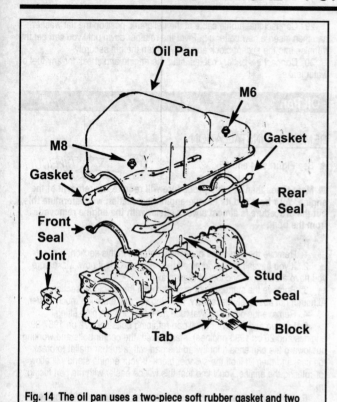

Fig. 14 The oil pan uses a two-piece soft rubber gasket and two seals—1987-89 engines

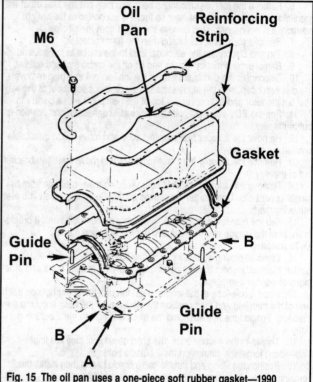

Fig. 15 The oil pan uses a one-piece soft rubber gasket—1990 engines

Oil Pump

The two-piece oil pump utilizes an inner and outer rotor and a pressure relief valve. A baffled pick-up tube is bolted to the body of the pump. The pump is driven via the auxiliary shaft which is itself driven by the crankshaft via a timing belt.

REMOVAL & INSTALLATION

◆ **See Figures 16 and 17**

1. Remove the oil pan as previously detailed. Remember that you probably need to remove the engine for this procedure.
2. Loosen and remove the pump pick-up tube brace bolt.
3. Loosen and remove the two pump mounting bolts and lift off the pump assembly. Pull out the pump shaft and take note of its orientation in the bore.
4. Insert the pump shaft in the same manner in which it came out.
5. Check that the pump and block mating surfaces are clean and then position the pump over the block so that the pump shaft slot is aligned correctly. Do not use a gasket or RTV sealant.
6. Tighten the pump mounting bolts to 14-21 ft. lbs. (19-28 Nm).
7. Position the pick-up tube brace and tighten the bolt to 60 inch lbs. (7 Nm).
8. Install the oil pan and engine.

DISASSEMBLY & ASSEMBLY

◆ **See Figure 17**

1. Remove the oil pan and oil pump.
2. Remove the mounting bolts (2) and pull off the pick-up tube.
3. Remove the 4 cover screws and lift off the cover.
4. Lift out the two rotors and sit them down carefully as they break very easily.

■ **The pump body, relief valve and spring are serviced as an assembly.**

To assemble:

5. Clean all components thoroughly in solvent and allow them to air dry completely—compressed air will speed the process if available. Clean the inside of the housing with a small toothbrush if necessary.
6. Check the cover, inner housing and both rotors for scratching, nicks or any other signs of wear. Replace as necessary.
7. Install the rotors so that the ID marks on each component are facing each other.
8. Measure the clearance between the outer rotor race and the inner bore of the housing. Clearance should be 0.001-0.0013 in. (0.03-0.033mm).
9. With the rotors still installed in the housing, lay a straight edge across the top of the assembly. Use a flat feeler gauge and measure the clearance between the straight edge and the rotors. Endplay must be 0.001-0.004 in. (0.03-0.10mm).
10. Check the bearing clearance by measuring the outer diameter of the pump shaft with a micrometer and then subtracting that figure from the inner diameter of the housing bearing. Clearance should be 0.0015-0.0029 in. (0.04-0.07mm).
11. Measurements outside of the specifications on any of the proceeding steps will dictate replacement of the associated component.
12. Install the housing cover and tighten the screws to 72-120 inch lbs. (8.1-13.6 Nm).
13. Connect the pick-up tube to the housing and tighten the 2 bolts to 14-21 ft. lbs. (19-28 Nm).
14. Install the oil pump and pan.

Engine Coupler And Flywheel

REMOVAL & INSTALLATION

◆ **See Figures 18, 19 and 20**

1. Remove the engine from the boat as detailed previously in this section.
2. Although not strictly necessary, we recommend removing the starter.
3. Cut the plastic tie that secures the housing drain hose (if equipped) and the pull the hose out of the fitting.

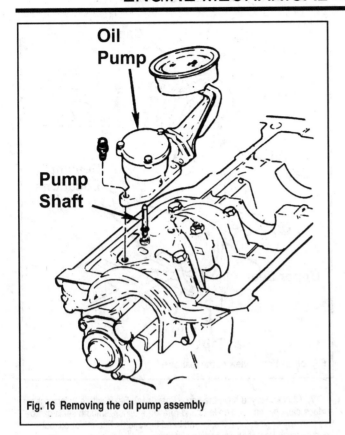

Fig. 16 Removing the oil pump assembly

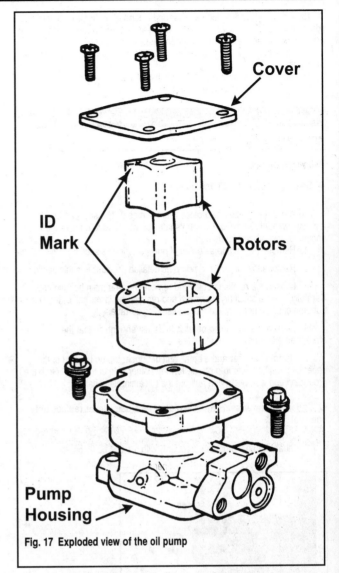

Fig. 17 Exploded view of the oil pump

4. Remove the 4 or 5 mounting bolts and slide out the lower flywheel housing cover.

5. Loosen and remove any connections attached to either of the ground studs on each side of the flywheel. Move the electrical leads out of the way.

6. Loosen the remaining retaining bolts (some of which were already removed with the shift bracket) for the flywheel housing and remove it. Take note of the positioning of any clamps so they may be installed in the same position.

7. Slide a wrench behind the coupler and loosen the six flywheel mounting nuts gradually and as you would the lug nuts on your car or truck—that is, in a diagonal star pattern.

8. Remove the coupler and then the flywheel.

To install:

9. Thoroughly clean the flywheel mating surface and check it for any nicks, cracks or gouges. Check for any broken teeth.

10. Install the flywheel over the dowel on the crankshaft.

■ **The 2.3L engine uses a completely different coupler than other OMC engines—make sure that you have the correct part if replacing with a new one.**

11. Slide the coupler over the studs so that it sits in the recess on the flywheel. Install new lock washers and tighten the mounting bolts to 14-17 ft. lbs. (19-23 Nm). Once again use the star pattern while tightening the bolts.

12. Position the flywheel housing and the shift bracket. Make sure that the clamps are in their original positions and insert the mounting bolts. Tighten them to 28-36 ft. lbs. (38-49 Nm).

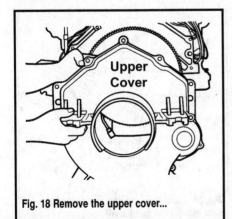

Fig. 18 Remove the upper cover...

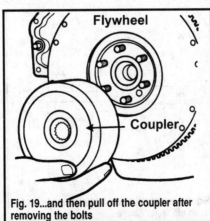

Fig. 19...and then pull off the coupler after removing the bolts

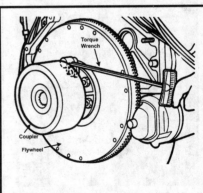

Fig. 20. Tighten the mounting bolts in a criss-cross pattern

13. Install the washer, lock washer and inner nut on the ground stud and tighten it to 20-25 ft. lbs. (27-34 Nm). Attach the electrical leads, install another lock washer and then tighten the outer nut securely.

14. Slide the lower flywheel cover into place and tighten the bolts to 60-84 inch lbs. (7-9 Nm).

15. Attach the drain hose with a new plastic tie.

16. Install the starter and engine.

Rear Main Oil Seal

REMOVAL & INSTALLATION

Two Piece Oil Seal

◆ See Figures 21, 22, 23 and 24

These engines utilize a two-piece rear main seal. The seal can be removed without removing the crankshaft. You will need to remove the engine for this procedure though.

1. Remove the engine as detailed previously in this section.

2. Remove the oil pan and pump as detailed previously in this section.

3. Loosen the retaining bolts and remove the rear main bearing cap. Carefully insert a small prybar or awl and remove the lower half of the seal from the cap. Do not damage the seal seating surface.

4. Loosen the remaining bearing bolts just enough so that the crankshaft drops down 1/16 in. at the rear.

5. Using a hammer and a small drift or punch, tap on the end of the upper seal until it starts to protrude form the other side of the race. Grab the protruding end with pliers and pull out the remaining seal half.

■ **Upper and lower seals must replaced as a pair. Never replace only one seal.**

6. Clean the seal grooves thoroughly with a small bottle brush.

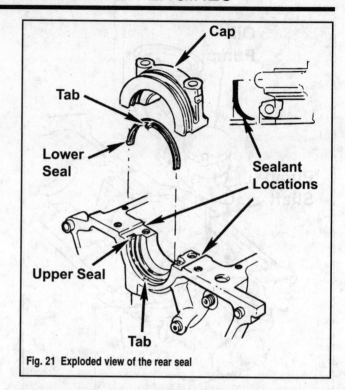

Fig. 21 Exploded view of the rear seal

7. Check that you have the correct new seal. Seals with a hatched inner surface can only be used on left hand rotation engines, smooth seals can be used on any engine. Coat the lip and bead thoroughly with motor oil, but keep oil away from the seal parting surfaces.

8. Your seal kit should come with an installation tool, if not, take a 0.004 in. feeler gauge and cut each side back about a half inch so that you're left with an 11/64 inch point. Bend the tool into the gap between the crankshaft and the seal seating surface. This will be your "shoe horn".

9. Position the upper half of the seal (lip facing the engine) between the crank and the tool so that the seal's bead is in contact with the tool tip. Roll the seal around the crankshaft using the tool as a guide until each end is flush with the cylinder block. Remove the tool.

■ **Make sure that the tab on the seal is facing the rear.**

10. Tighten up all bearing cap bolts (except the rear) to bring the crankshaft back up into position.

11. Insert the lower seal half into the main bearing cap with the lip facing the cap and the tab facing the rear. Start it so that one end is slightly below the edge of the cap and then use the tool to shimmy the seal all the way in until both edges are flush with the edge of the cap. Remove the tool.

12. Make sure that the cap/block mating surfaces and the seal ends are free of any oil and then apply a small amount of Perfect Seal to the block just behind where the upper seal ends are.

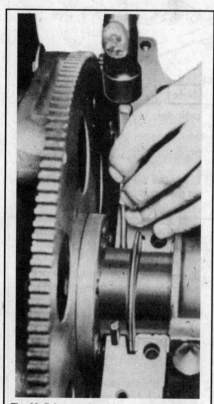

Fig. 22 Drive out the old upper seal with a small drift

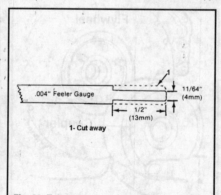

Fig. 23 Fabricate a seal installation tool out of an old feeler gauge...

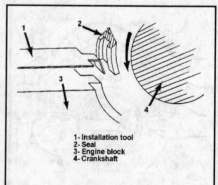

Fig. 24 ...and then use the tool to feed the seal around the crankshaft

✳✳ CAUTION

Do not get sealant on the seal ends.

13. Install the bearing cap and tighten the bolts on all caps to 80-90 ft. lbs. (109-122 Nm).
14. Install the oil pump and pan.
15. Install the engine.

One Piece Oil Seal

◆ See Figure 25

1. Remove the flywheel housing and cover as detailed in this section.
2. Remove the engine coupler and flywheel from the engine as detailed in this section.
3. Remove the engine and then the oil pan.
4. Using an awl, punch a hole into the metal surface of the seal between the inner lip and the cylinder block.
5. Install the threaded end of a slide hammer tool (# T77L-9533-B) into the end of the crankshaft and remove the seal.
6. Thoroughly clean the seal and mating surface with clean engine oil.
7. Position the seal on the installer tool (# T82L-6701-A) so that the spring side is facing the engine. Position the tool into place on the crankshaft, install the bolts and gradually (and alternately) tighten the bolts until the seal seats itself. The rear (outer) edge of the seal must be within 0.005 in. (0.13mm) of the rear edge of the cylinder block.

■ **If the bolts supplied with the installation tool are not available, flywheel/coupler bolts may be used.**

8. Remove the tool and install the oil pan.
9. Install the engine, flywheel and coupler.

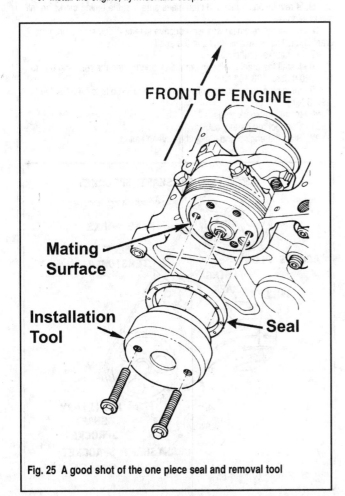

Fig. 25 A good shot of the one piece seal and removal tool

Front Cover

REMOVAL & INSTALLATION

◆ See Figure 26

■ **Any time the timing cover is removed, the timing belt should be inspected for wear or damage to determine if it should be replaced.**

1. Disconnect the negative battery cables.
2. Drain all water from the system as detailed in the Maintenance section.
3. Loosen the 4 water pump pulley bolts, but do not remove the pulley.
4. Remove the power steering pump and alternator drive belt.
5. Now remove the 4 bolts and pull off the water pump pulley.
6. Remove the crankshaft pulley bolt and pulley. If difficulty is encountered, you may need to obtain a flywheel holding tool to keep the crankshaft from turning while loosening the bolt. Another method is to presoak the area with penetrating lubricant and allow it to sit overnight. A bolt which has been pretreated in this fashion will usually break loose easier.
7. Remove the front cover retaining bolts (usually 4 Allen bolts and a Phillips screw) and remove the cover. Please note that the upper Allen bolt on the starboard side is linger than the others and must be installed in the same position.
To install:
8. Position the front cover on the engine. Install the retaining bolts/screw and tighten to 72-108 inch lbs. (8-12 Nm). Don't forget that the longer bolt goes in the top hole.
9. Install the crankshaft damper pulley and retaining bolt. Tighten the bolt to 100-120 ft. lbs. (136-163 Nm. It's not a bad idea to use a new washer here.
10. Install the water pump pulley and tighten the bolts to 14-21 ft. lbs. (19-28 Nm).
11. Install the drive belts and adjust the tension.
12. Connect the battery cables. Run the engine, check for leaks and check the ignition timing.

Timing Belt and Tensioner

CHECKING

1. Crank the engine without starting it until the timing mark on the crankshaft pulley lines up with the **TC** mark on the timing grid.
2. Pop out the rubber inspection plug found in the upper front cover and check that the camshaft sprocket timing marks are aligned. If not, rotate the engine another full turn and recheck it.

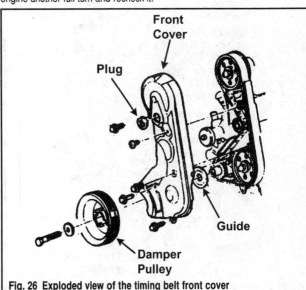

Fig. 26 Exploded view of the timing belt front cover

3. If the marks are aligned in the previous step, remove the distributor cap and confirm that the rotor is pointing to the No. 1 cylinder lead.

4. If the belt fails any of these quick check, you will need to remove it and retime the engine.

CORRECTING AN OUT-OF-TIME ENGINE

If the crankshaft pulley and camshaft sprocket timing marks were aligned in the Checking procedures, but the distributor rotor was not pointing at the No. 1 cylinder:

1. Scribe a mark on the side of the distributor housing indicating the exact positioning of the cap tower for the No. 1 cylinder's spark plug lead.

2. Pull out the high tension lead running from the ignition coil and ground it. Remove the distributor cap.

3. Remove the distributor hold-down clamp bolt and then slowly pull up on the distributor until you feel it disengage from the auxiliary shaft.

4. Rotate the rotor until it is 1/6th of a turn PAST the No.1 position. Insert the distributor back into the block until it is fully seated and engaged with the auxiliary shaft; you may have to wiggle the housing slightly to get everything to engage.

5. When the distributor is fully seated in the block, you should notice that the rotor rotated back around to the No. 1 position. If it did not, repeat this procedure again until it does.

6. Tighten the clamp bolt, install the cap and coil lead, and then start the engine. Check the ignition timing.

If, in the Checking procedure, you were unable to get the crankshaft and camshaft sprocket timing marks to align properly (and simultaneously), the timing belt has slipped. Refer to the Removal & Installation procedure and perform the steps.

REMOVAL & INSTALLATION

◆ See Figures 27, 28 and 29

1. Disconnect the negative battery cables.
2. Remove the front cover.
3. Slide the retaining washer for the belt off of the crankshaft—remember that the concave side faces outward.
4. Loosen the timing belt tensioner adjustment screw, position belt tensioner tool (# T74P-6254-A, or equivalent), on the tension spring roll pin and release the belt tensioner. Tighten the adjustment screw to hold the tensioner in the released position.

5. Remove the crankshaft pulley, hub and belt guide.
6. Remove the timing belt. If the belt is to be reused, mark the direction of rotation so it may be reinstalled in the same direction.

■ If the belt is to be reused, inspect it for wear or damage.

To install:

■ Removing the spark plugs will make rotating the crankshaft easier.

7. Rotate the engine and position the crankshaft sprocket to align the triangular mark on the sprocket with the TDC notch on the block.

8. Position camshaft sprocket so that the timing mark aligns with the camshaft timing pointer grid.

9. Remove the distributor cap and set the rotor to the No. 1 firing position by turning the auxiliary shaft.

10. Install the timing belt over the crankshaft sprocket and then counterclockwise over the auxiliary and camshaft sprockets. Align the belt fore-and-aft on the sprockets, but be very careful not to disturb any of the timing marks alignments.

11. Loosen the tensioner adjustment bolt to allow the tensioner to move against the belt. If the spring does not have enough tension to move the roller against the belt, it may be necessary to manually push the roller against the belt and tighten the bolt.

12. If you haven't already done so, remove a spark plug from each cylinder in order to relieve engine compression and to make sure the belt does not jump time during rotation in the next step.

13. Rotate the crankshaft two complete turns in the direction of normal rotation (clockwise when viewed from the front of the engine) to remove the slack from the belt.

14. Tighten the tensioner adjustment bolt to 28-40 ft. lbs. (40-55 Nm) and bracket bolt to 14-22 ft. lbs. (20-30 Nm).

15. Recheck the alignment of the timing marks. Rotate the engine two complete revolutions again and then make sure that the timing marks are all still in alignment.

16. Install the retaining washer (concave side out) and slide it onto the crankshaft. The washer is keyed to the shaft.

17. Install the front cover.

18. Install the crankshaft damper pulley and tighten the retaining bolt to 100-120 ft. lbs. (136-163 Nm).

19. Install the water pump pulley and tighten the bolts to 14-21 ft. lbs. (19-28 Nm).

20. Install the spark plugs.
21. Connect the battery cables.
22. Start the engine and check the ignition timing.

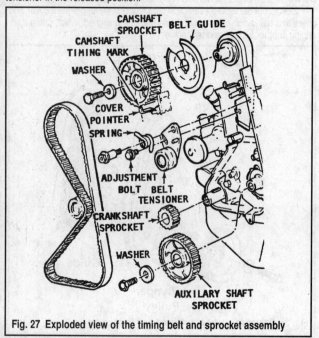

Fig. 27 Exploded view of the timing belt and sprocket assembly

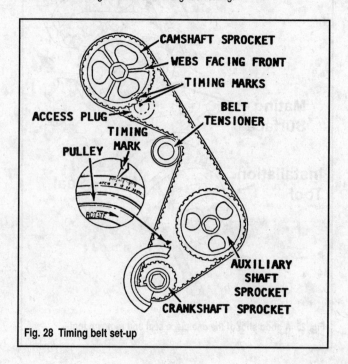

Fig. 28 Timing belt set-up

Timing Sprockets

REMOVAL & INSTALLATION

◆ **See Figures 27 and 30**

1. Disconnect the battery cables.
2. Remove the front cover and the timing belt.
3. Remove the camshaft and auxiliary shaft sprocket retaining bolts, as necessary. A holding tool (# T74P-6256-B) is extremely helpful here.
4. Remove the crankshaft, camshaft and auxiliary shaft sprockets, as necessary, using a suitable puller.

To install:

5. Install the crankshaft, camshaft and/or auxiliary shaft sprockets, as necessary.
6. While holding the sprocket from turning (using the holding tool), tighten the camshaft sprocket retaining bolt to 80-90 ft. lbs. (108-122 Nm) and/or the auxiliary sprocket retaining bolt to 28-40 ft. lbs. (38-54 Nm.
7. Install the timing belt and front cover.
8. Connect the battery cables.

Front Oil Seals

REMOVAL & INSTALLATION

◆ **See Figures 31, 32 and 33**

1. Disconnect the battery cables.
2. Remove the front cover and timing belt.
3. Remove the timing sprocket (crankshaft, camshaft and/or auxiliary shaft) under which the seal is being replaced.
4. Use seal remover tool (#T74P-6700-B, or an equivalent jawed seal puller), to remove the crankshaft, camshaft and/or auxiliary shaft seal(s). Position the tool so that the jaws are gripping the thin edge of the seal very tightly. Operate the jackscrew on the tool to remove the seal.

To install:

5. Lubricate the lips of the new seals with clean engine oil.
6. Use a threaded seal installer tool (# T74P-6150-A, or equivalent), to install the seals.
7. Install the timing sprocket(s).
8. Install the timing belt and front cover.
9. Connect the battery cables.

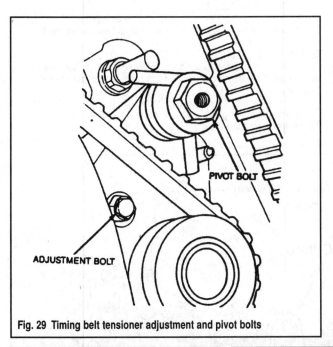

Fig. 29 Timing belt tensioner adjustment and pivot bolts

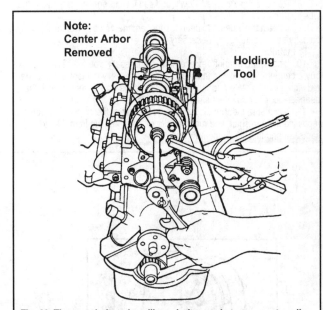

Fig. 30 The camshaft and auxiliary shaft sprockets are most easily removed or installed using a special holding tool

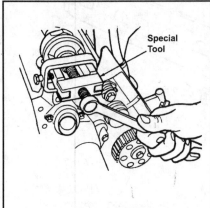

Fig. 31 Use a suitable tool to remove the camshaft seal . . .

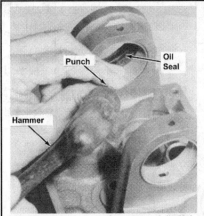

Fig. 32 ...or use a punch very carefully if the tool is not available

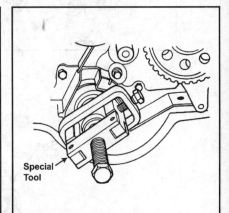

Fig. 33 You must use the tool on the crankshaft or auxiliary shaft seals

Camshaft

REMOVAL & INSTALLATION

◆ See Figures 32 thru 36

■ Depending on the available room/clearance in the engine compartment, it may be necessary to raise the engine or remove it entirely in order to gain sufficient clearance when pulling out the camshaft.

We recommend checking the camshaft lift prior to removing it from the cylinder block.
1. Disconnect the negative battery cable and drain the cooling system.
2. Remove the cylinder head cover.
3. Remove the front cover and the timing belt.
4. Compress the valve springs using valve spring compressor lever (# T47P-6565-A), or equivalent, and remove the cam followers (rocker arms).
5. Remove the camshaft sprocket retaining bolt and then remove the camshaft sprocket using a suitable puller.
6. Remove the camshaft seal using a seal removal tool.
7. Remove the two bolts pry off the camshaft rear retainer.
8. Remove the camshaft slowly and carefully to prevent damage to the lobes, journals or bearings.
9. Inspect the lobes and bearings for damage or wear. Replace the camshaft and bearings, as necessary.

To install:
10. If camshaft bearing replacement is necessary, you will have to rent or buy a bearing removal and installation tool. Essentially the tool is a threaded rod with a pulling plate and nut, along with various size drivers to push the bearings out of or pull the bearings into position.
11. Make sure the threaded plug is in the rear of the camshaft. If you are replacing the camshaft, you may have to remove the plug from the old camshaft and install it in the new one.

12. Coat the camshaft lobes with multi-purpose grease and lubricate the journals with heavy engine oil before installation. An engine assembly lube can also be used. Carefully slide the camshaft through the bearings.
13. Install the camshaft rear retainer and tighten the two bolts to 6-9 ft. lbs. (8-12 Nm).
14. Install a new camshaft seal using a suitable seal installer.
15. Install the camshaft sprocket and tighten the retaining bolt to 80-90 ft. lbs. (108-122 Nm).
16. Install the timing belt and front cover.
17. Install the cylinder head cover.
18. Connect the negative battery cable, then properly refill the engine cooling system.
19. Run the engine and check for leaks. Check the ignition timing.

Auxiliary Shaft

REMOVAL & INSTALLATION

◆ See Figure 37

1. Disconnect the battery cables.
2. Remove the front belt cover and the timing belt.
3. Remove the auxiliary shaft sprocket retaining bolt. Remove the sprocket using a puller.
4. Remove the distributor.
5. Remove the auxiliary shaft cover and the thrust plate.
6. Withdraw the auxiliary shaft from the block being careful not to damage the bearings.

To install:
7. Dip the auxiliary shaft in engine oil before installing. Slide the auxiliary shaft into the cylinder block, being careful not to damage the bearings.
8. Install the thrust plate and tighten the bolts to 6-9 ft. lbs. (8-12 Nm).
9. Position a new gasket and then install the auxiliary shaft cover. Tighten the cover screws to 6-9 ft. lbs. (8-12 Nm).

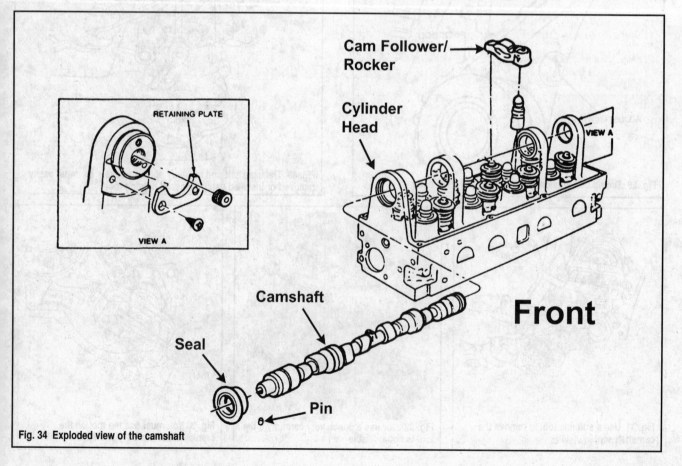

Fig. 34 Exploded view of the camshaft

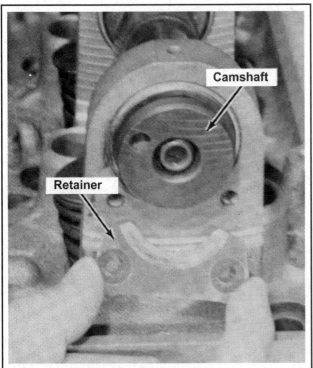

Fig. 35 Removing the rear retainer

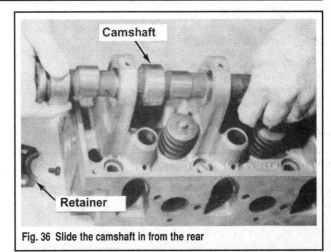

Fig. 36 Slide the camshaft in from the rear

Cylinder Head

REMOVAL & INSTALLATION

◆ See Figures 38, 39 and 40    DIFFICULT

1. Disconnect the battery cables and drain the cooling system.
2. Remove the flame arrestor.

■ Before removing the spark plug wires, check the configuration of your engine against the firing order diagrams found in the Maintenance section. If your engine is different in any way, make notations to the diagrams to assure proper installation. Also, be sure to tag all of the spark plug wires and the distributor cap or ignition coil terminals, as applicable.

3. Tag and remove the spark plug wires. Remove the spark plugs.
4. Remove the distributor assembly.
5. Tag and disconnect any vacuum hoses or electrical leads in the way of head removal.
6. Remove the intake manifold assembly.
7. Remove the camshaft cover from the cylinder head.
8. Remove the front cover retaining bolt(s) and remove the cover.
9. Remove the timing belt from the camshaft pulley and the auxiliary pulley. Refer to the Timing Belts section. If the belt is not going to be replaced, be sure to mark the current direction of rotation on the belt to assure proper installation.

■ The auxiliary shaft cover and cylinder front cover share a common gasket. Cut off the old gasket around the cylinder cover and use half of the new gasket on the auxiliary shaft cover.

10. Install the distributor.
11. Install the auxiliary shaft sprocket and tighten the bolt to 28-40 ft. lbs. (38-54 Nm).
12. Align the timing marks and install the timing belt and front cover.
13. Connect the battery cables.
14. Check the ignition timing.

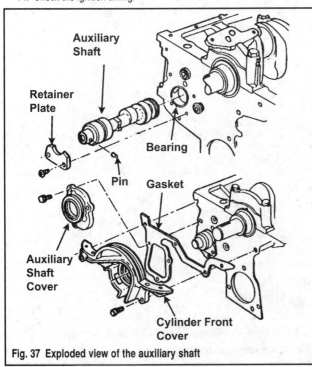

Fig. 37 Exploded view of the auxiliary shaft

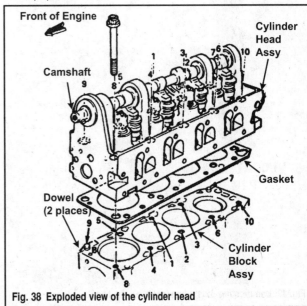

Fig. 38 Exploded view of the cylinder head

10. Remove the exhaust manifold.

11. Remove the timing belt idler and two bracket bolts. Remove the timing belt idler spring stop from the cylinder head.

12. Tag and disconnect the oil sending unit wire.

13. Remove the cylinder head bolts, then break the gasket seal and remove the cylinder head.

✳✳ WARNING

Do not pry the cylinder head to break the gasket seal unless it is absolutely necessary. Prying the cylinder head can deform the soft metal and compromise the gasket mating surfaces.

To install:

14. Clean all gasket mating surfaces and blow the oil out of the cylinder head bolt block holes.

15. Check the cylinder head for flatness using a straightedge and a feeler gauge. If the head gasket surface is warped greater than 0.006 in. (0.152mm), it must be resurfaced. Do not grind off more than 0.010 in. (0.254mm) from the cylinder head.

16. Position a new cylinder head gasket on the engine. Rotate the camshaft so that the locating pin is approximately 30 degrees to the right of the 6 o' clock position when facing the front of the cylinder head (this places the pin about at the five o'clock position) to avoid damage to the valves and pistons.

17. If they are available, cut the heads off of two old cylinder head bolts to use as guide studs. Thread the studs into the engine block at opposite corners to use as guides.

18. Make sure the gasket is properly fit to the block and then set the head into position. Apply a non-hardening gasket sealer to the bolt threads and install the bolts finger-tight.

19. Tighten the head bolts using 2 passes of the sequence as illustrated, first to 50-60 ft. lbs. (68-81 Nm) and then to 80-90 ft. lbs. (108-122 Nm).

20. Connect the oil sending unit wire.

21. Install the timing belt tensioner spring stop to the cylinder head.

22. Position the timing belt tensioner and tensioner spring to the cylinder head and install the retaining bolts. Rotate the tensioner against the spring with belt tensioner tool (# T74P-6254-A, or equivalent), and temporarily tighten.

23. Install the exhaust manifold using a new gasket. For details, please refer to the Exhaust Manifold procedure found earlier in this section.

24. Align the distributor rotor with the No. 1 plug location on the distributor cap and install it.

25. Install the timing belt over the sprockets. If the belt is not being replaced, make sure you have installed it in the same direction of rotation as it was before removal. Complete timing belt installation as detailed under the Timing Belt procedures located in this section. Be sure to check belt tension and alignment, then to properly tighten the tensioner and pivot bolts.

26. Install the front cover and tighten the retaining bolt(s) to 6-9 ft. lbs. (8-12 Nm).

27. Install the rocker arm cover and tighten the retaining bolts to 48-84 inch lbs. (5.5-9.5 Nm).

28. Install the intake manifold assembly. Tighten the bolts, in sequence, to 60-84 inch lbs. (6.8-9.5 Nm) and then retighten them all to 14-18 ft. lbs. (19-25 Nm).

29. Connect all remaining vacuum hoses and electrical leads as tagged during removal.

30. Position the alternator and install the drive belt.

31. Install the spark plugs, then install the spark plug wires as tagged and noted during removal.

32. Install the flame arrestor.

33. Connect the battery cables.

34. Bring the engine to normal operating temperature and check for leaks.

35. Check the ignition timing.

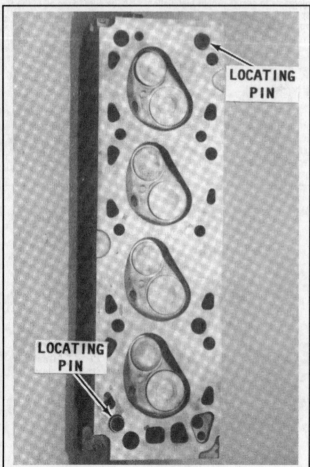

Fig. 39 Use locating pins when installing the head

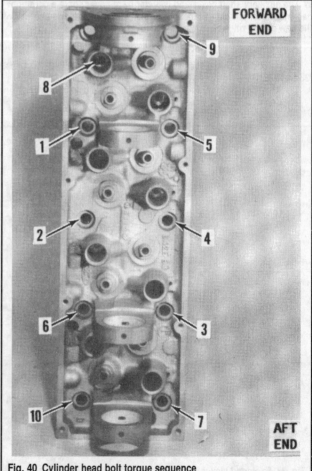

Fig. 40 Cylinder head bolt torque sequence

TORQUE SPECIFICATIONS

Component			ft. lbs.	inch lbs.	Nm
Auxiliary shaft		Sprocket bolt	28-40	-	38-54
		Thrust plate bolts	-	72-108	8-12
Camshaft		Cover	-	48-84	5.5-9.5
		Sprocket bolt	80-90	-	108-122
		Thrust plate (retainer)	-	72-108	8-12
Connecting rod cap			30-36	-	40-49
Crankshaft		Damper pulley bolt	100-120	-	136-163
		Main bearing cap	80-90	-	108-122
Cylinder head		Step 1	50-60	-	68-91
		Step 2	80-90	-	108-122
Cylinder head cover			-	48-84	5.5-9.5
Distributor clamp bolt			14-21	-	19-28
Engine coupler		Flywheel	54-64	-	73-87
		Flywheel housing	28-36	-	38-49
		Coupler nut	14-17	-	19-23
		Coupler ground stud	20-25	-	27-34
		Cover	-	60-84	7-9
Engine mounts	Front				
		Adjusting nut	100-120	-	136-163
		Bracket bolts	32-40	-	43-54
		Lower nut	60-75	-	81-102
	Rear				
		Center bolt 1987-88	18-20	-	24-27
		1989-90	44-52	-	60-71
		Lock nut	28-30	-	38-40
		Mounting bolts	20-25	-	27-34
Exhaust Elbow			12-14	-	16-19
Exhaust Manifold		Bolt	20-25	-	27-34
		Nuts	30-35		41-47
Exhaust Pipe		1987-89	10-12	-	14-16
		1990	27-34	-	27-34
Flywheel		Bolt	54-64	-	73-87
		Cover	-	60-84	7-9
		Housing	28-36	-	38-49
Front cover			-	72-108	8-12
Fuel pump			14-21	-	19-28
Intake manifold		Step 1	-	60-84	7-9
		Step 2	14-18	-	19-25
Main bearing cap			80-90	-	108-122
Oil pan		M6 bolts	-	72-96	8-12
		M8 bolts	-	96-120	11-14
Oil pump		Cover bolts	-	72-120	8-14
		Mounting bolts	14-21	-	19-28
		Pick-up Tube	14-21	-	19-28
Oil pressure sender			12-16	-	16-22
Oil withdrawal tube (flare)			15-18	-	20-24
Spark plug			5-10	-	6.8-13.6
Timing belt		Idler pulley	14-21	-	19-28
		Tensioner pivot	28-40	-	38-54
Valve cover			-	48-84	5.5-9.5
Water pump (circulating pump)			14-21	-	19-28
Water temperature sender			12-16		16-22

ENGINE SPECIFICATIONS

Component			Standard (in.) ①	Metric (mm) ①
Camshaft				
Clearance			0.001-0.003	0.025-0.076
	Wear limit		0.006	0.152
End Play			0.001-0.007	0.025-0.178
	Wear limit		0.009	0.229
Journal Diameter			1.7713-1.7720	44.99-45
Lobe Lift			0.2381 +/- 0.005	6.05 +/- 0.13
Out or Round			0.005	0.127
Runout			0.005	0.127
Connecting Rod				
Diameter			2.1720-2.1728	55.17-55.19
Out or Round			0.004	0.102
Piston pin bore			0.9104-0.9112	23.12-23-14
Side Clearance			0.0035-0.0105	0.089-0.267
	Wear limit		0.015	0.38
Out or Round			0.004	0.102
Taper (max)			0.004	0.102
Crankshaft				
Connecting rod journal diameter			2.0464-2.0472	51.98-52
Crankshaft End Play			0.004-0.008	0.102-0.203
Main Bearing	Journal Diameter		2.3982-2.3990	60.91-60.94
	Out or round		0.0006	0.015
	Runout		0.002	0.05
		Wear limit	0.005	0.127
	Taper (max)		0.0006	0.015
Piston				
Clearance			0.0014-0.0022	0.036
Diameter	Red		3.7780-3.7786	95.96-95.98
	Blue		3.7792-3.7798	95.99-96
Piston Pin				
Clearance			0.0002-0.0004	0.005
Diameter			0.9119-0.9124	23.16-23.18
Interference fit			Yes	Yes
Length			3.010-3.040	76.45-77.22
Piston Rings				
Compression	Gap		0.010-0.020	0.254
	Clearance		0.002-0.004	0.05-0.10
Oil	Gap		0.015-0.055	0.38-1.40
	Width		Snug	Snug

① Unless otherwise noted

ENGINE SPECIFICATIONS

Component			Standard (in.) ①	Metric (mm) ①
Valve system				
Face Angle			44 deg.	
Lifter			Hydraulic	Hydraulic
Rocker arm ratio			1.64:1	
Runout (max)			0.002	0.05
Seat Angle			45 deg.	
Seat Width	Intake		0.060-0.080	1.52-2.03
	Exhaust		0.070-0.090	1.78-2.29
Spring (Outer)	Free length		1.89	48
	Pressure	Intake	71-79 lbs. @ 1.56	316-351 N @ 39.62
		Exhaust	159-175 lbs. @ 1.16	707-778 N @ 29.46
Stem Clearance	Intake		0.0010-0.0027	0.025-0.069
	Exhaust		0.0015-0.0032	0.038-0.081

① Unless otherwise noted

4

ENGINE MECHANICAL GM 4 CYLINDER ENGINES

ENGINE MECHANICAL

General Information

The OMC 2.5L 153 cubic inch displacement engine and the 3.0L 181 cubic inch engine are manufactured by GMC and have been a favorite combination when mated to the OMC Cobra and SP stern drives.

This in-line four-cylinder powerplant uses a full pressure lubrication system with a disposable flow-thru oil filter cartridge. Oil pressure is furnished by a gear-type oil pump, driven by the distributor, which is driven by a helical gear on the camshaft. A regulator on the oil pump controls the amount of oil pressure output. The oil pump scavenges oil from the bottom of the oil pan and feeds it through the oil filter and then to the main oil gallery in the block. Drilled passages in the block and crankshaft distribute oil to the camshaft and crankshaft to lubricate the rod, main and camshaft bearings. The main oil gallery also feeds oil to the valve lifters, which pump oil up through the hollow pushrods to the rocker arms to lubricate the valve train in the cylinder head. Cylinder numbering and firing orders are identified in the illustrations at the end of the Maintenance section.

✳✳ CAUTION

NEVER, NEVER attempt to use standard automotive parts when replacing anything on your engine. Due to the uniqueness of the environment in which they are operated in, and the levels at which they are operated at, marine engines require different versions of the same part; even if they look the same. Stock and most aftermarket automotive parts will not hold up for prolonged periods of time under such conditions. Automotive parts may appear identical to marine parts, but be assured, OMC marine parts are specially manufactured to meet OMC marine specifications. Most marine items are super heavy-duty units or are made from special metal alloy to combat against a corrosive salt water atmosphere.

OMC marine electrical and ignition parts are extremely critical. In the United States, all electrical and ignition parts manufactured for marine application must conform to stringent U.S. Coast Guard requirements for spark or flame suppression. A spark from a non-marine cranking motor solenoid could ignite an explosive atmosphere of gasoline vapors in an enclosed engine compartment.

Engine Identification

ENGINE

◆ See Figures 1, 2 and 3

The engine serial numbers are the manufacturer's key to engine changes. These alpha-numeric codes identify the year of manufacture, the horsepower rating and various model/option differences. If any correspondence or parts are required, the engine serial number must be used for proper identification.

Remember that the serial number establishes the year in which the engine was produced, which is often not the year of first installation.

The engine specifications decal contains information such as the model number or code, the serial number (a unique sequential identifier given ONLY to that one engine) as well as other useful information.

An engine specifications decal can generally be found on top of the flame arrestor or on the side of the rocker arm cover - all pertinent serial number information can be found here—engine and drive designations, serial numbers and model numbers. Unfortunately this decal is not always legible on older boats and it's also quite difficult to find, so please refer to the following procedures for each individuals unit's serial number location.

■ **Serial numbers tags are frequently difficult to see when the engine is installed in the boat; a mirror can be a handy way to read all the numbers.**

The engine serial/model number is sometimes also stamped on the port rear side of the engine where it attaches to the bell housing; although on most later models it may instead be a metal plate attached in the same location. If your engine has a stamped number it will simply be the serial number; if you have a plate (and you should), it will always show a Model number and then the actual Serial number. Additionally, most models will also have this plate or sticker on the transom bracket.

• The first two characters identify the engine size in liters (L); **30** represents the 3.0L, **25** represents the 2.5L and so forth.
• The third character identifies the fuel delivery system; **2** designates a 2 bbl carburetor, **4** is a 4 bbl carburetor, and **F** is a fuel injected engine.
• The fourth character designates a major engine or horsepower change—it doesn't let you know what the change was, just that there was some sort of change. **A** means it is the first model released, **B** would be the second, and so forth
• The fifth character designates what type of steering system was used; **M** would be manual steering and **P** would be power steering.
• Now here's where it gets interesting; on 1986-87 engines and 1994-98 engines, the sixth, seventh and eighth characters designate the model year. The sixth and seventh actually show the model year, while the eighth is a random model year version code. **KWB** and **WXS** represent 1986; and **ARJ, ARF, FTC, SRC** or **SRY** show 1987. **MDA** is 1994, **HUB** is 1995, **NCA** is 1996, **LKD** is 1997 and **BYC** is 1998.
• On 1988-93 engines, the sixth character designates the direction of propeller rotation. **R** is right hand, **L** is left hand and **E** is either.
• Also on 1988-93 engines, the seventh, eighth and ninth characters designate the model year. The seventh and eighth actually show the model year, while the ninth is a random model year version code. **GDE** or **GDP** is 1988, **MED** or **MEF** is 1989, **PWC, PWR** or **PWS** is 1990, **RGD** or **RGF** is 1991, **AMH** or **AMK** is 1992 and **JVB** or **JVN** is 1993.

Any remaining characters are proprietary. So in example, a Model number on the ID plate that reads **302AMFTC** would designate a 1987 3.0L engine with a 2 bbl carburetor and manual steering, first model released.

Engine Model Designations

All engines covered here utilize unique identifiers assigned by OMC; surnames if you will—2.3, 3.0GL, 3.0GS, etc. Obviously the first two characters designate the engine size in litres (L). The second letter, a **G** or and **F** designate the engine manufacturer; General Motors (G) or Ford (F). The third through fifth letters can be found in different combinations, but the individual letter designates the same thing regardless of position. **L** designates limited output. **S** and **X** designate superior output. A 3.0GL will always have a lesser horsepower rating than a 3.0GS in a given model year; a 5.7GL will be less than a 5.7GS. **i** designates that the engine is fuel injected, if there is no **i** then you know the engine uses a carburetor.

Engine

REMOVAL & INSTALLATION

◆See Figures 4 thru 10

■ **Prior to removing the engine from your vessel, it is imperative to measure the engine height as detailed in the Determining Minimum Engine Height section on 1995-98 engines. DO NOT remove the engine until you have completed this procedure!**

1. Check the clearance between the front of the engine and the inside edge of the engine compartment bulkhead. If clearance is less than 6 in. (15.2mm), you will need to remove the stern drive unit because there won't be enough room to disengage the driveshaft from the engine coupler. More than 6 in. will provide enough working room to get the engine out without removing the drive, BUT, we recommend removing the drive anyway. If you intend on doing anything to the mounts or stringers, you will need to re-align the engine as detailed in the Engine Alignment section—which requires removing the drive, so remove the drive!
2. Remove the stern drive unit as detailed in the Drive Systems section.
3. Open or remove the engine hatch cover.
4. Disconnect the battery cables (negative first) at the battery and then disconnect them from the engine block and starter.

✳✳ CAUTION

Make sure that all switches and systems are OFF before disconnecting the battery cables.

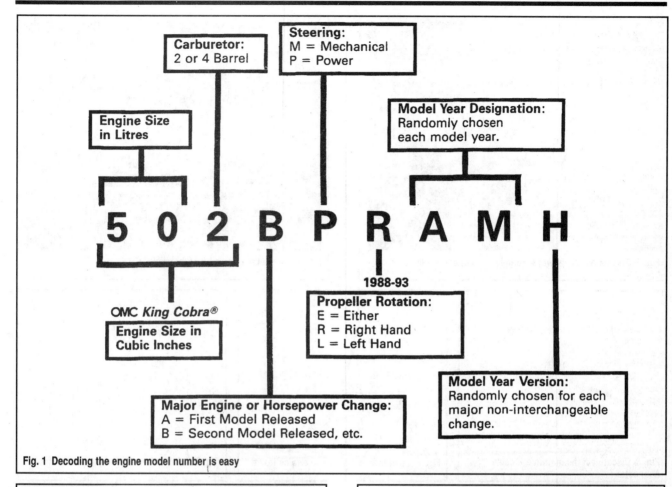

Carburetor:
2 or 4 Barrel

Steering:
M = Mechanical
P = Power

Model Year Designation:
Randomly chosen
each model year.

**Engine Size
in Litres**

5 0 2 B P R A M H

OMC *King Cobra*®
**Engine Size in
Cubic Inches**

1988-93
Propeller Rotation:
E = Either
R = Right Hand
L = Left Hand

Major Engine or Horsepower Change:
A = First Model Released
B = Second Model Released, etc.

Model Year Version:
Randomly chosen for each
major non-interchangeable
change.

Fig. 1 Decoding the engine model number is easy

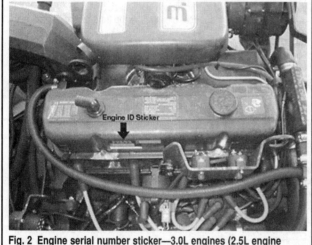

Fig. 2 Engine serial number sticker—3.0L engines (2.5L engine similar)

Fig. 3 You should also be able to find a plate on the transom bracket

5. Disconnect the two power steering hydraulic lines at the steering cylinder (models w/power steering). Carefully plug them and then tie them off somewhere on the engine, making sure that they are higher then the pump to minimize any leakage.

6. Disconnect the fuel inlet line at the fuel pump or filter (whichever comes first on your particular engine) and quickly plug it and the inlet— a clean golf tee and some tape works well in this situation. Make sure you have rags handy, as there will be some spillage.

7. Tag and disconnect the two-wire trim/tilt connector.

8. Pop the two-wire trim/tilt sender connector out of the retainer and then disconnect it. You may have to cut the plastic tie securing the cable in order to move it out of the way.

9. Locate the large rubber coated instrument cable connector (should be on the starboard side), loosen the hose clamp and then disconnect it from the bracket. Move it away from the engine and secure it. On early models, you will also need to unplug the three-wire trim/tilt cable connector just above it.

■Take note of your throttle arm attachment stud— is it a "push-to-close" or a "pull-to-close"?

10. Remove the cotter pin and washer from the throttle arm. Loosen the anchor block retaining nut and then spin the retainer away from the cable trunnion. Remove the throttle cable from the arm and anchor bracket. Be sure to mark the position of the holes that the anchor block was attached to.

Fig. 4 Disconnect the power steering lines at the cylinder

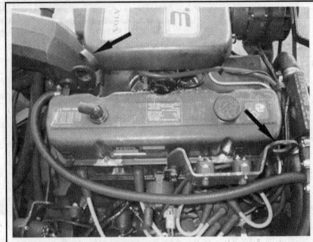

Fig. 5 Locate the engine lifting brackets...

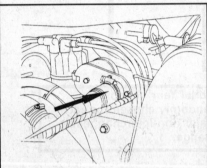

Fig. 6 Disconnect the instrument cable connector

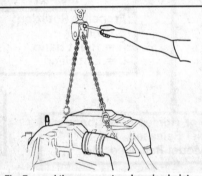

Fig. 7 ...and then connect and engine hoist

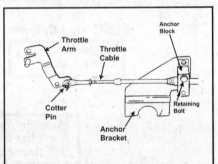

Fig. 8 Details of the throttle cable assembly

11. Loosen the hose clamps (4) and disconnect the exhaust elbow bellows from elbow. You may want to spray some WD-40 around the lip of the hose where it connects to the elbow, grasp it with both hands and wiggle it back and forth while pulling down on it.

12. Drain the cooling system as detailed in the Maintenance section.

13. Loosen the hose clamp on the water supply hose at the transom bracket and carefully slide it off the water tube. Attach the hose to the engine.

14. Disconnect the shift cables and position them out of the way.

15. Tag and disconnect any remaining lines, wires or hoses at the engine.

16. Attach a suitable engine hoist to the lifting eyes (usually front starboard side, rear port side) and take up any line slack until it is just taught.

■ **The engine hoist should have a capacity of at least 1500 lbs. (680 kg).**

17. Locate the rear engine mounts and remove both lock nuts and flat washers.

18. Locate the front engine mount and remove the two lag bolts.

19. If you listened to us at the beginning of the procedure, the drive unit should be removed. If so, slowly and carefully, lift out the engine. Try not to hit the power steering control valve, or any other accessories, while removing it from the engine compartment. If you didn't listen to us, and you had sufficient clearance in the engine compartment, the drive unit is probably still installed. Raise the hoist slightly until the weight is removed from the mounts and then carefully pull the engine forward until the driveshaft disengages from the coupler, now raise the engine out of the compartment.

To install:

20. Apply Engine Coupler grease to the splines of the coupler.

21. Slowly lower the engine into the compartment. If the drive unit was not removed, AND the crankshaft has not been rotated, insert the driveshaft into the coupler as you push the engine backwards until they engage completely and then lower the engine into position over the rear mounts until the front mounts just touch the stringers. If the shaft and coupler will not align completely, turn the crankshaft or driveshaft slightly until they mate correctly.

If the drive was removed, or the mounts were disturbed in any way, lower the engine into position over the rear mounts until the front mount just touches the stringer.

22. Install the two flat washers into the recess in the engine bracket side of the rear mounts and then install the two lock nuts. Tighten them to 28-30 ft. lbs. (38-40 Nm).

✱✱ CAUTION

Never use an impact wrench or power driver to tighten the locknuts.

23. Install the lag bolts into their holes on the front mounts and tighten each bolt securely.

24. If the drive was removed, the mounts were disturbed or the driveshaft/coupler didn't mate correctly, perform the engine alignment procedure detailed in this section. We think it's a good idea to do this regardless! After you've performed the alignment procedure, make sure that you hold the top adjusting nut and tighten the lower nut on the mount to 115-140 ft. lbs. (156-190 Nm).

25. Reconnect the exhaust bellows by sliding it up and over the elbow, position the clamps between the ribs in the hose and then tighten the clamps securely. Make sure you don't position the clamps into the expanding area.

26. Reconnect the water inlet hose. Lubricate the inside of the hose and wiggle it onto the inlet tube. Slide the clamp over the ridge and tighten it securely. This sounds like an easy step, but it is very important—if the hose, particularly the underside, is not installed correctly the hose itself may collapse or come off. Either scenario will cause severe damage to your engine, so make sure you do this correctly!

27. Carefully, and quickly, remove the tape and plugs so you can connect the power steering lines. Tighten the large fitting to 15-17 ft. lbs. (20-23 Nm) and the small fitting to 10-12 ft. lbs. (14-16 Nm). Don't forget to check the fluid level and bleed the system when you are finished with the installation.

28. Reconnect the trim/tilt connector so the two halves lock together.

29. Reconnect the trim position sender leads, the instrument cable, the engine ground wire, the battery cables and all other wires, lines of hoses that were disconnected during removal. Make sure you swab a light coat of grease around the fitting for the large engine/instrument cable plug.

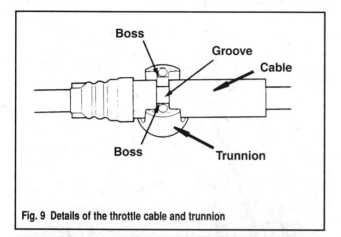

Fig. 9 Details of the throttle cable and trunnion

✳✳ CAUTION

Always make certain that all switches and systems are turned OFF before reconnecting the battery cables.

■ **Make sure all cables, wires and hoses are routed correctly before initially starting the engine.**

30. Unplug the fuel line and pump/filter fitting and reconnect them. Remember to check for leaks as soon as you start the engine.

31. Install and adjust the throttle cable. For complete details, please refer to the Fuel System section):

a. Remember we asked you to determine if you had a "push-to-close" or "pull-to-open" throttle cable (the throttle arm stud)? Position the remote control handle in Neutral—the propeller should rotate freely.

b. Turn the propeller shaft and the shifter into the forward gear detent position and then move the shifter back toward the Neutral position halfway.

c. Position the trunnion over the groove in the throttle cable so the internal bosses align and then snap it into the groove until it is fully seated.

d. Install the trunnion/cable into the anchor block so the open side of the trunnion is against the block. Position the assembly onto the bracket over the original holes (they should be the lower two of the four holes) and then install the retaining bolt and nut. When the nut is securely against the back of the bracket, tighten the bolt securely.

e. Install the connector onto the throttle cable and then pull the connector until all end play is removed from the cable. Turn it sideways until the hole is in alignment with the correct stud on the throttle arm. Slide it over the stud and install the washer and a new cotter pin. Make sure the cable is on the same stud that it was removed from. Tighten the jam nut against the connector

■ **On 1995-98 engines, the throttle arm connector nut must be installed on the cable with a minimum of 9 turns—meaning that at least 1/4 in. of thread should be showing between the end of the cable and the edge of the nut.**

32. Install and adjust the shift cables. Please refer to the Drive Systems section for further details.

33. Check and refill all fluids. Start the engine and check for any fuel or coolant leaks. Go have fun!

DETERMING MINIMUM ENGINE HEIGHT

1995-98 Engines

◆ See Figure 11

This procedure MUST be performed prior to removing the engine from the vessel.

1. With the engine compartment open, position a long level across the transom running fore and aft.

2. Have a friend or assistant steady the level while you measure from the bottom edge of the tool to the top of the exhaust elbow. Record the distance as "**1**".

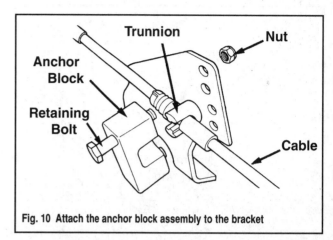

Fig. 10 Attach the anchor block assembly to the bracket

3. Now measure from the bottom of the level to the static water line on the drive unit. Record the distance as "**2**".

4. Subtract the elbow measurement (**1**) from the static waterline measurement (**2**). If the result is less than 9 in. (228mm), an exhaust elbow high rise kit must be installed (available from your local parts supplier).

ENGINE ALIGNMENT

◆ See Figures 12 and 13

Engine alignment is imperative for correct engine installation and also for continued engine and drive operation. It is a good idea to ensure proper alignment every time that the drive or engine have been removed. Engine alignment is checked by using OMC alignment tool #912273 and handle #311880 (Volvo #3851083-0 and #3850609-31). Engine alignment is adjusted by raising or lowering the front engine mount(s).

1. With the drive unit off the vessel, slide the alignment tool through the gimbal bearing and into the engine coupler. It should slide easily, with no binding or force. If not, check the gimbal bearing alignment as detailed in the Drive Systems section. If bearing alignment is correct, move to the next step.

2. If your engine utilizes a jam nut on the bottom of the mount bolt, loosen it and back it off at least 1/2 in.

3. Loosen the lock nut and back it off.

4. Now, determine if the engine requires raising or lowering to facilitate alignment—remember, the alignment tool should still be in position. Tighten or loosen the adjusting nut until the new engine height allows the alignment tool to slide freely.

5. Hold the adjusting nut with a wrench and then tighten the lock nut to 115-140 ft. lbs. (156-190 Nm). If your engine uses a jam nut, cinch it up against the lock nut.

6. Remove the alignment tool and handle.

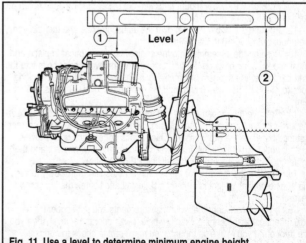

Fig. 11 Use a level to determine minimum engine height

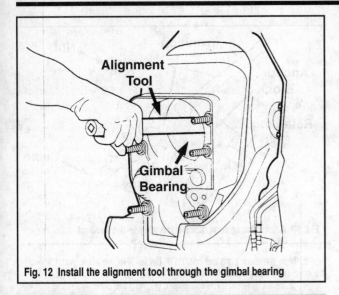

Fig. 12 Install the alignment tool through the gimbal bearing

Front Engine Mount

REMOVAL & INSTALLATION

◆ **See Figure 13**   ③ ⟨DIFFICULT

1. Position an engine hoist over the engine and hook it up to the two engine lifting eyes.
2. Remove the two lag bolts on each side of the mount where it rests on the stringer.
3. Raise the engine just enough to allow working room for removing the mount.
4. Remove the four mount-to-engine mounting bolts with their lock washers and lift out the mount. Don't forget the spacer on the port side.
5. Measure the distance between the bottom of the large washer on the lower side of the mounting bracket and the upper side of the smaller washer on top of the mount itself. Record it.
6. Position an open end wrench over the adjusting shaft just underneath the bracket to hold the shaft and then remove the top nut. Lift off the bracket and the spacer.
7. Remove the two bolts attaching the rubber mount to the upper bracket and remove it.
8. Remove the lower lock nut from the adjusting shaft and lift the shaft out of the lower mount. Some applications may have a jam nut under the locknut.

To install:

9. Drop the adjusting shaft into the lower mount and spin on the lower locknut until it is fingertight.
10. Slide the spacer over the top end of the shaft.
11. Position the rubber mount back into the upper bracket and tighten the bolts to 30-35 ft. lbs. (41-47 Nm).
12. Position the upper bracket over the adjusting shaft and then tighten the upper nut to 50-60 ft. lbs. (68-81 Nm).
13. Position the assembly onto the engine, hold the spacer in place and install the two long bolts on the port side. Install the two shorter bolts into the starboard side. We suggest using new lock washers here. Tighten the long bolts to 48-56 ft. lbs. (65-76 Nm) and the short bolts to 32-40 ft. lbs. (43-54 Nm).
14. Retrieve the measurement you took earlier and check it now. Loosen the lower locknut and rotate the adjusting shaft until you achieve the correct distance between the two washers. Once this is done, cinch up the lower locknut until it is tight against the lower surface of the lower mount and then tighten the adjusting nut to 115-140 ft. lbs. (156-190 Nm).
15. Position the mount over the lag screw holes and then slowly lower the engine until all weight is off the hoist. Install and tighten the lag screws securely.
16. If you're confident that your measurements and subsequent adjustment place the engine exactly where it was prior to removal, then you are through. If you're like us though, you may want to check the engine alignment before you fire up the engine.

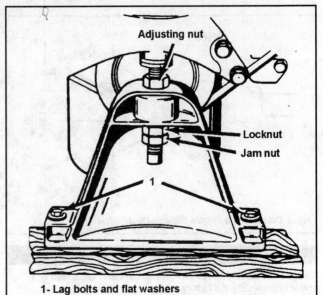

1- Lag bolts and flat washers

Fig. 13 A good view of the front engine mount—not all engines utilize the jam nut

Rear Engine Mounts

REMOVAL & INSTALLATION

③ ⟨DIFFICULT

1. Remove the engine as previously detailed.
2. Loosen the two bolts and remove the mount from the transom plate.
3. Hold the square nut with a wrench and remove the shaft bolt. Be sure to take note of the style and positioning of the two mount washers as you are removing the bolt. Mark them, lay them out, or write it down, but don't forget their orientation!!

To install

4. Slide the lower of the two washers onto the mount bolt, exactly as it came off.
5. Slide the bolt into the flat side of the rubber mount, install the remaining washer (as it came off!) and then spin on the square nut. Do not tighten it yet.

■ **Incorrect washer installation will cause excessive vibration during engine operation.**

6. Turn the assembly upside down and clamp the nut in a vise. Spin the assembly until the holes in the mounting plate are directly opposite any two of the flat sides on the nut. This is important, otherwise the slot on the engine pad will not engage the mount correctly. Secure the mount in this position and tighten the bolt to 18-20 ft. lbs. (24-27 Nm) on 1986-88 engines, or 44-52 ft. lbs. (60-71 Nm) on 1989-98 engines.
7. Remove the mount from the vise and position it on the transom plate. Install the bolts and washers and tighten each to 20-25 ft. lbs. (27-34 Nm).
8. Install the engine, making sure that the slot in the engine pad engages the square nut correctly. Install the two washers and locknut and tighten it to 28-30 ft. lbs. (38-41 Nm).

Cylinder Head (Valve) Cover

REMOVAL & INSTALLATION

② ⟨MODERATE

1. Open or remove the engine compartment hatch. Disconnect the negative battery cable.
2. Loosen the clamp and remove the crankcase ventilation hose at the cover (if equipped). Carefully move it out of the way.
3. Tag and disconnect the shift cut-out switch leads at the terminal block.
4. If your engine has a spark plug wire retainer attached to the cover, unclip the wires or remove the retainer.

5. Remove the fuel line at the pump and plug it, and the fitting, to avoid spills and system contamination. Remove the overflow hose also, plugging it and moving it out of the way. Have some rags nearby, as there will be some seepage either way. Move the line out of the way of the cover.

6. Remove the circuit breaker bracket and carefully position it out of the way.

7. Loosen the cover mounting bolts and lift off the cylinder head cover. Take note of any harness or hose retainers and clips that might be attached to certain of the mounting bolts; you need to make sure they go back in the same place.

To install:

8. Clean the cylinder head and cover mounting surfaces of any residual gasket material with a scraper or putty knife.

9. Position a new gasket on the cylinder head and then position the cover (don't forget the J-clips!). Tighten the mounting bolts to 45 inch lbs. (5 Nm) on 1986-93 engines, or 65 inch lbs. (7.3 Nm) on 1994-98 engines. Make sure any retainers or clips that were removed are back in their original positions.

10. Reconnect the fuel line to the pump. Check for leaks now, and after you restart the engine.

11. Connect the crankcase ventilation hose and the cut-out switch leads. Check that there were no other wires or hoses you may have repositioned in order to gain access to the cover.

12. Connect the battery cables.

Rocker Arms and Push Rods

REMOVAL & INSTALLATION

◆ See Figure 14

1. Open or remove the engine hatch cover and disconnect the negative battery cable. Remove the cylinder head cover as detailed previously.

2. Bring the piston in the No. 1 cylinder to TDC. If servicing only one arm, bring the piston in that cylinder to TDC. The No. 1 cylinder is the first cylinder at the front of the engine.

3. Loosen and remove the rocker arm nuts and lift out the balls. Lift the arm itself off of the mounting stud and pull out the pushrod. It is very important to keep each cylinder's component parts together as an assembly. We suggest drilling a set of holes in a 2X4 and positioning the pieces in the holes.

To install:

4. Clean and inspect the rocker assemblies.

5. Coat all bearing surfaces of the rocker assembly with engine oil.

6. Slide the push rods into their holes. Make sure that each rod seats in its socket on the lifter.

7. Position the rocker arm over the stud so that the cupped side rides on the push rod. Slide the ball over the stud, install the nut and tighten it securely, or until all play in the push rod is taken up.

8. Adjust the valves and install the cylinder head cover.

9. Connect the battery cable and check the idle speed.

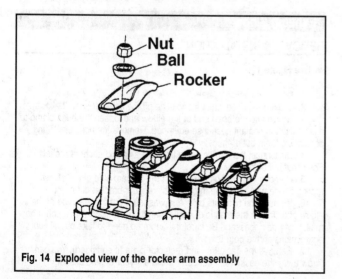

Fig. 14 Exploded view of the rocker arm assembly

VALVE ADJUSTMENT

◆ See Figure 15

✱✱ CAUTION

OMC recommends adjusting the valves while the engine is cold and while the engine is OFF.

These engines utilize hydraulic valve lifters, although there is no need for periodic valve adjustment, it is necessary to perform a preliminary adjustment after any work on the valve train/rocker assembly. All adjustment should be undertaken while the lifter is on the base circle of the camshaft lobe for that particular cylinder. This means the opposite side of the pointy part of each lobe.

1. Rotate the crankshaft, or bump the engine with the starter until the No. 1 cylinder is at TDC of the compression stroke— remember, this is the first cylinder at the front of the engine). Both lifters should be on the base circle of the camshaft. Note that the notch or mark on the damper pulley will be lined up with the **0** mark on the timing scale. Be careful here though, this could mean that either the No. 1 or the No. 4 piston is at TDC. Place your hand on the No. 1 cylinder's valve and check that it does not move as the mark on the pulley is approaching the **0** mark on the tab. If it does not move, you're ready to proceed; if it does move, you are on the No. 4 cylinder and need to rotate the engine an additional full turn (360°. This is important so make sure you've gotten it right!

Another method is to mark the position of the No. 1 cylinder's terminal on the distributor; do this on the body of the distributor because you're about to remove the cap. Remove the distributor cap and bump the engine over with the starter until rotor points to the mark you made previously for the No. 1 cylinder terminal. Once again, you're at TDC on the compression stroke.

2. Now that the No. 1 cylinder is at TDC, you can adjust both valves. Loosen the adjusting nut on the rocker until you can feel lash (play in the push rod) and then tighten the nut until the lash has been removed. Carefully jiggle the push rod while tightening the nut until it won't move anymore—this is zero lash. Tighten the nut an additional full turn to set the lifter and then you're done. Perform this procedure on each of the valves for this cylinder.

3. Perform the first two steps on the remaining cylinders in the proper order (2-4, front to back), setting the respective piston for each cylinder at TDC.

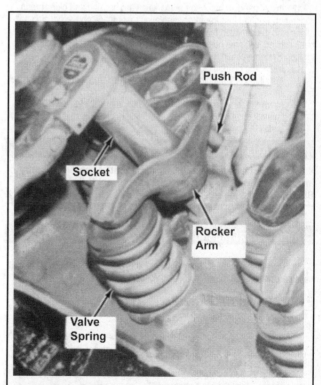

Fig. 15 Wiggle the push rod slowly while tightening the rocker adjusting nut

Hydraulic Valve Lifter

REMOVAL & INSTALLATION

MODERATE

1. Remove the cylinder head cover.
2. Using compressed air, thoroughly clean all dirt and grit from the cylinder head and related components.

✳✳ WARNING

If compressed air is not available, we highly recommend that you DO NOT proceed with this procedure. It is EXTREMELY important that no dirt gets into the lifter recesses before completing the installation.

3. Loosen the rocker arms and pivot the rockers off of the pushrods.
4. Disconnect the spark plug wires at the plugs.
5. Remove the high tension lead at the coil. Tag and disconnect both electrical connectors and then remove the coil and its bracket from the cylinder head.
6. Remove the distributor cap and mark the position of the rotor on the body of the distributor housing. Matchmark the distributor housing to the cylinder block and then remove the distributor, covering the hole carefully.
7. Remove the pushrod cover and gasket from the side of the cylinder block. On most later models, this will require removing the shift bracket assembly and moving it aside.
8. Remove the pushrods and lifters from the block, being very careful to keep track of where each one came from. Once again, we suggest using a 2x4 with holes drilled in it to store the rods and lifters; you'd be amazed at how quickly this job will fall apart if someone walks in and kicks the components that you have laid out on the floor!

To install:
9. Inspect the camshaft contact surface on the bottom of each lifter for excessive wear, galling or other damage. Discard the lifter if any of the above conditions are found. You'll probably also want to check out the camshaft lobe for damage also.
10. Coat the bottom of each lifter with Molykote® and then carefully install each one into its respective recess.
11. Install the pushrods and then install the pushrod cover with a new gasket. Tighten the retaining bolts securely. If you removed the shift bracket previously, install it now.
12. Install the distributor so the marks on the housing and block are in alignment. Confirm that the rotor orientation mark made on the housing still aligns with the direction that the rotor is pointing. If not, or if the engine has been rotated for some reason during this procedure, please refer to the Ignition System section for complete details.
13. Install the distributor cap.
14. Install the coil and its bracket. Reconnect all plug wires and electrical leads.
15. Move the rockers into position and then adjust the valve lash as detailed previously.
16. Install the cylinder head cover.

✳✳ CAUTION

If any, or all, of the lifters has been replaced with a new one it is very important that you add GM Engine Oil Supplement to the crankcase BEFORE starting the engine.

Combination Manifold

These engines incorporate the intake and exhaust manifolds into one unit called a combination manifold. It is serviced as a unit.

REMOVAL & INSTALLATION

◆ See Figure 16

MODERATE

1. Open or remove the engine compartment hatch. Disconnect the negative battery cable.
2. Drain all water from the engine, manifold and exhaust elbow as detailed in the Maintenance section.
3. Remove the flame arrestor and set it aside.

4. Disconnect the throttle cable from the carburetor, remove the anchor bolt and position the cable out of the way. Mark which anchor stud the cable was attached to for installation.
5. Disconnect the fuel line at the carburetor and fuel pump. Plug the pump inlet fitting. Make sure you have some rags handy to mop up the inevitable spill.
6. Remove the carburetor as detailed in the Fuel System section.
7. Remove the crankcase ventilation hose. Tag and disconnect any vacuum or electrical lines and then remove the carburetor.
8. Disconnect the shift cables.
9. Disconnect the water inlet line from the thermostat housing.
10. Remove the alternator and its mounting bracket as detailed in the Charging System section.

■ **On certain early engines without power steering, it may not be necessary to remove the alternator.**

11. Tag and disconnect any wires or hoses that may interfere with manifold removal.
12. Loosen the clamps and slide the exhaust pipe bellows off of the high rise elbow. Remove the mounting bolts (or nuts) and lift the elbow off of the manifold along with the throttle linkage plate.
13. If equipped, remove the oil cooler and mounting bracket.
14. Loosen and remove the manifold mounting bolts (4) and nuts (2) from the center outward and remove the manifold from the cylinder head—you may have to provide a little friendly persuasion.

To install:
15. Carefully clean all residual gasket material from the head, manifold and elbow mating surfaces with a scraper or putty knife. Inspect all gasket surfaces for scratches, cuts or other imperfections.
16. Position a new gasket on the cylinder head and install the manifold; making sure that everything is aligned properly. Tighten all bolts/nuts until they are just tight and then torque them to 20-25 ft. lbs. (27-34 Nm), starting in the center and alternately working your way out to the ends of the manifold.
17. Position a new gasket coated with sealer on the manifold, making sure that the indents line up and then install the elbow. You may also have an additional gasket and a restrictor plate on some models. Tighten the mounting bolts/nuts to 12-14 ft. lbs. (16-19 Nm). Connect the exhaust pipe/bellows and tighten the clamps securely.
18. Connect the water inlet line and the shift cables.
19. Install the carburetor with a new gasket and reconnect the throttle cable, choke wire and any vacuum lines. Remember which anchor stud the throttle was attached to.
20. Connect the fuel line to the carb and the fuel pump (remember to unplug the pump fitting) and then install the flame arrestor.
21. Install the alternator; tighten the alternator and bracket bolts to 26-30 ft. lbs. (35-41 Nm). Install the oil cooler if removed.
22. Make sure that any miscellaneous lines or hoses that you may have moved or disconnected during removal are reconnected and routed properly.
23. Fill the system with water, connect the battery cable and start the engine. When the engine reaches normal operating temperature, turn it off and re-torque the manifold bolts.

Exhaust Elbow

REMOVAL & INSTALLATION

◆ See Figure 16

1. Drain the cooling system.
2. Loosen the hose clamps and slide off the exhaust hose (bellows).
3. Reach into the open end of the elbow and loosen the hose clamp to disconnect the coolant hose if so equipped. Remove the hose and fitting, positioning them out of the way.
4. Disconnect the throttle cable from the arm and anchor bracket. Position it out of the way.
5. Remove the two bolts or nuts on the forward side of the elbow mating flange and swing the throttle cable anchor bracket out of the way.
6. Remove the two nuts/bolts and washers from the rear side of the elbow and lift it off the manifold. A little friendly persuasion with a soft rubber mallet may be necessary! Be careful though, no need to take out all your aggressions on the poor thing.
7. Remove the gasket(s) and restrictor plate (if equipped). You can throw away the gasket(s), but keep the plate if your engine uses it.

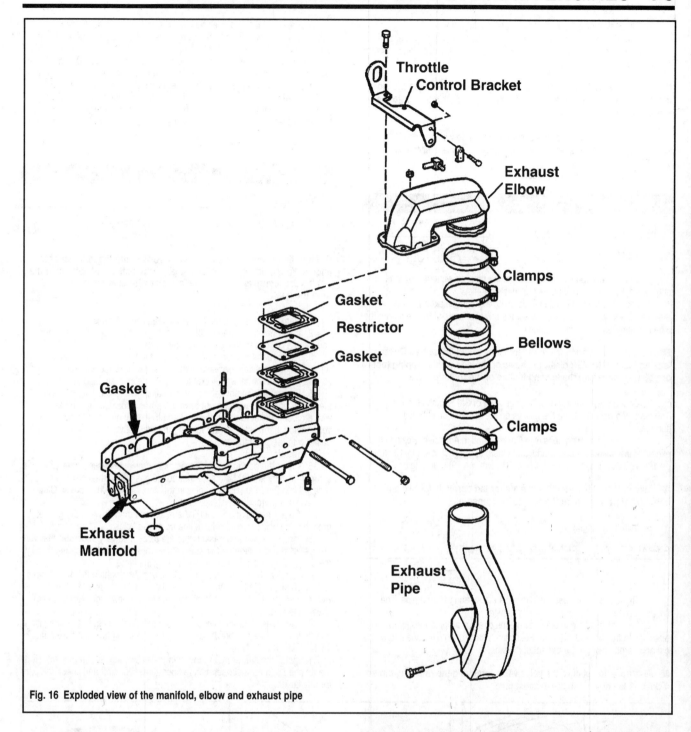

Fig. 16 Exploded view of the manifold, elbow and exhaust pipe

To install:

8. Clean the mating surfaces of the manifold and elbow thoroughly, coat both sides of a new gasket with Gasket Sealing compound and position it onto the manifold flange.

9. If your engine was so equipped, position the restrictor plate and/or second gasket onto the manifold. Don't forget to coat the gasket with sealing compound.

10. Position the elbow on the manifold, install the anchor bracket and screw the two bolts/nuts in finger tight. Install the remaining two bolts/nuts with new lock-washers and then tighten all four to 12-14 ft. lbs. (16-19 Nm).

11. Connect the coolant hose into the elbow and tighten the clamp.

12. Slide the exhaust hose, while wiggling it, all the way onto the elbow and tighten the clamp screws securely.

13. Reconnect the throttle cable.

Exhaust Hose (Bellows)

REMOVAL & INSTALLATION

◆ See Figures 16, 17, 18 and 19

1. Loosen all four hose clamps, two on top of the hose and two on the bottom.

2. Drizzle a soapy water solution over the top of hose where it mates with the exhaust elbow and let it sit for a minute.

3. Grasp the hose with both hands and wiggle it side-to-side while pulling down on it until it separates from the elbow.

4. Now wiggle it while pulling upwards until it pops off the exhaust pipe. Make sure that you secure the exhaust pipe while pulling off the hose.

5. Check the hose for wear, cracks and deterioration.
6. Coat the inside of the lower end of the hose with soapy water and wiggle it into position on the pipe. Remember to install the two clamps before sliding it over the end of the pipe.

■ **There are usually two ribs on one end of the hose—this is the side that attaches to the exhaust elbow. Do not position this end onto the exhaust pipe.**

7. Slide two clamps over the upper end (the side with two ribs!), lubricate the inside with soapy water and wiggle the hose over the elbow until it is in position.
8. Tighten all four clamps securely.

Exhaust Pipe

REMOVAL & INSTALLATION

◆ See Figure 16  MODERATE

1. It is unlikely you will be able to get the pipe off without removing the engine, so remove the engine as previously detailed.
2. Loosen the four retaining bolts at the transom shield and then remove the exhaust pipe. Carefully scrape any remnants of the seal from the pipe and transom mounting surfaces.

■ **The pipe mounting holes in the transom shield utilize Heli-Coil® locking inserts. NEVER clean the holes or threads with a tapping tool or you risk damaging the locking feature of the threads.**

3. Coat a new seal with OMC Adhesive M or 3M Scotch Grip Rubber Adhesive and position it into the groove on the transom shield mating surface.
4. Coat the mounting bolts with Gasket Sealing Adhesive. Position the exhaust pipe, insert the bolts and tighten them to 10-12 ft. lbs. (14-16 Nm) on 1986-89 engines; or 20-25 ft. lbs. (27-34 Nm) on 1990-98 engines.

■ **Make sure that the trim/tilt lines are routed correctly (above and behind) before tightening the pipe to the transom.**

5. Install the engine.

EXHAUST VALVE (FLAPPER) REPLACEMENT

 MODERATE

1. Remove the exhaust hose from the elbow and exhaust pipe. The flapper is located in the upper end of the pipe.
2. The valve is held in place by means of a pin running through two bushings in the sides of the pipe. Position a small punch over one end of the pin and carefully press the pin out of the pipe.

■ **Make sure you secure the valve while removing the retaining pin so it doesn't fall down into the exhaust pipe.**

3. Press out the two bushings and discard them. Coat two new bushings with Scotch Grip Rubber Adhesive and press them back into the sides of the pipe.
4. Position the new valve into the pipe with the long side DOWN. When looking at the valve, the molded retaining rings are off-center—the side with the rings should face the top of the pipe. When the valve is in place, coat the pin lightly with engine oil and slowly slide it through one of the bushings, through the two retaining holes on the valve and then through the opposite bushing. Make sure the pin ends are flush with the sides of the pipe on both sides.
5. Install the exhaust hose.

Oil Pan

REMOVAL & INSTALLATION

◆ See Figure 20  DIFFICULT

■ **More times than not this procedure will require the removal of the engine. Your boat and its unique engine installation will determine this, but the procedure is almost always easier with the engine removed from the boat.**

1. Remove the engine as previously detailed in this section.
2. If you haven't already drained the engine oil, do it now. Make sure you have a container and lots of rags available.
3. Remove the starter motor as detailed in the Electrical section. Although not absolutely necessary, this step will make the job much easier.
4. Remove the oil withdrawal tube from the oil drain plug fitting.
5. Loosen and remove the oil pan retaining bolts and nuts, starting with the center bolts and working out toward the pan ends. Lightly tap the pan with a rubber mallet to break the seal and then lift it off the cylinder block. If your engine stand will allow for rotating the engine, you'll find that this will be easier with the pan facing up.

To install:
6. Clean the pan mating surfaces of any residual gasket material with a scraper or putty knife. Make sure that no old gasket material has been pressed into the retaining bolt holes in the pan, block or front cover. Clean the pan itself thoroughly with solvent.
7. Coat a new gasket with RTV sealant and then position the gasket onto the pan being very careful to line up all the holes.
8. Move the pan and gasket onto the block. It is very important that you ensure all the holes line up correctly; sometimes a few bolts inserted through the pan and gasket will help the gasket stay in place.
9. Install all bolts and nuts finger tight and then tighten the 1/4-20 bolts to 80 inch lbs. (9 Nm) and the 5/16-18 bolts to 165 inch lbs. (19 Nm). Remember, on all engines, to start with the center bolts and work outward toward the ends of the pan.
10. Install the oil drain fitting and finger tighten the bolt. Rotate the fitting and then attach the withdrawal tube. Tighten the fitting bolt and flare nut to 15-18 ft. lbs. (20-24 Nm).
11. Install the starter motor and then install the engine (if removed). Run the engine up to normal operating temperature, shut it off and check the pan for any leaks.

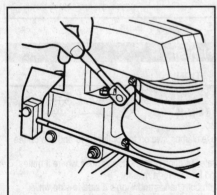

Fig. 17 Loosen the upper clamp screws and wiggle the bellows off of the elbow...

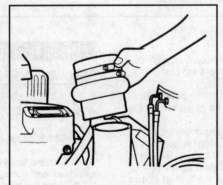

Fig. 18 ...loosen the lower clamp screws and then pull it off of the exhaust pipe

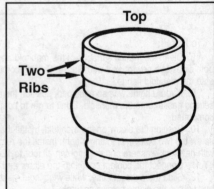

Fig. 19 Be sure that the bellows end with two ribs is on top

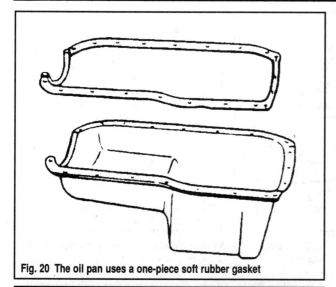

Fig. 20 The oil pan uses a one-piece soft rubber gasket

Oil Pump

The two-piece oil pump utilizes two pump gears and a pressure relief valve. A baffled pick-up tube is press-fit into the body of the pump. The pump is driven via the distributor shaft which is itself driven from a gear on the camshaft.

REMOVAL & INSTALLATION

◆ See Figure 21

1. Remove the oil pan as previously detailed. Remember that you probably need to remove the engine for this procedure.
2. Loosen and remove the pump pick-up tube support bracket bolt. The tube is pressed into the pump housing and should not be removed unless replacement is necessary.
3. Loosen and remove the two pump mounting bolts and lift off the pump assembly.
4. Check that the pump and block mating surfaces are clean and then position the pump over the block so that the pump drive shaft slot is aligned with the distributor tang. Make sure that the flange covers the alignment bushing. Do not use a gasket or RTV sealant.
5. Tighten the pump mounting bolts to 110-120 inch lbs. (12-14 Nm). Position the pick-up tube bracket and tighten the bolt to 60 inch lbs. (7 Nm).
6. Install the oil pan and engine.

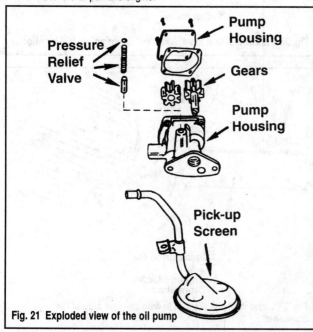

Fig. 21 Exploded view of the oil pump

Oil Filter Bypass Valve

After removing the oil filter, check the spring and small fiber valve for proper operation. Any signs of incorrect operation, or wear and deterioration will necessitate replacement.

REMOVAL & INSTALLATION

1. Drain the oil as detailed in the Maintenance section.
2. Remove the oil filter.
3. Using a small prybar, remove the valve.
4. Install a new valve and press it in by placing a 9/16 in. deep socket over it and tapping the socket lightly with a hammer.
5. Install the oil filter and refill the engine with the appropriate oil.

Engine Coupler And Flywheel

REMOVAL & INSTALLATION

◆ See Figures 22, 23 and 24

1. Remove the engine from the boat as detailed previously in this section.
2. Although not strictly necessary, we recommend removing the starter.
3. Remove the 4 or 5 mounting bolts and slide out the flywheel housing cover.
4. Loosen and remove any connections attached to either of the ground studs on each side of the flywheel. Move the electrical leads out of the way.
5. Remove the ESA module/shift bracket and position it out of the way on early models. On later models, remove the mounting bolts for the 10-pin connector and relay, and then move the bracket out of the way.
6. Cut the plastic tie that secures the housing drain hose (if equipped) and the pull the hose out of the fitting.
7. Loosen the remaining retaining bolts (some of which were already removed with the shift bracket) for the flywheel housing and remove it. Take note of the positioning of any clamps so they may be installed in the same position.
8. Slide a wrench behind the coupler and loosen the six flywheel mounting nuts gradually and as you would the lug nuts on your car or truck—that is, in a diagonal star pattern.
9. Remove the coupler and then the flywheel.
To install:
10. Thoroughly clean the flywheel mating surface and check it for any nicks, cracks or gouges. Check for any broken teeth.
11. Install the flywheel over the dowel on the crankshaft.
12. Slide the coupler over the studs so that it sits in the recess on the flywheel. Install new lock washers and tighten the mounting bolts to 40-45 ft. lbs. (54-61 Nm). Once again use the star pattern while tightening the bolts.
13. Position the flywheel housing and the shift bracket. Make sure that the clamps are in their original positions and insert the mounting bolts. Tighten them to 28-36 ft. lbs. (38-49 Nm).
14. Install the washer, lock washer and inner nut on the ground stud and tighten it to 20-25 ft. lbs. (27-34 Nm) on 1986-93 engines, or 35-40 ft. lbs. (47-54 Nm) on 1994-98 engines. Attach the electrical leads, install another lock washer and then tighten the outer nut securely.
15. Slide the lower flywheel cover into place and tighten the bolts to 60-84 inch lbs. (7-9 Nm).
16. Attach the drain hose with a new plastic tie.
17. Install the starter and engine.

Rear Main Oil Seal

REMOVAL & INSTALLATION

Two Piece Oil Seal (1986-90)

◆ See Figures 25, 26 and 27

These engines utilize a two-piece rear main seal. The seal can be removed without removing the crankshaft. You will need to remove the engine for this procedure though.
1. Remove the engine as detailed previously in this section.

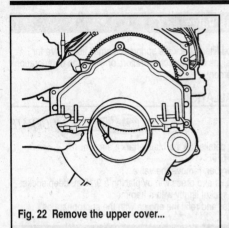

Fig. 22 Remove the upper cover...

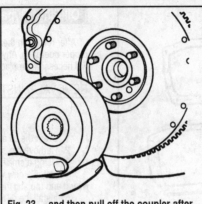

Fig. 23 ...and then pull off the coupler after removing the bolts

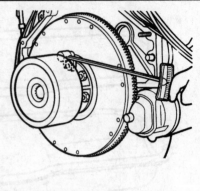

Fig. 24 Tighten the mounting bolts in a criss-cross pattern

2. Remove the oil pan and pump as detailed previously in this section.

3. Loosen the retaining bolts and remove the rear main bearing cap. Carefully insert a small prybar and remove the lower half of the seal from the cap. Do not damage the seal seating surface.

4. Using a hammer and a small drift, tap on the end of the upper seal until it starts to protrude form the other side of the race. Grab the protruding end with pliers and pull out the remaining seal half.

Fig. 25 Drive out the old upper seal with a small drift

■ **Upper and lower seals must replaced as a pair. Never replace only one seal.**

5. Check that you have the correct new seal. Seals with a hatched inner surface can only be used on left hand rotation engines, smooth seals can be used on any engine. Coat the lip and bead thoroughly with motor oil. Keep oil away from the seal parting surfaces.

6. Your seal kit should come with an installation tool, if not, take a 0.004 in. feeler gauge and cut each side back about a half inch so that you're left with an 11/64 inch point. Bend the tool into the gap between the crankshaft and the seal seating surface. This will be your "shoe horn".

7. Position the upper half of the seal (lip facing the engine) between the crank and the tool so that the seal's bead is in contact with the tool tip. Roll the seal around the crankshaft using the tool as a guide until each end is flush with the cylinder block. Remove the tool.

8. Insert the lower seal half into the main bearing cap with the lip facing the cap. Start it so that one end is slightly below the edge of the cap and then use the tool to shimmy the seal all the way in until both edges are flush with the edge of the cap. Remove the tool.

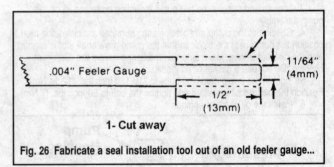

.004" Feeler Gauge

1 — 11/64" (4mm)

1/2" (13mm)

1- Cut away

Fig. 26 Fabricate a seal installation tool out of an old feeler gauge...

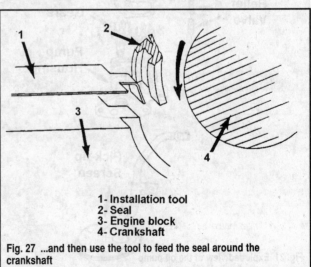

1- Installation tool
2- Seal
3- Engine block
4- Crankshaft

Fig. 27 ...and then use the tool to feed the seal around the crankshaft

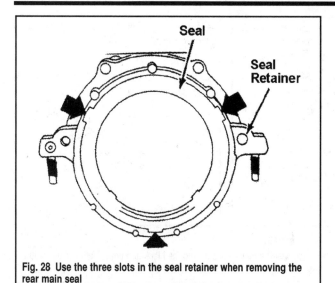

Fig. 28 Use the three slots in the seal retainer when removing the rear main seal

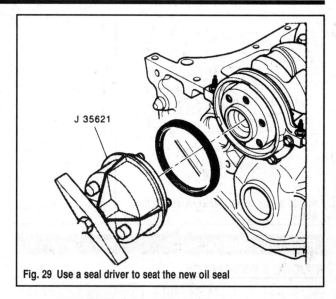

Fig. 29 Use a seal driver to seat the new oil seal

9. Make sure that the cap/block mating surfaces and the seal ends are free of any oil and then apply a small amount of Perfect Seal to the block just behind where the upper seal ends are.

10. Install the bearing cap and tighten the bolts to 10-12 ft. lbs. (14-16 Nm). Tap the end of the crankshaft forward and backward (as detailed in the appropriate section) and then tighten the bolts to 60-70 ft. lbs. (82-94 Nm).

11. Install the oil pump and pan.

12. Install the engine.

One Piece Oil Seal (1991-98)

◆ **See Figures 28 and 29**

It is not necessary to remove the engine or rear main bearing cap when removing the one-piece oil seal on these engines although you may find it easier to do just that.

1. Remove the flywheel housing and cover as detailed in this section.

2. Remove the engine coupler and flywheel from the engine as detailed in this section.

3. Remove the seal retainer. This is not absolutely necessary.

4. Insert a small prybar into one of the three slots in the edge of the seal retainer and slowly pry the seal out of the retainer. Be very careful not to nick or damage the sealing surface while prying out the seal.

5. Thoroughly clean the retainer surface and then install the retainer.

6. Spread a small amount of engine oil around the inside and outside edges of a new seal and position it over its slots in the retainer.

7. Position a seal driver (J-35621) over the seal and crankshaft and then thread the attaching screws into the holes in the crankshaft, tightening them securely. Turn the handle on the tool until it bottoms out—the seal is now in place.

8. Install the flywheel and engine coupler. Install the cover and flywheel housing.

Rear Main Oil Seal Retainer

REMOVAL & INSTALLATION

1991-98 Engines Only

◆ **See Figure 28**

It is not necessary to remove the engine or rear main bearing cap when removing the one-piece oil seal on these engines although you may find it easier to do just that.

■ **A new oil seal must be installed whenever the retainer is removed.**

1. Remove the oil pan.

2. Remove the flywheel.

3. Loosen the nuts/bolts and then lift out the retainer and gasket.

4. Replace the oil seal.

5. Clean all traces of old gasket material from the mating surfaces and position a new seal on the cylinder block.

6. Install the seal retainer, and a new gasket, and tighten the nuts/bolts to 135 inch lbs. (15 Nm).

7. Install the oil pan and flywheel as previously detailed.

Harmonic Balancer, Pulley And Hub

REMOVAL & INSTALLATION

◆ **See Figures 30 and 31**

1. If the engine is in the boat, install an engine hoist and tighten the chain so that the engine's weight is removed from the front engine mount. Remove the front engine mount.

2. Remove the seawater pump from the front of the crankshaft on models so equipped.

3. Remove the drive belts.

4. Loosen the bolts securing the pulley/harmonic balancer to the hub and remove the assembly—the 2.5L does not use the balancer.

5. Install special tool #J-6978-E onto the hub with two 3/8-24 x 2 in. bolts and a 5/16-24 x 2 in. bolt for the 2.5L, or three 3/8-24 x 2 in. bolts on the 3.0L. Tighten the tool press bolt and remove the hub. OMC suggests that you do not use a conventional gear puller for this procedure.

To install:

6. Coat the front cover oil seal lip with clean engine oil and then install the balancer/pulley over the crankshaft and key. You can use a rubber mallet to position it temporarily.

7. Install the installation tool #J-5590 onto the hub. Be sure that you thread the tool into the crankshaft at least 1/2 in. to protect the threads.

✳✳ CAUTION

The crankshaft actually extends slightly through the front of the assembly, so make sure to use the correct tool.

■ **In order that the timing mark can be properly positioned, there are two 3/8 in. holes and one 5/16 in. holes in the hub that must be matched.**

8. Install the drive belts and make sure that it is adjusted properly.

9. Install the seawater pump if removed.

10. Install the front mount and unhook the engine hoist.

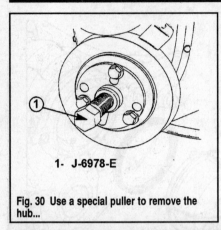

1- J-6978-E

Fig. 30 Use a special puller to remove the hub...

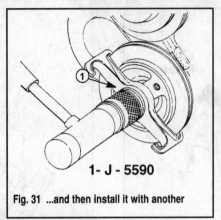

1- J - 5590

Fig. 31 ...and then install it with another

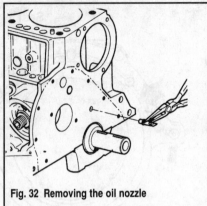

Fig. 32 Removing the oil nozzle

Front Cover And Oil Seal

REMOVAL & INSTALLATION

◆ See Figures 32, 33 and 34

■ This procedure may require engine removal, depending upon your particular boat. If necessary, remove the engine as detailed previously in this section.

1. Open the drain valves and drain the coolant from the block and exhaust manifold.
2. Loosen the alternator and power steering brackets to provide slack, and then remove the drive belts.
3. Remove the harmonic balancer/pulley/hub assembly.
4. Remove the water circulation pump.
5. Secure the engine with a hoist and remove the front engine mount if you haven't already removed the engine.
6. Drain the oil and remove the oil pan.
7. Loosen all of the front cover retaining screws and then pull off the cover and gasket. Clean all gasket material from the cover and block mating surfaces.
8. Pry the oil seal from the timing gear cover with a large drift or prybar.
9. Grasp the oil nozzle with a pair of pliers and pull it out of the cylinder block.

To install:

10. Insert a new oil nozzle into the block and drive it into position with a rubber mallet.

11. Coat both sides of a new oil seal lightly with grease and then position it into the cover so that the lip (open side) faces into the cover. With the cover on a clean flat surface, position a seal installer (#J-23042) over the seal and drive it into place with a hammer.

12. Clean all old gasket material from the cover and block mating surfaces. Coat a new gasket lightly with grease and stick it into position on the cylinder block.

13. The seal installer tool also acts as a centering tool. Install the tool onto the crankshaft and slide the cover over the tool and into position. Install the mounting screws and tighten them to 6-8 ft. lbs. (9-10 Nm) on 1986-92 engines, or 80 inch lbs. (9 Nm) on 1993-98 engines. Tighten the screws evenly and gradually.

14. Remove the centering tool.

15. Install the water circulation pump and the harmonic balancer/hub.

16. Install the oil pan.

17. Install all belts and check their tension.

18. Reinstall the engine mount; or if you removed the engine, install it now.

19. Refill the engine with oil and coolant.

Fig. 33 Use the special tool to drive the seal into the cover

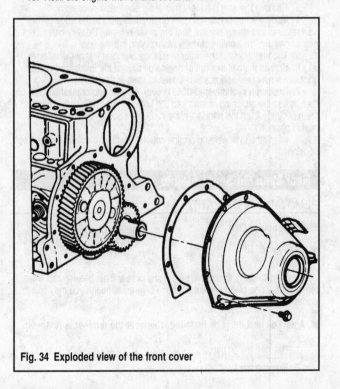

Fig. 34 Exploded view of the front cover

Camshaft, Bearings And Gear

We recommend checking the camshaft lift prior to removing it from the cylinder block.

CHECKING LIFT

◆ See Figure 35

1. Tag and disconnect the electrical connectors at the ignition coil.

2. Remove the cylinder head cover and rocker arms as detailed previously.

3. Using a special adaptor (#J-8520), connect a dial indicator so that its tip is positioned on the end of the pushrod—the adaptor should screw onto the end of the rocker stud.

4. Slowly rotate the crankshaft in the direction of engine rotation until the valve lifter is riding on the heel (back side of lobe) of the camshaft lobe. The pushrod should be at its lowest point when the lifter is on the heel.

■ **A remote starter works well for turning the engine over in this situation.**

5. Set the indicator to **0** and then rotate the engine until the pushrod is at the highest point of its travel. Camshaft lift should be 0.253 +/- 0.005 in. (6.426 +/- 0.127mm).

6. Continue rotating the engine until the pushrod is back at its lowest position—make sure that the indicator still reads **0**.

7. Repeat this procedure for the remaining pushrods.

8. Install the rocker arms and adjust the valve clearance.

9. Install the cylinder head cover and reconnect the coil leads.

REMOVAL & INSTALLATION

◆ See Figure 36 thru 41

■ **This procedure may require engine removal, depending upon your particular boat. If necessary, remove the engine as detailed previously in this section.**

1. Drain the engine oil.

2. Open the drain valves and drain the coolant from the cylinder block and exhaust manifold.

3. Remove the cylinder head cover and gasket. Loosen the rocker arm nuts just enough so that you can rotate the rockers off of the pushrod ends.

4. Mark the position of the distributor's No. 1 cylinder terminal on the housing of the distributor and then remove the cap. Matchmark the distributor and the cylinder block, loosen the hold down clamp and lift out the distributor.

5. Remove the ignition coil and side cover gasket. Take time to set up a system to keep the push rods and valve lifters in order, to ensure each will be installed back into the exact location from which it was removed. Withdraw each push rod and valve lifter in order.

6. Remove the harmonic balancer.

7. Remove the water circulation pump.

8. Secure the engine with a hoist and remove the front engine mount if you haven't already removed the engine.

9. Remove the oil pan.

10. Remove the front cover and gasket. Clean all gasket material from the cover and block mating surfaces.

11. Rotate the camshaft gear until the holes in the gear are aligned with the thrust plate screws. Remove the two screws. Carefully withdraw the camshaft gear and camshaft by pulling the gear straight forward and the shaft out of the block.

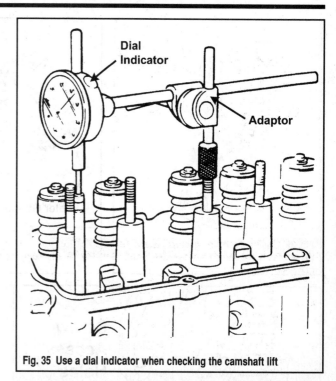

Fig. 35 Use a dial indicator when checking the camshaft lift

✳✳ CAUTION

Be very careful not to damage the camshaft bearings while removing the camshaft.

12. Check the gear and thrust plate end-play. This clearance should be 0.001-0.005 in. (0.025-0.127mm) max. If the decision is made to replace the camshaft gear, or the thrust plate, the gear must be pressed from the shaft. Gear removal from the camshaft requires the use of camshaft gear removal tool (#J-791) and an arbor press. Place the end of the removal tool onto the table of an arbor press, and then press the shaft free of the gear. The thrust plate must be positioned to prevent the woodruff key in the shaft from damaging the shaft when the shaft is pressed out of the gear. Also, be sure to support the end of the gear or the gear will be seriously damaged.

13. If the camshaft bearings are to be removed, you will need to remove the flywheel as previously detailed. Although it is not necessary, removing the crankshaft will also facilitate bearing removal, make sure that you move the connecting rods out of the way so they do not interfere with bearing removal.

14. Working from inside the block, drive out the rear cam bearing expansion plug (welch plug).

15. Slide the pilot tool into position in the inner bearing. Install a nut onto the puller screw so that the screw can be threaded into the tool with the nut still extending out the front of the cylinder block.

16. Install the remover onto the puller screw, slide the screw through the bore and thread it onto the pilot.

17. Hold the screw shaft with a wrench while turning the puller nut with another until the bearing comes out.

18. Now remove the pilot from the shaft and install it onto the drive handle so that the shoulder is against the handle. The front and rear bearings can now be driven out from the outside of the block.

To install:

19. Remove the handle from the pilot tool and install the inner bearing on the tool.

20. Position the tool and bearing to the rear of the inner bore. Install the screw shaft, with the remover, through the block and onto the pilot—from the front of the cylinder block.

21. Align the oil hole in the bearing with the oil gallery hole and then snug the puller nut up against the adaptor. Using two wrenches again, hold the screw shaft with one while turning the puller with the other until the bearing is in position.

■ **The oil hole is on the top of the bearing and will not be visible during installation. To make installation easier, align the two holes and then mark the opposite side of the bearing/block.**

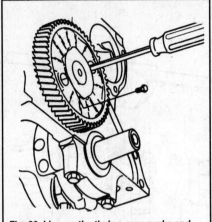

Fig. 36 Line up the timing gear marks and the access holes should rest over the thrust plate screws

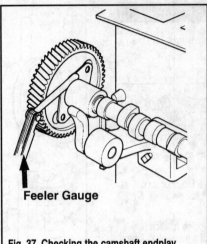
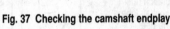

Feeler Gauge

Fig. 37 Checking the camshaft endplay

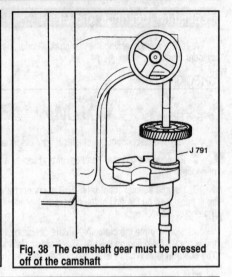

J 791

Fig. 38 The camshaft gear must be pressed off of the camshaft

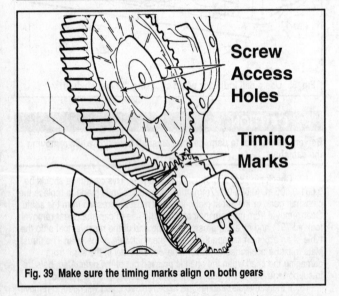

Screw Access Holes

Timing Marks

Fig. 39 Make sure the timing marks align on both gears

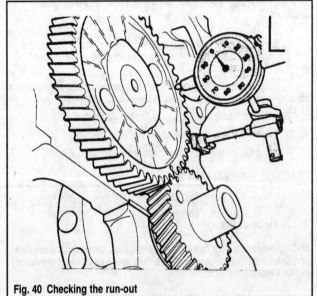

Fig. 40 Checking the run-out

22. Attach the drive handle to the tool and position the new front bearing onto the tool. Align the oil holes and drive the bearing in from the front (outside) of the cylinder block. Be sure that the bearing is driven in past the edge of the block at least 1/8 in. in order to expose the oil hole for the timing gear nozzle.

23. Repeat the last step for the rear bearing, but remember there is no oil nozzle hole in this bore so the bearing must be flush with the block.

24. Install a new expansion plug at the rear bearing.

25. To assemble the camshaft parts, first firmly support the camshaft at the back of the front journal in an arbor press. Next, place the gear spacer ring and the thrust plate over the end of the shaft, and install the Woodruff key in the shaft keyway. Install the camshaft gear and press it onto the shaft until it bottoms against the gear spacer ring. Check the end-play again.

26. If you removed the crankshaft, install it now.

27. Coat the camshaft lobes with G.M. Super Engine Oil Supplement and then pour the remainder of the can into the crankcase when you refill the engine with oil later in this procedure.

28. Slowly slide the camshaft and gear into the block, being careful not to damage the lobes or bearings. When the shaft is almost all the way in, rotate the camshaft and crankshafts until the timing marks on the teeth of each gear line up and then slide the camshaft in until the thrust plate meets the cylinder block and the holes in the plate line up with those in the block. Slowly rotate the crankshaft until the access holes in the cam gear are over the thrust plate holes. Install the two retaining screws and tighten them to 72-90 inch lbs. (8-10 Nm.)

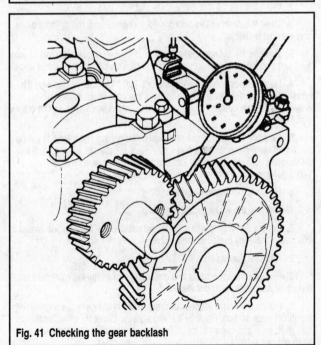

Fig. 41 Checking the gear backlash

29. Connect a dial indicator to the front of the cylinder block with an adaptor so that the needle on the indicator is in contact with the face of the camshaft gear. Rotate the engine 360° checking the run-out as you go. Repeat this procedure for the crankshaft gear. Camshaft runout should not exceed 0.004 in. (0.102mm). Crankshaft run-out should not exceed 0.003 in. (0.076mm). If run-out exceeds specification, remove the gear and check for burrs, otherwise replace the gear.

30. Move the dial indicator so that the needle is now riding on a gear tooth. Wiggle the shaft back and forth while checking the backlash reading on the indicator. Backlash should be 0.004-0.006 in. (0.102-0.152mm).

31. Install the oil pan. Install the front cover.

31. Install the harmonic balancer and water pump.

32. Installation of the remaining components is in the reverse order of removal. Don't forget to add the remainder of the Super Oil Supplement when refilling the engine with oil.

Cylinder Head

REMOVAL & INSTALLATION

◆ See Figures 42 and 43

 ② MODERATE

1. Drain the water from the cylinder block and manifold.

2. Remove the fuel line support brackets. Disconnect the fuel line at the carburetor and fuel pump, plug the fitting holes and remove the line.

3. Remove the combination manifold as previously detailed in this section; you can leave the carburetor attached if you like.

4. Disconnect the coolant hoses at the thermostat housing and move them out of the way. Have some rags available, as there will still be some coolant/water in the hoses.

5. Tag and disconnect the temperature sending unit lead at the thermostat housing, loosen the mounting bolts and then remove the housing and thermostat.

6. Tag and disconnect all wires at the ignition coil and then remove the mounting bracket bolt and lift off the coil.

7. Tag and disconnect the spark plug wires at the plugs; move them out of the way. Although not necessary, it's a good idea to remove the plugs themselves also.

8. Remove the circuit breaker bracket and then unbolt and remove the engine lifting bracket (if equipped).

9. Remove the cylinder head cover and rocker assemblies as detailed previously in this section. Remove the pushrods.

10. Loosen the cylinder head bolts, from the center bolts and working out to the ends of the head and then carefully lift the head off the block. You may need to persuade it with a rubber mallet—be careful! Set the head down carefully; do not sit it on cement.

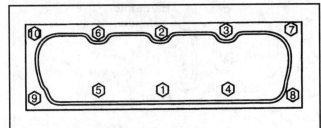
Fig. 42 Cylinder head tightening sequence

To install:

11. Carefully, and thoroughly, remove all residual head gasket material from the cylinder head and block mating surfaces with a scraper or putty knife. Check that the mating surfaces are free of any nicks or cracks. Make sure there is no dirt or old gasket material in any of the bolt holes. Refer to the Engine Rebuilding section for complete details on inspection and refurbishing procedures.

12. Position a new gasket over the cylinder block dowel pins. On models using a steel gasket, coat both sides of the gasket with a thin layer of Gasket Sealing Compound. Do not use too much sealing compound!!

13. Position the cylinder head over the dowels in the block. Coat the threads of the head bolts with Permatex® and install them finger tight. It never hurts to use new bolts, although it's not necessary. Tighten the bolts, a little at a time, in the sequence illustrated, until the proper tightening torque is achieved. 1986-93 engines have a final torque of 90-100 ft. lbs. (122-136 Nm), while 1994-98 engines have a final torque of 95 ft. lbs. (129 Nm)—make sure that you check the Torque Specifications chart for any preliminary steps/torques.

14. Install pushrods, rocker assemblies and the cylinder head cover.

15. Install the circuit breaker and engine lifting brackets. Install the spark plugs if they were removed and then connect the plug wires.

16. Install the coil and reconnect all the electrical leads.

17. Install the thermostat housing, the coolant hoses and the temperature sending lead.

18. Install the manifold and connect the fuel line. Don't forget to remove the fitting plugs.

19. Add coolant/water, connect the battery and check the oil. Start the engine and run it for a while to ensure that everything is operating properly. Keep an eye on the temperature gauge.

20. Re-tighten the cylinder head bolts after 20 hours of operation.

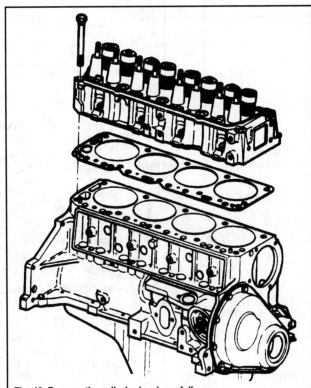
Fig. 43 Remove the cylinder head carefully

EXPLODED VIEWS

◆ See Figures 44, 45 and 46

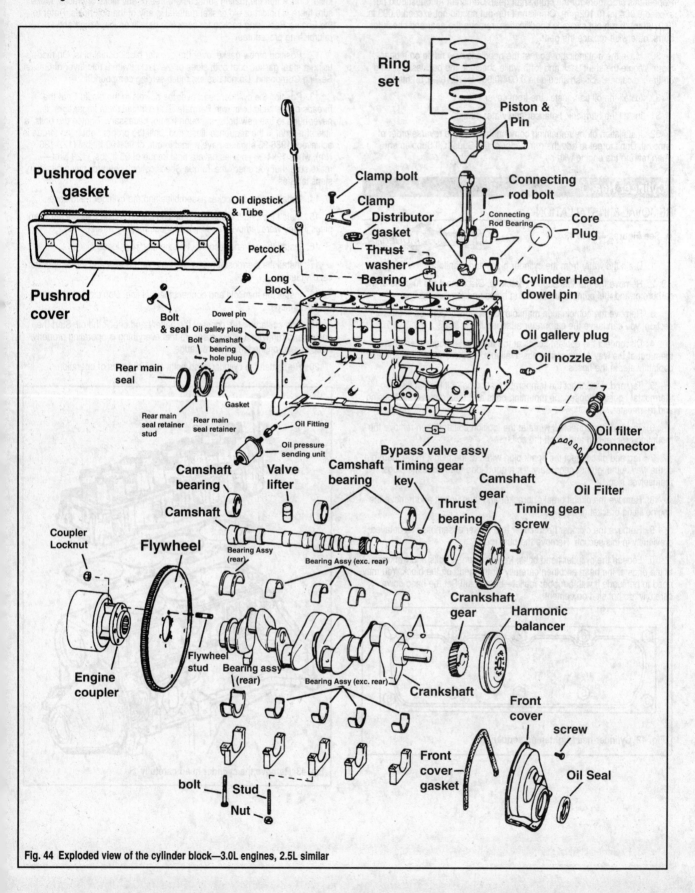

Fig. 44 Exploded view of the cylinder block—3.0L engines, 2.5L similar

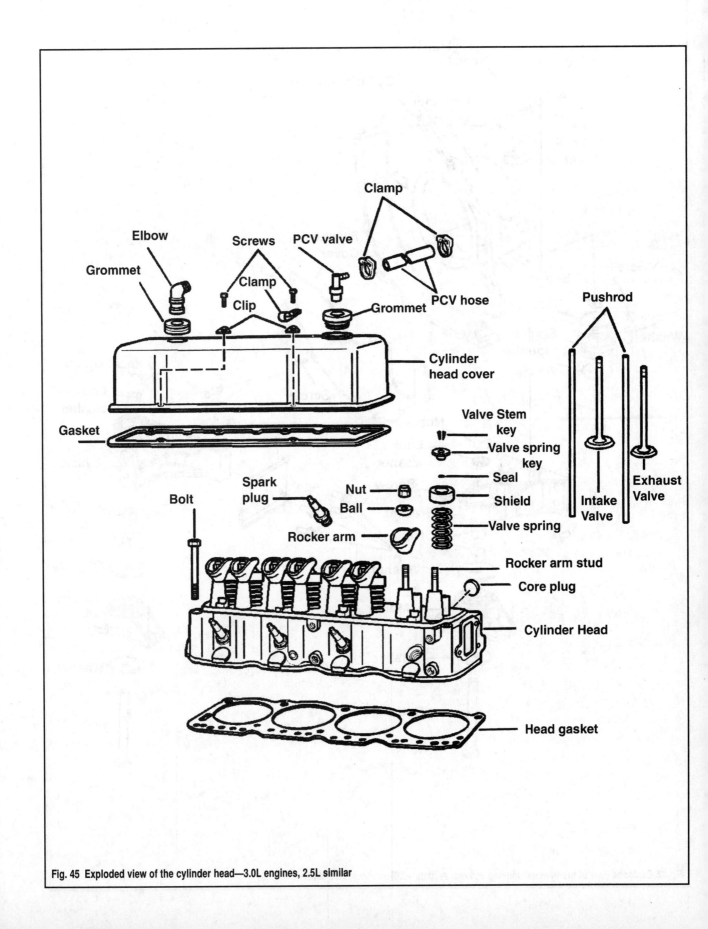

Fig. 45 Exploded view of the cylinder head—3.0L engines, 2.5L similar

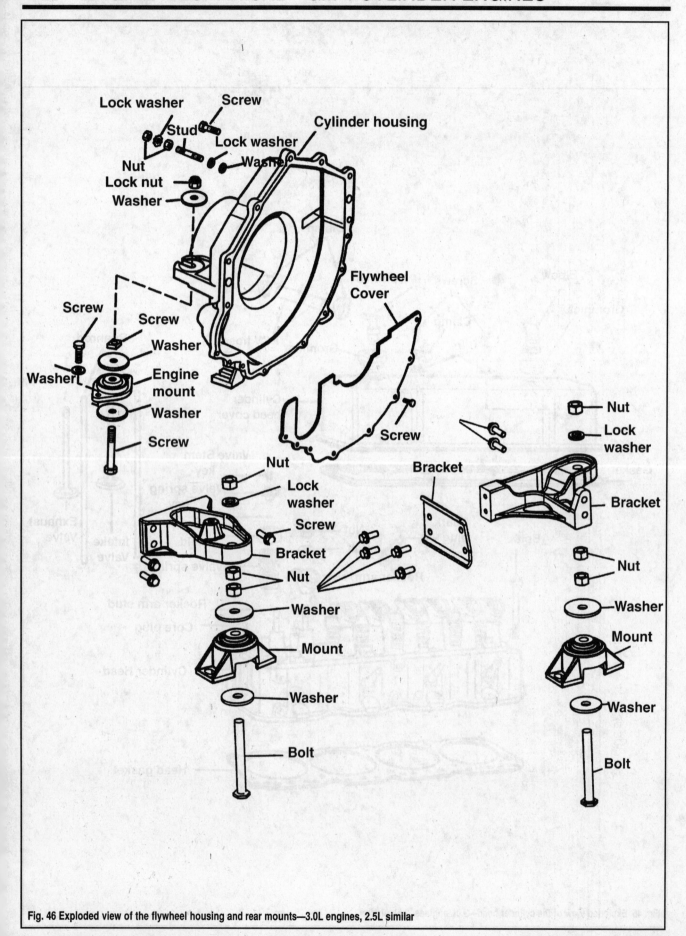

Fig. 46 Exploded view of the flywheel housing and rear mounts—3.0L engines, 2.5L similar

SPECIFICATIONS

TORQUE SPECIFICATIONS

Component		ft. lbs.	inch lbs.	Nm
Alternator bracket to engine		26-30	-	35-41
Camshaft sprocket	Sprocket	①	-	-
	Thrust plate	-	72-90	8-10
Combination manifold	Nut/bolt	20-25	-	27-34
Connecting rod cap		35	-	47
Crankshaft				
Main bearing cap (exc. rear)		60-70	-	81-95
Main bearing cap (rear)				
Step 1		10-12	-	14-16
Step 2 1986-90		60-70	-	82-94
Oil seal retainer 1991-98		-	135	15.3
Cylinder head				
1986-93		90-100	-	122-136
1994-98	Step 1	35	-	47
	Step 2	65	-	88
	Step 3	95	-	129
Cylinder head cover				
1986-93		-	45	5
1994-98		-	65	7.3
Distributor clamp bolt		20	-	27
Engine coupler				
Flywheel housing		28-36	-	38-49
Coupler nut		40-45	-	54-61
Coupler ground stud	1986-93	20-25	-	27-34
	1994-98	35-40	-	47-54
Cover		-	60-84	7-9
Engine mounts	Front			
Bracket bolts		30-35	-	41-47
Upper nut		50-60	-	68-81
Lower nut		115-140	-	156-190
Mounting bolts		48-56	-	65-76
	Port			
	Starboard	32-40	-	43-54
	Rear			
Center bolt	1986-88	18-20	-	24-27
	1989-98	44-52	-	60-71
Lock nut		28-30	-	38-40
Mounting bolts		20-25	-	27-34
Exhaust Elbow		12-14	-	16-19
Exhaust Pipe				
1986-89		10-12	-	14-16
1990-98		27-34	-	27-34
Front cover				
1986-92		6-8	-	9-10
1993-98		-	80	9
Main bearing cap				
Exc. rear		60-70	-	81-95
Rear	Step 1	10-12	-	14-16
	Step 2	60-70	-	81-95
Oil pan				
1/4 x 20 bolts		-	80	9
5/16 x 18 bolts		-	165	18.6
Drain retainer		15-18	-	20-24

① Sprocket is press-fit to camshaft

TORQUE SPECIFICATIONS

Component			ft. lbs.	inch lbs.	Nm
Oil pump	Cover bolts		-	65-75	7-8
	Mounting bolts		-	110-120	12-14
	Pick-up Tube		-	60	7
Oil pressure sender adapter			10-14	-	14-19
Oil seal retainer 1991-98			-	135	15.3
Oil withdrawal tube (flare)			15-18	-	20-24
Power steering pump					
	Bracket	Bolt	26-30	-	35-41
		Stud	13-15	-	18-20
	Hydraulic line fitting	Large	15-17	-	20-23
		Small	10-12	-	14-16
Raw water pump bracket			13-17	-	18-23
Timing chain cover					
1986-92			6-8	-	9-10
1993-98			-	80	9
Valve cover					
1986-93			-	45	5
1994-98			-	65	7.3
Water pump (circulating pump)			13-17	-	18-23
Water temperature sender			20	-	27

① Sprocket is press-fit to camshaft

ENGINE SPECIFICATIONS

Component			Standard (in.)①	Metric (mm)①
Camshaft				
Bearing Length			0.86	21.84
Bearing Outer Diameter	#1, #2		1.999-2.001	50.78-50.83
	#3		2.009-2.011	51.03-51.08
Journal Diameter	1986-93		1.8687	47.48
	1994-98		1.8692	47.47
Lobe Lift			0.253 +/- 0.005	6.43 +/- 0.13
Connecting Rod				
Bearing Clearance	2.5L		0.0007-0.0027	0.018-0.069
	3.0L		0.00085-0.00135	0.022-0.034
End Play			0.008-0.015	0.20-0.38
Length	2.5L		0.797-0.802	20.24-20.37
	3.0L		0.792-0.822	20.12-20.88
Crankshaft				
Crankpin	2.5L		1.999-2.000	50.78-50.80
	3.0L		2.099-2.100	53.32-53.34
Crankshaft End Play			0.002-0.006	0.05-0.15
Main Bearing	Clearance		0.0003-0.0029	0.008-0.074
	Journal Diameter		2.2983-2.2993	58.38-58.40
	Length	#1 thru 4	0.83 ②	21.08 ②
		#5	0.822	20.88
Piston				
Clearance	Limit at top land	2.5L	0.0245-0.0335	0.622-0.851
		3.0L	0.0255-0.0345	0.648-0.876
	Limit at skirt	2.5L	0.0005-0.0015	0.013-0.038
		3.0L	0.0025-0.0035	0.064-0.089
Groove Depth	Compression ring	2.5L	0.200-0.208	5.08-5.28
		3.0L	0.209-0.211	5.31-5.36
	Oil ring	2.5L	0.194-0.202	4.93-5.13
		3.0L	0.190-0.199	4.83-5.06
Piston Pin				
Clearance	Production		0.0003-0.0004	0.008-0.010
	Service Limit		0.001 Max	0.02 Max
Diameter			0.9270-0.9273	23.55-23.56
Interference fit			0.0008-0.0021	0.020-0.050
Length			2.990-3.010	75.95-76.45

① Unless otherwise noted
② 2.5L: 0.802 (20.37)
③ 3/4 to 1 full turn down from zero lash

ENGINE SPECIFICATIONS

Component				Standard (in.)①	Metric (mm)①
Piston Rings					
Compression	Gap	2.5L		0.010-0.020	0.25-0.51
		3.0L	Upper	0.010-0.025	0.25-0.51
			Lower	0.013-0.025	0.33-0.64
	Width			0.0775-0.0780	1.97-1.98
Oil	Gap			0.015-0.055	0.38-1.40
	Width	Groove		0.188-0.189	4.78-4.80
		Ring	2.5L	0.150-0.156	3.81-3.96
			3.0L	0.154-0.160	3.91-4.06
Valve system					
Face Angle				45 deg.	
Lash (hot)				③	③
Lifter				Hydraulic	Hydraulic
Rocker arm ratio				1.75:1	
Seat Angle				46 deg.	
Seat Width	Intake			0.0312-0.0625	0.79-1.59
	Exhaust			0.0625-0.0937	1.59-2.38
Spring (Outer)	Free length			2.08	52.83
	Pressure			78-86 lbs. @ 1.66	347-383 N @ 42.16
Stem Clearance	Intake	2.5L		0.0010-0.0027	0.025-0.069
		3.0L		0.0015-0.0032	0.038-0.081
	Exhaust			0.0010-0.0027	0.025-0.069

① Unless otherwise noted
② 2.5L: 0.802 (20.37)
③ 3/4 to 1 full turn down from zero lash

5

ENGINE MECHANICAL GM V6 AND V8 ENGINES

ENGINE MECHANICAL

General Information

NEVER, NEVER attempt to use standard automotive parts when replacing anything on your engine. Due to the uniqueness of the environment in which they are operated in, and the levels at which they are operated at, marine engines require different versions of the same part; even if they look the same. Stock and most aftermarket automotive parts will not hold up for prolonged periods of time under such conditions. Automotive parts may appear identical to marine parts, but be assured, OMC marine parts are specially manufactured to meet OMC marine specifications. Most marine items are super heavy-duty units or are made from special metal alloy to combat against a corrosive saltwater atmosphere.

OMC marine electrical and ignition parts are extremely critical. In the United States, all electrical and ignition parts manufactured for marine application must conform to stringent U.S. Coast Guard requirements for spark or flame suppression. A spark from a non-marine cranking motor solenoid could ignite an explosive atmosphere of gasoline vapors in an enclosed engine compartment.

V6 ENGINES

The OMC 4.3L, 262 cubic inch displacement V6 engine is manufactured by GMC. This engine is used in numerous models known as the 4.3, 4.3 HO, 4.3GL, 4.3GS and 4.3Gi.

The 4.3 and 4.3GL models are equipped with a 2-barrel carburetor, while the 4.3, 43 HO and 4.3GS models are equipped with a 4-barrel carburetor. Throttle body fuel injection was introduced on the 4.3Gi in 1996.

The lubrication systems and component locations on this engine are virtually identical to the larger V8 engines, except for having only three cylinders in each bank.

A balance shaft is mounted above the camshaft on all 1994 and later models and extends the entire length of the block and is supported on each end by a bearing. The balance shaft is driven by gears on the end of the camshaft and equalizes the dynamic forces known as harmonic vibrations minimizing engine vibration during routine operation of a V6 engine. Cylinder numbering and firing order is identified in the illustrations at the end of the Maintenance Section.

All 4.3L engines are left hand (counterclockwise) rotation when viewed from the stern of the boat. This does not necessarily indicate that your prop rotation is the same— always check them both!

V8 ENGINES

The OMC 5.0L, 305 cubic inch and 5.7L, 350 cubic inch displacement (small block) V8 engines are manufactured by GMC. These engines are used in the following configurations:

- 5.0 (2/4 bbl)
- 5.0GL (2 bbl)
- 5.7 (4 bbl)
- 5.7Gi (TBI)
- 5.7GL (2 bbl)
- 5.7GS (2/4 bbl)
- 5.7GSi (TBI)
- 5.7 LE (4 bbl)
- 350 (4 bbl)

The OMC 7.4L, 454 cubic inch and 8.2L, 502 cubic inch displacement (big block) V8 engines are also manufactured by GMC. These engines are used in the following configurations:

- 7.4 (4 bbl)
- 7.4 454 (4 bbl)
- 7.4 EFI (MPI)
- 7.4GL (4 bbl)
- 7.4Gi (MPI)
- 7.4GSi (MPI)
- 8.2GL (4 bbl)
- 8.2GSi (MPI)
- 454 (4 bbl)
- 454 HO (4 bbl)
- 502 (4 bbl)

The lubrication system is a force fed type where oil is supplied under full pressure to the crankshaft, main and connecting rod bearings, camshaft bearings and the valve lifters. Oil flow from the valve lifters is metered and pumped by the lifter through the hollow core pushrods to lubricate the rocker arms and valve train. All other components are lubricated by gravity and splash methods.

The oil pump is mounted on the rear main bearing cap and is driven by an extension shaft from the distributor—driven by the camshaft. Oil is drawn into the pump through the oil pick-up tube and screen. Should the screen become clogged, a relief valve in the screen will open and allow oil to be drawn into the pump.

Once oil reaches the pump, the pump forces oil through the lubrication system. A spring-loaded relief valve in the oil pump limits the maximum pump output pressure.

The pressurized oil flows out the pump through a full-flow disposable oil filter cartridge. On engines equipped with an oil cooler, the oil flows through the filter, out to the oil cooler via hoses and then returns to the block. Should the oil filter and/or cooler become clogged, a by-pass valve will open allowing the pressurized oil to by-pass the filter and cooler.

Some of the oil is then routed to the No. 5 crankshaft main bearing, the remainder of the oil pressure is routed to the main oil gallery. The main oil gallery is located above the camshaft and runs the full length of the block. Oil from the main gallery is routed through individual passages to the camshaft bearings, No's 1, 2, 3, and 4, crankshaft main bearings and the lifter galley's on each side of the block.

Holes in the camshaft bearings and crankshaft main bearings align with the holes in the block for oil flow. Grooves in the bearings allow oil to flow between the bearing and the component.

Oil in the lifter galleries is forced into each hydraulic lifter through a hole in the side of the lifter. Oil flowing through the lifter must pass through a metering valve in each of the lifters. The metered volume of oil then flows up through the hollow push-rods to the valve rockers. A small hole in the rocker arm allows oil to lubricate the valve train bearing surfaces. All excess oil drains back to the oil pan through oil return holes in the cylinder head.

A baffle plate or "splash pan" mounted below the main bearing caps prevents excess oil being thrown off the crankshaft from aerating the oil in the oil pan.

The distributor shaft and gear is lubricated by oil in the starboard lifter gallery. The timing chain and gears are lubricated with oil flowing out the front of the No. 1 main bearing journal. The mechanical fuel pump and pushrod is lubricated with oil thrown off the camshaft eccentric.

All V8 engines are left hand (counterclockwise) rotation when viewed from the stern of the boat except some inboards which are right hand (clockwise). This does not necessarily indicate that your prop rotation is the same— always check them both!

Cylinder numbering and firing order is identified in the illustrations at the end of the Maintenance Section.

Engine Identification

ENGINE

◆ See Figures 1, 2, 3 and 4

The engine serial numbers are the manufacturer's key to engine changes. These alpha-numeric codes identify the year of manufacture, the horsepower rating and various model/option differences. If any correspondence or parts are required, the engine serial number must be used for proper identification.

Remember that the serial number establishes the year in which the engine was produced, which is often not the year of first installation.

The engine specifications decal contains information such as the model number or code, the serial number (a unique sequential identifier given ONLY to that one engine) as well as other useful information.

An engine specifications decal can generally be found on top of the flame arrestor, on the side of the thermostat housing (early V6/V8 engines), or on the inner side of the rocker arm cover, usually near the breather/PCV line (port side on most models) - all pertinent serial number information can be found here—engine and drive designations, serial numbers and model numbers. Unfortunately this decal is not always legible on older boats and it's also quite difficult to find, so please refer to the following procedures for each individuals unit's serial number location.

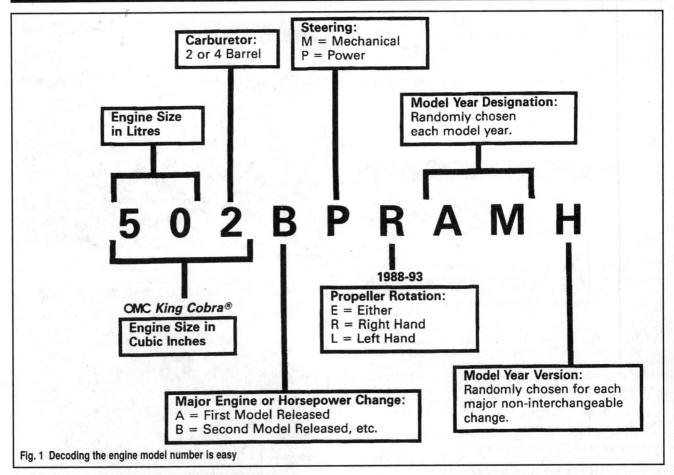

Fig. 1 Decoding the engine model number is easy

■ Serial numbers tags are frequently difficult to see when the engine is installed in the boat; a mirror can be a handy way to read all the numbers.

The engine serial/model number is sometimes also stamped on the port rear side of the engine where it attaches to the bell housing; although on most later models it may instead be a metal plate attached in the same location. If your engine has a stamped number it will simply be the serial number; if you have a plate (and you should), it will always show a Model number and then the actual Serial number. Additionally, most models will also have this plate or sticker on the transom bracket.

• The first two characters identify the engine size in liters (L); **43** represents the 4.3L, **50** represents the 5.0L and so forth.
• The third character identifies the fuel delivery system; **2** designates a 2 bbl carburetor, **4** is a 4 bbl carburetor, and **F** is a fuel injected engine.
• The fourth character designates a major engine or horsepower change—it doesn't let you know what the change was, just that there was some sort of change. **A** means it is the first model released, **B** would be the second, and so forth.

• The fifth character designates what type of steering system was used; **M** would be manual steering and **P** would be power steering.
• Now here's where it gets interesting; on 1986-87 engines and 1994-98 engines, the sixth, seventh and eighth characters designate the model year. The sixth and seventh actually show the model year, while the eighth is a random model year version code. **KWB** and **WXS** represent 1986; and **ARJ, ARF, FTC, SRC** or **SRY** show 1987. **MDA** is 1994, **HUB** is 1995, **NCA** is 1996, **LKD** is 1997 and **BYC** is 1998.
• On 1988-93 engines, the sixth character designates the direction of propeller rotation. **R** is right hand, **L** is left hand and **E** is either.
• Also on 1988-93 engines, the seventh, eighth and ninth characters designate the model year. The seventh and eighth actually show the model year, while the ninth is a random model year version code. **GDE** or **GDP** is 1988, **MED** or **MEF** is 1989, **PWC, PWR** or **PWS** is 1990, **RGD** or **RGF** is 1991, **AMH** or **AMK** is 1992 and **JVB** or **JVN** is 1993.

Any remaining characters are proprietary. So in example, a Model number on the ID plate that reads **574AMFTC** would designate a 1987 5.7L engine with a 4 bbl carburetor and manual steering, first model released. A number reading **74FAPRJVB** would designate a 1993 7.4L engine with fuel injection, power steering and a right hand propeller, first release; get the picture?

Fig. 2 Engine serial number sticker—4.3L engines

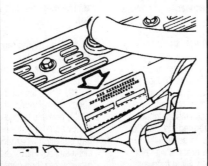

Fig. 3 Engine serial number sticker—V6 and V8 engines

Fig. 4 You should also be able to find a plate on the transom bracket

Engine Model Designations

Most engines covered here utilize unique identifiers assigned by OMC; surnames if you will—4.3GL, 5.0GL, 8.2GSi, etc. Obviously the first two characters designate the engine size in litres (L). The second letter, a **G** or and **F** designate the engine manufacturer; General Motors (G) or Ford (F). The third through fifth letters can be found in different combinations, but the individual letter designates the same thing regardless of position. **L** designates limited output. **S** and **X** designate superior output—a 4.3GL will always have a lesser horsepower rating than a 4.3GS in a given model year; a 5.7GL will be less than a 5.7GS. An **i** designates that the engine is fuel injected, if there is no **i** then you know the engine uses a carburetor.

Engine

REMOVAL & INSTALLATION

◆ **See Figures 5 thru 12**

■ **Prior to removing the engine from your vessel on all 1994-98 models, it is imperative to measure the engine height as detailed in the Determining Minimum Engine Height section. DO NOT remove the engine until you have completed this procedure!**

1. Check the clearance between the front of the engine and the inside edge of the engine compartment bulkhead. If clearance is less than 6 in. (15.2mm), you will need to remove the stern drive unit because there won't be enough room to disengage the driveshaft from the engine coupler. More than 6 in. will provide enough working room to get the engine out without removing the drive, BUT, we recommend removing the drive anyway. If you intend on doing anything to the mounts or stringers, you will need to re-align the engine as detailed in the Engine Alignment section—which requires removing the drive, so remove the drive!
2. Remove the stern drive unit as detailed in the Drive Systems section.
3. Open or remove the engine hatch cover.
4. Disconnect the battery cables (negative first) at the battery and then disconnect them from the engine block and starter.

✲✲ CAUTION

Make sure that all switches and systems are OFF before disconnecting the battery cables.

5. Disconnect the two power steering hydraulic lines at the steering cylinder (models w/ps). Carefully plug them and then tie them off somewhere on the engine, making sure that they are higher then the pump to minimize any leakage.
6. Disconnect the fuel inlet line at the fuel pump or filter (whichever comes first on your particular engine) and quickly plug it and the inlet— a clean golf tee and some tape works well in this situation. Make sure you have rags handy, as there will be some spillage.
7. Tag and disconnect the two-wire trim/tilt connector.
8. Pop the two-wire trim/tilt sender connector out of the retainer and then disconnect it. You may have to cut the plastic tie securing the cable in order to move it out of the way.
9. Locate the large rubber coated instrument cable connector (should be on the starboard side), loosen the hose clamp and then disconnect it from the bracket. Move it away from the engine and secure it. On early models, you will also need to unplug the three-wire trim/tilt cable connector just above it.

■ **Take note of you throttle arm attachment stud— is it a "push-to-close" or a "pull-to-close"? What hole is it on??**

10. Remove the cotter pin and washer from the throttle arm. Loosen the anchor block retaining nut and then spin the retainer away from the cable trunnion. Remove the throttle cable from the arm and anchor bracket. Be sure to mark the position of the holes that the anchor block was attached to.
11. Loosen the 4 hose clamps and disconnect the intermediate exhaust hose bellows from intermediate pipe. You may want to spray some WD-40 around the lip of the hose where it connects to the elbow, grasp it with both hands and wiggle it back and forth while pulling down on it. Slide it down over the lower pipe.

Fig. 5 Disconnect the power steering lines at the cylinder

Fig. 6 Disconnect the fuel inlet line

12. Drain the cooling system as detailed earlier in this section.
13. Loosen the hose clamp on the water supply hose at the transom bracket and carefully slide it off the water tube. Attach the hose to the engine. On big block engines, disconnect the water line at the oil cooler on the rear of the block.
14. Disconnect the shift cables and position them out of the way.
15. Tag and disconnect any remaining lines, wires or hoses at the engine.
16. Attach a suitable engine hoist to the lifting eyes and take up any line slack until it is just taught.

■ **The engine hoist should have a capacity of at least 1500 lbs. (680 kg).**

17. Locate the rear engine mounts and remove both lock nuts and flat washers.
18. Locate the front engine mounts and remove the two (per mount) lag bolts.
19. If you listened to us at the beginning of the procedure, the drive unit should be removed. If so, slowly and carefully, lift out the engine. Try not to hit the power steering control valve, or any other accessories, while removing it from the engine compartment. If you didn't listen to us, and you had sufficient clearance in the engine compartment, the drive unit is probably still installed. Raise the hoist slightly until the weight is removed from the mounts and then carefully pull the engine forward until the driveshaft disengages from the coupler, now raise the engine out of the compartment.

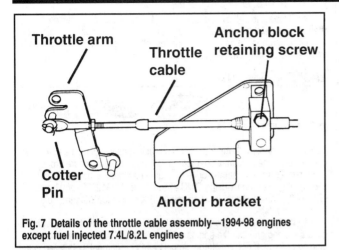

Fig. 7 Details of the throttle cable assembly—1994-98 engines except fuel injected 7.4L/8.2L engines

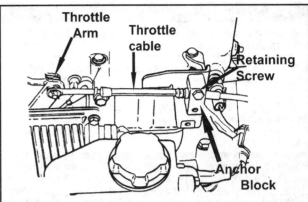

Fig. 8 Details of the throttle cable assembly—1994-98 7.4Gi/GSi and 8.2GSi engines

To install:

20. Apply Engine Coupler grease to the splines of the coupler.

21. Slowly lower the engine into the compartment. If the drive unit was not removed, AND the crankshaft has not been rotated, insert the driveshaft into the coupler as you push the engine backwards until they engage completely and then lower the engine into position over the rear mounts until the front mounts just touch the stringers. If the shaft and coupler will not align completely, turn the crankshaft or driveshaft slightly until they mate correctly.

If the drive was removed, or the mounts were disturbed in any way, lower the engine into position over the rear mounts until the front mount just touches the stringer.

22. Install the two flat washers into the recess in the engine bracket side of the rear mounts and then install the two lock nuts. Tighten them to 28-30 ft. lbs. (38-40 Nm).

✳✳ CAUTION

Never use an impact wrench or power driver to tighten the locknuts.

23. Install the lag bolts into their holes on the front mounts and tighten each bolt securely.

24. If the drive was removed, the mounts were disturbed or the driveshaft/coupler didn't mate correctly, perform the engine alignment procedure detailed in this section. We think it's a good idea to do this regardless!

25. Reconnect the exhaust bellows by sliding it up and over the pipe, position the clamps between the ribs in the hose and then tighten the clamps securely. Make sure you don't position the clamps into the expanding area.

26. Reconnect the water inlet hose. Lubricate the inside of the hose and wiggle it onto the inlet tube. Slide the clamp over the ridge and tighten it securely. This sounds like an easy step, but it is very important—if the hose, particularly the underside, is not installed correctly the hose itself may collapse or come off. Either scenario will cause severe damage to your engine, so make sure you do this correctly!

Fig. 9 Locate the engine lifting brackets...

27. Carefully, and quickly, remove the tape and plugs so you can connect the power steering lines. Tighten the large fitting to 15-17 ft. lbs. (20-23 Nm) and the small fitting to 10-12 ft. lbs. (14-16 Nm). Don't forget to check the fluid level and bleed the system when you are finished with the installation.

28. Reconnect the trim/tilt connector so the two halves lock together.

29. Reconnect the trim position sender leads, the instrument cable, the engine ground wire, the battery cables and all other wires, lines of hoses that were disconnected during removal. Make sure you swab a light coat of grease around the fitting for the large engine/instrument cable plug.

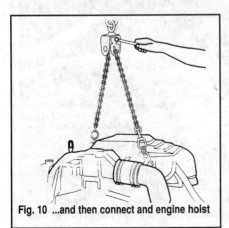

Fig. 10 ...and then connect and engine hoist

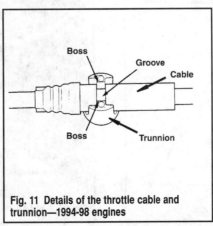

Fig. 11 Details of the throttle cable and trunnion—1994-98 engines

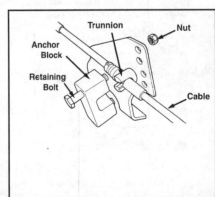

Fig. 12 Attach the anchor block assembly to the bracket—1994-98 engines

✳✳ CAUTION

Always make certain that all switches and systems are turned OFF before reconnecting the battery cables.

■ Make sure all cables, wires and hoses are routed correctly before initially starting the engine.

30. Unplug the fuel line and pump/filter fitting and reconnect them. Remember to check for leaks as soon as you start the engine.

31. Install and adjust the throttle cable. For complete details, please refer to the Fuel System section):

a. Remember we asked you to determine if you had a "push-to-close" or "pull-to-open" throttle cable (the throttle arm stud)? Position the remote control handle in Neutral—the propeller should rotate freely.

b. Turn the propeller shaft and the shifter into the forward gear detent position and then move the shifter back toward the Neutral position halfway.

c. Position the trunnion over the groove in the throttle cable so the internal bosses align and then snap it into the groove until it is fully seated.

d. Install the trunnion/cable into the anchor block so the open side of the trunnion is against the block. Position the assembly onto the bracket over the original holes (they should be the lower two of the four holes) and then install the retaining bolt and nut. When the nut is securely against the back of the bracket, tighten the bolt securely.

e. Install the connector onto the throttle cable and then pull the connector until all end play is removed from the cable. Turn it sideways until the hole is in alignment with the correct stud on the throttle arm. Slide it over the stud and install the washer and a new cotter pin. Make sure the cable is on the same stud that it was removed from. Tighten the jam nut against the connector

■ The throttle arm connector nut must be installed on the cable with a minimum of 9 turns—meaning that at least 1/4 in. of thread should be showing between the end of the cable and the edge of the nut.

32. Install and adjust the shift cables. Please refer to the Drive Systems section for further details.

33. Check and refill all fluids. Start the engine and check for any fuel or coolant leaks. Go have fun!

DETERMING MINIMUM ENGINE HEIGHT

1994-98 Engines Only

◆ See Figure 13

This procedure MUST be performed prior to removing the engine from the vessel.

1. With the engine compartment open, position a long level across the transom running fore and aft.

2. Have a friend or assistant steady the level while you measure from the bottom edge of the tool to the top of the exhaust elbow. Record the distance as "**1**".

3. Now measure from the bottom of the level to the static water line on the drive unit. Record the distance as "**2**".

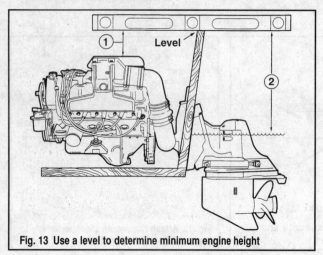

Fig. 13 Use a level to determine minimum engine height

4. Subtract the elbow measurement (**1**) from the static waterline measurement (**2**). If the result is less than 13 in. (330mm) on V6 engines or 14 in. (356mm) on V8 engines, an exhaust elbow high rise kit must be installed (available from your local parts supplier).

ENGINE ALIGNMENT

◆ See Figures 14 and 15

Engine alignment is imperative for correct engine installation and also for continued engine and drive operation. It is a good idea to ensure proper alignment every time that the drive or engine has been removed. Engine alignment is checked by using OMC alignment tool (#912273) and handle (#311880). Engine alignment is adjusted by raising or lowering the front engine mount(s).

1. With the drive unit off the vessel, slide the alignment tool through the gimbal bearing and into the engine coupler. It should slide easily, with no binding or force. If not, check the gimbal bearing alignment as detailed in the Drive Systems section. If bearing alignment is correct, move to the next step.

2. If your engine utilizes a jam nut on the bottom of the mount bolt, loosen it and back it off at least 1/2 in.

3. Loosen the lock nut and back it off.

4. Now, determine if the engine requires raising or lowering to facilitate alignment—remember, the alignment tool should still be in position. Tighten or loosen the adjusting nut until the new engine height allows the alignment tool to slide freely.

5. Hold the adjusting nut with a wrench and then tighten the lock nut to 100-120 ft. lbs. (136-163 Nm). If your engine uses a jam nut, cinch it up against the lock nut.

6. Remove the alignment tool and handle.

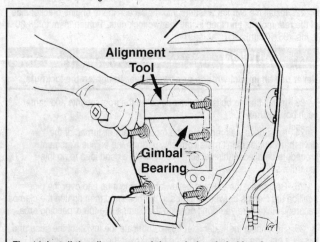

Fig. 14 Install the alignment tool through the gimbal bearing

Fig. 15 A good view of the front engine mount and its components

Front Engine Mounts

REMOVAL & INSTALLATION

◆ **See Figures 15 and 16**

 *DIFFICULT*

1. Position an engine hoist over the engine and hook it up to the two engine lifting eyes.

2. Remove the two lag screws/bolts on each side of the mount where it rests on the stringer.

3. Raise the engine just enough to allow working room for removing the mount.

4. Remove the three mount-to-engine mounting bolts with their lock washers and lift out the mount. One of the bolts is hidden under the bracket, behind the mount bolt.

■ **Some engines may use bolts of different lengths; mark them so you can replace them in their original locations.**

5. Measure the distance between the top of the large washer on the mount and the flat on the lower side of the mounting bracket. Record it.

6. Position a wrench over the bottom nut on the adjusting bolt, just underneath the mount, to hold the shaft and then remove the top nut. Lift off the bracket.

7. Remove the two lock nuts from the bolt and slide out the adjusting bolt. Remember which washer goes where.

To install:

8. Slide the adjusting bolt up through the mount and then position the small and large washers over the bolt—large over small.

9. Spin on the first (lower) lock nut and tighten it to 60-75 ft. lbs. (81-102 Nm).

10. Screw on the upper lock nut and then position the mount bracket over the bolt. Install the washer and adjusting nut and check the measurement taken in Step 5. Move the upper lock nut up or down until the correct specification is achieved and then tighten the adjusting nut to 100-120 ft. lbs. (136-163 Nm) on V6 and 5.0L/5.7L engines; 50-70 ft. lbs. (68-95 Nm) on 7.4L and 8.2L engines.

11. Spray the three mounting bolts with Loctite Primer N and allow them to air dry. Once dry, coat the bolts with Loctite or OMC Thread Sealing Agent and attach the mount to the engine. Tighten the bolts to 32-40 ft. lbs. (43-54 Nm).

12. Position the mount over the lag screw holes and then slowly lower the engine until all weight is off the hoist. Install and tighten the lag screws securely.

13. If you're confident that your measurements and subsequent adjustment place the engine exactly where it was prior to removal, then you are through. If you're like us though, you may want to check the engine alignment before you fire up the engine.

Rear Engine Mounts

REMOVAL & INSTALLATION

 *DIFFICULT*

◆ **See Figure 17**

1. Remove the engine as detailed previously.

2. Loosen the two bolts and remove the mount from the transom plate. Remember, you will already have removed the top lock nut and the washer during engine removal.

3. Hold the square nut with a wrench and remove the shaft bolt. Be sure to take note of the style and positioning of the two mount washers as you are removing the bolt. Mark them, lay them out, or write it down, but don't forget their orientation!!

To install:

4. Slide the lower of the two washers onto the mount bolt, exactly as it came off.

5. Slide the bolt into the flat (bottom) side of the rubber mount, install the remaining washer (as it came off!) and then spin on the square nut. Do not tighten it yet.

■ **Incorrect washer installation will cause excessive vibration during engine operation.**

6. Turn the assembly upside down and clamp the square nut in a vise. Spin the assembly until the holes in the mounting plate are directly opposite any two of the flat sides on the nut. This is important, otherwise the slot on the engine pad will not engage the mount correctly. Secure the mount in this position and tighten the bolt to 18-20 ft. lbs. (24-27 Nm) on 1986-88 engines, or 44-52 ft. lbs. (60-71 Nm) on 1989-98 engines.

7. Remove the mount from the vise and position it on the transom plate. Install the bolts and washers and tighten each to 20-25 ft. lbs. (27-34 Nm).

8. Install the engine, making sure that the slot in the engine pad engages the square nut correctly. Install the two washers and locknut and tighten it to 28-30 ft. lbs. (38-41 Nm).

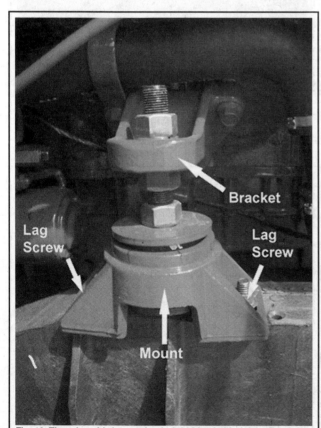

Fig. 16 There is a third mounting bolt hidden behind the adjusting bolt in this shot

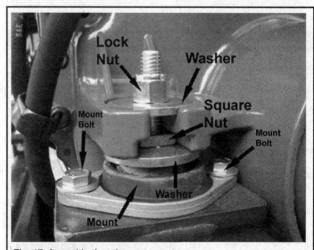

Fig. 17 A good look at the rear mount

Cylinder Head (Valve) Cover

REMOVAL & INSTALLATION

◆ See Figures 18, 19 and 20

■ In order to perform this procedure efficiently, we recommend removing the exhaust manifold in order to have sufficient working room to remove the cylinder head cover. Although not completely necessary, it's worth the extra effort to avoid the aggravation of working around the manifolds. Please refer to the manifold procedure later in this section.

1. Open or remove the engine compartment hatch. Disconnect the negative battery cable.
2. Loosen the clamp and remove the crankcase ventilation hose at the cover. Carefully move it out of the way.
3. Tag and disconnect any lines, leads or hoses that might be in the way of removal. In some instances you may be able to simply secure them out of the way without disconnecting them—you be the judge.
4. If your engine has a spark plug wire retainer attached to the cover, unclip the wires or remove the retainer and then tag and disconnect the plug wires at the spark plugs.
5. Loosen the cover mounting bolts (usually three on the V6, four on the 5.0L/5.7L and seven or eight on the 7.4L/8.2L) and lift off the cylinder head cover. Take note of any harness or hose retainers and clips that might be attached to certain of the mounting bolts; you need to make sure they go back in the same place.

To install:

6. Clean the cylinder head and cover mounting surfaces of any residual gasket material with a scraper or putty knife.
7. Position a new gasket on the cylinder head and then position the cover (don't forget the J-clips!). Tighten the mounting bolts to:
   • 100 inch lbs. (11.3 Nm) on 1986-93 4.3L engines
   • 62-115 inch lbs. (7-13 Nm) on 1994-96 4.3L engines
   • 106 inch lbs. (12 Nm) on 1997-98 4.3L engines
   • 45 inch lbs. (5 Nm) on 1986-96 5.0L/5.7L engines
   • 65 inch lbs. (7.3 Nm) on 1991 350 engines
   • 65 inch lbs. (7.3 Nm) on 1992 5.7 LE engines
   • 90 inch lbs. (10 Nm) on 1997 5.0L/5.7L engines (except the 1997 5.7GSi)
   • 106 inch lbs. (12 Nm) on 1997 5.7GSi and 1998 V8 engines
   • 115 inch lbs. (13 Nm) on 7.4L/8.2L engines, except 1996-97 and 1998 7.4Gi
   • 60-90 inch lbs. (7-10 Nm) on 1996-97 7.4L/8.2L engines
   • 72 inch lbs. (8 Nm) on 1998 7.4Gi engines
   Make sure any retainers or clips that were removed are back in their original positions.
8. Connect the crankcase ventilation hose and any other lines or hoses that may have been disconnected. Check that there were no other wires or hoses you may have repositioned in order to gain access to the cover.
9. Install the exhaust manifold.
10. Connect the battery cables.

Rocker Arms and Push Rods.

REMOVAL & INSTALLATION

1992-98 4.3L V6 and 1992-98 7.4L/8.2L V8 Engines

◆ See Figure 21

1. Open or remove the engine hatch cover and disconnect the negative battery cable. Remove the cylinder head cover as detailed previously.
2. Bring the piston in the No. 1 cylinder to TDC. If servicing only one arm, bring the piston in that cylinder to TDC. The No. 1 cylinder is the first cylinder at the port side of the engine.
3. Loosen and remove the rocker arm nut on the V6, shoulder bolt on the V8. Lift out the ball.
4. Lift the arm itself off of the mounting stud (or boss) and pull out the pushrod. On the V8, lift off the pushrod guide before removing the pushrod if it didn't come off with the rocker. It is very important to keep each cylinder's component parts together as an assembly. We suggest drilling a set of holes in a 2X4 and positioning the pieces in the holes.

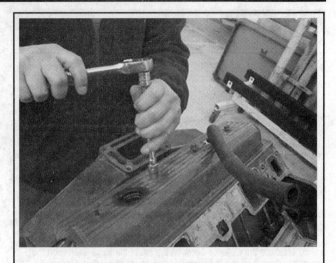

Fig. 18 Remove the bolts (5.7L shown)...

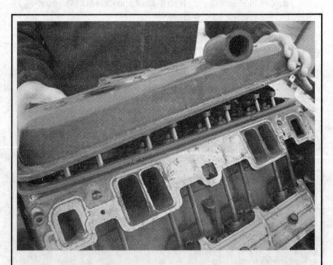

Fig. 19 ...and then lift off the cover

Fig. 20 Position a new gasket on the head

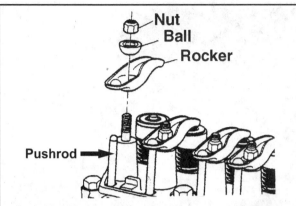

Fig. 21 Exploded view of the rocker arm assembly (most V8 engines use a shoulder bolt)

To install:

5. Clean and inspect the rocker assemblies. If any scuffing, wear or obvious deterioration is found replace the entire assembly (rocker, ball and pushrod, and guide if equipped). Roll each push rod across a flat, even surface (countertops work great for this); if it does not roll smoothly, replace it.

6. Coat all bearing surfaces of the rocker assembly with engine oil.

7. Slide the push rods into their holes. Make sure that each rod seats in its socket on the lifter.

8. On V8 engines, position the guide over the rod so the other side is sitting on the boss.

9. Position the rocker arm over the stud (boss) so that the cupped side rides on the push rod. Slide the ball over the stud, install the nut or bolt and tighten it to:

- 40 ft. lbs. (54 Nm) on 1992-96 V6 engines
- 20 ft. lbs. (27 Nm) on 1997-98 V6 engines
- 40 ft. lbs. (54 Nm) on 1992-98 V8 engines, except 1998 7.4Gi
- 45 ft. lbs. (61 Nm) on 1998 7.4Gi engines

10. No additional adjustment of the valves is necessary as the lash is set automatically when the rocker is tightened to specifications.

11. Install the cylinder head cover, connect the battery cable and check the idle speed.

1986-91 4.3L V6, 1986-98 5.0L/5.7L V8, 1990-91 7.4L V8 Engines

◆ **See Figures 21, 22 and 23**

1. Open or remove the engine hatch cover and disconnect the negative battery cable. Remove the cylinder head cover as detailed previously.

2. Bring the piston in the No. 1 cylinder to TDC. If servicing only one arm, bring the piston in that cylinder to TDC. The No. 1 cylinder on is the first cylinder at the port side of the engine.

3. Loosen and remove the rocker arm nuts and lift out the balls.

4. Lift the arm itself off of the mounting stud and pull out the pushrod. It is very important to keep each cylinder's component parts together as an assembly. We suggest drilling a set of holes in a 2X4 and positioning the pieces in the holes.

■ **On certain engines, the exhaust valve pushrods are longer then those for the intake valve.**

To install:

5. Clean and inspect the rocker assemblies, particularly the balls where they mate with the rockers.

6. Coat all bearing surfaces of the rocker assembly with engine oil.

7. Slide the push rods into their holes. Make sure that each rod seats in its socket on the lifter.

8. Position the rocker arm over the stud so that the cupped side rides on the push rod. Slide the ball over the stud, install the nut and tighten it until zero lash is present.

9. Adjust the valves as detailed in the Valve Adjustment section.

10. Install the cylinder head cover, connect the battery cable and check the idle speed.

VALVE ADJUSTMENT

1986-91 4.3L V6, 1986-98 5.0L/5.7L V8, 1988-91 7.4L V8 And 1998 7.4L/8.2L V8 Engines

◆ See Figure 24

These engines utilize hydraulic valve lifters, although there is no need for periodic valve adjustment, it is necessary to perform a preliminary adjustment after any work on the valve train/rocker assembly. All adjustment should be undertaken while the lifter is on the base circle of the camshaft lobe for that particular cylinder. This means the opposite side of the pointy part of each lobe.

1. Rotate the crankshaft, or bump the engine with the starter until the No. 1 cylinder is at TDC. Note that the notch or mark on the damper pulley will be lined up with the **0** mark on the timing scale. Be careful here though, this could mean that either the No. 1 or the No. 6 (No. 4, V6) piston is at TDC. Place your hand on the No. 1 cylinder's valve and check that it does not move as the mark on the pulley is approaching the **0** mark on the tab. If it does not move, you're ready to proceed; if it does move, you are on the No. 6 (No. 4) cylinder and need to rotate the engine an additional full turn. This is important so make sure you've gotten it right!

2. Now that the No. 1 cylinder is at TDC, you can adjust the following valves on V6 engines:

- No. 1 cylinder: intake and exhaust
- No. 2 cylinder: intake
- No. 3 cylinder: intake
- No. 5 cylinder: exhaust
- No. 6 cylinder: exhaust

Or these valves on V8 engines:

- No. 1 cylinder: intake and exhaust
- No. 2 cylinder: intake
- No. 3 cylinder: exhaust

Fig. 22 Wouldn't it be nice if your valve train looked this good?

Fig. 23 Close-up of the rocker assembly

- No. 4 cylinder: exhaust
- No. 5 cylinder: intake
- No. 7 cylinder: intake
- No. 8 cylinder: exhaust

3. Loosen the adjusting nut on the rocker until you can feel lash (play in the push rod) and then tighten the nut until the lash has been removed. Carefully jiggle the push rod while tightening the nut until it won't move anymore—this is zero lash. Tighten the nut an additional full turn (3/4 turn for the 1988-91 big block) to set the lifter and then you're done. Perform this procedure on each of the valves listed above.

4. Slowly rotate the engine an additional full turn and this will bring the No. 6 (No. 4 on V6) piston to TDC. The pulley notch/mark should once again be in line with the **0** on the timing tab. You can now adjust the remaining valves as you just did on the first pass:
- No. 2 cylinder: exhaust
- No. 3 cylinder: intake
- No. 4 cylinder: intake and exhaust
- No. 5 cylinder: intake
- No. 6 cylinder: intake

Or these valves on V8 engines:
- No. 2 cylinder: exhaust
- No. 3 cylinder: intake
- No. 4 cylinder: intake
- No. 5 cylinder: exhaust
- No. 6 cylinder: intake and exhaust
- No. 7 cylinder: exhaust
- No. 8 cylinder: intake

1992-98 4.3L V6, 1992-97 7.4L/8.2L V8 And 1998 7.4Gi V8 Engines

■ No initial valve lash adjustment is necessary on these engines; net lash is set when tightening the rocker arm nut or shoulder bolt to the proper torque. The rocker studs on engines with nuts are equipped with 'positive stop' shoulders.

Hydraulic Valve Lifter

REMOVAL & INSTALLATION

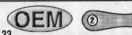

◆ See Figures 25 thru 33

■ Although all engines covered here are equipped with hydraulic lifters, some utilize conventional flat lifters, while others will be equipped with roller lifters. Procedures for each type are different, yet similar enough that we will cover them both within this procedure.

1. Using compressed air, thoroughly clean all dirt and grit from the cylinder head and related components.

✸✸ WARNING

If compressed air is not available, we highly recommend that you DO NOT proceed with this procedure. It is EXTREMELY important that no dirt gets into the lifter recesses before completing the installation.

2. Remove the cylinder head cover.
3. Remove the intake and exhaust manifolds.
4. Loosen the rocker arms and pivot the rockers off of the pushrods; or just remove them completely.
5. Remove the pushrods from the block, being very careful to keep track of where each one came from.
6. Remove the splash shield on 1998 7.4Gi engines.
7. On V6 and certain later V8 engines, loosen and remove the mounting bolts (two or four) and then lift off the lifter guide retainer, with the restrictors/guides. It is a good idea to mark the front side of the retainer for proper installation later on.

■ Not all engines utilize a guide retainer or restrictors/guides in all years, but it will be obvious whether or not your particular engine has them.

8. Remove the lifters. Although they should come right out, you may want to use a magnet, or in some cases on later engines you will need to use a lifter extractor tool (#J9290-1 or J-3049-A) to extract each lifter. On engines with roller lifters and lifter guides, it is important that the two match

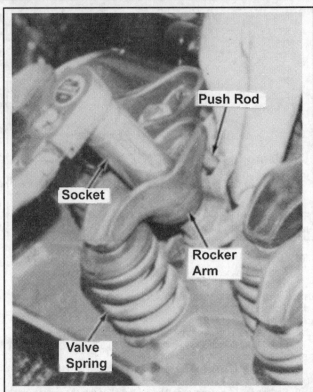

Fig. 24 Wiggle the push rod slowly while tightening the rocker arm adjusting nut

each other and that the roller rolls in the same direction as when it was removed; for this reason we think it's a good idea to match mark each lifter to its guide prior to removing it.

■ Once again, we suggest using a 2x4 with holes drilled in it to store the rods and lifters; you'd be amazed at how quickly this job will fall apart if someone walks in and kicks the components that you have laid out on the floor!

To install:

9. Clean all components thoroughly and let dry completely, use compressed air if at all possible.
10. Make sure that the push rod oil passages are clean and clear. Inspect the camshaft contact surface on the bottom of each lifter for excessive wear, galling or other damage. Discard the lifter if any of the above conditions are found. You'll probably want to check out the camshaft lobe for damage also.
11. Coat the bottom of each flat lifter with Molykote®, but on models with roller lifters use GM Engine Oil Supplement, and then carefully install each one into its respective recess.
12. Install the retainer, with restrictors/guides if equipped, and tighten the bolts to 12 ft. lbs. (16 Nm) on V6 engines, 18 ft. lbs. (25 Nm) on 5.0L/5.7L V8 engines and 20 ft. lbs. (26 Nm) on the big blocks. Make sure that the marks you made across the roller lifters and their guides match up and that the mark you made on the retainer is toward the front of the engine!!

■ On engines with restrictors/guides, the retainer MUST contact ALL guides. If bent, and it does not make contact, replace it with a new one. DO NOT attempt to bend it back into position.

13. Install the pushrods into the sockets on the lifters. Install the splash shield on the Gi.
14. Move the rockers into position and then tighten the nut as detailed in your engine's respective Rocker Arm procedure as detailed previously.
15. Install the intake and exhaust manifolds.
16. Install the cylinder head cover.

✸✸ CAUTION

If any, or all, of the lifters has been replaced with a new one it is very important that you add GM Engine Oil Supplement to the crankcase BEFORE starting the engine.

Fig. 25 A good look at the lifter, retainer and push rod—1986-96 4.3L V6 engines

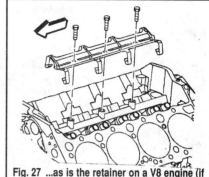

Fig. 26 The lifter retainer on 1997-98 V6 models is slightly different...

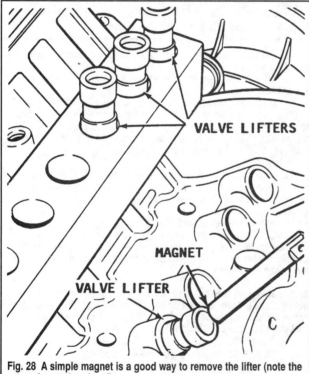

Fig. 27 ...as is the retainer on a V8 engine (if equipped)

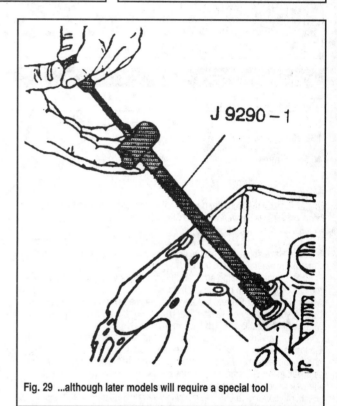

Fig. 28 A simple magnet is a good way to remove the lifter (note the set-up for storing them!)...

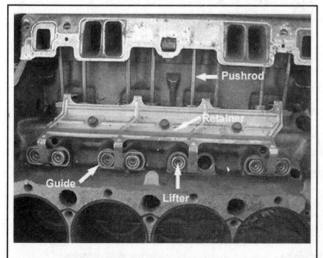

Fig. 29 ...although later models will require a special tool

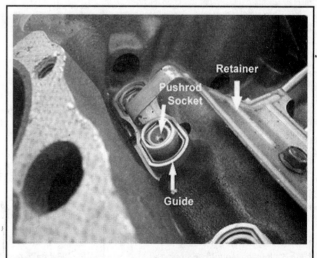

Fig. 30 Valve lifter retainer and guides on a V8 (V6 similar)

Fig. 31 Notice the guide and the pushrod socket in the lifter

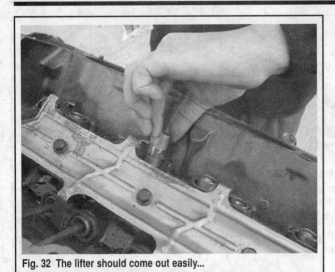

Fig. 32 The lifter should come out easily...

Fig. 33 ...and look like this (roller lifter shown)

Intake Manifold

REMOVAL & INSTALLATION

4.3L V6 Engines  MODERATE

◆ See Figures 34 thru 40

1. Open or remove the engine compartment hatch. Disconnect the negative battery cable.
2. Loosen the clamp and remove the crankcase ventilation or PCV hoses at the cylinder head covers. Carefully move them out of the way.
3. Drain all water from the cylinder block and manifolds.
4. Tag and disconnect all water hoses at the manifold and thermostat housing. Have some rags handy, since there will still be some water in them. Carefully move them out of the way.
5. Remove the flame arrestor and then disconnect the throttle cable at the carburetor/throttle body anchor block.
6. Disconnect the fuel line and plug the line end and the carb/throttle body. If you have a non-flexible line, disconnect it at the fuel pump also. Move both the cable and fuel line out of the way.
7. Tag and disconnect the lead at the temperature gauge sender unit. Do the same with the harness connected to the rear of the alternator. Tag and disconnect any other lines, leads or hoses that might be in the way of removal. In some instances you may be able to simply secure them out of the way without disconnecting them—you be the judge.
8. If your engine has a spark plug wire retainer attached to the cylinder head cover, unclip the wires or remove the retainer and then tag and disconnect the plug wires at the spark plugs.
9. Remove the distributor cap with the leads still connected. Mark the position of the distributor rotor to the distributor body. Scribe a matchmark across the distributor and the manifold. Loosen the clamp bolt and remove the distributor. Please refer to the Electrical section for further details on distributor removal. DO NOT turn the engine over once the distributor has been removed.
10. Remove the alternator and its mounting bracket.

11. Tag and disconnect any leads at the ignition module and move them out of the way.
12. Tag and disconnect the wire at the oil sending unit and then remove the unit itself.
13. Loosen and remove the manifold mounting bolts in the reverse order of the illustrated tightening sequence and then remove the manifold. Don't forget the solenoid bracket. There is a good likelihood you will need to pry the manifold off the block; be very careful that you don't scratch or mar the mating surfaces on the block, manifold or heads.
   To install:
14. Carefully remove all remaining gasket material from the manifold mating surfaces with a scraper or putty knife. Be careful that you don't accidentally drop any old gasket into the crankcase or intake ports on the cylinder head.
15. Inspect the manifold and all mating surfaces for any cracks or nicks.
16. On 1986-96 engines, position new seals on the cylinder block mating surfaces. Position new gaskets on the cylinder head mating surfaces. Use GM Silicone Rubber Sealer at all water passages and/or wherever a seal butts against a gasket.

■ Some engines may not have front or rear cylinder block seals. If yours is one of them, apply a 3/16 in. (5mm) bead of RTV sealant to the forward and aft edges of the cylinder block mating surface. Make sure that you run the bead at least a 1/2 in. up onto the gaskets.

17. On 1997-98 engines, coat the cylinder head side of the gasket with RTV sealant. DO NOT apply too much sealer—see the illustration. Position the gaskets onto the heads over the locater pins.
18. Apply a 1/2 in. (13mm) bead of RTV sealant to the forward and aft edges of the cylinder block mating surface. Make sure that you run the bead at least a 1/2 in. up onto the gaskets.
19. On all engines, install the manifold into place so that all the bolt holes line up, insert the bolts and tighten them to:
  • 30 ft. lbs. (41 Nm) on 1986-95 engines
  • 35 ft. lbs. (48 Nm) on 1996 engines
  • 11 ft lbs. (15 Nm) on 1997-98 engines. This is final torque, so be sure you check the Specifications chart for the first two stages.

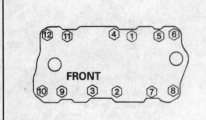

Fig. 34 Intake manifold tightening sequence—1986-95 engines

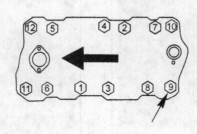

Fig. 35 Intake manifold tightening sequence—1996 engines

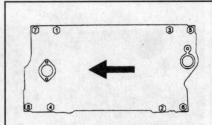

Fig. 36 Intake manifold tightening sequence—1997-98 engines

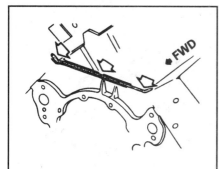

Fig. 37 Make sure to use new seals (or sealer) along the cylinder block-to-manifold mating surfaces—1986-96 engines

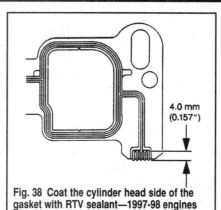

Fig. 38 Coat the cylinder head side of the gasket with RTV sealant—1997-98 engines

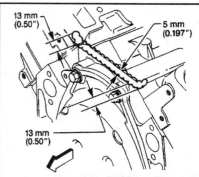

Fig. 39 Run a bead of sealant as shown on the forward block-to-manifold mating surface—1997-98 engines

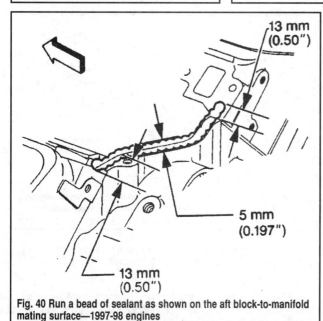

Fig. 40 Run a bead of sealant as shown on the aft block-to-manifold mating surface—1997-98 engines

Check the Torque Specifications chart for any further details on tightening needs. Tighten all bolts in the order shown in the illustration. On all 1997-98 engines, coat all eight bolts with thread locking sealant prior to installing them; remember also that these engines utilize a three step tightening process.

20. Install the oil sending unit and connect the wire.

21. Connect the ignition module leads and install the distributor as detailed in the Electrical section. Put the cap back on and reconnect the plug wires to the spark plugs.

22. Install all water hoses and tighten their clamps securely.

23. Install the fuel line at the carburetor/throttle body and fuel pump. Make sure you remove any plugs you may have inserted on removal.

24. Install the throttle cable and adjust it as detailed in the Fuel System section. Install the flame arrestor.

25. Connect all other lines, leads and hoses that may have been removed to facilitate manifold removal.

26. Install the alternator and mounting bracket. Reconnect the harness.

27. Connect the battery cable and start the engine. Check the ignition timing and idle speed. Check all hoses and seals for leaks.

5.0L And 5.7L V8 Engines

◆ See Figures 41, 42, 43, 44, 45 and 46

1. Open or remove the engine compartment hatch. Disconnect the negative battery cable.

2. Loosen the clamp and remove the crankcase ventilation or PCV hoses at the cylinder head covers. Carefully move them out of the way.

3. Drain all water from the cylinder block and manifolds.

4. Tag and disconnect all water hoses at the manifold and thermostat housing. Have some rags handy, since there will still be some water in them. Carefully move them out of the way.

5. Remove the flame arrestor and then disconnect the throttle cable at the carburetor/throttle body anchor block.

6. Disconnect the fuel line and plug the line end and the carb/throttle body/fuel rail fitting. If you have a non-flexible line, disconnect it at the fuel pump also. Move both the cable and fuel line out of the way.

7. Tag and disconnect the lead at the temperature gauge sender unit. Do the same with the harness connected to the rear of the alternator. Tag and disconnect any other lines, leads or hoses that might be in the way of removal. In some instances you may be able to simply secure them out of the way without disconnecting them—you be the judge.

8. If your engine has a spark plug wire retainer attached to the cylinder head cover, unclip the wires or remove the retainer and then tag and disconnect the plug wires at the spark plugs.

9. Remove the distributor cap with the leads still connected. Mark the position of the distributor rotor to the distributor body. Scribe a matchmark across the distributor and the manifold. Loosen the clamp bolt and remove the distributor. Please refer to the Electrical section for further details on distributor removal. DO NOT turn the engine over once the distributor has been removed.

10. Remove the alternator and its mounting bracket.

11. Tag and disconnect any leads at the ignition module and move them out of the way.

12. Tag and disconnect the wire at the oil sending unit and then remove the unit itself.

13. Loosen and remove the manifold mounting bolts in the reverse order of the illustrated tightening sequence and then remove the manifold. Don't forget the solenoid bracket. There is a good likelihood you will need to pry the manifold off the block; be very careful that you don't scratch or mar the mating surfaces on the block, manifold or heads.

To install:

14. Carefully remove all remaining gasket material from the manifold mating surfaces with a scraper or putty knife. Be careful that you don't accidentally drop any old gasket into the crankcase or intake ports on the cylinder head.

15. Inspect the manifold and all mating surfaces for any cracks or nicks.

16. On 1992-97 engines (exc. 1997 5.7GSi), position new seals on the cylinder block mating surfaces. Position new gaskets on the cylinder head mating surfaces. Use GM Silicone Rubber Sealer at all water passages and/or wherever a seal butts against a gasket.

■ Some engines may not have front or rear cylinder block seals. If yours is one of them, apply a 3/16 in. (5mm) bead of RTV sealant to the forward and aft edges of the cylinder block mating surface. Make sure that you run the bead at least a 1/2 in. up onto the gaskets.

17. On 1997 5.7GSi and all 1998 engines, coat the cylinder head side of the gasket with RTV sealant. DO NOT apply too much sealer—see the illustration. Position the gaskets onto the heads over the locater pins.

18. Apply a 1/2 in. (13mm) bead of RTV sealant to the forward and aft edges of the cylinder block mating surface. Make sure that you run the bead at least a 1/2 in. up onto the gaskets.

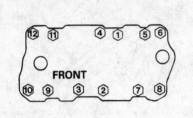

Fif. 41 Intake manifold tightening sequence—1992-97 5.7L engines except the 1997 5.7GSi

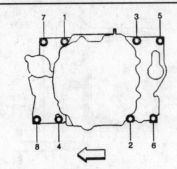

Fig. 42 Intake manifold tightening sequence—1997 5.7GSi and 1998 5.0L/5.7L engines

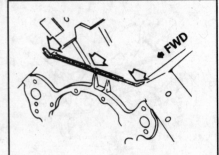

Fig. 43. Make sure to use new seals (or sealer) along the cylinder block-to-manifold mating surfaces—1992-97 engines except the 1997 5.7GSi

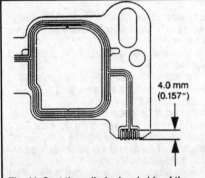

Fig. 44 Coat the cylinder head side of the gasket with RTV sealant—1997 5.7GSi and all 1998 engines

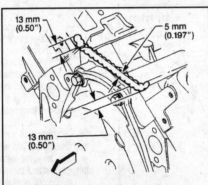

Fig. 45 Run a bead of sealant as shown on the forward block-to-manifold mating surface—1997 5.7GSi and all 1998 engines.

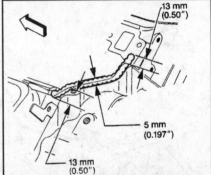

Fig. 46 Run a bead of sealant as shown on the aft block-to-manifold mating surface—1997 5.7GSi and all 1998 engines

19. On all engines, install the manifold into place so that all the bolt holes line up, insert the bolts and tighten them to:
- 30 ft. lbs. (41 Nm) on 1986-97 engines
- 11 ft lbs. (15 Nm) on 1997 GSi and 1998 engines. This is final torque, so be sure you check. This is final torque, so be sure you check the Specifications chart for the first two stages—27 inch lbs. (3 Nm) on the first pass, 106 inch lbs. (12 Nm) on the second.

Check the Torque Specifications chart for any further details on tightening needs. Tighten all bolts in the order shown in the illustration. On all 1998 engines (yes, this also goes for the 1997 5.7GSi), coat all eight bolts with thread locking sealant prior to installing them; remember also that these engines utilize a three step tightening process.

20. Install the oil sending unit and connect the wire.

21. Connect the ignition module leads and install the distributor as detailed in the Electrical section. Put the cap back on and reconnect the plug wires to the spark plugs.

22. Install all water hoses and tighten their clamps securely.

23. Install the fuel line at the carburetor/throttle body and fuel pump. Make sure you remove any plugs you may have inserted on removal.

24. Install the throttle cable and adjust it as detailed in the Fuel System section. Install the flame arrestor.

25. Connect all other lines, leads and hoses that may have been removed to facilitate manifold removal.

26. Install the alternator and mounting bracket. Reconnect the harness.

27. Connect the battery cable and start the engine. Check the ignition timing and idle speed. Check all hoses and seals for leaks.

7.4L And 8.2L V8 Engines (Except 1997-98 Gi/GSi)

◆ See Figures 47 thru 52    MODERATE

1. Open or remove the engine compartment hatch. Disconnect the negative battery cable.

2. Loosen the clamp and remove the crankcase ventilation or PCV hoses at the cylinder head covers. Carefully move them out of the way.

3. Drain all water from the cylinder block and manifolds.

4. Remove the flame arrestor cover. Remove the flame arrestor itself and then disconnect the throttle cable at the carburetor anchor block.

5. Disconnect the electrical leads and remove the alternator.

6. Tag and disconnect all water hoses at the manifold and thermostat housing. Have some rags handy, since there will still be some water in them. Carefully move them out of the way and then remove the thermostat housing.

7. Disconnect the fuel line and plug the line end and the carburetor fitting. If you have a non-flexible line, disconnect it at the fuel pump also. Move both the cable and fuel line out of the way.

8. Tag and disconnect the purple/white wire at the choke housing.

9. Unplug the main harness at the connector block. Disconnect it from the retaining clips and move it out of the way.

10. Tag and disconnect any other lines, leads or hoses that might be in the way of removal. In some instances you may be able to simply secure them out of the way without disconnecting them—you be the judge.

11. If your engine has a spark plug wire retainer attached to the cylinder head cover, unclip the wires or remove the retainer and then tag and disconnect the plug wires at the spark plugs.

12. Remove the distributor cap with the leads still connected. Mark the position of the distributor rotor to the distributor body. Scribe a matchmark across the distributor and the manifold. Loosen the clamp bolt and remove the distributor. Please refer to the Electrical section for further details on distributor removal. DO NOT turn the engine over once the distributor has been removed.

13. Tag and disconnect any leads at the ignition module and move them out of the way.

14. Loosen and remove the manifold mounting bolts in the reverse order of the illustrated tightening sequence and then remove the manifold; no need to separate the carburetor unless you are replacing the manifold. There is a good likelihood you will need to pry the manifold off the block; be very careful that you don't scratch or mar the mating surfaces on the block, manifold or heads.

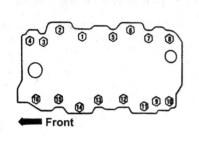

Fig. 47 Intake manifold tightening sequence—1988-90 engines

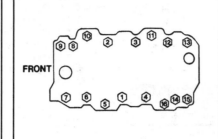

Fig. 48 Intake manifold tightening sequence—1991 engines

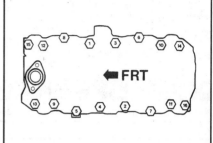

Fig. 49 Intake manifold tightening sequence—1992-95 engines

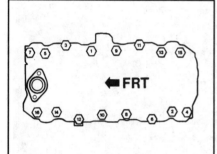

Fig. 50 Intake manifold tightening sequence—1996-97 engines

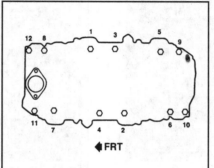

Fig. 51 Intake manifold tightening sequence—1998 engines

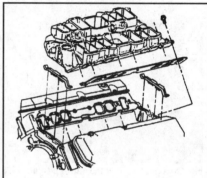

Fig. 52 Installing the intake manifold and gaskets

To install:

15. Carefully remove all remaining gasket material from the manifold mating surfaces with a scraper or putty knife. Be careful that you don't accidentally drop any old gasket into the crankcase or intake ports on the cylinder head.

16. Inspect the manifold and all mating surfaces for any cracks or nicks.

17. Position new seals on the cylinder block mating surfaces. Position new gaskets on the cylinder head mating surfaces—make sure the gaskets have the metal inserts covering the center exhaust port. Use GM Silicone Rubber Sealer at all water passages and/or wherever a seal butts against a gasket.

■ **Some engines may not have front or rear cylinder block seals. If yours is one of them, apply a 3/16 in. (5mm) bead of RTV sealant to the forward and aft edges of the cylinder block mating surface. Make sure that you run the bead at least a 1/2 in. up onto the gaskets.**

18. install the manifold into place so that all the bolt holes line up, insert the bolts and tighten them to:
- 30 ft. lbs. (41 Nm) on 7.4L engines.
- 35 ft. lbs. (47 Nm) on 8.2L engines

Check the Torque Specifications chart for any further details on tightening needs. Tighten all bolts in the order shown in the illustration.

19. Connect the ignition module leads and install the distributor as detailed in the Electrical section. Put the cap back on and reconnect the plug wires to the spark plugs.

20. Install all water hoses and tighten their clamps securely.

21. Install the fuel line at the carburetor and fuel pump. Make sure you remove any plugs you may have inserted on removal.

22. Install the throttle cable and adjust it as detailed in the Fuel System section. Install the flame arrestor and cover.

23. Connect all other lines, leads and hoses that may have been removed to facilitate manifold removal.

24. Install the alternator and mounting bracket. Reconnect the harness.

25. Connect the battery cable and start the engine. Check the ignition timing and idle speed. Check all hoses and seals for leaks.

1997-98 7.4L/8.2L Gi/GSi V8 Engines

Upper Manifold (Plenum)

◆ **See Figures 53 thru 59**

1. Open or remove the engine compartment hatch. Disconnect the negative battery cable.

2. Relieve the fuel system pressure as detailed in the Maintenance or Fuel System sections.

3. Loosen the clamp and remove the crankcase ventilation or PCV hoses at the cylinder head covers. Carefully move them out of the way.

4. Remove the flame arrestor cover. Remove the flame arrestor itself and then disconnect the throttle cable at the throttle body bell crank.

5. Tag and disconnect the electrical leads at the following:
- ECM **J1** and **J2**
- Knock sensor module
- Throttle position sensor
- IAC valve
- MAP sensor
- IAT sensor

6. Tag and disconnect the hose from the pressure regulator at the nipple on the port side of the plenum (1997 7.4Gi and 1998 GSi).

7. Tag and disconnect the vapor line at the pulse limiter on the port side, rear.

8. Loosen the mounting bolts (8 or 12) in the reverse order of the illustrated tightening sequence and lift off the plenum. Make sure that you take note of how the throttle and fuel line brackets were positioned. Remove and discard the eight O-rings.

9. Carefully cover the lower manifold port openings to prevent dirt from entering the engine.

✳✳ CAUTION

Upper intake plenums are made of aluminum—be extremely careful of how you handle it and where you set it down, particularly the mating surfaces.

To install:

10. Carefully remove any remaining gasket material from the plenum and manifold mating surfaces. Make sure the O-ring recesses in the lower manifold runners are free of grime and dirt.

11. Clean the inside of the plenum with Carburetor Cleaner or something similar. Do not use anything with methyl ethyl ketone!

12. Coat the new O-rings lightly with grease and insert them into the recesses on the lower manifold.

13. Position the plenum onto the manifold in such a way that you do not dislodge the O-rings. Coat the threads of the mounting bolts with Lubriplate® 777 (or similar) and install them. Tighten the bolts in sequence, in several steps, to 124 inch lbs. (14 Nm).

14. Connect the pulse limiter hose and tighten the clamp securely.

15. Reconnect all electrical leads previously removed.

16. Reconnect all vacuum lines previously removed.

17. Install the flame arrestor and reconnect the throttle cable. Check for correct throttle operation and adjust if necessary.

18. Connect the battery cables.

Fig. 53 A good look at the intake plenum (1998 7.4Gi)

Fig. 54 Disconnect the throttle linkage (7.4Gi)

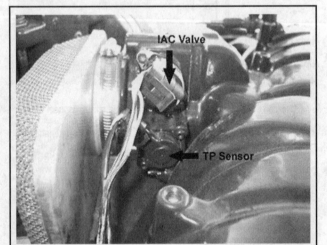

Fig. 55 Disconnect the sensor leads...

Fig. 56 ...and the ECM connectors

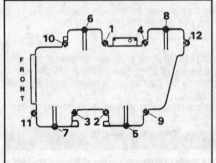

Fig. 57 Upper intake plenum tightening sequence—1997 7.4Gi and 1998 7.4GSi/8.2GSi

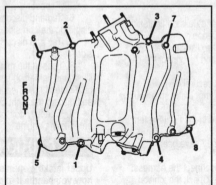

Fig. 58 Upper intake plenum tightening sequence—1998 7.4Gi

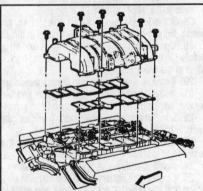

Fig. 59 Installing the intake plenum and gaskets

Lower Manifold

◆ **See Figures 52, 60 and 61**

1. Open or remove the engine compartment hatch. Disconnect the negative battery cable.
2. Drain all water from the cylinder block and manifolds.
3. Remove the upper intake manifold (plenum).
4. Disconnect the electrical leads and remove the alternator.
5. Tag and disconnect all water hoses at the manifold and thermostat housing. Have some rags handy, since there will still be some water in them. Carefully move them out of the way and then remove the thermostat housing.
6. Disconnect the fuel line and plug the line end and the throttle body fitting if not already done. If you have a non-flexible line, disconnect it at the fuel pump also. Move both the cable and fuel line out of the way.
7. Unplug the main harness at the connector block. Disconnect it from the retaining clips and move it out of the way.
8. Tag and disconnect any other lines, leads or hoses that might be in the way of removal. In some instances you may be able to simply secure them out of the way without disconnecting them—you be the judge.
9. If your engine has a spark plug wire retainer attached to the cylinder head cover, unclip the wires or remove the retainer and then tag and disconnect the plug wires at the spark plugs.
10. Remove the distributor cap with the leads still connected. Mark the position of the distributor rotor to the distributor body. Scribe a matchmark across the distributor and the manifold. Loosen the clamp bolt and remove the distributor. Please refer to the Electrical section for further details on distributor removal. DO NOT turn the engine over once the distributor has been removed.
11. Tag and disconnect any leads at the module and move them out of the way.
12. Loosen and remove the manifold mounting bolts in the reverse order of the illustrated tightening sequence and then remove the manifold; no need to separate the carburetor unless you are replacing the manifold. There is a good likelihood you will need to pry the manifold off the block; be very careful that you don't scratch or mar the mating surfaces on the block, manifold or heads.

To install:

13. Carefully remove all remaining gasket material from the manifold mating surfaces with a scraper or putty knife. Be careful that you don't accidentally drop any old gasket into the crankcase or intake ports on the cylinder head.
14. Inspect the manifold and all mating surfaces for any cracks or nicks.
15. Position new seals on the cylinder block mating surfaces. Position new gaskets on the cylinder head mating surfaces—make sure the gaskets have the metal inserts covering the center exhaust port. Use GM Silicone Rubber Sealer at all water passages and/or wherever a seal butts against a gasket.

■ **Some engines may not have front or rear cylinder block seals. If yours is one of them, apply a 3/16 in. (5mm) bead of RTV sealant to the forward and aft edges of the cylinder block mating surface. Make sure that you run the bead at least a 1/2 in. up onto the gaskets.**

16. On all engines, install the manifold into place so that all the bolt holes line up, insert the bolts and tighten them to:
   • 30 ft. lbs. (41 Nm) on 7.4L engines.
   • 35 ft. lbs. (47 Nm) on 8.2L engines
Check the Torque Specifications chart for any further details on tightening needs. Tighten all bolts in the order shown in the illustrations. On 1998 Gi engines, coat all bolts with thread locking sealant prior to installing them.
17. Connect the module leads and install the distributor as detailed in the Electrical section. Put the cap back on and reconnect the plug wires to the spark plugs.
18. Install all water hoses and tighten their clamps securely.
19. Install the fuel line at the carburetor/throttle body and fuel pump. Make sure you remove any plugs you may have inserted on removal.
20. Connect all other lines, leads and hoses that may have been removed to facilitate manifold removal.
21. Install the alternator and mounting bracket. Reconnect the harness.
22. Install the upper plenum.
23. Connect the battery cable and start the engine. Check the ignition timing and idle speed. Check all hoses and seals for leaks.

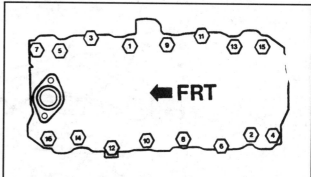

Fig. 60 Intake manifold tightening sequence— all 1997 7.4L engines

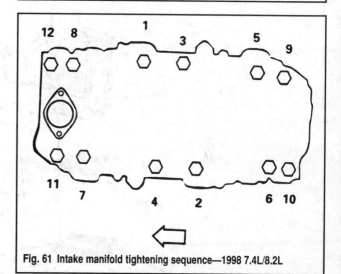

Fig. 61 Intake manifold tightening sequence—1998 7.4L/8.2L

Exhaust Manifold

REMOVAL & INSTALLATION

 MODERATE

◆ **See Figures 62, 63, 64, 65, 66 and 67**

1. Open or remove the engine compartment hatch. Disconnect the negative battery cable.
2. Drain all water and/or coolant from the engine, manifold and exhaust elbow as detailed in the Maintenance section.
3. Disconnect all water/coolant lines at the manifold and move them out of the way. It's a good idea to tag these so that you can ensure proper reconnection. Make sure that you have some rags handy, because there will probably be some spillage even though you've already drained the system.
4. Loosen the hose clamps (two/four) and then disconnect the exhaust pipe hose (bellows) at the elbow and move it out of the way. Grasp the hose with both hands and wiggle it back and forth while pulling it off the elbow. Take note of where all the hose clamps were situated.
5. If you intend to remove the starboard manifold, many engines will have the circuit breaker bracket attached to the elbow, loosen the mounting bolt(s) and remove the circuit breaker bracket (breaker still attached) and position it out of the way. Many later models will also have the ignition module bracket attached to the rear side as well; detach it and move it aside.
6. If you intend to remove the port elbow on EFI models, remove the fuel reservoir water hose and position it out of the way.
7. Also on the port elbow, all models with remote oil filters will have the filter bracket attached to the forward side of the elbow— be sure you support this carefully when removing the mounting bolts.
8. Remove, disconnect, or simply move out of the way, any hoses or wires which may be in the way of removal on your particular engine.
9. Loosen the manifold retaining bolts/nuts from the center outward and then pry off the manifold/elbow assembly.

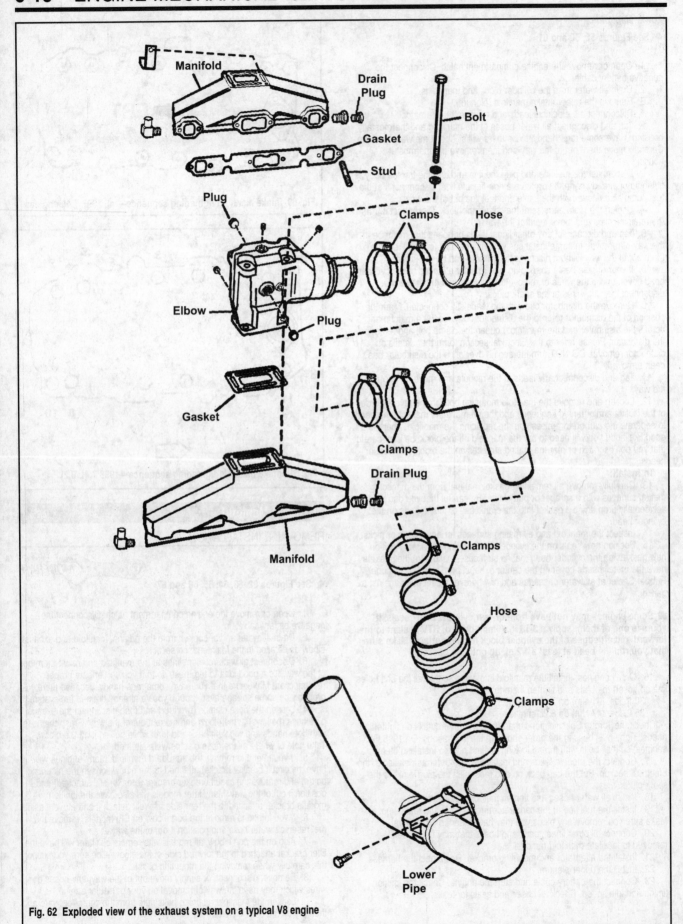

Fig. 62 Exploded view of the exhaust system on a typical V8 engine

Fig. 63 The port exhaust manifold uses 4 mounting nuts on the V6...

Fig. 64 ...as does the starboard side. Note that these engines did not use an elbow until 1991

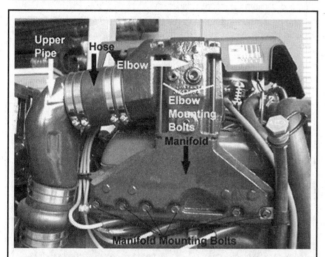

Fig. 65 Close-up of the manifold and related components—except 7.4L/8.2L engines

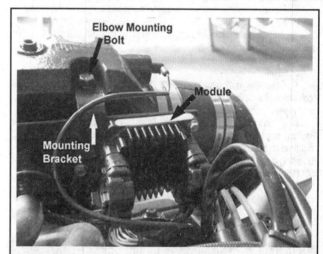

Fig. 66 The module bracket is held in place by one of the elbow mounting bolts on certain later models

10. If necessary, loosen the four retaining bolts and remove the exhaust elbow (except 1986-90 V6) from the manifold.

To install:

11. Carefully clean all residual gasket material from the head, manifold and elbow mating surfaces with a scraper or putty knife. Inspect all gasket surfaces for scratches, cuts or other imperfections.

12. Position a new gasket, without any sealant, on the cylinder head and install the manifold; making sure that everything is aligned properly. Tighten all nuts until they are just tight and then tighten them to 18-22 ft. lbs. (24-30 Nm) on the 1986-90 4.3L V6 and 20-26 ft. lbs. (27-35 Nm) on 1991-98 V6 and all V8 engines; starting in the center, and working your way out to the ends of the manifold. On 7.4L/8.2L engines, tighten any mounting bolts to 24-28 ft. lbs. (33-38 Nm).

13. If you removed the elbow, position a new gasket on the manifold, making sure that the indents line up (on seawater cooled engines) and then install the elbow. Tighten the mounting bolts to 10-12 ft. lbs. (14-16 Nm) on 1986-93 engines or 12-18 ft. lbs. (16-24 Nm) on 1994-98 engines; refer to the Exhaust Elbow procedures found later for more detail on installing the elbow.

14. Connect the exhaust pipe/bellows to the elbow and tighten the clamps securely.

15. If you removed the manifold or elbow plugs for some reason, make sure that the threads are coated with sealant before screwing them back in.

16. Connect the water/coolant hoses and tighten the clamps securely.

17. Make sure that any miscellaneous lines or hoses that you may have moved or disconnected during removal are reconnected and routed properly.

18. Fill the system with water or coolant, connect the battery cable and start the engine. When the engine reaches normal operating temperature, turn it off and re-torque the manifold bolts.

Fig. 67 On models with a remote oil filter, the filter bracket is secured by the two forward mounting bolts

High Rise Exhaust Elbow

REMOVAL & INSTALLATION

◆ See Figures 62 and 65  MODERATE

■ **1986-90 4.3L V6 engines did not use a high rise exhaust elbow.**

1. Drain the cooling system.
2. Loosen the two hose clamps and slide off the exhaust hose (bellows). Grasp it with both hands and pull it off while wiggling it from side to side. If it sticks, drip a little bit of soapy water around the lip.
3. If you intend to remove the starboard elbow, first loosen the mounting bolts and remove the circuit breaker bracket (if equipped) and position it out of the way. Many later models will also have the ignition module bracket attached to the rear side as well; detach it and move it aside.
4. If you intend to remove the port elbow on EFI models, remove the fuel reservoir water hose and position it out of the way.
5. Also on the port elbow, all models with remote oil filters will have the filter bracket attached to the forward side of the elbow—be sure you support this carefully when removing the mounting bolts.
6. Remove the four bolts, lock washers and washers from the elbow and lift it off the manifold. A little friendly persuasion with a soft rubber mallet may be necessary! Be careful though, no need to take out all your aggressions on the poor thing.
7. Remove the gasket and discard it.

To install:
8. Clean the mating surfaces of the manifold and elbow thoroughly, coat both sides of a new gasket with Gasket Sealing compound and position it onto the manifold flange.
9. Position the elbow on the manifold so the bolt holes are in alignment. Position any anchor brackets or mounting hardware for other components and then install the four mounting bolts and their washers. Tighten the mounting bolts to 10-12 ft. lbs. (14-16 Nm) on 1986-93 engines or 12-18 ft. lbs. (16-24 Nm) on 1994-98 engines.
10. Slide the exhaust hose, while wiggling it, all the way onto the elbow. Position the two clamps in their channels and tighten the clamp screws securely. Coating the inside of the hose with a soapy water solution will help you slip it on.

Exhaust Hoses (Bellows) And Intermediate Exhaust Pipe

REMOVAL & INSTALLATION

◆ See Figures 62, 68 and 69  MODERATE

■ **Not all engines will use an intermediate pipe between the upper bellows and the lower exhaust pipe.**

1. Starting with the upper hose, loosen all four hose clamps, two on top of the hose and two on the bottom.
2. Drizzle a soapy water solution over the top of the hose where it mates with the exhaust elbow and let it sit for a minute.
3. Grasp the hose with both hands and wiggle it side-to-side while pulling down on it until it separates from the elbow (or the top of the manifold on early V6 engines).
4. Now wiggle it while pulling upwards until it pops off the intermediate exhaust pipe.
5. The lower hose should be removed in the same manner as the upper.
6. Check the hose for wear, cracks and deterioration.
7. Coat the inside of the lower end of the lower hose with soapy water and wiggle it into position on the Y-pipe (lower). Remember to install the two clamps before sliding it over the end of the pipe.

■ **There is a step about 1 1/2 in. into the inside of one end on each the upper and lower hoses; upper only on the 7.4L/8.2L engines. The stepped side of BOTH hoses should fit over the intermediate pipe. This means that the stepped side on the upper hose faces DOWN, and the stepped side on the lower hose faces UP!**

8. Slide the two clamps over the upper end of the lower hose and then coat the inside with the soapy water solution and insert the bottom of the intermediate pipe into it fully until it seats on the step. Tighten the clamp screws securely.
9. Coat the inside of the lower end of the upper hose with soapy water and wiggle it into position on the intermediate pipe until it meets the step. Remember to install the two clamps before sliding it over the end of the pipe.
10. Slide two clamps over the upper end, lubricate the inside with soapy water and wiggle the hose over the elbow.
11. Tighten all four clamp screws securely.

Lower Exhaust Pipe (Y-Pipe)

REMOVAL & INSTALLATION

◆ See Figure 62  DIFFICULT

1. It is unlikely you will be able to get the pipe off without removing the engine, so remove the engine as previously detailed.
2. Loosen the four retaining bolts at the transom shield and then remove the exhaust pipe. Carefully scrape any remnants of the seal from the pipe and transom mounting surfaces.

■ **The pipe mounting holes in the transom shield utilize Heli-Coil® locking inserts. Never clean the holes or threads with a tapping tool or you risk damaging the locking feature of the threads.**

3. Coat a new seal with 3M Rubber Adhesive and position it into the groove on the transom shield mating surface.

Fig. 68 A good look at the upper hose...

Fig. 69 ...and the lower hose—not all engines will have the bulge in the lower hose like shown here

4. Coat the mounting bolts with Gasket Sealing Adhesive. Position the exhaust pipe, insert the bolts and tighten them to 10-12 ft. lbs. (14-16 Nm) on 1986-90 engines or 20-25 ft. lbs. (27-34 Nm) on 1991-98 engines.
5. Install the engine.

EXHAUST VALVE (FLAPPER) REPLACEMENT

 MODERATE

1. Remove the exhaust hoses and the intermediate exhaust pipe. The flapper is located in the upper end of the y-pipe.
2. The valve is held in place by means of a pin running through two bushings in the sides of the pipe. Position a small punch over one end of the pin and carefully press the pin out of the pipe.

■ **Make sure you secure the valve while removing the retaining pin so it doesn't fall down into the exhaust pipe.**

3. Press out the two bushings and discard them. Coat two new bushings with Scotch Grip Rubber Adhesive and press them back into the sides of the pipe.
4. Position the new valve into the pipe with the long side DOWN. When looking at the valve, the molded retaining rings are off-center—the side with the rings should face the top of the pipe. When the valve is in place, coat the pin lightly with engine oil and slowly slide it through one of the bushings, through the two retaining holes on the valve and then through the opposite bushing. Make sure the pin ends are flush with the sides of the pipe on both sides.
5. Install the exhaust hose.

Oil Pan

REMOVAL & INSTALLATION

1986-96 4.3L V6 and 1986-97 5.7L V8 Engines (Except 1997 5.7GSi)

 DIFFICULT

■ **More times than not, this procedure will require the removal of the engine. Your boat and its unique engine installation will determine this, but the procedure is almost always easier with the engine removed from the boat.**

1. Remove the engine as previously detailed in this section.
2. If you haven't already drained the engine oil, do it now. Make sure you have a container and lots of rags available.
3. Remove the oil dipstick and then remove the dipstick tube(s).
4. Remove the oil withdrawal tube from the oil drain plug fitting.
5. Loosen and remove the oil pan retaining bolts and nuts, starting with the center bolts and working out toward the pan ends. Lightly tap the pan with a rubber mallet to break the seal and then lift it off the cylinder block. If your engine stand will allow for rotating the engine, you'll find that this will be easier with the pan facing up.
To install:
6. Clean the pan mating surfaces of any residual gasket material with a scraper or putty knife. Make sure that no old gasket material has been pressed into the retaining bolt holes in the pan, block or front cover. Clean the pan itself thoroughly with solvent.
7. Position a new seal into the rear main bearing cap groove and press it into place; make sure that the ends are pressed into the small openings in the cylinder block.
8. Coat new side gaskets lightly with grease and position them so that the aft ends overlap the rear seal and the front ends are pressed into the groove between the timing cover and the cylinder block.
9. Install all bolts finger tight and then tighten the front and rear 5/16 in. bolts to 12-15 ft. lbs lbs. (16-20 Nm) and the remaining 1/4 in. bolts to 6-7.5 ft. lbs. (8-10 Nm). Remember to start with the center bolts and work outward toward the ends of the pan.
10. Install the oil drain fitting and finger tighten the bolt. Rotate the fitting and then attach the withdrawal tube. Tighten the fitting bolt and flare nut to 15-18 ft. lbs. (20-24 Nm).
11. Install the engine (if removed). Refill with all fluids. Run the engine up to normal operating temperature, shut it off and check the pan for any leaks.

1997-98 4.3L V6 Engines

 DIFFICULT

◆ See Figures 70, 71, 72, 73 and 74

■ **More times than not, this procedure will require the removal of the engine. Your boat and its unique engine installation will determine this, but the procedure is almost always easier with the engine removed from the boat.**

1. Remove the engine as previously detailed in this section.
2. If you haven't already drained the engine oil, do it now. Make sure you have a container and lots of rags available.
3. Remove the oil dipstick and then remove the dipstick tube(s).
4. Remove the oil withdrawal tube from the oil drain plug fitting.
5. Loosen and remove the oil pan retaining bolts and nuts, starting with the center bolts and working out toward the pan ends. Lightly tap the pan with a rubber mallet to break the seal and then lift it off the cylinder block. If your engine stand will allow for rotating the engine, you'll find that this will be easier with the pan facing up.
To install:
6. Clean the pan mating surfaces of any residual gasket material with a scraper or putty knife. Make sure that no old gasket material has been pressed into the retaining bolt holes in the pan, block or front cover. Clean the pan itself thoroughly with solvent.
7. Apply a small dab of RTV sealer to the joints on either side of the rear oil seal retainer and front cover, position a new pan gasket onto the pan being very careful to line up all the holes—do not use RTV sealant with this gasket other than where noted. Make sure that the sealer is applied at least 1 in. (25.4mm) in either direction from each of the four joints.
8. Move the pan and gasket onto the block; don't dawdle here because the RTV sealant applied in the previous step sets up very quickly. It is very important that you ensure all the holes line up correctly.

■ **On these engines, proper alignment of the rear edge of the pan and the block is imperative—if both surfaces are not flush, the engine-to-transom bracket alignment will not be rigid.**

9. Install the special straight edge (#J346773) at the rear and slide the pan back against it. Install the pan retaining bolts and nuts finger-tight.
10. Check the clearance between the three oil pan-to-bell housing contact points with a feeler gauge. If clearance exceeds 0.010 in. (0.254mm) at any of the three points you will need to remove the pan and start over again. Continue this until all three points are within specifications.
11. Now tighten the bolts to 18 ft. lbs. (25 Nm) and the nuts to 17 ft. lbs. (23 Nm). Remember to follow the torque sequence shown in the illustration.
12. On all engines again, install the oil drain fitting and finger tighten the bolt. Rotate the fitting and then attach the withdrawal tube. Tighten the fitting bolt and flare nut to 15-18 ft. lbs. (20-24 Nm).
13. Install the engine (if removed). Refill with all fluids. Run the engine up to normal operating temperature, shut it off and check the pan for any leaks.

1997 5.7GSi And 1998 5.0L/5.7L V8 Engines

 DIFFICULT

◆ See Figures 70, 71, 75 and 76

■ **More times than not, this procedure will require the removal of the engine. Your boat and its unique engine installation will determine this, but the procedure is almost always easier with the engine removed from the boat.**

1. Remove the engine as previously detailed in this section.
2. If you haven't already drained the engine oil, do it now. Make sure you have a container and lots of rags available.
3. Remove the oil dipstick and then remove the dipstick tube(s).
4. Remove the oil withdrawal tube from the oil drain plug fitting.
5. Loosen and remove the oil pan retaining bolts, studs and nuts, starting with the center fasteners and working out toward the pan ends. Lift off the reinforcing strips and then lightly tap the pan with a rubber mallet to break the seal and then lift it off the cylinder block. If your engine stand will allow for rotating the engine, you'll find that this will be easier with the pan facing up.

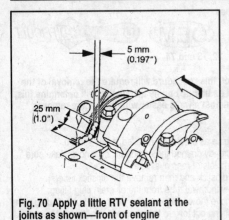

Fig. 70 Apply a little RTV sealant at the joints as shown—front of engine

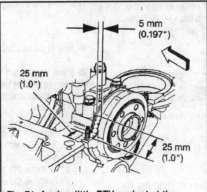

Fig. 71 Apply a little RTV sealant at the joints as shown—rear of engine

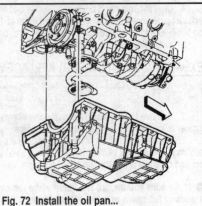

Fig. 72 Install the oil pan...

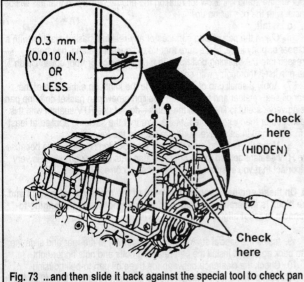

Fig. 73 ...and then slide it back against the special tool to check pan alignment

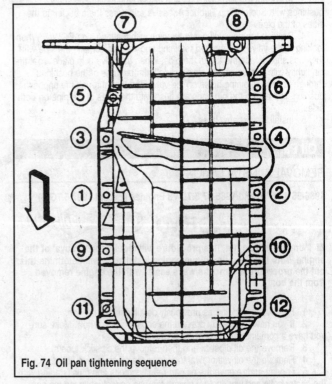

Fig. 74 Oil pan tightening sequence

To install:

6. Clean the pan mating surfaces of any residual gasket material with a scraper or putty knife. Make sure that no old gasket material has been pressed into the retaining bolt holes in the pan, block or front cover. Clean the pan itself thoroughly with solvent.

7. Apply a small dab (5mm) of GM Adhesive sealer to the joints on either side of the rear oil seal retainer and front cover, position a new pan gasket onto the pan being very careful to line up all the holes—do not use RTV sealant with this gasket other than where noted. Make sure that the sealer is applied at least 1 in. (25.4mm) in either direction from each of the four joints.

8. Move the pan and gasket onto the block; don't dawdle here because the RTV sealant applied in the previous step sets up very quickly. It is very important that you ensure all the holes line up correctly.

9. Position the reinforcing strips on each side and then tighten the bolts, stud or nuts on each corner to 15 ft. lbs. (20 Nm); tighten the remaining bolts and studs to 106 inch lbs. (12 Nm).

10. Install the oil drain plug or the oil drain fitting and finger tighten the bolt. Rotate the fitting and then attach the withdrawal tube. Tighten the fitting bolt and flare nut to 15-18 ft. lbs. (20-24 Nm).

11. Install the engine (if removed). Refill with all fluids. Run the engine up to normal operating temperature, shut it off and check the pan for any leaks.

1988-98 7.4L/8.2L V8 Engines

◆ See Figures 77, 78, 79 and 80

■ More times than not, this procedure will require the removal of the engine. Your boat and its unique engine installation will determine this, but the procedure is almost always easier with the engine removed from the boat.

1. Remove the engine as previously detailed in this section.
2. If you haven't already drained the engine oil, do it now. Make sure you have a container and lots of rags available.
3. Remove the oil dipstick and then remove the dipstick tube(s).
4. Remove the oil withdrawal tube from the oil drain plug fitting.
5. Loosen and remove the oil pan retaining bolts and nuts, starting with the center bolts and working out toward the pan ends. Lightly tap the pan with a rubber mallet to break the seal and then lift it off the cylinder block. If your engine stand will allow for rotating the engine, you'll find that this will be easier with the pan facing up.
6. Lift of the pan reinforcement strips on 1988-91 engines.

To install:

7. Clean the pan mating surfaces of any residual gasket material with a scraper or putty knife. Make sure that no old gasket material has been pressed into the retaining bolt holes in the pan, block or front cover. Clean the pan itself thoroughly with solvent.

On all but the 1998 7.4Gi:

8. Apply RTV sealer (Silicone Rubber Sealer) to the four corners (front and rear) of new side gaskets. Coat them with sealer and position them onto the cylinder block.

9. Position a new seal into the groove in the rear main bearing cap so that the ends mate evenly with the side gaskets.

10. Position a new front seal onto the bottom of the front cover and press the locating tips into the holes in the cover. Make sure that the ends of the seal mate evenly with the side gaskets.

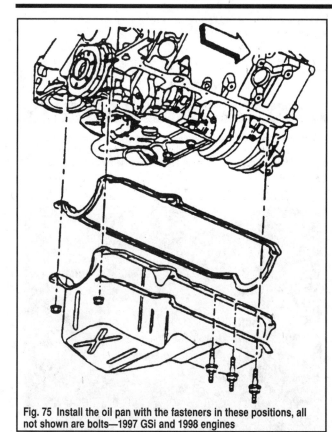

Fig. 75 Install the oil pan with the fasteners in these positions, all not shown are bolts—1997 GSi and 1998 engines

11. On 7.4L engines, install all bolts and nuts finger tight and then tighten the pan-to-front cover bolts to 70 inch lbs. (8 Nm) and the remaining bolts to 160 inch lbs. (18 Nm). On 8.2L engines, install all bolts and nuts finger tight and then tighten the pan-to-front cover bolts to 120 inch lbs. (13.6 Nm) and the remaining bolts to 200 inch lbs. (22.5 Nm). Refer to the tightening sequence illustration for 1988-91 engines; on 1992-93 engines, tighten them evenly and alternately from the center outward.

On the 1998 7.4Gi:

12. Apply RTV sealant to the four places that the front and rear main bearing caps mate with the cylinder block.

13. Position a new gasket and then install the oil pan. Thread in all mounting bolts finger tight and then tighten them to 18 ft. lbs. (25 Nm), alternately and from the center outward.

On all engines:

14. Install the oil drain fitting and finger tighten the bolt. Rotate the fitting and then attach the withdrawal tube. Tighten the fitting bolt and flare nut to 15-18 ft. lbs. (20-24 Nm); 20 ft. lbs. (28 Nm) on the 1998 7.4Gi..

15. Install the engine (if removed). Refill with all fluids. Run the engine up to normal operating temperature, shut it off and check the pan for any leaks.

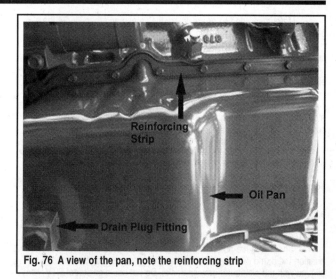

Fig. 76 A view of the pan, note the reinforcing strip

Oil Pump

The two-piece oil pump utilizes two pump gears and a pressure regulator valve enclosed in a two-piece housing. A baffled pick-up tube is press-fit into the body of the pump. The pump is driven via the distributor shaft which is itself driven from a gear on the camshaft. Oil passes through the pick-up screen, through the pump and then through the oil filter.

REMOVAL & INSTALLATION

◆ **See Figures 81, 82 and 83**

 DIFFICULT

1. Remove the oil pan as previously detailed. Remember that you probably need to remove the engine for this procedure.

2. Most V8 engines utilize an oil baffle (deflector) plate; remove the retaining nuts (three, and a bolt) and lift out the baffle. Certain V6 engines also use a baffle that is mounted to the pick-up screen.

3. Loosen and remove the pump mounting bolt(s) from the rear main bearing cap and lift off the pump assembly.

4. Pull out the driveshaft and retainer.

5. Loosen the four pump cover screws and lift off the cover with the screen attached.

■ The oil pump pick-up screen and pipe are press fit to the pump cover, they are serviced as an assembly so should not be separated; nor should they be replaced during normal service. If the assembly does require replacement, mount the pump in a vise and pull out the pipe. Do not reinstall the same assembly, use only a new replacement. Make sure that the screen is parallel with the bottom of the oil pan.

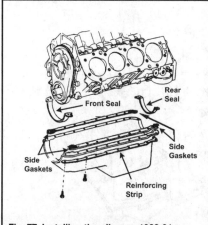

Fig. 77 Installing the oil pan—1988-91 engines

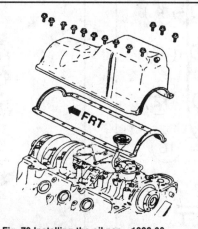

Fig. 78 Installing the oil pan—1992-98 engines (except 1998 7.4Gi)

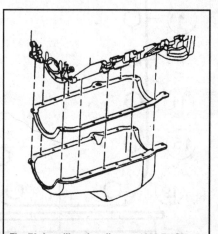

Fig. 79 Installing the oil pan—1998 7.4Gi

6. Matchmark the two gears where they mesh and then lift out the two gears. Remove the pressure regulator valve and it associated parts.

7. Clean all components in solvent and dry thoroughly. Inspect the pump body and gears for cracks, excessive wear or other damage

8. Install the pressure regulator valve into the housing cover with a new spring.

9. Install the retaining pin.

10. Install the drive gear into the housing. Install the idler gear so that the matchmarks mesh with the drive gear and the smooth side is toward the cover.

11. Install the pump cover and tighten the screws to:
- 6-9 ft. lbs. (8-12 Nm) on 1986-96 V6 engines
- 106 inch lbs. (12 Nm) on 1997-98 V6 engines
- 6-9 ft. lbs. (8-12 Nm) on 1986-97 5.0L/5.7L V8 engines (except the 1997 5.7GSi)
- 106 inch lbs. (12 Nm) on the 1997 5.7GSi and 1998 5.0L/5.7L V8 engines
- 80 inch lbs. (9 Nm) on all 7.4L/8.2L V8 engines (except the 1998 7.4Gi)
- 106 inch lbs. (12 Nm) on 1998 7.4Gi

12. Check that the pump and block mating surfaces are clean and then position the pump over the block so that the pump extension shaft tang is aligned with the distributor driveshaft slot. Do not use a gasket or RTV sealant. Tighten the pump mounting bolts to 65 ft. lbs. (90 Nm) on all but the 8.2L engine, and 70 ft. lbs. (95 Nm) on 8.2L engines

■ The pump should slide into place easily, if not recheck the alignment of the two shafts.

13. Install the oil baffle on engines so equipped and tighten the nuts to 30 ft. lbs. (40 Nm).

14. Install the oil pan and engine.

Oil Filter Bypass Valve

Almost all engines covered here utilize a bypass valve. After removing the oil filter, check the spring and small fiber valve for proper operation. Any signs of incorrect operation, or wear and deterioration will necessitate replacement.

Fig. 81 Oil pump and pick-up screen properly installed on a typical V8

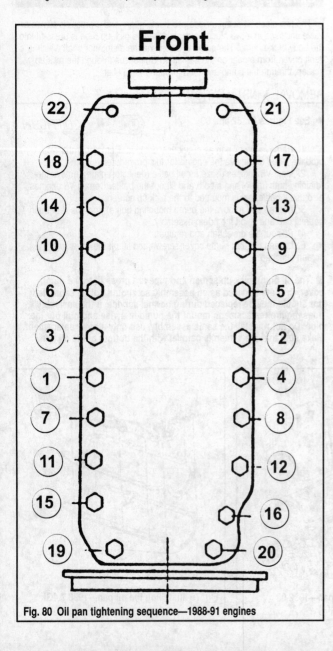

Fig. 80 Oil pan tightening sequence—1988-91 engines

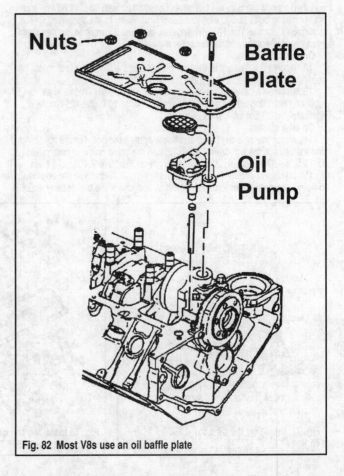

Fig. 82 Most V8s use an oil baffle plate

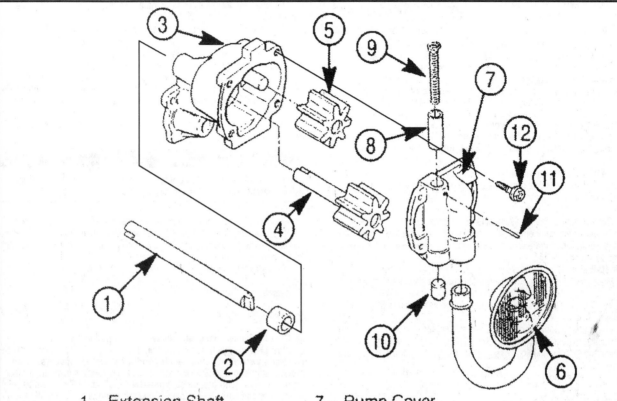

1 - Extension Shaft
2 - Shaft Coupling
3 - Pump Body
4 - Drive Gear and Shaft
5 - Idler Gear
6 - Pickup Screen and Pipe

7 - Pump Cover
8 - Pressure Regulator Valve
9 - Pressure Regulator Spring
10- Plug
11- Retaining Pin
12- Screws

Fig. 83 Exploded view of the oil pump

REMOVAL & INSTALLATION

◆ See Figure 84   ②   ◁MODERATE

1. Drain the oil as detailed in the Maintenance section.
2. Remove the oil filter.
3. Using a small prybar, remove the valve.
4. Install a new valve and press it in by placing a 9/16 in. deep socket over it and tapping the socket lightly with a hammer.
5. Install the oil filter and refill the engine with the appropriate oil.

Engine Coupler/Timing Ring And Flywheel

REMOVAL & INSTALLATION

■ Over the years, these engines were equipped in a variety of configurations—flywheel and torsional damper, flywheel, damper and coupler, flywheel and coupler, etc. And for a number of years, some models also had a timing ring attached. Early models are usually the ones equipped with a damper, while Cobra, SP and DP models usually were equipped with the coupler. King Cobra models are usually the ones that use the timing ring, although there were also some 4.3L engines that used it. Although we have attempted to narrow the applications in the following procedures, it was not always year or model specific. Please read the procedures FIRST in order to determine which one is appropriate for your particular engine's configuration.

Fig. 84 Oil filter bypass valve

Models With A Torsional Damper

■ Over the years, these engines were equipped in a variety of configurations—flywheel and torsional damper, flywheel, damper and coupler, flywheel and coupler, etc. And for a number of years, some models also had a timing ring attached. Early models are usually the ones equipped with a damper, while Cobra, SP and DP models usually were equipped with the coupler. King Cobra models are usually the ones that use the timing ring, although there were also some 4.3L engines that used it. Although we have attempted to narrow the applications in the following procedures, it was not always year or model specific. Please read the procedures FIRST in order to determine which one is appropriate for your particular engine's configuration.

1. Remove the engine as detailed previously.
2. Loosen the bolts and remove the flywheel housing and cover plate.
3. Remove the three mounting bolts and pull off the rear torsional damper.
4. Remove four of the six flywheel retaining bolts. Loosen the other two, any two is fine, and back them out about half way. Using the two bolts as a stop, carefully pull the flywheel off of the crankshaft flange. Remove the two bolts and flywheel.

To install:

5. Position the flywheel over the dowel pin on the flange and make sure that the holes line up correctly. Coat the bolt threads with engine oil, install them and tighten to 59 ft. lbs. (82 Nm). Tighten the six bolts in a diagonal star pattern.
6. Install the torsional damper and tighten the bolts securely.
7. Install the housing and plate. Coat all mounting bolts with engine oil and tighten the 9/16 in. housing bolts to 30 ft. lbs. (41 Nm), the 5/16 in. plate bolts securely.
8. Install the engine.

Models With A Timing Ring

◆ See Figures 86 and 87

■ Over the years, these engines were equipped in a variety of configurations—flywheel and torsional damper, flywheel, damper and coupler, flywheel and coupler, etc. And for a number of years, some models also had a timing ring attached. Early models are usually the ones equipped with a damper, while Cobra, SP and DP models usually were equipped with the coupler. King Cobra models are usually the ones that use the timing ring (350, 454, 502, 7.4L), although there were also some 4.3L engines that used it. Although we have attempted to narrow the applications in the following procedures, it was not always year or model specific. Please read the procedures FIRST in order to determine which one is appropriate for your particular engine's configuration.

1. Remove the engine from the boat as detailed previously in this section.
2. Loosen and remove any connections attached to either of the ground studs on each side of the flywheel. Move the electrical leads out of the way.
3. Loosen the 5 retaining bolts and slide out the lower housing cover.
4. Although not strictly necessary, we recommend removing the starter.
5. Locate the timing sensor cover on the housing, loosen the two nuts and lift off the cover. Remove the two nuts and washers and then pull out the sensor and position it out of the way.
6. Cut the plastic tie that secures the housing drain hose and the pull the hose out of the fitting.
7. Remove the flywheel housing retaining nuts/bolts and pull off the housing. Take note of the positioning of the oil cooler and its bracket.
8. Slide an offset wrench behind the coupler and loosen the six flywheel mounting nuts gradually and as you would the lug nuts on your car or truck—that is, in a diagonal star pattern. Remove the coupler.
9. Mark the dowel hole on the timing ring and pry it off the flywheel. Remove the flywheel.

To install:

10. Thoroughly clean the flywheel mating surface and check it for any nicks, cracks or gouges. Check for any broken teeth.
11. Install the flywheel over the dowel on the crankshaft.
12. Position the timing ring over the locating pin making sure the pin is in the correct hole—you did mark it, right? Press the ring into position.
13. Slide the coupler over the studs so that it sits in the recess on the flywheel. Install new lock washers and tighten the mounting nuts to 40-45 ft. lbs. (54-61 Nm). Once again use the star pattern while tightening the bolts.

14. Install the flywheel housing and attach the oil cooler. Tighten the nuts/bolts to 32-40 ft. lbs. (43-54 Nm).
15. Install the washer, lock washer and inner nut on the ground stud and tighten it to 15-20 ft. lbs. (20-27 Nm). Attach the electrical leads, install another lock washer and then tighten the outer nut securely.
16. Slide the drain hose in and attach it with a new plastic tie.
17. Position the timing sensor onto the mounting studs and press in on the spring tab so that the sensor seats itself correctly over the timing ring. Install the washers and nuts, with the tab still depressed, and tighten them to 48-64 inch lbs. (5-7 Nm). Once the nuts are tightened, press in on the spring tab again and confirm that the tab DOES NOT touch the timing ring. If it does, loosen the nuts and try it again.

✷✷ WARNING

If the sensor tab comes in contact with the teeth of the timing ring, it will be damaged on engine start-up.

18. Install the sensor and tighten the nuts to 48-64 inch lbs. (5-7 Nm).
19. Coat both sides of a new gasket with Gasket Sealing Compound and position it onto the cover. Slide the cover into position and tighten the bolts to 60-84 inch lbs. (7-9 Nm).
20. Install the engine.

Models With A Coupler

◆ See Figures 85, 86 and 87

■ Over the years, these engines were equipped in a variety of configurations—flywheel and torsional damper, flywheel, damper and coupler, flywheel and coupler, etc. And for a number of years, some models also had a timing ring attached. Early models are usually the ones equipped with a damper, while Cobra, SP and DP models usually were equipped with the coupler. King Cobra models are usually the ones that use the timing ring, although there were also some 4.3L engines that used it. Although we have attempted to narrow the applications in the following procedures, it was not always year or model specific. Please read the procedures FIRST in order to determine which one is appropriate for your particular engine's configuration.

1. Remove the engine from the boat as detailed previously in this section.
2. Loosen and remove any connections attached to either of the ground studs on each side of the flywheel. Move the electrical leads out of the way.
3. Although not strictly necessary, we recommend removing the starter.
4. Loosen the 4 or 5 retaining bolts and slide out the lower housing cover.
5. Cut the plastic tie that secures the housing drain hose and the pull the hose out of the fitting.
6. Remove the oil cooler and bracket, take note of the positioning of both and secure them out of the way.
7. Remove the flywheel housing.
8. Slide an offset wrench behind the coupler and loosen the six flywheel mounting nuts gradually and as you would the lug nuts on your car or truck—that is, in a diagonal star pattern.
9. Remove the coupler and then the flywheel.

To install:

10. Thoroughly clean the flywheel mating surface and check it for any nicks, cracks or gouges. Check for any broken teeth.
11. Install the flywheel over the dowel on the crankshaft.
12. Slide the coupler over the studs so that it sits in the recess on the flywheel. Install new lock washers and tighten the mounting nuts to 40-45 ft. lbs. (54-61 Nm) except on 1992 engines where the spec is 35-40 ft. lbs. (47-54 Nm). Once again use the star pattern while tightening the bolts.
13. Install the flywheel housing and attach the oil cooler. Tighten the bolts to 28-36 ft. lbs. (38-49 Nm) on 1986-91 engines or 32-40 ft. lbs. (43-54 Nm) on 1992-98 engines.
14. Install the washer, lock washer and inner nut on the ground stud and tighten it to 20-25 ft. lbs. (27-34 Nm) on 1986-91 engines or 15-20 ft. lbs. (20-27 Nm) on 1992-98 engines. Attach the electrical leads, install another lock washer and then tighten the outer nut securely.
15. Coat both sides of a new gasket with Gasket Sealing Compound and position it onto the cover. Slide the cover into position and tighten the bolts to 60-84 inch lbs. (7-9 Nm).
16. Slide the drain hose in and attach it with a new plastic tie.
17. Install the engine.

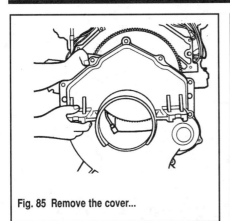

Fig. 85 Remove the cover...

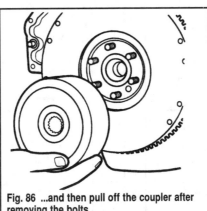

Fig. 86 ...and then pull off the coupler after removing the bolts

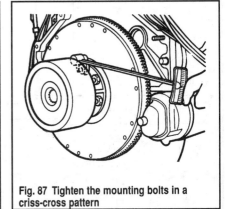

Fig. 87 Tighten the mounting bolts in a criss-cross pattern

Rear Main Oil Seal Retainer

REMOVAL & INSTALLATION

4.3L V6 And 5.0L/5.7L V8 Engines

◆ See Figure 88   ③ *DIFFICULT*

It is not necessary to remove the engine or rear main bearing cap when removing the one-piece oil seal on these engines although you may find it easier to do just that.

■ **A new oil seal must be installed whenever the retainer is removed.**

1. Remove the oil pan.
2. Remove the flywheel.
3. Loosen the nuts/bolts and then lift out the retainer and gasket.
4. Replace the oil seal.
5. Clean all traces of old gasket material from the mating surfaces and position a new seal on the cylinder block.
6. Install the seal retainer, with a new gasket, and tighten the nuts/bolts to 135 inch lbs. (15 Nm).
7. Install the oil pan and flywheel as previously detailed.

Rear Main Oil Seal

REMOVAL & INSTALLATION

4.3L V6 And 5.0L/5.7L V8 Engines

OEM ③ *DIFFICULT*

■ **Early engines may have been equipped with either a one piece or two piece oil seal.**

One Piece Seal

◆ See Figures 88, 89 and 90

It is not necessary to remove the engine or rear main bearing cap when removing the one-piece oil seal on these engines although you may find it easier to do just that.

1. Remove the flywheel housing and cover as detailed in this section.
2. Remove the engine coupler and flywheel from the engine as detailed in this section.
3. Remove the seal retainer. This is not absolutely necessary.
4. Insert a small prybar into one of the three slots in the edge of the seal retainer and slowly pry the seal out of the retainer. Be very careful not to nick or damage the sealing surface while prying out the seal.
5. Thoroughly clean the retainer surface and then install the retainer.
6. Spread a small amount of engine oil around the inside and outside edges of a new seal and position it into a Seal Driver (7J-35621).
7. Position the driver and seal over the crankshaft and then thread the attaching screws into the holes in the crankshaft, tightening them securely. Turn the handle on the tool until it bottoms out—the seal is now in place.
8. Install the flywheel and engine coupler. Install the cover and flywheel housing.

Two Piece Seal

◆ See Figures 91 and 92

These engines utilize a two-piece rear main seal. The seal can be removed without removing the crankshaft. You will need to remove the engine for this procedure though.

1. Remove the engine as detailed previously in this section.
2. Remove the oil pan and pump as detailed previously in this section.
3. Loosen the retaining bolts and remove the rear main bearing cap. Carefully insert a small prybar and remove the lower half of the seal. Do not damage the seal seating surface.

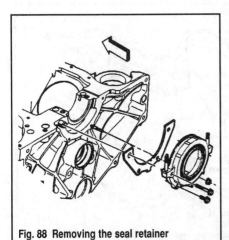

Fig. 88 Removing the seal retainer

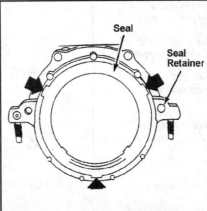

Fig. 89 Use the three slots in the seal retainer when removing the rear main seal

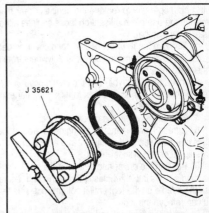

Fig. 90 Use a seal driver to seat the new oil seal

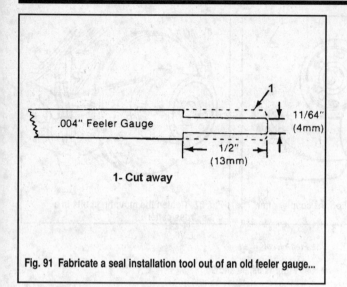

1- Cut away

Fig. 91 Fabricate a seal installation tool out of an old feeler gauge...

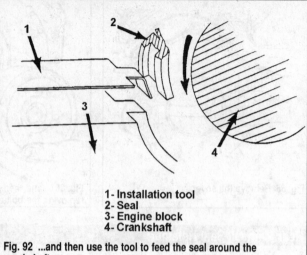

1- Installation tool
2- Seal
3- Engine block
4- Crankshaft

Fig. 92 ...and then use the tool to feed the seal around the crankshaft

4. Using a hammer and a small drift, tap on end of the upper seal until it starts to protrude form the other side of the race. Grab the protruding end with pliers and pull out the remaining seal half.

5. Check that you have the correct new seal. Seals with a hatched inner surface can only be used on left hand rotation engines, smooth seals can be used on any engine. Coat the lip and bead with motor oil. Keep oil away from the seal parting surfaces.

6. Your seal kit should come with an installation tool, if not, take a 0.004 in. feeler gauge and cut each side back about a half inch so that you're left with an 11/64 inch point. Bend the tool into the gap between the crankshaft and the seal seating surface. This will be your "shoe horn".

7. Position the upper half of the seal (lip facing the engine) between the crank and the tool so that the seal's bead is in contact with the tool tip. Roll the seal around the crankshaft using the tool as a guide until each end is flush with the cylinder block. Remove the tool.

8. Insert the lower seal half into the main bearing cap with the lip facing the cap. Start it so that one end is slightly below the edge of the cap and then use the tool to shimmy the seal all the way in until both edges are flush with the edge of the cap. Remove the tool.

9. Make sure that the cap/block mating surfaces and the seal ends are free of any oil and then apply a small amount of Perfect Seal to the block just behind where the upper seal ends are.

10. Install the bearing cap and tighten the bolts to the correct torque as detailed in the Torque Specifications chart.

11. Install the oil pump and pan.

12. Install the engine.

7.4L/8.2L V8 Engines

OEM ③ DIFFICULT

◆ See Figures 90 and 93

It is not necessary to remove the engine or rear main bearing cap when removing the one-piece oil seal on these engines although you may find it easier to do just that.

1. Remove the flywheel housing and cover as detailed in this section.

2. Remove the engine coupler and flywheel from the engine as detailed in this section.

3. Insert a small prybar into the seal groove and slowly pry the seal out of the groove. Be very careful not to nick or damage the sealing surface while prying out the seal.

4. Thoroughly clean the groove surface.

5. Spread a small amount of engine oil around the inside and outside edges of a new seal and position it into a Seal Driver (J-38841) so that the seal's lip faces the cylinder block.

6. Position the driver and seal over the crankshaft and then thread the attaching screws into the holes in the crankshaft, tightening them securely. Turn the handle on the tool until it bottoms out—the seal is now in place and you can remove the tool.

7. Install the flywheel and engine coupler. Install the cover and flywheel housing.

Harmonic Balancer, Pulley And Hub

REMOVAL & INSTALLATION

◆ See Figure 94   OEM ② MODERATE

1. Disconnect the battery cables.

2. Remove the bracket mounting bolts for the alternator and power steering pump.

3. Drain the engine and manifolds. Loosen the two mounting bolts and remove the seawater pump from the crankshaft.

4. Remove the drive or serpentine belt(s) as detailed in the Maintenance section.

5. Remove the three bolts and pull off the drive pulley attached to balancer.

6. Remove the balancer retaining bolt (if equipped) and install the special removal tool (#J-23523-03) onto the damper. Tighten the tool press bolt and remove the damper; don't lose the crankshaft key. OMC suggests that you do not use a conventional gear puller for this procedure.

To install:

7. Inspect the crank key and then install it into the shaft. Using a little GM Adhesive will make this easier.

8. Coat the front cover oil seal lip with clean engine oil and then install the damper with a proper installation tool (#J-23523-03, J-23523-E or J-39046). Be sure that you thread the tool into the crankshaft at least 1/2 in. to protect the threads. In a pinch you can use a block of wood and a plastic mallet, but be careful that the pulley does not shift on its mountings while you're hammering. Or, you can use a large washer and a 7/16-20 x 4 in. bolt, but we suggest the tool.

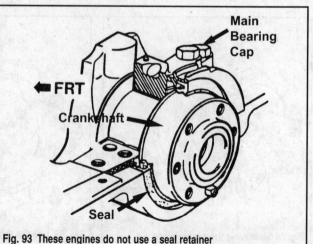

Fig. 93 These engines do not use a seal retainer

9. Install the plate, thrust bearing and nut on to the rod. Tighten the nut until it bottoms out and then remove the tool.

10. Install the retaining bolt (if equipped) and tighten it to:
• 60 ft. lbs. (81 Nm) on 1986-90 4.3L V6 and 5.0L/5.7L V8 engines (except LE and 350)
• 70 ft. lbs. (95 Nm) on 1990 5.7 LE and 350 V8 engines
• 70 ft. lbs. (95 Nm) on 1991-96 4.3L V6 and 5.7L V8 engines
• 74 ft. lbs. (100 Nm) on 1997-98 4.3L V6 and 5.7L V8 engines
• 85 ft. lbs. (115 Nm) on 7.4L V8 engines (except 1998 7.4Gi)
• 110 ft. lbs. (149 Nm) on 1998 7.4Gi V8 engines
• 90 ft. lbs. (122 Nm) on 8.2L V8 engines
Squirt a little RTV sealant into the crankshaft keyway to guard against oil seepage.

11. Install the drive pulley and tighten the bolts to 35 ft. lbs (48 Nm) on the V6, 43 ft. lbs. (58 Nm) on V8 engines.

12. Install the drive/serpentine belt and make sure that it is adjusted properly.

13. Install the seawater pump.

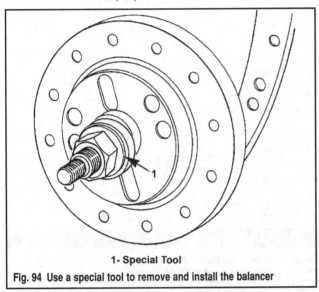

1- Special Tool

Fig. 94 Use a special tool to remove and install the balancer

Front Cover And Oil Seal

REMOVAL & INSTALLATION

All Except 1988-91 7.4L Engines

OEM ② MODERATE

◆ See Figures 95 and 96

■ This procedure may require engine removal, depending upon your particular boat. If necessary, remove the engine as detailed previously in this section.

1. Open the drain valves and drain the coolant from the block and exhaust manifold(s). Loosen the alternator and power steering brackets to provide slack, and then remove the drive belts.

2. Remove the oil pan on the 7.4L/8.2L V8, not always a bad idea on the other engines either, but OMC suggests that it must be removed on the big block engines. Although we can't recommend it, many people will leave the pan on and cut the seal as detailed in the procedures for the 1988-91 7.4L.

3. Remove the alternator and power steering pump bracket bolts.

4. Remove the water circulation pump.

5. Remove the harmonic balancer and pulley as detailed previously in this section.

6. Loosen the mounting bolts and remove the front cover.

7. If the oil seal needs replacement, pry it out from the front (outer) side of the cover with a small prybar.

8. Remove the front cover gasket.

To install:

9. Clean all gasket material from the cover and block mating surfaces with a scraper of putty knife. Be careful not to knock any pieces of gasket

into the timing assembly.

10. If you removed the oil seal, install a new one with the lip toward the inside of the cover. Position a support under the seal and cover and then press the seal into the cover with the proper tool (#J-23042 - 1986-93 engines, #J-35468 - 1992-98 7.4L/8.2L engines and 1994-98 4.3L/5.0L/5.7L engines).

11. Coat the lip of the oil seal with engine oil. Coat both sides of a new gasket with sealant and then position the gasket onto the engine. Install the cover so that all the bolt holes line up; there are dowel pins on the cylinder block that will help alignment. Tighten the bolts to:
• 80 inch lbs. (9 Nm) on 1986-96 V6 engines
• 106 inch lbs. (12 Nm) on 1997-98 V6 engines
• 80 inch lbs. (9 Nm) on 1986-97 5.0L/5.7L V8 engines (except 1992 5.7 LE and 1997 5.7Si)
• 100 inch lbs. (11 Nm) on 1992 5.7 LE engines
• 106 inch lbs. (12 Nm) on 1997 5.7FSi and all 1998 5.7L V8 engines
• 96 inch lbs. (11 Nm) on 1992-97 7.4L V8 engines
• 106 inch lbs. (12 Nm) on 1998 7.4L V8 engines
• 120 inch lbs. (14 Nm) on 8.2L V8 engines
Don't forget to use new cover on engines with a composite cover!

12. Install the oil pan.

13. Install the harmonic balancer and its pulley.

14. Install the circulation pump and its pulley. Pull the belts back on. Check their tension adjustment.

15. Install the engine if removed. Add oil and water/coolant, start the engine and check for any leaks.

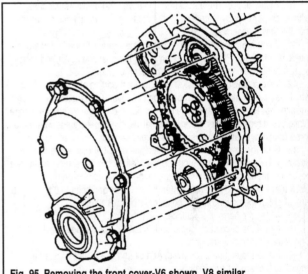

Fig. 95 Removing the front cover-V6 shown, V8 similar

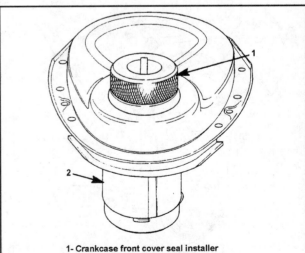

1- Crankcase front cover seal installer
2- Support (to prevent distorting cover)

Fig. 96 Make sure the front cover is supported securely before pressing in the oil seal

1988-91 7.4L Engines

◆ See Figure 97

■ **This procedure may require engine removal, depending upon your particular boat. If necessary, remove the engine as detailed previously in this section.**

1. Open the drain valves and drain the coolant from the block and exhaust manifold. Loosen the alternator and power steering brackets (or the idler pulley) to provide slack, and then remove the drive belts.

2. Drain the oil if you haven't already done so.

3. Remove the alternator and power steering pump bracket bolts.

4. Remove the water circulation pump.

5. Remove the harmonic balancer and pulley as detailed previously in this section.

6. Remove the two 7/16 in. oil pan-to-front cover bolts. Loosen the 3/8 in. cover mounting bolts and pry the front cover outward slightly. Insert a razor knife between the cover and block so that it is flush with the inside of the cover and then carefully cut the oil pan seal at each side of the cover.

7. Remove the cover.

8. If the oil seal needs replacement, pry it out from the front (outer) side of the cover with a small prybar.

9. Remove the front cover gasket.

To install:

10. Clean all gasket material from the cover and block mating surfaces with a scraper of putty knife. Be careful not to knock any pieces of gasket into the timing assembly.

11. If you removed the oil seal, install a new one with the lip toward the inside of the cover. Position a support under the seal and cover and then press the seal into the cover with the proper tool (#J-22102).

12. Take a new oil pan gasket and cut out a portion of it equal to that which you cut previously when removing the cover. Coat the oil pan mating surface with RTV sealer and carefully position the cut gasket into the flange. While you have the sealer out, run a small bead along the joints where the three (pan, cover and block) surfaces come in contact with one another.

13. Coat the lip of the oil seal with engine oil. Coat both sides of a new gasket with sealant and then position the gasket onto the engine. Install the cover so that all the bolt holes line up; there are dowel pins on the cylinder block that will help alignment. Tighten the cover-to-pan bolts to 8 ft. lbs. (10 Nm). Tighten the cover-to-block bolts to 6 ft. lbs. (8 Nm).

14. Install the harmonic balancer and its pulley.

15. Install the circulation pump and its pulley. Pull the belts back on. Check their tension adjustment.

16. Install the engine if removed. Add oil and water/coolant, start the engine and check for any leaks.

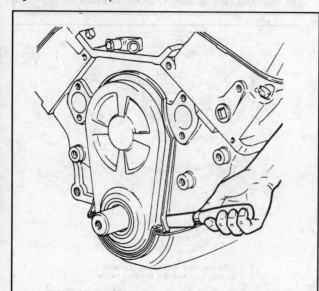

Fig. 97 Use a sharp knife to cut the oil pan seal

REMOVAL & INSTALLATION

◆ See Figures 98 and 99

1. Disconnect the battery cables.

2. Remove the mounting bolts for the alternator and power steering pump brackets.

3. Remove the crankshaft pulley and harmonic balancer as previously detailed in this section.

4. Remove the front cover as previously detailed in this section.

5. Rotate the camshaft slightly so that it creates tension on one side of the timing chain (either side is OK). Find a reference point on the same side of the cylinder block as the side that the timing chain is tight on and then measure from this point to the outer edge of the chain.

6. Rotate the camshaft in the opposite direction until the other side of the chain is tight. Press the inner side of the chain outward until it stops and then measure from your reference point on the cylinder block (obviously, do this from the same side of the chain as you did in the previous step) to the outer edge of the chain. This is timing chain deflection and it should be no more than 0.625 in. (16mm). If it is more than specification, the chain will require replacement.

7. Look carefully at the camshaft and crankshaft sprockets—you should notice a small indent on the front edge of one of the teeth on each sprocket. Bump the engine over until these two marks are in alignment as shown in the illustration—crank mark at 12:00 and cam mark at 6:00, a remote starter will work or you can screw the damper bolt back into the crankshaft and turn it with a wrench.

8. Dab a little paint across one of the chain links and the camshaft sprocket. Loosen the camshaft sprocket retaining bolts (three, although on the certain 1994-98 V6 engines, one will be a nut—don't be concerned if the stud comes out with the nut), grasp the sprocket on each side with the chain still attached and wiggle it off the shaft. It should come off readily, but if not, tap the bottom edge lightly with a rubber mallet.

⁑ WARNING

Never rotate the crankshaft once the timing chain has been removed. If moved, you'll risk damage to the pistons and/or valve train.

9. Mount a gear puller (#J-1619) over the crankshaft pulley and pull it off the shaft. Don't lose the key.

To install:

10. Clean the chain and sprockets in solvent and let them air dry. Check the chain for wear and damage, making sure there are no loose or cracked links. Check the sprockets for cracked or worn teeth.

11. Coat the woodruff key lightly with adhesive and position it on the crankshaft. Install the crankshaft sprocket onto the shaft with an installation tool (#J-22102).

12. Install the timing chain onto the camshaft sprocket so that the paint marks made during removal match up. If they do, and you haven't moved the engine, the timing marks on the two sprockets should also. Hold the sprocket/chain in both hands so the chain is hanging down, engage the chain around the crankshaft sprocket and then slide the cam sprocket/chain onto the camshaft making sure the dowel pin aligns with the hole in the sprocket. Do not force it! On 1994-98 V6 engines, make sure the balance shaft drive gear stud fits through the hole. Tighten the three mounting bolts/nut to:

- 20 ft. lbs. (27 Nm) on 1986-93 4.3L V6 and 5.0L/5.7L V8 (except 1992 5.7 LE) and all 7.4L V8 engines
- 100 inch lbs. (11 Nm) on 1992 5.7 LE
- 21 ft. lbs. (28 Nm) on 1994-98 4.3L V6 and 5.0L/5.7L V8 engines
- 25 ft. lbs. (34 Nm) on 8.2L V8 engines

⁑ CAUTION

Never force the camshaft sprocket onto the shaft or use a hammer, lest you loosen the rear welch plug.

13. Install the front cover, harmonic balancer and water pump pulley.

14. Install and adjust the drive belts.

15. Install the seawater pump and reconnect the alternator and power steering brackets.

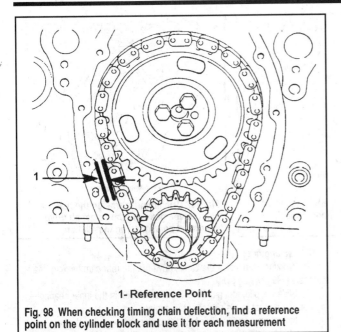

Fig. 98 When checking timing chain deflection, find a reference point on the cylinder block and use it for each measurement

1- Reference Point

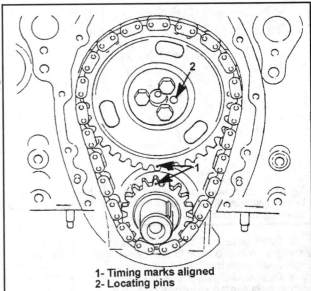

1- Timing marks aligned
2- Locating pins

Fig. 99 The two marks on each of the timing sprockets must be in alignment before removing or installing the timing chain

■ Remember that when the timing marks were aligned properly, the No. 4 (or No. 6, V8) cylinder was at TDC so if you removed the distributor for some reason, the rotor should be at the No. 4 (No. 6) post on the cap, NOT at the No. 1 post.

Balance Shaft

REMOVAL & INSTALLATION

1994-98 4.3L V6 Engines Only

◆ See Figures 100, 101 and 102

1. Remove the distributor cap and mark the position of the No. 1 cylinder terminal on the side of the housing. Matchmark the housing to the engine and then remove the distributor.
2. Remove the intake manifold.
3. Remove the drive belts and then remove the water circulating pump and its pulley.
4. Remove the harmonic balancer and its pulley.
5. Remove the oil pan.
6. Remove the front cover.
7. Fashion a small wedge out of an old piece of wood and jam it (carefully!) in between the teeth of the balance shaft driven gear and the camshaft drive gear (this is behind the sprocket) to hold the shafts from turning. Loosen, but do not remove, the driven gear bolt—although not on all applications, the chances are very good that this will be a Torx® head.
8. Loosen the three camshaft sprocket bolts/nut and then pull off the sprocket with the timing chain attached. Remove the camshaft drive gear.
9. Now you can remove the bolt loosened previously and pull off the balance shaft drive gear.
10. Remove the two mounting bolts from the balance shaft retainer and lift off the plate. You've probably already noticed that these will be Torx® fasteners also.
11. Insert a suitable prybar between the rear of the shaft and the cylinder block and carefully shimmy the shaft forward and out of the bearing housing in the rear of the block. Slide the entire shaft out through the front of the block. Be careful not to knock any debris into the inside of the cylinder block. You can also use a soft mallet and carefully tap the front bearing assembly to accomplish this step.
   To install:
12. Clean the balance shaft in solvent and let it dry completely. Inspect the rear bearing for wear or damage. Inspect the front bearing for excessive side play on the shaft, wear or damage. Check the bearing bore in the front of the cylinder block for scoring.

13. Lubricate the front and rear bearing with motor oil and then carefully slide the shaft through the front bore until it feeds into the rear bearing. OMC suggests using a Driver Handle (#J-8092) and Shaft Installer (#J-36996) to accomplish shaft installation, but many technicians simply tap the front edge of the shaft with a plastic mallet until the retaining ring on the front bearing seats against the cylinder block—you make the call as to which option you choose.
14. Install the thrust plate and tighten the two fasteners to 120 inch lbs. (14 Nm).
15. Position the driven gear onto the balance shaft with the bolt finger tight. Make sure that the timing mark is at the bottom of the gear.
16. Install the camshaft drive gear so that the mark aligns with the one on the balance shaft driven gear (it will be pointing UP). You may have to wiggle the two shafts until you get the marks to align correctly. Install the stud and tighten to 12 ft. lbs. (16 Nm).
17. Remove the balance shaft drive gear bolt again and coat the threads with Loctite. Screw it in and tighten it to 15 ft. lbs. (20 Nm); and then turn the bolt an additional 35°.
18. Install the timing chain and front cover.
19. Install the intake manifold.

Camshaft

CHECKING LIFT

◆ See Figures 103, 104 and 105

■ If the shaft is out of the engine, you can use a micrometer to take the heel-to-lobe measurement and the side-to-side measurement. Subtract the second measurement from the first to get lobe lift. Most times though, the shaft will still be in the cylinder block so perform the following procedure.

1. Tag and disconnect the electrical connectors at the ignition coil.
2. Remove the cylinder head cover and rocker arms as detailed previously.
3. Using a special adaptor (#J-8520), connect a dial indicator so that its tip is positioned on the end of the pushrod—the adaptor should screw onto the end of the rocker stud.
4. Slowly rotate the crankshaft in the direction of engine rotation until the valve lifter is riding on the heel (back side of lobe) of the camshaft lobe. The pushrod should be at its lowest point when the lifter is on the heel.

■ A remote starter works well for turning the engine over in this situation.

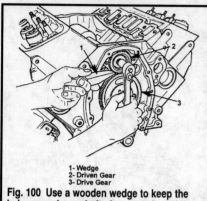

Fig. 100 Use a wooden wedge to keep the balance and camshafts from rotating while loosening the retaining bolt.

1- Wedge
2- Driven Gear
3- Drive Gear

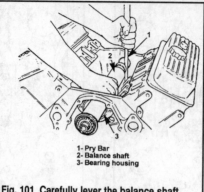

Fig. 101 Carefully lever the balance shaft forward and out of the cylinder block

1- Pry Bar
2- Balance shaft
3- Bearing housing

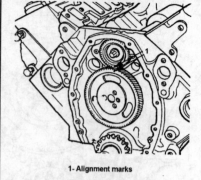

Fig. 102 The timing marks on the balance shaft driven gear and the camshaft drive gear must be aligned on installation

1- Alignment marks

5. Set the indicator to **0** and then rotate the engine until the pushrod is at the highest point of its travel. Camshaft lift should be as detailed in the Engine Specifications chart.

6. Continue rotating the engine until the pushrod is back at its lowest position—make sure that the indicator still reads **0**.

7. Repeat this procedure for the remaining pushrods.

8. Install the rocker arms and adjust the valve clearance.

9. Install the cylinder head cover and reconnect the coil leads.

REMOVAL & INSTALLATION

◆ **See Figures 106 and 107**

1. Remove the cylinder head covers and rocker assemblies as previously detailed in this section. Remove the pushrods.

2. Remove the intake manifold.

3. Remove the lifter restrictor mounting bolts and lift out the restrictors (if equipped). Matchmark the lifters to their bores and then lift out the valve lifters with a small magnet and store them in a rack with labels so they can be reinstalled in their original locations.

4. Remove the front cover and timing chain/sprockets as previously detailed in this section.

5. Remove the fuel pump and push rod on early V6 and small block V8 engines.

6. Remove the inner camshaft drive gear on 1994-98 V6 engines.

7. Remove the thrust plate/retainer.

8. Thread two (three on later V8s) 5/16-18 x 4 in. bolts into the camshaft bolt holes and carefully pull the camshaft out of the cylinder block. You may have to wiggle it back and forth a bit, so be sure you don't lean it up or down or else you could damage the bearings.

9. If the camshaft bearings are to be removed, you will need to remove the flywheel as previously detailed. Although it is not necessary, removing the crankshaft will also facilitate bearing removal, make sure that you move the connecting rods out of the way so they do not interfere with bearing removal.

1986-96 engines:

10. Working from inside the block, drive out the rear cam bearing expansion plug (welch plug).

11. Slide the pilot tool (#J-6098-01) into position in the inner bearing—the bearing closest to the center of the engine.

12. Install a nut and washer onto the puller screw so that the screw can be threaded into the tool with the nut still extending out the front of the cylinder block.

13. Index the pilot over the screw so that the open end is toward the nut on the puller screw.

14. Install the remover so the shoulder is facing the No. 3 bearing and it has sufficiently engaged the threads of the tool.

15. Hold the screw shaft with a wrench while turning the puller nut (front) with another until the bearing comes out. Repeat this procedure for the No. 2 bearing.

16. Now remove the pilot from the shaft, remove the tool and reassemble it on the rear of the engine to remove the No. 4 bearing.

17. Install the remover onto the drive handle so that the shoulder is against the handle. The front and rear bearings can now be driven out from the outside of the block.

1997-98 engines:

18. Working from inside the block, drive out the rear cam bearing expansion plug (welch plug).

19. Insert the camshaft bearing removal tool (#J-33049) through the first bearing (front of block) and into the bearing being removed. Make sure the collet is the correct one for your engine and turn the tool until it has seated into the bearing.

20. Push the centering cone up against the cylinder block and into the recess of the No. 1 bearing so that the tool centers itself. Drive the bearing out of the block.

21. Repeat the procedure to remove the remaining bearings; noting that the No. 1 bearing needs to be removed from the rear of the block to allow for proper centering of the tool.

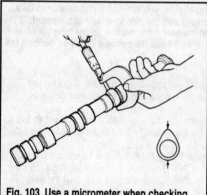

Fig. 103 Use a micrometer when checking the camshaft lift with the shaft out of the engine

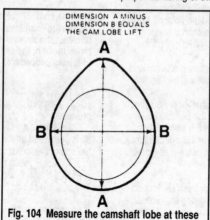

DIMENSION A MINUS
DIMENSION B EQUALS
THE CAM LOBE LIFT

Fig. 104 Measure the camshaft lobe at these two points with the micrometer

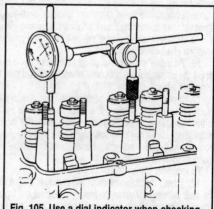

Fig. 105 Use a dial indicator when checking the camshaft lift with the shaft in the engine

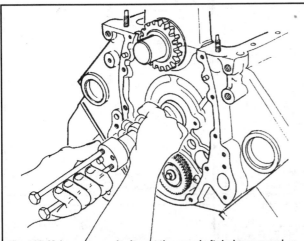

Fig. 106 Make sure you don't cant the camshaft during removal or installation. Note the two 4 in. bolts screwed into the end of the shaft (V6 shown)

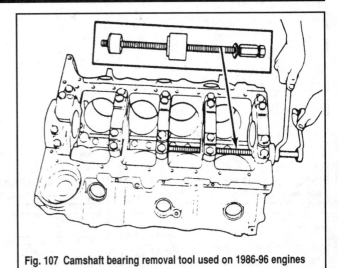

Fig. 107 Camshaft bearing removal tool used on 1986-96 engines

To install:

22. If you removed the bearings:
1986-96 engines:
23. Install the installer tool on the driver handle and then drive the front and rear bearings into the block from the outside toward the center (see the following Note on oil hole alignment).
24. Remove the handle from the pilot tool and install the inner bearing on the tool.
25. Position the tool and bearing to the rear of the inner bore. Install the screw shaft, with the remover, through the block and onto the pilot—from the front of the cylinder block.
26. Align the oil hole in the bearing with the oil gallery hole and then snug the puller nut up against the adaptor. Using two wrenches again, hold the screw shaft with one while turning the puller with the other until the bearing is in position.

■ **The oil hole(s) will not be visible during installation. To make installation easier, align the holes in the bearing with those in the block and then mark the opposite side of the bearing/block. To make it easier for you, the No.1 bearing should have the holes on equal sides of the 6 o'clock position; Nos. 2, 3 and 4 bearing holes should be positioned at the 5 o'clock position (toward the left side of engine) and even with the bottom of the cylinder bore; No. 5's oil hole should be at the 12 o'clock position.**

27. Repeat the last step for the remaining bearings.
28. Coat a new rear welch plug with sealant and install it so that it is flush with the surface of the cylinder block, or no more than 1/32 in. (0.792mm) deep when measured from the outer edge of its recess.
1997-98 engines:
29. Bearing installation on these engines is essentially the reverse of removal. Take note of the following though:
• Install the front and rear bearings first, from the outside toward the inside.
• Bearing bore sizes may vary, be sure you have the correct bearings when replacing them.
• Make sure that the oil holes in the bearing align with the holes in the block, during and after installation.
• Coat a new rear welch plug with sealant and install it so that it is flush with the surface of the cylinder block, or no more than 1/32 in. (0.792mm) deep when measured from the outer edge of its recess.
On all engines now:
30. Inspect the camshaft as detailed previously.
31. Coat the camshaft journals with engine oil. Coat the camshaft lobes and distributor drive gear teeth with GM Engine oil supplement or Molykote (in a pinch, just use engine oil).
32. Reinstall the long bolts and carefully insert the shaft into the cylinder block and slide it all the way in. Be very careful not to damage the bearings.
33. Remove the installation bolts and install the thrust plate/retainer (if equipped) and tighten the bolts to 106 inch lbs. (12 Nm) on 1986-96 V6 engines and all V8 engines, or 124 inch lbs. (14 Nm) on 1997-98 V6 engines.

34. Install the fuel pump and push rod on engines so equipped.
35. Install the drive gears, timing chain and front cover.
36. Drop the lifters back into their original bores so that they are aligned with the matchmarks and then install the restrictors. Install the push rods.
37. Install the intake manifold.
38. Install the rocker assemblies and the cylinder head covers.

Cylinder Head

REMOVAL & INSTALLATION

◆ **See Figures 108, 109, 110, 111, 112 and 113**

1. Drain the water from the cylinder block and manifold.
2. Remove the fuel line support brackets. Disconnect the fuel line at the carburetor/throttle body and fuel pump, plug the fitting holes and remove the line.
3. Remove the cylinder head cover and rocker assemblies as detailed previously in this section.
4. Remove the intake and exhaust manifolds as previously detailed; you can leave the carburetor/throttle body attached to the intake manifold if you like.
5. If you intend to remove the valve lifters, now is the time to do it. Either way, make sure that you cover the valley of the cylinder block carefully with plenty of rags to prevent dirt from entering any of the passages or settling on any components.
6. Tag and disconnect the spark plug wires at the plugs; move them out of the way. Although not necessary, it's a good idea to remove the plugs themselves also; plug the holes if you do.
7. Remove or relocate any components or connections that may interfere with the removal of an individual cylinder head.
8. Loosen the cylinder head bolts in the reverse order of the illustrated tightening sequence and then carefully lift the head off the block. You may need to persuade it with a rubber mallet—be careful! Set the head down carefully on two support blocks; do not sit it on cement.

■ **Its always a good idea to keep a record of which bolts came from which holes. It may sound silly, but on many engines they are different sizes and you wouldn't be the first person to break off a long bolt while tightening it in a short hole. Spend the extra few seconds and do this!**

To install:

9. Carefully, and thoroughly, remove all residual head gasket material from the cylinder head and block mating surfaces with a scraper or putty knife. Check that the mating surfaces are free of any nicks or cracks. Make sure there is no dirt or old gasket material in any of the bolt holes. Refer to the Engine Rebuilding section found for complete details on inspection and refurbishing procedures.

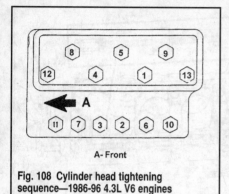

Fig. 108 Cylinder head tightening sequence—1986-96 4.3L V6 engines

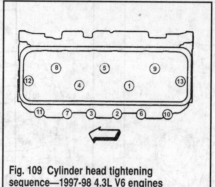

Fig. 109 Cylinder head tightening sequence—1997-98 4.3L V6 engines

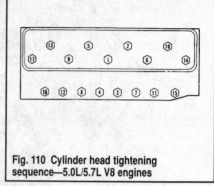

Fig. 110 Cylinder head tightening sequence—5.0L/5.7L V8 engines

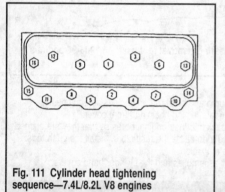

Fig. 111 Cylinder head tightening sequence—7.4L/8.2L V8 engines

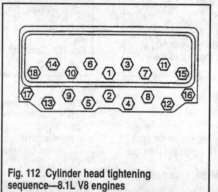

Fig. 112 Cylinder head tightening sequence—8.1L V8 engines

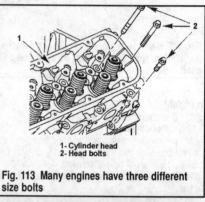

1- Cylinder head
2- Head bolts

Fig. 113 Many engines have three different size bolts

10. Position a new gasket over the cylinder block dowel pins. Do not use any sealer as the gasket is coated with a laquer which will provide the proper sealing effect once the engine gets to normal operating temperature the first time. DO NOT use automotive-type steel gaskets. Make sure the word **UP** is facing, you guessed it, up, if noted on your gasket.

11. Position the cylinder head over the dowels in the block. Coat the threads of the head bolts with Gasket Sealing compound and install them finger tight. It never hurts to use new bolts, although it's not necessary.

12. Tighten the bolts, a little at a time, in the sequence illustrated, until the proper tightening torque is achieved.

• 1986-97 4.3L V6 and 5.0L/5.7L V8 engines (except the 1997 5.7GSi: the first step should be 25 ft. lbs. (34 Nm), the second step should be 45 ft. lbs. (61 Nm), while the last step should be 65 ft. lbs. (88 Nm).

• 1998 4.3L V6 engines: tighten all the bolts, in sequence, to 22 ft. lbs. (30 Nm). Next, turn the short bolts (#11, 7, 3, 2, 6 & 10) an additional 55°; the medium bolts (#12 & 13) an additional 65°; and the long bolts (#1, 4, 8, 5 & 9) an additional 75°. Obviously you will need a torque/angle meter for these last passes.

• 1997 5.7GSi and all 1998 5.0L/5.7L V8 engines: tighten all the bolts, in sequence, to 22 ft. lbs. (30 Nm). Next, turn the short bolts (#3, 4, 7, 8, 11, 12, 15 & 16) an additional 55°; the medium bolts (#14 & 17) an additional 65°; and the long bolts (#1, 2, 5, 6, 9, 10 & 13) an additional 75°. Obviously you will need a torque/angle meter for these last passes.

• 7.4L/8.2L V8 engines: the first step should be 35 ft. lbs. (47 Nm), the second step should be 65 ft. lbs. (88 Nm), while the last step should be to 80 ft. lbs. (108 Nm) on 1988-91 engines or 85 ft. lbs. (115 Nm) on 1992-98 engines.

13. Install the rocker assemblies and the cylinder head cover. Don't forget the baffle plate and restrictors/retainers on engines so equipped.

14. Install the manifolds and connect the fuel line. Don't forget to remove the fitting plugs.

15. Install the spark plugs if they were removed and then connect the plug wires.

16. Install or connect any other components removed to facilitate getting the head off.

17. Add coolant/water, connect the battery and check the oil. Start the engine and run it for a while to ensure that everything is operating properly. Keep an eye on the temperature gauge.

18. It never hurts to re-tighten the cylinder head bolts again after 20 hours of operation.

Engine Circulating (Water) Pump

REMOVAL & INSTALLATION

◆ See Figure 114

 MODERATE

1. Disconnect the battery cables and then drain all water from the block and manifolds.

2. Drain all water/coolant from the cylinder block.

3. Loosen, but do not remove, the pump pulley mounting bolts.

4. Loosen the power steering pump and alternator bracket bolts and swivel them in until you are able to remove the belt(s). Different engines may have different systems, follow the belt back and loosen the appropriate components.

5. Now you can remove the pump pulley bolts along with the lock washers and clamping ring. Pull off the pulley.

6. Disconnect the water hoses at the pump.

7. Remove the mounting bolts and lift the pump off of the block.

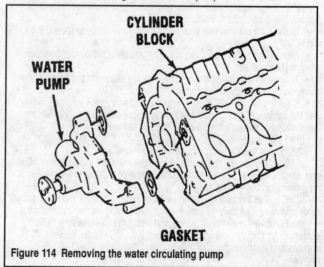

Figure 114 Removing the water circulating pump

To install:

8. Carefully scrape any old gasket material off both mounting surfaces. Inspect the pump for blockage, cracks or any other damage. Inspect the impeller for cracks. Replace either if necessary.

9. Coat both sides of a new gasket(s) with sealant and position on the cylinder block. Coat the threads of the pump mounting bolts with sealant, install the pump and tighten the bolts to:
- 30 ft. lbs. (41 Nm) on 4.3L V6 and 5.0L/5.7L/7.4L V8 engines
- 35 ft. lbs. (47 Nm) on 8.2L V8 engines.

10. Reconnect the water hoses and tighten the hose clamps securely.

11. Position the pump pulley and clamping ring on the boss. Screw the mounting bolts and lock washers in and tighten them securely.

12. Install the drive belt(s) and adjust them as detailed previously. Start the engine and check the system for leaks.

EXPLODED VIEWS

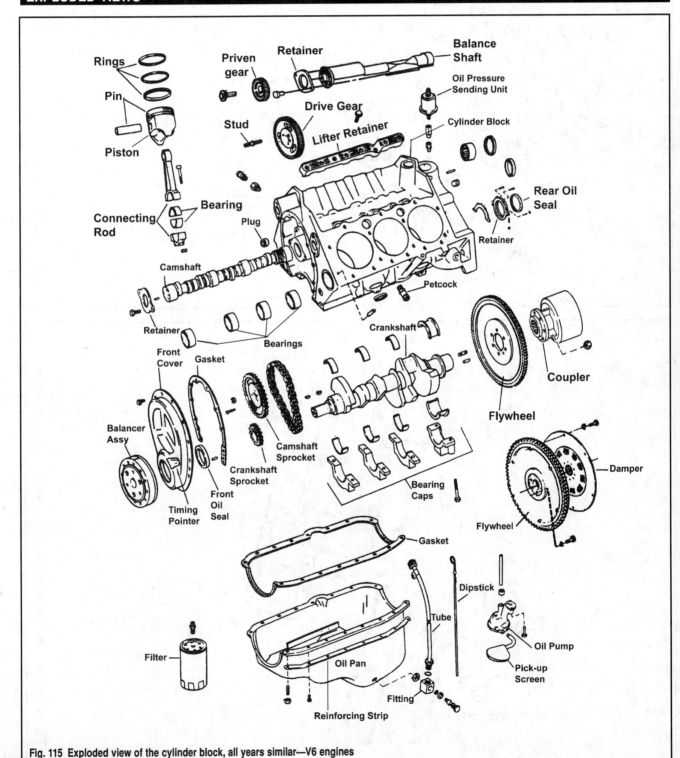

Fig. 115 Exploded view of the cylinder block, all years similar—V6 engines

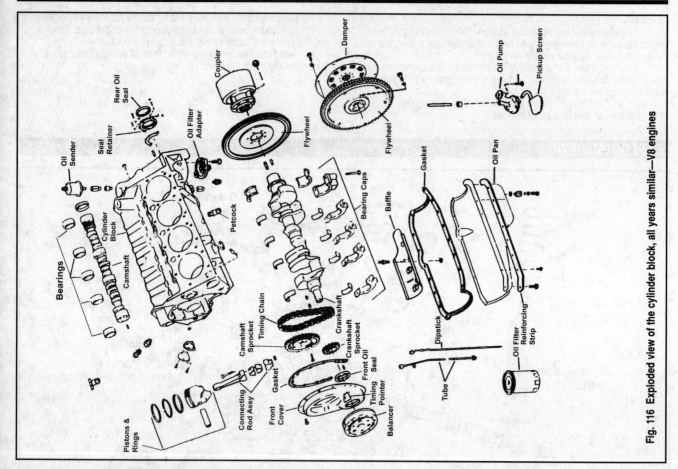

Fig. 116 Exploded view of the cylinder block, all years similar—V8 engines

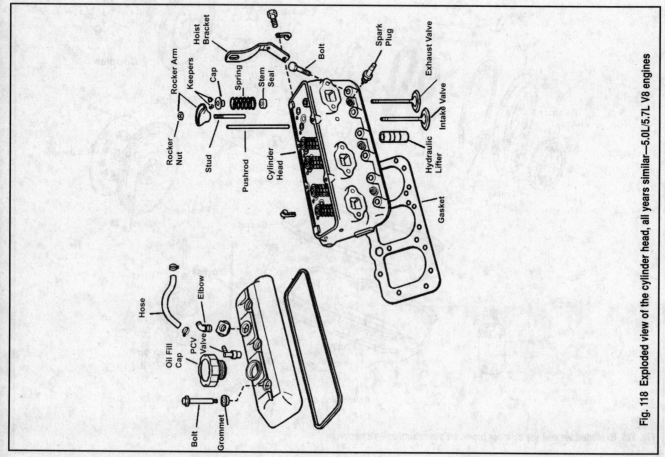

Fig. 118 Exploded view of the cylinder head, all years similar—5.0L/5.7L V8 engines

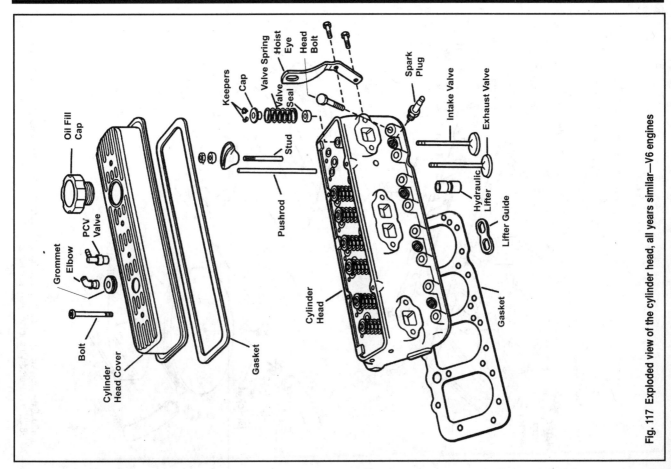

Fig. 117 Exploded view of the cylinder head, all years similar—V6 engines

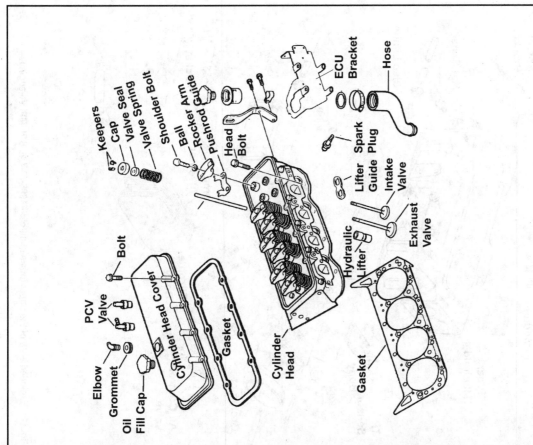

Fig. 119 Exploded view of the cylinder head, all years similar—7.4L/8.2L V8 engines

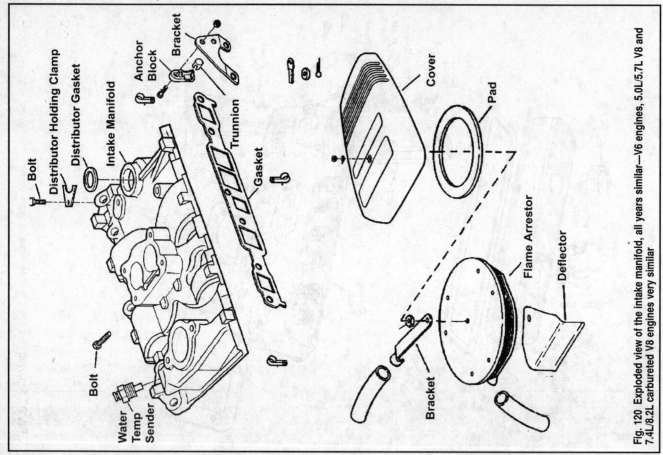

Fig. 120 Exploded view of the intake manifold, all years similar—V6 engines, 5.0L/5.7L V8 and 7.4L/8.2L carbureted V8 engines very similar

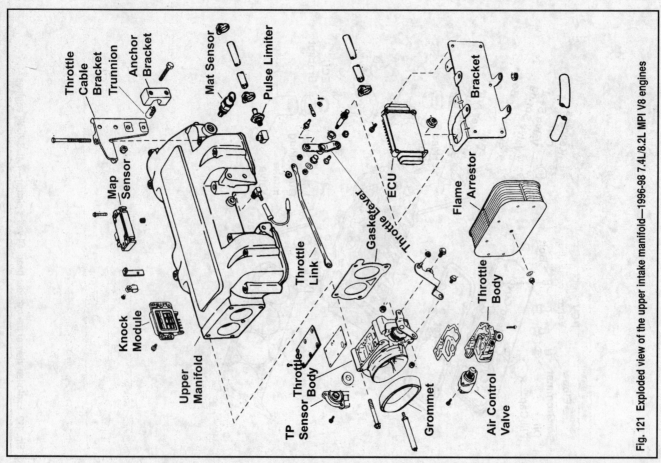

Fig. 121 Exploded view of the upper intake manifold—1996-98 7.4L/8.2L MPI V8 engines

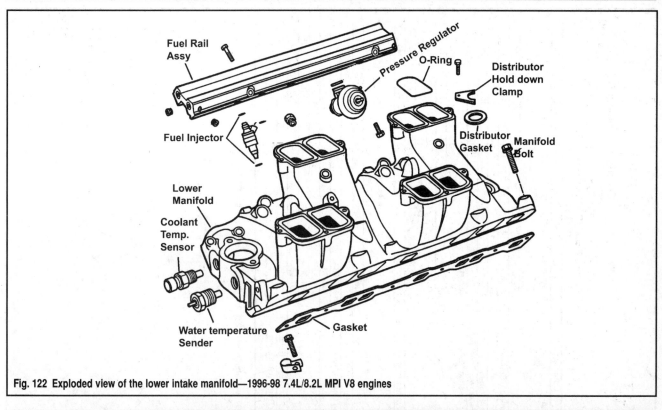

Fig. 122 Exploded view of the lower intake manifold—1996-98 7.4L/8.2L MPI V8 engines

SPECIFICATION CHARTS

TORQUE SPECIFICATIONS - 4.3L V6

Component	Detail	ft. lbs.	inch lbs.	Nm
Balance Shaft				
Driven Gear		15 + 35 deg.	-	20 + 35 deg.
Retainer (thrust plate)		-	120	14
Camshaft				
Sprocket	1986-93	20	-	27
	1994-98	21	-	28
Thrust plate	1986-96	-	106	12
	1997-98	-	124	14
Connecting Rod Cap	1986-96	45	-	60
	1997-98	20 + 70 deg.	-	27 + 70 deg.
Crankshaft				
Main Bearing Cap	1986-91	70	-	95
	1992-96	70	-	95
	1997-98			
	Outer on #2, #3, #4	80	-	110
	All others	77	-	105
Oil Seal Retainer		-	135	15.3
Cylinder Block				
Drain Plug		15	-	20
Oil Gallery Plug	Port	15	-	20
	Starboard	18	-	25
Cylinder Head				
Bolts	1986-97			
	Step 1	25	-	34
	Step 2	45	-	61
	Step 3	65	-	88
	1998	①	-	①
Cover	1986-93	-	100	11
	1994-96	-	62-115	7-13
	1997-98	-	106	12
Dipstick Tube		15-18	-	20-24
Distributor Clamp Bolt		20	-	27
Engine Mounts				
Front	To engine	32-40	-	43-54
	Adjusting nut	100-120	-	136-163
	Bolt	60-75	-	81-102
Rear	Locknut	28-30	-	38-40
	Mounting bolts	20-25	-	27-34
Exhaust Elbow	1986-93	10-12	-	14-16
	1994-98	12-18	-	16-24
Exhaust Manifold	1986-90	18-22	-	24-30
	1991-98	20-26	-	27-35

① 1st pass: 22 ft. lbs. (30 Nm); 2nd pass: short bolt - 55 deg., medium bolt - 65 deg., long bolt - 75 deg.

TORQUE SPECIFICATIONS - 4.3L V6

Component			ft. lbs.	inch lbs.	Nm
Exhaust Pipe	To transom	1986-90	10-12	-	14-16
		1991-98	20-25	-	27-34
Flywheel	Bellhousing	1986-91	28-36	-	38-39
		1992-98	32-40	-	43-54
	Coupler Nut		40-45	-	54-61
	Cover		-	60-84	7-9
	Stud	1986-91	20-25	-	27-34
		1992-98	15-20	-	20-27
Front Cover	Timing Sensor		-	48-64	5-7
		1986-96	-	80	9
		1997-98	-	106	12
Harmonic Balancer		1986-90	60	-	81
		1991-96	70	-	95
		1997-98	74	-	100
Ignition Coil	Clamp		-	7-10	0.8-1.13
Intake Manifold		1986-95	30	-	41
		1996	35	-	48
		1997-08	11	-	15
Main Bearing Cap		1986-91	70	-	95
	Outer on #2, #3, #4	1992-96	70	-	95
	All others		80	-	110
		1997-98	77	-	105
Manifolds	Exhaust	1986-90	18-22	-	24-30
		1991-98	20-26	-	27-35
	Intake	1986-95	30	-	41
		1996	35	-	48
		1997-08	11	-	15
Oil Filter Adaptor (Remote)	To block		20-25	-	27-34
Oil Filter Bypass Valve			20	-	27
Oil Pan	Bolts/nuts	1986-96			
	1/4 x 20 bolts		6-7.5	-	8-10
	5/16 x 18 bolts		12-15	-	16-20
		1997-98			
	Nuts		17	-	23
	Bolts/Studs		18	-	25
Dipstick Tube Fitting			15-18	-	20-24
Drain Retainer			15-18	-	20-24

TORQUE SPECIFICATIONS - 4.3L V6

Component			ft. lbs.	inch lbs.	Nm
Oil Pump	Cover bolts	1986-96	6-9	-	8-12
		1997-98	-	106	12
Oil Pressure Sender	Mounting Bolts		65	-	90
	Adaptor		10-14	-	14-19
	Sender		10-14	-	14-19
Oil Seal Retainer			-	135	15.3
Oil Withdrawal Tube (flare)			15-18	-	20-24
Power Steering Pump	Bracket	Bolt	26-30	-	35-41
		Stud/nut	13-15	-	18-20
	Lines	Large fitting	15-17	-	20-23
		Small fitting	10-12	-	14-16
Raw Water Pump Bracket			30	-	41
Rocker Arm	Nut	1986-96	40	-	54
		1997-98	20	-	27
	Stud		35	-	47
Spark Plugs			22 ①	-	30 ①
Timing Chain Cover		1986-96	-	80	9
		1997-98	-	106	12
Valve Cover		1986-93	-	100	11
		1994-96	-	62-115	7-13
		1997-98	-	106	12
Valve Lifter Guide Retainer			12	-	16
Water Pump (circulating pump)			30	-	41
Water Temperature Sender			18-22	-	24-30

① 1997 and later: New - 22 ft. lbs. (30 Nm); used - 15 ft. lbs. (20 Nm)

TORQUE SPECIFICATIONS - 5.0L/5.7L V8

Component			ft. lbs.	inch lbs.	Nm
Camshaft	Sprocket	1986-93	20	-	22
		1992 5.7 LE	-	100	11
		1994-98	21	-	28
	Retainer Bolt	1986-97	-	124	14
		1998	-	106	12
Connecting Rod Cap		1992-97 ①	45	-	60
		1998	20 +55 deg	-	27 +55 deg
Coolant Temp. Sender			15	-	20
Crankshaft	Main Bearing Cap	1986-91	70	-	95
		1992-97 ① Outer on #2, 3, 4	70	-	95
		All others	80	-	110
		1998	②	-	②
	Oil Seal Retainer		-	135	15.3
	Pulley Bolt		43	-	58
Cylinder Head	Bolts 1986-97 ①	Step 1	25	-	34
		Step 2	45	-	61
		Step 3	65	-	88
		1998	③	-	③
	Core Plug		15	-	20
	Cover	1986-96 ④	-	45	5
		1997 ①	-	90	10
		1998-03	-	106	12
Dipstick Tube			15-18	-	20-24
Distributor	Cap Bolt		-	21	2.4
	Clamp Bolt	1986-97 ①	20	-	27
		1998	18	-	25
Engine Mounts	Front	To engine	32-40	-	43-54
		Adjusting nut	100-120	-	136-163
		Bolt/Lower Nut	60-75	-	81-102
	Rear		28-30	-	38-40
Exhaust Elbow	To Manifold	1986-93	10-12	-	14-16
		1994-98	12-18	-	16-24
Exhaust Manifold			20-26	-	27-35
Exhaust Pipe	To transom	1986-90	10-12	-	14-16
		1991-98	20-25	-	27-34
Flame Arrestor	Cover		-	30-40	3.4-4.5
	Mounting studs		-	65-80	7-9
Flywheel	Bellhousing	1986-91	28-36	-	38-49
		1992-98	32-40	-	43-54
	Coupler nut		40-45	-	54-61
	Coupler stud	1986-91	20-25	-	27-34
		1992-98	15-20	-	20-27
Front Cover	Cover		-	60-84	7-9
		1986-97 ①	-	80	9
		1998	-	106	12

① 1997 5.7GSi: use 1998 specifications
② Inboard bolt: 74 ft. lbs. (100 Nm); Outboard bolt (4 bolt caps): 67 ft. lbs. (90 Nm)
③ 1st pass: 22 ft. lbs. (30 Nm); 2nd pass: short bolt - 55 deg., medium bolt - 65 deg., long bolt - 75 deg.
④ 1991 350 and 1992 5.7 LE: 65 inch lbs. (7 Nm)

TORQUE SPECIFICATIONS - 5.0L/5.7L V8

Component			ft. lbs.	inch lbs.	Nm
Harmonic Balancer	1986-90		60	-	81
	1991-96		70	-	95
	1997-98		74	-	100
Ignition Coil	Clamp		-	7-10	0.8-1.13
Intake Manifold	1986-97		30	-	41
	1998		⑥	-	⑥
Main Bearing Cap	1986-91		70	-	95
	1992-97 ①	Outer on #2, 3, 4	70	-	95
		All others	80	-	110
	1998		②	-	②
Manifolds	Exhaust		20-26	-	27-35
	Intake	1986-97	30	-	41
		1998	②	-	②
Oil Pan	1986-97 ①	1/4x20 bolts	6-7.5	-	8-10
		5/16x18 bolts	12-15	-	16-20
	1998	Corners	15	-	20
		All others	-	106	12
	Dipstick Tube Fitting		15-18	-	20-24
	Drain Plug		18	-	25
	Drain Retainer		15-18	-	20-24
Oil Pump	Cover bolts	1986-97 ①	6-9	-	8-12
		1998	-	106	12
	Mounting bolts		65	-	90
Oil Pressure Sender	Sender		10-14	-	14-19
	Adaptor		10-14	-	14-19
Oil Pressure Gauge Sensor			11	-	15
Oil Seal Retainer			-	135	15.3
Oil Withdrawal Tube (flare)			15-18	-	20-24
Power Steering Pump	Bracket	Bolt	26-30	-	35-41
		Stud/nut	13-15	-	18-20
	Lines	Large fitting	15-17	-	20-23
		Small fitting	10-12	-	14-16
Raw Water Pump Bracket			30	-	41
Rocker Arm	Nut		40	-	54
	Stud		35	-	47
Spark Plugs			22 ⑥	-	30 ⑥
Timing Chain Cover	1986-97 ①		-	80	9
	1998		-	106	12
Valve Cover	1986-96 ④		-	45	5
	1997 ①		-	90	10
	1998-03		-	106	12
Valve Lifter Guide Retainer			18	-	25
Water(Circulating) Pump			30	-	41
Water Temperature Sender			18-22	-	24-30
Windage Tray Nut			29	-	40

① 1997 5.7GSi: use 1998 specifications
② Inboard bolt: 74 ft. lbs. (100 Nm); Outboard bolt (4 bolt caps): 67 ft. lbs. (90 Nm)
③ 1991 350 and 1992 5.7 LE: 65 inch lbs. (7 Nm)
④ 1st pass: 27 inch lbs (3 Nm); 2nd pass: 106 inch lbs. (12 Nm); 3rd pass: 11 ft. lbs. (15 Nm)
⑥ 1998: New - 22 ft. lbs. (30 Nm); used - 11 ft. lbs. (15 Nm)

TORQUE SPECIFICATIONS - 7.4L/8.2L V8

Component				ft. lbs.	inch lbs.	Nm
Camshaft	Sprocket	7.4L		20	-	27
		8.2L		25	-	34
Connecting rod cap	1992-96	7.4L		48	-	65
		8.2L		73	-	99
	1997			48	-	65
	1998	7.4Gi		45	-	61
		7.4GSi		48	-	65
		8.2GSi		73	-	99
Coolant Temp. Sensor				15	-	20
Crankshaft	Main Bearing Cap	1988-91	Rear	70	-	95
			Others	110	-	149
		1992-98	7.4L	100	-	135
			8.2L	110	-	149
Cylinder Head	1988-91	Step 1		35	-	47
		Step 2		65	-	88
		Step 3		80	-	108
	1992-98	Step 1		35	-	47
		Step 2		65	-	88
		Step 3		85	-	115
	Cover	1988-95	Nuts	-	115	13
			Stud	-	60-90	7-10
		1996-97		-	60-90	7-10
		1998 (exc. Gi)	Nuts	-	115	13
			Stud	-	60-90	7-10
		1998 7.4Gi		-	72	8
Dipstick Tube				15-18	-	20-24
Distributor	Clamp Bolt			25	-	34
Engine Mounts	Front	To engine		32-40	-	43-54
		Adjusting nut		50-70	-	68-95
		Bolt/Lower Nut		60-75	-	81-102
	Rear			28-30	-	38-40
Exhaust Elbow	To Manifold	1988-93		10-12	-	14-16
		1994-98		12-18	-	16-24
Exhaust Manifold	Nut			20-26	-	27-35
	Bolt			24-28	-	33-38
Exhaust Pipe	To transom	1988-90		10-12	-	14-16
		1991-98		20-25	-	27-34
Flywheel	Bellhousing			32-40	-	43-54
	Bolt To Crank			70	-	95
	Coupler Nut			40-45	-	54-61
	Coupler Stud			20-25	-	20-27
	Cover			-	60-84	7-9
Front Cover	Timing sensor			-	48-64	5-7
	1988-91	Block bolts		6	-	8
		Pan bolts		8	-	10
	1992-97	7.4L		-	96	10.8
		8.2L		-	120	13.6
	1998	7.4L		-	106	12
		8.2L		-	120	13.6

TORQUE SPECIFICATIONS - 7.4L/8.2L V8

Component				ft. lbs.	inch lbs.	Nm
Fuel Rail	Bolt			-	89	10
	Stud			18	-	25
Harmonic Balancer	7.4L	Exc. 1998 Gi		85	-	115
		1998 7.4Gi		110	-	149
	8.2L			90	-	122
Intake Manifold	7.4L			30 ①	-	41 ①
	8.2L			35	-	47
Main Bearing Cap	1988-91	Rear		70	-	95
		Others		110	-	149
	1992-98	7.4L		100	-	135
		8.2L		110	-	149
Manifolds	Exhaust	Nut		20-26	-	27-35
		Bolt		24-28	-	33-38
	Intake	7.4L		30 ①	-	41 ①
		8.2L		35	-	47
MAP Sensor				18	-	25
Oil Filter Adaptor (Remote)				20-25	-	27-34
Oil Filter Bypass Valve				20	-	27
Oil Pan	All exc. 1998 7.4Gi	7.4L	To Block	-	160	18
			To Front Cover	-	70	8
		8.2L	To Block	-	200	22.5
			To Front Cover	-	120	14
	1998 7.4Gi			18	-	25
Oil Pump	Dipstick tube fitting			15-18	-	20-24
	Cover Bolts	Exc. 1998 7.4Gi		-	80	9
		1998 7.4Gi		-	106	12
	Mounting Bolts	7.4L		65	-	90
		8.2L		70	-	95
Power Steering Pump	Bracket	Bolt		26-30	-	35-41
		Stud/nut		13-15	-	18-20
	Lines	Large fitting		15-17	-	20-23
		Small fitting		10-12	-	14-16
Rocker Arm	Bolt	1992-98	Exc. 1998 Gi	40	-	54
			1998 7.4Gi	45	-	61
Spark Plugs				22 ②	-	30 ②
Timing Chain Cover	1988-91	Block bolts		6	-	8
		Pan bolts		8	-	10
	1992-97	7.4L		-	96	10.8
		8.2L		-	120	13.6
	1998	7.4L		-	106	12
		8.2L		-	120	13.6
Valve Cover	1988-95	Nuts		-	115	13
		Stud		-	60-90	7-10
	1996-97			-	60-90	7-10
	1998 (exc. Gi)	Nuts		-	115	13
		Stud		-	60-90	7-10
	1998 7.4Gi			-	72	8
Water (Circulating) Pump				30	-	41
Water Temperature Sender				18-22	-	24-30
Windage Tray Nut				29	-	40

① 7.4Gi: Lower manifold only; upper - 10 ft. lbs. (14 Nm)
② 1998: New - 22 ft. lbs. (30 Nm); used - 15 ft. lbs. (20 Nm)

ENGINE SPECIFICATIONS - 4.3L V6

Component				Standard (in.) ①	Metric (mm) ①
Piston Rings - Compression	Groove Clearance		1986-96	0.0012-0.0032	0.0305-0.0813
			1997-03	0.02-0.06	0.508-1.524
		Service Limit	1986-96	+0.001	+0.025
			1997-98	0.004 Max	0.10 Max
	Gap	Production	Top:	0.010-0.020	0.254-0.508
			2nd:	0.010-0.025	0.254-0.635
		Service Limit	1986-96	+0.010	+0.254
			1997-98	0.035 Max	0.89 Max
Piston Rings - Oil	Groove clearance	Production	1986-96	0.002-0.007	0.051-0.177
		Service Limit	1997-98	+0.001	+0.025
	Gap	Production	1986-96	0.008 MAX	0.203 MAX
			1997-98	0.015-0.055	0.381-1.397
			1997-03	0.015-0.050	0.381-1.27
		Service Limit	1986-96	+0.010	+0.254
			1997-98	0.065	1.651
Piston pin	Diameter		1986-96	0.9270-0.9273	23.546-23.553
			1997-98	0.927	23.545
	Clearance		Production	0.00025-0.00035	0.006-0.009
			Service Limit	0.001 Max	0.025 Max
	Interference fit			0.008-0.0016	0.0203-0.04060
Valve system	Face Angle			45 deg.	45 deg.
	Lash		1986-91	1 turn down	1 turn down
			1992-98	Net Lash	Net Lash
	Lift		1997-98 Intake	0.414	10.51
			Exhaust	0.428	10.87
	Lifter			Hydraulic	Hydraulic
	Rocker Arm Ratio			1.50:1	1.50:1
	Seat Angle			46 deg.	46 deg.
	Seat Width	Intake	1986-96	0.0312-0.0625	0.7937-0.1587
			1997-98	0.040-0.065	1.016-1.651
		Exhaust	1986-96	0.0625-0.0937	0.1587-0.2381
			1997-98	0.065-0.098	1.651-2.489
	Spring Free Length			2.03	51.6
	Spring Installed Ht.		1997-98	1.690-1.710	42.92-43.43
	Spring Pressure	Open	1986-96	194-206 lbs. @ 1.25	863-916 N @ 31.75
		Closed	1997-98	187-203 lbs. @ 1.27	832-903 N @ 32.25
				76-84 lbs. @ 1.70	338-374 N @ 43.16
	Stem-to-Guide Clear.	Production		0.0010-0.0027	0.0254-0.0686
		Service Limit	Intake	0.001 Max	0.025 Max
			Exhaust	0.002 Max	0.51 Max

① Unless otherwise noted

ENGINE SPECIFICATIONS - 4.3L V6

Component				Standard (in.) ①	Metric (mm) ①
Balance Shaft	Front Bearing Journal			2.1648-2.1654	55.985-55.001
	Rear Bearing Journal			1.4994-1.5000	38.084-38.100
	Rear Bearing Clearance			0.001-0.0036	0.0254-0.0914
Camshaft	End Play		1986-96	0.004-0.012	0.11-0.30
			1997-98	0.001-0.009	0.0254-0.2286
	Journal Diameter		1986-96	1.8682-1.8692	47.452-47.478
			1997-98	1.8677-1.8697	47.440-47.490
	Journal Out-of-Round			0.001 Max	0.025 Max
	Lobe Lift (+/- 0.002)	Intake	1986-96	0.234	5.943
			1997-98	0.2763	7.018
		Exhaust	1986-96	0.257	6.527
			1997-98	0.2855	7.025
	Runout			0.0015 Max	0.051 Max
Connecting Rod	Bearing Clearance	Production	1986-96	0.0013-0.0035	0.0330-0.0889
		Service Limit	1997-98	0.003 Max	0.07 Max
	Side Clearance		1992-96	0.0010-0.0030	0.0254-0.0762
			1997-98	0.006-0.014	0.152-0.356
Crankpin	Journal Diameter		1986-94	2.0988-2.0998	53.31-53.34
			1995-98	2.2487-2.2497	57.117-57.142
	Taper	Production		0.005 Max	0.0127 Max
		Service Limit		0.001 Max	0.02 Max
	Out-of-round	Production		0.0005 Max	0.0127 Max
		Service Limit		0.001 Max	0.02 Max
Crankshaft	Crankshaft End Play		1986-96	0.002-0.006	0.05-0.15
			1997-98	0.002-0.008	0.05-0.20
	Bearing Clearance	Production	#1	0.0008-0.0020	0.0203-0.0508
			#2, 3	0.0011-0.0023	0.0279-0.0584
			#4	0.0017-0.0032	0.043-0.081
		Service Limit	#1	0.0010-0.0015	0.03-0.06
			#2, 3	0.0010-0.0025	0.03-0.06
			#4	0.0025-0.0035	0.07-0.08
	Journal Diameter	1986-96	#1	2.4484-2.4493	62.1894-62.2122
			#2, 3	2.4481-2.4490	62.1817-62.2046
			#4	2.4479-2.4488	62.1767-62.1995
		1997-98	#1	2.4488-2.4495	62.199-62.217
			#2, 3	2.4485-2.4494	62.191-62.215
			#4	2.4479-2.4489	62.179-62.203
	Taper	Production		0.0002 Max	0.005 Max
		Service Limit		0.001 Max	0.02 Max
	Out-of-round	Production		0.0002 Max	0.005 Max
		Service Limit		0.001 Max	0.02 Max
Cylinder Bore	Diameter			4.0007-4.0017	101.618-101.643
	Out-of-round	Production		0.001 Max	0.025 Max
		Service Limit		0.002 Max	0.05 Max
	Taper			0.001 Max	0.025 Max
Cylinder Head	Gasket surface flatness			0.004 Max	0.10 Max
Piston	Clearance	Production	1986-96	0.0007-0.0017	0.0178-0.0431
			1997-98	0.0007-0.002	0.017-0.05
		Service Limit	1986-96	0.0027 Max	0.07 Max
			1997-98	0.0024 Max	0.60 Max

① Unless otherwise noted

ENGINE SPECIFICATIONS - 5.0L/5.7L V8

Component			Standard (in.) ①	Metric (mm) ①
Piston	Clearance	Production 1986-97 ②	0.0007-0.0017	0.0178-0.0431
		1998	0.0007-0.0021	0.018-0.053
		Service Limit	0.0027 Max	0.076 Max
Piston Rings - Compression	Groove Clearance	Production	0.0012-0.0032	0.031-0.081
		Service Limit	+0.001	+0.025
	Gap	Production Top: 1986-97 ②	0.010-0.020	0.254-0.508
		1998	0.010-0.026	0.254-0.51
		2nd: 1986-97 ②	0.010-0.025	0.254-0.635
		1998	0.018-0.026	0.46-0.66
Piston Rings - Oil	Groove clearance	Production	0.002-0.007	0.051-0.177
		Service Limit	+0.001	+0.025
	Gap	Production 1986-97 ②	0.015-0.055	0.381-1.397
		1998	0.010-0.030	0.25-0.76
		Service Limit 1986-97	+0.010	+0.254
		1998	+0.001	+0.0254
Piston Pin	Diameter	1986-97 ②	0.9270-0.9273	23.546-23.553
		1998	0.9270-0.9271	23.545-23.548
	Clearance	Production 1986-97 ②	0.00025-0.00035	0.006-0.009
		1998	0.0004-0.0008	0.010-0.020
		Service Limit	0.001 Max	0.025 Max
	Interference fit		0.0008-0.00016	0.0203-0.04060
Valve system	Face Angle		45 deg.	45 deg.
	Lash		1 turn dn. from zero	1 turn dn. from zero
	Lifter		Hydraulic	Hydraulic
	Rocker Arm Ratio		1.50:1	1.50:1
	Seat Angle		46 deg.	46 deg.
	Seat Width	Intake 1986-97 ②	0.0312-0.0625	0.7937-0.1587
		1998	0.045-0.065	1.14-1.65
		Exhaust 1986-97 ②	0.0625-0.0937	0.1587-0.2381
		1998	0.065-0.098	1.65-2.489
	Spring Free Length	1986-97 ②	2.03	51.6
		1998	2.02	51.3
	Spring Installed Ht.	1998	1.69-1.71	42.92-43.43
	Spring Pressure	Open 1986-97 ②	194-206 lbs. @ 1.25	863-916 N @ 31.75
		1998	187-203 lbs. @ 1.27	832-903 N @ 32.25
		Closed 1986-97 ②	76-84 lbs. @ 1.70	338-374 N @ 43.16
	Stem-to-Guide Clear.	Production	0.0010-0.0027	0.0254-0.0686
		Service Limit Intake	+0.001	+0.0254
		Exhaust	+0.002	+0.0508

① Unless otherwise noted
② 1997 5.7GSi: Use 1998 specifications

ENGINE SPECIFICATIONS - 5.0L/5.7L V8

Component			Standard (in.) ①	Metric (mm) ①
Camshaft	End Play		0.004-0.012	0.10-0.31
	Journal Diameter	1986-97 ②	1.8682-1.8692	47.452-47.478
		1998	1.8677-1.8697	47.440-47.490
	Journal Out-of-Round		0.001 Max	0.025 Max
	Lobe Lift (+/- 0.002)	Intake 1986-97 ②	0.269	6.833
		1998	0.274-0.278	6.97-7.07
		Exhaust 1986-97 ②	0.276	7.01
		1998	0.283-0.287	7.20-7.30
	Runout		0.0015 Max	0.051 Max
Connecting Rod	Bearing Clearance	1986-97 Production	0.0013-0.0035	0.0330-0.0889
		Service Limit	0.003 Max	0.076 Max
		1998	0.0013-0.0035	0.033-0.088
	Side Clearance	1986-97 ②	0.008-0.014	0.1520-0.356
		1998	0.006-0.014	0.16-0.35
Crankpin	Journal Diameter	1986-97	2.0986-2.0998	53.31-53.34
		1998	2.0986-2.0998	53.30-53.33
	Taper	Production 1986-97	0.0005 Max	0.0127 Max
		1998-03	0.0003 MAX	0.007 MAX
		Service Limit	0.001 Max	0.0254 Max
	Out-of-round	Production 1986-97	0.0005 Max	0.0127 Max
		1998-03	0.0003 MAX	0.007 MAX
		Service Limit	0.001 Max	0.0254 Max
Crankshaft	Crankshaft End Play	1986-97 ②	0.002-0.006	0.05-0.15
		1998	0.002-0.008	0.05-0.20
	Bearing Clearance 1986-97	Production #1	0.0008-0.0020	0.0203-0.0508
		#2, 3, 4	0.0011-0.0023	0.0279-0.0584
		#5	0.0017-0.0032	0.043-0.081
		Service Limit #1	0.0010-0.0015	0.025-0.064
		#2, 3, 4	0.0010-0.0025	0.025-0.064
		#5	0.0025-0.0035	0.064-0.089
	1998	Production #1	0.0007-0.0021	0.018-0.053
		#2, 3, 4	0.0009-0.0024	0.022-0.061
		#5	0.0010-0.0027	0.0254-0.069
		Service Limit #1	0.0010-0.0020	0.025-0.051
		#2, 3, 4	0.0010-0.0025	0.025-0.064
		#5	0.0015-0.003	0.038-0.076
	Journal Diameter	#1	2.4484-2.4493	62.1894-62.2122
		#2, 3, 4	2.4481-2.4490	62.1817-62.2046
		#5	2.4479-2.4488	62.1767-62.1995
	Taper	Production	0.0002 Max	0.005 Max
		Service Limit	0.001 Max	0.0254 Max
	Out-of-round	Production	0.0002 Max	0.005 Max
		Service Limit	0.001 Max	0.0254 Max
Cylinder Bore	Diameter	5.0L	3.735-3.737	94.88-94.92
		5.7L	4.0007-4.0017	101.618-101.643
	Out-of-round	Production	0.005 Max	0.012 Max
		Service Limit	0.002 Max	0.051 Max
	Taper		0.001 Max	0.025 Max
Cylinder Head	Gasket surface flatness		0.004 Max	0.102 Max

① Unless otherwise noted
② 1997 5.7GSi: Use 1998 specifications

ENGINE SPECIFICATIONS - 7.4L/8.2L V8

Component				Standard (in.) ①	Metric (mm) ①
Camshaft	End Play			0.004-0.012	0.10-0.31
	Journal Diameter	All Exc. 1998 7.4Gi		1.9482-1.9492	49.484-49.510
		1998 7.4Gi		1.9477-1.9497	49.471-49.522
	Journal Out-of-Round			0.001 Max	0.025 Max
	Lobe Lift (+/- 0.002)	Intake	7.4L		
			1998 7.4Gi	0.282	7.163
			All others	0.234	5.94
			8.2L	0.51	12.95
		Exhaust	7.4L		
			1998 7.4Gi	0.284	7.214
			All others	0.253	6.43
			8.2L	0.51	12.95
	Runout			0.002 Max	0.051 Max
Connecting Rod	Bearing Clearance	Production	1988-91	0.0009-0.0025	0.0229-0.0635
			1992-98	0.0011-0.0029	0.0279-0.0737
		Service Limit	1998 7.4Gi	0.0011-0.0039	0.028-0.099
			All others	0.0035 Max	0.076 Max
	Side Clearance	1988-91		0.015-0.021	0.033-0.584
		1992-98		0.0013-0.0230	0.033-0.584
Crankpin	Journal Diameter	7.4L		2.1988-2.1996	55.85-55.87
		8.2L		2.2	55.88
	Taper	Production		0.0005 Max	0.0127 Max
		Service Limit		0.001 Max	0.02 Max
	Out-of-round	Production		0.0005 Max	0.0127 Max
		Service Limit		0.001 Max	0.0254 Max
Crankshaft	Crankshaft End Play	1998 7.4Gi		0.005-0.011	0.127-0.279
		All others		0.006-0.010	0.152-0.254
	Bearing Clearance				
	1988-91	Production	#1, 2, 3, 4	0.0013-0.0025	0.033-0.064
			#5	0.0024-0.0040	0.061-0.102
		Service Limit	#1	0.002 Max	0.051 Max
			#2, 3, 4, 5	0.0035 Max	0.089 Max
	1992-98	Production	#1, 2, 3, 4	0.0017-0.0030	0.043-0.076
			#5	0.0025-0.0040	0.064-0.097
		Service Limit	#1, 2, 3, 4	0.0010-0.0030	0.025-0.076
			#5	0.0025-0.0040	0.064-0.102
	1998 7.4Gi	Production	#1	0.0017-0.0030	0.043-0.076
			#2, 3, 4	0.0011-0.0024	0.028-0.061
			#5	0.0025-0.0038	0.063-0.096
		Service Limit	#1, 2, 3, 4	0.0010-0.0030	0.025-0.076
			#5	0.0025-0.0040	0.064-0.102
	Journal Diameter			2.7482-2.7489	69.804-69.822
	Taper	Production	1998 7.4Gi	0.0004 Max	0.010 Max
			All others	0.0002 Max	0.005 Max
		Service Limit		0.001 Max	0.025 Max
	Out-of-round	Production	1998 7.4Gi	0.0004 Max	0.010 Max
			All others	0.0002 Max	0.005 Max
		Service Limit		0.001 Max	0.025 Max
Cylinder Bore	Diameter	7.4L		4.2500-4.2507	107.95-107.97
		8.2L		4.4655-4.4661	113.42-113.44
	Out-of-round			0.002 Max	0.05 Max
	Taper			0.001 Max	0.025 Max

① Unless otherwise noted

ENGINE SPECIFICATIONS - 7.4L/8.2L V8

Component				Standard (in.) ①	Metric (mm) ①
Cylinder Head	Gasket surface flatness			0.003 Max	0.08 Max
Piston	Clearance	Production	7.4L		
			1988-91	0.0008-0.0028	0.020-0.071
			1992-98	0.003-0.004	0.076-0.102
			1998 Gi	0.0018-0.0030	0.046-0.076
			8.2	0.004-0.005	0.102-0.127
		Service Limit		0.005 Max	0.127
Piston Rings - Compression	Groove Clearance	Production	1988-91	0.0017-0.0032	0.043-0.081
			1992-98	0.0012-0.0029	0.030-0.074
		Service Limit	1998 7.4Gi	0.0012-0.0039	0.031-0.099
			All others	+0.001	+0.025
	Gap	Production	1988-91	0.010-0.020	0.254-0.508
			1992-98		
			Top	0.010-0.018	0.254-0.457
			2nd	0.016-0.024	0.406-0.609
		Service Limit	1998 Gi		
			Top	0.010-0.028	0.254-0.711
			2nd	0.016-0.034	0.406-0.864
			All others	+0.010	+0.254
Piston Rings - Oil	Groove clearance	Production	1988-91	0.0005-0.007	0.013-0.178
			1992-98	0.0050-0.0065	0.127-0.165
		Service Limit	1998 7.4Gi	0.0050-0.0075	0.127-0.191
			All others	+0.001	+0.025
	Gap	Production	1988-91	0.015-0.055	0.381-1.397
			1992-98	0.010-0.030	0.254-0.762
		Service Limit	1998-00 Gi	0.010-0.040	0.254-1.016
			All others	+0.010	+0.254
Piston Pin	Diameter			0.9895-0.9897	25.133-25.138
	Clearance	Production	1988-91	0.00025-0.00035	0.006-0.009
			1992-98	0.0002-0.0007	0.005-0.018
		Service Limit		0.001 Max	0.025 Max
	Interference fit	1988-91		0.013-0.0021	0.330-0.053
		1992-96		0.0008-0.0016	0.020-0.0406
		1997-98		0.0021-0.0031	0.053-0.079
Valve System	Face Angle			45 deg.	45 deg.
	Lash	1988-91		3/4 turn down	3/4 turn down
		1992-97		Net Lash	Net Lash
		1998	7.4Gi	Net Lash	Net Lash
			All others	1 turn down	1 turn down
	Lifter			Hydraulic	Hydraulic
	Rocker Arm Ratio			1.70:1	1.70:1
	Seat Angle			46 deg.	46 deg.
	Seat Width	Intake	All exc. 98 Gi	0.0312-0.0625	0.7937-0.1587
			1998 7.4Gi	0.030-0.060	0.762-1.524
		Exhaust	All exc. 98 Gi	0.0625-0.0937	0.1587-0.2381
			1998 7.4Gi	0.060-0.095	1.524-2.413

① Unless otherwise noted

6

ENGINE MECHANICAL GM V8 ENGINES

ENGINE MECHANICAL

General Information

NEVER, NEVER attempt to use standard automotive parts when replacing anything on your engine. Due to the uniqueness of the environment in which they are operated in, and the levels at which they are operated at, marine engines require different versions of the same part; even if they look the same. Stock and most aftermarket automotive parts will not hold up for prolonged periods of time under such conditions. Automotive parts may appear identical to marine parts, but be assured, OMC marine parts are specially manufactured to meet OMC marine specifications. Most marine items are super heavy-duty units or are made from special metal alloy to combat against a corrosive saltwater atmosphere.

OMC marine electrical and ignition parts are extremely critical. In the United States, all electrical and ignition parts manufactured for marine application must conform to stringent U.S. Coast Guard requirements for spark or flame suppression. A spark from a non-marine cranking motor solenoid could ignite an explosive atmosphere of gasoline vapors in an enclosed engine compartment.

V8 ENGINES

The OMC 5.0L, 302 cubic inch and 5.8L, 351 cubic inch and 7.5L, 460 cubic inch displacement V8 engines are manufactured by Ford Motor Company. These engines are used in the following configurations:

- 5.0 (2/4 bbl)
- 5.0 EFI (EFI)
- 5.0Fi (EFI)
- 5.0FL (2 bbl)
- 5.0 HO (2 bbl)
- 5.8 (4 bbl)
- 5.8 EFI (EFI)
- 5.8Fi (EFI)
- 5.8FL (4 bbl)
- 5.8FSi (EFI)
- 5.8 HO (EFI)
- 5.8 LE (4 bbl)
- 7.5 (4 bbl)
- 351 (4 bbl)
- 351 EFI (EFI)
- 460 (4 bbl)

Cylinder numbering and firing order is identified in the illustrations at the end of the Maintenance Section.

Engine Identification

ENGINE

◆ See Figures 1 and 2

The engine serial numbers are the manufacturer's key to engine changes. These alpha-numeric codes identify the year of manufacture, the horsepower rating and various model/option differences. If any correspondence or parts are required, the engine serial number must be used for proper identification.

Remember that the serial number establishes the year in which the engine was produced, which is often not the year of first installation.

The engine specifications decal contains information such as the model number or code, the serial number (a unique sequential identifier given ONLY to that one engine) as well as other useful information.

An engine specifications decal can generally be found on top of the flame arrestor (early models), or on the inner side of the rocker arm cover, usually near the breather/PCV line (port side on most models); further, on fuel injected engine you can find it on the side of the TFI module bracket - all pertinent serial number information can be found here—engine and drive designations, serial numbers and model numbers. Unfortunately this decal is not always legible on older boats and it's also quite difficult to find, so please refer to the following procedures for each individuals unit's serial number location.

■ Serial numbers tags are frequently difficult to see when the engine is installed in the boat; a mirror can be a handy way to read all the numbers.

The engine serial/model number is usually stamped on the port rear side of the engine where it attaches to the bell housing; although on most later models it may instead be a metal plate attached in the same location. If your engine has a stamped number it will simply be the serial number; if you have a plate (and you should), it will always show a Product number, a Type/Model number and then the actual Serial number.

- The first two characters identify the engine size in liters (L); **50** represents the 5.0L, **58** represents the 5.8L and so forth.

- The third character identifies the fuel delivery system; **2** designates a 2 bbl carburetor, **4** is a 4 bbl carburetor, and **F** is a fuel injected engine.

- The fourth character designates a major engine or horsepower change—it doesn't let you know what the change was, just that there was some sort of change. **A** means it is the first model released, **B** would be the second, and so forth

- The fifth character designates what type of steering system was used; **M** would be manual steering and **P** would be power steering.

- Now here's where it gets interesting; on 1986-87 engines and 1994-98 engines, the sixth, seventh and eighth characters designate the model year. The sixth and seventh actually show the model year, while the eighth is a random model year version code. **KWB** and **WXS** represent 1986; and **ARJ, ARF, FTC, SRC** or **SRY** show 1987. **MDA** is 1994, **HUB** is 1995, **NCA** is 1996, **LKD** is 1997 and **BYC** is 1998.

- On 1988-93 engines, the sixth character designates the direction of propeller rotation. **R** is right hand, **L** is left hand and **E** is either.

- Also on 1988-93 engines, the seventh, eighth and ninth characters designate the model year. The seventh and eighth actually show the model year, while the ninth is a random model year version code. **GDE** or **GDP** is 1988, **MED** or **MEF** is 1989, **PWC, PWR** or **PWS** is 1990, **RGD** or **RGF** is 1991, **AMH** or **AMK** is 1992 and **JVB** or **JVN** is 1993.

Fig. 1 Engine serial number sticker

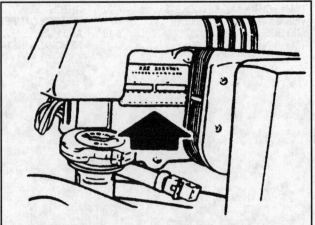

Fig. 2 On fuel injected engines, look on the TFI module bracket

Any remaining characters are proprietary. So in example, a Model number on the ID plate that reads **754APFTC** would designate a 1987 7.5L engine with a 4 bbl carburetor and power steering, first model released. A number reading **58FAPRJVB** would designate a 1993 5.8L engine with fuel injection, power steering and a right hand propeller, first release; get the picture?

Engine Model Designations

All engines covered here utilize unique identifiers assigned by OMC; surnames if you will—5.0FL, 5.8Fi, etc. Obviously the first two characters designate the engine size in litres (L). The second letter, a **G** or and **F** designate the engine manufacturer; General Motors (G) or Ford (F). The third through fifth letters can be found in different combinations, but the individual letter designates the same thing regardless of position. **L** designates limited output. **S** and **X** designate superior output—a 5.8Fi will always have a lesser horsepower rating than a 5.8FSi in a given model year. An **i** designates that the engine is fuel injected, if there is no **i** then you know the engine uses a carburetor.

Engine

REMOVAL & INSTALLATION

◆ **See Figures 3, 4, 5, 6, 7, 8 and 9**

■ **Prior to removing the engine from your 1994-96 vessel, it is imperative to measure the engine height as detailed in the Determining Minimum Engine Height section. DO NOT remove the engine until you have completed this procedure!**

1. Check the clearance between the front of the engine and the inside edge of the engine compartment bulkhead. If clearance is less than 6 in. (15.2mm), you will need to remove the stern drive unit because there won't be enough room to disengage the driveshaft from the engine coupler. More than 6 in. will provide enough working room to get the engine out without removing the drive, BUT, we recommend removing the drive anyway. If you intend on doing anything to the mounts or stringers, you will need to re-align the engine as detailed in the Engine Alignment section—which requires removing the drive, so remove the drive!

2. Remove the stern drive unit as detailed in the Drive Systems section.
3. Open or remove the engine hatch cover.
4. Disconnect the battery cables (negative first) at the battery and then disconnect them from the engine block and starter.

Make sure that all switches and systems are OFF before disconnecting the battery cables.

5. Disconnect the two power steering hydraulic lines at the steering cylinder (models w/ps). Carefully plug them and then tie them off somewhere on the engine, making sure that they are higher then the pump to minimize any leakage.
6. Disconnect the fuel inlet line at the fuel pump or filter (whichever comes first on your particular engine) and quickly plug it and the inlet— a clean golf tee and some tape works well in this situation. Make sure you have rags handy, as there will be some spillage.
7. Tag and disconnect the two-wire trim/tilt connector.
8. Pop the two-wire trim/tilt sender connector out of the retainer and then disconnect it. You may have to cut the plastic tie securing the cable in order to move it out of the way.
9. Locate the large rubber coated instrument cable connector (should be on the starboard side), loosen the hose clamp and then disconnect it from the bracket. Move it away from the engine and secure it. On early models, you will also need to unplug the three-wire trim/tilt instrument cable connector just above it.

■ **Take note of you throttle arm attachment stud— is it a "push-to-close" or a "pull-to-close"? What hole is it on??**

10. Remove the cotter pin and washer from the throttle arm. Loosen the anchor block retaining nut and then spin the retainer away from the cable trunnion. Remove the throttle cable from the arm and anchor bracket. Be sure to mark the position of the holes that the anchor block was attached to.
11. Loosen the hose clamps and disconnect the exhaust hose bellows from intermediate or exhaust pipe. You may want to spray some WD-40 around the lip of the hose where it connects to the elbow, grasp it with both hands and wiggle it back and forth while pulling down on it. Slide it down over the lower pipe.
12. Drain the cooling system as detailed in the Maintenance section.

Fig. 3 Disconnect the power steering lines at the cylinder

Fig. 4 Disconnect the instrument cable connector

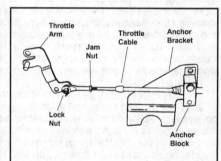

Fig. 5 Details of the throttle cable assembly—FL engines

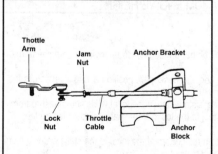

Fig. 6 Details of the throttle cable assembly—Fi and FSi engines

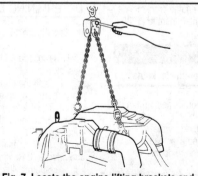

Fig. 7 Locate the engine lifting brackets and then connect and engine hoist

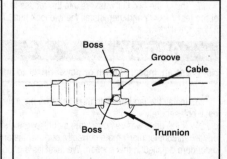

Fig. 8 Details of the throttle cable and trunnion

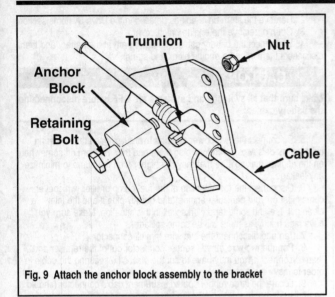

Fig. 9 Attach the anchor block assembly to the bracket

13. Loosen the hose clamp on the water supply hose at the transom bracket and carefully slide it off the water tube. Attach the hose to the engine.

14. Disconnect the shift cables and position them out of the way.

15. Tag and disconnect any remaining lines, wires or hoses at the engine.

16. Attach a suitable engine hoist to the lifting eyes and take up any line slack until it is just taught.

■ **The engine hoist should have a capacity of at least 1500 lbs. (680 kg).**

17. Locate the rear engine mounts and remove both lock nuts and flat washers.

18. Locate the front engine mounts and remove the two (per mount) lag bolts.

19. If you listened to us at the beginning of the procedure, the drive unit should be removed. If so, slowly and carefully, lift out the engine. Try not to hit the power steering control valve, or any other accessories, while removing it from the engine compartment. If you didn't listen to us, and you had sufficient clearance in the engine compartment, the drive unit is probably still installed. Raise the hoist slightly until the weight is removed from the mounts and then carefully pull the engine forward until the driveshaft disengages from the coupler, now raise the engine out of the compartment.

To Install:

20. Apply Engine Coupler grease to the splines of the coupler.

21. Slowly lower the engine into the compartment. If the drive unit was not removed, AND the crankshaft has not been rotated, insert the driveshaft into the coupler as you push the engine backwards until they engage completely and then lower the engine into position over the rear mounts until the front mounts just touch the stringers. If the shaft and coupler will not align completely, turn the crankshaft and driveshaft slightly until they mate correctly.

If the drive was removed, or the mounts were disturbed in any way, lower the engine into position over the rear mounts until the front mount just touches the stringer.

22. Install the two flat washers into the recess in the engine bracket side of the rear mounts and then install the two lock nuts. Tighten them to 28-30 ft. lbs. (38-40 Nm).

✳✳ CAUTION

Never use an impact wrench or power driver to tighten the locknuts.

23. Install the lag bolts into their holes on the front mounts and tighten each bolt securely.

24. If the drive was removed, the mounts were disturbed, or the driveshaft/coupler didn't mate correctly, perform the engine alignment procedure detailed in this section. We think it's a good idea to do this regardless!

25. Reconnect the exhaust bellows by sliding it up and over the pipe, position the clamps between the ribs in the hose and then tighten the clamps securely. Make sure you don't position the clamps into the expanding area.

26. Reconnect the water inlet hose. Lubricate the inside of the hose and wiggle it onto the inlet tube. Slide the clamp over the ridge and tighten it securely. This sounds like an easy step, but it is very important—if the hose, particularly the underside, is not installed correctly the hose itself may collapse or come off. Either scenario will cause severe damage to your engine, so make sure you do this correctly!

27. Carefully, and quickly, remove the tape and plugs so you can connect the power steering lines. Tighten the large fitting to 15-17 ft. lbs. (20-23 Nm) and the small fitting to 10-12 ft. lbs. (14-16 Nm). Don't forget to check the fluid level and bleed the system when you are finished with the installation.

28. Reconnect the trim/tilt connector so the two halves lock together.

29. Reconnect the trim position sender leads, the instrument cable, the engine ground wire, the battery cables and all other wires, lines of hoses that were disconnected during removal. Make sure you swab a light coat of grease around the fitting for the large engine/instrument cable plug.

✳✳ CAUTION

Always make certain that all switches and systems are turned OFF before reconnecting the battery cables.

■ **Make sure all cables, wires and hoses are routed correctly before initially starting the engine.**

30. Unplug the fuel line and pump/filter fitting and reconnect them. Remember to check for leaks as soon as you start the engine.

31. Install and adjust the throttle cable. For complete details, please refer to the Fuel System section):

a. Remember we asked you to determine if you had a "push-to-close" or "pull-to-open" throttle cable (the throttle arm stud)? Position the remote control handle in Neutral—the propeller should rotate freely.

b. Turn the propeller shaft and the shifter into the forward gear detent position and then move the shifter back toward the Neutral position halfway.

c. Position the trunnion over the groove in the throttle cable so the internal bosses align and then snap it into the groove until it is fully seated.

d. Install the trunnion/cable into the anchor block so the open side of the trunnion is against the block. Position the assembly onto the bracket over the original holes (they should be the lower two of the four holes) and then install the retaining bolt and nut. When the nut is securely against the back of the bracket, tighten the bolt securely.

e. Install the connector onto the throttle cable and then pull the connector until all end play is removed from the cable. Turn it sideways until the hole is in alignment with the correct stud on the throttle arm. Slide it over the stud and install the washer and a new cotter pin. Make sure the cable is on the same stud that it was removed from. Tighten the jam nut against the connector

■ **The throttle arm connector nut must be installed on the cable with a minimum of 9 turns—meaning that at least 1/4 in. of thread should be showing between the end of the cable and the edge of the nut.**

32. Install and adjust the shift cables. Please refer to the Drive Systems section for further details.

33. Check and refill all fluids. Start the engine and check for any fuel or coolant leaks. Go have fun!

DETERMINING MINIMUM ENGINE HEIGHT

◆ See Figure 10

This procedure MUST be performed prior to removing the engine from the vessel.

1. With the engine compartment open, position a long level across the transom running fore and aft.

2. Have a friend or assistant steady the level while you measure from the bottom edge of the tool to the top of the exhaust elbow. Record the distance as "1".

3. Now measure from the bottom of the level to the static water line on the drive unit. Record the distance as "2".

4. Subtract the elbow measurement (1) from the static waterline measurement (2). If the result is less than 14 in. (356mm), an exhaust elbow high rise kit must be installed (available from your local parts supplier).

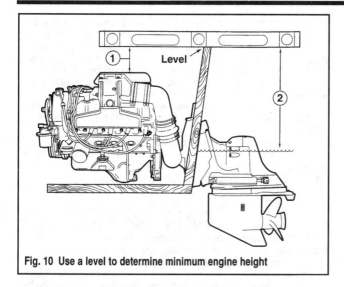

Fig. 10 Use a level to determine minimum engine height

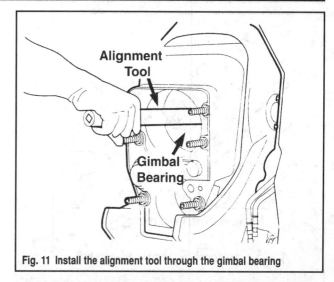

Fig. 11 Install the alignment tool through the gimbal bearing

ENGINE ALIGNMENT

5.0L/5.8L And 1990 7.5L V8 Engines

◆ **See Figures 11 and 12**

Engine alignment is imperative for correct engine installation and also for continued engine and drive operation. It is a good idea to ensure proper alignment every time that the drive or engine has been removed. Engine alignment is checked by using OMC alignment tool (#912273) and handle (#311880). Engine alignment is adjusted by raising or lowering the front engine mount(s).

1. With the drive unit off the vessel, slide the alignment tool through the gimbal bearing and into the engine coupler. It should slide easily, with no binding or force. If not, check the gimbal bearing alignment as detailed in the Drive Systems section. If bearing alignment is correct, move to the next step.

2. If your engine utilizes a jam nut on the bottom of the mount bolt, loosen it and back it off at least 1/2 in.

3. Loosen the lock nut and back it off.

4. Now, determine if the engine requires raising or lowering to facilitate alignment—remember, the alignment tool should still be in position. Tighten or loosen the adjusting nut until the new engine height allows the alignment tool to slide freely.

5. Hold the adjusting nut with a wrench and then tighten the lock nut to 40-50 ft. lbs. (54-60 Nm) on 1989-92 engines or 50-70 ft. lbs. (68-95 Nm) on 1993-96 engines. If your engine uses a jam nut, cinch it up against the lock nut.

6. Remove the alignment tool and handle.

1987-89 7.5L V8 Engines

◆ **See Figures 11 and 12**

Engine alignment is imperative for correct engine installation and also for continued engine and drive operation. It is a good idea to ensure proper alignment every time that the drive or engine has been removed. Engine alignment is checked by using OMC alignment tool (#912273) and handle (#311880). Engine alignment is adjusted by raising or lowering the front engine mount(s).

1. With the drive unit off the vessel, slide the alignment tool through the gimbal bearing and into the engine coupler. It should slide easily, with no binding or force. If not, check the gimbal bearing alignment as detailed in the Drive Systems section. If bearing alignment is correct, move to the next step.

2. There will be a slot on the lower mount providing access to the head of the mount bolt. Slide a 15/16 in. open end wrench into the slot and hold the bolt so that you can loosen the upper lock nut. It is not necessary to remove the nut.

Fig. 12 A good view of the front engine mount and its components

3. Now, determine if the engine requires raising or lowering to facilitate alignment—remember, the alignment tool should still be in position.

4. There should be a total of 4 slotted shims situated above and/or below the rubber mount; rearrange the shims until the new engine height allows the alignment tool to slide freely. The shims are slotted and will slide in and out of the mount easily. Never use more than 4 shims on an individual mount.

5. Install the nut and tighten it to 100-120 ft. lbs. (136-63 Nm).

6. Remove the alignment tool and handle.

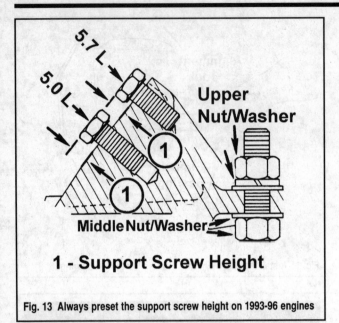

Fig. 13 Always preset the support screw height on 1993-96 engines

Front Engine Mounts

REMOVAL & INSTALLATION

5.0L/5.8L V8 Engines

◆ **See Figures 12 and 13**

1. Position an engine hoist over the engine and hook it up to the two engine lifting eyes (usually front starboard side, rear port side).
2. Measure the distance between the top of the large washer on the mount and the flat on the lower side of the mounting bracket. Record it.
3. Remove the two lag bolts on each side of the mount where it rests on the stringer.
4. Raise the engine just enough to allow working room for removing the mount.
5. Remove the two mount bracket-to-engine mounting bolts with their lock washers and lift out the mount/bracket.

■ **Some engines may use bolts of different lengths; mark them so you can replace them in their original locations.**

6. Carefully position the mount in a vise. Position a wrench over the head of the adjusting bolt, just underneath the mount to hold the shaft and then remove the upper nut and washer. Lift off the bracket.
7. Remove the two lock nuts from the bolt and slide out the adjusting bolt. Remember which washer goes where.

To Install:

8. Drop the small washer over the shaft and then slide the adjusting bolt up through the mount. Position the large washer over the bolt so it rests on the mount and then spin on the first (lower) lock nut and tighten it to 60-75 ft. lbs. (81-102 Nm).
9. Check that the mount support screw height is set to 1/2 in. (12.7mm) on the 1993 5.0L engine or 1/4 in. (6.35mm) on the 1994-96 5.0L. On the 1993 5.8L, set the screw to 7/16 in. (11.1mm), while on the 1994-96 5.8L engines set the screw to 3/16 in. (4.74mm). On all engines, make sure there is no paint on the head of the bolt, or on the mating surface at the block.

■ **1987-92 engines do not use a height support bolt.**

10. Spray the two mounting bolts with Loctite Primer N and allow them to air dry. Once dry, coat the bolts with Loctite or OMC Nut Lock and attach the mount bracket to the engine. Tighten the bolts to 32-40 ft. lbs. (43-54 Nm).
11. On engines with a height support bolt, tighten the correct support bolt for your engine until it just contacts the cylinder block. Now tighten it an additional 1/4 of a turn.

12. Screw on the middle lock nut and washer onto the mount bolt and then position the mount bracket over the bolt. Install the washer and upper adjusting nut and check the measurement taken previously. Move the upper and middle lock nuts up or down until the correct specification is achieved.
13. Position the mount over the lag screw holes and then slowly lower the engine until all weight is off the hoist. Install and tighten the lag screws securely.
14. Hold the middle nut with a wrench and then tighten the upper nut to 40-50 ft. lbs. (54-60 Nm) on 1989-92 engines or 50-70 ft. lbs. (68-95 Nm) on 1993-96 engines.
15. If you're confident that your measurements and subsequent adjustment place the engine exactly where it was prior to removal, then you are through. If you're like us though, you may want to check the engine alignment before you fire up the engine.

7.5L V8 Engines

1. Position an engine hoist over the engine and hook it up to the two engine lifting eyes (usually front starboard side, rear port side).
2. Remove the 2 lag bolts on each side of the mount where it rests on the stringer.
3. Raise the engine just enough to allow working room for removing the mount.
4. Remove the 3 mount bracket-to-engine mounting bolts with their lock washers and lift out the mount/bracket.

■ **Some engines may use bolts of different lengths; mark them so you can replace them in their original locations.**

5. Hold the mount bolt head and remove the upper lock nut. Note the positions of the adjusting shims and then remove them.
6. Pry out the rubber mounts.

To Install:

7. Coat the smaller side of a new rubber mount with soapy water and then press it into the top of the mount bracket. Repeat the procedure with the lower rubber mount.
8. Slide a small flat washer over the mount bolt and then insert the bolt into the lower mount foot and through the rubber mounts. Slide the larger washer over the end of the bolt and thread on the lock nut.
9. Insert the slotted shims around the bolt so that they are in the same position as when removed and then tighten the lock nut to 100-120 ft. lbs. (136-63 Nm).
10. Position the mount against the cylinder block, install the 3 mounting bolts and tighten them to 32-40 ft. lbs. (43-54 Nm).
11. Install the lag bolts and tighten them securely.
12. Remove the engine hoist.

Rear Engine Mounts

REMOVAL & INSTALLATION

◆ **See Figure 14**

1. Remove the engine as detailed previously.
2. Loosen the two bolts and remove the mount from the transom plate. Remember, you will already have removed the top lock nut and the washer during engine removal.
3. Hold the square nut with a wrench and remove the shaft bolt. Be sure to take note of the style and positioning of the two mount washers as you are removing the bolt. Mark them, lay them out, or write it down, but don't forget their orientation!!

To Install:

4. Slide the lower of the two washers onto the mount bolt, exactly as it came off.
5. Slide the bolt into the flat (bottom) side of the rubber mount, install the remaining washer (as it came off!) and then spin on the square nut. Do not tighten it yet.

■ **Incorrect washer installation will cause excessive vibration during engine operation.**

6. Turn the assembly upside down and clamp the square nut in a vise. Spin the assembly until the holes in the mounting plate are directly opposite any two of the flat sides on the nut. This is important, otherwise the slot on the engine pad will not engage the mount correctly. Secure the mount in this position and tighten the bolt to 44-52 ft. lbs. (60-71 Nm).

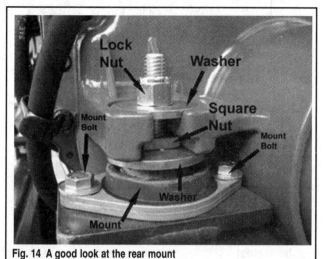

Fig. 14 A good look at the rear mount

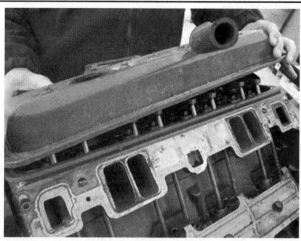

Fig. 15 Lift off the cylinder head cover

7. Remove the mount from the vise and position it on the transom plate. Install the bolts and washers and tighten each to 20-25 ft. lbs. (27-34 Nm).

8. Install the engine, making sure that the slot in the engine pad engages the square nut correctly. Install the two washers and locknut and tighten it to 28-30 ft. lbs. (38-41 Nm).

Cylinder Head (Valve) Cover

REMOVAL & INSTALLATION

◆ See Figures 15 and 16

 MODERATE

■ In order to perform this procedure efficiently, we recommend removing the exhaust manifold in order to have sufficient working room to remove the cylinder head cover. Although not completely necessary, it's worth the extra effort to avoid the aggravation of working around the manifolds. Please refer to the manifold procedure later in this section.

1. Open or remove the engine compartment hatch. Disconnect the negative battery cable.

2. Loosen the clamp and remove the crankcase ventilation hose at the cover (if equipped). Carefully move it out of the way.

3. Tag and disconnect any lines, leads or hoses that might be in the way of removal. In some instances you may be able to simply secure them out of the way without disconnecting them—you be the judge.

4. If your engine has a spark plug wire retainer attached to the cover, unclip the wires or remove the retainer and then tag and disconnect the plug wires at the spark plugs.

5. Loosen the cover mounting bolts and lift off the cylinder head cover. Take note of any harness or hose retainers and clips that might be attached to certain of the mounting bolts; you need to make sure they go back in the same place.

To Install:

6. Clean the cylinder head and cover mounting surfaces of any residual gasket material with a scraper or putty knife.

7. Position a new gasket onto the cover making sure that the tabs in the gasket engage the notches in the cover. Position the cover (don't forget the J-clips!) on the cylinder head and then tighten the mounting bolts to 60-72 inch lbs. (6.8-8.1 Nm) on 1987-93 engines or 36-60 inch lbs. (4-6 Nm) on 1994-96 engines. Make sure any retainers or clips that were removed are back in their original positions. On 1987-93 engines, wait 2 minutes and then retighten all of the bolts to the same specification once again.

8. Connect the crankcase ventilation hose and any other lines or hoses that may have been disconnected. Check that there were no other wires or hoses you may have repositioned in order to gain access to the cover.

9. Install the exhaust manifold.

10. Connect the battery cables.

Fig. 16 Always use a new gasket before installing the cover

Rocker Arms and Push Rods

REMOVAL & INSTALLATION

◆ See figures 17, 18, 19 and 20

1. Open or remove the engine hatch cover and disconnect the negative battery cable. Remove the cylinder head cover as detailed previously.

2. Bring the piston in the No. 1 cylinder to TDC. If servicing only one arm, bring the piston in that cylinder to TDC. The No. 1 cylinder is the first cylinder at the starboard side of the engine.

3. Loosen and remove the rocker arm bolt. Lift out the oil deflector on the 7.5L ad the fulcrum on all engines.

4. Lift the arm itself off of the mounting stud (or boss) and pull out the pushrod. Lift off the pushrod guide before removing the pushrod if it didn't come off with the rocker. It is very important to keep each cylinder's component parts together as an assembly. We suggest drilling a set of holes in a 2X4 and positioning the pieces in the holes.

To Install:

5. Clean and inspect the rocker assemblies. If any scuffing, wear or obvious deterioration is found replace the entire assembly (rocker, fulcrum and pushrod, and guide). Roll each push rod across a flat, even surface (countertops work great for this); if it does not roll smoothly, replace it.

6. Coat all bearing surfaces of the rocker assembly (valve stem tips, arm contact surfaces, fulcrum seat and push rod tip) with LubriPlate or Ford Multi-Purpose grease.

7. Slide the push rods into their holes. Make sure that each rod seats in its socket on the tappet.

8. Install a remote starter.

9. Bump the engine over until the No. 1 cylinder is at TDC after the compression stroke.

■ Valve arrangement running from front to back on the port side is: E-I-E-I-E-I-E-I, while on the starboard side it is: I-E-I-E-I-E-I-E.

10. Install the guide, rocker arm, fulcrum and bolt on the following valves.

5.0L and 7.5L engines
• No. 1: intake and exhaust
• No. 4: exhaust
• No. 5: exhaust
• No. 7: intake
• No. 8: intake

5.8L engines
• No. 1: intake and exhaust
• No. 3: exhaust
• No. 4: intake
• No. 7: exhaust
• No. 8: intake

11. Now bump the crankshaft around 180° from TDC and install the guide, rocker arm, fulcrum, deflector and bolt on the following valves:

5.0L and 7.5L engines
• No. 2: exhaust
• No. 4: intake
• No. 5: intake
• No. 6: exhaust

5.8L engines
• No. 2: exhaust
• No. 3: intake
• No. 6: exhaust
• No. 7: intake

12. Now bump the crankshaft around a further 90° (for a total of 270° from TDC where you started) on the 5.0L/5.8L engines. On the 7.5L engine, bump the crankshaft around a further 270° (that's a full 360° from TDC, plus an additional 90°). Install the guide, deflector, rocker arm, fulcrum and bolt on the following valves:

5.0L and 7.5L engines
• No. 2: intake
• No. 3: intake and exhaust
• No. 6: intake
• No. 7: exhaust
• No. 8: exhaust

5.8L engines
• No. 2: intake
• No. 4: exhaust
• No. 5: intake and exhaust
• No. 6: intake
• No. 8: exhaust

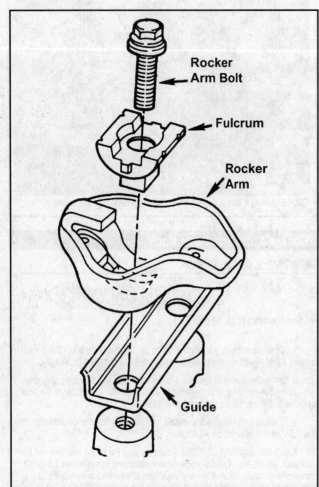

Fig. 17 Exploded view of the rocker arm assembly—5.0L/5.8L engines

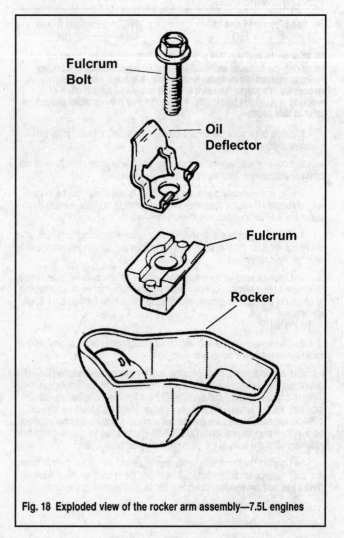

Fig. 18 Exploded view of the rocker arm assembly—7.5L engines

13. Now tighten all the fulcrum bolts to 18-25 ft. lbs (24-34 Nm).

■ **The fulcrum base MUST be inserted in the slot on the cylinder head BEFORE tightening any of the bolts.**

14. No additional adjustment of the valves is necessary as the lash is set automatically when the positive-stop rocker is tightened to specifications. To make sure that all components are operating properly though, it's a good idea to check the valve clearance as detailed following.

15. Install the cylinder head cover, connect the battery cable and check the idle speed.

VALVE CLEARANCE CHECK

◆ **See Figures 19 and 20**

Positive-stop rocker arm bolts virtually eliminate the need for a regular valve clearance adjustment, but wear, tear or damage will sometimes only be revealed by virtue of a preliminary clearance check with the hydraulic lifter in the fully collapsed position. Clearance figures for each engine are detailed in the Engine Specification chart and should be checked between the rocker arm and the tip of the valve stem with a feeler gauge.

■ **Valve arrangement running from front to back on the port side is: E-I-E-I-E-I-E-I, while on the starboard side it is: I-E-I-E-I-E-I-E.**

1. Locate the purple wire coming off of the **I** terminal on the starter solenoid, tag it and disconnect it—this will prevent the engine from starting when you install a remote starter. Install a remote starter.
2. Bump the engine over until the No. 1 cylinder is at TDC after the compression stroke.
3. Install a Lifter Compressor tool (#T71P-6513-B) onto the appropriate rocker arm and slowly apply enough pressure to bleed down the lifter until the rod is bottomed out. Check the clearance on the following valves:

5.0L and 7.5L engines
- No. 1: intake and exhaust
- No. 4: exhaust
- No. 5: exhaust
- No. 7: intake
- No. 8: intake

5.8L engines
- No. 1: intake and exhaust
- No. 3: exhaust
- No. 4: intake
- No. 7: exhaust
- No. 8: intake

4. Now bump the crankshaft around 180° from TDC, install the tool on the appropriate rocker arm, and check the clearance on the following valves:

5.0L and 7.5L engines
- No. 2: exhaust
- No. 4: intake
- No. 5: intake
- No. 6: exhaust

5.8L engines
- No. 2: exhaust
- No. 3: intake
- No. 6: exhaust
- No. 7: intake

5. Now bump the crankshaft around a further 90° (for a total of 270° from TDC where you started) on the 5.0L/5.8L engines. On the 7.5L engine, bump the crankshaft around a further 270° (that's a full 360° from TDC, plus an additional 90°). Install the tool and check the clearance on the following valves:

5.0L and 7.5L engines
- No. 2: intake
- No. 3: intake and exhaust
- No. 6: intake
- No. 7: exhaust
- No. 8: exhaust

5.8L engines
- No. 2: intake
- No. 4: exhaust
- No. 5: intake and exhaust
- No. 6: intake
- No. 8: exhaust

6. Allowable clearance on all valves should be:
- 5.0L—0.071-0.193 in. (1.8-4.9mm)
- 5.8L—0.098-0.198 in. (2.5-5.0mm)
- 7.5L—0.075-0.175 in. (1.9-4.4mm)

7. If clearance is not within specifications on 1986-93 engines, OMC makes replacement pushrods that are available either 0.060 in. longer or shorter than stock. If clearance is lower than spec, install the shorter rod; if higher than spec, install the longer pushrod. Valve clearance that is out of specification on 1994-96 engines will require replace components in the valvetrain.

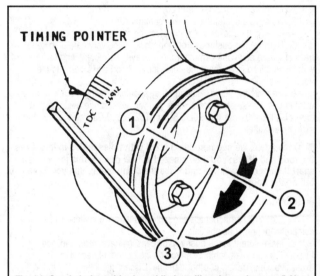

Fig. 19 Crankshaft positioning while installing the rockers on 5.0L and 5.8L engines. The No. 1 piston is at TDC after the compression stroke in position 1. Round two will have the crank moved 180° to position 2. While round three will move the crank an additional 90° to position 3

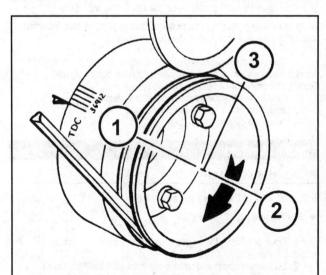

Fig. 20 Crankshaft positioning while installing the rockers on 7.5L engines. The No. 1 piston is at TDC after the compression stroke in position 1. Round two will have the crank moved 180° to position 2. While round three will move the crank an additional 270° to position 3

Hydraulic Valve Lifter

REMOVAL & INSTALLATION

■ If you have installed a new camshaft, always use new lifters.

1. Using compressed air, thoroughly clean all dirt and grit from the cylinder head and related components.

✳✳ WARNING

If compressed air is not available, we highly recommend that you DO NOT proceed with this procedure. It is EXTREMELY important that no dirt gets into the lifter recesses before completing the installation.

2. Remove the cylinder head cover.
3. Remove the intake and exhaust manifolds.
4. Loosen the rocker arms and pivot the rockers off of the pushrods.
5. Remove the pushrods from the block, being very careful to keep track of where each one came from. As always, we recommend using a 2x4 with holes drilled in it—mark the holes for each rod and insert into the appropriate hole.
6. Remove the lifters. Although they should come right out, you may want to use a magnet; or more likely, you will need to use a lifter extractor tool (#T70L-6500-A for the 5.0L/5.8L or #T52T-6500-DJD on the 7.5L)) to extract each lifter.

■ Once again, we suggest using a 2x4 with holes drilled in it to store the rods and lifters; you'd be amazed at how quickly this job will fall apart if someone walks in and kicks the components that you have laid out on the floor!

To Install:

7. Clean all components thoroughly and let dry completely, use compressed air if at all possible.
8. Make sure that the push rod oil passages are clean and clear. Inspect the camshaft contact surface on the bottom of each lifter for excessive wear, galling or other damage. Discard the lifter if any of the above conditions are found. You'll probably want to check out the camshaft lobe for damage also.
9. Coat each lifter, bore and camshaft lobe with clean engine oil or Multi-Purpose grease, and then carefully install each one into its respective recess.
10. Install the pushrods into the sockets on the lifters and coat their tips with Ford Multi-Purpose grease. Coat the valve stem ends as well.

■ When reusing existing components, it is imperative that the lifters and push rods are installed in their original bores.

11. Lubricate the rockers and fulcrum seats with multi-purpose grease. Move the rockers into position and then tighten the bolt as detailed in the Rocker Arm procedure as detailed previously.
12. Check the valve clearance as previously detailed.
13. Install the intake and exhaust manifolds.
14. Install the cylinder head cover.

Intake Manifold

REMOVAL & INSTALLATION

5.0L, 5.8L And 7.5L Engines W/Carburetor

◆ See Figures 21, 22, 23 and 24

1. Open or remove the engine compartment hatch. Disconnect the negative battery cable.
2. Loosen the clamp and remove the crankcase ventilation or PCV hoses at the cylinder head cover. Carefully move them out of the way.
3. Drain all water from the cylinder block and manifolds.
4. Tag and disconnect all water hoses at the manifold and thermostat housing. Have some rags handy, since there will still be some water in them. Carefully move them out of the way.

5. Remove the thermostat housing.
6. Remove the flame arrestor and then disconnect the throttle cable at the carburetor. Remove the throttle cable bracket from the manifold.
7. Disconnect the fuel line and plug the line end and the carburetor fitting. If you have a non-flexible line, disconnect it at the fuel pump also. Move both the cable and fuel line out of the way. There should also be a fuel pump vent line connected to the carburetor—tag it, disconnect it and move it out of the way.
8. Tag and disconnect the lead at the ignition coil.
9. Remove the distributor cap with the leads still connected and set it aside temporarily. Mark the position of the distributor rotor to the distributor body. Scribe a match mark across the distributor and the manifold. Loosen the clamp bolt and remove the distributor. Please refer to the Electrical section for further details on distributor removal. DO NOT turn the engine over once the distributor has been removed.
10. Tag and disconnect the electrical leads at the ignition coil. Remove the coil and its bracket.
11. Tag and disconnect the lead at the temperature gauge sender unit. Tag and disconnect any other lines, leads or hoses that might be in the way of removal. In some instances you may be able to simply secure them out of the way without disconnecting them—you be the judge.
12. If your engine has a spark plug wire retainer attached to the cylinder head cover, unclip the wires or remove the retainer and then tag and disconnect the plug wires at the spark plugs. Move the wires (and the cap) out of the way.
13. Loosen and remove the 12 manifold mounting bolts (16 on the 7.5L) in the reverse order of the illustrated tightening sequence and then remove the manifold. Leave the carburetor and decel valve attached. There is a good likelihood you will need to pry the manifold off the block; be very careful that you don't scratch or mar the mating surfaces on the block, manifold or heads.

To Install:

14. Carefully remove all remaining gasket material from the manifold mating surfaces with a scraper or putty knife. Be careful that you don't accidentally drop any old gasket into the crankcase or intake ports on the cylinder head.
15. Inspect the manifold and all mating surfaces for any cracks or nicks.
16. Apply a 1/8 in. (3.2mm) bead of RTV sealant to the forward and aft edges (4 corners) of the cylinder block mating surface where it meets the cylinder head. Make sure that you run the bead at least as wide as width of each seal.
17. Install the manifold gaskets onto the head. Position the cylinder block seals and run a 1/16 in, (1.6mm) bead of RTV sealant along the edge of each seal where it mates with the gaskets on the cylinder head.

■ The sealer sets up fairly quickly, so don't dawdle here!

18. Carefully install the manifold into place so that all the bolt holes line up. Run your finger along the edge of all seal areas to make certain that they are still positioned correctly; if not, remove the manifold and reposition it again. Insert the bolts, tighten them finger-tight and then tighten all bolts in sequence to 23-25 ft. lbs. (31-34 Nm) on the 5.0L/5.8L engines or 25-27 ft. lbs. (34-40 Nm) on the 7.5L. When you're through, repeat the tightening sequence one more time and then make sure to recheck at the end of the season.
   Check the Torque Specifications chart for any further details on tightening needs. Tighten all bolts in the order shown in the illustration.
19. Install the thermostat housing and adaptor.
20. Install the distributor making sure that the match marks align. Do the same with the distributor cap and reconnect the spark plug wires.
21. Install the coil and bracket. Reconnect the electrical leads and the distributor lead.
22. Unplug the fuel line(s) and reconnect them. Connect the vent hose and the PCV hose.
23. Install all water hoses and tighten their clamps securely.
24. Install the throttle cable bracket and then connect the cable itself. Adjust it as detailed in the Fuel System section. Install the flame arrestor.
25. Connect all other lines, leads and hoses that may have been removed to facilitate manifold removal.
26. Connect the battery cable and start the engine. Check the ignition timing and idle speed. Check all hoses and seals for leaks.

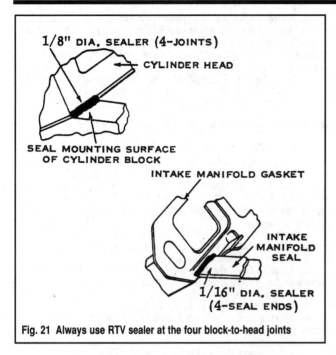

Fig. 21 Always use RTV sealer at the four block-to-head joints

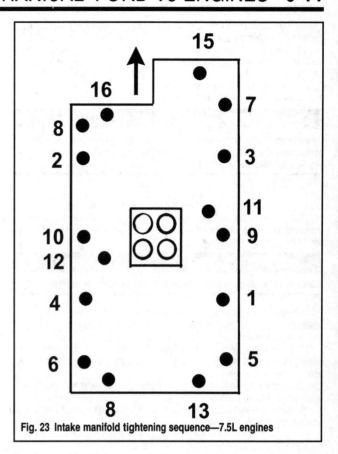

Fig. 23 Intake manifold tightening sequence—7.5L engines

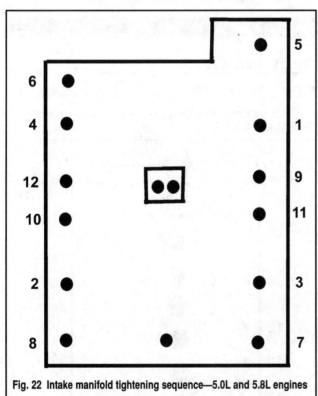

Fig. 22 Intake manifold tightening sequence—5.0L and 5.8L engines

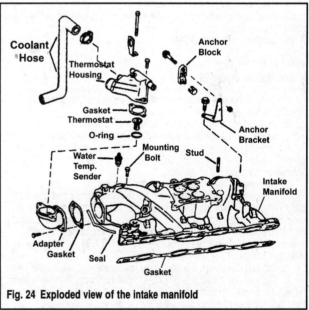

Fig. 24 Exploded view of the intake manifold

5.0 EFI, 5.0Fi And 5.8Fi/FSi Engines

Upper Intake Manifold (Plenum)

 MODERATE

■ **Please refer to the Exploded Views section for detailed illustrations of the upper manifold and related components.**

1. Open or remove the engine compartment hatch. Disconnect the negative battery cable.
2. Relieve the fuel system pressure as detailed in the Maintenance or Fuel System sections.

3. Loosen the 4 retaining screws and lift of the engine's plastic cover.
4. Tag and disconnect the ISC solenoid and Throttle Position sensor leads at the throttle body.
5. Remove the lock nut and disconnect the throttle cable at the lever on the throttle body. Remove the 2 manifold mounting bolts at the rear of the plenum that secure the throttle cable bracket. Position the cable and bracket out of the way.
6. Tag and disconnect the pressure regulator hose and the fuel reservoir line at the rear of the plenum.
7. Tag and disconnect the PCV hose at the valve.
8. On the forward starboard side, tag and disconnect the vacuum line at the MAP sensor. Move it and the filter out of the way.

9. Right next to the filter, remove the 2 forward plenum mounting bolts and lift off the shift bracket support.

10. At the module bracket next to the throttle body, tag and disconnect the high tension lead from the ignition coil, the primary lead also at the coil and the TFI-IV lead.

11. Remove the 2 remaining plenum screws and lift off the upper manifold with the throttle body attached.

✳✳ CAUTION

Be extremely careful of how you handle the and where you set it down, particularly the mating surfaces. You need also to safeguard against any foreign material entering the lower intake runners.

To Install:

12. Carefully remove any remaining gasket material from the plenum and manifold mating surfaces. Make sure that any vacuum lines running under the center of the plenum are plastic tied correctly and in a position so they don't get crimped on installation.

✳✳ WARNING

Never soak the plenum in commercial liquid cleaners or solvents.

13. Lay a new gasket onto the lower manifold runners. There are studs available to aid in positioning, but if you're careful and patient there should be no need for them.

14. Position the plenum onto the manifold in such a way that you do not dislodge the gasket. Install the 2 long mounting bolts and tighten them finger-tight.

15. Coat the terminal with Electrical Terminal grease and reconnect the 3 leads at the module bracket.

16. Install the shift support bracket with the next two manifold screws and screw them down finger-tight.

Install the throttle cable bracket with the 2 remaining manifold screws and then tighten all 6 mounting screws to 12-18 ft. lbs. (16-24 Nm).

17. Reconnect the remaining vacuum lines.

18. Reconnect the throttle cable. Check for correct throttle operation and adjust if necessary.

19. Connect the battery cables.

Lower Intake Manifold

◆ **See Figures 21 and 25**

1. Open or remove the engine compartment hatch. Disconnect the negative battery cable.

2. Drain all water from the cylinder block and manifolds.

3. Remove the upper intake manifold as detailed previously.

4. Tag and disconnect all water hoses at the manifold and thermostat housing. Have some rags handy, since there will still be some water in them. Carefully move them out of the way.

5. Remove the thermostat housing and adapter.

6. Remove the distributor cap with the leads still connected and set it aside temporarily. Mark the position of the distributor rotor to the distributor body. Scribe a match mark across the distributor and the manifold. Loosen the clamp bolt and remove the distributor. Please refer to the Electrical section for further details on distributor removal. DO NOT turn the engine over once the distributor has been removed.

7. Tag and disconnect the lead at the temperature gauge sender unit.

8. Tag and disconnect any other lines, leads or hoses that might be in the way of removal. In some instances you may be able to simply secure them out of the way without disconnecting them—you be the judge.

9. If your engine has a spark plug wire retainer attached to the cylinder head cover, unclip the wires or remove the retainer and then tag and disconnect the plug wires at the spark plugs. Move the wires (and the cap) out of the way.

10. Loosen and remove the 12 manifold mounting bolts in the reverse order of the illustrated tightening sequence and then remove the manifold. There is a good likelihood you will need to pry the manifold off the block; be very careful that you don't scratch or mar the mating surfaces on the block, manifold or heads.

To Install:

11. Carefully remove all remaining gasket material from the manifold mating surfaces with a scraper or putty knife. Be careful that you don't accidentally drop any old gasket into the crankcase or intake ports on the cylinder head.

12. Inspect the manifold and all mating surfaces for any cracks or nicks.

13. Apply a 1/8 in. (3.2mm) bead of RTV sealant to the forward and aft edges (4 corners) of the cylinder block mating surface where it meets the cylinder head. Make sure that you run the bead at least as wide as width of each seal.

14. Install the manifold gaskets onto the head. Position the cylinder block seals and run a 1/16 in, (1.6mm) bead of RTV sealant along the edge of each seal where it mates with the gaskets on the cylinder head.

■ **The sealer sets up fairly quickly, so don't dawdle here!**

15. Carefully install the manifold into place so that all the bolt holes line up. Run your finger along the edge of all seal areas to make certain that they are still positioned correctly; if not, remove the manifold and reposition it again. Insert the bolts, tighten them finger-tight and then tighten all bolts in sequence to 23-25 ft. lbs. (31-34 Nm). When you're through, repeat the tightening sequence one more time and then make sure to recheck at the end of the season.

Check the Torque Specifications chart for any further details on tightening needs. Tighten all bolts in the order shown in the illustration.

16. Install the thermostat housing and adaptor.

17. Install the distributor making sure that the match marks align. Do the same with the distributor cap and reconnect the spark plug wires.

18. Install the upper manifold.

19. Install all water hoses and tighten their clamps securely.

20. Connect all other lines, leads and hoses that may have been removed to facilitate manifold removal.

21. Connect the battery cable and start the engine. Check the ignition timing and idle speed. Check all hoses and seals for leaks.

Exhaust Manifold

REMOVAL & INSTALLATION

◆ **See Figures 26, 27 and 28**    ② *MODERATE*

1. Open or remove the engine compartment hatch. Disconnect the negative battery cable.

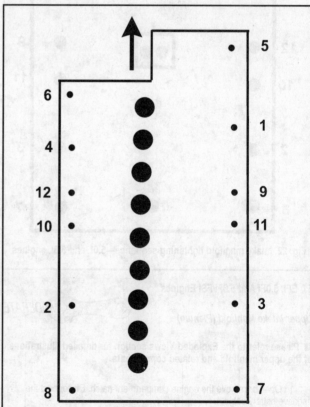

Fig. 25 Intake manifold (lower) tightening sequence—EFI engines

2. Drain all water and/or coolant from the engine, manifold and exhaust elbow as detailed in the Maintenance section.

3. Disconnect all water/coolant lines at the manifold and move them out of the way. It's a good idea to tag these so that you can ensure proper reconnection.

4. Loosen the hose clamps (two) and then squirt a little soapy water in between the hose and the elbow. Disconnect the exhaust pipe hose (bellows) at the elbow and move it out of the way. Grasp the hose with both hands and wiggle it back and forth while pulling it off the elbow. Take note of where all the hose clamps were situated.

■ **If you are unable to gain enough clearance to slide the hose off the elbow, remove it completely (as detailed elsewhere).**

5. If you intend to remove the starboard manifold, unsnap the electrical cable from the retaining clip and move it out of the way on FL models. On Fi and FSi models, remove the two bolts and lift off the ECA.

6. Carefully tag and disconnect the spark plug wires at the plugs and secure them above the manifold and out of the way.

7. Remove, disconnect, or simply move out of the way, any hoses or wires which may be in the way of removal on your particular engine.

8. Loosen the 8 manifold retaining nuts from the center outward and then pry off the manifold/elbow assembly.

9. If necessary, loosen the four retaining bolts and remove the exhaust elbow from the manifold.

To Install:

10. Carefully clean all residual gasket material from the head, manifold and elbow mating surfaces with a scraper or putty knife. Inspect all gasket surfaces for scratches, cuts or other imperfections.

11. Position a new gasket, without any sealant, on the cylinder head and install the manifold over the studs; making sure that everything is aligned properly. Tighten all nuts (with new lock washers) until they are just tight and then tighten them to 20-26 ft. lbs. (27-35 Nm). Starting in the center, and working your way out to the ends of the manifold.

12. If you removed the elbow, coat a new gasket with sealing compound and position it onto the manifold, making sure that everything lines up. Install the elbow. Tighten the mounting bolts to 12-14 ft. lbs. (16-20 Nm) on 1987-90 engines, or 12-18 ft. lbs. (16-24 Nm) on 1991-96 engines, refer to the Exhaust Elbow procedures found later for more detail on installing the elbow.

13. Connect the exhaust pipe/bellows to the elbow and tighten the clamps securely.

14. If you removed the manifold or elbow plugs for some reason, make sure that the threads are coated with sealant before screwing them back in.

15. Connect the water/coolant hoses and tighten the clamps securely.

16. Make sure that any miscellaneous lines or hoses that you may have moved or disconnected during removal are reconnected and routed properly.

17. Fill the system with water or coolant, connect the battery cable and start the engine. When the engine reaches normal operating temperature, turn it off and re-torque the manifold bolts.

Exhaust Elbow

REMOVAL & INSTALLATION

◆ **See Figures 26, 27 and 28**

 MODERATE

1. Drain the cooling system.

2. Loosen the two hose clamps and slide off the exhaust hose (bellows). Grasp it with both hands and pull it off while wiggling it from side to side. If it sticks, drip a little bit of soapy water around the lip.

3. If removing the starboard elbow on Fi and FSi models, remove the screws and disconnect the ECA—or ESA on carbureted engines

4. Remove the four bolts, lock washers and washers from the elbow and lift it off the manifold. A little friendly persuasion with a soft rubber mallet may be necessary! Be careful though, no need to take out all your aggressions on the poor thing.

5. Remove the gasket and discard it.

To Install:

6. Clean the mating surfaces of the manifold and elbow thoroughly, coat both sides of a new gasket with Gasket Sealing compound and position it onto the manifold flange.

7. Position the elbow on the manifold so the bolt holes are in alignment. Position any anchor brackets or mounting hardware for other components and then install the four mounting bolts and their washers. Tighten all four to 12-14 ft. lbs. (16-20 Nm) on 1987-90 engines, or 12-18 ft. lbs. (16-24 Nm) on 1991-96 engines.

8. Install the ECA/ESA on the starboard elbow and tighten the screws securely.

9. Slide the exhaust hose, while wiggling it, all the way onto the elbow. Position the two clamps in their channels and tighten the clamp screws securely. Coating the inside of the hose with a soapy water solution will help you slip it on.

Exhaust Hoses (Bellows) And Intermediate Exhaust Pipe

REMOVAL & INSTALLATION

◆ **See Figures 26, 29 and 30**

 MODERATE

1. Starting with the upper hose, loosen all four hose clamps, two on top of the hose and two on the bottom.

2. Drizzle a soapy water solution over the top of the hose where it mates with the exhaust elbow and let it sit for a minute.

3. Grasp the hose with both hands and wiggle it side-to-side while pulling down on it until it separates from the elbow.

4. Now wiggle it while pulling upwards until it pops off the intermediate exhaust pipe.

5. The lower hose should be removed in the same manner as the upper.

6. Check the hose for wear, cracks and deterioration.

7. Coat the inside of the lower end of the lower hose with soapy water and wiggle it into position on the Y-pipe (lower). Remember to install the two clamps before sliding it over the end of the pipe.

■ **There is a step about 1 1/2 in. into the inside of one end on each the upper and lower hoses. The stepped side of BOTH hoses should fit over the intermediate pipe. This means that the stepped side on the upper hose faces DOWN, and the stepped side on the lower hose faces UP!**

8. Slide the two clamps over the upper end of the lower hose and then coat the inside with the soapy water solution and insert the bottom of the intermediate pipe into it fully until it seats on the step. Tighten the clamp screws securely.

9. Coat the inside of the lower end of the upper hose with soapy water and wiggle it into position on the intermediate pipe until it meets the step. Remember to install the two clamps before sliding it over the end of the pipe.

10. Slide two clamps over the upper end, lubricate the inside with soapy water and wiggle the hose over the elbow.

11. Tighten all four clamp screws securely.

Lower Exhaust Pipe (Y-Pipe)

REMOVAL & INSTALLATION

◆ **See Figure 26**

 DIFFICULT

1. It is unlikely you will be able to get the pipe off without removing the engine, so remove the engine as previously detailed.

2. Loosen the retaining bolts at the transom shield and then remove the exhaust pipe. Carefully scrape any remnants of the seal from the pipe and transom mounting surfaces.

■ **The pipe mounting holes in the transom shield utilize Heli-Coil® locking inserts. Never clean the holes or threads with a tapping tool or you risk damaging the locking feature of the threads.**

3. Coat a new seal with 3M Rubber Adhesive and position it into the groove on the transom shield mating surface.

4. Coat the mounting bolts with Gasket Sealing Adhesive. Position the exhaust pipe, insert the bolts and tighten them to 20-25 ft. lbs. (27-34 Nm).

5. Install the engine.

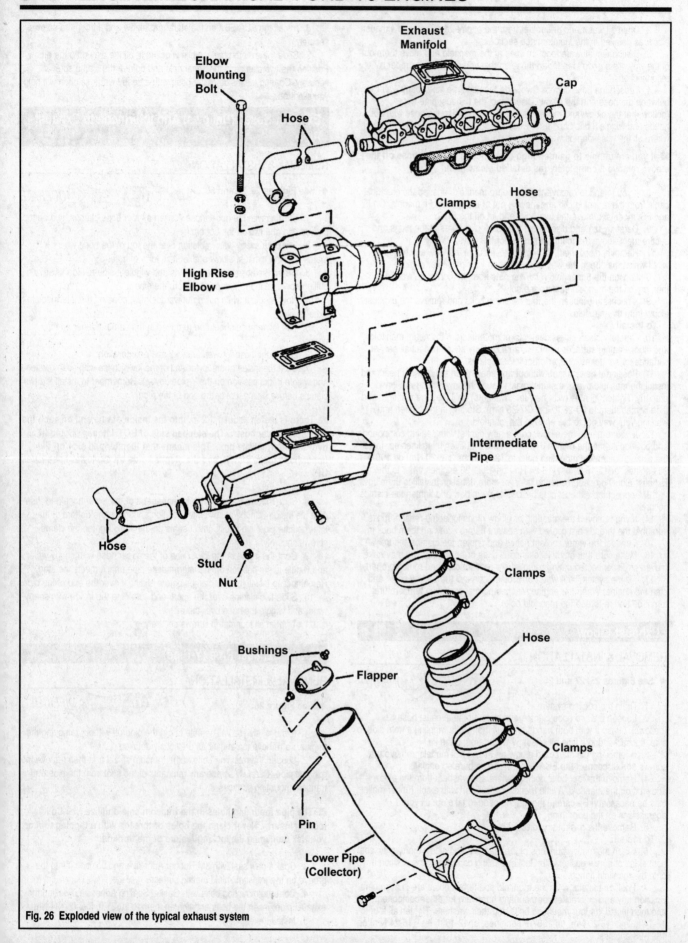

Fig. 26 Exploded view of the typical exhaust system

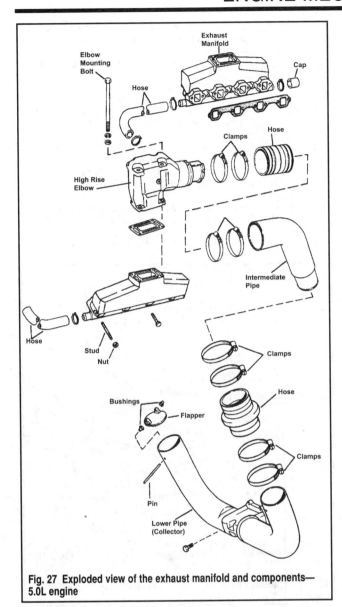

Fig. 27 Exploded view of the exhaust manifold and components— 5.0L engine

Fig. 28 Exploded view of the exhaust manifold and components— 7.5L engine

Fig. 29 A good look at the upper hose...

EXHAUST VALVE (FLAPPER) REPLACEMENT

6. Remove the exhaust hoses afnd the intermediate exhaust pipe. The flapper is located in the upper end of the y-pipe.

7. The valve is held in place by means of a pin running through two bushings in the sides of the pipe. Position a small punch over one end of the pin and carefully press the pin out of the pipe.

■ **Make sure you secure the valve while removing the retaining pin so it doesn't fall down into the exhaust pipe.**

8. Press out the two bushings and discard them. Coat two new bushings with Scotch Grip Rubber Adhesive and press them back into the sides of the pipe.

9. Position the new valve into the pipe with the long side DOWN. When looking at the valve, the molded retaining rings are off-center—the side with the rings should face the top of the pipe. When the valve is in place, coat the pin lightly with engine oil and slowly slide it through one of the bushings, through the two retaining holes on the valve and then through the opposite bushing. Make sure the pin ends are flush with the sides of the pipe on both sides.

10. Install the exhaust hose.

Fig. 30 ...and the lower hose—not all engines will have the bulge in the lower hose like shown here

Oil Pan

REMOVAL & INSTALLATION

◆ See Figure 31

 DIFFICULT

■ More times than not, this procedure will require the removal of the engine. Your boat and its unique engine installation will determine this, but the procedure is almost always easier with the engine removed from the boat.

1. Remove the engine as previously detailed in this section.
2. If you haven't already drained the engine oil, do it now. Make sure you have a container and lots of rags available.
3. Remove the oil dipstick and then remove the dipstick tube(s).
4. Remove the oil withdrawal tube from the oil drain plug fitting.
5. Loosen and remove the oil pan retaining bolts, studs and nuts, starting with the center fasteners and working out toward the pan ends. Lift off the pan braces and then lightly tap the pan with a rubber mallet to break the seal and then lift it off the cylinder block. If your engine stand will allow for rotating the engine, you'll find that this will be easier with the pan facing up.

To Install:

6. Clean the pan mating surfaces of any residual gasket material with a scraper or putty knife. Make sure that no old gasket material has been pressed into the retaining bolt holes in the pan, block or front cover. Clean the pan itself thoroughly with solvent.
7. Coat the mating surfaces of the oil pan and cylinder block with an oil-resistant sealant and then position a new gasket onto the block—make sure all the holes line up. In place of a one piece gasket like the other engines, the 7.5L uses 2 side gaskets and 2 seals—make sure that the tabs in the front cover seal and the rear main bearing seal are over the edges of each side gasket.
8. Move the pan onto the block; don't dawdle here because the RTV sealant applied in the previous step sets up very quickly. It is very important that you ensure all the holes line up correctly.
9. Position the pan braces on each side and then install a bolt at each corner to hold the pan in place. Install the rest of the bolts and tighten the 1/4-20 bolts to 84-108 inch lbs. (9.5-12.2 Nm) and the 5/16-18 bolts to 108-132 inch lbs. (12.2-14.9 Nm) from the center outward. On the fuel injected engines, tighten all bolts to 9-11 ft. lbs. (13-14 Nm); from the center outward.
10. Install the oil drain plug or the oil drain fitting and finger tighten the bolt. Rotate the fitting and then attach the withdrawal tube. Tighten the fitting bolt and flare nut to 15-18 ft. lbs. (20-24 Nm).
11. Install the engine (if removed). Refill with all fluids. Run the engine up to normal operating temperature, shut it off and check the pan for any leaks.

Oil Pump

The two-piece oil pump utilizes two pump rotors and a pressure regulator valve enclosed in a two-piece housing. A baffled pick-up tube is attached to the body of the pump. Oil passes through the pick-up screen, through the pump and then through the oil filter.

REMOVAL & INSTALLATION

◆ See Figures 31, 32 and 33

 DIFFICULT

1. Remove the oil pan as previously detailed. Remember that you probably need to remove the engine for this procedure.
2. Loosen the two inlet tube mounting bolts and the bracket bolt and then remove the pick-up tube/screen.
3. Loosen and remove the 2 pump mounting bolts and lift off the pump assembly along with the intermediate shaft.
4. Pull out the driveshaft and retainer.
5. Loosen the 4 cover mounting screws and lift off the pump cover.
6. Match mark the two rotors where they mesh and then lift out the outer rotor and shaft.
7. Pull out the cotter pin retaining the relief valve plug. To remove the plug itself, you will need to drill a small hole in the plug, insert a self-tapping screw and pull it out with a pair of pliers—don't lose the spring. Now you can remove the valve itself. Are you sure you want to do this?
8. Clean all components in solvent and dry thoroughly. Inspect the pump body and rotors for cracks, excessive wear or other damage

9. Install the relief valve into the housing cover with a new spring. Press in the plug until it seats itself and then install a new cotter pin.
10. Install the outer rotor and shaft into the housing so the earlier marks line up. Fill the housing with fresh engine oil, install the pump cover and tighten the screws to 72-120 inch lbs. (8-16 Nm).
11. Now fill the pump with engine oil through the inlet to prime it—wiggle the pump shaft to aid the oil in coating all internal parts.
12. Insert the intermediate shaft into the distributor socket until it seats itself fully. The stop on the shaft should just contact the crankcase, so position it until this is the case.
13. When the stop is positioned correctly, insert into the pump and install them as an assembly. Do NOT force anything into position; if it will not seat properly, the drive is probably not aligned with the distributor shaft correctly. Rotate the intermediate shaft slightly until they align. Tighten the oil pump bolts to 22-32 ft. lbs. (30-43 Nm).
14. Install the pick-up tube and screen. Tighten the tube-to-pump bolts to 10-15 ft. lbs. (14-20 Nm). Tighten the support bracket bolt to 22-32 ft. lbs. (30-43 Nm).
15. Install the oil pan and engine. Fill the engine with oil.

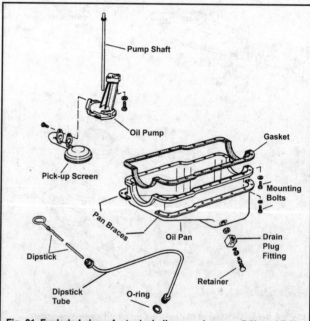

Fig. 31 Exploded view of a typical oil pan and pump—5.0L and 5.8L engines

Engine Coupler/Timing Ring And Flywheel

REMOVAL & INSTALLATION

■ Over the years, these engines were equipped in a variety of configurations—flywheel and torsional damper, flywheel, damper and coupler, flywheel and coupler, etc. And for a number of years, some models also had a timing ring attached. Early models are usually the ones equipped with a damper, while Cobra, SP and DP models usually were equipped with the coupler. King Cobra models are usually the ones that use the timing ring. Although we have attempted to narrow the applications in the following procedures, it was not always year or model specific. Please read the procedures FIRST in order to determine which one is appropriate for your particular engine's configuration.

Models With A Torsional Damper

 DIFFICULT

■ Over the years, these engines were equipped in a variety of configurations—flywheel and torsional damper, flywheel, damper and coupler, flywheel and coupler, etc. And for a number of years, some models also had a timing ring attached. Early models are usually the ones equipped with a damper, while Cobra, SP and DP models usually were equipped with the coupler. King Cobra models are usually the ones that use the timing ring. Although we have attempted to narrow

Fig. 32 Remove the pick-up tube...

Fig. 33 ...and then remove the pump itself

the applications in the following procedures, it was not always year or model specific. Please read the procedures FIRST in order to determine which one is appropriate for your particular engine's configuration.

1. Remove the engine as detailed previously.
2. Loosen the bolts and remove the flywheel housing and cover plate.
3. Remove the three mounting bolts and pull off the rear torsional damper.
4. Remove four of the six flywheel retaining bolts. Loosen the other two, any two is fine, and back them out about half way. Using the two bolts as a stop, carefully pull the flywheel off of the crankshaft flange. Remove the two bolts and flywheel.

To Install:

5. Position the flywheel over the dowel pin on the flange and make sure that the holes line up correctly. Coat the bolt threads with engine oil, install them and tighten to 59 ft. lbs. (82 Nm). Tighten the six bolts in a diagonal star pattern.
6. Install the torsional damper and tighten the bolts securely.
7. Install the housing and plate. Coat all mounting bolts with engine oil and tighten the 9/16 in. housing bolts to 30 ft. lbs. (41 Nm), tighten the 5/16 in. plate bolts securely.
8. Install the engine.

Models With A Timing Ring

◆ See Figures 34 and 35

 DIFFICULT

■ Over the years, these engines were equipped in a variety of configurations—flywheel and torsional damper, flywheel, damper and coupler, flywheel and coupler, etc. And for a number of years, some models also had a timing ring attached. Early models are usually the ones equipped with a damper, while Cobra, SP and DP models usually were equipped with the coupler. King Cobra models are usually the ones that use the timing ring. Although we have attempted to narrow the applications in the following procedures, it was not always year or model specific. Please read the procedures FIRST in order to determine which one is appropriate for your particular engine's configuration.

1. Remove the engine from the boat as detailed previously in this section.
2. Loosen and remove any connections attached to either of the ground studs on each side of the flywheel. Move the electrical leads out of the way.
3. Loosen the 5 retaining bolts and slide out the lower housing cover.
4. Although not strictly necessary, we recommend removing the starter.
5. Locate the timing sensor cover on the housing, loosen the two nuts and lift off the cover. Remove the two nuts and washers and then pull out the sensor and position it out of the way.
6. Cut the plastic tie that secures the housing drain hose and the pull the hose out of the fitting.
7. Remove the flywheel housing retaining nuts/bolts and pull off the housing. Take note of the positioning of the oil cooler and its bracket.
8. Slide an offset wrench behind the coupler and loosen the six flywheel mounting nuts gradually and as you would the lug nuts on your car or truck—that is, in a diagonal star pattern. Remove the coupler.
9. Mark the dowel hole on the timing ring and pry it off the flywheel. Remove the flywheel.

To Install:

10. Thoroughly clean the flywheel mating surface and check it for any nicks, cracks or gouges. Check for any broken teeth.
11. Install the flywheel over the dowel on the crankshaft.
12. Position the timing ring over the locating pin making sure the pin is in the correct hole—you did mark it, right? Press the ring into position.
13. Slide the coupler over the studs so that it sits in the recess on the flywheel. Install new lock washers and tighten the mounting nuts to 40-45 lbs. (54-61 Nm). Once again use the star pattern while tightening the bolts.
14. Install the flywheel housing and attach the oil cooler. Tighten the nuts/bolts to 32-40 ft. lbs. (43-54 Nm).
15. Install the washer, lock washer and inner nut on the ground stud and tighten it to 15-20 ft. lbs. (20-27 Nm). Attach the electrical leads, install another lock washer and then tighten the outer nut securely.
16. Slide the drain hose in and attach it with a new plastic tie.
17. Position the timing sensor onto the mounting studs and press in on the spring tab so that the sensor seats itself correctly over the timing ring. Install the washers and nuts, with the tab still depressed, and tighten them to 48-64 inch lbs. (5-7 Nm). Once the nuts are tightened, press in on the spring tab again and confirm that the tab DOES NOT touch the timing ring. If it does, loosen the nuts and try it again.

✸✸ WARNING

If the sensor tab comes in contact with the teeth of the timing ring, it will be damaged on engine start-up.

18. Install the sensor and tighten the nuts to 48-64 inch lbs. (5-7 Nm).
19. Coat both sides of a new gasket with Gasket Sealing Compound and position it onto the cover. Slide the cover into position and tighten the bolts to 60-84 inch lbs. (7-9 Nm).
20. Install the engine.

Models With A Coupler

◆ See Figures 34, 35 and 36

 DIFFICULT

■ Over the years, these engines were equipped in a variety of configurations—flywheel and torsional damper, flywheel, damper and coupler, flywheel and coupler, etc. And for a number of years, some models also had a timing ring attached. Early models are usually the ones equipped with a damper, while Cobra, SP and DP models usually were equipped with the coupler. King Cobra models are usually the ones that use the timing ring. Although we have attempted to narrow the applications in the following procedures, it was not always year or model specific. Please read the procedures FIRST in order to determine which one is appropriate for your particular engine's configuration.

1. Remove the engine from the boat as detailed previously in this section.
2. Loosen and remove any connections attached to either of the ground studs on each side of the flywheel. Move the electrical leads out of the way.
3. Although not strictly necessary, we recommend removing the starter.
4. Loosen the 4 or 5 retaining bolts and slide out the lower housing cover.

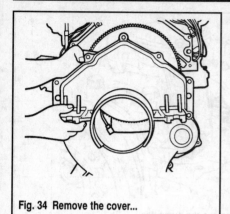

Fig. 34 Remove the cover...

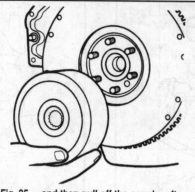
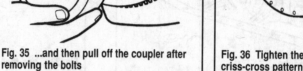

Fig. 35 ...and then pull off the coupler after removing the bolts

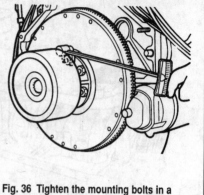

Fig. 36 Tighten the mounting bolts in a criss-cross pattern

5. Cut the plastic tie that secures the housing drain hose and the pull the hose out of the fitting.

6. Remove the oil cooler and bracket, take note of the positioning of both and secure them out of the way.

7. Remove the flywheel housing.

8. Slide an offset wrench behind the coupler and loosen the six flywheel mounting nuts gradually and as you would the lug nuts on your car or truck—that is, in a diagonal star pattern.

9. Remove the coupler and then the flywheel.

To Install:

10. Thoroughly clean the flywheel mating surface and check it for any nicks, cracks or gouges. Check for any broken teeth.

11. Install the flywheel over the dowel on the crankshaft.

12. Slide the coupler over the studs so that it sits in the recess on the flywheel. Install new lock washers and tighten the mounting nuts to 40-45 ft. lbs. (54-61 Nm) except on 1992 engines where the spec is 35-40 ft. lbs. (47-54 Nm). Once again use the star pattern while tightening the bolts.

4. Install the flywheel housing and attach the oil cooler. Tighten the bolts to 28-36 ft. lbs. (38-49 Nm) on 1987-91 engines or 32-40 ft. lbs. (43-54 Nm) on 1992-96 engines.

13. Install the washer, lock washer and inner nut on the ground stud and tighten it to 20-25 ft. lbs. (27-34 Nm) on 1987-91 engines or 15-20 ft. lbs. (20-27 Nm) on 1992-96 engines. Attach the electrical leads, install another lock washer and then tighten the outer nut securely.

14. Coat both sides of a new gasket with Gasket Sealing Compound and position it onto the cover. Slide the cover into position and tighten the bolts to 60-84 inch lbs. (7-9 Nm).

15. Slide the drain hose in and attach it with a new plastic tie.

16. Install the engine.

Rear Main Oil Seal

REMOVAL & INSTALLATION

5.0L And 5.8L Engines

◆ See Figures 37 and 38

OEM ③ ◁DIFFICULT

It is not necessary to remove the oil pan or rear main bearing cap when removing the one-piece oil seal on these engines although you may find it easier to do just that.

1. Remove the engine.

2. Remove the flywheel housing and cover as detailed in this section.

3. Remove the engine coupler and flywheel from the engine as detailed in this section.

4. Using a sharp awl or punch, carefully punch a hole into metal surface of the seal between the cylinder block and the lip.

5. Install a slide hammer (#T77L-9533-B), screw in the threaded end and remove the seal. Be very careful not to scratch or mar the seal or groove surfaces.

To Install:

6. Thoroughly clean the groove surface.

7. Coat the new seal with clean engine oil.

8. On 5.0L engines, position it into a Seal Driver (#T82L-6701-A) so that the seal's lip faces the cylinder block. Position the driver and seal over the crankshaft and then thread the attaching screws into the holes in the crankshaft. Tighten them securely, and alternately, until the seal is fully seated in the groove. The rear edge of the seal must be within 0.005 in. (0.127mm) of the outer edge of the cylinder block.

9. On 5.8L engines, insert the seal, lip facing forward, into a Seal Driver (#T65P-6701-A) and then position the tool over the centerline of the crankshaft. Keeping the tool perfectly straight, hit the drive end with a mallet until the leading edge of the tool comes into contact with the cylinder block.

10. On either engine, inspect the seal and groove surface after removing the tool to make sure nothing was damaged during installation.

11. Install the flywheel and engine coupler. Install the cover and flywheel housing.

12. Install the starter. Install the engine.

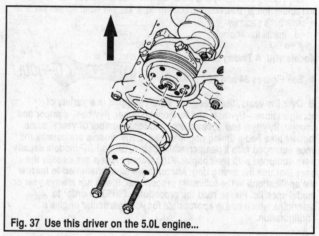

Fig. 37 Use this driver on the 5.0L engine...

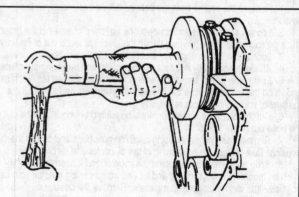

Fig. 38 ...or this one on the 5.8L engine

7.5L Engines

◆ See Figures 39, 40 and 41

These engines utilize a split-lip style rear main seal. The seal can be removed without removing the crankshaft. You will need to remove the engine for this procedure though.

1. Remove the engine as detailed previously in this section (although not strictly necessary, its highly recommended).

2. Remove the oil pan and pump as detailed previously in this section.

3. Loosen all of the main bearing cap bolts just enough so that the crankshaft drops down 1/32 in. at the rear.

4. Loosen the retaining bolts and remove the rear main bearing cap. Carefully insert a small prybar or awl and remove the lower half of the seal from the cap. Do not damage the seal seating surface.

5. Using a hammer and a small drift or punch, tap on the end of the upper seal until it starts to protrude form the other side of the race. Grab the protruding end with pliers and pull out the remaining seal half. OMC makes a seal removal tool for this purpose also, or many people will thread a small screw into one end of the seal and then pull it out that way—choose which way works best for you.

■ **Upper and lower seals must replaced as a pair. Never replace only one seal.**

6. Clean the seal grooves thoroughly with a small bottle brush.

7. Check that you have the correct new seals. Coat the entire seal half in clean engine oil.

8. Install the upper half of the seal (undercut side facing front of the engine) by rotating it on the journal until about 1/8 in. (3.2mm) is protruding from the parting surface. Be very careful here, as you need to make sure that no seal rubber has been shaved off by the bottom edge of the seal groove.

9. Tighten up all bearing cap bolts (except the rear) to 95-105 ft. lbs. (129-142 Nm) to bring the crankshaft back up into position.

10. Insert the lower seal half into the main bearing cap in the same manner as you did with the upper seal.

11. Make sure that the cap/block mating surfaces and the seal ends are free of any oil and then apply a small amount of Perfect Seal to the cap at the rear of the top mating surface. Do not apply sealer forward of the side seal groove.

12. Install the bearing cap and tighten the bolts on all caps to 95-105 ft. lbs. (129-142 Nm).

13. Install the oil pump and pan.

14. Install the engine.

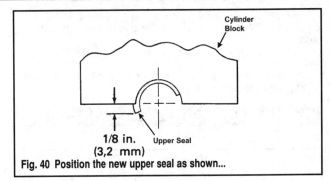

Fig. 40 Position the new upper seal as shown...

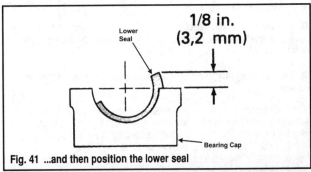

Fig. 41 ...and then position the lower seal

Vibration Damper, Pulley And Hub

REMOVAL & INSTALLATION

◆ See Figure 42

1. Disconnect the battery cables.

2. Remove the bracket mounting bolts for the alternator, raw water and power steering pump. Swivel them in to relieve belt tension.

3. Remove the drive belt(s) as detailed in the Maintenance section.

4. Drain the engine and manifolds.

5. Remove the three bolts and pull off the drive pulley attached to the balancer.

6. Remove the balancer retaining bolt and install special tool #T58P-6316-D onto the damper. Tighten the tool press bolt and remove the damper; don't lose the crankshaft key. OMC suggests that you DO NOT use a conventional gear puller for this procedure.

To Install:

7. Inspect the crank key and then install it into the shaft. Using a little adhesive will make this easier.

8. Coat the front cover oil seal lip with Ford Multi-Purpose grease and then install the damper so that the key and keyway align. Use a proper installation tool (#T79T-6316-A) to press the assembly into place on the crankshaft. Be sure that you thread the tool into the crankshaft at least 1/2 in. to protect the threads. In a pinch you can use a block of wood and a plastic mallet, but be careful that the pulley does not shift on its mountings while you're hammering. Or, you can use a large washer and a 7/16-20 x 4 in. bolt, but we suggest the tool.

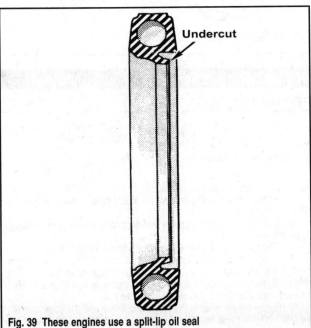

Fig. 39 These engines use a split-lip oil seal

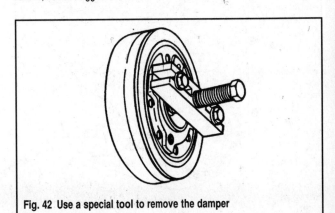

Fig. 42 Use a special tool to remove the damper

9. Install the retaining bolt and tighten it to 70-90 ft. lbs. (95-122 Nm). Squirt a little RTV sealant into the crankshaft keyway to guard against oil seepage.

10. Install the drive pulley and tighten the bolts to 22-32 ft. lbs (30-43 Nm).

11. Install the drive belts and make sure that they are adjusted properly.

12. Connect the battery cables.

Front Cover And Oil Seal

REMOVAL & INSTALLATION

◆ See Figures 43 and 44

■ This procedure may require engine removal, depending upon your particular boat. If necessary, remove the engine as detailed previously in this section.

■ It is not necessary to remove the front cover in order to remove the oil seal.

1. Open the drain valves and drain the coolant from the block and exhaust manifold. Loosen the alternator and power steering brackets (or the idler pulley) to provide slack, and then remove the drive belts.

2. Drain the oil if you haven't already done so.

3. Remove the alternator and power steering pump bracket bolts.

4. Remove the water circulation pump. The factory actually suggests that you can leave the pump on the cover, but we think removing it facilitates the process—either way will work, so you decide for yourself.

5. Remove the vibration damper and pulley as detailed previously in this section.

6. Remove the few oil pan-to-front cover bolts. Loosen the cover mounting bolts and pry the front cover outward slightly. Insert a razor knife between the cover and block so that it is flush with the inside of the cover and then carefully cut the oil pan gasket at each side of the cover.

7. Remove the cover.

8. If the oil seal needs replacement, position a seal remover tool (#T70P-6B070-B) over the seal and cover. Tighten the thru-bolts to force the puller under the flange and then tighten the puller bolts alternately until the seal is removed.

9. Remove the front cover gasket.

To Install:

10. Clean all gasket material from the cover and block mating surfaces with a scraper of putty knife. Be careful not to knock any pieces of gasket into the timing assembly.

11. If you removed the oil seal, coat a new one with Ford Multi-Purpose grease and insert it into the installation tool (#T70P-6B070-A). Position the seal and tool over the end of the crankshaft and push it forward until the seal starts seating itself into the cover. Screw the installation screw with its

washer and nut into the end of the crank and then tighten the nut against the washer and sleeve while forcing the seal into the cover. Remove the tool and inspect the front edge of the seal for damage.

12. Take a new oil pan gasket and cut out a portion of it equal to that which you cut previously when removing the cover. Coat the oil pan mating surface with RTV sealer and carefully position the cut gasket into the flange. While you have the sealer out, run a small bead along the joints where the three (pan, cover and block) surfaces come in contact with one another.

13. Coat the lip of the oil seal with engine oil. Coat both sides of a new gasket with sealant and then position the gasket onto the engine. Install the cover so that all the bolt holes line up and then use an alignment tool (#T61P-6019-B) to ensure proper cover-to-block alignment. Coat the mounting bolt threads with Ford Perfect Seal and screw them in finger-tight. Push in on the alignment tool and tighten the cover-to-pan bolts to 9-11 ft. lbs. (13-14 Nm). Tighten the cover-to-block bolts to 12-18 ft. lbs. (17-24 Nm). Remove the tool.

14. Install the vibration damper and its pulley.

15. Install the circulation pump and its pulley (or just the pulley if you chose earlier to leave the pump attached to the cover). Pull the belts back on. Check their tension adjustment.

16. Install the engine if removed. Add oil and water/coolant, start the engine and check for any leaks.

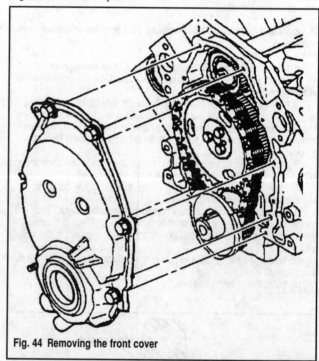

Fig. 44 Removing the front cover

Timing Chain And Sprockets

REMOVAL & INSTALLATION

◆ See Figures 45, 46 and 47

1. Disconnect the battery cables.

2. Remove the mounting bolts for the alternator and power steering pump brackets.

3. Remove the crankshaft pulley and harmonic balancer as previously detailed in this section.

4. Remove the front cover as previously detailed in this section.

5. Rotate the camshaft slightly so that it creates tension on one side of the timing chain (either side is OK). Find a reference point on the same side of the cylinder block as the side that the timing chain is tight on and then measure from this point to the outer edge of the chain.

6. Rotate the camshaft in the opposite direction until the other side of the chain is tight. Press the inner side of the chain outward until it stops and then measure from your reference point on the cylinder block (obviously, do

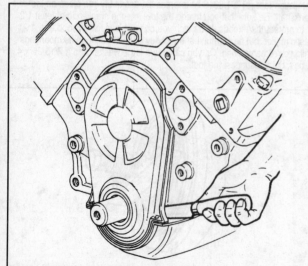

Fig. 43 Use a sharp knife to cut the oil pan seal

this from the same side of the chain as you did in the previous step) to the outer edge of the chain. This is timing chain deflection and it should be no more than 0.500 in. (12.7mm). If it is more than specification, the chain will require replacement.

7. Look carefully at the camshaft and crankshaft sprockets—you should notice a small indent on the front edge of one of the teeth on each sprocket (it may actually be at the base of the tooth on some sprockets). Bump the engine over until these two marks are in alignment as shown in the illustration—crank mark at 12:00 and cam mark at 6:00, a remote starter will work or you can screw the damper bolt back into the crankshaft and turn it with a wrench.

8. Dab a little paint across one of the chain links and the camshaft sprocket. Loosen the camshaft sprocket retaining bolt and remove it along with the fuel pump eccentrics. Grasp the two sprockets at top and bottom, with the chain still attached, and wiggle them off the shafts as an assembly. They should come off readily, but if not, tap the bottom edge lightly with a rubber mallet.

✳✳ WARNING

Never rotate the crankshaft once the timing chain has been removed. If moved, you'll risk damage to the pistons and/or valve train.

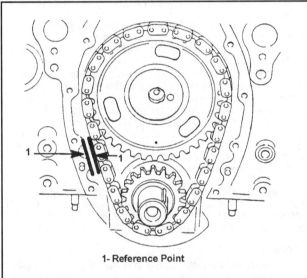

1- Reference Point

Fig. 45 When checking timing chain deflection, find a reference point on the cylinder block and use it for each measurement

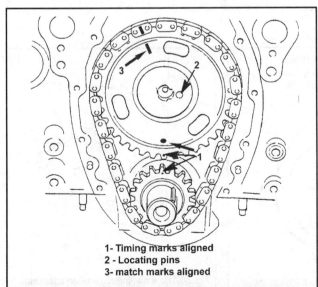

1- Timing marks aligned
2 - Locating pins
3- match marks aligned

Fig. 46 The two marks on each of the timing sprockets must be in alignment before removing or installing the timing chain

To Install:

9. Clean the chain and sprockets in solvent and let them air dry. Check the chain for wear and damage, making sure there are no loose or cracked links. Check the sprockets for cracked or worn teeth.

10. Install the timing chain onto the camshaft sprocket so that the paint marks made during removal match up. If they do, and you haven't moved the engine, the timing marks on the two sprockets should also. Hold the sprocket/chain in both hands so the chain is hanging down, engage the chain around the crankshaft sprocket making sure that the timing mark is pointing up, and then slide the assembly onto the two shafts making sure the dowel pin aligns with the hole in the camshaft sprocket. Do not force them!

11. Install the two-piece fuel pump eccentric and washer. Install the camshaft bolt and tighten it to 40-45 ft. lbs. (55-61 Nm).

✳✳ CAUTION

Never force either of the sprockets onto their shafts or use a hammer.

12. Install the front cover, vibration damper and water pump pulley.
13. Install and adjust the drive belts.
14. Reconnect the alternator and power steering brackets.

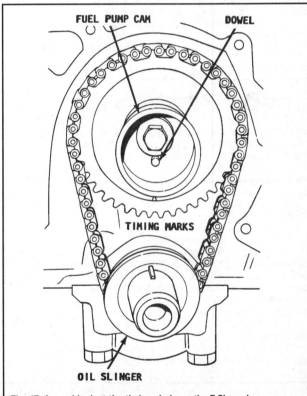

FUEL PUMP CAM   DOWEL

TIMING MARKS

OIL SLINGER

Fig. 47 A good look at the timing chain on the 7.5L engine

Camshaft

CHECKING LIFT

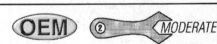

◆ See Figures 48, 49 and 50

■ If the shaft is out of the engine, you can use a micrometer to take the heel-to-lobe measurement and the side-to-side measurement. Subtract the second measurement from the first to get lobe lift. Most times though, the shaft will still be in the cylinder block so perform the following procedure.

■ Check each lobe consecutively, from front to back.

1. Tag and disconnect the electrical connectors at the ignition coil.
2. Remove the cylinder head cover and rocker arms as detailed previously.
3. Install a solid push rod into the bore, or use an adapter for ball end rods. Make sure that the rod is seated against the lifter, which is seated on the camshaft lobe.
4. Using a special adaptor, connect a dial indicator so that its tip is positioned on the end of the pushrod—the adaptor should screw onto the end of the rocker stud.
5. Slowly rotate the crankshaft in the direction of engine rotation until the valve lifter is riding on the heel (back side of lobe) of the camshaft lobe. The pushrod should be at its lowest point when the lifter is on the heel.

■ A remote starter works well for turning the engine over in this situation.

6. Set the indicator to **0** and then rotate the engine until the pushrod is at the highest point of its travel. Camshaft lift should be as detailed in the Engine Specifications chart.
7. Continue rotating the engine until the pushrod is back at its lowest position—make sure that the indicator still reads **0**.
8. Repeat this procedure for the remaining pushrods.
9. Install the rocker arms and adjust the valve clearance.
10. Install the cylinder head cover and reconnect the coil leads.

REMOVAL & INSTALLATION

1. Remove the cylinder head covers and rocker assemblies as previously detailed in this section. Remove the pushrods.
2. Remove the intake manifold.
3. Remove the lifters. Match mark the lifters to their bores and then lift them out and store them in a rack with labels so they can be reinstalled in their original locations.
4. Remove the front cover and timing chain/sprockets as previously detailed in this section.
5. Remove the fuel pump and push rod.
6. Remove the thrust plate/retainer.
7. Thread the cap bolt back into the camshaft bolt hole and carefully pull the camshaft out of the cylinder block. You may have to wiggle it back and forth a bit, so be sure you don't lean it up or down or else you could damage the bearings. Do not cant the shaft while removing it.
To Install:
8. Inspect the camshaft as detailed previously.
9. Coat the camshaft journals with engine oil. Coat the camshaft lobes with engine oil).
10. Carefully insert the shaft into the cylinder block and slide it all the way in. Be very careful not to damage the bearings.

11. Remove the installation bolt and install the thrust plate/retainer.
12. Install the sprockets, fuel pump eccentrics, timing chain and front cover.
13. Drop the lifters back into their original bores so that they are aligned with the match marks and then install the push rods.
14. Install the intake manifold.
15. Install the rocker assemblies and the cylinder head covers.

Cylinder Head

REMOVAL & INSTALLATION

◆ See Figure 51 and 52

1. Drain the water from the cylinder block and manifold.
2. Remove the fuel line support brackets. Disconnect the fuel line at the carburetor/throttle body and fuel pump, plug the fitting holes and remove the line.
3. Remove the cylinder head cover and rocker assemblies/push rods as detailed previously in this section.
4. Remove the intake and exhaust manifolds as previously detailed; you can leave the carburetor/throttle body attached to the intake manifold if you like.
5. If you intend to remove the valve lifters, now is the time to do it. Either way, make sure that you cover the valley of the cylinder block carefully with plenty of rags to prevent dirt from entering any of the passages or settling on any components.
6. Tag and disconnect the spark plug wires at the plugs; move them out of the way. Although not necessary, it's a good idea to remove the plugs themselves also; plug the holes if you do.
7. Remove or relocate any components or connections that may interfere with the removal of an individual cylinder head.
8. Loosen the cylinder head bolts in the reverse order of the illustrated tightening sequence and then carefully lift the head off the block. You may need to persuade it with a rubber mallet—be careful! Set the head down carefully on two support blocks; do not sit it on cement.

■ Its always a good idea to keep a record of which bolts came from which holes. It may sound silly, but on many engines they are different sizes and you wouldn't be the first person to break off a long bolt while tightening it in a short hole. Spend the extra few seconds and do this!

To Install:
9. Carefully, and thoroughly, remove all residual head gasket material from the cylinder head and block mating surfaces with a scraper or putty knife. Check that the mating surfaces are free of any nicks or cracks. Make sure there is no dirt or old gasket material in any of the bolt holes. Refer to the Engine Rebuilding section found for complete details on inspection and refurbishing procedures.

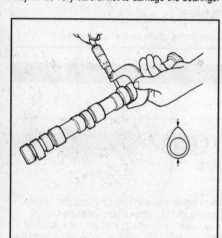

Fig. 48 Use a micrometer when checking the camshaft lift with the shaft out of the engine

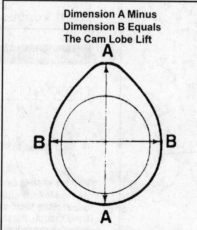

Dimension A Minus
Dimension B Equals
The Cam Lobe Lift

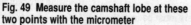

Fig. 49 Measure the camshaft lobe at these two points with the micrometer

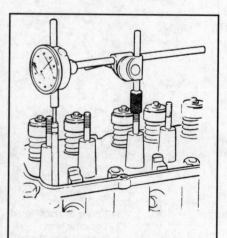

Fig. 50 Use a dial indicator when checking the camshaft lift with the shaft in the engine—typical

10. Inspect the head bolts on 5.0L engines. Engines may be equipped with one of two types of bolts. You may reuse all standard head bolts if this is what your engine has, but must replace any "torque-to-yield" bolts as shown in the accompanying illustration.

11. Position a new gasket over the cylinder block dowel pins. Do not use any sealer as the gasket is coated with a lacquer which will provide the proper sealing effect once the engine gets to normal operating temperature the first time. DO NOT use automotive-type steel gaskets. Many times the gasket will be marked, make sure the word **UP** is facing, you guessed it, 'up', if noted on your gasket.

12. Position the cylinder head over the dowels in the block. Install the head bolts finger tight. It never hurts to use new bolts, although it's not necessary (except as noted earlier).

13. Tighten the bolts, a little at a time, in the sequence illustrated, until the proper tightening torque is achieved.

- 1989-93 5.0L engines: the first step should be 50 ft. lbs. (68 Nm), the second step should be 60 ft. lbs. (81 Nm), the third step should be 65-70 ft. lbs. (88-95 Nm). In sequence.
- 1994-96 5.0L engines with standard bolts: the first step should be 55-65 ft. lbs. (75-88 Nm), the second step should be 65-72 ft. lbs. (88-98 Nm). In sequence.
- 1994-96 5.0L engines with "torque-to-yield" bolts: tighten all the bolts, in sequence, to 25-35 ft. lbs. (34-47 Nm). Next, tighten them to 45-55 ft. lbs. (61-75 Nm). Finally, tighten all bolts an additional 85°-95° (1/4 turn). Obviously you will need a torque/angle meter for these last passes.
- 1989-93 5.8L engines: the first step should be 90 ft. lbs. (122 Nm), the second step should be 100 ft. lbs. (136 Nm), the third step should be 112 ft. lbs. (152 Nm). In sequence.
- 1994-96 5.8L engines: the first step, in sequence, should be 95-105 ft. lbs. (129-143 Nm), the second step should be 105-112 ft. lbs. (143-151 Nm).
- engines: the first step should be 80 ft. lbs. (108 Nm), the second step should be 110 ft. lbs. (149 Nm), the third step should be 135 ft. lbs. (184 Nm). In sequence.

14. Install the rocker assemblies and the cylinder head cover.

15. Install the manifolds and connect the fuel line. Don't forget to remove the fitting plugs.

16. Install the spark plugs if they were removed and then connect the plug wires.

17. Install or connect any other components removed to facilitate getting the head off.

18. Add coolant/water, connect the battery and check the oil. Start the engine and run it for a while to ensure that everything is operating properly. Keep an eye on the temperature gauge.

19. It never hurts to re-tighten the cylinder head bolts again after 20 hours of operation.

Engine Circulating Pump

REMOVAL & INSTALLATION

◆ See Figure 53  ② MODERATE

1. Disconnect the battery cables and then drain all water from the block and manifolds.

2. Drain all water/coolant from the cylinder block.

3. Loosen, but do not remove, the pump pulley mounting bolts.

4. Loosen the power steering pump and alternator bracket bolts and swivel them in until you are able to remove the belt(s). Different engines may have different systems, follow the belt back and loosen the appropriate components.

5. Now you can remove the pump pulley bolts along with the lock washers and clamping ring. Pull off the pulley.

6. Remove the two power steering pump bracket bolts, the brace from the alternator bracket, the Torx® retaining bolt and the pump.

7. Disconnect the large water hose at the pump.

8. Tag and disconnect the electrical leads at the alternator and move them out of the way. Remove the bracket bolts and lift off the alternator.

9. Remove the mounting bolts and lift the pump off of the block. There's a good chance you will need to persuade the pump—tap it lightly with a rubber mallet.

To Install:

10. Carefully scrape any old gasket material off both mounting surfaces. Inspect the pump for blockage, cracks or any other damage. Inspect the impeller for cracks. Replace either if necessary.

11. Coat both sides of a new gasket with sealant and position on the cylinder block. Make sure that all of the holes are lined up.

12. Position the pump and screw in all bolts except those that attach the alternator and power steering pump brackets. Tighten the bolts to 12-18 ft. lbs. (17-24 Nm).

13. Reconnect the water hoses and tighten the hose clamps securely.

14. Position the spacer and the Torx® bolt behind the rear power steering pump bracket and tighten it securely. Attach the other bracket to the pump and tighten the 2 bolts.

15. Install the alternator and tighten the bracket-to-pump bolts to 18-20 ft. lbs. (24-27 Nm).

16. Attach the steering pump brace to the alternator and tighten the bolt securely.

17. Position the pump pulley and clamping ring on the boss. Screw the mounting bolts and lock washers in and tighten them to 14-20 ft. lbs. (19-27 Nm).

18. Install the drive belt(s) and adjust them as detailed previously. Start the engine and check the system for leaks

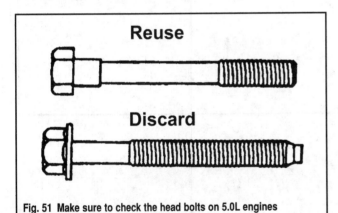

Fig. 51 Make sure to check the head bolts on 5.0L engines

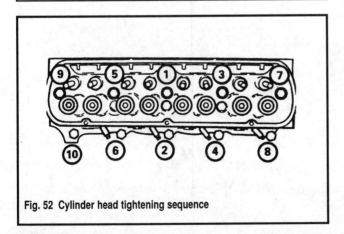

Fig. 52 Cylinder head tightening sequence

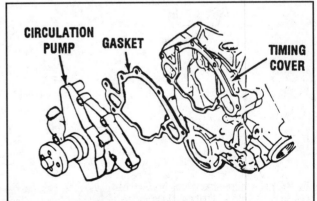

Fig. 53 Removing the water circulating pump

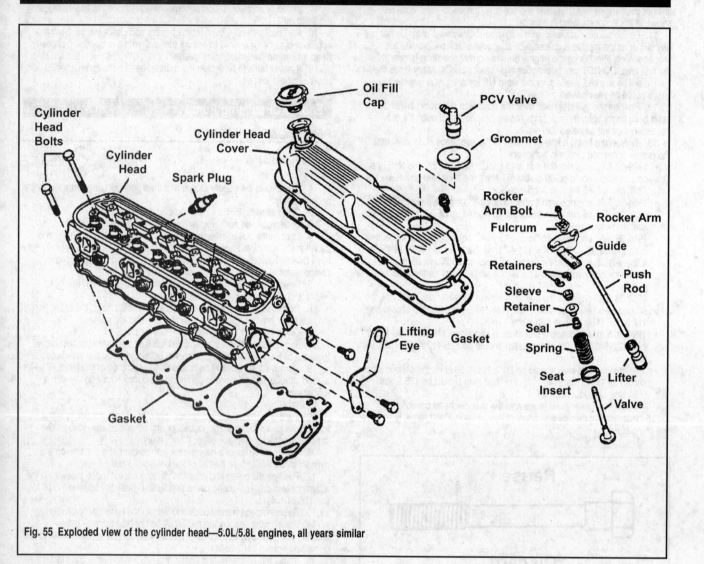

Fig. 55 Exploded view of the cylinder head—5.0L/5.8L engines, all years similar

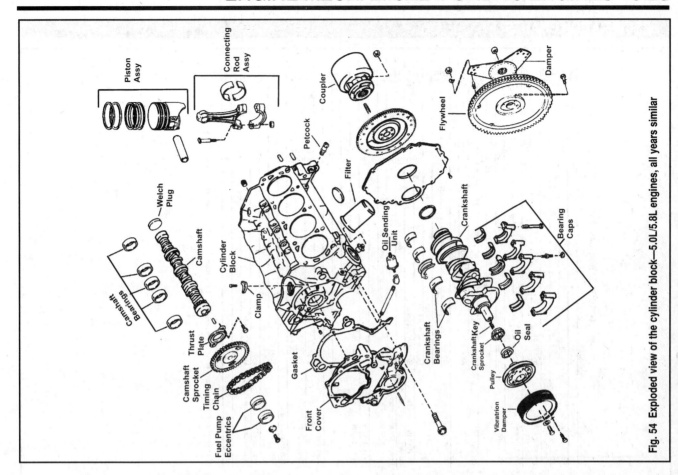

Fig. 54 Exploded view of the cylinder block—5.0L/5.8L engines, all years similar

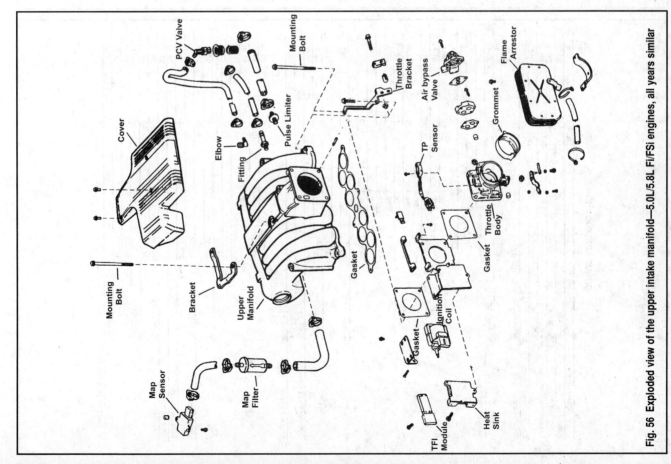

Fig. 56 Exploded view of the upper intake manifold—5.0L/5.8L FI/FSi engines, all years similar

SPECIFICATION CHARTS

TORQUE SPECIFICATIONS

Component			ft. lbs.	inch lbs.	Nm
Alternator	Bracket		18-20	-	24-27
Camshaft	Sprocket		40-45	-	55-61
Connecting Rod Cap	5.0L		19-24	-	26-33
	5.8L/7.5L		40-45	-	54-61
Coolant Temp Sender			18-22	-	24-30
Crankshaft	Main Bearing Cap	5.0L	60-70	-	81-95
		5.8L/7.5L	95-105	-	129-142
	Pulley Bolt		22-32	-	30-43
Cylinder Head	5.0L	1989-93 Step 1	50	-	68
		Step 2	60	-	81
		Step 3	65-70	-	88-95
		1994-96 Stretch Bolts Step 1	25-35	-	34-47
		Step 2	45-55	-	61-75
		Step 3	1/4 turn	-	1/4 turn
	5.8L	1989-93 Standard Bolts Step 1	55-65	-	75-88
		Step 2	65-72	-	88-98
		1994-96 Step 1	90	-	122
		Step 2	100	-	136
		Step 3	112	-	152
	7.5L	Step 1	95-105	-	129-143
		Step 2	105-112	-	143-151
		Step 1	80	-	108
		Step 2	110	-	149
		Step 3	135	-	184
Cylinder Head Cover	1987-93		-	60-72	7-8
	1994-96		-	36-60	4-6
Engine Mounts	Front 5.0L/5.8L	To engine	32-40	-	43-54
		Top nut 1989-92	40-50	-	54-60
		Top nut 1993-96	50-70	-	68-95
		Bolt/Lower Nut	60-75	-	81-102
	7.5L	To engine	32-40	-	43-54
		Top nut	100-120	-	136-163
	Rear	Locknut	28-30	-	38-40
		Mount	20-25	-	27-34
Exhaust Elbow	To Manifold	1987-90	12-14	-	16-20
		1991-96	12-18	-	16-24
Exhaust Manifold			20-26	-	27-35
Exhaust Pipe			20-25	-	27-34
Flywheel	Bellhousing	1987-91	28-36	-	38-49
		1992-96	32-40	-	43-54
	Coupler Nut		40-45①	-	54-61①
	Coupler Stud	1987-91	20-25	-	27-34
		1992-96	10-15	-	14-20
	Cover		-	60-84	7-9

① 1992: 35-40 ft. lbs. (47-50)

TORQUE SPECIFICATIONS

Component			ft. lbs.	inch lbs.	Nm
Front Cover	To Block		12-18	-	17-24
	To Oil Pan		9-11	-	13-14
Fuel Pump			18-20	-	24-27
Intake Manifold	5.0L/5.8L		23-25①	-	31-34①
	7.5L		25-27	-	34-40
Main Bearing Cap	5.0L		60-70	-	81-95
	5.8L/7.5L		95-105	-	129-142
Manifolds	Exhaust		20-26	-	27-35
	Intake	5.0L/5.8L	23-25②	-	31-34②
		7.5L	25-27	-	34-40
Oil Pan	Fi/FSi		9-11	-	13-14
	All others	1/4-20 Bolts	-	84-108	10-12
		5/16-18 Bolts	-	108-132	12-15
Oil Pump	Cover		-	72-120	6-16
	Inlet Tube To Bracket		10-15	-	14-20
	Inlet Tube Bracket		22-32	-	30-43
	Mounting Bolts		22-32	-	30-43
Oil Pressure Sender	Sender/Adaptor		10-14	-	14-19
Oil Withdrawal Tube (flare)			15-18	-	20-24
Power Steering Pump	Bracket		26-30	-	35-41
	Lines	Large Fitting	15-17	-	20-23
		Small Fitting	10-12	-	14-16
Rocker Arm			18-25	-	24-34
Spark Plugs	5.0L/5.8L	1989-90	15-20	-	20-27
		1991-95	5-10	-	7-13
		1996	10-15	-	14-20
	7.5L		5-10	-	7-13
Starter Motor	Bolts		24-30	-	33-41
	Bracket		16-19	-	22-26
Timing Chain Cover	To Block		12-18	-	17-24
	To Oil Pan		9-11	-	13-14
Torsional Damper			59	-	82
Valve Cover	1987-93		-	60-72	7-8
	1994-96		-	36-60	4-6
Vibration Damper			70-90	-	95-122
Water (Circulating) Pump	Mounting Bolts		12-18	-	17-24
	Pulley		14-20	-	19-27
Water Temperature Sender			18-22	-	24-30

① Fi and FSi models: figure is for lower manifold; upper manifold - 12-18 ft. lbs. (16-24 Nm)

ENGINE SPECIFICATIONS - 5.0L/5.8L

Component			Standard (in.) ⊕	Metric (mm) ⊕
Cylinder Bore	Diameter	5.0L	4.0004-4.0052	101.61-101.73
		5.8L	4.0000-4.0048	101.60-101.72
	Out-of-round	Production	0.0015 Max	0.038 Max
		Service Limit	0.005 Max	0.127 Max
	Taper	Production	0.001 Max	0.025 Max
		Service Limit	0.010 Max	0.254 Max
Cylinder Head	Surface Flatness		0.003 (in 6 in.)	0.076 (in 153mm)
Piston	Clearance		0.0018-0.0026	0.046-0.066
	Diameter 5.0L	Red	3.9984-3.9990	101.56-101.58
		Blue	3.9996-4.0002	101.59-101.61
		0.003 Over 1989-91	4.0008-4.0014	101.62-101.64
		1992-96	4.0020-4.0026	101.65-101.67
	5.8L	Red	3.9978-3.9984	101.54-101.56
		Blue	3.9990-3.9996	101.58-101.50
		0.003 Over 1989-91	4.0002-4.0008	101.61-101.62
		1992-96	4.0014-4.0020	101.64-101.65
Piston Pin	Clearance	Production	0.0002-0.0004	0.006-0.009
		Service Limit	0.0003-0.0005	0.0076-0.0127
	Diameter	Standard	0.9120-0.9123	23.164-23.172
		0.001 Over	0.9130-0.9133	23.190-23.197
		0.002 Over	0.9140-0.9143	23.216-23.223
Piston Rings - Compression	Side Clearance	Production	0.002-0.004	0.051-0.102
		Service Limit	+0.002 Max	+0.051 Max
	Gap		0.010-0.020	0.254-0.508
Piston Rings - Oil	Side clearance		Snug	Snug
	Gap		0.015-0.055	0.381-1.397
Valve system	Face Angle		44 deg.	44 deg.
	Lifter		Hydraulic	Hydraulic
	Clearance	Production	0.0007-0.0027	0.018-0.069
		Service Limit	0.005 Max	0.127
	Collapsed Gap	Production 5.0L	0.096-0.165	2.55 Max
		5.8L	0.123-0.173	3.12-4.39
		Service Limit 5.0L	0.071-0.193	1.80-4.90
		5.8L	0.098-0.198	2.49-5.03
	Push Rod	Diameter	0.8740-0.8745	22.199-22.213
		Run Out	0.015 Max	0.381 Max
	Rocker Arm Lift Ratio		1.61:1	1.61:1
	Seat Angle		45 deg.	45 deg.
	Seat Width		0.060-0.080	1.52-2.03
	Spring Free Length	Intake 5.0L (1989-91)	1.94	49.28
		5.0L (1992-96)	2.06	52.32
		5.0L (1995-96 EFI)	2.056	52.22
		5.8L (1989-91)	2.06	52.32
		5.8L (1992-96)	2.056	52.22

ENGINE SPECIFICATIONS - 5.0L/5.8L

Component			Standard (in.) ⊕	Metric (mm) ⊕
Camshaft	Bearing Clearance	Production	0.001-0.003	0.025-0.076
		Service Limit	0.006	0.152
	End Play	Production	0.001-0.007	0.025-0.178
		Service Limit	0.009	0.229
	Journal Diameter	#1	2.0805-2.0815	52.845-52.870
		#2	2.0655-2.0665	52.464-52.489
		#3	2.0505-2.0515	52.083-52.108
		#4	2.0355-2.0365	51.702-51.727
		#5	2.0205-2.0215	51.321-51.346
	Journal Out-of-Round		0.005 Max	0.127 Max
	Lobe Lift Intake	5.0L (1989-91)	0.2303	5.85
		5.0L (1992-93)	0.2375	6.033
		5.0L (1994-96)	0.26	6.604
		5.0L (1995-96 EFI)	0.3	7.62
		5.0 HO	0.26	6.604
		5.8L	0.278	7.061
		5.8L (1995-96 EFI)	0.3	7.62
	Lobe Lift Exhaust	5.0L (1989-91)	0.238	6.045
		5.0L (1992-93)	0.247	6.274
		5.0L (1994-96)	0.278	7.061
		5.0L (1995-96 EFI)	0.3	7.62
		5.0 HO	0.278	7.061
		5.8L	0.283	7.188
		5.8L (1995-96 EFI)	0.3	7.62
	Runout	Loss	0.005 Max	0.127 Max
	Timing Chain Deflection		0.5	12.7
Connecting Rod	Bearing Clearance	Production	0.0008-0.0015	0.020-0.038
		Service Limit	0.0008-0.0025	0.020-0.064
	Journal Diameter	5.0L	2.1228-2.1236	53.92-53.94
		5.8L	2.3103-2.3111	58.68-58.70
	Out-of-round		0.0006 MAX	0.015 Max
	Side Clearance	Production	0.010-0.020	0.25-0.51
		Service Limit	0.023	0.58
	Taper		0.0006 MAX	0.015 Max
Crankshaft	Bearing Clearance 5.0L	Production #1	0.0001-0.0015	0.0025-0.0381
		#2, 3, 4, 5	0.0005-0.0015	0.0127-0.0381
		Service Limit #1	0.0001-0.0020	0.0025-0.0508
		#2, 3, 4, 5	0.0005-0.0024	0.0127-0.0610
	5.8L	Production	0.0008-0.0015	0.0203-0.0381
		Service Limit	0.0008-0.0026	0.0203-0.0660
	Crankshaft End Play		0.004-0.008	0.102-0.203
	Journal Diameter	5.0L	2.2482-2.2490	57.10-57.13
		5.8L	2.9994-3.0002	76.19-76.21
	Out-of-round	Production	0.0006 Max	0.015 Max
		Total	0.002	0.051
	Taper		0.0006 Max	0.015 Max

ENGINE SPECIFICATIONS - 5.0L/5.8L

Component				Standard (in.) ①	Metric (mm) ①
Valve System (Cont'd)	Spring Free Length	Exhaust	5.0L (1989-91)	1.87	47.5
			5.0L (1992-96)	1.86	47.24
			5.0L (1995-96 EFI)	1.877	47.68
			5.8L (1989-91)	1.87	47.5
			5.8L (1992-96)	1.877	47.68
	Spring Assembled Ht.	Intake	5.0L (1989-91)	1.66-1.72	42.16-43.69
			5.0L (1992-93)	1.77-1.80	44.96-45.72
			5.0L (1994-96)	1.18	27.43
			5.0L (1995-96 EFI)	1.78	45.21
			5.8L (1989-91)	1.77-1.81	44.96-45.97
			5.8L (1992-93)	1.77-1.80	44.96-45.72
			5.8L (1994-96)	1.18	27.43
			5.8L (1995-96 EFI)	1.78	45.21
		Exhaust	5.0L (1989-91)	1.59-1.63	40.39-41.40
			5.0L (1992-93)	1.58-1.61	40.13-40.89
			5.0L (1994-96)	1.6	40.64
			5.8L (1989-91)	1.59-1.63	40.39-41.40
			5.8L (1992-93)	1.58-1.61	40.13-40.89
			5.8L (1994-96)	1.6	40.64
	Spring Pressure				
	Open	Intake	5.0L (1989-91)	190-210 lbs. @ 1.31	841-941 N @ 33.27
			5.0L (1992-93)	195-215 lbs. @ 1.15	867-956 N @ 29.21
			5.0L (1994-96)	194-214 lbs. @ 1.33	863-952 N @ 33.78
			5.0L (1995-96 EFI)	194-214 lbs. @ 1.33	863-952 N @ 33.78
			5.8L (1989-91)	190-210 lbs. @ 1.31	841-941 N @ 33.27
			5.8L (1992-93)	195-215 lbs. @ 1.15	867-956 N @ 29.21
			5.8L (1994-96)	208-228 lbs. @ 1.33	925-1014 N @ 33.78
			5.8L (1995-96 EFI)	194-214 lbs. @ 1.33	863-952 N @ 33.78
		Exhaust	5.0L (1989-91)	190-210 lbs. @ 1.31	841-941 N @ 33.27
			5.0L (1992-96)	195-215 lbs. @ 1.15	867-956 N @ 29.21
			5.8L (1989-91)	190-210 lbs. @ 1.31	841-941 N @ 33.27
			5.8L (1992-96)	195-215 lbs. @ 1.15	867-956 N @ 29.21
	Closed	Intake	5.0L (1989-91)	76-84 lbs. @ 1.69	340-376 N @ 42.92
			5.0L (1992-96)	74-82 lbs. @ 1.78	329-365 N @ 45.21
			5.8L (1992-96)	74-82 lbs. @ 1.78	329-365 N @ 45.21
		Exhaust	5.0L (1989-91)	76-84 lbs. @ 1.69	340-376 N @ 42.92
			5.0L (1992-96)	71-79 lbs. @ 1.60	318-353 N @ 40.64
			5.8L (1989-91)	71-79 lbs. @ 1.79	318-353 N @ 45.47
			5.8L (1992-96)	71-79 lbs. @ 1.60	318-353 N @ 40.64
		Service Limit		-10% @ Length	-10% @ Length
	Stem Diameter	Intake		0.3416-0.3423	8.677-8.694
		Exhaust		0.3411-0.3418	8.664-8.682
	Stem-to-Guide Clear.	Production	Intake	0.0010-0.0027	0.0254-0.0686
			Exhaust	0.0015-0.0032	0.038-0.081
		Wear Limit		0.0055	0.1397

ENGINE SPECIFICATIONS - 7.5L

Component			Standard (in.) ①	Metric (mm) ①
Camshaft	Bearing Clearance	Production	0.001-0.003	0.025-0.076
		Service Limit	0.006	0.152
	End Play	Production	0.001-0.006	0.025-0.152
		Service Limit	0.009	0.229
	Journal Diameter		2.1238-2.1248	53.95-53.97
	Journal Out-of-Round		0.0005 Max	0.0127 Max
	Lobe Lift	Intake	0.285	6.604
		Exhaust	0.29	7.061
		Loss	0.005 Max	0.127 Max
	Runout		0.005 Max	0.127 Max
	Timing Chain Deflection		0.5	12.7
Connecting Rod	Bearing Clearance	Production	0.0008-0.0015	0.020-0.038
		Service Limit	0.0008-0.0025	0.020-0.064
	Journal Diameter		2.4992-2.5000	63.48-63.50
	Out-of-round		0.0004 MAX	0.010 Max
	Side Clearance	Production	0.010-0.020	0.25-0.51
		Service Limit	0.023	0.58
	Taper		0.0004 MAX	0.010 Max
Crankshaft	Bearing Clearance	Production	0.0008-0.0015	0.020-0.038
		Service Limit #1	0.0008-0.0020	0.020-0.051
		#2,3,4,5	0.0008-0.0025	0.020-0.064
	Crankshaft End Play	Production	0.004-0.008	0.102-0.203
		Service Limit	0.012	0.31
	Journal Diameter		2.9994-3.0002	76.19-76.21
	Out-of-round	Production	0.0006 Max	0.015 Max
		Total	0.002	0.051
	Taper		0.0005 Max	0.012 Max
Cylinder Bore	Diameter		4.3600-4.3636	110-74-110.84
	Out-of-round	Production	0.0015 Max	0.038 Max
		Service Limit	0.005 Max	0.127 Max
	Taper	Production	0.001 Max	0.025 Max
		Service Limit	0.010 Max	0.254 Max
Cylinder Head	Surface Flatness		0.003 (in 6 in.)	0.076 (in 153mm)
Piston	Clearance		0.0022-0.0030	0.056-0.076
	Diameter	Red	4.3585-4.3591	110.71-110.72
Piston Pin	Clearance		0.0002-0.0005	0.006-0.0127
	Diameter	Standard	1.0398-1.0403	26.41-26.42
		0.001 Over	1.0410-1.0413	26.44-26.45
Piston Rings - Compression	Side Clearance	Production	0.002-0.004	0.051-0.102
		Service Limit	+0.006 Max	+0.152 Max
	Gap		0.010-0.020	0.254-0.508
Piston Rings - Oil	Side clearance		Snug	Snug
	Gap		0.010-0.035	0.254-0.889

ENGINE SPECIFICATIONS - 7.5L

Component				Standard (in.) ①	Metric (mm) ①
Valve system	Face Angle			44 deg.	44 deg.
	Lifter			Hydraulic	Hydraulic
		Clearance	Production	0.0007-0.0027	0.018-0.069
			Service Limit	0.005 Max	0.127
		Collapsed Gap			
			Production	0.075-0.175	1.905-4.445
			Service Limit	0.100-0.150	2.54-3.81
		Diameter		0.8740-0.8745	22.199-22.213
	Push Rod	Run Out		0.015 Max	0.381 Max
	Rocker Arm Lift Ratio			1.73:1	1.73:1
	Seat Angle			45 deg.	45 deg.
	Seat Width			0.060-0.080	1.52-2.03
	Spring Free Length			2.06	52.32
	Spring Assembled Ht.			1.80-1.83	45.72-46.48
	Spring Pressure	Open		300-330 lbs. @ 1.32	1334-1468 @ 33.74
		Closed		87-97 @ 1.82	387-432 @ 46.23
		Service Limit		-10% @ Length	-10% @ Length
	Stem Diameter	Intake		0.3416-0.3423	8.677-8.694
		Exhaust		0.3416-0.3423	8.677-8.694
	Stem-to-Guide Clear.	Production	Intake	0.0010-0.0027	0.0254-0.0686
			Exhaust	0.0010-0.0027	0.0254-0.0686
		Wear Limit		0.0055	0.1397

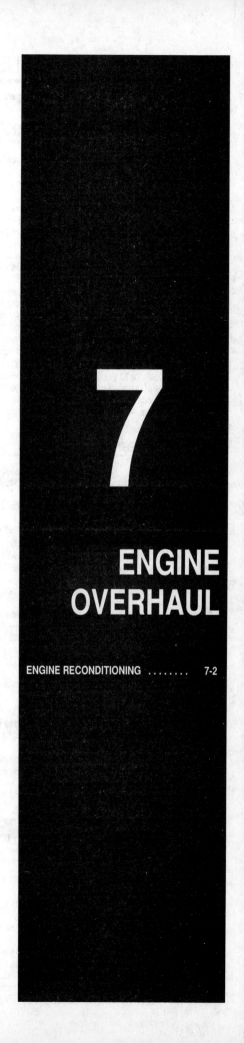

7

ENGINE OVERHAUL

ENGINE RECONDITIONING

Determining Engine Condition

Anything that generates heat and/or friction will eventually burn or wear out (for example, a light bulb generates heat, therefore its life span is limited). With this in mind, a running engine generates tremendous amounts of both; friction is encountered by the moving and rotating parts inside the engine and heat is created by friction and combustion of the fuel. However, the engine has systems designed to help reduce the effects of heat and friction and provide added longevity. The oiling system reduces the amount of friction encountered by the moving parts inside the engine, while the cooling system reduces heat created by friction and combustion. If either system is not maintained, a break-down will be inevitable. Therefore, you can see how regular maintenance can affect the service life of your engine. If you do not drain, flush and refill your cooling system at the proper intervals, deposits will begin to accumulate, thereby reducing the amount of heat it can extract from the coolant. The same applies to your oil and filter; if it is not changed often enough it becomes laden with contaminates and is unable to properly lubricate the engine. This increases friction and wear.

There are a number of methods for evaluating the condition of your engine. A compression test can reveal the condition of your pistons, piston rings, cylinder bores, head gasket(s), valves and valve seats. An oil pressure test can warn you of possible engine bearing, or oil pump failures. Excessive oil consumption, evidence of oil in the engine air intake area and/or bluish smoke from the exhaust may indicate worn piston rings, worn valve guides and/or valve seals.

COMPRESSION TEST

◆ See Figure 1

■ Please refer also to the Maintenance section for details specific to your engine.

A noticeable lack of engine power, excessive oil consumption and/or poor fuel mileage measured over an extended period are all indicators of internal engine wear. Worn piston rings, scored or worn cylinder bores, blown head gaskets, sticking or burnt valves, and worn valve seats are all possible culprits. A check of each cylinder's compression will help locate the problem.

■ A screw-in type compression gauge is more accurate than the type you simply hold against the spark plug hole. Although it takes slightly longer to use, it's worth the effort to obtain a more accurate reading.

1. Make sure that the proper amount and viscosity of engine oil is in the crankcase, then ensure the battery is fully charged.
2. Warm-up the engine to normal operating temperature, then shut the engine **OFF.**
3. Disable the ignition system.
4. Label and disconnect all of the spark plug wires from the plugs.
5. Thoroughly clean the cylinder head area around the spark plug ports, then remove the spark plugs.
6. Set the throttle plate to the fully open (wide-open throttle) position. You can block the throttle linkage open for this, or you can have an assistant operate the throttle lever in the boat.
7. Install a screw-in type compression gauge into the No. 1 spark plug hole until the fitting is snug.

✳✳ WARNING

Be careful not to cross-thread the spark plug hole.

8. According to the tool manufacturer's instructions, connect a remote starting switch to the starting circuit.
9. With the ignition switch in the **OFF** position, use the remote starting switch to crank the engine through at least five compression strokes (approximately 5 seconds of cranking) and record the highest reading on the gauge.
10. Repeat the test on each cylinder, cranking the engine approximately the same number of compression strokes and/or time as the first.
11. Compare the highest readings from each cylinder to that of the others. The indicated compression pressures are considered within specifications if the lowest reading cylinder is within 75 percent of the pressure recorded for the highest reading cylinder. For example, if your highest reading cylinder pressure was 150 psi (1034 kPa), then 75 percent of that would be 113 psi (779 kPa). So the lowest reading cylinder should be no less than 113 psi (779 kPa).

11. If a cylinder exhibits an unusually low compression reading, pour a tablespoon of clean engine oil into the cylinder through the spark plug hole and repeat the compression test. If the compression rises after adding oil, it means that the cylinder's piston rings and/or cylinder bore are damaged or worn. If the pressure remains low, the valves may not be seating properly (a valve job is needed), or the head gasket may be blown near that cylinder. If compression in any two adjacent cylinders is low, and if the addition of oil doesn't help raise compression, there is leakage past the head gasket. Oil and coolant in the combustion chamber, combined with blue or constant white smoke from the exhaust, are symptoms of this problem. However, don't be alarmed by the normal white smoke emitted from the exhaust during engine warm-up or from cold weather operation. There may be evidence of water droplets on the engine dipstick and/or oil droplets in the cooling system if a head gasket is blown.

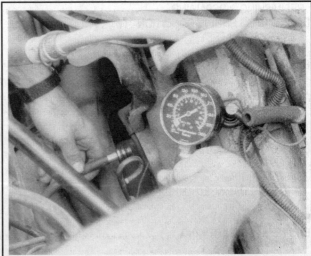

Fig. 1 A screw-in type compression gauge is more accurate and easier to use without an assistant

OIL PRESSURE TEST

Check for proper oil pressure at the sending unit passage with an externally mounted mechanical oil pressure gauge (as opposed to relying on a factory installed dash-mounted gauge). A tachometer may also be needed, as some specifications may require running the engine at a specific rpm.

1. With the engine cold, locate and remove the oil pressure sending unit.
2. Following the manufacturer's instructions, connect a mechanical oil pressure gauge and, if necessary, a tachometer to the engine.
3. Start the engine and allow it to idle.
4. Check the oil pressure reading when cold and record the number. You may need to run the engine at a specified rpm, so check the specifications.
5. Run the engine until normal operating temperature is reached.
6. Check the oil pressure reading again with the engine hot and record the number. Turn the engine OFF.
7. Compare your hot oil pressure reading to that given in the chart. If the reading is low, check the cold pressure reading against the chart. If the cold pressure is well above the specification, and the hot reading was lower than the specification, you may have the wrong viscosity oil in the engine. Change the oil, making sure to use the proper grade and quantity, then repeat the test.

Low oil pressure readings could be attributed to internal component wear, pump related problems, a low oil level, or oil viscosity that is too low. High oil pressure readings could be caused by an overfilled crankcase, too high of an oil viscosity or a faulty pressure relief valve.

Buy or Rebuild?

Now that you have determined that your engine is worn out, you must make some decisions. The question of whether or not an engine is worth rebuilding is largely a subjective matter and one of personal worth. Is the engine a popular one, or is it an obsolete model? Are parts available? Will it get acceptable gas mileage once it is rebuilt? Is the vessel it's being put into worth keeping? Would it be less expensive to buy a new engine, have your engine rebuilt by a pro, rebuild it yourself or buy a used engine? Or would it be simpler and less expensive to buy another boat? If you have considered all these matters and more, and have still decided to rebuild the engine, then it is time to decide how you will rebuild it.

■ **The editors at Seloc feel that most engine machining should be performed by a professional machine shop. Don't think of it as wasting money, rather, as an assurance that the job has been done right the first time. There are many expensive and specialized tools required to perform such tasks as boring and honing an engine block or having a valve job done on a cylinder head. Even inspecting the parts requires expensive micrometers and gauges to properly measure wear and clearances. Also, a machine shop can deliver to you clean, and ready to assemble parts, saving you time and aggravation. Your maximum savings will come from performing the removal, disassembly, assembly and installation of the engine and purchasing or renting only the tools required to perform the above tasks. Depending on the particular circumstances, you may save 40 to 60 percent of the cost doing these yourself.**

A complete rebuild or overhaul of an engine involves replacing all of the moving parts (pistons, rods, crankshaft, camshaft, etc.) with new ones and machining the non-moving wearing surfaces of the block and heads. Unfortunately, this may not be cost effective. For instance, your crankshaft may have been damaged or worn, but it can be machined undersize for a minimal fee.

So, as you can see, you can replace everything inside the engine, but, it is wiser to replace only those parts which are really needed, and, if possible, repair the more expensive ones. Later we will break the engine down into its two main components: the cylinder head and the engine block. We will discuss each component, and the recommended parts to replace during a rebuild on each.

Engine Overhaul Tips

Most engine overhaul procedures are fairly standard. In addition to specific parts replacement procedures and specifications for your individual engine, this is also a guide to acceptable rebuilding procedures. Examples of standard rebuilding practice are given and should be used along with specific details concerning your particular engine; which are found in the Engine Mechanical section.

Competent and accurate machine shop services will ensure maximum performance, reliability and engine life. In most instances it is more profitable for the do-it-yourself mechanic to remove, clean and inspect the component, buy the necessary parts and deliver these to a shop for actual machine work.

Much of the assembly work (crankshaft, bearings, piston rods, and other components) is well within the scope of the do-it-yourself mechanic's tools and abilities. You will have to decide for yourself the depth of involvement you desire in an engine repair or rebuild.

TOOLS

The tools required for an engine overhaul or parts replacement will depend on the depth of your involvement. With a few exceptions, they will be the tools found in a mechanic's tool kit. More in-depth work will require some or all of the following:
- A dial indicator (reading in thousandths) mounted on a universal base
- Micrometers and telescope gauges
- Jaw and screw-type pullers
- Scraper
- Valve spring compressor
- Ring groove cleaner
- Piston ring expander and compressor
- Ridge reamer
- Cylinder hone or glaze breaker
- Plastigage®
- Engine stand

The use of most of these tools is illustrated in the procedures. Many can be rented for a one-time use from a local parts jobber or tool supply house specializing in marine or automotive work.

Occasionally, the use of special tools is called for. See the information on Special Tools and the Safety Notice in the front of this manual before substituting another tool.

OVERHAUL TIPS

Aluminum has become extremely popular for use in engines, due to its low weight. Observe the following precautions when handling aluminum parts:
- Never hot tank aluminum parts (the caustic hot tank solution will eat the aluminum.
- Remove all aluminum parts (identification tag, etc.) from engine parts prior to the tanking.
- Always coat threads lightly with engine oil or anti-seize compounds before installation, to prevent seizure.
- Never over-tighten bolts or spark plugs especially in aluminum threads.

When assembling the engine, any parts that will be exposed to frictional contact must be pre-lubed to provide lubrication at initial start-up. Any product specifically formulated for this purpose can be used, but engine oil is not recommended as a pre-lube in most cases.

When semi-permanent (locked, but removable) installation of bolts or nuts is desired, threads should be cleaned and coated with Loctite® or another similar, commercial non-hardening marine sealant.

CLEANING

◆ See Figures 2, 3, 4 and 5

Before the engine and its components are inspected, they must be thoroughly cleaned. You will need to remove any engine varnish, oil sludge and/or carbon deposits from all of the components to insure an accurate inspection. A crack in the engine block or cylinder head can easily become overlooked if hidden by a layer of sludge or carbon.

Most of the cleaning process can be carried out with common hand tools and readily available solvents or solutions. Carbon deposits can be chipped away using a hammer and a hard wooden chisel. Old gasket material and varnish or sludge can usually be removed using a scraper and/or cleaning solvent. Extremely stubborn deposits may require the use of a power drill with a wire brush. If using a wire brush, use extreme care around any critical machined surfaces (such as the gasket surfaces, bearing saddles, cylinder bores, etc.). Use of a wire brush is NOT RECOMMENDED on any aluminum components. Always follow any safety recommendations given by the manufacturer of the tool and/or solvent. You should always wear eye protection during any cleaning process involving scraping, chipping or spraying of solvents.

An alternative to the mess and hassle of cleaning the parts yourself is to drop them off at a local marina or machine shop (or even an automotive garage). They will, more than likely, have the necessary equipment to properly clean all of the parts for a nominal fee.

Fig. 2 Use a gasket scraper to remove the old gasket material from the mating surfaces

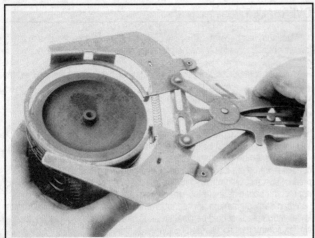

Fig. 3 Use a ring expander tool to remove the piston rings

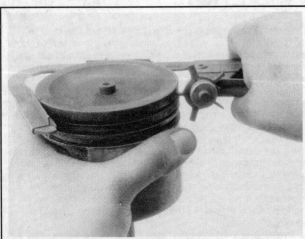

Fig. 4 Clean the piston ring grooves using a ring groove cleaner tool, or . . .

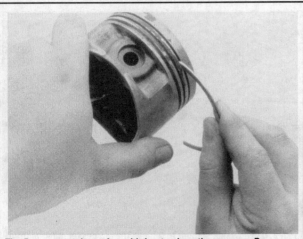

Fig. 5 . . . use a piece of an old ring to clean the grooves. Be careful, the ring can be quite sharp

Remove any oil galley plugs, freeze plugs and/or pressed-in bearings and carefully wash and degrease all of the engine components including the fasteners and bolts. Small parts such as the valves, springs, etc., should be placed in a metal basket and allowed to soak. Use pipe cleaner type brushes, and clean all passageways in the components. Use a ring expander and remove the rings from the pistons. Clean the piston ring grooves with a special tool or a piece of broken ring. Scrape the carbon off of the top of the piston. You should never use a wire brush on the pistons. After preparing all of the piston assemblies in this manner, wash and degrease them again.

When cleaning the cylinder head, remove carbon from the combustion chamber with the valves installed. This will avoid damaging the valve seats.

REPAIRING DAMAGED THREADS

◆ **See Figures 6, 7, 8, 9 and 10**

Several methods of repairing damaged threads are available. Heli-Coil® (shown here), Keenserts® and Microdot® are among the most widely used. All involve basically the same principle—drilling out stripped threads, tapping the hole and installing a pre-wound insert—making welding, plugging and oversize fasteners unnecessary.

Two types of thread repair inserts are usually supplied: a standard type for most inch coarse, inch fine, metric course and metric fine thread sizes and a spark lug type to fit most spark plug port sizes. Consult the individual tool manufacturer's catalog to determine exact applications. Typical thread repair kits will contain a selection of pre-wound threaded inserts, a tap (corresponding to the outside diameter threads of the insert) and an installation tool. Spark plug inserts usually differ because they require a tap equipped with pilot threads and a combined reamer/tap section. Most manufacturers also supply blister-packed thread repair inserts separately in addition to a master kit containing a variety of taps and inserts plus installation tools.

Before attempting to repair a threaded hole, remove any snapped, broken or damaged bolts or studs. Penetrating oil can be used to free frozen threads. The offending item can usually be removed with locking pliers or using a screw/stud extractor. After the hole is clear, the thread can be repaired, as shown in the series of accompanying illustrations and in the kit manufacturer's instructions.

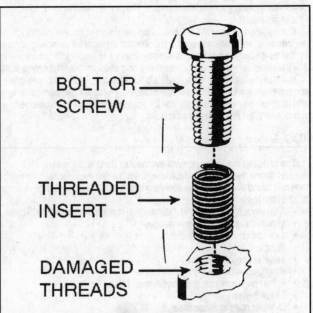

Fig. 6 Damaged bolt hole threads can be replaced with thread repair inserts

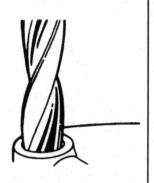

Fig. 7 Drill out the damaged threads with the specified size bit. Be sure to drill completely through the hole or to the bottom of a blind hole

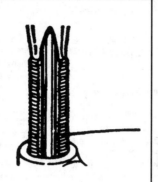

Fig. 8 Using the kit, tap the hole in order to receive the thread insert. Keep the tap well oiled and back it out frequently to clean out the chips

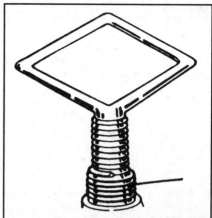

Fig. 9 Screw the insert onto the installer tool until the tang engages the slot. Thread the insert into the hole until it is 1/4–1/2 turn below the top surface, then remove the tool and break off the tang using a punch

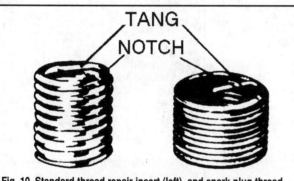

Fig. 10 Standard thread repair insert (left), and spark plug thread insert

Engine Preparation

To properly rebuild an engine, you must first remove it from the vessel, then disassemble and diagnose it. Ideally you should place your engine on an engine stand. This affords you the best access to the engine components. Follow the manufacturer's directions for using the stand with your particular engine. Remove the flywheel or coupler before installing the engine to the stand.

Now that you have the engine on a stand, and assuming that you have drained the oil and water/coolant from the engine, it's time to strip it of all but the necessary components. Before you start disassembling the engine, you may want to take a moment to draw some pictures, or fabricate some labels or containers to mark the locations of various components and the bolts and/or studs which fasten them. Modern day engines use a lot of little brackets and clips which hold wiring harnesses and such, and these holders are often mounted on studs and/or bolts that can be easily mixed up. The manufacturer spent a lot of time and money designing your engine/boat, and they wouldn't have wasted any of it by haphazardly placing brackets, clips or fasteners on the boat. If it's present when you disassemble it, put it back when you assemble, you will regret not remembering that little bracket which holds a wire harness out of the path of a rotating part.

You should begin by unbolting any accessories still attached to the engine, such as the water pump, power steering pump, alternator, etc. Then, unfasten any manifolds (intake or exhaust) which were not removed during the engine removal procedure. Finally, remove any covers remaining on the engine such as the rocker arm, front or timing cover and oil pan. Some front covers may require the balancer and/or crank pulley to be removed beforehand. The idea is to reduce the engine to the bare necessities (cylinder head(s), valve train, engine block, crankshaft, pistons and connecting rods), plus any other 'in block' components such as oil pumps, balance shafts and auxiliary shafts.

Finally, remove the cylinder head(s) from the engine block and carefully place on a bench. Disassembly instructions for each component follow later.

Cylinder Head

There are two basic types of cylinder heads used on today's engines: Overhead Valve (OHV) and the Overhead Camshaft (OHC). The latter can also be broken down into two subgroups: the Single Overhead Camshaft (SOHC) and the Dual Overhead Camshaft (DOHC). Generally, if there is only a single camshaft on a head, it is just referred to as an OHC head. An engine with an OHV cylinder head is also known as a pushrod engine.

Most cylinder heads these days are made of an aluminum alloy due to its light weight, durability and heat transfer qualities. However, cast iron was the material of choice in the past, and is still used on many engines today. Whether made from aluminum or iron, all cylinder heads have valves and seats. Most use two valves per cylinder, while the more hi-tech engines will utilize a multi-valve configuration using 3, 4 and even 5 valves per cylinder. When the valve contacts the seat, it does so on precision machined surfaces, which seals the combustion chamber. All cylinder heads have a valve guide for each valve. The guide centers the valve to the seat and allows it to move up and down within it. The clearance between the valve and guide can be critical. Too much clearance and the engine may consume oil, lose vacuum and/or damage the seat. Too little, and the valve can stick in the guide causing the engine to run poorly if at all, and possibly causing severe damage. The last component all cylinder heads have in common are valve springs. The spring holds the valve against its seat. It also returns the valve to this position when the valve has been opened by the valve train or camshaft. The spring is fastened to the valve by a retainer and valve locks (sometimes called keepers). Aluminum heads will also have a valve spring shim to keep the spring from wearing away the aluminum.

An ideal method of rebuilding the cylinder head would involve replacing all of the valves, guides, seats, springs, etc. with new ones. However, depending on how the engine was maintained, often this is not necessary. A major cause of valve, guide and seat wear is an improperly tuned engine. An engine that is running too rich, will often wash the lubricating oil out of the guide with gasoline, causing it to wear rapidly. Conversely, an engine which is running too lean will place higher combustion temperatures on the valves and seats allowing them to wear or even burn. Springs fall victim to the operating habits of the individual. A driver who often runs the engine rpm to the redline will wear out or break the springs faster then one that stays well below it. Unfortunately, 'hours of operation' takes it toll on all of the parts. Generally, the valves, guides, springs and seats in a cylinder head can be machined and re-used, saving you money. However, if a valve is burnt, it may be wise to replace all of the valves, since they were all operating in the same environment. The same goes for any other component on the cylinder head. Think of it as an insurance policy against future problems related to that component.

Unfortunately, the only way to find out which components need replacing, is to disassemble and carefully check each piece. After the cylinder head(s) are disassembled, thoroughly clean all of the components.

DISASSEMBLY

OHV Heads

◆ **See Figures 11 through 22**

Before disassembling the cylinder head, you may want to fabricate some containers to hold the various parts, as some of them can be quite small (such as keepers) and easily lost. Also keeping yourself and the components organized will aid in assembly and reduce confusion. Where possible, try to maintain a components' original location; this is especially important if there is not going to be any machine work performed on the components.

1. If you haven't already removed the rocker arms and/or shafts, do so now.
2. Position the head so that the springs are easily accessed.
3. Use a valve spring compressor tool, and relieve spring tension from the retainer.

■ **Due to engine varnish, the retainer may stick to the valve locks. A gentle tap with a hammer may help to break it loose.**

4. Remove the valve locks from the valve tip and/or retainer. A small magnet may help in removing the locks.
5. Lift the valve spring(s), tool and all, off of the valve stem.
6. Remove the valve seal from the stem and guide. If the seal is difficult to remove with the valve in place, try removing the valve first, then the seal. Follow the steps below for valve removal.
7. Position the head to allow access for withdrawing the valve.

■ **Cylinder heads that have seen a lot of hours and/or abuse may have mushroomed the valve lock grove and/or tip, causing difficulty in removal of the valve. If this has happened, use a metal file to carefully remove the high spots around the lock grooves and/or tip. Only file it enough to allow removal.**

8. Remove the valve from the cylinder head.
9. If equipped, remove the valve spring shim. A small magnetic tool or screwdriver will aid in removal.
10. Repeat Steps 3 though 9 until all of the valves have been removed

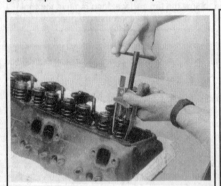

Fig. 11 When removing a valve spring, use a compressor tool to relieve the tension from the retainer...

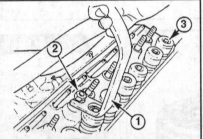

1- Valve spring compressor
2- Rocker arm nut
3- Valve locks

Fig. 12 ...you may also find a compressor that looks like this

Fig. 13 A small magnet will help in removal of the valve locks

Fig. 14 Be careful not to lose the small valve locks (keepers)

Fig. 15 Remove the valve seal from the valve stem—O-ring type seal shown

Fig. 16 Removing an umbrella/positive type seal

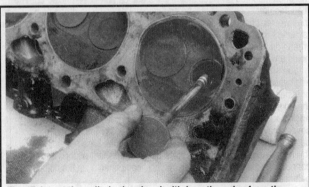

Fig. 17 Invert the cylinder head and withdraw the valve from the valve guide bore

INSPECTION

Now that all of the cylinder head components are clean, it's time to inspect them for wear and/or damage. To accurately inspect them, you will need some specialized tools:

- A 0–1 in. micrometer for the valves
- A dial indicator or inside diameter gauge for the valve guides
- A spring pressure test gauge

If you do not have access to the proper tools, you may want to bring the components to a shop that does.

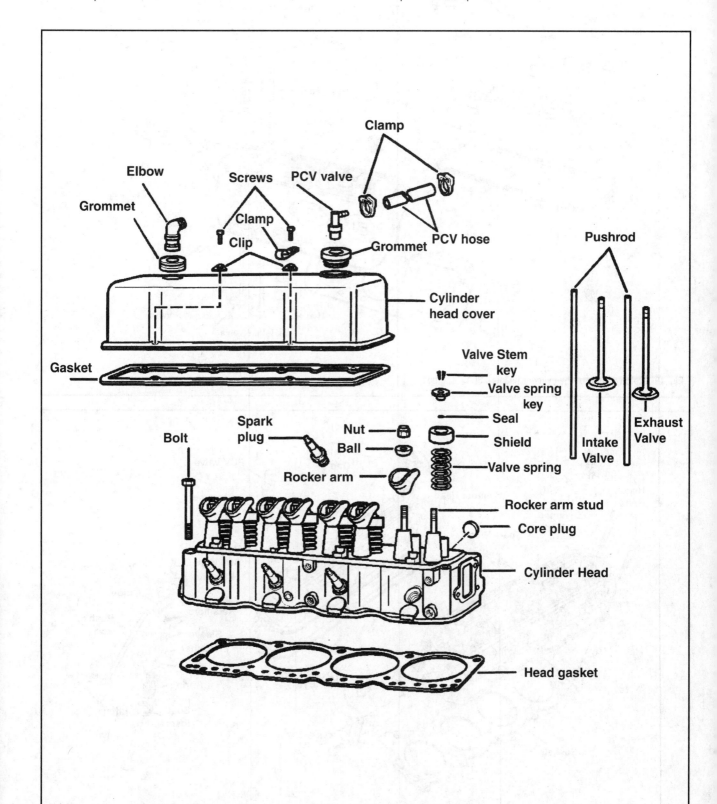

Fig. 18 Exploded view of the cylinder head—3.0L 4 cylinder engines

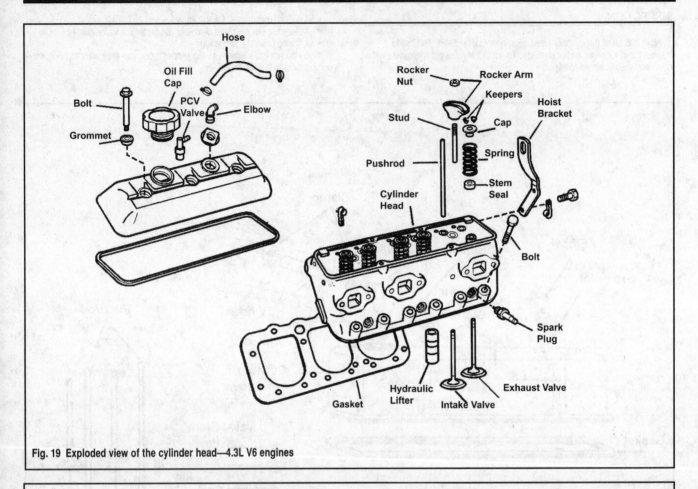

Fig. 19 Exploded view of the cylinder head—4.3L V6 engines

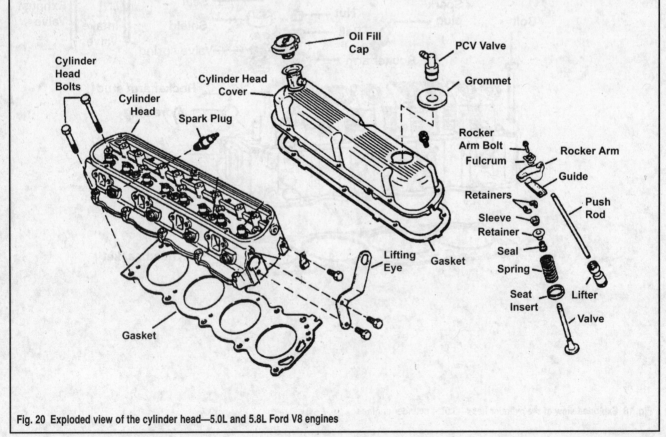

Fig. 20 Exploded view of the cylinder head—5.0L and 5.8L Ford V8 engines

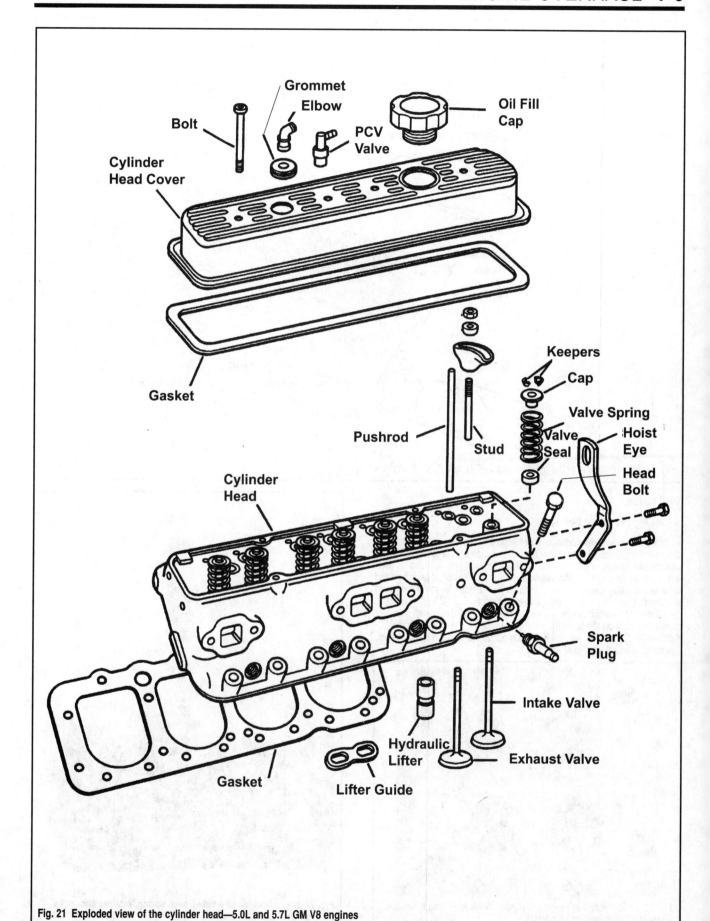

Fig. 21 Exploded view of the cylinder head—5.0L and 5.7L GM V8 engines

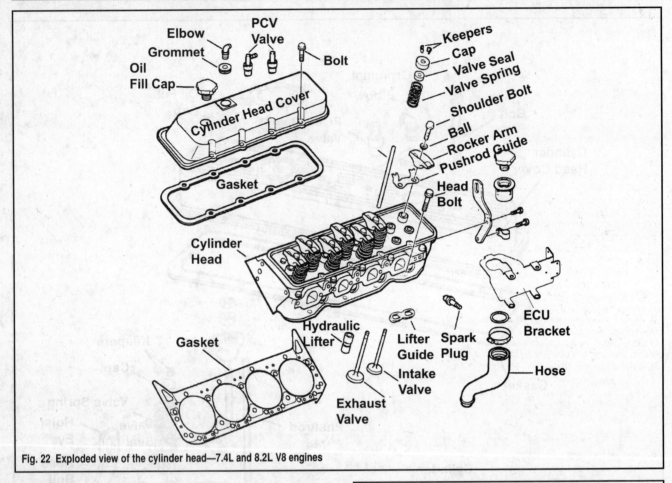

Fig. 22 Exploded view of the cylinder head—7.4L and 8.2L V8 engines

OHC Heads

◆ See Figures 23 and 24

Whether it is a single or dual overhead camshaft cylinder head, the disassembly procedure is relatively unchanged. One aspect to pay attention to is careful labeling of the parts on the dual camshaft cylinder head. There will be an intake camshaft and followers as well as an exhaust camshaft and followers and they must be labeled as such. In some cases, the components are identical and could easily be installed incorrectly. DO NOT MIX THEM UP! Determining which is which is very simple; the intake camshaft and components are on the same side of the head as was the intake manifold. Conversely, the exhaust camshaft and components are on the same side of the head as was the exhaust manifold.

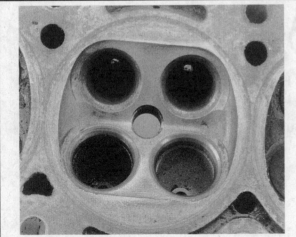

Fig. 23 Example of a multi-valve cylinder head. Note how it has 2 intake and 2 exhaust valve ports

Fig. 24 Exploded view of a valve, seal, spring, retainer and locks from an OHC cylinder head

Cup Type Camshaft Followers

◆ **See Figures 25, 26 and 27**

Most cylinder heads with cup type camshaft followers will have the valve spring, retainer and locks recessed within the follower's bore. You will need a C-clamp style valve spring compressor tool, an OHC spring removal tool (or equivalent) and a small magnet to disassemble the head.

1. If not already removed, remove the camshaft(s) and/or followers. Mark their positions for assembly.
2. Position the cylinder head to allow use of a C-clamp style valve spring compressor tool.

■ **It is preferred to position the cylinder head gasket surface facing you with the valve springs facing the opposite direction and the head laying horizontal.**

3. With the OHC spring removal adapter tool positioned inside of the follower bore, compress the valve spring using the C-clamp style valve spring compressor.
4. Remove the valve locks. A small magnetic tool or screwdriver will aid in removal.
5. Release the compressor tool and remove the spring assembly.
6. Withdraw the valve from the cylinder head.
7. If equipped, remove the valve seal.

■ **Special valve seal removal tools are available. Regular or needlenose type pliers, if used with care, will work just as well. If using ordinary pliers, be sure not to damage the follower bore. The follower and its bore are machined to close tolerances and any damage to the bore will effect this relationship.**

8. If equipped, remove the valve spring shim. A small magnetic tool or screwdriver will aid in removal.
9. Repeat Steps 3 through 8 until all of the valves have been removed.

ROCKER ARM TYPE CAMSHAFT FOLLOWERS

◆ **See Figures 28 through 36**

Most cylinder heads with rocker arm-type camshaft followers are easily disassembled using a standard valve spring compressor. However, certain models may not have enough open space around the spring for the standard tool and may require you to use a C-clamp style compressor tool instead.

1. If not already removed, remove the rocker arms and/or shafts and the camshaft. If applicable, also remove the hydraulic lash adjusters. Mark their positions for assembly.
2. Position the cylinder head to allow access to the valve spring.
3. Use a valve spring compressor tool to relieve the spring tension from the retainer.

■ **Due to engine varnish, the retainer may stick to the valve locks. A gentle tap with a hammer may help to break it loose.**

4. Remove the valve locks from the valve tip and/or retainer. A small magnet may help in removing the small locks.
5. Lift the valve spring, tool and all, off of the valve stem.
6. If equipped, remove the valve seal. If the seal is difficult to remove with the valve in place, try removing the valve first, then the seal. Follow the steps below for valve removal.
7. Position the head to allow access for withdrawing the valve.

■ Cylinder heads that have seen a lot of miles and/or abuse may have mushroomed the valve lock grove and/or tip, causing difficulty in removal of the valve. If this has happened, use a metal file to carefully remove the high spots around the lock grooves and/or tip. Only file it enough to allow removal.

8. Remove the valve from the cylinder head.
9. If equipped, remove the valve spring shim. A small magnetic tool or screwdriver will aid in removal.
10. Repeat Steps 3 though 9 until all of the valves have been removed.

INSPECTION

Now that all of the cylinder head components are clean, it's time to inspect them for wear and/or damage. To accurately inspect them, you will need some specialized tools:

- A 0-1 in. micrometer for the valves
- A dial indicator or inside diameter gauge for the valve guides
- A spring pressure test gauge

If you do not have access to the proper tools, you may want to bring the components to a shop that does.

Valves

◆ **See Figures 37 and 38**

MODERATE

The first thing to inspect are the valve he margin and face for any cracks, excessive wear or burning. The margin is the best place to look for burning. It should have a squared edge with an even width all around the diameter. When a valve burns, the margin will look melted and the edges rounded. Also inspect the valve head for any signs of tulipping. This will show as a lifting of the edges or dishing in the center of the head and will usually not occur to all of the valves. All of the heads should look the same, any that seem dished more than others are probably bad. Next, inspect the valve lock grooves and valve tips. Check for any burrs around the lock grooves, especially if you had to file them to remove the valve. Valve tips should appear flat, although slight rounding with high mileage engines is normal. Slightly worn valve tips will need to be machined flat. Last, measure the valve stem diameter with the micrometer. Measure the area that rides within the guide, especially towards the tip where most of the wear occurs. Take several measurements along its length and compare them to each other. Wear should be even along the length with little to no taper. If no minimum diameter is given in the specifications, then the stem should not read more than 0.001 in. (0.025mm) below the unworn area of the valve stem. Any valves that fail these inspections should be replaced.

Fig. 25 C-clamp type spring compressor and an OHC spring removal tool (center) for cup type followers

Fig. 26 Most cup type follower cylinder heads retain the camshaft using bolt-on bearing caps

Fig. 27 Position the OHC spring tool in the follower bore, then compress the spring with a C-clamp type tool

Fig. 28 Example of the shaft mounted rocker arms on some OHC heads

Fig. 29 Another example of the rocker arm type OHC head. This model uses a follower under the camshaft

Fig. 30 Before the camshaft can be removed, all of the followers must first be removed . . .

Fig. 31 . . . then the camshaft can be removed by sliding it out (shown), or unbolting a bearing cap (not shown)

Fig. 32 Compress the valve spring . . .

Fig. 33 . . . then remove the valve locks from the valve stem and spring retainer

Fig. 34 Remove the valve spring and retainer from the cylinder head

Fig. 35 Remove the valve seal from the guide. Some gentle prying or pliers may help to remove stubborn ones

Fig. 36 All aluminum and some cast iron heads will have these valve spring shims. Remove all of them as well

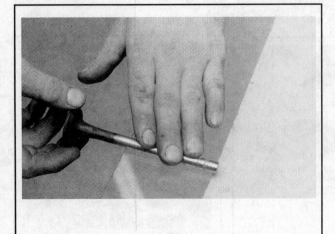

Fig. 37 Valve stems may be rolled on a flat surface to check for bends

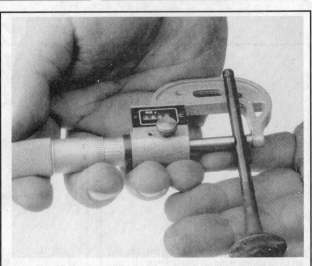

Fig. 38 Use a micrometer to check the valve stem diameter

Springs, Retainers and Valve Locks

◆ See Figures 39 and 40

The first thing to check is the most obvious, broken springs. Next check the free length and squareness of each spring. If applicable, insure to distinguish between intake and exhaust springs. Use a ruler and/or carpenter's square to measure the length. A carpenter's square should be used to check the springs for squareness. If a spring pressure test gauge is available, check each springs rating and compare to the specifications chart. Check the readings against the specifications given. Any springs that fail these inspections should be replaced.

The spring retainers rarely need replacing, however they should still be checked as a precaution. Inspect the spring mating surface and the valve lock retention area for any signs of excessive wear. Also check for any signs of cracking. Replace any retainers that are questionable.

Valve locks should be inspected for excessive wear on the outside contact area as well as on the inner notched surface. Any locks which appear worn or broken and its respective valve should be replaced.

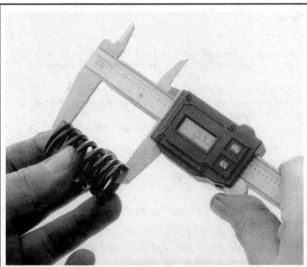

Fig. 39 Use a caliper to check the valve spring free-length

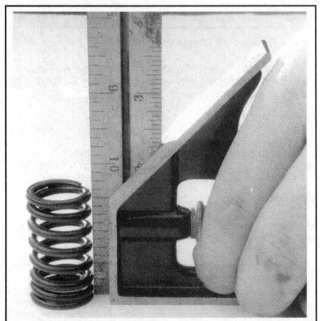

Fig. 40 Check the valve spring for squareness on a flat surface; a carpenter's square can be used

Cylinder Head

There are several things to check on the cylinder head: valve guides, seats, cylinder head surface flatness, cracks and physical damage.

Valve Guides

◆ See Figure 41

Now that you know the valves are good, you can use them to check the guides, although a new valve, if available, is preferred. Before you measure anything, look at the guides carefully and inspect them for any cracks, chips or breakage. Also if the guide is a removable style (as in most aluminum heads), check them for any looseness or evidence of movement. All of the guides should appear to be at the same height from the spring seat. If any seem lower (or higher) from another, the guide has moved. Mount a dial indicator onto the spring side of the cylinder head. Lightly oil the valve stem and insert it into the cylinder head. Position the dial indicator against the valve stem near the tip and zero the gauge. Grasp the valve stem and wiggle towards and away from the dial indicator and observe the readings. Mount the dial indicator 90 degrees from the initial point and zero the gauge and again take a reading. Compare the two readings for a out of round condition. Check the readings against the specifications given. An Inside Diameter (I.D.) gauge designed for valve guides will give you an accurate valve guide bore measurement. If the I.D. gauge is used, compare the readings with the specifications given. Any guides that fail these inspections should be replaced or machined.

Fig. 41 A dial gauge may be used to check valve stem-to-guide clearance; read the gauge while moving the valve stem

Valve Seats

A visual inspection of the valve seats should show a slightly worn and pitted surface where the valve face contacts the seat. Inspect the seat carefully for severe pitting or cracks. Also, a seat that is badly worn will be recessed into the cylinder head. A severely worn or recessed seat may need to be replaced. All cracked seats must be replaced. A seat concentricity gauge, if available, should be used to check the seat run-out. If run-out exceeds specifications the seat must be machined (if no specification is given use 0.002 in. or 0.051mm).

Cylinder Head Surface Flatness

◆ See Figures 42 and 43

After you have cleaned the gasket surface of the cylinder head of any old gasket material, check the head for flatness.

Place a straightedge across the gasket surface. Using feeler gauges, determine the clearance at the center of the straightedge and across the cylinder head at several points. Check along the centerline and diagonally on the head surface. If the warpage exceeds 0.003 in. (0.076mm) within a 6.0 in. (15.2cm) span, or 0.006 in. (0.152mm) over the total length of the head, the cylinder head must be resurfaced. After resurfacing the heads of a V-type engine, the intake manifold flange surface should be checked, and if necessary, milled proportionally to allow for the change in its mounting position.

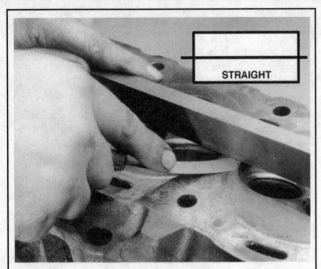

Fig 42 Check the head for flatness across the center of the head surface using a straightedge and feeler gauge

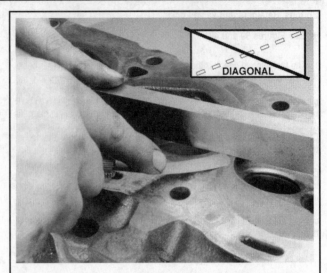

Fig. 43 Checks should also be made along both diagonals of the head surface

Cracks And Physical Damage

Generally, cracks are limited to the combustion chamber, however, it is not uncommon for the head to crack in a spark plug hole, port, outside of the head or in the valve spring/rocker arm area. The first area to inspect is always the hottest: the exhaust seat/port area.

A visual inspection should be performed, but just because you don't see a crack does not mean it is not there. Some more reliable methods for inspecting for cracks include Magnaflux®, a magnetic process or Zyglo®, a dye penetrant. Magnaflux® is used only on ferrous metal (cast iron) heads. Zyglo® uses a spray on fluorescent mixture along with a black light to reveal the cracks. It is strongly recommended to have your cylinder head checked professionally for cracks, especially if the engine was known to have overheated and/or leaked or consumed coolant. Contact a local shop for availability and pricing of these services.

Physical damage is usually very evident. For example, a broken mounting ear from dropping the head or a bent or broken stud and/or bolt. All of these defects should be fixed or, if not repairable, the head should be replaced.

Camshaft and Followers

Inspect the camshaft(s) and followers as described earlier.

REFINISHING & REPAIRING

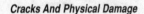

Many of the procedures given for refinishing and repairing the cylinder head components must be performed by a machine shop. Certain steps, if the inspected part is not worn, can be performed yourself inexpensively. However, you spent a lot of time and effort so far; why risk trying to save a couple bucks if you might have to do it all over again?

Valves

Any valves that were not replaced should be refaced and the tips ground flat. Unless you have access to a valve grinding machine, this should be done by a machine shop. If the valves are in extremely good condition, as well as the valve seats and guides, they may be lapped in without performing machine work.

It is a recommended practice to lap the valves even after machine work has been performed and/or new valves have been purchased. This insures a positive seal between the valve and seat.

Lapping The Valves

■ Before lapping the valves to the seats, read the rest of the cylinder head procedure to insure that any related parts are in acceptable enough condition to continue.

■ Before any valve seat machining and/or lapping can be performed, the guides must be within factory recommended specifications.

1. Invert the cylinder head.
2. Lightly lubricate the valve stems and insert them into the cylinder head in their numbered order.
3. Raise the valve from the seat and apply a small amount of fine lapping compound to the seat.
4. Moisten the suction head of a hand-lapping tool and attach it to the head of the valve.
5. Rotate the tool between the palms of both hands, changing the position of the valve on the valve seat and lifting the tool often to prevent grooving.
6. Lap the valve until a smooth, polished circle is evident on the valve and seat.
7. Remove the tool and the valve. Wipe away all traces of the grinding compound and store the valve to maintain its lapped location.

✳✳ WARNING

Do not get the valves out of order after they have been lapped. They must be put back with the same valve seat with which they were lapped.

Springs, Retainers and Valve Locks

There is no repair or refinishing possible with the springs, retainers and valve locks. If they are found to be worn or defective, they must be replaced with new (or known good) parts.

Cylinder Head

Most refinishing procedures dealing with the cylinder head must be performed by a machine shop. Read the procedures below and review your inspection data to determine whether or not machining is necessary.

Valve Guide

■ If any machining or replacements are made to the valve guides, the seats must be machined.

Unless the valve guides need machining or replacing, the only service to perform is to thoroughly clean them of any dirt or oil residue.

There are only two types of valve guides used on automobile engines: the replaceable-type (all aluminum heads) and the cast-in integral-type (most cast iron heads). There are four recommended methods for repairing worn guides.

- Knurling
- Inserts
- Reaming oversize
- Replacing

Knurling is a process in which metal is displaced and raised, thereby reducing clearance, giving a true center, and providing oil control. It is the least expensive way of repairing the valve guides. However, it is not necessarily the best, and in some cases, a knurled valve guide will not stand up for more than a short time. It requires a special knurlizer and precision reaming tools to obtain proper clearances. It would not be cost effective to purchase these tools, unless you plan on rebuilding several of the same cylinder head.

Installing a guide insert involves machining the guide to accept a bronze insert. One style is the coil-type which is installed into a threaded guide. Another is the thin-walled insert where the guide is reamed oversize to accept a split-sleeve insert. After the insert is installed, a special tool is then run through the guide to expand the insert, locking it to the guide. The insert is then reamed to the standard size for proper valve clearance.

Reaming for oversize valves restores normal clearances and provides a true valve seat. Most cast-in type guides can be reamed to accept an valve with an oversize stem. The cost factor for this can become quite high as you will need to purchase the reamer and new, oversize stem valves for all guides which were reamed. Oversizes are generally 0.003 to 0.030 in. (0.076 to 0.762mm), with 0.015 in. (0.381mm) being the most common.

To replace cast-in type valve guides, they must be drilled out, then reamed to accept replacement guides. This must be done on a fixture which will allow centering and leveling off of the original valve seat or guide, otherwise a serious guide-to-seat misalignment may occur making it impossible to properly machine the seat.

Replaceable-type guides are pressed into the cylinder head. A hammer and a stepped drift or punch may be used to install and remove the guides. Before removing the guides, measure the protrusion on the spring side of the head and record it for installation. Use the stepped drift to hammer out the old guide from the combustion chamber side of the head. When installing, determine whether or not the guide also seals a water jacket in the head, and if it does, use the recommended sealing agent. If there is no water jacket, grease the valve guide and its bore. Use the stepped drift, and hammer the new guide into the cylinder head from the spring side of the cylinder head. A stack of washers the same thickness as the measured protrusion may help the installation process.

Valve Seats

■ Before any valve seat machining can be performed, the guides must be within factory recommended specifications.

■ If any machining or replacements were made to the valve guides, the seats must be machined.

If the seats are in good condition, the valves can be lapped to the seats, and the cylinder head assembled. See the valves procedures for instructions on lapping.

If the valve seats are worn, cracked or damaged, they must be serviced by a machine shop. The valve seat must be perfectly centered to the valve guide, which requires very accurate machining.

Cylinder Head Surface

If the cylinder head is warped, it must be machined flat. If the warpage is extremely severe, the head may need to be replaced. In some instances, it may be possible to straighten a warped head enough to allow machining. In either case, contact a professional machine shop for service.

Cracks And Physical Damage

Certain cracks can be repaired in both cast iron and aluminum heads. For cast iron, a tapered threaded insert is installed along the length of the crack. Aluminum can also use the tapered inserts; however welding is the preferred method. Some physical damage can be repaired through brazing or welding. Contact a machine shop to get expert advice for your particular dilemma.

■ Any OHC cylinder head that shows excessive warpage should have the camshaft bearing journals align bored after the cylinder head has been resurfaced.

✷✷ WARNING

Failure to align bore the camshaft bearing journals could result in severe engine damage including but not limited to: valve and piston damage, connecting rod damage, camshaft and/or crankshaft breakage.

ASSEMBLY

The first step for any assembly job is to have a clean area in which to work. Next, thoroughly clean all of the parts and components that are to be assembled. Finally, place all of the components onto a suitable work space and, if necessary, arrange the parts to their respective positions.

OHV Engines

1. Lightly lubricate the valve stems and insert all of the valves into the cylinder head. If possible, maintain their original locations.
2. If equipped, install any valve spring shims which were removed.
3. If equipped, install the new valve seals, keeping the following in mind:
- If the valve seal presses over the guide, lightly lubricate the outer guide surfaces.
- If the seal is an O-ring type, it is installed just after compressing the spring but before the valve locks.
4. Place the valve spring and retainer over the stem.
5. Position the spring compressor tool and compress the spring.
6. Assemble the valve locks to the stem.
7. Relieve the spring pressure slowly and insure that neither valve lock becomes dislodged by the retainer.
8. Remove the spring compressor tool.
9. Repeat Steps 2 through 8 until all of the springs have been installed.

OHC Engines

◆ See Figure 44

CUP TYPE CAMSHAFT FOLLOWERS

To install the springs, retainers and valve locks on heads which have these components recessed into the camshaft follower's bore, you will need a small screwdriver-type tool, some clean white grease and a lot of patience. You will also need the C-clamp style spring compressor and the OHC tool used to disassemble the head.

1. Lightly lubricate the valve stems and insert all of the valves into the cylinder head. If possible, maintain their original locations.
2. If equipped, install any valve spring shims which were removed.
3. If equipped, install the new valve seals, keeping the following in mind:
- If the valve seal presses over the guide, lightly lubricate the outer guide surfaces.
- If the seal is an O-ring type, it is installed just after compressing the spring but before the valve locks.
4. Place the valve spring and retainer over the stem.
5. Position the spring compressor and the OHC tool, then compress the spring.
6. Using a small screwdriver as a spatula, fill the valve stem side of the lock with white grease. Use the excess grease on the screwdriver to fasten the lock to the driver.
7. Carefully install the valve lock, which is stuck to the end of the screwdriver, to the valve stem then press on it with the screwdriver until the grease squeezes out. The valve lock should now be stuck to the stem.
8. Repeat Steps 6 and 7 for the remaining valve lock.
9. Relieve the spring pressure slowly and insure that neither valve lock becomes dislodged by the retainer.
10. Remove the spring compressor tool.
11. Repeat Steps 2 through 10 until all of the springs have been installed.
12. Install the followers, camshaft(s) and any other components that were removed for disassembly.

ROCKER ARM TYPE CAMSHAFT FOLLOWERS

1. Lightly lubricate the valve stems and insert all of the valves into the cylinder head. If possible, maintain their original locations.

2. If equipped, install any valve spring shims which were removed.

3. If equipped, install the new valve seals, keeping the following in mind:

• If the valve seal presses over the guide, lightly lubricate the outer guide surfaces.

• If the seal is an O-ring type, it is installed just after compressing the spring but before the valve locks.

4. Place the valve spring and retainer over the stem.

5. Position the spring compressor tool and compress the spring.

6. Assemble the valve locks to the stem.

7. Relieve the spring pressure slowly and insure that neither valve lock becomes dislodged by the retainer.

8. Remove the spring compressor tool.

9. Repeat Steps 2 through 8 until all of the springs have been installed.

10. Install the camshaft(s), rockers, shafts and any other components that were removed for disassembly.

Engine Block

GENERAL INFORMATION

A thorough overhaul or rebuild of an engine block would include replacing the pistons, rings, bearings, timing belt/chain assembly and oil pump. For OHV engines also include a new camshaft and lifters. The block would then have the cylinders bored and honed oversize (or if using removable cylinder sleeves, new sleeves installed) and the crankshaft would be cut undersize to provide new wearing surfaces and perfect clearances. However, your particular engine may not have everything worn out. What if only the piston rings have worn out and the clearances on everything else are still within factory specifications? Well, you could just replace the rings and put it back together, but this would be a very rare example. Chances are, if one component in your engine is worn, other components are sure to follow, and soon. At the very least, you should always replace the rings, bearings and oil pump. This is what is commonly called a "freshen up".

Cylinder Ridge Removal

Because the top piston ring does not travel to the very top of the cylinder, a ridge is built up between the end of the travel and the top of the cylinder bore.

Pushing the piston and connecting rod assembly past the ridge can be difficult, and damage to the piston ring lands could occur. If the ridge is not removed before installing a new piston or not removed at all, piston ring breakage and piston damage may occur.

■ It is always recommended that you remove any cylinder ridges before removing the piston and connecting rod assemblies. If you know that new pistons are going to be installed and the engine block

will be bored oversize, you may be able to forego this step. However, some ridges may actually prevent the assemblies from being removed, necessitating its removal.

There are several different types of ridge reamers on the market, none of which are inexpensive. Unless a great deal of engine rebuilding is anticipated, borrow or rent a reamer.

1. Turn the crankshaft until the piston is at the bottom of its travel.

2. Cover the head of the piston with a rag.

3. Follow the tool manufacturers instructions and cut away the ridge, exercising extreme care to avoid cutting too deeply.

4. Remove the ridge reamer, the rag and as many of the cuttings as possible. Continue until all of the cylinder ridges have been removed.

DISASSEMBLY

◆ See Figures 45, 46 47 48 49 and 50

The engine disassembly instructions following assume that you have the engine mounted on an engine stand. If not, it is easiest to disassemble the engine on a bench or the floor with it resting on the bell housing or transmission mounting surface. You must be able to access the connecting rod fasteners and turn the crankshaft during disassembly. Also, all engine covers (timing, front, side, oil pan, whatever) should have already been removed. Engines which are seized or locked up may not be able to be completely disassembled, and a core (salvage yard) engine should be purchased.

On OHV engines, if not done during the cylinder head removal, remove the pushrods and lifters, keeping them in order for assembly. Remove the timing gears and/or timing chain assembly, then remove the oil pump drive assembly and withdraw the camshaft from the engine block. Remove the oil pick-up and pump assembly. If equipped, remove any balance or auxiliary shafts. If necessary, remove the cylinder ridge from the top of the bore. See the cylinder ridge removal procedure.

On OHC engines, if not done during the cylinder head removal, remove the timing chain/belt and/or gear/sprocket assembly. Remove the oil pick-up and pump assembly and, if necessary, the pump drive. If equipped, remove any balance or auxiliary shafts. If necessary, remove the cylinder ridge from the top of the bore. See the cylinder ridge removal procedure.

Rotate the engine over so that the crankshaft is exposed. Use a number punch or scribe and mark each connecting rod with its respective cylinder number. The cylinder closest to the front of the engine is always number 1. However, depending on the engine placement, the front of the engine could either be the flywheel or damper/pulley end. Generally the front of the engine faces the bow. Use a number punch or scribe and also mark the main bearing caps from front to rear with the front most cap being number 1 (if there are five caps, mark them 1 through 5, front to rear).

✳✳ WARNING

Take special care when pushing the connecting rod up from the crankshaft because the sharp threads of the rod bolts/studs will score the crankshaft journal. Insure that special plastic caps are installed over them, or cut two pieces of rubber hose to do the same.

Fig. 44 Once assembled, check the valve clearance and correct as needed

Fig. 45 Place rubber hose over the connecting rod studs to protect the crankshaft and cylinder bores from damage

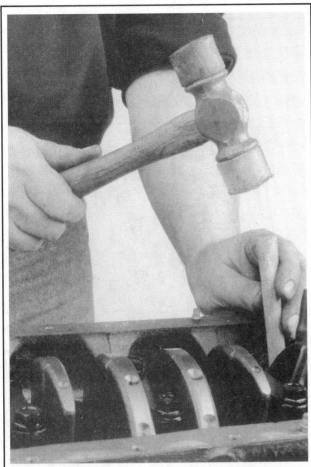

Fig. 46 Carefully tap the piston out of the bore using a wooden dowel

Again, rotate the engine, this time to position the number one cylinder bore (head surface) up. Turn the crankshaft until the number one piston is at the bottom of its travel, this should allow the maximum access to its connecting rod. Remove the number one connecting rods fasteners and cap and place two lengths of rubber hose over the rod bolts/studs to protect the crankshaft from damage. Using a sturdy wooden dowel and a hammer, push the connecting rod up about 1 in. (25mm) from the crankshaft and remove the upper bearing insert. Continue pushing or tapping the connecting rod up until the piston rings are out of the cylinder bore. Remove the piston and rod by hand, put the upper half of the bearing insert back into the rod, install the cap with its bearing insert installed, and hand-tighten the cap fasteners. If the parts are kept in order in this manner, they will not get lost and you will be able to tell which bearings came form what cylinder if any problems are discovered and diagnosis is necessary. Remove all the other piston assemblies in the same manner. On V-style engines, remove all of the pistons from one bank, then reposition the engine with the other cylinder bank head surface up, and remove that banks piston assemblies.

The only remaining component in the engine block should now be the crankshaft. Loosen the main bearing caps evenly until the fasteners can be turned by hand, then remove them and the caps. Remove the crankshaft from the engine block. Thoroughly clean all of the components.

INSPECTION

Now that the engine block and all of its components are clean; it's time to inspect them for wear and/or damage. To accurately inspect them, you will need some specialized tools:

- Two or three separate micrometers to measure the pistons and crankshaft journals
- A dial indicator

- Telescoping gauges for the cylinder bores
- A rod alignment fixture to check for bent connecting rods

If you do not have access to the proper tools, you may want to bring the components to a shop that does.

Generally, you shouldn't expect cracks in the engine block or its components unless it was known to leak, consume or mix engine fluids, it was severely overheated, or there was evidence of bad bearings and/or crankshaft damage. A visual inspection should be performed on all of the components, but just because you don't see a crack does not mean it is not there. Some more reliable methods for inspecting for cracks include Magnaflux®, a magnetic process or Zyglo®, a dye penetrant. Magnaflux® is used only on ferrous metal (cast iron). Zyglo® uses a spray on fluorescent mixture along with a black light to reveal the cracks. It is strongly recommended to have your engine block checked professionally for cracks, especially if the engine was known to have overheated and/or leaked or consumed coolant. Contact a local shop for availability and pricing of these services.

Engine Block

Engine Block Bearing Alignment

Remove the main bearing caps and, if still installed, the main bearing inserts. Inspect all of the main bearing saddles and caps for damage, burrs or high spots. If damage is found, and it is caused from a spun main bearing, the block will need to be align-bored or, if severe enough, replacement. Any burrs or high spots should be carefully removed with a metal file.

Place a straightedge on the bearing saddles, in the engine block, along the centerline of the crankshaft. If any clearance exists between the straightedge and the saddles, the block must be align-bored.

Align-boring consists of machining the main bearing saddles and caps by means of a flycutter that runs through the bearing saddles.

Deck Flatness

The top of the engine block where the cylinder head mounts is called the deck. Insure that the deck surface is clean of dirt, carbon deposits and old gasket material. Place a straightedge across the surface of the deck along its centerline and, using feeler gauges, check the clearance along several points. Repeat the checking procedure with the straightedge placed along both diagonals of the deck surface. If the reading exceeds 0.003 in. (0.076mm) within a 6.0 in. (15.2cm) span, or 0.006 in. (0.152mm) over the total length of the deck, it must be machined. Always check the Specification chart for your specific engine.

Cylinder Bores

◆ See Figure 51

The cylinder bores house the pistons and are slightly larger than the pistons themselves. A common piston-to-bore clearance is 0.0015–0.0025 in. (0.0381mm–0.0635mm). Refer to the Specification chart for your specific engine. Inspect and measure the cylinder bores. The bore should be checked for out-of-roundness, taper and size. The results of this inspection will determine whether the cylinder can be used in its existing size and condition, or a rebore to the next oversize is required (or in the case of removable sleeves, have replacements installed).

The amount of cylinder wall wear is always greater at the top of the cylinder than at the bottom. This wear is known as taper. Any cylinder that has a taper of 0.0012 in. (0.305mm) or more, must be re-bored. Measurements are taken at a number of positions in each cylinder: at the top, middle and bottom and at two points at each position; that is, at a point 90 degrees from the crankshaft centerline, as well as a point parallel to the crankshaft centerline. The measurements are made with either a special dial indicator or a telescopic gauge and micrometer. If the necessary precision tools to check the bore are not available, take the block to a machine shop and have them mike it. Also if you don't have the tools to check the cylinder bores, chances are you will not have the necessary devices to check the pistons, connecting rods and crankshaft. Take these components with you and save yourself an extra trip.

For our procedures, we will use a telescopic gauge and a micrometer. You will need one of each, with a measuring range which covers your cylinder bore size.

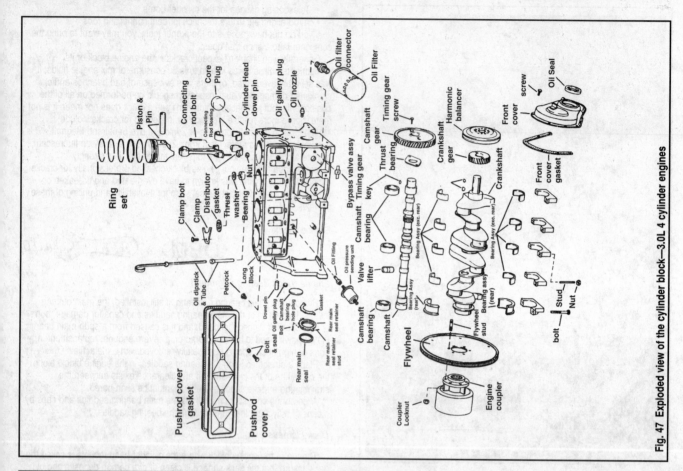

Fig. 47 Exploded view of the cylinder block—3.0L 4 cylinder engines

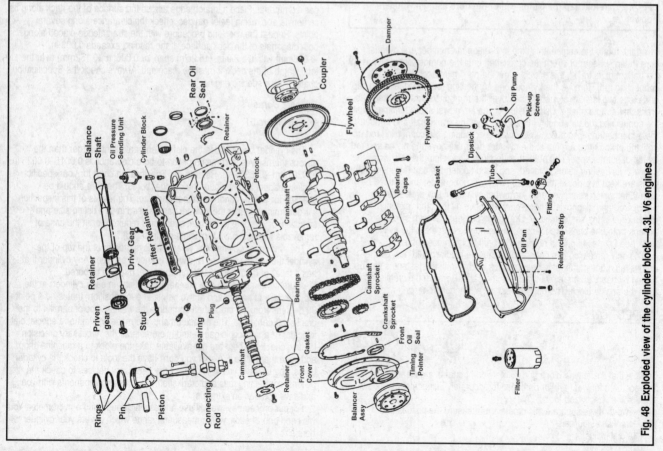

Fig. 48 Exploded view of the cylinder block—4.3L V6 engines

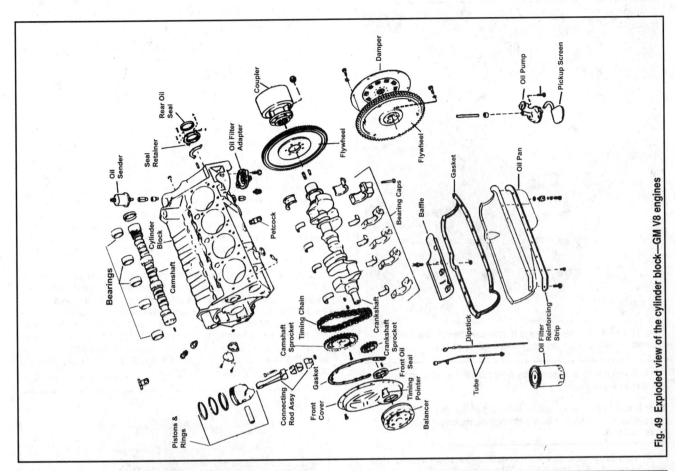

Fig. 49 Exploded view of the cylinder block—GM V8 engines

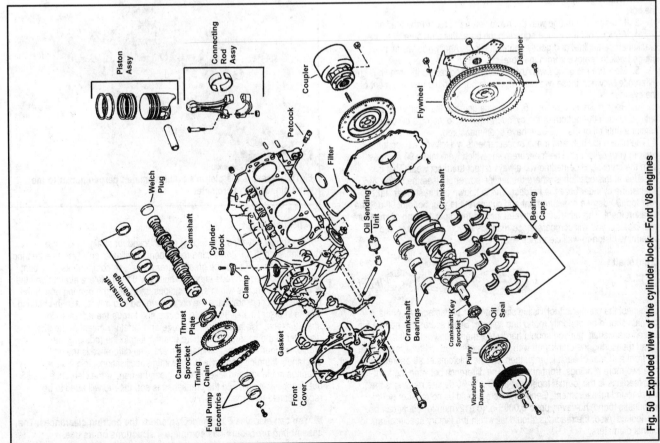

Fig. 50 Exploded view of the cylinder block—Ford V8 engines

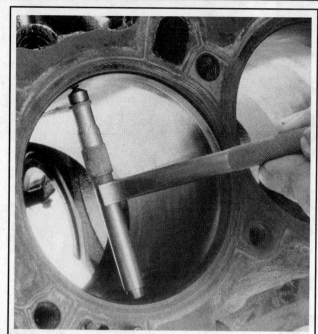

Fig. 51 Use a telescoping gauge to measure the cylinder bore diameter—take several readings within the same bore

1. Position the telescopic gauge in the cylinder bore, loosen the gauges lock and allow it to expand.

■ **Your first two readings will be at the top of the cylinder bore, then proceed to the middle and finally the bottom, making a total of six measurements.**

2. Hold the gauge square in the bore, 90 degrees from the crankshaft centerline, and gently tighten the lock. Tilt the gauge back to remove it from the bore.
3. Measure the gauge with the micrometer and record the reading.
4. Again, hold the gauge square in the bore, this time parallel to the crankshaft centerline, and gently tighten the lock. Again, you will tilt the gauge back to remove it from the bore.
5. Measure the gauge with the micrometer and record this reading. The difference between these two readings is the out-of-round measurement of the cylinder.
6. Repeat Steps 1 through 5, each time going to the next lower position, until you reach the bottom of the cylinder. Then go to the next cylinder, and continue until all of the cylinders have been measured.

The difference between these measurements will tell you all about the wear in your cylinders. The measurements which were taken 90 degrees from the crankshaft centerline will always reflect the most wear. That is because at this position is where the engine power presses the piston against the cylinder bore the hardest. This is known as thrust wear. Take your top, 90 degree measurement and compare it to your bottom, 90 degree measurement. The difference between them is the taper. When you measure your pistons, you will compare these readings to your piston sizes and determine piston-to-wall clearance.

Crankshaft

Inspect the crankshaft for visible signs of wear or damage. All of the journals should be perfectly round and smooth. Slight scores are normal for a used crankshaft, but you should hardly feel them with your fingernail. When measuring the crankshaft with a micrometer, you will take readings at the front and rear of each journal, then turn the micrometer 90 degrees and take two more readings, front and rear. The difference between the front-to-rear readings is the journal taper and the first-to-90 degree reading is the out-of-round measurement. Generally, there should be no taper or out-of-roundness found, however, up to 0.0005 in. (0.0127mm) for either can be overlooked. Also, the readings should fall within the factory specifications for journal diameters.

If the crankshaft journals fall within specifications, it is recommended that it be polished before being returned to service. Polishing the crankshaft insures that any minor burrs or high spots are smoothed, thereby reducing the chance of scoring the new bearings.

Pistons and Connecting Rods

Pistons

◆ See Figure 52

The piston should be visually inspected for any signs of cracking or burning (caused by hot spots or detonation), and scuffing or excessive wear on the skirts. The wrist pin attaches the piston to the connecting rod. The piston should move freely on the wrist pin, both sliding and pivoting. Grasp the connecting rod securely, or mount it in a vise, and try to rock the piston back and forth along the centerline of the wrist pin. There should not be any excessive play evident between the piston and the pin. If there are C-clips retaining the pin in the piston then you have wrist pin bushings in the rods. There should not be any excessive play between the wrist pin and the rod bushing. Normal clearance for the wrist pin is approx. 0.001–0.002 in. (0.025mm–0.051mm). Please refer to the Specification chart for your specific engine.

Use a micrometer and measure the diameter of the piston, perpendicular to the wrist pin, on the skirt. Compare the reading to its original cylinder measurement obtained earlier. The difference between the two readings is the piston-to-wall clearance. If the clearance is within specifications, the piston may be used as is. If the piston is out of specification, but the bore is not, you will need a new piston. If both are out of specification, you will need the cylinder rebored and oversize pistons installed. Generally if two or more pistons/bores are out of specification, it is best to rebore the entire block and purchase a complete set of oversize pistons.

Fig. 52 Measure the piston's outer diameter, perpendicular to the wrist pin, with a micrometer

Connecting Rod

You should have the connecting rod checked for straightness at a machine shop. If the connecting rod is bent, it will unevenly wear the bearing and piston, as well as place greater stress on these components. Any bent or twisted connecting rods must be replaced. If the rods are straight and the wrist pin clearance is within specifications, then only the bearing end of the rod need be checked. Place the connecting rod into a vice, with the bearing inserts in place, install the cap to the rod and torque the fasteners to specifications. Use a telescoping gauge and carefully measure the inside diameter of the bearings. Compare this reading to the rods original crankshaft journal diameter measurement. The difference is the oil clearance. If the oil clearance is not within specifications, install new bearings in the rod and take another measurement. If the clearance is still out of specifications, and the crankshaft is not, the rod will need to be reconditioned by a machine shop.

■ **You can also use Plastigage® to check the bearing clearances. The assembling procedure has complete instructions on its use.**

Camshaft

Inspect the camshaft and lifters/followers as described earlier.

Bearings

All of the engine bearings should be visually inspected for wear and/or damage. The bearing should look evenly worn all around with no deep scores or pits. If the bearing is severely worn, scored, pitted or heat blued, then the bearing, and the components that use it, should be brought to a machine shop for inspection. Full-circle bearings (used on most camshafts, auxiliary shafts, balance shafts, etc.) require specialized tools for removal and installation, and should be brought to a machine shop for service.

Oil Pump

■ **The oil pump is responsible for providing constant lubrication to the whole engine and so it is recommended that a new oil pump be installed when rebuilding the engine.**

Completely disassemble the oil pump and thoroughly clean all of the components. Inspect the oil pump gears and housing for wear and/or damage. Insure that the pressure relief valve operates properly and there is no binding or sticking due to varnish or debris. If all of the parts are in proper working condition, lubricate the gears and relief valve, and assemble the pump.

REFINISHING

◆ **See Figure 53**

Almost all engine block refinishing must be performed by a machine shop. If the cylinders are not to be re-bored, then the cylinder glaze can be removed with a ball hone. When removing cylinder glaze with a ball hone, use a light or penetrating type oil to lubricate the hone. Do not allow the hone to run dry as this may cause excessive scoring of the cylinder bores and wear on the hone. If new pistons are required, they will need to be installed to the connecting rods. This should be performed by a machine shop as the pistons must be installed in the correct relationship to the rod or engine damage can occur.

Pistons and Connecting Rods

◆ **See Figure 54**

Because of the tools necessary, only pistons with the wrist pin retained by C-clips are serviceable by the DIYer or technician. Press fit pistons require special presses and/or heaters to remove/install the connecting rod and should only be performed by a machine shop.

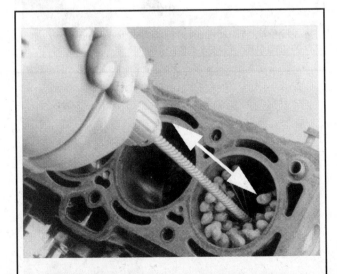

Fig. 53 Use a ball type cylinder hone to remove any glaze and provide a new surface for seating the piston rings

All pistons will have a mark indicating the direction to the front of the engine and the must be installed into the engine in that manner. Usually it is a notch or arrow on the top of the piston, or it may be the letter **F** cast or stamped into the piston.

C-Clip Type Pistons

1. Note the location of the forward mark on the piston and mark the connecting rod in relation.
2. Remove the C-clips from the piston and withdraw the wrist pin.

■ **Varnish build-up or C-clip groove burrs may increase the difficulty of removing the wrist pin. If necessary, use a punch or drift to carefully tap the wrist pin out.**

3. Insure that the wrist pin bushing in the connecting rod is usable and lubricate it with assembly lube.
4. Remove the wrist pin from the new piston and lubricate the pin bores on the piston.
5. Align the forward marks on the piston and the connecting rod and install the wrist pin.
6. The new C-clips will have a flat and a rounded side to them. Install both C-clips with the flat side facing out.
7. Repeat all of the steps for each piston being replaced.

ASSEMBLY

Before you begin assembling the engine, first give yourself a clean, dirt free work area. Next, clean every engine component again. The key to a good assembly is cleanliness.

Mount the engine block into the engine stand and wash it one last time using water and detergent (dishwashing detergent works well). While washing it, scrub the cylinder bores with a soft bristle brush and thoroughly clean all of the oil passages. Completely dry the engine and spray the entire assembly down with an anti-rust solution such as WD-40® or similar product. Take a clean lint-free rag and wipe up any excess anti-rust solution from the bores, bearing saddles, etc. Repeat the final cleaning process on the crankshaft. Replace any freeze or oil galley plugs which were removed during disassembly.

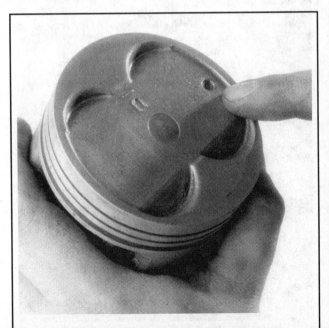

Fig. 54 Most pistons are marked to indicate positioning in the engine (usually a mark means the side facing the front)

Crankshaft

◆ **See Figures 55, 56, 57, 58 and 59**

1. Remove the main bearing inserts from the block and bearing caps.
2. If the crankshaft main bearing journals have been refinished to a definite undersize, install the correct undersize bearing. Be sure that the bearing inserts and bearing bores are clean. Foreign material under inserts will distort bearing and cause failure.
3. Place the upper main bearing inserts in bores with tang in slot.

■ **The oil holes in the bearing inserts must be aligned with the oil holes in the cylinder block.**

4. Install the lower main bearing inserts in bearing caps.
5. Clean the mating surfaces of block and rear main bearing cap.
6. Carefully lower the crankshaft into place. Be careful not to damage bearing surfaces.
7. Check the clearance of each main bearing by using the following procedure:
   a. Place a piece of Plastigage® or its equivalent, on bearing surface across full width of bearing cap and about 1/4 in. off center.
   b. Install cap and tighten bolts to specifications. Do not turn crankshaft while Plastigage® is in place.
   c. Remove the cap. Using the supplied Plastigage® scale, check width of Plastigage® at widest point to get maximum clearance. Difference between readings is taper of journal.
   d. If clearance exceeds specified limits, try a 0.001 in. or 0.002 in. undersize bearing in combination with the standard bearing. Bearing clearance must be within specified limits. If standard and 0.002 in. undersize bearing does not bring clearance within desired limits, refinish crankshaft journal, then install undersize bearings.

8. Install the rear main seal.
9. After the bearings have been fitted, apply a light coat of engine oil to the journals and bearings. Install the rear main bearing cap. Install all bearing caps except the thrust bearing cap. Be sure that main bearing caps are installed in original locations. Tighten the bearing cap bolts to specifications.
10. Install the thrust bearing cap with bolts finger-tight.
11. Pry the crankshaft forward against the thrust surface of upper half of bearing.
12. Hold the crankshaft forward and pry the thrust bearing cap to the rear. This aligns the thrust surfaces of both halves of the bearing.
13. Retain the forward pressure on the crankshaft. Tighten the cap bolts to specifications.
15. Measure the crankshaft end-play as follows:
   a. Mount a dial gauge to the engine block and position the tip of the gauge to read from the crankshaft end.
   b. Carefully pry the crankshaft toward the rear of the engine and hold it there while you zero the gauge.
   c. Carefully pry the crankshaft toward the front of the engine and read the gauge.
   d. Confirm that the reading is within specifications. If not, install a new thrust bearing and repeat the procedure. If the reading is still out of specifications with a new bearing, have a machine shop inspect the thrust surfaces of the crankshaft, and if possible, repair it.
16. Rotate the crankshaft so as to position the first rod journal to the bottom of its stroke.

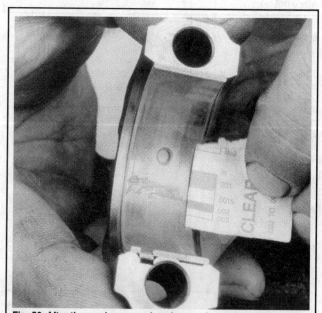

Fig. 56 After the cap is removed again, use the scale supplied with the gauging material to check the clearance

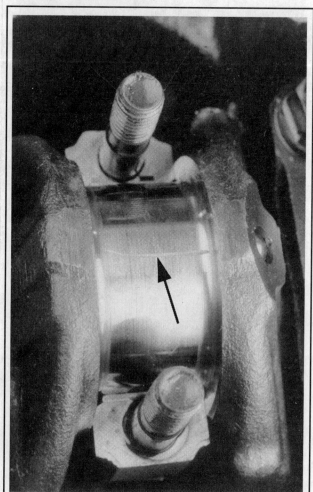

Fig. 55 Apply a strip of gauging material to the bearing journal, then install and torque the cap

Fig. 57 A dial gauge may be used to check crankshaft end-play...

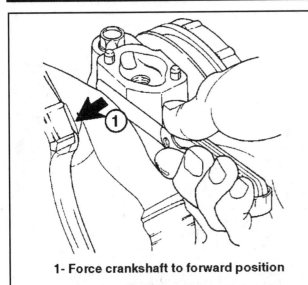

1- Force crankshaft to forward position

Fig. 58 ...or you can use a feeler gauge

Fig. 59 Carefully pry the crankshaft back and forth while reading the dial gauge for end-play

Pistons and Connecting Rods

 DIFFICULT

◆ See Figures 60, 61, 62, 63, 64, 65, 66 and 67

1. Before installing the piston/connecting rod assembly, oil the pistons, piston rings and the cylinder walls with light engine oil. Install connecting rod bolt protectors or rubber hose onto the connecting rod bolts/studs. Also perform the following:

   a. Select the proper ring set for the size cylinder bore.

   b. Position the ring in the bore in which it is going to be used.

   c. Push the ring down into the bore area where normal ring wear is not encountered.

   d. Use the head of the piston to position the ring in the bore so that the ring is square with the cylinder wall. Use caution to avoid damage to the ring or cylinder bore.

   e. Measure the gap between the ends of the ring with a feeler gauge. Ring gap in a worn cylinder is normally greater than specification. If the ring gap is greater than the specified limits, try an oversize ring set.

   f. Check the ring side clearance of the compression rings with a feeler gauge inserted between the ring and its lower land according to specification. The gauge should slide freely around the entire ring circumference without binding. Any wear that occurs will form a step at the inner portion of the lower land. If the lower lands have high steps, the piston should be replaced.

2. Unless new pistons are installed, be sure to install the pistons in the cylinders from which they were removed. The numbers on the connecting rod and bearing cap must be on the same side when installed in the cylinder bore. If a connecting rod is ever transposed from one engine or cylinder to another, new bearings should be fitted and the connecting rod should be numbered to correspond with the new cylinder number. The notch on the piston head goes toward the front of the engine.

3. Install all of the rod bearing inserts into the rods and caps.

4. Install the rings to the pistons. Install the oil control ring first, then the second compression ring and finally the top compression ring. Use a piston ring expander tool to aid in installation and to help reduce the chance of breakage.

5. Make sure the ring gaps are properly spaced around the circumference of the piston. Fit a piston ring compressor around the piston and slide the piston and connecting rod assembly down into the cylinder bore, pushing it in with the wooden hammer handle. Push the piston down until it is only slightly below the top of the cylinder bore. Guide the connecting rod onto the crankshaft bearing journal carefully, to avoid damaging the crankshaft.

6. Check the bearing clearance of all the rod bearings, fitting them to the crankshaft bearing journals. Follow the procedure in the crankshaft installation above.

7. After the bearings have been fitted, apply a light coating of assembly oil to the journals and bearings.

8. Turn the crankshaft until the appropriate bearing journal is at the bottom of its stroke, then push the piston assembly all the way down until the connecting rod bearing seats on the crankshaft journal. Be careful not to allow the bearing cap screws to strike the crankshaft bearing journals and damage them.

9. After the piston and connecting rod assemblies have been installed, check the connecting rod side clearance on each crankshaft journal.

10. Prime and install the oil pump and the oil pump intake tube.

11. Install the auxiliary/balance shaft/assembly if equipped.

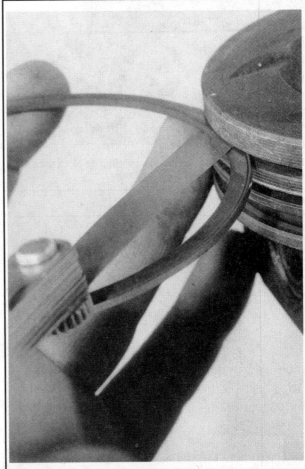

Fig. 60 Checking the piston ring-to-ring groove side clearance using the ring and a feeler gauge

Fig. 61 The notch on the side of the bearing cap matches the tang on the bearing insert

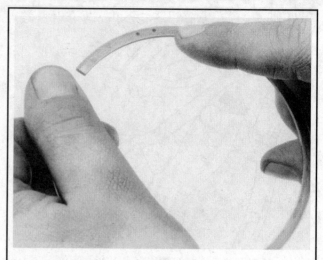

Fig. 62 Most rings are marked to show which side of the ring should face up when installed to the piston

Fig. 63 Install the piston and rod assembly into the block using a ring compressor and the handle of a hammer

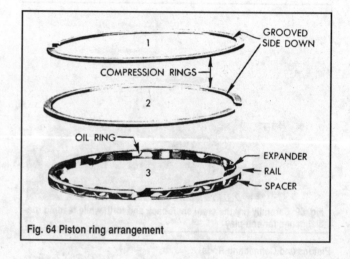

Fig. 64 Piston ring arrangement

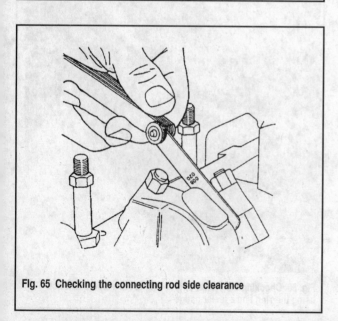

Fig. 65 Checking the connecting rod side clearance

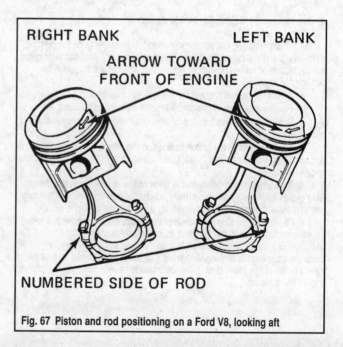

Fig. 67 Piston and rod positioning on a Ford V8, looking aft

Camshaft, Lifters And Timing Assembly

1. Install the camshaft.
2. Install the lifters/followers into their bores.
3. Install the timing gears/chain assembly.

Cylinder Head(S)

1. Install the cylinder head(s) using new gaskets.
2. Assemble the rest of the valve train (pushrods and rocker arms and/or shafts).

Engine Start-up and Break-in

STARTING THE ENGINE

Now that the engine is installed and every wire and hose is properly connected, go back and double check that all coolant and vacuum hoses are connected. Check that your oil drain plug is installed and properly tightened. If not already done, install a new oil filter onto the engine. Fill the crankcase with the proper amount and grade of engine oil. Fill the cooling system with a 50/50 mixture of coolant/water on models with a closed system.

1. Connect the battery.
2. Start the engine. Keep your eye on your oil pressure indicator; if it does not indicate oil pressure within 10 seconds of starting, turn the engine off.

✳✳ WARNING

Damage to the engine can result if it is allowed to run with no oil pressure. Check the engine oil level to make sure that it is full. Check for any leaks and if found, repair the leaks before continuing. If there is still no indication of oil pressure, you may need to prime the system.

3. Confirm that there are no fluid leaks (oil or other).
4. Allow the engine to reach normal operating temperature.
5. At this point you can perform any necessary checks or adjustments, such as checking the ignition timing.
6. Install any remaining components that were removed.

BREAKING IT IN

Make the first hours on the new engine, easy ones. Vary the speed but do not accelerate hard. Most importantly, do not lug the engine, and avoid sustained high speeds until at least 20 hours. Check the engine oil and coolant levels frequently. Expect the engine to use a little oil until the rings seat. Change the oil and filter at 20 hours and then follow the normal maintenance intervals from there out.

KEEP IT MAINTAINED

Now that you have just gone through all of that hard work, keep yourself from doing it all over again by thoroughly maintaining it. Not that you may not have maintained it before, heck you could have had a couple of thousand hours on it before doing this. However, you may have bought the vehicle used, and the previous owner did not keep up on maintenance. Which is why you just went through all of that hard work. See?

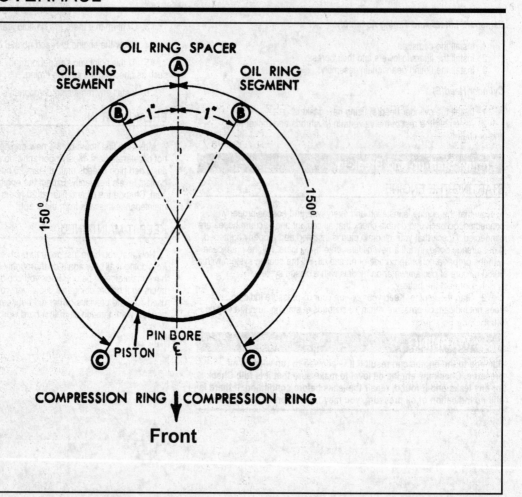

Fig. 66 Piston ring spacing

8

FUEL SYSTEMS
CARBURETORS

FUEL AND COMBUSTION

Fuel

Fuel recommendations have become more complex as the chemistry of modern gasoline changes. The major driving force behind the changes in gasoline chemistry is the search for additives to replace lead as an octane booster and lubricant. These new additives are governed by the types of emissions they produce in the combustion process. Also, the replacement additives do not always provide the same level of combustion stability, making a fuel's octane rating less meaningful.

In the 1960's and 1970's, leaded fuel was common. The lead served two functions. First, it served as an octane booster (combustion stabilizer) and second, in 4-stroke engines, it served as a valve seat lubricant. For 2-stroke engines, the primary benefit of lead was to serve as a combustion stabilizer. Lead served very well for this purpose, even in high heat applications.

Today, all lead has been removed from the refining process. This means that the benefit of lead as an octane booster has been eliminated. Several substitute octane boosters have been introduced in the place of lead. While many are adequate in an engine, most do not perform nearly as well as lead did, even though the octane rating of the fuel is the same.

OCTANE RATING

A fuel's octane rating is a measurement of how stable the fuel is when heat is introduced. Octane rating is a major consideration when deciding whether a fuel is suitable for a particular application. For example, in an engine, we want the fuel to ignite when the spark plug fires and not before, even under high pressure and temperatures. Once the fuel is ignited, it must burn slowly and smoothly, even though heat and pressure are building up while the burn occurs. The unburned fuel should be ignited by the traveling flame front, not by some other source of ignition, such as carbon deposits or the heat from the expanding gasses. A fuel's octane rating is known as a measurement of the fuel's anti-knock properties (ability to burn without exploding).

Usually a fuel with a higher octane rating can be subjected to a more severe combustion environment before spontaneous or abnormal combustion occurs. To understand how two gasoline samples can be different, even though they have the same octane rating, we need to know how octane rating is determined.

The American Society of Testing and Materials (ASTM) has developed a universal method of determining the octane rating of a fuel sample. The octane rating you see on the pump at a fuel dock is known as the pump octane number. Look at the small print on the pump. The rating has a formula. The rating is determined by the R+M/2 method. This number is the average of the research octane reading and the motor octane rating.

• The Research Octane Rating is a measure of a fuel's anti-knock properties under a light load or part throttle conditions. During this test, combustion heat is easily dissipated.

• The Motor Octane Rating is a measure of a fuel's anti-knock properties under a heavy load or full throttle conditions, when heat buildup is at maximum.

VAPOR PRESSURE

Fuel vapor pressure is a measure of how easily a fuel sample evaporates. Many additives used in gasoline contain aromatics. Aromatics are light hydrocarbons distilled off the top of a crude oil sample. They are effective at increasing the research octane of a fuel sample but can cause vapor lock (bubbles in the fuel line) on a very hot day. If you have an inconsistent running engine and you suspect vapor lock, use a piece of clear fuel line to look for bubbles, indicating that the fuel is vaporizing.

One negative side effect of aromatics is that they create additional combustion products such as carbon and varnish. If your engine requires high octane fuel to prevent detonation, de-carbon the engine more frequently with an internal engine cleaner to prevent ring sticking due to excessive varnish buildup.

ALCOHOL-BLENDED FUELS

When the Environmental Protection Agency mandated a phase-out of the leaded fuels in January of 1986, fuel suppliers needed an additive to improve the octane rating of their fuels. Although there are multiple methods currently employed, the addition of alcohol to gasoline seems to be favored because of its favorable results and low cost. Two types of alcohol are used in fuel today as octane boosters, methanol (wood alcohol) or ethanol (grain alcohol).

When used as a fuel additive, alcohol tends to raise the research octane of the fuel. There are, however, some special considerations due to the effects of alcohol in fuel.

• Since alcohol contains oxygen, it replaces gasoline without oxygen content and tends to cause the air/fuel mixture to become leaner.

• On older engines, the leaching affect of alcohol may, in time, cause fuel lines and plastic components to become brittle to the point of cracking. Unless replaced, these cracked lines could leak fuel, increasing the potential for hazardous situations.

• When alcohol blended fuels become contaminated with water, the water combines with the alcohol then settles to the bottom of the tank. This leaves the gasoline on a top layer.

■ **Modern fuel lines and plastic fuel system components have been specially formulated to resist alcohol leaching effects.**

HIGH ALTITUDE OPERATION

At elevated altitudes there is less oxygen in the atmosphere than at sea level. Less oxygen means lower combustion efficiency and less power output. As a general rule, power output is reduced three percent for every thousand feet above sea level.

On carbureted engines, re-jetting for high altitude does not restore lost power, it simply corrects the air-fuel ratio for the reduced air density and makes the most of the remaining available power. The most important thing to remember when re-jetting for high altitude is to reverse the jetting when returning to sea level. If the jetting is left lean when you return to sea level conditions, the correct air/fuel ratio will not be achieved and possible engine damage may occur.

RECOMMENDATIONS

All engine covered here are designed to run on unleaded fuel. Never use leaded fuel in your boat's engine. The minimum octane rating of fuel being used for your engine must be at least 87 AKI (outside the US: 90 RON), which means regular unleaded, but some engines may require higher octane ratings. OMC actually recommends the use of 89 AKI (93 RON) fuel as the ideal—in fact, anything less than this on many 4.3L/5.7L engines will require a change to the ignition timing. Fuel should be selected for the brand and octane that performs best with your engine. Check your owner's manual if in doubt. Premium unleaded is more stable under severe conditions but also produces more combustion products. Therefore, when using premium unleaded, more frequent de-carboning is necessary.

The use of a fuel too low in octane (a measure of anti-knock quality) will result in spark knock. Newer systems have the capability to adjust the engine's ignition timing to compensate to some extent, but if persistent knocking occurs, it may be necessary to switch to a higher grade of fuel. Continuous or heavy knocking may result in engine damage.

Combustion

In a high heat environment like an modern engine, the fuel must be very stable to avoid detonation. If any parameters affecting combustion change suddenly (the engine runs lean for example), uncontrolled heat buildup will occurs very rapidly.

The combustion process is affected by several interrelated factors. This means that when one factor is changed, the other factors also must be changed to maintain the same controlled burn and level of combustion stability.

• Compression—determines the level of heat buildup in the cylinder when the air-fuel mixture is compressed. As compression increases, so does the potential for heat buildup

• Ignition Timing—determines when the gasses will start to expand in relation to the motion of the piston. If the ignition timing is too advanced, gasses will be ignited and begin to expand too soon, such as they would during pre-ignition. The motion of the piston opposes the expansion of the gasses, resulting in extremely high combustion chamber pressures and heat. If the ignition timing is retarded, the gases are ignited later in relation to piston position. This means that the piston has already traveled back down the bore toward the bottom of the cylinder, resulting in less usable power.

• Fuel Mixture—determines how efficient the burn will be. A rich mixture burns slower than a lean one. If the mixture is too lean, it can't become explosive. The slower the burn, the cooler the combustion chamber, because pressure buildup is gradual.

• Fuel Quality (Octane Rating)—determines how much heat is necessary to ignite the mixture. Once the burn is in progress, heat is on the rise. The unburned poor quality fuel is ignited all at once by the rising heat instead of burning gradually as a flame front of the burn passing by. This action results in detonation (pinging).

There are two types of abnormal combustion—pre-ignition and detonation.

• Pre-ignition—occurs when the air-fuel mixture is ignited by some incandescent source other than the correctly timed spark from the spark plug.

• Detonation—occurs when excessive heat and or pressure ignites the air/fuel mixture rather than the spark plug. The burn becomes explosive.

In general, anything that can cause abnormal heat buildup can be enough to push an engine over the edge to abnormal combustion, if any of the four basic factors previously discussed are already near the danger point, for example, excessive carbon buildup raises the compression and retains heat as glowing embers.

CARBURETED FUEL SYSTEM

Troubleshooting

Troubleshooting fuel systems requires the same techniques used in other areas. A thorough, systematic approach to troubleshooting will pay big rewards. Build your troubleshooting checklist, with the most likely offenders at the top. Use your experience to adjust your list for local conditions. Everyone has been tempted to jump into the carburetor on a vague hunch. Pause a moment and review the facts when this urge occurs.

In order to accurately troubleshoot a carburetor or fuel system problem, you must first verify that the problem is fuel related. Many symptoms can have several different possible causes. Be sure to eliminate mechanical and electrical systems as the potential fault. Carburetion is the number one cause of most engine problems but there are other possibilities.

One of the toughest tasks with a fuel system is the actual troubleshooting. Several tools are at your disposal for making this process very simple. A timing light works well for observing carburetor spray patterns. Look for the proper amount of fuel and for proper atomization in the two fuel outlet areas (main nozzle and bypass holes). The strobe effect of the lights helps you see in detail the fuel being drawn through the throat of the carburetor. On multiple carburetor engines, always attach the timing light to the cylinder you are observing so the strobe doesn't change the appearance of the patterns. If you need to compare two cylinders, change the timing light hookup each time you observe a different cylinder.

Pressure testing fuel pump output can determine whether the fuel spray is adequate and if the fuel pump diaphragms are functioning correctly. A pressure gauge placed between the fuel pump(s) and the carburetor(s) will test the entire fuel delivery system. Normally a fuel system problem will show up at high speed where the fuel demand is the greatest. A common symptom of a fuel pump output problem is surging at wide open throttle but normal operation at slower speeds. To check the fuel pump output, install the pressure gauge and accelerate the engine to wide-open throttle. Observe the pressure gauge needle. It should always swing up the value indicated in the specification charts and remain steady. This reading would indicate a system that is functioning properly.

If the needle gradually swings down toward zero, fuel demand is greater than the fuel system can supply. This reading isolates the problem to the fuel delivery system (fuel tank or line). To confirm this, an auxiliary tank should be installed and the engine re-tested. Be aware that a bad anti-siphon valve on a built-in tank can create enough restriction to cause a lean condition and serious engine damage.

If the needle movement becomes erratic, suspect a ruptured diaphragm in the fuel pump.

To check for air entering the fuel system, install a clear fuel hose between the fuel screen and fuel pump. If air is in the line, check all fittings back to the boat's fuel tank.

Spark plug tip appearance is a good indication of combustion efficiency. The tip should be a light tan. A White insulator or small beads on the insulator indicate too much heat. A dark or oil fouled insulator indicates incomplete combustion. To properly read spark plug tip appearance, run the engine at the RPM you are testing for about 15 seconds and then immediately turn the engine OFF without changing the throttle position.

Reading spark plug tip appearance is also the proper way to test jet verifications in high altitude.

COMMON PROBLEMS

Fuel Delivery

Many times fuel system troubles are caused by a plugged fuel filter, a defective fuel pump or by a leak in the line from the fuel tank to the fuel pump. A defective choke may also cause problems. Would you believe, a majority of starting troubles which are traced to the fuel system are the result of an empty fuel tank or aged sour fuel.

Sour Fuel

Under average conditions (temperate climates), fuel will begin to break down in about four months. A gummy substance forms in the bottom of the fuel tank and in other areas. The filter screen between the tank and the carburetor and small passages in the carburetor will become clogged. The gasoline will begin to give off an odor similar to rotten eggs. Such a condition can cause the owner much frustration, time in cleaning components and the expense of replacement or overhaul parts for the carburetor.

Even with the high price of fuel, removing gasoline that has been standing unused over a long period of time is still the easiest and least expensive preventative maintenance possible. In most cases, this old gas can be used without harmful effects in an automobile using regular gasoline.

A gasoline preservative will keep the fuel fresh for up to twelve months. These products are available in most areas under various trade names.

Choke Problems

When the engine is hot, the fuel system can cause starting problems. After a hot engine is shut down, the temperature inside the fuel bowl may rise to 200°F and cause the fuel to actually boil. All carburetors are vented to allow this pressure to escape to the atmosphere. However, some of the fuel may percolate over the high-speed nozzle.

If the choke should stick in the open position, the engine will be hard to start. If the choke should stick in the closed position, the engine will flood, making it very difficult to start.

In order for this raw fuel to vaporize enough to burn, considerable air must be added to lean out the mixture. Therefore, the only remedy is to remove the spark plugs, ground the leads, crank the engine through about ten revolutions, clean the plugs, reinstall the plugs and start the engine.

If the needle valve and seat assembly is leaking, an excessive amount of fuel may enter the reed housing in the following manner. After the engine is shut down, the pressure left in the fuel line will force fuel past the leaking needle valve. This extra fuel will raise the level in the fuel bowl and cause fuel to overflow into the reed housing.

A continuous overflow of fuel into the reed housing may be due to a sticking inlet needle or to a defective float, which would cause an extra high level of fuel in the bowl and overflow into the reed housing.

Rough Engine Idle

If an engine does not idle smoothly, the most reasonable approach to the problem is to perform a tune-up to eliminate such areas as:
• Defective points
• Faulty spark plugs
• Timing out of adjustment
Other problems that can prevent an engine from running smoothly include:
• An air leak in the intake manifold
• Uneven compression between the cylinders
Of course any problem in the carburetor affecting the air/fuel mixture will also prevent the engine from operating smoothly at idle speed. These problems usually include:
• Too high a fuel level in the bowl
• A heavy float
• Leaking needle valve and seat
• Defective automatic choke
• Improper adjustments for idle mixture or idle speed

Excessive Fuel Consumption

Excessive fuel consumption can be the result of any one of four conditions or a combination of all.
• Inefficient engine operation.

- Faulty condition of the hull, including excessive marine growth.
- Poor boating habits of the operator.
- Leaking or out-of-tune carburetor.

If the fuel consumption suddenly increases over what could be considered normal, then the cause can probably be attributed to the engine or boat and not the operator.

Marine growth on the hull can have a very marked effect on boat performance. This is why sailboats always try to have a haul-out as close to race time as possible.

While you are checking the bottom, take note of the propeller condition. A bent blade or other damage will definitely cause poor boat performance.

If the hull and propeller are in good shape, then check the fuel system for possible leaks. Check the line between the fuel pump and the carburetor while the engine is running and the line between the fuel tank and the pump when the engine is not running. A leak between the tank and the pump many times will not appear when the engine is operating, because the suction created by the pump drawing fuel will not allow the fuel to leak. Once the engine is turned off and the suction no longer exists, fuel may begin to leak.

If a minor tune-up has been performed and the spark plugs, points (if equipped) and timing are properly adjusted, then the problem most likely is in the carburetor and an overhaul is in order.

Check the needle valve and seat for leaking. Use extra care when making any adjustments affecting the fuel consumption, such as the float level or automatic choke.

Engine Surge

If the engine operates as if the load on the boat is being constantly increased and decreased, even though an attempt is being made to hold a constant engine speed, the problem can most likely be attributed to the fuel pump or a restriction in the fuel line between the tank and the carburetor.

COMBUSTION RELATED PISTON FAILURES

When an engine has a piston failure due to abnormal combustion, fixing the mechanical portion of the engine is the easiest part of the equation. The hard part is determining what caused the problem in order to prevent a repeat failure. Think back to the four basic areas that affect combustion to find the cause of the failure.

Since you probably removed the cylinder head, inspect the failed piston and look for excessive deposit buildup that could raise compression or retain heat in the combustion chamber. Statically check the wide open throttle timing. Be sure that the timing is not over advanced. It is a good idea to seal these adjustments with paint to detect tampering.

Look for a fuel restriction that could cause the engine to run lean. Don't forget to check the fuel pump, fuel tank and lines, especially if a built in tank is used. Be sure to check the anti-siphon valve on built in tanks. If everything else looks good, the final possibility is poor quality fuel.

Operation

BASIC FUNCTIONS

◆ See Figure 1

Traditional carburetor theory often involves a number of laws and principles. The diagram illustrates several carburetor basics. If you blow air across a straw inserted into a container of liquid, a pressure drop is created in the straw column. As the liquid in the column is expelled, an atomized mixture (air and fuel droplets) is created. In a carburetor this is mostly air and a little fuel.

The actual ratio of air to fuel differs with engine conditions but is usually from 15 parts air to one part fuel at optimum cruise to as little as 7 parts air to one part fuel at full choke.

Using our example, what if the top of the container is covered and sealed around the straw, what will happen? No flow. This is typical of a clogged carburetor bowl vent. If the base of the straw is clogged or restricted what will happen? No flow or low flow. This represents a clogged main jet. If the liquid in the glass is lowered and you blow through the straw with the same force what will happen? Not as much fuel will flow. A lean condition occurs. If the fuel level is raised and you blow again at the same velocity what happens? The result is a richer mixture.

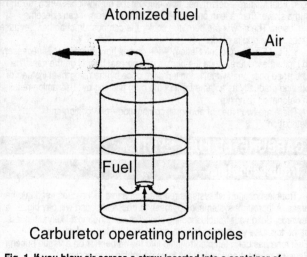

Carburetor operating principles

Fig. 1 If you blow air across a straw inserted into a container of liquid, a pressure drop is created in the straw column. As the liquid in the column is expelled, an atomized mixture (air and fuel droplets) is created

FUEL & AIR METERING

The carburetor is merely a metering device for mixing fuel and air in the proper proportions for efficient engine operation.

Float Systems

◆ See Figure 2

A small chamber in the carburetor serves as a fuel reservoir. A float valve admits fuel into the reservoir to replace the fuel consumed by the engine. If the carburetor has more than one reservoir, the fuel level in each reservoir (chamber) is controlled by identical float systems.

Fuel level in each chamber is extremely critical and must be maintained accurately. Accuracy is obtained through proper adjustment of the floats. This adjustment will provide a balanced metering of fuel to each cylinder at all speeds.

Following the fuel through its course, from the fuel tank to the combustion chamber of the cylinder, will provide an appreciation of exactly what is taking place. In order to start the engine, the fuel must be moved from the tank to the carburetor by a fuel pump installed in the fuel line.

After the engine starts, the fuel passes through the pump to the carburetor. All systems have some type of filter installed somewhere in the line between the tank and the carburetor. Most engines also have a filter as an integral part of the carburetor.

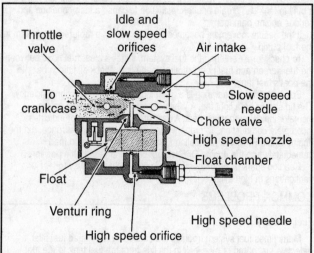

Fig. 2 Fuel flow through a venturi, showing principal and related parts controlling intake and outflow

At the carburetor, the fuel passes through the inlet passage to the needle and seat and then into the float chamber (reservoir). A float in the chamber rides up and down on the surface of the fuel. After fuel enters the chamber and the level rises to a predetermined point, a tang on the float closes the inlet needle and the flow entering the chamber is cut off. When fuel leaves the chamber as the engine operates, the fuel level drops and the float tang allows the inlet needle to move off its seat and fuel once again enters the chamber. In this manner, a constant reservoir of fuel is maintained in the chamber to satisfy the demands of the engine at all speeds.

A fuel chamber vent hole is located near the top of the carburetor body to permit atmospheric pressure to act against the fuel in each chamber. This pressure assures an adequate fuel supply to the various operating systems.

Air/Fuel Mixture

◆ **See Figure 3**

A suction effect is created each time the piston moves upward in the cylinder. This suction draws air through the throat of the carburetor. A restriction in the throat, called a venturi, controls air velocity and has the effect of reducing air pressure at this point.

The difference in air pressures at the throat and in the fuel chamber causes the fuel to be pushed out of metering jets extending down into the fuel chamber. When the fuel leaves the jets, it mixes with the air passing through the venturi. This fuel/air mixture should then be in the proper proportion for burning in the cylinders for maximum engine performance.

In order to obtain the proper air/fuel mixture for all engine speeds, some models have high and low speed jets. These jets have adjustable needle valves that are used to compensate for changing atmospheric conditions. In almost all cases, the high-speed circuit has fixed high-speed jets and is not adjustable.

A throttle valve controls the flow of air/fuel mixture drawn into the combustion chambers. A cold engine requires a richer fuel mixture to start and during the brief period it is warming to normal operating temperature. A choke valve is placed ahead of the metering jets and venturi. As this valve begins to close, the volume of air intake is reduced, thus enriching the mixture entering the cylinders. When this choke valve is fully closed, a very rich fuel mixture is drawn into the cylinders.

The throat of the carburetor is usually referred to as the barrel. Carburetors with single, double or four barrels have individual metering jets, needle valves, throttle and choke plates for each barrel. Single and two barrel carburetors are fed by a single float and chamber.

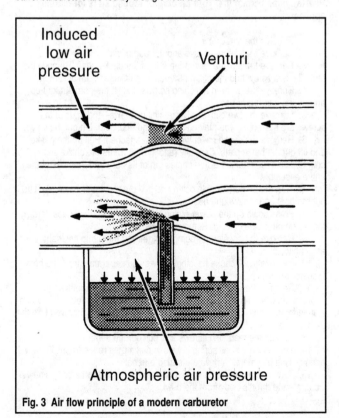

Fig. 3 **Air flow principle of a modern carburetor**

CARBURETOR CIRCUITS

The following section illustrates the circuit functions and locations of a typical marine carburetor.

Starting Circuit

◆ **See Figure 4**

The choke plate is closed, creating a partial vacuum in the venturi. As the piston rises, negative pressure in the crankcase draws the rich air-fuel mixture from the float bowl into the venturi and then on into the engine.

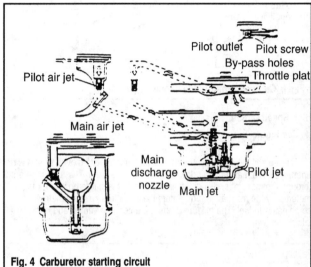

Fig. 4 **Carburetor starting circuit**

Low Speed Circuit

◆ **See Figure 5**

Zero–one-eighth throttle, when the pressure in the crankcase is lowered, the air-fuel mixture is discharged into the venturi through the pilot outlet because the throttle plate is closed. No other outlets are exposed to low venturi pressure. The fuel is metered by the pilot jet. The air is metered by the pilot air jet. The combined air-fuel mixture is regulated by the pilot air screw.

Mid-Range Circuit

◆ **See Figure 6**

One-eighth–three-eighths throttle, as the throttle plate continues to open, the air-fuel mixture is discharged into the venturi through the bypass holes. As the throttle plate uncovers more bypass holes, increased fuel flow results because of the low pressure in the venturi. Depending on the model, there could be two, three or four bypass holes.

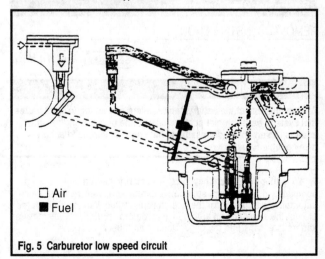

Fig. 5 **Carburetor low speed circuit**

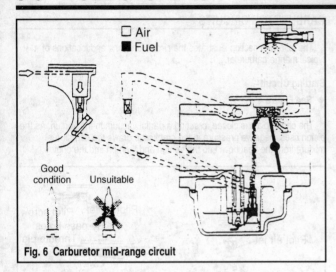

Fig. 6 Carburetor mid-range circuit

High Speed Circuit

◆ See Figure 7

Three-eighths–wide-open throttle, as the throttle plate moves toward wide open, we have maximum air flow and very low pressure. The fuel is metered through the main jet and is drawn into the main discharge nozzle. Air is metered by the main air jet and enters the discharge nozzle, where it combines with fuel. The mixture atomizes, enters the venturi and is drawn into the engine.

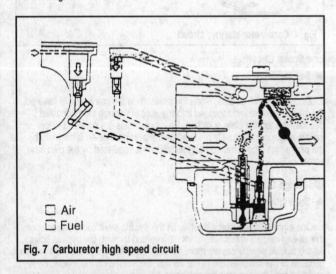

Fig. 7 Carburetor high speed circuit

Rochester 2bbl Carburetors

REMOVAL & INSTALLATION

✳✳ WARNING

Always disconnect the battery cables before attempting to work on the fuel system. Never smoke or allow open flame near the engine—this sounds like an obvious precaution, but you'd be surprised at how many people forget!

■ No matter how much they look alike, marine carburetors are completely different from automotive carburetors. Never substitute an automotive carburetor for the one on your engine! Venting procedures are not the same and an automotive carburetor could allow dangerous vapors to escape into the bilge. Don't even think about it.

1. Open the engine hatch or remove the covers and then disconnect the battery cables. Turn the fuel petcock **OFF** and/or shut down the fuel supply at the tank.
2. Remove the flame arrestor after disconnecting the vent hose. It's always a good idea to plug the throttle bores with a clean, lint-free cloth.
3. Disconnect the throttle cable at the carburetor and carefully move it out of the way.
4. Using two open-end wrenches, hold the fuel inlet nut at the carburetor securely and loosen the fuel line nut. Disconnect the two and carefully move the line out of the way. Plug both the inlet and line open ends to prevent fuel seepage.
5. Tag and disconnect the electric choke lead (2.3L only).
6. Tag and disconnect any other lines, leads or hoses that may be in the way of removal.
7. Loosen and remove the carburetor mounting nuts (early models) or bolts and washers and lift the unit off the manifold. Plug the opening with a clean lint-free cloth.
   To Install:
8. Clean the mating surfaces thoroughly of all residual gasket material.
9. Position a new gasket and then install the carburetor. On early models, tighten the nuts to 10-14 ft. lbs. (13-19 Nm). On all others, tighten the bolts to 12-15 ft. lbs. (16-20 Nm). Hopefully you remembered to remove the rag!
10. Reconnect the fuel line to the inlet line after removing the plugs and tighten it to 18 ft. lbs. (24 Nm). Don't forget to use two wrenches.
11. Connect the choke lead on the 2.3L engine.
12. Install and adjust the throttle cable as detailed in the Maintenance section. On the 2.5L and 3.0L engines, make sure that the distance between the centers of the two ball studs is 2 1/4 in. (57.2mm); on the 2.3L engines, the distance should be 2 3/32 in. 53.1mm).
13. Install the flame arrestor and reconnect the vent line.
14. Connect the battery cables. Start the engine and ensure there are no fuel leaks; shut down the engine immediately if there are. Check and adjust the idle speed and mixture.

DISASSEMBLY

◆ See Figures 8 thru 13

■ **Always be certain that your carburetor rebuild kit is for marine applications.**

1. Remove the carburetor.
2. Remove the idle vent valve and the pump link.
3. Remove the small screw at the end of the choke shaft. Remove the fast idle cam screw and then lift off the fast idle linkage as an assembly.
4. Remove the air horn attaching screws and lift the horn straight up and off the carburetor body.
5. Remove the two choke valve retaining screws—the ends of the screws are staked, so you may have to file the ends first. Lift out the valve.
6. Remove the choke cover with the coil attached. Pull up the gasket and lift out the baffle plate. Carefully (and slowly) rotate the choke shaft counterclockwise until the piston comes out of its bore and then lift out the entire assembly.
7. Turn the air horn upside down and pull out the float's hinge pin. Lift out the float and remove the needle.
8. Press down on the power piston shaft until the spring snaps, forcing the piston out of the casting. Remove the piston.
9. Remove the accelerator pump plunger and the pump lever/shaft assembly.
10. Remove the 2 choke housing screws and separate it from the horn. Discard the gasket.
11. Remove the pump plunger, return spring and main metering jets.
12. Remove the power valve from the body. Certain models will have an aluminum inlet ball in the bottom of the well—it will fall out when you turn the carb over, so don't lose it.
13. Loosen the mounting screws and pull out the venture cluster.
14. The discharge ball spring is held in place by a retainer; pull it out with needle-nose pliers and remove the spring and ball.
15. Turn the carburetor back over an remove the 3 throttle body screws. Separate the throttle body from the bowl.

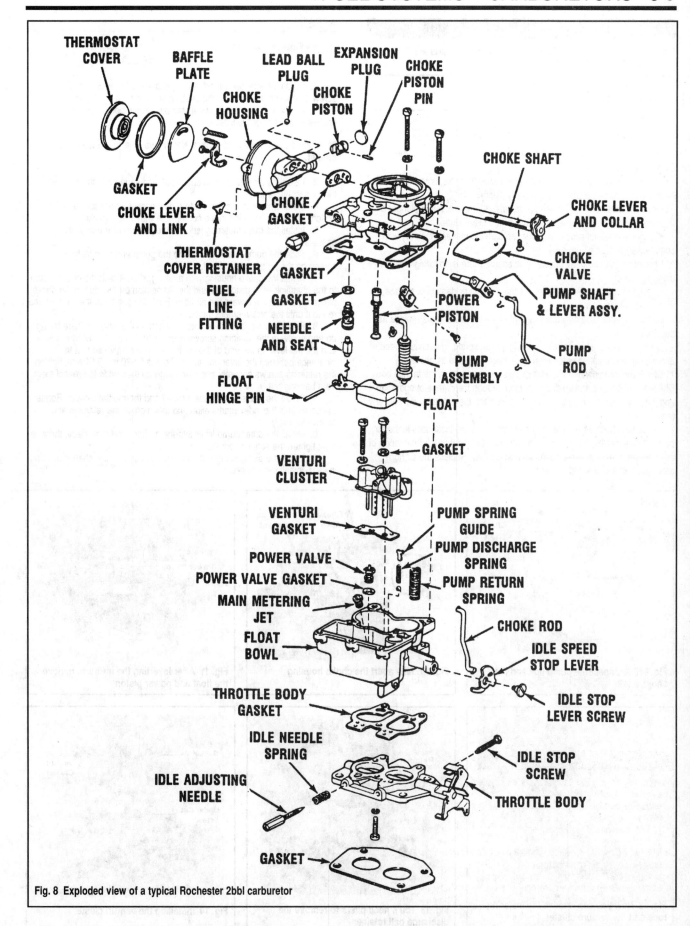

Fig. 8 Exploded view of a typical Rochester 2bbl carburetor

** WARNING

Do not disassemble the throttle body—replacement parts are NOT available.

CLEANING & INSPECTION

 MODERATE

Never use a wire brush or drill to clean jet passages or tubes in the carburetor.

Never allow the carburetor to soak in a cleaner bath for more than two hours. In fact, we recommend using spray cleaner.

Never immerse the float assembly, needle, accelerator pump plunger or fuel filter in cleaner. Wipe them carefully with a clean cloth.

Otherwise clean all allowable parts with carb cleaner and then dry with compressed air if at all possible.

Blow out and through all passages to ensure there is no foreign material clogging them.

Check the float needle and seat, if either is worn or damaged, replace them as a matched set.

Check the float assembly and hinge pin for wear or damage, replace as necessary.

Check the pump plunger, return spring, piston spring, idle mixture needle and all levers and linkages for wear or damage. Replace as necessary.

Check the throttle valve shaft for excessive looseness in the throttle body. Check that the valve and shaft do not bind through their range of operation and that the valve opens and closes fully. Replace the assembly if it fails any of these tests.

Check the choke valve lever and shaft for excessive looseness in the air horn. Check that the valve, lever and shaft do not bind through their range of operation and that the valve opens and closes fully. Replace the air horn assembly if it fails any of these tests.

ASSEMBLY

◆ See Figures 8 and 14

 MODERATE

1. Install the idle mixture adjusting needle into the throttle body and screw it down until it contacts the seat. Now back it out 1 1/2 turns.

2. Position a new gasket and install the throttle body. Tighten the screws to 50 inch lbs. (5.6 Nm).

■ **Always use a non-vent gasket. Never use a vented automotive gasket.**

3. Insert the steel discharge check ball and spring into the body and then insert the retainer.

4. Install the venture cluster with a new gasket. Make sure that the undercut screw is installed in the center hole with a new gasket.

5. Install the main metering jets and the power valve using new gaskets.

6. Slide the pump return spring into the pump and then install the inlet screen on the bottom of the bowl.

7. Install the choke housing with a new gasket. Attach the choke piston to the shaft/link—the piston pin and the flat section on the side of the choke piston must face out toward the air horn. Push the shaft into the air horn and rotate it until the piston enters the bore.

8. Position the choke valve onto the shaft so the letters **RP** are facing up and screw in the retaining screws finger-tight. Install the choke rod lever and the trip lever to the end of the shaft. Center the valve so that the clearance between the lever and the air horn is 0.020 in. (0.51mm), tighten the retaining screws securely and then rough up their ends to prevent them from backing out.

9. Install the baffle plate, gasket (new) and thermostatic cover. Rotate the cover until the index marks align and then tighten the retainers and screws securely.

10. Install the outer pump lever into the air horn and then install the inner arm; tighten the screw securely.

11. Install the pump plunger to the inner arm so that the shaft is pointing inward. Install the retainer.

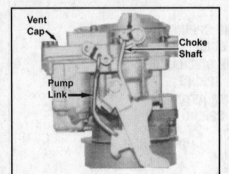

Fig. 9 Disconnect the pump link and the choke shaft

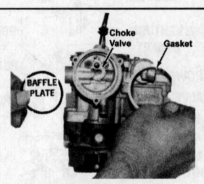

Fig. 10 Take apart the choke housing

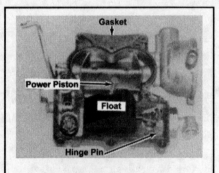

Fig. 11 After inverting the air horn, remove the float and power piston

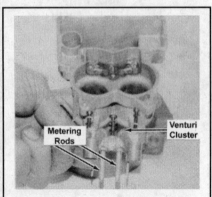

Fig. 12 Now you can remove the metering rods and the venture cluster

Fig. 13 You'll need pliers to remove the discharge ball retainer

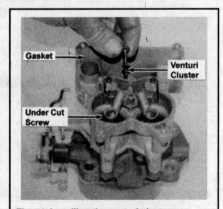

Fig. 14 Installing the venturi cluster

12. Position the screen on the float needle seat and then install the assembly onto the air horn.

13. Install the power piston into the vacuum cavity making sure that it can move without obstruction. Carefully stake the retainer to hold it in place.

14. Install a new air horn gasket and then attach the needle to the float. Carefully insert the needle into the float seat while, at the same time, guiding the float between the bosses. Insert the hinge pin and check the float level and drop.

15. Install the horn to the float bowl.

16. Install the carburetor.

ADJUSTMENT

Float Level

Floats With Vertical Seam

◆ See Figure 15

1. Following the disassembly procedure detailed previously, remove the bowl cover/air horn from the carburetor.

2. Turn the assembly upside down and check that the float pivots freely on the pin. Raise the float and let it fall—do not force it!

3. Using a standard carburetor gauge, measure the distance between the bottom of the air horn gasket (remember it's upside down, so this will be the top) and the top of the float; it should be within the specifications given in your rebuild kit. Carefully bend the float arm (at the rear of the float) with needle nose pliers to achieve the correct measurement. Make sure the float stays in alignment.

4. Its usually a good idea to check the float drop now.

5. Recheck your measurements one more time.

6. Reinstall the air horn and carburetor.

Floats With Horizontal Seam

◆ See Figure 16

1. Following the disassembly procedure detailed previously, remove the bowl cover/air horn from the carburetor.

2. Turn the assembly upside down and check that the float pivots freely on the pin. Raise the float and let it fall—do not force it!

3. Using a standard carburetor gauge, measure the distance between the bottom of the air horn gasket (remember it's upside down, so this will be the top) and the lower edge of the seam on the float; it should be within the specifications given in your rebuild kit. Carefully bend the float arm (at the rear of the float) with needle nose pliers to achieve the correct measurement. Make sure the float stays in alignment.

4. Its usually a good idea to check the float drop now.

5. Recheck your measurements one more time.

6. Reinstall the air horn while guiding the accelerator pump plunger into the well.

7. Slide the idle speed needle spring over the needle and screw it back into the throttle body.

8. Install the pump rod, choke rod and idle stop screw.

9. Install the carburetor.

Hollow Floats

◆ See Figure 17

1. Following the disassembly procedure detailed previously, remove the bowl cover/air horn from the carburetor.

2. Turn the assembly upside down and check that the float pivots freely on the pin. Raise the float and let it fall—do not force it!

3. Using a standard carburetor gauge, measure the distance between the bottom of the air horn gasket (remember it's upside down, so this will be the top) and the lip at the toe of the float; it should be within the specifications given in your rebuild kit. Carefully bend the float arm (at the rear of the float) with needle nose pliers to achieve the correct measurement. Make sure the float stays in alignment.

4. Its usually a good idea to check the float drop now.

5. Recheck your measurements one more time.

6. Reinstall the air horn and carburetor.

Float Drop

◆ See Figure 18

1. After checking the float level adjustment, turn the air horn over so it is right-side-up.

2. Allow the float to hang down freely and measure the distance between the bottom of the gasket and the lowest edge on the bottom of the float. On the 2.3L and 4.3L carburetor and later 3.0L carbs, measure to the gasket seam at the toe of the float.

3. If your measurement is not within the specification given in the kit, carefully bend the float tang with needle-nose pliers until it comes within spec.

4. Recheck your measurements one more time.

5. Reinstall the air horn and carburetor.

Idle Speed & Mixture

Please refer to the Maintenance section for adjustment procedures.

Pump Rod

◆ See Figure 19

1. Back out the idle stop screw and then close the throttle valves completely in their bores.

2. Position a pump gauge across the top of the air horn so that the leg is pointing down toward the top of the pump rod to the measurement listed in the Specifications chart.

3. If the lower edge of the gauge does not come in contact with the top of the rod, carefully bend the rod with needle-nose pliers until it does.

Choke Unloader

◆ See Figure 20

1. Remove the flame arrestor.

2. Move the linkage to position the throttle valves in the fully open position.

3. Insert the proper gauge (or drill bit) between upper edge of the plate and the inner side of the air horn wall. The gauge should just slide through—if not, bend the tang on the throttle lever until the measurement is correct.

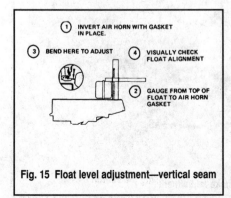

Fig. 15 Float level adjustment—vertical seam

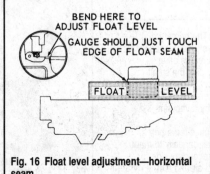

Fig. 16 Float level adjustment—horizontal seam

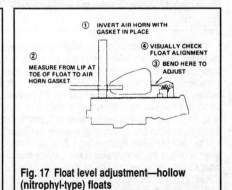

Fig. 17 Float level adjustment—hollow (nitrophyl-type) floats

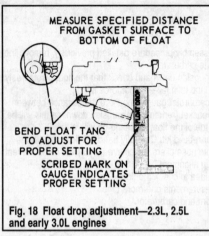

Fig. 18 Float drop adjustment—2.3L, 2.5L and early 3.0L engines

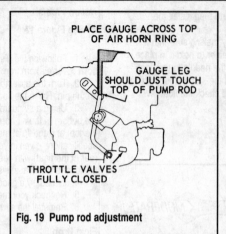

Fig. 19 Pump rod adjustment

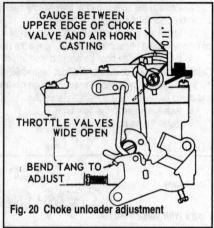

Fig. 20 Choke unloader adjustment

Intermediate Choke Rod—2.3L and 4.3L Engines

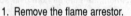 MODERATE

1. Remove the flame arrestor.
2. Remove the 3 mounting screws and lift off the thermostat cover, gasket and baffle.
3. Hold the choke valve closed and check that the edge of the actuator lever in the housing is lined up with the edge of the casting. If not, bend the choke rod with needle nose pliers at the bend just below the lever.

Vacuum Break—2.3L and 4.3L Engines

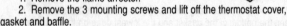 MODERATE

1. Remove the flame arrestor.
2. Stretch a small rubber band over the vacuum break diaphragm plunger so that the diaphragm is fully seated. Make sure that the rubber band is not interfering with the vacuum break link's movement.
3. Pull up on the intermediate choke lever until the choke valve is fully closed and then insert the proper sized gauge/bit between the inner wall of the air horn and the upper edge of the valve.
4. Bend the vacuum break link at the 90° bend below the unit.

Automatic Choke—2.5L, 3.0L and 4.3L Engines

◆ See Figure 21

 EASY

1. Remove the flame arrestor.
2. Loosen the 3 choke cover mounting screws and then rotate the cover until the mark on the cover is in alignment with the index line on the housing.
3. Tighten the screws.

Electric Choke—2.3L Engines

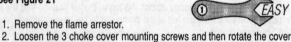 EASY

1. Remove the flame arrestor.
2. Loosen the choke cover mounting screws and then rotate the cover until the mark on the cover is in alignment with the index line on the housing.
3. If adjustment is necessary, move the cover one mark at a time, but never more than 2 marks in total from the index mark.
4. Tighten the screws.

Rochester 4 bbl Carburetor

DESCRIPTION

The Rochester Quadrajet carburetor has two stages. The primary (fuel inlet) side has small 1 3/8 in. bores with a triple venturi setup equipped with plain-tube nozzles. The carburetor operates much the same as other carburetors using the venturi principle. The triple venturi, plus the small primary bores, makes for a more stable and finer fuel control during idle and partial throttle operation. When the throttle is partially open, the fuel metering is accomplished with tapered metering rods, positioned by a vacuum-responsive piston and operating in specially designed jets.

The secondary side has two large, 2 1/4 in., bores. These large bores, when added to the primary side bores, provide enough air capacity to meet most engine requirements. The air valve is used in the secondary side for metering control and backs-up the primary bores to meet air and fuel demands of the engine.

The secondary air valve operates the tapered metering rods. These rods move in orifice plates and thus control fuel flow from the secondary nozzles in direct relation to the air flowing through the secondary bores.

The float bowl is designed to avoid problems of fuel spillage during sharp turns of the boat which could result in engine cutout and delayed fuel flow. The bowl reservoir is smaller than most four-barrel carburetors to reduce fuel evaporation during hot engine shut down.

The float system has one pontoon float and fuel valve which makes servicing much easier than on some other model carburetors. A fuel filter is located in the float bowl ahead of the float needle valve. This filter is easily removed for cleaning or replacement.

The throttle body is made of aluminum as part of a weight-reduction program and also to improve heat transfer away from the fuel bowl and prevent the fuel from "percolating" during hot engine shut down. A heat insulator gasket is used between the throttle body and bowl to help prevent "percolating".

✴✴ WARNING

Always disconnect the battery cables before attempting to work on the fuel system. Never smoke or allow open flame near the engine—this sounds like an obvious precaution, but you'd be surprised at how many people forget!

Fig. 21 Automatic choke adjustment

REMOVAL & INSTALLATION

■ **No matter how much they look alike, marine carburetors are completely different from automotive carburetors. Never substitute an automotive carburetor for the one on your engine! Venting procedures are not the same and an automotive carburetor could allow dangerous vapors to escape into the bilge. Don't even think about it.**

1. Open the engine hatch or remove the covers and then disconnect the battery cables. Turn the fuel petcock **OFF** and/or shut down the fuel supply at the tank.

2. Remove the flame arrestor after disconnecting the vent hose. It's always a good idea to plug the throttle bores with a clean, lint-free cloth.

3. Disconnect the throttle cable hardware from the throttle bracket and lever anchor studs. Remove the cable and carefully move it out of the way.

4. Remove the retaining clip and disconnect the choke spring rod.

5. Using two open end line nut wrenches, hold the fuel inlet nut at the carburetor securely and loosen the fuel line nut. Disconnect the two and carefully move the line out of the way. Plug both the inlet and line open ends to prevent fuel seepage.

6. Loosen and remove the carburetor mounting nuts/bolts (2 each) and washers, and lift the unit off the manifold. Plug the opening with a clean lint-free cloth. Remove the cable bracket.

To Install:

7. Clean the mating surfaces thoroughly of all residual gasket material, position a new gasket and then install the carburetor. Certain models may use an adaptor or wedge plate, install this first. Tighten the nuts to 10-14 ft. lbs. (13-19 Nm) and the bolts to 6-8 ft. lbs. (8-11 Nm). Hopefully you remembered to remove the rag!

8. Reconnect the fuel line to the inlet line after removing the plugs and tighten it to 18 ft. lbs. (24 Nm). Don't forget to use two wrenches.

9. Connect the choke spring rod and slide in the retaining clip.

10. Install and adjust the throttle cable as detailed later in this section.

11. Install the flame arrestor and reconnect the vent line.

12. Connect the battery cables. Start the engine and ensure there are no fuel leaks; shut down the engine immediately if there are. Check and adjust the idle speed and mixture.

DISASSEMBLY

◆ **See Figures 22 thru 30**

■ **Always be certain that your carburetor rebuild kit is for marine applications.**

1. Remove the carburetor.

2. Place the carburetor on the workbench in the upright position. If servicing an older carburetor, remove the idle vent valve attaching screw, and then remove the idle vent valve assembly. Remove the clip from the upper end of the choke intermediate rod, disconnect the choke rod from the upper choke shaft lever, and then remove the intermediate choke rod from the air horn.

3. Drive back the roll pin at the upper end of the accelerator pump lever just enough to remove the lever, and then disconnect the pump rod and lever.

4. Disconnect and remove the accelerator pump rod and lever from the horn.

5. Hold the air valve wide open and then remove the secondary metering rods by tilting and sliding the rods from the holes in the hanger.

6. Remove the nine air horn-to-bowl attaching screws, 2 of the screws are counter-sunk next to the primary venturi.

7. Remove the retaining clip from the vacuum-break link at the vacuum-break diaphragm. Disconnect the vacuum-break link from the vacuum-break assembly. Gently push apart the retaining ears of the bracket to release the vacuum-break canister.

8. Now, lift straight up on the air horn and remove it, taking care not to bend the accelerator pump and air bleed tubes sticking out from the air horn.

9. Remove the dashpot piston from the air-valve link by rotating the bend through the hole, and then remove the dashpot from the air horn by rotating the bend through the air horn. Further disassembly of the air horn is not necessary. Do not remove the air valves, air valve shaft, and secondary metering rod hangers because they are calibrated. Do not attempt to remove the high-speed air bleeds and accelerating well tubes because they are pressed into position.

10. Remove the accelerator pump piston from the pump well. Release the air horn gasket from the dowels on the secondary side of the bowl, and then pry the gasket from around the power piston and primary metering rods. Remove the pump return spring from the pump well.

11. Pull off the insert block, remove the plastic filler from over the float valve. Use a pair of needle-nosed pliers and pull straight up on the metering rod hanger directly over the power piston and remove the power piston and the primary metering rods. Disconnect the tension spring from the top of each rod and then rotate the rod and remove the metering rods from the power piston. Pull up just a bit on the float assembly hinge pin until the pin can be removed by sliding it toward the pump well.

12. Insert a screwdriver into the slots on the fuel inlet seat and remove the seat and gasket.

13. Remove both primary metering rod jets. Remove the pump discharge check ball retainer and the check ball.

14. Remove the baffle plates from the secondary side of the bowl.

15. Disconnect the vacuum hose from the tube connection on the bowl and from the vacuum break assembly. Remove the retaining screw, and then lift the assembly from the float bowl.

16. Remove the fuel inlet filter retaining nut, gasket, filter, and spring.

17. Remove the throttle body by taking out the two throttle body-to-bowl attaching screws, and then lift off the insulator gasket.

18. Remove the idle mixture screws and springs. Take care not to damage the secondary throttle valves.

CLEANING & INSPECTION

Never use a wire brush or drill to clean jet passages or tubes in the carburetor.

Never allow the carburetor to soak in a cleaner bath for more than two hours. In fact, we recommend using spray cleaner.

Never immerse the float assembly, needle, accelerator pump plunger or fuel filter in cleaner. Wipe them carefully with a clean cloth.

Otherwise clean all allowable parts with carb cleaner and then dry with compressed air if at all possible.

Blow out and through all passages to ensure there is no foreign material clogging them.

Check the float needle and seat, if either is worn or damaged, replace them as a matched set.

Check the float assembly and hinge pin for wear or damage, replace as necessary.

Check the pump plunger, return spring, piston spring, idle mixture needle and all levers and linkages for wear or damage. Replace as necessary.

Check the throttle valve shaft for excessive looseness in the throttle body. Check that the valve and shaft do not bind through their range of operation and that the valve opens and closes fully. Replace the assembly if it fails any of these tests.

Check the choke valve lever and shaft for excessive looseness in the air horn. Check that the valve, lever and shaft do not bind through their range of operation and that the valve opens and closes fully. Replace the air horn assembly if it fails any of these tests.

ASSEMBLY

◆ **See Figures 22, and 31 thru 42**

1. Turn the idle mixture adjusting screws in until they are barely seated, and then back them out two to three turns as a rough adjustment at this time. Never turn the adjusting screws down tight into their seats or they will be damaged.

2. Place a new insulator gasket on the bowl with the holes in the gasket indexed over the two dowels. Install, and then tighten the throttle body-to-bowl screws evenly.

3. Install the fuel inlet filter spring, filter, new gasket, and inlet nut. Tighten the nut.

4. Install the baffle plates in the secondary side of the bowl with the notches facing up.

5. Grasp the choke rod so that the intermediate choke lever is at the bottom. Suspend the assembly into the float bowl so that the hole with the flat side aligns with the hole in the side of the float bowl. Now, insert the choke shaft on the vacuum break through the hole until it engages the flat-sided hole in the choke lever. Install the screw and tighten it securely.

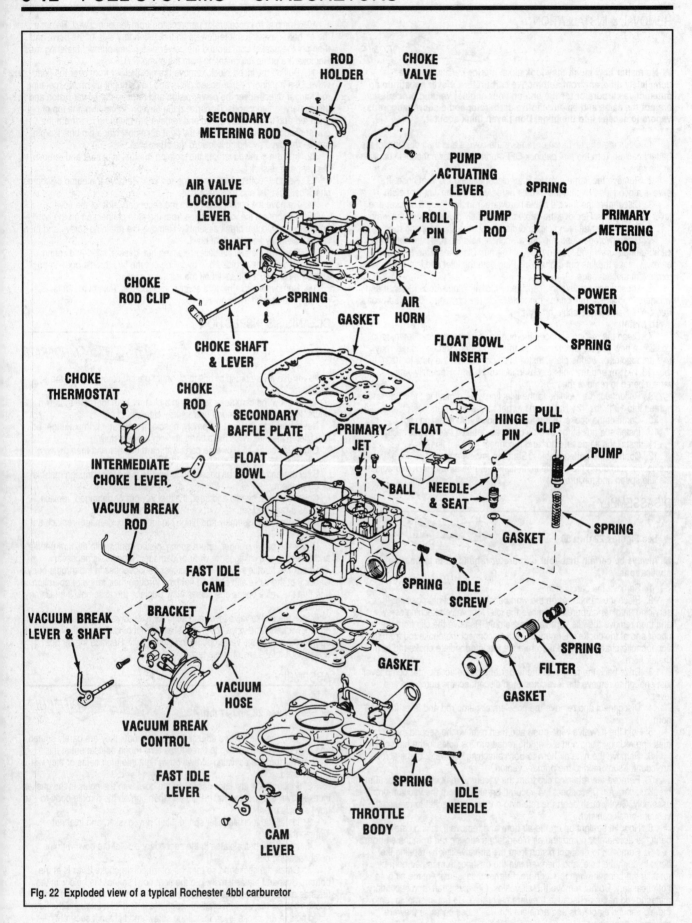

Fig. 22 Exploded view of a typical Rochester 4bbl carburetor

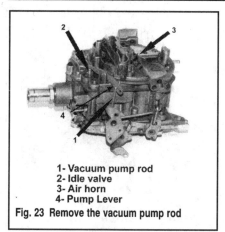

1- Vacuum pump rod
2- Idle valve
3- Air horn
4- Pump Lever

Fig. 23 Remove the vacuum pump rod

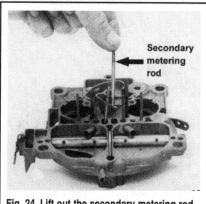

Fig. 24 Lift out the secondary metering rod

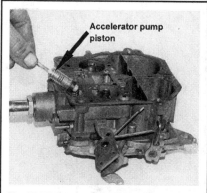

Fig. 25 Lift the accelerator pump piston and spring out of the recess

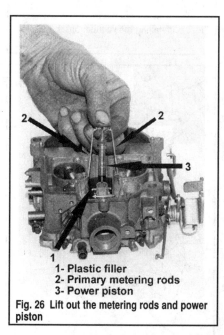

1- Plastic filler
2- Primary metering rods
3- Power piston

Fig. 26 Lift out the metering rods and power piston

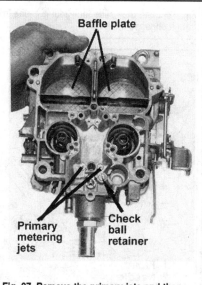

Fig. 27 Remove the primary jets and then remove the check ball retainer and ball

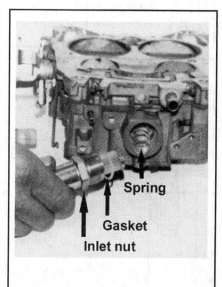

Fig. 28 Remove the fuel inlet nut and associated parts

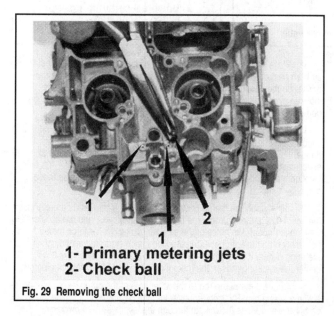

1- Primary metering jets
2- Check ball

Fig. 29 Removing the check ball

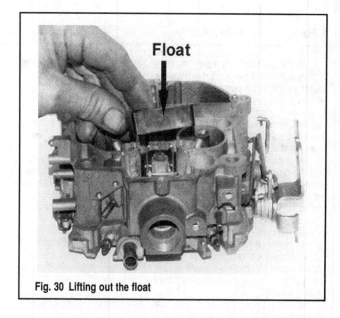

Fig. 30 Lifting out the float

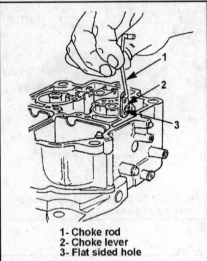

1- Choke rod
2- Choke lever
3- Flat sided hole

Fig. 31 Suspend the choke lever into the float bowl...

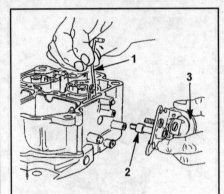

1- Choke rod
2- Choke shaft
3- Vacuum break control assembly

Fig. 32 ...and then slide the vacuum break in until it engages the lever

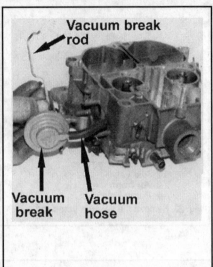

Vacuum break rod

Vacuum break   Vacuum hose

Fig. 33 Installing the vacuum break

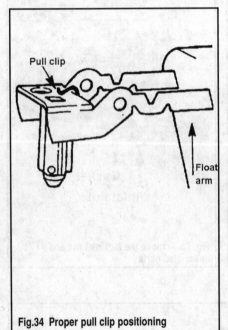

Pull clip

Float arm

Fig.34 Proper pull clip positioning

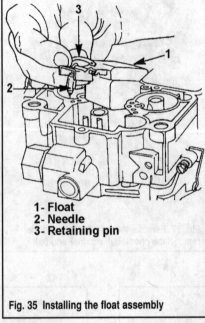

1- Float
2- Needle
3- Retaining pin

Fig. 35 Installing the float assembly

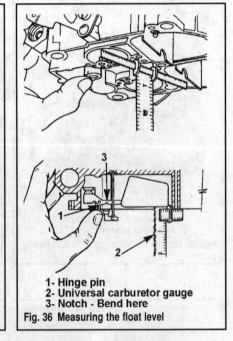

1- Hinge pin
2- Universal carburetor gauge
3- Notch - Bend here

Fig. 36 Measuring the float level

6. Install the primary main metering jets. Install a new float needle seat and gasket. Install the pump discharge check ball and retainer in the passage next to the pump well.

7. Install the pull clip on the needle with the open end toward the front of the bowl. Install the float by sliding the float lever under the pull clip from the front to the back. Hold the float assembly by the toe and with the float lever in the pull clip, install the retaining pin from the pump well side. Take care not to distort the pull clip.

8. Measure the distance from the top of the float bowl gasket surface, with the gasket removed, to the top of the float at the toe end. Check to be sure the retaining pin is held firmly in place and the tang of the float is seated on the float needle when making the measurement. Check that your measurement is 15/64 in. If not, carefully bend the float up or down until the correct measurement is reached.

9. Install the power piston spring in the power piston well. Install the primary metering jets, if they were removed during disassembly. Be sure the tension spring is connected to the top of each metering rod. Install the power-piston assembly in the well with the metering jets; the retainer should be flush with the casing. A sleeve around the piston holds the piston in place during assembly.

10. Install the plastic filler over the float needle. Press it down firmly until it is seated.

11. Place the accelerator pump return spring in the pump well. Install the air horn gasket around the primary metering rods and piston. Install the gasket on the secondary side of the bowl with the holes in the gasket indexed over the two dowels. Install the accelerating pump plunger in the pump well.

12. Install the secondary metering rods. Hold the air valve wide open and check to be sure the rods are positioned with their upper ends through the hanger holes and pointing toward each other. Hanger size may vary from the specifications chart due to inconsistencies in carburetor castings—the hole location may be different so make sure they are the same size as those originally installed.

13. Slowly position the air horn assembly on the bowl and carefully insert the secondary metering rods, the high-speed air bleeds, and the accelerating well tubes through the holes of the air horn gasket. Never force the air horn assembly onto the float bowl. Such action may distort the secondary metering plates. If the air horn assembly moves slightly sideways it will center the metering rods in the metering plates. Install the attaching screw.

14. Connect the pump rod to the inner hole of the pump lever and secure it with a spring clip. Position the lever into the casting mount on the carburetor and press the roll pin through the lever with a screwdriver.

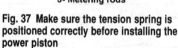

1- Power piston hanger
2- Tension spring
3- Metering rods

Fig. 37 Make sure the tension spring is positioned correctly before installing the power piston

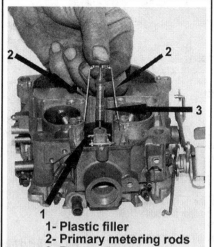

1- Plastic filler
2- Primary metering rods
3- Power piston

Fig. 38 Installing the power piston and metering rods

Filler

Fig. 39 Installing the float filler

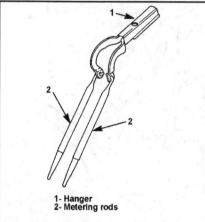

1- Hanger
2- Metering rods

Fig. 40 Positioning the secondary metering rods on the hanger

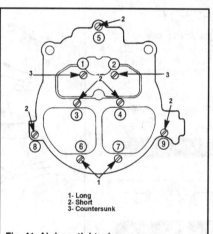

1- Long
2- Short
3- Countersunk

Fig. 41 Air horn tightening sequence

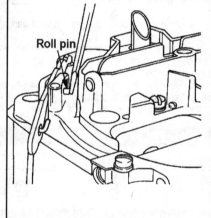

Roll pin

Fig. 42 Use a screwdriver to press the roll pin into place

15. Connect the bottom end of the choke shutter rod into the intermediate choke lever and secure it with a clip. Install the upper end of the rod into the choke blade lever and then press on the retaining clip.

16. Install the actuating rod into the air valve lever and then swivel the other end into the vacuum break arm and press on the retaining clip. Reconnect the vacuum line between the break and the carburetor.

17. Install the carburetor.

ADJUSTMENT

Accelerator Pump Rod

MODERATE

◆ See Figures 43 and 44

1. Back out the idle speed screw until it is no longer in contact with the throttle lever.

2. Move the pump rod to the inner hole on the lever if it's not already there.

3. Ensure that the throttle valves are closed fully and then measure the distance between the top of the carburetor (flame arrestor mounting surface) and the top of the pump plunger stem. Grasp the tip of the pump lever with a pair of needle nose pliers and bend it (carefully) until the distance is equal to that listed in the Specifications chart while supporting the lever.

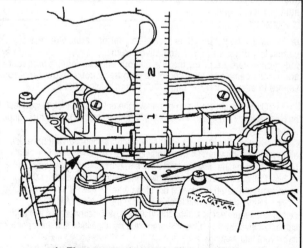

1- Flame arrestor mounting surface
2- Pump plunger stem

Fig. 43 Measure the distance between the top of the carburetor and the top of the plunger

Choke Coil Rod

◆ **See Figures 45 and 46**

1. Remove the flame arrestor.
2. Gently push the choke coil rod down into the housing.
3. The top of the choke coil rod must be even with the bottom of the hole in the vacuum break lever.
4. Bend the rod mid-way down at the kink to adjust it.

Float Level

◆ **See Figures 47 and 47a**

1. Press the float down lightly against the needle and hold the hinge pin in place securely.
2. Insert a float gauge (T-scale) into the hole near the flame arrestor stud. Measure from the top bowl casting to the top of the float at a point 3/16 in. (0.188mm) in from the end of the float at the toe.
3. If not within specifications, bend the float arm with needle nose pliers at the location shown in the illustration.

Idle Speed & Mixture

Please refer to the Maintenance section for adjustment procedures.

Vacuum Break

◆ **See Figures 48 and 49**

4. Press in on the vacuum break control diaphragm until it is fully seated—attaching a small rubber band will sometimes help here.
5. Rotate the vacuum break choke lever counterclockwise until the left tang comes in contact with the vacuum break rod. The choke rod must be at the bottom of the slot in the choke lever.
6. Bend the tang on the choke lever until you can achieve the correct gap in the choke valve when checked with a drill bit between the inner wall of the air horn and the upper side of the valve.

Holley 2300 2 bbl Carburetors

REMOVAL & INSTALLATION

✳✳ WARNING

Always disconnect the battery cables before attempting to work on the fuel system. Never smoke or allow open flame near the engine—this sounds like an obvious precaution, but you'd be surprised at how many people forget!

■ No matter how much they look alike, marine carburetors are completely different from automotive carburetors. Never substitute an automotive carburetor for the one on your engine! Venting procedures are not the same and an automotive carburetor could allow dangerous vapors to escape into the bilge. Don't even think about it.

1. Open the engine hatch or remove the covers and then disconnect the battery cables. Turn the fuel petcock **OFF** and/or shut down the fuel supply at the tank.
2. Remove the flame arrestor after disconnecting the vent hose. It's always a good idea to plug the throttle bores with a clean, lint-free cloth.
3. Unscrew the arrestor stud if the carburetor is to be replaced.
4. Pull out the cotter pin and disconnect the throttle cable at the carburetor and carefully move it out of the way.
5. Using two open-end wrenches, hold the fuel inlet nut at the carburetor securely and loosen the fuel line nut. Disconnect the two and carefully move the line out of the way. Plug both the inlet and line open ends to prevent fuel seepage.
6. Tag and disconnect the purple/white electric choke lead.
7. Tag and disconnect any other lines, leads or hoses that may be in the way of removal.
8. Loosen and remove the 4 carburetor mounting nuts/washers and lift the unit off the manifold. Plug the opening with a clean lint-free cloth.

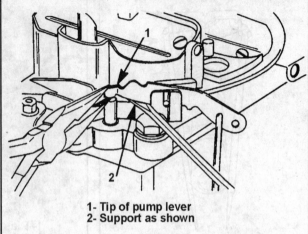

1- Tip of pump lever
2- Support as shown

Fig. 44 Bend the pump lever tip until the measurement is within specifications

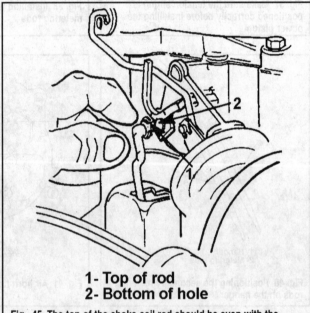

1- Top of rod
2- Bottom of hole

Fig. 45 The top of the choke coil rod should be even with the bottom of the hole in the choke lever

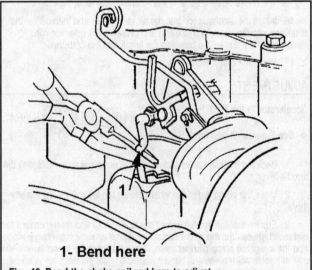

1- Bend here

Fig. 46 Bend the choke coil rod here to adjust

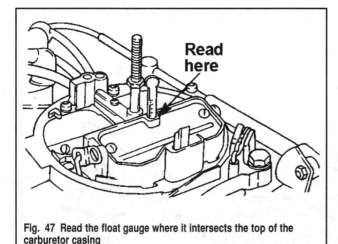

Fig. 47 Read the float gauge where it intersects the top of the carburetor casing

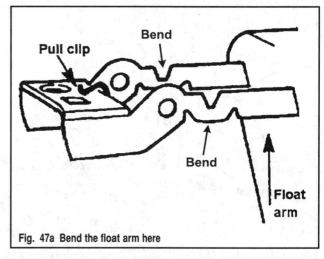

Fig. 47a Bend the float arm here

1- Tang
2- Choke rod
3- Vacuum break rod

Fig. 48 Rotate the lever until the tang comes in contact with the vacuum break rod

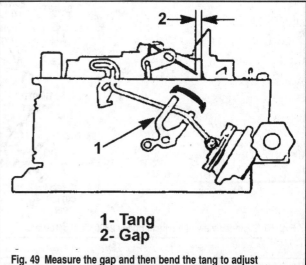

1- Tang
2- Gap

Fig. 49 Measure the gap and then bend the tang to adjust

■ Remember which stud the flat washer and ground wire (from choke) were attached to.

To Install:

9. Clean the mating surfaces thoroughly of all residual gasket material.
10. Position a new gasket and then install the carburetor. Position the flat washer and ground wire on the stud from which they were removed. Tighten the nuts to 10-14 ft. lbs. (13-19 Nm). Hopefully you remembered to remove the rag!
11. Reconnect the fuel line to the inlet after removing the plugs and tighten it to 11-13 ft. lbs. (15-18 Nm). Don't forget to use two wrenches.
12. Connect the choke lead.
13. Install and adjust the throttle cable.
14. Install the flame arrestor and reconnect the vent line.
15. Connect the battery cables. Start the engine and ensure there are no fuel leaks; shut down the engine immediately if there are. Check and adjust the idle speed and mixture.

DISASSEMBLY

◆ See Figures 50 thru 64

MODERATE

■ Disassembly can be made much easier by fabricating a holding fixture out of four 5/16 x 2 in. bolts and eight nuts. Thread a nut about halfway onto each of the bolts, insert each one through the mounting holes on the carburetor and then screw on the remaining nuts. Voila, four little legs to keep your valuable carburetor off of whatever surface you are working on.

✳✳ WARNING

Before even thinking about beginning this procedure, make certain that you have the correct rebuild kit for your make and model!

1. Remove the flame arrestor stud if you haven't already done this during removal.
2. Remove the fuel inlet line filter adapter from the float bowl. Lift out the filter screen and the O-ring—throw away the O-ring.
3. Grab the spring clip retainer at the end of the choke plate lever and pull it off with a pair of needle nose pliers.
4. Remove the 3 Phillips screws from the thermostatic cover on the choke housing and lift off the retaining ring, the cover and the gasket. Make sure that you disengage the cover spring from the tang on the choke lever.
5. Remove the 3 Phillips screws, this time holding the choke housing to the main body. Pull off the housing while making sure that the choke linkage arm completely disengages from the horn linkage lever.
6. Turn over the carburetor and remove the 4 accelerator pump cover retaining screws at the bottom of the fuel bowl. Lift off the cover, its diaphragm and the spring. Make sure the inlet check ball moves freely. With the bowl still upside down and the check ball seated in the retainer, check that the clearance between the ball and retainer is 0.011-0.013 in. (0.279-0.330mm). If out of spec, bend the wire retainer very carefully to bring clearance back into range, Remember that this is not a serviceable item, so don't bend it too far, and don't remove the ball; otherwise you will need to replace the entire bowl.
7. Remove the four 5/16 in. primary fuel bowl retaining bolts and lift off the bowl and metering block assemblies. Discard the gaskets. You will most likely need to tap the bowl/block assembly lightly with a small rubber mallet to break the seal.

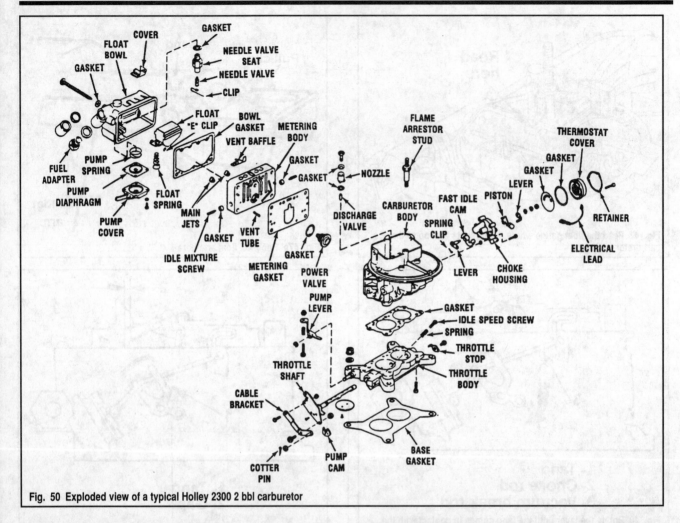

Fig. 50 Exploded view of a typical Holley 2300 2 bbl carburetor

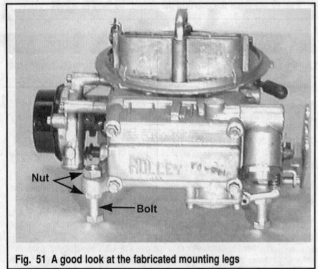

Fig. 51 A good look at the fabricated mounting legs

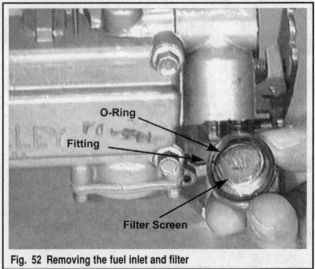

Fig. 52 Removing the fuel inlet and filter

■ Make sure that you take note of the positioning of the wire on the end of the fuel valve and the metal tab on the float itself. This wire must hook **UNDER** the tab on reassembly or the inlet valve will not open during operation.

8. On 3.0L engines, remove the fuel inlet valve assembly. Loosen and remove the lock nut on the outer end of the float pin and then remove the inner hex nut and the two washers. Slide out the pin. Loosen the 2 screws holding the float and then remove it and the bracket.

9. On all other engines, pull off the float retainer (E-clip) with a pair of needle nose pliers and slide the float off of the hinge pin. Remove the spring from the float (or pin) and then lift the baffle out of the bowl. Remove the inlet needle from its seat and then unscrew the seat, discarding the gasket.

10. When removing the float on all engines, the needle valve will fall into the bowl; make sure you remove it. Slide out the small plastic cover and unscrew the needle valve seat, pull out the gasket and discard it.

Fig. 53 Pull out the spring clip...

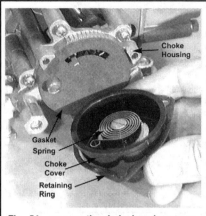
Fig. 54 ...remove the choke housing cover...

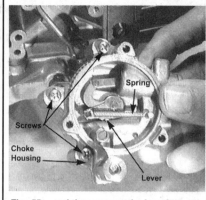
Fig. 55 ...and then remove the housing itself

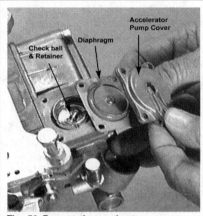
Fig. 56 Remove the accelerator pump cover, diaphragm and spring

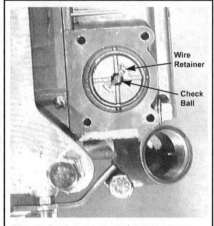
Fig. 57 Do not remove the check valve

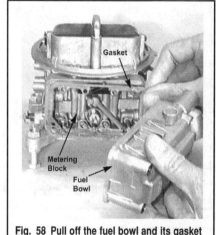
Fig. 58 Pull off the fuel bowl and its gasket

11. Remove the idle adjusting needles and their gaskets from the metering block (there should be one on each side of the block). It's never a bad idea to tighten them up while counting the turns until they make contact BEFORE you remove them as this will give you a decent benchmark for installation.

12. Remove the main jets with a screwdriver and then lift the vent baffle off of its pegs and throw away the two O-rings.

13. Unscrew the power valve with a 1 in. wrench. Discard the gasket.

14. Remove all old gasket material from the metering block and the float bowl mating surfaces.

15. Insert a small Phillips head screwdriver into the top of the carburetor and loosen the accelerator discharge nozzle so that you can pull out the nozzle. Make sure you get both gaskets out.

16. Turn the carburetor over (the discharge needle should fall out while you're doing this) and remove the 6 throttle body retaining screws and washers. Disconnect the vacuum hose from the choke and then lift off the throttle body.

17. Further disassembly should be unnecessary.

CLEANING & INSPECTION

Never use a wire brush or drill to clean jet passages or tubes in the carburetor.

Never allow the carburetor to soak in a cleaner bath for more than two hours. In fact, we recommend using spray cleaner.

Never immerse the throttle body, float assembly, needle, accelerator pump plunger or fuel filter in cleaner. Wipe them carefully with a clean cloth.

Otherwise clean all allowable parts with carb cleaner and then dry with compressed air if at all possible.

Blow out and through all passages to ensure there is no foreign material clogging them.

Check the float needle and seat, if either is worn or damaged, replace them as a matched set.

Check the float assembly and hinge pin for wear or damage, replace as necessary.

Check the pump plunger, return spring, piston spring, idle mixture needle and all levers and linkages for wear or damage. Replace as necessary.

Check the throttle valve shaft for excessive looseness in the throttle body. Check that the valve and shaft do not bind through their range of operation and that the valve opens and closes fully. Replace the assembly if it fails any of these tests.

Check the choke valve lever and shaft for excessive looseness in the air horn. Check that the valve, lever and shaft do not bind through their range of operation and that the valve opens and closes fully. Replace the air horn assembly if it fails any of these tests.

ASSEMBLY

◆ See Figures 50 thru 64 and 65

■ Please refer to the Disassembly section for additional illustrations of the assembly procedure.

1. Position a new gasket over the bottom of the main body so that all six holes line up correctly. Install the throttle body and screws and tighten them to 50 inch lbs. (5.6 Nm).

2. Drop the accelerator pump discharge needle into its bore in the body. Slide a new gasket over the nozzle screw, followed by the nozzle and another new gasket. Now install the nozzle so that the notches on its body align with those in the bore and tighten the screw to 15 inch lbs. (1.7 Nm).

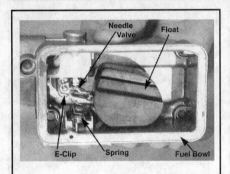

Fig. 59 A good look at the float assembly in the bowl

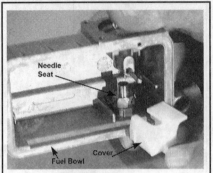

Fig. 60 Slide out the plastic cover

Fig. 61 Removing the metering block from the main body

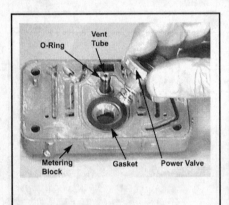

Fig. 62 Unscrewing the power valve

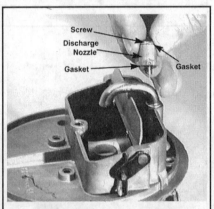

Fig. 63 Remove the accelerator discharge nozzle from inside the bore

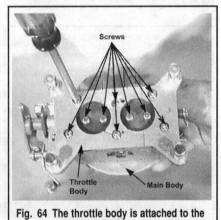

Fig. 64 The throttle body is attached to the bottom of the main body

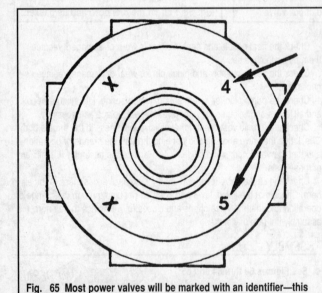

Fig. 65 Most power valves will be marked with an identifier—this one is a 4.5

3. Many rebuild kits will contain more than one power valve—check the number designated in the Carburetor Specifications chart for your engine and make sure you grab a valve with the corresponding number stamped on it. Now look at the fuel ports on the valve. A valve with multiple fuel passages will require a gasket with three internal projections; while a valve with only two passages will require a gasket without any internal projections and a round gasket. Make sure you choose the correct one. Install the valve into the metering block and tighten it to 100 inch lbs. (11.3 Nm).

■ In some cases, but not all, if you're working on a 4.3GS, you've probably already figured out by now that you don't have a power valve.

4. Slide new O-rings over each end of the vent baffle, coat them lightly with engine oil and insert one end into the block. Install the jets.

5. Install a new gasket onto each of the idle mixture needles. Screw them until they just seat on the block and then back them out an equivalent number of turns as noted on removal.

6. Coat the inner mating surface of the metering block with a very thin film of grease and position a new gasket over the dowels. Position the block and gasket onto the carburetor and press it into place.

7. Position a new gasket (metal) onto the needle valve seat and then tighten it into the fuel bowl to 10 inch lbs. (1.1 Nm). Slide the small cover into position. Place a new needle valve into the seat so that the spring clip faces out.

8. Slide the float onto the hinge pin, making sure that the clip rides over the float tab. Position the float spring into the grooves on the bowl and then install the small e-clip retainer onto the hinge pin.

9. Turn over the fuel bowl so the float tab is resting on top of the needle valve. If the float is not level (parallel to the bowl), carefully bend the tab until it is.

10. Position the diaphragm spring into the accelerator pump cavity. Lay a new diaphragm over the spring so the raised disc is facing up and resting on the pump lever. Install the cover over the diaphragm and tighten the screws securely.

11. Coat the outer mating surface of the metering block with a very thin film of grease and position a new gasket over the dowels. Position the bowl onto the block and press it into place. Install the mounting bolts, with new gaskets, until each is just finger-tight. Now tighten them again to 45 inch lbs. (5.1 Nm).

12. Check that the choke lever, spring and piston all move smoothly with no sign of binding, Squeeze a drop of light weight oil into the piston bore and actuate the piston several times. Choke components are not available individually— if you detect problems, replace the entire assembly.

13. Position the choke assembly on the main body so that the choke plate lever and the choke lever align. Tighten the 3 screws securely.

14. Insert a new cover gasket over the housing so the piston is protruding through the gasket. Now install the cover while guiding the lever into the spring end. Make sure that the lever engages the spring or the choke will be useless. Rotate the cover until the index mark is in the same place as noted during removal (or, check the Specifications chart). Slide on the retaining ring and then tighten the screws securely. The ring has raised tabs at each hole to aid in pulling the cover in tight; if after tightening the 3 screws you notice that the cover is still loose, remove the screws and flip the retaining ring over.

15. Install the spring clip over the choke plate lever, securing it to the choke lever.

16. Insert a new (or cleaned) fuel filter screen into the adapter, thread the assembly into the fuel bowl and tighten it to 100 inch lbs. (11.3 Nm).

17. Thread the flame arrestor stud into the top of the unit, tightening it securely.

18. Remove the fabricated legs if you haven't already and then install the carburetor. Check the throttle cable and idle speed adjustments.

ADJUSTMENT

Float Level

3.0L Engines

 MODERATE

■ If attempting to perform this procedure without removing the carburetor, please make certain that you first drain all fuel from the carburetor.

1. Remove the fuel bowl as detailed in the Carburetor Disassembly section.
2. Turn the bowl upside down and loosen the float adjustment screw.
3. Turn the adjustment nut until the float is sitting level in the bowl.
4. Hold the nut and tighten the screw securely.
5. Install the fuel bowl.

All Other Engines

◆ See Figure 66

■ If attempting to perform this procedure without removing the carburetor, please make certain that you first drain all fuel from the carburetor.

1. Remove the fuel bowl as detailed in the Carburetor Disassembly section.
2. Remove the float baffle and then turn the bowl upside down.
3. Carefully bend the curved float arm that comes in contact with the inlet needle until the float is level and parallel with the fuel bowl.
4. Install the fuel bowl.

Accelerator Pump Stroke

 EASY

Accelerator pump stroke is controlled by the position of the cam. The mounting screw must be positioned in the correct hole in the cam and throttle arm as detailed in the Carburetor Specifications chart.

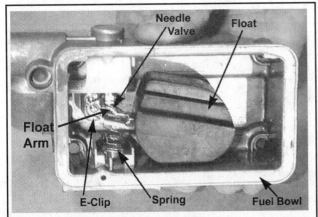

Fig. 66 Bend the arm to adjust the float level

Accelerator Pump Clearance

 EASY

1. Remove the flame arrestor.
2. Move the throttle plates to the wide open position and hold them there.
3. Depress the pump arm and then check the clearance between the arm (still depressed) and the head of the pump lever screw.
4. If clearance is not within the specifications detailed in the chart, rotate the adjusting screw IN to increase the clearance, or OUT to decrease it.
5. Check the pump operation; the pump should begin to move immediately when moving the throttle shaft. Any indication of hesitation on the pump's part will lead to hesitation or lag felt when using the throttle. Check for bent or worn components. Recheck you clearance adjustment and then check the pump stroke.
6. Install the flame arrestor.

Choke Vacuum

◆ See Figures 67 and 68

 MODERATE

1. Remove the flame arrestor.
2. Remove the 3 retaining screws and lift off the choke housing cover and retaining ring.
3. Using a paper clip or very small awl, press the choke piston down against the screw stop.
4. While holding the piston down against the stop, apply light pressure to the choke housing lever (as if you were closing it) until all free play is removed from the linkage.
5. Now insert the correct size drill bit into the choke opening and adjust it with the stop screw—tighten the screw to decrease the opening size and loosen it to increase the size of the opening.

✳✳ CAUTION

Be careful if you are increasing the size of the opening with the screw. Never back the screw out so far as to let the piston slip past it.

6. Reinstall the housing cover as detailed in the Carburetor Assembly procedures. Install the flame arrestor.

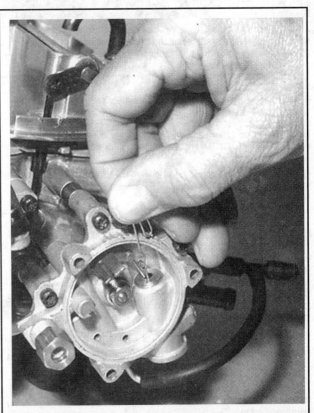

Fig. 67 Push the piston against the screw stop with a paper clip

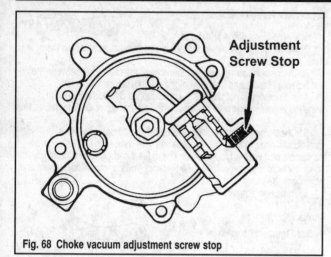

Fig. 68 Choke vacuum adjustment screw stop

Choke Unloader

1. Remove the flame arrestor.
2. With the throttle valves held in the wide open position, carefully insert the correct drill bit (see the Spec chart) between the lower edge of the plate and the inner wall of the air horn.
3. Now press lightly on the choke control lever while removing the bit. You should notice a very light drag on the bit as you are pulling it out. If not, or if you feel heavy pressure, use needle nose pliers to bend the tab on the throttle shaft kickdown lever (behind the choke housing) until you are satisfied with the drag.

Holley 4150/4160/4175 4 bbl Carburetors

REMOVAL & INSTALLATION

※※ WARNING
Always disconnect the battery cables before attempting to work on the fuel system. Never smoke or allow open flame near the engine—this sounds like an obvious precaution, but you'd be surprised at how many people forget!

■ No matter how much they look alike, marine carburetors are completely different from automotive carburetors. Never substitute an automotive carburetor for the one on your engine! Venting procedures are not the same and an automotive carburetor could allow dangerous vapors to escape into the bilge. Don't even think about it.

1. Open the engine hatch or remove the covers and then disconnect the battery cables. Turn the fuel petcock **OFF** and/or shut down the fuel supply at the tank.
2. Remove the flame arrestor after disconnecting the vent hose. It's always a good idea to plug the throttle bores with a clean, lint-free cloth.
3. Unscrew the arrestor stud if the carburetor is to be replaced.
4. Pull out the cotter pin and disconnect the throttle cable at the carburetor and carefully move it out of the way.
5. Using two open-end wrenches, hold the fuel inlet nut at the carburetor securely and loosen the fuel line nut. Disconnect the two and carefully move the line out of the way. Plug both the inlet and line open ends to prevent fuel seepage. On 5.8L Ford engines, cut the plastic tie and remove the vent hose.
6. Tag and disconnect the purple/white electric choke lead.
7. Tag and disconnect any other lines, leads or hoses that may be in the way of removal.
8. Loosen and remove the 4 carburetor mounting nuts/washers and lift the unit off the manifold. Plug the opening with a clean lint-free cloth.

■ Remember which stud the flat washer and ground wire (from choke) were attached to.

9. If replacing the carburetor, take note of the throttle lever bracket position and remove it.
To Install:
10. Clean the mating surfaces thoroughly of all residual gasket material.
11. Attach the throttle lever bracket from the old carburetor making sure its in the same position as on the original one.
12. Position a new gasket and then install the carburetor. Position the flat washer and ground wire on the stud from which they were removed. Tighten the nuts to 10-14 ft. lbs. (13-19 Nm). Hopefully you remembered to remove the rag!
13. Reconnect the fuel line to the inlet line after removing the plugs and tighten it to 11-13 ft. lbs. (15-18 Nm). Don't forget to use two wrenches. Don't forget to reinstall the vent hose on the 5.8L engine.
14. Connect the choke lead.
15. Install and adjust the throttle cable.
16. Install the flame arrestor stud and tighten it to 65-80 inch lbs. (7.3-9 Nm).
17. Install the flame arrestor and reconnect the vent line.
18. Connect the battery cables. Start the engine and ensure there are no fuel leaks; shut down the engine immediately if there are. Check and adjust the idle speed and mixture.

DISASSEMBLY

◆ See Figures 69 thru 77

■ Disassembly can be made much easier by fabricating a holding fixture out of four 5/16 x 2 in. bolts and eight nuts. Thread a nut about halfway onto each of the bolts, insert each one through the mounting holes on the carburetor and then screw on the remaining nuts. Voila, four little legs to keep your valuable carburetor off of whatever surface you are working on.

※※ WARNING
Before even thinking about beginning this procedure, make certain that you have the correct rebuild kit for your make and model!

1. Remove the flame arrestor stud if you haven't already done this during removal.
2. Remove the four primary fuel bowl retaining bolts and lift off the bowl and metering block assemblies. Discard the gaskets and lay any plates aside. You will most likely need to tap the bowl/block assembly lightly with a small rubber mallet to break the seal.
3. Pull the transfer tube out of the inner side of the carburetor mating surface and discard the O-rings.
4. Remove the idle adjusting needles and their gaskets from the metering block (there should be one on each side of the block). It's never a bad idea to tighten them up while counting the turns until they make contact BEFORE you remove them as this will give you a decent benchmark for installation.
5. Remove the main jets with a jet wrench (a screwdriver should work here as well) and then lift the vent baffle off of its pegs and throw away the two O-rings. Make sure that you take note of the jet numbers and which bore they came out of.

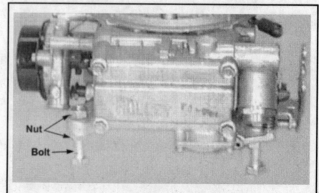

Fig. 69 A good look at the fabricated mounting legs

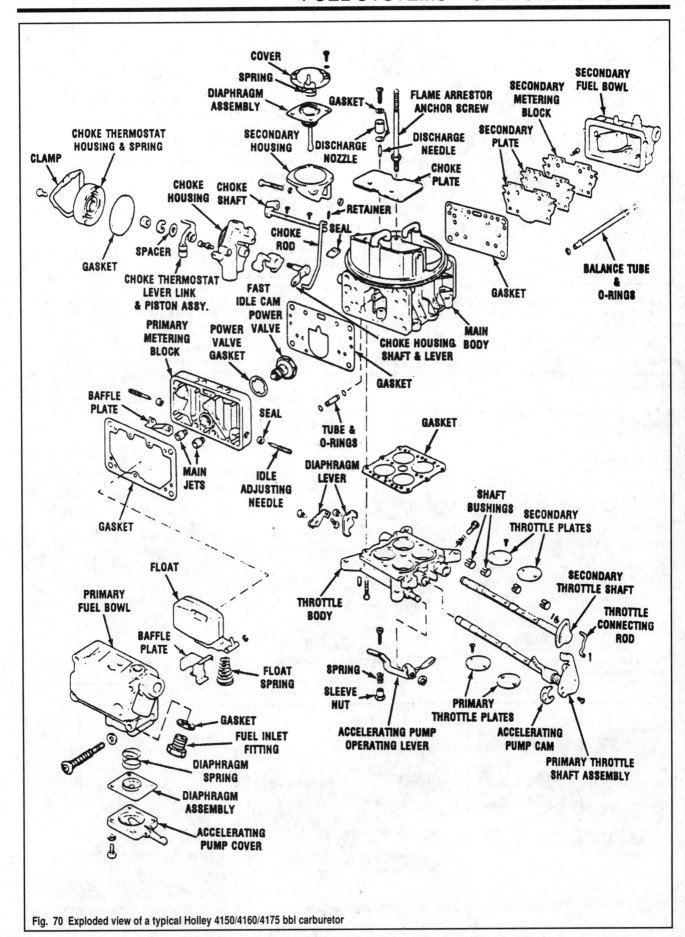

Fig. 70 Exploded view of a typical Holley 4150/4160/4175 bbl carburetor

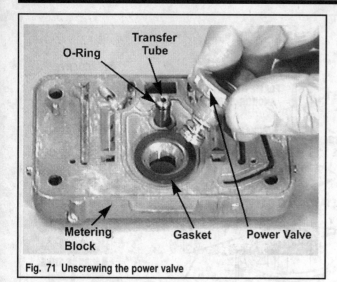

Fig. 71 Unscrewing the power valve

Labels: O-Ring, Transfer Tube, Metering Block, Gasket, Power Valve

6. Unscrew the power valve from the inner side of the metering block with a 1 in. wrench. Discard the gasket. Pull out the accelerator pump tube and discard the O-rings.

7. Working inside the primary bowl, pull off the float retainer (E-clip) with a pair of needle nose pliers and slide the float off of the hinge pin. Remove the spring from the float (or pin) and then lift the baffle (small plastic cover) out of the bowl. Remove the inlet needle from its seat and then unscrew the seat, discarding the gasket.

■ The needle valve will probably fall into the bowl; make sure you remove it.

8. Turn over the bowl and remove the 4 accelerator pump cover retaining screws at the bottom of the fuel bowl. Lift off the cover, its diaphragm and the spring. Make sure the inlet check ball moves freely. With the bowl still upside down and the check ball seated in the retainer, check that the clearance between the ball and retainer is 0.011-0.013 in. (0.279-0.330mm). If out of spec, bend the wire retainer very carefully to bring clearance back into range. Remember that this is not a serviceable item, so don't bend it too far, and don't remove the ball; otherwise you will need to replace the entire bowl.

9. Remove the fuel inlet line filter adapter from the float bowl. Lift out the filter screen and the O-ring—throw away the O-ring or gasket.

10. Remove any and all remaining gasket material from the bowl and block mating surfaces.

11. Remove the secondary fuel bowl and metering block by following the procedures above for the primary. Note that there will be no accelerator pump or power valve to remove here. Note also that if you are working on a V8 engine, you will probably need a 122 clutch-type screwdriver to remove the 4 bowl retaining screws.

12. Pull off the retaining clip at the bottom of the secondary vacuum diaphragm link rod.

13. Turn the carburetor over and remove the throttle body retaining screws and washers.

14. Pull the retaining clip off of the lower end of the choke rod with needle nose pliers. Remove the 3 retaining screws and lift off the choke housing cover and retaining ring. Now you can remove the 3 housing retaining screws and pull off the housing and vacuum passage gasket.

15. Remove the secondary vacuum housing from the main body—you should have disconnected the link rod in an earlier step. Remove the four cover screws and lift off the cover, spring, diaphragm and check ball from the housing.

16. Insert a small Phillips head screwdriver into the top of the carburetor and loosen the accelerator discharge nozzle so that you can pull out the nozzle. Make sure you get both gaskets out. Turn over the carburetor and let the needle fall into your hand.

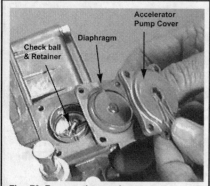

Fig. 72 Remove the accelerator pump cover, diaphragm and spring

Labels: Check ball & Retainer, Diaphragm, Accelerator Pump Cover

Fig. 73 Do not remove the check valve

Labels: Wire Retainer, Check Ball

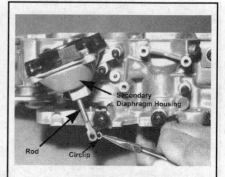

Fig. 74 Remove the secondary diaphragm rod circlip...

Labels: Secondary Diaphragm Housing, Rod, Circlip

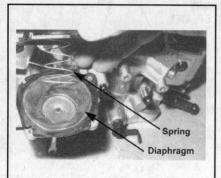

Fig. 75 ...the cover and diaphragm spring...

Labels: Spring, Diaphragm

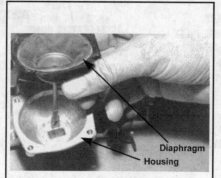

Fig. 76 ...and the diaphragm itself

Labels: Diaphragm Housing

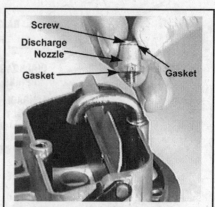

Fig. 77 Remove the accelerator discharge nozzle from inside the bore

Labels: Screw, Discharge Nozzle, Gasket, Gasket

■ Unless there is damage, do not remove the choke plate, shaft, rod or guide. The plate screws are staked and must be filed down to the shaft in order to remove anything.

17. Further disassembly should be unnecessary.

CLEANING & INSPECTION

◆ See Figures 78 and 79  ② *MODERATE*

Never use a wire brush or drill to clean jet passages or tubes in the carburetor.

Never allow the carburetor to soak in a cleaner bath for more than two hours. In fact, we recommend using spray cleaner.

Never immerse the throttle body, float assembly, needle, accelerator pump plunger or fuel filter in cleaner. Wipe them carefully with a clean cloth.

Otherwise clean all allowable parts with carb cleaner and then dry with compressed air if at all possible.

Blow out and through all passages to ensure there is no foreign material clogging them.

Check the float needle and seat, if either is worn or damaged, replace them as a matched set.

Check the float assembly and hinge pin for wear or damage, replace as necessary.

Check the pump plunger, return spring, piston spring, idle mixture needle and all levers and linkages for wear or damage. Replace as necessary.

Check the throttle valve shaft for excessive looseness in the throttle body. Check that the valve and shaft do not bind through their range of operation and that the valve opens and closes fully. Replace the assembly if it fails any of these tests.

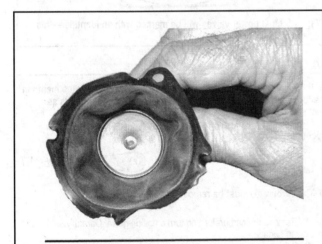

Fig. 79 Check the secondary vacuum diaphragm (top) and the accelerator pump diaphragm for tears, pin holes or other damage

Check the choke valve lever and shaft for excessive looseness in the air horn. Check that the valve, lever and shaft do not bind through their range of operation and that the valve opens and closes fully. Replace the air horn assembly if it fails any of these tests.

Always replace the seat valve assemblies. These components receive the most wear of any part in the carburetor and correct fuel level cannot be maintained if they are worn or damaged.

Check the secondary vacuum diaphragm by depressing the diaphragm stem and covering the vacuum fitting with your finger. Release the stem; if it moves out more than 1/16 in. in ten seconds, there is a leak and it will require replacement.

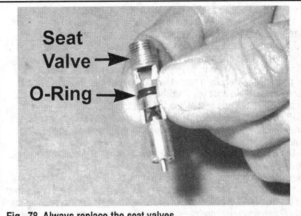

Fig. 78 Always replace the seat valves

ASSEMBLY

◆ See Figures 69 thru 77, 80 and 81  ③ *DIFFICULT*

■ Please refer to the Disassembly section for additional illustrations of the assembly procedure.

1. Drop the accelerator pump discharge needle into its bore in the throttle throat. Position the nozzle (and gasket) over the needle so that the nozzles are down and the slot is facing the air vent rib; install the retaining screw and tighten it securely.

2. Drop the check ball into the hole in the secondary vacuum diaphragm housing and then insert the diaphragm into the housing so that the vacuum hole lines up with the opening for the check ball. It's a good idea to squirt a drop of clean oil under each of the four corners on the diaphragm to keep it from moving when installing the cover. Place the spring onto the housing cover and then install the cover so that all screw and vacuum holes are in alignment. Tighten the cover screws securely while holding the diaphragm rod to keep the diaphragm from collapsing.

3. Position a new gasket over the secondary vacuum passage on the carburetor body and then install the housing onto the carburetor. Tighten the screws securely.

4. Position a new choke housing gasket on the carburetor. Install the housing while guiding the housing shaft lever onto the choke rod. The projection on the end of the choke rod must be under the fast idle cam so that the cam will be raised whenever the choke plate is closed. Tighten the mounting screws and lock washers and then slip the retaining clip onto the choke rod.

5. Install the choke housing cover with a new gasket. The small lever attached to the choke piston MUST index into the loop on the thermostatic coil. Position the retaining ring over the cover and then install the mounting screws finger-tight. Rotate the cover so that the index marks on the cover and housing line up as they were before removal; or check the Specifications chart. Tighten the screws.

6. Turn over the carburetor and install a new throttle body gasket. Position the throttle body while sliding the secondary vacuum diaphragm rod over operating lever. Tighten the mounting screws to 50 inch lbs. (5.6 Nm) and then press the retainer onto the diaphragm rod and operating lever.

7. Position a new gasket onto the fuel inlet nut, install the nut and tighten it securely.

8. Position the spring into the accelerator pump bore on the primary fuel bowl and then install the diaphragm so that the large end of the lever disc will be against the operating lever. Install the cover and tighten the screws securely. Move the operating lever so that it compresses the diaphragm to ensure that it has not been pinched and then tighten the cover screws to 5 inch lbs. (0.6 Nm).

9. On all models but the 5.8L Ford, insert the filter screen into the primary bowl and then install the inlet nut with a new gasket. On the 5.8L engine, install the inlet nut and filter with a new gasket.

10. Thread the fuel inlet needle seat into the bore with a new gasket. Drop in the needle and clip assembly and slide the plastic baffle into position.

11. Position the float into the bosses and then slide in the guide pin. Check the float setting as detailed in the Adjustments section.

12. Many rebuild kits will contain more than one power valve—check the number designated in the Carburetor Specifications chart for your engine and make sure you grab a valve with the corresponding number stamped on it. Now look at the fuel ports on the valve. A valve with multiple fuel passages will require a gasket with three internal projections; while a valve with only two passages will require a gasket without any internal projections and a round gasket. Make sure you choose the correct one. Install the valve into the metering block and tighten it to 100 inch lbs. (11.3 Nm).

13. Coat two new O-rings with clean engine oil and slide them onto the transfer tube. Press the tube into the primary metering block until it is fully seated.

14. Thread the main jets into their original bores. Remember that all jets should be marked with an individual number—refer to the Specifications chart for this number.

15. Install new gaskets on the idle mixture needles and then thread them into the block until they just come into contact with the surface and then back them out an equivalent number of turns to the figure recorded during the disassembly procedure; or check the Specification chart.

16. Coat the inner mating surface of the metering block with a very thin film of grease. Position a new gasket over the dowels. Position the block and gasket onto the carburetor and press it into place.

17. Coat the outer mating surface of the metering block with a very thin film of grease and position a new gasket over the dowels. Position the baffle plate. Make sure that the accelerator passages are not blocked and position the bowl onto the block and press it into place. Install the mounting bolts, with new gaskets, until each is just finger-tight. Now tighten them again to 45 inch lbs. (5.1 Nm).

18. Coat two new transfer/balance tube O-rings with Vaseline and slide them onto the tube(s) so they're against the flange on each end. Press one end into the recess in the bowl being careful not to disturb the O-ring.

19. Assemble the secondary fuel bowl and metering block by following the procedures above for the primary. Note that there will be no accelerator pump or power valve to assemble here so skip those steps.

20. After assembling all components associated with the secondary bowl and metering block, position a new gasket over the dowels on the inner side of the metering block. Now position the metering plate and another inner gasket on the block and attach the assembly to the carburetor. Install the retaining screws and tighten them securely.

21. Position the bowl over the block and then install it so that the fuel/balance tube slides into the recess. Install and tighten the retaining screws. Note also that if you are working on a V8 engine, you will need a #2 clutch-type screwdriver to remove the 4 bowl retaining screws.

22. Install the carburetor.

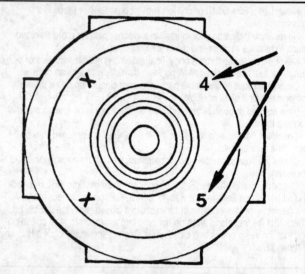

Fig. 81 Most power valves will be marked with an identifier—this one is a 4.5

ADJUSTMENT

■ If the carburetor has been rebuilt, please perform all adjustments in the following order; otherwise perform individual adjustments as necessary.

Secondary Throttle Valve Plate

◆ See Figure 82

② MODERATE

■ The carburetor must be removed for this adjustment.

1. Remove the carburetor and turn it upside down. Carefully set it down on a clean surface.

2. Back out the secondary throttle shaft stop screw until the secondary throttle plates are fully closed. This screw is imbedded in the throttle body, just underneath the secondary vacuum diaphragm rod.

3. Snap the throttle plates closed, turn the stop screw in until it just comes into contact with the secondary lever and then turn the screw in another 1/4 of a turn. The plates should not stick in the bores—if they do, turn in the screw slightly until they stop sticking.

4. Install the carburetor.

Float Level

◆ See Figure 83

② MODERATE

■ Primary and secondary fuel bowls are both adjusted in the same manner. Which one you start with is up to you.

1. Drain the fuel from the carburetor and then remove the fuel bowl and float assembly as detailed in the Disassembly procedures. Although not necessary, we recommend removing the carburetor.

2. Remove the float baffle and then turn the bowl upside down.

3. Carefully bend the curved float arm contacting the float inlet needle until the edge of the float is parallel with the edge of the fuel bowl.

4. Install the bowl with new gaskets. Install the carburetor if removed.

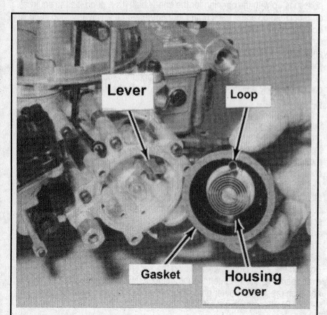

Fig. 80 The small lever MUST engage the loop in the housing cover

Secondary Throttle Stop Screw

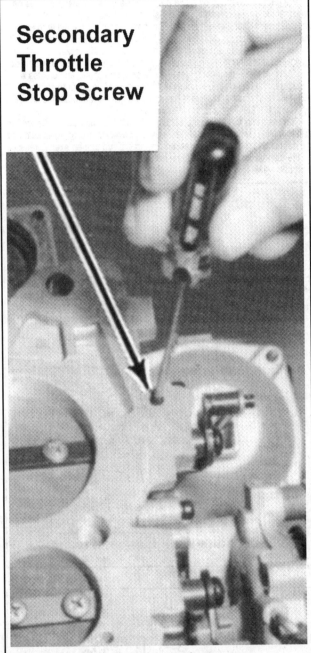

Fig. 82 Use a small screwdriver to turn the stop screw

Accelerator Pump Stroke

Accelerator pump stroke if fully controlled by the position of the pump cam—which is to say, the hole it is in on the throttle arm. Please refer to the Carburetor Specifications chart for the correct hole.

Accelerator Pump Clearance

1. Move the throttle to the wide open position and hold the pump lever down.

2. Insert a 0.0015 in. (0.038mm) flat feeler gauge between the bolt head under the pump lever and the pump arm. Turn the adjuster screw until you can obtain the correct clearance; turn it **in** to increase the clearance, or **out** to decrease the clearance.

3. The pump should begin to move as soon as the throttle shaft moves.

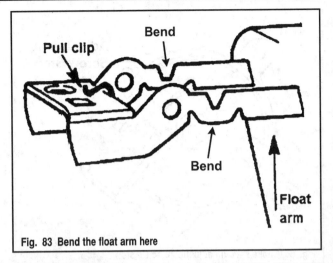

Fig. 83 Bend the float arm here

Choke Vacuum Qualification

◆ See Figures 84 and 85

1. Remove the choke housing and retaining ring.

2. Unbend a paper clip, insert it into the piston bore and push the piston down until it rests against the screw stop. Be careful that you do not scratch the piston or bore.

3. With the piston still against the stop, apply very light pressure to the choke housing lever in the closed position until all free-play has been removed from the linkage.

4. Refer to the Carburetor Specifications chart, and insert the correct size drill bit between the lower edge of the plate and the inner wall of the air horn. Tighten the screw stop to decrease clearance, or loosen the screw to increase clearance. The drill bit must slide through the opening with no drag.

✳✳ CAUTION

Never back the screw out so far as to allow the piston to drop past the screw.

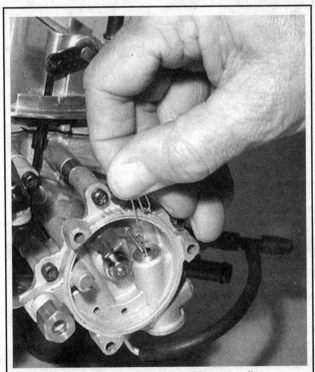

Fig. 84 Carefully force the piston down with a paper clip

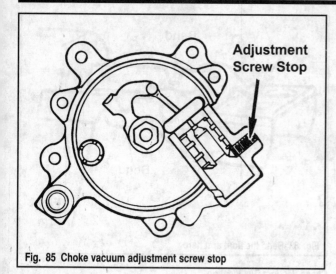

Fig. 85 Choke vacuum adjustment screw stop

Automatic Choke

◆ See Figure 86

The electric automatic choke has been set at the factory to give maximum performance under all operating and weather conditions. Index marks on the cover and housing should be aligned as specified in the Carburetor Specifications chart. A richer or leaner mixture can be obtained by loosening the cover screws and rotating the housing cover. Rotating the housing cover in a counterclockwise direction will richen the mixture, causing the choke to stay on for a longer time. Rotating the housing cover in the clockwise direction will lean out the mixture, causing the choke to kick off sooner.

It is a good idea to always keep alternative settings within two notched of the specified setting.

Fig. 86 Aligning the index marks will set the automatic choke for proper operation

Choke Unloader

Refer to the Carburetor Specifications chart for the proper size drill bit to use. Hold the throttle in the wide open position and insert the appropriate drill bit between the upper edge of the plate and the inner wall of the air horn. This will actually give the proper clearance at the lower edge of the plate.

Place you finger on the choke control lever and check that you can feel a very slight drag as the drill bit is moved through the opening. Carefully bend the tab on the kick down lever (under the housing) with needle nose pliers until the correct clearance is achieved and the drill bit passes through the opening with a slight drag.

Mechanical Fuel Pump

DESCRIPTION

◆ See Figures 87, 88, 89, 90 and 91

A mechanical-type fuel pump driven off the camshaft is utilized on all 2.3L, 2.5L and 3.0L engines, 1984-91 4.3L engines, 1986-92 5.0L/5.7L GM engines and all carbureted V8 Ford engines.

The fuel pump sucks gasoline from the fuel tank and delivers it to the carburetor in sufficient quantities, under pressure, to satisfy engine demands under all operating conditions.

The pump is operated by a two-part rocker arm. The outer part rides on an eccentric on the camshaft and is held in constant contact with the camshaft by a strong return spring. The inner part is connected to the fuel pump diaphragm by a short connecting rod. As the camshaft rotates, the rocker arm moves up and down. As the outer part of the rocker arm moves downward, the inner part moves upward, pulling the fuel diaphragm upward. This upward movement compresses the diaphragm spring and creates a vacuum in the fuel chamber below the diaphragm. The vacuum causes the outlet valve to close and permits fuel from the gas tank to enter the chamber by way of the fuel filter and the inlet valve.

Now, as the eccentric on the camshaft allows the outer part of the rocker arm to move upward, the inner part moves downward, releasing its hold on the connecting rod. The compressed diaphragm spring then exerts pressure on the diaphragm, which closes the inlet valve and forces fuel out through the outlet valve to the carburetor.

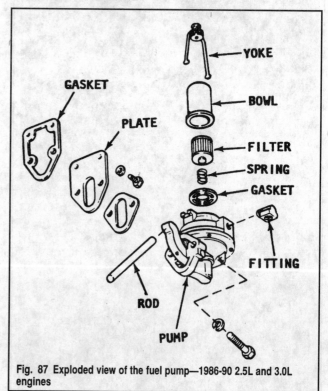

Fig. 87 Exploded view of the fuel pump—1986-90 2.5L and 3.0L engines

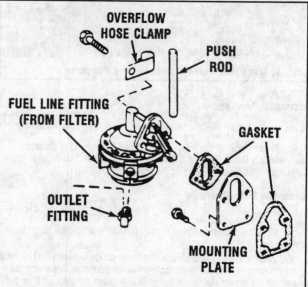

Fig. 88 Exploded view of the fuel pump—1991-92 3.0L engines, 2.3L engines and all GM V6/V8 engines

Fig. 89 A good shot of the pump on a 1993-94 3.0L engine

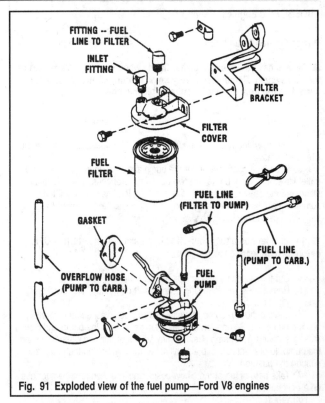

Fig. 91 Exploded view of the fuel pump—Ford V8 engines

2. Position a container under the pump and remove the fuel inlet and outlet lines. It is important to use two open-end wrenches; one to hold the fitting nut and the other to loosen the line nut. Plug the line with a golf tee or something similar to prevent additional fuel spillage. Take care not to spill fuel on a hot engine, because such fuel may ignite.

■ If your boat has a fuel shut-off valve, make sure it is turned OFF before disconnecting the fuel lines.

3. Disconnect the sight tube on engines so equipped.
4. Cut the plastic tie and pull off the vent/overflow line.
5. On the 3.0L engine, you may find that removing the starboard engine mount will be beneficial on many engines—you can be the judge by observing the working space.
6. Loosen the two pump mounting bolts/lock washers and carefully pull the pump out. Scrape any old gasket material from the pump and block mating surfaces. The pump pushrod may fall out of position when removing the pump; to prevent this, slip a small screwdriver in behind the pump and support the pushrod.
7. Take note of the pump elbow positioning and then remove it for use on a new pump.

To Install:

8. Check the pump pushrod for damage or wear.
9. Coat the threads of the elbow with Teflon® Pipe sealant—DO NOT use Teflon® tape. Install the elbow to the pump and tighten it to 48-60 inch lbs. (5.4-6.7 Nm).
10. Coat BOTH sides of the new gasket with Gasket Sealing compound and position it on block. Swab the end of the pushrod with grease and insert the pump into the cylinder block so that the rod is riding on the camshaft eccentric. Tighten the mounting bolts to 12-14 ft. lbs. (16-19 Nm) on 2.5L and 1986-90 3.0L engines; 20-25 ft. lbs. (27-34 Nm) on 2.3L, 4.3L, 5.0L and 5.7L engines and 1991-98 3.0L engines.
11. Connect the vent line and tie it into position.
12. Coat the threads of the fuel line pump fittings with Pipe Sealant, thread them into the pump base and then tighten to 15-18 ft. lbs. (20-24 Nm) while holding the inlet fittings with another wrench. Do not over-tighten and do not use Teflon tape.
13. Install the sight tube if removed.
14. Install the engine mount if removed. Don't forget to check the engine alignment.
15. Remove the container and connect the battery cables. Start the engine and check for fuel leaks.

Fig. 90 Exploded view of the fuel pump—1995-98 3.0L engines

Because the fuel pump diaphragm(s) is(are) moved downward only by the diaphragm spring, the pump delivers fuel to the carburetor only when the pressure in the outlet line is less than the pressure exerted by the diaphragm spring. This lower pressure condition exists when the carburetor float needle valve is unseated and the fuel passages from the pump into the carburetor float chamber are open.

When the needle valve is closed and held in place by the pressure of the fuel on the float, the pump builds up pressure in the fuel chamber until it overcomes the pressure of the diaphragm spring. This pressure almost stops movement of the diaphragm until more fuel is needed in the carburetor float bowl.

REMOVAL & INSTALLATION

All Except Ford V8 Engines

MODERATE

◆ See Figures 87 thru 90

■ Always have a fire extinguisher handy when working on any part of the fuel system. Remember, a very small amount of fuel vapor in the bilge, has the tremendous potential explosive power.

1. Open or remove the engine hatch/covers. Disconnect the battery cables.

5.0L/5.8L Ford Engines

◆ **See Figures 91 and 92**

■ **Always have a fire extinguisher handy when working on any part of the fuel system. Remember, a very small amount of fuel vapor in the bilge, has the tremendous potential explosive power.**

1. Open or remove the engine hatch/covers. Disconnect the battery cables.
2. Unscrew and remove the fuel inlet line at the fuel filter. Unscrew the filter and remove it.
3. Position a container under the pump and remove the fuel inlet and outlet lines. It is important to use two open-end wrenches; one to hold the fitting and the other to loosen the line nut. Plug the line with a golf tee or something similar to prevent additional fuel spillage. Take care not to spill fuel on a hot engine, because such fuel may ignite.

■ **If your boat has a fuel shut-off valve, make sure it is turned OFF before disconnecting the fuel lines.**

4. Cut the plastic tie and pull off the vent line.
5. Remove the two mounting bolts and lift off the fuel filter bracket, securing it out of the way.
6. Loosen the two pump mounting screws/lock washers and carefully pull the pump out. Scrape any old gasket material from the pump and block mating surfaces. The pump pushrod may fall out of position when removing the pump; to prevent this, slip a small screwdriver in behind the pump and support the pushrod.
7. Take note of the pump elbow positioning and then remove it for use on a new pump.

To Install:
8. Check the pump pushrod for damage or wear.
9. Coat the threads of the elbow with Teflon® Pipe sealant—DO NOT use Teflon® tape. Install the elbow to the pump and tighten it to 48-60 inch lbs. (5.4-6.7 Nm).
10. Coat BOTH sides of the new gasket with Gasket Sealing compound and position it on block. Swab the end of the pushrod with grease and insert the pump into the cylinder block so that the rod is riding on the camshaft eccentric. Tighten the mounting bolts to 18-20 ft. lbs. (24-27 Nm).
11. Install the filter bracket and tighten the bolts to 20-25 ft. lbs. (27-34 Nm).

Fig. 92 A good shot of the fuel pump

12. Connect the vent line and tie it into position.
13. Coat the threads of the fuel line pump fittings with Pipe Sealant, thread them into the pump base and then tighten to 15-18 ft. lbs. (20-24 Nm) while holding the inlet fittings with another wrench. Do not over-tighten and do not use Teflon tape.
14. Install the filter and then reconnect the inlet line.
15. Remove the container and connect the battery cables. Start the engine and check for fuel leaks.

7.5L Ford Engines

◆ **See Figures 91 and 93**

■ **Always have a fire extinguisher handy when working on any part of the fuel system. Remember, a very small amount of fuel vapor in the bilge, has the tremendous potential explosive power.**

1. Open or remove the engine hatch/covers. Disconnect the battery cables.
2. Unscrew and remove the fuel inlet and outlet lines at the fuel filter. On early models, you will need to drain the engine oil and then remove the oil filter to gain access here.
3. Unscrew and remove the filter fuel line at the fuel pump and remove it.
4. Remove the bolt, lock washer and plaster nut securing the dipstick clamp to the filter bracket. Remove the 2 mounting bolts and then lift off the bracket and filter. Carefully set it down upright as there will be fuel in the filter.
5. Position a container under the pump and remove the remaining fuel outlet line. It is important to use two open-end wrenches; one to hold the fitting and the other to loosen the line nut. Plug the line with a golf tee or something similar to prevent additional fuel spillage. Take care not to spill fuel on a hot engine, because such fuel may ignite.

■ **If your boat has a fuel shut-off valve, make sure it is turned OFF before disconnecting the fuel lines.**

6. Cut the plastic tie and pull off the vent line.
7. Remove the two mounting bolts and lift off the fuel filter bracket, securing it out of the way.
8. Loosen the two pump mounting bolts/lock washers and carefully pull the pump out. Scrape any old gasket material from the pump and block mating surfaces. The pump pushrod may fall out of position when removing the pump; to prevent this, slip a small screwdriver in behind the pump and support the pushrod.
9. Take note of the pump elbow positioning and then remove it for use on a new pump.

To Install:
10. Check the pump pushrod for damage or wear.
11. Coat the threads of the elbow with Teflon® Pipe sealant—DO NOT use Teflon® tape. Install the elbow to the pump and tighten it to 48-60 inch lbs. (5.4-6.7 Nm).
12. Coat BOTH sides of the new gasket with Gasket Sealing compound and position it on block. Swab the end of the pushrod with grease and insert the pump into the cylinder block so that the rod is riding on the camshaft eccentric. Tighten the mounting bolts to 20-25 ft. lbs. (27-34 Nm).
13. Connect the vent line and tie it into position.
14. Thread the pump outlet line in and tighten the line nut securely.
15. Install the filter and bracket and tighten the bolts finger-tight. Make sure that the long bolt goes in the upper hole.
16. Install the dipstick tube clamp and tighten it securely. Now go back and tighten the bracket mounting bolts to 20-25 ft. lbs. (27-34 Nm).
17. Coat the threads of the remaining fuel line fittings with Pipe Sealant, thread them into the pump base and then tighten to 15-18 ft. lbs. (20-24 Nm) while holding the fittings with another wrench. Do not over-tighten and do not use Teflon tape.
18. Install the oil filter if removed.
19. Remove the container and connect the battery cables. Start the engine and check for fuel leaks.

Fig. 93 A good shot of the fuel pump

PRESSURE TEST

■ **Always have a fire extinguisher handy when working on any part of the fuel system. Remember, a very small amount of fuel vapor in the bilge, has the tremendous potential explosive power.**

1. Open or remove the engine hatch/covers. Disconnect the battery cables.
2. Confirm that there is fuel in the tank. Also, make sure that the fuel filter and the carburetor inlet filter are clean and free of obstructions; we recommend replacing the filters before performing this test.
3. Position a container under the fuel pump and remove the outlet line (the line leading to the carburetor). It is important to use two open-end wrenches; one to hold the fitting elbow and the other to loosen the line nut. Plug the line with a golf tee or something similar to prevent additional fuel spillage.
4. Thread a 'tee' or a fuel pressure gauge connector into the fitting elbow on the pump and then reconnect the fuel line to the other end.
5. Connect a fuel pressure test gauge to the remaining fitting on the connector. Always follow the gauge manufacturer's instructions on set-up when installing the gauge.
6. Connect the battery cables, start the engine and let it idle. The fuel pressure should be 5 3/4-7 psi (39.6-48.3 kPa), except on the 7.5L Ford engine where it should be 5-6 psi (34.5-41.4 kPa).
7. Slowly run the engine up to 1,800 rpm and check that the pressure remains the same.
8. If pressure varies, or is off specification at all, replace the pump. All fuel pumps covered here are of the sealed variety; no adjustments or rebuilds are possible.
9. Turn off the engine, remove the pressure gauge and fuel line and then remove the connector. Reinstall the fuel line to the fitting and tighten to 18 ft. lbs. (24 Nm).

VACUUM TEST

■ **If experiencing slow starts or long cranking tests, perform this test after confirming that fuel pump pressure is acceptable.**

■ **Always have a fire extinguisher handy when working on any part of the fuel system. Remember, a very small amount of fuel vapor in the bilge, has the tremendous potential explosive power.**

1. Open or remove the engine hatch/covers. Disconnect the battery cables.
2. Confirm that there is fuel in the tank. Also, make sure that the fuel filter and the carburetor inlet filter are clean and free of obstructions.
3. Position a container under the fuel filter and remove the inlet line. It is important to use two open-end wrenches; one to hold the fitting elbow and the other to loosen the line nut. Plug the line with a golf tee or something similar to prevent additional fuel spillage.
4. Tag and disconnect all electrical leads at the (+) terminal of the ignition coil; tape the ends and position them out of the way. Reconnect the battery cables.
5. Connect a vacuum gauge to the remaining fitting on the connector. Always follow the gauge manufacturer's instructions on set-up when installing the gauge.
6. Crank the engine and note the reading on the gauge. The vacuum should be 9-10 in. Hg. (30-34 kPa).
7. Replace the fuel pump if the vacuum reading is outside the specified range. All fuel pumps covered here are of the sealed variety; no adjustments or rebuilds are possible.
8. Turn off the engine, remove the gauge and fuel line and then remove the connector. Reinstall the fuel line to the fitting and tighten to 18 ft. lbs. (24 Nm). Reconnect the ignition coil leads.

FUEL LINE TEST

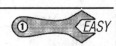

The fuel line, from the tank to the fuel pump, can be quickly tested by disconnecting the existing fuel line at the fuel pump and connecting a six-gallon portable tank and fuel line. This simple substitution eliminates the fuel tank and fuel lines in the boat. Now, start the engine and check the performance.

If the problem has been corrected, the fuel system between the fuel pump inlet and the fuel tank is at fault. This area includes the fuel line, the fuel pickup in the tank, the fuel filter, anti-siphon valve, the fuel tank vent, and excessive foreign matter in the fuel tank, and loose fuel fittings sucking air into the system. Improper size fuel fittings can also restrict fuel flow.

Possible cause of fuel line problems may be deterioration of the inside lining of the fuel line which may cause some of the lining to develop a blockage similar to the action of a check valve. Therefore, if the fuel line appears the least bit questionable, replace the entire line.

Another possible restriction in the fuel line may be caused by some heavy object lying on the line—a tackle box, etc.

Electric Fuel Pump

DESCRIPTION

◆ **See Figure 94**

All 1992-98 4.3L engines, 1993-98 5.0L/5.7L GM engines and all 7.4L/8.2L engines utilize an electric fuel pump with their carbureted fuel systems.

The electric fuel pump system includes the fuel tank/s, a water separating fuel filter, the electric fuel pump and a carburetor.

All pumps covered herein utilize a 12V motor running continuously at 4000 rpm during cranking or operation. The motor drives a conventional gear-type pump using incoming fuel as coolant. Fuel is supplied at varying pressures determined by both engine load and fuel consumption. A fuel pressure regulator eliminates the need for a return line.

Whenever the ignition switch is in the **START** position, voltage is supplied to the fuel pump relay via the **85** or **S** terminal on the starter relay. Once the relay is activated, voltage is applied to pump motor and then on to ground. This circuit is protected by a circuit breaker (12.5 amp on some models/years; 6 amp on others). Once the ignition switch is in the **RUN** position, voltage is supplied via the alternator "light" terminal.

Electric fuel pumps are not repairable, adjustable or rebuildable.

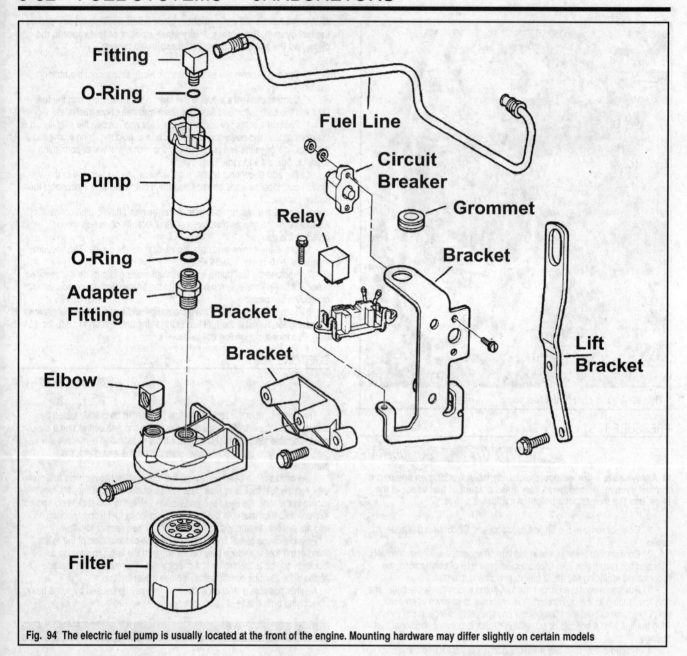

Fitting

O-Ring

Fuel Line

Pump

Circuit
Breaker

Relay

Grommet

O-Ring

Bracket

Adapter
Fitting

Bracket

Bracket

Elbow

Lift
Bracket

Filter

Fig. 94 The electric fuel pump is usually located at the front of the engine. Mounting hardware may differ slightly on certain models

REMOVAL & INSTALLATION

◆ See Figure 94

② MODERATE

■ Always have a fire extinguisher handy when working on any part of the fuel system. Remember, a very small amount of fuel vapor in the bilge, has the tremendous potential explosive power.

✱✱ CAUTION

All OMC fuel pumps are designed to meet U.S. Coast Guard fuel system regulations. Never substitute and automotive style pump.

1. Open or remove the engine hatch/covers. Disconnect the battery cables.
2. Position a container under the pump/filter assembly and remove the fuel inlet line at the filter bracket and the outlet line at the top of the pump. It is important to use two open-end wrenches; one to hold the fuel fitting nut or elbow and the other to loosen the line nut. Plug the line with a golf tee or something similar to prevent additional fuel spillage. Take care not to spill fuel on a hot engine, because such fuel may ignite.

3. Cut the plastic tie holding the electrical leads to the body of the pump. Unplug the electrical harness at the pump suppressor connector and move it out of the way.

4. Hold the pump end cap (just under the lower edge of the upper retaining bracket) with a wrench while loosening the lock nut (on top of bracket, under elbow). Unscrew the upper elbow.

✱✱ CAUTION

Not holding the end cap while loosening the lock nut will damage the internal O-ring, causing fuel leakage during operation.

5. Remove the 2 fuel filter bracket mounting bolts and lift off the pump/bracket/filter as an assembly, sliding the pump out of the upper grommet. Be careful of spilling fuel.

6. Lay the assembly down, or carefully mount the bracket in a vise. Hold the lower pump end cap with a wrench and unscrew the filter bracket-to-pump adapter.

✳✳ CAUTION

Not holding the end cap while loosening the lock nut will damage the internal O-ring, causing fuel leakage during operation.

To Install:

■ **Electric fuel pumps all look familiar, but will usually have different flow rates—always make sure that you have the correct pump for your engine.**

7. Slide a new O-ring over the threads of the adapter until it rests on the upper side of the hex nut.
8. Screw the adapter/bracket assembly into the pump while holding the lower end cap. Tighten the adapter to 8-10 ft. lbs. (11-14 Nm).
9. Slide the assembly up through the grommet in the bracket and position it so the mounting holes line up on the bracket. Tighten the filter bracket bolts to 20-25 ft. lbs. (27-34 Nm).
10. Thread the upper lock nut onto the elbow until it reaches the underside of the fitting. Slip on the washer and then a new O-ring. Thread the elbow into the top of the pump until it stops and then back it out just enough to align with the fuel line. Connect the line and tighten it securely.
11. Hold the upper end cap again and tighten the lock nut to 60-84 inch lbs. (6.8-9.5 Nm).
12. Reconnect the suppressor lead—push it in until the tabs click.

■ **If you notice any of the suppressor holes open, be sure to coat them with Liquid Neoprene to prevent water from entering.**

13. Install a new plastic tie around the pump body and electrical leads.
14. Reconnect the inlet line and tighten it securely.
15. Install the pump guard if removed.
16. Remove the container and connect the battery cables. Start the engine and check for fuel leaks.

PRESSURE TEST

◆ **See Figure 95**

■ **This test MUST be performed with the boat in the water and the engine running. It is imperative that an assistant be with you in the boat also to check the pressure readings.**

1. Open or remove the engine hatch/covers. Disconnect the battery cables.
2. Position a container under the fuel line connection at the carburetor (or the outlet side of the fuel pump) and remove the fuel line. It is important to use two open-end wrenches; one to hold the fitting nut/elbow and the other to loosen the line nut. Plug the line with a golf tee or something similar to prevent additional fuel spillage.

Engine Speed (rpm)	Fuel Pressure (psi)
Idle	4.9-8.5
1000	4.8-8.4
1500	4.7-8.3
2000	4.6-8.2
2500	4.5-8.1
3000	4.4-8.0
3500	4.3-7.9
4000	4.0-7.7
4500	--
4600	3.5-7.3

Fig. 95 Electric fuel pump pressure test specifications

3. Thread a 'tee' or a fuel pressure connector into the fitting nut/elbow on the carburetor or pump and then reconnect the fuel line to the other end.
4. Connect a fuel pressure test gauge to the remaining fitting on the connector. Always follow the gauge manufacturer's instruction when setting up the pressure gauge.
5. Connect the battery cables, start the engine and let it idle. The fuel pressure should be 4.9-8.5 psi.
6. Slowly run the engine up to 4600 rpm, checking the fuel pressure readings at each of the listed engine speed readings.

■ **New engines will almost always have lower operating pressures than engines that have been broken in,**

7. If pressure varies from specifications, replace the pump.
8. Turn off the engine, remove the pressure gauge and fuel line and then remove the connector. Reinstall the fuel line to the fitting and tighten to 18 ft. lbs. (24 Nm).

FUEL SYSTEM CHECKS

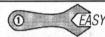

Check that the fuel filter is debris free and seals tightly against the mounting bracket and threads.
Check that the carburetor inlet filter screen is clean and free of debris.
Check for kinked, bent, crushed or damaged fuel lines.
Check for restricted anti-siphon valve; or incorrect valve.
Check for cracked, obstructed or damaged fuel pick-up screen in fuel tank.
Check for blocked or broken fuel tank vent.

COMPONENT REPLACEMENT

Circuit Breaker

1. Open or remove the engine hatch/covers. Disconnect the battery cables.
2. Locate the circuit breaker on the pump bracket. Tag and disconnect the electrical leads, remove the 2 mounting screws and lift off the breaker.
3. Install a new breaker and tighten the mounting bolts to 20-25 inch lbs. (2.3-2.8 Nm).
4. Attach the Red lead to one stud and the Red/Purple lead to the other. Tighten the nuts to 20-25 inch lbs. (2.3-2.8 Nm).
5. Coat both terminals with Liquid Neoprene. Reconnect the battery cables.

Relay And Bracket

1. Open or remove the engine hatch/covers. Disconnect the battery cables.
2. Cut the plastic tie, remove the wire retainer and then disconnect the Amphenol plug near the solenoid.
3. Tag and disconnect the two leads at the circuit breaker.
4. Tag and disconnect the Green electrical lead at the **Light** terminal on the rear of the alternator.
5. Cut the plastic tie holding the harness to the pump and then disconnect the connector at the pump suppressor.
6. Remove the two mounting screws and lift off the relay bracket. Unplug the relay.

To Install:

7. Pop the relay into the mounting bracket, position the bracket and tighten the screws to 20-25 inch lbs. (2.3-2.8 Nm).
8. Align the connector tabs and press the pump suppressor connector halves together until they click. Reattach the harness to the pump body with the plastic tie.
9. Reroute the green lead through the clamps on the front of the engine and connect it to the alternator.
10. Reattach the two circuit breaker leads and coat the terminals with Liquid Neoprene.
11. Reconnect the two halves of the Amphenol connector and press it into the wire retainer. Secure the connector to the retainer with a plastic tie.

Deceleration Valve

In order to prevent the engine from stalling during high vacuum conditions, a deceleration (gulp) valve is used on all carbureted Ford engines. Air is allowed to flow into the intake manifold, entering the air/fuel mixture so as to lean out the rich condition frequently created by high vacuum when the throttle valves are closed on rapid deceleration.

Sudden vacuum acts to draw down the diaphragm in the valve, opening the valve and allowing air to flow through the screen and into the manifold.

FUNCTION CHECK

1. Start the engine and allow it to reach normal operating temperature at idle.
2. Remove the small vacuum signal line from the vacuum source on the manifold.

3. Reconnect the vacuum line and listen carefully for air flowing through the screen and into the manifold. Confirm also that there is a noticeable drop in engine speed (rpm) when reconnecting the vacuum line.
4. If the air flow through the screen does not persist for at least a second (1), or, if the idle speed does not drop, check all hoses and lines connected to the valve for damage or blockage.
5. If the lines are OK, replace the valve.

REMOVAL & INSTALLATION

1. Open or remove the engine hatch/covers. Disconnect the battery cables.
2. Cut the plastic ties securing the vacuum lines and screen. Remove the 2 mounting screws and lift off the valve.
3. Install a new valve and tighten the screws securely.
4. Attach the two vacuum lines and the screen and secure them with plastic ties.

WIRING DIAGRAMS

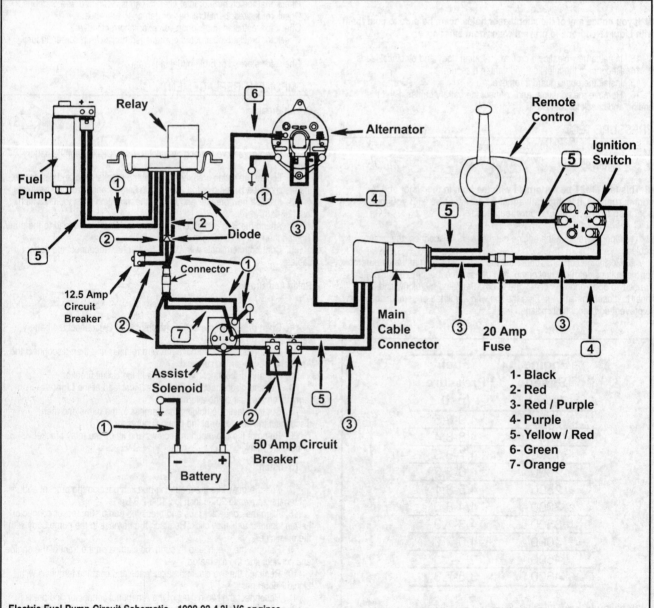

Electric Fuel Pump Circuit Schematic—1992-93 4.3L V6 engines

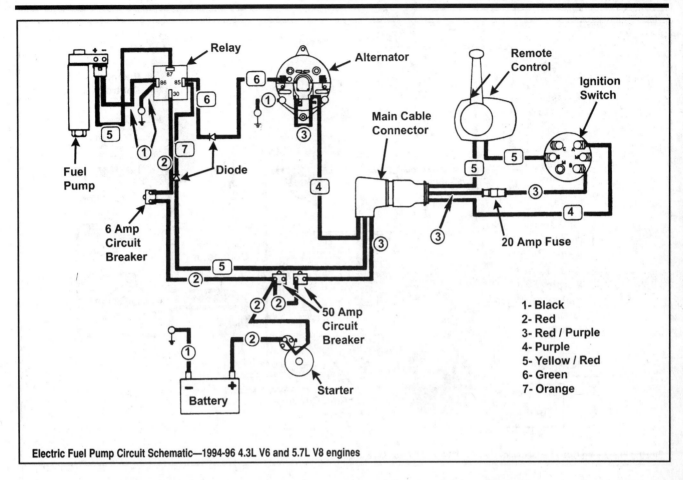

Legend (first schematic):
1- Black
2- Red
3- Red / Purple
4- Purple
5- Yellow / Red
6- Green
7- Orange

Electric Fuel Pump Circuit Schematic—1994-96 4.3L V6 and 5.7L V8 engines

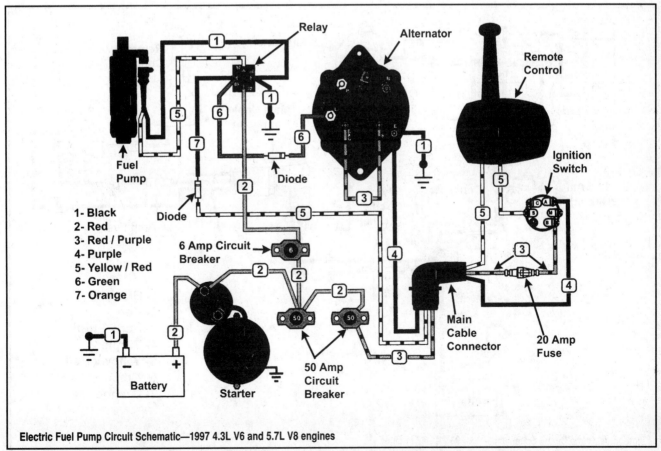

Legend (second schematic):
1- Black
2- Red
3- Red / Purple
4- Purple
5- Yellow / Red
6- Green
7- Orange

Electric Fuel Pump Circuit Schematic—1997 4.3L V6 and 5.7L V8 engines

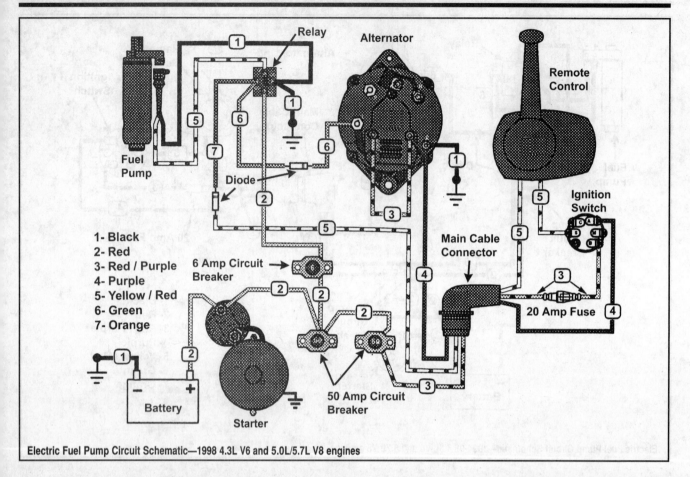

1- Black
2- Red
3- Red / Purple
4- Purple
5- Yellow / Red
6- Green
7- Orange

Electric Fuel Pump Circuit Schematic—1998 4.3L V6 and 5.0L/5.7L V8 engines

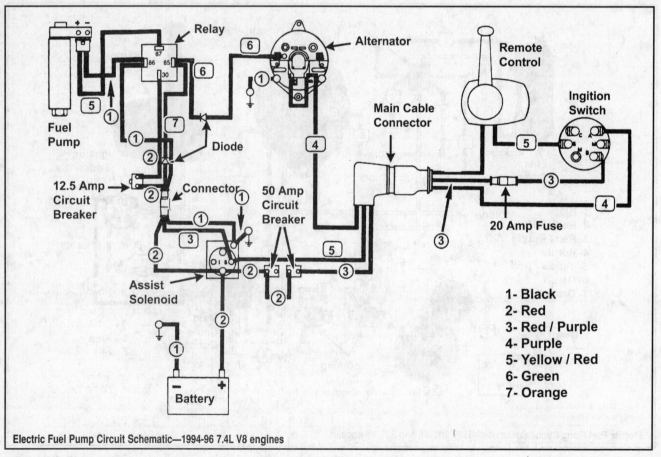

1- Black
2- Red
3- Red / Purple
4- Purple
5- Yellow / Red
6- Green
7- Orange

Electric Fuel Pump Circuit Schematic—1994-96 7.4L V8 engines

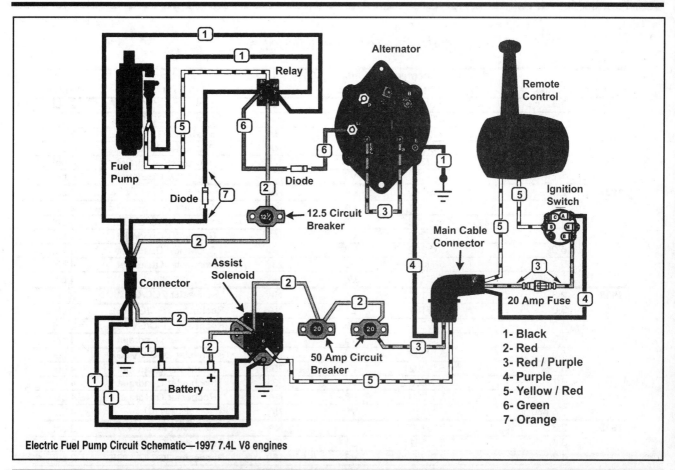

Electric Fuel Pump Circuit Schematic—1997 7.4L V8 engines

1- Black
2- Red
3- Red / Purple
4- Purple
5- Yellow / Red
6- Green
7- Orange

SPECIFICATIONS

TORQUE SPECIFICATIONS

Component			ft. lbs.	inch lbs.	Nm
Accelerator Pump Cover			-	5	0.6
Carburetor Mounting Nuts	Holley		10-14	-	13-19
	Rochester 2bbl	Early Models	10-14	-	13-19
		Later Models	12-15	-	16-20
	Rochester 4bbl	Nuts	10-14	-	13-19
		Bolts	6-8	-	8-11
Circuit Breaker			20-25	-	25-34
Fuel Bowl			-	45	5.1
Fuel Filter Bracket			20-25	-	25-34
Fuel Line Elbow			-	48-60	5.4-6.7
Fuel Line			15-18	-	20-24
Fuel Pump	Electric	Adapter	8-10	-	11-14
		Filter Bracket	20-25	-	25-34
		Lock Nut	-	60-84	6.8-9.5
	Mechanical	2.3L	20-25	-	25-34
		3.0L (1986-90)	12-14	-	16-19
		3.0L (1991-98)	20-25	-	25-34
		4.3L	20-25	-	25-34
		5.0L/5.7L (GM)	20-25	-	25-34
		5.0L/5.8L (Ford)	18-20	-	24-27
		7.5L	20-25	-	25-34
Inlet Needle seat			-	10	1.1
Power Valve			-	100	11.3

CARBURETOR APPLICATIONS

Year	Model	Displacement L/Cu. In.	Engine Type	HorsePower (At Prop.)	Carburetor Type	Fuel Delivery
1986	2.5	2.5/153	L4	NA	Rochester 2GC/GV /GE	2 bbl
	3.0	3.0/181	L4	NA	Rochester 2GC/GV /GE	2 bbl
	4.3	4.3/262	V6	NA	Rochester 2GC/GV /GE	2 bbl
	4.3	4.3/262	V6	NA	Rochester Quadrajet	4 bbl
	5.0	5.0/302	V8	185	Rochester Quadrajet	4 bbl
	5.7	5.7/350	V8	230	Rochester Quadrajet	4 bbl
1987	2.3	2.3/140	L4	NA	Rochester 2GC/GV ①	2 bbl
	3.0	3.0/181	L4	NA	Rochester 2GC/GV /GE	2 bbl
	4.3	4.3/262	V6	NA	Rochester 2GC/GV /GE	2 bbl
	4.3	4.3/262	V6	NA	Rochester Quadrajet	4 bbl
	5.0	5.0/302	V8	185	Rochester Quadrajet	4 bbl
	5.7	5.7/350	V8	230	Rochester Quadrajet	4 bbl
	7.5 460	7.5/460	V8	NA	Holley ①	4 bbl
1988	2.3	2.3/140	L4	NA	Rochester 2GC/GV ①	2 bbl
	3.0	3.0/181	L4	NA	Rochester 2GC/GV /GE	2 bbl
	4.3	4.3/262	V6	NA	Rochester 2GC/GV /GE	2 bbl
	4.3	4.3/262	V6	NA	Rochester Quadrajet	4 bbl
	5.0	5.0/302	V8	185	Rochester 2GC/GV /GE	2 bbl
	5.0	5.0/302	V8	200	Rochester Quadrajet	4 bbl
	5.7	5.7/350	V8	230	Rochester Quadrajet	4 bbl
	350	5.7/350	V8	NA	Rochester Quadrajet	4 bbl
	7.5 460	7.5/460	V8	NA	Holley ①	4 bbl
1989	2.3	2.3/140	L4	NA	Rochester 2GC/GV ①	2 bbl
	3.0	3.0/181	L4	120	Rochester 2GC/GV /GE	2 bbl
	4.3	4.3/262	V6	160	Rochester 2GC/GV /GE	2 bbl
	5.0	5.0/302	V8	185/200	Holley 2300	2 bbl
	5.7	5.7/350	V8	230	Rochester Quadrajet	4 bbl
	5.8	5.8/351	V8	235	Holley	4 bbl
	262	4.3/262	V6	NA	Rochester Quadrajet	4 bbl
	350	5.7/350	V8	NA	Rochester Quadrajet	4 bbl
	460	7.5/460	V8	NA	Holley ①	4 bbl
1990	2.3	2.3/140	L4	NA	Holley 2300 (350 cfm)	2 bbl
	3.0	3.0/181	L4	120	Holley 2300 (350 cfm) ②	2 bbl
	3.0 HO	3.0/181	L4	NA	Holley 2300 (500 cfm)	2 bbl
	4.3	4.3/262	V6	160	Holley 2300 (500 cfm)	2 bbl
	4.3 HO	4.3/262	V6	NA	Holley 4160 (575 cfm)	4 bbl
	5.0	5.0/302	V8	185	Holley 2300 (300 cfm)	2 bbl
	5.0 HO	5.0/302	V8	200	Holley 2300 (300 cfm)	2 bbl
	5.7	5.7/350	V8	230	Holley 4175 (650) ③	4 bbl
	5.7 LE	5.7/350	V8	NA	Holley 4175 (650) ③	4 bbl
	5.8	5.8/351	V8	235	Holley 4160 (600 cfm)	4 bbl
	350	5.7/350	V8	NA	Holley 4175 (650) ③	4 bbl
	454	7.4/454	V8	NA	Holley 4150 (850 cfm) ④	4 bbl
	460	7.5/460	V8	NA	Holley 4160 (750 cfm)	4 bbl

① Electric choke
② 3.0L PWC models used a Rochester 2 bbl carburetor
③ Spread bore
④ Dual feed

CARBURETOR APPLICATIONS

Year	Model	Displacement L/Cu. In.	Engine Type	HorsePower (At Prop.)	Carburetor Type	Fuel Delivery
1991	3.0	3.0/181	L4	120	Holley 2300 (350 cfm)	2 bbl
	3.0 HO	3.0/181	L4	NA	Holley 2300 (500 cfm)	2 bbl
	4.3	4.3/262	V6	160	Holley 2300 (500 cfm)	2 bbl
	4.3 HO	4.3/262	V6	NA	Holley 4160 (575 cfm)	4 bbl
	5.0	5.0/302	V8	185	Holley 2300 (300 cfm)	2 bbl
	5.0 HO	5.0/302	V8	200	Holley 2300 (490 cfm)	2 bbl
	5.7	5.7/350	V8	230	Holley 4175 (650) ①	4 bbl
	5.7 LE, 350	5.7/350	V8	NA	Holley 4175 (650) ①	4 bbl
	5.8	5.8/351	V8	235	Holley 4160 (600 cfm)	4 bbl
	7.4GL, 454	7.4/454	V8	300	Holley 4150 (750 cfm) ②	4 bbl
	454 HO	7.4/454	V8	390	Holley 4150 (850 cfm) ③	4 bbl
1992	3.0	3.0/181	L4	120	Holley 2300 (350 cfm)	2 bbl
	3.0 HO	3.0/181	L4	NA	Holley 2300 (500 cfm)	2 bbl
	4.3	4.3/262	V6	160	Holley 2300 (500 cfm)	2 bbl
	5.0	5.0/302	V8	185	Holley 2300 (300 cfm)	2 bbl
	5.0 HO	5.0/302	V8	200	Holley 2300 (490 cfm)	2 bbl
	5.7	5.7/350	V8	230	Holley 4175 (650) ①	4 bbl
	5.7 LE	5.7/350	V8	NA	Holley 4175 (650) ①	4 bbl
	5.8	5.8/351	V8	235	Holley 4160 (600 cfm)	4 bbl
	5.8 LE, 351	5.8/351	V8	235	Holley 4160 (600 cfm)	4 bbl
	7.4GL, 454	7.4/454	V8	300	Holley 4150 (750 cfm) ②	4 bbl
	8.2GL, 502, 454 HO	8.2/502	V8	390	Holley 4150 (850 cfm) ②	4 bbl
1993	3.0	3.0/181	L4	120	Holley 2300 (350 cfm)	2 bbl
	3.0 HO	3.0/181	L4	130	Holley 2300 (500 cfm)	2 bbl
	4.3	4.3/262	V6	160	Holley 2300 (500 cfm)	2 bbl
	4.3 HO	4.3/262	V6	NA	Holley 4160 (575 cfm)	4 bbl
	5.0	5.0/302	V8	185	Holley 2300 (300 cfm)	2 bbl
	5.0 HO	5.0/302	V8	200	Holley 2300 (490 cfm)	2 bbl
	5.8	5.8/351	V8	235	Holley 4160 (600 cfm)	4 bbl
	7.4GL	7.4/454	V8	300	Holley 4150 (750 cfm) ②	4 bbl
1994	3.0GL	3.0/181	L4	120	Holley 2300 (350 cfm)	2 bbl
	3.0GS	3.0/181	L4	135	Holley 2300 (500 cfm)	2 bbl
	4.3GL	4.3/262	V6	160	Holley 2300 (500 cfm)	2 bbl
	5.0FL	5.0/302	V8	190	Holley 2300 (490 cfm)	2 bbl
	5.7GL	5.7/350	V8	225	Holley 2300 (500 cfm)	2 bbl
	5.8FL	5.8/351	V8	230	Holley 4160 (600 cfm)	4 bbl
	7.4GL	7.4/454	V8	300	Holley 4150 (750 cfm) ②	4 bbl
1995	3.0GS	3.0/181	L4	130	Holley 2300 (500 cfm)	2 bbl
	4.3GL	4.3/262	V6	160	Holley 2300 (500 cfm)	2 bbl
	5.0FL	5.0/302	V8	190	Holley 2300 (490 cfm)	2 bbl
	5.8FL	5.8/351	V8	235	Holley 4160 (600 cfm)	4 bbl
	7.4GL	7.4/454	V8	300	Holley 4150 (750 cfm) ②	4 bbl
1996	3.0GS	3.0/181	L4	130	Holley 2300 (500 cfm)	2 bbl
	4.3GL	4.3/262	V6	160	Holley 2300 (500 cfm)	2 bbl
	4.3GS	4.3/262	V6	185	Holley 4160 (575 cfm)	4 bbl
	5.0FL	5.0/302	V8	190	Holley 2300 (490 cfm)	2 bbl
	5.8FL	5.8/351	V8	235	Holley 4160 (600 cfm)	4 bbl
	7.4GL	7.4/454	V8	300	Holley 4150 (750 cfm) ②	4 bbl

① Spread bore
② Single feed
③ Dual feed

CARBURETOR APPLICATIONS

Year	Model	Displacement L/Cu. In.	Engine Type	HorsePower (At Prop.)	Carburetor Type	Fuel Delivery
1997	3.0GS	3.0/181	L4	135	Holley 2300 (500 cfm)	2 bbl
	4.3GL	4.3/262	V6	190	Holley 2300 (500 cfm)	2 bbl
	4.3GS	4.3/262	V6	205	Holley 4160 (575 cfm)	4 bbl
	5.7GL	5.7/350	V8	215	Holley 2300 (500 cfm)	2 bbl
	5.7GS	5.7/350	V8	250	Holley 4175 (650 cfm) ①	4 bbl
	7.4GL	7.4/454	V8	300	Holley 4150 (750 cfm) ②	4 bbl
1998	3.0GS	3.0/181	L4	135	Holley 2300 (500 cfm)	2 bbl
	4.3GL	4.3/262	V6	190	Holley 2300 (500 cfm)	2 bbl
	4.3GS	4.3/262	V6	205	Holley 4160 (575 cfm)	4 bbl
	5.0GL	5.0/305	V8	220	Holley 2300 (500 cfm)	2 bbl
	5.7GS	5.7/350	V8	250	Holley 2300 (500 cfm)	2 bbl

① Spread bore
② Single feed

CARBURETOR SPECIFICATIONS - HOLLEY 2300 2BBL

Model			Choke Setting	Choke Unloader in. (mm) ①	Choke Vacuum in. (mm) ①	Float Setting	Intital Idle Mixture (Turns)	Main Jet	Power Valve	Accelerator Pump Lever in. (mm)	Pump Cam Position
300	5.0L	1989	1 notch lean	0.270-0.330 (6.8-8.4)	0.130-0.150 (3.3-3.9)	②	1 1/2 off seat	60	50	0.005-0.015 (0.130-0.381)	#1 Hole
		1990	1 notch lean	0.300 (7.6)	0.140 (3.6)	Level	1 1/2 off seat	60	50	0.010-0.015 (0.254-0.381)	#1 Hole
		1991	3 notches lean	0.300 (7.6)	0.228 (5.8)	②	1 1/8 off seat	61	3.5	0.010-0.015 (0.254-0.381)	#1 Hole
		1992-93	3 notches lean	0.300 (7.6)	0.228 (5.8)	②	1 1/8 off seat	61	3.5	0.010-0.015 (0.254-0.381)	Pink #1 Hole
350	3.0L, 3.0GL	1989-90	3 notches lean	0.380 (9.625)	0.250 (6.35)	Level	3/4 off seat	63	25	0.010-0.015 (0.254-0.381)	#2 Hole
		1991	3 notches lean	0.380 (9.625)	0.250 (6.35)	②	3/4 off seat	63	2.5	0.010-0.015 (0.254-0.381)	#1 Hole
		1992-94	3 notches lean	0.380 (9.625)	0.250 (6.35)	②	3/4 off seat	63	2.5	0.010-0.015 (0.254-0.381)	Orange #2 Hole
490	5.0 HO, 5.0FL	1992-96	3 notches lean	0.312 (7.93)	0.221 (5.6)	②	1 off seat	72	3.5	0.010-0.015 (0.254-0.381)	Orange #1 Hole
500	3.0 HO, 3.0GS	1990	5 notches lean	0.300 (7.62)	0.365 (9.3)	Level	③	75	25	0.010-0.015 (0.254-0.381)	#2 Hole
		1990-91	5 notches lean	0.300 (7.62)	0.365 (9.3)	②	③	75	2.5	0.010-0.015 (0.254-0.381)	#1 Hole
		1992-94	5 notches lean	0.300 (7.62)	0.365 (9.3)	②	③	75	2.5	0.010-0.015 (0.254-0.381)	Orange #2 Hole
		1995-98	5 notches lean	0.300 (7.62)	0.365 (9.3)	②	③	71	2.5	0.010-0.015 (0.254-0.381)	Orange #2 Hole
	4.3L, 4.3GL	1989-90	5 notches lean	3/8 (9.525)	0.250 (6.35)	Level	5/8 off seat	70	25	0.010-0.015 (0.254-0.381)	#1 Hole
		1991	5 notches lean	3/8 (9.525)	0.250 (6.35)	②	5/8 off seat	70	2.5	0.010-0.015 (0.254-0.381)	#1 Hole
		1992-98	5 notches lean	3/8 (9.525)	0.250 (6.35)	②	5/8 off seat	70	2.5	0.010-0.015 (0.254-0.381)	Red # 1 Hole
	5.7GL		5 notches lean	0.300 (7.62)	0.315 (8)	②	3/4-1 off seat	75	4.5	0.010-0.015 (0.254-0.381)	Red # 1 Hole
	5.7GS		5 notches lean	0.300 (7.62)	0.315 (8)	②	3/4-1 off seat	69	4.5	0.010-0.015 (0.254-0.381)	Yellow # 1 Hole

① Measured at lower edge of plate
② Parallel to fuel bowl when bowl inverted
③ 1 turn starboard to 1/2 turn port off seat

④ Level with fuel bowl inverted; Secondary: 1/8-1/4 turns in
⑤ 1998: Throttle lever side - 67; choke coil side - 68

CARBURETOR SPECIFICATIONS - HOLLEY 4BBL

Model	Cfm	Engine	Year	Choke Setting	Choke Unloader in. (mm) ①	Choke Vacuum in. (mm) ①	Float Setting	Intital Idle Mixture (Turns)	Main Jet	Power Valve	Accelerator Pump Lever in. (mm)	Accelerator Pump Cam Position	Secondary Throttle Plate (Turns)
		5.8L	1989	1 notch lean	0.270-0.330 (6.8-8.4)	0.130-0.150 (3.3-3.9)	②	1 1/2 Turns Out	73	65	0.005-0.015 (0.127-0.381)	#2 Hole	1/4 Beyond Contact
	750	7.5L	1987-88	1 notch rich	0.270-0.330 (6.8-8.4)	0.130-0.150 (3.3-3.9)	②	1 1/2 Turns Out	73	65	0.005-0.015 (0.127-0.381)	#2 Hole	1/4 Beyond Contact
			1989	1 notch lean	0.270-0.330 (6.8-8.4)	0.130-0.150 (3.3-3.9)	②	1 1/2 Turns Out	73	65	0.005-0.015 (0.127-0.381)	#1 Hole	1/4 Beyond Contact
4150	750	7.4L	1990	5 notches lean	1/4-1/2 (6.3-12.7)	0.415-0.425 (10.5-10.8)	②	1 Turn Out	88 (Pri.) 84 (Sec.)	6.5 (Pri.) 3.5 (Sec.)	0.010-0.015 (0.254-0.381)	#2 Hole	1/8 Beyond Contact
			1991-97	6 notches lean	0.344 (8.74)	0.312 (7.92)	②	1/4 off seat	73 (Pri.) 56 (Sec.)	2.5 (Pri.) 2.5 (Sec.)	0.010-0.015 (0.254-0.381)	Orange #1 Hole	1/8 Beyond Contact
	850	7.4L	1991-92	5 notches lean	1/4-1/2 (6.3-12.7)	0.415-0.425 (10.5-10.8)	②	1 Turn Out	88 (Pri.) 84 (Sec.)	6.5 (Pri.) 3.5 (Sec.)	0.010-0.015 (0.254-0.381)	Orange #1 Hole	1/8 Beyond Contact
4160	575	4.3 HO	1990	5 notches lean	0.300 (7.62)	0.250 (6.35)	②	3/4 Turn Out	67	25	0.010-0.015 (0.254-0.381)	#2 Hole	1/4 Beyond Contact
			1991	5 notches lean	0.300 (7.62)	0.250 (6.35)	②	3/4 Turn Out	67	2.5	0.010-0.015 (0.254-0.381)	#1 Hole	1/4 Beyond Contact
			1993	5 notches lean	0.300 (7.62)	0.250 (6.35)	②	3/4 Turn Out	67	2.5	0.010-0.015 (0.254-0.381)	Orange #2 Hole	1/4 Beyond Contact
		4.3GL, 4.3GS	1996-97	5 notches lean	0.300 (7.62)	0.250 (6.35)	②	3/4 off seat	67	2.5	0.010-0.015 (0.254-0.381)	Orange #1 Hole	1/4 Beyond Contact
			1998	5 notches lean	0.300 (7.62)	0.250 (6.35)	②	1 1/8 off seat	63	2.5	0.010-0.015 (0.254-0.381)	Green # 1 hole	1/8-1/4 Beyond Contact
	600	5.8L	1990	1 notch lean	0.300 (7.62)	0.140 (3.6)	②	1 1/2 Turns Out	73	65	0.010-0.015 (0.254-0.381)	#1 Hole	1/4 Beyond Contact
			1991-96	3 notches lean	0.300 (7.62)	0.228 (5.8)	②	7/8 off seat	67	2.5	0.010-0.015 (0.254-0.381)	Orange #1 Hole	1/8 Beyond Contact
		351	1992	3 notches lean	0.300 (7.62)	0.228 (5.8)	②	3/4 off seat	69	4.5	0.010-0.015 (0.254-0.381)	Orange #1 Hole	1/8 Beyond Contact
	750	7.5L	1990	1 notch lean	0.300 (7.62)	0.140 (3.6)	②	1 1/2 Turns Out	73	6.5	0.010-0.015 (0.254-0.381)	#1 Hole	1/4 Beyond Contact

① Measured at lower edge of plate
② Both parallel to fuel bowl when bowl is inverted

CARBURETOR SPECIFICATIONS - HOLLEY 4BBL

Model	Cfm	Engine	Year	Choke Setting	Choke Unloader in. (mm) ①	Choke Vacuum in. (mm) ①	Float Setting	Intital Idle Mixture (Turns)	Main Jet	Power Valve	Accelerator Pump Lever in. (mm)	Accelerator Pump Cam Position	Secondary Throttle Plate (Turns)
4175	650	5.7L	1990	Index	0.230-0.240 (5.9-6.1)	0.228 (5.8)	Level	1/4 off seat	60	50	0.010-0.015 (0.254-0.381)	#2 Hole	1/8 Beyond Contact
			1991	Index	0.230-0.240 (5.9-6.1)	0.228 (5.8)	②	1/4 off seat ③	60	5.0 ④	0.010-0.015 (0.254-0.381)	#1 Hole	1/8 Beyond Contact
			1992	Index	0.230-0.240 (5.9-6.1)	0.228 (5.8)	②	1/2 off seat	60	2.5	0.010-0.015 (0.254-0.381)	Orange #2 Hole	1/8 Beyond Contact
		5.7GL	1997	Index	0.230-0.240 (5.9-6.1)	0.228 (5.8)	②	1/2 off seat	60	2.5	0.010-0.015 (0.254-0.381)	Orange #2 Hole	1/8 Beyond Contact
		5.7 LE, 350	1990	Index	0.230-0.240 (5.9-6.1)	0.228 (5.8)	Level	1 off seat	60	5.0	0.010-0.015 (0.254-0.381)	#2 Hole	1/8 Beyond Contact
			1991	Index	0.230-0.240 (5.9-6.1)	0.228 (5.8)	②	1 off seat ⑤	60	5.0 ⑥	0.010-0.015 (0.254-0.381)	#1 Hole	1/8 Beyond Contact
			1992	Index	0.230-0.240 (5.9-6.1)	0.228 (5.8)	②	1/2 off seat	60	2.5	0.010-0.015 (0.254-0.381)	Orange #2 Hole	1/8 Beyond Contact

① Measured at lower edge of plate
② Both parallel to fuel bowl when bowl is inverted
③ For Part #986714. Part #987190: 1/2 turn off seat
④ For Part #986714. Part #987190: 2.5
⑤ Part #80390 is 1/2 turn off seat
⑥ Part #80390 and 80391 should be 2.5

CARBURETOR SPECIFICATIONS - ROCHESTER 2BBL

Year	Model	Engine	Float Level in. (mm)	Float Drop in. (mm)	Accelerator Pump Rod in. (mm)	Vacuum Break in. (mm)	Choke Unloader in. (mm)	Initial Choke Setting	Initial Idle Mixture (Turns)
1986	17086064	2.5L, 3.0L	1/2 (12.7)	1 3/4 (44.5)	7/8 (22.4)	-	5/64 (2.1)	Index	2 Out
	17083110	4.3L	17/32 (13.6)	1 3/4 (44.5)	1 15/64 (31.2)	3/16 (4.8)	5/16 (7.6)	Index	2 1/2 Out
1987	17086107	2.3L	17/32 (13.6)	1 3/4 (44.5)	1 15/64 (31.2)	3/16 (4.8)	5/16 (7.6)	Index	2 1/2 Out
	17086064	3.0L	1/2 (12.7)	1 3/4 (44.5)	7/8 (22.4)	-	5/64 (2.1)	Index	2 Out
	17085008	4.3L	17/32 (13.6)	1 3/4 (44.5)	1 15/64 (31.2)	3/16 (4.8)	5/16 (7.6)	Index	2 1/2 Out
1988	17086107	2.3L	17/32 (13.6)	1 3/4 (44.5)	1 15/64 (31.2)	3/16 (4.8)	5/16 (7.6)	Index	2 1/2 Out
	17086064	3.0L	1/2 (12.7)	1 3/4 (44.5)	7/8 (22.4)	-	5/64 (2.1)	Index	2 Out
	17085008	4.3L	17/32 (13.6)	1 3/4 (44.5)	1 15/64 (31.2)	3/16 (4.8)	5/16 (7.6)	Index	2 1/2 Out
	17059060	5.0L	17/32 (13.6)	1 3/4 (44.5)	1 15/64 (31.2)	3/16 (4.8)	5/16 (7.6)	Index	2 1/2 Out
1989	17086107	2.3L	17/32 (13.6)	1 3/4 (44.5)	1 15/64 (31.2)	3/16 (4.8)	5/16 (7.6)	Index	2 1/2 Out
	17086064	3.0L	1/2 (12.7)	1 3/4 (44.5)	7/8 (22.4)	-	5/64 (2.1)	Index	2 Out
	17085008	4.3L	17/32 (13.6)	1 3/4 (44.5)	1 15/64 (31.2)	3/16 (4.8)	5/16 (7.6)	Index	2 1/2 Out

CARBURETOR SPECIFICATIONS - ROCHESTER 4BBL

Year	Model	Engine	Float Level in. (mm)	Pump Rod Adjustment in. (mm)	Pump Rod Position	Vacuum Break in. (mm)	Choke Unloader in. (mm)	Initial Idle Mixture (Turns)
1986	17082515	4.3L	5/16 (7.77)	23/64 (9.14)	Inner Hole	1/8 (3.18)	15/64 (5.94)	3 3/4 Out
	17058286	5.0L, 5.7L	5/16 (7.77)	23/64 (9.14)	Inner Hole	1/8 (3.18)	15/64 (5.94)	3 3/4 Out
1987	17082515	4.3L	5/16 (7.77)	23/64 (9.14)	Inner Hole	1/8 (3.18)	15/64 (5.94)	3 3/4 Out
	17059286	5.0L, 5.7L	5/16 (7.77)	23/64 (9.14)	Inner Hole	1/8 (3.18)	15/64 (5.94)	3 3/4 Out
1988	17082515	4.3L	5/16 (7.77)	23/64 (9.14)	Inner Hole	1/8 (3.18)	15/64 (5.94)	3 3/4 Out
	17059286	5.0L, 5.7L	5/16 (7.77)	23/64 (9.14)	Inner Hole	1/8 (3.18)	15/64 (5.94)	3 3/4 Out
1989	17082515	4.3L, 262	5/16 (7.77)	23/64 (9.14)	Inner Hole	1/8 (3.18)	15/64 (5.94)	3 3/4 Out
	17059286	5.7L, 350	5/16 (7.77)	23/64 (9.14)	Inner Hole	1/8 (3.18)	15/64 (5.94)	3 3/4 Out

9

FUEL SYSTEMS FUEL INJECTION

FUEL AND COMBUSTION

FUEL SYSTEM APPLICATIONS

Year	Model	Displacement L/Cu. In.	Engine Type	Fuel Delivery
1993	5.0L	5.0/302	V8	MFI - FORD
	5.8L	5.8/351	V8	MFI - FORD
1994	5.0Fi	5.0/302	V8	MFI - FORD
	5.8Fi, 351	5.8/351	V8	MFI - FORD
1995	5.0 EFI	5.0/302	V8	MFI - FORD
	5.8 EFI	5.8/351	V8	MFI - FORD
	7.4 EFI	7.4/454	V8	MFI - GM
1996	4.3Gi	4.3/262	V6	TBI
	5.0Fi	5.0/302	V8	MFI - FORD
	5.7Gi	5.7/350	V8	TBI
	5.8Fi/Fsi	5.8/351	V8	MFI - FORD
	7.4Gi	7.4/454	V8	MFI - GM
1997	4.3Gi	4.3/262	V6	TBI
	5.7Gi/GSi	5.7/350	V8	TBI
	7.4Gi	7.4/454	V8	MFI - GM
1998	4.3Gi	4.3/262	V6	TBI
	5.7GSi	5.7/350	V8	TBI
	7.4Gi/GSi	7.4/454	V8	MFI - GM
	8.2GSi	8.2/502	V8	MFI - GM

Fuel

Fuel recommendations have become more complex as the chemistry of modern gasoline changes. The major driving force behind the changes in gasoline chemistry is the search for additives to replace lead as an octane booster and lubricant. These new additives are governed by the types of emissions they produce in the combustion process. Also, the replacement additives do not always provide the same level of combustion stability, making a fuel's octane rating less meaningful.

In the 1960's and 1970's, leaded fuel was common. The lead served two functions. First, it served as an octane booster (combustion stabilizer) and second, in 4-stroke engines, it served as a valve seat lubricant. For 2-stroke engines, the primary benefit of lead was to serve as a combustion stabilizer. Lead served very well for this purpose, even in high heat applications.

Today, all lead has been removed from the refining process. This means that the benefit of lead as an octane booster has been eliminated. Several substitute octane boosters have been introduced in the place of lead. While many are adequate in an engine, most do not perform nearly as well as lead did, even though the octane rating of the fuel is the same.

OCTANE RATING

A fuel's octane rating is a measurement of how stable the fuel is when heat is introduced. Octane rating is a major consideration when deciding whether a fuel is suitable for a particular application. For example, in an engine, we want the fuel to ignite when the spark plug fires and not before, even under high pressure and temperatures. Once the fuel is ignited, it must burn slowly and smoothly, even though heat and pressure are building up while the burn occurs. The unburned fuel should be ignited by the traveling flame front, not by some other source of ignition, such as carbon deposits or the heat from the expanding gasses. A fuel's octane rating is known as a measurement of the fuel's anti-knock properties (ability to burn without exploding).

Usually a fuel with a higher octane rating can be subjected to a more severe combustion environment before spontaneous or abnormal combustion occurs. To understand how two gasoline samples can be different, even though they have the same octane rating, we need to know how octane rating is determined.

The American Society of Testing and Materials (ASTM) has developed a universal method of determining the octane rating of a fuel sample. The octane rating you see on the pump at a fuel dock is known as the pump octane number. Look at the small print on the pump. The rating has a formula. The rating is determined by the R+M/2 method. This number is the average of the research octane reading and the motor octane rating.

• The Research Octane Rating is a measure of a fuel's anti-knock properties under a light load or part throttle conditions. During this test, combustion heat is easily dissipated.

• The Motor Octane Rating is a measure of a fuel's anti-knock properties under a heavy load or full throttle conditions, when heat buildup is at maximum.

VAPOR PRESSURE

Fuel vapor pressure is a measure of how easily a fuel sample evaporates. Many additives used in gasoline contain aromatics. Aromatics are light hydrocarbons distilled off the top of a crude oil sample. They are effective at increasing the research octane of a fuel sample but can cause vapor lock (bubbles in the fuel line) on a very hot day. If you have an inconsistent running engine and you suspect vapor lock, use a piece of clear fuel line to look for bubbles, indicating that the fuel is vaporizing.

One negative side effect of aromatics is that they create additional combustion products such as carbon and varnish. If your engine requires high-octane fuel to prevent detonation, de-carbon the engine more frequently with an internal engine cleaner to prevent ring sticking due to excessive varnish buildup.

ALCOHOL-BLENDED FUELS

When the Environmental Protection Agency mandated a phase-out of the leaded fuels in January of 1986, fuel suppliers needed an additive to improve the octane rating of their fuels. Although there are multiple methods currently employed, the addition of alcohol to gasoline seems to be favored because of its favorable results and low cost. Two types of alcohol are used in fuel today as octane boosters, methanol (wood alcohol) or ethanol (grain alcohol).

When used as a fuel additive, alcohol tends to raise the research octane of the fuel. There are, however, some special considerations due to the effects of alcohol in fuel.

• Since alcohol contains oxygen, it replaces gasoline without oxygen content and tends to cause the air/fuel mixture to become leaner.

• On older engines, the leaching affect of alcohol may, in time, cause fuel lines and plastic components to become brittle to the point of cracking. Unless replaced, these cracked lines could leak fuel, increasing the potential for hazardous situations.

• When alcohol blended fuels become contaminated with water, the water combines with the alcohol then settles to the bottom of the tank. This leaves the gasoline on a top layer.

■ **Modern fuel lines and plastic fuel system components have been specially formulated to resist alcohol leaching effects.**

RECOMMENDATIONS

All engine covered here are designed to run on unleaded fuel. Never use leaded fuel in your boat's engine. The minimum octane rating of fuel being used for your engine must be at least 87 AKI (outside the US: 90 RON), which means regular unleaded, but some engines may require higher octane ratings. OMC actually recommends the use of 89 AKI (93 RON) fuel as the ideal—in fact, anything less than this on many 4.3L/5.7L engines will require a change to the ignition timing. Fuel should be selected for the brand and octane that performs best with your engine. Check your owner's manual if in doubt. Premium unleaded is more stable under severe conditions but also produces more combustion products. Therefore, when using premium unleaded, more frequent de-carboning is necessary.

The use of a fuel too low in octane (a measure of anti-knock quality) will result in spark knock. Newer systems have the capability to adjust the engine's ignition timing to compensate to some extent, but if persistent knocking occurs, it may be necessary to switch to a higher grade of fuel. Continuous or heavy knocking may result in engine damage.

Combustion

In a high heat environment like an modern engine, the fuel must be very stable to avoid detonation. If any parameters affecting combustion change suddenly (the engine runs lean for example), uncontrolled heat buildup will occurs very rapidly.

The combustion process is affected by several interrelated factors. This means that when one factor is changed, the other factors also must be changed to maintain the same controlled burn and level of combustion stability.

• Compression—determines the level of heat buildup in the cylinder when the air-fuel mixture is compressed. As compression increases, so does the potential for heat buildup

• Ignition Timing—determines when the gasses will start to expand in relation to the motion of the piston. If the ignition timing is too advanced, gasses will be ignited and begin to expand too soon, such as they would during pre-ignition. The motion of the piston opposes the expansion of the gasses, resulting in extremely high combustion chamber pressures and heat. If the ignition timing is retarded, the gases are ignited later in relation to piston position. This means that the piston has already traveled back down the bore toward the bottom of the cylinder, resulting in less usable power.

• Fuel Mixture—determines how efficient the burn will be. A rich mixture burns slower than a lean one. If the mixture is too lean, it can't become explosive. The slower the burn, the cooler the combustion chamber, because pressure buildup is gradual.

• Fuel Quality (Octane Rating)—determines how much heat is necessary to ignite the mixture. Once the burn is in progress, heat is on the rise. The unburned poor quality fuel is ignited all at once by the rising heat instead of burning gradually as a flame front of the burn passing by. This action results in detonation (pinging).
There are two types of abnormal combustion—pre-ignition and detonation.

• Pre-ignition—occurs when the air-fuel mixture is ignited by some incandescent source other than the correctly timed spark from the spark plug.

• Detonation—occurs when excessive heat and or pressure ignites the air/fuel mixture rather than the spark plug. The burn becomes explosive.
In general, anything that can cause abnormal heat buildup can be enough to push an engine over the edge to abnormal combustion, if any of the four basic factors previously discussed are already near the danger point, for example, excessive carbon buildup raises the compression and retains heat as glowing embers.

FORD ELECTRONIC MULTI-POINT FUEL INJECTION (MFI)

Description And Operation

◆ **See Figures 1, 2, 3, 4 and 5**

The purpose of this section is to describe—in layman's terms whenever possible—the Multi-Port Fuel Injection (MFI) system installed on many Ford engines covered here.

Visual inspections, and simple tests that may be performed using only basic shop test equipment, will be given wherever possible. Again, we emphasize: specialized test equipment, hours of training and considerable experience is required to perform detailed service on a fuel injection system..

The first fuel injection system was introduced in Europe over 60 years ago—in 1932 on diesel truck engines. In the beginning, the system and individual components were quite expensive. Over the years, state-of-the-art microprocessors (commonly referred to as "computer chips"), have lowered the cost of electronically controlled fuel injection systems. Today, the price of EFI is getting close to the cost of modern carbureted systems.

An electronic fuel injection system is quite different from a carburetor system—even though they appear similar. The fuel injection system has a more efficient delivery of fuel to the cylinders than can be obtained with standard carburetor operation.

The EFI system provides a means of fuel distribution by precisely controlling the air/fuel mixture and under all operating conditions for, as near as possible, complete combustion.

On MFI systems, there is a throttle body located on the intake manifold, much like a carburetor except that only air is distributed and metered through the bores on the throttle body' Air for combustion is controlled by a throttle valve connected to the throttle linkage. Fuel is supplied by individual injectors mounted in the intake manifold and attached to a fuel rail assembly.

Each injector is "pulsed" or "timed" to open or close by an electronic signal from the ECM. While constantly receiving input from various sensors, the ECM performs high speed calculations of engine fuel requirements and then "pulses the injectors open or closed.

Ford's multi-point furl injection system is a pulse time/speed density fuel injection system. This is to say that fuel is metered into the intake air stream based on the engine's demand via 8 fuel injectors mounted on a tuned intake manifold.

Inputs from various engine sensors are fed into an on-board electronic engine control computer (EEC-IV) which computes the necessary fuel flow rates to maintain the prescribed air/fuel ratio at all times the engine is operating. The computer then sends an output signal to the injectors so they meter the correct quantity of fuel. The EEC-IV system is also capable of determining, and compensating for, engine age, uniqueness and altitude changes.

The fuel delivery system consists of two fuel pumps—a low pressure pump mounted on the engine, between the fuel tank and vapor separator; and a high pressure pump mounted in the separator reservoir, delivering fuel to the injectors. Also included are a vapor separator reservoir, a fuel filter and a fuel charging system,

The fuel charging system incorporates the electronic fuel injectors directly over each of the eight intake ports. When energized, the injectors spray a metered quantity of fuel directly into the intake air stream.

A pressure regulator located downstream from the injectors maintains constant pressure across the injector nozzles. Excess fuel supplied by the pump, but not needed by the engine, passes through the regulator and back to the separator reservoir.

On alternate revolutions of the crankshaft, one group of 4 injectors is energized simultaneously; with the remaining groups of four being energized on the next revolution. Injector "on time", or pulse width more accurately, is controlled by the EEC computer.

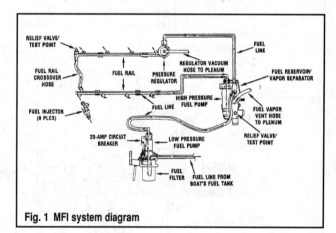

Fig. 1 MFI system diagram

FUEL CIRCUIT

Fuel from the fuel tank is delivered, through the fuel filter, to the vapor separator reservoir via the low pressure fuel pump. Once the ignition key is switched to the **ON** position, the pump will energize for approximately 2 seconds and then shut off until the electronic control assembly (ECA) receives a signal from the profile ignition pick-up (PIP) indicating that the engine is cranking or running.

The vapor separator reservoir is a fuel containment/vapor purge reservoir located at the rear of the engine, eliminating the need for fuel return lines. Fuel enters the reservoir from the bottom, passes through the high pressure pump on top of the assembly and then on to the fuel rail and injectors.

When fuel demand is less than the volume of fuel being supplied, line pressure will open an internal regulator in the low pressure pump allowing the fuel to circulate internally. Power for the delivery system and appropriate timing for the injectors is controlled by the ECA.

A vacuum/pressure operated diaphragm-type pressure regulator is located on the end of the fuel rail, downstream of the injectors. One side of the regulator senses fuel pressure while the other detects manifold vacuum; thus allowing the regulator to maintain the appropriate, and equal, fuel pressure at each injector.

Because of fuel vapors, the reservoir, and the fuel itself, are cooled by incoming seawater. Water enters passages in the reservoir from the transom shield water line and then continues on to the thermostat housing.

Inside the reservoir there is a float and needle mechanism, very similar to what you're familiar with on a carburetor, which is connected to a vacuum line from the air plenum. Any vapor will separate from the liquid fuel and rise to the top of the reservoir. As the amount of vapor increases, the fuel volume will decrease, causing the float to open the needle valve. Vacuum will then pull the vapor from the reservoir and into the intake plenum. A pulse limiter at the plenum prevents backfire from igniting these vapors. Once the vapor has been purged, the low pressure pump will feed more fuel into the reservoir, causing the level to rise and the float to close the needle valve.

Both the reservoir and the fuel rails are equipped with pressure relief valves.

ELECTRICAL CIRCUIT

As the ignition key is moved to the **ON** position, the EEC power relay is energized—providing power (current) to both the ECA and the fuel pump relay. Power for the pumps is supplied via 20 amp circuit breaker connected to the main 60 amp breakers on the engine.

As mentioned previously, if the switch is not moved past the **ON** position and into the **START** position, the ECA will shut off the pump after approximately 2 seconds. Once moved to the **START** and cranking commences, the ECA will receive a signal from the PIP and once again energize the pump. The pump will continue to run until the engine speed drops below 120 rpm.

COMPONENTS

Fuel

◆ See Figure 6

There are 8 electro-mechanical injectors attached to the fuel rail. They are charged with metering and atomizing fuel as delivered to the engine. Each injector body consists of a solenoid-actuated pintle and needle valve assembly.

A signal from the ECA activates the injector solenoid, causing the pintle to move inward and off the seat; thus allowing fuel to flow to the intake port. Fuel flow to the engine is regulated by the duration of solenoid energization, while atomization is obtained by contouring the pintle at the point where the fuel separates.

■ OMC utilizes a Deposit Resistant Injector (DRI) so that lean fuel delivery problems caused by bad fuel (low grade) have been eliminated.

These injectors have no pintle on the tip of the injector (as most injectors do); instead, fuel is metered via four small holes in a metering plate on the injector tip. As the inevitable fuel deposits coat the tip of the injector, the holes will not be affected; so metering characteristics remain correct.

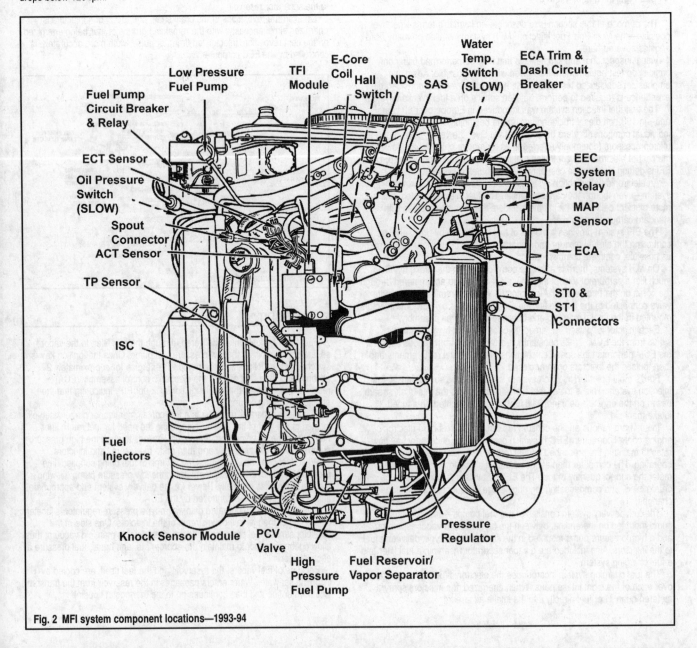

Fig. 2 MFI system component locations—1993-94

Fuel Pumps

As discussed previously, there are two electric fuel pumps utilized with this system. A low pressure pump attached to the fuel filter bracket and a high pressure pump mounted on top of the vapor separator reservoir.

Fuel Pressure Regulator

The fuel pressure regulator is mounted on the rear of the Starboard rail, downstream of the injectors.

The regulator is a simple diaphragm-type relief valve. One side of the diaphragm senses fuel pressure while the other side takes vacuum from the intake manifold. Nominal fuel pressure is established via the spring preload being applied to the diaphragm. Manifold pressure balance the other side of the diaphragm in order to maintain a constant pressure drop across all of the injectors. Excess fuel not needed by the engine bypasses the regulator and is returned to the separator reservoir.

Fuel Rail

The fuel rail delivers high pressure fuel from the vapor separator reservoir to each of the 8 injectors. The actual assembly consists of two hollow rails (one for each bank of cylinders) connected by a crossover. All fuel injectors and the pressure regulator are connected to the rail. All fuel rails are serviced as an assembly only, only the injectors and the regulator may be individually serviced.

Intake Manifold

The manifold assembly is actually made up of two separate manifolds; and upper and a lower. Runner lengths on the upper plenum are pre-tuned to optimize engine torque and power output.

Fuel injector pockets are machined into the lower manifold in order to prevent any air or fuel leakage. These pockets are also positioned so that they the injector fuel pattern sprays directly in front of each intake valve.

Throttle Body

The throttle body is mounted to the upper intake manifold, Port side and controls airflow to the engine by means of a butterfly-style valve. All throttle bodies are one-piece aluminum with a single bore. There is an air bypass channel around the throttle plate controlling all engine idle airflow, regulated by an electro-mechanical device controlled by the ECA. A linear actuator is incorporated into the throttle body which positions a variable area metering valve known as the Idle Speed Control (ISC) valve.

Vapor Separator Tank

The reservoir is located at the rear of the engine and consists of the assembly itself, the high pressure fuel pump, float assembly and various fuel and cooling lines. Operational details are discussed in the Description and Operation section.

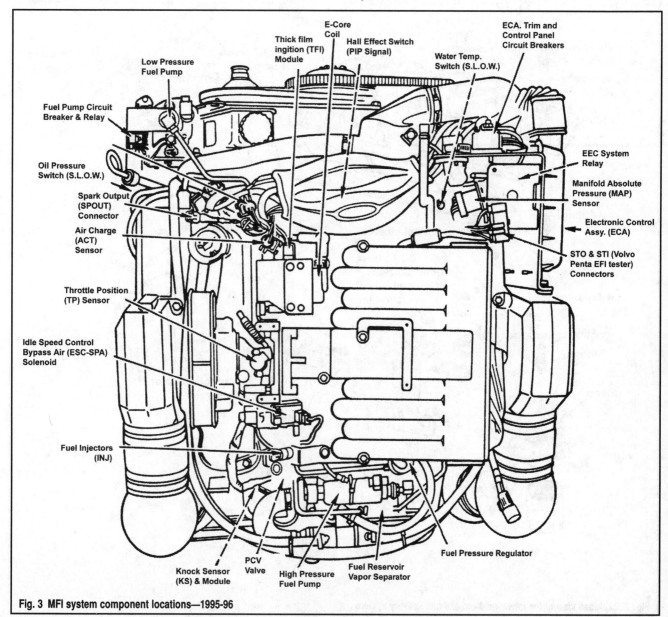

Fig. 3 MFI system component locations—1995-96

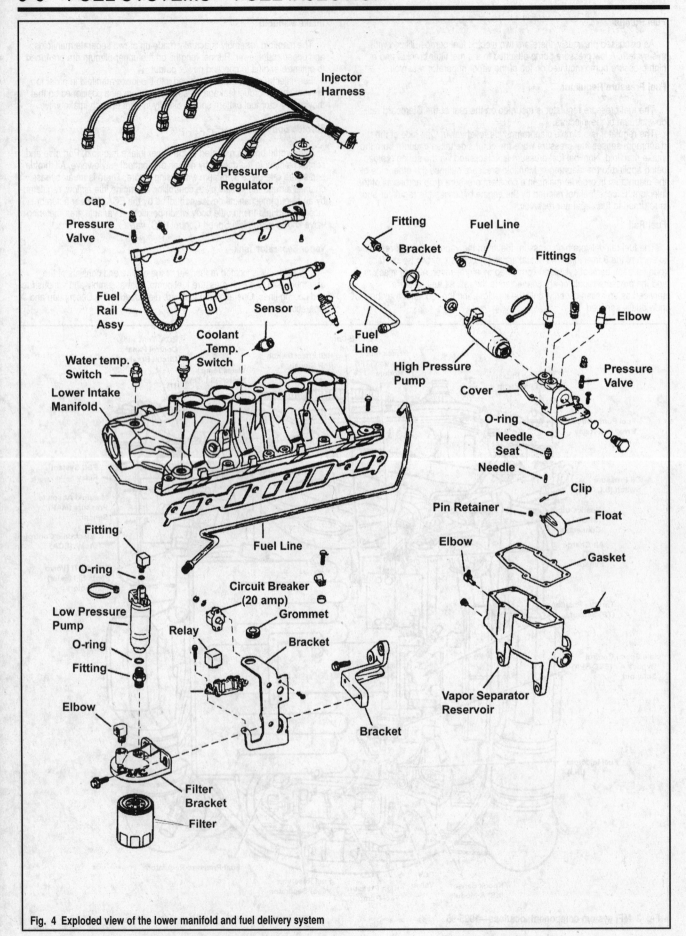

Fig. 4 Exploded view of the lower manifold and fuel delivery system

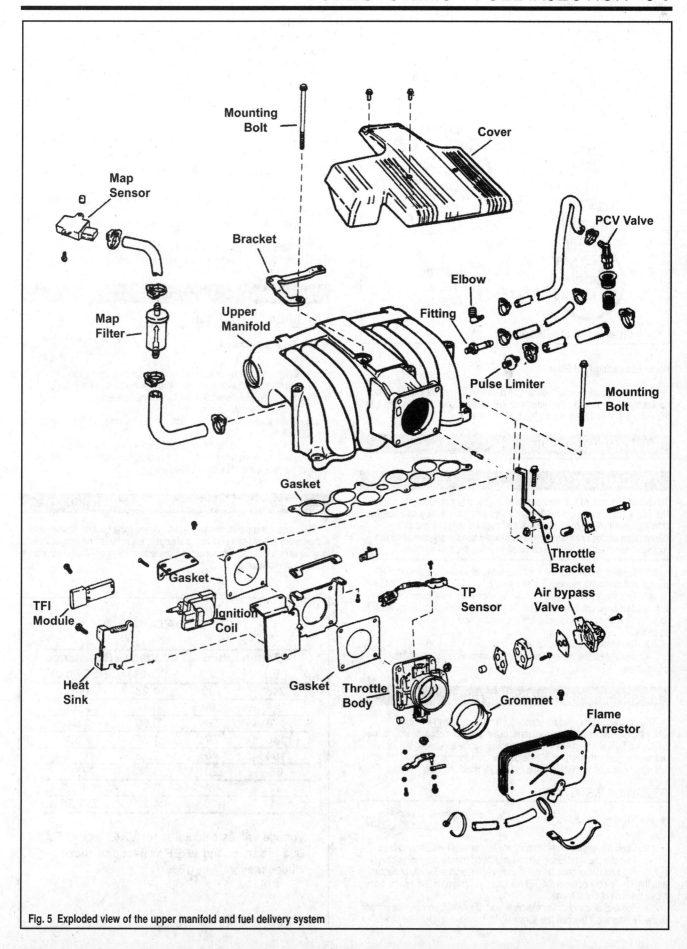

Fig. 5 Exploded view of the upper manifold and fuel delivery system

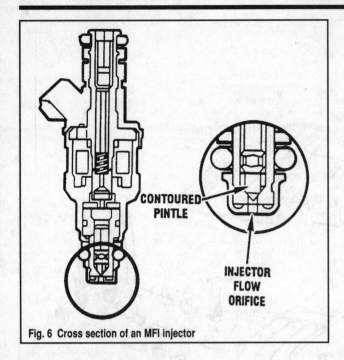

Fig. 6 Cross section of an MFI injector

Water Separating Fuel Filter

The water separator is a "typical" unit designed to prevent moisture from continuing on through the fuel lines and eventually through the injectors into the cylinders. It can be found on the front of the engine.

Relieving Fuel Pressure

 **CAUTION**

To reduce the risk of fire and personal injury, it is necessary to relieve the fuel system pressure before servicing any fuel system component. If this procedure is not performed, fuel may be sprayed out of a connection under extreme pressure. Always keep a dry chemical fire extinguisher near the work area when serving the fuel system.

1. Tag and disconnect the electrical leads at each of the electric fuel pumps and move the harness out of the way.
2. Turn the ignition key and crank the engine for ten seconds to relieve any residual pressure in the system. If the engine starts, don't worry, allow it to run until it dies out and then crank the engine again for a few more seconds.
3. Reconnect the fuel pump leads.
4. Disconnect the battery cables before commencing any additional procedures.

Air Charge Temperature Sensor (ACT)

The air charge temperature sensor (or air bypass valve) monitors incoming air temperature changes and relays this information directly to the ECA. The ECA will then advance spark timing so the idle speed will increase at lower engine temperatures. Since cold air contains more oxygen, the ECA will richen the fuel/air mixture when outside air is colder.

REMOVAL & INSTALLATION

◆ See Figures 2 and 3

1. Open the engine compartment and disconnect the battery cables.
2. Remove the 4 retaining screws and lift off the plastic engine cover.
3. Tag and disconnect the coil high tension lead, the E-coil connector and the TFI-IV connector at the TFI module, just forward of the throttle body. Move them all out of the way.
4. Tag and disconnect the electrical lead at the ACT sensor, just in front of the TFI module. Unscrew the sensor.

To Install:

5. Install the sensor and tighten it to 12-18 ft. lbs. (16.3-24.4 Nm). Coat the electrical lead with Terminal Grease and reconnect it.
6. Apply Terminal Grease to the lead and two connectors from the module and reconnect them. Confirm that each connection is secure.
7. Install the plastic cover and tighten the screws to 24-36 inch lbs. (2.7-4.1 Nm).
8. Connect the battery cables.

TESTING

◆ See Figures 7 and 8

1. With the ignition switch in the **OFF** position, disconnect the electrical lead at the ACT sensor.
2. Measure the resistance across the two terminals in the sensor connector.
3. Match your reading to that in the chart for the given temperature. If within specification, the sensor is OK, otherwise replace the sensor.

Circuit Breaker

REMOVAL & INSTALLATION

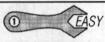

◆ See Figures 2, 3 and 4

1. Open the engine compartment and disconnect the battery cables.
2. Locate the circuit breaker on the fuel pump lift bracket (front of engine, Port side) and disconnect the two electrical leads.
3. Remove the screws/nuts and lift off the breaker.
4. Install a new circuit breaker and tighten the screws/nuts to 20-25 inch lbs. (2.3-2.8 Nm).
5. Connect the Red lead to one stud on the circuit breaker and the Red/Purple lead to the other. Tighten the nuts 20-25 inch lbs. (2.3-2.8 Nm). Coat the two terminals with Liquid Neoprene.

Engine Coolant Temperature Sensor (ECT)

This sensor is a thermistor immersed in the engine coolant passageway. A thermistor is a resistor capable of changing value based on temperature. Low coolant temperature produces a high resistance and a high temperature causes low resistance.

SENSOR DATA

Temperature (F)	(C)	Voltage (Volts)	Resistance (K Ohms)
248	120	0.28	1.18
212	100	0.47	2.07
176	80	0.80	3.84
140	60	1.35	7.60
104	40	2.16	16.15
68	20	3.06	37.30
32	0	3.87	94.98
-4	-20	4.33	271.20

Voltage values calculated for VREF equall 5V. Due to sensor and VREF variactions, these values may vary by 15%

Fig. 7 Sensor testing values

REMOVAL & INSTALLATION

◆ See Figures 2, 3, 9 and 10

 ① *EASY*

1. Open the engine compartment and disconnect the battery cables.
2. Remove the 4 retaining screws and lift off the plastic engine cover.
3. Tag and disconnect the coil high tension lead, the E-coil connector and the TFI-IV connector at the TFI module, just forward of the throttle body. Move them all out of the way.
4. Tag and disconnect the electrical lead at the ECT sensor, just in front of the TFI module. Unscrew the sensor.

To Install:
5. Install the sensor and tighten it to 10-15 ft. lbs. (13.6-20.3 Nm). Coat the electrical lead with Terminal Grease and reconnect it.
6. Apply Terminal Grease to the lead and two connectors from the module and reconnect them. Confirm that each connection is secure.
7. Install the plastic cover and tighten the screws to 24-36 inch lbs. (2.7-4.1 Nm).
8. Connect the battery cables.

TESTING

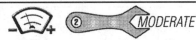 ② *MODERATE*

◆ See Figures 7 and 11

1. With the ignition switch in the **OFF** position, disconnect the electrical lead at the ECT sensor.
2. Measure the resistance across the two terminals in the sensor.
3. Start the engine and allow it to reach normal operating temperature. Check the resistance again.
4. Check the two readings against those in the chart and replace the sensor if they are incorrect.

EEC Power Relay

◆ See Figures 12 and 13

The EEC power relay receives direct battery power, protected by a circuit breaker and a fuse, via a red/purple electrical lead. Voltage from the ignition switch actuates the relay through a purple lead containing a diode to protect against reverse battery polarity. Finally, the purple/black lead carries battery voltage on the ECA at pins 37/57. All EEC-IV system components receive power via this lead.

REMOVAL & INSTALLATION

◆ See Figures 2, 3, 12 and 13

 ① *EASY*

1. Open the engine compartment and disconnect the battery cables.
2. Remove the ECA bracket and then disconnect the small vacuum hose to the MAP sensor.
3. Disconnect the electrical leads at the relay.
4. Cut the plastic tie, remove the retaining screw and lift off the relay.
5. Attach the relay to the bracket. Tighten the retaining screw to 24-36 inch lbs. (2.7-4.1 Nm) and install a new plastic tie.
6. Install the ECA bracket and reconnect the vacuum line.
7. Connect the battery cables.

Electronic Control Assembly (ECA)

REMOVAL & INSTALLATION

◆ See Figures 2 and 3

 ② *MODERATE*

1. Turn the ignition key to the **OFF** position. Do not perform any further steps in this procedure until you have done this.
2. Open the engine compartment and disconnect the battery cables.
3. Locate the ECA on the front side of the engine (Starboard) and cut the plastic tie around the rubber boot. Pull the boot back and remove the 10mm Hex screw holding the 60 pin connector on the front of the assembly. Unplug the connector and position it out of the way.
4. Remove the 4 cover retaining screws and then remove the plastic cover and the ECA.

To Install:
5. Position the assembly into the plastic cover so that the keyway lines up.
6. Attach the cover assembly to the bracket and tighten the screws to 24-36 inch lbs. (2.7-4.1 Nm).
7. Coat ALL terminal on the 60 pin connector with Terminal Grease and then plug it into the ECA. Tighten the Hex to 46-64 inch lbs. (5-7 Nm).
8. Stretch the rubber boot over the connector and install a new plastic tie.
9. Connect the battery cables.

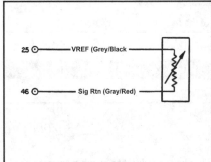

Fig. 8 ACT sensor schematic

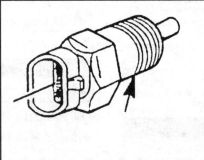

Fig. 9 Typical ECT sensor

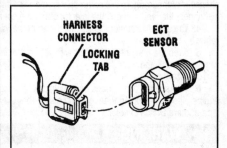

Fig. 10 The connector is held in place by a lock tab

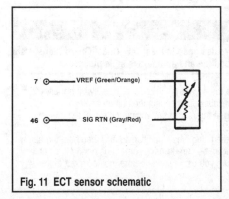

Fig. 11 ECT sensor schematic

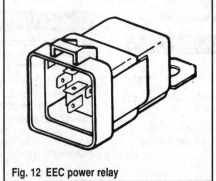

Fig. 12 EEC power relay

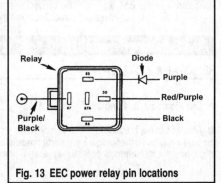

Fig. 13 EEC power relay pin locations

Fuel Injector

◆ See Figure 14

All EEC-IV EFI engine utilize an injector firing sequence that OMC calls 'bank-to-bank'—four injectors fire on one revolution of the crankshaft, while the remaining four fire on the next revolution of the shaft. All injectors receive power via the purple/black lead from the power relay at ECA pins 37/57. Ground circuits in the ECA control the actual firing; pin 58 fires the #1, 4, 5 and 8 injectors via the blue/yellow lead, while pin 59 fires the #2, 3, 6 and 7 injectors via the purple/white lead.

Fuel injector or circuit failure will NOT prompt the ECA to store a trouble code in its memory.

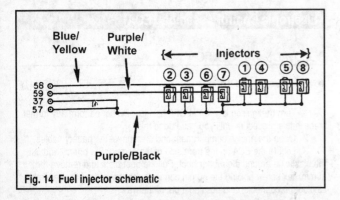

Fig. 14 Fuel injector schematic

REMOVAL & INSTALLATION

◆ See Figures 4 and 6    MODERATE

■ Always have a fire extinguisher handy when working on any part of the fuel system. Remember, a very small amount of fuel vapor in the bilge, has the tremendous potential explosive power.

1. Relieve the fuel pressure as previously detailed.
2. Open the engine compartment and disconnect the battery cables.
3. Remove the upper intake plenum and fuel rail. There is a good chance that the injectors will come up with the rail—if this happens, disconnect the lead and pull the injector out of the holder on the rail.
4. Tag and disconnect the electrical lead at the injector you intend to remove.
5. Grab the body of the injector with your hand and pull it upward while wiggling it back and forth.
To Install:
6. Replace the two O-rings on the injector, coating them with a light engine oil before slipping them on. Make sure that they are the fuel resistant type and are Brown in color.

✳✳ WARNING

Never use silicone grease on the injector O-ring.

7. Install the injector back into the rail the same way you removed it.
8. Install the fuel rail and reconnect the electrical lead to the injector. Make sure you coat both terminals and connector with Terminal grease.
9. Install the upper intake plenum and connect the battery cables.

Fuel Pressure Regulator

REMOVAL & INSTALLATION

◆ See Figures 2, 3, 4 and 15    MODERATE

■ Always have a fire extinguisher handy when working on any part of the fuel system. Remember, a very small amount of fuel vapor in the bilge, has the tremendous potential explosive power.

1. Relieve the fuel pressure as previously detailed.
2. Open the engine compartment and disconnect the battery cables.
3. Remove the upper intake plenum and fuel rail as previously detailed.
4. Remove the 3 Allen screws retaining the regulator and lift it off of the rail. Discard the gasket.
To Install:
5. Inspect the O-ring for cracks or other signs of deterioration, replace as necessary. We think it's a good idea to replace the O-ring regardless of condition. Either way, coat it with clean engine oil.

✳✳ WARNING

Never use silicone grease on the regulator O-ring.

6. Make sure that all residual gasket material is removed from the two mating surfaces and position a new gasket.
7. Install the regulator and tighten the screws to 27-40 inch lbs. (3.0-4.5 Nm).
8. Install the fuel rail and upper plenum.
9. Connect the battery cables.

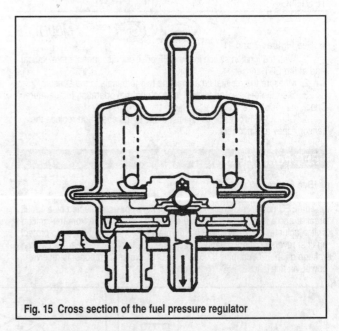

Fig. 15 Cross section of the fuel pressure regulator

Fuel Pump

REMOVAL & INSTALLATION

High Pressure

◆ See Figures 2, 3 and 4    MODERATE

■ Always have a fire extinguisher handy when working on any part of the fuel system. Remember, a very small amount of fuel vapor in the bilge, has the tremendous potential explosive power.

✳✳ CAUTION

All OMC fuel pumps are designed to meet U.S. Coast Guard fuel system regulations. Never substitute and automotive style pump.

1. Open or remove the engine hatch/cover and relieve the fuel system pressure as detailed previously. Disconnect the battery cable.
2. Locate the pump at the rear of the engine, on top of the fuel reservoir/vapor separator.
3. Make a note of the fitting elbow positioning and then hold the elbow while removing the fuel line; be sure to have some rags handy to catch any spilled fuel and make sure you plug the line before moving it out of the way.

4. With your wrench still on the elbow, back off the lock nut with another wrench. The lock nut is the nut immediately inside of the elbow; do not confuse it with the end cap nut on the end of the pump. Unscrew the elbow from the end of the pump and set it aside.

5. Remove the 2 screws securing the end bracket to the reservoir and then remove the bracket from the pump.

6. Remove the banjo nut from the end of the pump and then remove the pump. Discard all O-rings.

To Install:

7. Lubricate new O-rings with clean engine oil. Slide the small one onto the elbow fitting (right up against the flange) and the larger one onto the banjo nut so it sits in the groove on the inner side of the head.

8. Slide the banjo nut into the hole on the reservoir cover. Coat another new O-ring with clean engine oil and position it over the threads and up against the bracket.

9. Position the pump at a 45° angle, hold the end cap with a wrench and thread the banjo nut into the pump. Tighten it to 18-22 ft. lbs. (24.4-29.8 Nm). Be very careful to hold the end cap while tightening the banjo nut or you risk damaging the internal O-ring.

10. Position the support bracket over the other end of the pump and tighten the screws to 25-35 inch lbs. (2.8-4.0 Nm).

11. Screw the elbow fitting into the end of the pump so that it is in the same position as when removed. Cinch the lock nut down against the pump outlet and then tighten it to 60-84 inch lbs (6.8-9.5 Nm) while holding the pump end cap.

12. Install the fuel line and connect the battery cables.

Low Pressure

◆ See Figures 2, 3 and 4

■ **Always have a fire extinguisher handy when working on any part of the fuel system. Remember, a very small amount of fuel vapor in the bilge, has the tremendous potential explosive power.**

✳✳ CAUTION

All OMC fuel pumps are designed to meet U.S. Coast Guard fuel system regulations. Never substitute and automotive style pump.

1. Open or remove the engine hatch/cover and relieve the fuel system pressure as detailed previously. Disconnect the battery cable.

2. Position a container under the pump/filter assembly and remove the fuel inlet line at the filter bracket and the outlet line at the top of the pump. It is important to use two open-end wrenches; one to hold the fuel fitting nut or elbow and the other to loosen the line nut. Plug the line with a golf tee or something similar to prevent additional fuel spillage (OMC actually sells a plug for this purpose). Take care not to spill fuel on a hot engine, because such fuel may ignite.

3. Cut the plastic tie holding the electrical leads to the body of the pump. Unplug the electrical harness at the pump suppressor connector and move it out of the way.

4. Hold the pump end cap (just under the lower edge of the upper retaining bracket) with a wrench while loosening the lock nut (on top of bracket, under elbow). Unscrew the upper elbow.

✳✳ CAUTION

Not holding the end cap while loosening the lock nut will damage the internal O-ring, causing fuel leakage during operation.

5. Remove the 2 fuel filter bracket mounting bolts and lift off the pump/bracket/filter as an assembly, sliding the pump out of the upper grommet. Be careful of spilling fuel.

6. Lay the assembly down, or carefully mount the bracket in a vise. Hold the lower pump end cap with a wrench and unscrew the filter bracket-to-pump adapter.

✳✳ CAUTION

Not holding the end cap while loosening the lock nut will damage the internal O-ring, causing fuel leakage during operation.

To Install:

7. Slide a new O-ring over the threads of the adapter until it rests on the upper side of the hex nut.

8. Screw the adapter/bracket assembly into the pump while holding the lower end cap. Tighten the adapter to 8-10 ft. lbs. (11-14 Nm).

9. Slide the assembly up through the grommet in the bracket and position it so the mounting holes line up on the bracket. Tighten the filter bracket bolts to 20-25 ft. lbs. (27-34 Nm).

10. Thread the upper lock nut onto the elbow until it reaches the underside of the fitting. Slip on the washer and then a new O-ring. Thread the elbow into the top of the pump until it stops and then back it out just enough to align with the fuel line. Connect the line and tighten it securely.

11. Hold the upper end cap again and tighten the lock nut to 60-84 inch lbs. (6.8-9.5 Nm).

12. Reconnect the suppressor lead—push it in until the tabs click.

■ **If you notice any of the suppressor holes open, be sure to coat them with Liquid Neoprene to prevent water from entering.**

13. Install a new plastic tie around the pump body and electrical leads.

14. Reconnect the inlet line and tighten it securely.

15. Install the pump guard if removed.

16. Remove the container and connect the battery cables. Start the engine and check for fuel leaks.

Fuel Rail

REMOVAL & INSTALLATION

◆ See Figure 4

■ **Always have a fire extinguisher handy when working on any part of the fuel system. Remember, a very small amount of fuel vapor in the bilge, has the tremendous potential explosive power.**

1. Open or remove the engine hatch/cover and relieve the fuel system pressure as detailed previously. Disconnect the battery cable.

2. Remove the upper intake plenum/throttle body assembly as detailed previously.

3. Clean any and all debris from around the rail and injectors to prevent it from falling into the intake ports.

4. Carefully disconnect the fuel supply line at the rear Port side of the rail. Do the same with the return line on the Starboard side. Plug the lines and make sure you have plenty of rags available to mop up the inevitable spillage.

5. Make a note of the injector harness positioning on the rail and where it is secured. Cut the plastic ties securing the harness to the rails.

6. Remove the 4 rail retaining bolts (2 per side), carefully disengage the rails from each injector and lift off the rail. An alternate way is to simply remove the rail with the injectors attached, and then remove the injectors if necessary—we like this way better, particularly if you do not intend on replacing the rail or injectors. Cover the injector holes.

To Install:

7. If the injectors were removed, confirm that the caps are clean and debris-free.

8. Replace the two O-rings on each injector, coating them with a light engine oil before slipping them on. Make sure that the O-rings are the fuel resistant type and are Brown in color. No you don't have to replace the O-rings if they are in good condition, but why would you not do it while you've got the injectors out?? Cheap insurance, and can save you a lot of headache and time down the road.

✳✳ WARNING

Never use silicone grease on the injector O-ring.

9. If you left the injectors in position, seat each one into the rail cup, making sure the O-ring snaps into place. Otherwise, position the assembly over the ports and carefully pop the injectors into place. Once everything is in position, install the retaining bolts and tighten them to 70-105 inch lbs. (7.9-11.9 Nm).

10. Reroute the injector harness so that it approximates the original positioning and then install new plastic ties where you had removed the old ones. Make sure that all terminals and connections have been coated with Terminal Grease.

11. Install the manifold and throttle body assembly.

12. Connect the battery cables.

Fuel Reservoir/Vapor Separator

REMOVAL & INSTALLATION

◆ See Figures 2, 3 and 4

■ Always have a fire extinguisher handy when working on any part of the fuel system. Remember, a very small amount of fuel vapor in the bilge, has the tremendous potential explosive power.

1. Open or remove the engine hatch/cover and relieve the fuel system pressure as detailed previously. Disconnect the battery cable.
2. Loosen the hose clamps and then remove the water inlet and outlet hoses from the tank.
3. Remove the power steering cooler retaining bracket bolt and position the cooler out of the way. There is a special flat washer behind the cooler bracket—don't lose it when lifting off the assembly.
4. Clip the plastic tie securing the pump harness to the pump and then unplug the connector and move the harness out of the way.
5. Remove the small clamp and wiggle the vent hose from the nipple at the rear of the reservoir. Position it out of the way.
6. Remove the fuel inlet line from the low pressure pump. Remove the line running between the reservoir and pressure regulator. Remove the high pressure pump outlet line. Please refer to the Fuel Pump procedures for specific details on removing the fuel lines.
7. Loosen the clamp and remove the cooling outlet line (running to the thermostat housing) connected to the elbow on the Starboard side of the reservoir.
8. Remove the 2 mounting bolts at the bottom of the reservoir and lift off the assembly.

To Install:
9. Adjust the float level as detailed later in this section.
10. Position the reservoir onto the engine and tighten the mounting bolts to 24-36 ft. lbs. (38-49 Nm).
11. Reattach the coolant line to the elbow on the reservoir and tighten the clamp securely.
12. Connect the three fuel lines to the pump and reservoir as detailed in the Fuel Pump section.
13. Connect the vent line running from the upper manifold and tighten the clamp securely, but not so tight as to pinch the hose.
14. Coat the terminals on the fuel pump and pump harness connector with Terminal Grease and plug the connector into the pump. Secure the harness to the pump body with a new plastic tie.
15. Slide the flay washer over the mounting stud on the rear of the reservoir and then install the power steering cooler. Tighten the lock nut to 20-25 ft. lbs. (27-34 Nm).
16. Reconnect the cooling hoses and tighten their hose clamps securely.
17. Connect the battery cables.

FLOAT LEVEL

◆ See Figures 4 and 16

1. Disconnect all fuel lines and remove the high pressure fuel pump.
2. Remove the 8 reservoir cover retaining screws and lift off the cover and pump brackets. Discard the gasket.
3. Flip the cover assembly over and allow the float to drop down until it seats the float needle.
4. Measure the distance between the bottom edge of the inverted float (actually the top edge when installed in the reservoir) and the inner mounting surface of the reservoir cover.
5. Bend the float adjustment tab very carefully until the gap is 3/16 in. (4.76mm). Rotate the assembly once or twice and check that the float moves freely and that the gap has not changed.

■ Never force the float needle tip against the seat or you will damage the special tip.

6. Position a new cover gasket onto the reservoir and then carefully install the cover assembly. Tighten the retaining screws to 25-35 inch lbs. (2.8-4.0 Nm).
7. Install the fuel pump and reconnect the fuel lines.

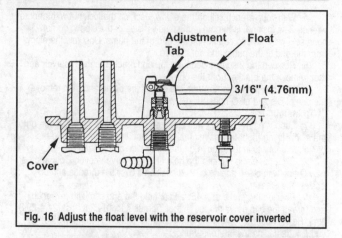

Fig. 16 Adjust the float level with the reservoir cover inverted

Idle Speed Control (ISC) Solenoid

◆ See Figures 17 and 18

The idle speed control by-pass air solenoid receives battery voltage via the purple/black lead from ECA pins 37/57. Ground is via the white/black ground wire from the ECA. This ground circuit serves to control voltage to the solenoid; the varying voltage then changes the position of the air by-pass valve. The greater the voltage, the more air will be allowed to bypass the throttle plate.

REMOVAL & INSTALLATION

◆ See Figures 2, 3 and 4

1. Open or remove the engine hatch/cover and relieve the fuel system pressure as detailed previously. Disconnect the battery cable.
2. Tag the 2-pin connector at the rear of the ISC solenoid and disconnect it from the solenoid. The solenoid is attached to the aft side of the throttle body.
3. Remove the 2 solenoid retaining screws and pull it off the adaptor. Discard the gasket.
4. Remove the 2 adaptor retaining screws and pull it off the throttle body. Discard the gasket. Pull the small muffler out of the adaptor body.

To Install:
5. Carefully wash the muffler in solvent and allow it to air dry.
6. Clean any residual gasket material from both sides of the adaptor, and from the throttle body and solenoid mating surfaces.
7. Insert the muffler into the adaptor so that is flush with, or below, the surface.
8. Position a new gasket onto the throttle body and then install the adaptor. Tighten the screws to 71-102 inch lbs. (8-11 Nm).
9. Position a new gasket onto the adaptor and install the solenoid. Tighten the screws to 71-102 inch lbs. (8-11 Nm).
10. Coat all terminals of the connector with Terminal Grease and reconnect the lead to the solenoid.
11. Install the engine cover and connect the battery cables.

Knock Sensor (KS) And Amplifier

◆ See Figures 19 and 20

At the first sign of spark knock, a pulsing signal is produced by the sensor and then sent through an amplifier to the ECA. After receiving this signal, the ECA will retard timing in three steps (but no more than 12° from the starting point) until knock is no longer sensed by the sensor. Once the MAP sensor detects a 3-4 in. Hg. change in engine vacuum, or the engine is shut off, the ECA will return the engine to its normal spark advance.

REMOVAL & INSTALLATION

◆ See Figure 19

1. Open or remove the engine hatch/cover and relieve the fuel system pressure as detailed previously. Disconnect the battery cable.

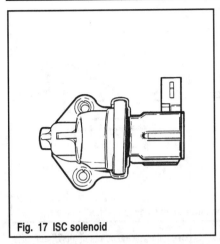

Fig. 17 ISC solenoid

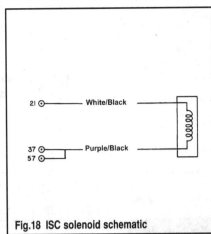

Fig.18 ISC solenoid schematic

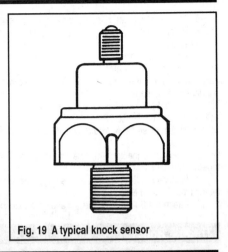

Fig. 19 A typical knock sensor

2. Locate the knock sensor on the rear of the cylinder block, just above the flywheel housing and behind the port side leg of the separator reservoir. Tag and disconnect the electrical lead at the sensor and then unscrew the sensor.

✳✳ CAUTION

The sensor is fragile and must be handled with care. If it is dropped, or banged in any manner, you must replace it.

3. Working right next to the sensor, cut the plastic tie around the amplifier and pop it out of its bracket. Pry off the retainer and then unplug the amplifier connector.

To Install:

4. Coat all terminals with Terminal Grease and plug in the amplifier connector. Press the retainer into position. Position the amplifier on the bracket and secure it with a new plastic tie.

5. Thoroughly clean the threads of the knock sensor and its hole. The sensor grounds through the threads, so there can be no residual paint or dirt. Carefully screw the sensor in and tighten it to 7-10 ft. lbs. (9.5-13.5 Nm).

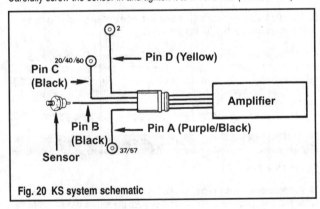

Fig. 20 KS system schematic

✳✳ CAUTION

DO NOT OVER-TIGHTEN THE KNOCK SENSOR.

6. Attach the sensor lead and connect the battery cables.

Manifold Absolute Pressure (MAP) Sensor

◆ **See Figure 21, 22 and 23**

The MAP sensor is a pressure transducer capable of measuring the changes in the intake manifold pressure. Pressure changes are the result of engine load and speed changes. MAP is the opposite of what is measured with a vacuum gauge.

When manifold pressure is high, vacuum is low— requires more fuel and of course the opposite is true. A low pressure—higher vacuum requires less fuel. The MAP sensor is also used to measure barometric pressure under certain conditions. This feature permits the ECM to automatically adjust for changes in operating altitude.

The ECM uses the MAP sensor to control fuel delivery and ignition timing. Reference voltage (VREF) is supplied to the sensor via a brown/yellow lead from the ECA. The sensor then outputs a frequency signal back to the ECA via the green/black lead and the circuit is completed with a signal return (SIG RTN) from the ECA to the sensor via the grey/red lead.

REMOVAL & INSTALLATION

◆ **See Figures 2, 3 and 5**

1. Open or remove the engine hatch/cover and relieve the fuel system pressure as detailed previously. Disconnect the battery cable.

2. Remove the Self-Test connector and cover from the top of the ECA bracket just forward of the Starboard exhaust elbow.

3. Remove the 3 ECA bracket mounting screws.

4. Loosen the small hose clamp and pull the vacuum line off the nipple on the sensor. Tag the line and position it out of the way. Remove the sensor filter at the other end of the line and clean it in a mild solvent.

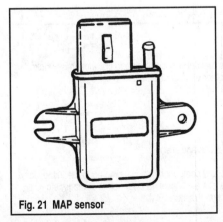

Fig. 21 MAP sensor

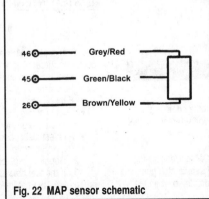

Fig. 22 MAP sensor schematic

Manifold Vacuum		Frequency
in. Hg	kPa	Hz
0	0	159
6	20.3	141
12	40.6	125
18	61	109
24	81.3	95
30	101.6	80

Fig. 23 MAP sensor data table

5. Tag the sensor lead and disconnect it from the sensor body.
6. Remove the 2 retaining screws and pull the sensor off the bracket.
To Install:
7. Attach the sensor to the ECA bracket so that the flat side is facing up (toward the bracket). Tighten the screws to 24-36 inch lbs (2.7-4.1 Nm). It is very important that the sensor is installed upside down to prevent condensation damage.
8. Install the sensor filter into the vacuum lines so that the arrow is pointing toward the sensor. Slide the vacuum line onto the sensor nipple and tighten all clamps securely. You may find it easier to connect the sensor terminal first.
9. Coat the connector terminals with Terminal Grease and snap together the connection.
10. Position the ECA bracket on the engine and tighten the screws to 60-84 inch lbs. (6.8-9.5 Nm).
11. Attach the Self-Test connector and cover. Connect the battery cables.

Throttle Body

REMOVAL & INSTALLATION

◆ **See Figures 5 and 24**

1. Open or remove the engine hatch/cover and relieve the fuel system pressure. Disconnect the battery cable.
2. Remove the 4 screws and lift off the plastic engine cover.
3. Remove mounting nut(s) and then remove the flame arrestor. Certain models will have a bracket holding the arrestor on.
4. Remove the throttle cable locknut at the arm on the throttle body and then disconnect the throttle linkage.
5. Tag and disconnect the wiring harness for the ISC solenoid and TP sensor.
6. Remove the 4 throttle body mounting nuts. Lift out the throttle body and set it down in a holding fixture to avoid damage to the valves. Stuff a clean, lint-free rag into the plenum opening.
7. Remove the TP sensor and ISC valve with their O-rings from the assembly if necessary.
8. Lift the coil bracket off of the studs.
To Install:
9. Carefully clean the throttle bore and valve. Do not use anything containing methyl ethyl ketone! If you didn't remove the TP sensor and ISC valve, make sure that you get no solvent on them during cleaning. CAREFULLY scrape any gasket material off the mating surfaces. Make sure all passages are free of dirt and completely dry before installation.

■ The inside of the throttle bore and the back (inner) side of the throttle valve have been coated with a special protective material which created a ridge at the point that the valve meets the bore—do not remove this ridge, or spray any kind of cleaner on the surfaces.

10. Slide a new gasket over the studs in the manifold so that the tab is at the bottom and facing rearward. Slide on the coil bracket.
11. Install the sensor and valve into the throttle body if they were removed.
12. Position the throttle body and new gasket (tab at bottom and facing the rear) over the studs and tighten the nuts to 12-18 ft. lbs. (16-24 Nm).
13. Connect the IAC and TP leads.
14. Connect the throttle cable and adjust as necessary.
15. Install the flame arrestor and engine cover.
16. Reconnect the battery cables.

Throttle Position Sensor (TP)

◆ **See Figures 25, 26 and 27**

The TP sensor is a potentiometer connected to the throttle shaft on the throttle body. One end of the sensor is connected to a 5-volt reference signal (VREF) from the ECA via a brown/yellow lead and the other end is connected to ECA ground.
A third wire (grey/white) is connected directly to the ECM to measure the voltage from the TP sensor. The voltage output of the TP sensor changes when the throttle valve angle is changed. The circuit is completed when the signal is returned (SIG RTN) via the grey/red wire.

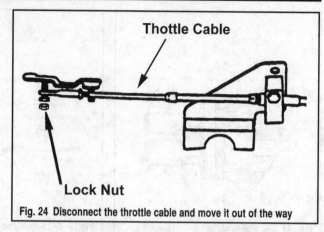
Fig. 24 Disconnect the throttle cable and move it out of the way

When the throttle position is closed, the voltage output of the TP sensor is low—about 1/2 volt. When the throttle valve is opened, the output increases and at WOT the output voltage should be close to 5 volts.
Therefore, the ECM can determine fuel delivery requirements based on throttle valve angle—operator demand.

REMOVAL & INSTALLATION

◆ **See Figures 2, 3 and 5**

1. Open or remove the engine hatch/cover and relieve the fuel system pressure. Disconnect the battery cable.
2. Remove the 4 screws and lift off the plastic engine cover.
3. Remove mounting nut(s) and then remove the flame arrestor.
4. Tag and disconnect the electrical connector at the sensor and move it out of the way.
5. Using a magic marker, draw a line across the body of the sensor and the throttle body to use as a reference mark on installation.
6. Unscrew the 2 mounting screws and remove the sensor. Clean the sensor with a dry cloth; make sure that the sensor is in good condition (no wear, cracks or damage), if not replace it.

■ If replacing the sensor with a new one, make sure to use the two new screws that came with the package.

7. Install the sensor so that the wiring harness would be laying across flame arrestor. Rotate the sensor clockwise until the reference mark made earlier lines up and tighten the screws to 18-26 inch lbs. (2-3 Nm).
8. Coat the connector terminals with Terminal Grease and then plug them together.
9. Install the flame arrestor and cover. Connect the battery cables.

System Diagnosis

◆ **See Figure 28**

■ The following system checks and tests MUST be followed in the order that they are presented so that system diagnosis can be properly exercised.

SYSTEM INTEGRITY CHECK

1. Inspect the entire fuel delivery system for leaks, looseness, kinks, corrosion, grounding or other damage. Be certain to check:
 • Fuel tank and lines
 • Fuel filter
 • Fuel pumps
 • Fuel injectors
 • Pressure regulator
 • All electrical connections and lines
 • Vapor Separator
2. Repair or replace any component not standing up to the inspection.
3. If the fuel delivery system shows no evidence of damage, etc., proceed to the Fuel Injection Pressure Test.

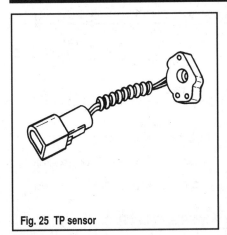

Fig. 25 TP sensor

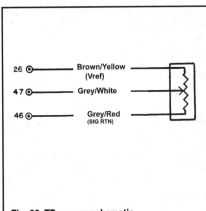

26 ⊙ — Brown/Yellow (Vref)
47 ⊙ — Grey/White
46 ⊙ — Grey/Red (SIG RTN)

Fig. 26 TP sensor schematic

Throttle Angle (deg.)	Voltage A (V)
0	0.50
10	0.97
20	1.44
30	1.90
40	2.37
50	2.84
60	3.31
70	3.78
80	4.24

A Voltage values calculated for VREF = 5V and may vary +/-15% due to sensor/VREF variations

Fig. 27 TP sensor data table

SYMPTOM CHART

SYMPTOM	Fuel System	Fuel Injectors	Throttle Body	Ignition System	Sensors	Electrical System	Cooling System	Engine Mechanical	Air Intake	Other
Cranking										
Will Not Crank		X				X		X		
Cranks But Does Not Run	X	X	X	X		X		X	X	
Hard To Start - Long Cranking Time	X	X		X	X			X	X	
Stalls After Start/Idle	X		X	X	X			X	X	
Idling										
Slow Return To Idle	X		X	X	X			X	X	
Rolling Idle/Runs Rough/Misses	X	X	X	X	X			X	X	
Fast Idle/Diesels	X		X	X			X	X	X	
Low Idle/Stalls/Quits After Decel			X			X			X	
Backfires While Idling	X			X	X			X		
Acceleration										
Stalls/Quits On Acceleration	X	X	X	X						
Bucks/Jerks During Acceleration	X	X	X	X					X	
Hesitation/Stumble/Surge	X	X	X	X	X				X	①
Backfires	X	X	X	X	X			X		
Lack Of/Low Power Output	X	X		X	X		X	X	X	②
Spark Knock	X			X	X		X	X	X	
Cruising										
Runs Rough/Cuts Out/Misses	X	X		X	X			X		
RPM Limited To 2700 Or Under	X	X		X	X		X	X		
Surges	X	X		X	X			X	X	
Backfires	X			X	X			X		
Low Power Output	X	X		X	X		X	X	X	②
Spark Knock	X			X	X		X	X	X	
Other Conditions										
Poor Fuel Economy	X			X			X	X	X	② ③
Strong Fuel Odors	X	X	X							④

① Check the throttle linkage and cable for binding
② Check stern drive oil. Possible drive bearing problems
③ Check hull for marine growth
④ Check fuel tank, vents and fuel lines for leaks

Fig. 28

FUEL INJECTION PRESSURE TEST

◆ **See Figure 29**

1. With the ignition switch in the **OFF** position, install a fuel pressure gauge at the pressure valve on the fuel rail.
2. Ground the Blue/Orange fuel pump lead at the STO (Self-Test Output) connector (pin #52).
3. Turn the ignition switch to the **ON** position (but do not start the engine) so that the fuel pumps are energized.
4. If the gauge reads 36-42 psi (248-290 kPa), proceed to the Fuel Pressure Leakdown Test.
5. If the gauge reads **0** or very low, proceed to Checking Voltage At FP Relay.

6. If the gauge shows the pressure to be high, proceed to Checking the Regulator For High Pressure.

FUEL PRESSURE LEAKDOWN TEST

◆ **See Figure 29**

1. With the ignition switch in the **OFF** position, install a fuel pressure gauge at the pressure valve on the fuel rail.
2. Connect a jumper wire to the Blue/Orange lead at the STO Self-Test connector (pin #52).
3. Turn the ignition switch to the **ON** position (but do not start the engine). Connect the jumper lead from the previous step to a suitable ground, run the fuel pumps for 30 seconds and then disconnect the jumper.

Take note of the fuel pressure reading on the gauge.

4. The pressure should remain within 3 psi of the running pressure for at least 3 minutes after the jumper is removed from ground.

5. If the pressure held, proceed to Pressure Regulator Diaphragm Check.

6. If the pressure did not remain constant, proceed to Pressure Regulator Valve Seat Check.

PRESSURE REGULATOR DIAPHRAGM CHECK

1. With the ignition switch in the **OFF** position, install a fuel pressure gauge at the pressure valve on the fuel rail.

2. Start the engine and allow it to run for 10 seconds. Turn it off for 10 seconds and then start it again and allow it to run for another 10 seconds.

3. Turn off the engine and disconnect the vacuum line at the regulator.

4. Examine the port and regulator body for signs of leakage.

5. If no fuel is present, proceed to Fuel Pressure Under Load Check.

6. If fuel leakage is present, replace the regulator and then perform the Fuel Injection Pressure Test again.

FUEL PRESSURE UNDER LOAD CHECK

1. With the ignition switch in the **OFF** position, install a fuel pressure gauge at the pressure valve on the fuel rail.

2. Start the engine and allow it to idle.

3. Disconnect the vacuum line at the pressure regulator. If the line has vacuum, plug it. Otherwise, determine why there is no vacuum, fix the problem, and then plug the line.

4. Run the engine up to full throttle and then back to idle at least 4 or 5 times while checking the fuel pressure gauge.

5. If the fuel pressure remains within 3 psi of itself across all readings, unplug the vacuum line and reconnect it.

6. If the pressure readings vary more than 3 psi, proceed to Fuel Filter Check.

FUEL PRESSURE REGULATOR CHECK

1. With the ignition switch in the **OFF** position, install a fuel pressure gauge at the pressure valve on the fuel rail.

2. Install a "T" in the vacuum line running from the regulator and the upper plenum. Install a vacuum gauge to the "T".

3. Start the engine and check both the pressure and vacuum gauges.

4. Increase the engine speed so that the vacuum level falls and check the pressure gauge.

5. If the fuel pressure increases as you accelerate and the vacuum decreases; and, if it decreases as you decelerate while the vacuum begins to increase, remove the gauges.

6. If the vacuum and pressure reading do not act inversely to one another, proceed to Vacuum Supply Check.

VACUUM SUPPLY CHECK

1. With the ignition switch in the **OFF** position, install a fuel pressure gauge at the pressure valve on the fuel rail.

2. Disconnect the vacuum line at the pressure regulator and plug it.

3. Connect a hand-operated vacuum pump to the regulator at the nipple.

4. Start the engine and check the fuel pressure while pumping the vacuum pump.

5. If the pressure reading changes as the vacuum changes, service the vacuum system and reconnect the vacuum line to the regulator.

6. If the fuel pressure does not change, replace the regulator.

FUEL PUMP RELAY VOLTAGE CHECK

◆ See Figure 29

1. With the ignition switch in the **OFF** position and the battery fully charged, ground the Blue/Orange fuel pump lead at the STO Self-Test connector (pin #52).

2. Connect a DVOM to the fuel pump relay.

3. Turn the ignition switch to the **ON** position (but do not start the engine). Measure the voltage at the relay.

4. If voltage is more than 10.5V, look for an open circuit between the relay and the ECA.

5. If voltage is less than 10.5V, proceed to Fuel Pump Relay Power Supply Check.

FUEL PUMP RELAY POWER SUPPLY CHECK

◆ See Figure 30

1. With the ignition switch in the **OFF** position and the battery fully charged, disconnect the electrical leads and remove the relay.

2. Turn the ignition switch to the **ON** position (but do not start the engine). Measure the voltage between the ground terminal in the relay connector (black lead, pin #86) and both **B+** pins (red/purple lead, pin #30 and purple/black lead, pin #85).

3. If voltage on both **B+** pins is higher than 10.5V, proceed to Fuel Pump Relay Operation Check.

4. If voltage is not higher than 10.5V, check for a crimped or torn wire between the pump relay and the EEC power relay or the battery (+).

FUEL PUMP RELAY OPERATION CHECK

◆ See Figure 30

1. Remove the relay from the bracket.

2. Ground terminal #86 with a jumper wire and then connect a B+ 12V power supply to #85. Measure the resistance between #30 and #87.

3. Resistance should be less than 1 ohm with the power source supplied and greater then 10,000 ohms when it is removed.

4. If not, replace the fuel pump relay.

FUEL PUMP OPERATION CHECK

◆ See Figure 29

1. With the ignition switch in the **OFF** position and the battery fully charged, ground the Blue/Orange fuel pump lead at the STO Self-Test connector (pin #52).

2. Turn the ignition switch to the **ON** position (but do not start the engine). You should be able to hear the fuel pumps and also feel a slight vibration to the touch.

3. If they are working, proceed to Fuel Filter Check.

4. If they are not working, check that the 20 amp circuit breaker is operating and the wiring is good. If everything is OK, proceed to Fuel Pump Voltage Check.

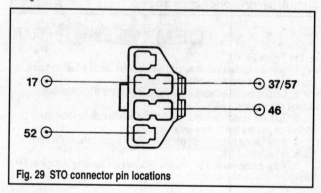

Fig. 29 STO connector pin locations

FUEL PUMP VOLTAGE CHECK

◆ **See Figure 29**

1. With the ignition switch in the **OFF** position and the battery fully charged, ground the Blue/Orange fuel pump lead at the STO Self-Test connector (pin #52).
2. Disconnect the electrical leads at the two fuel pumps.
3. Turn the ignition switch to the **ON** position (but do not start the engine).
4. Measure the voltage at the green/yellow power supply leads at each pump. If greater than 10.5V at each pump, proceed to Fuel Pump Ground Check.
5. If voltage is less than 10.5V, check the circuit between the ECA and the fuel pump connector and then proceed to Fuel Injection Pressure Test.

FUEL PUMP GROUND CHECK

1. With the ignition switch in the **OFF** position and the battery fully charged, disconnect the electrical leads at the two fuel pumps.
2. Measure the resistance between the black wire terminal and ground.
3. If each pump is less than 1 ohm, replace the pumps and then proceed to Fuel Injection Pressure Test.
4. If more than 1 ohm, check for an open wire to ground and then proceed to Fuel Injection Pressure Test.

FUEL FILTER CHECK

1. Remove the fuel filter and check its condition and also for any water.
2. If the filter requires replacement, proceed to Fuel Pump Ground Check.
3. If the filter is OK, reinstall it and repeat the Fuel Injection Pressure Test.

PRESSURE REGULATOR VALVE SEAT CHECK

1. Visually check the O-ring, gasket, mounting surfaces and regulator body for cracks or damage.
2. Connect a vacuum tester to the fuel outlet on the regulator and apply 20 in. Hg of vacuum.
3. If the vacuum drops below 10 in. Hg within 10 seconds, replace the regulator and then repeat the Fuel Injection Pressure Test.
4. If vacuum stays within specifications, proceed to Fuel Injector Function Check.

PRESSURE REGULATOR HIGH PRESSURE CHECK

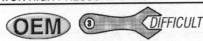

1. With the ignition switch in the **OFF** position and the battery fully charged, remove the fuel return line at the fuel rail and the vapor separator.
2. Install a clear, fuel resistant, hose between the rail and the reservoir.
3. Install a fuel pressure gauge at the pressure valve.

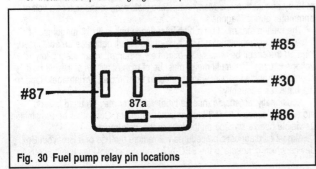

Fig. 30 Fuel pump relay pin locations

4. Ground the fuel pump lead as detailed in the Fuel Injection Pressure Check.
5. Turn the ignition switch to the **ON** position (but do not start the engine).
6. Note the fuel pressure reading and confirm that fuel is being returned to the reservoir through the clear tube.

✳✳ CAUTION

Do not keep the pump lead grounded for more than 10 seconds.

7. If the pressure is still higher than 36-42 psi (248-290 kPa), replace the pressure regulator and then repeat the Fuel Injection Pressure Test.
8. If pressure is no longer high, proceed to the Fuel Return System Check.

FUEL RETURN SYSTEM HIGH PRESURE CHECK

1. Check the fuel return lines for blockage or kinking.
2. Remove the return line at the pressure regulator and apply 3-5 lbs. of compressed air to the line.
3. If you hear air entering the reservoir, reconnect the line.
4. If no air can be heard entering the reservoir, proceed to the Fuel Return System Check.

FUEL RETURN SYSTEM CHECK

1. With the ignition switch in the **OFF** position and the battery fully charged, remove the fuel return line at the pressure regulator and the vapor separator reservoir.
2. Apply 3-5 lbs. of compressed air to the line at the regulator.
3. If air flows freely through the line, service the reservoir.
4. If air does not flow through the line, replace it or attempt to clear the blockage. Install the line and repeat the Fuel Injection Pressure Check.

FUEL INJECTOR FUNCTION CHECK

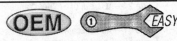

1. With the engine at normal operating temperature and idling, check that you can hear each injector pulsing with a mechanic's stethoscope.

■ **If the engine will not start, perform this step with the engine cranking.**

2. If you can detect that the injectors are operating, proceed to Fuel Injector Flow/Leakage Check.
3. If they are not working, proceed to Fuel Injector Resistance Check.

FUEL INJECTOR RESISTANCE CHECK

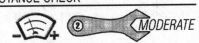

1. With the ignition switch in the **OFF** position and the battery fully charged, disconnect each injector connector individually.
2. Check that the resistance on each injector is 13-16 ohms.
3. If within range, proceed to Fuel Injector Continuity Signal Check.
4. If outside the range, replace the injector.

FUEL INJECTOR CONTINUITY CHECK

1. With the ignition switch in the **OFF** position and the battery fully charged, disconnect any injector lead on bank 1.
2. Connect a 12V test light to the B+ terminal in the connector and then touch the probe to the other connector terminal.
3. Start the engine and observe whether the test light blinks or not. Obviously, if the tester lights, then the circuit has been completed.

4. Repeat this procedure for bank 2.

5. If the circuit has been completed, proceed to Fuel Injector Flow/Leakage Check.

6. If the circuit does not indicate continuity on either bank, go back and test each injector connector on that bank. Service or replace the individual lead as necessary.

FUEL INJECTOR FLOW/LEAKAGE CHECK

1. Using a Fuel Injector Tester/Cleaner (#113-00001), clean and test the injectors. Now clean them once more. Match the color range on the tester to the top color of the injector and confirm that the flow rate is within specifications.

2. Check for any appreciable pressure drop while the special tool is connected to the system, either due to leakage or turning the unit off.

3. Check injector leak rate using the tool, anything over 1 drop/min. is too much.

4. If flow and leak rates were within specification, check all fuel lines for leaks. If all lines are OK, replace the fuel pump assembly and then repeat the Fuel Injection Pressure Check.

5. If either were out of specification, replace any defective injectors and then repeat this test and the Fuel Pressure Leakdown Check.

Diagnostic Trouble Codes—EEC IV System

- **No Code**—Loss of VREF
- **Code 10**—Vapor Separator
- **Code 11**—System Pass
- **Code 12**—RPM outside Self-Test upper band limit
- **Code 13**—RPM outside Self-Test lower band limit
- **Code 14**—PIP circuit fault
- **Code 15**—ROM test failed/KAM power in continuous (1993)
- **Code 15**—ROM test failed/KAM power interrupt (1994-96
- **Code 18**—Loss of TACH input to ECA/SPOUT circuit grounded
- **Code 19**—Failure in EEC reference voltage
- **Code 21**—ECT out of Self-test range
- **Code 22**—MAP out of Self-test range
- **Code 23**—TP out of Self-test range
- **Code 24**—ACT out of Self-test range
- **Code 25**—Inactive - Ignore code
- **Code 51**—40° indicated. ECT sensor circuit open
- **Code 52**—Shift assist circuit
- **Code 53**—TP circuit above maximum voltage
- **Code 54**—40° indicated. ACT sensor circuit open
- **Code 61**—254° indicated. ECT sensor circuit grounded
- **Code 63**—TP circuit below minimum voltage
- **Code 64**—254° indicated. ACT sensor circuit grounded
- **Code 67**—NDS circuit open
- **Code 72**—Insufficient MAP change - DYN RSP test
- **Code 73**— Insufficient TP change - DYN RSP test
- **Code 77**—Operator error - DYN RSP test
- **Code 87**—Fuel pump circuit failure
- **Code 95**—Fuel pump circuit open - ECA-to-motor ground
- **Code 96**—Fuel pump circuit open - BAT-to-relay
- **Code 98**—FMEM failure

GENERAL INFORMATION

Electronic Control Assembly (ECA)

One portion of the ECA is devoted to monitoring both input and output functions within the system. This ability forms the core of the self-diagnostic system. If a problem is detected within a circuit, the controller will recognize the fault, assign it an identification code, and store the code in memory. All codes are 2-digit numbers and may be retrieved during diagnosis.

While the EEC-IV system is capable of recognizing many internal faults, certain faults will not be recognized. Because the computer system sees only electrical signals, it cannot sense or react to mechanical or vacuum faults that may affect engine operation. Some of these faults though may affect another component which will set a code.

For example, the ECA monitors the output signal to the fuel injectors, but cannot detect a clogged injector. As long as the output driver is responding

correctly, the computer will read the system as functioning properly. However, the improper fuel will probably result in a lean mixture. This could, in turn, be detected by the knock sensor and then noticed by the ECA. Once the signal falls outside the pre-programmed limits, the ECA would notice a fault and then set a code.

Failure Mode Effects Management (FMEM)

The ECA contains back-up programs which allow the engine to operate if a sensor signal is lost. If a sensor input is seen to be out of range—either high or low—the FMEM program kicks in. The processor substitutes a fixed value for the missing sensor signal. The engine will continue to operate, although performance and driveability may be noticeably reduced. This function of the controller is sometimes referred to as the limp-home or fail-safe mode. If the missing sensor signal is restored, the FMEM system immediately returns the system to normal operation.

■ **When this cycle is in effect, the Self-Test function will show a service code 98.**

The error code associated with a particular fault will be stored in the Keep Alive Memory (KAM). If the fault is no longer detected, the engine will return to normal operating mode and the code will then be stored in Continuous Memory for the next 40 engine cycles before it is erased.

HAND-HELD SCAN TOOLS

◆ See Figure 31

Although codes may be read through the flashing of the light, the use of hand-held scan tools like the OMC MFI Tester (#500004) or an equivalent is highly recommended. There are alternate manufacturers of these tools so be very certain that the tools is correct for your intended use and application.

The scan tools allows any stored fault codes to be read from the ECA memory. Use of the scan tool also provides additional data during troubleshooting, but does not eliminate the use of charts. Scan tools make data collection much easier, but the data must be correctly interpreted by an operator familiar with the system.

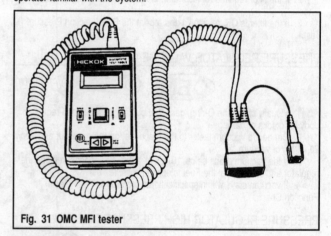

Fig. 31 OMC MFI tester

ELECTRICAL TOOLS

◆ See Figure 32

The most commonly required electrical diagnostic tool is the digital multi-meter (DVOM); allowing voltage, ohmage (resistance) and amperage to be read by one instrument. Many diagnostic tests require the use of a volt or ohmmeter during diagnosis.

The multi-meter must be a high impedance unit, with 10 megohms of impedance in the voltmeter. This type of meter will not place an additional load on the circuit being tested—extremely important when testing low voltage circuits. The multi-meter must be of high quality in all respects. It should be handled carefully and protected from impact or damage. Replace the batteries frequently.

Additionally, an analog (needle-type) voltmeter may be used to read stored fault codes if the MFI tester is not available. Codes will be transmitted as visible needle sweeps on the face of the instrument.

Almost all diagnostic procedures will require the use of a Breakout Box, a device which connects to the EEC-IV harness and provides testing ports for

Fig. 32 OMC Breakout Box

the 60 wires in the harness. Direct testing of the harness connectors at the terminals or by back-probing is not recommended; damage to the wiring and terminals is almost certain to occur.

Other necessary tools include a quality tachometer with inductive pick-up, a fuel pressure gauge with system adaptors and a vacuum gauge with an auxiliary source of vacuum.

READING CODES

Diagnosis of a problem requires attention to detail and following the diagnostic procedures in the correct order. Always resist the temptation to begin extensive testing before completing the preliminary diagnostic steps. The preliminary or visual inspection must be completed in detail before the diagnosis begins. In many cases this will shorten diagnostic time and often cure the problem without electronic testing.

Visual Inspection

This is possibly the most critical step of all diagnosis. A detailed examination of all connectors, wiring and vacuum hoses can often lead to a repair without further diagnosis. Performance of this step relies heavily on the skill of the person performing it; a careful inspector will inspect the underside of all hoses as well as the integrity of hard-to-reach hoses blocked by other components. All wiring should be checked carefully for any signs of strain, burning, crimping or terminal pull-out from a connector.

Checking connectors at components or in harnesses is required; usually pushing them together will reveal a loose fit. Pay particular attention to ground circuits, making certain they are not loose or corroded. Remember to check connectors and hose fittings at components not mounted on the engine. Any component or wiring in the vicinity of a fluid leak or spill should be given particular attention during inspection.

Additionally, inspect maintenance items such as belt condition and tension, battery charge and condition, and the radiator cap. Any of these items can easily set off a condition that will create a fault code.

Electronic Testing

If a code was set before a problem was self-corrected (such as a momentarily loose connector), the code will be erased if the problem does not re-occur within 40 engine cycles. Codes will be output and displayed as numbers on the hand-held tester.

In all cases, the code **11** is used to indicate PASS during testing. Please note that the PASS code may appear followed by other stored codes. These are codes from the Continuous Memory and may indicate intermittent faults, even though the system does not presently contain the fault. The PASS designation only indicates that the system passes all internal tests at the moment.

Key On Engine Off (KOEO) Test

■ **The engine must be a normal operating temperature before performing this test and all visual checks should be completed.**

1. Connect the MFI Tester to the STO and STI connectors on top of the ECA cover.
2. Set the Mode switch on the tester to the **FAST** position.
3. Set the SPKR switch on the tester to the **OFF** position.
4. Set the HOLD/TEST button to the **HOLD** position (raised).
5. Turn on the PWR switch. It should beep once and then briefly display **888** and then settle in at **000**.
6. Turn the ignition key to the **ON** position, but do not start the engine.
7. Press the HOLD/TEST button down to put the ECA into the Self-Test mode. After approximately 1 minute, any service codes should begin to appear on the screen.

✳✳ WARNING

Do not press the HOLD/TEST button again while the system is in Self-Test mode or you will erase all Continuous Memory codes.

8. Self-Test is complete when **CCRCVD** appears in the window. Press the HOLD/TEST button again to release it and lock all codes into the tester's memory.
9. Turn the ignition key to **OFF**.
10. You can use the **MEM/FWD** and **MEMREV** buttons to review the codes stored in the tool's memory. The first code will always be proceeded by a "beep" and the **1ST CD** prompt. Always review codes in the EXACT order in which they have been displayed.

Key On Engine Running (KOER) Test

■ **The engine must be a normal operating temperature before performing this test and all visual checks should be completed.**

1. Connect the MFI Tester to the STO and STI connectors on top of the ECA cover.
2. Set the Mode switch on the tester to the **FAST** position.
3. Set the SPKR switch on the tester to the **OFF** position.
4. Set the HOLD/TEST button to the **HOLD** position (raised).
5. Turn on the PWR switch. It should beep once and then briefly display **888** and then settle in at **000**.
6. Start the engine and allow it to idle for 10 seconds prior to moving on to the next step.
7. Press the HOLD/TEST button down to put the ECA into the Self-Test mode. After approximately 2 minutes, any service codes should begin to appear on the screen. For the first minute, the system will actuate the "High Idle" Self-test so you can confirm the ECA's control of timing.

■ **If a DYN RSP prompt appears, or a Code 10 appears toward the end of the "High Idle" test, ignore it.**

✳✳ WARNING

Do not press the HOLD/TEST button again while the system is in Self-Test mode or you will erase all Continuous Memory codes.

8. Self-Test is complete when **CCRCVD** appears in the window. Press the HOLD/TEST button again to release it and lock all codes into the tester's memory.
9. Turn the ignition key to **OFF**.
10. You can use the **MEM/FWD** and **MEMREV** buttons to review the codes stored in the tool's memory. The first code will always be proceeded by a "beep" and the **1ST CD** prompt. Always review codes in the EXACT order in which they have been displayed.

CLEARING CODES

Continuous Memory Codes

As mentioned previously, these codes are normally stored in the ECA memory for a duration of 40 engine warm-up cycles. To clear the codes for the purpose of testing or repair confirmation, perform the KOEO test. When the codes are just beginning to come up on the display, Press off the **HOLD/TEST** button and the codes will automatically be erased from the system's memory.

Do not disconnect the battery to clear the codes—they'll clear, but in the process set another code (19) for the ECA power loss.

VOLTAGE REFERENCE VALUES

Sensors

Sensor	DVOM Setting	Black Lead	Red Lead	Value	Notes
VREF	DCV	46	26	4.74-5.25V	-
TP	DCV	46	47	0.9-1.15V	Throttle Closed
				4.65V	WOT
ECT	DCV	46	7	0.87-1.17V	①
ACT	DCV	46	25	1.13-1.53V	①
MAP	Hz	46	45	159 Hz	②
PIP	DCV	46	56	0.0-0.3V	③
				VBAT	④
KS	DCV	46	2	0.3V	-
SAS (1993)	DCV	40	10	0	Not Actuated
				VBAT	Actuated
NDS (1993)	DCV	46	30	0	Neutral
				5	FWD & REV

① Measure the temperature with a pyrometer at the base of the sensor
② Based on 30 in.Hg pressure. Signal increases as pressure increases
③ Distributor cup opening in alignment with Hall Effect device
④ Distributor cup vane in alignment with Hall Effect device

Actuators

Sensor	DVOM Setting	Black Lead	Red Lead	Value
INJ Bank #1	DCV	40	58	VBAT
INJ Bank #2	DCV	40	59	VBAT
ISC	DCV	40	21	VBAT
FP	DCV	40	52	VBAT

Grounds

Sensor	DVOM Setting	Black Lead	Red Lead	Value
IGN	DCV	40/60	16	0
CSE	DCV	40/60	20	0
PWR	DCV	40/60	20	0

Power Supply

Sensor	DVOM Setting	Black Lead	Red Lead	Value	Notes
KAPWR	DCV	40/60	1	VBAT	Key On & Off
VPWR	DCV	40/60	37/57	VBAT	Key On only

VACUUM DIAGRAMS

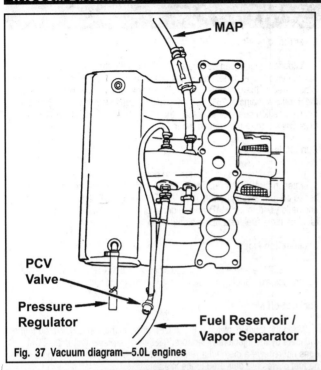

Fig. 37 Vacuum diagram—5.0L engines

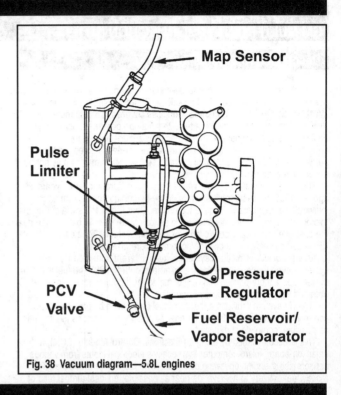

Fig. 38 Vacuum diagram—5.8L engines

PIN LOCATIONS

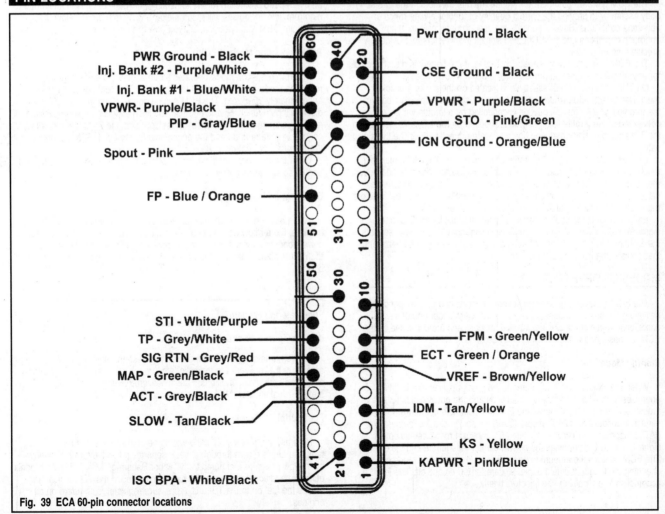

Fig. 39 ECA 60-pin connector locations

GM THROTTLE BODY FUEL INJECTION (TBI)
GM MULTI-POINT FUEL INJECTION (MFI)

Description And Operation

The purpose of this section is to describe—in layman's terms whenever possible—the Throttle Body Fuel Injection (TBI) and Multi-Port Fuel Injection (MFI) systems installed on many GM engines covered in this manual.

Visual inspections, and simple tests that may be performed using only basic shop test equipment. Again, we emphasize: specialized test equipment, hours of training and considerable experience is required to perform detailed service on a fuel injection system which may or may not be beyond the scope your skill level.

The first fuel injection system was introduced in Europe over 60 years ago—in 1932 on diesel truck engines. In the beginning, the system and individual components were quite expensive. Over the years, state-of-the-art microprocessors (commonly referred to as "computer chips"), have lowered the cost of electronically controlled fuel injection systems. Today, the price of EFI is getting close to the cost of modern carbureted systems.

An electronic fuel injection system is quite different from a carburetor system—even though they appear similar (particularly TBI). The fuel injection system has a more efficient delivery of fuel to the cylinders than can be obtained with standard carburetor operation.

The EFI system provides a means of fuel distribution by precisely controlling the air/fuel mixture and under all operating conditions for, as near as possible, complete combustion.

This is accomplished by using an Electronic Control Module (ECM), a small 'on-board' micro-computer that receives electrical inputs from various sensors about engine operating conditions. The ECM uses these inputs to modify fuel delivery to achieve, as near as possible, an ideal air/fuel ratio of 14.7:1.

The ECM program automatically signals the fuel injectors in the throttle body assembly to provide the correct quantity of fuel for a wide range of operating conditions. Several sensors are used to determine existing operating conditions and the ECM then signals the injectors to provide the precise amount of fuel.

The ECM has a "learning" capability. That is, if the battery is disconnected for any reason, the learning process has to begin all over again.

On TBI systems, the TBI assembly is located on the intake manifold, much like a carburetor, where air and fuel are distributed through a bore in the throttle body. Air for combustion is controlled by a throttle valve connected to the throttle linkage. Fuel is supplied by two injectors mounted in the TBI assembly; their metering tips are located directly above the throttle valve.

On MFI systems, there is a throttle body located on the intake manifold, much like a carburetor except that only air is distributed and metered through the bores on the throttle body. Air for combustion is controlled by a throttle valve connected to the throttle linkage. Fuel is supplied by individual injectors mounted in the intake manifold and attached to a fuel rail assembly.

Each injector is "pulsed" or "timed" to open or close by an electronic signal from the ECM. While constantly receiving input from various sensors, the ECM performs high speed calculations of engine fuel requirements and then "pulses the injectors open or closed.

MODES OF OPERATION

The ECM receives signals from several sensors, and then responds by delivering the proper amount of fuel to the cylinders under one of several conditions. These conditions are labeled "modes" and are controlled by the ECM, as described in the following short sections.

Starting Mode

When the ignition switch is rotated to the crank position, the ECM energizes the fuel pump relay for 2 seconds and the pump builds up pressure. The ECM then checks the Engine Coolant Temperature (ECT) sensor, Manifold Absolute Pressure (MAP) sensor, Intake Air Temperature (IAT) sensor and the Throttle Position (TP) sensor. From these incoming signals, the ECM determines the correct air/fuel ratio for starting the engine. The ECM controls the amount of fuel delivered in the starting mode by changing the length of time the injectors are turned on and off. This is accomplished by "pulsing" the injectors briefly.

Clear Flood Mode

A flooded engine can be cleared by opening the throttle to 75% of its travel. Once this is done, the ECM shuts down the fuel injectors and no fuel is delivered to the cylinders. The ECM will hold this injector rate as long as the throttle remains at 75% open and engine speed is below 400 rpm. If the throttle position changes to greater than 75% or slightly less than 75%, the ECM will return to the starting mode.

Run Mode

When the engine is first started and operating above 400 rpm, the system operates in the Run Mode. The ECM will calculate the desired air/fuel ratio based on rpm and input from the MAP, TP, IAT or ECT sensors. Higher engine load, from the MAP, and colder engine temperature, from the ECT requires more fuel, or a richer air/fuel ratio.

Acceleration Mode

If the ECM receives rapid change signals from the TP sensor, the ECM will provide extra fuel by increasing the injector pulse width.

Fuel Cut-off Mode

When the ignition switch is in the OFF position, no fuel is delivered to the cylinders by the injectors. Therefore, "dieseling" is prevented. If the ECM does not receive a distributor reference pulse—the engine is not operating—no fuel pulses are delivered to the injectors.

The fuel cut-off mode is also enabled at high engine rpm. This feature is an over-speed protection for the engine. Now, when cut-off is in effect due to high rpm, injector pulses will resume after engine rpm drops below the maximum OEM rpm specification (you now this as 'rev limit').

RPM Reduction Mode (SLOW)

The rpm reduction mode operates as an engine protection feature. The reduction mode permits normal fuel injection up to OEM specifications of approximately 2500 rpm. The ECM can recognize a change of state in a discrete input identifying an abnormal condition affecting engine operation. If a switch changes from its normal 'at rest' position, the ECM will sense the change in voltage and send the engine into Speed Limited Operational Warning (SLOW) safety mode.

Once the max rev rpm is hit, the system will disable half the injectors until engine speed drops below 1200 rpm.

Deceleration Mode

When backing off the throttle (decelerating) on certain late-model MFI engines, the air flow into the engine is reduced. The ECM recognizes the corresponding changes in the throttle position and manifold pressure and then shuts off fuel supply if the deceleration is very abrupt or for a long period.

SUBSYSTEMS

Fuel Metering System

◆ **See Figures 40 and 41**

As the name suggests, the fuel metering system "meters" the correct amount of fuel delivered to the cylinders through the intake manifold under all engine operating conditions. Fuel is delivered by the throttle body (TBI) or the individual injector (MFI) and is controlled by the ECM.

Fuel Supply

Naturally, the fuel supply will begin at the boat's fuel tank. From the fuel tank, the fuel is drawn through a water separating fuel filter and then moved by the low pressure fuel pump to the Vapor Separator Tank (VST). All models utilize an additional high pressure fuel pump found in the VST, which then moves the fuel on to the TBI unit. A pressure regulator is standard on all engines.

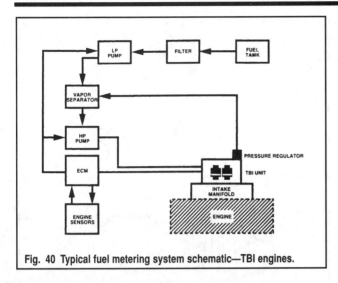

Fig. 40 Typical fuel metering system schematic—TBI engines.

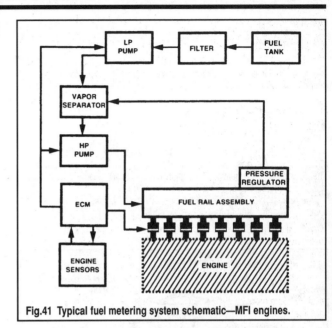

Fig.41 Typical fuel metering system schematic—MFI engines.

Water Separating Fuel Filter

The water separator is a "typical" unit designed to prevent moisture from continuing on through the fuel lines and eventually through the injectors into the cylinders.

Electric Fuel Pump

◆ See Figure 42

The low pressure fuel pump is located in line, just after the fuel separator assembly.

The high pressure fuel pump is located internally in the vapor separator tank.

When the ignition switch is moved to the **ON** position, the ECM will energize the fuel pump relay to ON, but only for a couple of seconds. The fuel pump almost instantly pressurizes the fuel system. As soon as the ignition switch is moved to the START position, the ECM energizes the fuel pump relay again and the fuel pump begins to operate.

Now, if the ECM fails to receive ignition reference pulses—indicating the engine is either cranking or actually operating—the ECM de-energizes the fuel pump relay and the fuel pump will stop.

An inoperative fuel pump relay can cause an "Engine Cranks, But Fails To Operate" condition.

The pump is capable of providing more fuel than is required by the injectors at WOT (wide open throttle). A pressure regulator is an integral part of the system maintaining fuel to the injectors at a predetermined regulated pressure. Excess fuel not required by the injectors is returned to the vapor separator tank by a separate fuel line.

Vapor Separator Tank/Fuel Reservoir

The vapor separator tank contains the internal electric fuel pump that pressurizes fuel to the throttle body. The tank also collects all excess fuel returned from the injectors through the pressure regulator.

A float valve on the fuel inlet line will keep the fuel level in the tank at a pre-determined height at all times. As the fuel is consumed by the engine and the level begins to drop, the float valve opens and allows fresh fuel to enter the tank. This float valve operates identically to a float in a carburetor.

Throttle Body

Models With TBI

The throttle body consists of the following assemblies:
• Fuel Meter Cover—also houses the pressure regulator on models with a vapor separator tank or fuel cell.
• Fuel Meter Body—two fuel injectors.
• Throttle Body.
• Throttle Valves—two throttle valves controlling air flow into the cylinders.
• IAC (Idle Air Control) valve.
• TP (Throttle Position) sensor.

From the above list of components, it can easily be appreciated why the throttle body is considered one of the most critical items in the injection system. If the proper amount of air and fuel are not injected into the cylinders, the engine will not operate efficiently or may even fail to start.

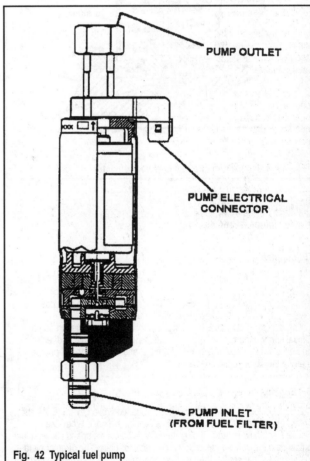

Fig. 42 Typical fuel pump

Models With MFI

The throttle body assembly is attached to the intake manifold or plenum and controls air flow to the engine, thus controlling engine output. Throttle valves within the assembly are controlled by the throttle cables. At idle, the valves are almost completely closed, while they open wider in proportion to the amount of throttle applied by the operator.

The TP sensor and IAC valve can also be found mounted to the throttle body.

Fuel Rail

Models With MFI

The fuel rail positions the injectors in the intake manifold, distributes fuel evenly to each injector and integrates the pressure regulator into the entire metering system.

Fuel Pressure Regulator

◆ See Figure 43

The pressure regulator is a diaphragm operated relief valve with fuel pump pressure on one side, and regulator spring pressure on the other side. The purpose of the regulator is to maintain a constant pressure differential across the injectors at all times under all conditions.

The regulator is located in the throttle body assembly on TBI models with a vapor separator tank or fuel cell, while on MFI models it is mounted on the fuel rail.

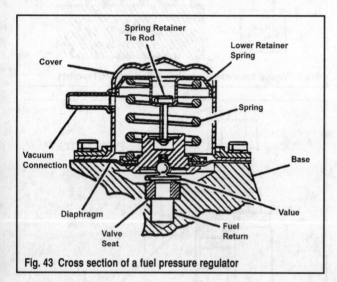

Fig. 43 Cross section of a fuel pressure regulator

Fuel Injectors

◆ See Figures 44 and 45

TBI Engines

Each injector is a solenoid-operated device, controlled by the ECM, that meters pressurized fuel into the intake manifold. When the injector's solenoid is energized by the ECM, a ball valve opens allowing fuel to flow past the valve and through a recessed flow director plate.

The director plate has 6 holes that control the fuel flow, generating a conical spray pattern of atomized fuel at the injector tip. Fuel is directed at the intake valve, becoming further atomized and then vaporized prior to entering the combustion chamber.

MFI Engines

Each injector is a solenoid-operated device, controlled by the ECM, that meters pressurized fuel into an individual engine cylinder. When the injector's solenoid is grounded by the ECM, a ball valve opens allowing fuel to flow past the valve and through a flow director plate. The director plate has six holes that control the fuel flow, generating a conical spray pattern of atomized fuel at the injector tip.

Intake Air Temperature Sensor (IAT)

This sensor, primarily used on many MFI engines, is a thermistor mounted on the underside of the intake plenum. A thermistor is a resistor capable of changing value based on temperature. Low coolant temperature produces a high resistance and a high temperature causes low resistance.

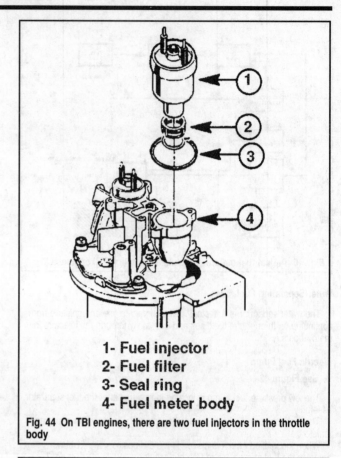

1- **Fuel injector**
2- **Fuel filter**
3- **Seal ring**
4- **Fuel meter body**

Fig. 44 On TBI engines, there are two fuel injectors in the throttle body

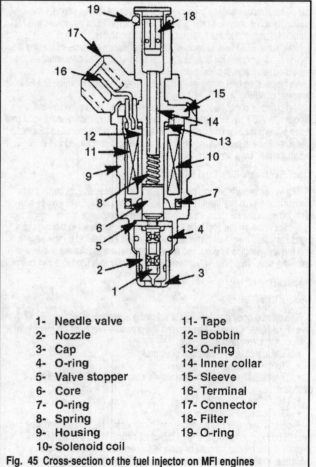

1-	Needle valve	11-	Tape
2-	Nozzle	12-	Bobbin
3-	Cap	13-	O-ring
4-	O-ring	14-	Inner collar
5-	Valve stopper	15-	Sleeve
6-	Core	16-	Terminal
7-	O-ring	17-	Connector
8-	Spring	18-	Filter
9-	Housing	19-	O-ring
10-	Solenoid coil		

Fig. 45 Cross-section of the fuel injector on MFI engines

Idle Air Control Valve (IAC)

◆ See Figure 46

The IAC valve assembly is mounted in the throttle body and controls engine idle speed. At the same time, the assembly prevents stalls due to changes in engine load.

Operation of the IAC can best be described as a device to control bypass air around the throttle valves. This feature is accomplished by moving a conical valve know as a "pintle" inward towards a seat—to decrease air flow or outward, away from the seat—to increase air flow. In this manner, a controlled amount of air moves around the throttle valve.

If rpm is too low, more air is bypassed around the throttle valve to increase rpm. If rpm is too high, less air is bypassed around the throttle valve to decrease rpm.

The ECM moves the IAC valve in small increments. These increments can only be measured using expensive special scan tool test equipment plugged into a DLC (Data Link Connector).

During engine idle speed, the proper position of the IAC valve is calculated by the ECM, and is based on coolant temperature and engine rpm. If the rpm drops below specification and the throttle valve is closed, the ECM senses a near stall condition and calculates a new valve position to prevent the stall.

Understand—engine idle speed is a function of total air flow into the cylinders based on IAC valve "pintle" position to maintain the desired idle speed for all engine operating conditions and loads.

The minimum idle air rate is set at the factory with a stop screw. This setting allows sufficient air flow by the throttle valves to cause the IAC valve "pintle" to be positioned a calibrated number of steps (counts) from the seat during "controlled" idle operation.

Throttle Position Sensor (TP)

◆ See Figure 47

The TP sensor is a potentiometer connected to the throttle shaft on the throttle body. One end of the sensor is connected to 5-volts from the ECM and the other end is connected to ECM ground.

A third wire is connected directly to the ECM to measure the voltage from the TP sensor. The voltage output of the TP sensor changes when the throttle valve angle is changed.

When the throttle position is closed, the voltage output of the TP sensor is low—about 1/2 V. When the throttle valve is opened, the output increases and at WOT the output voltage should be close to 5 V.

Therefore, the ECM can determine fuel delivery requirements based on throttle valve angle—operator demand.

Engine Coolant Temperature Sensor (ECT)

◆ See Figure 48

This sensor is a thermistor immersed in the engine coolant passageway. A thermistor is a resistor capable of changing value based on temperature. Low coolant temperature produces a high resistance and a high temperature causes low resistance.

Manifold Absolute Pressure Sensor (MAP)

◆ See Figure 48

The MAP sensor is a pressure transducer capable of measuring the changes in the intake manifold pressure. Pressure changes are the result of engine load and speed changes. MAP is the opposite of what is measured with a vacuum gauge.

When manifold pressure is high, vacuum is low— requires more fuel and of course the opposite is true. A low pressure—higher vacuum requires less fuel. The MAP sensor is also used to measure barometric pressure under certain conditions. This feature permits the ECM to automatically adjust for changes in operating altitude.

The ECM uses the MAP sensor to control fuel delivery and ignition timing.

Knock Sensor (KS)

◆ See Figure 50

The KS is usually mounted on the engine block drain "Y" fitting located on the lower starboard side of the engine block. On many other engines, the KS sensor is mounted on the block, next to the starter motor.

The ECM uses this signal in calculating ignition timing.

Knock Sensor Module

◆ See Figure 51

Used on 1994-98 engines, the KS module contains solid-state circuitry monitoring the KS AC voltage signal. If no spark knock is present an 8-10 volt signal is sent to the ECM. If a spark knock is present, the module will remove the 8-10 volt signal to the ECM.

Ignition Control (IC) Reference

The Hi Reference signal is supplied to the ECM by way of the "IC REF" line from the ignition module. The reference signal is used by the ECM to determine engine speed. This pulse type signal input creates the timing signal for pulsing of the fuel injections as well as the Ignition Control (IC) functions.

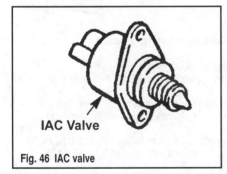

Fig. 46 IAC valve

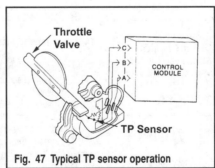

Fig. 47 Typical TP sensor operation

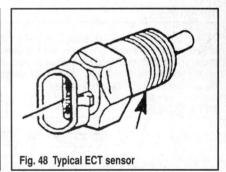

Fig. 48 Typical ECT sensor

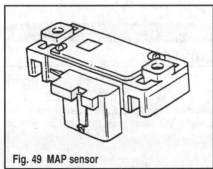

Fig. 49 MAP sensor

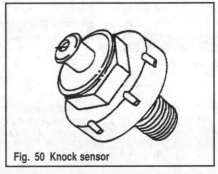

Fig. 50 Knock sensor

Fig. 51 Knock sensor and module system

Discrete Switch Inputs

Discrete switch inputs are utilized by the EFI system to identify abnormal conditions that could affect engine and stern drive operation. Normally these switches are at rest in an open position. When one of the discrete switches changes state from open to closed, the ECM will detect a change in the voltage value and responds by placing the engine into the power reduction mode.

The power reduction mode is an engine protection feature that allows the operator limited engine power up to 2500 rpm. Should the operator attempt to surpass this rpm limit, the ECM will reduce fuel and spark timing until the engine speed drops to approximately 1200 rpm. The power reduction mode will allow the operator maneuvering power but eliminates the possibility of high rpm engine damage, until the discrete switch fault condition is corrected.

Discrete switches are used to detect the critical engine and stern drive operations—namely Engine Oil Pressure, Oil Level, Emergency Stop, Water Flow and Stern Drive Fluid Level.

Trim Position Sensor

The trim position sensor provides the ECM with trim angle information which allows the ECM to determine if trim angle too high. When the angle reaches a pre-determined position, the system will move into an over-trim protection mode which is identical to the RPM Reduction Mode described previously.

Engine Control Module (ECM)

The ECM is the control center of the fuel injection system, monitoring input information from all sensors and then turning this information into system commands that affect engine performance.

Additionally, it will perform a diagnostic function check of the entire system; recognizing system problems and storing the resultant diagnostic trouble codes.

Relieving Fuel Pressure

✳✳ CAUTION

To reduce the risk of fire and personal injury, it is necessary to relieve the fuel system pressure before servicing any fuel system component. If this procedure is not performed, fuel may be sprayed out of a connection under extreme pressure. Always keep a dry chemical fire extinguisher near the work area when serving the fuel system.

TBI ENGINES

The throttle body utilizes a constant bleed feature that relieves fuel system pressure whenever the ignition switch is turned **OFF** and the engine is not running. No pressure relief procedure is required although prior to fuel system work, the negative battery cable should always be disconnected and the fuel tank filler cap should be loosened.

MFI ENGINES

1. Open the engine compartment and disconnect the battery cables.
2. Unscrew the relief valve cap on the high pressure fuel pump. Connect the special adapter (Owatonna #OTC-7272) to the valve and then connect a fuel pressure gauge (Owatonna #OTC-7211) to the adapter.

✳✳ CAUTION

Wrap a towel around the valve while connecting the special tools to avoid excessive fuel spillage.

3. Position the bleeder hose in an appropriate container and open the valve to bleed off all pressure (and any excess fuel!)
4. Remove the special tools and install the valve cap.

Circuit Breaker

REMOVAL & INSTALLATION

◆ See Figure 67

1. Open the engine compartment and disconnect the battery cables.
2. Locate the circuit breaker on the fuel pump lift bracket (front of engine) or the electrical bracket on later engines and disconnect the two electrical leads.

■ **Please refer to the Component Locations section.**

3. Remove the screws/nuts and lift off the breaker.
4. Install a new circuit breaker and tighten the screws/nuts to 20-25 inch lbs. (2.3-2.8 Nm).
5. Connect the Red lead to one stud on the circuit breaker and the Red/Purple lead to the other. Tighten the nuts 20-25 inch lbs. (2.3-2.8 Nm). Coat the two terminals with Liquid Neoprene.

Coolant Cover Assembly

REMOVAL & INSTALLATION

1996-98 MFI Engines Only (Except 1998 7.4Gi)

◆ See Figure 75

1. Remove the throttle body from the upper plenum as detailed elsewhere.
2. Remove the IAC valve.
3. Carefully turn the throttle body over and remove the coolant cover retaining screws. Lift off the cover and gasket.
4. Position a new gasket on the throttle body and then install the cover. Tighten the screws to 28 inch lbs. (3.2 Nm).
5. Install the IAC valve.
6. Install the throttle body.

Engine Coolant Temperature Sensor (ECT)

REMOVAL & INSTALLATION

◆ See Figures 52, 53, 71 and 72

■ **Please refer to the Component Locations section.**

1. Disconnect the battery cables.
2. Tag and disconnect the electrical connector at the sensor and move it out of the way.
3. Unscrew the ECT from the intake manifold (or housing) and remove it. Clean the sensor with a dry cloth; make sure to remove any excess sealant from the threads.
4. Coat the threads of the sensor with Teflon pipe sealant. Screw the sensor into the housing until it is hand tight and then tighten it to 108 inch lbs. (12 Nm).
5. Connect the electrical lead. Connect the battery cables.

Electronic Control Module (ECM)

REMOVAL & INSTALLATION

■ **Static electricity can severely damage the ECM! Take great care when removing the module that you never touch the connector pins on either side of the casing.**

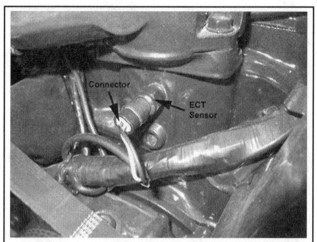

Fig. 52 The ECT sensor is located in the intake manifold on early TBI and MFI models (7.4L shown)...

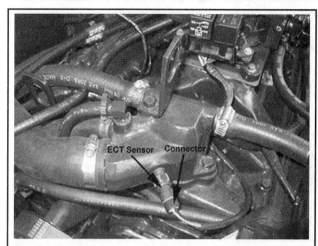

Fig. 53 ...or in the thermostat housing on later TBI and MFI models...

1. Open the engine compartment and disconnect the battery cables.
2. Locate the ECM and carefully disconnect the two harness leads (J1 and J2) going into each side of the module.

■ **Please refer to the Component Locations section.**

3. Loosen the 4 mounting screws and lift the ECM from the mounting bracket. Clean only with a clean, lint-free dry cloth. Check all connector pins for straightness and corrosion.

■ **If replacing the ECM, make sure that the new one has the same part and service number as the one being replaced.**

4. Install the ECM and tighten the mounting screws to 88-124 inch lbs. (10-14 Nm).
5. Reconnect the electrical lead at each end; the connectors and their sockets should be marked.

Fuel Injectors

REMOVAL & INSTALLATION

TBI Engines

◆ **See Figures 54, 55, 56, 57, 58 and 59**

✳✳ CAUTION

To reduce the risk of fire and personal injury, it is necessary to relieve the fuel system pressure before servicing any fuel system component. If this procedure is not performed, fuel may be sprayed out of a connection under extreme pressure. Always keep a dry chemical fire extinguisher near the work area when serving the fuel system.

1. Open or remove the engine hatch/cover and relieve the fuel system pressure. Disconnect the battery cable.
2. Remove the flame arrestor and then squeeze the plastic tabs on the injector electrical connectors and pull straight up until they come apart. Position the leads out of the way and make sure you tag them for correct reassembly. Many models will utilize a small grommet to secure the two harnesses to the side of the throttle body—just pop it out of the recess.
3. Loosen and remove the meter cover mounting screws and carefully lift the cover straight up and off of the throttle body assembly. Watch for the regulator seal and the outlet passage and cover gaskets—they may be hanging from the bottom of the cover.
4. Make sure that the old meter cover gasket is still in place. Lay a small metal rod across the throttle body and then insert a small pry bar under the injector lip and carefully pry the injector out and up using the metal rod as a fulcrum. On later models, you should be able to press on the injector tip with your finger to pop it out of the bore—try this first.

■ **Fuel injectors are serviced as an assembly. No disassembly is necessary or possible.**

To Install:

5. Inspect the injectors for damage or clogging. OMC recommends replacing damaged injectors. Always replace injectors with new ones that are the identical part number (as shown on the injector itself). All injectors are uniquely calibrated for flow rates so check the part number on the one being replaced and replace it with the same.
6. Install a new lower (small) O-ring after soaking it in automatic transmission fluid (1996), or Power Trim/Tilt & Power Steering Fluid (1997-98). Press the O-ring all the way up and against the injector fuel filter.
7. On certain 1996 engines, insert a new steel washer into the injector bore on the fuel meter body—do this only if your system had one there in the first place.
8. Soak a new upper O-ring in automatic transmission fluid (1996), or Power Trim/Tilt & Power Steering Fluid (1997-98). Install it into the injector bore on the fuel meter body. Make sure that it is seated correctly and flush with the top of the meter body surface.
9. Align the raised lug on the injector body with the notch in the meter cavity and install the injector, pushing it in until it is fully seated. The injector terminals should now be parallel with the throttle shaft.
10. Install the fuel meter cover and gasket. Reconnect the injector leads.
11. Install the flame arrestor and connect the battery cables. Turn the ignition switch to the **ON** position and check for leaks. Do not actually start the engine until confirming that there are no leaks at the throttle body or fuel lines.

MFI Engines

◆ **See Figures 60, 61, 71 and 72**

✳✳ CAUTION

To reduce the risk of fire and personal injury, it is necessary to relieve the fuel system pressure before servicing any fuel system component. If this procedure is not performed, fuel may be sprayed out of a connection under extreme pressure. Always keep a dry chemical fire extinguisher near the work area when serving the fuel system.

1. Relieve the pressure in the fuel system as detailed earlier.
2. Remove the upper intake plenum and fuel rail as detailed previously.
3. Slide the injector retaining clip to the side and pull the injector out of the rail.. Remove and discard the two O-rings from the ends of the injector.
4. Inspect the injector bores in the rail and the plenum for nicks, burrs or other damage. Clean the injector with a quality spray cleaner and dry completely with a lint-free cloth.

Fig. 54 Remove the connectors...

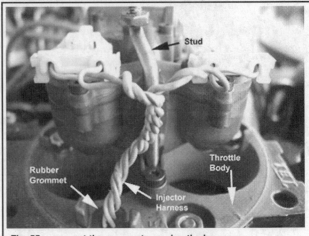

Fig. 55 ...pop out the grommet securing the harness...

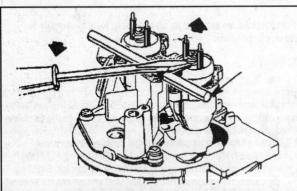

Fig. 56 ...and then carefully pry the injector out of the fuel meter body

✳✳ CAUTION

Fuel injectors are pre-coated with an anti-corrosion material— do not use an abrasive cleaner or wire brush when cleaning them.

5. Coat new O-rings with clean engine oil and slide them onto each end of the injector. Press a new injector clip over the injector. If replacing the injector, make absolutely certain that the new part carries the exact same number inscribed on the injector body.
6. Press the injector into the fuel rail so that the connector is facing away from the rail. Make sure that the new retainer clip snaps into the grooves on the rail.
7. Install the fuel rail and upper plenum.

Fuel Meter Body

REMOVAL & INSTALLATION

TBI Engines Only

◆ See Figure 62

■ The pressure regulator assembly is contained within the cover and is preset at the factory. Do not remove the four mounting screws for the pressure regulator installed in the fuel meter cover.

✳✳ CAUTION

To reduce the risk of fire and personal injury, it is necessary to relieve the fuel system pressure before servicing any fuel system component. If this procedure is not performed, fuel may be sprayed out of a connection under extreme pressure. Always keep a dry chemical fire extinguisher near the work area when serving the fuel system.

1. Open or remove the engine hatch/cover and relieve the fuel system pressure. Disconnect the battery cable.
2. Remove the flame arrestor and then squeeze the plastic tabs on the injector electrical connectors and pull straight up until they come apart. Position the leads out of the way and make sure you tag them for correct reassembly.
3. Remove the fuel meter cover assembly.
4. Remove the fuel injectors.
5. Disconnect the fuel inlet and outlet lines at the throttle body. Remove the fittings and their gaskets. Discard the gaskets. It may be necessary to remove the distributor cap on many models in order to gain access to the lines.

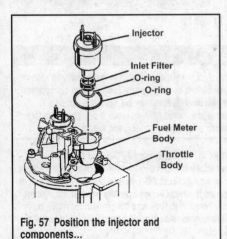

Fig. 57 Position the injector and components...

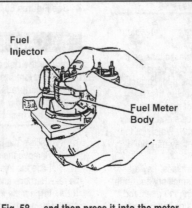

Fig. 58 ...and then press it into the meter body

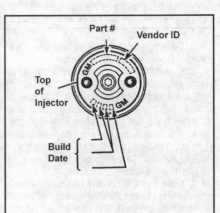

Fig. 59 Always check that the ID information on the new injector matches the old one

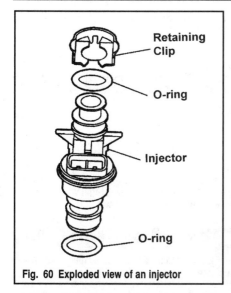

Fig. 60 Exploded view of an injector

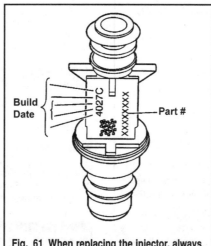

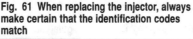

Fig. 61 When replacing the injector, always make certain that the identification codes match

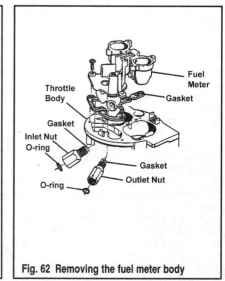

Fig. 62 Removing the fuel meter body

■ Although the fittings look similar, the inlet has a much larger bore than the outlet fitting so make sure you mark each of them for installation.

6. Remove the flame arrestor stud if you haven't already done so.

7. Remove the mounting screws and lift the meter body out of the throttle body. Remove the gasket and discard it.

To Install:

8. Position a new gasket on the throttle body mating surface. Make sure that the cut-outs in the gasket match the openings in the throttle body and then lower the meter body into place.

9. Coat the mounting screws with locking compound (Loctite 242) and then tighten them to 30 inch lbs. (4 Nm).

10. Position new gaskets and then spin in the two fuel fittings. Make sure you've got the right ones in position and then tighten the inlet to 30 ft. lbs. (40 Nm) and the outlet to 21 ft. lbs. (29 Nm).

11. Press new O-rings into the fittings and then screw the fuel lines in. Hold the fitting nut with a wrench while tightening the fuel line nut to 17 ft. lbs. (23 Nm).

12. Install the injectors and fuel meter cover as detailed in their respective procedures.

13. Install the flame arrestor stud.

14. Pop the injector lead connectors back on and connect the battery cables. Turn the ignition key to the **ON** position so that the fuel pump is energized and check for any fuel leaks. Do not actually start the engine until confirming that there are no leaks at the throttle body or fuel lines. Install the flame arrestor.

Fuel Meter Cover

REMOVAL & INSTALLATION

TBI Engines Only    MODERATE

◆ See Figure 63

■ The pressure regulator assembly is contained within the cover and is preset at the factory. Do not remove the four mounting screws for the pressure regulator installed in the fuel meter cover.

✳✳ CAUTION

To reduce the risk of fire and personal injury, it is necessary to relieve the fuel system pressure before servicing any fuel system component. If this procedure is not performed, fuel may be sprayed out of a connection under extreme pressure. Always keep a dry chemical fire extinguisher near the work area when serving the fuel system.

1. Open or remove the engine hatch/cover and relieve the fuel system pressure. Disconnect the battery cable.

2. Remove the flame arrestor and then squeeze the plastic tabs on the injector electrical connectors and pull straight up until they come apart. Position the leads out of the way and make sure you tag them for correct reassembly.

3. Loosen and remove the meter cover mounting screws (there are long and short screws so make sure you note which hole each one came out of) and carefully lift the cover straight up and off of the throttle body assembly. Watch for the damper seal and the outlet passage and cover gaskets—they may be hanging from the bottom of the cover.

■ On certain later models, you may have to loosen the flame arrestor stud and rotate it 180° to gain access to the screws.

4. Never immerse the cover in solvents or cleaner as damage to the regulator diaphragm could occur. Inspect the mating surfaces of the throttle body assembly and the regulator/cover for damage—use a magnifying glass if necessary! If any damage or scoring is found, replace the assembly.

5. Position a new regulator seal in the throttle body cavity and then position the outlet passage and cover gaskets on the assembly.

6. Drop the meter cover into position and then slip in the mounting screws. Make sure they are coated with locking compound and that the short screws go into the holes next to the injectors. Tighten all screws to 28 inch lbs. (3 Nm).

7. Pop the injector lead connectors back on and connect the battery cables. Turn the ignition key to the **ON** position so that the fuel pump is energized and check for any fuel leaks. Do not actually start the engine until confirming that there are no leaks at the throttle body or fuel lines. Install the flame arrestor.

Fuel Pressure Regulator

REMOVAL & INSTALLATION

TBI Engines

The pressure regulator is located in the throttle body fuel meter body and is not normally removed or serviced.

MFI Engines—Except 1998 7.4Gi   MODERATE

◆ See Figures 64 and 71

✳✳ CAUTION

To reduce the risk of fire and personal injury, it is necessary to relieve the fuel system pressure before servicing any fuel system component. If this procedure is not performed, fuel may be sprayed out of a connection under extreme pressure. Always keep a dry chemical fire extinguisher near the work area when serving the fuel system.

1. Open the engine cover and disconnect the battery cables. Relieve the fuel system pressure.

2. Remove the upper intake plenum and fuel rail as detailed previously.

3. Wiggle the vacuum line off of the regulator.

■ **Please refer to the Component Locations section.**

4. Hold the regulator body with a wrench and then disconnect the fuel return line.

5. Remove the mounting bolt and pull the regulator out of the rail. Discard the O-ring.

6. Use a small awl and carefully pry the regulator filter out of the fuel inlet port—this is the time to replace it.

To Install:

7. Inspect the regulator mounting surface for damage or irregularities. Clean the filter and blow it dry with compressed air if you have decided not to replace it.

8. Soak a new O-rings in clean engine oil and then position it into the regulator-to-rail bore and over the fuel line nipple.

9. Slide the regulator into position on the rail. Tighten the bolt to 84 inch lbs. (9.5 Nm).

10. Connect the outlet line to the regulator and spin in the fitting so it is finger-tight. Tighten the line nut to 13 ft. lbs. (17.5 Nm).

11. Connect the vacuum line and install the fuel rail and upper plenum. Connect the battery cables.

12. Turn the ignition switch to the **ON** position for 2 seconds and then turn it to the **OFF** position for 10 seconds. Now turn it back to the **ON** position and check for any fuel leaks before starting the engine.

MFI Engines—1998 7.4Gi

◆ **See Figures 65 and 72**

✳✳ CAUTION

To reduce the risk of fire and personal injury, it is necessary to relieve the fuel system pressure before servicing any fuel system component. If this procedure is not performed, fuel may be sprayed out of a connection under extreme pressure. Always keep a dry chemical fire extinguisher near the work area when serving the fuel system.

1. Open the engine cover and disconnect the battery cables. Relieve the fuel system pressure.

2. Remove the upper intake plenum and fuel rail as detailed previously.

■ **Please refer to the Component Locations section.**

3. Wiggle the vacuum line off of the regulator.

4. Remove the retaining ring at the rear of the regulator and then pull the unit from the housing along with the 2 O-rings, spacer and screen.

5. Coat new O-rings with clean engine oil and then install the spacer, large O-ring, screen and small O-ring onto the regulator in order. Hold the pieces on the regulator and slide it into the housing.

6. Press in the retaining ring and connect the vacuum line.

7. Connect the battery cables. Turn the ignition switch to the **ON** position for 2 seconds and then turn it back to the **OFF** position for 10 seconds. Turn the switch back **ON** and check for any fuel leaks before starting the engine.

Fuel Pump

REMOVAL & INSTALLATION

High Pressure

◆ **See Figure 66**

■ **Always have a fire extinguisher handy when working on any part of the fuel system. Remember, a very small amount of fuel vapor in the bilge, has the tremendous potential explosive power.**

✳✳ CAUTION

All OMC fuel pumps are designed to meet U.S. Coast Guard fuel system regulations. Never substitute and automotive style pump.

1. Open or remove the engine hatch/cover and relieve the fuel system pressure as detailed previously. Disconnect the battery cable.

2. Locate the pump at the rear of the engine, on top of the fuel reservoir/vapor separator.

■ **Please refer to the Component Locations section.**

3. Make a note of the fitting elbow positioning and then hold the elbow while removing the fuel line; be sure to have some rags handy to catch any spilled fuel and make sure you plug the line before moving it out of the way.

4. With your wrench still on the elbow, back off the lock nut with another wrench. The lock nut is the nut immediately inside of the elbow; do not confuse it with the end cap nut on the end of the pump. Unscrew the elbow from the end of the pump and set it aside.

5. Remove the 2 screws securing the end bracket to the reservoir and then remove the bracket from the pump.

6. Remove the banjo nut from the end of the pump and then remove the pump. Discard all O-rings.

To Install:

7. Lubricate new O-rings with clean engine oil. Slide the small one onto the elbow fitting (right up against the flange) and the larger one onto the banjo nut so it sits in the groove on the inner side of the head.

8. Slide the banjo nut into the hole on the reservoir cover. Coat another new O-ring with clean engine oil and position it over the threads and up against the bracket.

9. Position the pump at a 45° angle, hold the end cap with a wrench and thread the banjo nut into the pump. Tighten it to 18-22 ft. lbs. (24.4-29.8 Nm). Be very careful to hold the end cap while tightening the banjo nut or you risk damaging the internal O-ring.

10. Position the support bracket over the other end of the pump and tighten the screws to 25-35 inch lbs. (2.8-4.0 Nm).

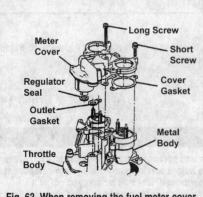

Fig. 63 When removing the fuel meter cover, keep track of the gaskets and seals

Fig. 64 The regulator is attached to the fuel rail

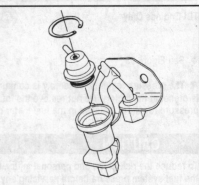

Fig. 65 The regulator is attached to a bracket at the rear of the fuel rail on 7.4Gi engine

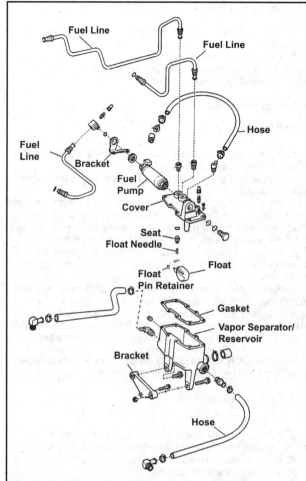

Fig. 66 Exploded view of the high pressure fuel pump and reservoir assembly (TBI shown, MFI similar)

11. Screw the elbow fitting into the end of the pump so that it is in the same position as when removed. Cinch the lock nut down against the pump outlet and then tighten it to 60-84 inch lbs (6.8-9.5 Nm) while holding the pump end cap.

12. Install the fuel line and connect the battery cables.

Low Pressure

◆ See Figures 67, 68 and 69

■ Always have a fire extinguisher handy when working on any part of the fuel system. Remember, a very small amount of fuel vapor in the bilge, has the tremendous potential explosive power.

✳✳ CAUTION

All OMC fuel pumps are designed to meet U.S. Coast Guard fuel system regulations. Never substitute and automotive style pump.

1. Open or remove the engine hatch/cover and relieve the fuel system pressure as detailed previously. Disconnect the battery cable.

■ Please refer to the Component Locations section.

2. Position a container under the pump/filter assembly and remove the fuel inlet line at the filter bracket and the outlet line at the top of the pump. It is important to use two open-end wrenches; one to hold the fuel fitting nut or elbow and the other to loosen the line nut. Plug the line with a golf tee or something similar to prevent additional fuel spillage (OMC actually sells a plug for this purpose). Take care not to spill fuel on a hot engine, because such fuel may ignite.

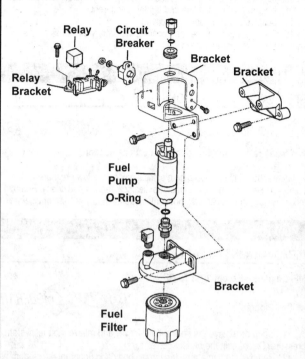

Fig. 67 Exploded view of the low pressure fuel pump and filter assembly (all engines similar)

3. Cut the plastic tie holding the electrical leads to the body of the pump. Unplug the electrical harness at the pump suppressor connector and move it out of the way.

4. Hold the pump end cap (just under the lower edge of the upper retaining bracket) with a wrench while loosening the lock nut (on top of bracket, under elbow). Unscrew the upper elbow.

✳✳ CAUTION

Not holding the end cap while loosening the lock nut will damage the internal O-ring, causing fuel leakage during operation.

5. Remove the 2 fuel filter bracket mounting bolts and lift off the pump/bracket/filter as an assembly, sliding the pump out of the upper grommet. Be careful of spilling fuel.

6. Lay the assembly down, or carefully mount the bracket in a vise. Hold the lower pump end cap with a wrench and unscrew the filter bracket-to-pump adapter.

✳✳ CAUTION

Not holding the end cap while loosening the lock nut will damage the internal O-ring, causing fuel leakage during operation.

To Install:

7. Slide a new O-ring over the threads of the adapter until it rests on the upper side of the hex nut.

8. Screw the adapter/bracket assembly into the pump while holding the lower end cap. Tighten the adapter to 8-10 ft. lbs. (11-14 Nm).

9. Slide the assembly up through the grommet in the bracket and position it so the mounting holes line up on the bracket. Tighten the filter bracket bolts to 20-25 ft. lbs. (27-34 Nm).

10. Thread the upper lock nut onto the elbow until it reaches the underside of the fitting. Slip on the washer and then a new O-ring. Thread the elbow into the top of the pump until it stops and then back it out just enough to align with the fuel line. Connect the line and tighten it securely.

11. Hold the upper end cap again and tighten the lock nut to 60-84 inch lbs. (6.8-9.5 Nm).

12. Reconnect the suppressor lead—push it in until the tabs click.

■ If you notice any of the suppressor holes open, be sure to coat them with Liquid Neoprene to prevent water from entering.

13. Install a new plastic tie around the pump body and electrical leads.
14. Reconnect the inlet line and tighten it securely.
15. Install the pump guard if removed.
16. Remove the container and connect the battery cables. Start the engine and check for fuel leaks.

Fuel Pump Relay

REMOVAL & INSTALLATION

◆ **See Figure 70**

■ **Please refer to the Component Locations section.**

Locate the relay in the engine electrical box on the low pressure fuel pump bracket. Tag and disconnect the electrical lead (if equipped). Remove the relay from the bracket. Do not soak the relay in any solvent of cleaner.

Fuel Rail

REMOVAL & INSTALLATION

MFI Engines—Except 1998 7.4Gi

◆ **See Figure 71**

1. Open or remove the engine hatch/cover and relieve the fuel system pressure. Disconnect the battery cable.
2. Remove the flame arrestor, throttle body and upper intake plenum as detailed elsewhere.
3. Tag and disconnect the electrical harness at each injector, moving all of them out of the way. Cut the plastic ties holding the injector harness to the rail and then carefully position the harness out of the way.
4. Making sure you have a container and plenty of rags available, remove the inlet and outlet fuel lines from the rail. Plug the lines and keep track of the O-rings.
5. Remove the 4 rail mounting screws and lift out the fuel rail. There's a good chance that some of the injectors are going to come out with the rail while others may stay in the manifold—don't worry about this.
6. Remove all injectors from the rail and the intake manifold.

To Install:

7. Inspect the injector seating recess in the fuel rail for pitting, nicks, burrs or any other irregularities. Replace the rail assembly if necessary.
8. Soak new injector O-rings in clean engine oil for a minute and then press them onto each side of the injector.
9. Press each injector (with a new retaining clip) into its recess on the rail until it clicks into place.
10. Position the rail/injector assembly onto the intake manifold and gently press down on the rail until the injectors seat fully into their bores. Install the rail mounting screws and tighten them to 88 inch lbs. (10 Nm).
11. Move the injector harness into position and reconnect all injectors. Secure the harness to the rail with new plastic ties.

12. Connect the two fuel lines, tightening the fittings to 13 ft. lbs. (17.5 Nm).
13. Install the plenum, throttle body and flame arrestor. Connect the battery cables.
14. Turn the ignition switch to the **ON** position for 2 seconds and then turn it to the **OFF** position for 10 seconds. Now turn it back to the **ON** position and check for any fuel leaks before starting the engine.

MFI Engines—1998 7.4Gi

◆ **See Figure 72**

1. Open or remove the engine hatch/cover and relieve the fuel system pressure. Disconnect the battery cable.
2. Remove the flame arrestor, throttle body and upper intake plenum as detailed elsewhere.
3. Tag and disconnect the electrical harness at each injector, moving all of them out of the way. Disconnect the 10-pin injector harness connector.
4. Remove the nut and retainer that secures the MAP sensor and then lift off the sensor—this is also one of the fuel rail retaining bolts.
5. Making sure you have a container and plenty of rags available, remove the inlet and outlet fuel lines from the rail and pressure regulator. Plug the lines.
6. Remove the 2 rail mounting bolts and the stud. Remove the 2 bolts holding down the fuel line bracket. Disconnect the vacuum line at the pressure regulator and then lift out the fuel rail and injector assembly. There's a good chance that some of the injectors are going to come out with the rail while others may stay in the manifold—don't worry about this.

■ **You will probably want to remove the distributor to provide better access to the two fuel line bracket mounting bolts.**

7. Remove all injectors from the rail.

To Install:

8. Inspect the injector seating recess in the fuel rail for pitting, nicks, burrs or any other irregularities. Replace the rail assembly if necessary.
9. Soak new injector O-rings in clean engine oil for a minute and then press them onto each side of the injector.
10. Press each injector (with a new retaining clip) into its recess on the rail until it clicks into place.
11. Position the rail/injector assembly onto the intake manifold and gently press down on the rail until the injectors seat fully into their bores. Install the rail mounting screws and tighten them to 88 inch lbs. (10 Nm). Tighten the stud to 18 ft. lbs. (25 Nm).
12. Move the injector harness into position and reconnect all injectors. Plug in the 10-pin harness connector.
13. Wiggle the vacuum line over the fitting on the regulator.
14. Connect the two fuel lines, tightening the fittings to 10-12 ft. lbs. (13.6-16.3 Nm). Make sure you use new O-rings and position them properly.
15. Install the MAP sensor.
16. Install the plenum, throttle body and flame arrestor. Connect the battery cables.

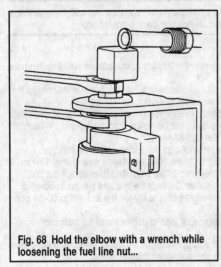

Fig. 68 Hold the elbow with a wrench while loosening the fuel line nut...

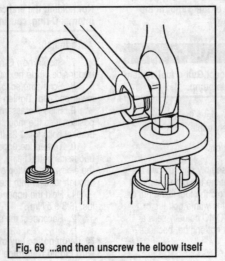

Fig. 69 ...and then unscrew the elbow itself

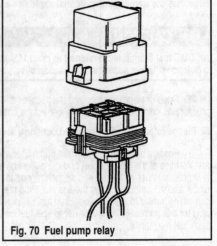

Fig. 70 Fuel pump relay

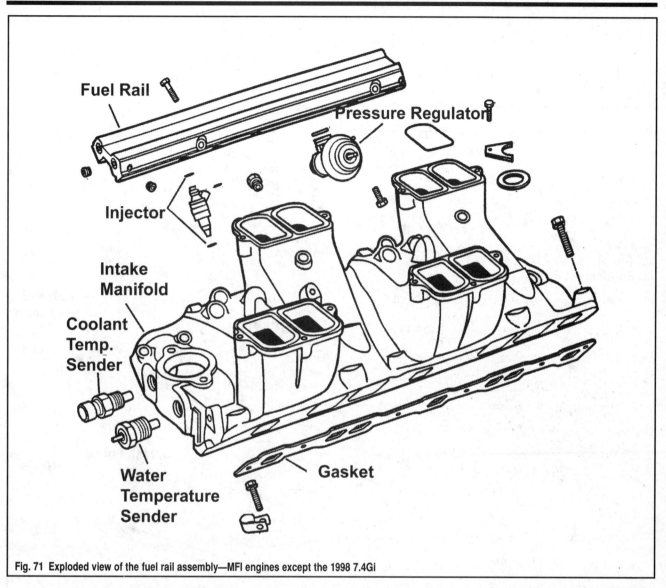

Fig. 71 Exploded view of the fuel rail assembly—MFI engines except the 1998 7.4Gi

17. Turn the ignition switch to the **ON** position for 2 seconds and then turn it to the **OFF** position for 10 seconds. Now turn it back to the **ON** position and check for any fuel leaks before starting the engine.

Idle Air Control Valve (IAC)

REMOVAL & INSTALLATION

◆ **See Figures 73 and 74**

1. Open the engine compartment and disconnect the battery cables.
2. Remove the flame arrestor.

■ **Please refer to the Component Locations section.**

3. Tag and disconnect the electrical connector at the valve and move it out of the way.
4. On flange-mounted valves, remove the 2 mounting screws and remove the valve. Remove the O-ring from the valve and throw it away.

To Install:

5. On either valve, clean the valve mating surfaces, pintle valve seat and air passage with carburetor cleaner— do not push or pull on the valve pintle. Do not soak the valve in liquid cleaner or solvent. Don't be concerned if you notice shiny spots on the pintle or seat. If you are planning to reinstall the original valve, there should be thread locking compound on the threads of the valve or mounting bolts; do not attempt to clean this compound.

■ If replacing the IAC valve, make sure that the new one has the same pintle shape and diameter. Measure the distance between the tip of the pintle and the valve mounting surface (A in the illustrations). If this measurement is greater than 1.102 in. (28mm), press the pintle in with your finger until it is within the specification. Measure the diameter of the pintle cone (B in the illustrations) and make sure that it is the same as the old valve.

6. Install a new O-ring or gasket onto the valve. Coat the O-ring with clean engine oil.

7. Install the valve and tighten the screws to 28 inch lbs. (3.2 Nm).

8. Reconnect the electrical lead.

9. Install the flame arrestor.

10. Connect the battery cables.

11. On 1996 TBI engines, turn the ignition key to the **ON** position for 10 seconds, start the engine and run it for 5 seconds, turn the key to the **OFF** position for 10 seconds. Start the engine and check that the idle is OK.

12. On all MFI engines and on 1997-98 TBI engines, start the engine and allow it to idle for 30 seconds, turn it Off for 10 seconds and then restart the engine and check that the idle is OK.

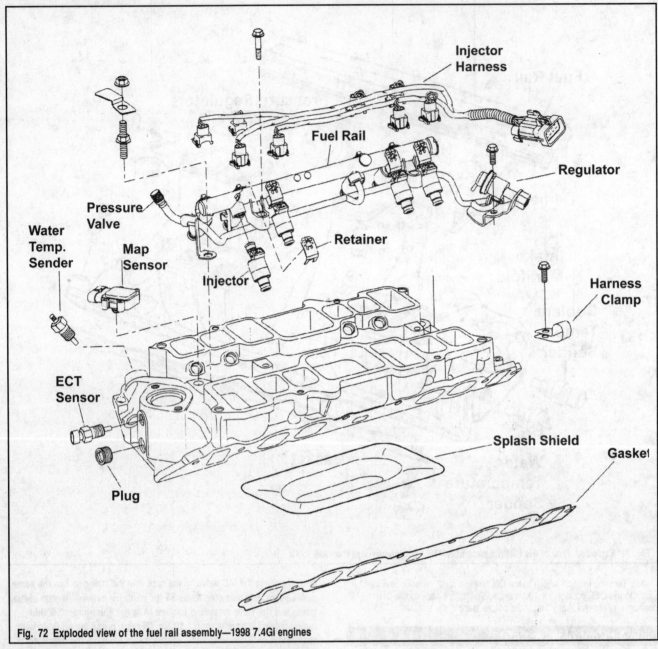

Injector Harness

Fuel Rail

Regulator

Pressure Valve

Water Temp. Sender

Map Sensor

Retainer

Injector

Harness Clamp

ECT Sensor

Splash Shield

Gasket

Plug

Fig. 72 Exploded view of the fuel rail assembly—1998 7.4Gi engines

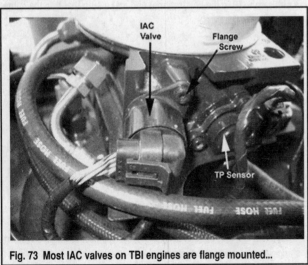

IAC Valve

Flange Screw

TP Sensor

Fig. 73 Most IAC valves on TBI engines are flange mounted...

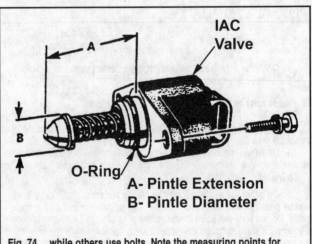

IAC Valve

A

B

O-Ring

A- Pintle Extension
B- Pintle Diameter

Fig. 74 ...while others use bolts. Note the measuring points for comparison

Intake Air Temperature Sensor (IAT)

REMOVAL & INSTALLATION

1. Open the engine compartment and disconnect the battery cables.

■ **Please refer to the Component Locations section.**

2. Tag and disconnect the electrical lead and unscrew the sensor from the upper intake plenum.
3. Clean the sensor with a dry cloth and inspect for signs of damage.
4. Coat the threads of the sensor with Teflon pipe sealant, install it into the plenum and tighten it to 11 ft. lbs. (15 Nm).
5. Connect the sensor lead and the battery cables.

Knock Sensor (KS)

REMOVAL & INSTALLATION

◆ **See Figure 75**

■ **Some engines will have two sensors (1998 7.4Gi), one on each side of the engine.**

1. Disconnect the battery cables.
2. Tag and disconnect the electrical connector at the sensor and move it out of the way.

■ **Please refer to the Component Locations section.**

3. Unscrew the sensor from the side of the cylinder block (or T-fitting) and remove the sensor. Clean the sensor with a dry cloth; especially the threads. If replacing the sensor, make sure that the new one is the same part number as the original.
4. Repeat Steps 2 and 3 for the second sensor on engines so equipped.
5. Install the sensor into the block or "T" and tighten it to 11-16 ft. lbs. (15-22 Nm). Do not use any thread sealer or locking solution unless it is positioned in a water jacket. Proper torque is imperative to ensure precise performance
6. Connect the electrical lead and the battery cables.

Knock Sensor Module

REMOVAL & INSTALLATION

■ **1998 7.4Gi engines will require removal of the oil fill tube for access to the module.**

1. Disconnect the battery cables.

■ **Please refer to the Component Locations section.**

2. Tag and disconnect the electrical connector at the module and move it out of the way.
3. Remove the 2 (or 4) module mounting bolts and lift off the module.

✳✳ WARNING

Never soak the module in liquid cleaner or solvent.

4. Install the module and tighten the mounting bolts securely.
5. Connect the electrical lead and the battery cables.

Manifold Absolute Pressure Sensor (MAP)

REMOVAL & INSTALLATION

◆ **See Figures 72, 76 and 77**

1. Open or remove the engine hatch/cover and disconnect the battery cable.
2. Locate the MAP sensor—on TBI engines it can be found on the ECM electrical bracket. On most MFI engines it is attached directly to the Starboard side of the upper intake plenum or directly to the intake manifold on 1998 7.4Gi engines.

■ **Please refer to the Component Locations section.**

3. Remove the upper intake plenum on the 1998 7.4Gi.
4. Tag and disconnect the electrical connector and the vacuum line (TBI only) at the sensor and move them out of the way.
5. Unscrew the mounting screw(s) and remove the sensor. On the 1998 7.4Gi, you must first remove the sensor retainer attached to the fuel rail. Clean the sensor with a dry cloth; make sure that the sensor seal is in good condition, if not replace it.
6. Install the sensor seal and then install the sensor and tighten the screws to 44-62 inch lbs. (5-7 Nm), except on the 1998 7.4Gi where you need to tighten the retainer nut to 18 ft. lbs. (24 Nm).
7. Install the upper plenum if removed.
8. Connect the electrical lead and the battery cables. Don't forget the vacuum line on TBI engines.

Throttle Body

REMOVAL & INSTALLATION

TBI Engines

◆ **See Figure 78, 79 and 80**

1. Open or remove the engine hatch/cover and relieve the fuel system pressure. Disconnect the battery cable.
2. Remove the flame arrestor and then squeeze the plastic tabs on the injector electrical connectors and pull straight up until they come apart. Position the leads out of the way and make sure you tag them for correct reassembly. Many models will utilize a rubber grommet securing the two harnesses to the side of the throttle body—just pop it out of the recess.
3. Disconnect the throttle cable and carefully move it out of the way.
4. Remove the distributor cap. Although not always necessary, this will make access much easier.

Fig. 75 The knock sensor is usually found just forward of the starter

Fig. 76 On all TBI engines, the MAP sensor is located on the ECM electrical bracket...

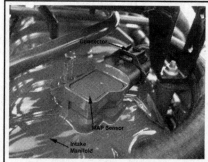

Fig. 77 ...or attached to the manifold on MFI engines

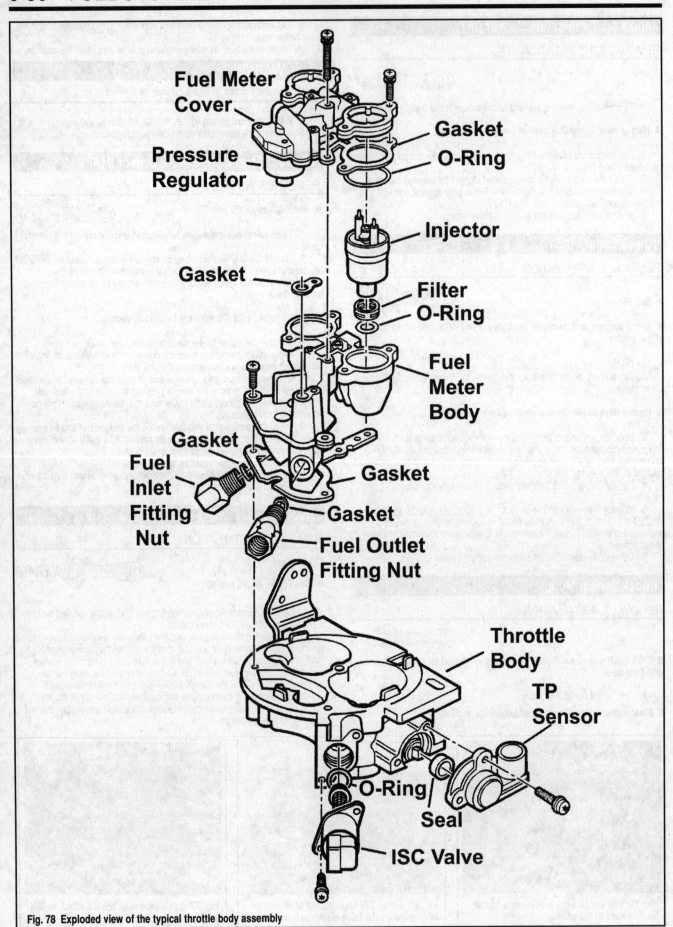

Fig. 78 Exploded view of the typical throttle body assembly

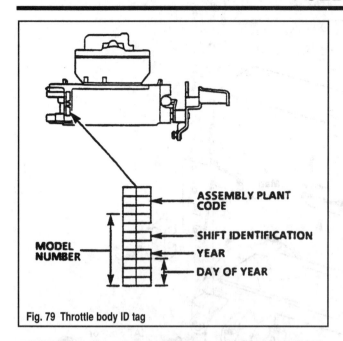

Fig. 79 Throttle body ID tag

ASSEMBLY PLANT CODE

SHIFT IDENTIFICATION

MODEL NUMBER

YEAR

DAY OF YEAR

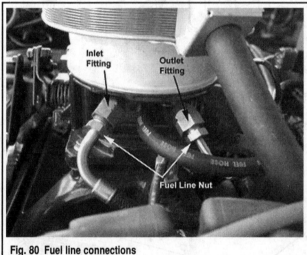

Fig. 80 Fuel line connections

Inlet Fitting

Outlet Fitting

Fuel Line Nut

5. Unplug the electrical connectors at the TP sensor and the IAC valve. Tag each lead and move them out of the way.

6. Disconnect the MAP sensor vacuum line at the throttle body (between the fuel lines on later engines).

7. Hold a container under the connections and then remove the fuel inlet and outlet lines at the throttle body. Have some rags available to wipe up any spills. Plug the fuel lines to prevent any further spillage. Don't forget to use a second flare wrench to hold the fitting nuts.

8. Remove the mounting bolts/nuts and lift the throttle body off of the adapter plate. Cover the manifold opening with a rag.

■ Be very careful of how, and where, you lay the unit down.

To Install:

9. Remove the TP sensor, IAC valve, fuel injectors and the meter cover and then thoroughly clean the throttle body assembly in an immersion type cleaner. Dry with compressed air if available and make certain all passages are clean.

10. Inspect all mating and casting surfaces for damage or cracks and scoring.

11. Install the sensor, valve, injectors and meter cover as detailed elsewhere in this section.

12. Position a new gasket on the adapter plate, install the throttle body and tighten the mounting nuts/bolts to 12 ft. lbs. (16 Nm).

13. Connect and adjust the throttle linkage.
14. Connect the MAP vacuum line if removed.
15. Connect the fuel lines and tighten the flange nuts to 17 ft. lbs. (23 Nm) using a back-up flare wrench on the fittings. Make sure that you install new O-rings prior to tightening the lines.
16. Connect the TP sensor and IAC valve electrical leads.
17. Press the harness grommet into the recess and then pop the injector lead connectors back on and connect the battery cables. Turn the ignition key to the **ON** position so that the fuel pump is energized and check for any fuel leaks. Do not actually start the engine until confirming that there are no leaks at the throttle body or fuel lines. Install the flame arrestor.
18. Start the engine and check for fuel leakage.

MFI Engines

◆ See Figure 81

1. Open or remove the engine hatch/cover and relieve the fuel system pressure. Disconnect the battery cable.
2. Remove mounting nut(s) and then remove the flame arrestor. Certain models will have a bracket holding the arrestor on.
3. Remove the throttle link nut and then separate the link from the throttle arm.
4. Tag and disconnect the wiring harness for the IAC valve and TP sensor.
5. Remove the 4 throttle body mounting bolts (some models use 3 nuts). Lift out the throttle body and set it down in a holding fixture to avoid damage to the valves. Stuff a clean, lint-free rag into the plenum opening and discard the gasket.
6. If necessary, remove the TP sensor and IAC valve with their O-rings from the assembly.

To Install:

7. Carefully clean the throttle bore and valves. Do not use anything containing methyl ethyl ketone and do not soak it in solvent—use a spray carb/choke cleaner! If you didn't remove the TP sensor and IAC valve, make sure that you get no solvent on them during cleaning. CAREFULLY scrape any gasket material off the mating surfaces. Make sure all passages are free of dirt and completely dry before installation.

8. Install the sensor and valve if removed. Position the throttle body and new gasket on the plenum and tighten the bolts to 11 ft. lbs. (15 Nm). If your engine uses nuts (1998 7.4Gi), tighten them to 89 inch lbs. (10 Nm).

9. Check that the throttle lever moves freely, snapping back to the closed position and then reconnect the throttle link to the lever. Tighten the nut securely. Make sure the rod end on 1998 7.4Gi engines has 0.5-0.75 in. (12.70-19.05mm) of available thread for correct engagement.

10. Connect the IAC and TP leads.
11. Install the flame arrestor.
12. Reconnect the battery cables.
13. Start the engine and allow it to idle for 30 seconds, turn it **OFF** for 10 seconds and then restart the engine and check that the idle is OK.

Throttle Body Adapter Plate

REMOVAL & INSTALLATION

TBI Engines Only

1. Open or remove the engine hatch/cover and relieve the fuel system pressure. Disconnect the battery cable.
2. Remove the flame arrestor and then squeeze the plastic tabs on the injector electrical connectors and pull straight up until they come apart. Position the leads out of the way and make sure you tag them for correct reassembly.
3. Remove the throttle body as detailed in the preceding procedure. Lift the gasket off the adapter if it hasn't already come up with the throttle body.
4. Remove the four mounting screws and lift off the adapter plate. Plug the intake manifold with a clean rag.
5. Install the adapter plate with a new gasket and tighten the mounting screw to 15 ft. lbs (19 Nm) on the V6 and 1998 V8, 30 ft. lbs. (40 Nm) on the 1996-97 V8.
6. Install the throttle body.

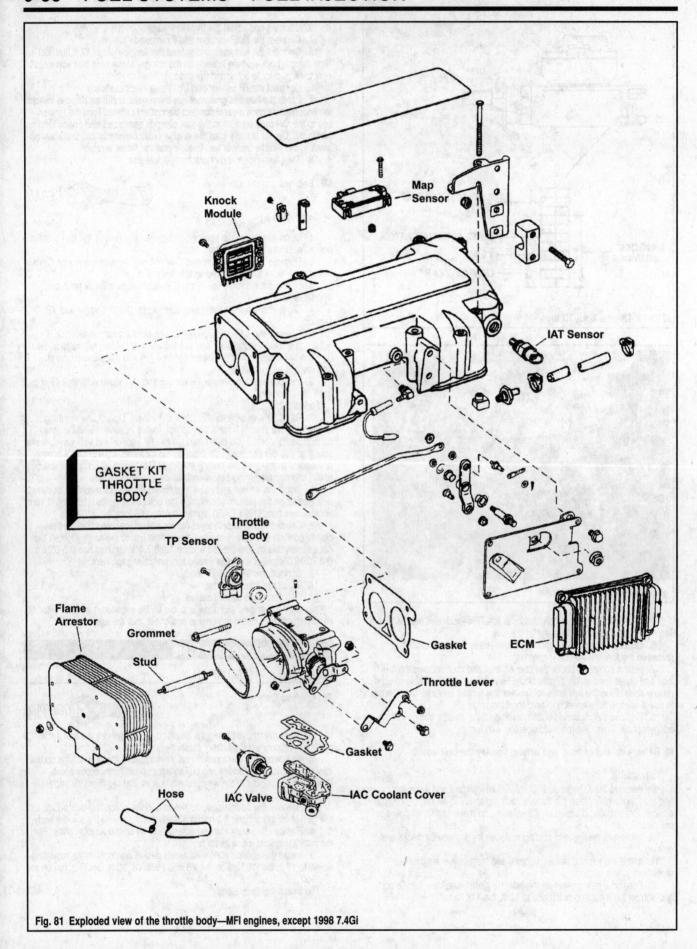

Fig. 81 Exploded view of the throttle body—MFI engines, except 1998 7.4Gi

Throttle Position Sensor (TP)

REMOVAL & INSTALLATION

◆ See Figure 82 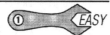 *EASY*

1. Open the engine compartment/hatch and disconnect the battery cables.
2. On the 1998 7.4Gi, remove the starboard high-rise exhaust elbow.
3. Remove the flame arrestor. Although not absolutely necessary, you may find it beneficial to remove the throttle body for better access to the sensor—although you must remove it on the 1998 7.4Gi.
4. Tag and disconnect the electrical connector at the sensor and move it out of the way.

■ Please refer to the Component Locations section.

5. Unscrew the 2 mounting screws and remove the sensor. Clean the sensor with a dry cloth; make sure that the sensor is in good condition (no wear, cracks or damage), if not replace it.

■ If replacing the sensor with a new one, make sure to use the two new screws that came with the package.

6. Position the sensor seal or O-ring over the throttle shaft.
7. With the throttle valve closed, install the sensor so it is positioned on the throttle shaft and tighten the screws to 18 inch lbs. (2 Nm). Coat the threads with Loctite® 242 (TBI engines) and don't forget the washers!
8. Install the throttle body and exhaust elbow if removed.
9. Connect the electrical lead. Connect the battery cables
10. Start the engine and check the sensor output voltage. It should be approximately 0.7V at idle and 5.0V at wide open throttle.
11. Install the flame arrestor.

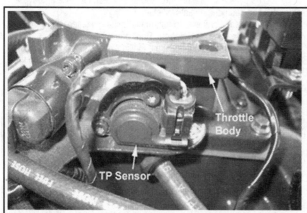

Fig. 82 The TP sensor is located on the lower edge of the throttle body

Vapor Separator Tank/Fuel Reservoir

REMOVAL & INSTALLATION

◆ See Figure 66  *MODERATE*

■ Certain early engines may not utilize this system.

■ Always have a fire extinguisher handy when working on any part of the fuel system. Remember, a very small amount of fuel vapor in the bilge, has the tremendous potential explosive power.

1. Open or remove the engine hatch/cover and relieve the fuel system pressure as detailed previously. Disconnect the battery cable.
2. Loosen the hose clamps and then remove the water inlet and outlet hoses from the tank.
3. Remove the power steering cooler retaining bracket nut and position the cooler out of the way. There is a special flat washer behind the cooler bracket—don't lose it when lifting off the assembly.

4. Clip the plastic tie securing the pump harness to the pump and then unplug the connector and move the harness out of the way.
5. Remove the small clamp and wiggle the vent hose from the nipple at the rear of the reservoir. Position it out of the way.
6. Remove the fuel inlet line coming from the low pressure pump. Remove the line running between the reservoir and pressure regulator. Remove the high pressure pump outlet line. Please refer to the Fuel Pump procedures for specific details on removing the fuel lines.
7. Loosen the clamp and remove the cooling outlet line (running to the thermostat housing) connected to the elbow on the Starboard side of the reservoir.
8. Remove the 2 mounting bolts at the bottom of the reservoir and lift off the assembly. The Port side mounting bolt may have a nut on the back of the bracket.

To Install:
9. Adjust the float level as detailed later in this section.
10. Position the reservoir onto the engine and tighten the mounting bolts to 28-36 ft. lbs. (38-49 Nm).
11. Reattach the coolant line to the elbow on the reservoir and tighten the clamp securely.
12. Connect the three fuel lines to the pump and reservoir as detailed in the Fuel Pump section.
13. Connect the vent line running from the upper manifold and tighten the clamp securely, but not so tight as to pinch the hose.
14. Coat the terminals on the fuel pump and pump harness connector with Terminal Grease and plug the connector into the pump. Secure the harness to the pump body with a new plastic tie.
15. Slide the flay washer over the mounting stud on the rear of the reservoir and then install the power steering cooler. Tighten the lock nut to 20-25 ft. lbs. (27-34 Nm).
16. Reconnect the cooling hoses and tighten their hose clamps securely.
17. Connect the battery cables. Turn the ignition key to the **ON** position so that the fuel pumps are energized and check for any fuel leaks. Do not actually start the engine until confirming that there are no leaks at the throttle body or fuel lines. Install the flame arrestor.
18. Start the engine and check for fuel leakage.

FLOAT LEVEL

◆ See Figure 83  *MODERATE*

1. Disconnect all fuel lines and remove the high pressure fuel pump.
2. Remove the 8 reservoir cover retaining screws and lift off the cover and pump brackets. Discard the gasket.
3. Flip the cover assembly over and allow the float to drop down until it seats the float needle.
4. Measure the distance between the bottom edge of the inverted float (actually the top edge when installed in the reservoir) and the inner mounting surface of the reservoir cover.
5. Bend the float adjustment tab very carefully until the gap is 3/16 in. (4.76mm). Rotate the assembly once or twice and check that the float moves freely and that the gap has not changed.

■ Never force the float needle tip against the seat or you will damage the special tip.

6. Position a new cover gasket onto the reservoir and then carefully install the cover assembly. Tighten the retaining screws to 25-35 inch lbs. (2.8-4.0 Nm).
7. Install the fuel pump and reconnect the fuel lines.

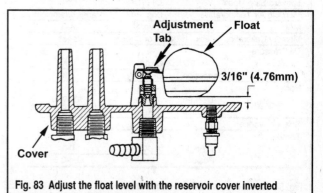

Fig. 83 Adjust the float level with the reservoir cover inverted

SYSTEM DIAGNOSIS

◆ **See Figure 84**

Prior to performing any diagnostics on the EFI system, a diagnostic circuit check must first be completed. This is an organized approach to identifying a problem created by a malfunction in the electronic engine control management system and can only be performed with the correct scan tool—although most tests can be accomplished with a less expensive code tool (CodeMate) and a digital multi-meter (DVOM). After hooking the tool up properly and finding that the on-board diagnostic system if functioning correctly and there are no codes displayed, the accompanying table should be used as a reference for what a normally functioning system should display. Use only the parameters listed in the chart.

If codes are displayed, move on to the next section.

PRECAUTIONS

◆ **See Figures 85 and 86**

Good shop practice, as well as good judgment requires all the following practices be observed.

1. The negative battery cable must be disconnected before any ECM system component is removed.

2. Always check to be sure the battery cables are securely connected, before starting the engine.

3. Never separate the battery from the on-board electrical system while the engine is operating.

4. Never separate the battery feed wire from the charging system while the engine is operating.

5. Disconnect the battery from the boat electrical system before starting to charge the battery.

6. Check to be sure all cable harnesses are securely connected and the battery terminals are clean and the cables securely connected.

7. Never connect or disconnect the wiring harness at the ECM when the ignition switch is in the on position.

8. Before any electric arc welding is attempted, disconnect the battery leads and the ECM connector/s.

9. Never direct a steam-cleaning nozzle at ECM components. Such action will cause damage or corrosion the component terminals.

10. Do not use any test equipment not specified in the diagnostic charts. Such equipment may give an incorrect reading and/or may actually damage good components.

11. A digital voltmeter with a rating of 10 megohms input impedance must be used when taking any voltage measurements.

12. A "low-amphere rated" test light must be used when a test light is specified for a test. Never use a high-amphere rated test light. Power the setup with the boat battery. If the ammeter indicates less than 3/10 amp current flow, the test light is safe to use.

A final word here. If a test light with 100 mA or less is used, a faint glow may show, when the test actually states "no light".

General Diagnostic Tests

■ **Please refer to wiring schematics at the end of this section for correct pin and circuit locations.**

ON-BOARD DIAGNOSTIC SYSTEM CHECK

1. Connect the tester to the diagnostic link connector (DLC)—either at the front of the engine near the thermostat housing, or on the Starboard side near the ECM bracket. Switch it to the **OFF** position.

2. With the ignition switch in the **ON** position and the engine OFF, the tester light should come on; if not, please refer to the following procedure for "No Code Tool Light or Data".

3. If the tester flashes a Code 12, check terminal **B** on the DLC for a short to ground—if no short is evident, the ECM will require replacement.

4. Switch the tester to the **ON** position and confirm that it flashes Code 12. If it doesn't, check the "Code Tester Light on Steady..." procedure following.

5. If DTC 51 is also present, proceed to that test. Otherwise, turn the tester OFF and see if the engine will start.

6. If the engine continues to run, move to the next step; if it won't run, move to the "Engine Cranks But Won't Run" procedure.

7. Turn off the engine but leave the ignition switch in the **ON** position. Switch the tester back to ON and check for any stored codes. Refer to the DTC charts at the end of this section.

Scan Position		Units	Data Value
BARO		Volts	3-5
Battery		Volts	12-14.5
Coolant Temperature		Deg. F/C	150-170 (66-77)
Engine Overtemp		OK/Overheat	OK
Fuel Consumption		GPH	1 to 2
Idle Air Control Valve			
	Follower	Counts (Steps)	0
	Minimum	Counts (Steps)	0-40
	Normal	Counts (Steps)	0-40
Injector			
	On Time Cranking	Msec	2.5-3.5
	Pulse Width	Msec	2-3
Knock Retard		Deg.	0
Lanyard Stop Mode		OFF/ON	OFF
Manifold Air Temperature		Deg. F/C	Varies w/ambient
MAP		Volts	1-3
Memory Calibration Check		Sum of Check	Varies
Oil Pressure			
	I/O	OK/LOW	OK
	Transmission	OK/LOW	OK
Rpm			
	Desired	Rpm	600-700
	Normal	Rpm	600-650
Spark Advance		Deg.	10-30
Throttle			
	Angle	0-100%	0-1
	Position	Volts	0.4-0.8

Fig. 84 Normal specifications for a scan tool

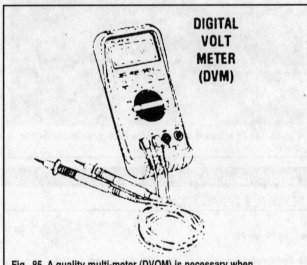

Fig. 85 A quality multi-meter (DVOM) is necessary when performing diagnostic tests...

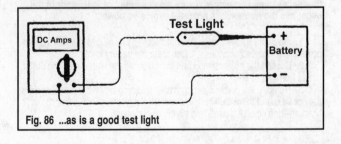

Fig. 86 ...as is a good test light

NO CT LIGHT OR NO DLC DATA

◆ **See Figure 88**

When a tester is connected to the DLC it should receive voltage via the circuit on terminal **F** and be grounded through the circuit on terminal **E**. There should always be a steady tester light when the ignition is **ON** (engine stopped). Obviously, always check for bad connections or frayed wires. It's also a good idea to check that the indicator bulb is not burned

■ **Please refer to wiring schematics for Tests A-1/A-2 in the Diagnostic Test Schematic section for correct pin and circuit locations.**

If The Engine Starts

1. Remove the Code Tool tester and turn the ignition switch to the **ON** position (engine OFF). Connect a test light to ground and touch the probe to terminal **F** of the DLC. If the light goes on move to Step 2; if it does not come on, the circuit between J1-32/J1-16 on the ECM and terminal **F** on the DLC is open or has a short to ground.
2. Now connect the test light to the battery positive and touch the probe to terminal **E** of the DLC. If the light goes on, move to the next step; if it does not come on, turn the ignition switch to the **OFF** position. Disconnect the J2 connector at the ECM and check the resistance between J2-31 and terminal **E** at the DLC. If close to 0 ohms, check the ECM connections or replace the ECM; otherwise, repair the open circuit.
3. Connect a test light between DLC terminals **F** and **E**. If the light goes on, replace the tool.

If The Engine Does Not Start

1. Check the ECM circuit breaker and replace if necessary. If the breaker is functioning properly, move to the next step.
2. With the ignition **OFF**, connect a test light to ground and probe J1-16 and J1-32 on the ECM connector. If the light goes on for each circuit, move to the next step. If it does not light, repair the open in that circuit.
3. Turn the ignition switch to the **ON** position (engine not running). Connect a test light to ground and probe ECM connector pin J1-11. If the light goes on, check the ECM grounds and then replace the ECM if they are OK. If it does not light, check the ECM/INJ circuit breaker. If it is OK, check the ECM/INJ relay. Otherwise, find and repair the ground in the J1-11 to breaker circuit.

CT LIGHT ON STEADY—WILL NOT FLASH DTC 12

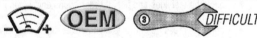

◆ **See Figure 88**

When the Code Tool tester is installed, it receives voltage through terminal **F** and is grounded through terminal **E**. When the ignition is **ON** and the engine OFF, there should always be a steady MIL. Obviously, always check for bad connections or frayed wires. It's also a good idea to check that the indicator bulb is not burned out.

■ **Please refer to wiring schematics for Tests A-1/A-2 in the Diagnostic Test Schematic section for correct pin and circuit locations.**

1. With the ignition switch in the **ON** position and the engine OFF, switch the tester ON. If the CT flashes DTC 12, look for bad connections in the circuit; if DTC 12 does not flash, go to Step 2.
2. Turn the ignition **OFF** and disconnect the J2 connector at the ECM. Turn the ignition switch back to the **ON** position. If the CT light comes ON, there is a short to ground between J2-31 and terminal **E** on the DLC connector; if the light does not come on, go to Step 3.
3. Turn the ignition **OFF**, disconnect the J1 connector at the ECM and connect a jumper wire between terminals **A** and **B** at the DLC. Connect a test light between terminal J1-7 and battery positive (B+). If the light goes ON, you'll need to verify that the tester is working properly by hooking it up to a known good system—we know its unlikely that you have another engine available, so all we can suggest is running down to your local dealer and asking them to confirm it is OK. If the light does not go on, check for an open circuit between terminal **A** on the DLC and ground; or between terminal **B** and J1-7 at the ECM

ENGINE CRANKS BUT WILL NOT RUN

The distributor ignition system and the fuel injector circuit are both supplied voltage via the EFI system relay. From the relay, circuit 902 delivers voltage to the injector/ECM fuse and to the coil. The following test assumes a properly functioning battery and adequate fuel supply. Is also a good idea to check all connections and wires before hand.

1994-96 Engines

1. Perform the On-Board Diagnostic System Check. Disconnect the TP sensor and attempt to start the engine. If it starts, replace the sensor, otherwise move to the next step.
2. Ensure that the ignition has been in the **OFF** position for at least 10 seconds and then switch it **ON**. If the fuel pumps run for 2 seconds and then stop, go to Step 3; if not, move on to the "Fuel System Diagnosis" procedure following.
3. Crank the engine for 3 second and listen for the fuel pumps to run while the engine is cranking. If they run, go to Step 5; if they do not run, go to Step 4.
4. Disconnect the connector at the ESA module and probe terminal **B** in the distributor harness connector while cranking the engine. If the meter needle fluctuates, go to the next step; if it does not, replace the IC module
5. Install a fuel pressure gauge to the pressure valve on the high pressure fuel pump; as per the manufacturer's instructions. Turn the ignition switch to the **ON** position. Check the fuel pressure while the pump is running. If 9-13 psi shows on TBI engines or 36-42 psi on MFI engines, move to the next step; if not, move on to the "Fuel System Diagnosis" procedure.
6. Check the secondary spark output. If spark strength and output are ok, move to the next step. Otherwise, disconnect the ESA module and try to start the engine. If it starts, replace the module; if it doesn't, perform the Ignition System diagnostic tests.
7. Disconnect the J1 connector at the ECM. Connect a DVOM and check for battery voltage (B+) at J1-16 and J1-32. If voltage is present at both terminals, move to the next step. If no voltage is present, check the 12.5 amp ignition circuit breaker. If the breaker is functioning properly and there is battery voltage present, repair the circuit between the two pins and the breaker; if not replace the breaker.
8. Turn the ignition switch to the **ON** position and use the DVOM to check for switched battery voltage at J1-11. If present, go to Step 11, if not, go to the next step.
9. Remove the ignition relay and check for battery voltage at the Red/Purple wire terminal. Turn the ignition switch **ON** and check for switched voltage at the Purple wire terminal. Check the relay ground circuit for continuity. If all three check out OK, replace the relay and check J1-11 again. If there is still no switched battery power, check the circuit between it and the relay.
10. In the previous step, if there was no battery voltage, there is a fault in the circuit going to the 12.5 amp circuit breaker. If there was no switched battery voltage, there is a fault in the circuit going to the ignition switch. If the there was a bad ground, fix it.
11. Disconnect one injector from each bank, turn the ignition switch **ON** and connect a DVOM. Probe the Pink/White wire terminal in each connector. If battery voltage is present, go to the next step; if not, check the circuit between the ignition relay and the injector.

■ **The ignition relay also supplies switched battery voltage to the knock sensor module—disconnect the module and try to start the engine, if it starts now, replace the module.**

12. Connect an injector test light to each of the injectors disconnected previously and then crank the engine. If both test lights blink brightly, the system is OK. If one (or both) lights stay on steadily, go to the next step. If they do not come on at all, go to Step 14.
13. Turn the ignition **OFF** and check the driver circuits for a short to ground. Disconnect the 2-pin connector at the ignition coil and then disconnect the J2 connector at the ECM. Attach a test light to the battery and probe terminals J2-5 and J2-21. If the light goes ON at either terminal, check for a fault in the ground circuit. If the light does not come ON, replace the ECM.

14. Turn OFF the ignition and check the ECM ground circuits. Disconnect both ECM connectors, connect a test light to the battery and then probe the following connectors: J2-15, J2-20, J1-14 and J1-30. If the light does not come ON, look for and repair a bad ground in the individual circuit. If it does come ON, turn off the ignition and check for an open in the driver circuits between the injectors and J2-15 or J2-21. If they check out OK, move to the next step; otherwise, repair the bad circuit.

15. Check the injectors for an open circuit—resistance across the injector terminals should be greater than 10 ohms. If it is, the injectors are OK and you will need to replace the ECM. If resistance is out of range, replace the injector.

1997-98 Engines

◆ **See Figures 89, 90 and 91**

■ **Please refer to wiring schematics for Test A-3 in the Diagnostic Test Schematic section for correct pin and circuit locations.**

1. Perform the On-Board Diagnostic System Check. Disconnect the TP sensor and attempt to start the engine. If it starts, replace the sensor, otherwise move to the next step.

2. Ensure that the ignition has been in the **OFF** position for at least 10 seconds and then switch it **ON**. If the fuel pumps run for 2 seconds and then stop, go to Step 3; if not, move on to the "Fuel System Diagnosis" procedure following.

3. Crank the engine for 3 seconds and listen for the fuel pumps to run. If they run, go to Step 5; if they do not run, go to Step 4.

4. With the ignition switch in the **OFF** position, disconnect the J2 connector at the ECM. Connect a DVOM to ground and probe the J2-8 pin while cranking the engine. If the DVOM reads 1-2V, confirm all ECM connections are good and then replace the ECM. If it does not read 1-2V, J2-8 is shorted to ground or the ignition module is bad.

5. Install a fuel pressure gauge to the pressure valve on the high pressure fuel pump; as per the manufacturer's instructions. Turn the ignition switch to the **ON** position. Check the fuel pressure while the pump is running. If 9-13 psi on 4.3L engines, 28-32 psi on V8 engines or 36-42 psi on 7.4L/8.2L engines, move to the next step; if not, move on to the "Fuel System Diagnosis" procedure.

6. Check the secondary spark output. If spark strength and output are ok, move to the next step. Otherwise, perform ignition system diagnostics.

7. Disconnect the J1 connector at the ECM. Connect a DVOM and check for battery voltage (B+) at J1-16 and J1-32. If voltage is present at both terminals, move to the next step. If no voltage is present, check the 12.5 amp ignition circuit breaker. If the breaker is functioning properly and there is battery voltage present, repair the circuit between the two pins and the breaker; if not replace the breaker.

8. Turn the ignition switch to the **ON** position and use the DVOM to check for switched battery voltage at J1-11. If present, go to Step 11, if not, go to the next step.

9. Remove the ignition relay and check for battery voltage at the Red/Purple wire terminal. Turn the ignition switch **ON** and check for switched voltage at the Purple wire terminal. Check the relay ground circuit for continuity. If all three check out OK, replace the relay and check J1-11 again. If there is still no switched battery power, check the circuit between it and the relay.

10. In the previous step, if there was no battery voltage, there is a fault in the circuit going to the 12.5 amp circuit breaker. If there was no switched battery voltage, there is a fault in the circuit going to the ignition switch. If the there was a bad ground, fix it.

11. Disconnect one injector from each bank, turn the ignition switch **ON** and connect a DVOM. Probe the Pink/White wire terminal in each connector. If battery voltage is present, go to the next step; if not, check the circuit between the ignition relay and the injector.

12. Connect an injector test light to each of the injectors disconnected previously and then crank the engine. If both test lights blink brightly, the system is OK. If one (or both) lights stay on steadily, go to the next step. If they do not come on at all, go to Step 14.

13. Turn the ignition **OFF** and check the driver circuits for a short to ground. Disconnect the 2-pin connector at the ignition coil and then disconnect the J2 connector at the ECM. Attach a test light to the battery and probe terminals J2-5 and J2-21. If the light goes ON at either terminal, check for a fault in the ground circuit. If the light does not come ON, replace the ECM.

14. Turn OFF the ignition and check the ECM ground circuits. Disconnect both ECM connectors, connect a test light to the battery and then probe the following connectors: J2-15, J2-20, J1-14 and J1-30. If the light does not come ON, look for and repair a bad ground in the individual circuit. If it does come ON, turn off the ignition and check for an open in the driver circuits between the injectors and J2-15 or J2-21. If they check out OK, move to the next step; otherwise, repair the bad circuit.

15. Check that the resistance across the injector terminals is less than 1.25-1.27 ohms. If it is, the injectors are OK and you will need to replace the ECM. If resistance is out of range, replace the injector.

FUEL SYSTEM DIAGNOSIS—TBI ENGINES

1. Connect a pressure gauge to the check valve for the high pressure pump. Turn the ignition switch **ON**. The pumps should run for 2 seconds and then shut off; while running the pressure should be 9-13 psi (27-31 psi on the 1998 5.7GSi). If it is, go to the next step. If below 9 psi (or 27), go to Step 6. If over 13 psi (or 31), go to Step 10.

2. Start the engine and allow it to idle until it reaches normal operating temperature. The fuel pressure should be within 1 psi of the previous reading. If it is, the system is OK; if not, go to the next step.

3. Check the fuel lines and filter for restrictions and repair or replace as necessary. If they are OK, go to the next step.

4. Disconnect the fuel line at the high pressure pump outlet and connect the pressure gauge to the pump (not the check valve). Turn the ignition **ON** and check that the pressure holds for 15-20 seconds after the pumps stop running. If pressure holds, go to the next step. If not, check for leaks; replacing the fuel pump if there are none found.

5. Disconnect the pressure gauge and reconnect the fuel line. Reconnect the gauge to the check valve. Disconnect the fuel return line at the vapor separator and connect another gauge to the line—do not reconnect the return line to the tank. Turn the ignition **ON** and confirm that the high pressure gauge reads 9-13 psi after the pumps stop, or 27-32 psi on the 1998 5.7GSi. If it does, the system is OK. If not, check both gauges; if equal, replace the regulator. If the high pressure gauge has dropped to near 0 psi and there is still pressure showing on the other gauge, check for a leaking injector or regulator.

6. If the fuel pressure in Step 1 was less than 9 psi (or 27 psi), check for any external fuel leaks. If any are found, fix them and go back to Step 1. If none are found, check that the battery is fully charged. Go to the next step if it is, otherwise charge it.

7. Check the fuel lines and filter for restrictions and water. Fix any problems you find, otherwise go to the next step.

8. Connect a pressure gauge to the low pressure check valve on the separator, turn the ignition **ON** and check that the pressure is 4-8 psi. If it is, go to the next step, otherwise replace the low pressure pump.

9. Disconnect the vent vacuum line at the separator and splice in a clear plastic hose between the end of the line and the tank. Turn the ignition **ON**. If fuel appears in the line, replace the vent valve in the tank. If none appears, check for leaking injectors and/or replace the regulator.

10. If the fuel pressure in Step 1 was greater than 13 psi (or 31 psi), disconnect the fuel return line at the separator tank and connect a pressure gauge between the line and the tank. Reconnect the line and turn the ignition **ON**. Check the pressure while the pumps are still running. If above 13 psi (or 31 psi), go to the next step, otherwise the system is OK.

11. Check the supply line from the filter back to the fuel tank to see if there has been another pump installed (or something producing pressure). If nothing is there, check for a restriction in the line between the regulator and the separator tank. Replace the regulator if nothing is found. If there is some sort of device installed between the filter and tank, remove it.

FUEL SYSTEM DIAGNOSIS—MFI ENGINES

◆ **See Figure 92**

■ **Please refer to wiring schematics for Test A-5 in the Diagnostic Test Schematic section for correct pin and circuit locations.**

1. Turn the ignition switch OFF. Install a fuel pressure gauge to the check valve for the high pressure pump. Turn the ignition switch to the ON position. The fuel pump should run for about 2 seconds and then shut off. Check the pressure while the pump is running and then after it stops; it should hold steady after the pump stops for at least 15-20 seconds. If the fuel pressure is 36-42 psi, go to Step 2.

If the pressure drops immediately to 0 psi after the pump shuts off, check the vacuum line at the pressure regulator for fuel—if fuel in line, replace the regulator, if no fuel in line, go to Step 4.

If the pressure was below 36 psi, go to Step 6; if above 42 psi, go to Step 11; if there was no pressure, or the pumps did not turn on, go to Step 13.

2. Start the engine and allow it to idle until it reaches normal operating temperature. If the fuel pressure is within 3-10 psi of the reading taken previously, the system if OK; if not, go to Step 3.

3. With the engine still idling, connect a vacuum gauge to the pressure regulator and apply 10 in. of vacuum to the pressure regulator. If the pressure drops by 3-10 psi, you have a vacuum leak at the regulator; if it doesn't drop by the above spec, replace the fuel pressure regulator.

4. Turn the ignition OFF, disconnect the fuel line at the high pressure pump outlet fitting and then connect a pressure gauge directly to the pump. Turn the ignition switch to the ON position. If the pressure holds for 15-20 seconds after the pump stops running, go to the next step; otherwise, check for leaks and then replace the pump.

5. With the ignition OFF, connect a pressure gauge to the high pressure pump test valve. Disconnect the fuel return line at the vapor separator and connect another pressure gauge directly to the line—do not reconnect the return line to the separator tank. Turn the ignition switch to the ON position. If the gauge connected to the high pressure test point holds at 36-42 psi, the system is OK.

If it doesn't hold, check both gauges. If they are each showing the same pressure (approximately), replace the regulator. If the high pressure gauge has dropped to 0 psi (or near 0) and there is still pressure left on the other gauge, check for leaking injector(s).

6. If the fuel pressure reading in Step 1 was less than 36 psi, check for fuel leaks. If leaks are found, repair them and restart the test. If no leaks are found, check that the battery is fully charged and then move to the next step.

7. Check to see if the fuel lines or filter are clogged. Look for water in the filter. Repair or replace any problems if found, otherwise, go to the next step.

8. Connect a pressure gauge to the low pressure test valve on the vapor separator tank. Turn the ignition switch to the ON position and check the fuel pressure. If 4-8 psi, go to the next step; otherwise, replace the low pressure fuel pump.

9. Disconnect the vacuum line at the pressure regulator. Connect a piece of clear plastic hose between the line and the regulator and then turn the ignition ON. If fuel appears in the hose, replace the regulator; otherwise, go to the next step.

10. Disconnect the vacuum line at the vapor separator. Connect a piece of clear plastic hose between the line and the separator and then turn the ignition ON. If fuel appears in the hose, replace the vent valve in the separator; otherwise, check for leaking fuel injectors and/or replace the pressure regulator.

11. If the pressure reading in Step 1 was over 42 psi, disconnect the fuel return line at the vapor separator and connect a pressure gauge between the line and tank. Turn the ignition ON and observe the pressure while the pumps are running. If above 42 psi, the system is OK; otherwise, move to the next step.

12. Check the supply line from the filter back to the fuel tank to see if there has been another pump installed (or something producing pressure). If nothing is there, check for a restriction in the line between the regulator and the separator tank. Replace the regulator if nothing is found. If there is some sort of device installed between the filter and tank, remove it.

13. If there was no pressure when the reading was taken in Step 1, perform the On-Board Diagnostic System Check.

14. Check the 20 amp circuit breaker at the fuel pump. Reset if necessary. Confirm that battery voltage is present at the breaker and then move to the next step. If no voltage is present, repair the open or ground in the circuit between the breaker and the battery.

15. Remove the fuel pump relay. Connect a test light to ground and probe terminals **30** and **87**. If the light comes ON, go to the next step; otherwise, check for an open circuit or bad connection back to the circuit breaker.

16. Connect a test light to the battery positive (B+) and probe terminal **85** on the relay socket while switching the ignition switch to the ON position. If the light comes ON for 2 seconds and then goes OFF, move to the next step; otherwise, go to Step 20.

17. Install a fused jumper wire between terminals **30** and **87** on the relay socket. If the fuel pumps run, replace the relay; otherwise, go to the next step.

18. Connect a test light to ground and then probe the terminal for the yellow/green wire on the fuel pump connector. If the light comes ON, go to the next step; otherwise, look for an open circuit or bad connection between the connector and terminal **87** on the relay socket. Remove the jumper and reinstall the relay.

19. Now connect the test light to power and probe the pump connector terminals with the black wire. If the light comes ON, replace BOTH fuel pumps; otherwise, look for an open or bad connection in the fuel pump ground circuits.

20. Disconnect the negative battery cable and then the J1 connector at the ECM. Connect an ohmmeter and check the continuity between terminal J1-23 and terminal **85** on the pump relay socket. If no continuity (0 ohms), replace the ECM; otherwise, look for an open circuit or bad connection.

IGNITION SYSTEM CHECK

Please refer to Section 10 for complete testing on the ignition system.

IDLE AIR CONTROL (IAC) VALVE FUNCTIONAL TEST

◆ **See Figure 93**

■ **Please refer to wiring schematics for Test A-7 in the Diagnostic Test Schematic section for correct pin and circuit locations.**

The ECM controls idle speed to a pre-set, desired rpm based on input from sensors and the actual engine rpm. Four separate circuits are used to move the IAC valve; said movement varying the amount of air allowed to bypass the throttle plates. Idle speed then is controlled, via the ECM, by the position of the IAC valve.

1. With the engine at normal operating temperature, turn it off, connect a tachometer and then restart it and allow the idle to stabilize. Record the idle speed and turn off the engine for 10 seconds. Disconnect the IAC harness connector, restart the engine and check the rpm. If the second recorded idle speed is higher than the first by 200 rpm or more, go to Step 2; if not, go to Step 3.

2. Reconnect the IAC harness. If the idle speed returns to within 75 rpm of the originally recorded speed in Step 1 within 30 seconds, the IAC circuit is functioning properly; if it does not, go to the next step.

3. Turn off the ignition for 10 seconds. Disconnect the IAC again and restart the engine. Connect a test light to ground and then touch the probe to each of the four IAC harness terminals. If the light blinks on each terminal, check for a bad IAC connection somewhere in the circuit, otherwise replace the valve; if it doesn't blink, look for an open or shorted circuit on the terminal(s) that didn't blink. Also, check for any bad connections at the ECM.

MANIFOLD ABSOLUTE PRESSURE (MAP) SENSOR OUTPUT CHECK

◆ **See Figures 87, 94 and 102**

■ **Please refer to wiring schematics for Test A-8 in the Diagnostic Test Schematic section for correct pin and circuit locations.**

The MAP sensor measures the change in intake manifold pressure resulting from engine load and rpm changes; converting these inputs into a voltage output. The ECM supplies the MAP sensor with a 5V reference signal—as manifold pressure changes so does the output voltage at the sensor.

1. Turn the ignition switch to the **ON** position and install the code tool to check for DTCs. If Code 33 is present, follow the procedures for Code 33 later in this section. If no Code 33 is present, check the MAP sensor voltage and compare this with that of a "known good" sensor; make sure they have the same color insert and 'hot stamp'. If the two voltage measurements are within 0.4V of each other, go to the next step; otherwise replace the sensor.

2. Remove the MAP sensor and then plug the vacuum port on the intake manifold.

3. Connect a hand vacuum pump to the port on the sensor, start the engine and note the sensor voltage.

4. Apply 10 inch Hg. (34 kPa) with the pump and note the change in voltage.

5. Subtract the second reading from the first. If greater than 1.5V, the sensor is OK and you should check the sensor port for a restriction or leaking seal. If less than 1.5V, check the sensor connection, replacing the sensor if the connection is OK.

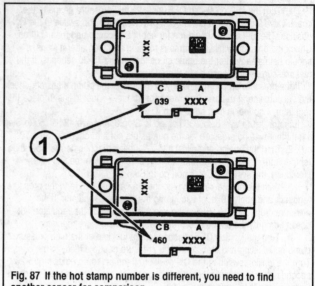

Fig. 87 If the hot stamp number is different, you need to find another sensor for comparison

Diagnostic Test Schematics

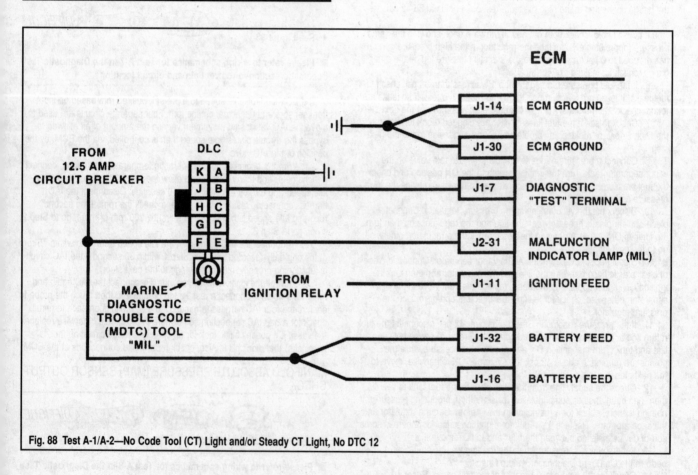

Fig. 88 Test A-1/A-2—No Code Tool (CT) Light and/or Steady CT Light, No DTC 12

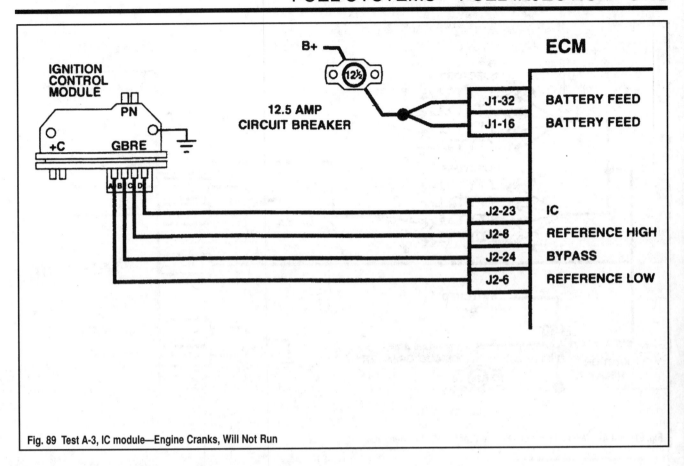

Fig. 89 Test A-3, IC module—Engine Cranks, Will Not Run

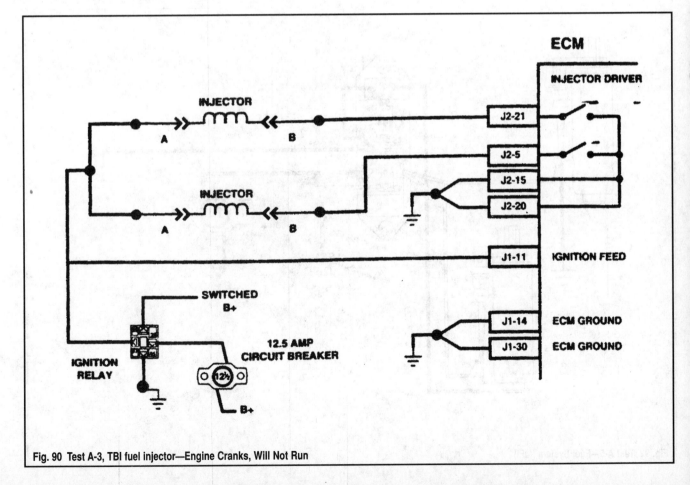

Fig. 90 Test A-3, TBI fuel injector—Engine Cranks, Will Not Run

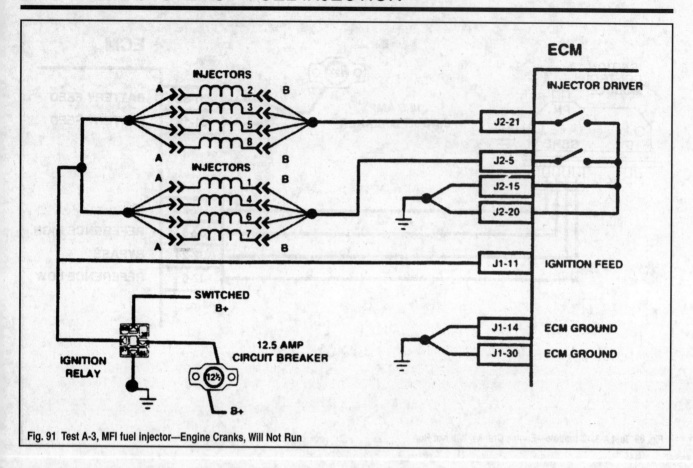

Fig. 91 Test A-3, MFI fuel injector—Engine Cranks, Will Not Run

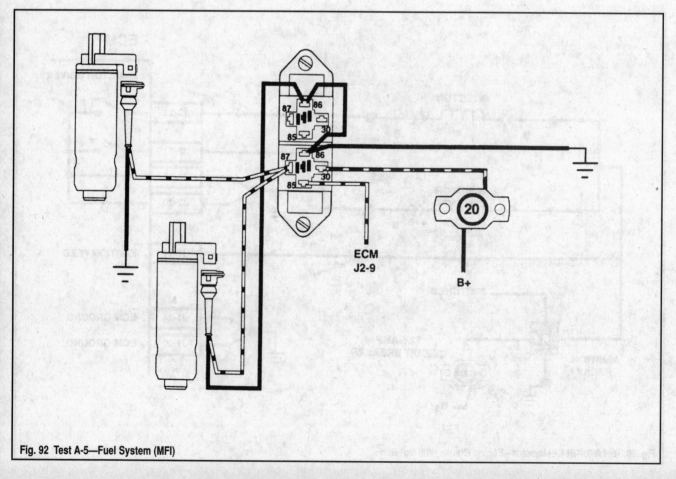

Fig. 92 Test A-5—Fuel System (MFI)

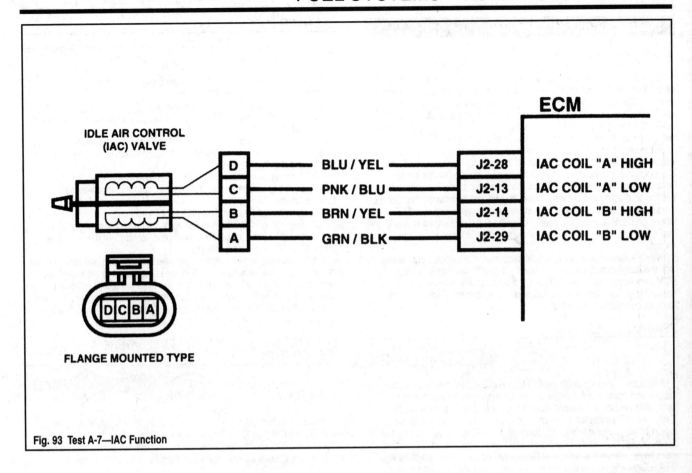

Fig. 93 Test A-7—IAC Function

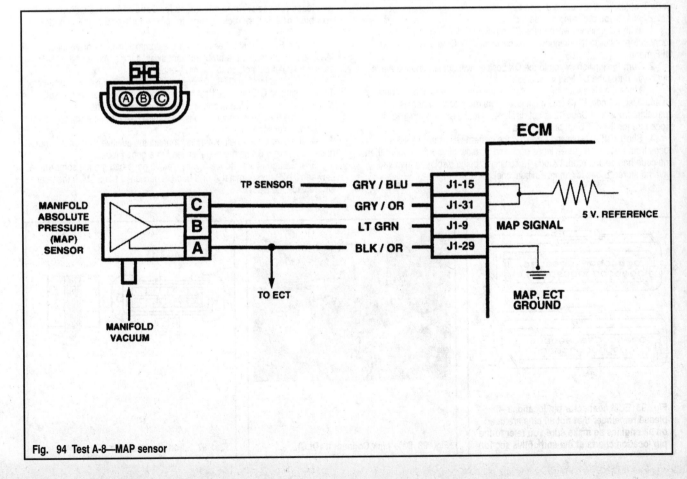

Fig. 94 Test A-8—MAP sensor

Diagnostic Trouble Codes (DTC)

◆ See Figures 95, 96 and 97

Operational problems during everyday engine operation are recognized by the ECM and a diagnostic trouble code is created to identify the particular problem. The ECM will retain this code, or a combination thereof, in its memory until such a time as you access it, read it and then clear it. If the reason for the code has not been repaired prior to clearing the code it will reset itself again shortly after the engine begins to operate.

Remember that DTCs do not necessarily indicate the specific problem or component, merely the area from which the problem is originating. Code charts for your particular engine can be found at the end of this section; while fault and function tests for each code are also found following the section on reading and clearing codes. You will find that more times than not, the source of the problem is a bad connection or frayed wire—particularly where intermittent codes are seen. Follow the steps in the tests carefully and you will find that diagnostics is not always as difficult as you may have thought.

Wiring schematics for all individual components or systems can be found with the fault/function test procedures. Complete system schematics and ECM symptoms charts can be found at the end of this section, while ECM connector pin locations are detailed here; remember there are two ECM harness connectors, J1 and J2.

READING CODES

All trouble codes stored in the ECM are displayed by means of a series of flashes and pauses; the number of flashes represents the number of the code, with the first and second digits being separated by a short pause. Each code will be flashed three times with a long pause separating the repeat of the code each time. For instance, DTC 12 would be represented as "flash, pause, flash-flash, long pause; and then it would repeat this cycle two more times. Count the number of flashes you observe and determine your DTC. Diagnostic trouble codes can be pulled with either a Scan tool or a Code Tool tester. When using a scan tool, simply follow the manufacturer's instructions. If you are using a code tool:

1. With the ignition switch in the **ON** position and the engine OFF, connect the tester to the diagnostic link connector (DLC) and switch it to the OFF mode.
2. Turn the ignition switch to the **ON** position without starting the engine and confirm that the CT light is on steady.
3. Move the CT switch to the **ON** position. The CT lamp should come on and flash a Code 12 (3 times), indicating the diagnostic system is operating properly. If Code 12 is not displayed, please refer to the following procedure for "No CT Light or DLC Data".
4. Shortly after displaying Code 12, the system will flash any stored codes from the system; three times each and in ascending order if more than one code has been stored. If Code 12 continues being displayed then you have no stored codes (or an intermittent one).

Once again, if using a scan tool, follow the tool manufacturer's instructions. If using a Code Tool tester:
1. Ensure that the battery is fully charged.
2. Connect the tester to the DLC and switch the ignition key to the **ON** position without starting the engine.
3. Select the ON position on the tester, disengage the remote controls shift function and then move the throttle through its full range, from idle to full throttle and then back to idle.
4. Turn the ignition switch to **OFF** for 20 seconds.
5. Switch the tester to the **OFF** position, start the engine and let it run for 20 seconds.
6. Turn the ignition switch to **OFF** for 20 seconds.
7. Turn the ignition switch to **ON** again (but do not start the engine). Switch the tester back to the **ON** position and confirm that the only code present is DTC 12. Remove the tool.
8. If there are codes present, check the battery again and then perform the procedure one more time. If codes are still apparent at the end of the second go-around (we're assuming that you fixed the code's problem before attempting to clear it!), refer to the appropriate troubleshooting or diagnostic charts. If the battery is not at full charge, you should hear the audio warning buzzer come on after engine start-up.

CODE 14—ENGINE COOLANT TEMPERATURE SENSOR (ECT)

◆ See Figure 98

The ECT sensor utilizes a thermistor to control the signal voltage being sent to the ECM. The ECM then applies specified voltage back through the **B** terminal to the sensor. When the coolant is cold, the sensor resistance is high. As the coolant warms, resistance lessens and voltage drops.

■ **Please refer to wiring schematics for Code 14 in the Diagnostic Trouble Code Schematics section for correct pin and circuit locations.**

1. The ECT and MAP sensor share a common circuit—disconnect the MAP sensor and see if the problem went away. If it did, move to the MAP sensor test; if it did not, go to the next step.
2. With the ignition **OFF**, disconnect the ECT sensor harness connector. Turn the ignition **ON** (but don't start the engine) and then connect a multimeter (DVOM) across the **A** and **B** sensor harness terminals.
3. If the voltage is above 4 volts, look for a bad electrical connection or replace the sensor.
4. If the voltage is below 4 volts, connect the positive lead on the multimeter to terminal **B** and the negative lead to a good ground.
5. If voltage is now above 4 volts, you've got a bad ground somewhere between J1-29 and terminal **A** or a bad connection at the ECM. If these are OK, replace the ECM.

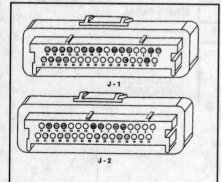

Fig. 95 ECM connector pin locations— please remember that not all pins are used on all engines so make sure you refer to the pin location charts at the end of this section

Fig. 96 Data Link Connector (DLC)...

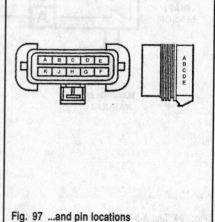

Fig. 97 ...and pin locations

6. If the voltage is not above 4 volts, remove the multi-meter and turn the ignition **ON**. Connect a test light to the positive battery terminal and then touch terminal **B**. If the light does not go ON, check for an open connection on the 5V circuit back to J1-2 or a bad connection at the ECM. If these are OK, replace the ECM.

7. If the light goes ON (1995-97), disconnect the J1/J2 connectors at the ECM. If the light stays ON, you've got a shorted circuit back to J1-2. If it goes out, circuit J1-2 is shorted to J1-29; otherwise, replace the ECM.

8. If the light goes ON (1998), disconnect the J1 connectors at the ECM and touch terminal **B** again. If the light stays ON, you've got a shorted circuit back to J1-2. If it goes out, disconnect the harness at the ECM and the sensor and check for continuity on both circuits between the sensor and the ECM. Each circuit should show continuity to its respective connecter terminal only—i.e. J1-2 to terminal **B** should show zero resistance, while J1-2 to terminal **A** should show infinity. If continuity readings on both circuits is correct, replace the ECM. If not, the circuit is shorted and should be repaired.

CODE 21—THROTTLE POSITION SENSOR (TP)

◆ **See Figure 99**

The TP sensor is a potentiometer and provides a voltage signal to the ECM that varies with the opening and closing of the throttles. Signal voltage will vary from approximately 0.7V (0.4-0.5V - 1998) at idle to slightly less than 4.9V at wide open throttle.

■ **Please refer to wiring schematics for Code 21 in the Diagnostic Trouble Code Schematics section for correct pin and circuit locations.**

9. The TP and MAP sensor share a common circuit—disconnect the MAP sensor and see if the problem went away. If it did, move to the MAP sensor test; if it did not, go to the next step.

10. With the ignition switch **OFF**, disconnect the TP sensor harness connector and then turn the ignition **ON** but do not start the engine. Connect a multi-meter (DVOM) between the **A** and **B** harness terminals.

11. If the reading in Step 2 is less than 4 volts, connect a multi-meter between harness connector **A** and a good engine ground. If the reading is over 4 volts, the circuit between J1-13 and terminal **B** is open to ground or you have a bad connection at the ECM; if both are OK, replace the ECM. If less than 4 volts, the 5V circuit for J1-15 is open or shorted to ground, or you have a bad connection at the ECM, or the MAP sensor 5V circuit from J1-31 is grounded; replace the ECM if they all check out OK.

12. If the reading in Step 2 is over 4 volts, connect the multi-meter between harness terminals **A** and **C**.

13. If the reading is under 4 volts, connect the multi-meter between terminal **C** and an engine ground. If the reading is over 4 volts, signal circuit J1-10 is shorted to circuit J1-15 or J1-31. If the reading is less than 4 volts, J1-10 is an open circuit or you have a bad connection at the ECM. If all are OK, replace the ECM.

14. If the reading is over 4 volts, turn the ignition **OFF**, remove the DVOM and connect a test light to the positive battery cable. Touch the probe to the **C** harness terminal. If the light doesn't come ON, replace the TP sensor. If the light comes on, disconnect the ECM and touch the harness terminal **C** with the probe. If the light comes ON, the circuit from J1-10 is shorted to ground; if the light doesn't come ON, replaced the ECM on 1995-97 engines. On 1998 engines, go to the next step.

15. Disconnect the harness at the ECM and the sensor and then check continuity on all three circuits between the TP and the ECM. Each circuit should only show continuity to its respective connector terminal—i.e. J1-13 to terminal **B** should show zero resistance, while J1-13 to terminals **A** and **C** should show infinity.

16. If all three circuits showed the correct continuity, replace the ECM. If not, check for shorts in each circuit.

CODE 23—INTAKE AIR TEMPERATURE (IAT) SENSOR

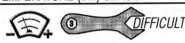

◆ **See Figure 100**

The IAT sensor, used only on MFI engines, utilizes a thermistor to control a voltage signal to the ECM. Resistance is high when intake air is cold so the ECM will see a high voltage. As the air warms, resistance lessens and voltage drops.

■ **Please refer to wiring schematics for Code 23 in the Diagnostic Trouble Code Schematics section for correct pin and circuit locations.**

1. Disconnect the IAT sensor and turn the ignition **ON** without starting the engine. Connect a multi-meter (DVOM) and check voltage across the sensor harness terminals. If over 4 volts, go to Step 2; if under 4 volts, go to Step 3.

2. Now check the resistance across the terminals. If less than 25,000 ohms, check the signal circuit for shorting to voltage; otherwise, check all circuits for bad connections or frayed wires. If over 25,000 ohms, replace the sensor.

3. Check the voltage between the harness connector signal circuit terminal **A** and ground. If above 4 volts, check for a bad sensor ground circuit (J1-13) or bad connections in the circuit; if OK, replace the ECM. If under 4 volts on 1995-97 engines, look for an open signal circuit or bad connections in the circuit; if OK, replace the ECM. If the reading is under 4 volts on 1998 engines, go to the next step.

4. Turn the ignition **OFF** and remove the DVOM. Connect a test light to B+ power source and then touch the probe to connector terminal **A**. If the light does not come ON, check for an open 5V circuit back to J1-24 or a bad connection at the ECM. If both are OK, replace the ECM. If the light comes ON, move to the next step.

5. Disconnect the ECM and then touch the test light probe to terminal **A** again. If the light comes ON, the 5V circuit to J1-24 is shorted to ground. If it doesn't come ON, go to the next step.

6. Disconnect the harness at the ECM and the sensor and then check continuity on both circuits between the IAT and ECM. Each circuit should show continuity to its respective connector terminal only—the circuit between J1-24 and terminal **A** should show zero resistance, while J1-24 to terminal **B** should show infinity. If the continuity on both circuits is correct, replace the ECM. If not, check for shorts in each circuit.

CODE 33—MANIFOLD ABSOLUTE PRESSURE SENSOR (MAP)

◆ **See Figure 101**

■ **Please refer to wiring schematics for Code 33 in the Diagnostic Trouble Code Schematics section for correct pin and circuit locations.**

The MAP sensor responds to changes in pressure in the manifold (vacuum). This information is sent to the ECM as signal voltage and will vary from 1.0-1.5 volts at idle to about 4.0-4.5 volts at full throttle. The ECM compensates for a failing Map sensor by defaulting to a MAP value program that varies with rpm. Circuit J1-31 provides a reference value of about 5 volts to the MAP sensor, while J1-29 is the ground. Circuit J1-9 carries the signal to the ECM.

1. With the ignition **OFF**, install a vacuum gauge to the vacuum port on the throttle body, start the engine and increase the idle to 1000 rpm. If the gauge reads steady at 14 in. Hg. or higher, go to Step 2; if it fluctuates or is lower, you have vacuum leak somewhere.

■ **On 1998 PFI engines, tee a vacuum gauge into the vacuum line running to the pressure regulator.**

2. Turn off the engine and remove the gauge. Disconnect the sensor electrical lead and connect a multi-meter (DVOM) between harness terminals A and C. Turn the ignition ON without starting the engine. If the reading is over 4 volts, go to Step 4. If under 4 volts, go to Step 3.

3. Connect the meter between harness terminal C and an engine ground. If the reading is over 4 volts, check circuit for an open or bad connection at the ECM or check if the ground circuit to J1-29 is open. If both are OK, replace the ECM. If under 4 volts, check circuit J1-31 for an open or short to ground, or bad connections at the ECM. If both are OK, replace the ECM.

4. Connect the meter between harness terminals **B** and **C** and then turn the ignition **ON** without starting the engine. If the reading is over 4 volts, go to Step 6 for 1995-97 engines or Step 7 for 1998 engines; if under 4 volts, go to Step 5.

5. Connect the meter between terminal **B** and an engine ground. If the reading is over 4 volts, circuit J1-9 is shorted to 5V circuits J1-15 or J1-31. If under 4 volts, circuit J1-9 is open or there's a bad connection at the ECM. If both are OK, replace the ECM.

6. Disconnect the terminals at the ECM and connect a test light to battery power (B+). Touch the probe to harness terminal **B**. If the light goes on, circuit J1-9 is shorted to ground. If not, replace the ECM.

7. With the ignition **OFF** and the DVOM removed, connect a test light to the battery positive (B+) source. Touch the probe to harness terminal **B**. If the light goes on, move to the next step. If it doesn't light, replace the sensor.

8. Disconnect the terminals at the ECM and touch the probe to harness terminal **B** again. If the light goes on, circuit J1-9 is shorted to ground. If not, move to the next step.

9. Disconnect the harness at the ECM and the sensor and then check continuity on all three circuits between the MAP and ECM. Each circuit should show continuity to its respective connector terminal only—the circuit between J1-15 and terminal **C** should show zero resistance, while J1-15 to terminals **A** and **B** should show infinity. If the continuity on all circuits is correct, replace the ECM. If not, check for shorts in each circuit.

CODE 42—IGNITION CONTROL (IC) CIRCUIT

◆ **See Figure 102**

■ **Please refer to wiring schematics for Code 42 in the Diagnostic Trouble Code Schematics section for correct pin and circuit locations.**

When the idle speed reaches the necessary rpm for IC and voltage is applied to circuit J2-24, circuit J2-23 should no longer be grounded. If J2-24 is open or grounded the module will not switch to IC mode, J2-23 voltage will be low and code 42 will then be set.

1. Clear all codes as detailed in this section. Start the engine and allow it to idle for 1 minute or until the MIL comes on. Turn off the engine but leave the ignition switch in the **ON** position. Switch the tester to Service mode and see if DTC 42 resets itself. If you get DTC 42, go to the next step; if no code, check the system circuits for bad connections or frayed wires.

2. Turn the ignition **OFF** and disconnect both ECM connectors. Connect a multi-meter (DVOM), set it to the Ohms scale and then check the circuit between J2-23 and ground. If resistance is 3000-6000 ohms, go to the next step; if under not in range, circuit J2-23 is open, grounded or there is a bad connection at the ignition module. If both are OK, replace the module.

3. With the DVOM still connected and grounding J2-23, connect a test light to the positive battery cable (B+) and touch the probe to circuit J2-24. The resistance should switch from over 3000 ohms to under 1000 ohms as the probe makes contact. If it does, go to the next step; if it doesn't drop, go to Step 5.

4. Reconnect the ECM connectors, start the engine and allow it to idle for 1 minute or until the MIL comes on. If it comes on and shows DTC 42, replace the ECM. If it doesn't come on, check the system circuits for bad connections or frayed wires, or a short to ground in circuits J2-23 or J2-24.

5. Connect a test light to a B+ battery power source and touch the probe to ECM harness terminal J2-24. If the light comes ON, go to the next step. If it doesn't come ON, check for poor connections or an open in circuit J2-24; if OK, replace the IC module.

6. Disconnect the 4-wire connector at the IC module and touch the test light probe to terminal J2-24. If the light comes ON, circuit J2-24 is shorted to ground. If it does not come ON, replace the IC module.

CODE 43—KNOCK SENSOR (KS)

◆ **See Figure 103**

■ **Please refer to wiring schematics for Code 43 in the Diagnostic Trouble Code Schematics section for correct pin and circuit locations.**

Detonation or spark knock is sensed by the module which sends a voltage signal to the ECM. As the sensor detects knock, the voltage drops, signaling the ECM to begin retarding timing. The ECM will retard timing whenever knock is detected, and rpm and idle speed are above a specified level.

1. Disconnect the 5-wire KS module connector. Connect a multi-meter (DVOM) and measure the resistance between terminal **E** and ground. If resistance is 99,500-100,500 ohms on the 1997-98 4.3L or 3300-4500 ohms on all other engines, go to Step 2. If not, the circuit between terminal **E** and the sensor is either open or shorted. If it's OK, replace the sensor.

2. Reconnect the 5-wire connector at the module and disconnect the connector at the sensor. Connect a test light to B+ battery power source and start the engine. Hold idle steady at 2500 rpm while touching the probe to the sensor harness terminal a few times. If you detect a noticeable drop in rpm, check the sensor terminal contacts and sensor threads, if they're OK, replace the sensor. If no drop in engine speed is noticeable, go to Step 3.

3. Turn the ignition **OFF** and disconnect the J1 connector at the ECM. Turn the ignition back **ON** and connect a multi-meter (DVOM) between harness terminal J1-1 and ground. If 8-10 volts are present, go to Step 4. If not, go to Step 6.

4. Allow voltage to stabilize and touch the test light probe to the harness connector terminal for the sensor. If the voltage changes, replace the ECM. If no change, go to the next step.

5. Disconnect the module connector again and touch the probe to its terminal **D**. If it lights, replace the sensor module. If it doesn't light, look for an open on the ground circuit for terminal **D**.

6. Connect a test light to ground, disconnect the 5-wire connector at the module and touch the probe to terminal **B**. If the light goes ON, circuit J1-1 is either open or shorted to ground. If OK, replace the KS module. If it doesn't go ON, the 12V circuit from the module to the ignition/injector relay is open or grounded.

CODE 51—ELECTRONIC CONTROL MODULE (ECM) CALIBRATION MEMORY FAILURE

■ **If DTC 51 has shown more than once but is intermittent, replace the ECM.**

Turn the ignition switch **ON** and clear the codes as detailed in this section. If the code resets itself, replace the ECM; if not, check all system circuits for bad connections and/or frayed wires.

Diagnostic Trouble Code Schematics

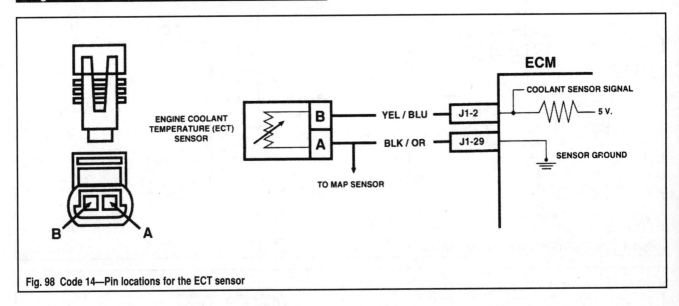

Fig. 98 Code 14—Pin locations for the ECT sensor

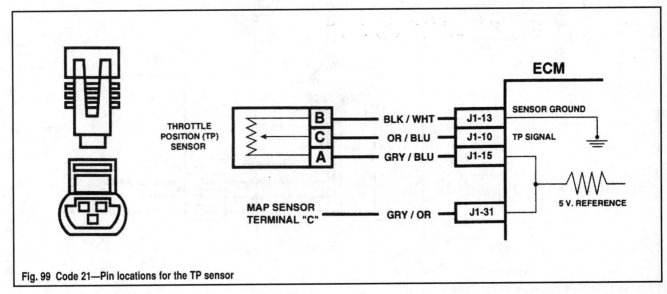

Fig. 99 Code 21—Pin locations for the TP sensor

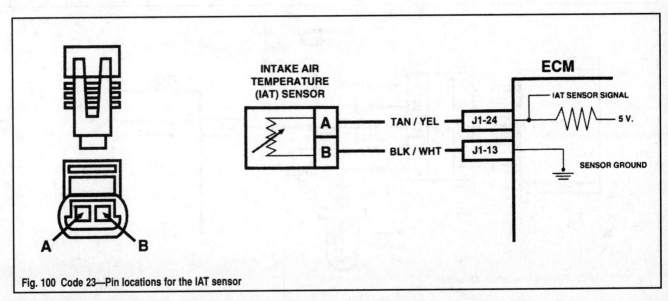

Fig. 100 Code 23—Pin locations for the IAT sensor

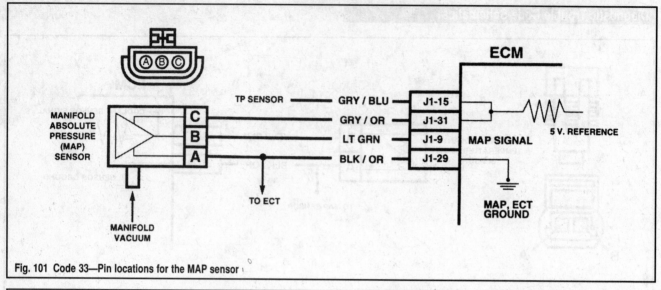

Fig. 101 Code 33—Pin locations for the MAP sensor

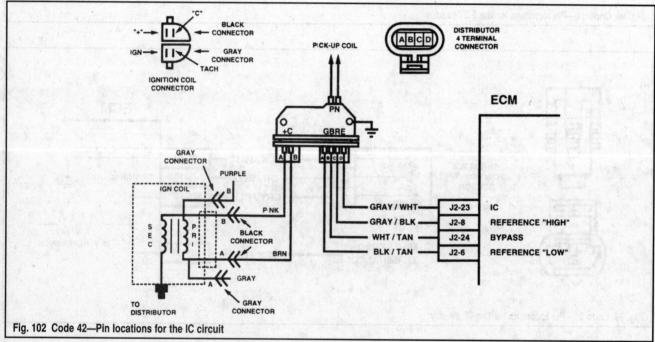

Fig. 102 Code 42—Pin locations for the IC circuit

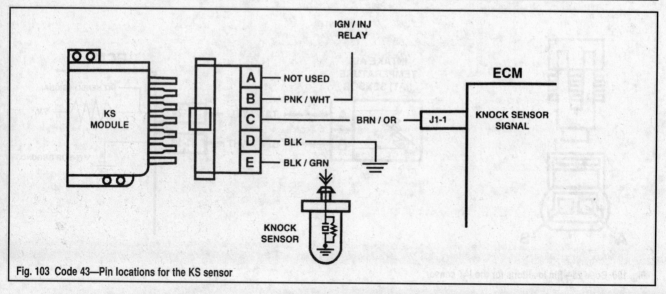

Fig. 103 Code 43—Pin locations for the KS sensor

WIRING SCHEMATICS

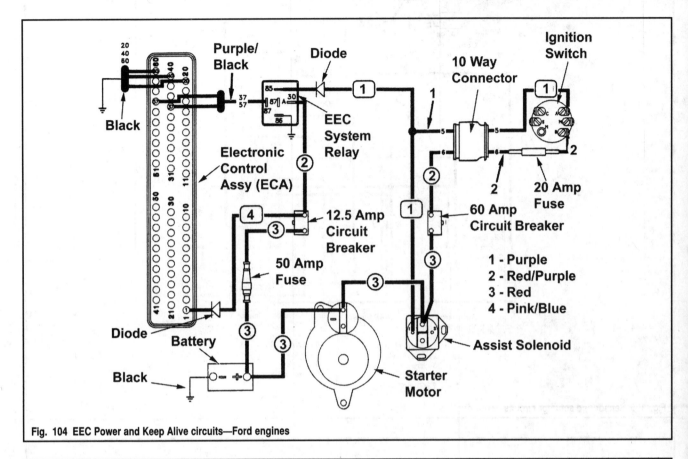

Fig. 104 EEC Power and Keep Alive circuits—Ford engines

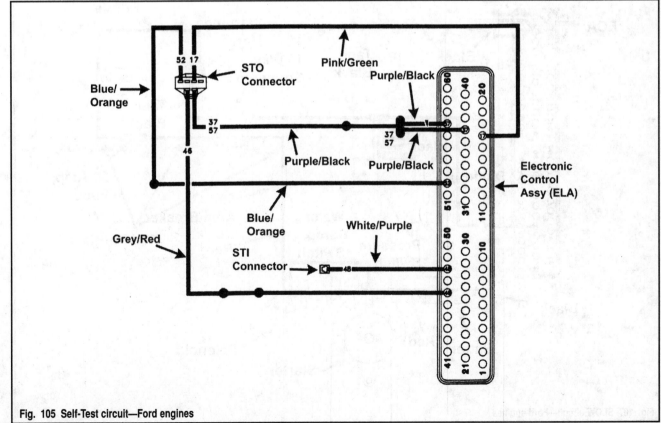

Fig. 105 Self-Test circuit—Ford engines

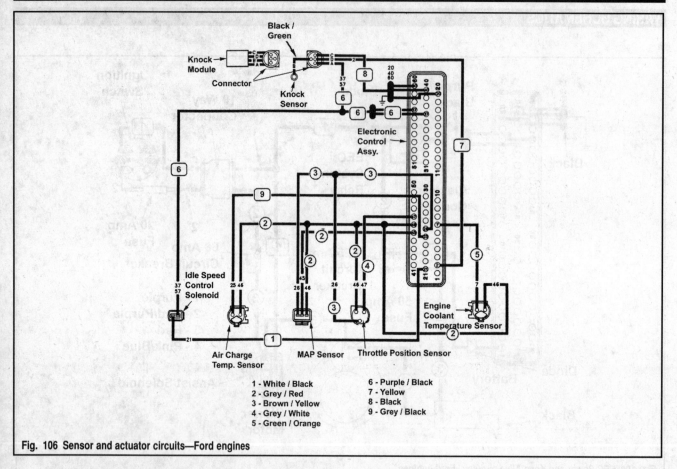

1 - White / Black
2 - Grey / Red
3 - Brown / Yellow
4 - Grey / White
5 - Green / Orange

6 - Purple / Black
7 - Yellow
8 - Black
9 - Grey / Black

Fig. 106 Sensor and actuator circuits—Ford engines

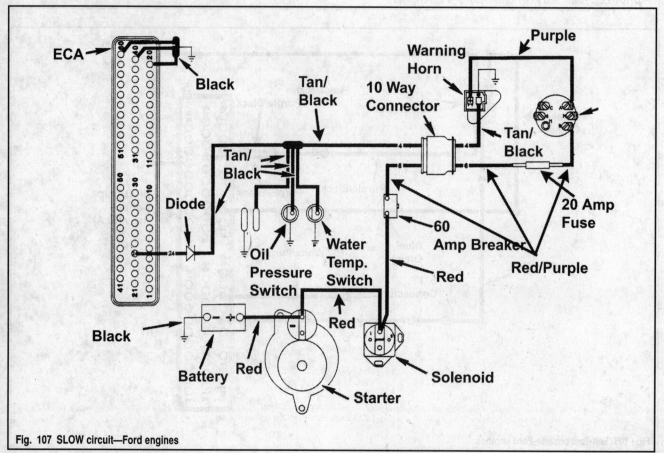

Fig. 107 SLOW circuit—Ford engines

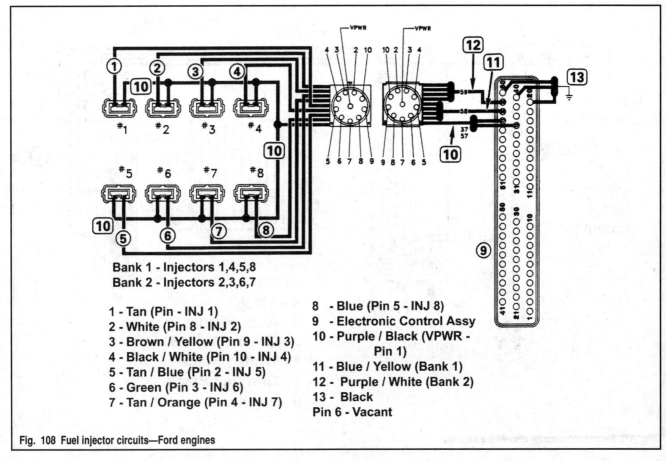

Bank 1 - Injectors 1,4,5,8
Bank 2 - Injectors 2,3,6,7

1 - Tan (Pin - INJ 1)
2 - White (Pin 8 - INJ 2)
3 - Brown / Yellow (Pin 9 - INJ 3)
4 - Black / White (Pin 10 - INJ 4)
5 - Tan / Blue (Pin 2 - INJ 5)
6 - Green (Pin 3 - INJ 6)
7 - Tan / Orange (Pin 4 - INJ 7)

8 - Blue (Pin 5 - INJ 8)
9 - Electronic Control Assy
10 - Purple / Black (VPWR - Pin 1)
11 - Blue / Yellow (Bank 1)
12 - Purple / White (Bank 2)
13 - Black
Pin 6 - Vacant

Fig. 108 Fuel injector circuits—Ford engines

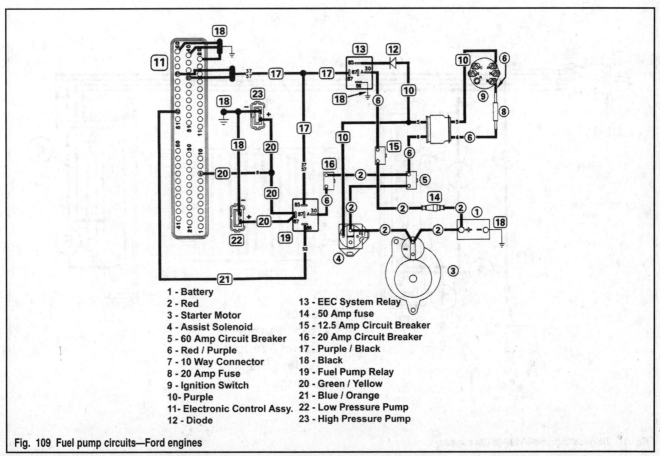

1 - Battery
2 - Red
3 - Starter Motor
4 - Assist Solenoid
5 - 60 Amp Circuit Breaker
6 - Red / Purple
7 - 10 Way Connector
8 - 20 Amp Fuse
9 - Ignition Switch
10- Purple
11- Electronic Control Assy.
12 - Diode

13 - EEC System Relay
14 - 50 Amp fuse
15 - 12.5 Amp Circuit Breaker
16 - 20 Amp Circuit Breaker
17 - Purple / Black
18 - Black
19 - Fuel Pump Relay
20 - Green / Yellow
21 - Blue / Orange
22 - Low Pressure Pump
23 - High Pressure Pump

Fig. 109 Fuel pump circuits—Ford engines

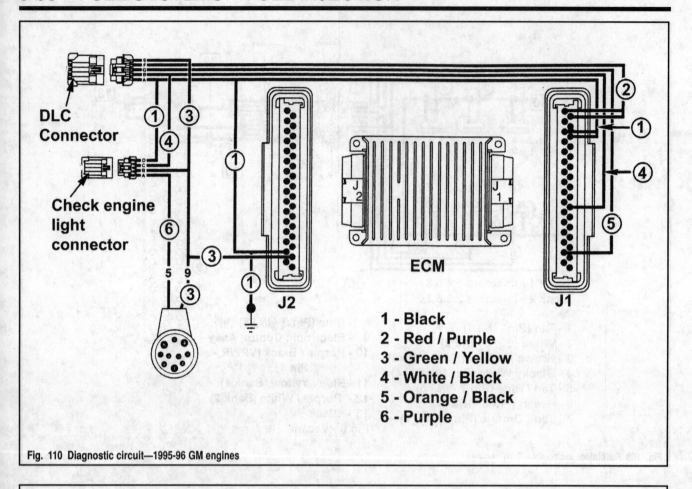

DLC Connector

Check engine light connector

ECM

J2

J1

1 - Black
2 - Red / Purple
3 - Green / Yellow
4 - White / Black
5 - Orange / Black
6 - Purple

Fig. 110 Diagnostic circuit—1995-96 GM engines

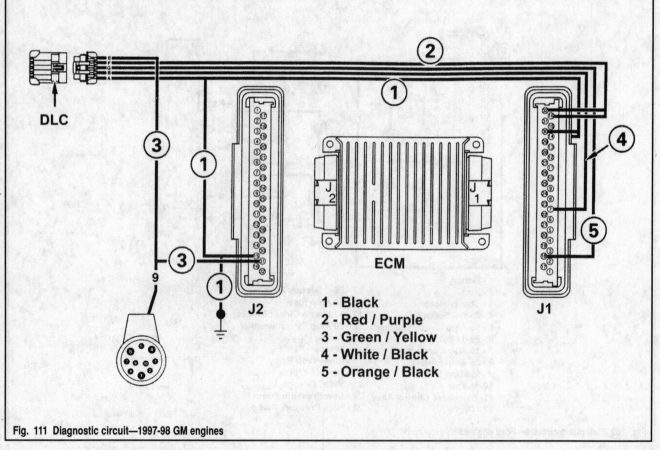

DLC

ECM

J2

J1

1 - Black
2 - Red / Purple
3 - Green / Yellow
4 - White / Black
5 - Orange / Black

Fig. 111 Diagnostic circuit—1997-98 GM engines

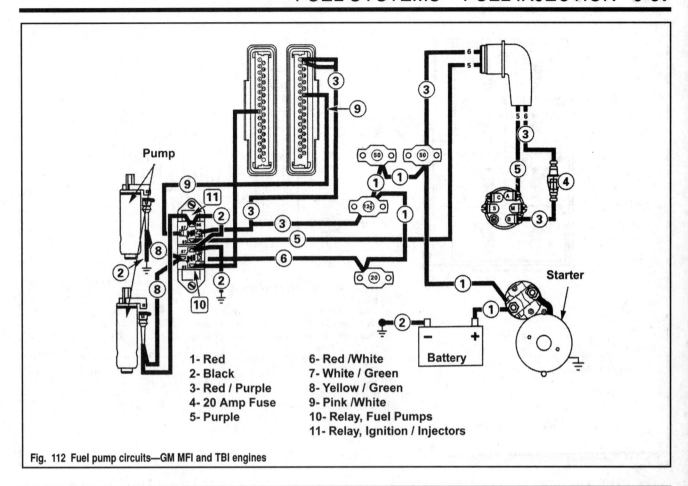

Pump

1- Red
2- Black
3- Red / Purple
4- 20 Amp Fuse
5- Purple
6- Red /White
7- White / Green
8- Yellow / Green
9- Pink /White
10- Relay, Fuel Pumps
11- Relay, Ignition / Injectors

Battery

Starter

Fig. 112 Fuel pump circuits—GM MFI and TBI engines

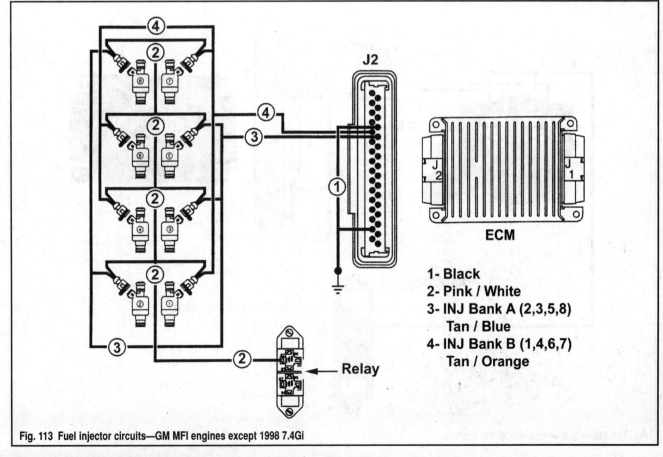

J2

ECM

Relay

1- Black
2- Pink / White
3- INJ Bank A (2,3,5,8)
   Tan / Blue
4- INJ Bank B (1,4,6,7)
   Tan / Orange

Fig. 113 Fuel injector circuits—GM MFI engines except 1998 7.4Gi

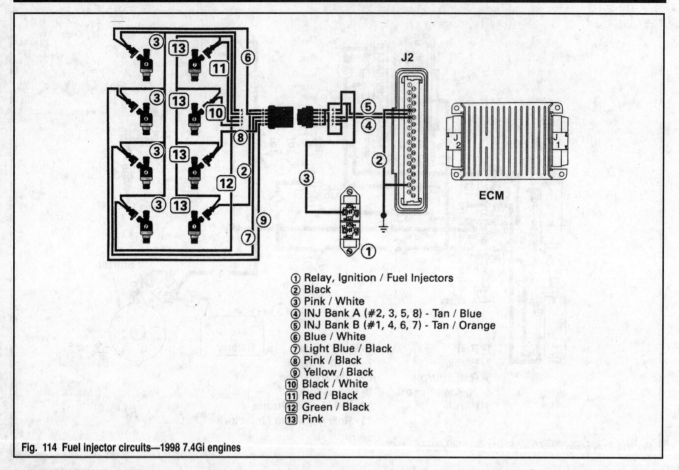

1. Relay, Ignition / Fuel Injectors
2. Black
3. Pink / White
4. INJ Bank A (#2, 3, 5, 8) - Tan / Blue
5. INJ Bank B (#1, 4, 6, 7) - Tan / Orange
6. Blue / White
7. Light Blue / Black
8. Pink / Black
9. Yellow / Black
10. Black / White
11. Red / Black
12. Green / Black
13. Pink

Fig. 114 Fuel injector circuits—1998 7.4Gi engines

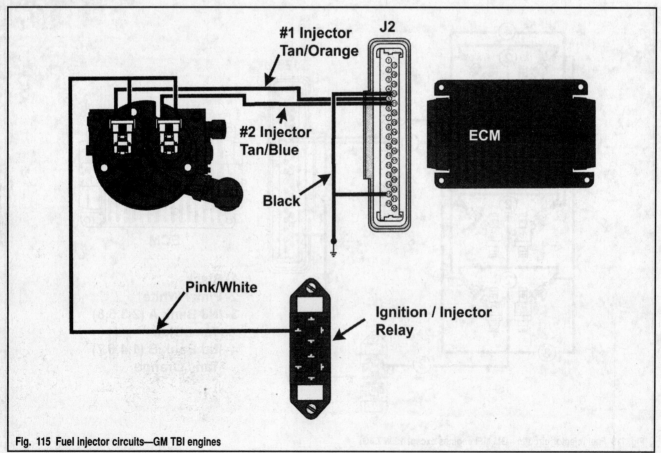

Fig. 115 Fuel injector circuits—GM TBI engines

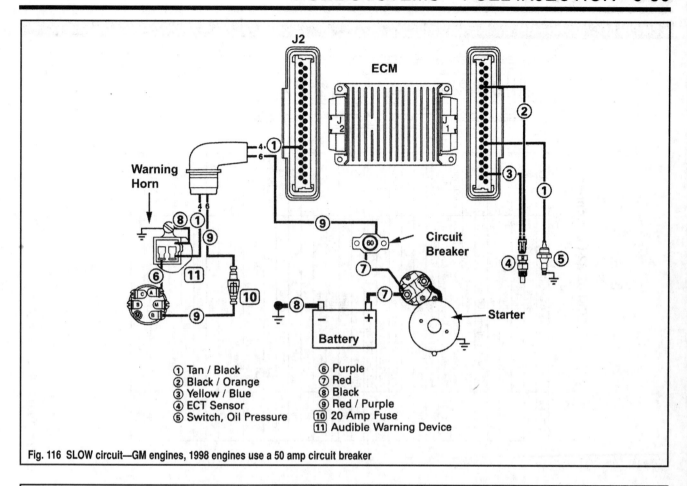

1. Tan / Black
2. Black / Orange
3. Yellow / Blue
4. ECT Sensor
5. Switch, Oil Pressure
6. Purple
7. Red
8. Black
9. Red / Purple
10. 20 Amp Fuse
11. Audible Warning Device

Fig. 116 SLOW circuit—GM engines, 1998 engines use a 50 amp circuit breaker

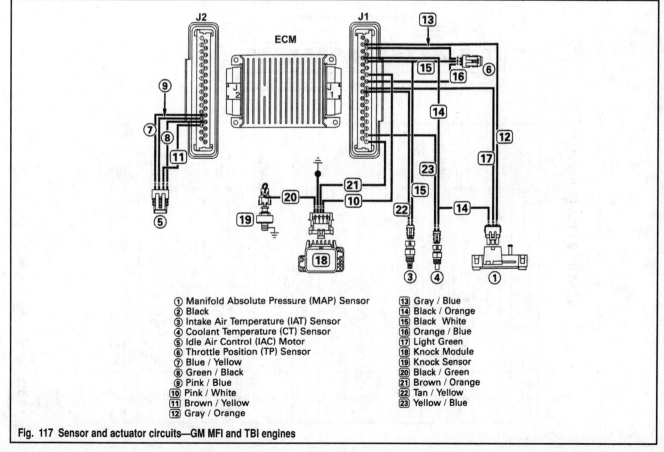

1. Manifold Absolute Pressure (MAP) Sensor
2. Black
3. Intake Air Temperature (IAT) Sensor
4. Coolant Temperature (CT) Sensor
5. Idle Air Control (IAC) Motor
6. Throttle Position (TP) Sensor
7. Blue / Yellow
8. Green / Black
9. Pink / Blue
10. Pink / White
11. Brown / Yellow
12. Gray / Orange
13. Gray / Blue
14. Black / Orange
15. Black White
16. Orange / Blue
17. Light Green
18. Knock Module
19. Knock Sensor
20. Black / Green
21. Brown / Orange
22. Tan / Yellow
23. Yellow / Blue

Fig. 117 Sensor and actuator circuits—GM MFI and TBI engines

ECM PIN LOCATIONS—GM ENGINES

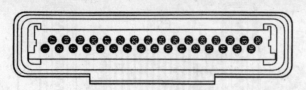

ECM J2 32-Pin Connector

J2 Pin No.	Circuit Description	Wire Color
1	Blank	
2	Blank	
3	Blank	
4	Blank	
5	Cyl. 1-7-4-6 Injector Driver	Tan/Orange
6	Distributor Low Reference and ESA	Black/Tan
7	Fuel Injector Jumper	Purple/White
8	Distributor High Reference and ESA	Gray/Black
9	Fuel Pump Relay Driver	White/Green
10	Blank	
11	Blank	
12	S.L.O.W. Warning Horn	Tan/Black
13	IAC Coil "A" Low	Pink/Blue
14	IAC Coil "B" High	Brown/Yellow
15	Ground	Black
16	Blank	
17	Blank	
18	Blank	
19	Blank	
20	Ground	Black
21	Cyl. 2-8-3-5 Injector Driver	Tan/Blue
22	Fuel Injector Jumper	Purple/White
23	(ESA) Electronic Shift Assist and (IC) Electronic Spark Timing	White
24	IC Bypass	White/Tan
25	Blank	
26	Blank	
27	Blank	
28	IAC Coil "A" High	Blue/Yellow
29	IAC Coil "B" Low	Green/Black
30	MIL / Check Engine Lamp Output	Green/Yellow
31	Blank	
32		

Fig. 119 J2 pin locations—1995 engines

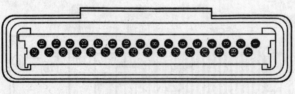

ECM J1 32-Pin Connector

J1 Pin No.	Circuit Description	Wire Color
1	Spark Retard Signal (ESC)	Brown/Orange
2	ECT Sensor	Yellow/Blue
3	Blank	
4	Blank	
5	Master/Slave (ECM) Twin Engine Only	Yellow/Gray
6	Oil Pressure Switch - S.L.O.W.	Tan/Black
7	Diagnostic "Test" (Service Mode)	White/Black
8	Blank	
9	MAP Sensor Signal	Light Green
10	TP Sensor Signal	Orange/Blue
11	Ignition 12V	Pink/White
12	Blank	
13	IAT and TP Sensor Ground	Black/White
14	ECM Ground	Black
15	TP 5 Volt Reference	Gray/Blue
16	Battery 12V	Red/Purple
17	Blank	
18	Serial Data (Scan Tool Communication)	Orange/Black
19	Blank	
20	Blank	
21	Not Used	Pink
22	Blank	
23	Blank	
24	IAT Sensor Signal	Tan/Yellow
25	Blank	
26	Blank	
27	Blank	
28	Blank	
29	MAP and ECT Sensor Ground	Black/Orange
30	ECM Ground	Black
31	MAP Sensor 5V Reference	Gray/Orange
32	Battery 12V	Red/Purple

Fig. 118 J1 pin locations—1995 engines

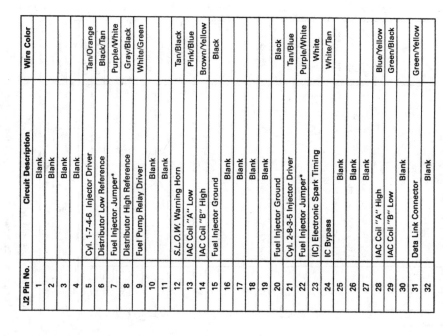

ECM J2 32-Pin Connector

J2 Pin No.	Circuit Description	Wire Color
1	Blank	
2	Blank	
3	Blank	
4	Blank	
5	Cyl. 1-7-4-6 Injector Driver	Tan/Orange
6	Distributor Low Reference	Black/Tan
7	Fuel Injector Jumper*	Purple/White
8	Distributor High Reference	Gray/Black
9	Fuel Pump Relay Driver	White/Green
10	Blank	
11	Blank	
12	S.L.O.W. Warning Horn	Tan/Black
13	IAC Coil "A" Low	Pink/Blue
14	IAC Coil "B" High	Brown/Yellow
15	Fuel Injector Ground	Black
16	Blank	
17	Blank	
18	Blank	
19	Blank	
20	Fuel Injector Ground	Black
21	Cyl. 2-8-3-5 Injector Driver	Tan/Blue
22	Fuel Injector Jumper*	Purple/White
23	(IC) Electronic Spark Timing	White
24	IC Bypass	White/Tan
25	Blank	
26	Blank	
27	Blank	
28	IAC Coil "A" High	Blue/Yellow
29	IAC Coil "B" Low	Green/Black
30	Blank	
31	Data Link Connector	Green/Yellow
32	Blank	

* MFI models only; not used on TBI models

Fig. 121 J2 pin locations—1996-97 engines

ECM J1 32-Pin Connector

J1 Pin No.	Circuit Description	Wire Color
1	Spark Retard Signal (ESC)	Brown/Orange
2	ECT Sensor, S.L.O.W.	Yellow/Blue
3	Blank	
4	Blank	
5	ECM Master/Slave (not used)	Yellow/Gray
6	Oil Pressure Switch - S.L.O.W.	Tan/Black
7	Diagnostic "Test" (Service Mode)	White/Black
8	Blank	
9	MAP Sensor Signal	Light Green
10	TP Sensor Signal	Orange/Blue
11	Ignition 12V	Pink/White
12	Blank	
13	IAT* and TP Sensor Ground	Black/White
14	ECM Ground	Black
15	TP 5 Volt Reference	Gray/Blue
16	Battery 12V	Red/Purple
17	Blank	
18	DLC, Master/Slave Communication	Orange/Black
19	Blank	
20	Blank	
21	Blank	
22	Blank	
23	Blank	
24	IAT Sensor* Signal	Tan/Yellow
25	Blank	
26	Blank	
27	Blank	
28	Blank	
29	MAP and ECT Sensor Ground, S.L.O.W.	Black/Orange
30	ECM Ground	Black
31	MAP Sensor 5V Reference	Gray/Orange
32	Battery 12V	Red/Purple

* MFI models only; not used on TBI models

Fig. 120 J1 pin locations—1996-97 engines

ECM J2 32-Pin Connector

J2 Pin No.	Circuit Description	Wire Color
1	Blank	
2	Blank	
3	Blank	
4	Blank	
5	Injector Driver #	Tan/Orange
6	IC Reference "LOW"	Black/Tan
7	Fuel Injector Jumper*	Purple/White
8	IC Reference "HIGH"	Gray/Black
9	Fuel Pump Relay Driver	White/Green
10	Blank	
11	Blank	
12	S.L.O.W.™ Warning Horn	Tan/Black
13	IAC Coil "A" Low	Pink/Blue
14	IAC Coil "B" High	Brown/Yellow
15	Fuel Injector Ground	Black
16	Blank	
17	Blank	
18	Blank	
19	Blank	
20	Fuel Injector Ground	Black
21	Injector Driver +	Tan/Blue
22	Fuel Injector Jumper*	Purple/White
23	IC Signal	White
24	IC Bypass	White/Tan
25	Blank	
26	Blank	
27	Blank	
28	IAC Coil "A" High	Blue/Yellow
29	IAC Coil "B" Low	Green/Black
30	Blank	
31	Data Link Connector	Green/Yellow
32	Blank	

* MFI models only; not used on TBI models
# MFI models: Cylinders 1-4-6-7   TBI models: Injector No. 1
- MFI models: Cylinders 2-3-5-8   TBI models: Injector No. 2

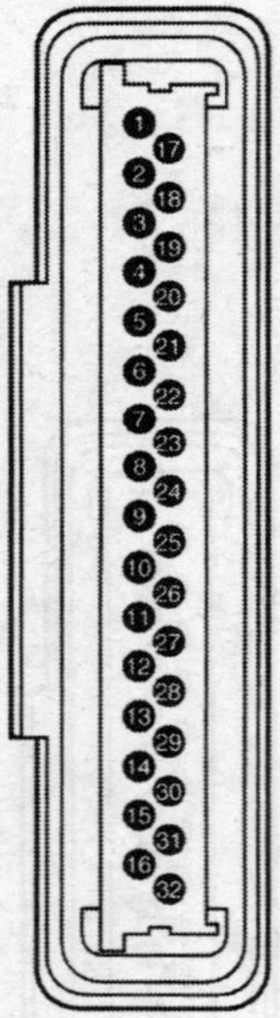

Fig. 123 J2 pin locations—1998 engines

ECM J1 32-Pin Connector

J1 Pin No.	Circuit Description	Wire Color
1	Spark Retard Signal (ESC)	Brown/Orange
2	ECT Sensor, S.L.O.W.™	Yellow/Blue
3	Blank	
4	Blank	
5	ECM Master/Slave (not used)	Yellow/Gray
6	Oil Pressure Switch - S.L.O.W.™	Tan/Black
7	Diagnostic "Test" (Service Mode)	White/Black
8	Blank	
9	MAP Sensor Signal	Light Green
10	TP Sensor Signal	Orange/Blue
11	Ignition 12V	Pink/White
12	Blank	
13	IAT* and TP Sensor Ground	Black/White
14	ECM Ground	Black
15	TP 5 Volt Reference	Gray/Blue
16	Battery 12V	Red/Purple
17	Blank	
18	DLC, Master/Slave Communication	Orange/Black
19	Blank	
20	Blank	
21	Blank	
22	Blank	
23	Blank	
24	IAT Sensor* Signal	Tan/Yellow
25	Blank	
26	Blank	
27	Blank	
28	Blank	
29	MAP and ECT Sensor Ground, S.L.O.W.™	Black/Orange
30	ECM Ground	Black
31	MAP Sensor 5V Reference	Gray/Orange
32	Battery 12V	Red/Purple

* MFI models only; not used on TBI models

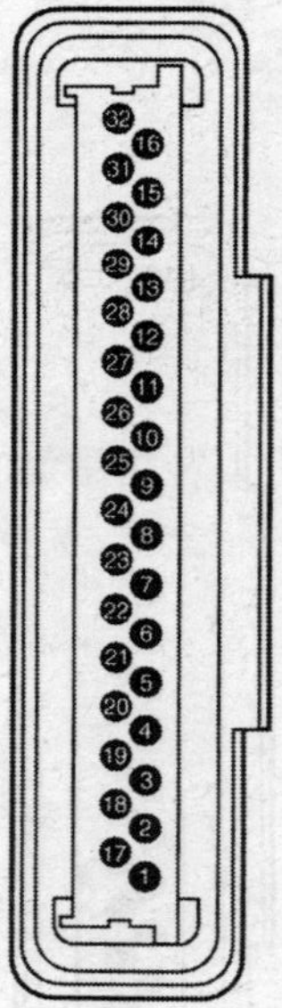

Fig. 122 J1 pin locations—1998 engines

This chart is to further aid in diagnosis of symptoms. These voltages were derived from a known good engine. The voltages shown were done with the electrical system completely hooked up and operational. The voltages are to help identify what voltage is needed to operate the different circuits. **NEVER ATTEMPT TO OBTAIN THESE VOLTAGES BY PROBING WIRES OR CONNECTORS. Serious damage could result to wiring, or connectors, with the loss of engine operation.** The voltages you may get may vary due to low battery charge or other reasons, but they should be close.

The "B+" symbol indicates a system voltage.
THE FOLLOWING CONDITIONS MUST BE MET BEFORE TESTING:
● Engine operating temperature ● Engine idling (for "Engine Operating" column)
● Test terminal not grounded

ECM J1 Input Connector and Symptoms Identification

PIN	PIN FUNCTION	WIRE COLOR	COMPONENT CONNECTOR	NORMAL VOLTAGE IGNITION "ON"	NORMAL VOLTAGE ENGINE OPERATING	DTC(s) AFFECTED	POSSIBLE SYMPTOMS
1	KS SIGNAL	BROWN/ORANGE	KNOCK SENSOR	9.5V	9.5V	43	POOR FUEL ECONOMY, POOR PERFORMANCE, DETONATION
2	ECT SIGNAL	BLUE	ECT SENSOR	1.95V (2)	1.95V (2)	14	POOR PERFORMANCE EXHAUST ODOR, ROUGH IDLE, RPM REDUCTION
3	NOT USED	-	-	-	-	-	-
4	NOT USED	-	-	-	-	-	-
5	MASTER SLAVE	YELLOW/GRAY	IN-LINE BOAT HARNESS	B+	B+	NONE	LACK OF DATA FROM OTHER ENGINE
6	OIL PRESSURE SWITCH	TAN/BLACK	10 PIN CONNECTOR	(5)	B+	NONE	RPM REDUCTION MODE
7	DIAGNOSIS TEST TERMINAL	WHITE/BLACK	DLC	B+	B+	NONE	INCORRECT IDLE, POOR PERFORMANCE
8	NOT USED	-	-	-	-	-	-
9	MAP SIGNAL	LIGHT GREEN	MAP SENSOR	4.9V	1.46V (3)	33	POOR PERFORMANCE SURGE, POOR FUEL ECONOMY, EXHAUST ODOR
10	TP SIGNAL	ORANGE/BLUE	TP SENSOR	0.62V (4)	0.62V (4)	21	POOR PERFORMANCE AND ACCELERATION, INCORRECT IDLE
11	IGNITION FUSED	PINK/WHITE	SPLICE	B+	B+	NONE	CT INOPERABLE, NO START
12	NOT USED	-	-	-	-	-	-
13	TP & IAT GROUND	BLACK/WHITE	TP SENSOR	(5)	(5)	21	HIGH IDLE, ROUGH IDLE, POOR PERFORMANCE
14	ECM GROUND	BLACK	ENGINE BLOCK	(5)	(5)	NONE	NO CHANGE
15	TP 5V REF	GRAY/BLUE	TP, MAP SENSOR	5V	5V	21, 33	LACK OF POWER, IDLE HIGH
16	BATTERY FEED	RED/PURPLE	SPLICE	B+	B+	NONE	NO START

(1) BATTERY VOLTAGE FOR FIRST TWO SECONDS, THEN 0 VOLTS
(2) VARIES WITH TEMPERATURE
(3) VARIES WITH MANIFOLD VACUUM
(4) VARIES WITH THROTTLE MOVEMENT
(5) LESS THAN 0.5 VOLT (500Mv)

Fig. 124 ECM connector and symptoms identification—1995 J1 (1 of 2)

ECM J1 Input Connector and Symptoms Identification

PIN	PIN FUNCTION	WIRE COLOR	COMPONENT CONNECTOR	NORMAL VOLTAGE IGNITION "ON"	NORMAL VOLTAGE ENGINE OPERATING	DTC(s) AFFECTED	POSSIBLE SYMPTOMS
17	NOT USED	-	-	-	-	-	-
18	SERIAL DATA	ORANGE/BLACK	DLC	5V	5V	NONE	NO SERIAL DATA
19	NOT USED	-	-	-	-	-	-
20	NOT USED	-	-	-	-	-	-
21	E-STOP	PINK	BULLET CONNECTOR	B+	B+	NONE	RPM REDUCTION MODE
22	NOT USED	-	-	-	-	-	-
23	NOT USED	-	-	-	-	-	-
24	IAT SENSOR	TAN/YELLOW	IAT SENSOR	3.2V (2)	3.2V (2)	23	ROUGH IDLE
25	NOT USED	-	-	-	-	-	-
26	NOT USED	-	-	-	-	-	-
27	NOT USED	-	-	-	-	-	-
28	NOT USED	-	-	-	-	-	-
29	MAP & ECT GROUND	BLACK/ORANGE	MAP AND ECT SENSOR	(5)	(5)	14 & 33	LACK OF PERFORMANCE EXHAUST ODOR, STALL
30	ECM GROUND	BLACK	ENGINE BLOCK	(5)	(5)	NONE	NO CHANGE
31	MAP, 5V REFERENCE	GRAY/ORANGE	MAP SENSOR	5V	5V	33	LACK OF POWER, SURGE, ROUGH IDLE, EXHAUST ODOR
32	BATTERY FEED	RED/PURPLE	SPLICE	B+	B+	NONE	NO START

(1) BATTERY VOLTAGE FOR FIRST TWO SECONDS, THEN 0 VOLTS
(2) VARIES WITH TEMPERATURE
(3) VARIES WITH MANIFOLD VACUUM
(4) VARIES WITH THROTTLE MOVEMENT
(5) LESS THAN 0.5 VOLT (500Mv)

Fig. 125 ECM connector and symptoms identification—1995 J1 (2 of 2)

ECM J2 Output Connector and Symptoms Identification

This chart is to further aid in diagnosis of symptoms. These voltages were derived from a known good engine. The voltages shown were done with the electrical system completely hooked up and operational. The voltages are to help identify what voltage is needed to operate the different circuits. NEVER ATTEMPT TO OBTAIN THESE VOLTAGES BY PROBING WIRES OR CONNECTORS. Serious damage could result to wiring, or connectors, with the loss of engine operation. The voltages you may get may vary due to low battery charge or other reasons, but they should be close.

The "B+" symbol indicates a system voltage.
THE FOLLOWING CONDITIONS MUST BE MET BEFORE TESTING:
• Engine operating temperature • Engine idling (for "Engine Operating" column)
• Test terminal not grounded

PIN	PIN FUNCTION	WIRE COLOR	COMPONENT CONNECTOR	NORMAL VOLTAGE		DTC(s) AFFECTED	POSSIBLE SYMPTOMS
				IGNITION "ON"	ENGINE OPERATING		
1	NOT USED	-	-	-	-	-	-
2	NOT USED	-	-	-	-	-	-
3	NOT USED	-	-	-	-	-	-
4	NOT USED	-	-	-	-	-	-
5	CYL 1-7-4-6 INJECTOR DRIVER	TAN/ORANGE	INJECTOR	B+	B+	NONE	ROUGH IDLE, LACK OF POWER, STALL
6	IC REF LOW & ESA	BLACK/TAN	IC MODULE	(5)	(5)	NONE	NO CHANGE
7	JUMPER	PURPLE/WHITE	ECM	2.9V	2.9V	NONE	NO CHANGE
8	IC REF HIGH & ESA	GRAY/BLACK	IC MODULE	5V	1.6V	NONE	NO RESTART
9	FUEL PUMP RELAY DRIVER	WHITE/GREEN	FUEL PUMP RELAY	(1)(5)	B+	NONE	NO START
10	NOT USED	-	-	-	-	-	-
11	NOT USED	-	-	-	-	-	-
12	RPM REDUCE WARNING HORN	TAN/BLACK	10 PIN CONNECTOR	(5)	(5)	NONE	LOSS OF RPM OVER 2500
13	IAC "A" LOW	PINK/BLUE	IAC VALVE	NOT USABLE	NOT USABLE	NONE	ROUGH UNSTABLE OR INCORRECT IDLE
14	IAC "B" HIGH	BROWN/YELLOW	IAC VALVE	NOT USABLE	NOT USABLE	NONE	ROUGH UNSTABLE OR INCORRECT IDLE
15	FUEL INJECTOR GROUND	BLACK	ENGINE GROUND	(5)	(5)	NONE	NO CHANGE
16	NOT USED	-	-	-	-	-	-

(1) BATTERY VOLTAGE FOR FIRST TWO SECONDS, THEN 0 VOLTS
(2) VARIES WITH TEMPERATURE
(3) VARIES WITH MANIFOLD VACUUM
(4) VARIES WITH THROTTLE MOVEMENT
(5) LESS THAN 0.5 VOLT (500Mv)

Fig. 126 ECM connector and symptoms identification—1995 J2 (1 of 2)

ECM J2 Output Connector and Symptoms Identification

PIN	PIN FUNCTION	WIRE COLOR	COMPONENT CONNECTOR	NORMAL VOLTAGE		DTC(s) AFFECTED	POSSIBLE SYMPTOMS
				IGNITION "ON"	ENGINE OPERATING		
17	NOT USED	-	-	-	-	-	-
18	NOT USED	-	-	-	-	-	-
19	NOT USED	-	-	-	-	-	-
20	FUEL INJECTOR GROUND	BLACK	ENGINE GROUND	(5)	(5)	NONE	NO CHANGE
21	CYL 2-8-3-5 INJECTOR DRIVER	TAN/BLUE	INJECTOR	B+	B+	NONE	ROUGH IDLE, LACK OF POWER, STALL
22	PIN 7	PURPLE/WHITE	-	B+	B+	NONE	ROUGH IDLE, LACK OF POWER, STALL
23	IC or ESA SIGNAL	WHITE	IC MODULE	(5)	1.2V	42	STALL, WILL RESTART IN BYPASS MODE, LACK OF POWER
24	IC BYPASS	WHITE/TAN	IC MODULE	(5)	4.5V	42	LACK OF POWER, FIXED TIMING
25	NOT USED	-	-	-	-	-	-
26	NOT USED	-	-	-	-	-	-
27	NOT USED	-	-	-	-	-	-
28	IAC "A" HIGH	BLUE/YELLOW	IAC VALVE	NOT USABLE	NOT USABLE	NONE	ROUGH UNSTABLE OR INCORRECT IDLE
29	IAC "B" LOW	GREEN/BLACK	IAC VALVE	NOT USABLE	NOT USABLE	NONE	ROUGH UNSTABLE OR INCORRECT IDLE
30	NOT USED	-	-	-	-	-	-
31	MALFUNCTION INDICATOR LAMP	GREEN/YELLOW	DLC	(5)	(5)	NONE	LAMP INOPERABLE
32	NOT USED	-	-	-	-	-	-

(1) BATTERY VOLTAGE FOR FIRST TWO SECONDS, THEN 0 VOLTS
(2) VARIES WITH TEMPERATURE
(3) VARIES WITH MANIFOLD VACUUM
(4) VARIES WITH THROTTLE MOVEMENT
(5) LESS THAN 0.5 VOLT (500Mv)

Fig. 127 ECM connector and symptoms identification—1995 J2 (2 of 2)

ECM J1 Connector and Symptoms Identification

This chart is to further aid in diagnosis of symptoms. These voltages were derived from a known good engine. The voltages shown were done with the electrical system completely hooked up and operational. The voltages are to help identify what voltage is needed to operate the different circuits. **NEVER ATTEMPT TO OBTAIN THESE VOLTAGES BY PROBING WIRES OR CONNECTORS.** Serious damage could result to wiring, or connectors, with the loss of engine operation. The voltages you may get may vary due to low battery charge or other reasons, but they should be close.

The "B+" symbol indicates a system voltage.
THE FOLLOWING CONDITIONS MUST BE MET BEFORE TESTING:
● Engine at operating temperature ● Engine idling (for "Engine Operating" column)
● Test terminal not grounded

PIN	PIN FUNCTION	WIRE COLOR	COMPONENT CONNECTOR	NORMAL VOLTAGE		DTC(s) AFFECTED	POSSIBLE SYMPTOMS
				IGNITION "ON"	ENGINE OPERATING		
1	KS SIGNAL	BROWN/ORANGE	KNOCK SENSOR	9.5V	9.5V	43	POOR FUEL ECONOMY, POOR PERFORMANCE, DETONATION
2	ECT SIGNAL S.L.O.W.	YELLOW/BLUE	ECT SENSOR	1.95V (2)	1.95V (2)	14	POOR PERFORMANCE, EXHAUST ODOR, ROUGH IDLE, RPM REDUCTION
3	NOT USED	-	-	-	-	-	-
4	NOT USED	-	-	-	-	-	-
5	MASTER/SLAVE (not used)	YELLOW/GRAY	IN-LINE BOAT HARNESS	B+	B+	NONE	LACK OF DATA FROM OTHER ENGINE
6	OIL PRESSURE SWITCH, S.L.O.W.	TAN/BLACK	10 PIN CONNECTOR	(5)	B+	NONE	RPM REDUCTION MODE
7	DATA LINK CONNECTOR	WHITE/BLACK	DLC	B+	B+	NONE	INCORRECT IDLE, POOR PERFORMANCE
8	NOT USED	-	-	-	-	-	-
9	MAP SIGNAL	LIGHT GREEN	MAP SENSOR	4.9V	1.46V (3)	33	POOR PERFORMANCE SURGE, POOR FUEL ECONOMY, EXHAUST ODOR
10	TP SIGNAL	ORANGE/BLUE	TP SENSOR	0.62V (4)	0.62V (4)	21	POOR PERFORMANCE AND ACCELERATION, INCORRECT IDLE
11	ECM SWITCHED 12V	PINK/WHITE	SPLICE	B+	B+	NONE	CT INOPERABLE, NO START
12	NOT USED	-	-	-	-	-	-
13	TP & IAT* GROUND	BLACK/WHITE	TP, IAT SENSOR	(5)	(5)	21	HIGH IDLE, ROUGH IDLE, POOR PERFORMANCE
14	ECM GROUND	BLACK	ENGINE BLOCK	(5)	(5)	NONE	NO CHANGE
15	TP 5V REF	GRAY/BLUE	TP, MAP SENSOR	5V	5V	21, 33	LACK OF POWER, IDLE HIGH
16	ECM B+	RED/PURPLE	SPLICE	B+	B+	NONE	NO START

(1) BATTERY VOLTAGE FOR FIRST TWO SECONDS, THEN 0 VOLTS
(2) VARIES WITH TEMPERATURE
(3) VARIES WITH MANIFOLD VACUUM
(4) VARIES WITH THROTTLE MOVEMENT
(5) LESS THAN 0.5 VOLT (500Mv)
* MFI models only; not used on TBI models

Fig. 128 ECM connector and symptoms identification—1996-97 J1 (1 of 2)

ECM J1 Connector and Symptoms Identification

PIN	PIN FUNCTION	WIRE COLOR	COMPONENT CONNECTOR	NORMAL VOLTAGE		DTC(s) AFFECTED	POSSIBLE SYMPTOMS
				IGNITION "ON"	ENGINE OPERATING		
17	NOT USED	-	-	-	-	-	-
18	MASTER/SLAVE (not used)	ORANGE/BLACK	DLC, IN-LINE BOAT HARNESS	5V	5V	NONE	NO SERIAL DATA
19	NONE	-	-	-	-	-	-
20	NOT USED	-	-	-	-	-	-
21	NONE	-	-	-	-	-	-
22	NOT USED	-	-	-	-	-	-
23	NOT USED	-	-	-	-	-	-
24	IAT SENSOR*	TAN/YELLOW	IAT SENSOR*	3.2V (2)	3.2V (2)	23	ROUGH IDLE
25	NOT USED	-	-	-	-	-	-
26	NOT USED	-	-	-	-	-	-
27	NOT USED	-	-	-	-	-	-
28	NOT USED	-	-	-	-	-	-
29	MAP/ECT GROUND, S.L.O.W.	BLACK/ORANGE	MAP AND ECT SENSOR	(5)	(5)	14 & 33	LACK OF PERFORMANCE EXHAUST ODOR, STALL
30	ECM GROUND	BLACK	ENGINE BLOCK	(5)	(5)	NONE	NO CHANGE
31	MAP, 5V REFERENCE	GRAY/ORANGE	MAP SENSOR	5V	5V	33	LACK OF POWER, SURGE, ROUGH IDLE, EXHAUST ODOR
32	ECM B+	RED/PURPLE	SPLICE	B+	B+	NONE	NO START

(1) BATTERY VOLTAGE FOR FIRST TWO SECONDS, THEN 0 VOLTS
(2) VARIES WITH TEMPERATURE
(3) VARIES WITH MANIFOLD VACUUM
(4) VARIES WITH THROTTLE MOVEMENT
(5) LESS THAN 0.5 VOLT (500Mv)
* MFI models only; not used on TBI models

Fig. 129 ECM connector and symptoms identification—1996-97 J1 (2 of 2)

This chart is to further aid in diagnosis of symptoms. These voltages were derived from a known good engine. The voltages shown were done with the electrical system completely hooked up and operational. The voltages are to help identify what voltage is needed to operate the different circuits. NEVER ATTEMPT TO OBTAIN THESE VOLTAGES BY PROBING WIRES OR CONNECTORS. Serious damage could result to wiring, or connectors, with the loss of engine operation. The voltages you may get may vary due to low battery charge or other reasons, but they should be close.

The "B+" symbol indicates a system voltage.
THE FOLLOWING CONDITIONS MUST BE MET BEFORE TESTING:
• Engine at operating temperature • Engine idling (for "Engine Operating" column)
• Test terminal not grounded

ECM J2 Connector and Symptoms Identification

PIN	PIN FUNCTION	WIRE COLOR	COMPONENT CONNECTOR	NORMAL VOLTAGE IGNITION "ON"	NORMAL VOLTAGE ENGINE OPERATING	DTC(s) AFFECTED	POSSIBLE SYMPTOMS
1	NOT USED	-	-	-	-	-	-
2	NOT USED	-	-	-	-	-	-
3	NOT USED	-	-	-	-	-	-
4	NOT USED	-	-	-	-	-	-
5	INJECTOR DRIVER BANK B	TAN/ORANGE	INJECTOR	B+	B+	NONE	ROUGH IDLE, LACK OF POWER, STALL
6	IC REF LOW	BLACK/TAN	IC MODULE	(5)	(5)	NONE	NO CHANGE
7	INJECTOR JUMPER*	PURPLE/WHITE	ECM	2.9V	2.9V	NONE	NO CHANGE
8	IC REF HIGH	GRAY/BLACK	IC MODULE	5V	1.6V	NONE	NO RESTART
9	FUEL PUMP RELAY DRIVER	WHITE/GREEN	FUEL PUMP RELAY	(1)(5)	B+	NONE	NO START
10	NOT USED	-	-	-	-	-	-
11	NOT USED	-	-	-	-	-	-
12	S.L.O.W. WARNING HORN	TAN/BLACK	10 PIN CONNECTOR	(5)	(5)	NONE	LOSS OF RPM OVER 2500
13	IAC "A" LOW	PINK/BLUE	IAC VALVE	NOT USABLE	NOT USABLE	NONE	ROUGH UNSTABLE OR INCORRECT IDLE
14	IAC "B" HIGH	BROWN/YELLOW	IAC VALVE	NOT USABLE	NOT USABLE	NONE	ROUGH UNSTABLE OR INCORRECT IDLE
15	FUEL INJECTOR GROUND	BLACK	ENGINE GROUND	(5)	(5)	NONE	NO CHANGE
16	NOT USED	-	-	-	-	-	-

(1) BATTERY VOLTAGE FOR FIRST TWO SECONDS, THEN 0 VOLTS
(2) VARIES WITH TEMPERATURE
(3) VARIES WITH MANIFOLD VACUUM
(4) VARIES WITH THROTTLE MOVEMENT
(5) LESS THAN 0.5 VOLT (500Mv)
* MFI models only; not used on TBI models

Fig. 130 ECM connector and symptoms identification—1996-97 J2 (1 of 2)

ECM J2 Connector and Symptoms Identification

PIN	PIN FUNCTION	WIRE COLOR	COMPONENT CONNECTOR	NORMAL VOLTAGE IGNITION "ON"	NORMAL VOLTAGE ENGINE OPERATING	DTC(s) AFFECTED	POSSIBLE SYMPTOMS
17	NOT USED	-	-	-	-	-	-
18	NOT USED	-	-	-	-	-	-
19	NOT USED	-	-	-	-	-	-
20	FUEL INJECTOR GROUND	BLACK	ENGINE GROUND	(5)	(5)	NONE	NO CHANGE
21	INJECTOR DRIVER BANK A	TAN/BLUE	INJECTOR	B+	B+	NONE	ROUGH IDLE, LACK OF POWER, STALL
22	INJECTOR JUMPER*	PURPLE/WHITE	ECM	B+	B+	NONE	ROUGH IDLE, LACK OF POWER, STALL
23	IC SIGNAL	WHITE	IC MODULE	(5)	1.2V	42	STALL, WILL RESTART IN BYPASS MODE, LACK OF POWER
24	IC BYPASS	WHITE/TAN	IC MODULE	(5)	4.5V	42	LACK OF POWER, FIXED TIMING
25	NOT USED	-	-	-	-	-	-
26	NOT USED	-	-	-	-	-	-
27	NOT USED	-	-	-	-	-	-
28	IAC "A" HIGH	BLUE/YELLOW	IAC VALVE	NOT USABLE	NOT USABLE	NONE	ROUGH UNSTABLE OR INCORRECT IDLE
29	IAC "B" LOW	GREEN/BLACK	IAC VALVE	NOT USABLE	NOT USABLE	NONE	ROUGH UNSTABLE OR INCORRECT IDLE
30	NOT USED	-	-	-	-	-	-
31	DLC	GREEN/YELLOW	DLC	(5)	(5)	NONE	-
32	NOT USED	-	-	-	-	-	-

(1) BATTERY VOLTAGE FOR FIRST TWO SECONDS, THEN 0 VOLTS
(2) VARIES WITH TEMPERATURE
(3) VARIES WITH MANIFOLD VACUUM
(4) VARIES WITH THROTTLE MOVEMENT
(5) LESS THAN 0.5 VOLT (500Mv)
* MFI models only; not used on TBI models

Fig. 131 ECM connector and symptoms identification—1996-97 J2 (2 of 2)

ECM J1 Connector and Symptoms Identification

This chart is to further aid in diagnosis of symptoms. These voltages were derived from a known good engine. The voltages shown were done with the electrical system completely hooked up and operational. The voltages are to help identify what voltage is needed to operate the different circuits. NEVER ATTEMPT TO OBTAIN THESE VOLTAGES BY PROBING WIRES OR CONNECTORS. Serious damage could result to wiring, or connectors, with the loss of engine operation. The voltages you may get may vary due to low battery charge or other reasons, but they should be close.

The "B+" symbol indicates a system voltage.
THE FOLLOWING CONDITIONS MUST BE MET BEFORE TESTING:
- Engine at operating temperature • Engine idling (for "Engine Operating" column)
- Test terminal not grounded

PIN	PIN FUNCTION	WIRE COLOR	COMPONENT CONNECTOR	NORMAL VOLTAGE		DTC(s) AFFECTED	POSSIBLE SYMPTOMS
				IGNITION "ON"	ENGINE OPERATING		
1	KS SIGNAL	BROWN/ORANGE	KNOCK SENSOR	9.5V	9.5V	43	POOR FUEL ECONOMY, POOR PERFORMANCE, DETONATION
2	ECT SIGNAL S.L.O.W."	YELLOW/BLUE	ECT SENSOR	1.95V (2)	1.95V (2)	14	POOR PERFORMANCE, EXHAUST ODOR, ROUGH IDLE, RPM REDUCTION
3	NOT USED	-		-	-	-	-
4	NOT USED	-		-	-	-	-
5	MASTER/SLAVE (not used)	YELLOW/GRAY	IN-LINE BOAT HARNESS	B+	B+	NONE	LACK OF DATA FROM OTHER ENGINE
6	OIL PRESSURE SWITCH, S.L.O.W."	TAN/BLACK	10 PIN CONNECTOR	(5)	B+	NONE	RPM REDUCTION MODE
7	DATA LINK CONNECTOR	WHITE/BLACK	DLC	B+	B+	NONE	INCORRECT IDLE, POOR PERFORMANCE
8	NOT USED					-	
9	MAP SIGNAL	LIGHT GREEN	MAP SENSOR	4.9V	1.46V (3)	33	POOR PERFORMANCE, SURGE, POOR FUEL ECONOMY, EXHAUST ODOR
10	TP SIGNAL	ORANGE/BLUE	TP SENSOR	0.4-0.5V (4)	0.4-0.5V (4)	21	POOR PERFORMANCE AND ACCELERATION, INCORRECT IDLE
11	ECM SWITCHED 12V	PINK/WHITE	SPLICE	B+	B+	NONE	CT INOPERABLE, NO START
12	NOT USED			-	-	-	-
13	TP & IAT GROUND	BLACK/WHITE	TP, IAT SENSOR	(5)	(5)	21	HIGH IDLE, ROUGH IDLE, POOR PERFORMANCE
14	ECM GROUND	BLACK	ENGINE BLOCK	(5)	(5)	NONE	NO CHANGE
15	TP 5V REF	GRAY/BLUE	TP, MAP SENSOR	5V	5V	21, 33	LACK OF POWER, IDLE HIGH
16	ECM B+	RED/PURPLE	SPLICE	B+	B+	NONE	NO START

(1) BATTERY VOLTAGE FOR FIRST TWO SECONDS, THEN 0 VOLTS
(2) VARIES WITH TEMPERATURE
(3) VARIES WITH MANIFOLD VACUUM
(4) VARIES WITH THROTTLE MOVEMENT
(5) LESS THAN 0.5 VOLT (500Mv)

* MFI models only; not used on TBI models

Fig. 132 ECM connector and symptoms identification—1998 J1 (1 of 2)

ECM J2 Connector and Symptoms Identification

PIN	PIN FUNCTION	WIRE COLOR	COMPONENT CONNECTOR	NORMAL VOLTAGE		DTC(s) AFFECTED	POSSIBLE SYMPTOMS
				IGNITION "ON"	ENGINE OPERATING		
17	NOT USED	-	-	-	-	-	-
18	NOT USED	-	-	-	-	-	-
19	NOT USED	-	-	-	-	-	-
20	FUEL INJECTOR GROUND	BLACK	ENGINE GROUND	(5)	(5)	NONE	NO CHANGE
21	INJECTOR DRIVER BANK A	TAN/BLUE	INJECTOR	B+	B+	NONE	ROUGH IDLE, LACK OF POWER, STALL
22	INJECTOR JUMPER*	PURPLE/WHITE	ECM	B+	B+	NONE	ROUGH IDLE, LACK OF POWER, STALL
23	IC SIGNAL	WHITE	IC MODULE	(5)	1.2V	42	STALL, WILL RESTART IN BYPASS MODE, LACK OF POWER
24	IC BYPASS	WHITE/TAN	IC MODULE	(5)	4.5V	42	LACK OF POWER, FIXED TIMING
25	NOT USED	-	-	-	-	-	-
26	NOT USED	-	-	-	-	-	-
27	NOT USED	-	-	-	-	-	-
28	IAC "A" HIGH	BLUE/YELLOW	IAC VALVE	NOT USABLE	NOT USABLE	NONE	ROUGH UNSTABLE OR INCORRECT IDLE
29	IAC "B" LOW	GREEN/BLACK	IAC VALVE	NOT USABLE	NOT USABLE	NONE	ROUGH UNSTABLE OR INCORRECT IDLE
30	NOT USED	-	-	-	-	-	-
31	DLC	GREEN/YELLOW	DLC	(5)	(5)	NONE	-
32	NOT USED	-	-	-	-	-	-

(1) BATTERY VOLTAGE FOR FIRST TWO SECONDS, THEN 0 VOLTS
(2) VARIES WITH TEMPERATURE
(3) VARIES WITH MANIFOLD VACUUM
(4) VARIES WITH THROTTLE MOVEMENT
(5) LESS THAN 0.5 VOLT (500Mv)

* MFI models only; not used on TBI models

Fig. 131 ECM connector and symptoms identification—1996-97 J2 (2 of 2)

ECM J2 Connector and Symptoms Identification

This chart is to further aid in diagnosis of symptoms. These voltages were derived from a known good engine. The voltages shown were done with the electrical system completely hooked up and operational. The voltages are to help identify what voltage is needed to operate the different circuits. **NEVER ATTEMPT TO OBTAIN THESE VOLTAGES BY PROBING WIRES OR CONNECTORS.** Serious damage could result to wiring, or connectors, with the loss of engine operation. The voltages you may get may vary due to low battery charge or other reasons, but they should be close.

The "B+" symbol indicates a system voltage.
THE FOLLOWING CONDITIONS MUST BE MET BEFORE TESTING:
• Engine at operating temperature • Engine idling (for "Engine Operating" column)
• Test terminal not grounded

PIN	PIN FUNCTION	WIRE COLOR	COMPONENT CONNECTOR	NORMAL VOLTAGE		DTC(s) AFFECTED	POSSIBLE SYMPTOMS
				IGNITION "ON"	ENGINE OPERATING		
1	NOT USED	-	-	-	-	-	-
2	NOT USED	-	-	-	-	-	-
3	NOT USED	-	-	-	-	-	-
4	NOT USED	-	-	-	-	-	-
5	INJECTOR DRIVER BANK B	TAN/ORANGE	INJECTOR	B+	B+	NONE	ROUGH IDLE, LACK OF POWER, STALL
6	IC REF LOW	BLACK/TAN	IC MODULE	(5)	(5)	NONE	NO CHANGE
7	INJECTOR JUMPER*	PURPLE/WHITE	ECM	2.9V	2.9V	-	NO CHANGE
8	IC REF HIGH	GRAY/BLACK	IC MODULE	5V	1.6V	NONE	NO RESTART
9	FUEL PUMP RELAY DRIVER	WHITE/GREEN	FUEL PUMP RELAY	(1)(5)	B+	NONE	NO START
10	NOT USED	-	-	-	-	-	-
11	NOT USED	-	-	-	-	-	-
12	S.L.O.W." WARNING HORN	TAN/BLACK	10 PIN CONNECTOR	(5)	(5)	NONE	LOSS OF RPM OVER 2500
13	IAC "A" LOW	PINK/BLUE	IAC VALVE	NOT USABLE	NOT USABLE	NONE	ROUGH UNSTABLE OR INCORRECT IDLE
14	IAC "B" HIGH	BROWN/YELLOW	IAC VALVE	NOT USABLE	NOT USABLE	-	ROUGH UNSTABLE OR INCORRECT IDLE
15	FUEL INJECTOR GROUND	BLACK	ENGINE GROUND	(5)	(5)	NONE	NO CHANGE
16		-	-	-	-	-	-

(1) BATTERY VOLTAGE FOR FIRST TWO SECONDS, THEN 0 VOLTS
(2) VARIES WITH TEMPERATURE
(3) VARIES WITH MANIFOLD VACUUM
(4) VARIES WITH THROTTLE MOVEMENT
(5) LESS THAN 0.5 VOLT (500Mv)
* MFI models only; not used on TBI models

Fig. 134 ECM connector and symptoms identification—1998 J2 (1 of 2)

ECM J1 Connector and Symptoms Identification

PIN	PIN FUNCTION	WIRE COLOR	COMPONENT CONNECTOR	NORMAL VOLTAGE		DTC(s) AFFECTED	POSSIBLE SYMPTOMS
				IGNITION "ON"	ENGINE OPERATING		
17	NOT USED	-	-	-	-	-	-
18	MASTER/SLAVE (not used)	ORANGE/BLACK	DLC, IN-LINE BOAT HARNESS	5V	5V	NONE	NO SERIAL DATA
19	NONE	-	-	-	-	-	-
20	NOT USED	-	-	-	-	-	-
21	NONE	-	-	-	-	-	-
22	NOT USED	-	-	-	-	-	-
23	NOT USED	-	-	-	-	-	-
24	IAT SENSOR*	TAN/YELLOW	IAT SENSOR*	3.2V (2)	3.2V (2)	23	ROUGH IDLE
25	NOT USED	-	-	-	-	-	-
26	NOT USED	-	-	-	-	-	-
27	NOT USED	-	-	-	-	-	-
28	NOT USED	-	-	-	-	-	-
29	MAP/ECT GROUND, S.L.O.W."	BLACK/ORANGE	MAP AND ECT SENSOR	(5)	(5)	14 & 33	LACK OF PERFORMANCE EXHAUST ODOR, STALL
30	ECM GROUND	BLACK	ENGINE BLOCK	(5)	(5)	NONE	NO CHANGE
31	MAP, 5V REFERENCE	GRAY/ORANGE	MAP SENSOR	5V	5V	33	LACK OF POWER, SURGE, ROUGH IDLE, EXHAUST ODOR
32	ECM B+	RED/PURPLE	SPLICE	B+	B+	NONE	NO START

(1) BATTERY VOLTAGE FOR FIRST TWO SECONDS, THEN 0 VOLTS
(2) VARIES WITH TEMPERATURE
(3) VARIES WITH MANIFOLD VACUUM
(4) VARIES WITH THROTTLE MOVEMENT
(5) LESS THAN 0.5 VOLT (500Mv)
* MFI models only; not used on TBI models

Fig. 133 ECM connector and symptoms identification—1998 J1 (2 of 2)

ECM J2 Connector and Symptoms Identification

PIN	PIN FUNCTION	WIRE COLOR	COMPONENT CONNECTOR	NORMAL VOLTAGE		DTC(s) AFFECTED	POSSIBLE SYMPTOMS
				IGNITION "ON"	ENGINE OPERATING		
17	NOT USED	–	–	–	–	–	–
18	NOT USED	–	–	–	–	–	–
19	NOT USED	–	–	–	–	–	–
20	FUEL INJECTOR GROUND	BLACK	ENGINE GROUND	(5)	(5)	NONE	NO CHANGE
21	INJECTOR DRIVER BANK A	TAN/ BLUE	INJECTOR	B+	B+	NONE	ROUGH IDLE, LACK OF POWER, STALL
22	INJECTOR JUMPER*	PURPLE/ WHITE	ECM	B+	B+	NONE	ROUGH IDLE, LACK OF POWER, STALL
23	IC SIGNAL	WHITE	IC MODULE	(5)	1.2V	42	STALL, WILL RESTART IN BYPASS MODE, LACK OF POWER
24	IC BYPASS	WHITE/ TAN	IC MODULE	(5)	4.5V	42	LACK OF POWER, FIXED TIMING
25	NOT USED	–	–	–	–	–	–
26	NOT USED	–	–	–	–	–	–
27	NOT USED	–	–	–	–	–	–
28	IAC "A" HIGH	BLUE/ YELLOW	IAC VALVE	NOT USABLE	NOT USABLE	NONE	ROUGH UNSTABLE OR INCORRECT IDLE
29	IAC "B" LOW	GREEN/ BLACK	IAC VALVE	NOT USABLE	NOT USABLE	NONE	ROUGH UNSTABLE OR INCORRECT IDLE
30	NOT USED	–	–	–	–	–	–
31	DLC	GREEN/ YELLOW	DLC	(5)	(5)	NONE	–
32	NOT USED	–	–	–	–	–	–

(1) BATTERY VOLTAGE FOR FIRST TWO SECONDS, THEN 0 VOLTS
(2) VARIES WITH TEMPERATURE
(3) VARIES WITH MANIFOLD VACUUM
(4) VARIES WITH THROTTLE MOVEMENT
(5) LESS THAN 0.5 VOLT (500Mv)

* MFI models only; not used on TBI models

Fig. 135 ECM connector and symptoms identification—1998 J2 (2 of 2)

DIAGNOSTIC TROUBLE CODE (DTC) CHARTS

DTC CHART
Ford MFI

Code	Description
No Code	Loss of VREF
10	Vapor Separator
11	System Pass
12	RPM outside Self-Test upper band limit
13	RPM outside Self-Test lower band limit
14	PIP circuit fault
15	ROM test failed/KAM power in continuous (1993)
15	ROM test failed/KAM power interrupt (1994-96)
18	Loss of TACH input to ECA/SPOUT circuit grounded
19	Failure in EEC reference voltage
21	ECT out of Self-test range
22	MAP out of Self-test range
23	TP out of Self-test range
24	ACT out of Self-test range
25	Inactive – Ignore
51	-40 deg. indicated. ECT sensor circuit open
52	Shift assist circuit
53	TP circuit above maximum voltage
54	-40 deg. indicated. ACT sensor circuit open
61	254 deg. indicated. ECT sensor circuit grounded
63	TP circuit below minimum voltage
64	254 deg. indicated. ACT sensor circuit grounded
67	NDS circuit open
72	Insufficient MAP change - DYN RSP test
73	Insufficient TP change - DYN RSP test
77	Operator error - DYN RSP test
87	Fuel pump circuit failure
95	Fuel pump circuit open – ECA-to-motor ground
96	Fuel pump circuit open – BAT-to-relay
98	FMEM failure

DTC CHART
GM Engines

Code	Description
14	Engine coolant temperature (ECT) sensor circuit
21	Throttle position (TP) sensor circuit
23	Intake air temperature (IAT) sensor circuit (MFI Only)
33	Manifold absolute pressure (MAP) sensor circuit
42	Ignition control (IC) system
43	Knock sensor (KS) circuit
51	ECM EEProm fault

Fig. 136 and 137

FUEL FLOW DIAGRAMS

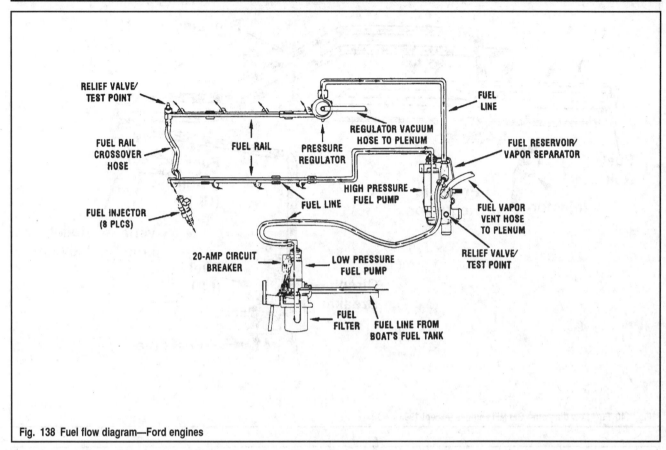

Fig. 138 Fuel flow diagram—Ford engines

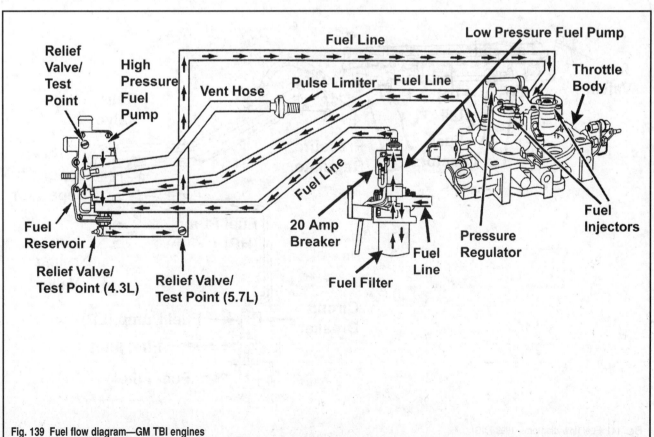

Fig. 139 Fuel flow diagram—GM TBI engines

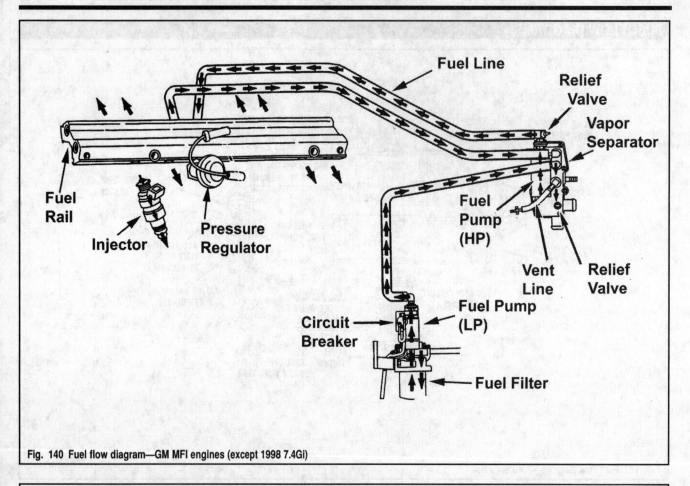

Fig. 140 Fuel flow diagram—GM MFI engines (except 1998 7.4Gi)

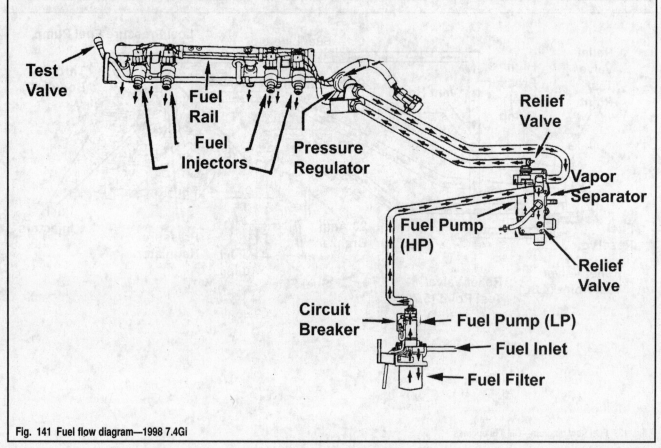

Fig. 141 Fuel flow diagram—1998 7.4Gi

COMPONENT LOCATIONS

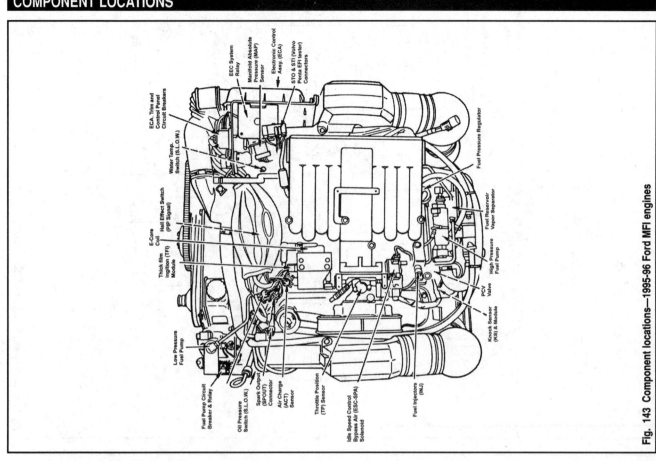

Fig. 143 Component locations—1995-96 Ford MFI engines

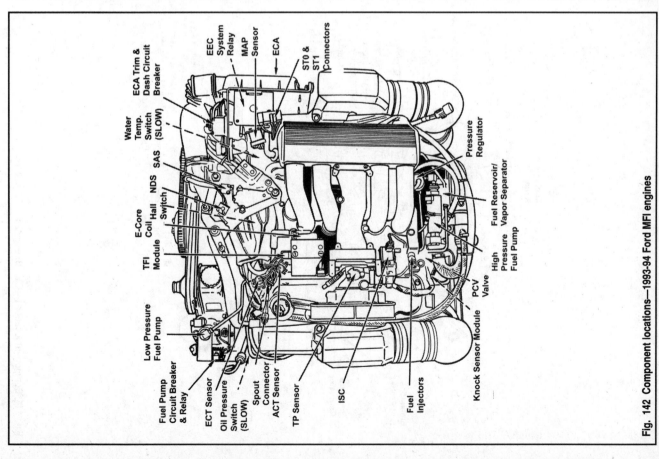

Fig. 142 Component locations—1993-94 Ford MFI engines

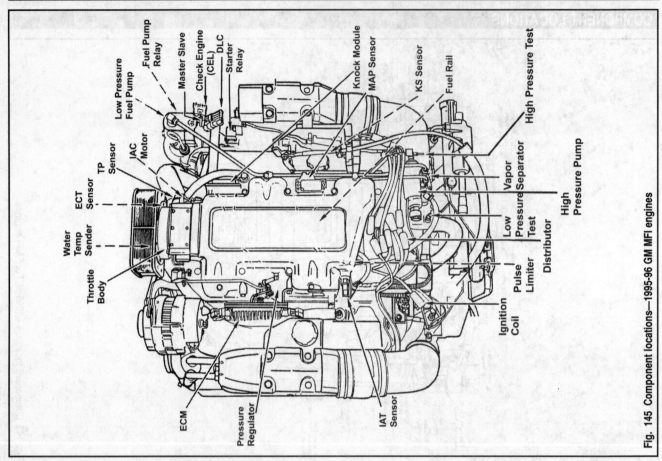

Fig. 145 Component locations—1995-96 GM MFI engines

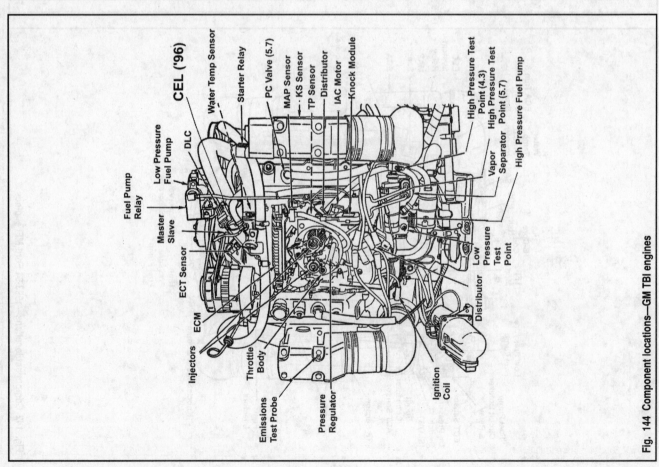

Fig. 144 Component locations—GM TBI engines

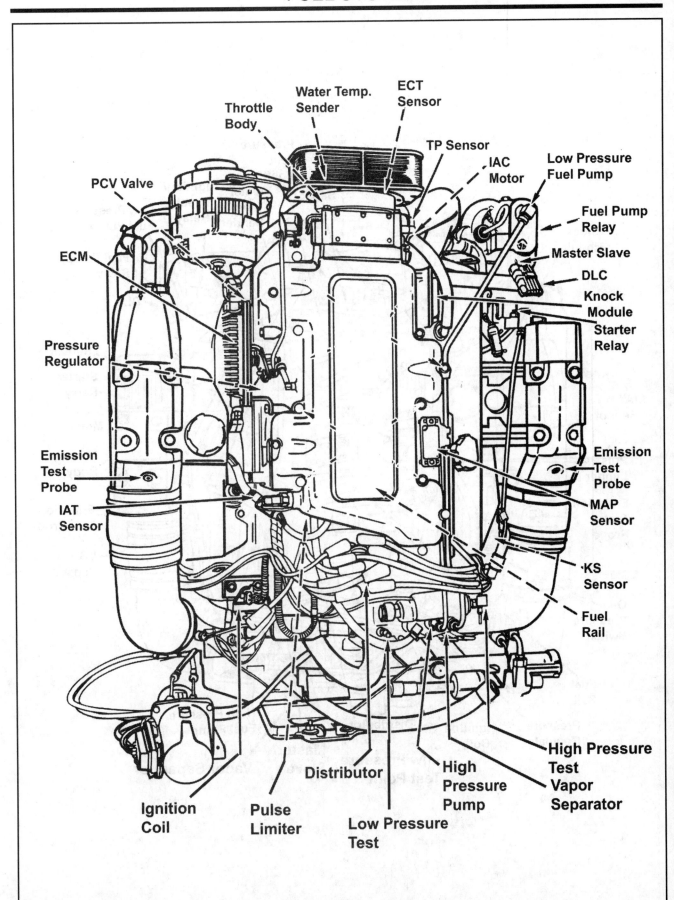

Fig. 46 Component locations—1997-98 GM MFI engines (except 1998 7.4Gi)

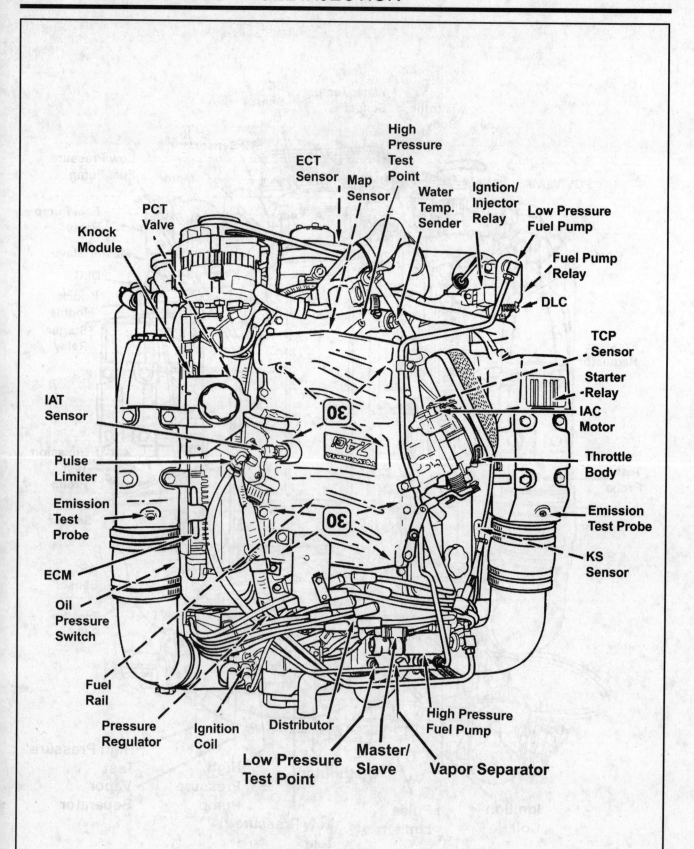

Fig. 47 Component locations—1998 7.4Gi

TORQUE SPECIFICATIONS
Ford Multi-Point Fuel Injection (MFI)

Component	ft. lbs.	inch lbs.	Nm
ACT Sensor	12-18	-	16-24
Circuit Brreaker	-	20-25	2.3-2.8
ECA Bracket-to-Engine	-	60-84	7-10
ECA Cover	-	24-36	2.7-4.1
ECA Module-to-Bracket	-	24-36	2.7-4.1
ECA Connector (60 Pin)	-	46-64	5-7
ECT Sensor	10-15	-	13.6-20.3
EEC Relay-to-Bracket	-	24-36	2.7-4.1
Elbow Lock Nut	-	60-84	7-10
Flame Arrestor	-	24-48	2.7-5.4
Fuel Line Fittings	10-12	-	13.6-16.3
Fuel Pressure Regulator	-	27-40	3.0-4.5
Fuel Pump (Lower) Adapter	8-10	-	11-14
Fuel Pump Banjo Nut	18-22	-	24.4-29.8
Fuel Pump (Lower) Bracket	20-25	-	27-34
Fuel Rail-to-Manifold	-	70-105	7.9-11.7
Fuel Reservoir Cover	-	25-35	2.8-4.0
Fuel Reservoir-to-Engine	24-36	-	38-49
ISC Solenoid	-	71-102	8-11
ISC Solenoid-to-Adaptor	-	71-102	8-11
Knock Sensor	7-10	-	9.5-13.5
MAP Sensor	-	24-36	2.7-4.1
Power Steering Cooler-to-Engine	20-25	-	27-34
Shift Bracket			
Large Nut	20-25	-	27-34
Small Nut	-	60-84	7-10
Throttle Body Adaptor	-	71-102	8-11
Throttle Body-to-Manifold	12-18	-	16-24
Throttle Position Sensor (TPS)	-	18-26	2-3
Upper-to-Lower Manifold	12-18	-	16-24

TORQUE SPECIFICATIONS
GM Fuel Injection (MFI/TBI)

Component		ft. lbs.	inch lbs.	Nm
Circuit Breaker				
	Nut	-	20-25	2.3-2.8
	Screw	-	20-25	2.3-2.8
Coolant Cover (MFI Only)		-	28	3.2
ECM		-	88-124	10-14
ECT Sensor		-	108	12
Filter Bracket Inlet Elbow (MFI Only)		12	-	16
Fuel Filter Bracket (MFI Only)		20-25	-	27-34
Fuel Line Nuts				
	MFI	10-12	-	13.6-16.3
	TBI	17	-	23
Fuel Meter Body (TBI Only)		-	30	4
Fuel Meter Cover (TBI Only)		-	28	3
Fuel Meter Inlet Nut (TBI Only)		30	-	40
Fuel Meter Outlet Nut (TBI Only)		21	-	29
Fuel Pressure Regulator (MFI Only)		-	84	9.5
Fuel Pump (LP) Adapter (MFI Only)		8-10	-	11-14
Fuel Pump (LP) End Cap (MFI Only)		-	60-84	6.8-9.5
Fuel Pump (HP) Bracket (MFI Only)		-	27-35	2.8-4.0
Fuel Pump Banjo Nut (MFI Only)		18-22	-	24.7-29.8
Fuel Rail (MFI Only)		-	88	10
Fuel Reservoir				
	Cover	-	25-35	2.8-4.0
	To Block	28-36	-	38-49
IAC Valve		-	28	3.2
IAT Sensor (MFI Only)		11	-	15
Intake Plenum (MFI Only)		15	-	20
Knock Sensor		11-16	-	15-22
MAP Sensor				
	1998 7.4Gi	18	-	24
	All Others	-	44-62	5-7
Outlet Elbow (HP Pump)		-	60-84	6.8-9.5
Throttle Body				
	1998 7.4Gi	-	89	10
	All Other MFI	11	-	15
	All TBI	12	-	16
Throttle Body Adapter Plate (TBI Only)				
	V6	15	-	19
	1996-97 V8	30	-	40
	1998 V8	15	-	19
Throttle Position Sensor (TP)		-	18	2

10

IGNITION AND ELECTRICAL SYSTEMS

UNDERSTANDING AND TROUBLESHOOTING ELECTRICAL SYSTEMS

Basic Electrical Theory

◆ See Figure 1

For any 12 volt, negative ground, electrical system to operate, the electricity must travel in a complete circuit. This simply means that current (power) from the positive terminal (+) of the battery must eventually return to the negative terminal (-) of the battery. Along the way, this current will travel through wires, fuses, switches and components. If for any reason the flow of current through the circuit is interrupted, the component(s) fed by that circuit will cease to function properly.

Perhaps the easiest way to visualize a circuit is to think of connecting a light bulb (with two wires attached to it) to the battery—one wire attached to the negative (-) terminal of the battery and the other wire to the positive (+) terminal. With the two wires touching the battery terminals, the circuit would be complete and the light bulb would illuminate. Electricity would follow a path from the battery to the bulb and back to the battery. It's easy to see that with longer wires on our light bulb, it could be mounted anywhere. Further, one wire could be fitted with a switch so that the light could be turned on and off.

The normal marine circuit differs from this simple example in two ways. First, instead of having a return wire from each bulb to the battery, the current travels through a single ground wire, which handles all the grounds for a specific circuit. Secondly, most marine circuits contain multiple components, which receive power from a single circuit. This lessens the amount of wire needed to power components.

HOW ELECTRICITY WORKS: THE WATER ANALOGY

Electricity is the flow of electrons—the sub-atomic particles that constitute the outer shell of an atom. Electrons spin in an orbit around the center core of an atom. The center core is comprised of protons (positive charge) and neutrons (neutral charge). Electrons have a negative charge and balance out the positive charge of the protons. When an outside force causes the number of electrons to unbalance the charge of the protons, the electrons will split off the atom and look for another atom to balance out. If this imbalance is kept up, electrons will continue to move and an electrical flow will exist.

Many people have been taught electrical theory using an analogy with water. In a comparison with water flowing through a pipe, the electrons would be the water and the wire is the pipe.

The flow of electricity can be measured much like the flow of water through a pipe. The unit of measurement used is amps, frequently abbreviated as amps (a). You can compare amperage to the volume of water flowing through a pipe. When connected to a circuit, an ammeter will measure the actual amount of current flowing through the circuit. When relatively few electrons flow through a circuit, the amperage is low. When many electrons flow, the amperage is high.

Water pressure is measured in units such as pounds per square inch (psi); electrical pressure is measured in units called volts (v). When a voltmeter is connected to a circuit, it is measuring the electrical pressure. When electrical pressure is low, then voltage is considered to be low. When electrical pressure is high, then voltage is considered to be high.

The actual flow of electricity depends not only on voltage and amperage but also on the resistance of the circuit—the higher the resistance, the higher the force necessary to push the current through the circuit. The standard unit for measuring resistance is an ohm (Ω). Resistance in a circuit varies depending on the amount and type of components used in the circuit and the overall condition of the components and wires. If we assume that everything in our circuit is new, then, the main factors which determine resistance are:
• Material—some materials have more resistance than others. Those with high resistance are said to be insulators. Rubber materials (or rubber-like plastics) are some of the most common insulators used, as they have a very high resistance to electricity. Very low resistance materials are said to be conductors. Copper wire is among the best conductors. Silver is actually a superior conductor to copper and is used in some relay contacts but its high cost prohibits its use as common wiring. Most marine wiring is made of copper.
• Size—the larger the wire size being used, the less resistance the wire will have. This is why components that use large amounts of electricity usually have large wires supplying current to them.

• Length—for a given thickness of wire, the longer the wire, the greater the resistance. The shorter the wire, the less the resistance. When determining the proper wire for a circuit, both size and length must be considered to design a circuit that can handle the current needs of the component.
• Temperature—with many materials, the higher the temperature, the greater the resistance (positive temperature coefficient). Some materials exhibit the opposite trait of lower resistance with higher temperatures (negative temperature coefficient). These principles are used in many of the sensors on the engine.

OHM'S LAW

There is a direct relationship between current, voltage and resistance that can be summed up by a statement known as Ohm's law.
• Voltage (E) is equal to amperage (I) times resistance (R): $E = I \times R$
• Other forms of the formula are $R = E/I$ and $I = E/R$
In each of these formulas, E is the voltage in volts, I is the current in amps and R is the resistance in ohms. The basic point to remember is that as the resistance of a circuit goes up, the amount of current that flows in the circuit will go down, if voltage remains the same.

The amount of work that the electricity can perform is expressed as power. A unit of power is known as a watt (W). There is a direct relationship between power, voltage and current that can be summed up by the following formula:
• Power (W) is equal to amperage (I) times voltage (E): $W = I \times E$

■ This formula is only true for direct current (DC) circuits, The alternating current formula is a tad different but since the electrical circuits in most boats are DC type, we need not get into AC circuit theory.

Electrical Components

POWER SOURCE

◆ See Figure 2

Power is supplied to the boat by two devices: The battery and the alternator. The battery supplies electrical power during starting or during periods when the current demand of the boat's electrical system exceeds the output capacity of the alternator. The alternator supplies electrical current when the engine is running. The alternator does not just supply the current needs of the boat but it also recharges the battery.

In most modern boats, the battery is a lead/acid electrochemical device consisting of six 2 volt subsections (cells) connected in series, so that the unit is capable of producing approximately 12 volts of electrical pressure. Each subsection consists of a series of positive and negative plates held a short distance apart in a solution of sulfuric acid and water.

The two types of plates are of dissimilar metals, thus setting up a chemical reaction inside the battery case. It is this reaction that produces current flow from the battery when its positive and negative terminals are connected to an electrical load. The alternator, restoring the battery to its original chemical state replaces the power removed from the battery.

The following is a brief description of how the system works:
The battery stores electricity and acts as a "sponge" for the whole system. It mops up all generated
current until it's fully charged and it release s energy on demand.

Permanent magnets that create the moving magnetic field. If your engine has good spark, you can take it for granted that the magnets are in working order because the ignition and charging systems share the same magnets.

The alternator windings are the stationary coils of wire that the magnets rotate around. They produce the electrical charge. Simply put, the more windings in your stator, the greater the potential output in amps your charging system you'll have.

The rectifier consists of a series of diodes or electrical one-way valves. The rectifier overcomes one of the disadvantages of a current-generating system using permanent magnets and stator windings, which is that the current produced within the windings is alternating current (AC). You can't use AC to charge batteries. They accept only direct current (DC). So the rectifier is designed to convert AC current to a usable form of DC current simply called "rectified AC."

All engines covered here utilize a voltage regulator; either combined with the rectifier or standing alone. The regulator automatically reduces the output of generated current as the battery becomes fully charged.

GROUND

All boats use some sort of a ground return circuit. Direct ground components are grounded to an electrically conductive metal component through their mounting points. These electrically conductive metal components are then grounded to the battery.

All other components use some sort of ground wire which leads directly back to the battery. The electrical current runs through the ground wire and returns to the battery through the ground (-) cable. If you look, you'll see that the battery ground cable connects between the battery and a heavy gauge ground wire.

■ **It should be noted that a good percentage of electrical problems can be traced to bad or loose grounds.**

PROTECTIVE DEVICES

◆ **See Figure 3**

It is possible for large surges of current to pass through the electrical system of your boat. If this surge of current were to reach components in the circuit, the surge could burn them out or severely damage them. Surges can also overload the wiring, causing the harness to get hot and melt the insulation. To prevent this, fuses, circuit breakers and/or fusible links are connected into the supply wires of the electrical system. These items are nothing more than a built-in weak spot in the system. When an abnormal amount of current flows through the system, these protective devices work as follows to protect the circuit:

• Fuse—when an excessive electrical current passes through a fuse, the fuse "blows" (the conductor melts) and opens the circuit, preventing the passage of current.

• Circuit Breaker—a circuit breaker is basically a self-repairing fuse. It will open the circuit in the same fashion as a fuse but when the surge subsides, the circuit breaker can be reset and does not need replacement.

• Fusible Link—a fusible link (fuse link or main link) is a short length of special, high temperature insulated wire that acts as a fuse. When an excessive electrical current passes through a fusible link, the thin gauge wire inside the link melts, creating an intentional open to protect the circuit.

To repair the circuit, the link must be replaced. Some newer type fusible links are housed in plug-in modules, which are simply replaced like a fuse, while older type fusible links must be cut and spliced if they melt. Since this link is very early in the electrical path, it's the first place to look if nothing on the boat works, yet the battery seems to be charged and is properly connected.

✳✳ CAUTION

Always replace fuses, circuit breakers and fusible links with identically rated components. Under no circumstances should a component of higher or lower amperage rating be substituted.

SWITCHES & RELAYS

◆ **See Figure 4**

Switches are used in electrical circuits to control the passage of current. The most common use is to open and close circuits between the battery and the various electric devices in the system. Switches are rated according to the amount of amperage they can handle. If a switch rated for the sufficient amperage is not used in a circuit, the switch could overload and cause damage.

Some electrical components which require a large amount of current to operate use a special switch called a relay. Since these circuits carry a large amount of current, the thickness of the wire in the circuit is also greater. If this large wire were connected from the load to the control switch, the switch would have to carry the high amperage load and the space needed for wiring in the boat would be twice as big to accommodate the increased size of the wiring harness. To prevent these problems, a relay is used.

Relays are composed of a coil and a set of contacts. When the coil has a current passed though it, a magnetic field is formed and this field causes the contacts to move together, completing the circuit. Most relays are normally open, preventing current from passing through the circuit but they can take any electrical form depending on the job they are intended to do. Relays can be considered "remote control switches." They allow a smaller current to operate devices that require higher amperages. When a small current operates the coil, a larger current is allowed to pass by the contacts. Some common circuits that may use relays are horns, lights, starter, electric fuel pumps and other high draw circuits.

LOAD

Every electrical circuit must include a "load" (something to use the electricity coming from the source). Without this load, the battery would attempt to deliver its entire power supply from one pole to another. This would result in a "short circuit" of the battery. All this electricity would take a short cut to ground and cause a great amount of damage to other components in the circuit by developing a tremendous amount of heat. This condition could develop sufficient heat to melt the insulation on all the surrounding wires and reduce a multiple wire cable to a lump of plastic and copper.

WIRING & HARNESSES

The average boat contains miles of wiring, with hundreds of individual connections. To protect the many wires from damage and to keep them from becoming a confusing tangle, they are organized into bundles, enclosed in plastic or taped together and called wiring harnesses. Different harnesses serve different parts of the boat. Individual wires are color coded to help trace them through a harness where sections are hidden from view.

Marine wiring can either be single strand wire, multi-strand wire or printed circuitry. Single strand wire has a solid metal core and is usually used inside such components as alternators, motors, relays and other devices. Multi-strand wire has a core made of many small strands of wire twisted together into a single conductor. Most of the wiring in a marine electrical system is made up of multi-strand wire, either as a single conductor or grouped

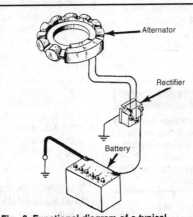

Fig. 1 **This example illustrates a simple circuit. When the switch is closed, power from the positive (+) battery terminal flows through the fuse and the switch, and then to the light bulb. The light illuminates and the circuit is completed through the ground wire back to the negative (-) battery terminal**

Fig. 2 **Functional diagram of a typical charging circuit showing the relationship of the stator, solid-state rectifier and the battery**

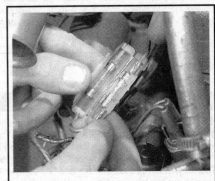

Fig. 3 **Fuses protect the vessel's electrical system from abnormally high amounts of current flow**

together in a harness. All wiring is color coded on the insulator, either as a solid color or as a colored wire with an identification stripe. A printed circuit is a thin film of copper or other conductor that is printed on an insulator backing. Occasionally, a printed circuit is sandwiched between two sheets of plastic for more protection and flexibility. A complete printed circuit, consisting of conductors, insulating material and connectors is called a printed circuit board. Printed circuitry is used in place of individual wires or harnesses in places where space is limited, such as behind instrument panels.

Since marine electrical systems are very sensitive to changes in resistance, the selection of properly sized wires is critical when systems are repaired. A loose or corroded connection or a replacement wire that is too small for the circuit will add extra resistance and an additional voltage drop to the circuit.

The wire gauge number is an expression of the cross-section area of the conductor. Boats from countries that use the metric system will typically describe the wire size as its cross-sectional area in square millimeters—with this method, the larger the wire, the greater the number. Another common system for expressing wire size is the American Wire Gauge (AWG) system. As gauge number increases, area decreases and the wire becomes smaller. An 18 gauge wire is smaller than a 4 gauge wire. A wire with a higher gauge number will carry less current than a wire with a lower gauge number. Gauge wire size refers to the size of the strands of the conductor, not the size of the complete wire with insulator. It is possible, therefore, to have two wires of the same gauge with different diameters because one may have thicker insulation than the other.

It is essential to understand how a circuit works before trying to figure out why it doesn't. An electrical schematic shows the electrical current paths when a circuit is operating properly. Schematics break the entire electrical system down into individual circuits. In a schematic, usually no attempt is made to represent wiring and components as they physically appear on the boat, switches and other components are shown as simply as possible. Face views of harness connectors show the cavity or terminal locations in all multi-pin connectors to help locate test points.

CONNECTORS

◆ See Figures 5, 6, 7 and 8

Weatherproof connectors are most commonly used where the connector is exposed to the elements. Terminals are protected against moisture and dirt by sealing rings that provide a weather tight seal. All repairs require the use of a special terminal and the tool required to service it.

Unlike standard blade type terminals, these weatherproof terminals cannot be straightened once they are bent. Make certain that the connectors are properly seated and all of the sealing rings are in place when connecting leads.

Test Equipment

Pinpointing the exact cause of trouble in an electrical circuit is most times accomplished by the use of special test equipment. The following sections describe different types of commonly used test equipment and briefly explain how to use them in diagnosis. In addition to the information covered below, the tool manufacturer's instruction manual (provided with most tools) should be read and clearly understood before attempting any test procedures.

JUMPER WIRES

◆ See Figure 9

✳✳ CAUTION

Never use jumper wires made from a thinner gauge wire than the circuit being tested. If the jumper wire is of too small a gauge, it may overheat and possibly melt. Never use jumpers to bypass high resistance loads in a circuit. Bypassing resistances, in effect, creates a short circuit. This may, in turn, cause damage and fire. Jumper wires should only be used to bypass lengths of wire or to simulate switches.

Jumper wires are simple, yet extremely valuable, pieces of test equipment. They are basically test wires that are used to bypass sections of a circuit. Although jumper wires can be purchased, they are usually fabricated from lengths of standard marine wire and whatever type of connector (alligator clip, spade connector or pin connector) that is required

for the particular application being tested. In cramped, hard-to-reach areas, it is advisable to have insulated boots over the jumper wire terminals in order to prevent accidental grounding.

It is also advisable to include a standard marine fuse in any jumper wire. This is commonly referred to as a "fused jumper". By inserting an in-line fuse holder between a set of test leads, a fused jumper wire can be used for bypassing open circuits. Use a 5-amp fuse to provide protection against voltage spikes.

Jumper wires are used primarily to locate open electrical circuits. If an electrical component fails to operate, connect the jumper wire between the component and a good ground. If the component operates only with the jumper installed, the ground circuit is open.

If the ground circuit is good but the component does not operate, the circuit between the power feed and component may be open. By moving the jumper wire successively back from the component toward the power source, you can isolate the area of the circuit where the open is located. When the component stops functioning or the power is cut off, the open is in the segment of wire between the jumper and the point previously tested.

You can sometimes connect the jumper wire directly from the battery to the "hot" terminal of the component but first make sure the component uses 12 volts in operation. Some electrical components, such as fuel injectors or sensors are designed to operate on about 4 to 5 volts and running 12 volts directly to these components will cause damage.

TEST LIGHTS

◆ See Figure 10

The test light is used to check circuits and components while electrical current is flowing through them. It is used for voltage and ground tests. To use a 12-volt test light, connect the ground clip to a good ground and probe wherever necessary with the pick. The test light will illuminate when voltage is detected. This does not necessarily mean that 12 volts (or any particular amount of voltage) is present, it only means that some voltage is present.

■ **It is advisable before using the test light to touch its ground clip and probe across the battery posts or terminals to make sure the light is operating properly.**

✳✳WARNING

Do not use a test light to probe electronic ignition, spark plug or coil wires. Never use a pick-type test light to probe wiring on electronically controlled systems unless specifically instructed to do so. Any wire insulation that is pierced by the test light probe should be taped and sealed with silicone after testing.

Like the jumper wire, the 12V test light is used to isolate opens in circuits. But, whereas the jumper wire is used to bypass the open to operate the load, the 12V test light is used to locate the presence of voltage in a circuit. If the test light illuminates, there is power up to that point in the circuit, if the test light does not illuminate, there is an open circuit (no power). Move the test light in successive steps back toward the power source until the light in the handle illuminates. The open is between the probe and a point that was previously probed.

The self-powered test light is similar in design to the 12V test light but contains a 1.5V penlight battery in the handle. It is most often used in place of a multi-meter to check for open or short circuits when power is isolated from the circuit (continuity test).

The battery in a self-powered test light does not provide much current. A weak battery may not provide enough power to illuminate the test light even when a complete circuit is made (especially if there is high resistance in the circuit). Always make sure that the test battery is strong. To check the battery, briefly touch the ground clip to the probe, if the light glows brightly, the battery is strong enough for testing.

✳✳WARNING

A self-powered test light should not be used on any electronically controlled system or component. The small amount of electricity transmitted by the test light is enough to damage many electronic marine components.

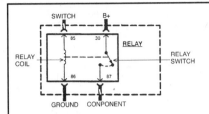

Fig. 4 Relays are composed of a coil and a switch. These two components are linked together so that when one operates, the other operates at the same time. The large wires in the circuit are connected from the battery to one side of the relay switch (B+) and from the opposite side of the relay switch to the load (component). Smaller wires are connected from the relay coil to the control switch for the circuit and from the opposite side of the relay coil to ground

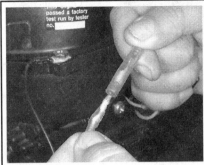

Fig. 5 Bullet connectors are some of the more common electrical connectors found on an engine

Fig. 6 A typical weatherproof electrical connector

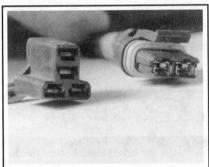

Fig. 7 Hard shell (left) and weatherproof (right) connectors have replaceable terminals

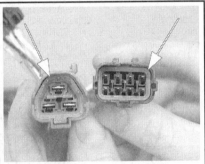

Fig. 8 The seals on weatherproof connectors must be kept in good condition to prevent the terminals from corroding

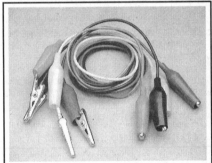

Fig. 9 Jumper wires are simple, yet extremely valuable, pieces of test equipment

MULTI-METERS

◆ See Figure 11

Multi-meters are an extremely useful tool for troubleshooting electrical problems. They can be purchased either in analog or digital form and have a price range to suit any budget. A multi-meter is a voltmeter, ammeter and multi-meter (along with other features) combined into one instrument. It is often used when testing solid-state circuits because of its high input impedance (usually 10 mega-ohms or more). A brief description of the multi-meter main test functions follows:

• Voltmeter—the voltmeter is used to measure voltage at any point in a circuit or to measure the voltage drop across any part of a circuit. Voltmeters usually have various scales and a selector switch to allow the reading of different voltage ranges.

The voltmeter has a positive and a negative lead. To avoid damage to the meter, always connect the negative lead to the negative (-) side of the circuit (to ground or nearest the ground side of the circuit) and connect the positive lead to the positive (+) side of the circuit (to the power source or the nearest power source).

■ The negative voltmeter lead will always be Black and the positive voltmeter will always be some color other than Black (usually Red).

• Multi-meter—the multi-meter is designed to read resistance (measured in ohms) in a circuit or component. Most multi-meters will have a selector switch which permits the measurement of different ranges of resistance (usually the selector switch allows the multiplication of the meter reading by 10, 100, 1,000 and 10,000). Some multi-meters are "auto-ranging" which means the meter itself will determine which scale to use.

Since an internal battery powers the meters, the multi-meter can be used like a self-powered test light. When the multi-meter is connected, current from the multi-meter will flow through the circuit or component being tested. Since the multi-meter's internal resistance and voltage are known values, the amount of current flow through the meter depends on the resistance of the circuit or component being tested.

The multi-meter can also be used to perform a continuity test for suspected open circuits. In using the meter for making continuity checks, do not be concerned with the actual resistance readings. Zero resistance (or any ohm reading) indicates continuity in the circuit. Infinite resistance indicates an opening in the circuit. A high resistance reading where there should be none indicates a problem in the circuit.

Checks for short circuits are made in the same manner as checks for open circuits, except that the circuit must be isolated from both power and normal ground. Infinite resistance indicates no continuity, while zero resistance indicates a dead short.

✱✱WARNING

Never use a multi-meter to check the resistance of a component or wire while there is voltage applied to the circuit.

• Ammeter—an ammeter measures the amount of current flowing through a circuit in units called amps (amps). At normal operating voltage, most circuits have a characteristic amount of amps, called "current draw" which can be measured using an ammeter. By referring to a specified current draw rating, then measuring the amps, and comparing the two values; one can determine what is happening within the circuit to aid in diagnosis.

For example, an open circuit will not allow any current to flow, so the ammeter reading will be zero. A damaged component or circuit will have an increased current draw, so the reading will be high.

The ammeter is always connected in series with the circuit being tested. All of the current that normally flows through the circuit must also flow through the ammeter, if there is any other path for the current to follow, the ammeter reading will not be accurate. The ammeter itself has very little resistance to current flow and, therefore, it will not affect the circuit but it will measure current draw only when the circuit is closed and electricity is flowing. Excessive current draw can blow fuses and drain the battery, while a reduced current draw can cause motors to run slowly, lights to dim and other components to not operate properly.

Troubleshooting the Electrical System

When diagnosing any electrical problem organized troubleshooting is a must. The complexity of electrical systems on modern boats and their power plants demands that you approach any problem in a logical organized manner. There are certain troubleshooting techniques, which are standard:

• **Establish when the problem occurs**—Does the problem appear only under certain conditions? Were there any noises, odors or other unusual symptoms?

• **Check for obvious problems**—Problems such as broken wires and loose or dirty connections can cause major problems. Always check the obvious before assuming something complicated (or expensive) is the cause.

■ **Experience has shown that most problems tend to be the result of a fairly simple and obvious cause, such as loose or corroded connectors, bad grounds or damaged wire insulation, which causes a short. This makes careful visual inspection of components during testing essential to quick and accurate troubleshooting.**

• **Isolate the problem area**—Make some simple tests and observations, then eliminate the systems that are working properly. Test for problems systematically to determine the cause once the problem area is isolated. Are all the components functioning properly? Is there power going to electrical switches and motors. Performing careful, systematic checks will often turn up most causes on the first inspection, without wasting time checking components that have little or no relationship to the problem.

• **Verify all systems after repairs are completed**—Some causes can be traced to more than one component, so a careful verification of repair work is important in order to pick up additional malfunctions that may cause a problem to reappear or a different problem to arise. A blown fuse, for example, is a simple problem that may require more than another fuse to repair. If you don't look for a problem that caused a fuse to blow, a shorted wire (for example) may go undetected.

VOLTAGE

◆ See Figure 12

This test determines voltage available from the battery and should be the first step in any electrical troubleshooting procedure after visual inspection. Many electrical problems, especially on electronically controlled systems, can be caused by a low state of charge in the battery. Excessive corrosion at the battery cable terminals can cause poor contact that will prevent proper charging and full battery current flow.

1. Set the voltmeter selector switch to the 10-volt position.
2. Connect the multi-meter negative lead to the battery's negative (-) terminal and the positive lead to the battery's positive (+) terminal.
3. Turn the battery (ignition) switch **ON** to provide a load.
4. A well-charged battery should register over 12 volts. If the meter reads below 11.5 volts, the battery power may be insufficient to operate the electrical system properly.
5. Charge the battery and retest.

VOLTAGE DROP

◆ See Figure 13

When current flows through a load, the voltage beyond the load drops. This voltage drop is due to the resistance created by the load and also by small resistances created by corrosion at the connectors and damaged insulation on the wires. The maximum allowable voltage drop under load is critical, especially if there is more than one load in the circuit, since all voltage drops are cumulative.

1. Set the voltmeter selector switch to the 10-volt position.
2. Connect the multi-meter negative lead to the battery negative (-) terminal or another good ground.
3. Touch the multi-meter positive lead to the battery's positive (+) terminal to determine battery voltage.
4. Operate the circuit and check the voltage prior to the first component (load).
5. There should be little or no voltage drop (from battery voltage) in the circuit prior to the first component. If a voltage drop exists, the wire or connectors in the circuit are suspect.
6. While operating the first component in the circuit, probe the ground-side of the component with the positive (+) meter lead and observe the voltage readings. A small voltage drop should be noticed. The resistance of the component causes this voltage drop.
7. Repeat the test for each component (load) down the circuit.
8. If a large voltage drop is noticed, the preceding component, wire or connector is suspect.

RESISTANCE

◆ See Figures 14 and 15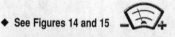

✳✳WARNING

Never use a multi-meter with power applied to the circuit. The multi-meter is designed to operate on its own power supply. The normal 12-volt electrical system voltage can damage the meter!

1. Isolate the circuit from the boat's power source.
2. Ensure that the battery (ignition) switch is **OFF**.
3. Isolate at least one side of the circuit to be checked, in order to avoid reading parallel resistances. Parallel circuit resistances will always give a lower reading than the actual resistance of either of the branches.
4. Connect the meter leads to both sides of the circuit (wire or component) and read the actual resistance measured in ohms on the meter scale. Make sure the selector switch is set to the proper ohm scale for the circuit being tested, to avoid misreading the multi-meter test value.
5. Compare this reading to the resistance specification for the component or formulate the theoretical resistance using Ohms Law.

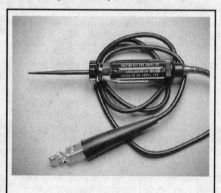

Fig. 10 A test light is used to detect the presence of voltage in a circuit

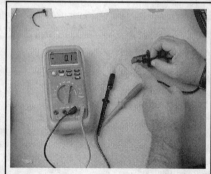

Fig. 11 Multi-meters are essential for diagnosing faulty wires, switches and other electrical components

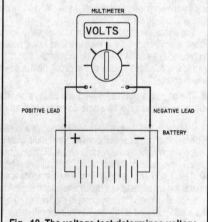

Fig. 12 The voltage test determines voltage available from the battery and should be the first step in any electrical troubleshooting procedure after visual inspection

OPEN CIRCUITS

◆ **See Figures 16 and 17**

This test already assumes the existence of an open in the circuit and it is used to help locate the open portion.

1. Isolate the circuit from power and ground.
2. Connect the self-powered test light or multi-meter ground clip to the ground-side of the circuit and probe sections of the circuit sequentially.
3. If the light is out or there is infinite resistance, the open is between the probe and the circuit ground.
4. If the light is on or the meter shows continuity, the open is between the probe and the end of the circuit toward the power source.

SHORT CIRCUITS

◆ **See Figure 18**

■ **Never use a self-powered test light to perform checks for opens or shorts when power is applied to the circuit under test. The test light can be damaged by outside power.**

1. Isolate the circuit from power and ground.
2. Connect the self-powered test light or multi-meter ground clip to a good ground and probe any easy-to-reach point in the circuit.
3. If the light comes on or there is continuity, there is a short somewhere in the circuit.

4. To isolate the short, probe a test point at either end of the isolated circuit (the light should be on or the meter should indicate continuity).
5. Leave the test light probe engaged and sequentially open connectors or switches; remove parts, etc. until the light goes out or continuity is broken.
6. When the light goes out, the short is between the last two circuit components, which were opened.

Wire and Connector Repair

Almost anyone can replace damaged wires, as long as the proper tools and parts are available. Wire and terminals are available to fit almost any need. Even the specialized weatherproof, molded and hard shell connectors are now available from aftermarket suppliers.

Be sure the ends of all the wires are fitted with the proper terminal hardware and connectors. Wrapping a wire around a stud is never a permanent solution and will only cause trouble later. Replace wires one at a time to avoid confusion. Always route wires in the same manner of the manufacturer.

When replacing connections, make absolutely certain that the connectors are certified for marine use. Automotive wire connectors may not meet United States Coast Guard (USCG) specifications.

■ **If connector repair is necessary, only attempt it if you have the proper tools. Weatherproof connectors require special tools to release the pins inside the connector. Attempting to repair these connectors with conventional hand tools will damage them.**

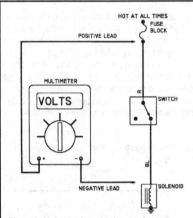

Fig. 13 Voltage drop is due to the resistance created by the load and also by small resistances created by corrosion at the connectors and damaged insulation on the wires

Fig. 14 Using a multi-meter to check resistance on the secondary side of the ignition coil

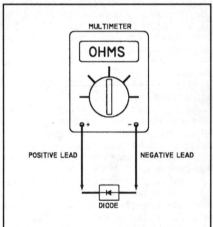

Fig. 15 When performing resistance tests, always isolate the circuit from power

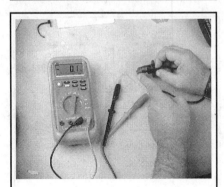

Fig. 16 The infinite display on this multi-meter (0.1) indicates that the circuit is open

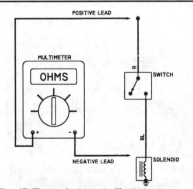

Fig. 17 The easiest way to illustrate an open circuit is to consider an example circuit with a switch. When the switch is turned OFF, power does not flow through the circuit to the load. Thus, the circuit is open

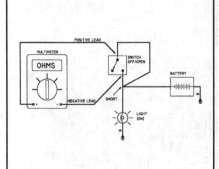

Fig. 18 In this illustration, the circuit between the battery and light should be open because the switch is turned OFF. However, battery voltage is reaching the light at the point of the short; this could possibly be caused by chaffed wires

BATTERY

The battery is one of the most important parts of the electrical system. In addition to providing electrical power to start the engine, it also provides power for operation of the running lights, radio and electrical accessories.

Because of its job and the consequences (failure to perform in an emergency), the best advice is to purchase a well-known brand, with an extended warranty period, from a reputable dealer.

The usual warranty covers a pro-rated replacement policy, which means the purchaser is entitled to consideration for the time left on the warranty period if the battery should prove defective before the end of the warranty.

Many manufacturers have specifications on the size and type of battery to use for their engines. If in doubt as to how large a battery the boat requires, make a liberal estimate and then purchase the one with the next higher amp rating.

■ Please refer to the Maintenance section for all procedures on cleaning, testing, storage and maintenance.

BATTERY CONSTRUCTION

◆ See Figure 19

A battery consists of a number of positive and negative plates immersed in a solution of diluted sulfuric acid. The plates contain dissimilar active materials and are kept apart by separators. The plates are grouped into elements. Plate straps on top of each element connect all of the positive plates and all of the negative plates into groups.

The battery is divided into cells holding a number of the elements apart from the others. The entire arrangement is contained within a hard plastic case. The top is a one-piece cover and contains the filler caps for each cell. The terminal posts protrude through the top where the battery connections for the boat are made. Each of the cells is connected to its neighbor in a positive-to-negative manner with a heavy strap called the cell connector.

MARINE BATTERIES

◆ See Figure 20

Because marine batteries are required to perform under much more rigorous conditions than automotive batteries, they are constructed differently than those used in automobiles or trucks. Therefore, a marine battery should always be the No. 1 unit for the boat and other types of batteries used only in an emergency.

Marine batteries have a much heavier exterior case to withstand the violent pounding and shocks imposed on it as the boat moves through rough water and in extremely tight turns. The plates are thicker and each plate is securely anchored within the battery case to ensure extended life. The caps are spill proof to prevent acid from spilling into the bilge when the boat heels to one side in a tight turn or is moving through rough water. Because of these features, the marine battery will recover from a low charge condition and give satisfactory service over a much longer period of time than any type intended for automotive use.

✳✳WARNING

Never use a maintenance-free battery with an engine that is not voltage regulated. The charging system will continue to charge as long as the engine is running and it is possible that the electrolyte could boil out if periodic checks of the cell electrolyte level are not done.

BATTERY RATINGS

◆ See Figure 21

Three different methods are used to measure and indicate battery electrical capacity:
• Amp/hour rating
• Cold cranking performance
• Reserve capacity

The amp/hour rating of a battery refers to the battery's ability to provide a set amount of amps for a given amount of time under test conditions at a constant temperature. Therefore, if the battery is capable of supplying 4 amps of current for 20 consecutive hours, the battery is rated as an 80 amp/hour battery. The amp/hour rating is useful for some service operations, such as slow charging or battery testing.

Cold cranking performance is measured by cooling a fully charged battery to 0°F (-17°C) and then testing it for 30 seconds to determine the maximum current flow. In this manner the cold cranking amp rating is the number of amps available to be drawn from the battery before the voltage drops below 7.2 volts.

The illustration depicts the amount of power in watts available from a battery at different temperatures and the amount of power in watts required of the engine at the same temperature. It becomes quite obvious—the colder the climate, the more necessary for the battery to be fully charged.

Reserve capacity of a battery is considered the length of time, in minutes, at 80°F (27°C), a 25 amp current can be maintained before the voltage drops below 10.5 volts. This test is intended to provide an approximation of how long the engine, including electrical accessories, could operate satisfactorily if the stator assembly or lighting coil did not produce sufficient current. A typical rating is 100 minutes.

■ If possible, the new battery should have a power rating equal to or higher than the unit it is replacing.

BATTERY LOCATION

◆ See Figure 22

Every battery installed in a boat must be secured in a well protected, ventilated area. If the battery area lacks adequate ventilation, hydrogen gas, which is given off during charging, is very explosive. This is especially true if the gas is concentrated and confined.

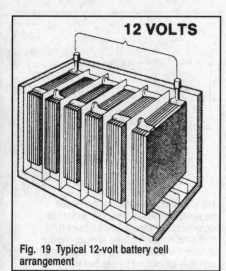

Fig. 19 Typical 12-volt battery cell arrangement

Fig. 20 Cut-away look at a typical marine battery

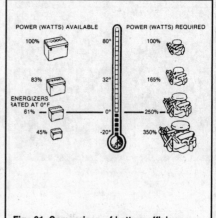

Fig. 21 Comparison of battery efficiency and engine demands at various temperatures

DUAL BATTERY INSTALLATION

◆ See Figure 23, 24, 25 and 26

Three methods are available for utilizing a dual-battery hook-up.

1. A high-capacity switch can be used to connect the two batteries. The accompanying illustration details the connections for installation of such a switch. This type of switch installation has the advantage of being simple, inexpensive, and easy to mount and hookup. However, if the switch is forgotten in the closed position, it will let the convenience loads run down both batteries and the advantage of the dual installation is lost. However, the switch may be closed intentionally to take advantage of the extra capacity of the two batteries, or it may be temporarily closed to help start the engine under adverse conditions.

2. A relay, can be connected into the ignition circuit to enable both batteries to be automatically put in parallel for charging or to isolate them for ignition use during engine cranking and start. By connecting the relay coil to the ignition terminal of the ignition-starting switch, the relay will close during the start to aid the starting battery. If the second battery is allowed to run down, this arrangement can be a disadvantage since it will draw a load from the starting battery while cranking the engine. One way to avoid such a condition is to connect the relay coil to the ignition switch accessory terminal. When connected in this manner, while the engine is being cranked, the relay is open, but when the engine is running with the ignition switch in the normal position, the relay is closed, and the second battery is being charged at the same time as the starting battery.

3. A heavy duty switch installed as close to the batteries as possible can be connected between them. If such an arrangement is used it must meet the standards of the American Boat and Yacht Council, or the Fire Protection Standard for Motor Craft, N.F.P.A. No. 302.

BATTERY CHARGERS

◆ See Figure 27

Before using any battery charger, consult the manufacturer's instructions for its use. Battery chargers are electrical devices that change Alternating Current (AC) to a lower voltage of Direct Current (DC) that can be used to charge a marine battery. There are two types of battery chargers—manual and automatic.

A manual battery charger must be physically disconnected when the battery has come to a full charge. If not, the battery can be overcharged and possibly fail. Excess charging current at the end of the charging cycle will heat the electrolyte, resulting in loss of water and active material, substantially reducing battery life.

■ As a rule, on manual chargers, when the ammeter on the charger registers half the rated amperage of the charger, the battery is fully charged. This can vary and it is recommended to use a hydrometer to accurately measure state of charge.

Automatic battery chargers have an important advantage—they can be left connected (for instance, overnight) without the possibility of overcharging the battery. Automatic chargers are equipped with a sensing device to allow the battery charge to taper off to near zero as the battery becomes fully charged. When charging a low or completely discharged battery, the meter will read close to full rated output. If only partially discharged, the initial reading may be less than full rated output, as the charger responds to the condition of the battery. As the battery continues to charge, the sensing device monitors the state of charge and reduces the charging rate. As the rate of charge tapers to zero amps, the charger will continue to supply a few milliamps of current—just enough to maintain a charged condition.

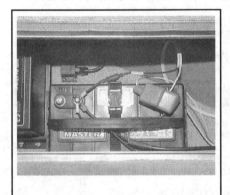

Fig. 22 A good example of a secured, well protected and ventilated battery area

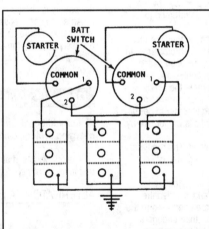

Fig. 23 Schematic diagram for a typical three battery, two engine hookup.

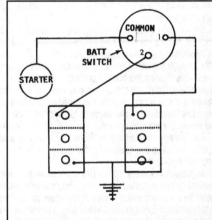

Fig. 24 Schematic diagram for a typical two battery, one engine hookup.

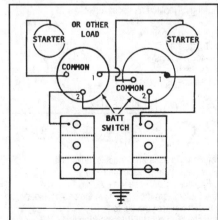

Fig. 25 Schematic diagram for a typical two battery, two engine hookup.

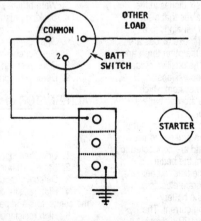

Fig. 26 Schematic diagram for a typical single battery, one engine hookup.

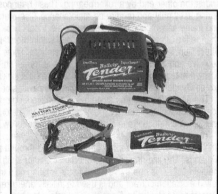

Fig. 27 Automatic chargers, such as the Battery Tender® from Deltran, are equipped with a sensing device to allow the battery charge to taper off to near zero as the battery becomes fully charged

BATTERY CABLES

Battery cables don't go bad very often but like anything else, they can wear out. If the cables on your boat are cracked, frayed or broken, they should be replaced.

When working on any electrical component, it is always a good idea to disconnect the negative (-) battery cable. This will prevent potential damage to many sensitive electrical components

Always replace the battery cables with one of the same length or you will increase resistance and possibly cause hard starting. Smear the battery posts with a light film of dielectric grease or a battery terminal protectant-spray once you've installed the new cables. If you replace the cables one at a time, you won't mix them up.

■ **Anytime you disconnect the battery cables, it is recommended that you disconnect the negative (-) battery cable first. This will prevent you from accidentally grounding the positive (+) terminal when disconnecting it, thereby preventing damage to the electrical system.**

Before you disconnect the cable(s), first turn the ignition to the **OFF** position. This will prevent a draw on the battery that could cause arcing. When the battery cable(s) are reconnected (negative cable last), be sure to check all electrical accessories are all working correctly.

CHARGING SYSTEM

■ **For additional, and model specific, electrical schematics of the charging system, please refer to the Wiring Diagrams section.**

General Information

The charging system provides electrical power for operation of the vessel's ignition system, starting system and all electrical accessories. The battery serves as an electrical surge or storage tank, storing (in chemical form) the energy originally produced by the engine driven alternator. The system also provides a means of regulating output to protect the battery from being overcharged and to avoid excessive voltage to the accessories.

The storage battery is a chemical device incorporating parallel lead plates in a tank containing a sulfuric acid/water solution. Adjacent plates are slightly dissimilar, and the chemical reaction of the two dissimilar plates produces electrical energy when the battery is connected to a load such as the starter motor. The chemical reaction is reversible, so that when the alternator is producing a voltage (electrical pressure) greater than that produced by the battery, electricity is forced into the battery, and the battery is returned to its fully charged state.

Newer engines use alternating current alternators, because they are more efficient, can be rotated at higher speeds, and have fewer brush problems. In an alternator, the field usually rotates while all the current produced passes only through the stator winding. The brushes bear against continuous slip rings. This causes the current produced to periodically reverse the direction of its flow. Diodes (electrical one way valves) block the flow of current from traveling in the wrong direction. A series of diodes is wired together to permit the alternating flow of the stator to be rectified back to 12 volts DC for use by the vessel's electrical system.

The voltage regulating function is performed by a regulator. The regulator is often built into the alternator; this system is termed an integrated or internal regulator.

An alternator differs from a DC shunt generator in that the armature is stationary, and is called the stator, while the field rotates and is called the rotor. The higher current values in the alternator's stator are conducted to the external circuit through fixed leads and connections, rather than through a rotating commutator and brushes as in a DC generator. This eliminates a major point of maintenance.

The rotor assembly is supported in the drive end frame by a ball bearing and at the other end by a roller bearing. These bearings are lubricated during assembly and require no maintenance. There are six diodes in the end frame assembly. These diodes are electrical check valves that also change the alternating current developed within the stator windings to a Direct Current (DC) at the output (BAT) terminal. Three of these diodes are negative and are mounted flush with the end frame while the other three are positive and are mounted into a strip called a heat sink. The positive diodes are easily identified as the ones within small cavities or depressions.

The alternator charging system is a negative (-) ground system which consists of an alternator, a regulator, a charge indicator, a storage battery and wiring connecting the components, and fuse link wire.

The alternator is belt-driven from the engine. Energy is supplied from the alternator/regulator system to the rotating field through two brushes to two slip-rings. The slip-rings are mounted on the rotor shaft and are connected to the field coil. This energy supplied to the rotating field from the battery is called excitation current and is used to initially energize the field to begin the generation of electricity. Once the alternator starts to generate electricity, the excitation current comes from its own output rather than the battery.

The alternator produces power in the form of alternating current. The alternating current is rectified by 6 diodes into direct current. The direct current is used to charge the battery and power the rest of the electrical system.

When the ignition key is turned to the **ON** position, current flows from the battery, through the charging system indicator light, to the voltage regulator, and on to the alternator. When the engine is started, the alternator begins to produce current. As the alternator turns and produces current, the current is divided in two ways: part to the battery (to charge the battery and power the electrical components of the vessel), and part is returned to the alternator (to enable it to increase its output). In this situation, the alternator is receiving current from the battery and from itself. A voltage regulator is wired into the current supply to the alternator to prevent it from receiving too much current, which would cause it to put out too much current. Conversely, if the voltage regulator does not allow the alternator to receive enough current, the battery will not be fully charged and will eventually go dead.

The battery is connected to the alternator at all times, whether the ignition key is turned **ON** or not. If the battery were shorted to ground, the alternator would also be shorted. This would damage the alternator. To prevent this, a fuse link is installed in the wiring between the battery and the alternator. If the battery is shorted, the fuse link melts, protecting the alternator.

An alternator is better that a conventional, DC shunt generator because it is lighter and more compact, because it is designed to supply the battery and accessory circuits through a wide range of engine speeds, and because it eliminates the necessary maintenance of replacing brushes and servicing commutators.

General System Troubleshooting

■ **For additional, and model specific, electrical schematics of the charging system, please refer to the Wiring Diagrams section.**

The following symptoms and possible corrective actions will be helpful in restoring a faulty charging system to proper operation.

ALTERNATOR FAILS TO CHARGE

1. Drive belt loose or broken. Replace and/or adjust drive belt.
2. Corroded or loose wires or connection in the charging circuit. Inspect, clean, and tighten.
3. Warn brushes or slip rings. Replace as necessary.
4. Sticking brushes.
5. Open field circuit. Trace and repair.
6. Open charging circuit. Trace and repair.
7. Open circuit in the stator windings.
8. Open rectifier. Replace the unit.
9. Defective regulator. Replace as necessary.

ALTERNATOR CHARGES LOW OR UNSTEADY

1. Drive belt loose or broken. Replace and/or adjust drive belt.
2. Battery charge too low. Charge or replace the battery.
3. Defective regulator. Replace as necessary.
4. High resistance at the battery terminals. Remove the cables, clean the connectors and battery posts, replace and tighten.
5. High resistance in the charging circuit. Trace and repair.
6. Open stator winding. Replace alternator.
7. High resistance in the ground circuit. Trace and repair.

ALTERNATOR OUTPUT LOW & LOW BATTERY

1. Drive belt slipping. Adjust or replace belt.
2. High resistance in the charging circuit. Trace and repair.
3. Shorted or open circuit in rectifier. Replace as necessary.
4. Grounded stator windings. Replace alternator.

ALTERNATOR OUTPUT TOO HIGH—BATTERY OVERCHARGED

1. Faulty voltage regulator. Replace as necessary.
2. Regulator base not grounded properly. Correct condition to make good ground.
3. Faulty ignition switch. Replace switch.

ALTERNATOR TOO NOISY

1. Worn, loose, or frayed drive belt. Replace belt and adjust properly.
2. Alternator mounting loose. Tighten all mounting hardware securely.
3. Worn alternator bearings. Replace.
4. Interference between rotor and stator leads, or rectifier leads. Check and correct.
5. Rotor or fan damaged. Replace.
6. Open or shorted rectifier. Replace.
7. Open or shorted winding in stator. Replace.

AMMETER FLUCTUATES CONSTANTLY

1. High resistance connection in the alternator or voltage regulator circuit. Trace and repair.

Alternator

■ **For additional, and model specific, electrical schematics of the charging system, please refer to the Wiring Diagrams section.**

PRECAUTIONS

Several precautions must be observed when performing work on alternator equipment.
• If the battery is removed for any reason, make sure that it is reconnected with the correct polarity. Reversing the battery connections may result in damage to the one-way rectifiers.
• Never short across or ground any of the alternator terminals unless specifically mentioned in a particular test procedure.
• Never operate the alternator with the main circuit broken. Make sure that the battery, alternator, and regulator leads are not disconnected while the engine is running.
• Never attempt to polarize an alternator.
• When charging a battery that is installed in the vessel, disconnect the battery cables.
• When arc (electric) welding is to be performed on any part of the vessel, disconnect the negative battery cable and alternator leads.
• Never disconnect the battery cables while the engine is running

CHARGING SYSTEM INSPECTION

1. Make sure the battery connections are clean and tight and that the battery is in good condition and fully charged.
2. Check the drive belt for damage or looseness.
3. Check the wiring harness at the alternator. The harness connector should be tight and latched. Make sure that the output terminal of the alternator is connected to the vessel battery positive lead.
4. Verify that all charging system related fuses and electrical connections are tight and free of damage.

5. Check the mounting bolts for proper torque.
The alternator does not require periodic lubrication. The rotor shaft is mounted on bearings at the drive end and the slip ring end. Each bearing contains its own permanent grease supply.

TESTING

Many times the alternator is suspected of being defective when the battery is not receiving a charge and is constantly being depleted of its energy. Most of the time a heavily corroded wire terminal, broken wire or worn out battery is the actual problem. Perform the preliminary checks listed above to eliminate any problem areas in the charging circuitry before performing the output tests.
If the battery is constantly undercharged, verify all accessories are being switched off when the engine is not running. Check to see if a new accessory (fish finder, live bait tank, etc.) has been added which will place a heavy ampere draw on the battery when operating at low speeds. The battery may be drained when operating at slow speeds for long periods of time.
Check the physical condition and charge state of the battery. The battery should be 75% (1.230 specific gravity) of a full charge. If one or more cells in the battery are defective, the battery should be replaced.
Inspect the alternator system wiring for corroded or loose terminals, damaged or frayed wiring and/or loose wire harness connectors. Check the drive belt for physical condition and proper tension.
If the charging system has passed all the above visual checks, perform the output tests to determine if the alternator is defective.
The following tests require the use of a voltmeter/multi-meter capable of reading 0-20 volts DC. These tests will determine if the alternator and other components within the alternator circuit are in satisfactory working condition.

Alternator Output Test

◆ **See Figures 28, 29 and 30**

■ **Always perform the following test prior to any others so that you can first confirm that a charging problem exists.**

1. Slide off the rubber boot and disconnect the orange lead at the **POS+** terminal (1986-96), or **B** terminal (1997-98), on the rear of the alternator. Connect and ammeter in series between the terminal and the orange lead—positive probe to terminal and negative probe to lead.
2. Connect a multi-meter between the **POS+**, or **B**, terminal and an engine ground—positive probe to terminal and negative probe to ground.
3. Connect a carbon pile (or Stevens LB-85 Load Bank) across the battery.
4. Start the engine and allow it to idle. Run the engine to 2000 rpm (1986), 650 and 1700 rpm (1991-92), or 650, 1500 and 2000 or higher rpm (1987-90 and 1993-98). Check the meter readings at each benchmark. Compare them to the minimum output and voltage range numbers given for your alternator in the Alternator Test Specifications at the end of this section.
5. If a battery overcharge condition is indicated by excessive water usage, the meter's showing consistent charge or the alternator output at idle (no load) exceeding 14.7V (15.2V on 1986, 14.8V for the 1993-96 65 amp alternator and 15 V on 1997-98); check voltage at the purple lead connected to the **S** terminal on the alternator—low voltage, or no voltage, will cause overcharging. If voltage is OK, there is a short at the voltage regulator and it should be replaced.
6. If a battery undercharge condition is indicated (or no charge at all), perform the following tests in order.

Battery Charging Circuit Test

◆ **See Figures 31, 32 and 33**

1. Slide the rubber boot off of the **POS+** (1987-96), or **B** terminal (1997-98), terminal. Connect the positive lead of a multi-meter to the orange lead (still connected) and the negative lead to an engine ground. The meter should show normal battery voltage (approx. 12V).

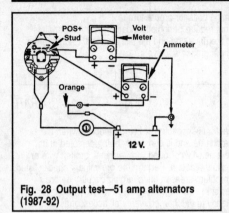

Fig. 28 Output test—51 amp alternators (1987-92)

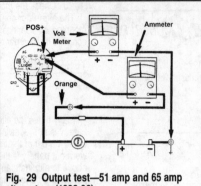

Fig. 29 Output test—51 amp and 65 amp alternators (1993-96)

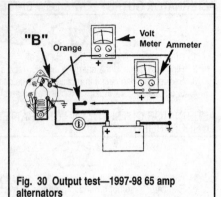

Fig. 30 Output test—1997-98 65 amp alternators

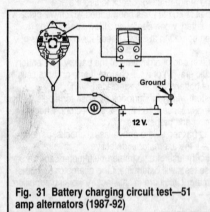

Fig. 31 Battery charging circuit test—51 amp alternators (1987-92)

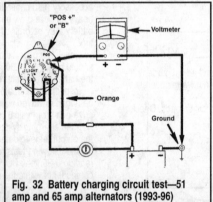

Fig. 32 Battery charging circuit test—51 amp and 65 amp alternators (1993-96)

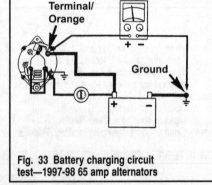

Fig. 33 Battery charging circuit test—1997-98 65 amp alternators

2. If voltage is not within 1V of battery voltage, check the circuit between the orange lead and the circuit breaker, and then the red lead to the starter assist solenoid and then repeat the Output Test.

3. If no voltage is shown, check for an open circuit and repeat the Output Test.

4. If the reading is within specification, move to the next test if working on a 1987-96 51/65 amp alternator, or the Ignition Circuit Test if working on the 1997-98 65 amp alternator.

Red/Purple "EX" Or "EXC" Lead Test

1987-96 51/65 Amp Alternators Only

◆ **See Figures 34 and 35**

1. Disconnect the red/purple lead at the **EX** or **EXC** terminal on the back of the alternator.

2. Connect a multi-meter between the lead and an engine ground—positive probe to the lead and negative probe to ground.

3. Turn the ignition switch to the **ON** position but do not start the engine. The meter should show normal battery voltage (approx. 12V). If it does, move on to the next test.

4. If the meter indicates less than battery voltage, check the purple wire circuit between the alternator and the ignition switch for corrosion or bad connections. Repair and then repeat the Output Test.

5. If the meter indicates an open circuit (0V), check and repair the purple wire circuit and then repeat the Output Test.

6. If the meter reading is 0V and the fuse or circuit breaker blows each time the ignition switch is turned **ON**, the purple wire is grounded against another wire or component. Repair the ground and then repeat the Output Test.

Purple "S" Lead Test

1986 42 Amp Alternators Only

1. Disconnect the purple lead at the terminal on the back of the alternator.

2. Connect a multi-meter between the lead and an engine ground—positive probe to the lead and negative probe to ground.

3. Turn the ignition switch to the **ON** position but do not start the engine. The meter should show normal battery voltage (approx. 12V). If it does, check the battery.

4. If the meter indicates less than battery voltage, check the purple circuit between the alternator and the ignition switch for corrosion or bad connections. Repair and then repeat the Output Test.

5. If all wiring is OK, check the battery again; perform a specific gravity test and a load test. A fully charges battery, or a sulfated one, would make the alternator show normal voltage but still be low on amperage.

1987-96 51/65 Amp Alternators Only

◆ **See Figures 36 and 37**

1. Disconnect the purple lead at the **S** terminal on the back of the alternator.

2. Connect a multi-meter between the lead and an engine ground—positive probe to the lead and negative probe to ground.

3. Turn the ignition switch to the **ON** position but do not start the engine. The meter should show normal battery voltage (approx. 12V). If it does, move on to the Excite/Sense Circuit Test.

4. If the meter indicates less than battery voltage, check the purple circuit between the alternator and the ignition switch for corrosion or bad connections. Repair and then repeat the Output Test.

5. If the meter indicates an open circuit (0V), check and repair the circuit and then repeat the Output Test.

6. If the meter reading is 0V and the fuse or circuit breaker blows each time the ignition switch is turned **ON**, the purple wire is grounded against another wire or component. Repair the ground and then repeat the Output Test.

7. If all wiring is OK, check the battery again; perform a specific gravity test and a load test. A fully charges battery, or a sulfated one, would make the alternator show normal voltage but still be low on amperage.

8. If everything is still OK, move to the Excite/Sense Circuit Test.

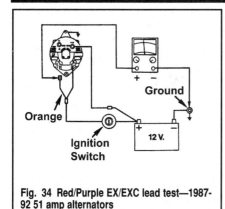

Fig. 34 Red/Purple EX/EXC lead test—1987-92 51 amp alternators

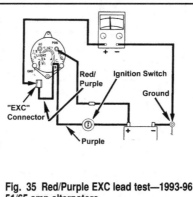

Fig. 35 Red/Purple EXC lead test—1993-96 51/65 amp alternators

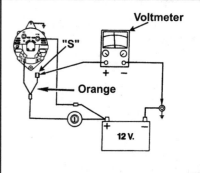

Fig. 36 Purple "S" lead test—1987-92 51 amp alternators

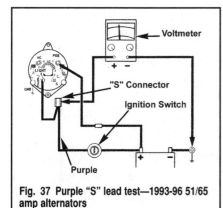

Fig. 37 Purple "S" lead test—1993-96 51/65 amp alternators

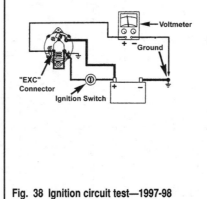

Fig. 38 Ignition circuit test—1997-98

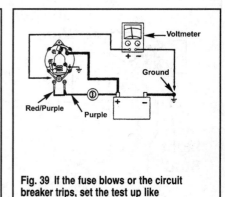

Fig. 39 If the fuse blows or the circuit breaker trips, set the test up like this—1997-98

Ignition Circuit Test

 MODERATE

1997-98 Alternators Only

◆ See Figures 38 and 39

1. Connect a multi-meter between the **EXC** lead and an engine ground—positive probe to the lead and negative probe to ground.
2. Turn the ignition switch to the **ON** position, but do not start the engine. The meter should show normal battery voltage (approx. 12V). If it does, move on to the Excite/Sense Circuit Test.
3. If the meter indicates less than battery voltage (1 or more volts less), disconnect the purple and red/purple leads, connect and ohmmeter and check continuity back to the ignition switch. Repair and then repeat the Output Test.
4. If the meter indicates an open circuit (0 V), disconnect the purple and red/purple leads and repeat the voltage check.
5. If the meter reading is 0V and the fuse or circuit breaker blows each time the ignition switch is turned **ON**, disconnect the purple and red/purple leads and repeat the voltage check in Steps 1-2. If the fuse/breaker doesn't blow (or trip), the circuit is grounding through the alternator and you will need to replace it. If they do blow, there's a problem in the ignition switch circuit—find it and correct it. Repeat the Output Test.
6. If all wiring is OK, check the battery again; perform a specific gravity test and a load test. A fully charged battery, or a sulfated one, would make the alternator show normal voltage but still be low on amperage.
7. If everything is still OK, move to the Excite/Sense Circuit Test.

Excite/Sense Circuit Test

 MODERATE

1987-98 Engines Only

◆ See Figures 40, 41 and 42

1. Connect a multi-meter between the **Light** stud (1987-96), or **L2** terminal (1997-98), on the rear of the alternator and ground—positive lead to the stud, negative lead to ground. There should not be a wire connected to the stud.

■ On many models (particularly those with a mechanical fuel pump), you will need to unscrew the protective cap on the stud.

2. Turn the ignition switch to the **ON** position but do not start the engine. The meter should read 1.5-3.0 V on 1987-96 models, or 1-2 V on 1997-98 models. If it does, move to the Diode-Trio Test on 1987-96 models; if not, replace the voltage regulator.

3. If on replacing the regulator, the alternator continues to fail to produce the specified voltage and amperage, disassemble the alternator and perform the internal tests detailed later.

Diode-Trio Test

 MODERATE

1987-96 Alternators Only

◆ See Figures 43 and 44

1. Bypass the diode-trio by connecting a jumper wire between the **POS+** terminal and the **LIGHT** stud on the back of the alternator.

■ On models with a mechanical fuel pump, you will need to unscrew the protective cap on the Light stud.

2. Connect a multi-meter between the **POS+** terminal stud and an engine ground—positive lead to the stud, negative to the ground.

3. Start the engine and allow it to idle. If you detect charging voltage at the **POS+** terminal stud, replace the diode-trio.

4. If your alternator passed both the Excite/Sense Test and this test, replace the voltage regulator. If on replacing the regulator, the alternator continues to fail to produce the specified voltage and amperage, disassemble the alternator and perform the internal tests detailed later.

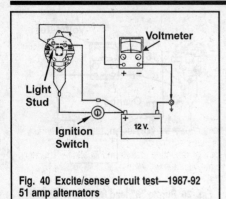

Fig. 40 Excite/sense circuit test—1987-92 51 amp alternators

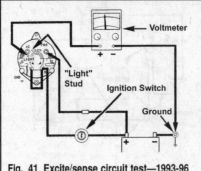

Fig. 41 Excite/sense circuit test—1993-96 51/65 amp alternators

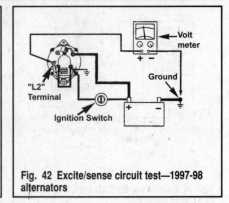

Fig. 42 Excite/sense circuit test—1997-98 alternators

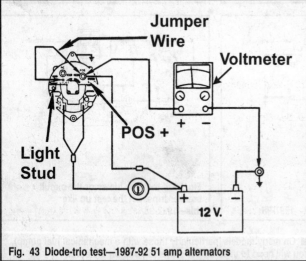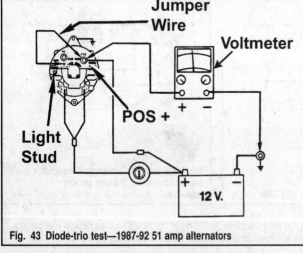

Fig. 43 Diode-trio test—1987-92 51 amp alternators

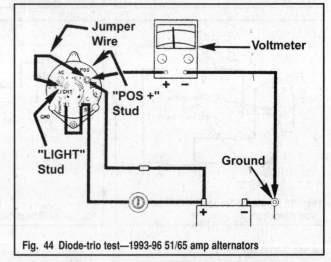

Fig. 44 Diode-trio test—1993-96 51/65 amp alternators

REMOVAL & INSTALLATION

◆ **See Figures 45 and 46** ② MODERATE

1. Disconnect both the negative and positive battery cables from the battery terminal posts. On the back of the alternator, tag and disconnect all electrical leads from the terminals at the rear of the alternator. Although you can always refer to the wiring schematics on installation, we strongly suggest making good notes of which terminal each lead was connected to.

2. Loosen the alternator pivot and belt adjustment bolt(s). Pivot the alternator inward and remove the alternator drive belt from the pulley.

3. Remove the alternator mounting bolt(s) and nut(s)—drive belt models use a mounting bolt and a pivot bolt. Lift the alternator free of the engine.

To Install:

4. Install the alternator into the mounting bracket with the bolt, washers, spacer (if equipped) and nuts. Do not tighten the bolts yet; just leave them finger tight.

5. Reconnect all leads to their original terminals. Coat the purple and red/purple leads with Liquid Neoprene. Make sure that the rubber boots or plastic caps are positioned correctly so they cover the connection fully.

6. Install the drive belt and adjust it as detailed in the Maintenance section and then tighten the bolts securely.

7. Reconnect the battery cables.

DISASSEMBLY

■ As expressed earlier in this manual, some times the equipment, small parts and experience required to repair the alternator will suggest that it is more cost effective if the alternator is taken to a shop which specializes in alternator rebuilding. Be sure to inform the shop that the alternator is for a marine application, to ensure the terminal ends and covers are installed properly.

If a specialty shop is not available, use the following instructions and exploded diagram to disassemble the alternator.

1986 Alternators ③ DIFFICULT

1. Remove the alternator.

2. Carefully position the alternator into a soft-jawed vise and then scribe or paint reference marks across the two housing halves.

3. Loosen and remove the 4 housing bolts. Pry the metal tabs away from the rubber plug and then lift off the end screen.

4. Remove the small rubber connector from the housing.

5. Position a 5/16 in. Allen wrench in vise. Slide the alternator shaft onto the wrench and then remove the end nut with a 15/16 in. wrench. Remove the washer, pulley, fan and collar from the rotor shaft.

6. Carefully pull the rotor assembly up and off the end of the slip ring frame, being very careful to not lose the brush springs. Its not a bad idea to cover the end frame with a piece of tape to keep any dirt out of the bearing.

7. Remove the diode trio screw and insulating washer holding it to the regulator. Now remove the 3 nuts securing the trio to the rectifier bridge and lift out the trio and the stator.

Fig. 45 Disconnect all electrical leads at the rear of the alternator...

Fig. 46 ...and then remove the mounting and pivot bolt (typical installation shown)

1987-96 Alternators

◆ See Figures 47, 48 and 49

1. Remove the alternator.
2. Run an oversize drive belt around the pulley and then clamp it into a vise so that the belt protects the pulley/alternator. Loosen the pulley nut.
3. Lift the assembly out of the vise and place it carefully on a clean workbench. Remove the pulley nut, pulley, fan, woodruff key and the spacer.
4. If equipped, remove the protective cap from the **LIGHT** stud. Remove any nuts, washers and insulators from the remaining studs or terminals; remember to make note of which hardware came from which stud.
5. Remove the two screws in the center and pry off the rear cover.
6. Remove the two nuts holding down the brush holder, but do not pull it off yet.

Fig. 47 Disconnect the three stator wires and the strap

7. Tag and disconnect the three stator wires on top of the rectifier bridge. Unscrew the locknut on the **POS+** terminal and remove the capacitor lead (if equipped). Lift off the small bracket connecting the left stator terminal to the **AC** stud. Remove the 4 mounting screws, straighten out the **POS+** strap and lift off the diode trio/rectifier assembly.

■ **The rectifier bridge will most likely be stuck to the rear housing; carefully pry it loose.**

8. Slide the brush holder out of the housing, making sure that the brushes don't snap out as they clear the slip rings.
9. Remove the 2 regulator mounting screws and lift off the regulator. Make sure you don't lose the two fiber washers that most units have underneath the regulator!
10. Scribe (or paint) a mark across the front and rear housings, loosen the four screws and separate the pieces. Inserting 2 small prybars at points 180° apart (there should be slots) and using them carefully as levers will make this much easier. Make sure that you don't insert the bars any more than 1/16 in. into the slots
11. If the rotor/stator assembly didn't come out when you separated the housings, use the prybars again to lever the assembly up and out of the housing.

1997-98 Alternators

◆ See Figure 50

1. Remove the alternator.
2. Run an oversize drive belt around the pulley and then clamp it into a vise so that the belt protects the pulley/alternator. Loosen the pulley nut.
3. Lift the assembly out of the vise and place it carefully on a clean workbench. Remove the pulley nut, pulley, fan, woodruff key and the spacer.
4. Remove the protective cap from the **L2** terminal.
5. Remove the 2 flange nuts and 1 screw and then lift off the regulator cover.
6. Remove the two mounting screws and lift out the regulator/brush holder assembly.
7. Disconnect and remove the capacitor (1 nut).
8. Remove the insulators from the three remaining terminals (**L2, P** and **B**).
9. Scribe a mark on both housings and matching marks on the stator, as an aid to properly assemble the alternator frame later on.
10. Remove the four thru-bolts. Separate the rear housing and stator from the front housing and rotor by carefully inserting 2 small prybars no more than 1/16 in. into the slots on opposite sides of the stator frame; UNDER the stator but not into the windings. Pry the rear housing stator up and separate them. Never pry anywhere except at the slot or the castings will be damaged.
11. Install the rotor/front housing assembly in a soft-jawed vise; clamping it very carefully around the rotor armature. Grab the front housing and pull it upwards while tapping the end of the rotor shaft with a plastic mallet. Lift out the spacer.

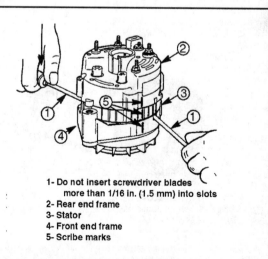

1- Do not insert screwdriver blades more than 1/16 in. (1.5 mm) into slots
2- Rear end frame
3- Stator
4- Front end frame
5- Scribe marks

Fig. 48 Use two screwdrivers to separate the frames

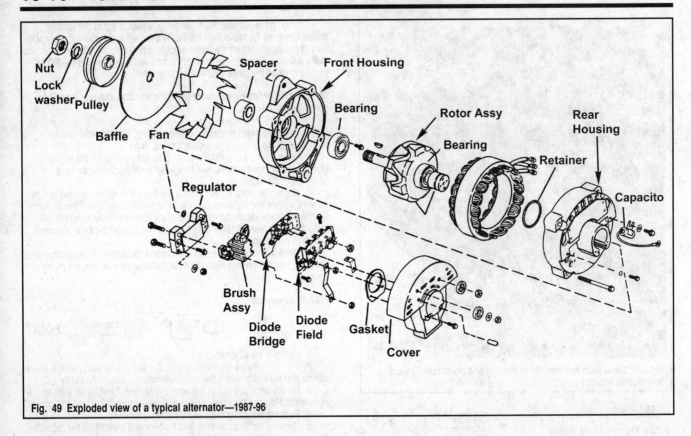

Fig. 49 Exploded view of a typical alternator—1987-96

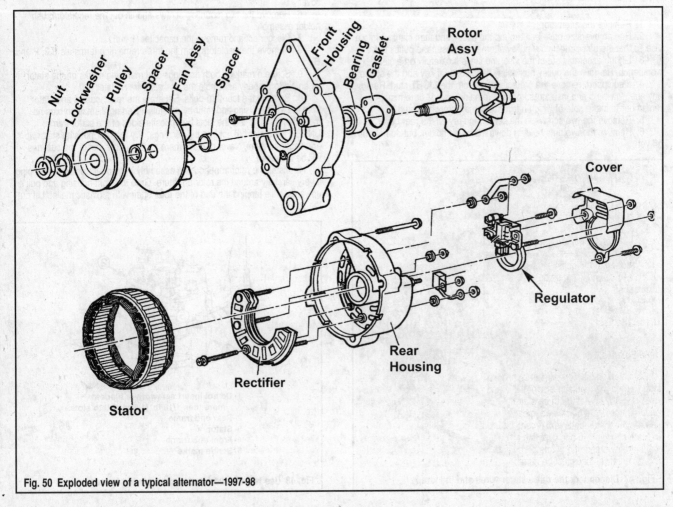

Fig. 50 Exploded view of a typical alternator—1997-98

12. Remove the 3 bearing retainer screws from the front housing and press out the bearing.

13. Lay the stator/rear housing assembly on its back and unsolder the 3 stator leads while holding the corresponding rectifier lead with needle nose pliers to dissipate the heat. Remove the stator from the housing.

14. Remove the center mounting screw and lift out the rectifier.

CLEANING & INSPECTION

The following components are to be visually inspected for serviceability. Electrical checks and tests are detailed previously in this section.

Use a clean soft and wipe any dirt or debris off the parts. Do not clean electrical components with solvent, because such action may damage the item.

Brush

Inspect the brush casing for cracks or damaged brush leads. Check the brush leads for poor solder connections and damaged leads. Check for broken springs and excessive brush wear. If the brushes are worn to less than 3/16 in. (5mm).

Rotor

Inspect the end of the rotor shaft for stripped or damaged threads. Check the pole piece fingers for damage caused by worn bearings. Inspect the rear bearing for smooth quiet rolling action. If the threads cannot be repaired or the bearing is defective, the entire rotor assembly must be replaced.

Clean the rotor slip rings with 400 grit sand paper. Blow off any dust using compressed air and inspect the rotor slip rings for grooves, pits, flat spots or out-of-round more than 0.002 in. (0.051mm). If any of the above conditions are found, the rotor assembly must be replaced.

Stator

Inspect the insulating enamel for heat discoloration because discoloration is an indication of a shorted diode or grounded winding. Check the stator for damaged insulation and wires from contact with the rotor. Replace the stator if any of the above defects are obvious.

End Frames (Housings)

Inspect the end frames for cracks, distortion, stripped threads or worn bearing bore (bearing seized on shaft and spinning in bore). If any of the above damage is obvious the end frame must be replaced.

Fan and Pulley

Inspect the fan for bent or cracked fins, broken welds or worn mounting hole. Check the pulley sheaves for trueness, excessive wear, grooves, pits, nicks and corrosion. If the pulley sheave damage cannot be repaired with a file or wire brush, the pulley must be replaced or drive belt wear will be accelerated.

Diode Trio Test

1986 Engines

Obtain an ohmmeter and select the Rx1000 scale. Attach one ohmmeter lead (red) to the regulator arm of the diode. Attach the other meter lead (black) to any one of the three stator terminals. Observe the meter reading, and then move the black lead to each of the 2 remaining terminals. Now reverse the meter leads and repeat the procedure. The ohmmeter should indicate continuity in one direction and a high resistance, when the meter leads are reversed. If any of the three terminals fail the test, the diode trio, as an assembly, must be replaced.

1987-96 Engines

Obtain an ohmmeter and select the Rx1000 scale. Attach one ohmmeter lead to any one of the three small tabs. Attach the other meter lead to the **LIGHT** terminal (the terminal with the metal strap). Observe the meter reading, and then reverse the meter leads and again observe the reading. The ohmmeter should indicate continuity in one direction and a high resistance, when the meter leads are reversed. Test the other two small tabs in the same manner. If any one of the three tabs fail the test, the diode trio, as an assembly, must be replaced.

1997-98 Engines

◆ See Figure 51

1. Connect an ommeter. Connect one meter lead to the common side of the diode. Connect the other meter lead to the opposite end of the diode and note the meter reading.

2. Reverse the meter leads and again note the meter reading. The meter should indicate a high resistance in one direction and continuity in the other when the leads are reversed.

3. Repeat this procedure for the remaining diodes.

4. Each diode must show continuity and high resistance in the same direction as the others. If not, replace the rectifier assembly.

Rectifier Bridge Positive Diode Test

1986 Engines

Connect an ohmmeter and set it to the Rx100 scale. Connect one lead to the grounded heat sink and the other to one of the 3 terminals. Note the reading on the meter, reverse the leads and note the reading again. If both readings are zero, or very high, replace the rectifier bridge. If one reading is zero and the other is very high, the bridge is OK.

Now repeat the tests between the grounded heat sink and the other 2 terminals.

1987-96 Engines

Attach one ohmmeter lead to the **POS+** terminal strap. Attach the other meter lead to one of the three stator attachment terminals. As in the previous test, the ohmmeter should indicate continuity in one direction and a high resistance when the meter leads are reversed. Test the other two stator attaching terminals in the same manner. If any one of the three stator attaching terminals fails the test, the entire rectifier bridge assembly must be replaced.

1997-98 Engines

◆ See Figure 52

1. Obtain an ohmmeter and connect the negative lead of the meter to the rectifier diode heat sink. Connect the other meter lead to one of the rectifier diode terminal and note the reading. If the meter does not indicate continuity, the diode is open and the rectifier must be replaced. Repeat the procedure for each of the diodes.

2. Now reverse the probes and check each diode in the opposite direction. The meter should indicate no continuity; if any continuity is present, replace the rectifier.

Rectifier Bridge Negative Diode Test

1987-96 Engines

Attach one ohmmeter lead to the rectifier bridge body. Attach the other meter lead to one of the three stator attachment terminals. As in the previous two tests, the ohmmeter should indicate continuity in one direction and a high resistance when the meter leads are reversed. Test the other two stator attaching terminals in the same manner. If any one of the three stator attaching terminals fail the test, the entire rectifier bridge assembly must be replaced.

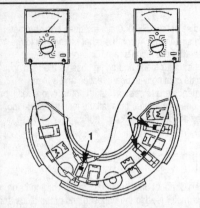

1- Common side of diode circuit board
2- Repeat test for these two diodes

Fig. 51 Testing the diode trio—1997-98

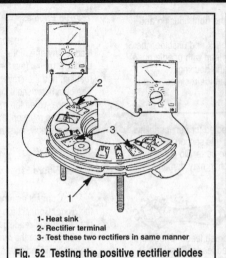

1- Heat sink
2- Rectifier terminal
3- Test these two rectifiers in same manner

Fig. 52 Testing the positive rectifier diodes

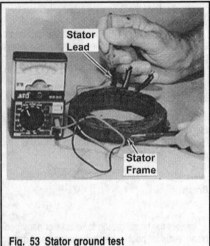

Fig. 53 Stator ground test

1997-98 Engines

◆ **See Figure 50**

1. Obtain an ohmmeter and connect the positive lead of the meter to the rectifier diode heat sink. Connect the other meter lead to one of the rectifier diode terminal and note the reading. If the meter does not indicate continuity, the diode is open and the rectifier must be replaced. Repeat the procedure for each of the diodes.

2. Now reverse the probes and check each diode in the opposite direction. The meter should indicate no continuity; if any continuity is present, replace the rectifier.

Stator Ground Test

◆ **See Figure 53**

Select the Rx1000 scale on the ohmmeter. Make contact with one meter lead to the stator frame. Make contact with the other meter lead to one of the stator leads. The meter must indicate infinity, or no continuity. Repeat this test for the other two stator leads. If any one lead fails the test, the stator assembly must be replaced.

Stator Open Circuit Test

◆ **See Figure 54**

Select the lowest possible scale on the ohmmeter. Make contact with the meter leads to the first and second stator leads (from the left). The meter should register approximately 0.1 ohms on 1987-96 alternators, or 0.2 ohms on 1997-98 alternators. Repeat this test with the first and third stator leads, and then the second and third stator leads. In each test, the resistance value should be the same. If the results vary, the stator assembly must be replaced.

Rotor Ground Test

◆ **See Figure 55**

Select the Rx1000 scale on the ohmmeter. Make contact with one meter lead to any one of the two slip rings. Make contact with the other meter lead to the rotor body. The meter should indicate infinity, or no continuity. Any other reading would indicate the rotor assembly is grounded and therefore must be replaced.

Rotor Open/Short Circuit

◆ **See Figure 56**

Select the Rx10 scale on the ohmmeter. Make contact with one meter lead to one slip ring, and the other slip ring with the second meter lead. The meter should indicate 2.6 ohms on 19986 engines, 3.5-4.7 ohms on 1987-92 engines or 4.1-4.7 ohms on 1993-96 engines. A lower reading indicates shorted windings. A higher reading indicates an excessive resistance. An infinite reading indicates an open circuit. If the reading is not within specification, the rotor assembly must be replaced.

ASSEMBLY

1986 Engines

1. Clean the brushes with a soft cloth and a mild solvent.

2. Press both brushes into the holder and then insert a toothpick through the hole in the rear of the alternator housing to hold them in place during the rest of the procedure.

✳✳WARNING

Never use a metal pick to hold the brushes in place.

3. Check to see if there is a small dab of sealer in the recess on the edge of the stator frame. If not, squeeze a small amount of RTV silicone sealer into the recess.

4. Position the stator into the housing and rotate it until the notches in the frame are aligned with the holes in the housing.

5. Position the diode trio and install the special insulated screw.

6. Move the stator terminal onto the rectifier bridge, thread on the 3 nuts and tighten them securely.

7. Clean the rotor slip rings with mild solvent. Line up the reference marks on the housings and connect them so that they fit snugly over the stator frame.

8. Remove the toothpick.

9. Press the rubber plug into the voltage regulator until it is snug.

10. Slide the screen over the unit and then install the housing bolts. Tighten the bolts to 35-50 inch lbs. (46-68 Nm).

1987-96 Engines

◆ **See Figure 47**

1. Position the stator/rotor assembly into the front housing so that the three terminals are just under the upper mounting boss on the housing. Make sure that the indents in the frame of the stator line up over each of the thru-bolt holes.

2. Position the rear housing over the assembly and then feed the stator leads through the slots in the back of the housing. Set the cover onto the stator assembly so that the alignment marks made during disassembly line up, install the 4 bolts and tighten them to 50-60 inch lbs. (5.6-6.8 Nm).

■ **Spin the rotor shaft to ensure that the rotor rotates freely within the assembly.**

3. Lay the fiber washers over the regulator mounting bolt holes and install the regulator. Tighten the screws securely.

4. Position the brush holder over the regulator so that the banded surface is facing away from the regulator. Squeeze the brushes with your fingers and slide the holder into the grooves at the of the housing. Tighten the right nut finger tight and leave the left one loose.

5. Straighten the **B+** strap from the rectifier, swab a little heat sink compound over the mounting surface and place the diode/rectifier bridge in position on the rear housing.

6. Feed the strap through the slot in the diode trio body and position the unit onto the bridge. Tighten the mounting screws securely.

7. Bend the strap into place so that it fits against base of the diode trio output stud. Slide the condenser lead over the output stud and tighten the lock nut securely.

8. Slide the small bracket into position over the **AC** stud and the terminal and then reconnect the 3 stator leads to the diode trio.

9. Make sure the strap from the **LIGHT** terminal is positioned over the left stud on the brush holder and then install the final holder mounting nut. Tighten both nuts securely.

10. Position a new felt gasket into the slip ring pocket on the inside of the rear cover. Position the cover over the regulator/brush holder, install the screws and tighten them securely.

11. Reinstall all hardware to the rear terminals and studs.

12. Install the spacer, woodruff key, fan and pulley and then tighten the nut finger tight. Place the assembly back in a vise with the drive belt around the pulley and tighten the nut to 35-50 ft. lbs. (48-68 Nm).

13. Install the alternator and connect the electrical leads.

1997-98 Engines

◆ **See Figure 50**

1. Position a new bearing into the front housing bearing bore. Position the front end frame into an arbor press. Using the appropriate size mandrel, press the bearing into the bore until the bearing contacts the end frame. Place the bearing retainer plate over the bearing and secure with 3 screws and washers.

2. Position the rotor into a soft jawed vise with the pulley shaft upwards. Install the front housing over the shaft. Slide the short spacer over the shaft, the fan (blades down), the tall spacer and then the pulley. Install the washer and nut and tighten it to 47 ft. lbs. (63 Nm).

3. Position the rectifier into the rear housing and tighten the retaining screw securely. Slide in the stator and then re-solder the three electrical leads.

4. Position the rear housing assembly over the rotor so that the terminal studs on the rear are on the upper pivot boss side and the marks made earlier are in alignment. Press the frames together and then install the screws, tightening them securely.

5. Push the brushes down so they are flush with the top of the holder and then insert a #54 (0.050 in./1mm) drill bit into the hole in the holder so they stay compressed. Position the brush/regulator in the cavity on the back of the rear end frame and tighten the two mounting screws to 42 inch lbs. (4.2 Nm). Pull out the drill bit so that the brushes release against the slip rings.

6. Install the strap to the **L2** stud and tighten the nut securely.

7. Install the cover. Tighten the 2 nuts to 51 inch lbs. (5.8 Nm) and the screw to 30 inch lbs. (3.4 Nm). Pop on the insulator caps.

8. Install the alternator.

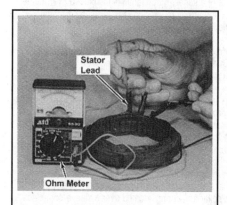

Fig. 54 Stator open circuit test

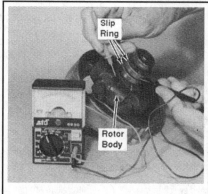

Fig. 55 Performing the rotor ground test.

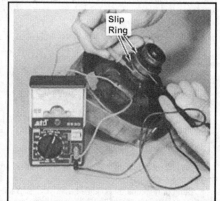

Fig. 56 Rotor open/short circuit test

STARTER CIRCUIT

Description and Operation

The cranking system used all OMC engines consists of a 12 V DC cranking motor (starter), a starter-mounted or field housing-mounted solenoid, an assist solenoid or starter relay, a neutral start (safety) switch, the ignition switch, 50 or 60 amp circuit breaker, 20 amp inline fuse and all necessary wiring to complete the circuit.

All engines utilize an electric starter motor coupled with a mechanical gear mesh between the starter motor and the flywheel, similar to the method used to crank an automobile engine.

As the name implies, the sole purpose of the starter motor circuit is to control operation of the starter motor to crank the engine until it is operating. The circuit includes a relay or magnetic switch (solenoid) to connect or disconnect the motor from the battery. The operator controls the switch with a key switch.

A neutral safety switch is installed into the circuit to permit operation of the starter motor only if the shift control lever is in neutral. This switch is a safety device to prevent accidental engine start when the engine is in gear.

The starter motor is a series-wound electric motor that draws a heavy current from the battery. It is designed to be used only for short periods of time to crank the engine for starting. To prevent overheating the motor, cranking should not be continued for more than 30 seconds without allowing

the motor to cool for at least three minutes. Actually, this time can be spent in making preliminary checks to determine why the engine fails to start.

Power is transmitted from the starter motor to the flywheel through a Bendix drive. This drive has a pinion gear mounted on screw threads. When the motor is operated, the pinion gear moves upward and meshes with the teeth on the flywheel ring gear.

When the engine starts, the pinion gear is driven faster than the shaft and as a result, it screws out of mesh with the flywheel. A rubber cushion is built into the Bendix drive to absorb the shock when the pinion meshes with the flywheel ring gear. The parts of the drive must be properly assembled for efficient operation. If the screw shaft assembly is reversed, it will strike the splines and the rubber cushion will not absorb the shock.

The sound of the motor during cranking is a good indication of whether the starter motor is operating properly or not. Naturally, temperature conditions will affect the speed at which the starter motor is able to crank the engine. The speed of cranking a cold engine will be much slower than when cranking a warm engine. An experienced operator will learn to recognize the favorable sounds of the engine cranking under various conditions.

The job of the starter motor relay (or starter assist solenoid) is to complete the circuit between the battery and starter motor. It does this by closing the starter circuit electro-magnetically, when activated by the key switch. This is a completely sealed switch, which meets SAE standards for marine applications. Do not substitute an automotive-type relay for this application. It is not sealed and gasoline fumes can be ignited upon starting the engine. The relay consists of a coil winding, plunger, return spring, contact disc and four externally mounted terminals. The relay is installed in series with the positive battery cables mounted to the two larger terminals. The smaller terminals connect to the neutral switch and ground.

To activate the relay, the shift lever is placed in neutral, closing the neutral switch. Electricity coming through the ignition switch goes into the relay coil winding which creates a magnetic field. The electricity then goes on to ground in the engine. The magnetic field surrounds the plunger in the relay, which draws the disc contact into the two larger terminals. Upon contact of the terminals, the heavy amperage circuit to the starter motor is closed and activates the starter motor. When the key switch is released, the magnetic field is no longer supported and the magnetic field collapses. The return spring working on the plunger opens the disc contact, opening the circuit to the starter.

When the armature plate is out of position or the shift lever is moved into forward or reverse gear, the neutral switch is placed in the open position and the starter control circuit cannot be activated. This prevents the engine from starting while in gear.

Troubleshooting the Starting System

If the starter motor spins but fails to crank the engine, the cause is usually a corroded or gummy Bendix drive. The drive should be removed, cleaned and given an inspection.

1. **Before wasting too much time troubleshooting the starter motor circuit, the following checks should be made. Many times, the problem will be corrected.**
- Battery fully charged.
- Shift control lever in neutral.
- Main 20-amp fuse is good (not blown).
- All electrical connections clean and tight.
- Wiring in good condition, insulation not worn or frayed.

2. **Starter motor cranks slowly or not at all.**
- Faulty wiring connection
- Short-circuited lead wire
- Shift control not engaging neutral (not activating neutral start switch)
- Defective neutral start switch
- Starter motor not properly grounded
- Faulty contact point inside ignition switch
- Bad connections on negative battery cable to ground (at battery side and engine side)
- Bad connections on positive battery cable to magnetic switch terminal
- Open circuit in the coil of the magnetic switch (relay)
- Bad or run-down battery
- Excessively worn down starter motor brushes
- Burnt commutator in starter motor
- Brush spring tension slack
- Short circuit in starter motor armature

3. **Starter motor keeps running.**
- Melted contact plate inside the magnetic switch
- Poor ignition switch return action

4. **Starter motor picks up speed, put pinion will not mesh with ring gear.**
- Worn down teeth on clutch pinion
- Worn down teeth on flywheel ring gear

Starter Motor

DESCRIPTION & OPERATION

Direct Drive—1987-92 Ford Engines

Direct drive starters consist of a set of field coils positioned over pole pieces, which are attached to the inside of a heavy iron frame. An armature and over-running clutch drive mechanism are included inside the iron frame.

The armature consists of a series of iron laminations placed over a steel shaft, a commutator, and the armature winding. The windings are heavy copper ribbons assembled into slots in the iron laminations. The ends of the windings are soldered or welded to the commutator bars. These bars are electrically insulated from each other and from the iron shaft.

An over-running clutch drive arrangement is installed near one end of the starter shaft. This clutch drive assembly contains a pinion that is made to move along the shaft by means of a shift lever to engage the engine ring gear for cranking. The relationship between the pinion gear and the ring gear on the engine flywheel provides sufficient gear reduction to meet cranking requirement speed for starting.

The over-running clutch drive has a shell and sleeve assembly, which is splined internally to match the spiral splines on the armature shaft. The pinion is located inside the shell. Spring-loaded rollers are also inside the shell and they are wedged against the pinion and a taper inside the shell. Some starters use helical springs and others use accordion type springs. Four rollers are used. A collar and spring, located over a sleeve completes the major parts of the clutch mechanism.

When the starter is not in use, one of the field coils is connected directly to ground through a set of contacts. When the starter is first connected to the battery, a large current flows through the grounded field coil, actuating a moveable pole shoe. The pole shoe is attached to the starter drive plunger lever and thus the drive is forced into engagement with the flywheel.

When the pole shoe is seated fully, it opens the grounding contacts on the field coil and the starter then moves to normal operation. A holding coil is utilized to keep pole shoe in position during the time that the starter is cranking the engine.

Direct Drive—1986-94 GM Engines

Direct drive starters consist of a set of field coils positioned over pole pieces, which are attached to the inside of a heavy iron frame. An armature, an over-running clutch drive mechanism, and a solenoid are included inside the iron frame.

The armature consists of a series of iron laminations placed over a steel shaft, a commutator, and the armature winding. The windings are heavy copper ribbons assembled into slots in the iron laminations. The ends of the windings are soldered or welded to the commutator bars. These bars are electrically insulated from each other and from the iron shaft.

An overrunning clutch drive arrangement is installed near one end of the starter shaft. This clutch drive assembly contains a pinion that is made to move along the shaft by means of a shift lever to engage the engine ring gear for cranking. The relationship between the pinion gear and the ring gear on the engine flywheel provides sufficient gear reduction to meet cranking requirement speed for starting.

The overrunning clutch drive has a shell and sleeve assembly, which is splined internally to match the spiral splines on the armature shaft. The pinion is located inside the shell. Spring-loaded rollers are also inside the shell and they are wedged against the pinion and a taper inside the shell. Some starters use helical springs and others use accordion type springs. Four rollers are used. A collar and spring, located over a sleeve completes the major parts of the clutch mechanism.

When the solenoid is energized and the shift lever operates, it moves the collar endwise along the shaft. The spring assists movement of the pinion into mesh with the ring gear on the flywheel. If the teeth on the pinion fail to mesh for just an instant with the teeth on the ring gear, the spring compresses until the solenoid switch is closed; current flows to the armature; the armature rotates; the spring is still pushing on the pinion; the pinion teeth mesh with the ring gear; and cranking begins.

Torque is transferred from the shell to the pinion by the rollers, which are wedged tightly between the pinion and the taper cut into the inside of the shell. When the engine starts, the ring gear drives the pinion faster than the armature; the rollers move away from the taper; the pinion overruns the shell; the return spring moves the shift lever back; the solenoid switch is opened; current is cutoff to the armature; the pinion moves out of mesh with the ring gear; and the cranking cycle is completed. The start switch should be opened immediately when the engine starts to prevent prolonged overrun.

Gear Reduction—1993-96 Ford Engines And 1995-98 GM Engines

Gear reduction starter motors are small and light weight for the amount of work produced. These starters have small permanent magnets mounted inside the field frame. The placement and strength of these magnets differ between models. Therefore the field frames are not interchangeable.

The permanent magnets take the place of the large current-carrying iron core field-coil magnets previously used on the larger direct drive starter motors. The motor armature is supported on both ends by a permanently lubricated roller or ball bearing assembly to reduce the drag and friction created by the motor speed of approximately 7,000 rpm.

A planetary gear reduction unit is mated between the drive motor armature and the Bendix drive gear. This planetary gear drive results in an over all gear reduction of approximately 4:1. Through the gear reduction, the conversion of motor high speed, low torque is converted to a high torque, low speed gear drive output. The gear reduction results in a final engine cranking speed of approximately 1,750 rpm.

The starter is designed to operate under heavy loads and produce high power for only short periods of time. Therefore, never crank the engine for more than 30 seconds without allowing a minimum cooling off time of two minutes, before attempting to crank the engine again.

As with all marine installations, a safety switch is installed in the remote control shift box. This switch is designed to open the starter circuit to prevent the engine from starting, if the shift lever is in any position other than the neutral position.

The unit has a Bendix follow-thru type drive designed to overcome disengagement of the flywheel ring gear when engine speed exceeds cranking motor speed.

A helical cut shaft is designed to quickly engage and disengage the Bendix drive. An internal one-way clutch in the Bendix allows the motor to drive the Bendix. If the engine should start, and the flywheel begin to drive the Bendix, the one-way clutch will release and allow the drive to overrun and release the cranking motor armature. The helical splined shaft will then disengage the Bendix from the flywheel.

■ As of 1995, OMC suggests that their GM starters are no longer serviceable. No replacement parts are available. Check with your local parts supplier.

PRECAUTIONS

1. Always make sure that each battery cable is connected to the correct terminal on the battery.
2. Never disconnect the battery while the engine is running.
3. When using a battery charger or booster, always make sure that the positive battery cable on the charger is connected to the positive terminal on the battery. The same goes for the negative side.
4. Always make sure that both battery cables are disconnected prior to connecting a charger or booster.
5. Always make sure the battery is in good operating condition.
6. Always make sure the battery leads and terminals are clean.

TROUBLESHOOTING

◆ See Figures 57 and 58
Regardless of how or where the solenoid is mounted, the basic circuits of the starting system on all makes of cranking motors are the same and similar tests apply. In the following testing and troubleshooting procedures, the differences are noted.

✳✳ CAUTION

Always take time to vent the bilge when making any of the tests as prevention against igniting any fumes accumulated in that area. As a further precaution, remove the high-tension wire from the center of the distributor cap and ground it securely to prevent sparks.

All cranking motor problems fall into one of three areas:
1. The cranking motor fails to rotate.
2. The cranking motor spins rapidly, but does not crank the engine.
3. The cranking motor cranks the engine, but too slowly to affect engine start.

The following paragraphs provide a logical sequence of tests designed to isolate a problem in the cranking system.

Battery

1. Turn on several of the cabin lights (or any accessories). Turn the ignition switch to the **ON** position and note the effect on the brightness of the lights. With a properly functioning electrical system, the lights will dim slightly and the starter will crank the engine at a reasonable rate. If the lights dim considerably and the engine does not turn over, one of several causes may be at fault.

2. If the lights go out completely, or dim considerably, the battery charge is low or almost dead. The obvious remedy is to charge the battery; switch over to a secondary battery if one is available; or to replace it with a known fully charged one.

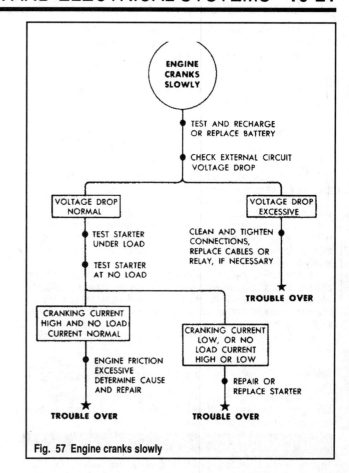

Fig. 57 Engine cranks slowly

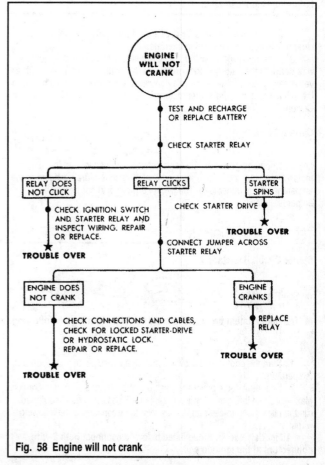

Fig. 58 Engine will not crank

3. If the starting relay clicks, sounding similar to a machine gun firing, the battery charge is too low to keep the starting relay engaged when the starter load is brought into the circuit.

4. If the starter spins without cranking the engine, the drive is broken. The starter will have to be removed for repairs.

5. If the lights do not dim, and the starter does not operate, then there is an open circuit. Proceed to the Cable Connection Test.

Cable Connection

1. If the starter fails to operate and the lights do not dim when the ignition switch is turned to start, the first area to check is the connections at the battery, starting relay, starter and neutral-safety switch.

2. First, remove the cables at the battery; clean the connectors and posts; replace the cables; and tighten them securely.

3. Now, try the starter. If it still fails to crank the engine, try moving the shift box selector lever from neutral to forward to determine if the neutral-safety switch is out of adjustment or the electrical connections need attention.

4. Sometimes, after working the shift lever back-and-forth and perhaps a bit sideways, the neutral-switch connections may be temporarily restored and the engine can be started. Disconnect the leads; clean the connectors and terminals on the switch; replace the leads; and tighten them securely at the first opportunity.

5. If the starter still fails to crank the engine, move on to the Solenoid Test.

Solenoid

1. The solenoid, commonly called the starter relay, is checked by directly bridging between the terminal from the battery (the large heavy one) to the terminal from the ignition switch.

✳✳ CAUTION

Take every precaution to ensure there is no gasoline fumes in the bilge before making these tests.

2. If a bilge blower is installed, operate it for at least five minutes to clear any fumes accumulated in the bilge.

3. Turn the ignition switch to the **ON** position. Now, connect a jumper wire between the battery lead terminal and the ignition lead terminal with a very heavy piece of wire. If the relay operates, the trouble is in the circuit to the ignition switch. If the starter motor still fails to operate, continue with the Current Draw Test.

Current Draw

Lay an amperage gauge on the cable between the battery and the starter motor. Attempt to crank the engine and note the current draw reading of the amperage gauge under load. The current draw should not exceed 190 amperes.

TESTING—GM ENGINES

Starter Circuit Resistance

◆ See Figure 59

■ **The battery must be fully charged prior to performing the following test.**

1. Start the engine and allow it to run until it reaches normal operating temperature.

2. Shut the engine down and ensure that it will not start during cranking. Make sure that the ignition switch is in the **OFF** position, and, if equipped, run the bilge pump to clear any fumes prior to disconnecting the following wires:

• **All models except those listed below: disconnect both 2-wire connectors at the ignition coil.**

• **1993 4.3L and 1994 5.7L models (carburetor): disconnect both primary wires at the coil terminals**

• **1993-97 7.4L and 8.2L carbureted models: disconnect the engine harness 14-pin connector at the ignition module**

3. Connect a voltmeter and connect the positive lead to the lower terminal on the solenoid. Touch the negative terminal to the starter ground so that it grounds.

4. Turn the ignition switch **ON** and crank the engine observing the initial meter reading. If the meter reads more than 9 V at normal cranking speed, the starter and switch are OK. If the cranking speed is low and the meter reads less than 9 V, check your battery, battery terminals and/or the solenoid. Perform the Solenoid Contacts Test.

✳✳ CAUTION

Never operate the starter for more than 10 seconds at a time without allowing the motor to cool for at least 3 min.

Solenoid Contacts Test

◆ See Figure 60

1. Connect a voltmeter and switch it to the high scale. Connect the negative lead to the lower solenoid terminal and the positive lead to the upper (BAT) terminal.

2. Turn the ignition switch to the **ON** position and crank the engine. As soon as you begin cranking the engine, flip the meter back to the low scale and take note of the reading. Turn the meter back to the high scale and shut off the engine.

3. If the solenoid switch contacts are OK, the meter reading should have been 1/10 V or less. Anything above this will necessitate switch repair or replacement.

Solenoid Amperage Test

◆ See Figure 61

1. Remove the starter from the engine.

2. Remove the screw from the solenoid motor terminal and then bend the field leads slightly until they are clear of the terminal.

3. Connect a heavy gauge jumper wire to the Motor terminal on the solenoid and ground it against the starter body.

4. Connect an ammeter and a carbon pile in series between the **S** terminal on the solenoid (small one) and the positive terminal on a battery. Connect a heavy gauge jumper wire between the starter body and the negative post on the battery.

5. Connect a voltmeter between the base of the solenoid and the **S** terminal on the solenoid.

6. Adjust the resistance on the carbon pile (slowly!) until the voltmeter is showing 10 volts and then check the reading on the ammeter. With the solenoid at room temperature, the reading should be 47-55 amps; showing current draw of both windings in parallel.

7. Remove the jumper between the Motor terminal and the starter body. Readjust the carbon pile until the voltmeter is reading 10 V again. The ammeter should read 15-20 amps; showing the current draw for hold-in winding alone.

8. Any readings outside the ranges given above will necessitate replacement of the solenoid.

TESTING—FORD ENGINES

Starter Circuit Resistance

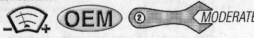

1987-92 Engines

◆ See Figure 62

1. With the ignition switch in the **OFF** position, disconnect the distributor primary wire at the ignition coil (negative terminal).

2. Connect a voltmeter and set it to the low scale. Connect the negative lead to the starter terminal and the positive lead to the positive battery terminal. Turn the ignition switch to the **ON** position; the meter should read 0.5V.

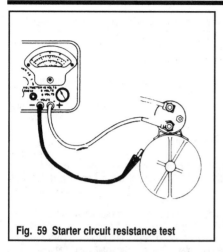

Fig. 59 Starter circuit resistance test

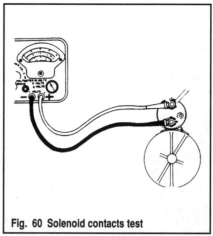

Fig. 60 Solenoid contacts test

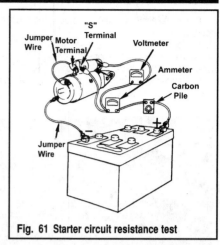

Fig. 61 Starter circuit resistance test

3. Now move the negative lead to the starter terminal on the solenoid and turn the ignition switch back to the **ON** position; the meter should read 0.3V.

4. Now move the negative lead to the battery terminal on the solenoid and turn the ignition switch back to the **ON** position; the meter should read 0.1V.

5. Now move the negative lead to the negative battery terminal on the solenoid and move the positive lead to an engine ground. Turn the ignition switch back to the **ON** position; the meter should read 0.1V.

1993-96 Engines

■ The battery must be fully charged prior to performing the following test.

1. Start the engine and allow it to run until it reaches normal operating temperature.

2. Shut the engine down and ensure that it will not start during cranking. Make sure that the ignition switch is in the **OFF** position, and, if equipped, run the bilge pump to clear any fumes prior to disconnecting the following wires:

- **EFI models: disconnect the 3-wire connector at the E-coil.**
- **Carbureted models: disconnect both primary wires at the ignition coil terminals**

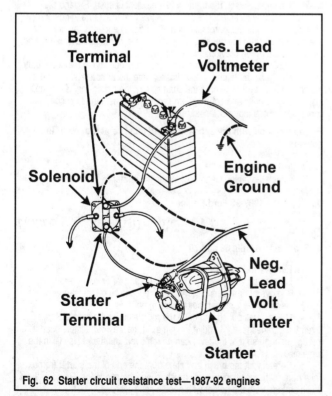

Fig. 62 Starter circuit resistance test—1987-92 engines

3. Connect a voltmeter between the **B+** terminal on the motor and the housing.

4. If battery voltage is not present, check all wires and connections between the battery and the solenoid. Also check all grounds for opens or shorts.

5. If battery voltage is present, connect a remote starter switch between **B+** and **S** terminals. Turn the remote starter ON. If the starter does not crank, replace it. If it does crank, check the wires and connections between the solenoid and assist solenoid for opens or shorts.

** CAUTION

Never operate the starter for more than 10 seconds at a time without allowing the motor to cool for at least 3 min.

1993-96 Engines

1. Connect a voltmeter between an engine ground and the **S** terminal on the assist solenoid. Turn the ignition switch to the **START** position and confirm that the meter shows battery voltage.

2. If no voltage is shown, check the assist solenoid ground and then repeat Step 1. If there is still not battery voltage present, check the circuit from the terminal back through the cable connector, neutral safety switch, ignition switch, fuse and circuit breaker—correct any problems found.

3. If voltage is found to be present in Step 1, reconnect the voltmeter to the solenoid output terminal and check for battery voltage. If none is indicated, replace the assist solenoid.

REMOVAL & INSTALLATION

◆ See Figure 63

1. Disconnect the battery cables.

2. Tag and disconnect all wires at the starter solenoid. Move them out of the way.

3. Cut the plastic tie holding the drain hose from the flywheel housing.

4. On GM engines, remove the nut from the support bracket. Remove the ring gear guard at the leading edge of the flywheel housing.

5. Loosen the two mounting bolts and then pull them out of the starter housing. Carefully lift out the starter. Mounting bolts on GM engine run vertically, while mounting bolts on Ford engine run horizontally.

■ Be aware that some applications will utilize a shim(s) between the motor and cylinder block. Do not lose this shim!

To Install:

6. Position the shim (if equipped) and then install the starter. Slide in the thru-bolts and tighten them to 30-36 ft. lbs (41-49 Nm) on GM engines, 20-25 ft. lbs. (27-34 Nm) on 1987-92 Ford engines, or 15-20 ft. lbs. (21-27 Nm) on 1993-96 Ford engines.

7. On GM engines, install the ring gear plate and tighten the bolts to 5-7 ft. lbs. (7-9 Nm).

8. On GM engines, install the support bracket nut and tighten it securely.

9. Reconnect all electrical leads, tightening their nuts securely and coating the solenoid connections with Liquid Neoprene. On Ford engines, tighten the battery cable to 80-120 inch lbs. (9-13 Nm).

10. Reattach the drain hose to the solenoid with a new plastic tie.

11. Connect the battery cables.

Fig. 63 Starter mounting bolts

DISASSEMBLY & ASSEMBLY

■ Before disassembling begins, read these few words first on a time and money saving idea. Most dealers do not find it economical to stock all the small parts or spend the time to service a starter motor. In most cases the labor cost alone will exceed the cost of a new or rebuilt unit.

Throughout the country, specialty shops may be found specializing in and servicing starter motors, alternators and other electrical assemblies. A customer can usually obtain a new or rebuilt starter for a modest cost usually on the spot or same day service. Some marine dealers even stock new and rebuilt cranking motors ready for installation. This means the repair can be completed and the boat back in service before the end of the day.

If the decision is made to disassemble the unit, check with the marine dealer first for parts availability, order time, and cost compared to a new or rebuilt unit off the shelf. These units usually come with some type of warranty. Therefore, if it fails within the warranty period, additional cost is not involved.

Direct Drive—1986-94 GM Engines

◆ See Figures 64, 65 and 66

1. Remove the starter.
2. Remove the solenoid.
3. Scribe or paint match-marks across the commutator end plate and frame; and across the frame and drive housing.
4. Loosen and remove the 2 thru-bolts and pull off the end plate.
5. Tap the field frame where it mates with the drive housing a few times with a rubber mallet and then pull off the frame along with the field coil.
6. Pull the brush holder pivot pins out and lift the holders and springs out of the frame. Remove the brush screws and pull the brushes from the holder.
7. Remove the 2 bolts holding the bearing plate to the drive housing and then lift the armature assembly out of the housing.
8. Slide the thrust collar off of the pinion end of the armature shaft and then remove the thrust washer from the opposite end of the armature shaft..
9. Locate a small piece of pipe (1/2 in. I.D) and slide it over the shaft. Tap it lightly with a hammer until you drive the pinion stop retainer in toward the armature. Remove the snap-ring from its groove and then slide the retainer and armature assembly off of the shaft.

10. Remove the retaining pin from the brush support and holders and then remove them all after disconnecting the leads.

11. Pop off the small retaining ring and pull the pivot pin (lever shaft) out of the drive housing. Lift the lever and plunger assembly from the housing.

To assemble:

12. Clean all components thoroughly and inspect them. Perform the previous tests.

13. Coat the shift lever linkage lightly with silicone grease. Position it into the drive housing so the open side is facing out and then slide in the pivot pin. Snap on the retaining ring.

14. Lubricate the drive end of the armature shaft with a thin coating of 10W motor oil or silicone lubricant. Install the bearing plate so the raised side faces the armature and then install the fiber washer. Slide the assist spring and clutch assembly onto the armature shaft so that the pinion is facing outward. Slide the retainer onto the shaft with the cupped surface facing the end of the shaft (away from the pinion).

15. Stand the armature on its end on a wooden surface with the commutator end down. Position the snap ring on the upper end of the shaft and hold it in place with a block of wood. Now, tap on the wood block with a hammer to force the snap ring over the end of the shaft. Slide the snap ring down into the groove. Assemble the thrust collar onto the shaft with the shoulder next to the snap ring.

16. Place the armature flat on the work bench, and then position the retainer and thrust collar next to the snap ring. Next, using two pair of pliers at the same time (one pair on each side of the shaft), grip the retainer and thrust collar and squeeze until the snap ring is forced into the retainer.

17. Lubricate the drive housing bushing with a thin coating of silicone lubricant. Make sure the thrust collar is in place against the snap ring and the retainer.

18. Squirt a few drops of 10W oil into the drive housing bushing and then slide the armature and clutch assembly into place in the drive housing while engaging the shift lever with the clutch.

19. Swivel the assembly until the holes in the center bearing plate line up with the holes in the housing and then install the 2 screws. Tighten them securely.

20. Install the brushes into the brush holders. Assemble the insulated (black) and grounded brush holders together with the V-spring and position them as a unit on the support pin. Push the holders and springs to the bottom of the support, and then rotate the spring to engage the center of the V-spring in the support.

21. Attach the ground wire to the grounded brush and the field lead wire to the insulated brush.

22. Position the field frame over the armature and apply a thin coating of liquid neoprene between the frame and the solenoid case. Expand the brushes until they just clear the commutator. Place the frame in position against the drive housing, rotating it until the dowel engages the hole and seats the frame into the housing. Take care not to damage the brushes and making sure that the previous marks line up.

23. Apply a coating of silicone lubricant to the bushing in the commutator end frame. Place the leather brake washer onto the armature shaft and slide the commutator end plate onto the shaft so that the match-marks line up. Install the thru-bolts and tighten them securely. Connect the field coil connectors to the motor solenoid terminal.

■ The thru-bolt with the exposed thread must be at the top of the housing.

24. Install the solenoid.
25. Check the pinion clearance. Install the starter.

Direct Drive—1987-92 Ford Engines

◆ See Figure 67, 68, 69 and 70

1. Remove the starter motor.
2. Loosen the clamp screw and remove the starter cover retaining clamp. Remove the remaining mounting screw and lift off the starter cover.
3. Remove the 2 starter motor through bolts.
4. Use a small punch and drive the roll pin out of the starter plunger lever. It is not necessary to drive the pin all of the way out; simply push it through enough so that the lever can be removed, but the pin is still in the remaining housing hinge hole.
5. Carefully lift the brush end plate from the end of the starter frame. Take note of the positioning of all brushes, springs and leads. Remove the 4 brush springs from the holder with needle nose pliers.

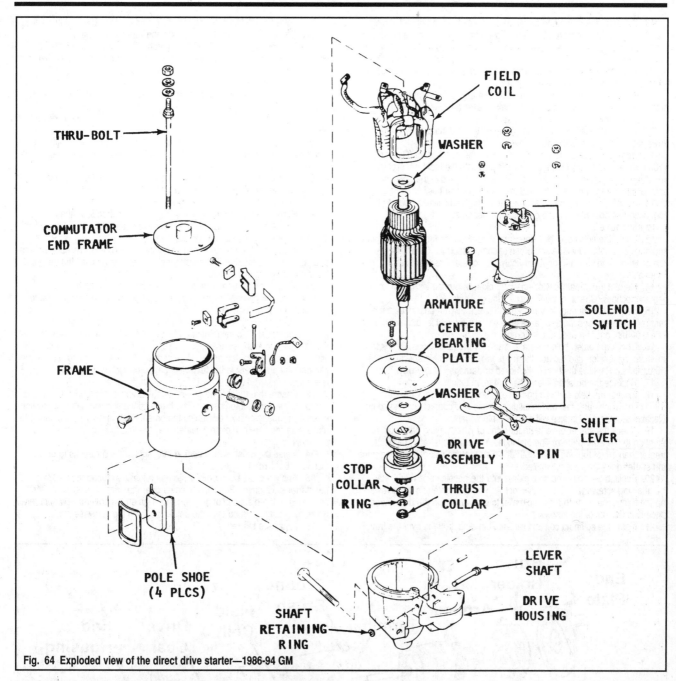

THRU-BOLT

COMMUTATOR
END FRAME

FRAME

POLE SHOE
(4 PLCS)

FIELD
COIL

WASHER

ARMATURE

CENTER
BEARING
PLATE

WASHER

DRIVE
ASSEMBLY

STOP
COLLAR

RING

THRUST
COLLAR

SOLENOID
SWITCH

SHIFT
LEVER

PIN

LEVER
SHAFT

DRIVE
HOUSING

SHAFT
RETAINING
RING

Fig. 64 Exploded view of the direct drive starter—1986-94 GM

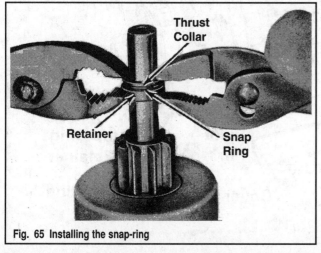

Thrust
Collar

Retainer

Snap
Ring

Fig. 65 Installing the snap-ring

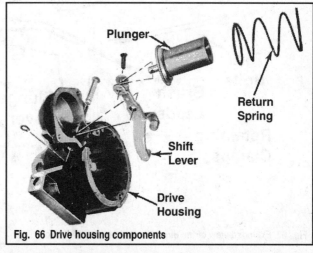

Plunger

Return
Spring

Shift
Lever

Drive
Housing

Fig. 66 Drive housing components

6. Separate the frame and armature assembly from the drive housing.

7. Remove the plunger lever return spring from its seat in the housing. Lift out the plunger and lever assembly off of the armature and out of the frame. Remove the armature.

8. Remove the thrust washer, retainer and stop ring from the end of the armature. Slide the starter drive assembly off.

9. Reach under the brush holder at a point that is 180° from the grommet and carefully lift it out of the frame. Never remove the field coils or brushed unless you intend to replace them!

10. Locate the field coil retaining sleeve on the coil that operates the drive gear, bend up the tab and remove the sleeve.

11. Remove the 3 coil retaining screws with the special tool (Owatonna #10044-A) and a press (which actually just prevents the tool from slipping out of the screw head). Unsolder the coil leads and remove the pole shoes and coils from the frame—a 300 watt iron works well for this.

12. Unsolder (or cut) the brush leads from the field coils as close to the connection as possible. Cut the ground brush leads at the frame.

To assemble:

13. Clean the field coils, armature, armature shaft, commutator, end plate and drive end housing with a soft brush or compressed air. Clean all other components in solvent and dry thoroughly. Remove any residual sealing material.

14. Inspect the armature windings for broken or burned insulation and/or bad connections. Check for open circuits and grounds.

15. Set the commutator in V-blocks and check the run-out. Inspect the shaft and bearing for scoring or excessive wear. If the unit is rough or more than 0.005 in. out-of-round, turn it or replace it.

16. Check the brush holder for any cracks or broken springs. Check the brush spring tension and replace them if tension is not within specifications (40 oz. min.). Replace the holder if cracked or damaged in any way.

17. Replace the brushes if less than 1/4 in. in length.

18. Inspect the field coils for burned or cracker insulation. Check the brush connections and the lead insulation. Brush and contact kits should be available, but everything else will require replacement.

19. Check the starter drive teeth for signs of wear, paying particular attention to the wear pattern. The pinion teeth should penetrate to a depth greater than 1/2 of the ring gear tooth depth so that the risk of early ring gear and starter drive failure is minimized.

20. Position the coils and pole pieces into the housing and then install the retaining screws with the special tool. As you are tightening the screws, tap the frame sharply with a plastic mallet to seat and align the pole shoes. Once finished, stake the screws.

21. Install the solenoid coil and retainer, bending the tabs over the frame.

22. Solder the field coil leads with rosin solder and a 300 watt iron. Always check for continuity before continuing with the assembly.

23. Position new field coil brush leads onto the coil wire and install the clip (provided with the brush). Solder the lead, clip and terminal together.

24. Secure the solenoid coil ground wire under the tab on top of the frame.

25. Solder the ground brush leads to the frame at the point where they were originally cut off.

26. Install the brush springs in the holder so that the flat side is down and they are fully seated.

27. Insert the holder into the frame so that the large lug and notch are located next to the grommet. Make sure that you don't pinch the brush leads between the holder and the frame

28. Coat the armature splines lightly with Extreme Pressure Grease and slide the starter drive onto the armature. Attach a new stop ring and then slide on a new retainer.

29. Remove all grease from the commutator and then slide the armature into the frame. Install the plunger and lever assembly so that the fingers on the lever engage the collar behind the drive.

30. Coat the drive housing bushing lightly with Extreme Pressure Grease and position the armature thrust washer on the bushing. Install the plunger lever return spring into its seat.

31. Install the armature and frame assembly carefully so that the projection on the plunger lever engages the return spring before the frame seats into the drive housing.

32. Insert a punch into the housing hinge and lever holes and wiggle it until the holes are all in alignment and then drive the pin into place. Depress the plunger and then release it; confirm that the plunger arm makes contact with the copper band, the drive gear extends fully, and the spring returns the plunger and drive gear to the fully disengaged position without any hesitation.

33. Ensure that there is no grease on the brushes. Hold back the spring and push the spring into position. The brush spring should snap into position when it's properly positioned behind the brush. Push down on each spring to make sure they are completely seated. Lay the brush leads into the correct channels.

34. Install the brush end plate and then tighten the thru-bolts to 55-75 inch lbs. (6-8 Nm).

35. Apply a small bead of RTV silicone adhesive around the flange on the starter drive cover. Position the cover over the drive plunger so that it engages the 2 projections on the housing. Tighten the cover screw securely. Slide the retaining clamp over the unit and tighten the screw securely.

36. Install the starter.

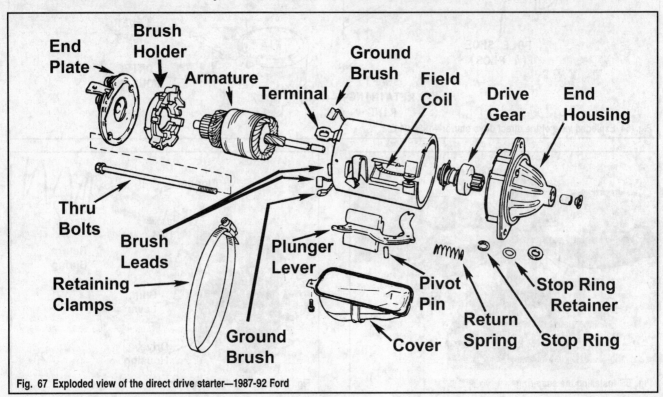

Fig. 67 Exploded view of the direct drive starter—1987-92 Ford

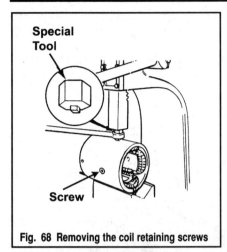

Fig. 68 Removing the coil retaining screws

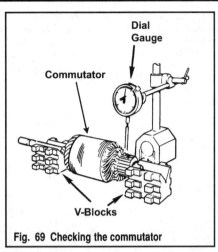

Fig. 69 Checking the commutator

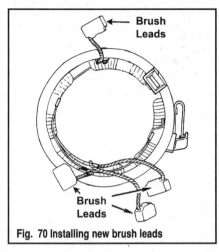

Fig. 70 Installing new brush leads

Gear Reduction—1993-96 Ford Engines And 1995-98 GM Engines

◆ See Figures 71 thru 85

■ OMC provides no facility for rebuilding the GM starter motor after 1995. Although the following procedure will be very similar for a 1995-98 GM starter motor, we strongly recommend checking with your local OMC retailer to confirm parts availability prior to beginning disassembly on a starter installed on a 1995 and later GM engine.

1. Remove the starter.
2. Disconnect the wire brush lead from the brush block to the solenoid **M** terminal. Position it out of the way.
3. Remove the 2 solenoid retaining bolts from the inner side of the drive housing. Pull the solenoid out of the housing slowly while unhooking the end of the plunger from the shift lever.
4. Place a scribe mark or a strip of tape (or paint) on the drive housing, field frame and end plate. The tape or scribe marks will assist in the alignment of the drive housing and field frame during the assembling procedures.
5. Remove the two long thru bolts on each side of the motor case. Slide the bolts out from the end plate. Make sure you remove the bolts with the larger heads, the smaller ones are the brush plate screws and should not be removed yet. Pull the drive housing off of the field frame making sure that you save the seal assembly.
6. Carefully pry the ends of the shift lever from the pins on the drive assembly and remove the lever.
7. Remove the retainer and stop ring from the end of the driveshaft and pull the starter drive assembly from the shaft.
8. Press the snap ring off of the driveshaft. Slide off the washer and then remove the gear assembly from the shaft.
9. Tap the retainer off of the stationary gear in such a way that you do not break the 2 pins on the gear. Lift out the 3 planetary gears from the shaft inside the stationary gear. Rotate the assembly so the shaft is vertical and then tap the gear on a hard surface until the armature thrust ball comes free inside the shaft.
10. Mount the end of the shaft in a vise with soft jaws and then use a small prybar to remove the shaft retainer. Slide the thrust washer off of the shaft and then slide the shaft out of the stationary gear.
11. Remove the two bolts securing the end plate to the brush holder. Pull the positive brush lead seal out of the notch in the frame, slide the holder off the end of the commutator and then pull the armature assembly out of the field frame.

■ The armature will be held in the field frame by permanent magnets.

To assemble:

12. Clean the armature, shaft and brush holder with a brush and compressed air. Do not use a grease dissolving solvent to clean electrical components, planetary gears, or Bendix drive. Solvent will damage wiring insulation and washout the lubricant in bearings and drive gears.

13. Verify the brush holder is not damaged and the brushes are held firmly against the commutator.
14. Inspect the armature commutator for grooves or out of round condition. If the commutator is worn it can be turned down on a lathe. Check the armature for shorts using a growler or test light.
15. Check the commutator for run-out. Inspect the armature shaft and two bearings for scoring or excessive wear with a dial indicator. If the commutator is rough, or more than 0.005 in. (0.12mm) out of round, it must be replaced.
16. Check the roller bearings for wear and rough spots. If there is any roughness or the bearing is stiff, replace the bearing.
17. Check the planetary gear set and the gear teeth for broken or missing teeth. Check to be sure the gears mesh and roll freely without binding or rough spots. If any of these conditions are found, replace the entire planetary gear set.
18. Check the springs in the brush holder to be sure none are broken. Check the spring tension and replace if the tension is not 64 oz. Check the insulated brush holder for cracks that could cause a short. Check the length of each brush; if worn to 0.250 in. (6.35mm) or less, replace them.
19. Coat each end of the armature shaft/splines with a light coating of wheel bearing grease. Slide the armature into the field frame assembly.
20. Install the brush holder assembly into the end of the frame. A brush holder service tool which fits over the end of the commutator while installing the assembly can be fabricated from a piece of scrap wood or metal as shown in the accompanying illustration. Place this tool over the end of the commutator and then slide the holder assembly down over it. As it slides down the tool, the brushes are retracted into the holder. When the holder makes contact with the commutator and is in position, pull out the tool.

If you can't, or won't, fabricate this tool, position the holder against the commutator so that the positive lead and grommet are aligned with the field frame. Use a small screwdriver to push each brush into the holder while tilting the holder against the commutator; as each brush clears the commutator, remove the screwdriver and the brush will release against the commutator. Guide the lead and grommet into the notch in the field frame.

21. Dab some wheel bearing grease into the bearing bore on the end plate and then install the plate onto the end of the frame. Install the small brush screws and tighten them to 20-30 inch lbs. (2.3-3.4 Nm).
22. Apply a coating of wheel bearing grease to the shaft assembly splines and bearing surfaces. to the stationary gear and planetary gears. Slide the gear shaft through the stationary gear and then position the planetary gears onto the posts on the top of the shaft. Rotate the gears until they are fully seated. Squeeze some grease into the center recess of the shaft assembly and install the thrust ball.
23. Position the gear retainer onto the top of the gear assembly so that is rides in the two pins securely.
24. Slide the armature thrust washer over the other end of the shaft and secure it with the snap ring.
25. Slide the drive assembly over the shaft and into position so that it engages the gear. Slide on the stop ring until it sits in the groove on the shaft, follow it with the retainer—press the retainer down until it rides over the stop ring.
26. Stretch the shift lever over the pins on the side of the drive.

27. Coat the drive housing bushing lightly with bearing grease. Slide the drive assembly into the drive housing slightly and then insert the seal, small end down, into the housing notch behind the lever. Position the plastic keeper over the small tab on the gear assembly—the longer side fits down into the drive housing.

28. Carefully slide field frame down and into the drive housing so that the match- marks made previously all line up. Feed the two long thru-bolts through and tighten them to 45-84 inch lbs. 5.1-9.5 Nm).

29. Insert the solenoid into the drive housing, attaching the plunger to the shift lever. Rotate the unit so that the **M** terminal is facing the starter. Tighten the mounting bolts to 45-85 inch lbs. (5.1-9.6 Nm).

30. Attach the positive brush connector to the solenoid terminal and tighten the nut to 80-120 inch lbs. (9-13 Nm).

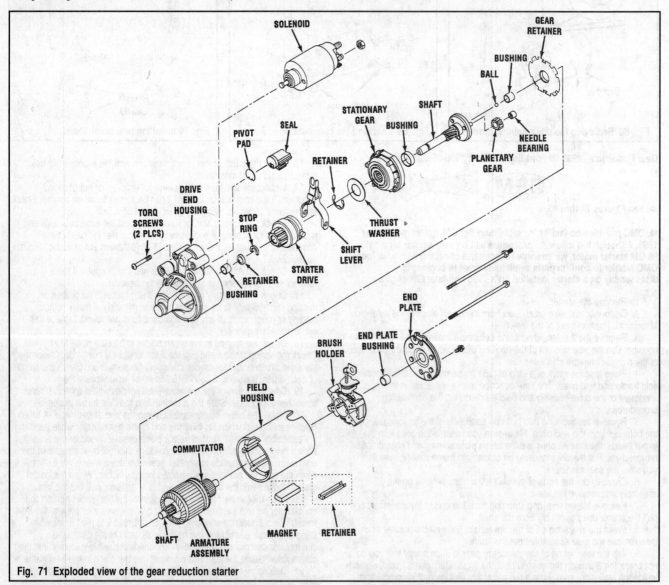

Fig. 71 Exploded view of the gear reduction starter

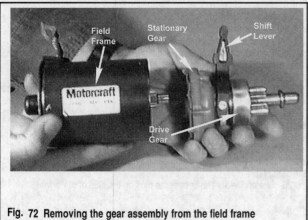

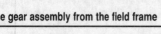

Fig. 72 Removing the gear assembly from the field frame

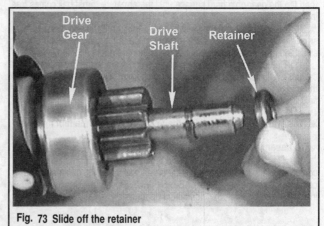

Fig. 73 Slide off the retainer

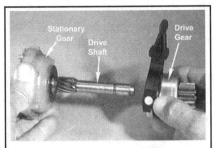

Fig. 74 Pull the drive assembly off of the shaft...

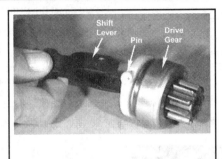

Fig. 75 ...and then remove the shift lever

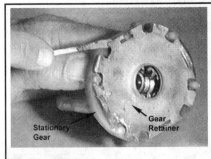

Fig. 76 Pry off the gear retainer...

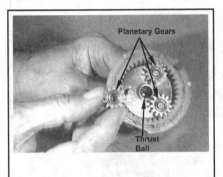

Fig. 77 ...and then remove the planetary gears and thrust ball

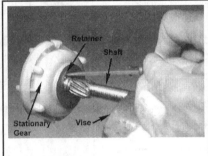

Fig. 78 Pry up the retainer...

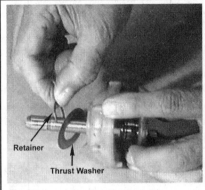

Fig. 79 ...and then remove it and the washer

Fig. 80 Lift out the brush holder...

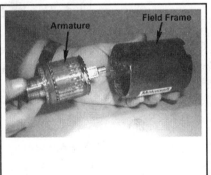

Fig. 81 ...and then remove the armature assembly

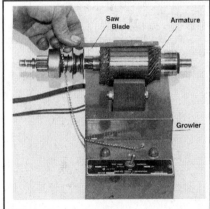

Fig. 82 Testing the armature

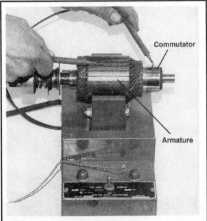

Fig. 83 Testing the armature

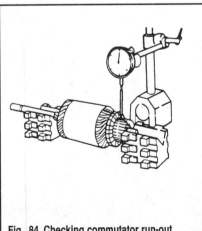

Fig. 84 Checking commutator run-out

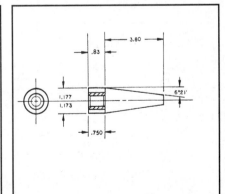

Fig. 85 Fabricating a brush holder installation tool

PINION CLEARANCE

Direct Drive Only—1986-94 GM Engines

◆ **See Figure 86**

Pinion clearance should always be checked after reassembly of the starter to insure proper adjustment.

1. Connect a 6V power source between the **S** terminal on the solenoid and ground.

✳✳ CAUTION

Never use more than 6V or the starter motor my actually operate.

2. Connect a heavy gauge jumper wire to the solenoid motor terminal and quickly touch the other end to the body of the starter so the solenoid energizes and the pinion moves into the cranking position.

3. Push the pinion back toward the commutator end until there is no slack. Measure the gap between the pinion and the retainer. If not 0.010-0.159 in. (0.25-4.04mm) on direct drive models, the solenoid linkage or shift lever yoke buttons are worn and will require replacement; or the shift lever mechanism has been assembled improperly. No adjustment is possible.

Solenoid

REMOVAL & INSTALLATION

GM Direct Drive (1986-94)

1. Remove the starter motor.
2. Remove the screw/washer from the starter connector strap terminal.
3. Remove the two solenoid mounting screws, twist it clockwise until you can remove the flange key from the slot in the starter and then lift off the solenoid. Don't lose the return spring.

To Install:

4. Slide the return spring over the plunger.
5. Position the solenoid at the starter and rotate it until the flange key engages the keyway.
6. Install the two mounting screws and tighten them securely.
7. Install the connector strap and then install the starter.

Ford Direct Drive (1987-92)

1. Disconnect the battery cables.
2. Tag and disconnect the 4 electrical leads at the solenoid. Move them out of the way.
3. Remove the mounting bolts and lift out the solenoid.
4. Install the solenoid and tighten the mounting bolts securely.
5. Reconnect the solenoid leads and the battery cables.

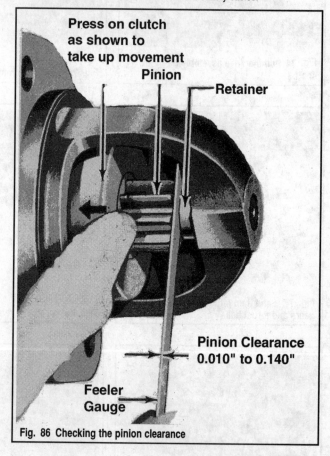

Fig. 86 Checking the pinion clearance

IGNITION SYSTEM

OMC engines covered in this manual utilize one of six different ignition systems: conventional breaker points, Delco Electronic Spark Timing System (EST), Prestolite Breakerless Inductive Distributor (BID), OMC Spitfire/Electronic Engine Management (EMM), Ford Thick Film Integrated Ignition System (TFI-IV), GM High Energy Ignition System (HEI) and

Distributorless Electronic Ignition (DEI). Please refer to the Application chart for details on your specific engine.

Descriptions of each system, and complete repair procedures are detailed under each of the respective system heads.

IGNITION SYSTEM APPLICATIONS

Year	Model	Displacement L/Cu. In.	Engine Type	Ignition System
1986	2.5L	2.5/153	L4	Breaker Point
	3.0L (All)	3.0/181	L4	Breaker Point
	4.3L (All)	4.3/262	V6	Breaker Point
	5.0L (All)	5.0/305	V8	Breaker Point
	5.7L (All)	5.7/350	V8	Breaker Point
	7.5L	7.5/460	V8	Breaker Point
1987	2.3L	2.3/140	L4	Breaker Point
	3.0L (All)	3.0/181	L4	Breaker Point
	4.3L (All)	4.3/262	V6	Breaker Point
	5.0L (All)	5.0/305	V8	Breaker Point
	5.7L (All)	5.7/350	V8	Breaker Point
	7.5L	7.5/460	V8	Breaker Point
1988	2.3L	2.3/140	L4	Breaker Point
	3.0L (All)	3.0/181	L4	Breaker Point
	4.3L (All)	4.3/262	V6	Breaker Point
	5.0L (All)	5.0/305	V8	Breaker Point
	5.7L (All)	5.7/350	V8	Breaker Point
	7.4L 454	7.4/454	V8	Breaker Point
	7.5L 460	7.5/460	V8	Breaker Point
1989	2.3L	2.3/140	L4	Breaker Point
	3.0L (All)	3.0/181	L4	Breaker Point
	4.3L (All)	4.3/262	V6	Breaker Point
	5.0L (All)	5.0/302	V8	Breaker Point
	5.7L (All)	5.7/350	V8	Breaker Point
	7.5L	7.5/460	V8	Breaker Point
1990	2.3L	2.3/140	L4	Breaker Point
	3.0L (All)	3.0/181	L4	HEI
	4.3L	4.3/262	V6	Breaker Point
	4.3L HO	4.3/262	V6	Spitfire/EEM
	5.0L (All)	5.0/302	V8	Breaker Point
	5.7L	5.7/350	V8	Breaker Point
	5.7L LE	5.7/350	V8	Breaker Point
	5.8L	5.8/351	V8	Breaker Point
	350	5.7/350	V8	Spitfire/EEM
	454	7.4/454	V8	Spitfire/EEM
	460	7.5/460	V8	Spitfire/EEM
1991	3.0L (All)	3.0/181	L4	EST
	4.3L	4.3/262	V6	BID
	4.3L HO	4.3/262	V6	Spitfire/EEM
	5.0L (All)	5.0/302	V8	BID
	5.7L (All)	5.7/350	V8	BID
	5.8L (All)	5.8/351	V8	BID
	350	5.7/350	V8	Spitfire/EEM
	7.4GL	7.4/454	V8	Spitfire/EEM
1992	3.0L (All)	3.0/181	L4	EST
	4.3L (All)	4.3/262	V6	BID
	5.0L (All)	5.0/302	V8	BID
	5.7L (All)	5.7/350	V8	BID
	5.8L (All)	5.8/351	V8	BID
	5.8 LE, 351	5.8/351	V8	Spitfire/EEM
	7.4GL, 454	7.4/454	V8	Spitfire/EEM
	8.2GL, 502	8.2/502	V8	Spitfire/EEM

BID	Breakerless Inductive Distributor
EEM	Electronic Engine Management
EST	Electronic Spark Timing
HEI	High Energy Ignition

Fig. 87

IGNITION SYSTEM APPLICATIONS

Year	Model	Displacement L/Cu. In.	Engine Type	Ignition System
1993	3.0L (All)	3.0/181	L4	EST
	4.3L (All)	4.3/262	V6	BID
	5.0L	5.0/302	V8	BID
	5.0 EFI	5.0/302	V8	TFI-IV
	5.8L	5.8/351	V8	BID
	5.8 LE, 351	5.8/351	V8	Spitfire/EEM
	5.8 EFI	5.8/351	V8	TFI-IV
	454, 454 H(	7.4/454	V8	Spitfire/EEM
	8.2GL, 502	8.2/502	V8	Spitfire/EEM
1994	3.0GL/GS	3.0/181	L4	EST
	4.3GL	4.3/262	V6	EST
	5.0FL	5.0/302	V8	BID
	5.0Fi	5.0/302	V8	TFI-IV
	5.7GL	5.7/350	V8	BID
	5.8FL	5.8/351	V8	BID
	5.8Fi	5.8/351	V8	TFI-IV
	351 EFI	5.8/351	V8	TFI-IV
	7.4GL	7.4/454	V8	Spitfire/EEM
1995	3.0GS	3.0/181	L4	EST
	4.3GL	4.3/262	V6	EST
	5.0FL	5.0/302	V8	BID
	5.0 EFI	5.0/302	V8	TFI-IV
	5.8FL	5.8/351	V8	BID
	5.8 EFI/HO	5.8/351	V8	TFI-IV
	7.4Gi	7.4/454	V8	EST-DI
	7.4GL	7.4/454	V8	Spitfire/EEM
1996	3.0GS	3.0/181	L4	EST
	4.3GL/GS	4.3/262	V6	EST
	4.3Gi	4.3/262	V6	EST-DI
	5.0FL	5.0/302	V8	BID
	5.0Fi	5.0/302	V8	TFI-IV
	5.7Gi	5.7/350	V8	EST-DI
	5.8FL	5.8/351	V8	BID
	5.8Fi/FSi	5.8/351	V8	TFI-IV
	7.4Gi	7.4/454	V8	EST-DI
	7.4GL	7.4/454	V8	Spitfire/EEM
1997	3.0GS	3.0/181	L4	EST
	4.3GL/GS	4.3/262	V6	EST
	4.3Gi	4.3/262	V6	EST-DI
	5.7GL/GS	5.7/350	V8	BID
	5.7Gi/GSi	5.7/350	V8	EST-DI
	7.4Gi	7.4/454	V8	EST-DI
	7.4GL	7.4/454	V8	Spitfire/EEM
1998	3.0GS	3.0/181	L4	EST
	4.3GL/GS	4.3/262	V6	EST
	4.3Gi	4.3/262	V6	EST-DI
	5.0GL	5.0/305	V8	BID
	5.7GL	5.7/350	V8	BID
	5.7GSi	5.7/350	V8	EST-DI
	7.4Gi/GSi	7.4/454	V8	EST-DI
	8.2GSi	8.2/502	V8	EST-DI

BID	Breakerless Inductive Distributor
EEM	Electonic Engine Management
EST	Electronic Spark Timing
EST-DI	Electronic Spark Timing - Distributor Ignition
TFI-IV	Thick Film Integrated - IV

Fig. 88

BREAKER POINT IGNITION SYSTEM

■ Please refer to Maintenance section for complete breaker point gap, dwell and timing adjustment procedures.

Description

Used primarily on all engines through 1989, and a few engines in 1990; this ignition system consists of a primary and a secondary circuit. The low-voltage current of the ignition system is carried by the primary circuit. Components of the primary circuit include the ignition switch, ballast resistor, neutral-safety switch, primary winding of the ignition coil, contact points in the distributor, condenser and the low-tension wiring.

The secondary circuit carries the high-voltage surges from the ignition coil that result in a high-voltage spark between the electrodes of each spark plug. The secondary circuit includes the secondary winding of the coil, coil-to-distributor high-tension lead, distributor rotor and cap, ignition cables, and the spark plugs.

When the contact points are closed and the ignition switch is **ON**, current from the battery or from the alternator flows through the primary winding of the coil, through the contact points to ground. The current flowing through the primary winding of the coil creates a magnetic field around the coil windings and energy is stored in the coil. Now, when the contact points are opened by rotation of the distributor cam, the primary circuit is broken. The current attempts to surge across the gap as the points begin to open, but the condenser absorbs the current. In so doing, the condenser creates a sharp break in the current flow and a rapid collapse of the magnetic field in the coil. This sudden change in the strength of the magnetic field causes a voltage to be induced in each turn of the secondary windings in the coil.

The ratio of secondary windings to the primary windings in the coil increases the voltage to about 20,000 volts. This high voltage travels through a cable to the center of the distributor cap, through the rotor to an adjacent distributor cap contact point, and then on through one of the ignition wires to a spark plug.

When the high-voltage surge reaches the spark plug it jumps the gap between the insulated center electrode and the grounded side electrode. This high voltage jump across the electrodes produces the energy required to ignite the compressed air/fuel mixture in the cylinder.

The entire electrical build-up, breakdown, and transfer of voltage is repeated as each lobe of the distributor cam passes the rubbing block on the contact breaker arm, causing the contact points to open and close. At high engine rpm operation, the number of times this sequence of actions takes place is staggering.

Beginning at the key switch, current flows to the ballast resistor and then to the positive side of the coil. When the resistor is cold its resistance is approximately one ohm. The resistance increases in proportion to the resistor's rise in temperature.

While the engine is operating at idle or slow speed, the cam on the distributor shaft revolves at a relatively slow rate. Therefore, the breaker points remain closed for a slightly longer period of time. Because the points remain closed longer, more current is allowed to flow and this current flow heats the ballast resistor and increases its resistance to cut down on current flow thereby reducing burning of the contact points.

During high rpm engine operation, the reduced current flow allows the resistor to cool enough to reduce resistance, thus increasing the current flow and effectiveness of the ignition system for high-speed performance.

The voltage drops about 25% during engine cranking due to the heavy current demands of the starter. These demands reduce the voltage available for the ignition system. In order to reduce the problem of less voltage, the ballast resistor is by-passed during cranking. This releases full battery voltage to the ignition system.

The shift cut-out switch, sometimes referred to as an interrupt switch, is connected between the primary side of the ignition coil and ground. This switch is normally open. The function of this switch is to ground the ignition system during a shift to neutral. By grounding the ignition system, gear pressure is released and the shift is made much easier. Obviously, if the ignition system is grounded, the cylinders will not fire during this period. In actual practice only a few cylinders fail to fire and it is usually not noticeable. The shift cutout switch is mounted on the transom and is activated by the remote control shift cable. If this switch is not adjusted properly, or if it shorts out (is grounded) then the primary side of the ignition coil will be grounded and the engine will not start.

In order to obtain the maximum performance from the engine, the timing of the spark must vary to meet operating conditions. For idle, the spark advance should be as low as possible. During high-speed operation, the spark must occur sooner, to give the air/fuel mixture enough time to ignite, burn, and deliver power to the piston for the power stroke.

Manual setting and centrifugal advance are the two methods of obtaining the constantly changing demands of the engine. The manual setting is made at idle speed. This setting allows the contact points to open at a specified position of the piston in the same manner as with conventional ignition systems.

Troubleshooting

Any problem in the ignition system must first be localized to the primary or secondary circuit before the defective part can be identified.

GENERAL TESTS

◆ See Figure 89

Disconnect the ignition coil wire from the center of the distributor cap and hold it about 1/4 in. from a good engine ground. Turn the ignition switch to the **ON** position, and crank the engine with the starter. If you observe a good spark, go to the Secondary Circuit Test. If you do not have a good spark, go to the Primary Circuit Test.

Primary Circuit Test

◆ See Figure 90

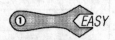

Remove the distributor cap; lift off the rotor; and then turn the crankshaft until the contact points close. Turn the ignition switch to the **ON** position and open and close the contact points using a small screwdriver or a non-metallic object. Hold the high-tension coil wire about 1/4 in. from a good engine ground. If you notice a good spark jump from the wire to the ground, the primary circuit checks out; proceed to the Secondary Circuit Test. If there is no spark, proceed to the Contact Point Test.

Contact Point Test

◆ See Figure 91

Remove the distributor cap and rotor. Rotate the crankshaft until the points are open, and then insert some type of insulator between the points. Now, hold the high-tension coil wire about 1/4 in. from a good engine ground, and at the same time move a small screwdriver up and down with the screwdriver shaft touching the moveable point and the tip making intermittent contact with the contact point base plate. In this manner, the screwdriver is being used for a set of contact points. If a spark is observed from the high-tension wire to ground, then the problem is in the contact points. Replace the set of points. If there is no spark from the high-tension wire to ground, the problem is either a defective coil or condenser. To test the condenser, see the Condenser Test.

Condenser Test

◆ See Figure 91

Condensers seldom cause a problem. However, there is always the possibility one may short out and ground the primary circuit. Before testing the condenser, check to be sure one of the primary wires or connections inside the distributor has not shorted out to ground.

The most accurate method of testing a condenser is with an instrument manufactured for that purpose. However, seldom is one available, especially during an emergency. Therefore, the following procedure is outlined for emergency troubleshooting the condenser and the primary circuit insulation for a short.

First, remove the condenser from the system. Take care that the metallic case of the condenser does not touch any part of the distributor. Next, insert a piece of insulating material between the contact points. Now, move the blade of a small screwdriver up and down with the shaft of the screwdriver making contact with the movable contact point and the tip making and breaking contact with the contact point base plate. Observe for a low-tension

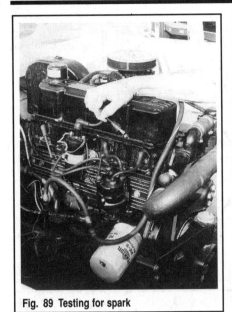

Fig. 89 Testing for spark

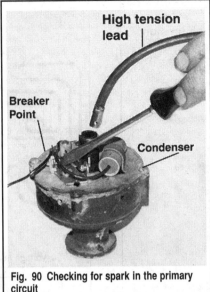

Fig. 90 Checking for spark in the primary circuit

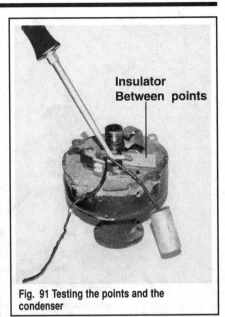

Fig. 91 Testing the points and the condenser

spark between the tip of the screwdriver and the contact point base plate as you make and break the contact with the screwdriver tip. You should observe a spark during this test and it will prove the primary circuit complete through the neutral-safety switch, the primary side of the ignition coil, the ballast resistor, the shift cut-out switch, and the primary wiring inside the distributor. If you have a spark, reconnect the condenser and again make the same test with the screwdriver. If you do not get a spark, either the condenser is defective and should be replaced, or the shift control switch should be adjusted or replaced.

If you were unable to get a spark with the condenser disconnected, it means no current is flowing to this point, or there is a short circuit to ground. Use a continuity tester and check each part in turn to ground in the same manner as you did at the movable contact point. If you get a spark, indicating current flow, at one terminal of the part, but not at the other, then you have isolated the defective unit.

Secondary Circuit Test

 MODERATE

◆ See Figure 92

The secondary circuit should not be tested using emergency troubleshooting procedures unless the primary circuit has been tested and proven satisfactory, or any problems discovered in the primary circuit have been corrected.

If the primary circuit tests are satisfactory, use the same procedures as outlined in the Primary Circuit Test to check the secondary circuit. Hold the high-tension coil lead about 1/4 in. from a good ground and at the same time, use a small screwdriver to open and close the contact points. A spark at the high-tension lead proves the ignition coil is good. However, if the engine still fails to start and the problem has been traced to the ignition system, then the defective part or the problem must be in the secondary circuit.

The distributor cap, the rotor, high-tension leads, or the plugs may require attention or replacement. To test the rotor, go to the Rotor Test. If you were unable to observe a spark during the Secondary Circuit test just described, the ignition coil is defective and must be replaced.

Rotor Test

EASY

◆ See Figure 93

With the distributor cap removed and the rotor in place on the distributor shaft, hold the high-tension coil lead about 1/4 in. from the rotor contact spring, and at the same time crank the engine with the ignition switch turned on. If a spark jumps to the rotor, it means the rotor is shorted to ground and must be replaced. If there is no spark to the rotor, it means the insulation is good and the problem is either in the distributor cap (check it for cracks), in the high-tension leads (check for poor insulation or replace it), or in the spark plugs (replace them).

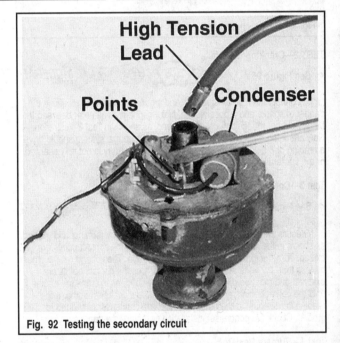

Fig. 92 Testing the secondary circuit

IGNITION VOLTAGE TESTS

Many times hard starting and mis-firing problems are caused by defective or corroded connections. Such a condition can lower the available voltage to the ignition coil. Therefore, make voltage tests at critical points to isolate such a problem. Move the voltmeter test probes from point-to-point in the following order.

Test 1—Voltage Loss Across Entire Ignition Circuit

 MODERATE

◆ See Figure 94

Connect a voltmeter between the battery side of the ignition coil and the positive post of the battery, as shown in the Test 1 illustration. Crank the engine until the contact points are closed. Turn the ignition switch to the **ON** position. The voltage loss should not exceed 3.2 volts. This figure allows for a 0.2 loss across each of the connections in the circuit, plus a calibrated 2.4 volt drop through the ballast resistor. If the total voltage loss exceeds 3.2 volts, then it will be necessary to isolate the corroded connection in the circuit of the key resistor, the wiring, or at the battery.

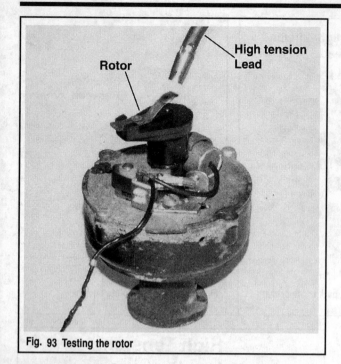

Fig. 93 Testing the rotor

TEST 2—Cranking System

◆ See Figure 95

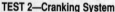

Disconnect the high tension wire and ground it securely to minimize the danger of sparks and a possible fire. Next, connect a voltmeter between the battery side of the ignition coil and ground, as shown in Test 2 illustration. Now crank the engine and check the voltage. A normal system should have a reading of 8.0 volts. If the voltage is lower, the battery is not fully charged or the starter is drawing too much current.

Test 3—Contact Points and Condenser

◆ See Figure 96

Measure the voltage between the distributor primary terminal and ground, as shown in the illustration. Crank the engine until the contact points are closed. Turn the ignition switch to the **ON** position. The voltage reading must be less than 0.2 volts. A higher reading indicates the contact points are oxidized and must be replaced. To check the condenser, crank the engine until the contact points are open, and then take a voltage reading. If the reading is not equal to the battery voltage, the condenser is shorted to ground. Check the condenser installation or replace the condenser.

Test 4—Primary Resistor

◆ See Figure 97

Disconnect the battery wire at the primary resistor to prevent damage to the ohmmeter. Connect an ohmmeter across the terminals of the resistor, as shown in the illustration. The specified resistance is between 1.3 and 1.4 ohms. If the reading does not fall within this range, replace the resistor.

Test 5—Voltage Loss In The Ignition Switch, Ammeter, Or Battery Cable

◆ See Figure 98

Crank the engine until the points are closed. Connect the voltmeter to the battery post (not the cable terminal) and to the load side of the ignition switch, as shown in the illustration. Now, turn the ignition switch to the **ON** position and note the voltage reading. The meter reading should not be more than 0.8 volts. A 0.2 volt drop across each of the connections is permitted.

If the voltage drop is more than 0.8 volts, move the test probe to the "hot" side of the ammeter. If the reading is 0.4 volts, the ignition switch is satisfactory. Once the corroded connection has been located, remove the nut, clean the wire terminal and connector, and then tighten the connection securely.

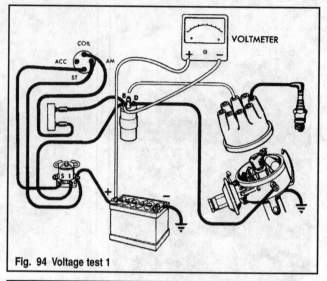

Fig. 94 Voltage test 1

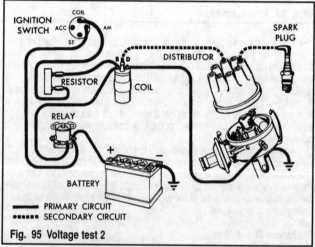

Fig. 95 Voltage test 2

Test 6—Distributor Condition

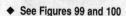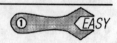

The condition of the distributor can be quickly and conveniently checked with a timing light.

Under normal timing light procedures the trigger wire from the timing light is attached to the spark plug wire of the No. 1 cylinder. In this test, connect the trigger wire to the fifth cylinder in the firing order for a V8 engine.

The timing mark and the pointer should align in the same position as it did with the number one cylinder. If there is a variation of a few degrees, the distributor shaft bushings or cam lobes may be worn and the condition will have to be corrected.

Before setting the timing, make sure the point dwell is correct. Take care to aim the timing light straight at the mark. Sighting from an angle may cause an error of two or three degrees.

Distributor Cap And Rotor

REMOVAL & INSTALLATION

◆ See Figures 99 and 100

■ It is not necessary to remove the spark plug and ignition coil leads to remove the cap. If you do remove them, be sure to carefully mark each lead and tag it for reinstallation in the proper cap socket.

1. Loosen the two cap retaining screws and carefully lift off the cap. It never hurts to paint a small mark across the cap and distributor body to aid in installation. There are a few models out there using only one retaining screw.

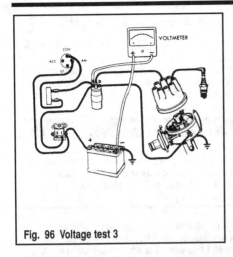

Fig. 96 Voltage test 3

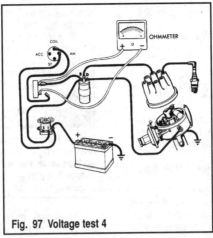

Fig. 97 Voltage test 4

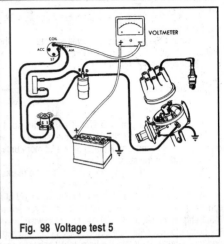

Fig. 98 Voltage test 5

2. Paint a small mark on the distributor body at the end of the rotor and then pull the rotor up and off the shaft.

3. If you have removed the electrical leads from the cap, clean it in warm, soapy water and allow it to dry completely. Check the cap for cracks or other damage. Marine caps should have brass contacts—if yours has aluminum contacts, replace it as someone has previously installed an automotive cap.

4. Check the rotor for cracks and other damage.

5. Install the rotor onto the shaft so that the end is pointing to the mark on the distributor base made earlier.

6. Install the cap onto the distributor. There should be locating tabs cut into the distributor housing to aid installation; you've also got the mark you made previously. Tighten the retaining screws securely.

Distributor

REMOVAL

■ It is not necessary to remove the spark plug leads to remove the cap. If you do remove them, be sure to carefully mark each lead and tag it for reinstallation in the proper cap socket.

1. Disconnect the distributor primary lead at the ignition coil terminal and then remove the cap and rotor.

2. Bump the engine over so that the No. 1 cylinder's piston is at TDC.

3. Scribe or paint a small alignment mark on the distributor body at the rotor notch on the distributor shaft. Scribe or paint a small mark across the distributor housing and cylinder block.

4. Loosen and remove the distributor clamp bolt and then slowly lift the distributor out of the cylinder block.

■ The distributor will actually rotate slightly as you lift it out, so don't be surprised or concerned.

INSTALLATION

Engine Not Disturbed

1. Install the rotor onto the shaft so that the marks made earlier align.

2. Install a new gasket on the housing.

3. Grasp the rotor and spin the shaft about 1/8 of a turn, counterclockwise, past the alignment mark (if you are working on a Ford 7.5L (460) engine, position the rotor 1/8 turn clockwise before the mark). Position the distributor over the hole in the block so that the alignment marks match and then work it down into the hole until it engages completely with the oil pump shaft.

■ It is OK to wiggle the rotor slightly during installation so that the distributor gear meshes properly with the oil pump. It is imperative though that the rotor ends up pointing to the alignment mark made during removal when installation is complete.

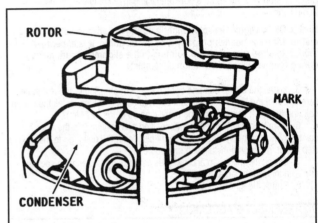

Fig. 99 Paint a small mark on the distributor housing for rotor alignment

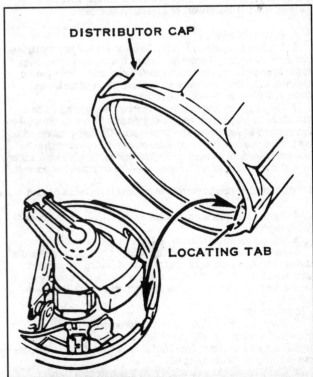

Fig. 100 Most models will have locating tabs cut into the housing to aid in installing the cap

4. Install the clamp bolt and tighten it securely, but no more than 20 ft. lbs. (27 Nm).

5. Check the point gap and dwell as detailed in the Maintenance section and then install the cap and coil lead.

6. Check the ignition timing.

Engine Disturbed

1. Set the No. 1 piston to TDC. Remove the No. 1 spark plug, put your thumb over the hole and slowly crank the engine until you feel compression on the No. 1 cylinder. Crank the engine a bit more until the pointer lines up with the mark on the scale.

2. Install the rotor onto the shaft so that the marks made earlier align.

3. Install a new gasket on the housing.

4. Grasp the rotor and spin the shaft about 1/8 of a turn, counterclockwise, past the alignment mark (if you are working on a Ford 7.5L (460) engine, position the rotor 1/8 turn clockwise before the mark). Position the distributor over the hole in the block so that the alignment marks match and then work it down into the hole until it engages completely with the oil pump shaft.

■ **It is OK to wiggle the rotor slightly during installation so that the distributor gear meshes properly with the oil pump. It is imperative though that the rotor ends up pointing to the alignment mark made during removal when installation is complete.**

5. Install the clamp bolt and tighten it finger-tight. Rotate the distributor carefully until the breaker points just begin to open and then install the clamp bolt and tighten it securely, but no more than 20 ft. lbs. (27 Nm).

6. Install the distributor cap into position and confirm that the rotor points to the terminal for the No. 1 spark plug lead. If it does, go to the next step; if not, repeat the installation procedure.

7. Check the point gap and dwell as detailed in Maintenance section and then install the cap and coil lead.

8. Check the ignition timing.

DISASSEMBLY & ASSEMBLY

◆ **See Figure 101, 102 and 103**

■ **Engines using a breaker point ignition system may have been equipped with a Delco, Mallory or a Prestolite distributor.**

1. Remove the distributor.

2. Disconnect the primary wire and the condenser lead from the breaker point assembly terminal. Remove the breaker point assembly by removing the two attaching screws. Remove the condenser attaching screw and the condenser. On Mallory distributors, you will have to remove the terminal screw running through the side of the housing.

3. Pull the primary lead through the opening in the housing. Scribe a mark across the breaker plate and housing and then remove the two breaker plate attaching screws. Remove the breaker plate. On Mallory models, make sure that you take note of the relationship between the curved notch in the plate and the inside of the housing. On Delco models, there is a small rubber grommet in the plate holding the primary wire— do not disturb this grommet.

■ **Further disassembly of the distributor body is not recommended.**

4. Remove the thrust washer installed only on certain counterclockwise rotating engines.

5. Remove the weight retainers, and then remove the weights.

6. Remove the distributor cap clamps. Scribe a mark on the gear and a matching mark on the distributor shaft as an aid in locating the pin holes during assembly. Place the distributor shaft in a V-block, and then use a drift punch to remove the roll pin. Remove the gear from the shaft. Remove the shaft collar roll pin.

To assemble:

7. Never wash the distributor cap, rotor, condenser, or breaker plate assembly of a distributor in any type of cleaning solvent. Such compounds may damage the insulation of these parts or, in the case of the breaker plate assembly, saturate the lubricating felt.

8. Check the shaft for wear and fit in the distributor body bushings. If either the shaft or the bushings are worn, replace the shaft and distributor body as an assembly. Use a set of V-blocks and check the shaft alignment

with a dial gauge. If the run-out is more than 0.002 in., the shaft and body must be replaced.

9. Inspect the breaker plate assembly for damage and replace it if there are signs of excessive wear.

10. Check to be sure the advance weights fit free on their pins and do not have any burrs or signs of excessive wear. Check the cam fit on the end of the shaft. The cam should not fit loose but it should still be free without binding.

✳✳WARNING

Marine distributors have a corrosion resistant coating applied to the return spring on top of the breaker plate and on the two small weight springs under the plate. For this reason, as well as other good reasons, automotive parts—point sets, springs, or other distributor parts—should never be used as a replacement.

11. Always replace the points with a new set during a distributor overhaul. The condenser seldom gives trouble, but good shop practice dictates a new condenser with a new set of points. Some point sets still include a condenser in the package. If you have paid for a new condenser, you might as well install it and be free of concern over that part.

12. Lubricate the distributor shaft with crankcase oil, and then slide it into the distributor body. Slide the collar onto the shaft; align the holes in the collar with the hole in the shaft; and then install a new pin.

13. Install the distributor cap clamps. Left-hand rotating engine distributor assemblies may have an additional thrust washer between the collar and the base. Use a feeler gauge between the collar and the distributor base to check the shaft end-play. The end-play should be between 0.001-0.010 in.

14. Install the gear onto the shaft with the marks, on the gear and the shaft made during disassembly, aligned. The holes through the gear and the shaft should be aligned after the gear is installed. Install the gear roll pin.

15. Fill the grooves in the weight pivot pins with distributor cam lubricant. Position the weights in the distributor with the weight you identified with a mark during disassembly matched with the marked pivot pin. Secure the weights in place with the retainers. Slide the thrust washer onto the shaft. Fill the upper distributor shaft grooves with distributor cam lubricant.

16. Install the cam assembly with the marked spring bracket near the marked spring bracket on the stop plate. If a new cam assembly is being installed, take care to be sure the cam is installed with the hypalon-covered stop in the correct cam plate control slot. The proper slot can be determined by measuring the length of the slot used on the old cam' and then using the corresponding slot on the new cam. Some new cams will have the size of the slot stamped in degrees near the slot. If the wrong slot is used, the maximum advance will not be correct.

17. Coat the distributor cam lobes with a light film of distributor cam lubricant. Install the retainer and wick. Use a few drops of SAE 10W engine oil on the wick. Install the weight springs, with the spring and bracket marked during disassembly, matched.

18. Place the breaker plate in position and then secure with the attached screws. Remember on Mallory models that the curved notch should be spanning the hole where the terminal screw goes through the housing. On Delco models, make sure that the small rubber grommet in seated correctly in the plate and the notch in the housing.

19. Push the primary wire through the opening in the distributor. Place the breaker point assembly and the condenser in position and secure them in place with the attaching screws. Don't forget the terminal screw on the Mallory! Connect the primary wire and the condenser lead to the breaker point primary terminal.

20. Adjust the point gap and dwell as outlined in the following procedures, then install the distributor.

Ignition Coil

REMOVAL & INSTALLATION

1. Disconnect the negative battery cable.
2. Disconnect the high tension lead coming from the distributor.
3. Tag and disconnect the coil electrical leads.
4. Remove the coil mounting bolts/screws and lift out the coil.
5. Install the coil and tighten the mounting hardware securely.
6. Connect the electrical leads and the high tension lead.
7. Connect the battery cable.

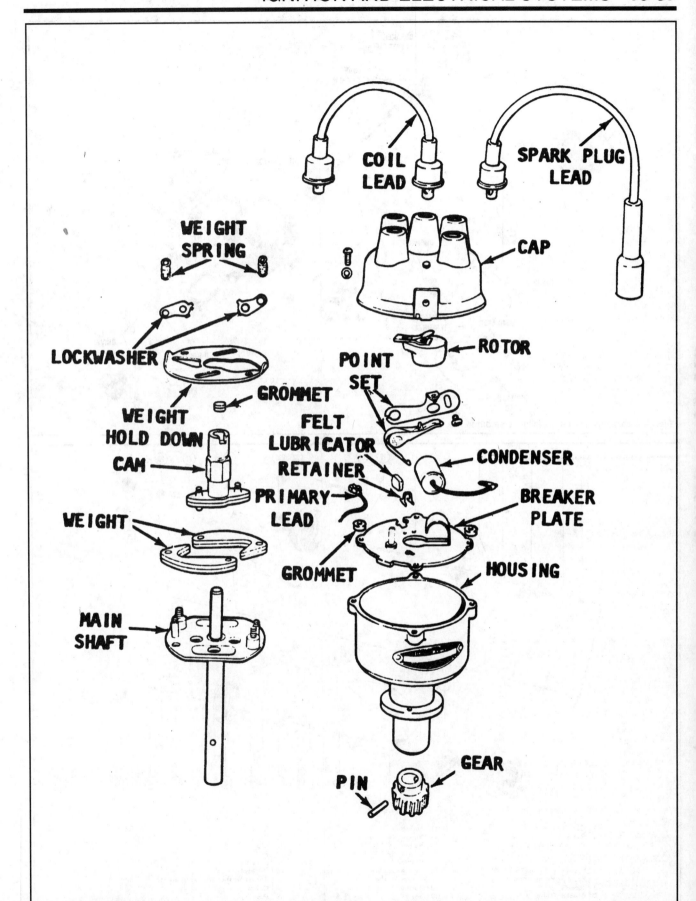

Fig. 101 Exploded view of a Delco distributor

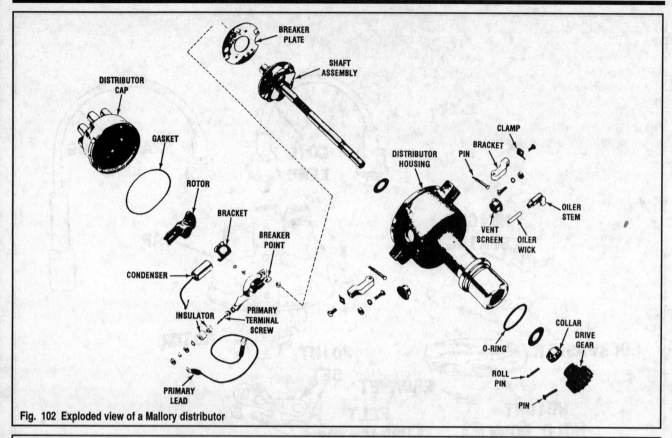

Fig. 102 Exploded view of a Mallory distributor

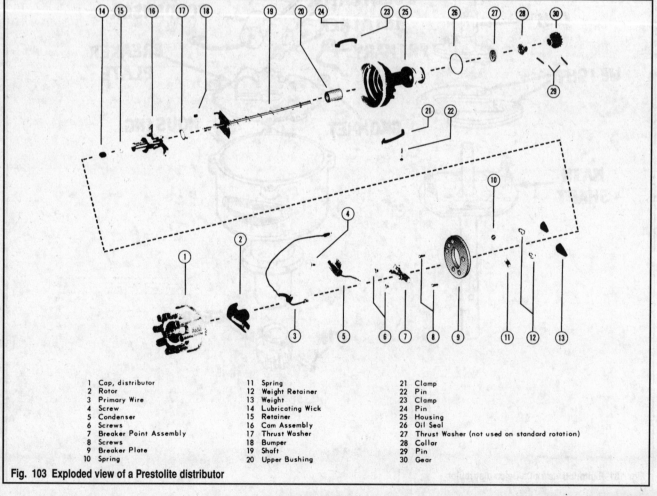

1	Cap, distributor	11	Spring	21	Clamp
2	Rotor	12	Weight Retainer	22	Pin
3	Primary Wire	13	Weight	23	Clamp
4	Screw	14	Lubricating Wick	24	Pin
5	Condenser	15	Retainer	25	Housing
6	Screws	16	Cam Assembly	26	Oil Seal
7	Breaker Point Assembly	17	Thrust Washer	27	Thrust Washer (not used on standard rotation)
8	Screws	18	Bumper	28	Collar
9	Breaker Plate	19	Shaft	29	Pin
10	Spring	20	Upper Bushing	30	Gear

Fig. 103 Exploded view of a Prestolite distributor

ELECTRONIC SPARK TIMING (EST) SYSTEM

DISTRIBUTOR IGNITION (DI) SYSTEM

Description

◆ See Figures 104, 105 and 106

Basic components of this system include a pointless distributor, a remote ignition coil, an electronic ignition module and a pick-up coil. The distributor does not contain breaker points, a condenser or a centrifugal advance arrangement. As the name of the system implies, spark advance and ignition timing are handled electronically. There are little to no differences between the two systems; EST is used on carbureted engines, while DI is used on EFI engines.

The distributor utilizes an internal magnetic pick-up assembly consisting of a permanent magnet, a pole piece with internal teeth and the actual pick-up coil. The pick-up coil itself is sealed against moisture and electro-mechanical interference. When the rotating teeth of the timer core (which is attached to the top of the distributor shaft) line up with the internal teeth on the pole piece, voltage is induced in the pick-up coil. The coil then sends a signal to the ignition module; thus triggering the primary ignition circuit. Current flow in the primary circuit is interrupted and voltage as high as 35,000 volts is induced in the secondary windings of the ignition coil; this voltage is then through the secondary ignition circuit to fire the spark plugs.

The distributor is connected to the ignition coil by a high tension secondary wire and 2 low voltage primary wires. Because of this high voltage, a special thermoplastic, injection molded, glass-reinforced distributor cap is used.

A timer core (trigger wheel) is mounted near the upper end of the distributor shaft. This "wheel" has "teeth"—the number corresponding to the number of engine cylinders. An ignition module is mounted close to the wheel.

The ignition coil produces greater spark voltage, longer spark, and operates at a much higher rpm than normal. The secondary windings are wrapped around the primary windings, which are in turn wrapped around an iron core. The coil generates a very high secondary voltage of up to 35,000 V when the primary circuit has been broken. There are a pair of 2-wire connectors used for:

- Battery voltage input
- Primary voltage to the ignition module
- Ignition module trigger signal input
- Tachometer output signal

The ignition module is located inside the distributor and is a solid state unit that uses transistorized relays and switches to control any associated circuits. The module serves the following functions:

- Changing the voltage signal from the pick-up coil to a "square" digital signal

- Sending the digital signal as a reference signal (REF HI) for ignition control
- Providing a ground reference signal (REF LO)
- Providing a means to control spark advance (Ignition mode and/or Module mode)
- Providing a trigger signal for the ignition coil

The pick-up coil assembly consists of the stationary pole piece and a pick-up coil and magnet; which rides inside the pole piece. The coil produces an alternating voltage signal as the teeth pass the magnet. A signal is produced for each cylinder of the engine during one complete revolution of the engine. The coil assembly is connected to the ignition module with a 2-wire connector.

Troubleshooting

◆ See Figure 107 and 108

✷✷WARNING

This Delco EST breakerless ignition system requires the use of a jumper wire—OMC 986662—or one may be fabricated using a six inch piece of 16AWG wire and a male bullet terminal on each end. This jumper is required to shunt (turn off), the electronic spark advance function in the module while setting initial timing.

✷✷ CAUTION

Always ground the high tension lead of the coil any time it is disconnected from the distributor. Failure to disconnect the lead could result in fire or an explosion, if gas vapors are present. Reason: A high voltage discharge in the secondary circuit of the coil may occur when the ignition switch is turned to the on and off positions. This high voltage discharge may occur even if the engine is not cranked.

Before spending too much time and money attempting to trace a problem to the ignition system, a compression check of the cylinders should be made. If the cylinders do not have adequate compression, troubleshooting and attempted service of the ignition or fuel system will fail to give the desired results of satisfactory engine performance.

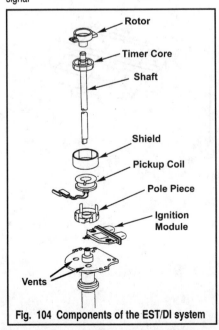

Fig. 104 Components of the EST/DI system

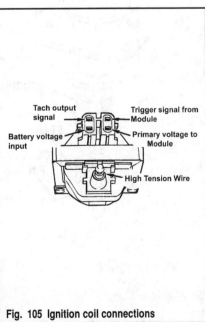

Fig. 105 Ignition coil connections

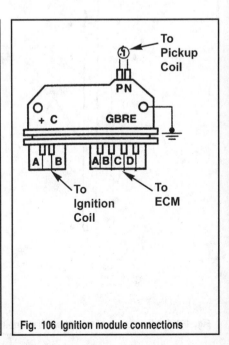

Fig. 106 Ignition module connections

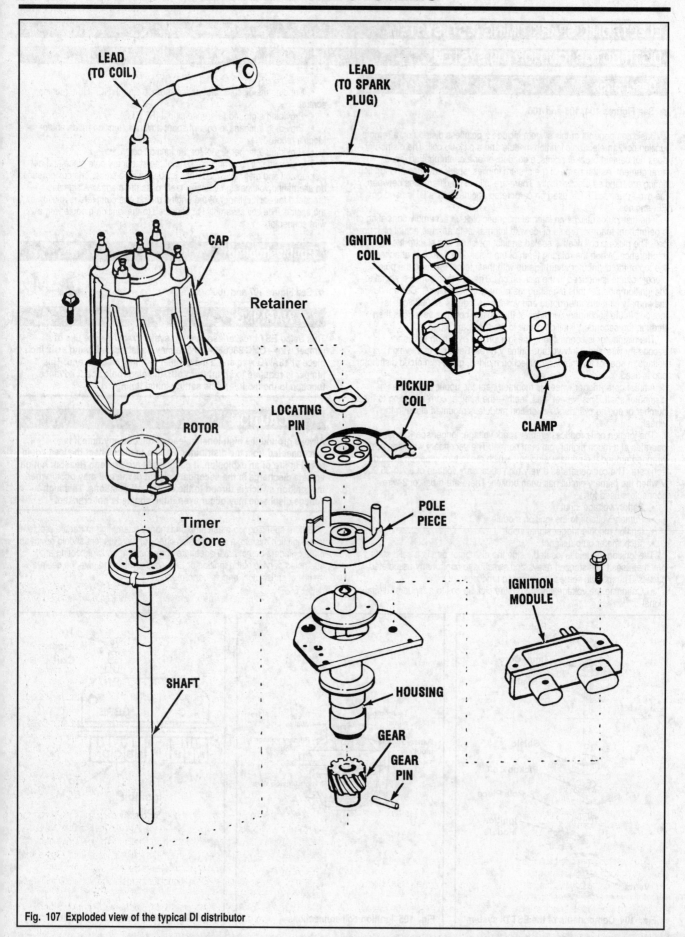

Fig. 107 Exploded view of the typical DI distributor

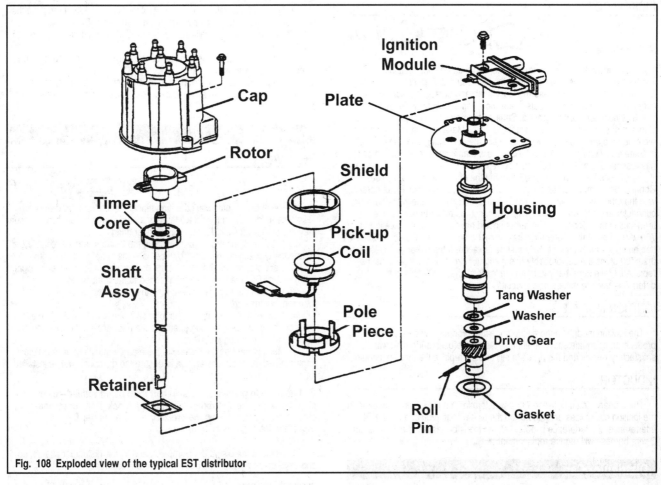

Fig. 108 Exploded view of the typical EST distributor

VISUAL CHECKS

Check the coil tower for carbon tracking. Check the terminals for secure connections. Verify the polarity is correct. Check the coil nipple to be sure it is sealed and insulated. If flashover should occur at this location, the engine will fail to start.

Check the distributor cap for carbon tracking. Clean the cap if it is dirty. Moisture and dirt make a good path for flashover. If a carbon track has started, the cap must be replaced. Check the rotor for carbon tracking and cleanliness. Again, if carbon track has started, the rotor must be replaced.

Check the high tension leads to the spark plugs for burning, cracking and any type of deterioration. Check the spark plug for fouling, proper gap and possible cracked insulator.

12 VOLT (B+)

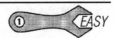

◆ See Figure 109

1. Connect a multi-meter as per the manufacturers instructions.
2. Disconnect the purple and gray wire connector at the ignition coil.
3. Connect the positive lead on the meter to the terminal for the purple wire in the connector; connect the negative lead to ground.
4. Turn the ignition switch to the **ON** position and check that the meter reads at least 8 V.
5. Plug the connector back into the coil and disconnect the pink and brown connector at the distributor.
6. Connect the positive lead on the meter to the terminal for the pink wire in the connector; connect the negative lead to ground.
7. Turn the ignition switch to the **ON** position and check that the meter reads at least 8 V.

IGNITION COIL

◆ See Figure 109

✱✱ CAUTION

Disconnect both connectors from the coil before performing the following test.

1. Check for short to ground: Connect one meter lead to the frame as a ground. Make contact with the other meter lead to the Purple wire terminal on the coil. The meter should indicate an infinite reading. If the meter indicates less than an infinite reading, the ignition coil has a short to ground. The coil must be replaced.
2. Check for an open or short in the Primary circuit: Set the ohmmeter to the low scale. Connect one meter lead to the Purple wire terminal. Make contact with the other meter lead to the Gray wire terminal for the tach lead. The meter should indicate 0.35-0.45 ohms. If the meter indicates higher ohms, the primary circuit probably has an open. If the meter reading is low, the primary circuit has a short. Either condition requires that the coil must be replaced.
3. Check for an open or short in the Secondary circuit. Set the meter to the high scale. Connect one meter lead to the Purple wire terminal. Make contact with the other meter lead to the high tension terminal. The meter should indicate 7500-9000 ohms. If the meter indicates less than 7500 ohms, the circuit has a short. If the meter indicates more than 9000 ohms, the circuit probably has an open. Either condition demands that the coil must be replaced.
4. Install the black coil connector (distributor harness) first and then install the gray connector (engine harness).

PICK-UP COIL

◆ **See Figure 110**

1. Connect a multi-meter as per the manufacturers instructions.
2. Remove the screws securing the distributor cap, and then remove the cap and the rotor. Release the locking tab and unplug the pick-up coil connector. On almost all models these two wires should be white and green.
3. Check for short to ground: Set the ohmmeter to the high scale. Connect one meter lead to the distributor body. Make contact with the other meter lead to either of the wire terminals in the connector. The meter should indicate an infinite reading. A reading of less than infinite, indicates the coil has a shorted circuit. The coil must be replaced.
4. Check for open or shorted coil: Set the ohmmeter to the high scale. Connect one meter lead to the green wire pickup coil terminal. Make contact with the other meter lead to the white wire coil terminal. If the wires on your connector are a different color, simply connect the meter to each of the terminals in the connector. The meter should have a constant indication of 700-900 ohms. If the meter indicates more than 900 ohms, the coil probably has an open circuit and the coil must be replaced. If the meter indicates less than 700 ohms the coil probably has a shorted circuit and the coil must be replaced. Make sure that you bend the leads slightly while testing to determine any intermittent open circuits.

IGNITION MODULE

The ignition module in the distributor either produces a spark or it fails to produce an adequate spark. If all other tests have been performed with satisfactory results and the problem still exists—replace the ignition module.

INDUCTOR

The primary circuit between the 2-wire terminal on the ignition module and the ignition coil utilizes an inductor on the pink wire to protect against RFI interference. If interference is detected, the inductor has failed and the entire 2-wire harness will require replacement.

Distributor Cap

REMOVAL & INSTALLATION

◆ **See Figure 111 and 112**

■ **It is not necessary to remove the spark plug leads at the cap when removing it. If you do remove them, be sure to tag each of them to ensure correct installation.**

1. Disconnect the battery cables.
2. Remove the distributor cap mounting screws and lift off the cap. Some caps may have a vent connection; if yours does, unplug the hose.

3. Clean the cap with warm soapy water—if you have not removed the spark plug wires, you must, obviously, do it at this time. Blow it dry with compressed air; or if not available, make sure it air dries completely prior to installation.
4. Check the cap contacts for excessive burning, wear or corrosion. Check the cap for cracks or wear.
5. Install the cap by fitting the tab into the notch in the distributor housing and tighten the screws securely. Reconnect the plug wires if disconnected.

Distributor

REMOVAL

1. Begin by disconnecting the high tension leads from the distributor. Take time to identify and tag each lead as an aid during installation to ensure they are properly installed for correct cylinder firing.
2. Lift the locking tabs and disconnect the 2-and 4-terminal connectors. Rotate the crankshaft until No. 1 cylinder is in the firing position—both valves for No. 1 are closed and the timing mark on the harmonic balancer is aligned with the 0° (TDC) mark on the grid attached to the front cover.
3. Remove the two screws securing the distributor cap in place, and then remove the cap.
4. Now, note the position of the rotor tip. Take time to make a reference mark on the distributor housing to enable the rotor and the housing to be properly aligned during installation.
5. Make a mark on the distributor base and a matching mark on the engine as an aid during installation to ensure the distributor will be installed back in its original position.

■ **Take care to prevent the crankshaft from being rotated—even slightly—while the distributor is out of the block. If the crankshaft should be rotated—follow the procedures listed under Engine Disturbed later in this section.**

6. Remove the distributor clamp bolt and lift the distributor straight up and clear of the engine. The gasket should be discarded.

INSTALLATION

Engine Not Disturbed

1. Perform the following procedures if the distributor and engine were marked prior to removal and there is no evidence to indicate that the crankshaft was rotated while the distributor was removed. If the distributor was not marked, as instructed in the removal procedures, or if the crankshaft was rotated—even slightly—while the distributor was out, proceed to the Engine Disturbed section.

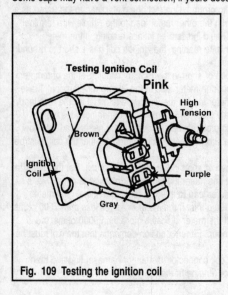

Testing Ignition Coil

Pink

High Tension

Brown

Ignition Coil

Purple

Gray

Fig. 109 Testing the ignition coil

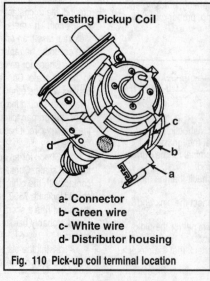

Testing Pickup Coil

a- Connector
b- Green wire
c- White wire
d- Distributor housing

Fig. 110 Pick-up coil terminal location

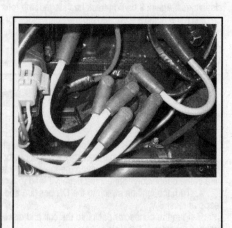

Fig. 111 Remove the cap with the wires attached (3.0L shown)

2. Install the rotor and rotate it approximately 1/8 turn counterclockwise from the reference mark made prior to disassembling. Rotor rotation at this time is necessary because as the distributor shaft gear indexes with the camshaft gear, the shaft will rotate clockwise about 1/8 turn—the rotor will then be back where it should be for the No. 1 cylinder to fire.

3. Position a new distributor gasket in place on the engine. A reference mark should have been made on the distributor and the block prior to removal, as instructed in the removal procedures. Install the distributor into the block with the mark on the distributor roughly aligned with the mark on the block. Push the distributor fully into the block until the housing is seated.

✳✳WARNING

If necessary, rotate the rotor slightly to permit the gear on the lower end of the distributor shaft to index with the gear on the camshaft. The "spade" on the lower end of the distributor shaft must engage the slot for the oil pump. You can also insert a small screwdriver into the opening and turn the pump shaft. Failure to index would result in no engine oil circulation—disaster. The distributor shaft collar should now be fully seated on the block. However, once the distributor is in place the rotor reference marks and the distributor housing marks should both be aligned.

4. Secure the distributor in place with the hold down clamp and bolt. Tighten the bolt to the proper specification as detailed in the Torque Specifications chart for your particular engine (snug, but not more than 20 ft. lbs.).

5. Connect the 2-and 3-wire leads into the distributor.

6. Install the cap and tighten the screws securely. The cap can only be installed properly—one way. Lubricate the sockets in the distributor cap with wheel bearing grease, or the equivalent. Install the spark plug high tension leads (if they were removed) using the identification made during removal to ensure proper cylinder firing.

7. Check the ignition timing.

Engine Disturbed

The following procedures are to be performed if reference marks were not made for the rotor, the distributor housing and the engine block or if the crankshaft was rotated while the distributor was out of the block.

1. Rotate the crankshaft until the No. 1 cylinder is ready to fire—both valves are closed and the timing mark on the harmonic balancer is aligned with the 0° mark on the grid attached to the front cover.

2. Position a new gasket in place on the engine block. Install the distributor into the block.

■ **If it is not possible to fully seat the distributor in place on the block, press down lightly on the distributor housing and at the same time rotate the rotor slightly. This action will permit the gear on the lower end of the distributor shaft to index with the camshaft gear. The "spade" on the lower end of the distributor shaft should engage the slot for the oil pump. The distributor shaft collar should now be fully seated on the block.**

Fig. 112 A good look at the cap on the 7.4Gi

3. Once the distributor is fully seated, install the clamp and bolt, but leave it just loose enough to permit rotating the distributor with strong hand pressure. At this point, the rotor must be in position to fire the No. 1 cylinder.

4. Just place the cap in position on the distributor. Scribe a mark on the distributor housing aligned with the No. 1 spark plug terminal. Now, remove the cap and verify the rotor is aligned with the mark just made for the No. 1 spark plug terminal. If the rotor does not align with the mark, rotate the distributor housing—with difficulty, because the clamp was not tightened—right or left to align the rotor with the mark for the No. 1 terminal. If it is not possible to align the rotor with the No. 1 mark, the distributor must be removed the entire procedure started over again.

5. Tighten the distributor clamp bolt to the proper specification as detailed in the Torque Specifications chart for your particular engine (snug, but not more than 20 ft. lbs.).

6. Check all high tension leads and connect them in the proper sequence for correct cylinder firing.

7. Connect the 2-wire engine harness connector to the distributor.

8. Check the ignition timing.

DISASSEMBLY

◆ See Figures 107, 108, and 113 thru 119

Remove the distributor from the engine, as outlined in the previous section. The following procedures pick up the work after the distributor has been removed and is on a suitable work surface.

1. Pull the rotor free from the upper end of the distributor shaft.

2. Disconnect the connector at the ignition module, pry the retainer free and then lift out the pick-up coil.

3. Disconnect the 2 ignition module leads. Remove the two screws securing the module to the distributor, and then remove the module. It may be necessary to carefully pry the module loose from the distributor.

4. Make a mark on the drive tang at the lower end of the distributor shaft and another mark on the collar of the gear aligned with the first mark as an aid to installing the gear back in its original position. After the marks have been made, drive the roll pin free of the gear with a small (4.5mm) punch.

5. Slide the gear free of the shaft, and then separate the shaft by pulling it up through the top of the pick-up coil—the timer core will come with it.

6. Carefully pry off the retainer on the end of the housing securing the shield, coil and pole piece. Slide them all off and discard the retainer.

Fig. 113 A good view of the EST distributor

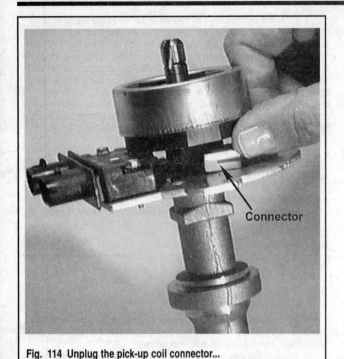

Fig. 114 Unplug the pick-up coil connector...

CLEANING AND INSPECTING

Obtain a clean cloth and wipe the distributor cap clean. Inspect the cap for chips, cracks and any sign of a carbon path. If such a path is discovered, the cap must be replaced because such a path would permit high tension leakage.

Clean loose corrosion from the terminal segments inside the cap. Do not use emery cloth or sandpaper. If the segments are deeply grooved, a new cap should be installed.

Inspect the terminal sockets for corrosion. Clean the sockets using a stiff wire brush to loosen the corrosion. After the sockets are clean, lubricate them with wheel bearing grease, or equivalent.

Inspect the rotor for cracks. Check to be sure the tip of the rotor is not badly burned. Such a condition demands the rotor be replaced.

Inspect the trigger wheel for any sign of contact with the sensor. If the distributor has been removed from the block, make an attempt to check the distributor shaft for excessive wear between the shaft and the bushings in the housing.

ASSEMBLY

◆ **See Figures 107, 108 and 113 thru 120**

The following procedures pickup the work after disassembled parts have been cleaned and replacement items have been obtained and are on hand.

1. Align the pole piece with the housing and slide it onto the end of the housing. Align the tab on the coil with the hole in the base of the housing.

Fig. 115 ...and remove the ignition module

Fig. 116 Driving out the roll pin

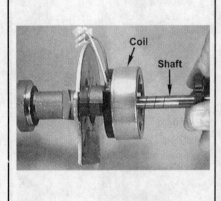

Fig. 117 Close up of the pin and the gear

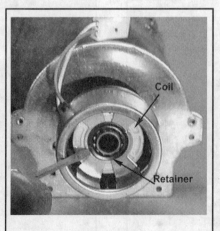

Fig. 118 Pull the shaft up through the coil. Note that the timer core stays with the shaft

Fig. 119 Pry up the retainer

Fig. 120 Coat the plate with grease

2. Install the shield and a new retainer over the end of the housing. Position a 5/8 in. socket over the retainer and lightly tap it into place over the end of the housing—make sure it is seated firmly into the groove.

3. Coat the distributor shaft with oil and slide into and through the center of the housing. Spin it several times to lubricate it and the housing.

4. Slide the star washer onto the shaft so that the teeth are pointing up toward the top of the housing. Slide the flat washer and the gear onto the shaft so that the alignment marks made earlier line up. Insert the roll pin and drive it down until it is flush.

5. Make sure the surface of the distributor plate is clean and then spread an even coat of Thermal Conductive grease onto the bottom of the module. Position the module on the plate and tighten the two screws securely; clean any excess grease from the plate and module.

6. Connect the lead from the coil to the back of the module and lock it in place.

7. Align the rotor with the notch in the distributor shaft, and then press it to secure it in place.

Ignition Coil

REMOVAL & INSTALLATION

② MODERATE

1. Disconnect the negative battery cable.
2. Disconnect the high tension lead coming from the distributor.
3. Tag and disconnect the coil electrical leads (two 2-pin connectors).
4. Remove the two coil mounting bolts and lift out the coil (with the bracket attached, if equipped).
5. Install the coil and tighten the mounting bolts to 20-25 ft. lbs. (27-34 Nm).
6. Connect the electrical leads and the high tension lead.
7. Connect the battery cable.

PRESTOLITE INTEGRAL BREAKERLESS INDUCTIVE DISTRIBUTOR (BID) SYSTEM

Description

◆ See Figures 121, 122 and 123

Basic components of this system include a pointless distributor and a remote ignition coil. The distributor has the external appearance of a conventional distributor—the cap and rotor will appear quite familiar, but the remainder of the internal items are completely different. A trigger "wheel" is mounted near the upper end of the distributor shaft. This "wheel" has "teeth"—the number corresponding to the number of engine cylinders. A horseshoe shaped sensor/electronic ignition module is mounted around the distributor shaft and "wheel". The sensor portion of the module and the "wheel" replace the breaker points and condenser from earlier systems. This sensor/module controls ignition timing and ignition coil saturation.

The BID system contains two separate circuits—a primary circuit—and a secondary circuit. The primary circuit handles the low voltage current supplied by the battery or by the alternator. Major components of this circuit are the primary winding of the ignition coil, the electronic control module including the sensor and the low tension wiring. The secondary circuit handles the high voltage surges produced by the ignition coil. This high voltage is routed to each spark plug through the high tension leads. Components of this circuit include the secondary winding of the ignition coil, the high tension leads between the coil and the distributor and between the distributor, the high tension leads to each spark plug, the distributor rotor, and the distributor cap.

The sensor is a coil of very fine wire molded into a plastic housing. The housing is mounted on the base plate in the distributor and is connected directly to, and considered an integral part of, the ignition module. The sensor is not serviceable and must be replaced as an assembly along with the module.

The electronic ignition module is a self-contained solid state unit encapsulated with potting compound. This compound provides a protective barrier against vibration and moisture. The module contains an oscillator and a transistor. It is not serviceable and must be replaced along with the sensor.

The BID system operates on the principle of a metal detector. The detector is the sensor in the module and the detected metal is the "tooth" in the trigger "wheel".

As the "wheel" near the upper end of the distributor shaft rotates, the sensor detects the metal "tooth"; the oscillator in the ignition module is at a low level; the transistor is OFF; and primary current does not flow. If this were a breaker point system—the points would be open.

After the "tooth" of the "wheel" passes the sensor, the oscillator is at a high level, the transistor is ON; and current flows in the primary winding. With a breaker point system—the points would be closed.

The high voltage surge developed in the secondary winding of the ignition coil is transferred to the proper spark plug through the cable; the center of the distributor cap; the distributor rotor; to the proper cap contact point when the rotor aligns with the point; and finally through the high tension lead to the spark plug in the cylinder.

This sequence is repeated as each "tooth" on the trigger "wheel" passes the sensor.

Initial manual setting of the distributor is made when a tooth of the trigger wheel passes the sensor at the instant the No. 1 piston is at an exact position in the cylinder—namely TDC. This position is indicated by the timing mark on all GM engines, or the pointer on Ford engines, and the timing grid.

At high engine rpm, the spark must occur earlier than at idle speed. This advance spark gives the fuel/air mixture time to ignite, burn, and deliver its power to the piston as it starts down on the power stroke. Therefore, the spark advance is governed by engine speed. A centrifugal advance arrangement consisting of an advance cam, two weights, two springs and a weight base plate is installed in the distributor to accomplish the advance timing.

When the engine is at idle, the springs hold the advance weights so there is no advance. Therefore, spark timing occurs as set by the initial manual setting of the distributor.

When engine speed increases, centrifugal force causes the weights to move outward and push against the advance cam, thus rotating the weight plate. As the plate rotates, the teeth of the trigger wheel will pass the sensor earlier in the compression stroke, thus advancing the spark.

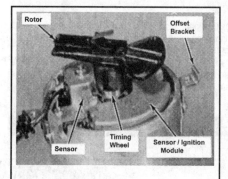

Fig. 121 A close-up of the BID distributor with the cap removed...

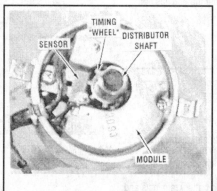

Fig. 122 ...and a better view with the rotor removed

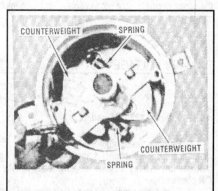

Fig. 123 Centrifugal weight arrangement

Troubleshooting

** CAUTION

Always ground the high tension lead of the coil any time it is disconnected from the distributor. Failure to disconnect the lead could result in fire or an explosion, if gas vapors are present. Reason: A high voltage discharge in the secondary circuit of the coil may occur when the ignition switch is turned to the on and off positions. This high voltage discharge may occur even if the engine is not cranked.

Before spending too much time and money attempting to trace a problem to the ignition system, a compression check of the cylinders should be made. If the cylinders do not have adequate compression, troubleshooting and attempted service of the ignition or fuel system will fail to give the desired results of satisfactory engine performance.

VISUAL CHECKS

Check the coil tower for carbon tracking. Check the terminals for secure connections. Verify the polarity is correct. Check the coil nipple to be sure it is sealed and insulated. If flashover should occur at this location, the engine will fail to start.

Check the distributor cap for carbon tracking. Clean the cap if it is dirty. Moisture and dirt make a good path for flashover. If a carbon track has started, the cap must be replaced. Check the rotor for carbon tracking and cleanliness. Again, if carbon track has started, the rotor must be replaced.

Check the high tension leads to the spark plugs for burning, cracking and any type of deterioration. Check the spark plug for fouling, proper gap and possible cracked insulator.

SENSOR AIR GAP

◆ See Figure 124

All models equipped with BID ignition systems require an air gap adjustment between the sensor wheel and the trigger wheel; particularly after the distributor has been removed. It is also a good idea to ensure that the air gap is within specification prior to performing any ignition timing procedures.

1. Open or remove the engine hatches.
2. Remove the distributor cap and pull off the rotor as detailed later in this section.
3. Connect a remote starter switch as detailed in the manufacturer's instructions.
4. Check the sensor coil, trigger wheel and teeth for signs of wear or other obvious damage. Using the remote starter, crank the engine around until a tooth on the wheel aligns with the contact on the coil.
5. Insert a non-metallic flat-bladed feeler gauge between the coil and tooth. The gap should be 0.008 in. (0.20mm).
6. If not within specification, loosen the small screw (A) on the trigger coil slightly and move the coil carefully until the gap is within spec. Tighten the screw securely—but not too tight— and recheck that the gap has not changed.
7. Disconnect the remote starter.
8. Install the rotor and distributor cap.

PRIMARY VOLTAGE TEST

◆ See Figures 125, 126 and 127

1. Remove the distributor cap and connect a voltmeter as per the manufacturer's instructions. Bump the engine over until the sensor is between any two teeth on the trigger wheel. Turn the ignition switch to the **ON** position.
2. Connect the meter leads to their corresponding battery terminals and confirm that 12-13 V is present at the battery. Charge or replace the battery if voltage is less than suggested.

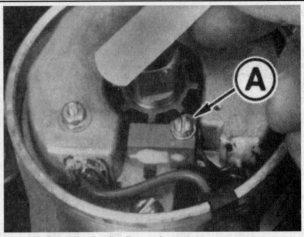

Fig. 124 Checking the distributor air gap

3. Connect the negative lead on the meter to the negative battery terminal or a good ground. Connect the positive lead to the positive terminal on the ignition coil. The meter should indicate battery voltage as shown in Step 2 (within 1V); if less, move to the Voltage Drop Test.
4. With the negative lead still connected as in the previous step, connect the positive lead to negative terminal on the coil. If the meter indicates 4-8V, ground the high tension lead from the coil and position a screwdriver in front of the sensor face, discharging the coil. The meter should now indicate 12-13V. If less than suggested, the coil or the sensor/module will require replacement. Replace the coil as necessary and then perform the test with the screwdriver again—if still no spark, replace the sensor and reset the air gap.

If the meter indicated less than 4V, disconnect the meter lead at the coil and remove the wire at the negative terminal on the coil. Now connect the meter lead to the negative terminal again and confirm that the reading is 12-13V. If so, there is a short in the sensor assembly and you will need to replace the unit. If the reading is only 0-4V, the coil primary circuit has an open and you'll need to replace the coil.

If the meter originally showed more than 8V in the first test of this step, check for an open distributor ground circuit or a shorted coil primary circuit. If the distributor ground is bad, repair the circuit. If it is OK, replace the sensor assembly. If coil primary circuit is showing a short, replace the coil.

VOLTAGE DROP TEST

1. Remove the distributor cap and connect a voltmeter as per the manufacturer's instructions. Bump the engine over until the sensor is between any two teeth on the trigger wheel. Turn the ignition switch to the **ON** position.
2. Connect the positive meter lead to the positive terminal on the battery. Connect the negative meter lead to the positive terminal on the ignition coil. If the meter indicates less than 1V, check all connections. If the meter fluctuates, check for a loose connection. Repair either option prior to moving to the next step.
3. Connect the negative meter lead to the negative terminal on the battery. Connect the positive meter lead to the housing on the sensor assembly. If the meter indicates more than 1V, check and tighten all ground connections and inspect the negative battery terminal. Confirm that the sensor mounting screws are tight. Repeat the test, if the reading is above 1V again, replace the battery ground cable.
4. Check the advance mechanism in the distributor and the ignition timing.

Distributor Cap

REMOVAL & INSTALLATION

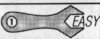

■ It is not necessary to remove the spark plug leads at the cap when removing it. If you do remove them, be sure to tag each of them to ensure correct installation.

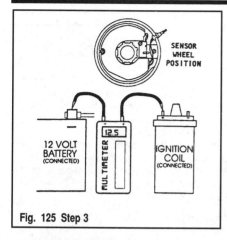

Fig. 125 Step 3

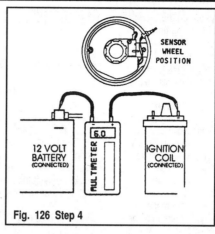

Fig. 126 Step 4

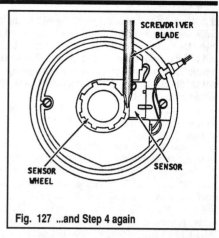

Fig. 127 ...and Step 4 again

1. Disconnect the battery cables.

2. Remove the distributor cap mounting screws and lift off the cap. Some caps may have a vent connection; if yours does, unplug the hose.

3. Clean the cap with warm soapy water—if you have not removed the spark plug wires, you must, obviously, do it at this time. Blow it dry with compressed air; or if not available, make sure it air dries completely prior to installation.

4. Check the cap contacts for excessive burning, wear or corrosion. Check the cap for cracks or wear.

5. Install the cap by fitting the tab into the notch in the distributor housing and tighten the screws securely. Reconnect the plug wires if disconnected.

Distributor

REMOVAL

1. Begin by disconnecting the high tension leads from the distributor. Take time to identify and tag each lead as an aid during installation to ensure they are properly installed for correct cylinder firing.

2. Rotate the crankshaft until No. 1 cylinder is in the firing position—both valves for No. 1 are closed and the timing mark (or pointer) on the harmonic balancer is aligned with the 0° mark on the grid attached to the front cover.

3. Remove the two screws securing the distributor cap in place, and then remove the cap.

4. Now, note the position of the rotor tip. Take time to make a reference mark on the distributor housing to enable the rotor and the housing to be properly aligned during installation.

5. Make a mark on the distributor base and a matching mark on the engine as an aid during installation to ensure the distributor will be installed back in its original position.

■ **Take care to prevent the crankshaft from being rotated—even slightly—while the distributor is out of the block. If the crankshaft should be rotated—follow the procedures listed under Engine Disturbed later in this section.**

6. Remove the distributor clamp bolt and lift the distributor straight up and clear of the engine. The gasket should be discarded.

INSTALLATION

Engine Not Disturbed

■ **When installing a new distributor on a Ford engine, make sure that you add Ford Engine Oil Supplement to the oil before starting the engine for the first time.**

1. Perform the following procedures if the distributor and engine were marked prior to removal and there is no evidence to indicate that the crankshaft was rotated while the distributor was removed. If the distributor

was not marked, as instructed in the removal procedures, or if the crankshaft was rotated—even slightly—while the distributor was out, proceed to the Engine Disturbed section.

2. Install the rotor and rotate it approximately 1/8 turn counterclockwise from the reference mark made prior to disassembling on GM engines; if the engine you're working on is a Ford, rotate the rotor 1/8 turn clockwise. Rotor rotation at this time is necessary because as the distributor shaft gear indexes with the camshaft gear, the shaft will rotate clockwise about 1/8 turn—the rotor will then be back where it should be for the No. 1 cylinder to fire.

3. Position a new distributor gasket in place on the engine. A reference mark should have been made on the distributor and the block prior to removal, as instructed in the removal procedures. Install the distributor into the block with the mark on the distributor roughly aligned with the mark on the block. Push the distributor fully into the block until the housing is seated.

✳✳WARNING

If necessary, rotate the rotor slightly to permit the gear on the lower end of the distributor shaft to index with the gear on the camshaft. The "spade" on the lower end of the distributor shaft must engage the slot for the oil pump. Failure to index would result in no engine oil circulation—disaster. The distributor shaft collar should now be fully seated on the block. However, once the distributor is in place the rotor reference marks and the distributor housing marks should both be aligned.

4. Secure the distributor in place with the hold down clamp and bolt. Tighten the bolt to the proper specification as detailed in the Torque Specifications chart for your particular engine (snug, but no more than 20 ft. lbs.).

5. Install the cap and tighten the screws securely. The cap can only be installed properly—one way. Lubricate the sockets in the distributor cap with Extreme Pressure grease, or the equivalent. Install the spark plug high tension leads (if they were removed) using the identification made during removal to ensure proper cylinder firing.

6. Check the ignition timing.

Engine Disturbed

■ **When installing a new distributor on a Ford engine, make sure that you add Ford Engine Oil Supplement to the oil before starting the engine for the first time.**

The following procedures are to be performed if reference marks were not made for the rotor, the distributor housing and the engine block or if the crankshaft was rotated while the distributor was out of the block.

1. Rotate the crankshaft until the No. 1 cylinder is ready to fire—both valves are closed and the timing mark (or pointer) on the harmonic balancer is aligned with the 0° mark on the grid attached to the front cover.

2. Position a new gasket in place on the engine block. Install the distributor into the block.

■ If it is not possible to fully seat the distributor in place on the block, press down lightly on the distributor housing and at the same time rotate the rotor slightly. This action will permit the gear on the lower end of the distributor shaft to index with the camshaft gear. The "spade" on the lower end of the distributor shaft should engage the slot for the oil pump. The distributor shaft collar should now be fully seated on the block.

3. Once the distributor is fully seated, install the clamp and bolt, but leave it just loose enough to permit rotating the distributor with strong hand pressure. At this point, the rotor must be in position to fire the No. 1 cylinder.

4. Just place the cap in position on the distributor. Scribe a mark on the distributor housing aligned with the No. 1 spark plug terminal. Now, remove the cap and verify the rotor is aligned with the mark just made for the No. 1 spark plug terminal. If the rotor does not align with the mark, rotate the distributor housing—with difficulty, because the clamp was tightened slightly—right or left to align the rotor with the mark for the No. 1 terminal. If it is not possible to align the rotor with the No. 1 mark, the distributor must be removed the entire procedure started over again.

5. Tighten the distributor clamp bolt to the proper specification as detailed in the Torque Specifications chart for your particular engine (snug, but no more than 20 ft. lbs.).

6. Check all high tension leads and connect them in the proper sequence for correct cylinder firing.

7. Check the ignition timing.

DISASSEMBLY

◆ See Figures 123, 128, 129, and 130

Remove the distributor from the engine, as outlined in the previous section. The following procedures pick up the work after the distributor has been removed and is on a suitable work surface.

1. Pull the rotor free from the upper end of the distributor shaft.

2. Loosen the sensor screw (near the trigger wheel, remove the 2 mounting screws and lock-washers and then lift out the sensor/module assembly. Pull the grommet out of the notch in the distributor housing and then disconnect the 2 leads at the ignition coil.

3. No further disassembly is necessary.

CLEANING AND INSPECTING

Obtain a clean cloth and wipe the distributor cap clean. Inspect the cap for chips, cracks and any sign of a carbon path. If such a path is discovered, the cap must be replaced because such a path would permit high tension leakage.

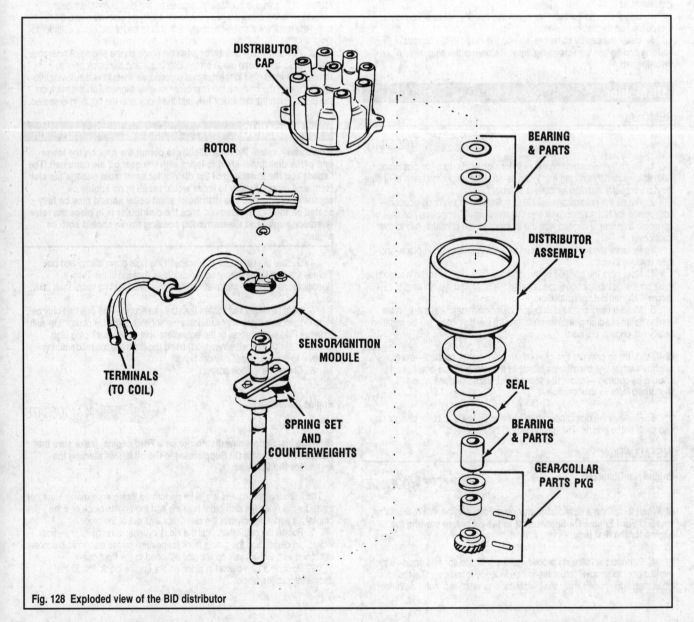

Fig. 128 Exploded view of the BID distributor

Clean loose corrosion from the terminal segments inside the cap. Do not use emery cloth or sandpaper. If the segments are deeply grooved, a new cap should be installed.

Inspect the terminal sockets for corrosion. Clean the sockets using a stiff wire brush to loosen the corrosion. After the sockets are clean, lubricate them with wheel bearing grease, or equivalent.

Inspect the rotor for cracks. Check to be sure the tip of the rotor is not badly burned. Such a condition demands the rotor be replaced.

Inspect the trigger wheel for any sign of contact with the sensor. If the distributor has been removed from the block, make an attempt to check the distributor shaft for excessive wear between the shaft and the bushings in the housing.

ASSEMBLY

◆ See Figures 123, 128, 129 and 130

The following procedures pickup the work after disassembled parts have been cleaned and replacement items have been obtained and are on hand.

1. Coat the bottom side of the sensor housing with Heat Sink compound and then position the assembly into the distributor. Tighten the 2 mounting screws (with new lock-washers) securely. Bump the engine around until a tooth on the trigger wheel lines up with the mark on the sensor and then adjust the air gap.

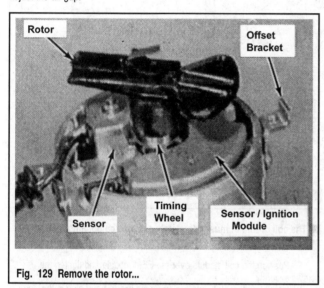

Fig. 129 Remove the rotor...

2. Connect the lead from the assembly to the coil and press the grommet back into the edge of the housing.

3. Align the rotor with the notch in the distributor shaft, and then press it to secure it in place.

4. Install the distributor.

Ignition Coil

REMOVAL & INSTALLATION

1. Disconnect the negative battery cable.

2. Disconnect the high tension lead coming from the distributor.

3. Tag and disconnect the coil electrical leads (two 2-pin connectors).

4. Remove the two coil mounting bolts and lift out the coil (with the bracket attached, if equipped).

5. Install the coil and tighten the mounting bolts to 20-25 ft. lbs. (27-34 Nm).

6. Connect the electrical leads and the high tension lead.

7. Connect the battery cable.

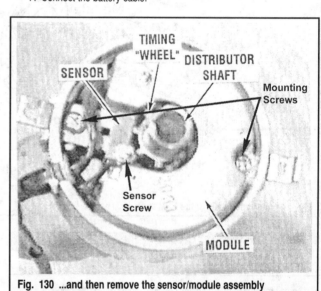

Fig. 130 ...and then remove the sensor/module assembly

SPITFIRE/ELECTRONIC ENGINE MANAGEMENT (EMM) SYSTEM

Description & Operation

◆ See Figure 131

The Electronic Engine Management system is an electrical control system offering automatic monitoring and maintenance of a number of various circuits including the SLOW system (Speed Limiting Operational Warning), an engine speed limiter, a knock sensor, and variable spark advance.

The SLOW system is continually monitoring oil and water temperatures—emitting an audible warning when it detects low oil pressure or high water temperature. After sounding the alarm, the system will automatically, and gradually, reduce the engine speed by 200 rpm per second until it reaches 2800 rpm. Low oil pressure or overheating that occurs below 2800 rpm will set off the alarm, but NOT activate the system. The system is programmed with an automatic 5 second delay prior to activation in order to account for momentary loss of pressure during normal operation. The system will automatically shut itself off and reset upon engine shut-down.

The engine speed limiter, much like a governor, will shut of the signal to the ignition system when it detects an engine speed of 200 rpm above the programmed maximum limit. As soon as the engine speed drops back down into the allowable range, the system will de-activate and allow for normal engine operation.

A cylinder head-mounted knock sensor monitors for detonation or pre-ignition, gradually retarding spark advance until the condition is eliminated. Once activated, the system will maintain the retarded advance until the engine speed changes by more than 50 rpm. Once this happens, the system will gradually increase advance until the normal spark curve is achieved for the engine speed and operating conditions.

The system program provides two separate variable spark advance curves for the ignition module. The normal curve is for 89 AKI octane fuel and it uses the knock sensor to protect the engine and modify the curve in the case of lower octane fuels. There is also a default curve for 86 AKI octane fuel that works to protect the engine when lower octane fuel is detected and/or in the event of a defective knock sensor.

The system utilizes a breakerless electronic ignition system which requires no adjustments and almost no maintenance.

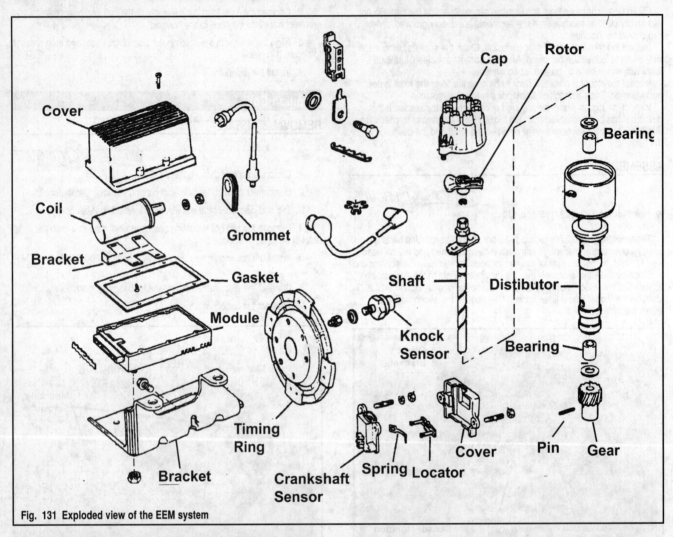

Fig. 131 Exploded view of the EEM system

System Components

COIL MODULE

The coil module is a large black housing, mounted on the inside of the Port exhaust high rise. It actually sits on top of the EEM module. The module contains one conventional 12V ignition coil.

1. Locate the module assembly and remove the screw securing the 14-pin connector on the side of the EEM module.

2. Remove the coil high tension lead and then pop the two modules out of their retaining bracket.

3. Separate the coil module from the EEM module by removing the six T25 Torx® screws. Carefully lay the coil housing to the side and remove the 4 coil bracket screws. Tag and disconnect the EEM electrical leads running to the coil. Slide the coil tower grommet out of the housing and off the tower. Remove the coil.

4. Connect an ohmmeter as per the manufacturer's instructions and set it to the low scale. Attach the meter leads to the two coil terminals. Check that the meter reads 1.26-1.54 ohms.

5. Now switch the meter to the high scale and attach one of the meter leads to either of the two coil terminals. Connect the other meter lead to the inside of the high tension tower. Correct resistance is 9400-11000 ohms.

6. Meter readings outside of the suggested ranges will necessitate replacement of the coil.

7. Reconnect the EEM module leads to the coil— if you neglected to tag them, the blue lead goes to the positive terminal and the yellow lead goes to the negative terminal.

8. Slide the grommet over the high tension tower and then position the coil into the housing while slipping in the grommet. Install the mounting bracket and tighten the screws securely.

■ **The coil grommet must completely seal off the opening in the housing when installed.**

9. Position the coil module over the EEM module and tighten the screws securely.

10. Install the assembly into the mounting bracket and plug in the connector. Install the retaining screw and tighten it securely.

DISTRIBUTOR

1. Begin by disconnecting the high tension leads from the distributor. Take time to identify and tag each lead as an aid during installation to ensure they are properly installed for correct cylinder firing.

2. Rotate the crankshaft until No. 1 cylinder is in the firing position— both valves for No. 1 are closed and the timing mark (or pointer) on the harmonic balancer is aligned with the 0° mark on the grid attached to the front cover.

3. Remove the two screws securing the distributor cap in place, and then remove the cap.

4. Now, note the position of the rotor tip. Take time to make a reference mark on the distributor housing to enable the rotor and the housing to be properly aligned during installation.

5. Make a mark on the distributor base and a matching mark on the engine/intake manifold as an aid during installation to ensure the distributor will be installed back in its original position.

■ **Take care to prevent the crankshaft from being rotated—even slightly—while the distributor is out of the block.**

6. Remove the distributor clamp bolt and lift the distributor straight up and clear of the engine. The gasket should be discarded.

7. Install the rotor and rotate it approximately 1/8 turn counterclockwise from the reference mark made prior to disassembling on GM engines; if the engine you're working on is a Ford, rotate the rotor 1/8 turn clockwise. Rotor rotation at this time is necessary because as the distributor shaft gear indexes with the camshaft gear, the shaft will rotate clockwise about 1/8 turn—the rotor will then be back where it should be for the No. 1 cylinder to fire.

8. Position a new distributor gasket in place on the engine. A reference mark should have been made on the distributor and the block prior to removal, as instructed in the removal procedures. Install the distributor into the block with the mark on the distributor roughly aligned with the mark on the block. On GM engines, the mark should be at the 10:00 position; while on Ford engines, the mark should be at the 8:00 position. Push the distributor fully into the block until the housing is seated.

9. If necessary, rotate the distributor housing slightly until the marks on the rotor and housing are in exact alignment.

10. Secure the distributor in place with the hold down clamp and bolt. Tighten the bolt to the proper specification as detailed in the Torque Specifications chart for your particular engine (snug, but no more than 20 ft. lbs.).

11. Install the cap and tighten the screws securely. The cap can only be installed properly—one way. Lubricate the sockets in the distributor cap with Extreme Pressure grease, or the equivalent. Install the spark plug high tension leads (if they were removed) using the identification made during removal to ensure proper cylinder firing.

EEM MODULE

The EEM module is located underneath the coil module on a bracket attached to the Port exhaust riser. It monitors and controls all EEM functions as previously described.

1. Locate the module assembly and remove the screw securing the 14-pin connector on the side of the EEM module.

2. Remove the coil high tension lead and then pop the two modules out of their retaining bracket.

3. Separate the coil module from the EEM module by removing the six T25 Torx® screws. Carefully lay the coil housing to the side and then tag and disconnect the EEM electrical leads running to the coil. Position the coil module to the side.

4. Reconnect the EEM module leads to the coil— if you neglected to tag them, the blue lead goes to the positive terminal and the yellow lead goes to the negative terminal.

5. Position the coil module over the EEM module and tighten the screws securely.

6. Install the assembly into the mounting bracket and plug in the connector. Install the retaining screw and tighten it securely.

KNOCK SENSOR

The knock sensor is mounted at the rear of the cylinder head. When removing, disconnect the electrical lead and carefully unscrew the sensor—never use an impact wrench. Since the sensor grounds it self to the engine through its threads, make certain that the threads on both the sensor and the head are clean and debris-free. Tighten to 7-10 ft. lbs. (9.5-13.5 Nm), being VERY careful not to over-tighten it.

TIMING RING

A timing ring is mounted on the crankshaft between the flywheel and the engine coupler. The ring has four outer vanes which are used to notify the EEM module of the correct timing for firing of each cylinder. An inner ring is also utilized as a 'start-up' reference for the No. 1 cylinder. Removal and installation procedures are contained with the Engine Mechanical section.

TIMING SENSOR

The timing sensor is mounted on the upper flywheel housing, on the Port side. This is a very simple electronic switching device, receiving power and ground through the EEM module. Two Hall Effect switches inside the sensor sense the timing ring movement and send an electrical impulse back to the module which uses these signals to control ignition system operation. Removal and installation of the sensor is a snap, but is also detailed in the appropriate Engine Mechanical section (usually in the Flywheel/Coupler procedures).

FORD THICK FILM INTEGRATED IGNITION SYSTEM (TFI-IV)

Description & Operation

◆ **See Figure 132 and 133**

The TFI-IV system, part of a greater system called Electronic Engine Control System 4th generation (EEC-IV) is used on all fuel injected Ford engines. This is an electronic engine control system consisting of a network of electronic and electro-mechanical components that continuously vary engine operation to meet pre-programmed operating parameters.

The heart of the system is a "black box" called and electronic control assembly (ECA). The ECA is linked to all of the system components by the wiring harness and is sealed against dirt and moisture. The ECA monitors and controls all engine operating conditions through various sensor inputs and via its own command outputs. All operating parameters are pre-programmed and dependant on engine size and operating conditions.

The TFI-IV ignition system used with the EEC-IV system utilizes the following major components: distributor, ECA, TFI module and coil.

The distributor is driven by a drive gear on the camshaft and a driven gear attached to the lower end of the distributor shaft. At the upper end of the distributor, a rotary vane cup rotates with the distributor. Electrically, the distributor utilizes a Hall Effect vane switched On and Off by the ECA, The vane assembly is a Hall sensor on one side and a permanent magnet on the other. The rotary vane cup, made of a ferrous metal, triggers the signals. When an opening in the vane cup moves in front of the magnet and sensor, a magnetic flux field is completed from the magnet, through the Hall sensor, and back to the magnet.

As the vane cup rotates and closes the window opening, the flux lines are shunted through the vane and back to the magnet. During this time, voltage is produced as the vane passes through the window opening. When the vane clears the opening, the window edge causes the signal to go to 0 volts. This pulsed signal is called the Profile Ignition Pickup signal, or PIP signal. One window of the vane cup is a different size from the others and is used for crankshaft positioning and engine speed information by the ECA. This signal is known as the "Signature PIP". The distributor uses no centrifugal weights or vacuum advance motors since spark is controlled by the ECA. With the exception of a base timing adjustment, no adjustments are necessary.

The two PIP signals provide the ECA with engine speed and crankshaft positioning information. This information is processed by the processor within the ECA and sends out a Spark Output, abbreviated SPOUT, signal to the TFI module.

The TFI-IV module composed of a custom integrated circuit, a Darlington output device and associated thick film integrated components. The module is mounted remotely on a heat sink next to the ignition coil. When commended by the SPOUT signal from the ECA, it opens and closes the ignition coil primary circuit to produce the secondary spark output from the coil to the distributor and on to the spark plugs.

The module is controlled, and spark timing adjustment is made, by the timed SPOUT signal from the ECA. In the event of no SPOUT signal, such as during base timing adjustment or an ECA failure, a second PIP signal from the Hall Effect switch in the distributor is sent to the module and will provide a base timing specification for engine operation. The SPOUT circuit includes a connector on the wiring harness which is removed when setting base timing. The TFI module also incorporates an Ignition Diagnostic Monitor (IDM) circuit which provides the ECA with an rpm reference signal. Failure of the IDM circuit will store a service code in the memory of the ECA.

The ignition coil is remotely mounted next to the TFI module and is commonly referred to as the E-Coil or E-Core coil. The coil consists of an externally wound lamination around an iron core and is epoxy-filled to resist heat, moisture and vibration. The coil is small in size but capable of delivering 38Kv at maximum output.

Power for the E-coil is provided from the ECA through the wiring harness connecting to the positive terminal on the coil. The negative side of the coil is connected to the TFI module; thus completing the primary side of the ignition circuit. The TFI module will turn the primary current flow ON when there is no

SPOUT signal. Current flow is turned OFF as commanded by the SPOUT signal. When the primary current field collapses, a high voltage output is generated in the secondary windings, delivered to the distributor and then on to the spark plugs.

The knock sensor is simply a signal generating device that detects engine detonation (knock) and converts this frequency signal into voltage through the use of a piezo-element mounted in a threaded metal housing. Special construction makes the element only sensitive to the particular engine vibrations associated with spark knock.

When spark knock occurs, the knock sensor produces a low voltage pulsing electrical signal that is then amplified in the knock module. The signal is then sent to the ECA which will then retard the ignition timing in three steps until the knock is no longer sensed, or the retard reaches a maximum of 8°. Normal spark advance will return after the MAP sensor detects a 3-4 in. Hg. change in engine vacuum.

The SLOW operational system is designed to protect the engine from damage should it lose oil pressure or experience excessively high coolant temperature. This system will activate automatically without operator action. The condition can be verified by checking either the oil or temperature gauges. In the event of a high coolant temperature (200°F and above), or low oil pressure (less than 5 psi), the coolant temperature switch or the oil pressure switch will ground a low voltage signal in the ECA. The ECA will then begin to alter the fuel injector firing sequence, causing the engine to run rough and eventually bringing the engine speed below 2700 rpm. Once below 2700 rpm, the engine will run smoothly but be limited to a max of 2700 rpm by the ECA. The engine will remain in this SLOW operational mode until such a time as the problem is corrected or the temperature/psi moves back into the normal range. If the vessel is equipped with an Audible Alarm option, a warning horn will sound as soon as the engine enters the SLOW mode.

Engine speed limiter operation is controlled by, you guessed it, the ECA. Sensor operation, such as the rpm data from the PIP signal, is also monitored by the ECA. If engine speed exceeds the programmed maximum by 400 rpm, the speed limiter circuit will be activated, effectively halting all fuel injector firing until the engine speed falls back within the allowable limits. Once engine speed is back below the limit, the speed limiting circuit will be de-activated and injector firing will resume.

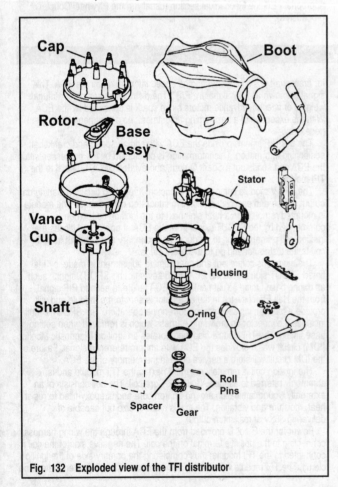

Fig. 132 Exploded view of the TFI distributor

Fig. 133 Major components of the TFI-IV system

Distributor Cap

REMOVAL & INSTALLATION

◆ See Figure 134

■ It is not necessary to remove the spark plug leads at the cap when removing it. If you do remove them, be sure to tag each of them to ensure correct installation.

1. Disconnect the battery cables.
2. Remove the plastic engine cover and then remove the rubber boot covering the distributor.
3. Remove the distributor cap mounting screws and lift off the cap. Some caps may have a vent connection; if yours does, unplug the hose.
4. Clean the cap with warm soapy water—if you have not removed the spark plug wires, you must, obviously, do it at this time. Blow it dry with compressed air; or if not available, make sure it air dries completely prior to installation.
5. Check the cap contacts for excessive burning, wear or corrosion. Check the cap for cracks or wear.
6. Install the cap and note the square alignment locator. Tighten the screws to 18-23 inch lbs. (2.0-2.6 Nm). Reconnect the plug wires if disconnected.
7. Install the rubber boot and the plastic engine cover. Reconnect the battery cables.

Distributor

REMOVAL & INSTALLATION

1. Begin by disconnecting the high tension leads from the distributor. Take time to identify and tag each lead as an aid during installation to ensure they are properly installed for correct cylinder firing.
2. Rotate the crankshaft until No. 1 cylinder is in the firing position—both valves for No. 1 are closed and the timing mark (or pointer) on the harmonic balancer is aligned with the 0° mark on the grid attached to the front cover.
3. Remove the two screws securing the distributor cap in place, and then remove the cap.
4. Now, note the position of the rotor tip. Take time to make a reference mark on the distributor housing to enable the rotor and the housing to be properly aligned during installation.
5. Push in on the tab and disconnect the 8-pin connector from the side of the housing.
6. Make a mark on the distributor base and a matching mark on the engine as an aid during installation to ensure the distributor will be installed back in its original position.

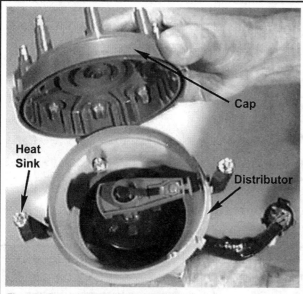

Fig. 134 Removing the distributor cap

■ Take care to prevent the crankshaft from being rotated—even slightly—while the distributor is out of the block. If the crankshaft should be rotated—follow the procedures listed under Engine Disturbed later in this section.

7. Remove the distributor clamp bolt and lift the distributor straight up and clear of the engine. The gasket should be discarded.
To Install:

■ When installing a new distributor, make sure that you add Ford Engine Oil Supplement to the oil before starting the engine for the first time.

8. Insure that the No. 1 piston is at TDC of the compression stroke—make sure that the timing pointer is aligned with the TDC mark on the damper timing scale.
9. Rotate the shaft of the distributor so that the tip of the rotor is aligned with the mark on the housing made previously.
10. Lubricate the distributor drive gear lightly with engine oil and install the unit into the engine. Rotate the distributor shaft as you are inserting the unit so that the leading edge of a vane is centered in the stator assembly.
11. Rotate the distributor in the block again to ensure that the vane is centered on the stator and the rotor is pointing directly at the mark. If you cannot get the vane and stator to align correctly, lift the unit out of the block just enough so that the drive gear disengages with the camshaft gear, rotate the shaft slightly so that it engages with another tooth, and then reinsert the unit back into the block and repeat the alignment process.

■ If you have difficulty getting the oil pump shaft to engage with the distributor shaft, drop a little dawb of grease on the end of the oil pump shaft after positioning it in the center of the opening and then reinstall the distributor.

12. Install the hold-down clamp and tighten the bolt finger-tight.
13. Coat the 8-pin connector with Electric Terminal grease and then plug it into the housing.
14. Install the distributor cap and slide on the rubber boot.
15. Position the shift lever in the **NEUTRAL** position and install a variable advance inductive timing light as per the manufacturer's instructions.
16. Start the engine and allow it to reach normal operating temperature.
17. Remove the shorting bar from the 2-wire SPOUT connector located in the wiring harness next to the distributor.
18. With the engine idling, shine the timing light at the marks and slowly rotate the distributor until initial timing is 5° BTDC.
19. Coat the terminals with Electric Terminal grease and reinstall the shorting bar to the SPOUT connector. Confirm that the timing is advanced past the original setting.
20. Remove the timing light and tighten the clamp bolt to 17-25 ft. lbs. (23-34 Nm).
21. Recheck the timing and adjust as necessary.

◆ **See Figures 132 and 135 thru 141**

1. Remove the distributor cap and pull off the rotor.
2. Remove the distributor.
3. Loosen the two captive screws that secure the base adapter to the distributor housing and lift off the base assembly.
4. Use a magic marker or a little paint and mark the rotary vane cup and the distributor gear to confirm their orientation during installation.
5. Hold the distributor gear and remove the 2 vane cup retaining screws. Lift off the vane cup.
6. Lay the distributor on its side and support the shaft end with a small piece of wood. Use a blunt-end punch and drive out the roll pins holding the distributor gear and the spacer to the shaft.
7. Install a universal bearing separator over the gear and install the inverted assembly into an arbor press. Install Bearing Removal Tool (#D84L-950-A) or the equivalent and then press the distributor shaft out of the gear—make sure to hold the unit as it's coming out of the gear.
8. Clean all oil residue and dirt from the shaft. Use a piece of emery cloth to polish the shaft and remove any burrs. Slide the spacer off of the shaft if possible, otherwise, install the assembly back into the press and drive it off as you did the gear. Make sure that the bearing separator is between the distributor housing and the spacer.
9. Slide the distributor shaft out of the housing. Remove the screw from the stator lead and retainer and lift out the stator retainer.
10. Carefully pry the stator lead out of the cut-out in the side of the distributor housing and then lift the stator assembly up and free of the center flange.
To Install:

✳✳ CAUTION

Never clean any electrical components in a petroleum-based solvent.

11. Inspect the bushing in the distributor base for signs of wear or damage—the bushing is not a separate component, so the entire distributor assembly must be replaced.
12. Inspect the O-ring for wear or damage, replace as necessary.
13. Clean all electrical components with contact cleaner.
14. Inspect the stator for damaged wires, loose pins and/or cracked connectors. Replace the assembly if damaged or defective.
15. Position the stator over the center flange bushing and pres it into place. Slide the harness into its cut-out in the housing and then install the retainer. Tighten the screws to 15-35 inch lbs. (1.7-4.0 Nm).
16. Lubricate the bushing and shaft seal with clean engine oil and then slide the shaft down and through the housing.
17. Slide the spacer onto the bottom of the shaft so the flange is facing the housing. Align the roll pin hole in the spacer with the hole in the shaft and then press it into place with a suitable sized mandrel. Make sure that the two holes are lined up and then press in the roll pin.
18. Slide the drive gear onto the bottom of the shaft so the flange is facing the spacer. Align the roll pin hole in the gear flange with the hole in the shaft and then press it into place with a suitable sized mandrel. Make sure that the two holes are lined up and then press in the roll pin.

■ If the holes in the spacer or gear do not line up with the holes in the shaft, DO NOT attempt to force then into alignment by inserting a punch or similar instrument. Remove them and then start over.

19. Turn the distributor upright and position the rotary vane cup onto the upper end of the distributor shaft so the notches in the cup align with those in the shaft flange. Install the retaining screws and tighten them to 25-35 inch lbs. (2.8-4.0 Nm). Spin the distributor shaft a few times to check for smooth rotation and a lack of binding. Make sure that the vane cup does not come in contact with the stator—replace the distributor if it does.
20. Align the notch on the base adapter with the notch in the housing, install the screws and tighten them securely.
21. Slide the rotor into position so that the index leg of the rotor engages the notch in the rotary vane cup.
22. Install the distributor cap and then install the distributor. Check and adjust the timing.

Fig. 135 Remove the rotor...

Fig. 136 ...and then remove the adapter base

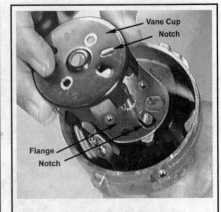

Fig. 137 Lifting out the rotary vane cup

Fig. 138 Drive out the roll pins with a hammer and drift

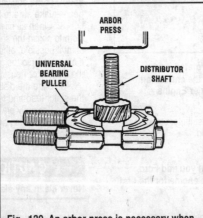

Fig. 139 An arbor press is necessary when removing the drive gear.

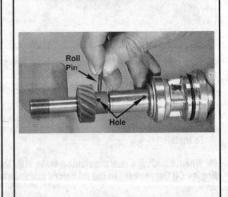

Fig. 140 Make sure that the holes are aligned before pressing in the roll pin

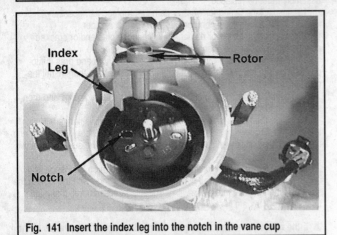

Fig. 141 Insert the index leg into the notch in the vane cup

E-Core Coil

REMOVAL & INSTALLATION

 MODERATE

1. Disconnect the battery cables.
2. Remove the plastic engine cover and then wiggle the high tension lead off of the coil.
3. Insert a very small screwdriver under the lock clip on the side of the coil and pry it out to release the electrical connector.
4. Remove the 2 large bolts (on top) securing the coil to the TFI bracket. Remove the 2 small bolts (on the side) and lift off the coil and coil bracket.

To Install:

5. Position the coil and bracket onto the TFI bracket and tighten the small bolts to 24-36 inch lbs. (2.7-4.1 Nm). Install the larger bolts through the top and tighten them to 60-84 inch lbs. (6.8-9.5 Nm).
6. Coat the connector terminals with Electrical Terminal grease and plug in the connector until the locking latch snaps into place.
7. Coat the coil lead and terminal lightly with wheel bearing grease and connect the lead to the terminal.
8. Install the engine cover, connect the battery cables, start the engine and check for proper operation.

TESTING

MODERATE

◆ See Figures 142, 143 and 144

1. Connect a multi-meter as per the manufacturer's instructions.
2. Connect the meter leads between the positive coil terminal and the secondary terminal. Secondary circuit resistance should be 7600-9400 ohms. Test **A**.
3. Connect the meter leads between the negative coil terminal and the secondary terminal. Secondary circuit resistance should be 7600-9400 ohms. Test **A**.
4. Connect the meter leads between the negative coil terminal and the positive terminal. Primary circuit resistance should be 0.39-0.42 ohms.
5. Connect the meter leads between the positive terminal and the coil case ground. The meter should show infinity. Test **B**.
6. Connect the meter leads between the secondary terminal and the coil case ground. The meter should show infinity. Test **C**.
7. Any readings not within the ranges given will indicate the need to replace the coil.

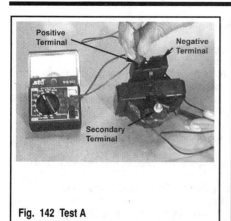

Fig. 142 Test A

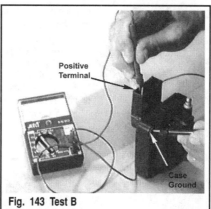

Fig. 143 Test B

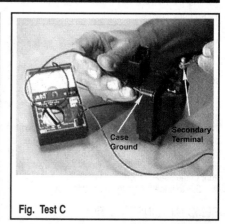

Fig. Test C

TFI Module

REMOVAL & INSTALLATION

◆ See Figure 145

1. Disconnect the battery cables.
2. Remove the plastic engine cover and then wiggle the high tension lead off of the coil.
3. Remove the E-coil.
4. Grasp the sides of the module connector, push in on each side clip while pulling on the body of the connector and then remove the connector.
5. Remove the 2 mounting bolts and lift off the module assembly. Remove the 2 bolts securing the module to the heat sink and separate the two.

To Install:
6. Coat the back of the module with a 1/32 in. (0.79mm) thick layer of Heat Sink Compound, position the module to the heat sink, and then tighten the screws to 24-36 inch lbs. (2.7-4.1 Nm).
7. Position the assembly on the bracket and tighten the bolts to 60-84 inch lbs. (6.8-9.5 Nm).
8. Coat the connector terminals with Electric Terminal grease and push the two connectors together so that the latches lock into position.
9. Install the E-coil. Install the engine cover and connect the battery cables.
10. Start the engine and adjust the initial timing if necessary.

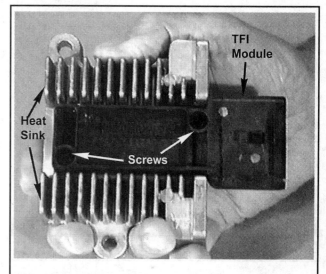

Fig. 145 Close-up of the module and heat sink

INSTRUMENTS AND GAUGES

Oil And Temperature Gauges

TROUBLESHOOTING

◆ See Figure 146

 MODERATE

The body of oil and temperature gauges must be grounded and they must be supplied with 12 volts. Many gauges have a terminal on the mounting bracket for attaching a ground wire. A tang from the mounting bracket makes contact with the gauge. Check to be sure the tang does make good contact with the gauge.

Ground the wire to the sending unit and the needle of the gauge should move to the full right position indicating the gauge is in serviceable condition.

Check the sender unit for a defective temperature warning system.

If a problem arises on a boat equipped with water, temperature, and oil pressure lights, the first area to check is the light assembly for loose wires or burned-out bulbs.

When the ignition key is turned to the **ON** position, the light assembly is supplied with 12 volts and grounded through the sending unit mounted on the engine. When the sending unit makes contact because the water temperature is too hot or the oil pressure is too low, the circuit to ground is completed and the lamp should light.

To check the bulb: turn the ignition switch to the **ON** position. Disconnect the wire at the sending unit, and then ground the wire. The lamp on the dash should light. If it does not light, check for a burned-out bulb or a break in the wiring to the light.

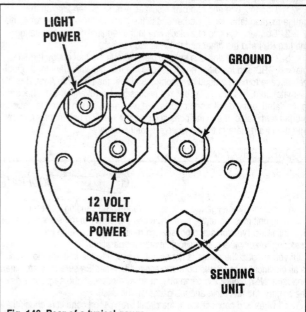

Fig. 146 Rear of a typical gauge

TESTING

1. With the ignition switch in the **OFF** position, remove the wire to the sending unit.

2. Turn the ignition switch to the **RUN** position and confirm that the gauge needle is seated on the post at the left side; or all the way to the left of the scale if there is no post.

3. Turn the ignition switch OFF again and connect a jumper wire between the ground terminal and the sending unit terminal.

4. Turn the ignition switch back to the **RUN** position and confirm that the gauge needle is seated on the post at the right side; or all the way to the right of the scale if there is no post.

5. Replace the gauge if anything fails.

REMOVAL & INSTALLATION

1. Disconnect the battery cables.

2. Tag and disconnect the electrical leads at the back of the gauge.

3. Disconnect the light socket, remove the holding strap and lift out the gauge.

To Install:

4. Position the gauge into the mounting hole, install the strap and tighten the nuts securely.

■ **Be careful not to tighten the holding strap too tightly or you risk distorting the gauge casing.**

5. Connect the ground wire and then connect the remaining leads.

6. Install the light socket and then coat all terminal connections with liquid neoprene or equivilent.

7. Connect the battery cables.

Fuel Gauge

HOOKUP

The Boating Industry Association recommends the following color coding be used on all fuel gauge installations:
- Black—for all grounded current-carrying conductors.
- Pink—insulated wire from the fuel gauge sending unit to the gauge.
- Red—insulated wire for a connection from the positive side of the battery to any electrical equipment.

1. Connect one end of a pink insulated wire to the terminal on the gauge marked tank and the other end to the terminal on top of the tank unit.

2. Connect one end of a black wire to the terminal on the fuel gauge marked **IGN** and the other end to the ignition switch.

3. Connect one end of a second black wire to the fuel gauge terminal marked **GRD** and the other end to a good ground. It is important for the fuel gauge case to have a good common ground with the tank unit. Aboard an all-metal boat, this ground wire is not necessary. However, if the dashboard is insulated, or made of wood or plastic, a wire must be run from the gauge ground terminal to one of the bolts securing the sending unit in the fuel tank, and then from there to the negative side of the battery.

TROUBLESHOOTING

In order for the fuel gauge to operate properly the sending unit and the receiving unit must be of the same type and preferably of the same make. The following symptoms and possible corrective actions will be helpful in restoring a faulty fuel gauge circuit to proper operation.

If you suspect the gauge is not operating properly, the first area to check is all electrical connections from one end to the other. Be sure they are clean and tight. Next, check the common ground wire between the negative side of the battery, the fuel tank, and the gauge on the dash.

If all wires and connections in the circuit are in good condition, check the sending unit.

If the pointer does not move from the empty position one of four faults could be to blame:
1. The dash receiving unit is not properly grounded.
2. No voltage at the dash receiving unit.
3. Negative meter connections are on a positive grounded system.
4. Positive meter connections are on a negative grounded system.

If the pointer fails to move from the full position, the problem could be one of three faults.
5. The tank sending unit is not properly grounded.
6. Improper connection between the tank sending unit and the receiving unit on the dash.
7. The wire from the gauge to the ignition switch is connected at the wrong terminal.

If the pointer remains at the 3/4 full mark, it indicates a six-volt gauge is installed in a 12-volt system.

If the pointer remains at about 3/8 full, it indicates a 12-volt gauge is installed in a six-volt system.

Erratic Fuel Gauge Readings

Inspect all of the wiring in the circuit for possible damage to the insulation or conductor. Carefully check:
1. Ground connections at the receiving unit on the dash.
2. Harness connector to the dash unit.
3. Body harness connector to the chassis harness.
4. Ground connection from the fuel tank to the trunk floor pan.
5. Feed wire connection at the tank sending unit.

Gauge Always Reads Full—When Ignition Switch Is In ON Position

1. Check the electrical connections at the receiving unit on the dash; the body harness connector to chassis harness connector; and the tank unit connector in the tank.

2. Make a continuity check of the ground wire from the tank to the tank floor pan.

3. Connect a known good tank unit to the tank feed wire and the ground lead. Raise and lower the float and observe the receiving unit on the dash. If the dash unit follows the arm movement, replace the tank sending unit.

Gauge Always Reads Empty—When Ignition Switch Is In ON Position

Disconnect the tank unit feed wire and do not allow the wire terminal to ground. The gauge on the dash should read full.

If Gauge Reads Empty

1. Connect a spare control unit into the control unit harness connector and ground the unit. If the spare unit reads full, the original unit is shorted and must be replaced.

2. A reading of empty indicates a short in the harness between the tank sending unit and the gauge on the control panel.

If Gauge Reads Full

1. Connect a known good tank sending unit to the tank feed wire and the ground lead.

2. Raise and lower the float while observing the gauge on the control panel. If the control panel gauge follows movement of the float, replace the tank sending unit.

Gauge Never Reads Full

This test requires shop test equipment.

1. Disconnect the feed wire to the tank unit and connect the wire to ground thru a variable resistor or thru a spare tank unit.

2. Observe the control panel gauge reading. The reading should be full when resistance is increased to about 90 ohms. This resistance would simulate a full tank.

3. If the check indicates the control panel gauge is operating properly, the trouble is either in the tank sending unit rheostat being shorter, or the float is binding. The arm could be bent, or the tank may be deformed. Inspect and correct the problem.

TESTING

1. With the ignition switch in the **OFF**, remove the wire to the sending unit (lower right, looking at the back of the gauge).

2. Turn the ignition switch to the **RUN** position and confirm that the gauge needle is seated on the post at the left side; or all the way to the left of the scale if there is no post.

3. Turn the ignition switch back to the **OFF** position and connect a jumper wire between the ground terminal (upper right from the back of gauge) and the sending unit terminal.

4. Turn the ignition switch to the **RUN** position again and confirm that the gauge needle is seated on the post at the right side or all the way to the right of the scale if there is no post.

5. Replace the gauge if anything fails.

REMOVAL & INSTALLATION

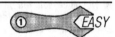

1. Disconnect the battery cables.
2. Tag and disconnect the electrical leads at the back of the gauge.
3. Disconnect the light socket, remove the holding strap and lift out the gauge.

To Install:

4. Position the gauge into the mounting hole, install the strap and tighten the nuts securely.

■ **Be careful not to tighten the holding strap too tightly or you risk distorting the gauge casing.**

5. Connect the ground wire and then connect the remaining leads.
6. Install the light socket and then coat all terminal connections with liquid neoprene or equivilent.
7. Connect the battery cables.

Battery Gauge

TESTING

1. Make sure that the battery is fully charged and then turn the ignition switch to the **RUN** position. The gauge should read battery voltage, otherwise replace it.

2. If the gauge is out of the dash, turn the ignition switch OFF and connect a jumper wire between the ground terminal (upper right looking at the back of the gauge) and the negative battery terminal. Connect another jumper between the power terminal (lower left looking at the back of the gauge) and the positive battery terminal. The gauge should register battery voltage, otherwise replace it.

REMOVAL & INSTALLATION

1. Disconnect the battery cables.
2. Tag and disconnect the electrical leads at the back of the gauge.
3. Disconnect the light socket, remove the holding strap and lift out the gauge.

To Install:

4. Position the gauge into the mounting hole, install the strap and tighten the nuts securely.

■ **Be careful not to tighten the holding strap too tightly or you risk distorting the gauge casing.**

5. Connect the ground wire and then connect the remaining leads.
6. Install the light socket and then coat all terminal connections with liquid neoprene or equivalent.
7. Connect the battery cables.

Speedometer

TESTING

■ **An air compressor is necessary for this procedure. It is imperative that its pressure gauge is extremely accurate.**

Disconnect the hose at the back of the gauge and then connect a line from an air compressor. Apply 5.3 psi of air pressure to the speedometer and check that the gauge reads 20 mph (+/- 1 mph). Now apply 27.8 psi pressure and check that the gauge reads 45 mph (+/- 1 mph). If either of the readings is not as specified, replace the speedometer.

REMOVAL & INSTALLATION

1. Disconnect the battery cables.
2. Tag and disconnect the electrical leads and the hose at the back of the gauge.
3. Disconnect the light socket, remove the holding strap and lift out the gauge.

To Install:

4. Position the gauge into the mounting hole, install the strap and tighten the nuts securely.

■ **Be careful not to tighten the holding strap too tightly or you risk distorting the gauge casing.**

5. Connect the ground wire and then connect the remaining leads. Connect the hose.
6. Install the light socket and then coat all terminal connections with liquid neoprene or equivilent.
7. Connect the battery cables.

Tachometer

TESTING

◆ **See Figure 147**

Connect a tachometer as per the manufacturer's instructions. Start the engine and check a few different engine speeds on the boat's tachometer against the service tach. If using a 6000 rpm service tach, variations of +/- 150 rpm are acceptable; if using an 8000 rpm tach, variations of +/- 200 rpm are acceptable. Replace the tachometer if not within specifications.

All tachometers supplied with OMC engines should have a small switch on the back of the unit to program it for the particular engine that it is connected to. Switch calibration is as follows:

• 4 cylinder engines—arrow points to position **2**
• V6 engines—arrow points to position **3**
• V8 engines—arrow points to position **4**

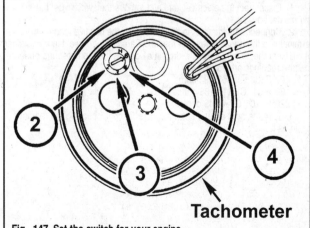

Fig. 147 Set the switch for your engine

REMOVAL & INSTALLATION

1. Disconnect the battery cables.

2. Tag and disconnect the electrical leads at the back of the gauge.

3. Disconnect the light socket, remove the holding strap and lift out the gauge.

To Install:

4. Position the gauge into the mounting hole, install the strap and tighten the nuts securely.

■ **Be careful not to tighten the holding strap too tightly or you risk distorting the gauge casing.**

5. Connect the ground wire and then connect the remaining leads. .
6. Install the light socket and then coat all terminal connections with liquid neoprene or equivalent.
7. Connect the battery cables.

SENDING UNITS AND SWITCHES

Oil Pressure Sending Unit

TESTING

1986-91 Engines

Start the engine. Allow it to idle until it reaches normal operating temperature and then run the engine up to full throttle and back again. Observe the gauge and replace the sender if unsatisfactory. If the reading is still unsatisfactory after doing the sender, check the gauge.

1992-98 Engines

■ **This test should be performed only after confirming that the oil pressure gauge is operating properly.**

The oil pressure sending unit is located on the either side of the engine block, depending on your engine.
Connect a multi-meter as per the manufacturer's instructions. Connect the positive lead of the meter to the sender terminal and the negative lead to the hex nut on the back of the sender housing. With the engine not running, the meter should register continuity as shown in the accompanying chart.
Start the engine and check the sender at the following oil pressure readings:
- 0 psi—227-257 ohms
- 40 psi—92-114 ohms
- 80 psi—21.5-49.5 ohms
If any readings vary, replace the unit.

Water Temperature Sender

TESTING

The water temperature sending unit is usually located in the thermostat housing.
1. Disconnect the electrical lead and remove the switch from the thermostat housing.
2. Connect a multi-meter to the switch with the positive lead on the terminal and the negative lead on the hex. Carefully immerse it in a container of oil. Heat the oil very carefully, drop in a cooking thermometer and observe that the readings are as follows:
- 100°F (38°C): 403-493 ohms
- 160°F (71°C): 118-138 ohms
- 220°F (105°C): 44-49 ohms
3. Turn the heat off and allow the water to cool, checking the readings again as it cools.

Audio Warning System

TESTING

■ **Not all vessels will be equipped with this system which will sound an alarm when the oil pressure or water temperature reach a specified level. Under normal engine operating conditions, the horn will sound when the ignition switch is turned to the ON position, and continue sounding until the engine is started and oil pressure reaches at least 2-6 psi. The system will also when it senses water temperature above 200°F (93°C).**

Oil Pressure Switch

As mentioned previously, the audible warning oil pressure switch is calibrated to "make" or "break" contact at 4 +/-2 psi.
1. Connect a multi-meter as per the manufacturer's instructions. Connect the positive lead to the switch terminal and the negative lead to a good ground. The meter should indicate full continuity (0 ohms).
2. Start the engine and run it at idle. The meter should indicate no continuity (infinity).
3. Replace the switch if either reading is incorrect.

Water Temperature Switch

The audible warning switch is calibrated to "make" or "break" contact when temperature exceeds 195-205°F (88-98°C).
1. Connect a multi-meter as per the manufacturer's instructions. Connect the positive lead to the switch terminal and the negative lead to a good ground.
2. Carefully lower the switch into a container of oil with a cooking thermometer. Carefully heat the container (no flames please) until the thermometer reads above 200°F (93°C).
3. When the oil is below the above temperature, the meter should read no continuity (infinity). At, or above, the designated temperature, the meter should now show full continuity (0 ohms).

Warning Horn

1. Turn the ignition switch to the **ON** position but do not start the engine. The buzzer should sound and then turn off.
2. If the buzzer does not sound, disconnect the electrical lead at the water temperature sender and touch it briefly to engine ground. If the buzzer sounds, you have a problem in the wiring circuit; if it does not sound, replace it.
3. Disconnect the electrical lead at the oil pressure sender and touch it briefly to engine ground. If the buzzer sounds, you have a problem in the wiring circuit; if it does not sound, replace it.

Ignition Switch

TESTING

◆ **See Figure 148**

■ Before performing this test, please ensure that all fuses and the starter motor are in good condition.

1. Disconnect the battery cables and connect a multi-meter as per the manufacturer's instructions.
2. Tag and disconnect all leads at the back of the switch.
3. With the ignition switch in the **OFF** position, test for continuity across all of the terminals on the back of the switch—there should be none (the meter should show infinity). Note that on stern drives, terminals **M** and **C** are not used.
4. Move the ignition switch to the **ON** position. Check for continuity between terminals **A** and **B**.
5. Move the ignition switch to the **START** position. There should be continuity between terminals **S** and **B**.
6. Reconnect all leads and coat them with Liquid Neoprene. Disconnect the meter and connect the battery cables.

Fig. 148 Rear of a typical ignition switch

WIRING DIAGRAMS

INDEX

CODE	WIRE COLOR
B	BLACK
BL	BLUE
BN	BROWN
GN	GREEN
GR	GREY
LBL	LIGHT BLUE
LBN	LIGHT BROWN
LTGN	LIGHT GREEN
OR	ORANGE
P	PINK
PU	PURPLE
R	RED
SB	SOLID BLACK
T	TAN
VO	VIOLET
W	WHITE
Y	YELLOW

NOTE: Main wire color is the first color, while the stripe/trace color is the second color on multi-colored wires.

T/B = Tan wire with Black stripe

R/PU = Red wire with Purple stripe

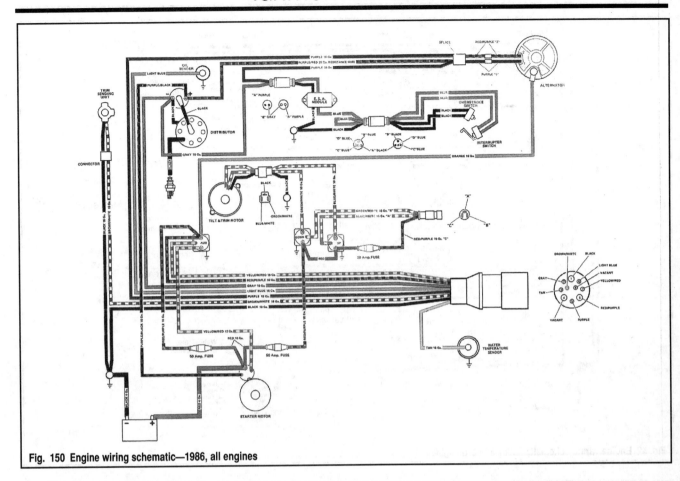

Fig. 150 Engine wiring schematic—1986, all engines

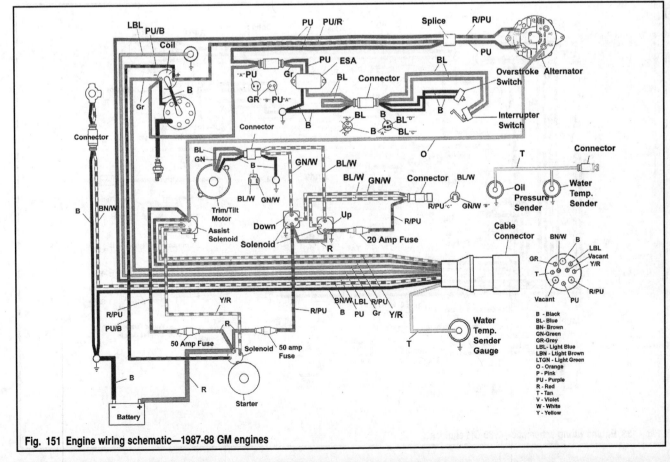

Fig. 151 Engine wiring schematic—1987-88 GM engines

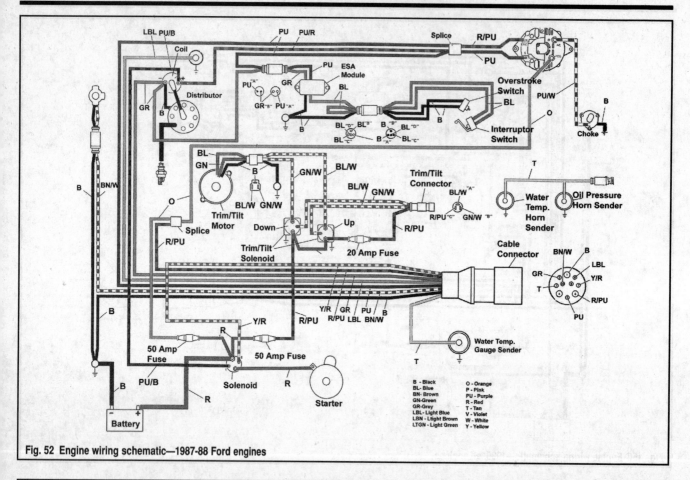

Fig. 52 Engine wiring schematic—1987-88 Ford engines

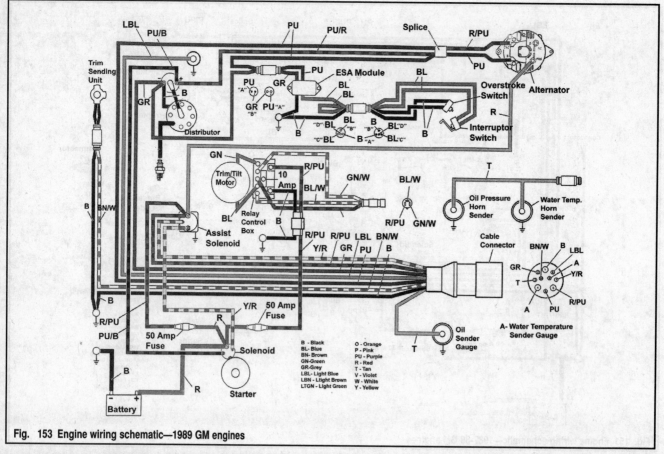

Fig. 153 Engine wiring schematic—1989 GM engines

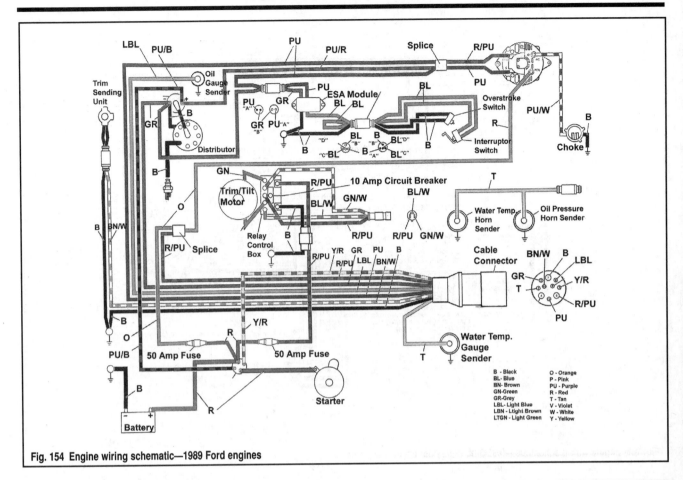

Fig. 154 Engine wiring schematic—1989 Ford engines

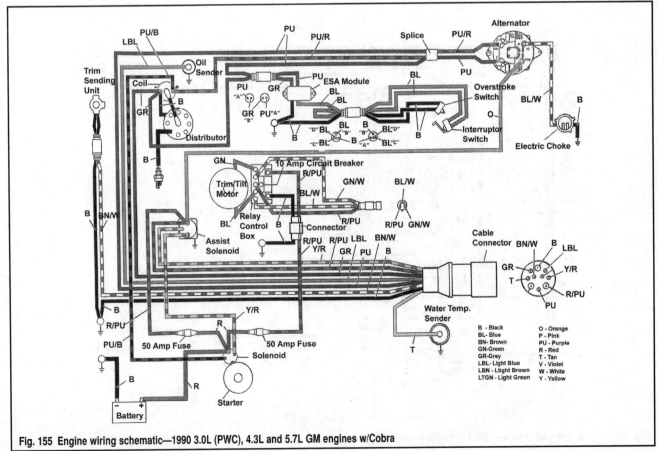

Fig. 155 Engine wiring schematic—1990 3.0L (PWC), 4.3L and 5.7L GM engines w/Cobra

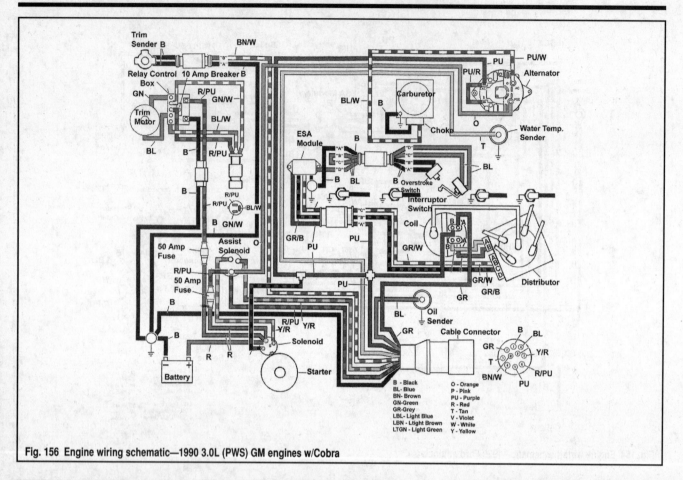

Fig. 156 Engine wiring schematic—1990 3.0L (PWS) GM engines w/Cobra

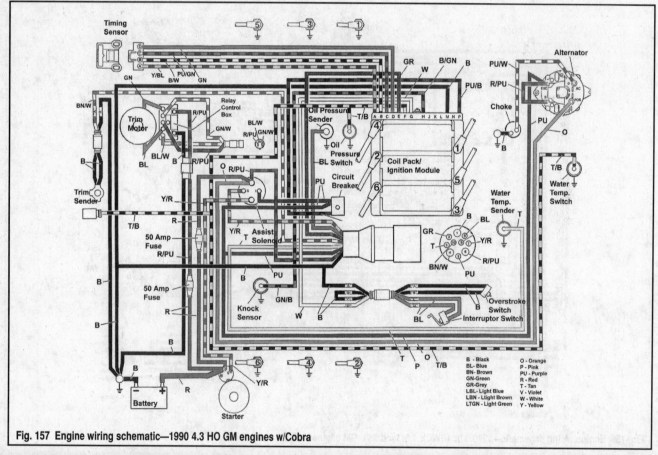

Fig. 157 Engine wiring schematic—1990 4.3 HO GM engines w/Cobra

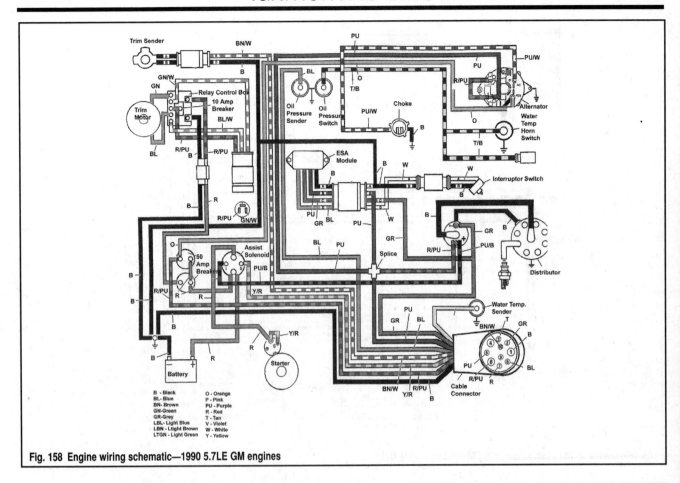

Fig. 158 Engine wiring schematic—1990 5.7LE GM engines

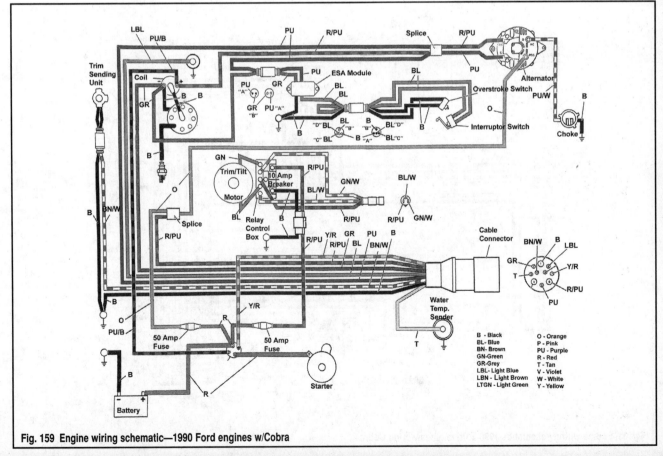

Fig. 159 Engine wiring schematic—1990 Ford engines w/Cobra

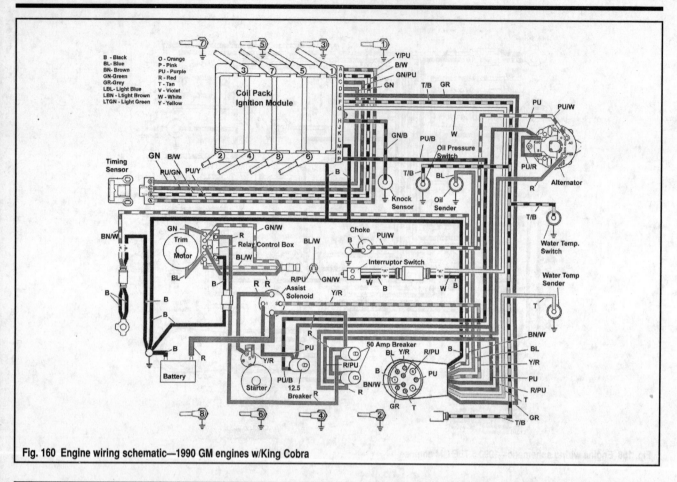

Fig. 160 Engine wiring schematic—1990 GM engines w/King Cobra

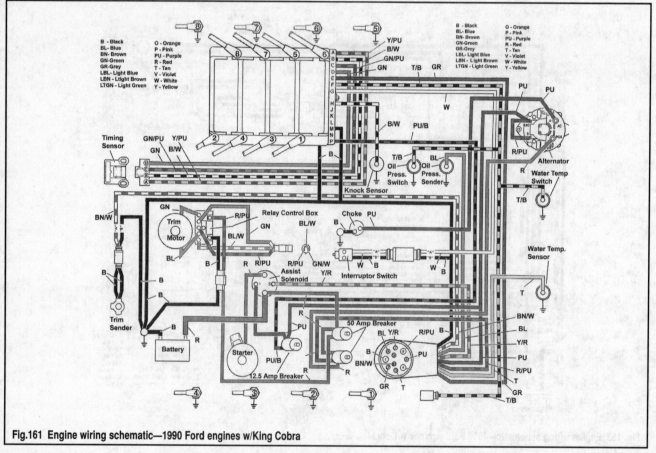

Fig.161 Engine wiring schematic—1990 Ford engines w/King Cobra

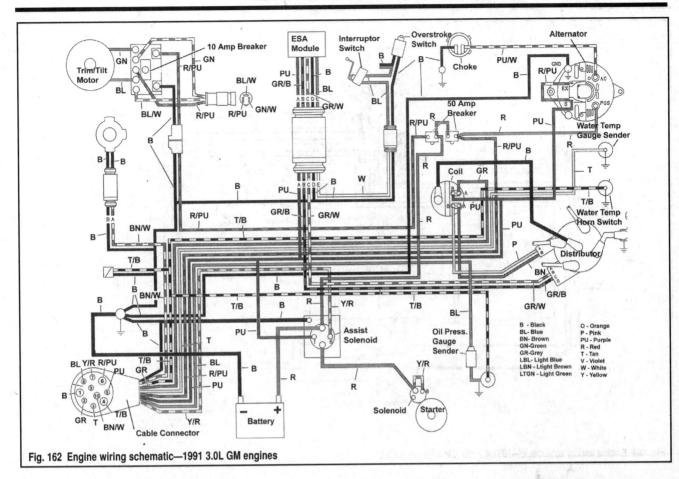

Fig. 162 Engine wiring schematic—1991 3.0L GM engines

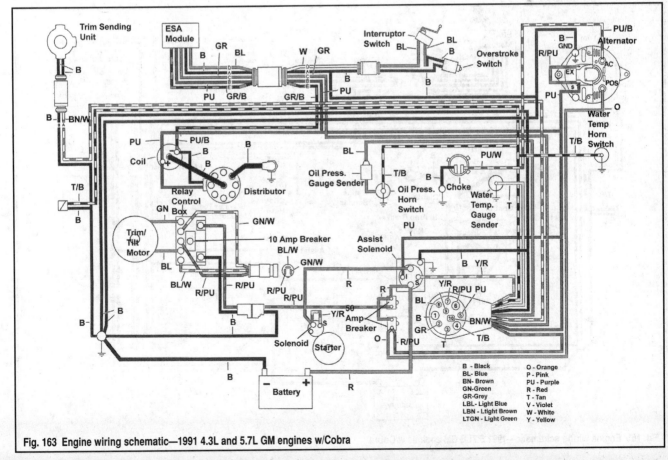

Fig. 163 Engine wiring schematic—1991 4.3L and 5.7L GM engines w/Cobra

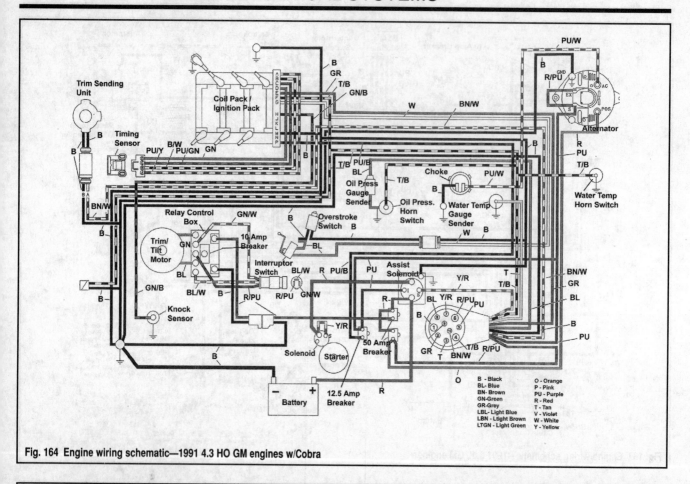

Fig. 164 Engine wiring schematic—1991 4.3 HO GM engines w/Cobra

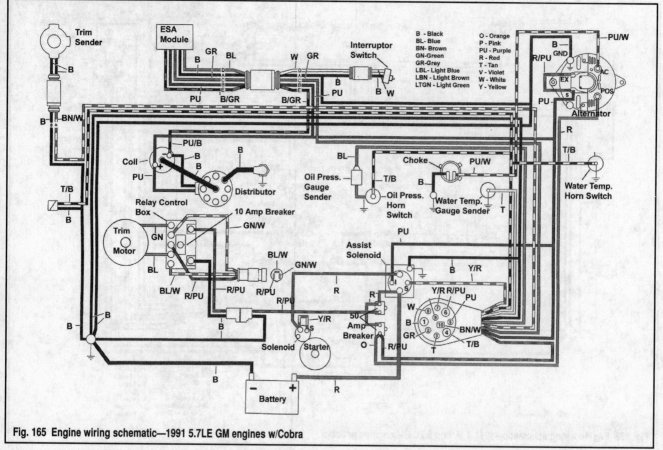

Fig. 165 Engine wiring schematic—1991 5.7LE GM engines w/Cobra

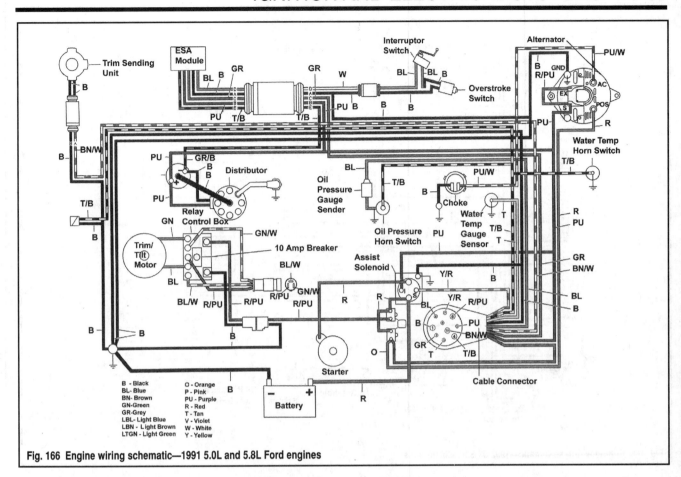

Fig. 166 Engine wiring schematic—1991 5.0L and 5.8L Ford engines

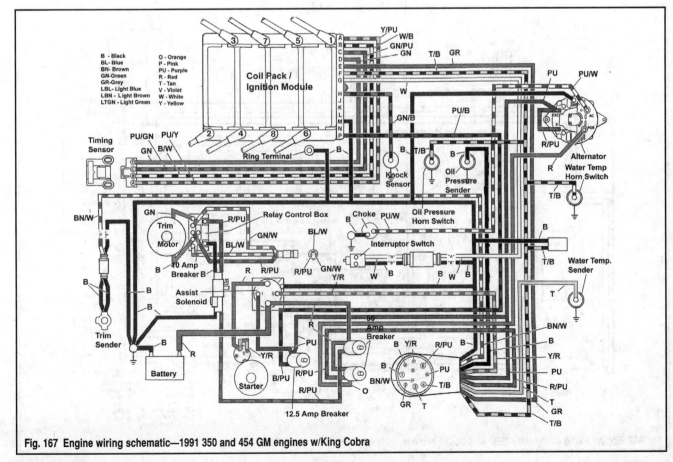

Fig. 167 Engine wiring schematic—1991 350 and 454 GM engines w/King Cobra

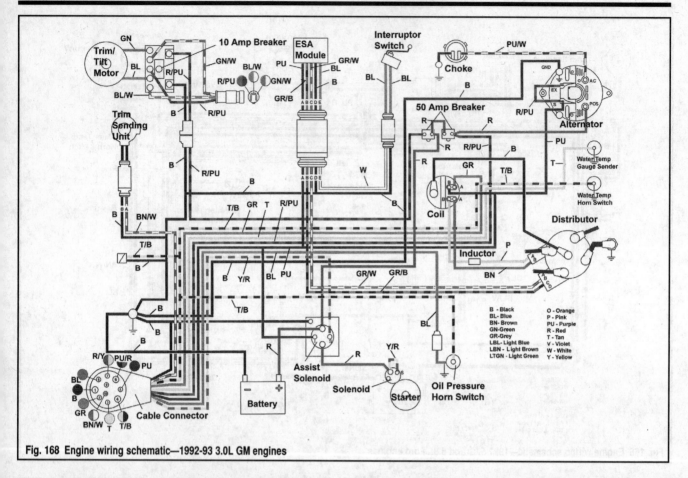

Fig. 168 Engine wiring schematic—1992-93 3.0L GM engines

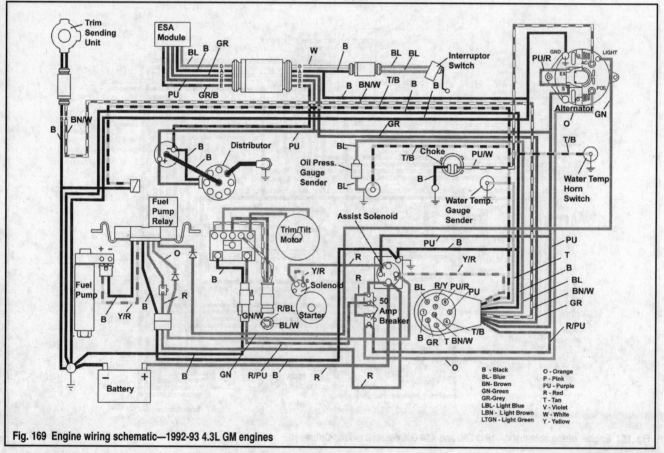

Fig. 169 Engine wiring schematic—1992-93 4.3L GM engines

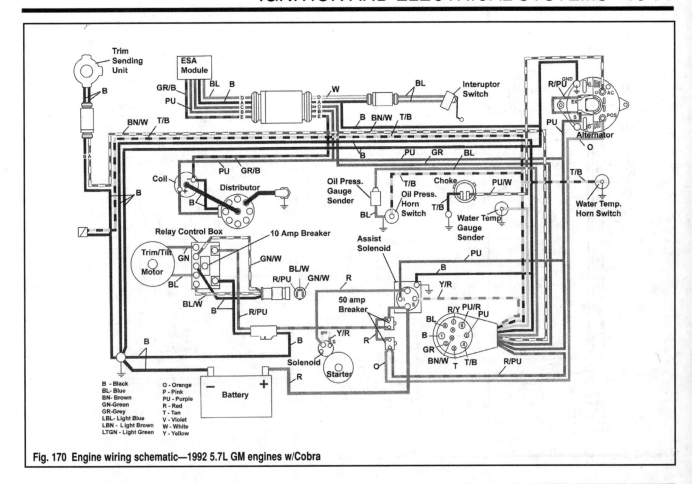

Fig. 170 Engine wiring schematic—1992 5.7L GM engines w/Cobra

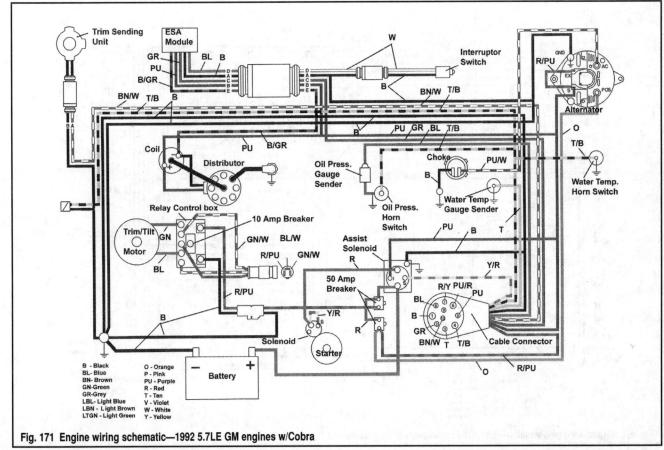

Fig. 171 Engine wiring schematic—1992 5.7LE GM engines w/Cobra

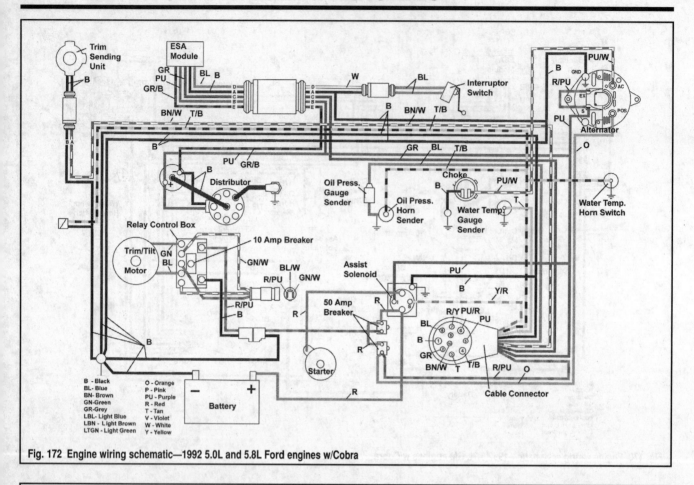

Fig. 172 Engine wiring schematic—1992 5.0L and 5.8L Ford engines w/Cobra

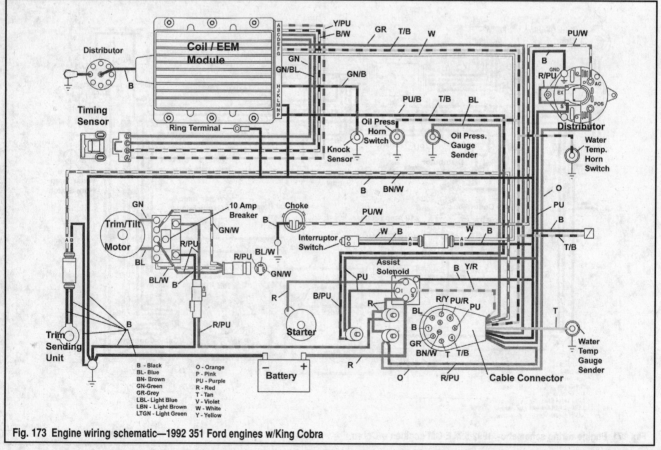

Fig. 173 Engine wiring schematic—1992 351 Ford engines w/King Cobra

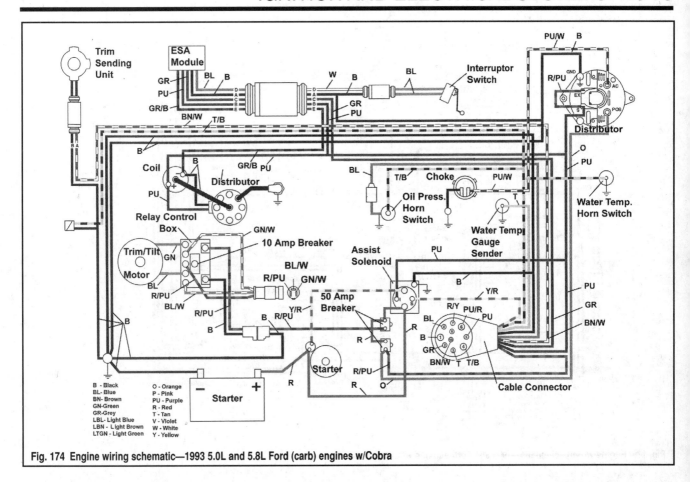

Fig. 174 Engine wiring schematic—1993 5.0L and 5.8L Ford (carb) engines w/Cobra

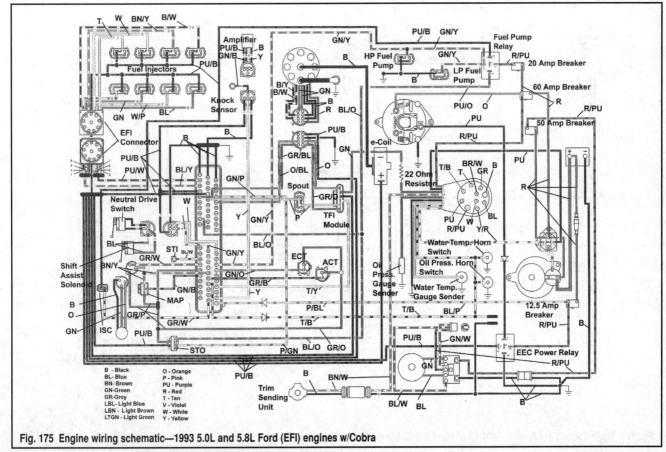

Fig. 175 Engine wiring schematic—1993 5.0L and 5.8L Ford (EFI) engines w/Cobra

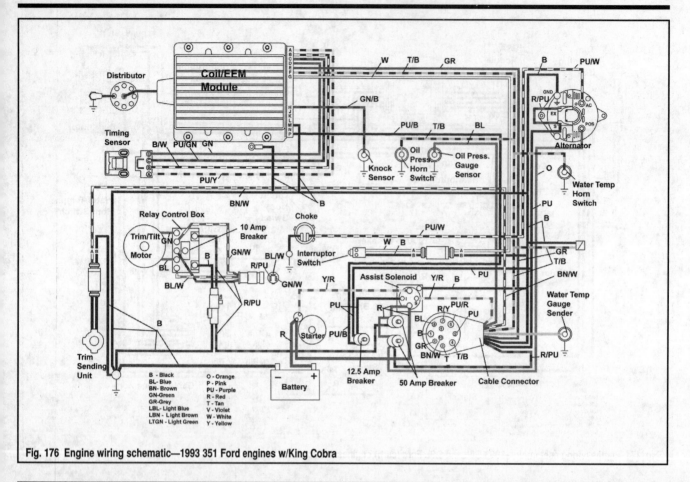

Fig. 176 Engine wiring schematic—1993 351 Ford engines w/King Cobra

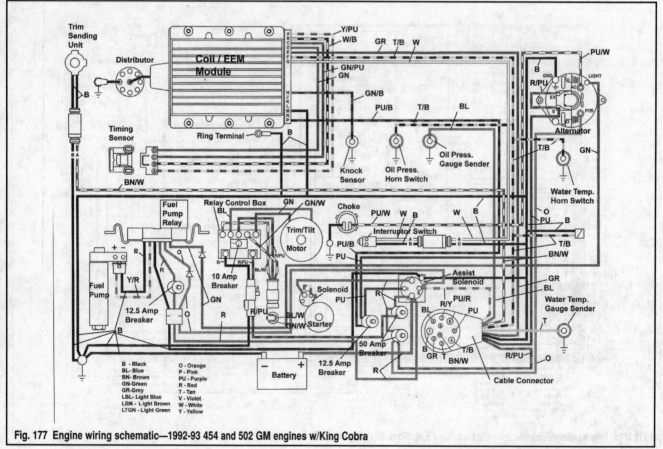

Fig. 177 Engine wiring schematic—1992-93 454 and 502 GM engines w/King Cobra

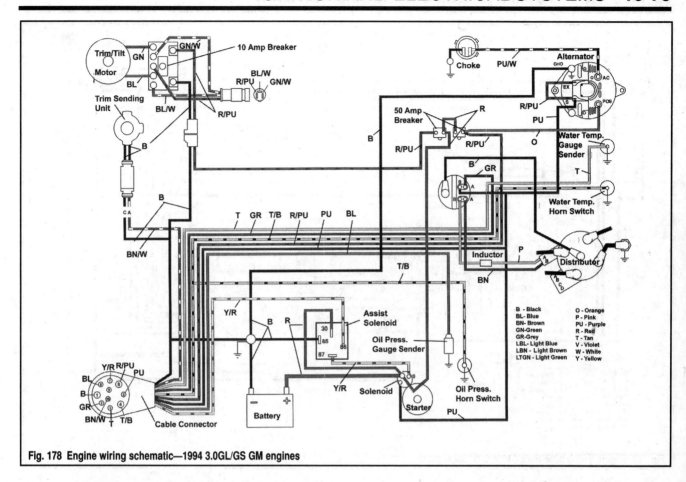

Fig. 178 Engine wiring schematic—1994 3.0GL/GS GM engines

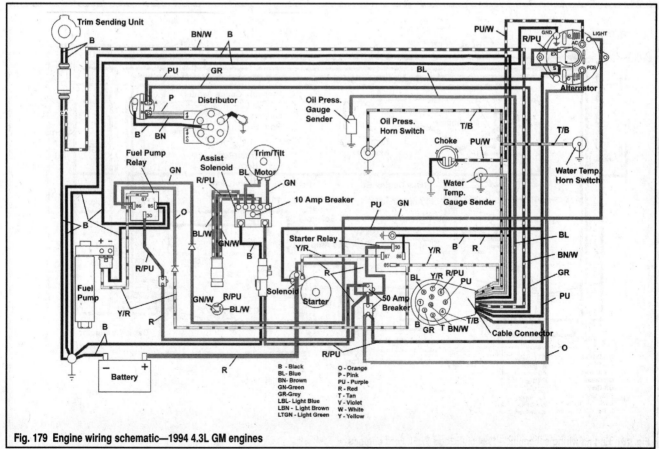

Fig. 179 Engine wiring schematic—1994 4.3L GM engines

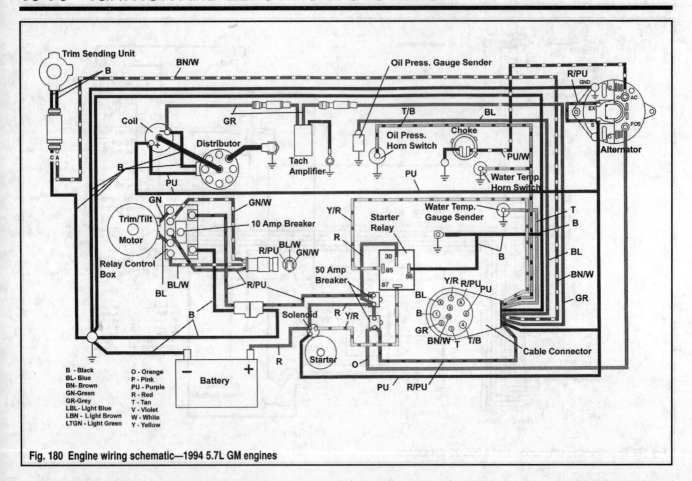

Fig. 180 Engine wiring schematic—1994 5.7L GM engines

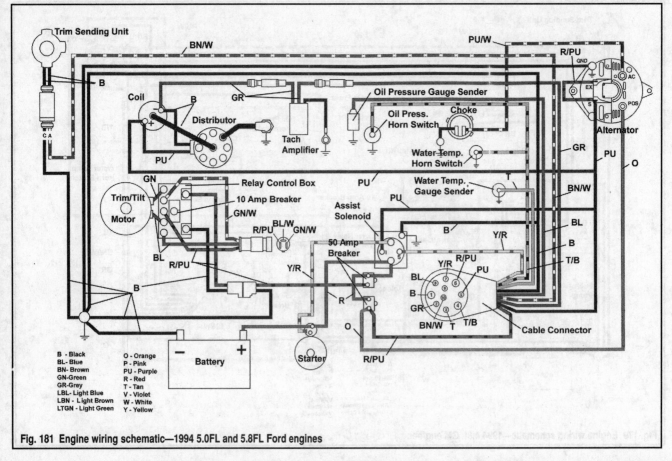

Fig. 181 Engine wiring schematic—1994 5.0FL and 5.8FL Ford engines

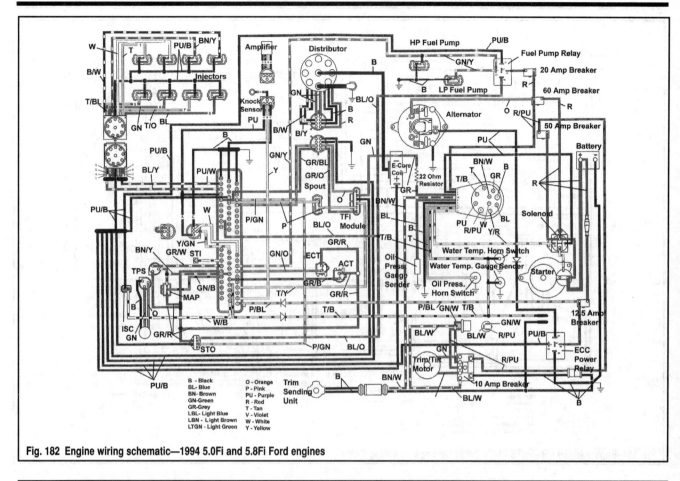

Fig. 182 Engine wiring schematic—1994 5.0Fi and 5.8Fi Ford engines

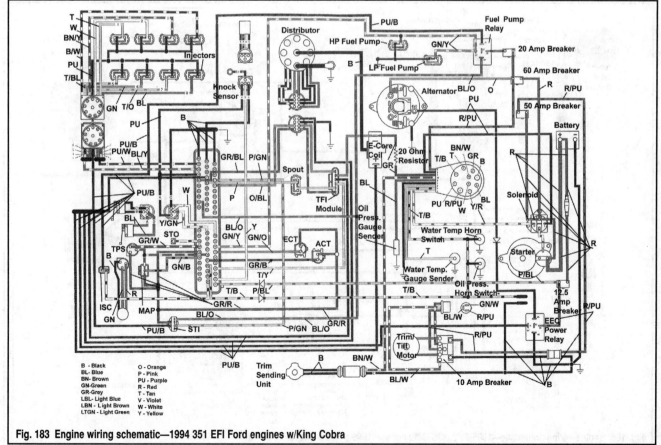

Fig. 183 Engine wiring schematic—1994 351 EFI Ford engines w/King Cobra

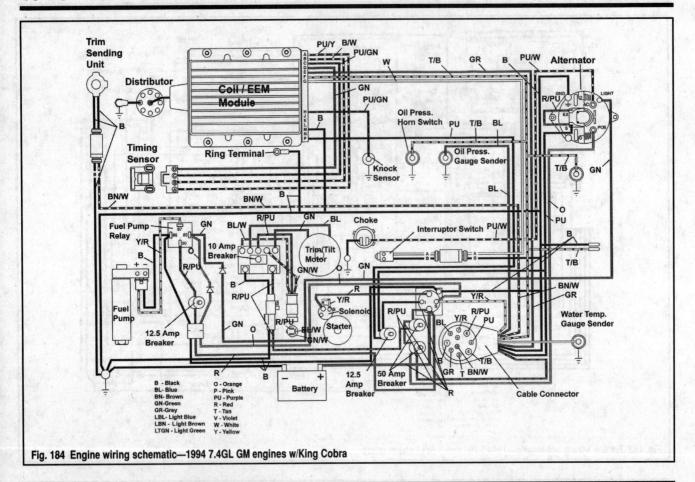

Fig. 184 Engine wiring schematic—1994 7.4GL GM engines w/King Cobra

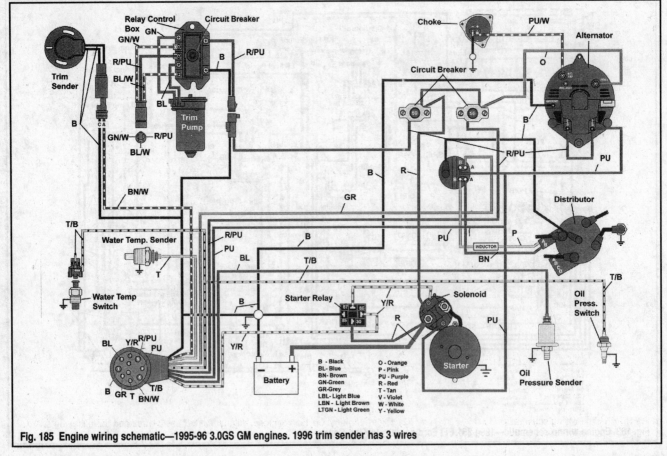

Fig. 185 Engine wiring schematic—1995-96 3.0GS GM engines. 1996 trim sender has 3 wires

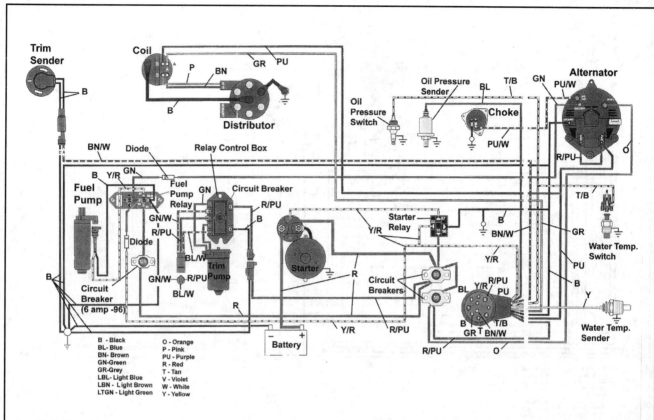

Fig. 186 Engine wiring schematic—1995-96 4.3GL/GS GM engines. On 1996 engines, the trim sender has 3 wires and there may be a 10 amp fuse on the R/PU wire at the relay control box

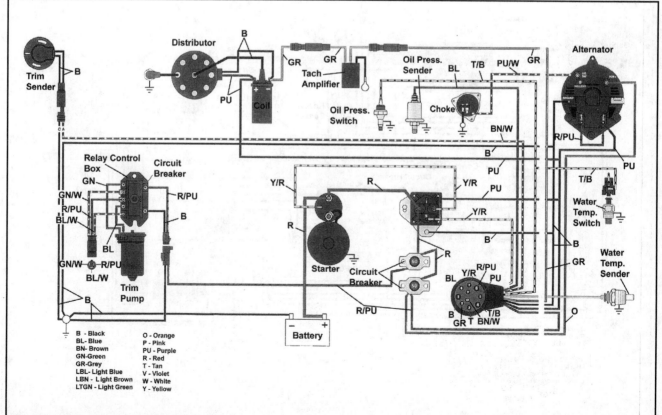

Fig. 187 Engine wiring schematic—1995-96 5.0FL and 5.8FL Ford engines. On 1996 engines, the trim sender has 3 wires and there may be a 10 amp fuse on the R/PU wire at the relay control box

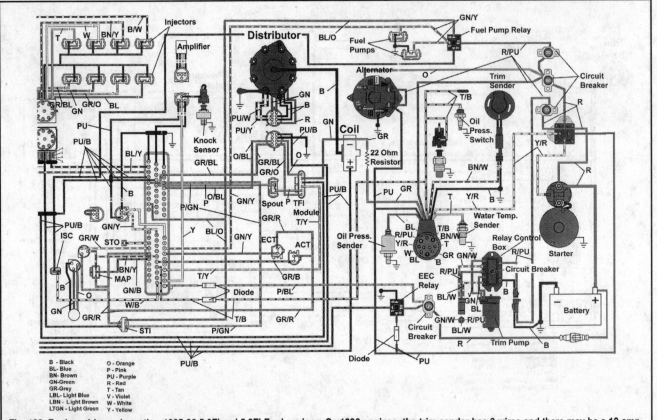

Fig. 188 Engine wiring schematic—1995-96 5.0Fi and 5.8Fi Ford engines. On 1996 engines, the trim sender has 3 wires and there may be a 10 amp fuse on the R/PU wire at the relay control box

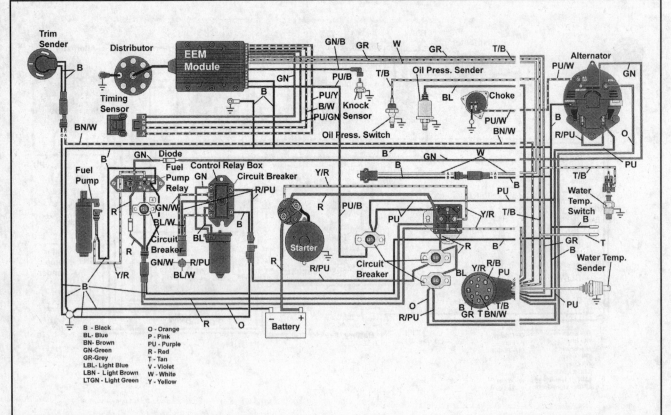

Fig. 189 Engine wiring schematic—1995-96 7.4GL GM engines w/King Cobra. On 1996 engines, the trim sender has 3 wires and there may be a 10 amp fuse on the R/PU wire at the relay control box

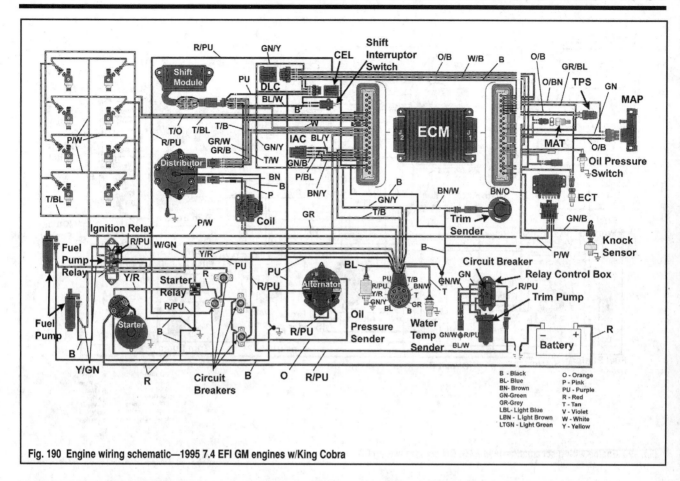

Fig. 190 Engine wiring schematic—1995 7.4 EFI GM engines w/King Cobra

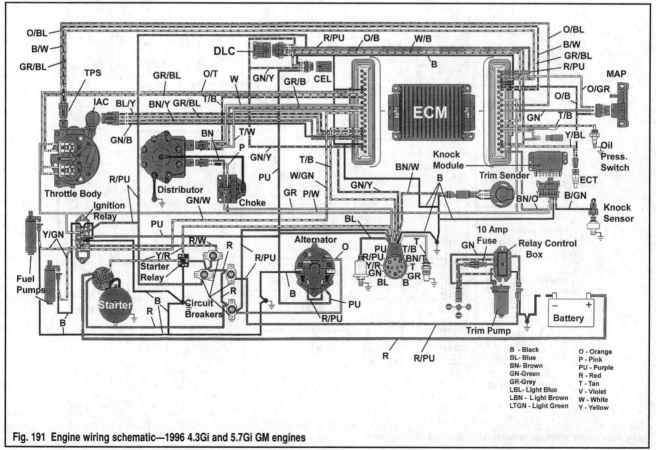

Fig. 191 Engine wiring schematic—1996 4.3Gi and 5.7Gi GM engines

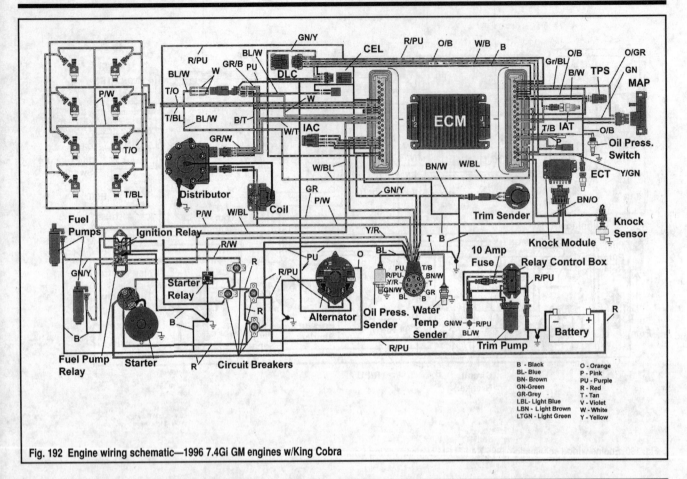

Fig. 192 Engine wiring schematic—1996 7.4Gi GM engines w/King Cobra

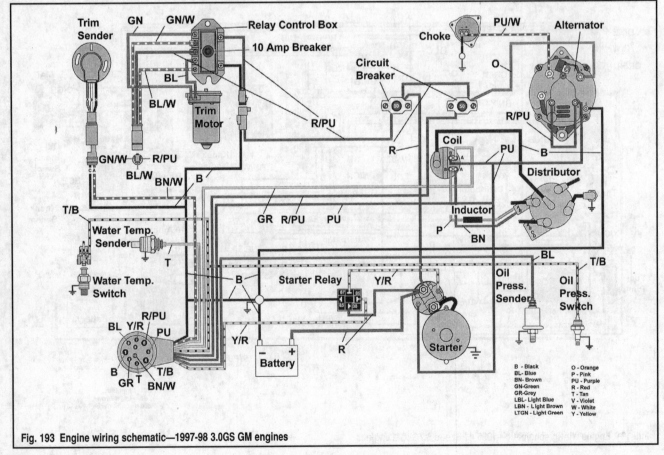

Fig. 193 Engine wiring schematic—1997-98 3.0GS GM engines

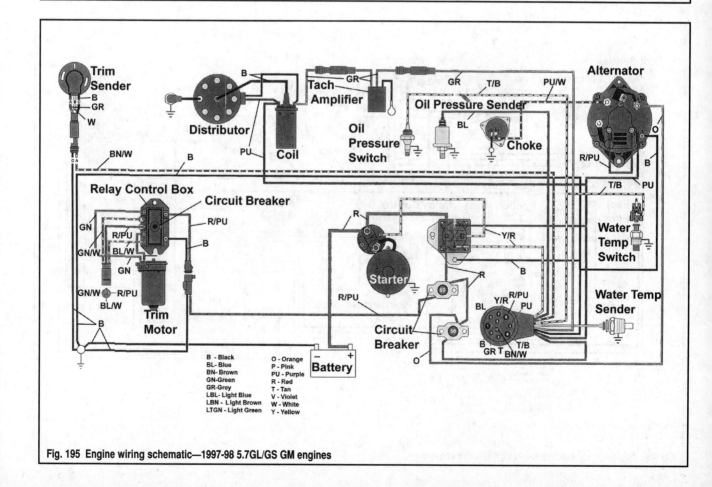

Fig. 194 Engine wiring schematic—1997-98 4.3GL/GS GM engines

Fig. 195 Engine wiring schematic—1997-98 5.7GL/GS GM engines

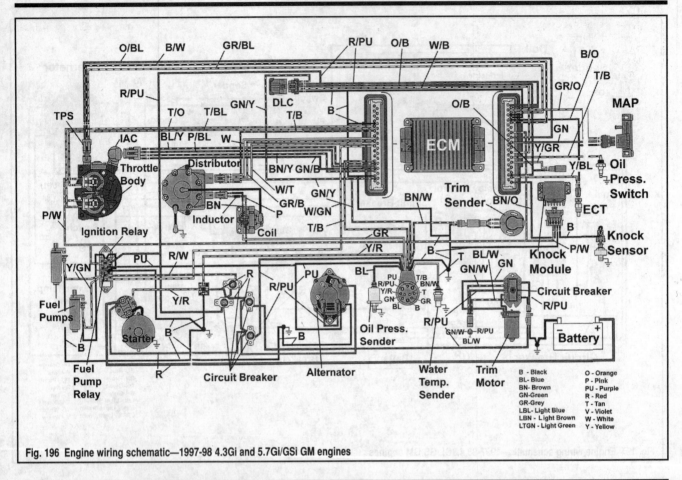

Fig. 196 Engine wiring schematic—1997-98 4.3Gi and 5.7Gi/GSi GM engines

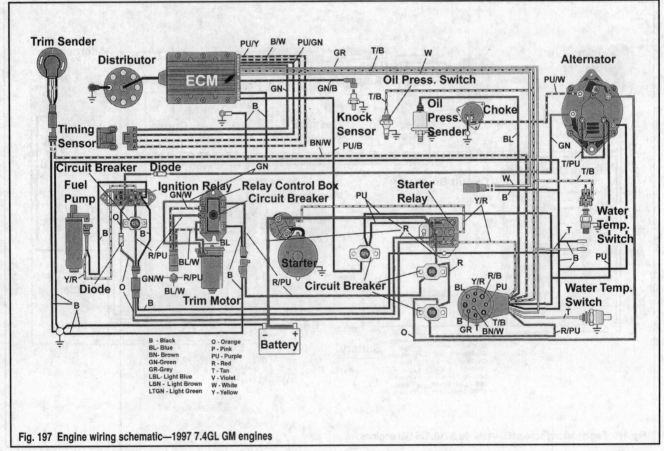

Fig. 197 Engine wiring schematic—1997 7.4GL GM engines

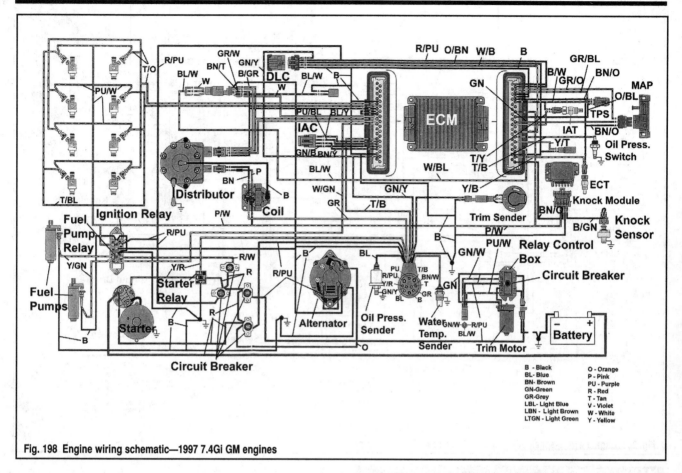

Fig. 198 Engine wiring schematic—1997 7.4Gi GM engines

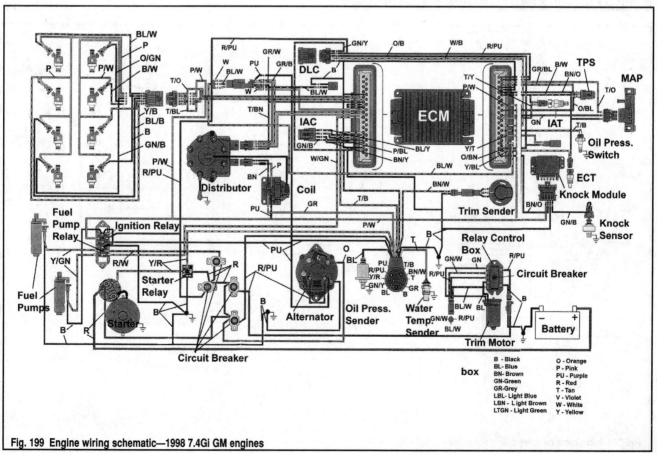

Fig. 199 Engine wiring schematic—1998 7.4Gi GM engines

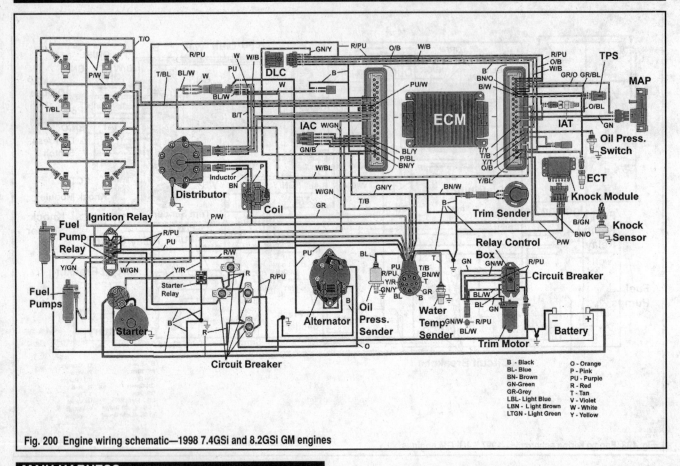

Fig. 200 Engine wiring schematic—1998 7.4GSi and 8.2GSi GM engines

MAIN HARNESS

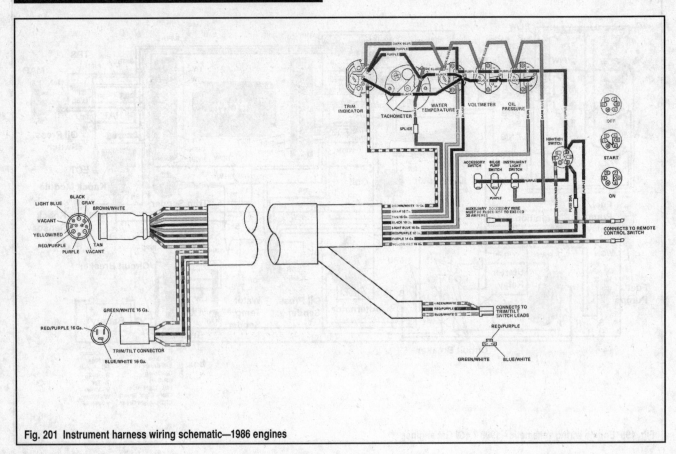

Fig. 201 Instrument harness wiring schematic—1986 engines

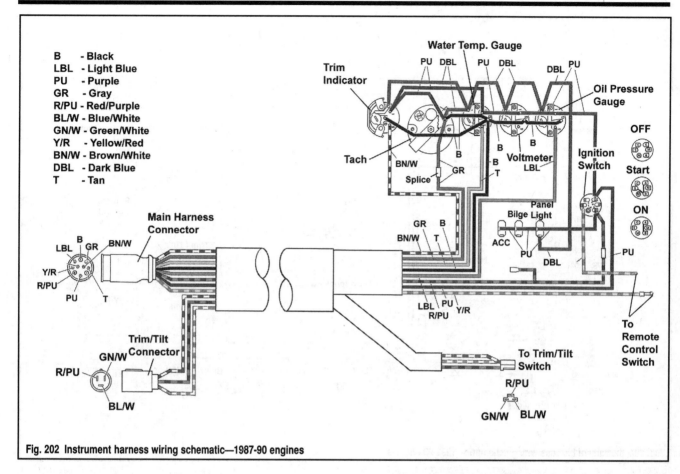

Fig. 202 Instrument harness wiring schematic—1987-90 engines

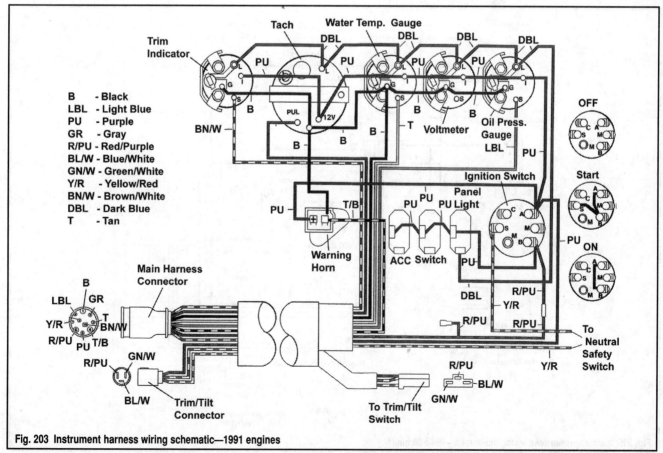

Fig. 203 Instrument harness wiring schematic—1991 engines

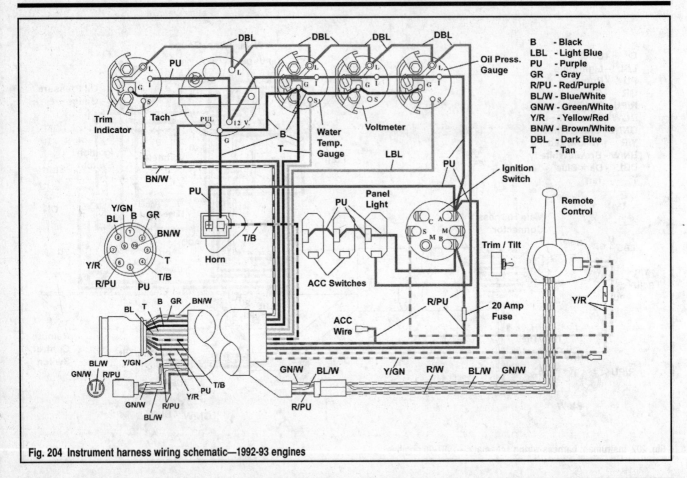

Fig. 204 Instrument harness wiring schematic—1992-93 engines

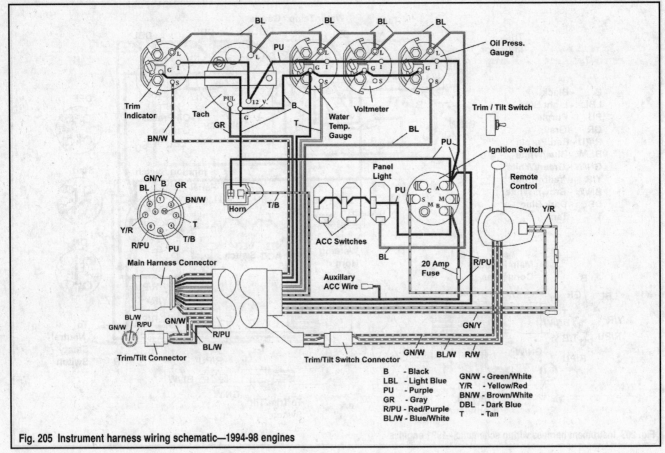

Fig. 205 Instrument harness wiring schematic—1994-98 engines

ALTERNATOR TEST SPECIFICATIONS
1986

Model	Voltage Range (Volts)	Output (Amps)	Regulator	Belt Tension (in.)	Max. Output @ 600 rpm (Amps)	Max. Output @ 2000 rpm (Amps)	AC Stud (Volts)	Condenser Capacity (Microfareds) [1]
Delcotron 42	13.9-15.5	42	Internal	1/4-1/2	18	37	7-8	+/- 0.1

[1] Variation to the number stamped on the capacitor base

ALTERNATOR TEST SPECIFICATIONS
1987-90

Model	Voltage Range (Volts)	Output (Amps)	Regulator	Belt Tension (in.)	Max. Output @ 650 rpm (Amps)	Max. Output @ 1500 rpm (Volts)	Max. Output @ 3000 rpm (Amps)	AC Stud (Volts)	Condenser Capacity (Microfareds)
51	13.9-14.7	51	Internal	1/4-1/2	20	13.9-14.7	45	7-8	0.4-0.6

ALTERNATOR TEST SPECIFICATIONS
1991-92

Model	Min. Engine Speed @ Output (rpm)	Voltage Range (Volts)	Output (Amps)	Regulator	Belt Tension (in.)	Max. Output @ 650 rpm (Amps)	Max. Output @ 1700 rpm (Amps)	Light Stud (Volts)	AC Stud (Volts)	Condenser Capacity (Microfareds)
51	400	13.9-14.7	51	Internal	1/4-1/2	20	51	13.9-14.7	7-8	0.4-0.6

ALTERNATOR TEST SPECIFICATIONS
1993-96

Model	Min. Engine Speed @ Output (rpm)	Voltage Range (Volts)	Output (Amps)	Regulator	Belt Tension (in.)	Max. Output @ 650 rpm (Amps)	Max. Output @ 1500 rpm (Amps)	Max. Output @ 2000 + rpm (Amps)	Light Stud (Volts)	AC Stud (Volts)	Condenser Capacity (Microfareds)
51	400	13.9-14.7	51	Internal	1/4-1/2	20	47	51	13.9-14.7	7-8	0.4-0.6
65	400	13.5-14.8	65	Internal	1/4-1/2	20	53	56	13.5-14.8	7-8	0.4-0.6

ALTERNATOR TEST SPECIFICATIONS
1997-98

Model	Min. Engine Speed @ Output (rpm)	Voltage Range (Volts)	Output (Amps)	Regulator	Belt Tension (in.)	Max. Output @ 650 rpm (Amps)	Max. Output @ 1500 rpm (Amps)	Max. Output @ 2000 + rpm (Amps)	L2 Terminal (Volts)	P Terminal (Volts)	Condenser Capacity (Microfareds)
Prestolite 65	400	14.0-14.7	65	Internal	1/4-1/2	20	53	56	13-14	6.5-7.5	0.4-0.6

STARTER SPECIFICATIONS
GM Engines

Model	No-Load Test					Load Test			Solenoid Current Draw	
	Min. (Amps)①	Max. (Amps)①	Volts	Min. (RPM)	Max. (RPM)	Volts	Min. (Amps)	Max. (Amps)	Hold-In Winding (Amps @ V)	Both Windings In Parallel (Amps @ V)
1986-87 3.0L & 5.0L	60	85	9	6,800	10,300	4.8	440	490	15-20 @ 10	47-55 @ 10
1986-87 4.3L & 5.7L	65	95	9	6,800	10,300	4.3	490	560	15-20 @ 10	47-55 @ 10
1988-98	65	95	9	6,800	10,300	4.3	490	560	15-20 @ 10	47-55 @ 10

① Figure includes the solenoid

STARTER SPECIFICATIONS
Ford Engines

Model	Brushes		Spring Tension (Oz.)	Current Draw Normal Load (Max. Amps)	Cranking Speed (rpm)	Cranking Power (V)	Motor	Max. Load (Amps)	No Load (Amps)
	Length In. (mm)	Wear Limit In. (mm)					Min. Stall Torque @ 5V Ft, Lbs, (Nm)		
1986-87	0.50 (12.7)	0.25 (6.35)	40	150	110	8	9 (12)	460	70
1988 2.3L	0.50 (12.7)	0.25 (6.35)	40	150	110	8	9 (12)	460	70
7.5L	0.50 (12.7)	0.25 (6.35)	40	150	110	8	10 (12)	475	85
1989-90 2.3L, 5.0L, 5.8L	0.50 (12.7)	0.25 (6.35)	40	150	180	8	9 (12)	460	70
7.5L	0.50 (12.7)	0.25 (6.35)	40	150	110	8	10 (12)	475	85
1991-92	0.50 (12.7)	0.25 (6.35)	40	150	180	8	9 (12)	460	70
1993-94	0.66 (16.8)	0.25 (6.35)	64	140-200	200-250	8	11 (15)	800	60-80

11

COOLING SYSTEM

COOLING SYSTEMS

Description & Operation

◆ **See Figures 1 and 2**

Cooling water is a critical phase of engine operation. Cooling water passes through the cylinder heads and block, drawing off heat generated by fuel combustion and engine friction. Proper operation of the cooling system is critical to maintaining satisfactory engine operation and performance.

All OMC engines are cooled by means of one of two systems: an external-water, raw water system; or a Closed system, which actually incorporates the raw water system into a closed automotive-style anti-freeze system. Engines covered here may come equipped with either of the two systems. Please refer to the flow diagrams at the end of this section.

RAW WATER SYSTEM

■**Please refer to the Water Flow Diagrams section for detailed views of your engine's cooling system.**

As implied by the name, this system utilizes water from outside the boat to cool the engine and certain related components—this system is also frequently called a seawater system and the names can (and probably will) be used interchangeably. All versions of this system utilize two water pumps, hoses and a thermostat.

All 1986-93 engines and all King Cobras through 1995 utilize a water pump (impeller) mounted in the top of the upper gear housing to draw the water up through the drive unit and then on to the engine circulating pump attached to the front of the cylinder block—although different in construction, this pump is quite similar to a typical water pump you would find on your car or truck. From here the water is circulated through the engine block, cylinder heads and exhaust manifold(s); and then expelled back into the body of water where it originated.

All 1994-95 Cobra models and all 1996-98 models use a belt-driven raw water pump mounted on the lower starboard side of the engine to draw the water up through the drive unit and then on to the engine circulating pump attached to the front of the cylinder block—although different in construction, this pump is quite similar to a typical water pump you would find on your car or truck. From here the water is circulated through the engine block, cylinder heads and exhaust manifold(s); and then expelled back into the body of water where it originated.

The raw water pump is a self-priming impeller-style pump. Special compound nylon/rubber blades give the blades the ability to flex. The inside surface of the pump is eccentric, causing the blades to move (flex) from a larger volume area to a smaller volume area as they rotate within the pump body.

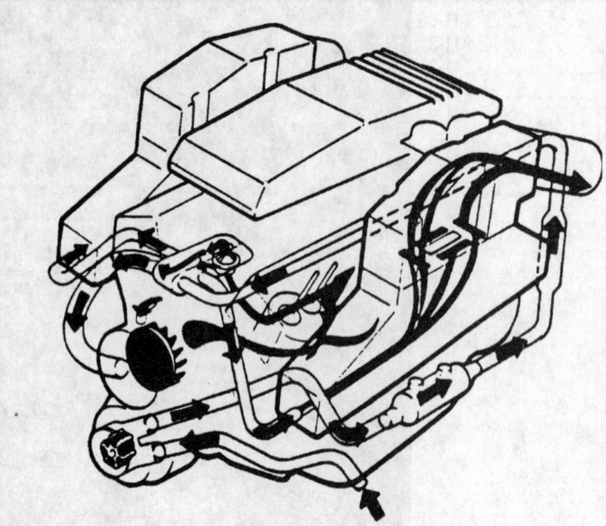

During periods of low speed operation there will be very little resistance to water flow within the pump. The impeller blades will flex along the eccentric surface, thus providing a positive displacement action. As operational speed increases, resistance will begin to build within the pump; causing the ends of the impeller blades to begin flexing inward, slowly changing the pump action from positive displacement to that of a circulation-style pump.

The belt-driven engine circulating pump is mounted on the front of the cylinder block on all models<as mentioned previously, this pump looks identical and is positioned similar to that which you may be familiar with on your car or truck. A pulley is bolted to the forward edge of the pump shaft hub and is driven via a belt from the crankshaft pulley. The pump shaft and permanently lubricated bearing assembly is pressed into the pump cover—a seal pressed into the cover prevents any cooling water from escaping. Pump bearings are specifically manufactured for use in marine applications; never be tempted to substitute an automotive style pump!

Raw water is picked up through screened openings on either side of the lower unit. Water is then drawn through the lower and upper gear housing of the drive unit and through the transom plate. A line then carries the water to an oil or power steering cooler and then on to the water pump, or directly to the pump—some models do not use an oil cooler, while others position the cooler upstream of the water pump. After exiting the water pump, the cooling water then enters the inlet side of the thermostat housing and moves in a number of directions. Some water is routed through the hoses directly to the exhaust manifold(s) for cooling, while the remaining water moves down to the engine circulating pump where it is driven into the water jackets surrounding each cylinder. After working its way through all of the cylinders, the water is forced upward through two passages and into the cylinder head(s) where it cools the combustion chambers. Once at the forward side of the head(s), the water enters the thermostat housing; where, if sufficiently warm, it will cause the thermostat to open and be diverted partially back to the circulating pump and/or on to the exhaust manifolds where it will be introduced into the exhaust gas flow and out of the boat. If the water is not yet warm enough to cause the thermostat to open, it will be diverted entirely back to the circulating pump.

CLOSED SYSTEM

■ **Please refer to the Water Flow Diagrams section for detailed views of your engine's cooling system.**

This system is actually two systems in one—a closed freshwater system and a raw or seawater system. All versions of this system utilize two water pumps, hoses, a thermostat and a heat exchanger.

In the closed portion of the system, a mixture of freshwater and anti-freeze is circulated through-out the engine block, cylinder head and a heat exchanger (similar to your car's radiator). The pressurized water/coolant never leaves the system and is thermostatically controlled.

Lacking a radiator and fan like an automobile, it is necessary to find another means of keeping the fluid in the closed system from boiling and this is where the second portion of this system comes into play. Raw water from outside the vessel is drawn in by means of the belt-driven engine-mounted seawater pump. Unlike true raw water models as described previously, the seawater is pumped into the heat exchanger rather than through the engine.

Fig. 1 A simple drawing of a raw water cooling system

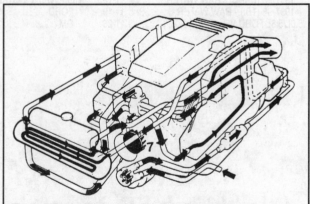

Fig. 2 A simple drawing of a closed cooling system

In fact, it bypasses the engine circulating pump altogether. Once in the exchanger, the seawater is routed through tubes surrounding the closed system tubes, thus cooling the fresh water/anti-freeze mix as it passes through the exchanger. The heated seawater is then routed through the exhaust manifold(s) and then back overboard.

The raw water portion of this system operates exactly like detailed in the previous section, with one exception—after exiting the seawater pump, the cooling water moves to (and through) the heat exchanger and then on to the exhaust manifolds. Raw water never enters the engine to cool the block and heads, since this has already been taken care of by the coolant in the closed portion of the system.

Troubleshooting

The following paragraphs list troubles encountered in the various portions of the system with accompanying probable causes of the problem. The causes are given in a logical order of checking until the problem is corrected.

ENGINE OVERHEATS—RAW WATER SYSTEM

- Loose or broken engine circulating pump belt or pick-up pump belt. Check belt condition and/or tension.
- Inaccurate temperature gauge or sender.
Check ground wire connection at gauge. Disconnect sender wire and make contact with a good engine ground—the gauge should climb to full scale.
- An accessory or barnacles in front of water intakes on drive causing turbulence. Check screens for debris and perform the Raw Water Test.
- Defective raw water pump.
- Loose hose connections between pick-up and pump—sucking air.
- Pump fails to hold prime due to air leaks. Perform Raw Water Flow Test.
- Ice in water passages.
- Defective engine circulating pump. Check pump seals and gasket for leakage.
- Defective thermostat. Test thermostat as detailed in the Maintenance section.

MAINTENANCE AND TESTING

First, the most important words in this manual: THE ENGINE CANNOT BE OPERATED FOR EVEN FIVE SECONDS WITHOUT WATER MOVING THROUGH THE WATER PICK-UP/RAW WATER PUMP OR THE PUMP IMPELLER WILL BE DAMAGED. THEREFORE, NEVER START THE ENGINE, EVEN FOR TESTING PURPOSES, WITHOUT THE BOAT BEING IN THE WATER, OR PROVISION HAVING BEEN MADE FOR WATER TO PASS THROUGH THE SEAWATER PICK-UP PUMP.

■ **Marine thermostats are generally rated at 140-160°. An automotive-type thermostat must never be used because of the higher temperature ratings. Such a high rating would cause the engine to run much hotter than normal.**

Cooling System

The cooling system, raw water or closed, should be cleaned and flushed at least once every two years; more often if possible. Please refer to the Maintenance section for detailed flushing procedures.

DRAINING, FLUSHING & FILLING

The cooling system should be drained, cleaned, and refilled each season, although OMC's recommendations are for every two years on normal anti-freeze systems. We think its cheap insurance to do it every season, but you certainly can't go wrong by following the factory's suggestion. The bow of the boat must be higher than the stern to properly drain the cooling system. If the bow is not higher than the stern, water will remain in the cylinder block and in the exhaust manifold. Insert a piece of wire into the drain holes, but not in the petcock, to ensure sand, silt, or other foreign material is not blocking the drain opening.

ENGINES OVERHEATS—CLOSED SYSTEM

■ **In addition to items detailed under Raw Water:**

- Closed cooling system reservoir level low.
- Plugged heat exchanger cores.
- Too much anti-freeze in closed system.
- Fresh water kit not installed properly.
- Exhaust elbow dump fittings "bottomed" on inner water jacket.

ENGINE OVERHEATS—MECHANICAL

- Incorrect ignition timing.
- Spark plug wires crossed.
- Lean air/fuel mixture.
- Pre-ignition—spark plugs are wrong heat range.
- Engine laboring—engine rpm below specifications at WOT.
- Poor lubrication.
- Water in cylinders—due to warped cylinder head.
- Distributor not functioning properly.
- Clogged exhaust elbows.
- Exhaust flappers stuck closed.
- Cabin hot water heater incorrectly connected to engine.

WATER IN CYLINDERS

- Backwash through exhaust system.
- Loose cylinder head bolts.
- Blown cylinder head gasket.
- Warped cylinder head.
- Cracked or corroded exhaust manifold.
- Cracked block in valve lifter area.
- Improper engine or exhaust hose installation.
- Cracked cylinder wall.

WATER IN OIL

- Backwash through exhaust system.
- Water seeping past piston rings from flooded combustion chamber.
- Thermostat stuck open or missing—condensation forms because engine is operating too cool.
- Cracked cylinder block.
- Intake manifold water passage leak.

If the engine is not completely drained for winter storage, trapped water can freeze and cause severe damage. The water in the oil cooler—if so equipped—must also be drained.

■ **For complete details, procedures and illustrations on draining, filling and/or flushing of the cooling system, please refer to the Maintenance section.**

DRIVE BELTS

■ **For complete details, procedures and illustrations on drive belt removal and adjustment, please refer to the Maintenance section.**

FLUID LEVEL CHECK

■ **For complete details, procedures and illustrations on checking the fluid level in the closed cooling system, please refer to the Maintenance section.**

PRESSURE TEST

■ **For complete details, procedures and illustrations covering pressure testing the cooling system, please refer to the Maintenance section.**

PRESSURE CAP TEST

■ **For complete details, procedures and illustrations covering testing the closed cooling system pressure cap, please refer to the Maintenance section.**

RAW WATER COOLING TEST

If you suspect that the raw water system is taking in air, perform the following test.

✱✱ WARNING

This test MUST be performed with the vessel in the water. DO NOT perform this test with the vessel out of water with a flushing attachment attached.

1. Disconnect the water hose at the raw water pump outlet and then disconnect the same hose at the inlet on the thermostat housing.
2. Connect a clear vinyl hose of the same diameter between the pump and the housing, Tighten the clamps securely.
3. Start the engine and drive the boat at the lowest rpm at which the overheating problem has been observed. Have an assistant (do not do this yourself—stay at the helm!) check the clear hose for evidence of air bubbles being drawn through the system.
4. If air bubbles are present, air is being sucked into the system somewhere between the pump and the inlet holes on the lower unit. Check all hoses, clamps and fittings for damage or other signs of deterioration. Check the lower unit water tube, guide, seal, grommet and water passage cover gasket for damage, signs of deterioration or leakage. Check the pump impeller plate gasket and housing O-ring for damage and/or leakage.
5. If no bubbles are observed in Step 3, remove the clear hose and reconnect the original water hose.

RAW WATER FLOW TEST

If you suspect that a low cooling water supply is causing an overheat condition, perform the following test.

✱✱ WARNING

This test MUST be performed with the vessel in the water. DO NOT perform this test with the vessel out of water with a flushing attachment attached.

1. Disconnect the water hose at the thermostat housing inlet. Hold the hose vertically so that the open end is not higher than the top of the engine.
2. Start the engine and allow it to idle. A column of water should rise from the hose approximately 2-4 in. high. Less than 2 in. indicates a likely restriction in the supply line, or the raw water pump is defective. Check for a blocked inlet screen at the drive or a crimped/broken pivot housing at the gimbal housing/transom plate. Also, confirm that the pump impeller is not worn or broken.

✱✱ CAUTION

Don't forget to shut the engine down immediately after observing the height of water coming out of the hose!

3. Reconnect the hose to the housing and tighten the clamp securely.

Raw Water Pump (Impeller)

REMOVAL & INSTALLATION

Stern Drive Unit

All Models Except 1990-95 King Cobra

◆See Figures 3 thru 11

■ As noted previously, all 1986-93 models and 1994-95 King Cobra models utilize a stern drive mounted raw water pump. There is no engine-mounted external water supply pump on these engines.

1. Loosen the 3 rear cover mounting bolts at the rear of the upper gear housing and lift off the cover.

2. Disconnect the small drain hose at the port side of the water pump.
3. Remove the 3 pump housing mounting bolts. Carefully tap the side of the housing (very lightly) with a rubber mallet and then pull of the housing. The seal, liner and impeller should all come off with the housing, but if not, remove them separately.
4. Insert a small prybar between the impeller plate and the adapter, twist it and remove the plate and gasket.
5. The adapter itself is held in place via 2 O-rings and some sealer. Removal requires driving it out from the other side with a 1 3/8 in. pipe. Obviously, access will require removal of the drive and then the U-joint shaft assembly (as detailed in the Drive Unit section). Once this is accomplished, remove the 2 adapter mounting bolts and then press the adapter out through the U-joint shaft opening on the front of the gear housing.

■ Occasionally, if you are lucky, you can pull the adapter out from the rear with a pair of vise grip pliers. If you attempt it this way, be very careful that you do not damage the adapter with the pliers.

To Install:

6. Clean all components in solvent and allow to dry thoroughly.
7. Clean all sealer from the adapter surfaces. Inspect the O-ring grooves and gasket surfaces for nicks or burrs.
8. Clean all sealer from the pump housing seal recess and the liner cavity. Make sure that the vent passage is clear. Check the housing for cracks, damage or obvious signs of wear.
9. Check the contact surfaces on the impeller plate for signs of wear, scoring or distortion. If worn, you can turn the plate over and reuse it when installing the pump.
10. The impeller blades should not be set in a bent position or have flat edges where they contact the liner.
11. If the you removed the liner from the pump housing, or if it just came out on removal, swab a light coating of Gasket Sealing Compound on the inside of the housing and then press the liner into the housing until it seats itself. Wipe off any excess sealer that may have been squeezed into the liner through the ports.
12. Lubricate the impeller with OMC HiVis Gearcase Lube, position the impeller over the liner so the drive pocket is facing out and twist the impeller into the liner in a clockwise direction.
13. If the adapter was removed, coat both sides of a new adapter gasket with Gasket Sealing Compound and position it onto the rear of the adapter. Coat both new O-rings with the same sealant and install them. Slide a Seal Protector (# 913501) over the end of the water pump shaft and then press the adapter into position—you may have to tap it into place with a rubber mallet. Coat the threads of the mounting bolts with Gasket Sealing Compound, install them and tighten to 12-14 ft. lbs. (16-19 Nm). Remove the tool.
14. Coat both side of a new impeller plate gasket with Sealing

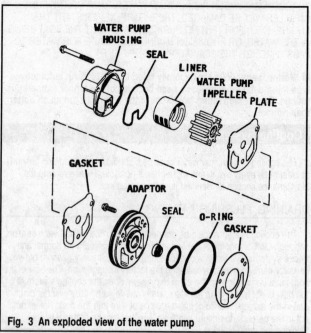

Fig. 3 An exploded view of the water pump

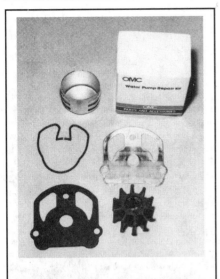

Fig. 4 Contents of a standard water pump repair kit

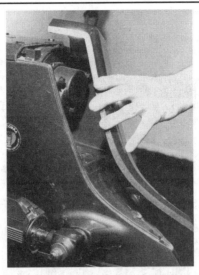

Fig. 4 Remove the rear cover...

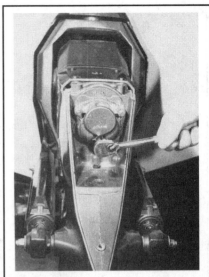

Fig. 5 ...and then remove the three pump mounting bolts

Compound and position it on the adapter. Make sure that the relief slot in the adapter remains clear.

15. Install the impeller plate, remembering that you can flip it over if one of the sides was worn.

16. Cover a new seal ring with Sealing Compound and position it into the groove in the pump housing. Reach in with a small screwdriver and rotate the impeller (clockwise only!) until the narrow end of the drive pocket is pointing at the lower bolt hole in the housing.

17. Now rotate the U-joint shaft until the narrow side of the drive wedge is pointing down at the lower mounting bolt hole in the impeller plate.

18. Position the housing assembly over the shaft until it seats on the impeller plate. Coat the threads of the mounting bolts with Gasket Sealing Compound and then tighten them to 108-132 inch lbs. (12-15 Nm).

19. Reconnect the vent hose to the pump housing.

20. Install the rear gear housing cover and then tighten the bolts to 108-132 inch lbs. (12-15 Nm).

Fig. 7 Remove the pump housing (with the impeller and liner)...

Fig. 8 ...and then pry off the impeller plate

Fig. 9 Remove the two adapter mounting bolts...

Fig. 10 ...and then carefully pull it out of the housing

Fig. 11 Make sure that the wedge on the pump shaft is pointing down

1990-95 King Cobra

◆ See Figure 12

■ As noted previously, all 1986-93 models and 1994-95 King Cobra models utilize a stern drive mounted raw water pump. There is no engine-mounted external water supply pump on these engines.

1. Loosen the 3 rear cover mounting bolts at the rear of the upper gear housing and lift off the cover.
2. Disconnect the small drain hose at the port side of the water pump (center on later King Cobra).
3. Remove the 3 pump housing mounting bolts. Carefully tap the side of the housing (very lightly) with a rubber mallet and then pull of the housing. The seal, liner and impeller should all come off with the housing, but if not, remove them separately.
4. Remove the 3 water pump plate bolts and carefully pry the plate off the adapter housing.
5. Remove the 4 adapter housing bolts and pry off the housing slowly, one corner at a time.

To Install:

6. Clean all components in solvent and allow to dry thoroughly.
7. Clean all sealer from the adapter surfaces. Inspect the O-ring grooves and gasket surfaces for nicks or burrs.
8. Clean all sealer from the pump housing seal recess and the liner cavity. Make sure that the vent passage is clear. Check the housing for cracks, damage or obvious signs of wear.
9. Check the contact surfaces on the impeller plate for signs of wear, scoring or distortion. If worn, you can turn the plate over and reuse it when installing the pump.
10. The impeller blades should not be set in a bent position or have flat edges where they contact the liner.
11. If the you removed the liner from the pump housing, or if it just came out on removal, swab a light coating of OMC Scotch Grip Rubber Adhesive 1300 on the inside of the housing and then press the liner into the housing until it seats itself. Wipe off any excess sealer that may have been squeezed into the liner through the ports.
12. Lubricate the impeller with OMC Ultra-HPF Gearcase Lube, position the impeller over the liner so the drive pocket is facing out (toward you) and twist the impeller into the liner in a counterclockwise direction.

13. If the adapter housing was removed, coat the housing bore with OMC Ultra-HPF Gearcase Lube and then position the unit onto the gearcase housing. Rotate the U-joint slightly until the gear teeth engage and then install the mounting bolts; the long ones go in the lower holes. Tighten all bolts to 10-12 ft. lbs. (14-16 Nm).
14. Coat both sides of a new adapter gasket with Gasket Sealing Compound and position it onto the adapter housing. Make sure that the slot in the gasket is facing down and to the left (when looking from the rear of the drive) and that there is no sealing compound in the slot. Install the pump plate over the gasket and tighten the bolts to 10-12 ft. lbs. (14-16 Nm).
15. Cover a new seal ring with OMC Scotch Grip Rubber Adhesive 1300 and position it into the groove in the pump housing. Reach in with a small screwdriver and rotate the impeller (counterclockwise only!) until the narrow end of the drive pocket is pointing at the lower bolt hole in the housing.
16. Now rotate the U-joint shaft until the narrow side of the drive wedge is pointing down at the lower mounting bolt hole in the impeller plate.
17. Position the housing assembly over the shaft until it seats on the plate. Coat the threads of the mounting bolts with Gasket Sealing Compound and then tighten them to 9-11 inch lbs. (12-15 Nm).
18. Reconnect the vent hose to the pump housing.
19. Install the rear gear housing cover and then tighten the bolts to 9-11 inch lbs. (12-15 Nm).

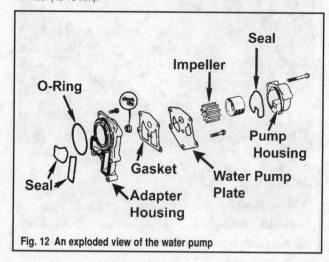

Fig. 12 An exploded view of the water pump

Belt-Driven

◆ See Figure 13

 MODERATE

If performing the following while boat is in the water, close the seacock. If your boat is not equipped with a seacock, disconnect and plug the seawater inlet line to prevent water from entering the system.

■ **The engine should be OFF and cool.**

1. Drain the engine completely as detailed in the Maintenance section.
2. Loosen the hose clamps on the two water lines at the back of the pump. Wiggle the hoses off of their fittings and set them aside. Remember that the inside hose (closest to the engine) is the supply hose and goes to the inlet.
3. Loosen the pump mounting bolts at the cylinder block and brace. Swivel the pump in until you can remove the drive belt and then pull the bolts out and remove the pump assembly.

To Install:
4. Inspect the pump assembly for cracks, damage or other signs of wear.
5. Position the pump in the mounting bracket so that the hose fittings are horizontal (inlet fitting on the inside) and the cam screw side of the housing facing upward—this is very important, reversing the positions will cause overheating and damage to the engine.
6. Attach the entire assembly to the cylinder block, screwing in the bolts only finger-tight.
7. Swivel the pump in until you can fit the drive belt over the pulley and then adjust the belt tension, tightening the mounting bolts securely.
8. Reconnect the water hoses, supply on the inside, and tighten the hose clamps securely.
9. Start the engine and check for leaks.

DISASSEMBLY & ASSEMBLY

Stern Drive Unit

■ **Due to the nature of the removal process, disassembly procedures for the stern drive-mounted water pump are included in the Removal & Installation procedures.**

Belt-Driven

All 1994 Engines And 1995 3.0L Engines

◆ See Figure 14  MODERATE

1. Remove the pump.
2. Remove the 3 bearing housing-to-impeller housing mounting bolts and their washers and separate the two housings. The easiest way to accomplish this is to stand the assembly on the pump pulley and hold the pulley shaft while lifting and turning the impeller housing counterclockwise.

3. The impeller should stay on the shaft, but if it sticks in the housing, carefully grasp it with pliers and pull it out. Remove the O-ring and discard it.
4. Remove the cam screw and its washer. Remove the cam only if necessary; otherwise leave it and the screw in place.
5. If the impeller remained on the pump shaft after splitting the cases, lift it off and pull out the shaft key.
6. Lift off the endplate and its gasket. Use #1 retaining ring pliers and pinch out the ring securing the seal assembly. Lift out the seal assembly. No further disassembly is necessary—the pulley, shaft, bearings and housing are serviced as an assembly only.

To assemble:
7. Remove all residual gasket material and then clean all parts with solvent. Allow them to dry completely, or use compressed air.
8. Inspect the impeller housing and cam for wear, cracks or any signs of deterioration; replacing as necessary.
9. Inspect the end plate for wear, scoring or distortion, particularly the side that comes in contact with the impeller. If only wear is evident, turn the plate over when you install it; otherwise replace it.
10. Inspect the impeller for bent, cracked or broken blades. Also check out the blade contact surfaces to see that they are not flat. Replace as necessary.
11. Install the ceramic seal into the housing, pressing it and the retaining ring in until the ring snaps into the groove. Do not use any grease on the seal.
12. Coat the end plate gasket lightly with Gasket Sealing Compound and position it onto the housing, Drop the end plate over the shaft and into position on the gasket.
13. Install the shaft key in the slot and then lightly press the impeller onto the shaft.
14. If you replaced the cam, position it into the housing and then slide a new washer onto the screw. Coat the threads lightly with Sealing Compound and install the screw, tightening it securely.
15. Coat the impeller surfaces and the O-ring groove lightly with OMC Triple Guard grease. Press in a new O-ring and then install the impeller housing over the assembly, rotating it counterclockwise while lightly pressing it downward. Once in position, rotate the housing slowly until the assembly screw next to the boss is aligned with the cam screw. Install the three retaining bolts and tighten them securely.

1995 V6/V8 Engines And All 1996-98 Engines

1. Remove the pump.
2. Position the pump on a clean, flat surface with the pulley facing down. Remove the 3 end cover retaining screws and lift off the cover and gasket. Throw away the gasket.
3. Carefully grab the impeller with needle nose pliers and pull it out of the housing.

To assemble:
4. Remove all residual gasket material and then clean all parts with solvent. Allow them to dry completely, or use compressed air.
5. Inspect the impeller housing and cam for wear, cracks or any other signs of deterioration; replacing as necessary.

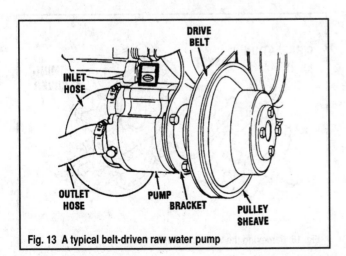

Fig. 13 A typical belt-driven raw water pump

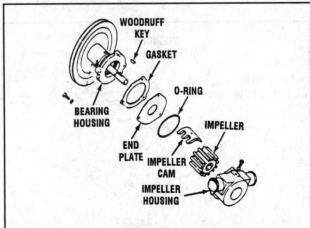

Fig. 14 Exploded view of a belt-driven raw water pump, most models will not have the hose fittings opposed as shown

6. Inspect the end cover for wear, scoring or distortion, particularly the side that comes in contact with the impeller; replacing as necessary.

7. Inspect the impeller for bent, cracked or broken blades. Also check out the blade contact surfaces to see that they are not flat. Replace as necessary.

8. Coat the impeller surfaces lightly with OMC Triple Guard grease and slide it into the housing until it is flush with the upper housing gasket surface.

9. Position a new gasket (no sealant) and then install the end cover. Tighten the screws securely.

10. Install the pump.

Thermostat

For all removal and testing procedures, please refer to the Maintenance section.

Water (Engine) Circulating Pump

REMOVAL & INSTALLATION

GM Engines

◆ See Figure 15

1. Disconnect the battery cables and then drain all water from the block and manifolds.

2. Drain all water/coolant from the cylinder block.

3. Loosen, but do not remove, the pump pulley mounting bolts.

4. Loosen the power steering pump and alternator bracket bolts and swivel them in until you are able to remove the belt(s). Different engines may have different systems, follow the belt back and loosen the appropriate components.

5. Now you can remove the pump pulley bolts along with the lockwashers and clamping ring. Pull off the pulley.

6. Disconnect the water hoses at the pump.

7. Remove the mounting bolts and lift the pump off of the block.

To Install:

8. Carefully scrape any old gasket material off both mounting surfaces. Inspect the pump for blockage, cracks or any other damage. Inspect the impeller for cracks. Replace either if necessary.

9. Coat both sides of a new gasket(s) with sealant and position on the cylinder block. Coat the threads of the pump mounting bolts with sealant, install the pump and tighten the bolts to:
- 13-17 ft. lbs. (18-23 Nm) on 4 cyl. engines
- 30 ft. lbs. (41 Nm) on 4.3L V6 and 5.0L/5.7L/7.4L V8 engines
- 35 ft. lbs. (47 Nm) on 8.2L V8 engines

10. Reconnect the water hoses and tighten the hose clamps securely.

11. Position the pump pulley and clamping ring on the boss. Screw the mounting bolts and lock washers in and tighten them securely.

12. Install the drive belt(s) and adjust them as detailed previously. Start the engine and check the system for leaks.

Ford Engines

◆ See Figure 16

1. Disconnect the battery cables and then drain all water from the block and manifolds.

2. Drain all water/coolant from the cylinder block.

3. Loosen, but do not remove, the pump pulley mounting bolts.

4. Loosen the power steering pump (if equipped) and alternator bracket bolts and swivel them in until you are able to remove the belt(s). Different engines may have different systems, follow the belt back and loosen the appropriate components.

5. Now you can remove the pump pulley bolts along with the lock washers and clamping ring. Pull off the pulley.

6. Remove the two power steering pump bracket bolts, the brace from the alternator bracket, the Torx® retaining bolt and the pump.

7. Disconnect the large water hose at the pump.

8. Tag and disconnect the electrical leads at the alternator and move them out of the way. Remove the bracket bolts and lift off the alternator.

9. Remove the mounting bolts and lift the pump off of the block. There's a good chance you will need to persuade the pump—tap it lightly with a rubber mallet.

To Install:

10. Carefully scrape any old gasket material off both mounting surfaces. Inspect the pump for blockage, cracks or any other damage. Inspect the impeller for cracks. Replace either if necessary.

11. Coat both sides of a new gasket with sealant and position on the cylinder block. Make sure that all of the holes are lined up.

12. Position the pump and screw in all bolts except those that attach the alternator and power steering pump brackets. Tighten the bolts to 14-21 ft. lbs. (19-28 Nm) on the 2.3L, or 12-18 ft. lbs. (17-24 Nm) on the V8s.

13. Reconnect the water hoses and tighten the hose clamps securely.

14. Position the spacer and the Torx® bolt behind the rear power steering pump bracket and tighten it securely. Attach the other bracket to the pump and tighten the 2 bolts.

Install the alternator and tighten the bracket-to-pump bolts to 18-20 ft. lbs. (24-27 Nm).

15. Attach the steering pump brace to the alternator and tighten the bolt securely.

16. Position the pump pulley and clamping ring on the boss. Screw the mounting bolts and lock washers in and tighten them to 14-20 ft. lbs. (19-27 Nm).

17. Install the drive belt(s) and adjust them as detailed previously. Start the engine and check the system for leaks.

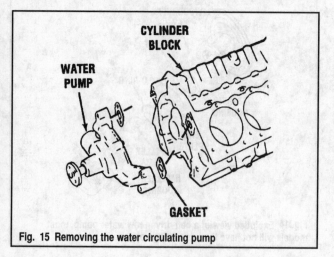

Fig. 15 Removing the water circulating pump

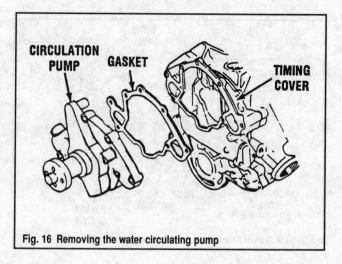

Fig. 16 Removing the water circulating pump

FLOW DIAGRAMS

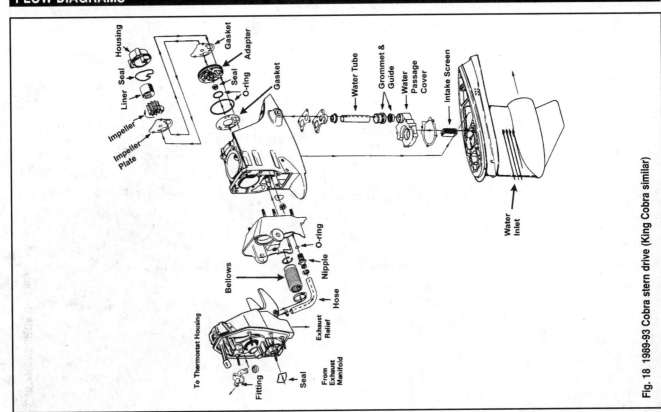

Fig. 18 1989-93 Cobra stern drive (King Cobra similar)

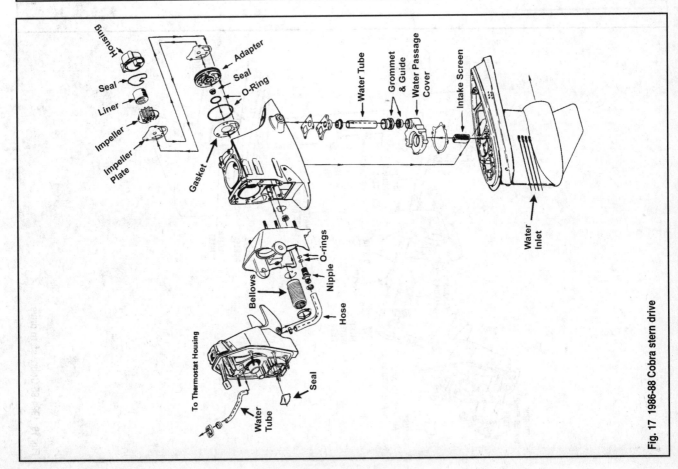

Fig. 17 1986-88 Cobra stern drive

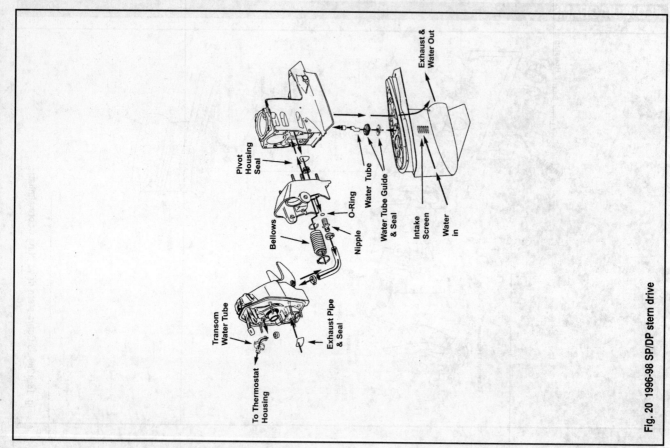

Fig. 20 1996-98 SP/DP stern drive

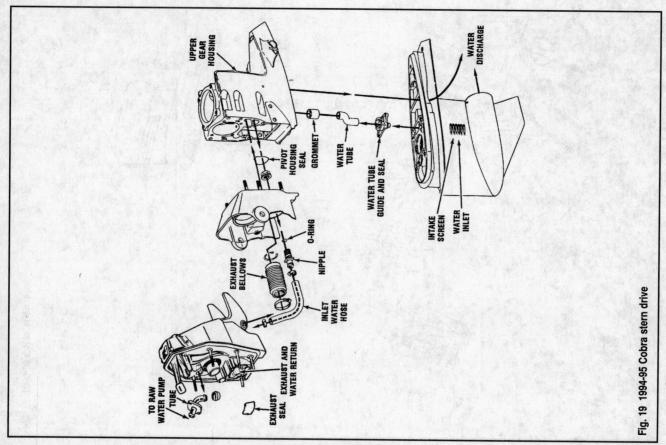

Fig. 19 1994-95 Cobra stern drive

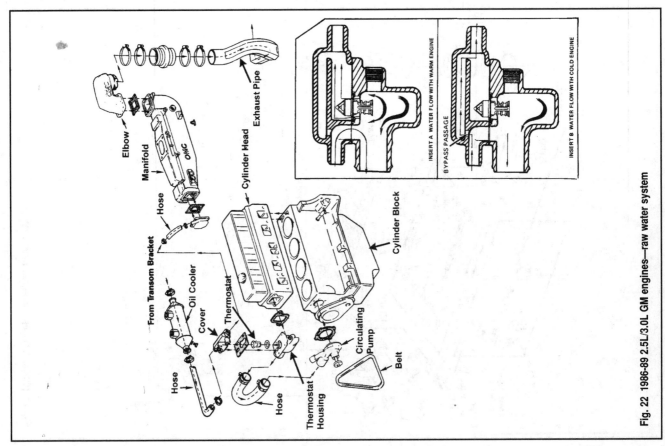

Fig. 22 1986-89 2.5L/3.0L GM engines—raw water system

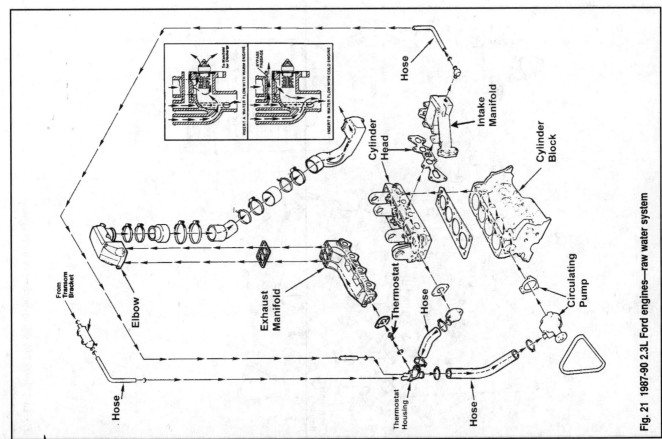

Fig. 21 1987-90 2.3L Ford engines—raw water system

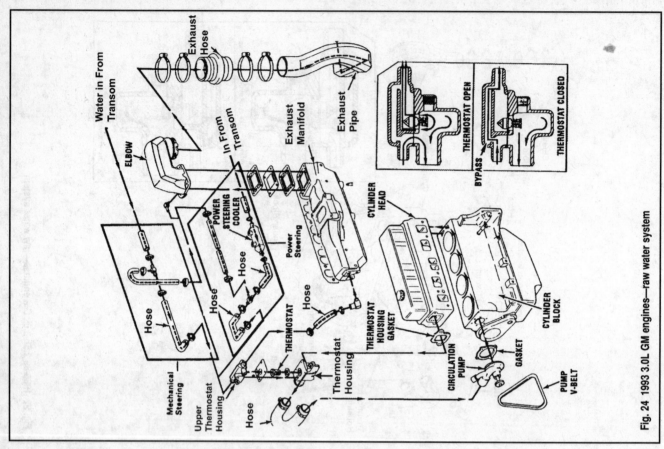

Fig. 24 1993 3.0L GM engines—raw water system

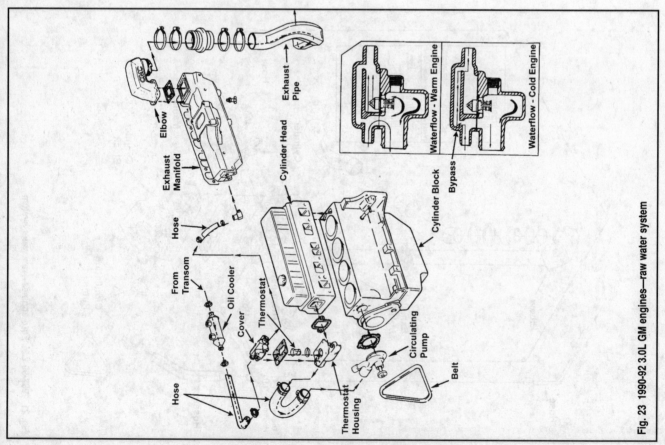

Fig. 23 1990-92 3.0L GM engines—raw water system

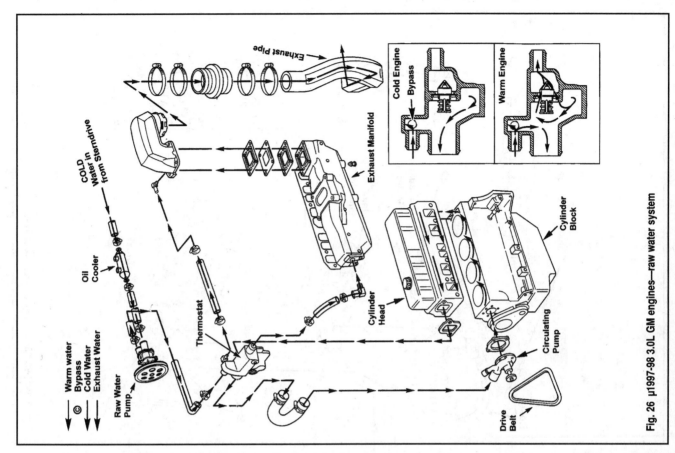

Fig. 26 µ1997-98 3.0L GM engines—raw water system

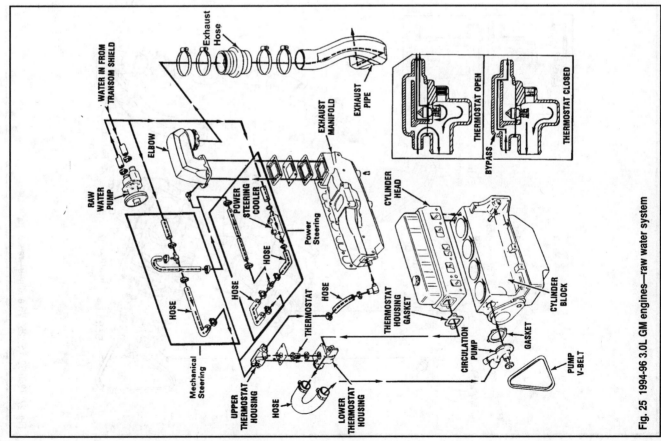

Fig. 25 1994-96 3.0L GM engines—raw water system

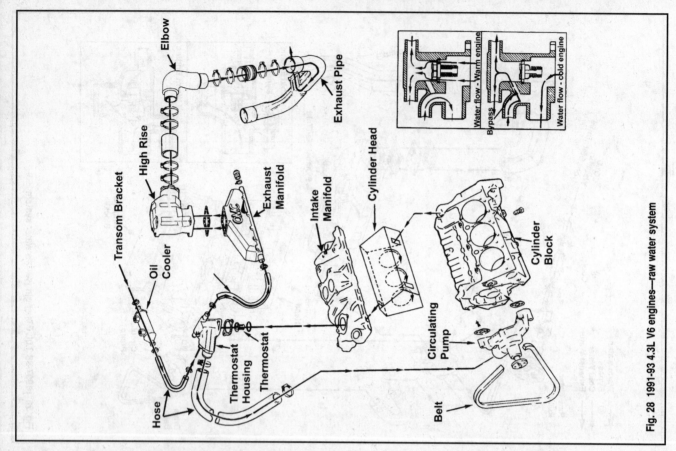

Fig. 28 1991-93 4.3L V6 engines—raw water system

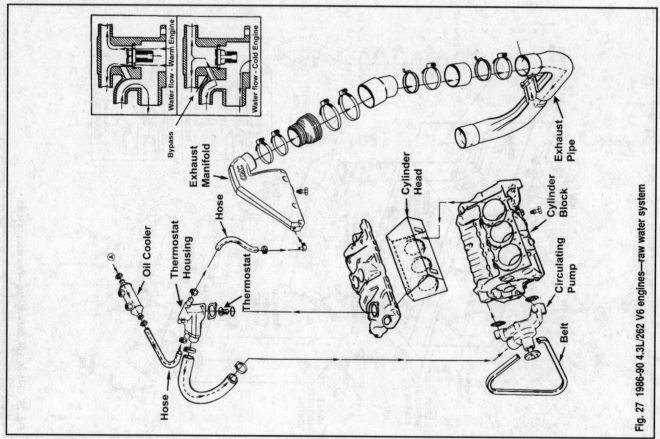

Fig. 27 1986-90 4.3L/262 V6 engines—raw water system

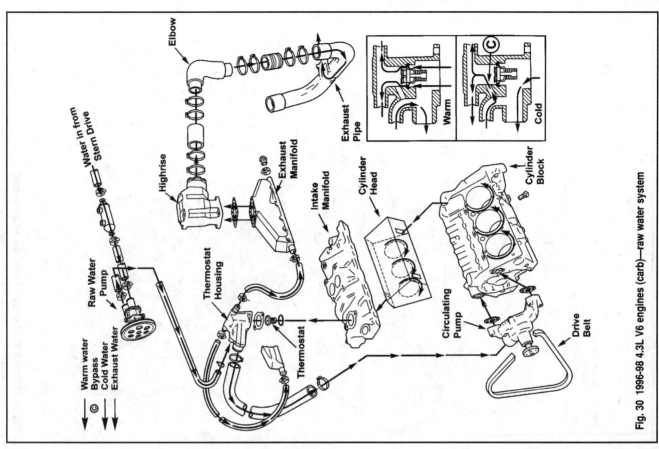

Fig. 30 1996-98 4.3L V6 engines (carb)—raw water system

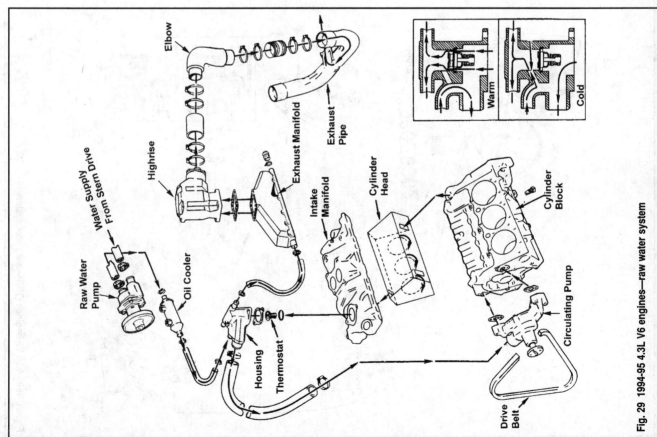

Fig. 29 1994-95 4.3L V6 engines—raw water system

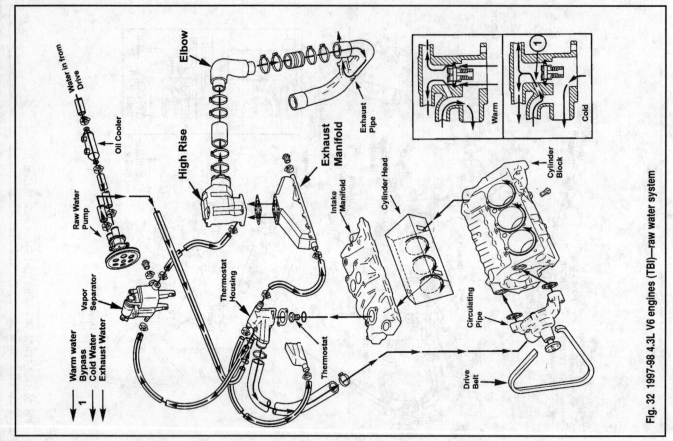

Fig. 32 1997-98 4.3L V6 engines (TBI)—raw water system

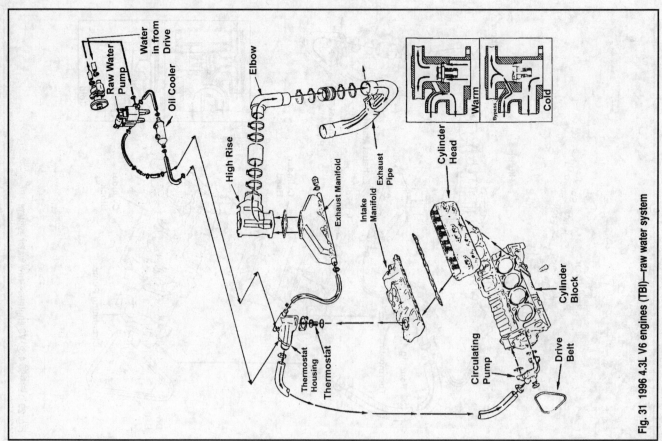

Fig. 31 1996 4.3L V6 engines (TBI)—raw water system

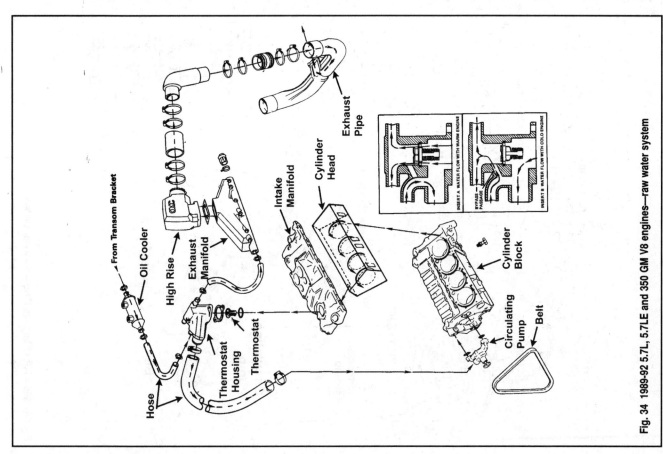

Fig. 34   1989-92 5.7L, 5.7LE and 350 GM V8 engines—raw water system

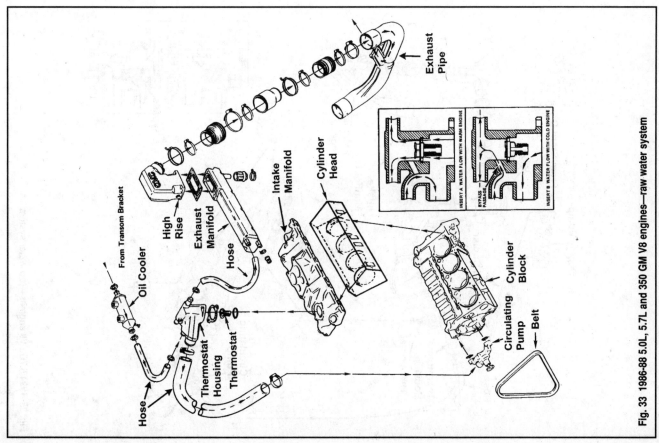

Fig. 33   1986-88 5.0L, 5.7L and 350 GM V8 engines—raw water system

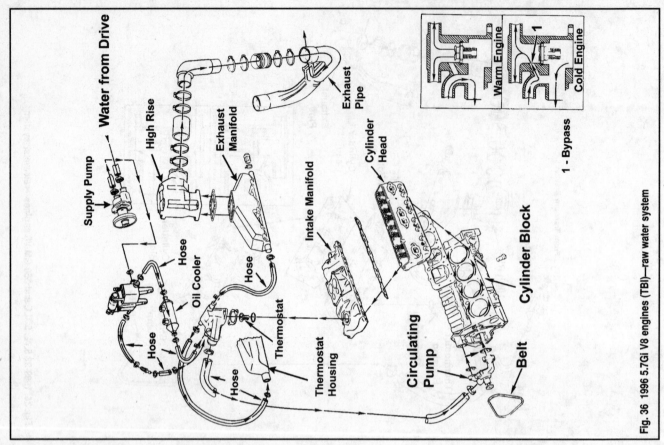

Fig. 36 1996 5.7Gi V8 engines (TBI)—raw water system

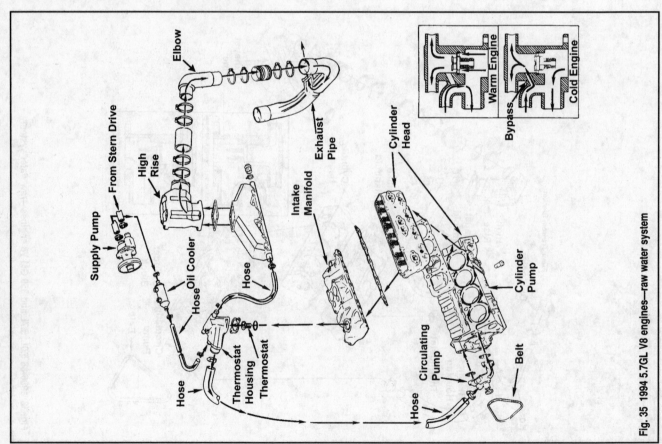

Fig. 35 1994 5.7GL V8 engines—raw water system

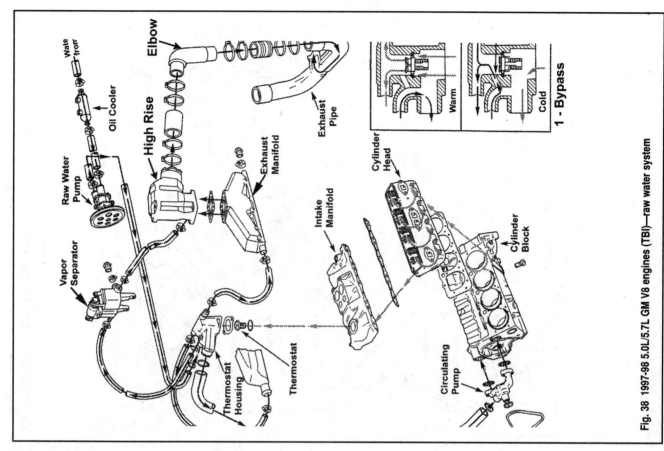

Fig. 38 1997-98 5.0L/5.7L GM V8 engines (TBI)—raw water system

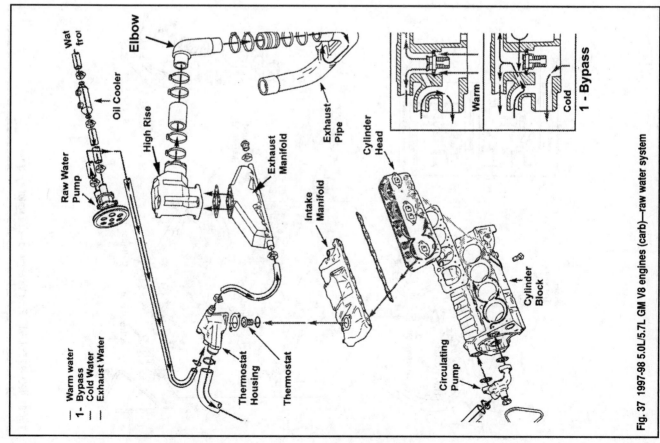

Fig. 37 1997-98 5.0L/5.7L GM V8 engines (carb)—raw water system

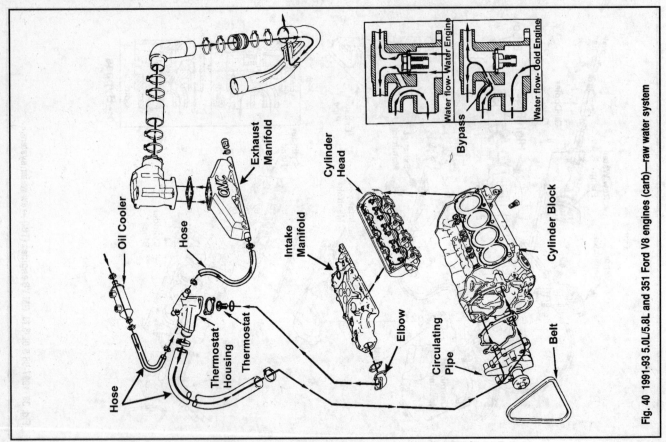

Fig. 40 1991-93 5.0L/5.8L and 351 Ford V8 engines (carb)—raw water system

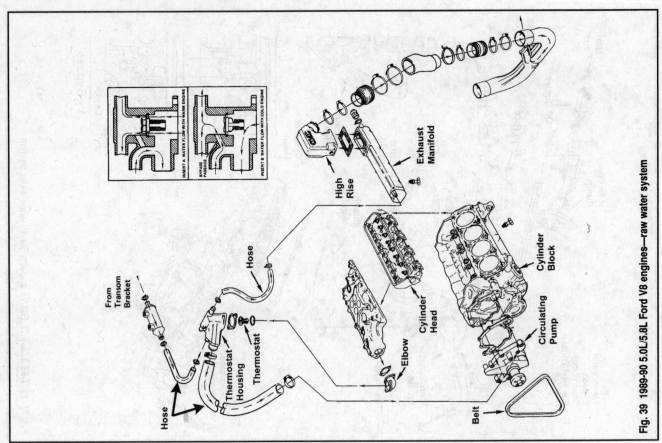

Fig. 39 1989-90 5.0L/5.8L Ford V8 engines—raw water system

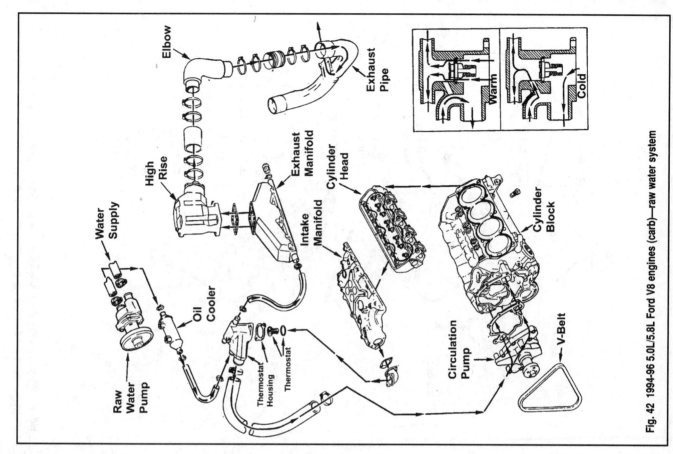

Fig. 42 1994-96 5.0L/5.8L Ford V8 engines (carb)—raw water system

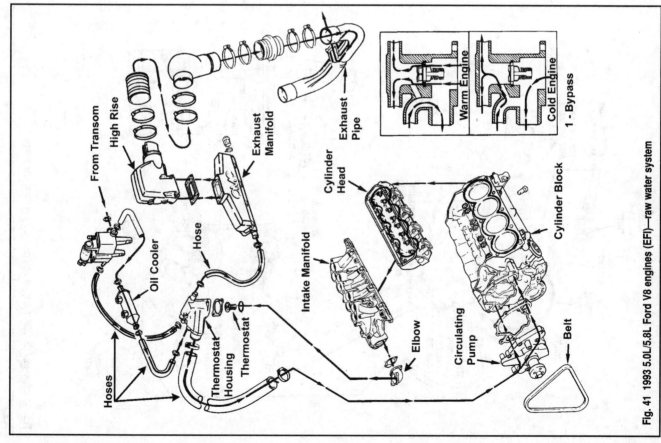

Fig. 41 1993 5.0L/5.8L Ford V8 engines (EFI)—raw water system

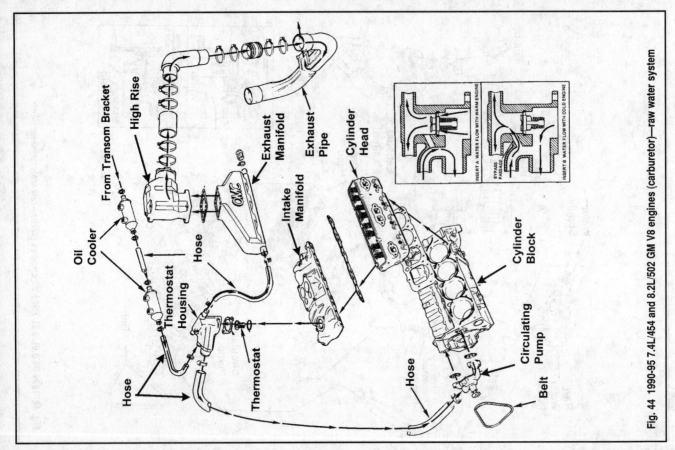

Fig. 44 1990-95 7.4L/454 and 8.2L/502 GM V8 engines (carburetor)—raw water system

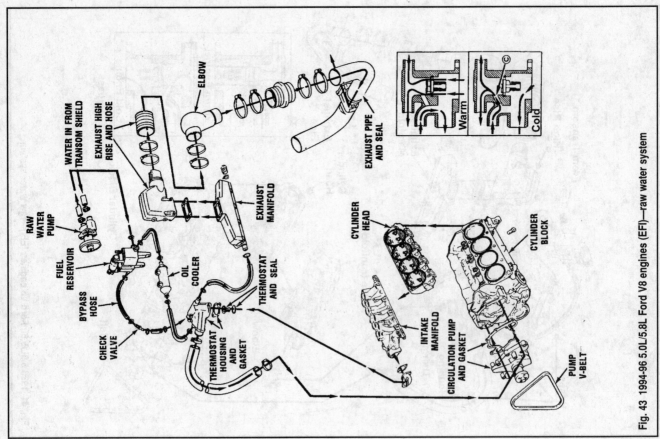

Fig. 43 1994-96 5.0L/5.8L Ford V8 engines (EFI)—raw water system

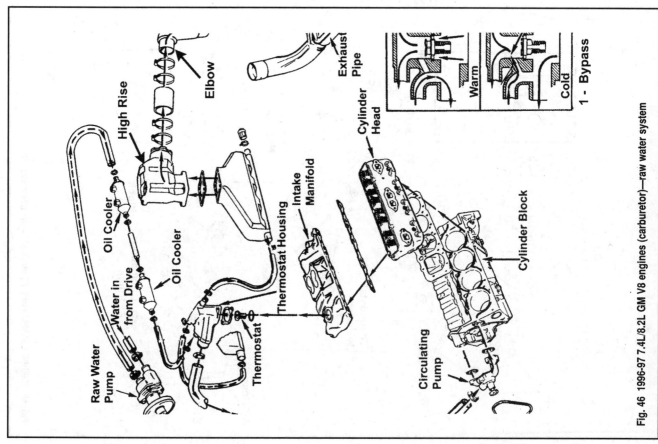

Fig. 46 1996-97 7.4L/8.2L GM V8 engines (carburetor)—raw water system

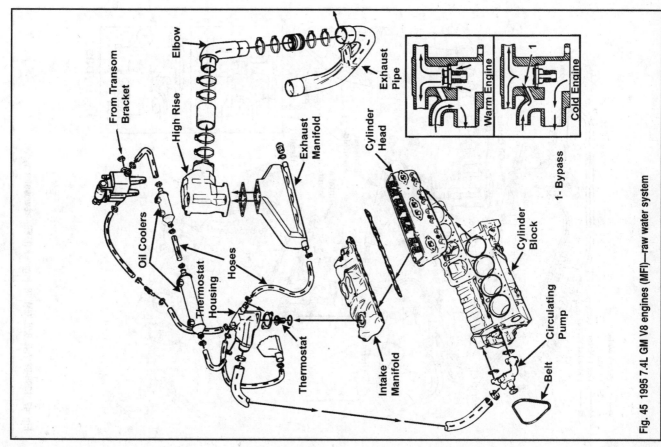

Fig. 45 1995 7.4L GM V8 engines (MFI)—raw water system

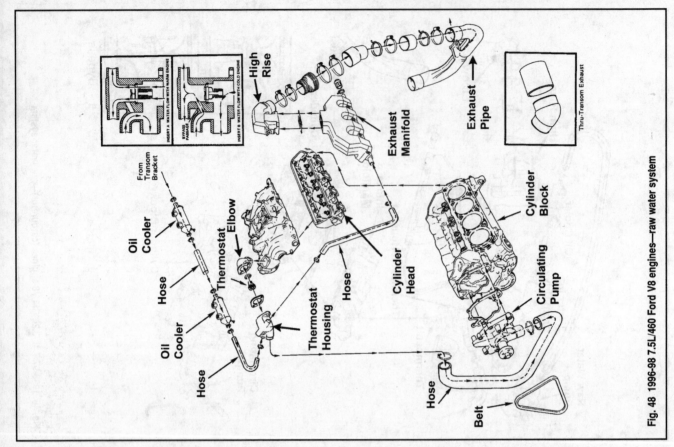

Fig. 48 1996-98 7.5L/460 Ford V8 engines—raw water system

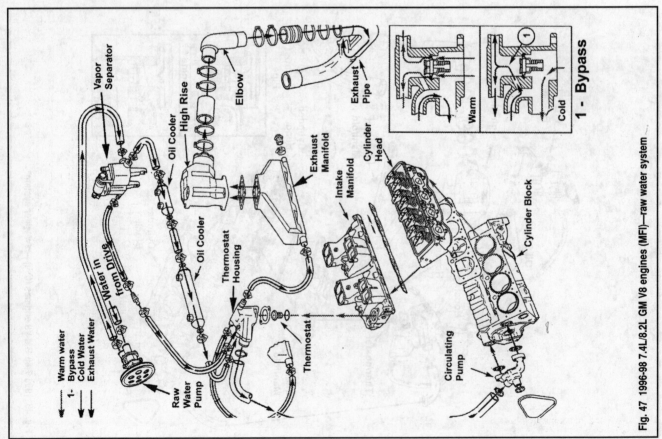

Fig. 47 1996-98 7.4L/8.2L GM V8 engines (MFI)—raw water system

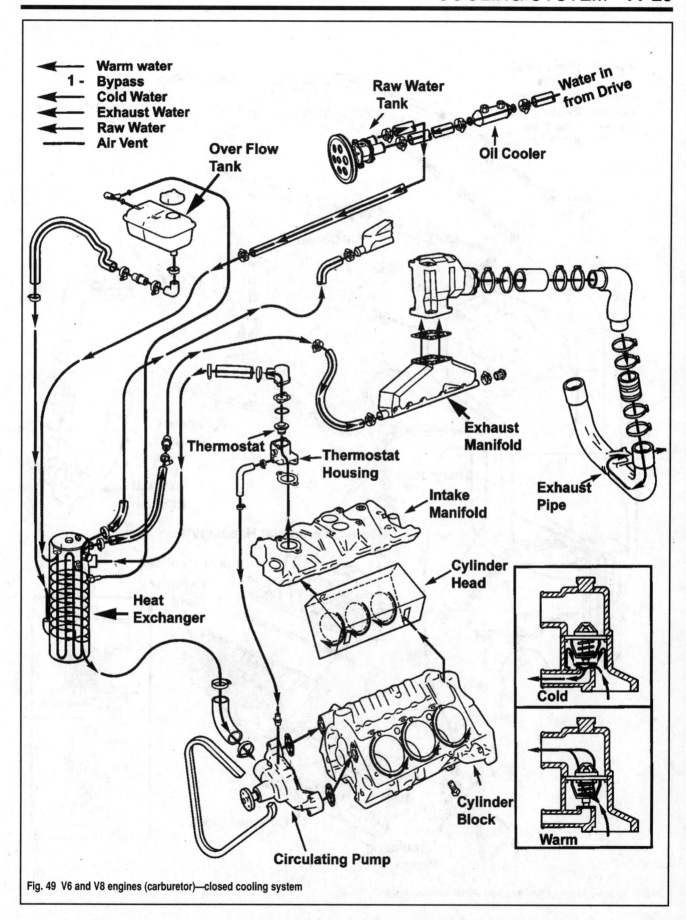

Fig. 49 V6 and V8 engines (carburetor)—closed cooling system

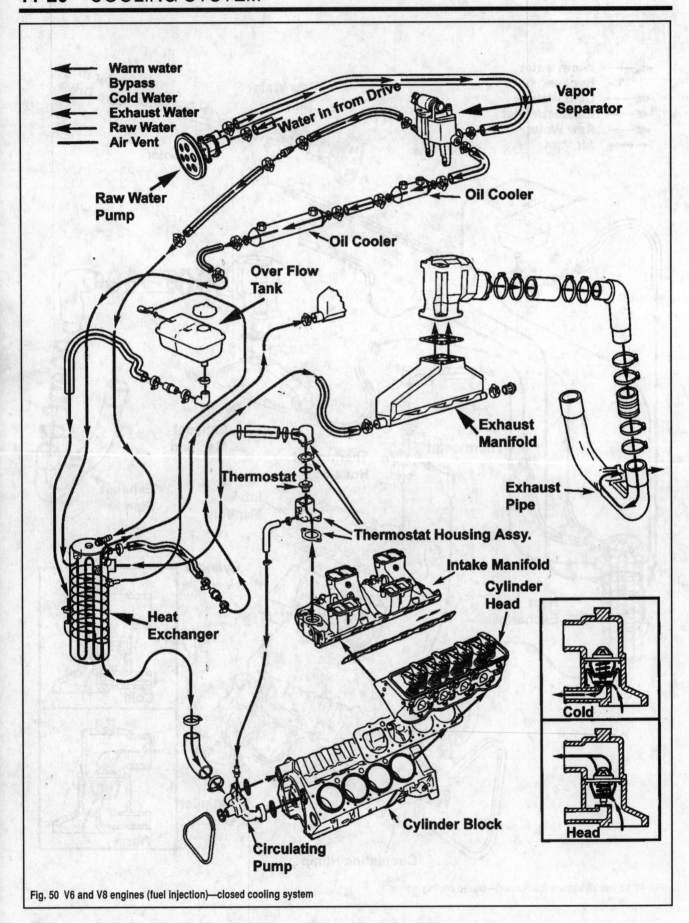

Warm water
Bypass
Cold Water
Exhaust Water
Raw Water
Air Vent

Water in from Drive

Vapor Separator

Raw Water Pump

Oil Cooler

Oil Cooler

Over Flow Tank

Exhaust Manifold

Exhaust Pipe

Thermostat

Thermostat Housing Assy.

Intake Manifold
Cylinder Head

Heat Exchanger

Cold

Head

Cylinder Block

Circulating Pump

Fig. 50 V6 and V8 engines (fuel injection)—closed cooling system

12

STERN DRIVE SP/SX/SX COBRA, COBRA AND KING COBRA, DP/DP DUOPROP

STERN DRIVE UNIT—COBRA, KING COBRA, SP, SX, SX COBRA, DP AND DP DUOPROP MODELS

Description

■ Single propeller drives may have been labeled as Cobra, King Cobra, SP, SX or SX Cobra models. Dual propeller drives may have been labeled as King Cobra, DP or DP DuoProp models. For the purposes of this section, we will frequently refer to all single propeller models as SP and all duo prop models as DP.

That which we refer to as the stern drive is actually a number of individual components attached and working together to transfer the power of the engine into a viable propulsion system for your boat. All stern drive units can be broken down into their component assemblies.

The transom assembly consists of an inner transom plate, gimbal housing, gimbal ring, and a pivot housing—all attached to the transom of the vessel. The inner transom plate is, obviously, attached to the inner side of the transom.

On the other side of the transom, and attached to the transom plate, are the gimbal housing, gimbal ring and pivot housing. The pivot housing is attached to the gimbal ring via roller bearings and is what allows for the up and down (trim) movement of the stern drive unit itself. The gimbal ring is also attached to the gimbal housing via roller bearings and is what allows for side-to-side movement of the unit, or, steering.

The stern drive unit, or at least that thing that is most visible when viewing the stern of the boat, is made up of two component assemblies: the upper gear housing and the lower gear housing.

The upper gear housing, frequently called the driveshaft housing or simply the upper unit is attached to the pivot housing at the top and the lower gear housing at the bottom. Power from the engine, brought through the transom assembly via the splined driveshaft and universal joint is transferred via a pinion gear to the forward and reverse gears on the vertical shaft leading to the lower unit by means of a set of drive and driven gears. The pinion, forward and reverse gears are helical cut and are only available as a matched set.

A cone-type clutch (on all but the Cobra) is splined to the vertical shaft and engages the forward and reverse gears whenever the shift lever is moved to the appropriate position

The lower gear housing, or lower unit, is attached to the bottom of the upper housing. Power, or propulsion, comes through the vertical shaft from the upper unit, is transferred to the propeller shaft via a pinion gear, and causes the propeller to rotate.

On duo prop units, there is a matched set of three gears and dual propeller shafts. The lower pinion gear is splined to the vertical shaft and is in constant mesh with the two other gears; one on the inner and one on the outer propeller shafts. The forward gear is splined to the inner propeller shaft, while the aft gear is a machined part of the outer propeller shaft. As the vertical shaft rotates, the forward gear and inner shaft is driven in one direction while the aft gear and outer shaft are driven in the opposite direction. The aft gear is attached to the same pinion gear as the forward gear, but 180° opposite, so each shaft and propeller rotates at the same speed, but in opposite directions.

The horizontal and vertical driveshafts are both mechanically connected. Therefore, anytime the engine is operating, the horizontal and vertical driveshafts are constantly rotating with engine rpm. A double yoke universal joint assembly in the horizontal driveshaft allows the stern drive to be raised or lowered to a required trim/tilt position (within limits), while the engine is operating.

The trim/tilt system consists of an electrically driven hydraulic pump with dual hydraulic cylinders mounted between the transom assembly and the lower unit and the controls necessary for efficient operation.

Troubleshooting

The following are a list of potential drive unit problems and their possible causes:

GEAR HOUSING NOISE

- Metal particles in the unit oil supply.
- Propeller incorrectly installed.
- Propeller or propeller shaft bent.
- Incorrect drive gear shimming—gear housing back-lash or pinion gear height.
- Worn or damaged gears or bearings.

DRIVESHAFT HOUSING NOISE

- Steering lever may be contacting the transom cut-out edges when turning.
- Flywheel housing on the engine coming in contact with the inner transom plate or the exhaust pipe.
- Bad propeller.
- U-joint cross and bearing assembly O-rings of incorrect size or installed wrong.
- U-joint cross and bearing assemblies have excessive side-play.
- Bearing caps on the U-joint are coming in contact with the center socket or the driveshaft housing bearing retainer.
- Rough or scored U-joint cross and bearings.
- Worn or missing O-rings on the U-joint shaft rattling against the gimbal bearing.
- Worn or damaged splines on the driveshafts and/or couplers.
- Incorrect engine alignment.
- Damaged, rough, worn or loose gimbal bearing.
- Gimbal bearing seated incorrectly.
- Incorrect clearance between the gimbal housing and plate.
- Bell housing and gimbal housing incorrectly aligned.
- Weak or flexing transom.
- Weak, missing or mis-aligned rear engine mount.

DRIVE UNIT WILL NOT SLIDE INTO BELL HOUSING

- U-joint and engine coupler splines not aligned.
- Unit not in Forward gear.
- Incorrectly aligned shift shaft coupler.
- Engine out of alignment.
- Incorrectly installed gimbal bearing.
- Damaged or worn splines on the driveshaft or coupler.

DRIVE UNIT WILL NOT SHIFT—SHIFT HANDLE MOVES

- Shift cables not adjusted properly.
- Shift cables not connected.
- Inner wire on cable broken.
- Gear housing crank installed incorrectly.

DRIVE UNIT WILL NOT SHIFT—SHIFT HANDLE DOES NOT MOVE

- Remote control box/assembly installed incorrectly.
- Broken, worn or damaged linkage.
- Shift shaft or lever stuck.
- Controls or cables installed or adjusted incorrectly.

HARD SHIFTING

- Shift cables out of adjustment.
- Shift cut-out switch broken or improperly adjusted.
- Shift cable too short, or too long.
- Corroded cables.
- Shift shaft bushings corroded or damaged.
- Shift crank or clutch actuating spool worn or damaged.
- Shaft shaft damaged, worn or broken.
- Shift cable attaching nuts too tight.

JUMPS OUT OF GEAR

- Incorrectly adjusted shift cables.
- Worn or damaged clutch or gears.

Stern Drive Unit

REMOVAL & INSTALLATION

1986-93 Cobra And 1988-95 King Cobra Models

◆ See Figures 1 thru 7

■ All special tool part numbers listed below are OMC factory tool numbers—we realize that if you reading this after 2001, they may be difficult, if not impossible, to obtain. With the exception of some of the earliest tools, the vast majority of these tools should be available from a Volvo Penta representative. Although the part numbers themselves may be different in many cases, the tool should be the same.

■ The vessel must be out of the water to perform this procedure.

1. Disconnect the battery cables.
2. Lower the drive unit into the full DOWN or IN position. Drain the stern drive unit oil as detailed in the Maintenance section—make sure that you remove the oil dipstick. Check the drained lubricant carefully to determine if it contains any water or metal particles. Rub some of the lubricant between your fingers. Any metal particles in the lubricant will thus be evident. Do not be mislead if the color of the lubricant is metal colored. This is not a harmful condition and is caused by the lubricant used in the unit the first time after manufacture.
3. Remove the propeller(s) if you have not already done so. Although not strictly necessary, we strongly suggest removing the propellers as a precaution.
4. Support the stern drive before detaching it from the transom. Use either an engine/drive hoist or a dolly. If using the hoist, make sure it is capable of a 500 lb. capacity. Screw a lifting eye (1/2 x 13 thread, available at any home center) into the oil dipstick hole as far as possible and then attach the hoist chain to the eye. Tighten the hoist chain just enough to relieve the weight of the drive. If using a dolly, slide it under the anti-cavitation plate and tighten the two clamps.
5. Remove the plastic cap on the drive end of each trim/tilt cylinder. Remove the lock nut and flat washer. Support the hydraulic cylinder and lightly tap the pivot rod out of the cylinders with a rubber mallet. Carefully lower the cylinders down until they rest against the spray plate.
6. Loosen and remove the 6 mounting nuts (3 per side) securing the upper unit to the transom bracket. Throw the nuts away.
7. Pull the drive unit backward and away from the transom bracket, making sure that you support the U-joints and shaft as they come out of the pivot housing bellows.
   On the 1990-95 King Cobra, pull the unit back just far enough to expose the shift cable connectors. Pull up the latch, remove the connector pin and then pull the drive off all the way.
8. Support the drive in a suitable holding fixture.

To install:

The engine must be properly aligned or the input and output shaft splines will be destroyed after a short time of engine operation. If the engine has not been removed while the drive is out the alignment should still be satisfactory, but you must still use the proper alignment tools.

■ If you've gone through the trouble of removing the stern drive, we recommend this as a good time to perform some preventative, although not necessary, maintenance—remove the U-joint and exhaust bellows and replace them as detailed in the Gimbal Housing section; why not do the water hose also? Cheap insurance, so spend a few bucks and do it!

9. Inspect the condition of the water sealing seal and replace with a new seal if any doubts as to its condition.
10. Check the water hose and its fitting. Make sure the clamp nut is tightened to 96-120 inch lbs. (11-14 Nm).
11. Install a new pivot housing gasket and then coat the lip of the U-joint bellows lightly with Triple Guard grease.
12. Slide an alignment tool (# 912273), splined end first, through the bellows and gimbal bearing and into the engine coupler. The tool should slide into the coupler very easily; if not, the engine is out of adjustment and will have to be moved as detailed in the Engine Alignment procedures given in the appropriate section for your engine.

■ If you are not sure of which way that the engine needs to be adjusted, coat the splines of the tool with grease and reinsert it again until it binds and then remove it very carefully. The pattern left on the alignment tool should tell you which way to adjust the engine.

13. Once the engine is aligned correctly, remove the tool and shift the drive into Neutral; making sure that the shift rod in the drive is also in the Neutral position. Coat the shift lever pin liberally with Triple Guard grease.
14. Coat the 6 mounting studs with Drive Sealing compound.
15. Coat the splines on the driveshaft liberally with Moly Grease. Lubricate 2 new shaft O-rings with clean engine oil and slide them onto the shaft until they are seated in their grooves. Do not get grease in the grooves.
16. Use a grease gun on the U-joint zerk fitting and top out the joints. Coat the surface of the U-joint coupler with grease. Coat the mounting stud threads with Gasket Sealing compound.
17. On 1990-95 King Cobras, move the remote control shift handle protruding from the pivot housing up and into the Forward position. On the port side of the drive (next to the U-joint yoke), pull the shift linkage out and flip the connector latch up.
18. Raise the unit off of the holding fixture with your hoist and move it into position in front of the transom. Carefully slide the unit forward while guiding the driveshaft into the engine coupler and the drive over the mounting studs. Wiggle the propeller slightly once the shaft splines make contact with the coupler until they engage properly and then slide the unit all the way forward until the case mates with the pivot housing.

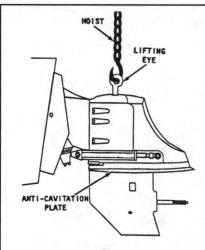

Fig. 1 Screw in a lifting eye and then attach a hoist

Fig. 2 Remove the plastic cap

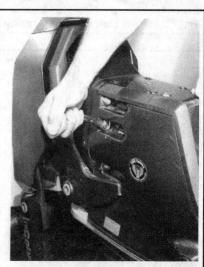

Fig. 3 Loosen the mounting bolts...

On the 1990-95 King Cobras, stop the drive just at the point where its possible to connect the shift handle to linkage. Line up the holes, insert the pin and press the latch down until it snaps into position. Now finish mating the drive to the housing.

18. Install 6 new locknuts and tighten them with a 5/8 in. swivel socket to 25 ft. lbs. (34 Nm) starting with the center nuts and then working back and forth in a criss-cross pattern. Now repeat the same sequence until they are tightened to 50 ft. lbs. (68 Nm).

19. Press the trim/tilt cylinder bushing into the boss on each side of the upper unit. Install bushings into the outer side of each cylinder, align the cylinder with the boss and press in the pivot pin very carefully. You may have to give it a few taps with a rubber mallet.

20. On cylinders with lock nuts, install the washers and then the nuts (new!). Tighten the nuts until there is an equal amount of threads exposed on each side of the pin and then tighten both nuts to 32-34 ft. lbs. (43-46 Nm), except on the 1990-95 King Cobras where it is 10-12 ft. lbs. (14-16 Nm). Replace the plastic caps.

21. Detach the lifting hoist. Install the oil level indicator and tighten it to 48-72 inch lbs. (5-8 Nm).

22. Adjust the trim sending unit and the shift mechanism.

23. Install the shift cover and tighten the bolt(s) securely. Fill the unit with oil, screw in the dipstick, install the propeller(s) and connect the battery cables.

1994-98 Single Propeller Models (SP/SX/SX Cobra)

1996-98 Dual Propeller Models (DP/DP DuoProp)

◆ **See Figures 1, 2, 6, 7, 8, 9, 10 and 11**

■ All special tool part numbers listed below are OMC factory tool numbers—we realize that if you reading this after 2001, they may be difficult, if not impossible, to obtain. With the exception of some of the

Fig. 4 ...and then separate the drive from the housing

earliest tools, the vast majority of these tools should be available from a Volvo Penta representative. Although the part numbers themselves may be different in many cases, the tool should be the same.

■ **The vessel must be out of the water to perform this procedure.**

1. Disconnect the battery cables.

2. Lower the drive unit into the full DOWN or IN position. Drain the stern drive unit oil as detailed in the Maintenance section—make sure that you remove the oil dipstick. Check the drained lubricant carefully to determine if it contains any water or metal particles. Rub some of the lubricant between your fingers. Any metal particles in the lubricant will thus be evident. Do not be mislead if the color of the lubricant is metal colored. This is not a harmful condition and is caused by the lubricant used in the unit the first time after manufacture.

3. Remove the propeller(s) if you have not already done so. Although not strictly necessary, we strongly suggest removing the propellers as a precaution.

4. Support the stern drive before detaching it from the transom. Use either an engine/drive hoist or a dolly. If using the hoist, make sure it is capable of a 500 lb. capacity. Screw a lifting eye (1/2 x 13 thread, available at any home center) into the oil dipstick hole as far as possible and then attach the hoist chain to the eye. Tighten the hoist chain just enough to relieve the weight of the drive. If using a dolly, slide it under the anti-cavitation plate and tighten the two clamps.

5. Remove the 3 bolts securing the shift cover to the rear of the unit and lift off the cover. Remove the jam nut from the shift cable, pull out the cotter pin holding the shift cable end and remove the washer. Remove the pivot cube from the end of the cable and unscrew the brass nut if equipped.

6. Loosen the shift cable anchor clamp screw on the starboard side of the drive, slide the anchor over to relieve the cable and then turn the drive hard to point before removing the shift cable.

7. Remove the plastic cap on the drive end of each trim/tilt cylinder. Remove the elastic lock nut and flat washer; on certain 1998 units, the nut may have been replaced with an e-clip so remove the clip and flat washer from the rod.

8. Support the hydraulic cylinder and lightly tap the pivot rod out of the cylinders with a rubber mallet. Carefully lower the cylinders down until they rest against the spray plate.

■ **If the bushings come out with the pivot rod, make sure that you keep the two grounding clips.**

9. Loosen and remove the 6 elastic mounting nuts (3 per side) securing the upper unit to the transom bracket. Throw the nuts away.

10. Pull the drive unit backward and away from the transom bracket, making sure that you support the U-joints and shaft as they come out of the pivot housing bellows.

11. Support the drive in a suitable holding fixture.

To install:

The engine must be properly aligned or the input and output shaft splines will be destroyed after a short time of engine operation. If the engine has not been removed while the drive is out the alignment should still be satisfactory, but you must still use the proper alignment tools.

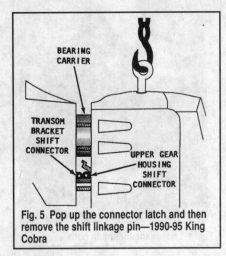

Fig. 5 Pop up the connector latch and then remove the shift linkage pin—1990-95 King Cobra

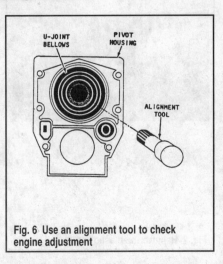

Fig. 6 Use an alignment tool to check engine adjustment

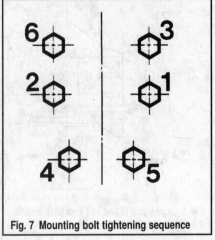

Fig. 7 Mounting bolt tightening sequence

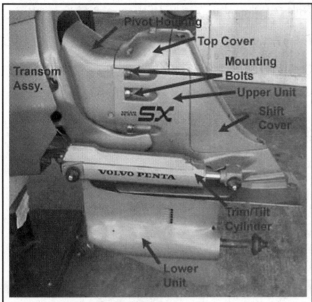

Fig. 8 A good shot of the drive unit

■ If you've gone through the trouble of removing the stern drive, we recommend this as a good time to perform some preventative, although not necessary, maintenance—remove the U-joint and exhaust bellows and replace them as detailed in the Gimbal Housing section; why not do the water hose also? Cheap insurance, so spend a few bucks and do it!

12. Inspect the condition of the water sealing seal and replace with a new seal if any doubts as to its condition.

13. Check the water hose and its fitting. Make sure the clamp nut is tightened to 96-120 inch lbs. (11-14 Nm).

14. Apply a light coating of OMC Triple Guard grease to the tapered end of the pinion bearing carrier. Make sure that the rubber sealing ring is seated properly in the groove of the pivot housing and then coat the edge of the U-joint bellows with grease.

15. Slide an alignment tool (# 3851083), splined end first, through the bellows and gimbal bearing and into the engine coupler. The tool should slide into the coupler very easily; if not, the engine is out of adjustment and will have to be moved as detailed in the Engine Alignment procedures given in the appropriate section for your engine.

■ If you are not sure of which way that the engine needs to be adjusted, coat the splines of the tool with grease and reinsert it again until it binds and then remove it very carefully. The pattern left on the alignment tool should tell you which way to adjust the engine.

16. Once the engine is aligned correctly, remove the tool and shift the drive into gear (either one) by moving the eccentric piston lever up or down.

17. Coat the 6 mounting studs with Drive Sealing compound.

18. Coat the splines on the driveshaft liberally with Moly Grease. Lubricate 2 new shaft O-rings with clean engine oil and slide them onto the shaft until they are seated in their grooves. Do not get grease in the grooves.

19. Use a grease gun on the U-joint zerk fitting and top out the joints. Coat the surface of the U-joint coupler with grease.

20. Raise the unit off of the holding fixture with your hoist and move it into position in front of the transom. Carefully slide the unit forward while guiding the driveshaft into the engine coupler and the drive over the mounting studs. Wiggle the propeller slightly once the shaft splines make contact with the coupler until they engage properly and then slide the unit all the way forward until the case mates with the pivot housing.

21. Install 6 new locknuts and tighten them with a 5/8 in. swivel socket to 25 ft. lbs. (34 Nm) starting with the center nuts and then working back and forth in a criss-cross pattern. Now repeat the same sequence until they are tightened to 50 ft. lbs. (68 Nm).

22. Press the trim/tilt cylinder bushing into the boss on each side of the upper unit. Install bushings into the outer side of each cylinder, align the cylinder with the boss and press in the pivot pin very carefully. You may have to give it a few taps with a rubber mallet.

23. On cylinders with lock nuts, install the washers and then the nuts (new!). Tighten the nuts until there is an equal amount of threads exposed on each side of the pin and then tighten both nuts to 10-12 ft. lbs. (14-16 Nm). Replace the plastic caps.

24. Detach the lifting hoist.

25. Turn the drive over hard to port. Coat the end of the shift cable casing with grease and then slide it through the transom sleeve. Make sure the large and small seals are still in place on the cable and slide it into the housing. Swivel the anchor clamp into place in the groove on the cable and tighten the bolt securely.

26. Thread the cube onto the end of the shift cable about half way. Move the remote control handle to Neutral and then position the remote control lever on the eccentric piston in its neutral detent (lever should be level. Now turn the cube in or out until it is in alignment with the center of the slot in the crank.

27. Push the cube pin through the hole and then install the washer and a new cotter pin. Install the jam nut and tighten it securely against the cube.

28. Adjust the trim sending unit.

29. Install the shift cover and tighten the bolt(s) securely. Fill the unit with oil, screw in the dipstick, install the propeller(s) and connect the battery cables.

Fig. 9 Remove the 3 shift cover bolts...

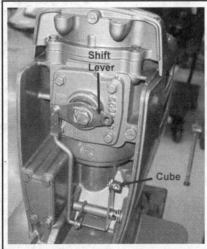

Fig. 10 ...and then disconnect the shift cable

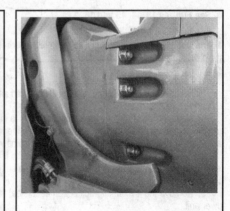

Fig. 11 There are 3 nuts on each side

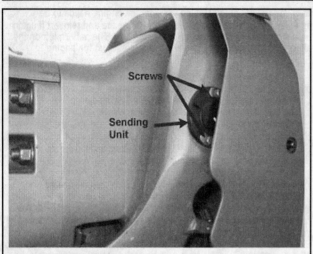

Fig. 12 The trim sending unit

TRIM SENDING UNIT ADJUSTMENT

◆ See Figure 12   MODERATE

   1. Move the trim switch to the DOWN or IN position and confirm that the drive unit is in the full down position.
   2. Turn the wheel hard over to port and then loosen the sending unit bolts on the starboard side of the gimbal ring.
   3. Disconnect the sending unit plug in the harness, connect an ohmmeter and rotate the sending unit until the meter shows a reading of 10-12 ohms.
   4. Tighten the sending units screws to 24 inch lbs. (2.7 Nm) and reconnect the plug.

UPPER GEAR HOUSING

Description

   Output power from the engine is connected to the stern drive through a horizontal driveshaft. A coupler is bolted to the flywheel and has a splined hub in the center. The end of the horizontal driveshaft indexes with, and slides into, the center of the hub. Power from the engine is then transmitted through the horizontal driveshaft to a pinion gear set where power direction is changed from horizontal to vertical.
   The upper driven gear is pressed onto the outside diameter of the upper driveshaft. The upper driveshaft is splined on the lower end. When the upper gear housing is mated to the lower unit, the end of the lower unit driveshaft indexes into the splined end of the upper driveshaft. Engine power is then transferred down into the lower gear unit.
   The horizontal and vertical driveshafts are both mechanically connected. Therefore, anytime the engine is operating, the horizontal and vertical driveshafts are constantly rotating with engine rpm. A double yoke universal joint assembly in the horizontal driveshaft allows the stern drive to be raised, lowered or turned side-to-side to a required trim/tilt position (within limits), while the engine is operating.

Upper Gear Housing

REMOVAL & INSTALLATION

1986-93 Cobra Models

◆ See Figures 13, 14, 15 and 16  MODERATE

■ **The upper gear housing may be removed after removing the entire stern drive unit from the vessel.**

   1. Disconnect the battery cables.
   2. Remove the stern drive as previously detailed and position it in a support fixture.
   3. Make a note of the of where the index mark is in relationship to the grid on the bottom of the trim tab and then remove the trim tab bolt and the tab. Make sure that you write the tab mark down somewhere!
   4. Remove the bolt that was partially covered by the trim tab and then remove second bolt forward of this one. The middle bolt is for the anode and does not need to be removed.
   5. Remove the 4 mounting bolts (2 on each side) from the sides of the lower unit.
   6. Lift the housing off of the lower unit until the shift rod disengages the guide pin and then swivel it 90 degrees to port while lifting it completely off. Move it to an appropriate stand.
   To install:
   7. Thoroughly clean the upper case-to-lower housing mating surfaces.

   8. Make sure that the nylon plug is positioned at the base of the splines inside the lower driveshaft. Coat the splines at both ends of the intermediate shaft with clean engine oil and insert the shaft into the lower driveshaft.
   9. Rotate the shift rod head 90 degrees to port.
   10. Install the plastic water tube into the recess in the bottom of the upper unit.
   11. Position the upper unit over the lower unit and then slowly lower it while guiding the water tube and shaft into place. As the shift rod head begins to appear in the exhaust opening, swivel it forward until it can engage the guide pin. Wiggle the propeller shaft until the shaft splines engage those in the coupler and then lower the unit fully until it sits on the lower unit.

■ **There may be a yellowish-green substance in the threads of the mounting bolts—this helps the bolts from coming loose and if not evident, they will need to be replaced with new ones.**

   12. Coat the threads of all bolts with Gasket Sealing compound. Install the bolts, 4 short ones (on the sides) and 2 longer ones (in the rear). Tighten them all to 22-24 ft. lbs. (30-33 Nm).
   13. Position the trim tab so that the index mark lines up with the correct mark on the grid and then tighten the bolt to 28-32 ft. lbs. (38-42 Nm).
   14. Install the drive unit.

1988-95 King Cobra Models

◆ See Figure 17  MODERATE

■ **The upper gear housing may be removed after removing the entire stern drive unit from the vessel.**

   1. Disconnect the battery cables.
   2. Remove the stern drive as previously detailed and position it in a support fixture.
   3. Remove the 6 mounting bolts (3 on each side) from the sides of the lower unit and the 1 bolt at the rear of the plate.
   4. Lift the housing off of the lower unit and move it to an appropriate stand. The water tube and/or shaft coupler may come off (or fall out) with the upper unit, so watch for them and reinstall them in the lower unit if necessary.
   To install:
   5. Thoroughly clean the upper case-to-lower housing mating surfaces.
   6. Slide the small nylon support ring over the top of the driveshaft and into position in the recess.
   7. Coat the new O-rings for the driveshaft retainer and oil passage with OMC Hi-Vis oil and then press them into their respective grooves.
   8. Make sure that the tabs on the driveshaft retainer are aligned and in place with the nearest slots.
   9. Install the intermediate shaft onto the driveshaft. Press the water supply tube into the grommet on the bottom of the housing.

Fig. 13 Remove the side bolts...

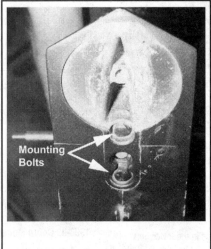

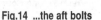

Fig.14 ...the aft bolts

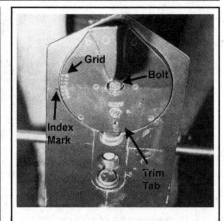

Fig. 15 Don't forget to note the trim tab position before removing it

10. Position the upper unit over the lower unit and then slowly lower it while guiding the water tube and intermediate shaft into place. Feed the impeller housing vent tube into the exhaust chamber on the lower unit. Wiggle the propeller shaft until the upper shaft splines engage those in the coupler and then lower the unit fully until it sits on the lower unit.

11. Coat the threads of all bolts with Gasket Sealing compound. Install the bolts and tighten them to 32-40 ft. lbs. (43-54 Nm).

12. Install the drive unit.

1994-98 Single Propeller Models (SP/SX/SX Cobra)

1996-98 Dual Propeller Models (DP/DP DuoProp)

⊘② ⊳MODERATE

◆ See Figures 8, 11, 12, 13, 14, 15 and 18 thru 22

■ The upper gear housing may be removed after removing the entire stern drive unit from the vessel.

1. Disconnect the battery cables.

2. Remove the stern drive as previously detailed and position it in a support fixture.

3. Remove the 4 mounting bolts (2 on each side) from the sides of the lower unit and the 2 bolts at the rear of the plate. DP drives units may use studs on the sides, so they will have nuts if this is the case.

4. Lift the housing off of the lower unit and move it to an appropriate stand. The water tube and/or shaft coupler may come off (or fall out) with the upper unit, so watch for them and reinstall them in the lower unit if necessary.

To install:

5. Thoroughly clean the upper case-to-lower housing mating surfaces.

6. Clean the oil screen thoroughly. Coat a new O-ring lightly with clean engine oil and press it into the groove. Install the oil screen.

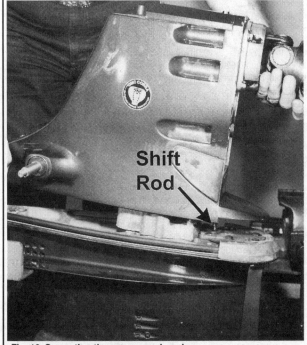

Fig. 16 Separating the upper gear housing

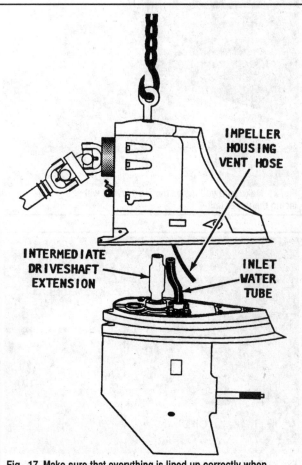

Fig. 17 Make sure that everything is lined up correctly when mating the upper and lower units

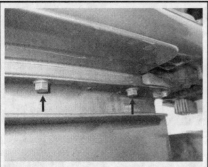
Fig. 18 Remove the two bolts on each side...

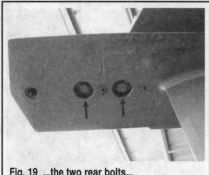
Fig. 19 ...the two rear bolts...

Fig. 20 ...and then separate the cases and lift off the upper unit

7. Make sure that the intermediate shaft coupler is installed correctly on the shaft with the groove facing upward.

8. Install the plastic water tube guide into the recess in the lower unit. Check the positioning of the upper and lower grommets on the tube, coat the lower grommet with sealing compound and position it into the recess and over the guide. Certain early models may use a retainer that is held in place with 4 screws, if you have one, tighten them to 10-12 ft. lbs. (14-16 Nm).

9. Replace any O-rings.

10. Position the upper unit over the lower unit and then slowly lower it while guiding the water tube and shaft coupler into place. Wiggle the propeller shaft until the upper shaft splines engage those in the coupler and then lower the unit fully until it sits on the lower unit.

11. Coat the threads of all bolts (or studs) with sealing compound. Install the bolts (or nuts) and tighten the 4 short ones (on the sides) to 22-24 ft. lbs. (30-33 Nm) and the 2 longer ones (aft) to 32-40 ft. lbs. (43-54 Nm).

12. Install the drive unit.

Fig. 21 Make sure you install the shaft sleeve with the groove facing the upper unit

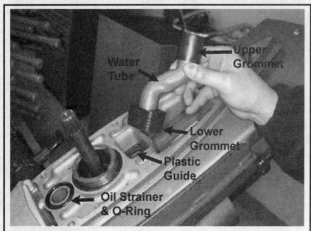
Fig. 22 Position the water tube before moving the upper unit into place

DISASSEMBLY & ASSEMBLY

1986-93 Cobra Models

1988-89 King Cobra Models

◆ See Figures 23 thru 50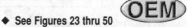

■ All special tool part numbers listed below are OMC factory tool numbers—we realize that if you reading this after 2001, they may be difficult, if not impossible, to obtain. With the exception of some of the earliest tools, the vast majority of these tools should be available from a Volvo Penta representative. Although the part numbers themselves may be different in many cases, the tool should be the same.

■ OMC makes a special spline socket (# 311875) and holding fixture (# 912278) to hold and support the upper housing—we strongly recommend using these tools, as the unit must be held securely throughout the disassembly process.

1. Remove the upper housing and install it into an appropriate support fixture.

2. Remove the 3 mounting bolts and lift off the rear cover. Remove the 4 upper cover bolts and lift off the cover. Throw away the Large O-ring, but make sure that you keep the 4 bolts separate as they are the longest bolts in the upper unit and are important for securing the cover during drive lifting operations.

3. Remove the water pump and adapter as detailed in the Cooling System section.

4. Loosen and remove the 4 bearing carrier bolts. Insert 2 small prybars between the carrier and the housing and then lever the U-joint/carrier assembly out of the housing—have a friend or assistant help you by pulling out on the driveshaft. Save any shim material that is not damaged. Remove the 2 O-rings on the shaft and the one on the inner flange of the carrier and throw them away. Please refer to the U-Joints section for service procedures on the U-joints.

■ Be very careful not to press or strike the impeller shaft while removing the assembly.

5. Separate the U-joint shaft from the driveshaft at the yoke closest to the bearing carrier (see U-Joints section).

6. Clamp the flat side of a universal bearing separator below the small bearing on the shaft and position the assembly into a press. Slide a bearing remover (# 912269) over the impeller shaft so the small side is toward the bearing and then press off the bearing. Make sure you catch the shaft.

7. Cut a 2-7/8 in. length of 1 in. diameter pipe and press the piece through the yokes on the end of the shaft. Position the yoke (with support pipe) into a vise so the shaft is pointing up. Install a spanner wrench (# 912272) over the spanner nut, attach a breaker bar and remove the nut. This nut is secured with Ultra-Lock adhesive, so it may be necessary to heat the nut in order to break it loose.

8. Install a remover tool (# 912269) over the impeller shaft so the small end is toward the gear. Support the bearing carrier and press the shaft out. If the large bearing requires replacement, install a bearing separator between the gear and bearing so the curved side is facing the bearing. Insert a cup remover (# 912291) into the gear and remove the race.

■ **Removal of the impeller shaft will damage the shaft—never remove the shaft unless necessary due to damage or wear.**

9. The impeller shaft is pressed into the driveshaft. Remove requires drilling a hole through the shaft and installing a puller. On replacement, coat the new shaft lightly with engine oil, position the large end of an installer (# 912269) over the U-joint shaft and insert the new shaft into the tool. Press the shaft in until it is flush with the end of the tool.

10. Remove the O-ring from the bearing carrier if you didn't already do it while removing the assembly. Position the forward face of the carrier (the side the U-joint comes out of) down on a press. Insert the nose of an installer (# 912287) through the seal and then press it out of the carrier. Make sure that the carrier is face down, or you will damage the carrier during the removal process.

11. Sit the recessed side of a holding block on the top side of the carrier. Lay the base plate on the block and then install the puller (see pinion bearing carrier steps for part numbers) and remove the bearing race. Make sure that this is necessary, because you will damage the race on removal.

12. Insert a spanner tool (# 984330) into the upper bore and secure the arm to the mating surface with 2 of the upper cover bolts. Tighten them alternately until the tool is secure and seated in the retainer ring around the upper driven gear. Attach a breaker bar and loosen the ring; once loosened sufficiently, remove the tool and unscrew the ring. Remember that the tool will be under considerable tension so loosen the bolts a little at a time.

13. Wrap a wooden dowel or broom stick with a rag and insert it into the bore from the bottom (you may have to turn the unit over). Carefully tap the dowel with a rubber mallet until the pinion bearing carrier comes loose; remove the carrier. Remove all shim material from the lip of the carrier and inside the housing.

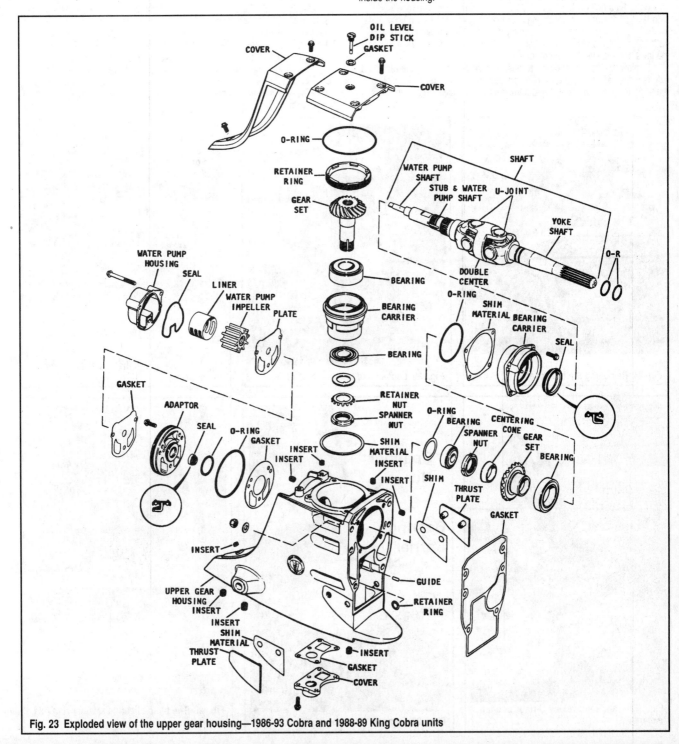

Fig. 23 Exploded view of the upper gear housing—1986-93 Cobra and 1988-89 King Cobra units

14. Turn the carrier over and bend back the locking tab holding the spanner nut in position. Clamp a spline socket (# 314438) in a vise and then slide the carrier onto it, gear side first. Position a spanner wrench (# 912272) over the nut, connect a 1/2 x 3/4 in. adapter to a 1/2 in. breaker bar and remove the nut, retainer and washer.

15. The gear and shaft are one piece and are pressed into the lower bearing. Position a holding block (# 912278) over 2 plates on your press so the recessed side is facing up. Position the retainer ring over the block and then sit the carrier on the assembly with the gear facing down. Press the shaft and gear free of the carrier.

16. If the pinion shaft bearing requires replacement, position the shaft into a universal bearing separator so the curved side of the tool is between the bearing and the gear. Place the assembly in a press and remove the bearing.

17. Position the carrier on a flat surface so the small bearing race is facing up. Sit a holding block (# 912278) on the carrier with the recessed side facing up. Lay the retainer ring over the block and then lay a base plate on the ring. Assemble a bridge assembly (#432127) and puller jaws (# 432129), install it on the components so the jaws are under the race, expand the jaws and remove the bearing race.

18. Now flip the carrier over and position a holding block so the recessed side is facing down. Lay a base plate on the block. Install the puller tool so the jaws are under the race, expand the jaws and remove the bearing race.

To assemble:

19. Replace all O-rings, gaskets and seals with new ones whenever possible. Clean all parts (except rubber or plastic) in solvent and dry them with compressed air.

20. Coat all metallic components with clean engine oil.

21. Check all gears and gear teeth carefully to be sure they are not chipped, cracked, nicked or otherwise damaged in any manner. Replacement gears always come in matched sets, so when you find a bad one, replace them all.

22. Inspect all bearings and bearing races for pits, grooves, burns or signs of uneven wear. Replace as necessary.

23. Check all shafts and shaft splines carefully to be sure they are not chipped, cracked, nicked or otherwise damaged in any manner.

24. Check the thrust plate dimensions again as detailed previously. Add shim material as necessary.

25. Coat both sides of a new water tube cover gasket with Gasket Sealing compound and then position into the housing so that the round opening is on the port side as you look at the housing from the rear. Install the cover and tighten the bolts to 12-14 ft. lbs. (16-19 Nm). Finally, press a new seal into the cover and then twist the water tube in until it seats.

26. Set the pinion bearing carrier on a press so the wider side is facing up. Position a new bearing race into the carrier and then lay an installer (# 912276) over the race so the **A** side is facing up. Press the race in until it seats completely.

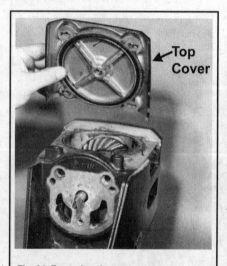

Fig. 24 Removing the top cover

Fig. 25 Loosen and remove the 4 bolts...

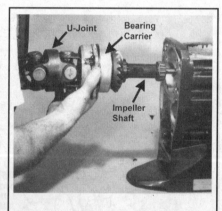

Fig. 26 ...pry or pull the carrier away from the housing...

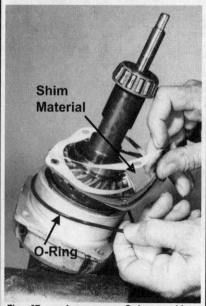

Fig. 27 ...and remove any O-rings or shim material

Fig. 28 Remove the spanner nut

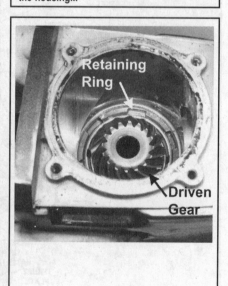

Fig. 29 The pinion bearing retainer ring has notches...

27. Turn the carrier over and repeat the previous step for the smaller bearing race.

28. If you removed the large pinion bearing, lay the **A** side of the installer on a press. Coat the inside of the bearing with oil, position it on the installer and insert the shaft. Press in the shaft until the gear seats itself on the bearing. If you removed the smaller bearing, turn the carrier over and repeat the procedure for this one.

29. Thread in the spline socket used during disassembly and mount the carrier assembly in a vise. Spray the shaft and spanner nut threads with Locquic Primer and allow them to dry completely. Slide the washer and tab retainer over the shaft. Apply OMC Ultra-Lock adhesive to the threads of the nut and then screw it on so the beveled edge is facing the carrier. Finger-tight only at this point please.

30. Install the spanner wrench and breaker bar and then tighten the nut while rotating the carrier body until there is no shaft end play evident. Remove the unit, install a holding block (# 912278) into the vise and insert the carrier with the gear side facing up. Connect a torque wrench to the top of the spline socket and check that the rolling torque is 3-9 inch lbs. (0.3-1.0 Nm) with 6 inch lbs. (0.7 Nm) being the ideal. If rolling torque is too low, tighten the spanner nut until a recess in the nut has lined up with the next locking tab. Recheck the torque and continue this procedure until it falls within specifications. If the torque reading is too high, remove the nut,

retainer and washer. Turn the assembly over and position it in press so that you can exert enough pressure to move the shaft (just barely) slightly. Re-install the hardware and repeat the procedure until you can bring the rolling torque measurement into range.

31. Bend at least one locking tab over and into a recess on the spanner nut.

32. If you removed the drive gear bearing on the upper carrier, a new one must be installed. Coat the inside of the bearing lightly with oil and position it onto the gear. Move the assembly to a press and press the bearing onto the gear using an installer (#912276) with side **A** facing the bearing. If you also remove the race, coat the outer side of a new one with oil and press it in using the same procedure.

33. Install a new seal onto the stepped side an installer (# 912286) so that the protruding lip is facing away. Coat the metal casing of the seal with Gasket Sealing compound and then press the seal in until it seats fully. Coat the lip with Triple Guard grease. Once this is done, insert the carrier into the housing as a quick check for out-of-roundness—it should slide in freely, otherwise it will require replacement.

34. Sit the gear/bearing assembly on a press (gear side down) and position the carrier onto the bearing. Coat the U-joint shaft splines and the rollers with engine oil and then insert the shaft into the gear, pressing it in until fully seated. Carrier movement of up to 1/8 in. is not unusual, so don't be concerned if the bearing is not tight in the carrier.

Fig. 30 ...insert the tool so it seats in them...

Fig. 31 ...and then loosen the ring...

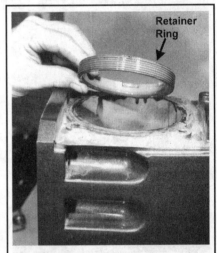

Fig. 32 ...until you can remove it

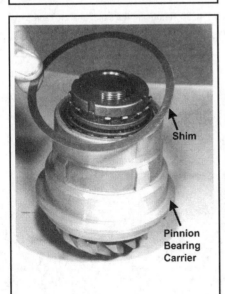

Fig. 33 Once the pinion bearing carrier is out, remove the shim material

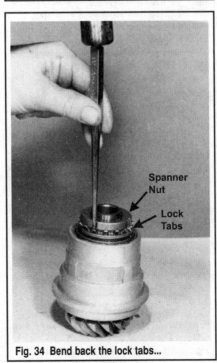

Fig. 34 Bend back the lock tabs...

Fig. 35 ...and remove the spanner nut

35. Spray the threads of the spanner nut and shaft with OMC Locquic Primer and allow to air dry completely.

36. Find the piece of pipe you cut earlier and reinstall it into the yoke so that you can clamp the assembly in a vise with the gear side facing up. Slide the centering cone down over the shaft and set fully into the gear with a drift and plastic mallet.

37. Coat the exposed shaft threads evenly with OMC Ultra Lock adhesive and then thread on the spanner nut so the beveled edge is against the gear. Install the spanner wrench and a 90 ° torque wrench and tighten the nut to 200 ft. lbs. (271 Nm).

38. Install a bearing installer (# 912269) into the base of a press so the large opening is facing up. Coat the inside of the impeller bearing with engine oil and sit it in the opening on the tool. Now coat the impeller shaft lightly with Extreme Pressure grease and press the assembly into the bearing until it seats itself on the shaft.

39. Determining shimming needs for the carrier assembly will require a shim gauge kit (# 984329). Remove the O-ring from the carrier if you haven't already done so, slide the race over the rear bearing and then insert the kit's center adaptor. Do not install any of the original shims.

40. Position the assembly into the fixture frame, rotating the carrier until the bolt holes align and then install the 3 bolts provided with the kit. Tighten them securely. Tighten the pre-load screw on the end of the fixture until it just reaches the reference groove in the adaptor, thus setting the preload. Tighten the lock nut securely and then rotate the shaft at least 4 times to set the rollers.

41. Clamp the frame in a vise so the gear is pointing up. Check the carrier for a stamped gear ratio rating and then select the appropriate gauge rod from the kit. Attach the rod to the inner hole in the center adaptor and tighten the bolt securely.

42. Measure the clearance between the rod and a flat on top of one of the gear teeth; subtract this figure from 0.020 in. to determine the shim thickness needed behind the rear bearing race. Now remove the gauge rod and install the long gauge rod in the outer hole on the adaptor. Measure the clearance between the end of the rod and the carrier flange, add this figure to the shim thickness you decided on for the rear bearing and then SUBTRACT 0.020 in. This will provide you with the correct shim thickness for bearing pre-load.

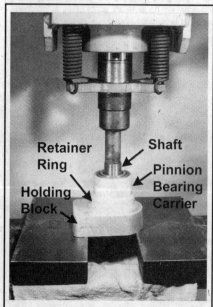

Fig. 36 Press out the gear and shaft...

Fig. 37 ...and then separate the bearing from the gear

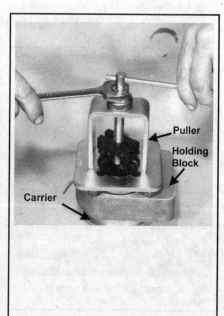

Fig. 38 Removing the small bearing race from the pinion carrier

Fig. 39 Forcing out the impeller shaft bearing race

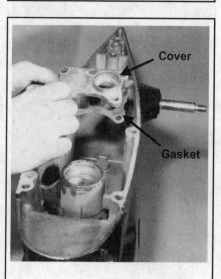

Fig. 40 Installing the water tube cover

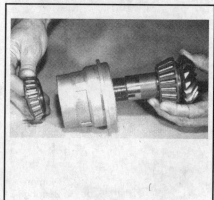

Fig. 41 A good shot of the pinion bearing carrier

43. OK, now that we've assembled all of the parts, let's put it all back together! Position a shim gauge onto the gear teeth flats on the pinion bearing carrier and measure the clearance between the arm of the gauge and the underside of the carrier rim. Subtract, you guessed it, 0.020 in. from this measurement. To determine the shim thickness needed for the carrier.

44. Install the appropriate amount of shims (previous step) onto the carrier seat in the housing and then install the carrier assembly. There are lugs on the side of the carrier housing which must align with the locating slots in the housing bore—the oil circulation slot visible inside the carrier is directly in line with one of the lugs so it makes for a good reference while sliding the carrier into the housing bore. Turn the housing on its side and then insert the carrier using the intermediate shaft to push it into place until the gear teeth tops are 3.10 in. (78.7mm) below the cover surface of the gear housing. You should also be able to see all of the retainer ring threads.

45. Spin the housing back into the upright position and then coat the retainer ring threads lightly with oil before tightening it down by hand. Once in place, install the spanner tool used during disassembly and tighten the ring to 145-165 ft. lbs. (197-224 Nm).

46. Slide a 5/8 in. ID flat washer onto a 5/8 in. x 3 in. bolt. Position the correct amount of shim material (as determined previously) into the bearing race seat, coat the rear bearing race with oil and then install it into the seat. Position a puller head (# 307636) into the water pump adapter housing recess so the flat side is against the case. Now position the installer (# 314429) into the race, insert the bolt through the puller and thread it into the installer. Tighten the bolt until the race seats itself fully.

47. Assemble the U-joints to the inner yoke and then slide the correct amount of shim material (as determined previously) over the shaft and against the inner flange of the carrier so the notches are all in alignment.

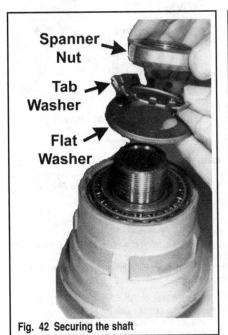

Fig. 42 Securing the shaft

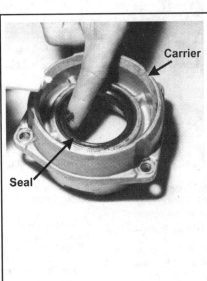

Fig. 43 Be sure to coat the forward bearing carrier seal with grease

Fig. 44 Pressing in the U-joint shaft

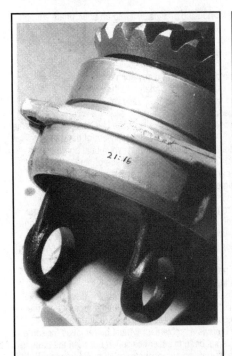

Fig. 45 The gear ratio should be etched on the carrier housing

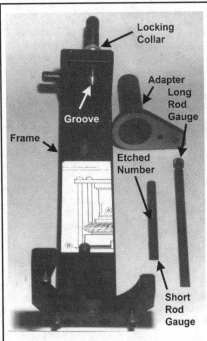

Fig. 46 A good shot of the Shim Gauge Kit

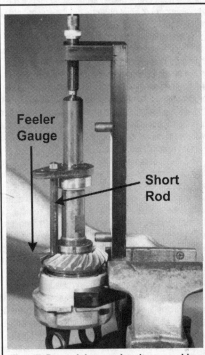

Fig. 47 Determining rear bearing race shim thickness

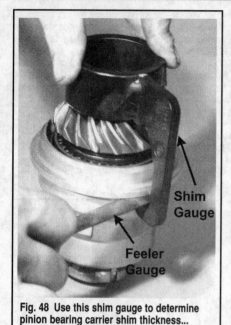

Fig. 48 Use this shim gauge to determine pinion bearing carrier shim thickness...

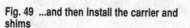

Fig. 49 ...and then install the carrier and shims

Fig. 50 Checking backlash with a dial indicator

Coat a new O-ring with Gasket Sealing compound and install it into the groove on the carrier.

48. Slide the entire assembly into the housing until the O-ring seats itself. Coat the threads of the 4 mounting bolts with Gasket Sealing Compound, install them and tighten to 12-14 ft. lbs. (16-19 Nm) in a crosswise pattern. Rotate the shafts at least 4 times to seat the bearing. Endplay must be Zero (0); if not, recheck all shimming measurements.

49. Attach a dial indicator to a stand of the front of the housing so that the indicator pin is resting against one of the forward gear teeth at a radius of about 1.580 in. from the centerline. Slowly rotate the shaft back and forth so that the teeth on the two gears just make contact in one direction and then again in the other direction. The dial indicator should read 0.009-0.015 in. (0.23-0.38mm). If outside this range, an equal amount of shim material will have to be added or removed at the rear bearing race, the upper bearing carrier and the pinion bearing carrier. Subtract the desired lash (0.012 in. is the mid-point of the range) from your actual lash reading (0.020 in.) to arrive at the amount of shim change that it necessary (0.020 - 0.012 = 0.008). Half of this figure (0.008 divided by 2 = 0.004) is the shim thickness that will have to be added or removed from the three locations. If you wish to decrease the lash, you would remove 0.004 in. of shims from the rear bearing race and the upper bearing carrier, and then you would add 0.004 in. of shim material to the pinion bearing carrier. If you had to increase the lash, you would add the figure to the first two locations and then remove that amount from the pinion bearing carrier. Always remember that no matter how you need to change the lash (increase or decrease) the upper carrier and the rear bearing race always have the same action, while the pinion bearing carrier will always be the opposite of those two—add to the first two, remove from the last one, or vice versa.

50. Almost done! Rotate the housing in the stand so the front edge is now down and the U-joint shafts are hanging (the bottom of the unit is now on the side). Insert the intermediate shaft into the bottom of the pinion bearing carrier. Slide on the spline socket and attach a torque wrench. Rotate the driveshaft fast enough so that you can get a steady reading on the torque wrench and check that it is 6-21 inch lbs. (0.7-2.4 Nm). This is the rolling torque; if it is out of the range, check the backlash again. If backlash is still OK, change the shim thickness under the upper bearing carrier; adding shims lower the rolling torque, while removing them will raise the figure.

51. Install the water pump and adapter.

52. Coat a new O-ring with Gasket Sealing compound, press it into the groove in the top cover and then install the cover. Coat the threads of the mounting bolts (remember these are unique!) with Gasket Sealing compound and then tighten them to 12-14 ft. lbs. (16-19 Nm).

53. Install the rear cover and tighten the bolts to 108-132 inch lbs. (12-15 Nm).

54. Install the upper housing to the lower unit.

✳ **1990-95 King Cobra Models** ✳

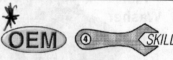

◆ **See Figures 51 thru 74**

■ All special tool part numbers listed below are OMC factory tool numbers—we realize that if you reading this after 2001, they may be difficult, if not impossible, to obtain. With the exception of some of the earliest tools, the vast majority of these tools should be available from a Volvo Penta representative. Although the part numbers themselves may be different in many cases, the tool should be the same.

■ OMC makes a special spline socket (# 3850598) and holding fixture (# 3850605) to hold and support the upper housing—we strongly recommend using these tools, as the unit must be held securely throughout the disassembly process.

1. Remove the upper housing and install it into an appropriate support fixture.

2. Loosen and remove the 4 u-joint bearing carrier bolts. Insert 2 small prybars between the carrier and the housing (there should be 2 slots) and lever the carrier assembly out of the housing—have a friend or assistant help you by pulling out on the driveshaft. Save any shim material that is not damaged. Remove the 2 O-rings on the shaft and the one on the inner flange of the carrier and throw them away. Please refer to the U-Joints section for service procedures on the U-joints.

3. Disconnect the bearing carrier yoke from the U-joint as detailed in the U-Joint section. Position the shaft and joint assembly out of the way.

4. Obtain a 1 in. diameter piece of pipe from your local home center and cut a piece 2 7/8 in. (73mm) long. Slide this piece of pipe through the holes in the end of the carrier yoke and then install the assembly in a vise so that the jaws are bearing on the ends of the pipe and the yoke.

5. The retaining bolt is a E16 Torx bolt and will be difficult to break loose. Once loosened, remove the bolt and retainer and then lift the carrier and pinion gear assembly off of the yoke. You may have to use a press and bearing separator if the yoke shaft does not come out freely.

6. Position a support tube (# 914747) into a press and stuff a small rag down into the tube. Lay the carrier on top of the tube so the gear is facing into the tube, insert the hex end of a bearing installer (# 914701) into the gear hub and press out the gear and bearing.

7. Examine the roller bearing and the race. Check for discoloration, chipped of damaged rollers, gouges or rough spots in the race. If both the bearing and race are in good shape, there is no need to separate the bearing and the gear. If there is a problem, or it must be removed for some other reason, use a universal bearing puller positioned between the bearing and gear. Put the unit in a press and press the gear off of the bearing. Be warned that you will probably damage the bearing during the removal process, so do not remove it unless necessary.

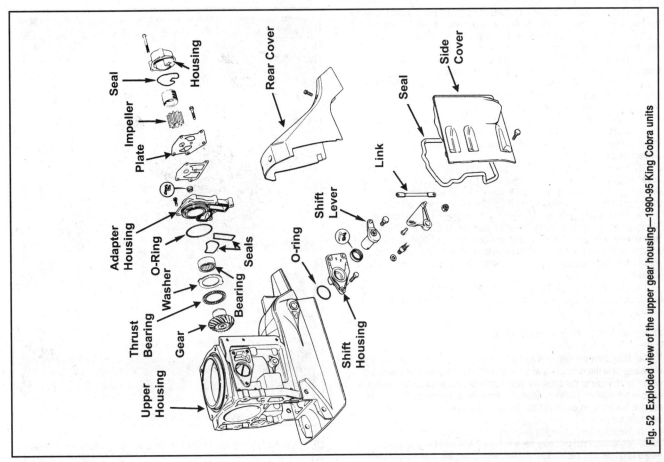

Fig. 52 Exploded view of the upper gear housing—1990-95 King Cobra units

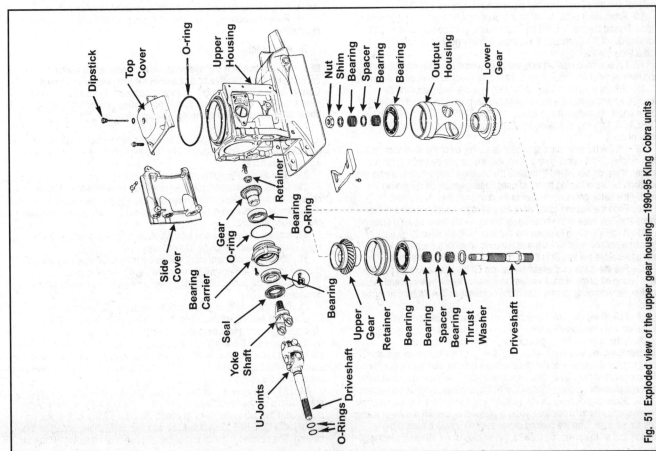

Fig. 51 Exploded view of the upper gear housing—1990-95 King Cobra units

8. Reach inside the carrier with a pair of snap ring pliers and remove the retaining ring. Position the carrier in a support tool (#914683) and place them on a press. Lay the large side of the removal tool (# 914748) on the bearing and press it and the seal out of the carrier.

9. Using a conventional 3-jawed puller, remove the inner bearing race from the u-joint side of the carrier. Now turn the carrier over, position it in the earlier support tool (# 914747) and place them on a press. Insert the larger side of a removal tool (# 914685) into the carrier so that it rests on the remaining race and then press out the race.

10. Remove the 3 mounting bolts and lift off the rear cover. Carefully pull the small drain hose off of the water pump housing nipple and tuck it out of the way. Remove the water pump and adaptor as detailed in the Cooling System section.

11. Remove the 6 mounting bolts for the port side cover and lift off the cover. Reach in with needle nosed pliers and pull out the cotter pin holding the shift lever link to the lever. Pull the link off of the lever, remove the 3 mounting bolts and carefully pull out the shift lever housing assembly.

12. Pull the shift shoe and spring out of the inner side of the housing. Remove and discard the O-ring. Carefully grab the roll pin and pull it out of the housing, then pull the roller and shaft tube out as well. Now you should be able to remove the shift lever. You should be able to pry out the inner seal with an awl, but if necessary, use the special bridge (# 431227) and puller (# 432129).

13. Pivot the bellcrank (below the shifter bore) forward as far as it will go without forcing it, lift up the clasp and remove the linkage pin. Pry the clasp off of the shift lever.

■ **Not all models will use the linkage pin.**

14. Now pivot the bellcrank as far back as it will go; rotate the bellcrank clockwise until the pivot pin is at the 6 o'clock position and pry it (carefully!) off the stud. Pull out the cotter pin and disconnect the link from the bellcrank, remove the washer. Remove the spring and retainer from the stud. Finally, remove the pivot pin cotter pin and disconnect the shift lever from the bellcrank.

15. Remove the 4 mounting bolts and lift off the upper housing top cover—again, you will probably have to tap it a few times with a rubber mallet and then pry it apart from the housing. Save any undamaged shim material and the oil deflector, but discard the cover seal (flat O-ring).

16. Fabricate a t-shaped removal tool using 3/8-16 x 1 1/2 in. threaded stock. Thread the tool into the top of the output gear housing and pull the housing out of the gearcase. If you have difficulty getting the housing out, use a slide hammer.

17. Lift out the driveshaft extension tube and then reach into the bore and remove all shim material; save them if they look to be in good shape.

18. Secure a spline socket (# 914699) in a vise and then slide the housing assembly into the socket. Attach a six-point socket and breaker bar to the upper retaining nut and remove the nut. THIS NUT HAS LEFT HAND THREADS, so turn it clockwise to loosen it. Lift off the nut and any shims underneath it.

29. Lift the housing and upper gear assembly off of the driveshaft and then reach into the upper gear and remove the needle bearings (2) and spacer. If you do not intend to replace the bearings, make sure you note their positions before removing them, as used bearings must be reinstalled in exactly the same positions relative ot the driveshaft and each other.

20. Slide the bearing collar off of the end of the shaft and remove the shims from the base of the housing (or the top of the lower bearing retainer).

21. If you are going to reuse the clutch cone, it will need to go back in the same position from which it was removed so mark the top of the cone with an indelible marker. Spin the clutch cone up and off of the shaft.

22. Pull the lower gear assembly up and off the shaft and then spin the cone spring off of the shaft. Remove the 2 lower needle bearings and thrust washer, remembering to take note of their orientation if you intend to reuse them.

23. Pull off the remaining bronze thrust washer and remove the driveshaft from the spline socket.

24. If the upper or lower gear/bearing assemblies are to be disassembled, install a support tool (# 914747) in an arbor press and stuff a small rag into the tube. Position the assembly into the tube so the gear is facing down and then lay a removal tool (# 914707) onto the gear hub with the large diameter opening facing DOWN. Press the gear out of the bearing.

25. Some bearings are slip-fit, so see if you can remove the bearing from the retainer by hand. If not, then yours is pressed in and will require pressing it off for removal. Turn the bearing carrier over and place it back in the support tool on the press. So that the curved side of the retainer is facing

Fig. 53 A cross section of the King cobra upper gear housing

UP. Position a removal tool (# 914693) on the bearing and press the bearing out of the retainer. Remember that you will most likely damage the bearing during this process, so don't do it unless absolutely necessary.

26. Repeat the previous two steps if upper gear and bearing removal is necessary.

To assemble:

■ **It is imperative that the upper and lower gears are installed in the correct positions, relative to one another if they are being reused. Correct backlash adjustment and tooth contact will spread the load over a larger area of the gear tooth; preventing breakdown and abnormal wear and also ensuring quieter operation.**

27. Replace all O-rings, gaskets and seals with new ones whenever possible. Clean all parts (except rubber or plastic) in solvent and dry them with compressed air.

28. Coat all metallic components with clean engine oil.

29. Check all gears and gear teeth carefully to be sure they are not chipped, cracked, nicked or otherwise damaged in any manner. Replacement gears always come in matched sets, so when you find a bad one, replace them all.

30. Inspect all bearings and bearing races for pits, grooves, burns or signs of uneven wear. Replace as necessary.

31. Check all shafts and shaft splines carefully to be sure they are not chipped, cracked, nicked or otherwise damaged in any manner.

32. Check the cone clutch/sliding sleeve for wear, heat or any other damage.

33. Coat the outside of a new lowers bearing's case and the inner surface of the bearing retainer liberally with Ultra-HPF Gearcase Lube. If your bearing is a slip-fit, simply press it into the retainer so the side with a notch is facing UP. If press-fit, set the retainer on the base of a press and lay the bearing into it so the side with a notch is facing UP. Place the installer tools (# 914692 and # 914693) on top of the bearing (in that order) and then press it into the retainer.

34. Lay the recessed side of the special tool (#914691) on the base of a press. Turn over the assembly and place it on the tool with the bearing side down. Coat the hub of the lower gear and the inner surface of the bearing liberally with Ultra-HPF Gearcase Lube and then position the gear over the bearing. The beveled (closed) end of the retainer must be facing UP toward

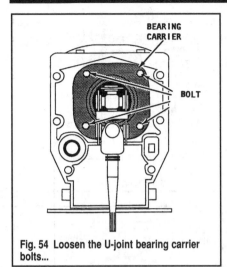

Fig. 54 Loosen the U-joint bearing carrier bolts...

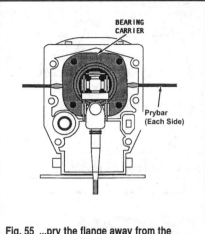

Fig. 55 ...pry the flange away from the housing...

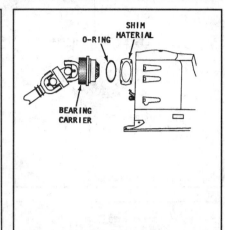

Fig. 56 ...and remove the assembly from the housing

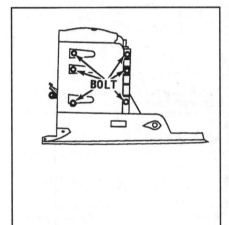

Fig. 57 Remove the port side cover to access the shift linkage

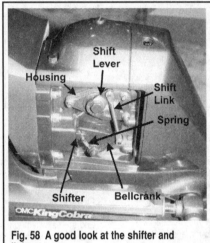

Fig. 58 A good look at the shifter and linkage

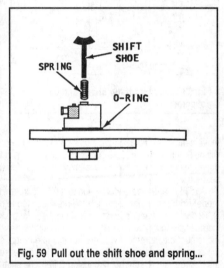

Fig. 59 Pull out the shift shoe and spring...

the bottom of the gear. Lay the larger end of the installer tool (# 914748) over the gear and press it into the bearing. Perform the Output Gear Shimming procedures.

35. Sit the bearing housing on the base of a press so the tapered side is facing UP. Coat the outside of a new upper bearing case and the inside of the housing bore liberally with Ultra-HPF Gearcase Lube and position the bearing on top of the housing so the side with the small notch is facing UP. Now lay the installer plate (# 914692) on the bearing so the stepped side is against the bearing and sit the other tool (# 914693) on top. Carefully press the bearing in until it just seats in the housing being careful not to apply too much pressure with the press.

36. Place the tool (# 914691), recessed side down, on the base of a press. Turn the bearing housing over so that the bearing is resting on the tool. Coat the gear hub and inner diameter of the bearing liberally with Ultra-HPF Gearcase Lube and position the gear in the bearing. Set an installer (# 914748) on the gear and press it in until it is fully seated. Perform the Output Gear Shimming procedures.

37. Clamp the spline socket in the vise again, insert the driveshaft and slide the bottom bronze washer over the shaft. Lubricate the lower needle bearing with Ultra-HPF Gearcase Lube and slide the lower bearing onto the shaft, followed by the small spacer and the upper bearing. Remember if you are using the old bearings that they must be installed in the same positions as they were removed; new bearings may be installed in any order you wish.

38. Position the tapered end of the cone spring over the top of the shaft and then pull out on the lower end while feeding the spring down and over the shaft until the bottom end is sitting in the recess above the bearing. Do this slowly and it will go much easier.

39. Slide the lower bearing assembly onto the shaft and slowly lower it down over the spring, retainer side down. Lift the bearing slightly while wiggling it back and forth until the spring is visible through the top of the bearing—once in place, the spring should lift and spin freely.

40. Install the clutch cone over the shaft. If reusing the old one, make sure that the marks made during disassembly are correct. If using a new clutch, there should be a **T** stamped in one of the valleys and this should be on the top.

41. Install the correct amount of shims onto the rim of the retainer as determined in the Upper Output Gear Shimming section. Slide the bearing collar onto the shaft and then lower the bearing housing into place so it rests on the shims.

42. Install the 2 upper needle bearings, separating them with the small spacer. Be careful that they are not damaged by catching the shaft shoulder. Remember if you are using the old bearings that they must be installed in the same positions as they were removed; new bearings may be installed in any order you wish. Use a small screwdriver and press down around the edge of the upper bearing to make sure they are fully seated.

43. Remove the arm from a shift fixture (# 986631) and position it upside down on top of the bearing. Measure the depth from the top of the arm to the top of the gear hub with a micrometer and record it as **A**. Next, measure the depth from the top of the arm to the top of the bearing collar. Record this as **B**.

44. Subtract dimension **A** from dimension **B** and record the new figure as dimension **C**. Compare your figure to the accompanying chart to determine the needed shim thickness. Coat the shim with Ultra-HPF Gearcase Lube and stick it to the bottom of the bearing retaining nut. Install the nut and shim, tightening to 140-160 ft. lbs. (190-217 Nm). If the lock patch on the nut has worn away, make sure you use a new nut. THE NUT IS LEFT HAND THREAD, so be sure that you tighten it in a counterclockwise direction.

45. Check your shimming by placing the bearing housing, splined shaft down, in a press support. Hold the retaining nut with just enough force to keep the shaft from turning. Attach a dial indicator in a manner that will allow the needle access to the upper bearing's case. Force the housing down until the indicator zeroes and then lift up on the housing while checking vertical

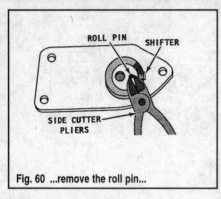

Fig. 60 ...remove the roll pin...

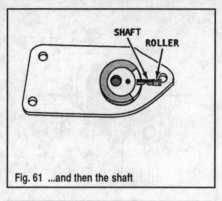

Fig. 61 ...and then the shaft

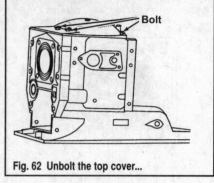

Fig. 62 Unbolt the top cover...

Fig. 63 ...and lift off the shims and oil deflector

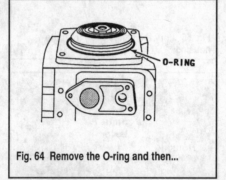

Fig. 64 Remove the O-ring and then...

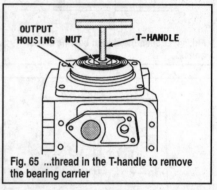

Fig. 65 ...thread in the T-handle to remove the bearing carrier

movement. If you figured your shims correctly in the previous steps, there will be 0.005-0.010 in. (0.13-0.25mm) of movement. If this is the case, you're done. If your check shows vertical movement outside the range, perform the next step.

46. To change the endplay, subtract 0.0075 in. (midpoint of the desired range) from the actual reading in the previous step to determine dimension **C**. Reducing endplay requires increasing the shim thickness, while increasing endplay requires decreasing shim thickness. If your vertical movement measurement was greater than the range, add dimension **C** to the shim thickness you originally installed. If the measurement was less than the desired range, subtract dimension **C** from the shim thickness you originally installed. Either way, check this figure against the chart to find the correct new shim, remove the retaining nut and original shim, and then install the nut with the new shim.

47. Install the driveshaft extension into the housing bore and then thread your fabricated tool into the top of the output gear housing assembly. Coat the housing case with Ultra-HPF Gearcase Lube and then lower the assembly into the bore so that the notches at the front are in alignment after you have engaged the driveshaft tube. The housing is a slip-fit in the bore and should not be forced, when correctly seated and aligned, the shift access hole on the port side of the housing will be lined up with the corresponding hole in the gearcase housing. Remove the T-handle.

48. Install the deflector on top of the output gear housing and then position the correct amount of shim material as determined in the Top Cover Shimming procedure. Coat a new top cover O-ring with Ultra-HPF Gearcase Lube and insert it into the groove in the top of the housing. Install the cover and tighten the bolts to 32-40 ft. lbs. (43-54 Nm), alternately and in a crosswise pattern.

49. Coat the hub of the pinion gear with Ultra-HPF Gearcase Lube. Support the bearing in the tool (# 914687), position the gear in the bearing and press it in with the installer tool (# 914708) until fully seated.

50. Position the nose of the U-joint bearing carrier in a support tool (# 914683) on the base of a press. Coat the outer case of the large bearing race with gear lube and place it on the carrier. Insert the small end of the installation tool (# 914690) into the cup and press in the race until fully seated. Turn the carrier over, coat the outer surface of the small bearing race with gear lube, and then place it into the carrier. Insert the smaller side of the installation tool into the race and press it into the carrier until fully seated.

51. Thread a nut onto an appropriately sized bolt, insert it through one of the flange bolt holes in the carrier (it doesn't matter which one) and then thread on another nut. Cinch the nuts against each side of the flange until the bolt is secure.

52. Position a support tool (# 914707) on the base of a press and then insert the gear and bearing assembly into the tool, gear side down. Coat the pinion gear hub with gear lube. Now drop the carrier onto the bearing and insert the yoke bearing so that it is resting on the end of the end of the gear inside the carrier. Lay the installation tool (#914687) over the top of the bearing and then cover the tool with a flat plate or something similar.

53. Connect a torque wrench to the bolt that you installed on the carrier and then slowly begin to press on the bearing until you get a rolling torque figure of 1-4 inch lbs. (0.11-0.45 Nm). Relieve the pressure from the press, hold the gear and recheck rolling torque. Although it is normal for the torque to decrease slightly when pressure is released, the rolling torque figure should still stay within the suggested range. If it does, remove the assembly and remove the bolt; if out of range on the low side, press the bearing again and recheck.

■ **Never press the bearing too far. If your rolling torque measurement is outside the range on the high side, remove the bearing, reinstall it and repeat the procedure.**

54. Insert a depth gauge into the carrier and measure the distance between the back of the bearing and the end of the gear. This measurement will determine the correct amount of shims to use behind the U-joint coupler shaft; measure each shim separately and make sure that you use at least one, but not more than seven.

55. Coat the inner lip and outer case of a new oil seal with gear lube and insert it onto the installer (# 914689) so the side with the silver insert is facing up. Place the carrier, gear side down, onto a support tool (# 914707), insert the seal and installer and then press in the seal until it bottoms in the carrier.

56. Place the pinion bearing carrier on a workbench or other flat surface so the gear is facing upward. Install a shim gauge (# 914688) over the carrier so it is resting on the gear.

57. Use a flat blade feeler gauge and measure the gap between the gauge and the carrier. This figure is to be used when adding shims behind the carrier.

■ **Do not take measurements at the corners of the carrier.**

58. Coat the U-joint coupler shims lightly with oil and slide them over the shaft and up against the back of the coupler. Coat the shaft splines liberally with gear lube.

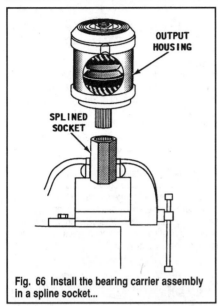

Fig. 66 Install the bearing carrier assembly in a spline socket...

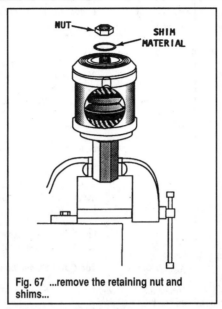

Fig. 67 ...remove the retaining nut and shims...

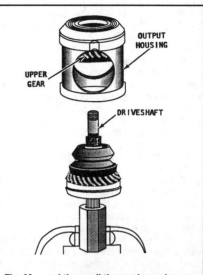

Fig. 68 ...and then pull the carrier and upper gear off of the lower gear and shaft

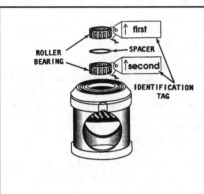

Fig. 69 Remove the needle bearings and make sure that you identify them for reuse...

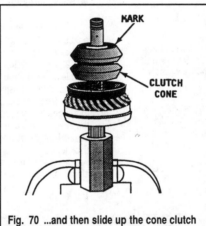

Fig. 70 ...and then slide up the cone clutch

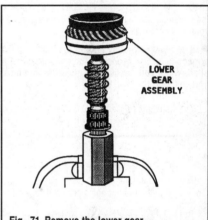

Fig. 71 Remove the lower gear...

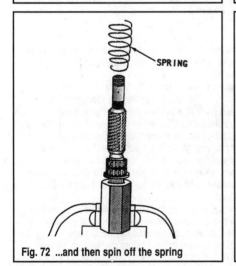

Fig. 72 ...and then spin off the spring

End Play Shim Chart

Dimension C (in.)	Shim #	Size (in.)
0.002-0.005	915167	0.063
0.006-0.009	915168	0.067
0.010-0.013	915103	0.071
0.014-0.017	915104	0.075
0.018-0.021	915105	0.079
0.022-0.025	915106	0.083
0.026-0.029	915107	0.087
0.030-0.033	915108	0.091
0,034-0.037	915109	0.095
0.038-0.041	915110	0.099

Fig. 73

Shifter Shim Chart

Clearance Measurement	Number of Shims R'qd.
0.000-0.008	2
0.009-0.018	3
0.019-0.028	4
0.029-0.038	5
0.039-0.048	6
0.049-0.058	7

Fig. 74

59. Place a support tool on the base of a press, rest the carrier on the tool with the gear down and then insert the coupler shaft into the gear hub. Press the shaft in until it seats fully.

60. Slide the support tube that you created previously in the disassembly procedure into the yoke holes and clamp the yoke in a vise. Slip the retainer into the end of the gear and then install the bolt, tightening it to 80-90 ft. lbs. (108-122 Nm).

61. Assemble and install the U-joints and drive shaft.

62. Slide the correct amount of shims over the pinion gear and up against the carrier flange so that the notches in the shims and housing are lined up. Coat the assembly lightly with oil and then slide it into the housing. The bolt holes are not symmetrical, so use the oil circulation holes in the carrier and the housing to line everything up. Tap the carrier lightly with a rubber mallet to seat everything and then install the mounting bolts. Tighten them to 12-14 ft. lbs. (16-19 Nm) in cross-wise pattern.

63. Coat the metal case of a new shift housing seal with Gasket Sealing compound and place the seal on the large end of an installer tool (# 914685) so the open side of the tool is facing away from the tool. Rest the housing in

a support on the base of a press and then position the seal and tool over the housing bore. Place a socket of sufficient size on the installer tool and press the seal in until the tool seats itself.

64. Insert the shift shaft all of the way into the housing and then rotate the lever until the tip is pointing at the upper index mark on the housing. Press in a new roll pin until it is protruding about 1/8 in. (3.2mm) above the shifter. Coat a new O-ring with gear lube and slide it into position around the base of the housing. Insert the spring and then the shift shoe into the shift shaft, making sure that the spring is fully seated inside the bore.

65. Install the water pump adaptor and water pump as detailed in the Cooling System section. Install the rear cover and tighten the bolts to 9-11 ft. lbs. (12-15 Nm).

66. Check the thrust plate dimensions as detailed in the Shimming section found later. Add shim material as necessary.

67. Remove the upper gear housing from the support tool and then install is on the lower unit.

68. Make sure that the small reference dot on the shift shoe is facing upward and the shift arm is at the upper index mark. Slide the housing into the bore and tighten the bolts to 12-14 ft. lbs. (16-19 Nm).

69. Wedge a small pry bar between the shift lever and one of the housing mounting bolts. Remove the lever bolt and any shim material. Remove the pry bar and move the lever until the end is pointing at the lower index mark on the housing.

70. Install the bolt and then slowly begin to tighten it as you rotate the propeller shaft. When you just begin to feel the eccentric high spot on the clutch cone, stop tightening the bolt and then check the clearance between the bolt head and shifter housing with a flat bladed feeler gauge. Record your measurement.

■ **If you have difficulty finding the high spot on the clutch cone, connect a torque wrench to the propeller shaft and use it to turn the shaft. When you notice the highest reading on the torque wrench, you've found the high spot.**

71. Shims are available in 0.010 in. (0.25mm) size only so compare your measurement to the accompanying chart to determine the correct amount of shims. You must use a minimum of 2 shims, but no more than 7. Rotate the propeller shaft until the clutch cone is off of the high spot and then remove the bolt.

72. Clean the threads of the bolt thoroughly, slide on the shim material and then coat the threads with Teflon pipe sealant. Thread in the bolt, secure the lever with the pry bar again and then tighten the bolt to 15-20 ft. lbs. (20-27 Nm).

73. Make sure that the lever is still pointing at lower index mark and rotate the propeller shaft again. If you can still feel the eccentric high spot on the clutch, the bolt will require more shims.

74. Coat the slot on the shifter bellcrank and the pivot pin lightly with gear lube grease and then insert the pin through the shift lever and bellcrank so that the flat head of the pin will be on the gearcase side of the bellcrank when installed. Install the washer and cotter pin.

75. Connect the spring to the retainer and then to the pivot pin. Coat the linkage pivot pin with grease and then connect the shift link to the bellcrank with the pin and a new cotter pin.

76. Coat the small bronze bushing in the bellcrank with Tripleguard grease. Insert the shift lever into the housing with the bellcrank linkage pin pointing straight down and slide the bellcrank onto the stud.

77. Slide the spring retainer over the stud and tighten the nut so that the retainer tab engages the notch in the bellcrank. Pivot the bellcrank all of the way forward and press the clasp onto the end of the shift lever, making sure that it snaps into position on both sides and pivots freely.

78. Dab a little grease on the bellcrank and shift lever pivot pins and also the overstroke slot. Connect the shift link to the shift lever with the washer and a new cotter pin.

79. Coat the inner edge of the side cover and a new foam seal with Scotch Grip Rubber Adhesive and then position the seal into the cover. Press the seal into place firmly so that the ribs on the seal and cover engage fully. Install the cover and tighten the bolts to 9-11 ft. lbs. (12-15 Nm) in a cross-wise pattern.

80. Install the drive unit.

Backlash And Shim Adjustments

The following procedures cover backlash and shim adjustments for the upper gear housing on 1990-95 King Cobra models only. The sequence of instructions is divided into sections and they all must be performed in the order given to ensure efficient and long life operation of the upper gear housing.

Output Gear Shimming

1. Position the lower gear/bearing assembly on a shim fixture (# 986631) so that the bearing/retainer is down and the end of the fixture arm is over the top of the gear teeth. Tighten the knob on the fixture securely.

2. Measure the clearance between the bottom of the arm and the top of the gear teeth with a flat bladed feeler gauge. This figure is the thickness of shim material to be positioned on the seat inside the housing.

3. Lay the upper gear/bearing assembly on a flat surface with the bearing/retainer facing down. Position a shim gauge (# 914721) over the gear.

4. Make sure that all previous shim material has been removed from the edge of the retainer and the lip inside the lower edge of the output gear housing. Position the gear housing over the gauge so that it is resting on the bearing retainer and then measure the clearance between the top of the shim gauge and the gear teeth. Note this measurement and then subtract it from 0.040 in.—this is the amount of shims you will need, gather them and set aside until called for in the assembly procedure.

Top Cover Shimming

◆ **See Figures 75, 76 and 77**

1. Position the deflector on top of the output gear housing and then lay a 0.096 in. (2.44mm) lower gearcase shim ring (# 913865) on top of the deflector.

2. Install the top cover (no bolts), thread a washer and nut onto your T-handle, insert it through the oil level hole in the top of the cover and thread it into the output housing shaft. Tighten the nut to 3-5 inch lbs. (0.3-0.6 Nm).

3. Measure the clearance between the bottom of the cover and the top of the gear housing at: a) the front of the housing as close to the center of the cover as you can, and b) the rear of the housing at the center of the cover. Take the average of these two readings and subtract it from the thickness of the lower gearcase shim ring (0.096 in.); call it **A**. Now add a pre-determined preload figure of 0.004 (0.10mm) to the result (**A**). This is the shim material thickness required to pre-load the output gear housing.

Backlash

◆ **See Figures 78, 79, 80 and 81**

This test will determine the relative clearance between the pinion and drive gears. We assume that you have already assembled the components as detailed up to this point in the Assembly procedures and that the unit is still in the holding fixture.

1. Mount a dial indicator holding bracket onto the support plate (or upper unit itself) and install a dial indicator so that the indicator needle is resting directly on the mid-point of a tooth on the lower gear. Zero the indicator and then wiggle the gear very slowly while observing the reading on the gauge. Do not wiggle the gear far enough so that the pinion gear begins to rotate the U-joint; just far enough so that the tooth on the lower gear comes in contact with the corresponding tooth on the pinion gear from either direction. Move the gauge away from the tooth, rotate the gear 180° and then set the indicator needle up as you did the last time. Wiggle the gear again and observe the reading.

2. Pull the gauge back away from the gear and slide it up the holding bracket so that you can perform the same sequence of measurements on the reverse (upper) gear. Observe and record each reading.

3. Backlash should be is equal to the number stamped on the gear sleeve, plus or minus 0.002 in. (0.05mm). Remember that lash differences between the two gears are OK as long as they fall within the appropriate range. Shim calculations should be determined by subtracting the mid-point of the range (this is the number stamped on the gear) from the reading taken with the gauge; or, by subtracting the reading from the midpoint. Lash measurements that are out of specification will require multiple shimming changes—please refer to the Drive Gear Backlash Table when determining how to correct any shimming discrepancies. When using the table, pay close attention to the notes and remember that you must always follow the order of the positions as indicated.

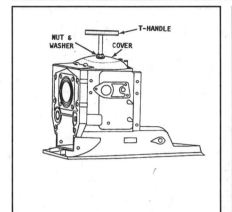

Fig. 75 Use the T-handle to load the top cover...

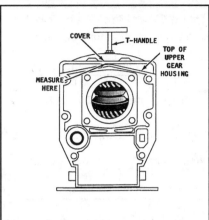

Fig. 76 ...and then take a measurement at the front...

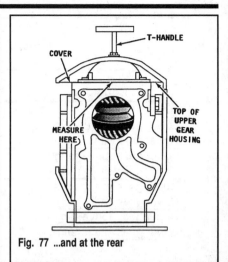

Fig. 77 ...and at the rear

Fig. 78 Check backlash on the lower gear...

Fig. 79 ...and then the upper gear

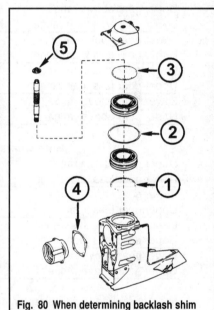

Fig. 80 When determining backlash shim changes, refer to this illustration for positioning when using the chart

■ The shim thickness-to-backlash ratio is approximately 2:1, therefore each 0.002 in. of backlash will be offset by 0.004 in. of shim material.

■ Any change to the bearing carrier shim (position 4) will affect lash on both drive gears. Any change to the upper bearing shim (position 2) will only impact lash on the upper bearing/gear, while any changes to the lower bearing shims will impact the lower both gears.

Thrust Plate Shimming

◆ See Figures 82 and 83

1. Use a vernier caliper and measure the width across the thrust plates on the sides of the forward portion of the housing. Make sure there are not gaps between the thrust plates and housing before taking this measurement. Total width of this dimension should be 5.591-5.620 in. (14.20-14.27 cm). If your measurement is within this range, there is no reason to remove the thrust plates; if not, the thrust plates will need to be removed and shims material added or removed.

2. Pry out the retaining rings (2/side) and lift off the thrust plates. There should be a minimum of one shim under each plate and no more than four on either side. If there are more than 4 shims per side (or your measurement calls for adding more than 4), the thrust plates have been worn and will require replacement.

3. Shim material must be equal on each side—when adding or removing from one side, you must do the same on the opposite side. Shims are only available in thicknesses of 0.015 in. (0.38mm) and since they must be installed or removed in pairs (1/side), you will always be impacting the dimension by 0.030 in. (0.76mm).

4. Never exceed the maximum width (5.620 in.) as this will lead to interference between the upper housing and the transom assembly.

5. Once you have added (or removed) the correct amount of shim material, slip the thrust plate into position so the tab goes through the housing and then install the retainer rings over the stud with the concave sides facing the housing. Hold the inner side of the ring with a 1/2 in. wrench and then tap the plate with a rubber mallet until everything seats into the housing.

Drive Gear Backlash Table
(Please refer to the illustration for shim positions)

Condition	Positions Requiring Shim Changes				
	1	2	3	4	5
One Gear In Spec, One Gear Out Of Spec ①					
Lower gear less than range, Upper gear within range	Remove shims	Add shims	NC	NC	Reshim
Lower gear above range, Upper gear within range	Add shims	Remove shims	NC	NC	Reshim
Upper gear less than range, Lower gear within range	NC	Add shims	Remove shims	NC	Reshim
Upper gear above range, Lower gear within range	NC	Remove shims	Add shims	NC	Reshim
Both Gears Out Of Spec By Same Amount ②③					
Upper gear above range, Lower gear less than range	Remove shims	NC	Add shims	NC	NC
Lower gear above range, Upper gear less than range	Add shims	NC	Remove shims	NC	NC
Lower gear less than range, Upper gear less than range	NC	NC	NC	Add shims	NC
Lower gear above range, Upper gear above range	NC	NC	NC	Remove shims	NC
Both Gears Out Of Spec By Different Amounts ③④					
Upper gear above range, Lower gear less than range	Remove shims	A/R	Reshim	NC	Reshim
Lower gear above range, Upper gear less than range	Add shims	A/R	Reshim	NC	Reshim
Lower gear less than range, Upper gear less than range	Remove shims	Add shims	Reshim	NC	Reshim
Lower gear above range, Upper gear above range	Add shims	Remove shims	Reshim	NC	Reshim

NC No change to shims required

A/C Add or remove shims as necessary

① Add or remove the same amount of shim material at each position indicated

② Add or remove the same amount of shim material at each position indicated

③ A change to shim 1 affects lash at both gears. A change to shim 2 affects lash at upper gear only.

④ Add or remove different amounts of shim material at each position indicated

Fig. 81

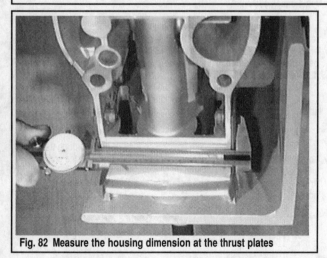

Fig. 82 Measure the housing dimension at the thrust plates

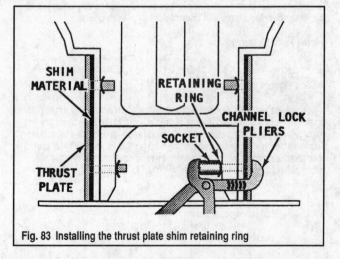

Fig. 83 Installing the thrust plate shim retaining ring

1994-98 Single Propeller Models (SP/SX/SX Cobra)

1996-98 Dual Propeller Models (DP/DP DuoProp)

◆ See Figures 84 thru 125

■ All special tool part numbers listed below are OMC factory tool numbers—we realize that if you reading this after 2001, they may be difficult, if not impossible, to obtain. With the exception of some of the earliest tools, the vast majority of these tools should be available from a Volvo Penta representative. Although the part numbers themselves may be different in many cases, the tool should be the same.

■ OMC makes a special spline socket (# 3850598) and holding fixture (# 3850605) to hold and support the upper housing—we strongly recommend using these tools, as the unit must be held securely throughout the disassembly process.

1. Remove the upper housing and install it into an appropriate support fixture.

2. Disconnect the shift link at the eccentric lever and then remove the shift mechanism as detailed in the Shift Mechanism section. Make sure that you take note of the position of the shift shoe as you are pulling out the assembly because you will need to know this during assembly.

3. Loosen and remove the 4 pinion bearing carrier bolts. Insert 2 small prybars between the carrier and the housing and lever the U-joint/carrier assembly out of the housing—have a friend or assistant help you by pulling out on the driveshaft. Save any shim material that is not damaged. Remove the 2 O-rings on the shaft and the one on the inner flange of the carrier and throw them away. Please refer to the U-Joints section for service procedures on the U-joints.

4. Remove the 4 mounting bolts and lift off the upper housing top cover—again, you will probably have to tap it a few times with a rubber mallet and then pry it apart from the housing. The aft bolts may be difficult to remove because of the lip on top of the cover. Save any undamaged shim material, but discard the cover seal (flat O-ring).

5. Secure the upper shaft if the unit is not already in the special fixture and then remove the upper shaft nut with a 30mm wrench. Remember, this nut is a LEFT HAND THREAD, so it is turned CLOCKWISE to loosen it—there should be an arrow with the loosening direction embossed in the top of the nut also.

■ This upper shaft nut also acts as a shim, thus it is available in 4 different sizes indicating the shoulder thickness of the particular nut.

6. Use a magic marker and mark a reference line across the top of the needle bearing cage and upper bearing so they may be installed with identical orientation. Pull out the needle bearing and then lift out the upper gear/bearing; you may want to reach into the bearing carrier bore and push it up from the bottom. Pull out any shim material and set it aside if it is undamaged.

7. Reach into the top of the housing and pull out the cone clutch and spring.

8. Removing the bearing retainer ring cannot be done without the use of OMC/Volvo's special tool (# 3850604); a spanner wrench. Insert the tool into the top of the housing, over the shaft, and onto the retainer ring, rotating it slowly until the flange indexes with the cut-outs along the perimeter of the ring. Rotate the top flange to line up its holes with two on the top of the housing and then secure it with 2 top cover bolts, alternately tightening each bolt two turns at a time until it is tight.

9. Position a socket over the top of the tool and use a good sized breaker bar to loosen the retainer ring. Once loosened, loosen the top flange bolts slowly and alternately, and then remove the tool. Reach in and lift out the ring.

10. Now reach in and pull out the upper shaft and lower gear/bearing assembly, setting it aside temporarily. Use a small pick and pry out any remaining shim material left in the bore.

11. The 2 lower bearing C-clips and spacer may have fallen out when you lifted out the shaft; if they did, remove them. If not, turn the assembly upside down and push up on the shaft until you can pop out the clips and then slide off the spacer.

12. Use a magic marker and mark reference lines across the needle bearing and roller bearing, and across the gear and bearing where they mate—if being re-used, they must be re-installed at the same orientation. Remove the needle bearing and the roller bearing/lower gear from the shaft.

13. If the upper or lower gear/bearing assemblies are to be disassembled, install a bearing tool (# 3850606) in an arbor press with the large diameter opening facing UP. Position the bearing into the tool so it rests on the ring of the bearing sleeve and the gear is facing UP. Position the drift tool (# 884263) and press the sleeve off of the bearing.

14. Pull out the bearing assembly and position it back into the tool so the gear is facing down. Position the drift tool again and now press the gear off of the bearing. Repeat the procedure for the other gear/bearing assembly.

15. Examine the spring retainer on the upper shaft. If everything is in good shape, skip this step. Otherwise, remove the circlip with a pair of circlip pliers and slide the retainer off the shaft.

16. Disconnect the pinion bearing carrier yoke from the U-joint as detailed in the U-Joint section. Position the shaft and joint assembly out of the way.

17. Obtain a 1 in. diameter piece of pipe from your local home center and cut a piece 2 7/8 in. long. Slide this piece of pipe through the holes in the end of the carrier yoke and then install the assembly in a vise so that the jaws are bearing on the ends of the pipe and the yoke.

18. The retaining bolt is a #50 Torx bolt and will be difficult to break loose. Once loosened, remove the bolt and retainer and then lift the carrier and pinion gear assembly off of the yoke.

19. Turn the carrier over and remove the C-clip with a pair of snap-ring pliers. Hold a cloth over the opening while compressing the ring, as it can cause damage if it slips off the pliers accidentally.

20. Position the special sleeve (# 884938) in a press with the larger diameter side facing upward. Position the carrier into the sleeve so the gear is facing down, place the driver tool (# 884266) on top of the gear and then press out the gear and roller bearing. Remove and discard the crush sleeve.

21. Examine the roller bearing and the race inside the carrier. Check for discoloration, chipped of damaged rollers, gouges or rough spots in the race. If both the bearing and race are in good shape, there is no need to separate the bearing and the gear. If there is a problem, or it must be removed for some other reason, use a universal bearing puller positioned between the bearing and gear. Put the unit in a press and press the gear off of the bearing.

22. Remove the small oil screen from the housing.

23. Although you would not normally remove the water passage plate unless looking for blockage or debris causing a restricted coolant flow, remove the 4 bolts and pry off the plate.

To assemble:

■ It is imperative that the drive and shift gears are installed in the correct positions, relative to one another. Correct backlash adjustment and tooth contact will spread the load over a larger area of the gear tooth; preventing breakdown and abnormal wear and also ensuring quieter operation.

24. Replace all O-rings, gaskets and seals with new ones whenever possible. Clean all parts (except rubber or plastic) in solvent and dry them with compressed air.

25. Coat all metallic components with clean engine oil.

26. Check all gears and gear teeth carefully to be sure they are not chipped, cracked, nicked or otherwise damaged in any manner. Replacement gears always come in matched sets, so when you find a bad one, replace them all.

27. Inspect all bearings and bearing races for pits, grooves, burns or signs of uneven wear. Replace as necessary.

28. Check all shafts and shaft splines carefully to be sure they are not chipped, cracked, nicked or otherwise damaged in any manner.

29. Check the cone clutch/sliding sleeve for wear, heat or any other damage.

30. Clean the oil screen thoroughly and press it into the recess until it seats completely and is flush with the housing surface.

31. Inspect the water passage for signs of blockage or debris; check the cover for distortion or corrosion and then replace the cover with a new gasket. Coat both sides of the gasket with sealing compound and tighten the screws to 60-84 inch lbs. (6.8-9.5 Nm).

32. Check the thrust plate dimensions as detailed in the Shimming section found later. Add shim material as necessary.

33. Position the bearing carrier into the special tool (#884938) with the large diameter upward. Lightly coat the outside surface of the small bearing race with synthetic gear lubricant and then position it into the carrier. Install the driver handle (# 9991801 and mandrel (# 884932) with the small side down and press the race into the carrier.

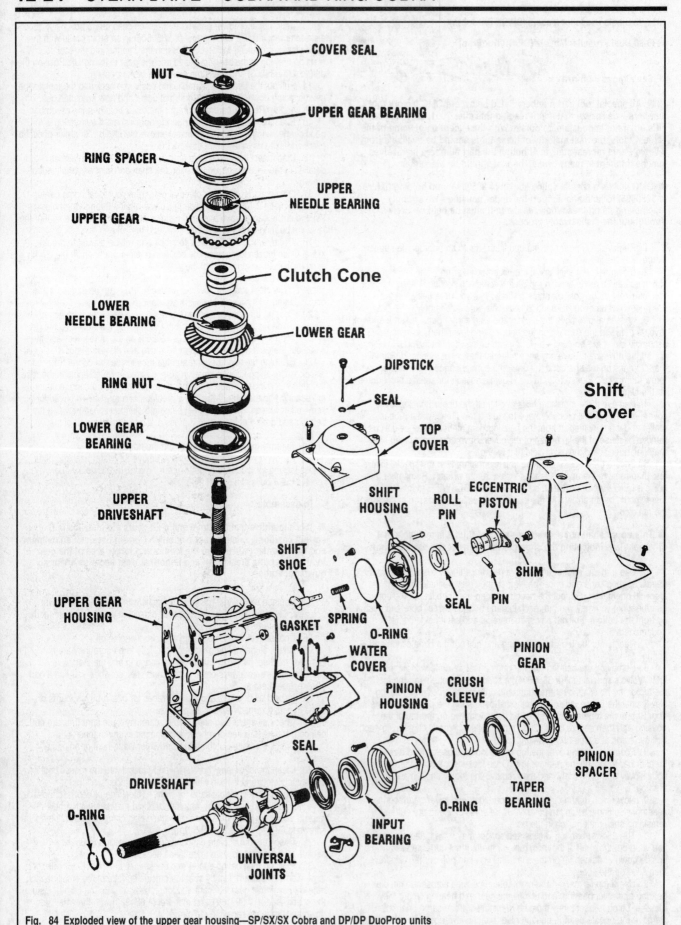

Fig. 84 Exploded view of the upper gear housing—SP/SX/SX Cobra and DP/DP DuoProp units

34. Turn the carrier over and install the bearing race for the large bearing. Coat the outside surface with synthetic gear oil and then press it into position with the tools from the previous step— this time though, the large side should face down.

35. Lubricate the outer diameter of the pinion gear hub with GL5 synthetic gear oil and then set it in a press. Position a new roller bearing over the hub, taper facing upward, and press it onto the gear with the installer tool (# 884263) until it seats against the gear.

36. With the gear/bearing still in place, slide a new crush sleeve over the input shaft. Lubricate both bearings again with the synthetic lube and then position the carrier (large opening up) over the gear and bearing.

37. Position the small bearing into the carrier and over the shaft. Use the driver (# 884263) and press it into place until there is approximately 0.040-0.060 in. (1.0-1.5mm) of end play in the bearing box. Be very careful not to press the bearing on so far as to distort the crush sleeve.

38. Coat the outside circumference of a new carrier seal with Gasket Sealing compound and then position the seal into the installation tool (# 3850607) so that the protruding lip fits into the recess in the tool. Press the seal into the carrier until it fully seats itself.

39. Carefully squeeze the retaining clip with circlip pliers and position it over the seal and into the recessed groove. Be very careful when releasing the ring from the pliers because if it slips, or is not in the groove, it will come out with extreme force. Make sure you are wearing eye protectors! Coat the exposed edge of the seal with water resistant grease

40. Position the yoke shaft back into the vise so the shaft is pointing up and the vise jaws are clamped on the pipe support. Bring the carrier assembly down and over the shaft while feeding the shaft through the seal and into the splines on the pinion gear.

41. Once the shaft and gear splines are engaged correctly, coat the threads of a new pinion gear retainer bolt with Loctite Primer and allow it to dry completely. Install the retainer into the recess on the gear and then coat the bolt threads with Loctite. Thread the bolt in until it is finger-tight—its now time to check rolling torque. DO NOT HURRY THESE NEXT STEPS! Figure on at least a half hour and 6-8 separate checks before you get it right. If you go too fast, or too tight, you will distort the crush sleeve and have to rebuild the entire assembly again.

a.) With the assembly still secured in the vise by the yoke, tighten the pinion bolt while turning the carrier to seat the bearings. Do not turn the bolt more than 1/16 of a turn at any one time!

b.) Remove the assembly from the vise and carefully clamp the vise jaws over the carrier body.

c.) Install a torque wrench on the pinion bolts and check to see at what reading the yoke shaft begins to rotate as you apply torque to the bolt. Optimum rolling torque should be 9-14 inch lbs. (1.0-1.6 Nm), although on new assemblies, we'd like to see it up around 13-14 inch lbs.

d.) Loosen the vise jaws and remount the assembly so the vise has the yoke again. Tighten the bolt another 1/16th of a turn (or less) while rotating the carrier. Reposition the assembly in the vise and check the rolling torque again. Repeat these few steps, slowly, until you are confident the proper torque has been achieved and the crush sleeve has not been prematurely distorted. Bearing pre-load is now set.

42. Install the driveshaft/U-joint assembly into the yoke as detailed in the U-Joint section.

43. Position the lower gear into a press or holding tool so the gear is facing down. Coat the inside of the lower bearing with GL-5 synthetic gear lube and slide it over the flange and onto the gear so the numbered side is facing UP. Use the special drift (#884168) and press the bearing onto the flange until it is fully seated.

44. Repeat the previous step for the upper gear and bearing.

45. Place the support tool (# 884938) into a press so the large diameter opening is facing upward. Coat the inside of the adapter ring with GL-5 synthetic gear lube and sit the ring on top of the support so the flange is facing upward. Now position the upper gear into the tools so the gear is facing down, position a piece of flat stock over the assembly and then drive the sleeve into place with the installer tool (# 3850606).

46. Repeat the previous step for the lower gear.

■ **Although they look identical, the upper and lower bearing sleeves are actually different thicknesses. Do not mix the two sleeves up.**

Fig. 85 Remove the mounting bolts and pull out the shift mechanism

Fig. 86 Loosen and remove the 4 bolts...

Fig. 87 ...pry the carrier away from the housing...

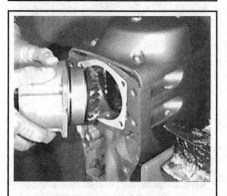

Fig. 88 ...and remove the assembly

Fig. 89 Remove the 4 mounting bolts...

Fig. 90 ...and lift off the top cover

Fig. 91 Save any undamaged shims...

Fig. 92 ...but discard the flat O-ring seal

Fig. 93 The shaft nut should be marked with a shoulder size code and with a directional arrow

Fig. 94 Remove the shaft nut

Fig. 95 Lift out the needle bearing...

Fig. 96 ...and the upper bearing/gear assembly

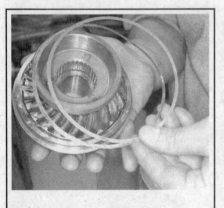

Fig. 97 Remove any old shim material...

Fig. 98 ...lift out the cone clutch...

Fig. 99 ...and then the spring

Fig. 100 Notice the cut-outs on the retainer ring flange

Fig. 101 Insert the special tool...

Fig. 102 Tighten the top flange with bolts from the cover

Fig. 103 Remove the retainer ring

Fig. 104 Lift out the lower gear and bearing...

Fig. 105 ...and pry out the shims

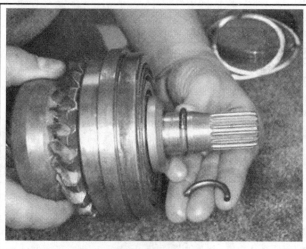

Fig. 106 Remove the C-clip and spacer...

Fig. 107 ...and then slide the needle bearing and lower gear assembly from the upper shaft

Fig. 108 Press off the bearing ring...

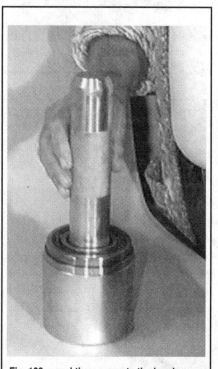

Fig. 109 ...and then separate the bearing from the gear

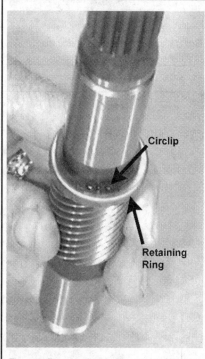

Fig. 110 Remove the circlip and retaining ring

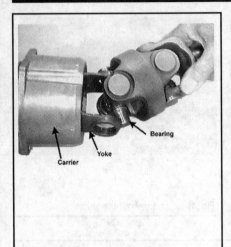

Fig. 111 Disconnect the U-joint assembly from the yoke...

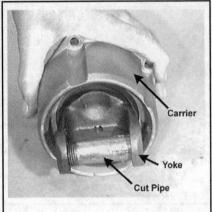

Fig. 112 ...and then install the piece of pipe before clamping the assembly in a vise

Fig. 113 Loosen the retaining bolt...

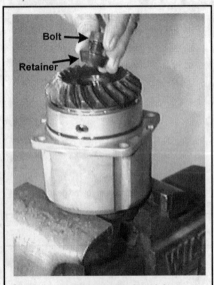

Fig. 114 ...and then remove it and the retainer

Fig. 115 Lift off the carrier assembly...

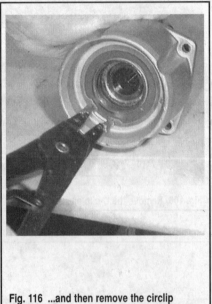

Fig. 116 ...and then remove the circlip

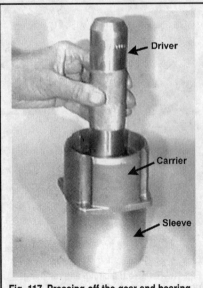

Fig. 117 Pressing off the gear and bearing

Fig. 118 Pull out the small oil screen

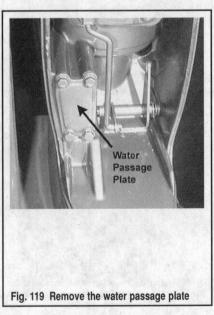

Fig. 119 Remove the water passage plate

47. Slide the spring retainer onto the splined end of the vertical shaft so that the recessed side is facing the helix (spiral on the shaft). Use a pair of snap-ring pliers and install the circlip in the groove above the retainer. Make sure you wear safety glasses and do not release the pliers until the circlip is fully seated in the groove.

48. Slide the splined end of the vertical shaft through the lower gear from the gear side, making sure that the groove in the shaft is on the side of the bearing. Coat needle bearings lightly with GL5 synthetic gear lube and then slide them onto the shaft and into the bearing and gear assembly. New bearings can be installed without concern for their orientation to one another; but if you are re-using the old ones, make sure the marks made during disassembly are in alignment.

49. Slide the ring spacer over the splined end of the shaft until is seats itself on the top of the lower bearing. Install the 2 circlips into the groove on the shaft and then pull the shaft down and through the gear/bearing until the clips seat into the recess in the ring. Hold the shaft on the helix side and shake it lightly a few times to confirm that the clips are seated correctly in the ring.

50. Reach into the top of the upper housing bore and clean off the shim ledge at the bottom of the bore. Position the correct amount of shim material on the ledge as determined in the Drive Gear Shimming section.

51. Insert the shaft and gear assembly through the bore, splined end first, until the bearing is sitting on the shim material. There is a chance that the vertical shaft can slide out of the gear, so adjust your support fixture so that it is supporting the shaft. Even a slight downward slide of the shaft will release the circlips, so be careful about not letting the shaft fall out.

52. Coat the bearing retaining ring with GL5 synthetic gear lube and then thread the ring in until it is hand-tight and fully seated—the top of the ring must be below the lower edge of the pinion bearing carrier bore. Install the spanner tool (# 3850604) into the bore so it mates with the ring, position the tension arm over two of the top cover holes and tighten it securely with 2 of the top cover bolts. Tighten the ring to 145-165 ft. lbs. (197-224 Nm) and then remove the tool.

53. Slide the clutch spring over and down the shaft until it is seated on the ring spacer. Slide the cone clutch over the shaft so that the **TOP** marking is facing upward. When the clutch hits the helix, rotate it clockwise so that it spins down the shaft until it is resting on the spring.

54. Position the correct amount of shim material (as determined in the Drive Gear Shimming procedures) onto the top mating surface of the upper housing so that it fits around the opening of the shaft bore. Position the upper gear and bearing assembly over the top of the vertical shaft and then press it into the housing bore until it seats on the top of the housing and over the shim material. Be very careful not to disturb the shims while centering the bearing in the housing bore; in fact, you may wish to loosen the support fixture so that centering the shaft assembly and bearing is easier.

55. Coat needle bearings lightly with GL5 synthetic gear lube and then slide them onto the shaft and into the bearing and gear assembly. New bearings can be installed without concern for their orientation to one another; but if you are re-using the old ones, make sure the marks made during disassembly are in alignment.

59. Install the shaft nut and tighten it to 96-110 ft. lbs. (130-149 Nm). Remember that the vertical shaft end play is controlled by the size of the shoulder on this nut, so perform the Vertical Shaft End Play procedure prior to installing the nut. Remember, this is a LEFT HAND thread nut and is tightened by turning it COUNTERCLOCKWISE.

57. Install the correct amount of shim material (as determined in the Pinion Bearing Carrier Shimming procedure) over the pinion gear and against the carrier flange. Install the carrier into the upper housing and tighten the bolts to 12-14 ft. lbs. (16-19 Nm). Do not install the O-ring at this time.

■ The shim material will only fit onto the carrier so that they align with the mounting bolt holes in one direction. Do not force them, or allow mis-alignment of the holes.

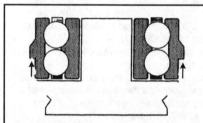

Fig. 120 Make sure the square step on the lower bearing faces the lower gear on 1998 and later drives

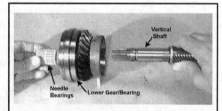

Fig. 121 Slide the vertical shaft through the lower bearing and then press in the needle bearings

Fig. 122 Slide the spacer onto the shaft...

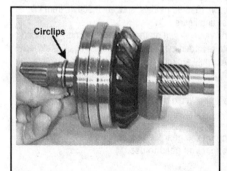

Fig. 123 ...and then install the circlips

Fig. 124 Position the lower gear shims in the upper housing...

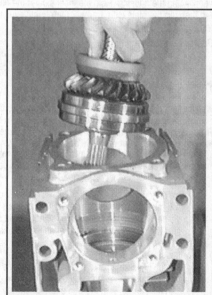

Fig. 125 ...and then insert the shaft and gear assembly

58. Coat a new cover gasket with Gasket Sealing compound and position it on the upper unit mating surface so the tab with the hole in it is over the bolt hole with the oil circulating passage. Install the correct amount of shim material as determined in the Top Cover Shimming procedure. Install the top cover and tighten the cover bolts, alternately, to 16-18 ft. lbs. (22-24 Nm). Make sure that you coat the bolt threads with sealing compound also.

59. Loosen the mounting bolts and carefully pull the pinion bearing carrier out of the upper housing. Coat a new O-ring with GL5 synthetic gear lube and slide it onto the carrier flange so that it seats into the groove. Make sure that the shim material is positioned correctly and then install the assembly into the housing. Coat the mounting bolt threads with sealing compound and then tighten them to 12-14 ft. lbs. (16-19 Nm).

60. Install the shift mechanism.

Backlash And Shim Adjustments

The following procedures cover backlash and shim adjustments for the upper gear housing on 1994-98 Single Propeller Models and 1996-98 Dual Propeller Models only. The sequence of instructions are divided into sections and they all must be performed in the order given to ensure efficient and long life operation of the upper gear housing.

Thrust Plate Shimming

◆ See Figure 126

1. Use a vernier caliper and measure the width across the thrust plates on the sides of the forward portion of the housing. Make sure there are not gaps between the thrust plates and housing before taking this measurement. Total width of this dimension should be 5.007-5.036 in. (12.72-12.79 cm). If your measurement is within this range, there is no reason to remove the thrust plates; if not, the thrust plates will need to be removed and shims material added or removed.

2. Pry out the retaining rings (2/side) and lift off the thrust plates. There should be a minimum of one shim under each plate and no more than four on either side. If there are more than 4 shims per side (or your measurement calls for adding more than 4), the thrust plates have been worn and will require replacement.

3. Shim material must be equal on each side—when adding or removing from one side, you must do the same on the opposite side. Shims are only available in thicknesses of 0.015 in. (0.38mm) and since they must be installed or removed in pairs (1/side), you will always be impacting the dimension by 0.030 in. (0.76mm).

4. Never exceed the maximum width (5.036 in.) as this will lead to interference between the upper housing and the transom assembly.

5. Once you have added (or removed) the correct amount of shim material, slip the thrust plate into position so the tab goes through the housing and then install the retainer rings over the stud with the concave sides facing the housing. Hold the inner side of the ring with a 1/2 in. wrench and then tap the plate with a rubber mallet until everything seats into the housing.

Output Gear Shimming

◆ See Figures 127 and 128

■ All special tool part numbers listed below are OMC factory tool numbers—we realize that if you reading this after 2001, they may be difficult, if not impossible, to obtain. With the exception of some of the earliest tools, the vast majority of these tools should be available from a Volvo Penta representative. Although the part numbers themselves may be different in many cases, the tool should be the same.

1. Both the upper and lower gears will have a factory etching on the collar which must be used in determining the proper shim thicknesses. This etching will have a + (plus) or - (minus) sign followed by a number—this number must be read in thousandths of an inch. In example, a +12 should be converted to 0.012 in., while a -5 should be read a -0.005 in. Check each gear for this number and mark it down for reference.

2. Lay the lower gear onto a clean, level surface so that the gear if facing down. Install a shim tool (# 3850701) over the gear, making sure that the tool is in full contact with the end of the gear. Use a flat blade feeler gauge and measure the gap between the outer edge of the bearing and the tool. Record this measurement.

3. The value shown on the etching must now be added or subtracted from the gap recorded in the previous step. A positive etched marking should be subtracted, while a negative number should be added. If our measured gap was 0.021 in. and the etching was +12, the shim thickness would be 0.009 in. (0.021 in. - 0.012 in. = 0.009 in.). If the etched mark was -5, the shim thickness would be 0.026 in. (0.021 in. + 0.005 in. = 0.026 in.). When selecting your shim material, make sure that it is clean and undamaged; and remember that you must use at least one shim and cannot use more than four shims.

■ If the etched marking on the gear's collar is 0, then the measured gap will be the shim thickness that you need to add. Sounds obvious, but you'd be surprised...

4. Repeat the previous two steps on the upper gear, taking your measurement this time between the tool and the lower edge of the adapter ring.

Pinion Bearing Carrier Shimming

◆ See Figure 129

■ All special tool part numbers listed below are OMC factory tool numbers—we realize that if you reading this after 2001, they may be difficult, if not impossible, to obtain. With the exception of some of the earliest tools, the vast majority of these tools should be available from a Volvo Penta representative. Although the part numbers themselves may be different in many cases, the tool should be the same.

1. Just like the drive gears, the pinion gear has a number etched into it which must be used in determining the proper shim thicknesses. This etching will have a + (plus) or - (minus) sign followed by a number—this number must be read in thousandths of an inch. In example, a +12 should be converted to 0.012 in., while a -5 should be read a -0.005 in. Check each gear for this number and mark it down for reference.

2. Position the pinion bearing housing/carrier in a vise and secure it carefully with the gear facing upward.

3. On 1994-96 drives, use a depth gauge and measure the distance from the top of the teeth on the gear to the flange on the housing—take a few measurements in different positions and then average each of the readings to get your actual figure. Record this figure.

4. On 1997 and later drives, install a shimming fixture tool (# 3850600) onto the top of the gear and then use a depth gauge and measure the distance from the top of the tool to the flange on the housing—take a few measurements in different positions and then average each of the readings to get your actual figure. On 1995-98 models, subtract 0.500 in. from this figure and record the resulting figure.

5. On all models, the value shown on the etching must now be added or subtracted from the depth recorded in the previous step. A positive etched marking should be added, while a negative number should be subtracted. If our measured depth (minus the 0.500) was 2.045 in. and the etching was +12, the figure would be 2.057 in. (2.045 in. + 0.012 in. = 2.057 in.). If the etched mark was -5, the figure would be 2.040 in. (2.045 in. - 0.005 in. = 2.040 in.).

■ If the etched marking on the gear's collar is 0, then the measured depth will be the figure that you need to use.

6. Now, subtract the factory's nominal figure from the figure arrived at in the previous step. Use 1.898 in. for all drives with 1.97:1 ratios; all other ratios should use a nominal figure of 1.971 in. If our measured depth was 2.057 in. and the nominal figure being used was 1.971 in., the shim thickness would be 0.086 in. (2.057 in. - 1.971 in. = 0.086 in.). If the nominal figure was 1.898 in., the shim thickness would be 0.159 in. (2.057 in. - 1.898 in. = 0.159 in.).

■ Shims should only be placed between the bearing carrier and the upper housing. When selecting your shim material, make sure that it is clean and undamaged; and remember that you must use at least one shim and cannot use more than five shims.

Fig. 126 Measure the housing dimension at the thrust plates

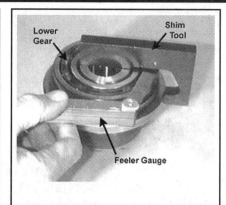

Fig. 127 Measure the distance between the tool and the top of the bearing...

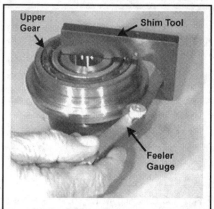

Fig. 128 ...and then between the edge of the adapter ring and the tool

Vertical Shaft End Play

DIFFICULT

◆ See Figures 130 and 131

Vertical movement (end play) of the upper shaft must fall within certain parameters and is controlled by the size of the shoulder on the upper driveshaft nut. There are 4 nuts available, each with a different thickness shoulder on the bottom of the nut. Each nut will have a number etched into the top surface that corresponds to the shoulder thickness. **0** has no shoulder; **2** has a shoulder thickness of 0.008 in.; **4** is 0.016 in. thick; and **6** is 0.024 in. thick. Another thing to remember is that this nut is a LEFT HAND THREAD, meaning it is tightened in a COUNTERCLOCKWISE direction. You should notice an arrow indicating this etched into the top of the nut also.

1. Lubricate the threads on the shaft with GL5 synthetic gear oil and spin the nut on in a counterclockwise direction. Make sure that the lower end of the shaft is clamped in the holding fixture and then tighten the nut to 96-110 ft. lbs. (130-149 Nm).

2. Once the nut is tight, push up on the shaft from the lower end and then insert a feeler gauge between the bottom of the nut's shoulder and the top of the upper bearing. Clearance should be 0.002-0.010 in. (0.05-0.25mm) and is adjusted by means of changing the nut. If the clearance is greater than specified, change to a nut with a thicker shoulder; if clearance is too low, install a nut with a thinner shoulder. After determining the correct nut, remove the old one and install a new one, tightening it to the correct torque figure.

3. If you are unable to bring the clearance into specifications with one of the 4 nuts, confirm that the upper gear, bearing and support ring are assembled correctly. If they are, the bearing is shot and will need to be replaced.

Top Cover Shimming

MODERATE

◆ See Figure 132

1. Position 0.040 in. of shim material onto the top of the upper bearing and then install the cover being careful not to disturb the shims. Do not install a new cover gasket at this time.

2. Press down lightly, and evenly, on the cover while using a flat blade feeler gauge to check the clearance between the bottom of the cover and the top of the upper bearing. Take this measurement in several places and record the average of your measurements.

3. Top cover clearance should be 0.002-0.004 in. (0.05-0.10mm); take the midpoint (0.003 in.) and add it to the initial shim thickness that you started with (0.040 in.). Record this figure (0.043 in.).

4. Now, subtract the figure arrived at in Step 2 from the figure in Step 3 and this will be the shim thickness to add to the existing shim material under the cover. You must always use at least one shim, and never more than four.

5. Lift off the cover and position the new shim material on top of the existing one(s). Reinstall the cover without disturbing the shims and tighten the bolts to 16-18 ft. lbs. (22-24 Nm).

Backlash

OEM
DIFFICULT

◆ See Figures 133, 134, 135 and 136

This test will determine the relative clearance between the pinion and drive gears. We assume that you have already assembled the components as detailed up to this point in the Assembly procedures and that the unit is still in the holding fixture.

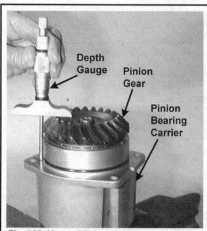

Fig. 129 Use a depth gauge to measure the depth. Later models must be measured from the top of a special tool

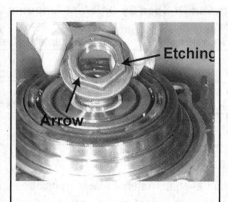

Fig. 130 The shaft nut is etched with a number corresponding to shoulder thickness

Fig. 131 Checking the vertical shaft end play

Fig. 132 Checking the top cover clearance

Fig. 133 Check backlash on the lower gear...

Fig. 134 ...and then the upper gear

Drive Gear Backlash Table

(Please refer to the illustration for shim positions)

Condition	Positions Requiring Shim Changes				
	1	2	3	4	5
One Gear In Spec, One Gear Out Of Spec ①					
Lower gear less than range, Upper gear within range	Remove shims	NC	NC	NC	Reshim w/new nut
Lower gear above range, Upper gear within range	Add shims	NC	NC	NC	Reshim w/new nut
Upper gear less than range, Lower gear within range	NC	Add shims	Reshim	NC	Reshim w/new nut
Upper gear above range, Lower gear within range	NC	Remove shims	Reshim	NC	Reshim w/new nut
Both Gears Out Of Spec By Same Amount ②					
Upper gear above range, Lower gear less than range	Remove shims	Remove shims	Reshim	NC	Reshim w/new nut
Lower gear above range, Upper gear less than range	Add shims	Add shims	Reshim	NC	Reshim w/new nut
Lower gear less than range, Upper gear less than range	NC	NC	NC	Add shims	NC
Lower gear above range, Upper gear above range	NC	NC	NC	Remove shims	NC
Both Gears Out Of Spec By Different Amounts ③					
Upper gear above range, Lower gear less than range	Remove shims	Remove shims	Reshim	NC	Reshim w/new nut
Lower gear above range, Upper gear less than range	Add shims	Add shims	Reshim	NC	Reshim w/new nut
Lower gear less than range, Upper gear less than range	Remove shims	Add shims	Reshim	NC	Reshim w/new nut
Lower gear above range, Upper gear above range	Add shims	Remove shims	Reshim	NC	Reshim w/new nut

NC No change to shims required
① Add or remove the same amount of shim material at each position indicated
② Add or remove the same amount of shim material at each position indicated
③ Add or remove different amounts of shim material at each position indicated

Fig. 135

1. Mount a dial indicator holding bracket onto the support plate (or upper unit itself) and install a dial indicator so that the indicator needle is resting directly on the mid-point of a tooth on the lower gear. Zero the indicator and then wiggle the gear very slowly while observing the reading on the gauge. Do not wiggle the gear far enough so that the pinion gear begins to rotate the U-joint; just far enough so that the tooth on the lower gear comes in contact with the corresponding tooth on the pinion gear from either direction. Move the gauge away from the tooth, rotate the gear 180° and then set the indicator needle up as you did the last time. Wiggle the gear again and observe the reading.

2. Pull the gauge back away from the gear and slide it up the holding bracket so that you can perform the same sequence of measurements on the reverse (upper) gear. Observe and record each reading.

3. Backlash should be 0.006-0.011 in. (0.15-0.28mm) on each gear. Remember that lash differences between the two gears are OK as long as they fall within the appropriate range. Shim calculations should be determined by subtracting the mid-point of the range (0.0085 in.) from the reading taken with the gauge; or, by subtracting the reading from the midpoint. Lash measurements that are out of specification will require multiple shimming changes—please refer to the Drive Gear Backlash Table when determining how to correct any shimming discrepancies. When using the table, pay close attention to the notes and remember that you must always follow the order of the positions as indicated.

■ **The shim thickness-to-backlash ratio is approximately 2:1, therefore each 0.002 in. of backlash will be offset by 0.004 in. of shim material.**

■ **Any change to the bearing carrier shim (position 4) will affect lash on both drive gears. Any change to the upper bearing shim (position 2) will only impact lash on the upper bearing/gear, while any changes to the lower bearing shims will only impact the lower bearing/gear.**

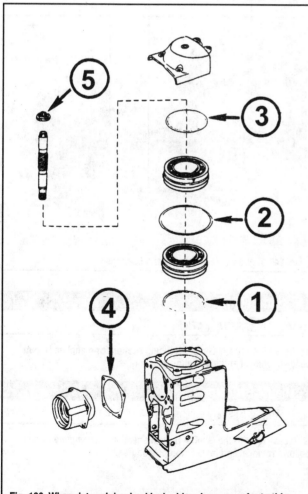

Fig. 136 When determining backlash shim changes, refer to this illustration for positioning when using the chart

U-Joint/Drive Shaft Assembly

REMOVAL & INSTALLATION

■ **U-joint and driveshaft removal and installation procedures are an integral part of the Upper Gear Housing procedures. Please refer to this section for complete details on removal and installation of these components.**

DISASSEMBLY & ASSEMBLY

◆ **See Figures 137 thru 140**

■ **All special tool part numbers listed below are OMC factory tool numbers—we realize that if you reading this after 2001, they may be difficult, if not impossible, to obtain. With the exception of some of the earliest tools, the vast majority of these tools should be available from a Volvo Penta representative. Although the part numbers themselves may be different in many cases, the tool should be the same.**

Although it is not necessary to remove the pinion bearing carrier/U-joint assembly for service, you may find it more convenient to remove it before performing the following procedures. Additionally, we recommend using the special tool provided by OMC (# T74-4635) when performing this procedure because it really is easier with the tool. That being said, it is certainly not necessary, so we have provided instructions with and without the tool.

1. Clean the assembly in solvent and allow it to dry thoroughly.

2. Inspect the splines on both yokes for wear or cracking, replacing anything if a problem is found.

3. Inspect the joints themselves for wear, knocking or too much side play. If problems are found, separate the joints.

4. On the inside of each U-joint arm there is a locking ring (circlip); drive each locking ring off the U-joint cross bearing with a small punch and hammer. Discard all rings as they are removed—do not consider re-using them! The locking rings may come off with considerable force, so hold a bunched-up shop cloth over them as you are pressing each one off the bearing and don't forget to wear eye protection.

Without the special tool:

5. Set the pinion carrier on a vise with the jaws open far enough that the bearing cap will be able to pass through.

6. Position a ball peen hammer with the ball on the top bearing cap and then tap it with another hammer until the lower bearing cap begins to come out of the yoke. Grab it with a pair of pliers and pull it all the way out of the yoke.

7. Rotate the U-joint assembly 180° and repeat the previous step until you can remove the cross-member (spider) from the yoke. Make sure that you take note of the positioning of the grease fitting in relation to the yoke.

8. Press the other bearings out in the same manner, as described in the previous steps.

With the special tool:

9. Clamp the U-joint in a vise so that one of the bearings is running north-south.

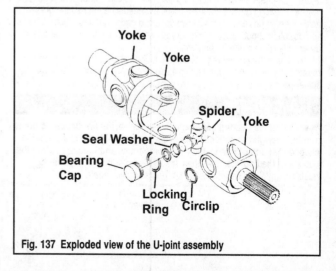

Fig. 137 Exploded view of the U-joint assembly

Fig. 138 Drive off the locking ring

10. Install the open adapter from the special tool kit into the open end of the removal tool. Position the tool over the bearing/caps and press the joint out of its yoke.

■ **Never press a bearing cap all the way through the yoke. As soon as the pressure-side cap is most of the way through the yoke, remove the tool and pull off the cap with a pair of pliers.**

11. Rotate the joint 180° and then re-clamp it in the vise. Install the tool again and repeat the previous step. Remove the bearing and cap. Make sure that you take note of the positioning of the grease fitting in relation to the yoke.

12. Press the other bearings out in the same manner, as described in the previous steps.

To assemble:

■ **Spiders and needle bearings should only be replaced as a set— never install new bearing onto a used spider or vise-versa.**

13. Install new seal washers onto each end of the spider and then position it into the yoke. Certain models may have spiders with grease fittings; make sure that they are facing positioned at a 45° angle to any two lobes so they will be accessible after installation. Now is a good time to use the fitting and fill it with grease until visible at each lobe.

14. Lubricate the U-joint bearing caps with a liberal amount of water-resistant bearing grease and then install the needle bearing so the grease holds them in position. Position two opposing caps and press them in as far as possible without forcing them; all you need to do is get it in far enough to allow installation of the locking ring.

15. Once the groove is visible, press in the new locking ring and then repeat the procedure on the remaining bearings. Always make sure that the spider pivots properly in the bearing caps and is not hanging up.

✳✳ CAUTION

Never force the bearing caps onto the spider. If you are unable to press it in far enough to expose the locking ring groove, there is a good chance that one of the needles has come loose and lodged itself in the bottom of the cap. Remove the caps and spider, check the bearing positioning in the cap and start the process over again.

16. Install the bearing carrier if you removed it.

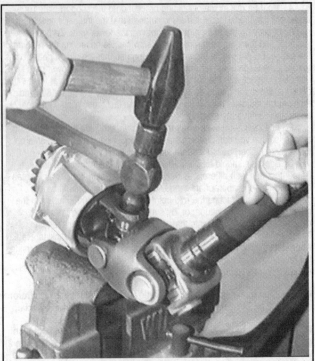

Fig. 139 Tap the bearing through until you can pull off the bearing cap

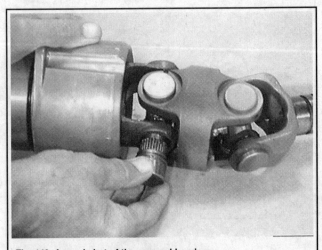

Fig. 140 A good shot of the cap and bearings

U-Joint Bellows

REMOVAL & INSTALLATION

■ **Please refer to the Gimbal Housing section for complete U-joint bellows removal and installation procedures.**

Exhaust Bellows

REMOVAL & INSTALLATION

■ **Please refer to the Gimbal Housing section for complete U-joint bellows removal and installation procedures.**

Shift Mechanism

REMOVAL & INSTALLATION

1994-98 Single Propeller Models (SP/SX/SX Cobra)

1996-98 Dual Propeller Models (DP/DP DuoProp)

◆ **See Figures 141, 142 and 143**

1. Using a pair of needle nose pliers, remove the cotter pin securing the shift link to the lever on the eccentric piston and move the link out of the way. Don't loose the washer.

2. Loosen and remove the 4 shift mechanism mounting bolts and then remove the assembly. Take note of the position of the shift shoe as you remove the assembly and make sure that you write it down to retrieve on assembly.

3. Remove and discard the O-ring.

4. Coat a new O-ring with GL5 synthetic gear lube and press it back into the groove on the mechanism.

5. Install the shift mechanism into the upper housing, making sure that the shift shoe is in the same position, and tighten the bolts to 60-84 inch lbs. (6.8-9.5 Nm).

6. Reconnect the shift lever with a new cotter pin.

DISASSEMBLY, ASSEMBLY & SHIMMING

1994-98 Single Propeller Models (SP/SX/SX Cobra)

1996-98 Dual Propeller Models (DP/DP DuoProp)

◆ **See Figures 144, 145, 146 and 147**

1. Remove the shift cover.

2. Remove the shift mechanism. It is not necessary to remove the drive unit, or even the upper housing to remove the mechanism.

** CAUTION

Be careful that the spring-loaded shift shoe does not fall out during removal.

3. Lift out the shift shoe and spring, making sure that you take note of the positioning of each.

4. Carefully position the assembly in a vise with soft jaws and then drive the tension pin into the piston with a small punch until you are able to pull out the piston pin with a pair of pliers.

5. Mark the piston positioning in the housing and then pull the eccentric piston out of the housing.

6. Turn the piston over and tap it a few times until the tension pin pops out.

7. Remove the O-ring from its recess on the inner side of the housing. Pry out the seal ring from the other side.

Fig. 141 Disconnect the shift link...

Fig. 142 ...remove the mounting bolts...

Fig. 143 ...and pull out the shift mechanism

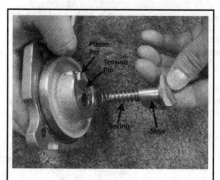

Fig.144 Remove the shift shoe and spring

Fig. 145 Pry out the O-ring

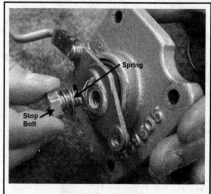

Fig. 146 Coat the stop bolt threads with sealant before installing it

To assemble:

8. Clean all components thoroughly and allow to dry completely. Check the piston for signs of scoring or other wear.

9. Coat a new sealing ring on the outside perimeter with Gasket Sealing compound and then press it into position in the housing so that the side with the smaller groove faces into the housing. Although it may be possible by hand, we suggest using a press and a drift (# 884259); position the drift (large diameter down) and press the seal in until the tool seats on the housing. Now you can coat the lips of the seal with Triple Guard grease.

10. Position the piston, through the seal and bushing, into the bore so that the offset screw is on the side of the housing with the notch in the top (starboard).

11. Insert the piston pin into the its hole so that the hole in the pin lines up with the tension pin hole. Insert the tension pin and drive it in until it is flush with the piston surface.

12. Slide the spring over the shift shoe shaft and then insert the shaft into the piston. Coat a new O-ring with GL5 synthetic gear lube and press it into its groove.

13. Coat the shift mechanism and housing mating surfaces with Permatex and then install the entire mechanism into the housing so that the stop bolt hole is on the starboard side and the longer side of the shoe is also on the starboard side. Tighten the mounting bolts securely.

14. Support the piston arm while removing the shift shoe stop bolt along with any shims underneath it.

15. Remove the upper and lower mounting bolts on the starboard side and then install the special bracket (#3850057) with the bolts that came with it. Tighten the 2 bolts to 60-84 inch lbs. (6.8-9.5 Nm).

16. Position the shift system for shimming by moving the piston lever until the starboard hole aligns with the hole in the bracket. Install the shoulder bolt supplied with the bracket and tighten it securely.

17. Reinstall the stop bolt (without shims!) and tighten it until you cannot rotate the vertical shaft. Now loosen the bolt very slowly, a little at a time, until the shaft just begins to turn freely without drag.

18. Insert a flat bladed feeler gauge between the underside of the bolt head and the piston lever. Record the measurement. Remove the bolt, coat

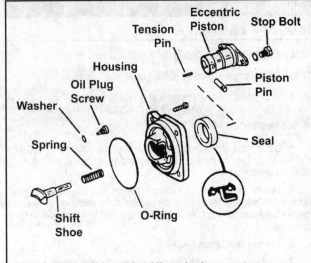

Fig. 147 Exploded view of the shift mechanism

the threads with Teflon pipe sealant, and add as many shims as it takes to get as close to this measurement WITHOUT going over it. You must use at least one shim, but not more than 4 shims. Install the bolt and tighten it to 10-12 ft. lbs. (14-16 Nm) and confirm that the vertical shaft turns freely; if not, repeat the procedure.

■ **Shift shoe stop bolt shims come in only one size.**

19. Reconnect the shift rod and install the shift cover.

LOWER GEARCASE (LOWER UNIT)

Description

Power from the engine is transferred through a splined driveshaft (sometimes called the output shaft) and U-joint assembly to the upper gear housing. Splines on the forward end of the U-joint index with matching splines of the shaft connected to the engine in the flywheel housing.

From the U-joint, power is transferred through a pinion gear to the forward (lower) and reverse (upper) gears which are in constant mesh with the pinion gear. The pinion, forward and reverse gears are all helical cut and sold as a matched set.

As previously mentioned, a cone clutch is splined to the vertical driveshaft (except on the Cobra where it is on the propeller shaft) and engages the forward and reverse gear whenever the shift lever is moved to the appropriate position. When the clutch engages either gear, the power and rotational direction is then transferred to the vertical driveshaft. The forward and reverse gears change the direction of power 90° through the vertical driveshaft and intermediate coupling down to the lower unit.

On single propeller units, a pinion gear on the lower end of the vertical driveshaft transfers power again 90° through a matching gear on the propeller shaft. The pinion gear on the lower end of the shaft and the driven gear on the prop shaft may be changed, thus affecting final gear ratio. These gears are sold as matching sets only and must always be installed this way. Engine power is then transmitted through the propeller shaft and out to the propeller.

Duo-prop units operate similarly to the singles except that they have a matched set of three gears and dual propeller shafts. The lower pinion gear is splined to the vertical shaft and is in constant mesh with the two prop shaft gears. The forward gear is splined to the inner propeller shaft, while the aft gear is a machined part of the outer prop shaft. As the vertical shaft rotates, the forward gear and inner shaft is driven in one direction, while the aft gear and outer shaft is driven in the opposite direction (because it is engaged to the same pinion gear, but 180° opposite the inner shaft's gear). The final result? Each prop shaft and propeller rotate at the same speed, but in opposite directions.

Dual propellers provide a larger blade area than a stern drive with a single propeller. The counter-rotating propellers also provide a much higher torque load and a self-canceling "roll" force.

Lower Gear Housing (Lower Unit)

REMOVAL & INSTALLATION

The lower unit may be separated and removed from the upper gear housing while the stern drive remains attached to the boat or after the stern drive has been removed.

If you plan on disassembling the unit, remove the propeller from the lower gear housing as detailed elsewhere in this section.

If the unit is still on the boat, drain the drive oil as detailed in the Maintenance section. Allow the lubricant to drain into the container; as the lubricant drains, check the color. Dark brown to black indicates normal old lubricant, while a chalky white to cream color indicates the presence of water. The presence of any water in the gear lubricant is bad news. The unit must be completely disassembled, inspected, the cause of the problem determined and corrected. Close attention should be given to the seals on the propeller shaft, driveshaft and the bearing carrier O-ring. Examine the magnet on the end of the fill/drain plug for evidence of metal particles. The presence of tiny small metal dust like shavings indicates normal wear of the gears, bearings and shafts within the lower unit. Large metal chips or heavy fillings indicate extensive internal damage is taking place and the lower unit must be completely disassembled and inspected. All worn and/or damaged components must be replaced.

1986-93 Cobra Models

◆ See Figures 13, 14, 15 and 16

■ **The lower gear housing may be removed without removing the entire stern drive unit from the vessel.**

If the drive has been completely remove from the boat, simply follow the upper gear housing removal procedures and lift it off the lower unit, otherwise:

1. Move the drive to the full down position and the drain the oil.
2. Now move the drive to the full tilt up position.
3. Position an appropriate support fixture around or under the lower unit.
4. Disconnect the battery cables.
5. Make a note of the of where the index mark is in relationship to the grid on the bottom of the trim tab and then remove the trim tab bolt and the tab. Make sure that you write the tab mark down somewhere!
6. Remove the bolt that was partially covered by the trim tab and then remove second bolt forward of this one. The middle bolt is for the anode and does not need to be removed.
7. Remove the 4 mounting bolts (2 on each side) from the sides of the lower unit.
8. Carefully lower the housing down and away from the upper unit until the shift rod disengages the guide pin and then swivel it 90 degrees to port while lowering it completely away. Move it to an appropriate stand.

To install:

9. Thoroughly clean the upper case-to-lower housing mating surfaces.
10. Make sure that the nylon plug is positioned at the base of the splines inside the lower driveshaft. Coat the splines at both ends of the intermediate shaft with clean engine oil and insert the shaft into the lower driveshaft.
11. Rotate the shift rod head 90 degrees to port.
12. Install the plastic water tube into the recess in the bottom of the upper unit.
13. Position the lower unit under the upper unit and then slowly raise it while guiding the water tube and shaft into place. As the shift rod head begins to appear in the exhaust opening, swivel it forward until it can engage the guide pin. Wiggle the propeller shaft until the shaft splines engage those in the coupler and then raise the unit fully until it mates with the upper unit.

■ **There may be a yellowish-green substance in the threads of the mounting bolts—this helps the bolts from coming loose and if not evident, they will need to be replaced with new ones.**

14. Coat the threads of all bolts with Gasket Sealing compound. Install the bolts,4 short ones (on the sides) and 2 longer ones (in the rear). Tighten them all to 22-24 ft. lbs. (30-33 Nm).
15. Position the trim tab so that the index mark lines up with the correct mark on the grid and then tighten the bolt to 28-32 ft. lbs. (38-42 Nm).
16. Install the drive unit if removed.

1988-95 King Cobra Models

◆ **See Figure 17**

 MODERATE

■ **The lower gear housing may be removed without removing the entire stern drive unit from the vessel.**

If the drive has been completely remove from the boat, simply follow the upper gear housing removal procedures and lift it off the lower unit, otherwise:

1. Move the drive to the full down position and the drain the oil.
2. Now move the drive to the full tilt up position.
3. Position an appropriate support fixture around or under the lower unit.
4. Disconnect the battery cables.
5. Remove the 6 mounting bolts (3 on each side) from the sides of the lower unit and the 1 bolt at the rear of the plate.
6. Lower the housing away from the upper unit and move it to an appropriate stand. The water tube and/or shaft coupler may come off (or fall out), so watch for them and reinstall them in the lower unit if necessary.

To install:

7. Thoroughly clean the upper case-to-lower housing mating surfaces.
8. Slide the small nylon support ring over the top of the driveshaft and into position in the recess.
9. Coat the new O-rings for the driveshaft retainer and oil passage with OMC Hi-Vis oil and then press them into their respective grooves.
10. Make sure that the tabs on the driveshaft retainer are aligned and in place with the nearest slots.
11. Install the intermediate shaft onto the driveshaft. Press the water supply tube into the grommet on the bottom of the upper housing.

12. Position the lower unit under the upper unit and then slowly raise it while guiding the water tube and intermediate shaft into place. Feed the impeller housing vent tube into the exhaust chamber on the lower unit. Wiggle the propeller shaft until the upper shaft splines engage those in the coupler and then raise the unit fully until it mates with the upper unit.
13. Coat the threads of all bolts with Gasket Sealing compound. Install the bolts and tighten them to 32-40 ft. lbs. (43-54 Nm).
14. Install the drive unit if removed.

1994-98 Single Propeller Models (SP/SX/SX Cobra)

1996-98 Dual Propeller Models (DP/DP DuoProp)

 MODERATE

◆ **See Figures 11, 12, 13, 14, 22 and 148**

■ **The lower gear housing may be removed without removing the entire stern drive unit from the vessel.**

If the drive has been completely remove from the boat, simply follow the upper gear housing removal procedures and lift it off the lower unit, otherwise:

1. Move the drive to the full down position and the drain the oil.
2. Position an appropriate support fixture around or under the lower unit.
3. Now move the drive to the full tilt up position.
4. Remove the 4 mounting bolts (2 on each side) from the sides of the lower unit and the 2 bolts at the rear of the plate. DP drives units may use studs on the sides, so they will have nuts if this is the case.
5. If your unit utilizes a trim tab, mark its position with a magic marker, remove the mounting bolt and then remove the tab.
6. Lower the lower unit and move it and the stand away from the drive. The water tube and/or shaft coupler may come off (or fall out) when the units are separated, so watch for them and reinstall them in the lower unit if necessary.

■ **If the drive unit was already off of the boat, simply remove the attaching bolts/nuts and lift off the upper unit.**

To install:

7. Thoroughly clean the upper case-to-lower housing mating surfaces.
8. Clean the oil screen thoroughly. Coat a new O-ring lightly with clean engine oil and press it into the groove. Install the oil screen.
9. Make sure that the intermediate shaft coupler is installed correctly on the shaft with the groove facing upward.
10. Install the plastic water tube guide into the recess in the lower unit. Check the positioning of the upper and lower grommets on the tube, coat the lower grommet with sealing compound and position it into the recess and over the guide. Certain early models, usually 1994, may use a retainer that is held in place with 4 screws, if you have one, tighten them to 10-12 ft. lbs. (14-16 Nm).
11. Replace any O-rings after first coating them with GL5 synthetic gear lube.

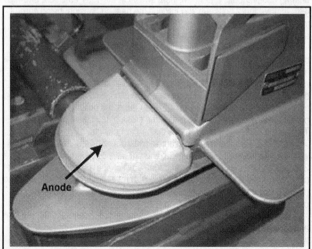

Fig. 148 Make sure you install the anode before mating the two cases

12. If removed, install the forward anode and tighten the 2 bolts to 60-84 inch lbs. (6.8-9.5 Nm),

13. Position the lower unit under the upper unit and then slowly raise it while guiding the water tube and shaft coupler into place. Wiggle the propeller shaft until the upper shaft splines engage those in the coupler and then lower the unit fully until it sits on the lower unit.

14. Coat the threads of all bolts (or studs) with Gasket Sealing compound. Install the bolts (or nuts) and tighten the 4 short ones (on the sides) to 22-24 ft. lbs. (30-33 Nm) and the 2 longer ones (aft) to 32-40 ft. lbs. (43-54 Nm).

15. Install the trim tab, if you removed it, so that the earlier marks are in alignment and tighten the bolt to 14-16 ft. lbs. (19-22 Nm).

16. Install the drive unit.

DISASSEMBLY & ASSEMBLY

1986-93 Cobra And 1988-89 King Cobra Models - Standard Rotation

◆ See Figures 149 thru 182

■ All special tool part numbers listed below are OMC factory tool numbers—we realize that if you reading this after 2001, they may be difficult, if not impossible, to obtain. With the exception of some of the earliest tools, the vast majority of these tools should be available from a Volvo Penta representative. Although the part numbers themselves may be different in many cases, the tool should be the same.

■ DO NOT ATTEMPT THIS PROCEDURE WITHOUT THE SPECIAL TOOLS!! We know that most of us, whether professional technicians or shade-tree mechanics, hate spending money on special tools. Most of us don't want to spend the money, even if we had it, so we find ways to fabricate our own version of special tools. Be warned in advance that you will NOT be able to create these tools on your own. Please read through the following steps beforehand and make a list of the tools that you will need to acquire—if you are unwilling, or unable, to get your hands on them, take the drive to someone who already has them.

1. Remove the lower unit and support it in an appropriate stand.
2. Grasp the intermediate shaft and pull it out of the housing.
3. Pull out the plastic water tube guide; remove the water tube and seal if they came out with the guide.
4. Remove the 5 locknuts on the water passage housing, throw away the nuts but keep the washers. The housing will need to be persuaded, so insert a small prybar between the housing and mating surface near one of the two forward studs and then lever the housing up while wiggling it back and forth. Once loose, pull it up and over the driveshaft housing. Throw away the gasket and O-ring.
5. Remove the 4 driveshaft bearing housing mounting bolts, but do not remove the housing yet.
6. Remove the propeller if you haven't already done so. On all models but the 460 King Cobra and 1991-93 Cobra, remove the 4 propeller shaft bearing housing bolts and throw them away. On the 1991-93 Cobra and the 460, remove the 2 bolts running through the bearing housing, being careful to catch the retainers on the inside of the housing lip as they fall off of the end of the bolts.
7. On 2.3L and 3.0L engines, connect 2 puller bolts (# 316982) to a puller head (# 307636), insert the bolts through the head and then thread them into the bearing housing. Tighten the large center bolt in the puller head so that it presses against the propeller shaft and then continue tightening it to remove the bearing carrier/housing. On all other 1986-90 engines, attach 2 puller legs (# 330278) to the same puller head and then position it so the legs are underneath the cross bars in the carrier housing. Tighten the center bolt and remove the carrier housing.

On 1991-93 engine/drives, assemble two 10 in. 5/16-18 threaded rods with flat washers and nuts to a puller body (# 307636). Attach a press bolt (# 307637) and handle (# 307638) and then position the assembly so that you can threads the other end of the long rods into the bearing housing. With the press bolt on the end of the propeller shaft, tighten it against the shaft, pulling the carrier out of the housing.

8. Inspect the bearing carrier anode for signs of deterioration. If the anode appears to be less than 50% in size, replace it. Why not replace it while you've got the carrier out anyway? Cheap insurance.

9. Carrier housing bearing should be replaced only if necessary. Insert a small prybar or awl and remove the 2 oil seals. Use the special tools (on 1986-90 drives, #391259 and #391010 - small bearing, or #391012 - large bearing; on 1991-93 drives, #432127 and # 432130) if you don't want to damage the bearings. But if you followed our advice at the start of this step, you wouldn't be removing the bearings unless they were already damaged, so the easy way is to insert a large drift or punch and drive each bearing out of the carrier.

10. Reach in and remove the 2 retaining rings with circlip pliers (# 331045) and then lift out the retaining plate, reverse gear, thrust bearing and thrust washer. Note that the 460 King Cobra and 1991-1993 Cobra do not utilize the 2 retaining rings or the retaining plate.

11. Getting a wrench on the pinion nut is tough—rotate the propeller shaft and at the same time pull upward on the shift rod, this will engage forward gear and provide a bit more clearance. Now reinsert the intermediate shaft into the top of the unit and install a spline socket (# 311875) over it. Insert a pinion nut holder (# 334455) into the propeller shaft bore and over the nut, connect a breaker bar to the spline socket and then turn the driveshaft to loosen the pinion nut. Remove the nut, socket and intermediate shaft. Lift off the driveshaft bearing housing (you removed the nuts already).

12. Remove all of the driveshaft housing O-rings and throw them away. Insert your finger and rotate the needle bearings to ensure freedom of movement. Check for any damage or corrosion. If you notice a problem of any kind, replace the housing as an assembly.

13. Wiggle the lower driveshaft back and forth while pulling it up and out of the gearcase; the thrust bearing, washer and shims should come out with it but if not, remove them also.

14. Removal of the lower driveshaft bearing will destroy the bearing, so refrain from this step unless you suspect a problem with the bearing. Remove the small Phillips screw from the starboard side of the gearcase near the water inlets on all but the 1990-93 Cobra. On the 1990-93 Cobra, the bearing is held in place by a small retaining ring that rides in a groove above the bearing; insert a small prybar through the looped end of the ring and pop it out of the groove. Assemble the special tool (# 391257) as detailed in the illustrations and then drive the bearing down and out the bottom on the 2.3L and 3,0L engines, or pull the bearing up and out the top on all other engines.

■ Make sure that all loose needles are properly positioned in the bearing cage BEFORE attempting to remove the lower bearing.

15. Reach into the propeller shaft bore and remove the pinion gear and nut. Check to see if any needle bearings may also have fallen loose and remove them too.

16. Back on top of the case, remove the 2 anode bolts and lift it off of the case. Press down on the shift rod to allow the detent lever clearance inside the gearcase when the propeller shaft is removed. Unscrew the shift rod until it is free, but do not remove it from the cover yet. Remove the 6 shift rod cover mounting bolts and then lift off the cover and rod as an assembly. Remove and throw away the gasket. Do not remove the shift rod from the cover unless you intend to replace the small O-ring and wiper as they will be damaged by the rod's threads while pulling it through.

17. Pull out the propeller shaft with the forward gear and bearing housing assembly. Make sure that the fluid drain plug is removed or you may break the plug's tip during removal of the shaft.

18. Insert a small awl under one end of the clutch dog spring and pry it out of the groove while feeding it up and over the top of the clutch dog. Push the small pin out of the clutch dog, allowing the bearing housing assembly to separate from the clutch and both assemblies to come off of the propeller shaft.

19. If it hasn't already fallen off, pull the forward gear, thrust bearing and washer out of the bearing housing. Push down on the detent shifter (arm) until it is all the way down and then pull out the shifter shaft and cradle. Now pull back up on the detent shifter until it pops back into the Neutral position. Push the pin out of the housing with a small pick and then reach into the front of the housing, remove the shifter from the detent and then remove the shifter through the bearing opening. Turn the detent shifter 90° in either direction and remove it from the housing

■ On 2.3L and 3.0L engines, rotate the detent shifter 90° to either side and then lift the shifter out through the top of the housing. Remove the shift lever through the same opening

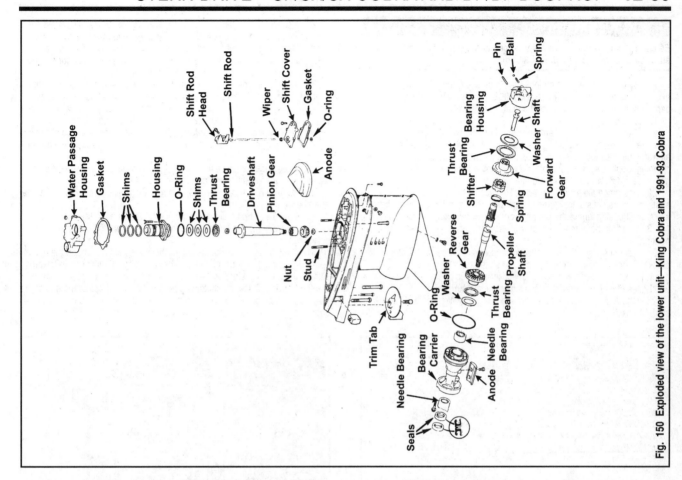

Fig. 150 Exploded view of the lower unit—King Cobra and 1991-93 Cobra

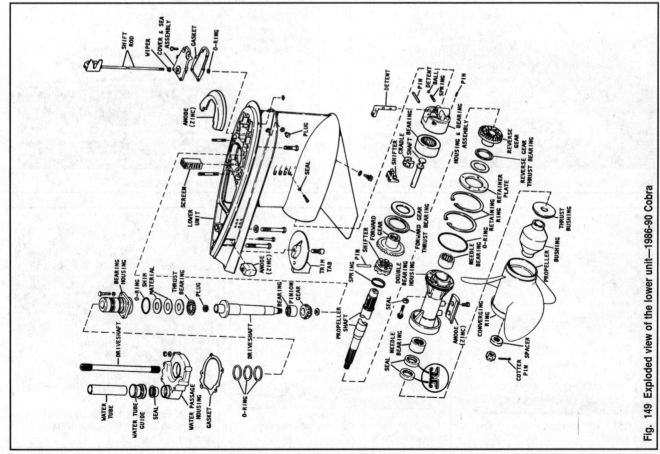

Fig. 149 Exploded view of the lower unit—1986-90 Cobra

✳✳ CAUTION

Be careful, the detent ball and spring can come out of the housing with surprising force once the detent is removed.

20. By now, the needle bearings have probably fallen out on all units except those mated with the 2.3L and 3.0L engine (because they utilize captive needles)—make sure you have all 25 of them!

To assemble:

21. Wash all parts in solvent and blow them dry with compressed air. Remove all traces of seal and gasket material from all mating surfaces. Blow all water, oil passageways and screw holes clean with compressed air. After cleaning, apply a light coating of gear lubricant to the bright surfaces of all gears, bearings, and shafts as prevention against rusting and corrosion.

22. Use a fine file to remove burrs. Replace all O-rings, gaskets, and seals to ensure satisfactory service from the unit. Clean the corrosion from inside the housing where the bearing carrier was removed.

23. Check to be sure the water intake is clean and free of any foreign material.

24. Inspect the gear case, housings, and covers inside and out for cracks. Check carefully around screw and shaft holes. Check for burrs around machined faces and holes. Check for stripped threads in screw holes and traces of gasket material remaining on mating surfaces.

25. Check O-ring seal grooves for sharp edges, which could cut a new seal. Check all oil holes.

26. Inspect the bearing surfaces of the shafts, splines, and keyways for wear and burrs. Look for evidence of an inner bearing race turning on the shaft. Check for damaged threads. Measure the run-out on all shafts to detect any bent condition. If possible, check the shafts in a lathe for out-of-roundness.

27. Inspect the gear teeth and shaft holes for wear and burrs. Hold the center race of each bearing and turn the outer race. The bearing must turn freely without binding or evidence of rough spots. Never spin a ball bearing with compressed air or it will be ruined. Inspect the outside diameter of the outer races and the inside diameter of the inner races for evidence of turning in the housing or on a shaft. Deep discoloration and scores are evidence of overheating.

28. Inspect the thrust washers for wear and distortion. Measure the washers for uniform thickness and flatness.

29. Replace ALL seals, O-rings, and gaskets to ensure maximum service after the work is completed.

30. Coat the detent ball and spring lightly with Needle Bearing grease, insert the spring into the forward gear bearing housing, followed by the detent ball.

31. On 2.3L and 3.0L models, insert the detent shifter through the detent opening and into the housing. On all others, position the shifter end toward either side of the housing, depress the ball and spring and press the shifter into the housing. Once the shifter is past the detent ball, rotate the end of the lever to the back of the housing and then move it until the Neutral position is engaged.

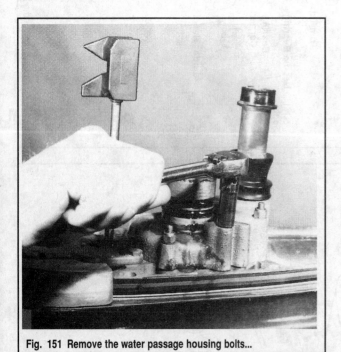

Fig. 151 Remove the water passage housing bolts...

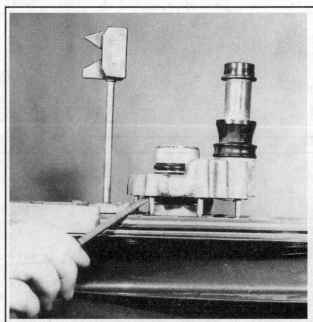

Fig. 152 ...carefully pry the housing loose...

Fig. 153 ...and then remove the housing and gasket

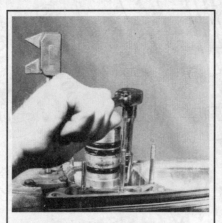

Fig. 154 Remove the driveshaft bearing housing bolts

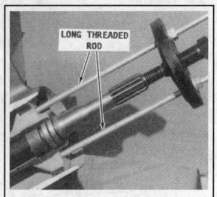

Fig. 155 Removing the bearing carrier housing on the 2.3L and 3.0L (others similar)

■ Do not push the detent shifter all of the way in at this time or you will jam it in the housing.

32. On models with loose needles, coat all 25 bearings with grease and install them into the cage if the bearing has not been replaced with a new one (in which case the new one is most likely a caged bearing with no loose needles).

33. Insert the shift lever so the side with the narrow arms is facing the detent shifter. Attach it to the neck on the detent shifter, align the pivot holes with those in the housing and then insert the pivot pin.

34. On 2.3L and 3.0L engines, rotate the detent shifter 90° to either side of the housing and then insert the shifter through the top of the housing. Install the shift lever and pivot pin as detailed in the previous step.

35. Press the detent shifter to its full down position, lay the cradle on the shifter shaft and then attach them both to the wider arms on the shift lever. While still holding the shift shaft in place, pull the detent shifter back up into the Neutral position.

36. Lay the thrust bearing on the back side of the forward gear and then position the washer on the bearing. Carefully insert the assembly into the bearing housing without disturbing the thrust bearing or any of the needle bearings.

37. Take a look at the clutch dog and identify the end with the part number embossed on it (it should also have "Prop end" etched on this same side); this is the end of the dog that must face the rear of the propeller shaft, that is to say it must face the propeller. Once located, slide this end onto the inner (forward) end of the propeller shaft so that the splines index properly and the retaining pin holes line up. Now insert the shift shaft into the propeller shaft until the retaining pin hole lines up with the other two holes and then press in the pin. Slide a new retaining spring down over the propeller shaft and position it so that at least 3 coils are covering each end of the retaining pin WITHOUT overlapping one another.

38. Obtain a driveshaft holding socket (# 314438) and clamp it in a vise. Slide the driveshaft into the socket and then set the pinion gear on the end of the shaft. Tighten the nut to 70-75 ft. lbs. (95-102 Nm), except on the 460 King Cobra models where it should be 100-110 ft. lbs. (136-149 Nm). Remove the shaft from the holder.

39. Position the thrust bearing, washer and driveshaft bearing housing on the other end of the driveshaft without any shims. Face the stepped side of an alignment plate (part of the shim gauge kit # 984329) away from the bearing housing, slide the plate over the shaft and seat it against the housing. Move the entire assembly into a shim fixture (# 984329). Tighten the tension screw on the fixture against the end of the driveshaft until the screw sleeve reaches the groove in the pointed end of the screw. Tighten the tension screw lock nut on the outside of the fixture until it is tight against the frame.

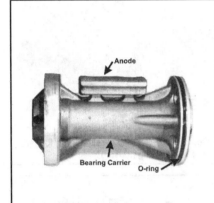

Fig. 156 A good look at the bearing carrier

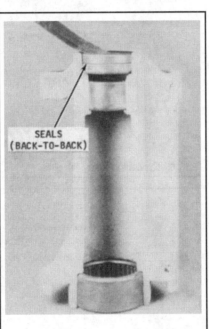

Fig. 157 Prying out the two oil seals

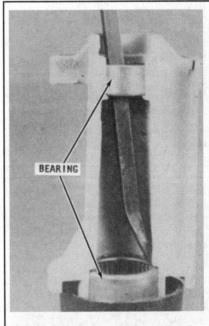

Fig. 158 Drive the bearings out with a drift

Fig. 159 Remove the retaining rings (except on the 460 and 1991-93 Cobra)

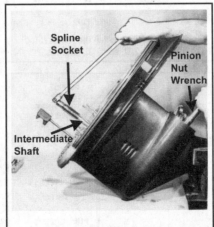

Fig. 160 Rotate the driveshaft to remove the pinion nut

Fig. 161 Lift off the bearing housing...

Fig. 162 ...and pull out the lower driveshaft

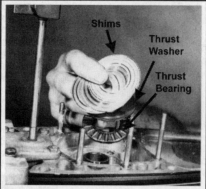

Fig. 163 The shims and thrust bearing should come out with it

Fig. 164 Reach in and remove the pinion gear

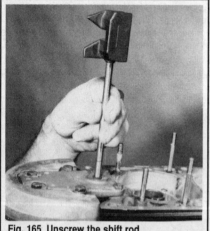

Fig. 165 Unscrew the shift rod...

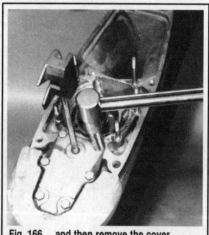

Fig. 166 ...and then remove the cover

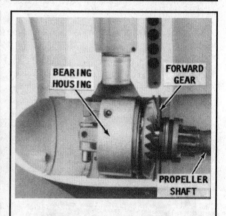

Fig. 167 Pull out the propeller shaft, forward gear and bearing housing

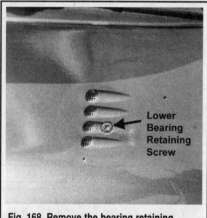

Fig. 168 Remove the bearing retaining screw (except 1990-93 Cobra)...

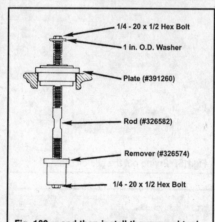

Fig. 169 ...and then install the removal tool like this on the 2.3L and 3.0L engines...

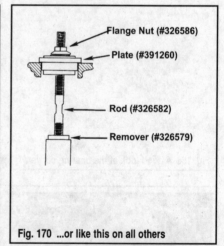

Fig. 170 ...or like this on all others

Fig. 171 Remove the pin from the clutch dog

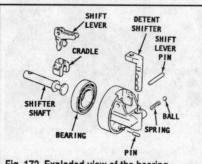

Fig. 172 Exploded view of the bearing carrier shift assembly

Fig. 173 Press down on the detent shifter to remove the shift shaft...

40. Tilt the entire fixture back until it is resting on the support lugs. Select the following gauge bars depending upon your particular application:
- 2.3L and 3.0L engines—# 328366
- V6 and V8 engines except the 460 King Cobra—# 328367
- 460 King Cobra—# 330224

41. Position the correct gauge into the fixture so the curved head is up and behind the pinion gear and the base is resting against the bearing housing. Hold the gauge tightly against the bearing housing and measure the clearance between the other end and the pinion gear. Make sure that you rotate the shaft a few times and take a few readings in different places. Take an average of your readings and then subtract this figure from 0.020 in. to determine the correct amount of shims required to properly position the pinion gear. On 460 King Cobra models, use 0.030 in. as the figure you subtract from.

42. Check your measuring accuracy by installing the shims and then checking the clearance again. If you come up with 0.020 in. (0.030 in. for the 460), then you measured correctly. Otherwise, repeat the last step and try it again. Once correct, remove the shaft from the fixture and disassemble the individual components.

43. If you removed the lower driveshaft bearing, use the same group of tools to press it back into the housing bore. Coat the outside of the bearing case lightly with Needle Bearing grease and then slide the bearing onto the tool so the side with letters on it is facing the installer. Position the assembly into the bore and drive it in until the driving rod is seated on the spacer.

44. On 1990-93 units, slide the retaining ring into the bore and make sure that it is fully seated into the recess. On all others, slide a new O-ring over the small bearing retaining bolt and spray the threads with LocQuic Primer. Once dry, coat them with Nut Lock, install the bolt and tighten it to 48-84 inch lbs. (5-9 Nm).

45. Press down on the detent lever in the forward gear bearing housing until it engages the Reverse position (this is the lowest position in its travel). Tilt the top of the lower unit forward so that the propeller shaft bore is raised slightly and then insert the propeller shaft assembly into the bore so that the locating pin on the front side of the bearing housing engages the recess in the nose of the gear housing. Once the pin is properly fit into the recess, the detent lever will be centered (front to back) on the shift rod cavity.

46. Coat both sides of a new shift rod cover gasket with Gasket Sealing compound and lay it on the gear housing mating surface. Take a look down the bore to gauge the distance to the detent lever and slide the shift rod through the cover far enough so that you will be able to begin threading it into the detent without having the cover touch the new gasket. Position the cover and rod over the bore and thread the rod into the detent 4-5 full turns. Now you can position the cover on the gasket! Coat the threads of the mounting bolts with Gasket Sealing compound and tighten them alternately to 60-84 inch lbs. (7-9 Nm).

47. Slowly rotate the propeller shaft while pulling up on the shift rod so that you engage Forward gear. Look down through the driveshaft bore and confirm that the clutch dog is fully engaged with the lugs on the forward gear.

48. Install the anode and tighten the bolts securely.

49. Reach into the propeller shaft bore and hold the pinion gear under the lower driveshaft bearing while an assistant inserts the driveshaft into the gear from the top. Install the intermediate shaft and then slide a spline

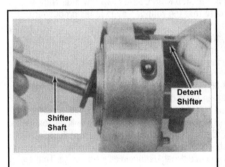
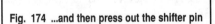

Fig. 174 ...and then press out the shifter pin

Fig. 175 Remove the detent shifter

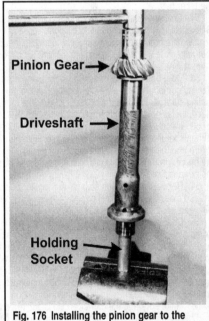

Fig. 176 Installing the pinion gear to the driveshaft

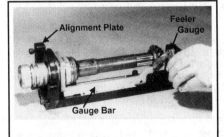

Fig. 177 Checking shim thickness on the pinion

Fig. 178 Make sure that the locating pin lines up with the hole in the nose of the housing

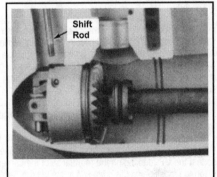

Fig. 179 Center the detent under the shift rod bore

Fig. 180 Hold the pinion nut while turning the driveshaft from the top

Fig. 181 Install the reverse gear and bearing...

Fig. 182 ...and then the retainer plate (not all models)

socket (# 311875) over the top splines. Position the pinion nut into a nut holder (# 334455) and hold it in place while rotating the shafts with a torque wrench attached to the spline socket. Tighten the nut to 70-75 ft. lbs. (95-102 Nm) except the 460 King Cobra which must be tightened to 100-110 ft. lbs. (136-149 Nm). Remove the socket and intermediate shaft.

50. Position the thrust bearing, thrust washer and shim material over the top of the driveshaft. Coat the small (new) O-ring with oil and slide it over the base of the driveshaft bearing housing. Next, coat the 3 larger exterior rings with Gasket Sealing compound and slide them over the top of the housing and into their recesses; position the upper ring first and then work down with the other two.

51. Coat the entire exterior surface (very lightly) of the housing with Gasket Sealing compound to reduce corrosion. Position the housing over the shaft so the bolt holes line up properly and then install the bolts and tighten them alternately to 14-16 ft. lbs. (19-22 Nm). Don't forget to coat the bolt threads with sealing compound.

52. Coat both sides of a new water passage cover gasket with Gasket Sealing compound and slide the gasket over the studs and into place on the mating surface. Position the cover itself over the studs and lightly tap it into place. Install new lock nuts and tighten them to 9-11 ft. lbs. (12-15 Nm).

53. If you replaced the large bearing on the propeller shaft bearing carrier, coat the outside of the case with oil. Position the housing in a holding fixture and then slide the bearing onto an installer (# 326562) so the side of the bearing with lettering is facing the large diameter side of the tool. Press the bearing into the housing until it is fully seated.

54. Repeat the previous step for the smaller bearing, but face the lettered side toward the smaller end of the tool. On V6 and V8 engines, you will need another installer (# 321517).

55. Coat the outer casing of 2 new seals with Gasket Sealing Compound. Position the inner seal on the stepped side of the installation tool so the lip is facing the housing. Turn the tool over and install the outer seal so the lip is facing away from the housing. Use # 326551 for the 2.3L and 3.0L models, and # 326690 or # 336311 (1990-93 Cobra) for all other engines. Make sure that the seal lips and inner cavity are coated with grease.

56. Make sure that the carrier anode is tightened to 5-7 ft. lbs. (7-9 Nm).

57. Lay the thrust bearing and then the washer over the hub in the back of the reverse gear. Slide the gear over the propeller shaft and into the bore. Don't forget the smaller forward washer on 1990-93 which rides on the face of the gear.

58. On all models except the 460 King Cobra and 1990-93 Cobra, slide the retainer plate over the propeller shaft and up against the reverse gear. Carefully grab a retaining ring with the appropriate pliers, position it over the inner groove and then release it so it seats fully in the groove. Repeat this with the second ring. Choose two holes in the retainer plate that are 180° apart and then thread in 2 guide pins (# 383175)—1/4 in. threaded rods of sufficient length will also work here.

59. On all models, install new O-rings coated with Triple Guard grease into their respective grooves on the carrier and then coat the O-ring flanges with Gasket Sealing compound.

60. On all models except the 460 King Cobra and the 1990-93 Cobra, slide the carrier housing onto the 2 guide pins and into the gear housing bore so that the **UP** marking on the rear flange of the carrier is, you guessed it...on top, positioning the drain slot at the bottom. Tap the carrier softly with a rubber mallet until it is seated below the hydrostatic rings. Coat the threads (and seals) of 4 new carrier bolts with Gasket Sealing compound, install 2 of them, remove the guide pins and the install the remaining 2 bolts. Tighten all four, alternately, to 10-12 ft. lbs (14-16 Nm).

61. On 460 King Cobra models, slide the carrier over the shaft and into the bore so the anode is on the bottom and the mounting bolt holes are in the vertical position. Tap the carrier lightly with a rubber mallet to seat the O-rings. Position a retainer behind the bolt hole and hold it there while threading in the bolt; repeat for the remaining bolt. The large end of the retainer must fit into the recess inside the propeller shaft bore, while the smaller end should fit between the bosses on the carrier housing. Tighten the bolts to 18-20 ft. lbs. (24-27 Nm).

■ If the bolt retainers on the 460 King Cobra have one side with rounded edges, make sure that the rounded edge is facing outward.

62. On 1990-93 Cobra models, slide the carrier over the shaft and into the bore so the anode is on the bottom and the mounting bolt holes are in the vertical position. Tap the carrier lightly with a rubber mallet to seat the O-rings. Install the 2 retaining tabs, thread in the bolts and then tighten the bolts to 20-24 ft. lbs. (27-33 Nm). Make sure that you coat the bolt threads with Locquic Primer and then UltraLock before installing them.

63. Position a shift rod height gauge (# 912277) on the upper mating surface of the gear housing. Now screw the rod (in or out) until the top of the shift rod head just contacts the upper arm of the gauge (about 7-13/64 in.). The notched side of the head must be facing forward, so it is acceptable if you have to turn the rod and additional 1/2 turn in either direction. That's it.

64. Make sure that the small nylon plug is in place inside the lower driveshaft. Coat the splines on each end of the intermediate shaft with gear oil and then insert the shaft into the top of the lower driveshaft.

65. Spin the shift rod head 90° to the port side and then install the water tube into the guide on top of the gear case.

66. Install the upper gear housing as detailed in the appropriate section and then install the drive unit if it was removed. Fill the drive unit with oil.

1986-93 Cobra And 1988-89 King Cobra Models - Counter-Rotation

◆ See Figures 151, 152, 153, 154, 160, 161, 162, 163, 164, 165, 166, 170, 176, 177 and 183

■ Procedures have changed on the 1993 Cobra lower unit, please refer to 1994-98 Single Propeller Unit section for disassembly and assembly on this unit.

■ All special tool part numbers listed below are OMC factory tool numbers—we realize that if you reading this after 2001, they may be difficult, if not impossible, to obtain. With the exception of some of the earliest tools, the vast majority of these tools should be available from a Volvo Penta representative. Although the part numbers themselves may be different in many cases, the tool should be the same.

■ DO NOT ATTEMPT THIS PROCEDURE WITHOUT THE SPECIAL TOOLS!! We know that most of us, whether professional technicians or shade-tree mechanics, hate spending money on special tools. Most of us don't want to spend the money, even if we had it, so we find ways to fabricate our own version of special tools. Be warned in advance that you will NOT be able to create these tools on your own. Please read through the following steps beforehand and make a list of the tools that you will need to acquire—if you are unwilling, or unable, to get your hands on them, take the drive to someone who already has them.

1. Remove the lower unit and support it in an appropriate stand.
2. Grasp the intermediate shaft and pull it out of the housing.
3. Pull out the plastic water tube guide; remove the water tube and seal if they came out with the guide.
4. Remove the 5 locknuts on the water passage housing, throw away the nuts but keep the washers. The housing will need to be persuaded, so insert a small prybar between the housing and mating surface near one of the two forward studs and then lever the housing up while wiggling it back and forth. Once loose, pull it up and over the driveshaft housing. Throw away the gasket and O-ring.
5. Remove the 4 driveshaft bearing housing mounting bolts, but do not remove the housing yet.
6. Remove the propeller if you haven't already done so. Remove the 2 bolts running through the bearing housing, being careful to catch the retainers on the inside (460) or outside (Cobra) of the housing lip as they fall off of the end of the bolts.
7. Thread an adapter (# 432398) onto a slide hammer (# 432128) until it is fully seated and then thread the other end of the adapter onto the end of the propeller shaft. Use the hammer to back the bearing carrier and propeller shaft out of the housing bore.
8. Inspect the bearing carrier anode for signs of deterioration. If the anode appears to be less than 509 in size, replace it. Why not replace it while you've got the carrier out anyway? Cheap insurance.
9. Pull the O-rings off of the bearing carrier and carefully clamp the carrier in a vise. Fit a spanner wrench (# 432400 - Cobra, # 432399 - 460 King Cobra) into the small recesses and loosen the forward gear assembly.

■ **The forward assembly is serviced as a unit only, no disassembly is possible.**

10. Push the propeller shaft out of the carrier (from the aft side) and then reach in and remove the thrust bearing and washer if it did not come out with the shaft. Discard the thrust bearing.
11. Position a puller (# 432129) and bridge (#432127) over the carrier so that the jaws of the puller are underneath the lower oil seal and then remove the two seals. Obviously, if you intend to replace the seals anyway, you can get away without the special tools and just pry them out (carefully!).
12. Neither bearing should be removed unless you intend to replace them, as they will be damaged during the removal process. A bridge (# 432127) and puller (# 432129) are the special tools required, but if the bearing is already worn or damaged...just drive it out with a drift.
13. Getting a wrench on the pinion nut is tough—rotate the propeller shaft and at the same time pull upward on the shift rod, this will engage forward gear and provide a bit more clearance. Now reinsert the intermediate shaft into the top of the unit and install a spline socket (# 311875) over it. Insert a pinion nut holder (# 334455) into the propeller shaft bore and over the nut, connect a breaker bar to the spline socket and then turn the driveshaft to loosen the pinion nut. Remove the nut, socket and intermediate shaft. Lift off the driveshaft bearing housing (you removed the nuts already).
14. Remove all of the driveshaft housing O-rings and throw them away. Insert your finger and rotate the needle bearings to ensure freedom of movement. Check for any damage or corrosion. If you notice a problem of any kind, replace the housing as an assembly.
15. Wiggle the lower driveshaft back and forth while pulling it up and out of the gearcase; the thrust bearing, washer and shims should come out with it but if not, remove them also.
16. Removal of the lower driveshaft bearing will destroy the bearing, so refrain from this step unless you suspect a problem with the bearing. Remove the small Phillips screw from the starboard side of the gearcase near the water inlets. Assemble the special tool (# 391257) as detailed in the illustrations and then pull the bearing up and out the top on all other engines.

■ **Make sure that all loose needles are properly positioned in the bearing cage BEFORE attempting to remove the lower bearing.**

17. Reach into the propeller shaft bore and remove the pinion gear and nut. Check to see if any needle bearings may also have fallen loose and remove them too.
18. Back on top of the case, remove the 2 anode bolts and lift it off of the case. Press down on the shift rod to allow the detent lever clearance inside the gearcase when the propeller shaft is removed. Unscrew the shift rod until it is free, but do not remove it from the cover yet. Remove the 6 shift rod cover mounting bolts and then lift off the cover and rod as an assembly. Remove and throw away the gasket. Do not remove the shift rod from the cover unless you intend to replace the small O-ring and wiper as they will be damaged by the rod's threads while pulling it through.

19. Insert a long piece of stock (or a handle of some sort) into the shift rod bore from the top until you can depress the shift detent. Once the detent is held down sufficiently, grasp the clutch shaft and pull it (and the bearing housing) out of the propeller shaft bore.
20. Insert a small awl under one end of the clutch dog spring and pry it out of the groove while feeding it up and over the top of the clutch dog. Push the small pin out of the clutch dog, allowing the bearing housing assembly to separate from the clutch and both assemblies to come off of the clutch shaft.

■ **There are detent balls in the end of the shaft—these are not serviceable, and if damaged will necessitate replacement of the entire shaft.**

21. Remove the reverse gear, thrust bearing and washer from the end of the bearing housing. There is a small bearing inside of the reverse gear which is not serviceable and should not be removed.
22. On all units but the 460 King Cobra, press the shifter pivot pin out of the housing and then pull out the shift shaft, cradle and shifter. Remove the shifter and cradle from the shift shaft.
On the 460, remove the shift shaft and cradle from the housing. Pull out the pivot pin and then lift out the shifter.
23. Remove all 25 needles from the bearing housing and then use a small awl or pick to press the detent out of the housing. Remove the detent ball and spring. On the 460, wrap a rag around the housing, rotate the detent 90° in either direction and then pull it out of the housing. Be careful here, as the detent ball and spring can pop out of the housing with considerable force while removing the detent.

To assemble:

24. Wash all parts in solvent and blow them dry with compressed air. Remove all traces of seal and gasket material from all mating surfaces. Blow all water, oil passageways and screw holes clean with compressed air. After cleaning, apply a light coating of gear lubricant to the bright surfaces of all gears, bearings, and shafts as prevention against rusting and corrosion.
25. Use a fine file to remove burrs. Replace all O-rings, gaskets, and seals to ensure satisfactory service from the unit. Clean the corrosion from inside the housing where the bearing carrier was removed.
26. Check to be sure the water intake is clean and free of any foreign material.
27. Inspect the gear case, housings, and covers inside and out for cracks. Check carefully around screw and shaft holes. Check for burrs around machined faces and holes. Check for stripped threads in screw holes and traces of gasket material remaining on mating surfaces.
28. Check O-ring seal grooves for sharp edges, which could cut a new seal. Check all oil holes.
29. Inspect the bearing surfaces of the shafts, splines, and keyways for wear and burrs. Look for evidence of an inner bearing race turning on the shaft. Check for damaged threads. Measure the run-out on all shafts to detect any bent condition. If possible, check the shafts in a lathe for out-of-roundness.
30. Inspect the gear teeth and shaft holes for wear and burrs. Hold the center race of each bearing and turn the outer race. The bearing must turn freely without binding or evidence of rough spots. Never spin a ball bearing with compressed air or it will be ruined. Inspect the outside diameter of the outer races and the inside diameter of the inner races for evidence of turning in the housing or on a shaft. Deep discoloration and scores are evidence of overheating.
31. Inspect the thrust washers for wear and distortion. Measure the washers for uniform thickness and flatness.
32. Replace ALL seals, O-rings, and gaskets to ensure maximum service after the work is completed.
33. Coat the detent ball and spring lightly with Needle Bearing grease, insert the spring into the reverse gear bearing housing, followed by the detent ball.
34. On 460 models, position the detent arm toward either side of the housing and then depress the detent ball and spring with a punch while pushing the detent into the housing. Once past the ball, rotate the lever toward the rear of the housing and then move it until the Neutral position is engaged. Insert the shifter so the side with the narrow arms is facing the bearing opening, reach inside and connect the shifter to the neck of the detent. Align the pivot holes with those in the housing and then insert the pivot pin.
35. On all other models, insert the detent into the housing and press it in until it is next to the detent ball—making sure that the ball is resting on the spring. Insert a small pick into the other end of the detent bore and press in the detent ball while pushing the detent in until it pops into the Neutral

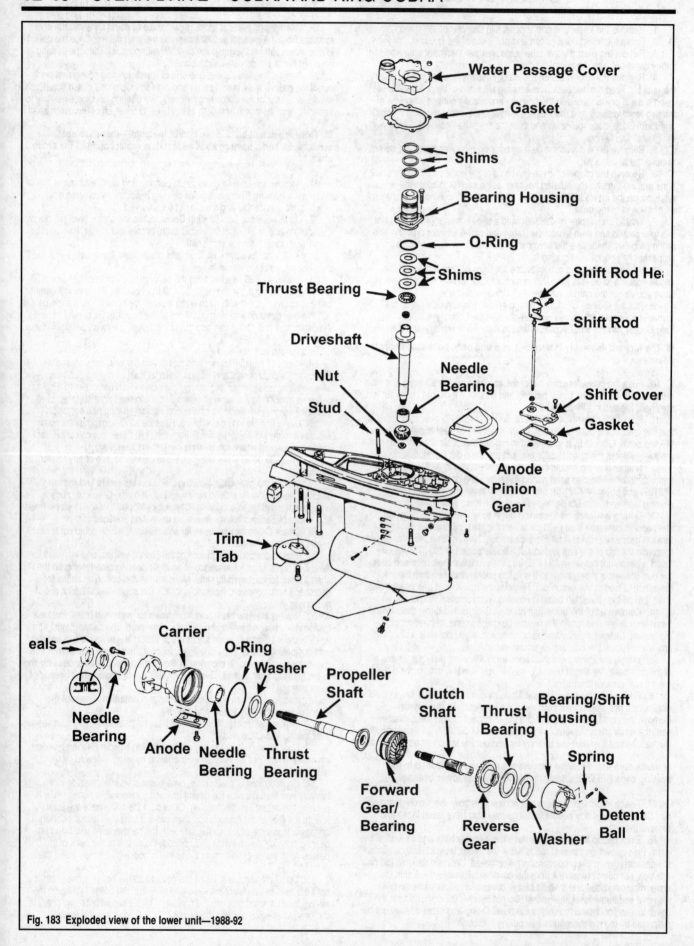

Fig. 183 Exploded view of the lower unit—1988-92

position. Position the cradle onto the shift shaft and lay the shifter in place on the cradle. Slide the shaft into the housing and insert the shifter tip into the pocket on the detent. Align the pivot holes with those in the housing and then insert the pivot pin.

36. Coat all 25 bearings with grease and install them into the cage if the bearing has not been replaced with a new one (in which case the new one is most likely a caged bearing with no loose needles).

37. On the 460, confirm that the detent is still in Neutral, rest the cradle on the shift shaft and then connect the cradle to the shifter arms. Depress the detent until it holds the shaft and cradle in place.

38. Insert the shift lever so the side with the narrow arms is facing the detent shifter. Attach it to the neck on the detent shifter, align the pivot holes with those in the housing and then insert the pivot pin.

39. Lay the thrust bearing on the back side of the reverse gear and then position the washer on the bearing. Carefully insert the assembly into the bearing housing without disturbing the thrust bearing or any of the needle bearings.

40. Take a look at the clutch dog and identify the end with the part number embossed on it (it should also have "Prop end" etched on this same side); this is the end of the dog that must face the rear of the clutch shaft, that is to say it must face the propeller. Once located, slide this end onto the inner (forward) end of the shaft so that the splines index properly and the retaining pin holes line up. Now insert the shift shaft into the clutch shaft until the retaining pin hole lines up with the other two holes and then press in the pin. Slide a new retaining spring down over the propeller shaft and position it so that at least 3 coils are covering each end of the retaining pin WITHOUT overlapping one another.

41. Obtain a driveshaft holding socket (# 314438) and clamp it in a vise. Slide the driveshaft into the socket and then set the pinion gear on the end of the shaft. Tighten the nut to 70-75 ft. lbs. (95-102 Nm), except on the 460 King Cobra models where it should be 100-110 ft. lbs. (136-149 Nm). Remove the shaft from the holder.

42. Position the thrust bearing, washer and driveshaft bearing housing on the other end of the driveshaft without any shims. Face the stepped side of an alignment plate (part of the shim gauge kit # 984329) away from the bearing housing, slide the plate over the shaft and seat it against the housing. Move the entire assembly into a shim fixture (# 984329). Tighten the tension screw on the fixture against the end of the driveshaft until the screw sleeve reaches the groove in the pointed end of the screw. Tighten the tension screw lock nut on the outside of the fixture until it is tight against the frame.

43. Tilt the entire fixture back until it is resting on the support lugs. Select the following gauge bars depending upon your particular application:
- V6 and V8 engines except the 460 King Cobra—# 328367
- 460 King Cobra—# 330224

44. Position the correct gauge into the fixture so the curved head is up and behind the pinion gear and the base is resting against the bearing housing. Hold the gauge tightly against the bearing housing and measure the clearance between the other end and the pinion gear. Make sure that you rotate the shaft a few times and take a few readings in different places. Take an average of your readings and then subtract this figure from 0.020 in. to determine the correct amount of shims required to properly position the pinion gear. On 460 King Cobra models, use 0.030 in. as the figure you subtract from.

45. Check your measuring accuracy by installing the shims and then checking the clearance again. If you come up with 0.020 in. (0.030 in. for the 460), then you measured correctly. Otherwise, repeat the last step and try it again. Once correct, remove the shaft from the fixture and disassemble the individual components.

46. If you removed the lower driveshaft bearing, use the same group of tools to press it back into the housing bore. Coat the outside of the bearing case lightly with Needle Bearing grease and then slide the bearing onto the tool so the side with letters on it is facing the installer. Position the assembly into the bore and drive it in until the driving rod is seated on the spacer. Note that the 460 King Cobra uses 19 needle bearings, while all others use 18.

47. Slide a new O-ring over the small bearing retaining bolt and spray the threads with LocQuic Primer. Once dry, coat them with Nut Lock, install the bolt and tighten it to 48-84 inch lbs. (5-9 Nm).

48. Press down on the detent lever in the reverse gear bearing housing until it engages the Reverse position (this is the lowest position in its travel). Tilt the top of the lower unit forward so that the propeller shaft bore is raised slightly and then insert the shift shaft assembly into the bore so that the locating pin on the front side of the bearing housing engages the recess in the nose of the gear housing. Once the pin is properly fit into the recess, the detent lever will be centered (front to back) on the shift rod cavity.

49. Coat both sides of a new shift rod cover gasket with Gasket Sealing compound and lay it on the gear housing mating surface. Take a look down the bore to gauge the distance to the detent lever and slide the shift rod through the cover far enough so that you will be able to begin threading it into the detent without having the cover touch the new gasket. Position the cover and rod over the bore and thread the rod into the detent 4-5 full turns. Now you can position the cover on the gasket! Coat the threads of the mounting bolts with Gasket Sealing compound and tighten them alternately to 60-84 inch lbs. (7-9 Nm).

50. Slowly rotate the clutch shaft while pushing down on the shift rod. Look down through the driveshaft bore and confirm that the clutch dog is fully engaged with the lugs on the reverse gear.

51. Install the anode and tighten the bolts securely.

52. Reach into the propeller shaft bore and hold the pinion gear under the lower driveshaft bearing while an assistant inserts the driveshaft into the gear from the top. Install the intermediate shaft and then slide a spline socket (# 311875) over the top splines. Position the pinion nut into a nut holder (# 334455) and hold it in place while rotating the shafts with a torque wrench attached to the spline socket. Tighten the nut to 70-75 ft. lbs. (95-102 Nm) except the 460 King Cobra which must be tightened to 100-110 ft. lbs. (136-149 Nm). Remove the socket and intermediate shaft.

53. Position the thrust bearing, thrust washer and shim material over the top of the driveshaft. Coat the small (new) O-ring with oil and slide it over the base of the driveshaft bearing housing. Next, coat the 3 larger exterior rings with Gasket Sealing compound and slide them over the top of the housing and into their recesses; position the upper ring first and then work down with the other two.

54. Coat the entire exterior surface (very lightly) of the housing with Gasket Sealing compound to reduce corrosion. Position the housing over the shaft so the bolt holes line up properly and then install the bolts and tighten them alternately to 14-16 ft. lbs. (19-22 Nm). Don't forget to coat the bolt threads with sealing compound.

55. Coat both sides of a new water passage cover gasket with Gasket Sealing compound and slide the gasket over the studs and into place on the mating surface. Position the cover itself over the studs and lightly tap it into place. Install new lock nuts and tighten them to 9-11 ft. lbs. (12-15 Nm).

56. If you replaced the large bearing on the propeller shaft bearing carrier, coat the outside of the case with oil. Position the housing in a holding fixture and then slide the bearing onto an installer (# 432401) so the side of the bearing with lettering is facing the large diameter side of the tool. Press the bearing into the housing until tool is seated.

57. Repeat the previous step for the smaller bearing, but face the lettered side toward the smaller end of the tool.

58. Coat the outer casing of 2 new seals with Gasket Sealing Compound. Position the inner seal on the stepped side of the installation tool so the lip is facing the housing. Turn the tool over and install the outer seal so the lip is facing away from the housing. Use tool # 326690. Make sure that the seal lips and inner cavity are coated with grease.

59. Secure the carrier housing in a vise and then insert the thrust washer and a new thrust bearing into place. Coat the bearing surfaces of the propeller shaft with gearcase oil and then slide it into and through the housing until it is resting on the thrust bearing.

60. Coat the threads of the forward gear lightly with oil and then thread it into the housing as far as you can by hand. Use a spanner (# 432400, or # 432399 on the 460) and tighten the gear to 100 ft. lbs. (136 Nm).

61. Make sure that the carrier anode is tightened to 60-84 inch lbs. (7-9 Nm).

62. Install new O-rings (3 on the 460 and 1 on the others) coated with Triple Guard grease into their respective grooves on the carrier and then coat the O-ring flanges with Gasket Sealing compound.

63. Install the intermediate shaft and then shift the gearcase into Reverse, making sure that the clutch dog and gear are fully engaged.

64. Slide the carrier/shaft into the bore so the anode is on the bottom and the mounting bolt holes are in the vertical position. Once the carrier assembly seats in the bore, hold the intermediate shaft and then wiggle the propeller shaft until the internal splines on the two shafts engage. Pres the carrier in further by hand until the retaining tab slots are flush with those in the gearcase bore. In stall the retaining tabs and then tighten the bolts to 18-20 ft. lbs. (24-27 Nm).

65. Position a shift rod height gauge (# 912277) on the upper mating surface of the gear housing. Now screw the rod (in or out) until the top of the shift rod head just contacts the upper arm of the gauge (about 7-13/64 in.). The notched side of the head must be facing forward, so it is acceptable if you have to turn the rod and additional 1/2 turn in either direction. That's it.

66. Remove the intermediate shaft and make sure that the small nylon plug is in place inside the lower driveshaft. Coat the splines on each end of the intermediate shaft with gear oil and then insert the shaft into the top of the lower driveshaft.

67. Spin the shift rod head 90° to the port side and then install the water tube into the guide on top of the gear case.

68. Install the upper gear housing as detailed in the appropriate section and then install the drive unit if it was removed. Fill the drive unit with oil.

1990-95 King Cobra And 1994-98 SP/SX/SX Cobra Single Propeller Models

◆ See Figures 184 thru 206

■ Please note that although not mentioned in the above head, the 1993 Cobra counter-rotating unit is also covered in this section.

■ All special tool part numbers listed below are OMC factory tool numbers—we realize that if you reading this after 2001, they may be difficult, if not impossible, to obtain. With the exception of some of the earliest tools, the vast majority of these tools should be available from a Volvo Penta representative. Although the part numbers themselves may be different in many cases, the tool should be the same.

■ DO NOT ATTEMPT THIS PROCEDURE WITHOUT THE SPECIAL TOOLS!! We know that most of us, whether professional technicians or shade-tree mechanics, hate spending money on special tools. Most of us don't want to spend the money, even if we had it, so we find ways to fabricate our own version of special tools. Be warned in advance that you will NOT be able to create these tools on your own. Please read through the following steps beforehand and make a list of the tools that you will need to acquire—if you are unwilling, or unable, to get your hands on them, take the drive to someone who already has them.

1. Remove the propeller if not already removed. Install the lower unit in an appropriate holding fixture.

2. Reach into the propeller shaft opening and remove the retainer screw on the upper starboard side. Lift out the retainer securing the bearing housing to the lower unit.

3. Use a T27 Torx bit and remove the propeller shaft bearing housing set screw on the upper port side of the aft end of the propeller shaft housing.

4. Remove the 3 Torx bolts securing the anode retainer to the propeller shaft bearing housing—the small bolt is a T30 bit and the 2 larger bolts are T40 bits. Grasp the retainer and pull it and the internal anode out of the housing. Replace the anode if it has deteriorated by more than a third.

5. Connect a bearing housing removal tool (# 914684) to the bearing housing and turn it counterclockwise, using a ratchet and breaker bar, until the housing is loosened and you can spin it all the way out of the case. Once out of the lower unit, remove the tool and carefully pull off any shim material. Remove the O-ring and throw it away.

6. Use a small punch or drift and pry out the small oil seal in the end of the bearing housing. The seal will be damaged during removal, so be sure that you throw it away now.

7. Turn the housing over and install the special bearing puller (# 432129), rod (# 432127) and guide plate (# 914700). Slide the guide plate over the rod so the stepped side facing upward and insert the assembly into the housing. Expand the jaws of the puller so they are tightly behind and under the bearing cup (race) and then pull out the cup.

8. Carefully pull the propeller shaft out of the housing bore. Reinstall the propeller nut to protect the threads and position the shaft/bearing into a bearing knife. Press off the bearing.

9. If you did not remove the intermediate shaft sleeve during case separation, remove it now and inspect the splines. If you did not already remove it, and you can't find it, look in the bottom of the upper housing where is sometimes get stuck. Same thing with the water tube, grommets and retainer; if you haven't already removed them, do it now.

■ Early SP models and most King Cobra models may use a retainer/guide that is attached with 4 small screws.

10. Remove and discard any remaining O-rings on the top of the lower unit mating surface.

11. Install the special spanner tool (# 3850601- SP, # 914696 - KC) over the driveshaft and onto the retainer so that it indexes with the slots in the edge of the retainer. Back out the retainer, disconnect the tool and throw away the O-ring.

12. Install the special spline socket (# 3850598 - SP, # 914699) onto the upper end of the driveshaft. Slide the special handle (# 311880) through an alignment plate (# 914686) and then thread a pinion nut holder (# 914694 1993-94 SP or # 3854864 for 1995-98 SP and all KC) onto the end of the handle. Insert the assembly into the propeller shaft bore while wiggling the driveshaft back and forth until the holder slides onto the pinion nut. Turn the driveshaft counterclockwise with a wrench on the spline socket until the pinion nut is completely unthreaded from the bottom of the shaft.

13. Remove the tools and the pinion nut. Reach into the housing bore and put your hand under the pinion gear while someone else pulls upward on the shaft—you'll want to catch the gear and needle bearings while they are removing the shaft. The upper bearing race will probably come out with the bearing.

If you have long arms, you can probably accomplish this step alone; but it's really much easier with an assistant.

14. Reach into the bore again and remove the propeller shaft gear/bearing and needle bearings (which will fall out if they didn't already when you removed the pinion).

15. Insert a 3-armed puller into the housing bore and remove the bearing race (cup) at the back of the bore. Reach in and remove any shim material; which will have been damaged, so discard it.

> ✱✱ **WARNING**
>
> Do not remove the propeller shaft bearing and oil slinger from the gear unless it is defective and you plan on replacing it.

16. Position the assembly in a universal bearing separator tool so the tool edge is between the bearing and the oil slinger. Place the separator onto a suitable support and press off the bearing with a bearing remover (# 914708) and handle (# 378737).

17. Repeat the previous procedure to remove the oil slinger, positioning the tool edge between the slinger and the gear. Remember the earlier Warning: if you do not intend to replace the oil slinger, do not remove it.

18. On 1993-94 SP drives and all King Cobra drives, assemble a 2-armed puller (# 432129) into a deck bridge (# 432127) and position them over the driveshaft bore so that it sits on a suitable support. Insert the arms of the puller through the lower driveshaft bearing race and under the shim material. Expand the arms out and under the race and then tighten the puller nut to remove the race (cup) and shims. Check the bore to be sure there are no additional shims left; throw away all shims, as they will have been damaged in the removal process.

On 1995-98 SP drives, assemble the race remover tip (# 3855859), the rod, nut and washer (# 3855860) and the guide plate (# 914700). Insert the remover tip into the driveshaft bore until the guide plate seats itself, tighten the nut and remove the lower bearing race and shims. Check the bore to be sure there are no additional shims left; throw away all shims, as they will have been damaged in the removal process.

19. Coat the 19 pinion bearing needles with Needle Bearing Grease and then assemble them back into the bearing case. Assemble a bearing puller (# 914706), guide plate (# 914700) and remover (# 914704) and then insert the tools into the top of the driveshaft bore until the remover rests on the top of the bearing and the guide plate is seated in the bore recess. Lay a small shop towel under the bearing in the propeller shaft bore and then tap the top of the special tool with a hammer until it forces the bearing out of its seat. Remove the tools, rag, bearing and needles.

> ✱✱ **CAUTION**
>
> Do not attempt to remove the pinion bearing without re-installing the needles. Without the needle bearing in place, the cage will distort during the removal process and require replacement.

> ✱✱ **WARNING**
>
> Do not remove the driveshaft bearings unless it/they are defective and you plan on replacing them.

20. Lift the loose upper bearing race off of the driveshaft. Position a universal bearing separator into an arbor press so the curved face is facing up and then insert the driveshaft so the lower bearing is sitting on the tool. Carefully press the shaft down and through the bearing until they are loose and then lift the bearings off of the shaft. Be sure that you take note of which bearing went in which position; they look alike but are not and cannot be interchanged.

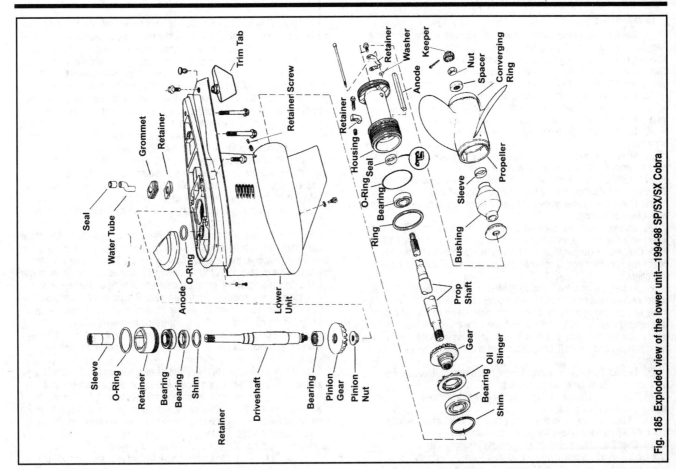

Fig. 185 Exploded view of the lower unit—1994-98 SP/SX/SX Cobra

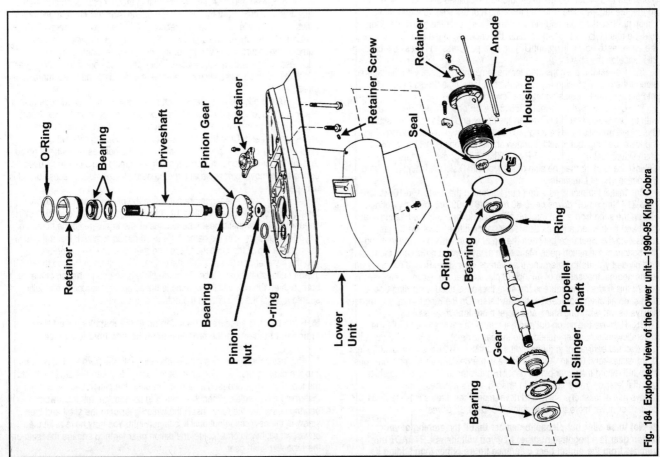

Fig. 184 Exploded view of the lower unit—1990-95 King Cobra

To assemble:

21. Wash all parts in solvent and blow them dry with compressed air. Remove all traces of seal and gasket material from all mating surfaces. Blow all water, oil passageways and screw holes clean with compressed air. After cleaning, apply a light coating of gear lubricant to the bright surfaces of all gears, bearings, and shafts as prevention against rusting and corrosion.

22. Use a fine file to remove burrs. Replace all O-rings, gaskets, and seals to ensure satisfactory service from the unit. Clean the corrosion from inside the housing where the bearing carrier was removed.

23. Check to be sure the water intake is clean and free of any foreign material.

24. Inspect the gear case, housings, and covers inside and out for cracks. Check carefully around screw and shaft holes. Check for burrs around machined faces and holes. Check for stripped threads in screw holes and traces of gasket material remaining on mating surfaces.

25. Check O-ring seal grooves for sharp edges, which could cut a new seal. Check all oil holes.

26. Inspect the bearing surfaces of the shafts, splines, and keyways for wear and burrs. Look for evidence of an inner bearing race turning on the shaft. Check for damaged threads. Measure the run-out on all shafts to detect any bent condition. If possible, check the shafts in a lathe for out-of-roundness.

27. Inspect the gear teeth and shaft holes for wear and burrs. Hold the center race of each bearing and turn the outer race. The bearing must turn freely without binding or evidence of rough spots. Never spin a ball bearing with compressed air or it will be ruined. Inspect the outside diameter of the outer races and the inside diameter of the inner races for evidence of turning in the housing or on a shaft. Deep discoloration and scores are evidence of overheating.

28. Inspect the thrust washers for wear and distortion. Measure the washers for uniform thickness and flatness.

29. Replace ALL seals, O-rings, and gaskets to ensure maximum service after the work is completed.

30. All bearings utilize an assembly notch, this notch must always be facing AWAY from the respective gear when installed.

31. Coat the inside of the smaller driveshaft bearing with GL5 synthetic gear lube. Thread the pinion nut into the lower end of the driveshaft for protection and then slide the bearing onto the splined end (top) of the shaft so that the tapered side is facing the bottom of the shaft (the end with the pinion nut). Install a bearing installer (# 914698) over the end of the shaft so that the raised lip on the tool contacts the backside of the bearing. Now rest the entire assembly (pinion nut UP) on an arbor press and press the bearing into place on the shaft.

32. Repeat the previous step for the remaining bearing, with the tapered side of the bearing facing the top of the shaft this time. Press the bearing into position until it seats against the lower bearing.

33. Clamp a spline socket (# 3850598 - SP or # 914699 - KC) in a vise with the spline end facing UP. Insert the upper end (splined) of the shaft into the socket and slide the bearing race down over the smaller bearing—this is the lower bearing, but it will be on top at the moment. Now slide a pinion shim fixture tool (# 3855098 - SP or # 986617 - KC) over the shaft and against the race so that the side of the tool with 3 slots is facing up toward the pinion end of the shaft.

34. Install the pinion gear and nut, tightening the nut to 150-160 ft. lbs. (203-217 Nm) on SP drives or 60-80 ft. lbs. (81-109 Nm) on King Cobras. Rotate the shim fixture tool through a few revolutions until the bearing seats itself and then insert a flat bladed feeler gauge into each of the 3 slots, measuring the clearance between the top of the inner hub on the tool and the bottom of the pinion gear. Mark each measurement down, add them up and divide by three to determine the average clearance. Round off this figure to the nearest thousandth of an inch (3 places after the decimal point) and record this figure. This figure is the shim thickness needed on King Cobra drives; on all others, proceed to the next step. On the King Cobra, you must always use at least one shim, and never more than three shims.

35. Remove the pinion nut and gear; throw away the nut. Locate the shim allowance number etched onto the backside of the gear and convert it to a decimal equivalent in thousandths of an inch—**+5** translates to 0.005 in., **-2** translates to -0.002 in. Add this figure to the one arrived at in the previous step and record it as your pinion gear shim thickness. When adding shim material later on, remember that in arriving at this thickness, you must always use at least one shim, and never more than three shims (1993-94 SP drives), or never more than five shims (1995-98 SP drives).

■ Not to be silly, but please remember that if the etching of your pinion gear is a negative number (-2), you will then SUBTRACT this number from the pinion gear clearance figure rather than adding it.

36. Assemble the correct amount of shim material for the pinion gear and position it onto the bearing shoulder in the top of the driveshaft bore.

37. Coat the outer surface of the bearing race with GL5 synthetic gear lube and position it squarely into the driveshaft bore so that the taper, or large opening, is facing upward. Now assemble the race driver (# 914706), guide plate (# 914700) and installer (# 914703) so the stepped end of the guide plate faces the threads on the driver and the stepped end of the installer is also facing the threads. Insert the tool into the bore and push down on the plate until it indexes with the bore. Tap the end of the driver with a hammer until the races seats against the shim material on the shoulder.

38. Coat the propeller shaft gear and oil slinger lightly with GL5 synthetic gear lube. Place the gear (teeth down) on a universal bearing separator for support and slide the oil slinger over the hub so that the pins align with the holes in the gear. Install a support plate (# 914698) and bearing installer (# 914687) over the hub and press the slinger into place on the gear. Make sure that the pins and holes are aligned BEFORE pressing the slinger into place.

39. With the gear and oil slinger still sitting in the separator tool, coat the inner race of the bearing with GL5 synthetic gear lube and slide the bearing over the gear hub so the tapered side is facing UP. Press the bearing into place with the installer tool (# 914707) until it is seated against the oil slinger.

40. On the King Cobra, place the gear and bearing on the race and rotate the gear to properly position the rollers. Set the bearing in the groove of a shim fixture tool (# 986631). Rotate the arm of the fixture over the gear and tighten the set screw. Slide a feeler gauge between the bottom of the arm and the top of the gear. Record your measurement as this is your shim thickness. You must always use at least one shim, and never more than three.

41. On SP drives, with the gear and bearing assembly still sitting in the separator tool, slide the bearing race into position over the bearing and then turn it a few revolutions until the bearings are fully seated. Position a shimming fixture (# 3850600) onto the bearing so its recessed side faces the bearing. Install a depth gauge over the shim fixture and measure the depth from the top of the fixture to the end of the gear hub; subtract 0.5 in. (thickness of the fixture) and record the resulting figure.

42. Locate the shim allowance number etched onto the backside of the gear and convert it to a decimal equivalent in thousandths of an inch—**+5** translates to 0.005 in., **-2** translates to -0.002 in. Add this figure to the one arrived at in the previous step and record it. OMC suggests a nominal dimension of 0.106 in., so subtract the figure you just recorded from this number and record the new number as your propeller shaft gear shim thickness. When adding shim material later on, remember that in arriving at this thickness, you must always use at least one shim, and never more than three.

43. Turn the lower unit over so that it can rest on its forward edge (use a piece of wood to support it). Select the correct number of shims and insert them into the propeller shaft bore so that they sit in the bearing race recess at the forward end. Coat the outer surface of the bearing race with GL5 synthetic gear lube and position squarely into the bore as far as it will go so that the larger opening is facing out or aft. Assemble the same tools used during the removal procedure and the press the race into the recess until it is fully seated.

44. If the needle bearings are not still in the pinion bearing cage, coat them with grease and re-install all 19 of them...again! Coat the outside of the cage with GL5 synthetic gear lube and install the bearing into the bottom of the driveshaft bore with the same tools used during removal. Insert the tool through the top of the driveshaft bore and then position the pinion bearing into the installer and thread the installer onto the bottom of the rod through the propeller shaft bore. Once the guide plate is seated in the top of the bore, tighten the nut until the bearing is pulled up into its recess and fully seated against the shoulder in the bottom of the bore.

■ If the bearing appears to be spinning in the installer, insert the pinion nut holder into the propeller shaft bore to hold it in place.

45. Install the propeller shaft gear/bearing into the bore so that is seats into the bearing race with the gear teeth facing aft. Insert the driveshaft into the top of the bore and feed it carefully through the pinion bearing without disturbing the needles. When the shaft is in far enough that the lower bearing seats into the race, install the pinion gear onto the shaft and then screw in the new pinion nut until it is finger-tight. You may have to lift the driveshaft slightly in order to get the pinion gear teeth to engage the teeth on the propeller shaft gear.

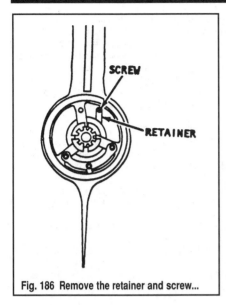

Fig. 186 Remove the retainer and screw...

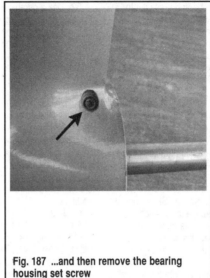

Fig. 187 ...and then remove the bearing housing set screw

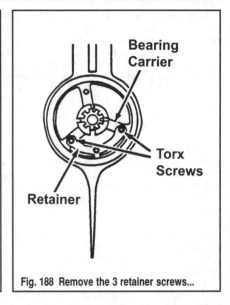

Fig. 188 Remove the 3 retainer screws...

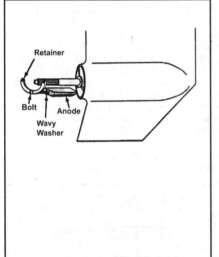
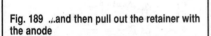

Fig. 189 ...and then pull out the retainer with the anode

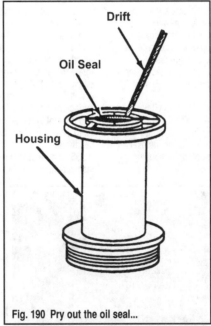

Fig. 190 Pry out the oil seal...

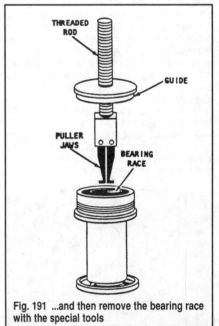

Fig. 191 ...and then remove the bearing race with the special tools

Fig. 192 If you haven't done it already, remove the shaft sleeve...

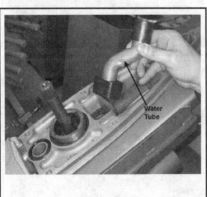

Fig. 193 ...and the water tube (some models may have a different retainer)

Fig. 194 After removing the driveshaft retainer...

46. Coat the upper bearing race with synthetic gear lube and press it into place until it seats over the bearing. Install a new O-ring (coated with clean oil) and then thread in the shaft retainer. Install the spanner tool over the retainer and screw it in all the way until it seats on the upper bearing race.

47. Install the spline socket over the top of the driveshaft. Carefully remove the pinion nut and coat the threads with Loctite Primer. Once the primer has dried, coat the threads again, this time with Loctite locking compound and then thread the nut into the shaft. Install the nut holder tools and then tighten the nut to 150-160 ft. lbs. (203-217 Nm) on SP drives or 175-185 ft. lbs. (237-251 Nm) on King Cobra drives by turning the driveshaft.

48. With the spline socket still installed on the upper end of the driveshaft, turn the lower unit over so that it can rest on its forward edge (use a piece of wood to support it). Attach a breaker bar to the socket and rotate the driveshaft a few times to ensure that the bearings are seated fully. Remove the bar and install a dial torque wrench; pull on the wrench and observe the gauge reading as the driveshaft just begins to rotate. The driveshaft rolling torque should be 2-4 inch lbs. (0.22-0.45 Nm) on SP drives or 1-4 inch lbs. (0.11-0.45 Nm) on King Cobras; if not, remove the wrench and install the spanner wrench so you can tighten or loosen the driveshaft retainer until the rolling torque comes into specification. Do this slowly, a little at a time, checking the torque in between each time you tighten or loosen the retainer. Record the actual setting.

49. Sit the propeller shaft bearing housing on a flat surface with the aft side facing down. Assemble the special seal installer tool (bolt # 914706, guide # 914700, and installer # 915970) so that the tapered side of the guide is facing the threads. Slide the bearing housing seal onto the bolt so that the lip is facing the guide. Coat the outside casing (metal) of the seal with Gasket Sealing compound and then position the assembly into the housing so that the guide seats itself into the housing. Tap the end of the bolt until the seal is seated fully. Remove the tool and coat the seal lip with grease.

50. Coat the outside of the housing bearing race with GL5 synthetic gear lube and then press it into the housing using the same tools as for the seal except that the race must use a different installer (# 914703).

51. Coat the threads of the propeller shaft bearing housing with GL5 synthetic gear lube and then coat a 0.160 in. shim ring (# 914720) with Needle Bearing grease. Position the shim ring on the housing and thread the housing into the bore. Do not use an O-ring yet. Connect a housing installer (# 914684) so that the slots align with the ribs and then tighten the housing to 80 ft. lbs. (108.5 Nm).

52. Use a depth gauge and measure the depth from the edge of the propeller shaft housing bore to the top of each of the 3 ribs on the bearing housing. Make sure that the gauge pointer is well away from the hole in each rib to avoid a false reading from the slight indentation around each hole. Add the 3 measurements together and divide them by three—round the results off to thousandths of an inch. Record this figure and then add 0.160 in. to it (thickness of the shim ring). Record the final figure for the initial propeller shaft bearing preload shim thickness; although it is not necessarily the actual shim thickness you will end up using, read on.

53. Remove the bearing housing again and pull off the shim ring. Install the propeller shaft into the bore and then re-install the bearing housing again (without the shim ring or the O-ring) until it is hand-tight. Install the housing installer tool and re-install the spline socket to the top of the driveshaft if you had already removed it. In order to set the final rolling torque, you'll need to turn the driveshaft with the torque wrench while turning the housing installer tool with a breaker bar at the same time, until the torque wrench shows a reading of:

• 14-23 inch lbs. (1.58-2.60 Nm) on SP drives with drive ratios of 1.43:1, 1.51:1, 1.60:1, 1.79:1 and 1.89:1

• 11-18 inch lbs. (1.24-2.03 Nm) on SP drives with gear ratios of 1.66:1, 1.85:1, 1.95:1, 1.97:1 and 2.18:1

It's a little different on the King Cobra, final rolling torque should be set based on your figure taken earlier for the initial rolling torque. If your initial rolling torque figure was:

• 1 inch lb. - set the final rolling torque to 2-5 inch lbs. (0.23-0.57 Nm)
• 2 inch lb. - set the final rolling torque to 3-6 inch lbs. (0.34-0.68 Nm)
• 3 inch lb. - set the final rolling torque to 4-7 inch lbs. (0.45-0.79 Nm)
• 4 inch lb. - set the final rolling torque to 5-8 inch lbs. (0.57-0.90 Nm).

54. Install your propeller without the thrust washer. Install the brass pacer and then tighten the propeller nut BY HAND until the propeller will no longer rotate freely. Install a gear lash extension (# 986616 - KC, # 3850602 - all others) over the upper end of the driveshaft. Now rig a support bracket with a dial indicator so that the indicator needle runs parallel to the case mating surface and contacts the groove in the arm on the special tool. Grab the tool and attempt to rotate it while confirming that the indicator reads 0.009-0.013 in. (0.23-0.33mm) on the King Cobra, or 0.005-0.012 in. (0.13-0.30mm) on all others.

55. If gear lash is within specifications, remove the tool and gauge. Gear lash is controlled by the amount of shim material under the lower bearing

Fig. 195 ...and the O-rings

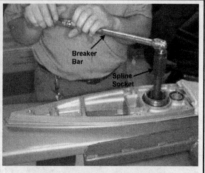

Fig. 196 Turn the spline socket and...

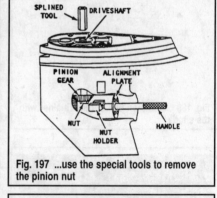

Fig. 197 ...use the special tools to remove the pinion nut

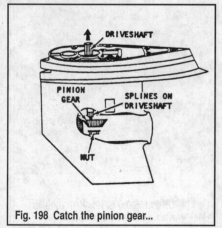

Fig. 198 Catch the pinion gear...

Fig. 199 ...while removing the driveshaft

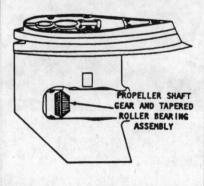

Fig. 200 Pull out the propeller shaft gear and bearing...

race on the driveshaft and behind the propeller gear bearing race; if backlash is less then the minimum figure in the range or more than the maximum figure in the range, ALL shimming procedures MUST be performed again. Recheck all of your shimming measurements; if you find differences in your calculations, recheck the backlash. If it is still out of specification, or if all of your shim measurements came out the same the second time around, move to the next two steps.

56. If backlash was less than 0.009 in. on King Cobra drives, subtract your actual figure from 0.011 in. (which is the upper limit of correct backlash) and then divide the result by two and round it off to thousandths of an inch. THIS figure is the shim thickness that will have to be ADDED under the lower bearing race, and REMOVED from behind the propeller shaft bearing race.

57. If backlash was less than 0.005 in. on 1993-94 SP drives, subtract your actual figure from 0.012 in. (which is the upper limit of correct backlash) and then divide the result by two and round it off to thousandths of an inch. THIS figure is the shim thickness that will have to be ADDED under the lower bearing race, and REMOVED from behind the propeller shaft bearing race.

58. If backlash was less than 0.005 in. on 1995-98 SP drives, subtract your actual figure from 0.0085 in. (which is the optimum backlash) and then divide the result by two and round it off to thousandths of an inch. THIS figure is the shim thickness that will have to be ADDED under the lower bearing race, and REMOVED from behind the propeller shaft bearing race.

59. If backlash was more than 0.013 in. on King Cobra drives, subtract 0.011 in. (which is the lower limit of correct backlash) from your actual figure and then divide the result by two and round it off to thousandths of an inch. THIS figure is the shim thickness that will have to be REMOVED from under the lower bearing race, and ADDED behind the propeller shaft bearing race.

60. If backlash was more than 0.012 in. on 1993-94 SP drives, subtract 0.005 in. (which is the lower limit of correct backlash) from your actual figure and then divide the result by two and round it off to thousandths of an inch. THIS figure is the shim thickness that will have to be REMOVED from under the lower bearing race, and ADDED behind the propeller shaft bearing race.

61. If backlash was more than 0.012 in. on 1995-98 SP drives, subtract 0.0085 in. (which is the optimum backlash) from your actual figure and then divide the result by two and round it off to thousandths of an inch. THIS figure is the shim thickness that will have to be REMOVED from under the lower bearing race, and ADDED behind the propeller shaft bearing race.

62. Reinstall the propeller shaft bearing housing. Use a depth gauge and measure the depth from the edge of the propeller shaft housing bore to the top of each of the 3 ribs on the bearing housing. Make sure that the gauge pointer is well away from the hole in each rib to avoid a false reading from the slight indentation around each hole. Add the 3 measurements together and divide them by three—round the results off to thousandths of an inch. Record this figure and then subtract it from the figure determined when calculating the initial propeller shaft bearing preload shim thickness a few steps ago. The result will be the thickness of shims that need to be added to the forward edge of the housing; lightly coat it with needle bearing grease and then position it on the housing.

63. Coat the housing threads and a new O-ring with GL5 synthetic gear lube, install the O-ring and then thread the assembly into the bore.

64. Install the housing installation tool again, making sure the slots and ribs line up and then connect a torque wrench at a 90° angle to the tool. Tighten the housing to 200-225 ft. lbs. (271-305 Nm) and then re-check the rolling torque.

a. If the final rolling torque is too HIGH (as per the specifications detailed previously), remove the bearing housing and install the next size LARGER shim ring.

b. If the final rolling torque is too LOW (as per the specifications detailed previously), remove the bearing housing and install the next size SMALLER shim ring.

c. Re-install the housing (again!!) and confirm that rolling torque is now within specification.

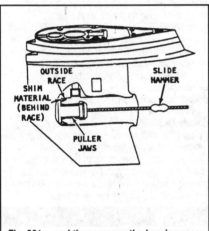

Fig. 201 ...and then remove the bearing race

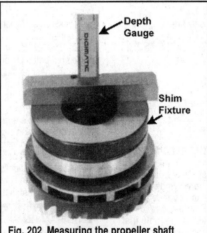

Fig. 202 Measuring the propeller shaft gear/bearing shim thickness

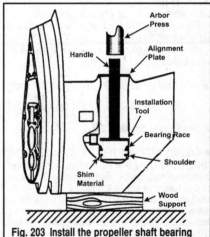

Fig. 203 Install the propeller shaft bearing race...

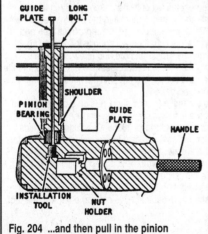

Fig. 204 ...and then pull in the pinion bearing

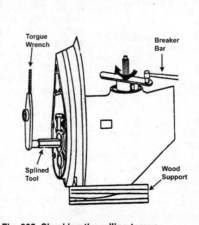

Fig. 205 Checking the rolling torque

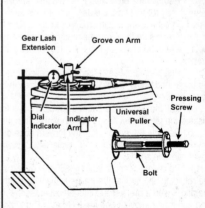

Fig. 206 Checking the backlash. If a puller is not available, install the propeller

65. Coat the threads of the bearing housing set screw with Loctite Primer. After the primer dries, coat the threads with Loctite locking compound (or OMC Ultra Lock) and install the screw into the lower unit casing. Tighten it to 42-60 inch lbs. (4.8-6.8 Nm), do not tighten the screw more than recommended or you will distort the lip of the bearing housing.

66. Coat the threads of the housing retainer screw with Gasket Sealing compound. Position the thicker edge of the retainer toward the handle of the retainer installation tool (# 3850603) while compressing the spring-loaded tip. Insert the tool into the exhaust bore on top of the lower unit and lower it until it lines up with the hole in the housing. Use a 1/4 in. drive ratchet with a 3/8 in. socket and an extension to install the retainer screw. Tighten it to 20-25 ft. lbs. (27-34 Nm). Push the tool toward the rear of the chamber until it pops off the retainer and then remove it.

67. Install the water tube retainer into the recess on top of the unit; remember that some drives may secure this retainer with 4 small screws—if yours is one of these, tighten the screws to 10-12 ft. lbs. Coat the grommet with gasket sealing compound and position it into the recess over the retainer.

68. Coat new driveshaft retainer and oil passage O-rings with synthetic gear lube and install them into their respective grooves.

69. Install the anode and tighten the 2 bolts to 60-84 inch lbs. (6.8-9.5 Nm).

Install the intermediate driveshaft sleeve over the driveshaft so that the groove is facing UP.

70. Install the water tube and upper seal into the lower grommet. Make sure that the oil filter screen is clean and debris-free, and then install it.

71. Move the lower unit underneath the drive and raise it into position while guiding in the water tube and indexing the sleeve to the upper housing. You may have to wiggle the propeller shaft to get the sleeve to index with the upper shaft splines. Coat the threads of the mounting bolts with gasket sealing compound and then install the 4 short ones on the side and the 2 long ones at the rear. On SP drives, tighten the side (short) bolts to 22-24 ft. lbs. (30-33 Nm) and the aft (long) bolts to 32-40 ft. lbs. (43-54 Nm). On King Cobra drives, tighten all 7 bolts to 32-40 ft. lbs. (43-54 Nm).

72. Install the trim tab so that the marks made previously are in alignment and then tighten the bolt to 14-16 ft. lbs. (19-22 Nm).

73. Install the propeller as detailed elsewhere in this section. Fill the unit with lubricant.

1996-98 Dual Propeller Models (DP/DP DuoProp)

◆ See Figures 207 thru 226

■ **DO NOT ATTEMPT THIS PROCEDURE WITHOUT THE SPECIAL TOOLS!! We know that most of us, whether professional technicians or shade-tree mechanics, hate spending money on special tools. Most of us don't want to spend the money, even if we had it, so we find ways to fabricate our own version of special tools. Be warned in advance that you will NOT be able to create these tools on your own. Please read through the following steps beforehand and make a list of the tools that you will need to acquire—if you are unwilling, or unable, to get your hands on them, take the drive to someone who already has them.**

1. Remove the propellers if not already removed. The outer propeller bearing may remain on the outer shaft; if so, make sure that you remove it also. Install the lower unit in an appropriate holding fixture.

2. We're sure that you've already drained the oil, but if you happened to replace the drain plug, you'll need to remove it again. Use an 8mm Allen wrench.

3. Pull the shaft sleeve (sometimes called the intermediate shaft) off of the upper end of the driveshaft; if not there, look in the bottom end of the upper housing.

4. Pull the water tube out of the recess in the top of the lower unit. Discard the upper seal and pry out the retainer in the bottom of the recess.

5. Remove the anode from the forward edge of the unit. If more than 1/3 of the anode has deteriorated, replace it. Certain models may utilize a cover over the anode.

6. Position a spanner tool (# 3855877) over the propeller shafts and on the retaining ring so that the waves in the tool match those on the inside of the ring. Insert a breaker bar into the cut-out in the tool, loosen the retaining ring and then remove it and the 2 O-rings. Throw away the O-rings.

7. Lift out the bearing housing retainer ring and then pull out the O-ring and throw it away.

8. Thread a puller tool (# 884789) onto the end of the outer propeller shaft. Tighten the 5/8-11 in. bolt on the end of the tool against the inner shaft and slowly let the tool remove the outer shaft and bearing housing. Remove the tool once the assembly is out of the lower unit and then press the outer shaft out of the housing. Remove the 2 O-rings on the housing and throw them away.

9. Assemble the pusher tip (# 3855921) to the tube (# 3855922) and tighten the set screw securely. Thread a shaft adaptor (# 3855919) into the end of the inner propeller shaft and then slide the tool over the adapter and shaft until the notch in the end of the pusher slides over the pinion nut and seats on the forward gear—you will probably need to wiggle the driveshaft slightly until the nut and notch line up correctly.

10. Install a spline socket (# 3850598) onto the upper end of the driveshaft. Insert a breaker bar into the socket and turn the shaft a few times to loosen (but NOT remove!) the pinion nut, which you are holding with the pusher tool.

11. Coat the shaft adaptor in the end of the inner shaft with wheel bearing grease and then thread on the special nut (# 3855920). Position a wrench on the flats at the end of the tube and then tighten the special nut (while holding the tube) until the inner shaft pulls out of the forward gear. Remove the tools and pull the shaft out of the tube.

12. Loosen the screw and remove the pusher tip from the tube. Set the pusher tip in a press so that the notch is facing down and then insert the propeller shaft into the top of the tip with the tapered side of the bearing facing up. Thread an old propeller nut onto the end of the shaft, to protect the threads, and then press the bearing off of the shaft, Remove any shim material.

13. Reach into the propeller shaft bore and unscrew the pinion nut; do not throw it away yet.

14. Install a spanner wrench (# 3850601) over the upper end of the driveshaft and onto the retainer so that the ridges in the tool index with the notches in the retainer. Loosen and remove the retainer. If the O-ring does not come out with the retainer, reach in and pull it out; either way, throw it away. While you're at it, pull out the O-ring for the oil screen also.

15. Slide a driveshaft puller (# 3855923) over the upper end of the shaft and tighten the 2 Allen screws on the sides until it clamps securely onto the shaft. Now tighten the 2 bolts alternately and a little at a time until the shaft is pulled out of the pinion gear. Make sure that you have a rag in the bore under the gear so it doesn't damage the case when it drops. Remove the tool and pull the shaft out of the bore.

16. Reach into the propeller shaft bore and remove the pinion and forward gears.

17. Assemble a race remover (# 3855859) to one end of a special rod (# 3855860), slide on a guide plate (# 914700) so the stepped side is facing the remover and then install a washer and nut on top of the plate. Insert the tool into the driveshaft bore so the remover is under the lip of the lower bearing race. Slide the guide plate down the rod so it sits in the recess at the top of the bore and then tighten the nut slowly until the race breaks free. Remove any shim material that does not come out with the tool and race.

18. If the pinion bearing needles fell out when removing the gear and shaft, coat them with grease and re-install them into the bearing case. Remove the race remover tip from the tool used in the previous step and insert the rod down through the driveshaft bore, carefully feeding it through the pinion bearing. Reach into the propshaft bore and thread a pinion removal tip (# 3855868) onto the end of the rod. Seat the guide plate into the recess at the top of the bore and tighten the nut until the pinion bearing breaks free. Remove the tool and bearing.

19. Carefully turn the lower unit over and support is on a piece of wood so that the propeller shaft bore is facing upward. Assemble a race remover (# 3855862) onto the end of a special rod (# 3855860)), slide on a guide plate (# 3855863) so the stepped side is facing the remover and then install a washer and nut on top of the plate. Insert the tool into the propshaft bore and then push the remover into the forward bearing race until it locks into place behind the race. Slide the guide plate down the rod so it sits in the recess at the top of the bore and then tighten the nut slowly until the race breaks free. Remove any shim material that does not come out with the tool and race.

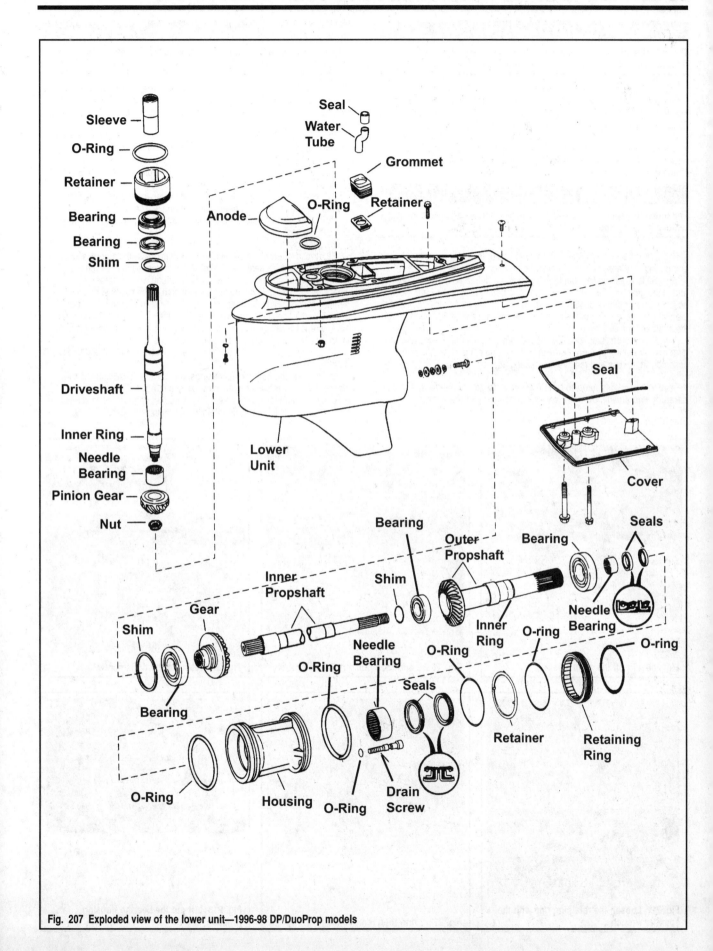

Fig. 207 Exploded view of the lower unit—1996-98 DP/DuoProp models

The propeller bearing will be damaged during removal so do not remove it from the housing unless you intend to replace it.

20. Lay a support base (# 3855926) on a press so the notched side is facing up and then position the bearing housing into the support so the aft side is down. Insert a seal and bearing removal tool (# 3855924) into the housing and press out the bearing and seals. Throw them all away.

21. Push a race remover (# 3855862) into the housing so that it snaps into place behind the bearing race. Sit the housing back into the support base with the aft side facing up and then use a puller (# 884789) to press out the race.

The outer propeller bearing will be damaged during removal so do not remove it unless you intend to replace it.

22. Thread a puller (# 884789) onto the end of the outer propeller shaft and position the assembly in a press with the gear facing up.

23. Slide a dismantling tool (# 884803) and a drift (# 884143) into open end of the gear and then press the seals and bearing out of the gear. Now position a puller tool (# 884832) into the gear opening so that the flanges on the tool grip the inner lip of the bearing race, insert a drift (# 884143) through the shaft and remove the race.

24. Clamp a universal bearing separator in between the bearing and the gear and thread the puller (# 884789) onto the end of the shaft. Support the assembly in a press and remove the bearing and the sleeve.

■ You will probably need to heat the sleeve to get it off—do this carefully and slowly. Do not create any hot spots on the shaft.

The forward gear and/or driveshaft bearings will be damaged during removal so do not remove it unless you intend to replace it.

25. Install a universal bearing separator between the forward gear and bearing and then position the unit is a press with the gear side down. Position a suitably sized mandrel on the gear hub and press the gear out of the bearing.

26. Thread the pinion nut into the end of the driveshaft and then install a universal bearing separator between the two bearings. Position the assembly on a press with the nut facing upward and then press off the two bearing. Move the separator to the sleeve and repeat the procedure, paying attention to the previous Note with regard to applying heat.

27. Reach into the propshaft bore and loosen the mounting bolt for the magnets. Remove the 2 magnets and 3 spring washers.

28. If necessary, remove the 2 bolts and pull down the exhaust cover. Discard the seal.

To assemble:

29. Wash all parts in solvent and blow them dry with compressed air. Remove all traces of seal and gasket material from all mating surfaces. Blow all water, oil passageways and screw holes clean with compressed air. After cleaning, apply a light coating of gear lubricant to the bright surfaces of all gears, bearings, and shafts as prevention against rusting and corrosion.

30. Use a fine file to remove burrs. Replace all O-rings, gaskets, and seals to ensure satisfactory service from the unit. Clean the corrosion from inside the housing where the bearing carrier was removed.

31. Check to be sure the water intake is clean and free of any foreign material.

32. Inspect the gear case, housings, and covers inside and out for cracks. Check carefully around screw and shaft holes. Check for burrs around machined faces and holes. Check for stripped threads in screw holes and traces of gasket material remaining on mating surfaces.

Fig. 208 Lift off the shaft sleeve...

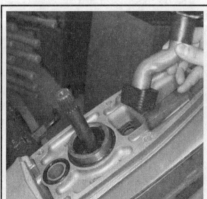

Fig. 209 ...and then pull out the water tube

Fig. 210 Remove the drain screw

Fig. 211 Loosen the retaining ring with the spanner wrench...

Fig. 212 ...and then remove the ring

Fig. 213 Lift out the bearing housing retainer

33. Check O-ring seal grooves for sharp edges, which could cut a new seal. Check all oil holes.

34. Inspect the bearing surfaces of the shafts, splines, and keyways for wear and burrs. Look for evidence of an inner bearing race turning on the shaft. Check for damaged threads. Measure the run-out on all shafts to detect any bent condition. If possible, check the shafts in a lathe for out-of-roundness.

35. Inspect the gear teeth and shaft holes for wear and burrs. Hold the center race of each bearing and turn the outer race. The bearing must turn freely without binding or evidence of rough spots. Never spin a ball bearing with compressed air or it will be ruined. Inspect the outside diameter of the outer races and the inside diameter of the inner races for evidence of turning in the housing or on a shaft. Deep discoloration and scores are evidence of overheating.

36. Inspect the thrust washers for wear and distortion. Measure the washers for uniform thickness and flatness.

37. Replace ALL seals, O-rings, and gaskets to ensure maximum service after the work is completed.

38. Coat the inside of the smaller driveshaft bearing with GL5 synthetic gear lube. Thread the pinion nut into the lower end of the driveshaft for protection and then slide the bearing onto the splined end (top) of the shaft so that the tapered side is facing the bottom of the shaft (the end with the pinion nut). Install a bearing installer (# 3850617) over the end of the shaft so that the raised lip on the tool contacts the backside of the bearing. Now rest the entire assembly (pinion nut UP) on an arbor press and press the bearing into place on the shaft.

39. Repeat the previous step for the remaining bearing, with the tapered side of the bearing facing the top of the shaft this time. Press the bearing into position until it seats against the lower bearing.

40. Clamp a spline socket (# 3850598) in a vise with the spline end facing UP. Insert the upper end (splined) of the shaft into the socket and slide the bearing race down over the smaller bearing—this is the lower bearing, but it will be on top at the moment. Now slide a pinion shim fixture tool (# 3855870) over the shaft and against the race so that the side of the tool with 3 slots is facing up toward the pinion end of the shaft.

41. Install the pinion gear and nut, tightening the nut to 72-87 ft. lbs. (98-118 Nm). Rotate the shim fixture tool through a few revolutions until the bearing seats itself and then insert a flat bladed feeler gauge into each of the 3 slots, measuring the clearance between the top of the inner hub on the tool and the bottom of the pinion gear. Mark each measurement down, add them up and divide by three to determine the average clearance. Round off this figure to the nearest thousandth of an inch (3 places after the decimal point) and record this figure.

42. Loosen the pinion nut a few turns (don't remove it), clamp the upper portion of the shim fixture in a vice and then strike the nut and shaft with a mallet until the pinion comes loose from the shaft.

43. Remove the pinion nut and gear; throw away the nut. Locate the shim allowance number etched onto the backside of the gear and convert it to a decimal equivalent in thousandths of an inch—**+5** translates to 0.005 in., **-2** translates to -0.002 in. Add this figure to the one arrived at previously and record it as your pinion gear shim thickness. When adding shim material later on, remember that in arriving at this thickness, you must always use at least one shim, and never more than five shims.

Fig. 214 Use the tool to pull out the outer shaft and housing..

Fig. 215 ...and then press the shaft out of the housing

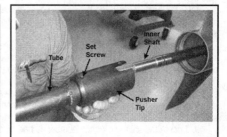

Fig. 216 Slide the special tool over the inner shaft...

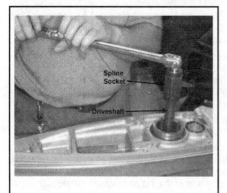

Fig. 217 ...and then turn the driveshaft to loosen the pinion nut

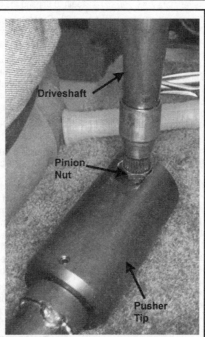

Fig. 218 See how the notch must fit around the pinion nut?

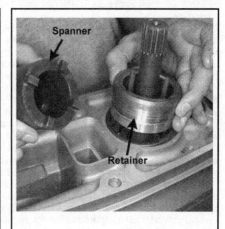

Fig. 219 Remove the retainer...

■ Not to be silly, but please remember that if the etching of your pinion gear is a negative number (-2), you will then SUBTRACT this number from the pinion gear clearance figure rather than adding it.

44. Assemble the correct amount of shim material for the pinion gear and position it onto the bearing shoulder in the top of the driveshaft bore.

45. Coat the outer surface of the bearing race with GL5 synthetic gear lube and position it squarely into the driveshaft bore so that the taper, or large opening, is facing upward. Now assemble the race driver (# 3850623), guide plate (# 3850619) and installer (# 385021) so the stepped end of the guide plate faces the threads on the driver and the stepped end of the installer is also facing the threads. Insert the tool into the bore and push down on the plate until it indexes with the bore. Tap the end of the driver with a hammer until the races seats against the shim material on the shoulder.

46. If the needle bearings are not still in the pinion bearing cage, coat them with grease and re-install all of them. Coat the outside of the cage with GL5 synthetic gear lube and install the bearing into the bottom of the driveshaft bore with the same tools used during removal. Insert the tool through the top of the driveshaft bore and then position the pinion bearing into the installer and thread the installer onto the bottom of the rod through the propeller shaft bore. Once the guide plate is seated in the top of the bore, tighten the nut until the bearing is pulled up into its recess and fully seated against the shoulder in the bottom of the bore.

■ If the bearing appears to be spinning in the installer, insert the pinion nut holder into the propeller shaft bore to hold it in place.

47. Coat the forward gear and the inner race of the bearing lightly with GL5 synthetic gear lube. Place the gear (teeth down) on a universal bearing separator for support and slide the bearing over the gear hub so the tapered side is facing UP. Press the bearing into place with the installer tool (# 3855861) until it is seated.

48. Remove the separator tool and sit the gear/bearing on a flat surface with the gear teeth facing down. Slide the bearing race into position over the bearing and then turn it a few revolutions until the bearings are fully seated. Position a shimming fixture (# 3850600) onto the bearing so its recessed side faces the bearing. Install a depth gauge over the shim fixture and measure the depth from the top of the fixture to the end of the gear hub; subtract 0.5 in. (thickness of the fixture) and record the resulting figure.

49. Locate the shim allowance number etched onto the backside of the gear and convert it to a decimal equivalent in thousandths of an inch—+5 translates to 0.005 in., -2 translates to -0.002 in. Add this figure to the one arrived at in the previous step and record it. OMC suggests a nominal dimension of 0.055 in., so subtract the figure you just recorded from this number and record the new number as your forward gear shim thickness. When adding shim material later on, remember that in arriving at this thickness, you must always use at least one shim, and never more than three shims.

50. Thread a puller (# 884789) onto the end of the outer shaft to protect the threads and position the unit onto a press with the puller side resting on the press surface. Coat the inner surface of the bearing race with GL5 synthetic gear lube, position it over the end of the shaft and then press it into position with a race installer (# 3855865) until it seats fully in the gear.

51. Remove the shaft and tool and then position the race installer on the on the press with the gear side of the shaft sitting in it. Lightly coat the inner surface of the bearing with synthetic gear lube and slide it over the end of the shaft so the tapered side is facing upward. Press the bearing into place with an installer tool (# 3855866) until it seats fully on the back of the gear.

52. Coat the inside of the sleeve and the area on the shaft where it rides with Loctite Primer; once dry, coat the shaft area with Loctite 609. Slide the sleeve over the shaft and then use the installer to press it into position.

53. Coat the outside of the needle bearing cage with GL5 synthetic gear lube and position the bearing into the end of the shaft. Press the bearing all of the way into the shaft with a bearing installer (# 3855928).

54. Position the double lipped seal into the installer tool so that the rubber case faces the stepped side on the installer, coat the outside surface with gasket sealing compound and then press it into the shaft until the tool seats on the upper edge of the shaft. Position the single lip seal into the installer so the open side is facing the tool and then press it into the shaft until the tool seats itself. Coat the seal lips with water repellant grease.

55. Position a bearing installer (# 3855861) into a press so that the recessed side is facing upward. Center the forward gear on top of the tool so the bearing is facing down. Lubricate the splined end of the inner shaft with synthetic gear lube and slip the shaft into the gear. Thread a propeller nut into the end of the shaft and lightly tap the shaft into the gear until it is seated.

56. Position a shim fixture tool (# 3855872) onto a flat surface and then place the forward gear bearing race into the tool so the tapered side faces up and the race rests on the 3 feet in the tool.

57. Insert the shaft and gear into the tool so the bearing is resting in the race. Slide a different shim fixture (# 3855871) down and over the shaft until it seats on the bearing flange. Now slide the outer gear bearing over the shaft until it seats on the top of the fixture tool.

58. Position a protection sleeve (# 884976) over the seals in the end of the outer shaft and then slide the shaft down and over the inner shaft and into the outer fixture tool. Remove the seal protector.

59. Rotate each of the shaft a few times to ensure that the gears are seated. Press down on the assembly to load it and then tighten each of the 3 set screws to stabilize the assembly. Use a depth gauge to measure the distance between the top edge of the shim fixture and the small ridge on the back of the outer gear; take three measurements, each one over the set screws in the tool. Round your measurements off to thousandths of an inch, add them up and divide by three to reach an average measurement and record the figure.

60. Subtract the forward gear shim figure arrived at earlier from the average figure in the previous step and record this figure. Locate the shim allowance number etched onto the backside of the outer gear (or on the outside of the outer shaft) and convert it to a decimal equivalent in thousandths of an inch—+5 translates to 0.005 in., -2 translates to -0.002 in. Add this figure to the one arrived at in the previous step and record it as your outer gear shim thickness. When adding shim material later on, remember that in arriving at this thickness, you must always use the least amount of shim material possible to get to the figure.

61. Turn the lower unit over so that it can rest on its forward edge (use a piece of wood to support it). Select the correct number of shims (as determined earlier) and insert them into the propeller shaft bore so that they sit in the forward gear bearing race recess at the forward end of the bore. Coat the outer surface of the bearing race with GL5 synthetic gear lube and position squarely into the bore as far as it will go. Assemble the same tools used during the removal procedure and the press the race into the recess until it is fully seated.

62. Lightly coat the forward gear bearing with GL5 synthetic gear lube and then install it into the bore so that is seats into the bearing race with the gear teeth facing aft. Insert the driveshaft into the top of the bore and feed it carefully through the pinion bearing without disturbing the needles. When the shaft is in far enough that the lower bearing seats into the race, install the pinion gear onto the shaft and then screw in the new pinion nut until it is finger-tight. We recommend using a pinion nut installation tool (# 3855930). You may have to lift the driveshaft slightly in order to get the pinion gear teeth to engage the teeth on the forward gear.

63. Coat the upper bearing race with synthetic gear lube and press it into place until it seats over the bearing. Thread in the shaft retainer hand-tight after coating its threads with gear lube.

64. Thread a propeller shaft adaptor onto the end of the inner shaft and then slide a pusher tip and tube over the shafts so that the notch lines up around the pinion nut. Thread the special nut (# 3855920) onto the adaptor and tighten it just enough to secure the tools to the shaft, but not so much as to pull the shaft out of the forward gear.

65. Install the spline socket over the top of the driveshaft and then tighten the nut to 72-87 ft. lbs. (98-118 Nm) by turning the driveshaft.

66. With the spline socket still installed on the upper end of the driveshaft, turn the lower unit over so that it can rest on its forward edge (use a piece of wood to support it). Attach a beaker bar to the socket and rotate the driveshaft a few times to ensure that the bearings are seated fully. Remove the bar and install a dial torque wrench; pull on the wrench and observe the gauge reading as the driveshaft just begins to rotate. The driveshaft initial rolling torque should be 2-4 inch lbs. (0.22-0.45 Nm); if not, remove the wrench and install the spanner wrench so you can tighten or loosen the driveshaft retainer until the rolling torque comes into specification. Do this slowly, a little at a time, checking the torque in between each time you tighten or loosen the retainer. Record the actual setting.

67. Select the correct amount of shim material for the outer gear, remembering that you must use as few as possible to arrive at the proper thickness and that the thinnest shims should be positioned in between the thickest ones. Slide the shims over the inner shaft and into position.

68. Coat the inner race of the outer gear bearing lightly with GL5 synthetic gear lube and slide it over the inner shaft and up against the shims

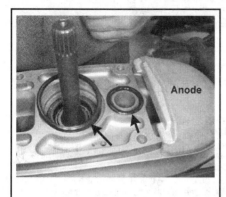
Fig. 220 ...and remove these two O-rings

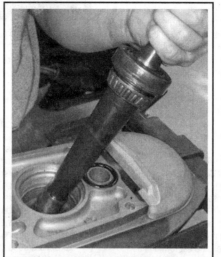
Fig. 221 Lift out the driveshaft

Fig. 222 A nice look at how the shafts and gears are oriented in the case

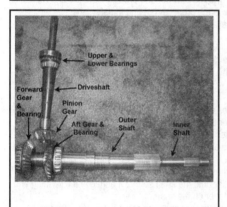
Fig. 223 Installing the inner propeller shaft

Fig. 224 Installing the pinion nut

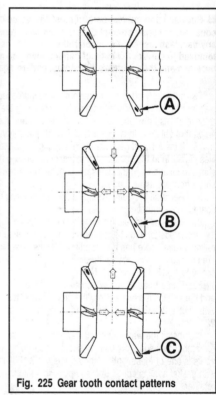
Fig. 225 Gear tooth contact patterns

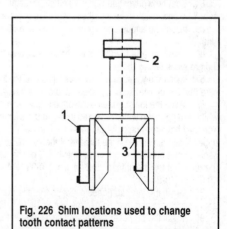
Fig. 226 Shim locations used to change tooth contact patterns

69. Thread the shaft adaptor onto the inner shaft again and then slide the special tube (# 3855922) over the shaft until it is resting on the outer gear bearing. Coat the threads on the end of the tube with wheel bearing grease and thread the special nut (# 3855920) onto the adaptor. Tighten the nut slowly until the bearing seats itself.

70. Coat the gear teeth on the outer gear with marking dye. Install a seal protector (# 884976) onto the end of the outer shaft and then slide the shaft into place over the inner shaft. Remove the protector.

71. Install another seal protector sleeve (# 884807) over the bearing housing seals and then slide the housing into position over the shafts—no O-rings yet. Seat the housing in the bore so that the oil drain hole is facing the skeg on the bottom of the case. Install the bearing housing retainer over the housing so that the tab on the retainer fits into the hole in the side of the bore.

72. Coat the threads of the retaining ring with synthetic gear lube and then spin it into the case and against the retainer; once again, no O-rings yet.

73. Reinstall the spline socket to the top of the driveshaft and the spanner tool to the retaining ring and set the final rolling torque by turning the driveshaft while tightening the retaining ring. Set the rolling torque to 18-27 inch lbs. (2-3 Nm) on 1.95:1 and 2.30:1 drives, or 27-35 inch lbs. (3-4 Nm) on 1.68:1 and 1.78:1 drives. Record your actual measurement.

■ Before attempting to check backlash, you must first verify that there is 0 (zero) endplay at any of the shafts.

74. Install the driveshaft puller tool (# 3855923) and lock it down on the shaft to prevent it from turning. Install a lash tool (# 3855873) on the inner propeller shaft (but not on the splines) and then connect a dial indicator so that the tip is resting on the outer line on the arm of the tool. Wiggle the inner shaft back and forth while checking the indicator gauge—forward gear lash should be 0.006-0.012 in. 0.15-0.30mm). Record the actual measurement.

75. Reposition the lash tool so that it is on the outer shaft, position the needle of the dial indicator on the inner line and then wiggle the outer shaft while observing the indicator gauge—outer gear lash should be 0.006-0.014 in. (0.15-0.35mm). Record the actual measurement.

76. If backlash is within specifications, remove the tool and gauge. If not within the correct range, you must first confirm that the gear teeth contact patterns are correct.

77. Wrap a thick rag around the propeller shafts and hold the shafts tightly while turning the driveshaft slightly to load the gears. Remove the bearing housing and outer shaft as detailed previously and check the tooth contact pattern made in the dye you applied during installation against the illustration.

    a. Correct contact should look like **A** in the illustration.

    b. If the contact pattern looks like **B** in the illustration, you will need to remove shims from locations **1** and **2** in the illustration. You will also need to add shims to location **3** in the illustration. There is no set formula, so you'll have to try it a few times in small increments until the patterns look good.

    c. If the contact pattern looks like **C** in the illustration, you will need to add shims to locations **1** and **2** in the illustration. You will also need to remove shims from location **3** in the illustration. There is no set formula, so you'll have to try it a few times in small increments until the patterns look good.

■ **Backlash and contact pattern must always be corrected in consecutive steps; contact pattern must be correct BEFORE finalizing any backlash issues. If it is determined that shims need to be added or removed to correct the contact pattern, then you must recheck the backlash and then recheck the contact pattern again.**

78. If the forward gear backlash must be increased (your lash reading was less than the specifications) without changing the contact pattern, remove shims from location **1** in the illustration and then add an equal amount of shims to locations **2** and **3** in the illustration.

79. If the forward gear backlash must be decreased (your lash reading was higher than the specifications) without changing the contact pattern, add shims to location **1** in the illustration and then remove an equal amount of shims from locations **2** and **3** in the illustration.

80. Reinstall the outer shaft and bearing housing using new O-rings coated with synthetic gear lube.

81. Reinstall the retaining ring and recheck the final rolling torque—use the spline socket and spanner wrench to tighten the ring until the actual rolling torque figure has been achieved. Make sure that the oil drain holes are aligned and then install the drain screw with a new O-ring. Tighten the screw to 10-15 ft. lbs. (14-20 Nm).

82. Run a bead of yellow Loctite around the exhaust cover seal groove and then install a new seal into the groove. Coat the mounting bolt threads with Loctite 242, install the cover and tighten the bolts to 14-17 ft. lbs. (19.5-22.5 Nm).

83. Install the water tube retainer into the recess on top of the unit; remember that some drives may secure this retainer with 4 small screws—if yours is one of these, tighten the screws to 10-12 ft. lbs. Coat the grommet with gasket sealing compound and position it into the recess over the retainer.

84. Coat new driveshaft retainer and oil passage O-rings with synthetic gear lube and install them into their respective grooves.

85. Install the anode and tighten the 2 bolts to 60-84 inch lbs. (6.8-9.5 Nm).

86. Install the intermediate driveshaft sleeve over the driveshaft so that the groove is facing UP.

87. Install the water tube and upper seal into the lower grommet. Make sure that the oil filter screen is clean and debris-free, and then install it.

88. Move the lower unit underneath the drive and raise it into position while guiding in the water tube and indexing the sleeve to the upper housing. You may have to wiggle the propeller shaft to get the sleeve to index with the upper shaft splines. Coat the threads of the mounting bolts/studs with gasket sealing compound and then install the 4 nuts on the side and the 2 bolts at the rear. Tighten the nuts to 22-24 ft. lbs. (30-33 Nm) and the aft bolts to 32-40 ft. lbs. (43-54 Nm).

89. Install the trim tab (if equipped) so that the marks made previously are in alignment and then tighten the bolt to 14-16 ft. lbs. (19-22 Nm).

90. Install the propeller as detailed elsewhere in this section. Fill the unit with lubricant.

Propeller

REMOVAL & INSTALLATION

SP Single Prop Models

◆ **See Figures 227, 228 and 229**

All SP drives are shipped from the factory as standard right hand rotation propellers; which is to say that they rotate clockwise when viewed from the behind the vessel. As mentioned elsewhere in this section, it is possible to switch the unit to a left hand rotation. A different propeller is required depending on the direction of rotation—you can't just switch it back and forth at your whim! In fact, OMC suggests that a counter-rotation propeller should be used ONLY in twin engine configurations, and ONLY on the Port engine. Most propellers will be marked with an **R** or an **L**.

1. Disconnect the negative battery cable to ensure that the engine will not start while working on the propeller.

2. Bend around the ends of the cotter pin and pull it out. Remove the keeper over the propeller nut.

3. Wedge a block of wood between the propeller and the anti-ventilation plate. Loosen and remove the propeller nut and thrust washer.

4. Grasp the propeller and pull it from the shaft. Lift off the thrust bushing.

■ **The rubber hub inside the propeller is serviceable, but only by an authorized OMC facility.**

To install:

5. Clean the splines and threads on the propeller shaft thoroughly. Check for nicks, burrs or other obvious signs of damage.

6. Check for and remove any fishing line that may be entangled.

7. Coat the splines on the propeller shaft lightly with OMC Triple Guard grease. Slide the thrust bushing onto the shaft so that inner taper is toward the gearcase and matches the taper on the shaft.

8. Slide the propeller onto the shaft so that it engages the bushing and then seats against the taper on the shaft.

9. Slide on the thrust washer and then move the control lever to the FORWARD position to lock up the propeller shaft. Install the nut and tighten it to 70-80 ft. lbs. (95-108 Nm),

10. Position the keeper over the nut so that the cotter pin holes line up and then push in a new pin and bends the ends back against the keeper.

11. Shift the control into NEUTRAL and confirm that the propeller turns freely.

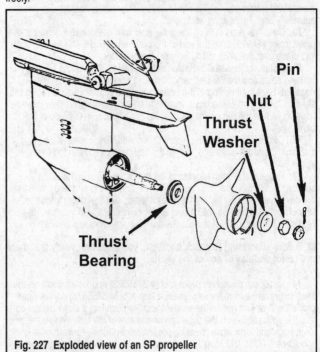

Fig. 227 Exploded view of an SP propeller

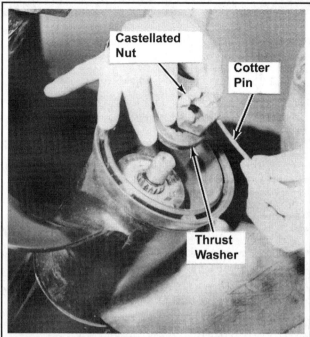

Fig. 228 Remove the propeller nut and washer (keeper not shown)...

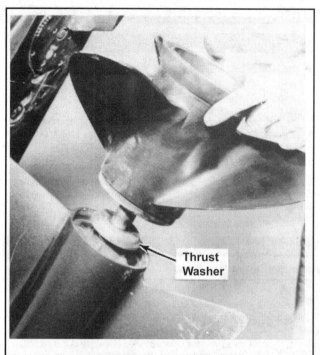

Fig. 229 ...and then pull off the propeller and bushin2

DP Duo Prop Models

OEM ② MODERATE

◆ See Figure 230

1. Disconnect the negative battery cable to prevent accidental starting of the engine while working on the propeller.

2. Position a small block of wood between the drive and aft propeller to keep it from turning. Move the shift lever into FORWARD and then remove the propeller nut with a 30mm wrench.

3. Grasp the propeller and slide it aft and off of the shaft. If the propeller will not slide off easily, you are going to need to apply heat from a torch—we do not recommend persuading it with a mallet. Use your torch sparingly, heating the hub only so much as to allow removal. There is a good chance that when it finally comes off, the rubber hub will stay with the shaft; if so, cut the hub off of the shaft with a saw and replace it with a new one. In fact, we highly recommend replacing the rubber hub after using a torch even if it does come off with the propeller.

■ OMC suggests that although the hub is serviceable, is may only be serviced by an authorized OMC facility, so try to avoid using the torch if at all possible.

4. Now wedge the block of wood in between the forward propeller and the housing and move the remote control lever to the REVERSE position. Remove the inner nut with the special tool (# 3855876) and then pull off the propeller, paying attention to the comments in the previous step regarding 'stuck' propellers.

■ Most later models utilize a plastic nut on the forward propeller—we suggest replacing it whenever you remove it.

■ You can fabricate the special tool for removing the inner nut in less than an hour. Purchase a 12 in. length of pipe (2 in. inner diameter) at your local building supplies store. Cut three 1/2 in. deep slots into one end of the pipe with a metal-cutting hacksaw blade—each slot should be 120° apart, although you can position the pipe over the nut and mark the location of each rib on the nut. Slide the pipe over the shafts and onto the inner nut and then turn it with a plumber's pipe wrench.

To install:

■ The two propellers on these units are a matched set. The part number and size coding will be marked on the sides of each propeller; D for an aluminum propeller or a F for a stainless steel propeller. This letter will be followed by a number which indicates the pitch; as the number increases, so does the pitch. Propellers for these units come only in matched sets, propellers with different part numbers, pitch of construction materials should NEVER be used together.

5. Clean the splines and threads on the propeller shafts thoroughly. Check for nicks, burrs or other obvious signs of damage. Check for any entangled fishing line and remove it.

6. Coat the entire length of both propeller shafts with Triple Guard propeller shaft grease. Move the control lever into the FORWARD position.

7. Slide the forward propeller and nut onto the shaft and then tighten the nut to 45 ft. lbs. (60 Nm).

8. Move the control lever to the REVERSE position and slide on the aft propeller and nut. Tighten the nut to 50 ft. lbs. (70 Nm).

9. Reconnect the battery cable.

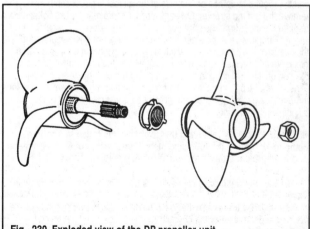

Fig. 230 Exploded view of the DP propeller unit

TRANSOM ASSEMBLY

Description

◆ See Figures 231, 232, 233 and 234

The transom assembly consists of four main components: the pivot housing, gimbal ring, gimbal housing and the inner transom plate.

The pivot housing connects to the upper housing of the stern drive unit at its aft side and rides in the gimbal ring via pivot bolts. The pivot housing also serves as the aft terminus of the exhaust bellows, U-joint bellows and water tube. Tilt capability, via the trim/tilt cylinders is enabled by the pivot housing.

The gimbal ring, aside from holding the pivot housing, rides in the gimbal housing and allows for side-to-side motion of the drive assembly. The steering arm and a lower pivot shaft control this motion.

The gimbal housing is attached to the vessels transom and the inner transom plate.

Pivot Housing

REMOVAL & INSTALLATION

 DIFFICULT

◆ See Figures 231 thru 241

1. Remove the stern drive unit as detailed elsewhere in this section.
2. Turn the steering wheel so the transom assembly is hard over to Port. Loosen the shift cable anchor bracket bolt and then slide out the bracket, disengaging the shift cable. Push (or pull) the shift cable through the pivot and gimbal housings and then pull it all the way through the inner sleeve and connector tube. Remove the seals.

■ If you only intend to remove the pivot housing and have no need for shift cable service, simply push the cable through the pivot housing and then secure it out of the way.

3. Use a magic marker and draw a line across the trim sender and the gimbal ring. Remove the 2 mounting screws, but do not pull out the sender unit.
4. Reach into the housing and disengage the U-joint bellows from the housing lip and then push the bellows out of the housing.
5. On all models except the King Cobra, insert a pair of snap-ring pliers into the exhaust bore and remove the exhaust bellows retaining ring—wear safety glasses and make sure you are very careful when releasing the ring from the pliers. Push the bellows out of the housing.
6. Disconnect the water hose if you haven't already and then the water line nipple nut. Press the nipple as far through the case as you able at this time.
7. Turn the steering wheel hard over to Starboard this time and then find the ground strap bolt on the side of the gimbal ring. Loosen the bolt and remove the ground strap(s). On early models, there may be two bolts; if so, remove the strap from the upper bolt.
8. Straighten out the steering wheel so the housing is centered and then remove the pivot pin from each side of the gimbal ring using a 1/2 in. hex drive and ratchet. Although it is unlikely that the housing will fall out of the ring on its own, support it while removing the pins just to be safe. Otherwise, grasp the bottom of the housing and pull up on it while swiveling it free of the gimbal ring. Pull the friction and thrust washers off of the pivot pin lugs in the housing.

■ If you are simply removing the pivot housing to get to another component, ignore the following steps if you wish, but we recommend performing them either way—once again, cheap insurance!

9. Pick the water passage seal (O-ring) out of its recess and throw it away.
10. Unscrew the small water passage drain plug on the port side of the housing and throw away the O-ring.
To install:
11. Clean the housing thoroughly in solvent and allow it to air dry completely.
12. Carefully remove any sealant or adhesive from the coolant passage and bellows openings. Clean all threads and drain holes thoroughly.

13. Check the ground strap connectors for frayed or loose ends. Inspect the friction and thrust washers for any damage and replace them if necessary, although it's not a bad idea to do it anyway while you have the unit apart—your call.
14. Coat a new drain screw O-ring with sealing compound, slide it onto the screw and then thread in the screw. Tighten it to 50-60 inch lbs. (6-7 Nm).
15. Run a bead of 3M adhesive around the water passage recess and then press in a new seal.
16. Install the friction and pivot washers onto each lug. Coat the thrust washer with grease.
17. Position the housing into the gimbal ring while guiding the water line nipple into the housing. Make sure that each bellows is in place directly behind the opening in the housing.
18. Align one side of the housing so that the hole lines up with the hole in the gimbal ring. Look through the hole to ensure that the neither of the washers has moved and is blocking the hole. Insert the pivot pin and screw it in until it just seats itself, but do not tighten it yet.
19. Repeat the previous step for the other side of the housing. When the pin is in place, tilt the housing up and down with your hand to ensure freedom of movement. If OK, tighten each pin to 105-120 ft. lbs. (142-163 Nm). Make sure that you hold the housing while tightening the pins, because it will have a tendency to swivel in the direction that you are tightening.
20. Push, and pull, the nipple all the way through the housing. Make sure that the drain hole is facing down and then install the nut and tighten it to 96-120 inch lbs. (11-14 Nm). Ensure that the nipple does not spin while tightening the nut and make sure that the nipple threads extend past the end of the nut. If they do not, loosen the nut and check the positioning of the square nipple collar on the other side of the housing.
21. Coat the entire surface of the V-shaped lip for the U-joint bellows with Gasket Sealing compound. Pull the bellows through the opening in the housing and press it into the groove until it seats around its entire circumference. Any flat spots on the bellows will indicate that it has not completely engaged the V-groove.
22. Reach into the exhaust bore and pull the top of the bellows into the bore. Tilt the housing, reach around and push the rest of the bellows into the bore. Make sure that it seats in the bellows channel around the lip of the housing bore and there are no flat spots indicating that the bellows has not engaged the lip fully. Reinstall the retaining clip with snap-ring pliers so the opening is at the top of the bellows. Don't forget to wear safety glasses.
23. Reconnect the ground strap(s) on the port side of the gimbal ring.
24. Feed the shift cable through its opening and install the retainer. Install the stern drive.

Gimbal Ring

REMOVAL & INSTALLATION

 OEM DIFFICULT

◆ See Figures 231, 232, 233 and 242

1. Remove the stern drive unit and pivot housing as detailed elsewhere in this section. If you left the trim/tilt cylinders connected to the gimbal ring when removing the drive, remove them now.
2. Turn the steering wheel hard over to port. Use a magic marker and draw a small line across the trim sender and gimbal ring on the starboard side of the ring. Remove the 2 mounting screws, but do not yet remove the unit itself.
3. Remove the bolt, washer and ground wire on the side of the ring (if equipped).
4. Reach under the gimbal housing and remove the 3 mounting bolts (4 on the King Cobra) for the steering support bracket. Disconnect any ground straps that may be attached to the bolts and then pry down the bracket—you may have to persuade it with a few taps on a drift so be careful.
5. Working inside the boat, pull the cotter pin out of the steering arm pin and then remove the pin. Loosen the bolt on the steering arm and disconnect the ground strap.
6. Remove the 2 bolts and lift off the gimbal housing cover (1998 models may use 4 bolts). Now remove the 2 bolts and lift off the steering cavity cap, throwing away the small foam gasket.

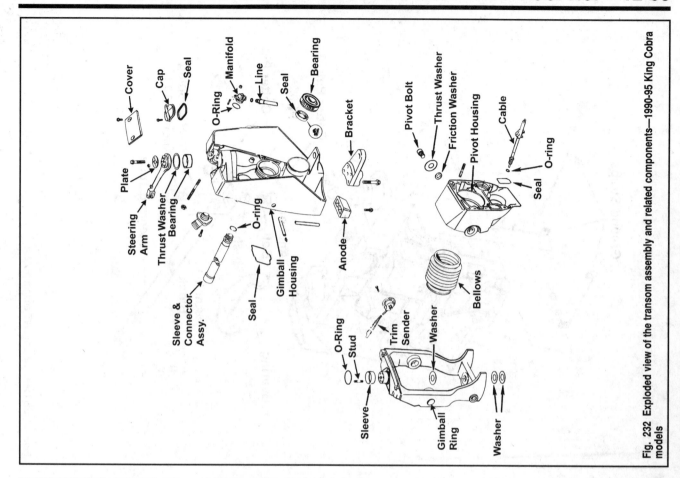

Fig. 232 Exploded view of the transom assembly and related components—1990-95 King Cobra models

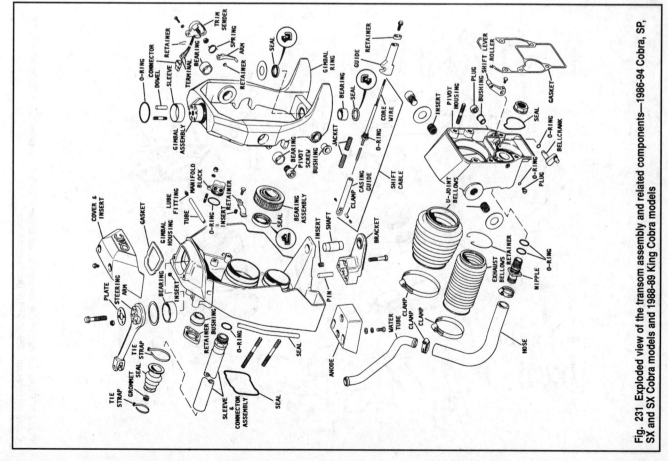

Fig. 231 Exploded view of the transom assembly and related components—1986-94 Cobra, SP, SX and SX Cobra models and 1988-89 King Cobra models

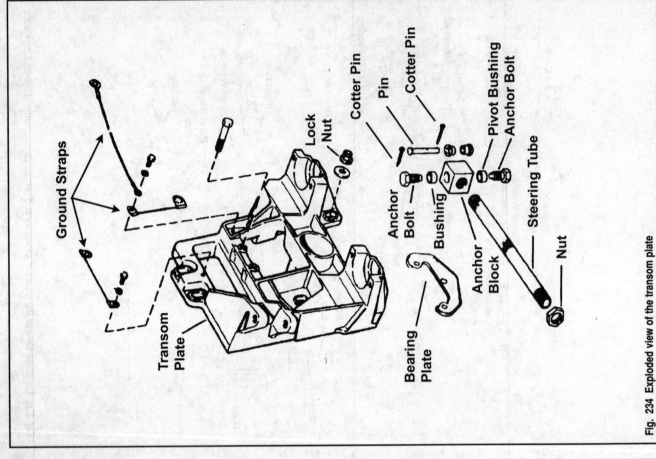

Fig. 234 Exploded view of the transom plate

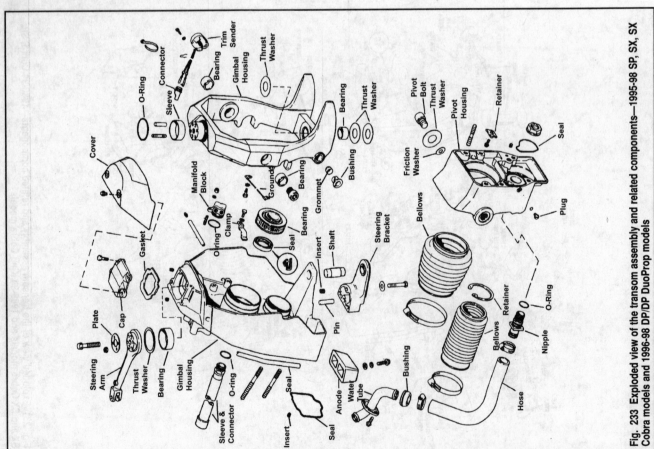

Fig. 233 Exploded view of the transom assembly and related components—1995-98 SP, SX, SX Cobra models and 1996-98 DP/DP DuoProp models

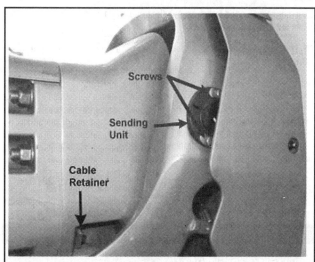

Fig. 235 The shift cable retainer is located on the starboard side of the unit

■ 1998 models may not use the inner steering cavity cap, so just pull out the inner sealing ring.

7. Remove the 4 locknuts and lift the retaining plate off of the steering arm. Remove the center steering arm bolt. Position a steering arm puller (# 984146) over the 2 dowels inside the arm and then thread in the removal bolt. Tighten the bolt slowly while holding the gimbal ring; when the ring comes free, remove the tool and the steering arm.

8. Lift the thrust washer out of the top of the gimbal housing.

■ If you are simply removing the gimbal ring to get to another component, ignore the following steps if you wish, but we recommend performing them either way—once again, cheap insurance!

9. Remove the O-ring from the upper steering post and throw it away.

10. Insert the large diameter end of a removal tool (# 912281) into either of the tilt bearing holes on the sides of ring and drive out the bearing. Repeat the procedure on the opposite side.

11. Use a slide hammer and puller tool (# 432130) to pull out the trim/tilt cylinder nylon bushing on each side of the lower end of the ring.

12. Insert the smaller end of the removal tool used earlier into the lower steering support shaft bearing and drive out the pivot bearing.

13. If the upper steering post bushing appears to be damaged, use a small chisel and split one side of it until you can remove it.

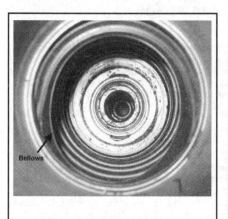

Fig. 236 Disengage the U-joint bellows from the lip and then push it through

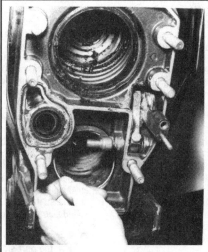

Fig. 237 Next, remove the exhaust bellows retaining clip (except on King Cobras)

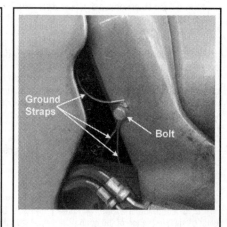

Fig. 238 Later models may have more than one ground strap attached here

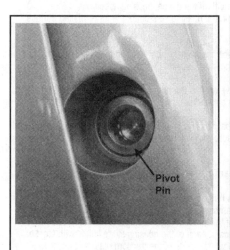

Fig. 239 Find the pivot pin...

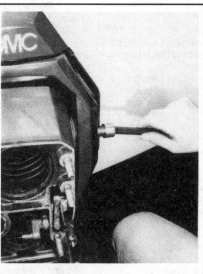

Fig. 240 ...and then remove it with a hew wrench

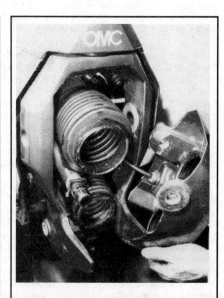

Fig. 241 Removing the pivot housing

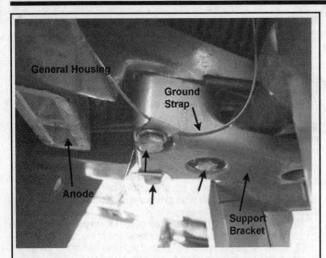

Fig. 242 A good look at the lower steering support (except King Cobra)

To install:

14. Coat the outer surface of a new lower pivot bearing with clean engine oil and position it into the bore on the bottom of the ring. Use the removal tool and drive it into the bore from the bottom of the ring.

15. Coat the outer surface of two lower bushings and then press them into the ring until they seat with the removal tool used earlier. Make sure you coat the inside of each bushing with grease.

16. Coat the outer surface of each new pivot pin bearing lightly with clean engine oil and position the bearings into the outside of the gimbal ring. Insert the blunt end of the removal tool into the bearing and drive it into the ring with a rubber mallet. The bearing must seat somewhere between being flush with the INNER edge of the gimbal ring and 0.010 in. (0.254mm) inside the edge.

17. If you had to cut off the upper steering post bushing, lubricate the inside of a new one with grease and then position it over the post. Thread the center steering arm and bolt in and tighten it until the bushing seats itself. Remove the arm and bolt. Coat a new O-ring with grease and install it over the post and into the recess at the base.

18. Coat the thrust washer with grease and position it over the upper pivot bearing in the top of the gimbal housing. Move the steering arm into position over the washer from inside the boat. Guide the gimbal ring up and into the gimbal housing until the post engages the steering arm and then install the center bolt so that it is finger-tight.

■ A good way to ensure that the ring is centered correctly in the upper bearing is to observe the gimbal studs—when the center bolt is still only finger-tight, the top of the studs should either flush or above the upper surface of the steering arm. If this is case, DO NOT tighten the center bolt or you will damage the bearing.

19. After confirming correct positioning, tighten the center bolt to 64-72 ft. lbs. (88-98 Nm). Position the plate over the arm and studs and then thread on 4 new locknuts. Tighten them to 13-15 ft. lbs. (18-20 Nm).

20. On models using a cap, position a new foam gasket into the cap so that the notches align correctly. Install the cap and tighten the bolts to 60-84 inch lbs. (7-9 Nm). If you have a drive that does not use a cap, run a bead of 3M adhesive along the seal groove and then press in a new sealing ring.

21. Install the gimbal housing cover and tighten the bolts to 60-84 inch lbs. (7-9 Nm). On models with only 2 mounting bolts, make sure that the dowels under the cover align with the holes in the gimbal housing BEFORE tightening the bolts.

22. Working inside the boat, connect the ground strap to the steering arm and tighten the bolt securely. Slide the hydraulic ram assembly over the steering arm so that the holes line up. Coat the steering pin with grease and insert it down through the top of the arm. Install a new cotter pin and bend back the arms around the pin.

23. Coat a new nylon washer and the lower pivot post with grease. Slide the washer between the gimbal ring and housing, ensure that all three holes line up and then slide the pivot post up through the holes until it is seated.

24. Now position the lower steering support bracket and seat it by tapping it a few times with a rubber mallet. Apply a light upward pressure to the bracket and the gimbal ring and then use the thrust washers to estimate

how many you'll need to fit between the upper surface of the bracket and the lower surface of the ring.

■ The support bracket will have lip that protrudes from its upper surface that fits over the pivot pin. This is where the thrust washer estimation must be done, anywhere else will lead to an incorrect estimate and the wrong preload.

25. Add one extra thrust washer to your estimate from the previous step and then slide them into position, never using less than one or more than five.

26. Connect the ground straps to the steering support mounting bolts and then tighten the bolts (with new lock washers) to 18-20 ft. lbs. (24-27 Nm).

27. Adjust the trim sender as detailed in the Trim/Tilt section and then tighten the screws to 18-24 inch lbs. (2-3 Nm).

28. Install the pivot housing and stern drive.

Gimbal Housing And Transom Plate

REMOVAL & INSTALLATION

◆ **See Figures 231, 232, 233, 243 and 244**

■ It is not necessary to remove the pivot housing and gimbal ring when removing the gimbal housing/transom plate assembly, although you may find it easier if you do. Because the housing is attached to the trim lines and pump, you'll need to decide whether to disconnect the lines at the manifold or to disconnect the pump and feed it through— neither way is preferred.

1. Remove the engine, drive unit, lower exhaust pipe and the steering cylinder as detailed in their appropriate sections.

2. Remove the bolt and washer attaching the ground strap to the steering arm (if equipped).

3. Working inside the boat, remove the 2 bolts and 4 nuts (and washers) and then lift the transom plate off the studs. If you are doing anything other than laying the plate aside, disconnect the shift cable sleeve as detailed in the next procedure. On 1990-95 King Cobra models and 1998 7.4L/8.2L models, there are two additional bolts and washers.

■ Remember that the hydraulic lines will be routed ABOVE the upper stud on the port side and through a recess in the inner plate.

4. Pry out the exhaust bore O-ring seal and throw it away.

5. Remove the 2 remaining nuts and their washers and then pull off the transom bearing plate. Lift off the gimbal housing assembly.

6. Pry the large gimbal housing seal out of the inner mating surface recess.

To install:

7. Clean the outer transom-to-gimbal housing mating surfaces. Coat the entire seal groove in the inner edge of the housing with 3M Scotch Grip Rubber Adhesive 1300. Start at the top of the housing and work the seal into the groove as you move around the housing. When you get back to the top, cut the seal to fit and coat each end with adhesive before pressing them together.

8. Install the gimbal housing onto the transom. If you left the pump and lines attached, make sure that you feed them through the cut-out in the transom first.

■ Do not apply sealer between the transom and the housing seal.

9. If the trim lines had been disconnected, reinstall them at this time as detailed in the Disassembly procedure and paying attention to the Note in the Removal portion of this procedure regarding line routing.

10. Coat the threads of each of the housing studs lightly with Gasket Sealing Compound. Coat the inside of the transom plate alignment tube lightly with grease.

11. Make sure that the two hydraulic lines are positioned over the upper port stud and then install the inner transom plate.

12. Position the transom bearing plate over the two lowest studs so the flat side is against the transom. Install the washers and new locknuts;

tightening the nuts until they just make contact with the plate.

13. Install the remaining 4 washers with new locknuts; tightening the nuts until they just make contact with the plate. Don't forget the ground strap on the upper port stud!

14. Coat the threads of the 2 upper alignment bolts lightly with Gasket Sealing Compound and screw them in until they are finger-tight.

15. Tighten the 6 locknuts, in the sequence shown, to 20-25 ft. lbs. (27-34 Nm). Now tighten the 2 upper bolts to 12-14 ft. lbs. (16-19 Nm). On 1998 7.4/8.2L models, use the same torque figures, but refer to the second illustration for the tightening sequence.

16. Install the ground strap to the steering arm on models so equipped and tighten the bolt securely.

17. Run a bead of the 3M Scotch Grip around the exhaust bore seal recess and then install a new O-ring, lightly pressing it into the groove all the way around the bore.

18. Install all remaining components.

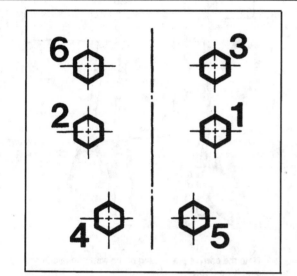

Fig. 243 Transom plate tightening sequence—all except 1998 7.4/8.2L models

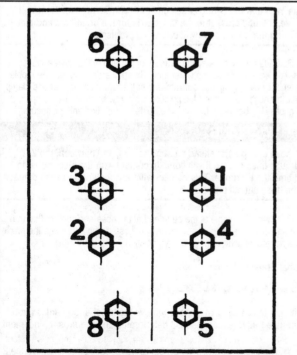

Fig. 244 Transom plate tightening sequence—1990-95 King Cobra models and 1998 7.4/8.2L model

SHIFT CABLE SLEEVE REPLACEMENT

1. Using a crows-foot adapter on your wrench, unscrew the sleeve connector at the transom plate, pull away the sleeve and throw away the O-ring.

2. Coat a new O-ring with Gasket Sealing compound and then slide it onto the connector until it sits against the inner side of the hex head.

3. Coat the connector threads lightly with sealing compound, thread it into the plate and tighten it to 46-52 ft. lbs. (62-71 Nm).

DISASSEMBLY & ASSEMBLY

◆ **See Figures 231, 232, and 233**

Disconnect the battery cables. Remove the trim/tilt cylinders, pivot housing and gimbal ring as previously detailed. Once these procedures are accomplished, move through the following procedures as necessary to remove or disassemble what is necessary.

Disassembly of the gimbal housing really consists of removing those remaining components that are attached to it and we suggest that you perform each of these procedures in the order that they are shown here; reverse the order when assembling everything.

U-Joint & Exhaust Bellows

◆ **See Figures 231, 232, 233, 245 and 246**

■ **1990-95 King Cobra models have no exhaust bellows.**

1. Loosen the hose clamp screws and pull out the ground strap.
2. Pull off the two bellows.
To install:
3. Coat the outer surface of the gimbal bearing flange with Gasket Sealing Compound.

■ **It is not necessary to coat the exhaust flange with sealer.**

4. Slide a clamp ring over the smaller side of the U-joint bellows and then slip the bellows over the bearing flange so that the rib on the inner surface of the bellows end rides in the groove on the surface of the flange.

5. Insert the ground strap under the clamp ring at the 9 o'clock position and then spin the clamp until the screw mechanism is between the 1 and 2 o'clock position; with the screw head facing down. Tighten the clamp ring securely.

6. The larger end of the exhaust bellows also has a rib on the inner side; slide a clamp ring over this end and then push the bellows onto the flange until the rib seats into the groove on the flange. Make sure that the exhaust relief cut-out is facing down.

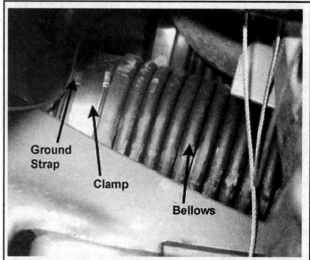

Fig 245 A close view of the bellows

Fig. 246 Make sure that you slide the clamp on first

7. Insert the ground strap under the clamp ring at the 9 o'clock position and then spin the clamp until the screw mechanism is at the 3 o'clock position; with the screw head facing upward. Tighten the clamp ring securely.

Water Hose And Tube

◆ See Figures 231, 232, 233 and 247

1. Remove the exhaust and U-joint bellows.
2. Working through the two hydraulic lines attached to the trim manifold, loosen the water line hose clamp and wiggle the end of the hose off of the water tube in the housing. Don't forget to pull off the ground strap and position it out of the way.
3. Working from inside the boat loosen the 2 (or 3) water tube-to-transom plate mounting bolts and then pull the tube and its grommet out of the housing and through the transom. If you have trouble with the grommet, push it through from the gimbal housing side.
   To install:
4. Check the grommet on the water tube for any cracks or other obvious signs of deterioration or wear. If necessary to replace it, slide a new one onto the tube so that the tapered end is facing the hooked side of the water tube.
5. Push the water tube and grommet through the transom plate and into the gimbal housing so that the hooked end goes in first and is pointing upward. Hold the tube in place and insert the 2 bolts, tightening them to 120-144 inch lbs. (14-16 Nm). On the 1990-95 King Cobra, tighten the 3 bolts to 60-84 inch lbs. (7-9 Nm).
6. Check the water hose for any cracks or other obvious signs of deterioration or wear; pay particular attention to the hose clamp seating areas on each end.
7. Check the nipple in the aft end of the hose; clean it thoroughly and then pull off the O-ring and replace it with a new one. Coat the O-ring with grease and make sure that it seats into the groove closest to the square on the inner side of the nipple. If the nipple requires replacement, simply loosen the hose clamp and pull it from the end of the hose.

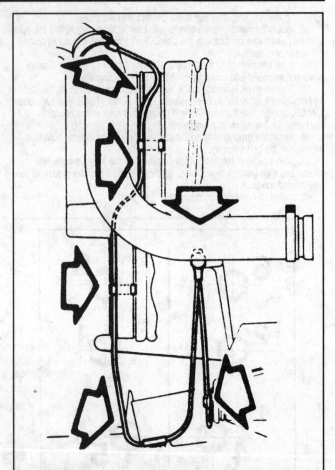

Fig. 247 Note the correct positioning of the water hose and ground strap (except King Cobra)

✳✳ CAUTION

Make sure that the O-ring is seated in the correct groove. The outer groove, the one closest to the threads, is for the drain hole and placing an O-ring into this groove will cause big problems.

8. Slide a new hose clamp over the upper end of the hose and then press the hose end onto the water tube. Swivel the hose so that the nipple (aft) end is pointing directly outward and to the rear. Slide the ground clamp under the hose clamp and then position the screw head on the clamp so that it is accessible between the 2 hydraulic lines before tightening it securely.

✳✳ WARNING

The hose clamp screw head must be behind, and between, the 2 hydraulic lines or else it will come in contact with the pivot housing and limit the tilt range. This contact will, obviously, also cause damage to the hose and clamp.

9. Check the nipple at the outer end of the hose and confirm that the drain hole is facing down. If not, loosen the hose clamp and rotate the nipple in the hose until achieving the correct position.

Gimbal Housing Anode

◆ See Figures 231, 232, 233 and 248

■ It should not be necessary to remove the lower anode unless you determine that it is less than 2/3 deteriorated and requires replacement.

1. Working underneath the gimbal housing, reach up and remove the 2 mounting bolts and their washers. Pull down the anode. 1998 7.4L and 8.2L engines may use two smaller anodes with 4 bolts here.

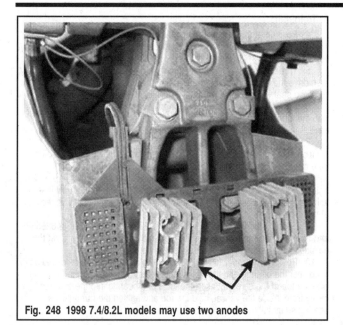

Fig. 248 1998 7.4/8.2L models may use two anodes

2. Clean the mounting surface thoroughly and then install a new anode(s). Use new lock washers and tighten the mounting bolts to 12-14 ft. lbs. (16-19 Nm) except on the 1998 7.4L/8.2L applications using two anodes where they should be tightened to 5-7 ft. lbs. (6-9 Nm).

Trim Sender

◆ **See Figures 231, 232, 233 and 249**

Remember that you removed the trim sender during gimbal ring removal but, did not disconnect it.

1. Working inside the boat, disconnect the trim sender harness connector at the engine.

2. Take note of the wire positioning inside of the plug and then use a socket removal tool (# 322699) and push the wire terminals out of the connector plug.

3. Now, back outside the vessel, locate the harness grommet where it feeds through the transom assembly—just above the water tube in the gimbal housing. Carefully pull off the retaining clip with a pair of needle nosed pliers.

4. Grasp the grommet and pull backward on it while prying it out with a small screwdriver. Once free of the housing, pull the wires all the way through the transom assembly. Cut any plastic ties securing the harness and then remove the trim assembly.

To install:

5. Coat the harness grommet with grease and then feed the leads through the hole in the gimbal housing and press the grommet into place until it is seated. Pop the retaining clip into position over the base of the grommet.

✳✳ WARNING

Make sure that you do not route the harness leads under the extension tube or hydraulic lines.

6. Coat the connector plug with alcohol or solvent and then use the special tool and press the leads back into the connector plug until they seat—make sure they go back in the same positions as they were prior to removal. On MOST models, the ribbed black wire goes into the **A** terminal, the white wire into the **B** terminal and the smooth black wire into the **C** terminal.

7. Reconnect the harness at the engine.

8. Use a new plastic tie and secure the harness to one of the hydraulic lines.

9. Install the sender unit into the gimbal ring when you install the ring.

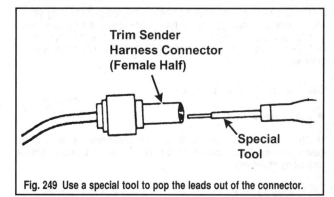

Fig. 249 Use a special tool to pop the leads out of the connector.

Hydraulic Lines & Manifold

◆ **See Figures 231, 232 and 233**

1. Loosen and remove the hydraulic line retaining clamps on each side of the gimbal housing.

2. Although not strictly necessary, we recommend removing the grease extension tube in order to gain access to the starboard hydraulic lines. Clamp a pair of vise grips onto the tube (not too tight please) as close to the gimbal bearing bore as possible. Insert a small prybar between the pliers and the bearing flange and then pry the tube out with the bar while rotating back and forth with the pliers.

3. Loosen each of the hydraulic line nuts at the manifold and carefully pull each line out of its recess. Remove each O-ring and throw them away. It's a good idea to plug the line ends with a golf tee or something similar to minimize fluid leakage.

4. Move inside the vessel and remove the 2 lines connected to the inner side of the manifold. Mark them for installation, plug them and secure them out of the way.

5. Back outside the boat, loosen and remove the mounting nut on the upper side of the manifold from the stud. The manifold is a tight fit in the housing with an O-ring—you may be able to pull it right out, but you'll probably need to position a small drift on the back (inner) side and force it through.

To install:

6. Coat a new manifold O-ring with Gasket Sealing compound and slide it onto the manifold flange. Coat 2 new O-rings for the inner hydraulic lines with hydraulic fluid and slide them over the line tips.

7. Pull the 2 inside lines through the bore in the housing and then position the manifold near the bore so that the inner side is facing to Starboard. Route the lines to the manifold so they are parallel and then tighten the fittings to 14-18 ft. lbs. (19-24 Nm). It is imperative that the lines get reinstalled on the port from which they were detached! As a check, once installed, the lower line should be attached to the **UP** inlet on the trim pump valve body.

■ **Yes, we know it is unlikely that you will be able to get a torque wrench on the line nut—do it by feel, its about as tight as a spark plug, just don't over-tighten them.**

8. Feed the lines and through the hole in the housing until the manifold seats on it's mounting stud. Install a new locknut onto the stud and tighten it to 108-132 inch lbs. (12-15 Nm).

9. Coat 4 new O-rings with hydraulic fluid and install them on each of the outer (gimbal housing side) hydraulic line tips. Attach the lower of the 2 Port lines and then the upper. The shorter of the 2 Starboard lines should be connected to the lower of the two fittings on the starboard side of the manifold and then connect the remaining (longer) line to the upper fitting. Tighten all lines to 14-18 ft. lbs. (19-24 Nm)—see the previous Note.

10. Route each pair of lines together and then install the retaining clamp on each side. Tighten the bolts securely.

11. If you removed the grease extension tube, pull the grease fitting off of the outer end of the tube. Coat the last 1/4 in. of the tube on the fitting side with Gasket Sealing compound and then insert the tube through the outer wall of the housing.

12. Spray the tapered end of the tube and the hole in the housing with Loctite primer and allow it to dry completely. Once dry, coat the tapered end with Loctite and then drive the end of the tube into the bearing flange. Wipe off any excess Loctite and then install the grease fitting to the opposite end of the tube.

Gimbal Bearing/Seal & Steering Bearing

◆ See Figures 231, 232 and 233

■ **Gimbal and steering bearing removal is not necessary to remove the gimbal housing and transom plate. Do not remove these bearing unless absolutely necessary.**

1. Insert a standard 3-jawed puller (Owatonna # OEM-4184 or the equivalent) into the gimbal bearing bore and position the jaws between the bearing and the inner seal. Expand the jaws tightly and then remove the bearing. The bearing will be damaged during removal so throw it away.

2. Reinsert the puller into the bearing bore again and position the jaws under the seal this time. Tighten the jaws and remove the seal.

3. Remove the thrust washer from the steering bearing in the top of the gimbal housing. Install a driver tool (# 912270) into a remover tool (# 912287) and position the tools underneath the bearing. Work the driver and press the bearing up and out of the housing bracket.

To install:

4. Position a new gimbal bearing seal onto the installer tool (# 912279) so the open end is facing the tool. Coat the metal casing with Gasket Sealing compound.

5. Thread the installer (and seal) onto a driver handle (# 311880) and then drive the seal into the bore until it seats. Remove the tools and coat the lip of the seal with grease.

6. Rotate the outer band of the gimbal bearing until the slot is aligned with the lubrication hole and scribe a reference mark across the bearing case.

7. Coat the exterior of the bearing with clean engine oil and then position it into the bore so that the reference mark is facing the extension tube opening.

8. Thread the raised side of a gimbal housing seal installer (# 912279) onto the driver handle (# 311880) and then drive the bearing in until it seats fully. Lubricate the bearing with grease through the grease fitting on the end of the extension tube—starboard side of the gimbal housing.

9. Slide a steering bearing installer (# 912287) onto the rod (# 326582) and then position the bearing over the rod and onto the installer. Coat the outside of the bearing lightly with clean engine oil.

10. Position a stopper (# 912286) over the bearing so the recessed side is against the bearing and the rod is threaded into the tool so that it's flush with the top of the stopper. Insert the tools and bearing into the steering cavity from inside the vessel, hold the rod and tighten the nut until the bearing seats fully.

TORQUE SPECIFICATIONS

Component			Ft. Lbs.	Inch Lbs.	Nm
			Torque		
Anode	Gimbal Housing	All Except	12-14	-	16-19
	Lower Unit		-	60-84	6.8-9.5
Gimbal Ring	Bolts To Pivot Housing		105-120	-	142-163
	Support Bracket		18-20	-	24-27
Lower Unit - DP/DP DuoProp					
	Drain Plug		10-15	-	14-20
	Exhaust Plate		14-17	-	19.5-22.5
	Fill Plug		4-6	-	5.4-8.1
	Pinion Nut		72-87	-	98-118
	Propeller Nut				
		Front	45	-	60
		Rear	50	-	70
	Upper Unit Bolts				
		3/8	22-24	-	30-33
		7/16	32-40	-	43-54
	Upper Unit Nuts		22-24	-	30-33
Lower Unit - SP, SX SX Cobra, King Cobra	Anode		-	60-84	6.8-9.5
	Anode Retainer		12-14	-	16-19
	Drain Screw		-	60-84	6.8-9.5
	Fill Screw		-	60-84	6.8-9.5
	Oil Level Gauge (Dipstick)		-	48-72	5.4-8.1
	Pinion Nut	King Cobra	175-185	-	237-251
		All Others	150-160	-	203-217
	Propshaft Bearing Housing		200-225	-	271-305
	Propshaft Bearing Housing Retainer		20-25	-	27-34
	Propeller Nut		70-80	-	95-108
	Upper Unit Bolts - King Cobra		32-40	-	43-54
	Upper Unit Bolts - SP/SX/SX Cobra				
		3/8-16	22-24	-	30-33
		7/16-14	32-40	-	43-54
	Set Screw		-	42-60	4.8-6.8
	Trim Tab		14-16	-	19-22
	Water Tube Guide		10-12	-	14-16
Manifold Nut			-	108-132	12-15
Pivot Housing	Bolts To Gimbal Ring		105-120	-	142-163
	Drain		-	50-60	6-7
Shift Cable Sleeve	1994-97		46-52	-	62-71
Steering Arm	Bolt		65-72	-	88-98
	Cover		-	60-84	7-9
	Studs		13-15	-	18-20
Transom Plate	Nuts		20-25	-	27-34
	Bolts		12-14	-	16-19
Transom Shield Cover			-	60-84	7-9
Trim Line Nut	1990-95 King Cobra		14-18	-	19-24
	All Others		-	84-108	9.5-12.2
Trim Sender			-	18-24	2.0-2.7

TORQUE SPECIFICATIONS (Cont'd)

Component			Ft. Lbs.	Inch Lbs.	Nm
				Torque	
Trim/Tilt	Pivot Pin		-	120-144	14-16
Upper Housing	Bearing Retainer Ring		145-165	-	197-224
	Bellcrank	King Cobra			
	Drain Screw				
		DP/DP DuoProp	10-15	-	14-20
		All Others	-	60-84	6.8-9.5
	Fill Screw		-	60-84	6.8-9.5
	Impeller Plate		10-12	-	14-16
	Lower Unit Bolts - DP/DP DuoProp				
		3/8	23	-	31
		7/16	36	-	49
	Lower Unit Bolts - King Cobra		32-40	-	43-54
	Lower Unit Bolts - SP/SX/SX Cobra				
		3/8-16	22-24	-	30-33
		7/16-14	32-40	-	43-54
	Lower Unit Nuts - DP/DP DuoProp		23	-	31
	Oil Level Gauge/Plug		-	48-72	5.4-8.1
	Output Gear Housing	King Cobra	140-160	-	190-217
	Pinion Bearing Carrier		12-14	-	6.8-9.5
	Pinion Gear	Cobra/King Cobra	80-90	-	108-122
	Rear Cover		-	108-132	12-15
	Shift Shoe Stop Screw	King Cobra	15-20	-	20-27
		All Others	10-12	-	14-16
	Shift Housing	King Cobra	9-11	-	12-15
	Side Cover	King Cobra	12-14	-	16-19
	Top Cover	King Cobra	32-40	-	43-54
		All Others	16-18	-	22-24
	Upper Driveshaft Nut		96-110	-	130-149
	Water Passage Cover		-	60-84	6.8-9.5
	Water Pump Adapter	Cobra/King Cobra	10-12	-	14-16
	Water Pump Housing	Cobra/King Cobra	9-11	-	12-15
Water Nipple			-	96-120	11-14
Water Outlet	1990-95 King Cobra		-	60-84	7-9
	All Others		-	120-144	14-16

DRIVE MODEL APPLICATIONS

Year	Engine	Model	Model Number	Gear Ratio
1986	2.5, 3.0	Cobra	983841	1.71:1
	4.3	Cobra	983842	1.68:1
	5.0	Cobra	983843	1.50:1
	5.7	Cobra	983844	1.41:1
1987	2.3	Cobra	984540, 985193	2.00:1
	3.0	Cobra	984042, 985194	1.71:1
	4.3	Cobra	984541, 985195	1.68:1
	5.0	Cobra	984542	1.50:1
	5.7	Cobra	984543, 985196	1.41:1
	7.5	Cobra	984544, 985197	1.43:1
1988	2.3	Cobra	985267	2.00:1
	3.0	Cobra	985268	1.71:1
	4.3	Cobra	985269, 985279	1.68:1
	5.0	Cobra	985271, 985272	1.50:1
	5.7	Cobra/King	985273, 985274	1.41:1
	7.4 454	King Cobra	985275, 985276	1.43:1
	7.5 460	King Cobra	984275, 985276	1.43:1
1989	2.3	Cobra	985683	2.00:1
	3.0	Cobra	985684	1.71:1
	4.3	Cobra	985685, 985686	1.68:1
	5.0	Cobra	985687, 985688	1.59:1
	5.7	Cobra	985689, 985690	1.41:1
	5.8	Cobra	985689, 985690	1.41:1
	262	King Cobra	985685, 985686	1.68:1
	350	King Cobra	985689, 985690	1.41:1
	460	King Cobra	985691, 985692	1.43:1
1990	2.3	Cobra	986030	2.00:1
	3.0, 3.0 HO	Cobra	986321, 986322	1.86:1
	4.3, 4.3 HO	Cobra	986032, 986033	1.68:1
	5.0, 5.0 HO	Cobra	986035, 986036	1.59:1
	5.7	Cobra	986037, 986038	1.41:1
	5.7 LE	King Cobra	986043	1.42:1
	5.8	Cobra	986037, 986038	1.41:1
	350, 454, 460	King Cobra	986043	1.43:1
1991	3.0, 3.0 HO	Cobra	986696	1.86:1
	4.3, 4.3 HO	Cobra	986698, 986699	1.68:1
	5.0, 5.0 HO	Cobra	986700, 986701	1.59:1
	5.7, 5.8	Cobra	986702, 986703	1.41:1
	5.7 LE, 350	King Cobra	986704	1.43:1
	7.4GL, 454, 454 HO	King Cobra	986704	1.43:1

DRIVE MODEL APPLICATIONS

Year	Engine	Model	Model Number	Gear Ratio
1992	3.0, 3.0 HO	Cobra	987120	1.86:1
	4.3	Cobra	987122, 987123	1.68:1
	5.0, 5.0 HO	Cobra	987124, 987125	1.59:1
	5.7, 5.8	Cobra	987126, 987127	1.41:1
	5.7/5.8 LE, 351	King Cobra	987526	1.50:1
	7.4GL, 454, 454 HO	King Cobra	987128	1.43:1
	8.2GL, 502	King Cobra	987128	1.43:1
1993	3.0, 3.0 HO	Cobra	987343	1.86:1
	4.3, 4.3 HO	Cobra	987344, 987345	1.68:1
	5.0, 5.0 HO	Cobra	987346, 987347	1.59:1
	5.0 EFI	Cobra	987683, 987684, 988103, 988104	1.59:1
	5.8	Cobra	987999, 988000	1.41:1
	5.8 EFI	Cobra	987683, 987684	1.50:1
	5.8 LE, 351	King Cobra	987351, 987685	1.49:1/1.63:1
	7.4GL, 454, 454 HO	King Cobra	987350, 987351	1.49:1/1.63:1
	8.2GL, 502	King Cobra	987350	1.42:1
1994	3.0GL, 3.0GS	Cobra	3868029, 3868102	1.85:1
	4.3GL	Cobra	3868041, 3868076, 3868107, 3868095	1.66:1
	5.0FL, 5.0Fi	Cobra	3868076, 3868095	1.66:1
	5.7GL	Cobra	3868065, 3868114	1.60:1
	5.8FL, 5.8Fi	King Cobra	3868052, 3868096	1.51:1
	351 EFI	King Cobra	3868129	1.63:1
	7.4GL	King Cobra	3868074, 3868122	1.63:1
		King Cobra	3868121, 3868073	1.42:1
1995	3.0GS	Cobra	3868155	1.85:1
	4.3GL, 5.0FL	Cobra	3868156	1.66:1
	5.0 EFI	Cobra	3868157	1.60:1
	5.8FL, 5.8 EFI	Cobra	3868158	1.51:1
	5.8 HO EFI	King Cobra	3868207	1.43:1
	7.4GL	King Cobra	3868212	1.63:1
	7.4 EFI	King Cobra	3868211	1.42:1
1996	3.0GS	SP	3868391	1.97:1
	4.3Gi, 4.3GL, 4.3GS	SP	3868464	1.85:1
		SP	3868464	1.85:1
		SP	3868389	1.66:1
		DP	3868440	2.32:1
	5.0Fi	SP	3868389	1.66:1
		SP	3868388	1.60:1
		DP	3868441	1.95:1
	5.0FL	SP	3868389	1.66:1
		DP	3868441	1.95:1

SP Single Prop
DP Duo Prop

DRIVE MODEL APPLICATIONS

Year	Engine	Model	Model Number	Gear Ratio
1996 (cont'd)	5.7GL	SP	3868388	1.60:1
		DP	3868441	1.95:1
	5.7Gi, 5.8Fi, 5.8FL	SP	3868387	1.51:1
		DP	3868441	1.95:1
	5.8FSi	SP	3868387	1.51:1
		DP	3868441	1.95:1
		DP	3868442	1.78:1
	7.4Gi	SP	3868386	1.43:1
		DP	3868443	1.68:1
	7.4GL	SP	3868386	1.43:1
		DP	3868442	1.78:1
1997	3.0GS	SP	3868391	2.18:1
	4.3Gi, 4.3GL, 4.3GS	SP	3868464	1.85:1
		DP	3868389	1.66:1
		DP	3868440	2.32:1
			3868441	1.95:1
	5.7GL	SP	3868388	1.60:1
		DP	3868441	1.95:1
	5.7Gi, 5.7GS	SP	3868387	1.51:1
		DP	3868441	1.95:1
	5.7GSi	SP	3868387	1.51:1
		DP	3868386	1.43:1
		DP	3868441	1.95:1
	7.4Gi	SP	3868386	1.43:1
		DP	3868443	1.68:1
	7.4GL	SP	3868386	1.43:1
		DP	3868442	1.78:1

SP Single Prop
DP Duo Prop

DRIVE MODEL APPLICATIONS

Year	Engine	Model	Model Number	Gear Ratio
1998	3.0GS	SP	3868814	2.18:1
			3868722	1.97:1
	4.3GL	SP	3868721	1.89:1
			3868720	1.79:1
		DP	3868688	2.32:1
			3868708	1.95:1
	4.3GS, 4.3Gi	SP	3868720	1.79:1
		DP	3868688	2.32:1
			3868708	1.95:1
	5.0GL	SP	3868856	1.60:1
		DP	3868708	1.95:1
	5.7GS	SP	3868718	1.51:1
			3868717	1.43:1
		DP	3868708	1.95:1
	5.7GSi	SP	3868717	1.43:1
		DP	3868708	1.95:1
	7.4Gi	SP	3868709	1.78:1
			3868717	1.43:1
		DP	3868708	1.95:1
			3868709	1.78:1
	7.4GSi	DP	3868710	1.68:1
			3868709	1.78:1
	8.2GSi	DP	3868710	1.68:1
			3868709	1.78:1
			3868710	1.68:1

SP Single Prop
DP Duo Prop

13

TRIM AND TILT

SELECT (POWER) TRIM AND TILT—1986-88 COBRA AND KING COBRA

Description

◆ See Figure 1

■ **Not all models use all components in all years.**

This power trim/tilt system is a hydro-mechanical, electrically controlled system allowing the stern drive to be raised when moving through the water at idle speed, when approaching the beach or shore, and when trailering the boat. The drive can also be adjusted through-out its trim range while the boat is underway and at any speed. Thus enabling the helmsman to maintain the best bow position for maximum speed, efficiency and ride for any water or throttle conditions.

An electric motor, controlled by a switch and relays, provides the power for a hydraulic pump. The pump, in turn, forces fluid through a set of hydraulic hoses to a set of hydraulic cylinders. Depending on the motor's direction, fluid will flow into one end of the cylinders to raise or lower the drive unit.

A control switch may be mounted on the remote control handle or on the control panel. When activated, the **UP** or **DOWN** relay is energized. Current to the motor rotates the motor in one of two directions, thus raising or lowering the drive unit.

The power trim/tilt system consists of a valve body (manifold), hydraulic reservoir, electric motor, up and down solenoids, trim/tilt switch, trim sending unit, trim indicator gauge, trim limit switch, two hydraulic trim cylinders, and the necessary hydraulic lines and electrical harnesses for the system to function efficiently.

The manifold, hydraulic reservoir, electric motor, and up/down solenoids make up a single unit known as the pump assembly that is mounted inside the boat. Two hydraulic trim cylinders are mounted, one on each side, of the stern drive unit. One end of the cylinder is connected to the gimbal housing and the opposite end is attached to the drive housing.

The trim sender unit is mounted on the starboard side of the gimbal ring and is connected to the trim gauge on the control panel through the wiring harness. When the stern drive is trimmed **IN** or **OUT** the gauge moves to indicate the stern drive's position.

A trim limit switch may be connected to the **UP** solenoid on the trim/tilt hydraulic pump. This limit switch allows the stern drive to be trimmed out only to a preset position. If the stern drive is moved out past the support brackets under high throttle setting could impose high stress loads on the boat transom and possibly loss of steering control. There may also be a 10 amp circuit breaker or fuse serving the same purpose.

The relationship of the various units in the trim/tilt system are discussed in the following paragraphs.

VALVE BODY/MANIFOLD

◆ See Figure 1

The valve body, or manifold, is the central mounting point for the pump, valves, reservoir and electric motor. The hydraulic pump is capable of creating pressure up to 3000 psi. Internal pressure relief valves set at the factory control the output pressure in the UP and DOWN circuits. A system thermal relief valve protects the pump, valves, lines, and cylinder packing from thermal expansion of the hydraulic fluid in the system when operating or sitting in the hot sun. The pump, pressure and thermal relief valves are all set at the factory and adjustments are not possible. If any of these components are found to be defective during troubleshooting or disassembly, they must be replaced with identical items. A replaceable filter is mounted on the inlet opening to the pump.

All models utilize a manual release valve. If the pump or motor fails, manually raising or lowering the stern drive would require the appropriate hydraulic line be loosened on the manifold external fitting.

RESERVOIR

All models utilize an integrated reservoir. A large screw-on cap and opening provides for easy servicing of the reservoir. The cap vents the reservoir to atmosphere pressure when the fill neck seal is removed inside the cap.

HYDRAULIC PUMP

The hydraulic fluid pump (valve body) is a gear- type pump, very similar to an engine oil pump. The pump is attached to the manifold with four fasteners and cannot be disassembled. If the pump fails, it must be replaced as an assembly. The reservoir must be drained and removed to gain access to the pump.

MANUAL RELEASE VALVE

◆ See Figure 2

The manual release (or relief) valve is used on all models and is located on the base of the manifold. This valve allows the drive unit to be raised or lowered manually without the use of electric or hydraulic power. By rotating the valve counterclockwise, the drive unit will lower under its own weight; closing the valve will stop the process. The valve must be fully open to raise the unit, and it must be tightly closed in order to hold a trim or tilt position while static or moving forward.

TRIM-IN/TILT-DOWN CHECK VALVE

◆ See Figure 1

The trim-in/tilt-down check valve is located in the valve body. This valve opens under pressure to allow hydraulic fluid from the pump to the aft chambers during the Down cycle. The aft chamber with the rod is identified as the chamber behind the cylinder.

The valve also allows fluid to return from the aft chambers to the pump during the Up cycle, after the valve has been opened by the pump control piston.

In addition to the above, the valve prevents the drive unit from raising during operation in Reverse, acting as a reverse lock; and prevents fluid flow from the high volume side of the cylinders during a high pressure impact condition.

PUMP CONTROL PISTON

◆ See Figure 1

The pump control piston is located directly behind the trim-in/tilt-down check valve inside the valve body. A spring at both ends of the piston tends to keep the piston centered in the valve body bore. The piston is moved via pressure acting on the trim-in/tilt-down check valve as the drive is moved UP or DOWN.

TRIM-OUT/TILT-UP CHECK VALVE

◆ See Figure 1

This valve is located externally at the bottom of the valve body. It opens under pressure to allow fluid to flow from the forward chambers (in front of the piston) back to the pump. The valve serves to maintain oil pressure in both cylinders so that the drive unit will hold a particular trim or tilt position while moving forward. It will return fluid to the pump during the Down cycle after having been opened by the pump control piston.

TILT-UP RELIEF VALVE

◆ See Figure 1

This valve is located externally on the back of the valve body. It functions to control maximum pressure in the circuit during the UP cycle. It also limits pressure at the full tilt-up 'stall' position to a maximum of 1200-1600 psi.

TRIM-IN RELIEF VALVE

◆ See Figure 1

This valve controls pressure in the system during the DOWN cycle. Due to rod volume, more fluid returns from the high volume side of the trim cylinders than the low volume side can handle; excess fluid passes through the valve which will keep pressure on the Down side at 400-800 psi.

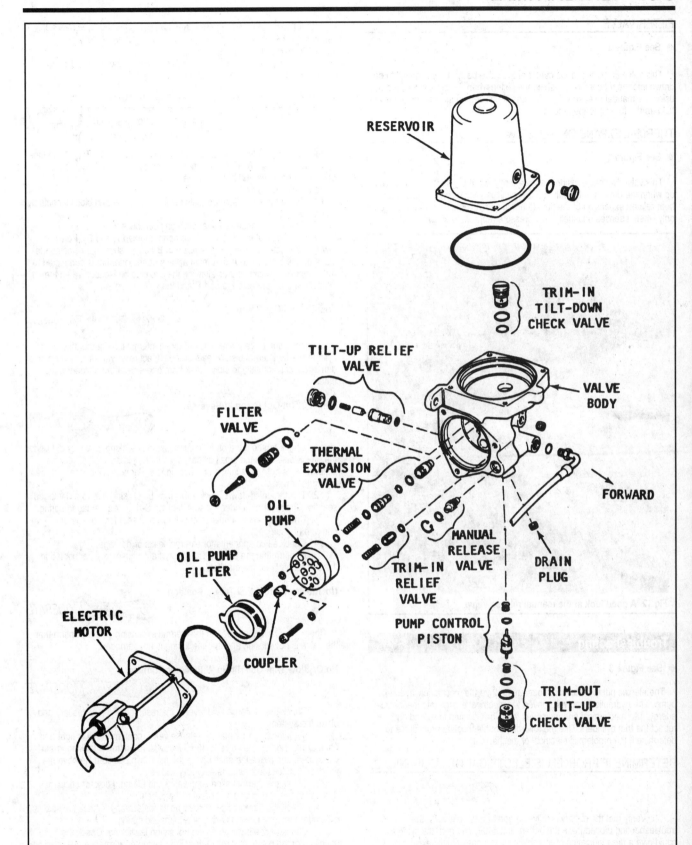

Fig. 1 Exploded view of the trim/tilt motor and pump assembly—most models similar

FILTER VALVE

◆ **See Figure 1**

This valve is located in the center of the valve body. It permits fluid to be drawn into the aft chambers behind the piston when the drive is lowered by using the manual release valve. The check ball behind the filter prevents any fluid from returning to the reservoir.

THERMAL EXPANSION RELIEF VALVE

◆ **See Figure 1**

This valve is located inside the valve body and it prevents hydraulic lock-up when the drive is in the full TILT UP position, which can be caused by heat-related expansion of the fluid (thermal expansion). The valve will open only when it senses unusually high pressure (approx. 2250-2750 psi).

Fig. 2 A good look at the manual release valve

Troubleshooting

◆ **See Figure 3**

The electric pump motor must operate to drive the mechanical hydraulic pump. The hydraulic pump will convert the electrical energy into mechanical energy. The mechanical energy moves the hydraulic fluid to extend and retract the trim cylinders. Use the following troubleshooting procedures to determine if the problem is electrical or mechanical.

DETERMINE IF PROBLEM IS ELECTRICAL OR MECHANICAL

1. Verify that the battery is fully charged before continuing the troubleshooting procedure. If the battery is not fully charged, the system could give a false indication of a problem when none actually exists.
2. Verify the fluid level in the reservoir is up to the **MAX** mark on the side of the reservoir; or the bottom of the filler hole on earlier models.
3. Depress the **UP** or **DOWN** button on the control panel and listen for the pump motor to operate or the solenoid to click.
4. If the pump motor does not operate, and the solenoids do not "click", the problem is electrical.

5. If the pump motor operates, but the stern drive does not move, the problem is most likely a hydraulic or mechanical problem. Refer to hydraulic/mechanical troubleshooting.
6. On early models, check the manual release valve— it MUST be closed and tightened to 45-55 inch lbs. (5-6 Nm).

HYDRAULIC AND MECHANICAL PROBLEMS

All testing procedures should be performed using a OMC Trim/Tilt In-Line Pressure Tester (# 983977) consisting of a pressure gauge, adapter, tester body, shut-off valve and hydraulic line.

To install the tester:
1. Raise the stern drive up to its full tilt-up position and block it securely with wood.
2. Open the manual release valve one complete turn.
3. The body of the tester has two ports (marked **A** and **B**). Port **A** is always connected to the gauge, while port **B** can be shut off by closing the valve on the tester. The **A** port must always be the one that is connected to the component or circuit thus allowing the circuit to be isolated when the vessel's line is connected to port **B** and the valve is closed.

Trim-In/Trim-Out Creep

This condition is frequently caused by an internal leak at the trim cylinders—the internal piston seals or the impact relief valves may be dirty or damaged, causing fluid to slowly leak from one side of the piston to the other.

Trim-Out/Tilt-Out Creep In Reverse

1. Check to see if the pump control piston is sticking. Look for dirt under the check balls or in the small oil passages in the piston itself. Check the piston and its bore for scuffing or other signs of damage. Check the ball seat for signs of wear.
2. The trim-in/tilt-down check valve may be sticking due to a damaged seal tip. Check for dirt under the seal, leaking O-ring or a damaged spring.
3. Check the manual release check ball and seat for leakage or damage (if equipped).
4. Check all components for external leaks or damage.
5. Confirm that the UP and DOWN hydraulic hoses are connected to the correct ports on the manifold.

No Tilt-Down From Full Tilt-Up Position

Check the O-ring and seal on the thermal expansion valve. Confirm that the valve is not sticking and that the spring is not damaged.

No Or Slow Trim-Out/Tilt-Up With Motor Running

1. Check the hydraulic fluid level in the reservoir. Fill if necessary and bleed the system.
2. If equipped with a manual release valve, confirm that it is tight and not leaking. Tighten the valve to 45-55 inch lbs. (5-6 Nm). If leaking, drain the system and move the drive unit to the full Down position. Remove the valve and check the plastic seal, O-ring and seat.
3. Check the thermal expansion valve and O-ring. Look for sticking, damage to the spring or dirt under the check ball.
4. Check the trim-out/tilt-up check valve for damage. Check the seal tip, dirt under the seal, leaking O-ring or a damaged spring.
5. Check to see if the pump control piston is sticking. Check the condition of the piston and the O-ring. Poor piston performance may also be affected by an internal leak in the trim cylinders—the cylinder piston seals or impact relief valves may be damaged.
6. The trim-in/tilt-down check valve may be sticking due to a damaged seal tip. Check for dirt under the seal, leaking O-ring or a damaged spring.
7. Check the tilt-up relief valve for sticking, damaged spring or a dirty seal tip. Check that the manifold O-ring is not distorted.

Power Trim Diagnostic Chart
1986-88

Symptom/Problem	Possible Cause
Static leak-down and/or trim-in creep in Forward gear	③ ⑤ ④ ⑥ ⑰
Trim-out/tilt-upcreep in Reverse gear (no reverse lock)	⑧ ⑨ ⑥ ⑰
Leaking, Up and Down (trim-in/out creep)	⑦
Slow or no trim-out/tilt-up while motor running	① ③ ④ ⑤ ⑥ ⑦ ⑧ ⑪ ⑬ ⑭
No trim-out/tilt-up, motor does not run	⑯ ⑬ ⑭ ⑮
Slow or no trim-in/tilt-down while motor running	① ⑤ ⑥ ⑧ ⑨ ⑩
No trim-in/tilt-down, motor does not run	⑯ ⑬ ⑭ ⑮
No trim/tilt Up or Down, motor runs	① ⑬ ⑭
No trim/tilt Up or Down, motor runs fast	⑫
No trim/tilt Up or Down, motor does not run	⑬ ⑭ ⑮ ⑯
No tilt-down from full tilt-up position, motor stalls	④
Erratic operation, jumpy motion, pulsing/hissing fluid sounds	②
Continual use or loss of hydraulic fluid	⑰

① Air or low/no oil in reservoir. Fill reservoir and cycle pump
② Air in hydraulic system. Cycle pump through range 3-5 times. Fill reservoir
③ Manual release valve loose or leaking. Check seal ring, O-ring and seats. Tighten valve to 45-55 inch lbs. (5-6 Nm).
④ Thermal expansion valve sticking or damaged spring.Dirt under check ball.
⑤ Trim-out/tilt-up check valve sticking, damaged seal tip, dirt under seal, leaking O-ring, damaged spring
⑥ Pump control piston sticking, dirt under check balls or in small oil passages.
⑦ Trim/tilt cylinder internal leak. Check for damaged piston seal impact relief valves, or dirt under relief valves
⑧ Trim-in/tilt-down check valves sticking. Damaged seal tip, dirt under seal, leaking O-ring or bad spring
⑨ Manual release check ball leaking. Check for damaged or dirty ball or seat.
⑩ Trim-in relief valve leaking. Check seal tip, seat or spring. Check for sticking or dirt.
⑪ Tilt-up relief valve leaking. Check sticking valve, seal tip, seal or spring. Check valve body O-ring.
⑫ Pump coupling broken.
⑬ 10 or 50 amp circuit breaker tripped
⑭ Motor inoperative, replace motor.
⑮ Relay control box malfunction. Replace box.
⑯ Switch circuit malfunction.
⑰ External oil leak. Check all lines, connections, cylinders and valve body.
⑱ Pump-to-manifold oil line attachment reversed.

8. Check the condition of the oil pump. Dirt can be in the pump or the coupling may have failed.

9. Check that the motor output is up to specification.

No Or Slow Trim-Out/Tilt-Down With Motor Running

1. Check the hydraulic fluid level in the reservoir. Fill if necessary and bleed the system.

2. Check the trim-out/tilt-up check valve for damage. Check the seal tip, dirt under the seal, leaking O-ring or a damaged spring.

3. Check to see if the pump control piston is sticking. Check the condition of the piston and the O-ring. Poor piston performance may also be affected by an internal leak in the trim cylinders—the cylinder piston seals or impact relief valves may be damaged.

4. The trim-in/tilt-down check valve may be sticking due to a damaged seal tip. Check for dirt under the seal, leaking O-ring or a damaged spring.

5. The filter valve, or the check valve behind it, may be damaged or leaking.

6. Check the trim-in relief valve for sticking, damaged spring or a dirty seal tip.

7. Check the condition of the oil pump. Dirt can be in the pump or the coupling may have failed.

8. Check that the motor output is up to specification.

No Trim/Tilt, Motor Will Not Run

More times than not this condition is caused by an electrical problem in the switch circuit, relay/solenoid circuit, or with the motor. Please refer to the Electrical Problems section.

No Trim/Tilt, Motor Runs

1. Check the hydraulic fluid level in the reservoir. Fill if necessary and bleed the system.

2. If equipped with a manual release valve, confirm that it is tight and not leaking. Tighten the valve to 45-55 inch lbs. (5-6 Nm). If leaking, drain the system and move the drive unit to the full Down position. Remove the valve and check the plastic seal, O-ring and seat.

3. Check the thermal expansion valve located in the end of the manual release valve. Look for sticking, damage to the spring or dirt under the check ball.

4. Check the condition of the oil pump. Dirt can be in the pump or the coupling may have failed.

ELECTRICAL PROBLEMS

■ Please refer to the electrical schematics later in this section.

Before performing any electrical tests, always verify that the battery has a full charge. A low battery can always give a false indication that some other problem in the system exists.

Check the circuit breakers in the main engine electrical box and in the trim/tilt relay control box have not been tripped for some reason. Push in the button and reset the breaker.

Trim/Tilt Switch

During switch operation, electrical current flows from the input terminal (red/purple wire) to the respective output terminals.

1. Check all electrical connections. Check the trim/tilt fuse or circuit breaker and the trim/tilt pump fuse/breaker. Check the battery.

2. Connect a test light or a multi-meter as per the manufacturer's instructions and touch the probe to the red/purple wire terminal. The tester should light or the meter should indicate battery voltage; if not, recheck the fuses or circuit breakers and/or check any related electrical connections.

3. Move the probe to the green/white wire terminal and move the switch to the Down position. The tester should light or the meter should indicate battery voltage; if not, confirm that all electrical connections are OK and then replace the switch.

4. Now move the probe to the blue/white wire terminal and move the switch to the Up position. The tester should light or the meter should indicate battery voltage; if not, confirm that all electrical connections are OK and then replace the switch.

5. If all test prove to be good and the pump will still not operate, refer to the Pump Relay test later in this section.

Trim/Tilt Motor

◆ See Figures 4 and 5

A stall-torque test is the only reliable means of checking motor performance. Trim/tilt motors are not serviceable and must be replaced as a unit.

✳✳ WARNING

Never operate the trim/tilt motor for any longer than it takes to complete the following tests. Always allow the motor to cool sufficiently between each test. Performing the two successive tests on a hot motor will lead to inaccurate stall-torque readings.

1. Check all electrical connections. Check the trim/tilt circuit breaker or fuse and the trim/tilt pump fuse/breaker. Check the battery.

2. Remove the pump/motor assembly and then remove the motor from the valve body. Install it carefully in a soft-jawed vise.

3. Connect an ammeter or multi-meter in series with the **B+** terminal on the battery (make sure it's fully charged!) and the blue lead at the motor. Next connect a voltmeter (or DVOM) so the positive lead is at the motor's blue lead and the negative lead is at the motor's green lead.

4. Install an inch lb. torque wrench with a suitable adapter to the armature shaft on the pump. Have someone study the wrench while you move to the next step.

5. Now complete the circuit by running a jumper wire between the negative battery cable and the green lead and then take note of all of the readings. The ammeter should show 185 amps, the voltmeter should show 12 volts (min.) and the torque wrench should show a stall-torque of not less than 20.6 inch lbs.

6. Disconnect the 2 meters and the jumper and then reconnect the ammeter in series between the battery positive terminal and the green lead at the motor.

7. Next, make sure that the motor has cooled down and then complete the circuit by connecting the jumper wire between the negative terminal on the battery and the blue lead. This will allow you a stall-torque reading in the opposite direction; both the stall-torque reading and the ammeter reading should be the same as in the earlier step.

8. A low current reading coupled with a low torque reading indicates high resistance in the internal connections or in the brush contacts. Conversely, a high current reading and a low torque reading indicates a bad armature or field windings. Either way, the motor will have to be replaced.

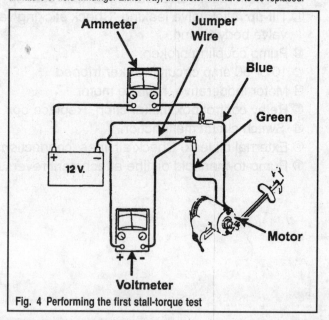

Fig. 4 Performing the first stall-torque test

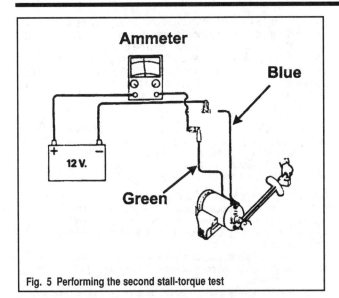

Fig. 5 Performing the second stall-torque test

Trim Sending Unit

1. Check all electrical connections. Check the trim/tilt circuit breaker or fuse and the trim/tilt pump breaker/fuse. Check the battery.
2. Tag and disconnect the sending unit harness at the connector.
3. Install an ohmmeter, or multi-meter, as per the manufacturer's instructions. Connect the probes to the 2 sender leads and then move the sender arm through a full Up and Down cycle while observing the meter.
4. If the meter does not read 0 to 190-210 ohms through out the cycle, the unit must be replaced.

Service Precautions

The following nine points should always be observed when installing, testing, or servicing any part of the power trim/tilt system.
1. Coat the threads of all fittings and O-rings with Power Trim/Tilt or Steering Fluid, or an equivalent product.
2. Use only clean, new, Power Trim/Tilt and Steering fluid when filling the hydraulic fluid reservoir. If the Power Trim and Steering fluid is not available, substitute with SAE 10W-30 or 10W-40 motor oil in an emergency and then drain the system as soon as possible.
3. The stern drive must be in the full UP/OUT position to relieve hydraulic pressure when checking the fluid level or adding fluid to the reservoir. Normal fluid level should be maintained between the **MIN** and **MAX** marks on the side of the reservoir. If the quantity is low, add fluid until the level is up to the **MAX** mark on the side of the reservoir. Obviously, on models without the plastic reservoir, the level should be to the bottom of the filler hole.
4. If the pump stops during long use, allow the pump motor to cool at least five minutes before starting again. An internal thermal circuit protects the pump motor from overheating. If the pump still will not operate, the system breaker may have tripped or the fuse may have blown.
5. Keep the work area clean when servicing or disassembling parts. The smallest amount of dirt or lint can cause failure of the pump to operate.
6. The manifold, pump assembly, pressure relief valves, and thermal relief valves must be replaced as individual assemblies. There are no piece parts available for these components except for the O-rings that are sold in a kit.
7. The motor is protected from internal fluid leaks and external moisture by seals, O-rings, and a grommet on the field winding harness. If the motor fails due to hydraulic fluid, clean the motor thoroughly and replace the pump internal seal and O-ring. If the motor fails from moisture as evidenced by internal corrosion, the wire harness grommet is damaged or deteriorated, and the pump motor should be replaced. Sometimes liquid neoprene can be used on the wire harness grommet to seal minor cracks. Use liquid neoprene with care to prevent contaminating the electric motor or hydraulic valves.

8. Keep the trim cylinders attached to the forward anchor pin during servicing and repairs. Do not allow the trim cylinders to hang by their hydraulic hoses. Such practice may kink and damage the hoses. Use care when handling trim cylinders during removal/installation of the stern drive unit. Rough treatment of the hoses could result in a weakened hose, partial separation at the fittings, or bending of the metal tubing, any one of which could restrict the flow of fluid to the trim cylinders.
9. Always hold both fittings with a wrench when tightening or loosening hydraulic hoses and fittings. When installing flex hoses, be sure there are no twists or sharp bends in the flex hose.

System Testing

High pressure testing of hydraulic components requires expensive special gauges, tools and highly trained personnel to accurately interpret test results. A danger to the operator and others in the area always exists during the testing process. Therefore, it is highly recommended that the boat, with the trim/tilt system installed, be taken to a qualified repair facility having the necessary equipment to conduct the testing properly and safely.

Do not attempt to work on hydraulic hose connections without the proper tools designed for that specific purpose. The use of a common box end wrench will quite likely result in "rounding off" the flats on a hydraulic fitting. It is imperative that you have a set of "flare" or "line" wrenches that will almost guarantee the fitting will not to be damaged.

System Bleeding

TRIM CYLINDERS

Before installing a new or rebuilt trim cylinder:
1. Compress the trim cylinders so they are in their fully retracted (Down) position.
2. Unscrew the end caps on each cylinder (use the special spanner wrench as detailed in the Disassembly & Assembly procedure) and fill the low-volume chamber with OMC Power Trim/Tilt & Steering Fluid. Install the end caps and tighten to 25-30 ft. lbs. (34-41 Nm).
3. Install the cylinders and connect the hydraulic lines.

TRIM MOTOR & PUMP

■ **Please refer to the Maintenance section for procedures on filling the pump reservoir.**

Verify that the manual release valve is in the fully closed position and that the drive is in the full **UP** position.

✱✱ WARNING

Never remove the pump reservoir fill plug unless the drive is in the full UP position.

Remove the reservoir plug and fill the reservoir with fluid until the level reaches the bottom of the hole. Replace the plug. Run the drive unit through at least 5 Up/Down cycles to purge any residual air in the system, stop the drive in the full Up position and check the fluid level one last time; topping off if necessary.

Trim Sender

ADJUSTMENT

■ **Before adjusting the trim sender, the sender must be installed in the gimbal ring and the drive must be installed on the boat.**

1. Turn the drive over hard to port and set it to full negative (Down) trim. Turn the ignition switch to the **ON** position; if the trim gauge shows the drive to be in the Full Down position there is no need to proceed with the adjustment, if not, move to the next step.

2. Loosen the 2 trim sender mounting screws, turn the ignition switch back to the **ON** position and then slowly rotate the sender until the trim gauge registers a Full Down position. Obviously, you'll need an assistant for this step unless you've got a long neck and great eyes.

3. Tighten the sender mounting screws to 18-24 inch lbs. (2.0-2.7 Nm) and then recheck the gauge.

Trim/Tilt Cylinder

■ If troubleshooting procedures have isolated a problem to the trim/tilt cylinders, for example: a leaking oil scraper seal around the rod or a defective impact valve, it is strongly recommended that the cylinder(s) be removed and replaced with new ones, rather than attempting to disassemble the cylinders. This recommendation is based on the fact considerable difficulty may be encountered in removing the end cap from the cylinder. Even with the aid of the special tool for this purpose, the task is most difficult.

REMOVAL & INSTALLATION

◆ See Figures 6 thru 12

 MODERATE

1. Remove the plastic cap on the aft (drive) end of each trim/tilt cylinder. Remove the elastic lock nut and flat washer.

2. Support the hydraulic cylinder and lightly tap the pivot rod out of the cylinders with a rubber mallet—tap on one side while pulling on the other side. Carefully lower the cylinders down.

■ If the bushings come out with the pivot rod, make sure that you keep the two grounding clips.

3. Grab the pivot ends of each cylinder and pull them out to the fully extended position. Loosen the manual release valve on models so equipped and then remove the filler plug at the reservoir to relieve any residual pressure in the system. Close the valve and install the cap.

4. Working at the forward end of the cylinder, unscrew the bolt and disconnect the ground wire. Remove the cylinder cover retainer and lift off the cover. These models will also have a small screw and line retainer over the hydraulic line running to the aft end of the cylinder; remove them.

5. With the stern drive still in the full DOWN position, place a suitable drain pan under the hydraulic lines, tag and then disconnect the two hydraulic hoses from the end of the cylinder (cover with rag) using the correct size "Flare Nut" or "Line Wrench". These wrenches will prevent damaging the hex flats on the line or hose fittings if they are extremely tight and/or a slight amount of corrosion has built up on the fitting. Most standard open-end wrenches will flex under high torque loads, causing the wrench to slip, damaging the hex fitting on the hydraulic line.

6. Remove the O-rings on the end of each line fitting. Install plugs into the hoses and/or fittings to prevent draining the trim/tilt hydraulic system any more than necessary.

7. Pry off (or unscrew) the plastic pivot pin cap (if equipped) on the end of the forward pin. Remove the locknut from the end of the pin and then slide off the flat washer and then the bushing from the pivot pin.

8. Pull the pivot pin out of the cylinder and gimbal ring, or grasp the cylinder on the inside of each pin and pull it off the anchor pins. Remove the flat washer and bushing on the inside surface of the cylinder ends or the anchor pins.

To Install:

■ Certain models may come equipped with limited tilt cylinders. They will contain a spacer that reduces total tilt rod travel and can be identified by the letter L that is stamped on the end of the tilt rod eye. Limited tilt cylinders must always be installed in pairs and never mixed with standard rods providing full travel.

9. Press the trim/tilt cylinder bushing into the boss on each side of the gimbal ring. Install bushings into the outer side of each cylinder. Although it is not necessary to replace the bushings unless they are worn excessively or damaged, you may want to consider replacing them anyway while you have everything apart.

10. Clean the forward pivot pin and then coat it with grease. Apply a light coat of Gasket Sealing Compound to the threaded ends of the pin and then center it in the ring.

11. Align the cylinder and then press it on to the pivot pin very carefully. You may have to give it a few taps with a rubber mallet.

12. Install the washers and then the lock nuts (new!). Tighten the nuts until there is an equal amount of thread exposed on each side of the pin and then tighten both nuts to 32-34 ft. lbs. (43-46 Nm). Replace the plastic caps.

13. Coat new O-rings with hydraulic fluid and then install them into the grooves on the end of each hydraulic line fitting. Remove the plugs and attach each line to its respective fitting on the cylinder. Tighten each fitting to 14-18 ft. lbs. (19-24 Nm) with a flare wrench.

14. Install the aft line retainer and tighten the bolt securely. Also on these models, position and install the cylinder cover.

15. Install the retainer; and then on all models, install the ground strap and tighten the bolt securely.

16. Clean the aft pivot pin thoroughly and then coat it with grease.

17. Make sure that the grounding clips are in position and then press the bushings into the end of each cylinder and into the bore boss in the drive unit. As we did with the forward end bushings, we suggest replacing the bushing as preventative maintenance, although it is perfectly acceptable to reuse the old bushing as long as they are in good shape.

18. Align the cylinder ends with the bores on each side of the drive and then insert the pivot pin into one end and press it through until it comes out the other side—you will probably have to give it a few taps with a mallet.

19. Install the flat washers and lock nuts. Tighten the nuts until there is an equal amount of thread showing on each side and then tighten each nut to 32-34 ft. lbs. (43-46 Nm). Install the plastic caps.

20. Fill the reservoir with Power Trim/Tilt & Steering fluid and run the drive through its full range of motion at least 5 times to purge any air in the system.

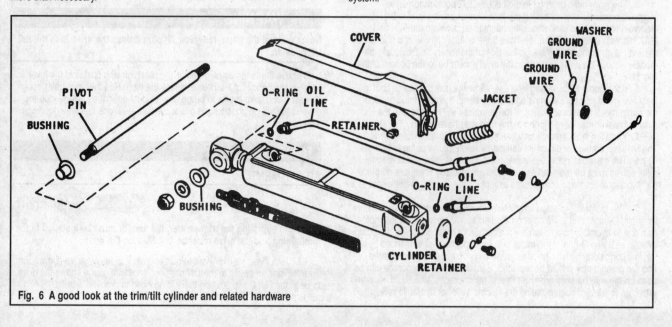

Fig. 6 A good look at the trim/tilt cylinder and related hardware

Fig. 7 Removing the aft nuts

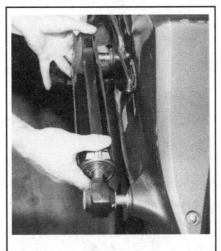

Fig. 8 Carefully pull the cylinder off of the drive...

Fig. 9 ...and then remove the rear nut

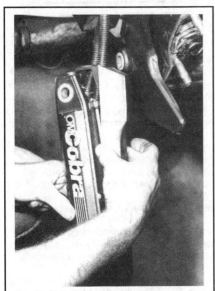

Fig. 10 Pull of the cylinder cover...

Fig. 11 ...remove the retainer...

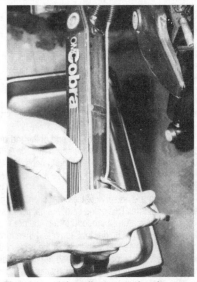

Fig. 12 ...and then disconnect the aft hydraulic line

DISASSEMBLY & ASSEMBLY

◆ See Figures 6 thru 19   OEM ③ DIFFICULT

■ If troubleshooting procedures have isolated a problem to the trim/tilt cylinders, for example: a leaking oil scraper seal around the rod or a defective impact valve, it is strongly recommended that the cylinder(s) be removed and replaced with new ones, rather than attempting to disassemble the cylinders. This recommendation is based on the fact considerable difficulty may be encountered in removing the end cap from the cylinder. Even with the aid of the special tool for this purpose, the task is most difficult.

1. Remove the trim cylinder(s) as previously detailed.
2. Hold the cylinder over a drain pan with the hydraulic line ports facing down and into the pan; extend, and then retract the cylinder 2-3 times to remove all fluid from the cylinder.
3. Remove the pivot pin bushings if you intend to replace them.
4. Place the cylinder in a vise equipped with soft jaws and carefully tighten it.

5. Obtain a spanner wrench (# 912084) or an equivalent tool. Removal of the end cap is difficult using the special tool so exercise care. Insert the tangs of the spanner into the holes in the end cap. If needed, slide a long breaker bar onto the tool so the bar is in the same plane as the tool. This position will provide maximum mechanical advantage. Remove the end cap. Tap the breaker bar, if necessary, but bear in mind—if the holes in the end of the cap become damaged (elongated), the cylinder might as well be given the "deep six". Continue to unscrew the end cap until it is held by a single thread. Extend the rod, and then continue to remove the end cap and piston assembly.
6. Clean and degrease the piston rod thoroughly and then install it in a vise using a rod holder (# 983213). Carefully insert the pivot pin through the hole in the end piece and spin it off the end of the rod. If you are unable to loosen the end piece, heat the shank very carefully with a torch or heat gun and then try it again.

■ If you find that the piston rod is slipping in the holder tool while trying to spin off the end piece, wrap a piece of paper around the rod before inserting it into the tool.

Fig. 13 Drain the cylinder into a container

Fig. 14 Remove the end cap...

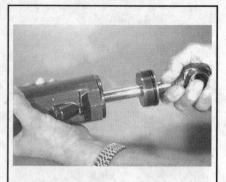

Fig. 15 ...and then pull the piston rod out of the cylinder

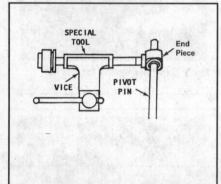

Fig. 16 Use the pivot pin to remove the end piece

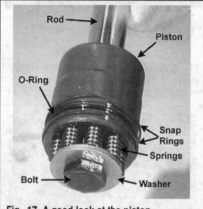

Fig. 17 A good look at the piston

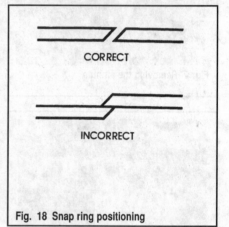

Fig. 18 Snap ring positioning

7. Pull the end cap off of the end of the rod and then pry out the scraper with a small awl. Do the same with the 2 O-rings, one inside and one outside. Discard the scraper and the O-rings.

8. Move to the end piston and remove the 2 split rings and the O-ring; throw them all away.

9. There are no serviceable components in the assembly, so there is no service procedure. This being said though, if you suspect impact valve leakage there is a quick procedure to try before replacing everything. Install the piston rod in the holder tool again and then loosen the large bolt on the end of the piston just enough to allow the relief check balls to unseat themselves; but not too much. Clean the piston components with solvent and then blow everything dry with compressed air. Tighten the bolt securely until the balls seat themselves.

To Assemble:

10. Make an effort to keep the work area clean, because any contamination on the piston could lead to a malfunction. Inspect the interior of the cylinder for any signs of scoring or roughness. Clean all surfaces with safety solvent and blow them dry with compressed air.

11. If the piston has any signs of corrosion or damage it must be replaced with a new unit.

12. Clean the end cap threads with a wire brush until you have removed all traces of old sealant.

13. Lubricate all O-rings and interior components with Power Trim/Tilt and Steering Fluid.

14. Position a new scraper over the end cap so that the lip is facing upward and then use a small seal installer (# 326545), or an appropriately sized socket, to press the scraper into the cap.

15. Coat new O-rings with fluid and then install them into their respective grooves in side and outside the end cap.

16. Coat a new piston O-ring with fluid and roll it into place in the recess in the piston. Position new split ring retainers on each side of the O-ring so that their ends match but do not overlap. These rings are quite fragile, so be careful when installing them into the recess.

17. Install a seal protector tool (# 326005) over the end of the piston rod and slide on the end cap. Make sure that the rod, O-ring and cap are thoroughly lubricated with hydraulic fluid.

18. Make sure that the piston rod threads are free of all old sealant and then spray some Loctite Primer on the threads. While waiting for it to dry, secure the rod in the holding fixture again and then coat the threads with Loctite. Screw on the end piece and tighten it securely. Remove the rod and tool from the vise.

19. Re-clamp the cylinder in the vise and coat the cylinder bore and piston thoroughly with fluid. Grab the piston body so that you compress the split rings with your fingers and then feed it into the cylinder bore. Push the piston and rod all of the way into the cylinder. Once the piston is completely in, cap the hydraulic line ports and fill the cylinder with as much hydraulic fluid as possible.

20. Slide the end cap down the rod and thread it into the cylinder. Install the spanner wrench and tighten it to 25-30 ft. lbs. (34-41 Nm). Once there, continue rotating the end cap until the two holes are parallel with the pivot pin hole in the end piece.

21. Attach the cylinders, check the fluid level in the reservoir and then run the drive through its full range of motion 5 times to purge any remaining air in the system.

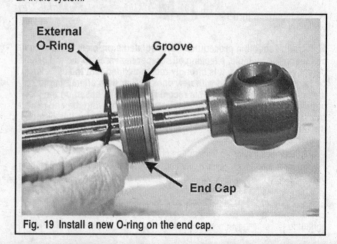

Fig. 19 Install a new O-ring on the end cap.

Trim/Tilt Pump Assembly

The pump motor is a 12-volt DC bi-directional motor. The armature of the motor rests inside the field winding and is energized through brushes contacting the commutator on the end of the armature. When the up or down switches are closed, current is directed through the motor field in one of two directions causing the motor to run in one desired direction.

The opposite end of the armature is mechanically splined to the hydraulic pump. The gear rotor action of the pump causes the fluid to flow in either direction.

The motor case is sealed with O-rings and a shaft seal to prevent the entry of water or hydraulic fluid into the motor cavity.

Before going directly to the pump as a source of trouble, check the battery to be sure it is up to a full charge. Inspect the wiring for loose connections, corrosion and damaged wires. Take a good look at the control switches and connection for evidence of a problem.

The control switches may be quickly eliminated as a source of trouble by connecting the pump directly to the battery for testing purposes.

REMOVAL & INSTALLATION

◆ See Figures 20, 21, 22 and 23

■ Integrated reservoir models are identifiable because they have the motor on the bottom and the reservoir on the top. There may be some models in this year span that came through with the newer style pump; motor on top and clear plastic reservoir on the bottom.

1. Using the trim switch, raise the stern drive to the full UP position and then support the bottom of the unit with a piece of wood. Remember that these models, you can raise the drive even if the system is not working by loosening the manual release valve.
2. Disconnect the battery cables from the battery terminals.
3. Relieve the pressure in the system by loosening the manual release valve one complete turn counterclockwise.
4. Mark or tag the hydraulic hoses (top or bottom) before disconnecting them from the pump. The inlet and outlet ports on the manifold should also be stamped (usually) with **UP** or **DN** markings identifying each port. Place a suitable container under the fittings and then use a line nut wrench to back the fitting out of the manifold port. Cap the hoses to prevent fluid leakage into the boat and dirt from entering the system. OMC actually sells line and port caps if you are interested, otherwise you can improvise with something from your shop like taping a baggie around them.
5. Disconnect the trim harness plug and the battery.
6. Remove the 3 lag bolts securing the trim/tilt pump assembly to the boat. Lift the assembly out and free of the boat.
7. Try to keep the assembly standing upright while out of the boat.

To Install:

8. Position the trim/tilt pump assembly in the boat. Install the lag bolts through the pump support bracket and into the transom. Tighten the bolts securely.
9. Connect the hydraulic hoses to their ports in the manifold—make sure they thread into the correct port. Tighten the line nuts securely.
10. Fill the reservoir with OMC Power Trim/Tilt and Steering Fluid to the bottom of the filler hole.
11. Connect the trim connector plug. Connect the battery cables and then cycle the stern drive Up and Down several (at least 5) times to bleed all air from the system. Check the reservoir level after each cycle and service, if needed.

Fig. 20 Disconnect the hydraulic hoses at the pump...

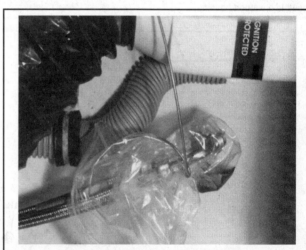

Fig. 21 ...and them plug or bag them to keep out any dirt

Fig. 22 Unplug the trim connector

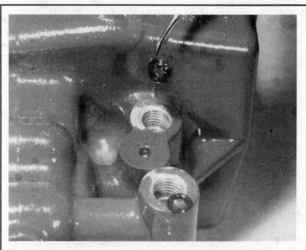

Fig. 23 Make sure that you remove the hydraulic line O-rings

DISASSEMBLY & ASSEMBLY

◆ **See Figures 24 thru 40**

1. Remove the pump assembly.

2. Drain all the remaining fluid in the reservoir into a suitable container. Discard the fluid in accordance with any local hazardous waste ordinances.

3. Remove the 4 reservoir-to-manifold mounting screws (Phillips) and then hold the assembly over a suitable container while you separate the two units. Remove the O-ring seal from the bottom of the reservoir and throw it away.

4. Turn the remaining assembly over and remove the 3 motor-to-manifold mounting screws (Phillips). Separate the motor from the manifold and then remove the O-ring and discard it.

5. Remove the oil pump filter screen from the pump surface. Clean it thoroughly and set it aside for assembly.

6. Pull the drive coupling out of the pump body and check the slotted and hex ends for wear or damage—replace as necessary. Check the rubber ball in the end of the coupling.

7. Loosen the 3 pump-to-valve body Allen bolts alternately and only ONE TURN at a time until you can lift off the gear housing (pump). There are 2 short bolts and 1 long bolt. If the pump needs to be flushed, use only new hydraulic fluid—never use solvent. Do not remove the pump assembly screws; the pump is serviced as an assembly only.

✳✳ WARNING

The pump housing contains pistons with springs that exert an upward pressure on the unit; it is imperative that when loosening the mounting screws, they only be turned one complete revolution at a time, moving on to the next screw diagonally across from it each time until all spring pressure has been eliminated. Failure to follow this procedure will result in a warped housing.

■ The mating surfaces of the pump housing and manifold are precision machined and do not use a gasket; extreme care should be exercised during removal. Be very careful no to gouge, scratch or mar either of these services.

8. Working in the valve body, scrape away any visible staked metal and then pull out the filter valve and washer. Confirm that the filter is clean and undamaged. Thread in a No. 6 screw or a tap and pull up the screw to remove the valve seat. Use a small magnet to remove the check ball and check its seat for any damage. Remove the O-rings and throw them away.

9. Use a small awl or pick and pull out the 2 oil pump O-rings in the valve body.

10. Unscrew and remove the trim-in relief valve spring/valve. Check the spring, valve and seal tip for visible signs of damage. If necessary, replace the seal tip by pressing on a new tip so the flat side is against the valve and the ribbed side is against the seat.

11. Pull out the thermal expansion valve core and spring. Removing the valve body will require fabricating a special tool from a small length of extra control cable core wire (or a paper clip). Bend the wire into the shape illustrated and make sure that the end is smooth and unfrayed. Insert the tool into the valve body and pull on it until the valve pops out.

12. Use a small (No. 1) pair of circlip pliers to remove the manual release valve retaining clip and then unscrew the valve from the valve body. Pull out the O-ring and throw it away. Check the sealing surface on the tip of the valve.

13. Remove the tilt-up relief valve plug on the back side of the valve body and then pull out the spring and valve core. Removing the valve body itself will require the special tool made earlier; insert the tool (or a bent pick) into the cross-drilled holes in the body of the valve and pull out firmly.

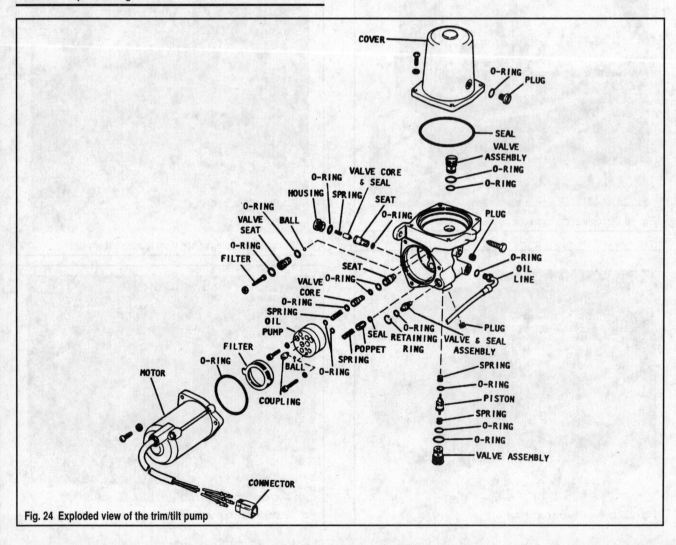

Fig. 24 Exploded view of the trim/tilt pump

14. Insert a drag link socket (or equivalent) into the trim-in/tilt-down check valve and unthread it from the valve body. Pull out the upper spring in the bore and then pull out the pump control piston with needle nose pliers. Don't forget the lower spring. Leave the 2 O-rings on the check valve for now.

15. Remove the trim-out/tilt-up check valve from the bottom (outside) of the valve body.

16. Use a small pin punch to check each check valve for freedom of operation—insert the punch into the hole in the bottom of the valve and press down on the spring, making sure that it returns freely.

To Assemble:

17. Thoroughly clean all components in solvent and blow them dry with compressed air. Pay particular attention to drying all manifold passages. Keep everything clean and lint-free.

18. Inspect the pump drive coupling for excessive wear or damage. Check all valves and bores for any sign of scoring or corrosion. Check the machined mating surfaces of the pump housing and valve body for scratches, grooves or other marring.

19. Coat all components thoroughly with Power Trim/Tilt & Steering fluid before installation—do not assemble any component dry!

20. Inspect the tilt-up relief valve seal tip and seat. Coat new O-rings with hydraulic fluid and install them on the valve seat and cap. Press the seat into the bore on the back of the valve body, insert the spring and valve core and then install the cap. Tighten to 50-60 inch lbs. (5.5-7.0 Nm).

21. Install new O-rings on the trim-out/tilt-up check valve and then thread the valve into the valve body. Tighten to 50-60 inch lbs. (5.5-7.0 Nm).

22. Insert the lower control piston spring into the bore in the valve body. Slide a new O-ring onto the piston and position it into the valve body so the end with the O-ring is down. Slide a new O-ring onto the trim-in/tilt-down check valve and then threads it into the piston assembly bore after inserting the upper piston spring. Tighten the check valve to 50-60 inch lbs. (5.5-7.0 Nm).

23. Take a look at the small O-rings on the thermal expansion valve body and core, if they were removed or look damaged in any way, replace them. Install the valve body, core and spring into the main valve body.

24. Inspect the sealing tip on the trim-in relief valve. If you haven't already replaced it, we suggest doing it now as described in the disassembly step. Install the valve and spring into the valve body.

25. Press 2 new O-rings into the filter valve body and press it into the main valve body. Install the filter and washer.

26. Position the oil pump onto the valve body mating surface without disturbing the O-rings and insert each of the screws. Tighten the screws to 25-35 inch lbs. (2.8-4.0 Nm), one full rotation at a time, in an alternating pattern.

27. Insert the drive coupling into the drive gear and rotate it slowly by hand to make sure there is no binding evident. If you are unable to turn it by hand, loosen the pump housing screws a little (remember, one turn at a time) and then try it again. If the coupling now turns, you've got a problem in the housing; check all components and start over again.

Fig. 25 Remove the motor mounting bolts...

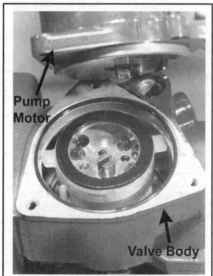

Fig. 26 ...and then lift off the motor

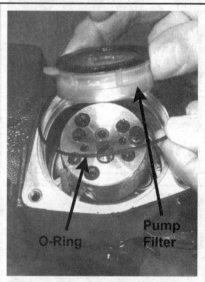

Fig. 27 Remove the oil pump filter and O-ring...

Fig. 28 ..and then lift out the pump coupling

Fig. 29 Remove the 3 pump mounting bolts...

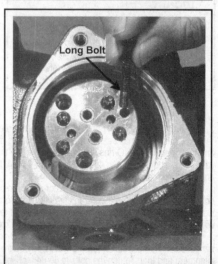

Fig. 30 ...remember that one is longer than the other two...

Fig. 31 ...and then lift the pump from the valve body

Fig. 32 Remove the filter valve and then the 2 O-rings

Fig. 33 Remove the trim-in relief valve and thermal expansion springs from the valve body

Fig. 34 Use a small pick or the special tool to remove the thermal expansion valve body

Fig. 35 Remove the release valve retaining clip

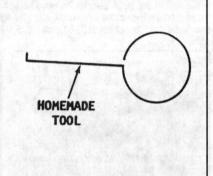

Fig. 36 Fabricate a small removal tool out of a paper clip

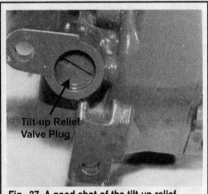

Fig. 37 A good shot of the tilt-up relief valve plug

Fig. 38 Unscrew the trim-in/tilt-down check valve...

Fig. 39 ...and then remove the upper control piston spring

28. Install the round oil pump filter screen onto the pump housing.

29. Check the small seal on the inner end of the manual release valve and then install a new O-ring. Thread the valve into the bore on the outer valve body and tighten it to 45-55 inch lbs. (5-6 Nm). Install a new retaining ring with circlip pliers.

30. Install a new O-ring onto the motor and then position it to the valve body so the mounting holes line up correctly. Now rotate the coupler until the slot lines up with the blade on the motor shaft and then install the motor (it will only go on one way). Tighten the bolts to 60-84 inch lbs. (7-9 Nm).

31. Turn the assembly over and install a new O-ring on the reservoir flange. Position the reservoir on the valve body so that the bolt holes are in alignment and then tighten the bolts to 60-84 inch lbs. (7-9 Nm).

32. Install the assembly in the boat, fill with fluid and test for leaks.

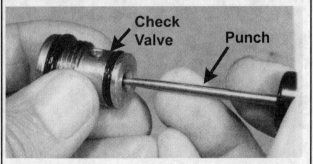

Fig. 40 Insert a small punch into the check valve to check return spring action

POWER TRIM AND TILT—1989-98 COBRA, KING COBRA, SP(SX) AND DP

Description

◆ **See Figure 41**

■ **Not all models use all components in all years.**

This power trim/tilt system is a hydro-mechanical, electrically controlled system allowing the stern drive to be raised when moving through the water at idle speed, when approaching the beach or shore, and when trailering the boat. The drive can also be adjusted through-out its trim range while the boat is underway and at any speed. Thus enabling the helmsman to maintain the best bow position for maximum speed, efficiency and ride for any water or throttle conditions.

An electric motor, controlled by a switch and relays, provides the power for a hydraulic pump. The pump, in turn, forces fluid through a set of hydraulic hoses to a set of hydraulic cylinders. Depending on the motor's direction, fluid will flow into one end of the cylinders to raise or lower the drive unit.

A control switch may be mounted on the remote control handle or on the control panel. When activated, the **UP** or **DOWN** relay is energized. Current to the motor rotates the motor in one of two directions, thus raising or lowering the drive unit.

The power trim/tilt system consists of a valve body (manifold), hydraulic reservoir, electric motor, up and down solenoids, trim/tilt switch, trim sending unit, trim indicator gauge, trim limit switch, two hydraulic trim cylinders, and the necessary hydraulic lines and electrical harnesses for the system to function efficiently.

The manifold, hydraulic reservoir, electric motor, and up/down solenoids make up a single unit known as the pump assembly that is mounted inside the boat. Two hydraulic trim cylinders are mounted, one on each side, of the stern drive unit. One end of the cylinder is connected to the gimbal housing and the opposite end is attached to the drive housing.

The trim sender unit is mounted on the starboard side of the gimbal ring and is connected to the trim gauge on the control panel through the wiring harness. When the stern drive is trimmed **IN** or **OUT** the gauge moves to indicate the stern drive's position.

A trim limit switch may be connected to the **UP** solenoid on the trim/tilt hydraulic pump. This limit switch allows the stern drive to be trimmed out only to a preset position. If the stern drive is moved out past the support brackets under high throttle setting could impose high stress loads on the boat transom and possibly loss of steering control. There may also be a 10 amp circuit breaker or fuse serving the same purpose.

The relationship of the various units in the trim/tilt system are discussed in the following paragraphs.

VALVE BODY/MANIFOLD

◆ **See Figure 41**

The valve body, or manifold, is the central mounting point for the pump, valves, reservoir and electric motor. The hydraulic pump is capable of creating pressure up to 3000 psi. Internal pressure relief valves set at the factory control the output pressure in the UP and DOWN circuits. A system thermal relief valve protects the pump, valves, lines, and cylinder packing from thermal expansion of the hydraulic fluid in the system when operating or sitting in the hot sun. The pump, pressure and thermal relief valves are all set at the factory and adjustments are not possible. If any of these components are found to be defective during troubleshooting or disassembly, they must be replaced with identical items. Each of the valves is color coded to ensure the correct valve is installed in the correct port. A replaceable filter is mounted on the inlet opening to the pump.

Models through 1996 utilize a manual release valve. Models after 1997 offer no provisions for a manual relief valve on the manifold. If the pump or motor fails, manually raising or lowering the stern drive would require the appropriate hydraulic line be loosened on the manifold external fitting.

■ **There may be some 1996 models without a manual relief valve, and there will be some 1997 models with a manual relief valve—checking your manifold is the only way to tell for sure.**

RESERVOIR

All models utilize an integrated reservoir, while some later models may use a different reservoir which is a translucent 1 1/2 quart reservoir that attaches to the bottom of the pump manifold. A **MIN** and **MAX** mark cast into the reservoir makes it easy to check the fluid quantity. A large screw-on cap and opening provides for easy servicing of the reservoir. The cap vents the reservoir to atmosphere pressure when the fill neck seal is removed inside the cap.

HYDRAULIC PUMP

The hydraulic fluid pump is a gear-type pump, very similar to an engine oil pump. The pump is attached to the manifold with four fasteners and cannot be disassembled. If the pump fails, it must be replaced as an assembly. The reservoir must be drained and removed to gain access to the pump.

MANUAL RELEASE VALVE

1989-96 Models Only

◆ **See Figures 41 and 42**

■ **There may be some 1997 models that are still utilizing the manual release valve; conversely, there may be some 1996 models that are not equipped with this component. Simply check the valve body to be sure.**

The manual release (or relief) valve is used on 1989-96 models and is located on the port side of the base of the manifold. This valve allows the drive unit to be raised or lowered manually without the use of electric or hydraulic power. By rotating the valve counterclockwise, the drive unit will lower under its own weight; closing the valve will stop the process. The valve must be fully open to raise the unit, and it must be tightly closed in order to hold a trim or tilt position while static or moving forward.

MANUAL RELEASE CHECK BALL & RETAINER

1989-96 Models Only

◆ **See Figure 41**

■ **There may be some 1997 models that are still utilizing the manual release valve; conversely, there may be some 1996 models that are not equipped with this component. Simply check the valve body to be sure.**

The manual release check ball and retainer is positioned in a seat next to the trim-in/tilt-down check valve. The retainer limits the check ball movement. The check ball allows fluid to flow directly from the reservoir into the manifold when the pump is not operating. This function is required to enable fluid to be drawn into the hydraulic cylinders when raising the stern drive unit manually.

TRIM-IN/TILT-DOWN CHECK VALVE

◆ **See Figure 41**

The trim-in/tilt-down check valve is located in the manifold. This valve opens under pressure to allow hydraulic fluid from the pump to the low volume side of the cylinders during the Down cycle. The cylinder chamber with the rod is identified as the low volume side.

The valve also allows fluid to return from the low volume side of the cylinder to the pump during the Up cycle, after the valve has been opened by the pump control piston.

In addition to the above, the valve prevents the drive unit from raising during operation in Reverse, acting as a reverse lock; and prevents fluid flow from the high volume side of the cylinders during a high pressure impact condition.

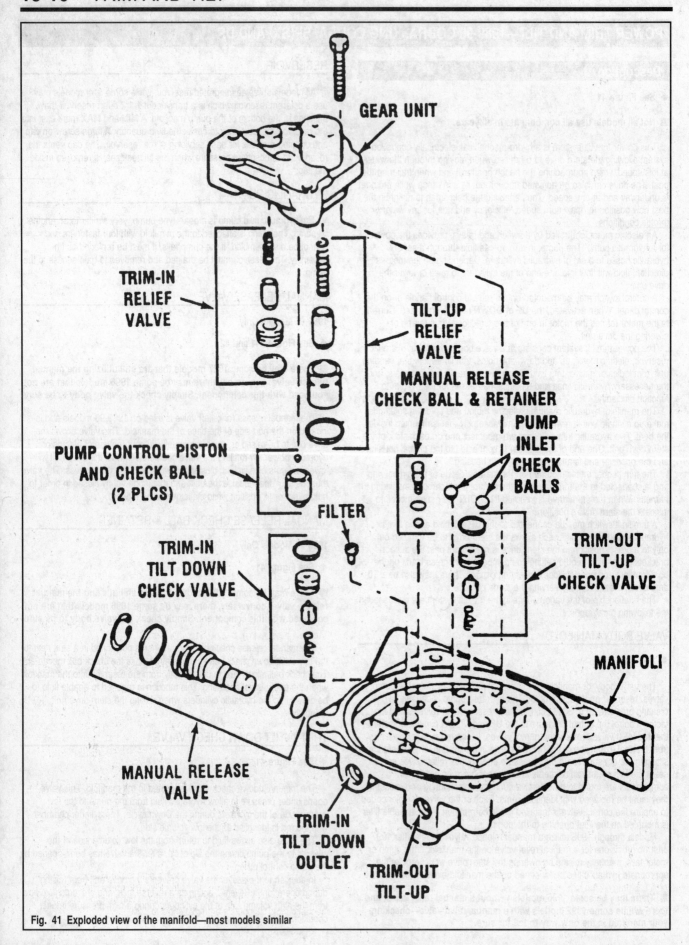

Fig. 41 Exploded view of the manifold—most models similar

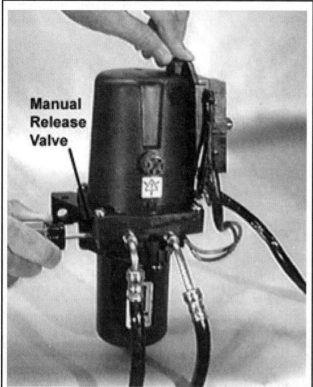

Fig. 42 Early models were equipped with a manual release valve

TRIM-IN/TILT-DOWN OUTLET PORT

◆ See Figure 41

The outlet port on the manifold is connected to the rod end (low volume) side of the trim cylinders be a combination of hydraulic hoses and lines. It is directly in line with the manual relief valve.

Hydraulic fluid, under pressure, flows through this port to both cylinders during cylinder retraction. Return fluid from both cylinders, during cylinder extension, will flow back through this port.

TRIM-OUT/TILT-UP CHECK VALVE

◆ See Figure 41

This valve is located in a seat connected to the UP port in the manifold. It opens under pressure to allow fluid flow to the high volume side of the trim cylinders during UP operation. The valve serves to maintain oil pressure in the high volume chamber of cylinder so that the drive unit will hold a particular trim or tilt position while moving forward. It will return fluid to the pump during the Down cycle after having been opened by the pump control piston.

TRIM-OUT/TILT-UP OUTLET PORT

◆ See Figure 41

This outlet port on the manifold is marked with the word **UP** and is connected to the cylinder end (high volume side) of the trim cylinders. Hydraulic fluid, under pressure, flows through this port to both cylinders during trim cylinder extension. Return fluid from each cylinder, during retraction, flows back through the port to the pump.

THERMAL EXPANSION RELIEF VALVE

◆ See Figure 41

This valve is located inside the manual release valve (1989-96) or on the gear housing (1997-98) and it prevents hydraulic lock-up when the drive is in the full tilt-up position, which can be caused by heat-related expansion of the fluid (thermal expansion). The valve will open only when it senses unusually high pressure (approx. 2250-2750 psi).

INLET CHECK BALLS

◆ See Figure 41

These check balls rest in seats on either side of the oil pump gears, allowing fluid from the reservoir into the pump when they open. When closed, obviously, they prevent fluid in the pump from returning to the reservoir.

PUMP CONTROL PISTON

◆ See Figure 41

Oil pressure in the pump moves the control pistons to open the trim-in/tilt-down check valve during the UP cycle and the trim-out/tilt-up check valve during the DOWN cycle. There are two checks balls that either allow or restrict fluid movement through the pistons depending on the cycle selection.

PUMP FILTER & COUPLER

◆ See Figure 41

A serviceable filter is located between the pump motor and the reservoir; filtering all fluid flowing into the pump from the reservoir. A coupler is used to connect the motor shaft to the oil pump drive gear.

FILTER SCREEN

1989-96 Models Only

◆ See Figure 41

Certain models may be equipped with a serviceable filter screen between the manifold and gear housing that filters fluid returning to the reservoir from the manual relief valve passage.

TILT-UP RELIEF VALVE

◆ See Figure 41

This valve is located in the gear housing, next to the Trim-in relief valve; it is noticeably larger than its companion. It functions to control maximum pressure in the circuit during the UP cycle. It also limits pressure at the full tilt-up 'stall' position to a maximum of 1200-1600 psi.

TRIM-IN RELIEF VALVE

◆ See Figure 41

As mentioned previously, this valve is located in the gear housing next to the tilt-up relief valve and is noticeably smaller than its companion. It controls pressure in the system during the DOWN cycle. Due to rod volume, more fluid returns from the high volume side of the trim cylinders than the low volume side can handle; excess fluid passes through the valve which will keep pressure on the Down side at 400-800 psi.

Troubleshooting

◆ See Figures 43 and 44

The electric pump motor must operate to drive the mechanical hydraulic pump. The hydraulic pump will convert the electrical energy into mechanical energy. The mechanical energy moves the hydraulic fluid to extend and retract the trim cylinders. Use the following troubleshooting procedures to determine if the problem is electrical or mechanical.

DETERMINE IF PROBLEM IS ELECTRICAL OR MECHANICAL

1. Verify that the battery is fully charged before continuing the troubleshooting procedure. If the battery is not fully charged, the system could give a false indication of a problem when none actually exists.
2. Verify the fluid level in the reservoir is up to the **MAX** mark on the side of the reservoir; or the bottom of the filler hole on earlier models.
3. Depress the **UP** or **DOWN** button on the control panel and listen for the pump motor to operate or the solenoid to click.

Power Trim Diagnostic Chart 1997-98

Symptom/Problem	Possible Cause
Static leak-down and/or trim-in creep in Forward gear	③④⑮
Trim-out/tilt-upcreep in Reverse gear (no reverse lock)	⑤⑦⑮
Leaking, Up and Down (trim-in/out creep)	⑥
Slow or no trim-out/tilt-up while motor running	①③④⑤⑥⑦⑨⑯⑯
No trim-out/tilt-up, motor does not run	①③④⑤⑥⑯⑰⑱
Slow or no trim-in/tilt-down while motor running	①③④⑤⑥⑦⑧⑩
No trim-in/tilt-down, motor does not run	⑪⑫⑬⑭
No trim/tilt Up or Down, motor runs	①③⑫
No trim/tilt Up or Down, motor runs fast	⑩
No trim/tilt Up or Down, motor does not run	⑪⑫⑬⑭
No tilt-down from full tilt-up position, motor stalls	③
Erratic operation, jumpy motion, pulsing/hissing fluid sounds	②
Continual use or loss of hydraulic fluid	⑮

① Air or low/no oil in reservoir. Fill reservoir and cycle pump
② Air in hydraulic system. Cycle pump through range 3-5 times. Fill reservoir
③ Thermal expansion valve sticking or damaged spring. Dirt under check ball.
④ Trim-out/tilt-up check valve sticking, damaged seal tip, dirt under seal, leaking O-ring, damaged spring
⑤ Pump control piston sticking, dirt under check balls or in small oil passages.
⑥ Trim/tilt cylinder internal leak. Check for damaged piston seal impact relief valves, or dirt under relief valves
⑦ Trim-in/tilt-down check valves sticking. Damaged seal tip, dirt under seal, leaking O-ring or bad spring
⑧ Trim-in relief valve leaking. Check seal tip, seat or spring. Check for sticking or dirt.
⑨ Tilt-up relief valve leaking. Check sticking valve, seal tip, seal or spring. Check valve body O-ring.
⑩ Pump coupling broken.
⑪ 10 or 50 amp circuit breaker tripped
⑫ Motor inoperative, replace motor.
⑬ Relay control box malfunction. Repllace box.
⑭ Switch circuit malfunction.
⑮ External oil leak. Check all lines, connections, cylinders and valve body.
⑯ Pump-to-manifold oil line attachment reversed.

Power Trim Diagnostic Chart 1989-96

Symptom/Problem	Possible Cause
Static leak-down and/or trim-in creep in Forward gear	③④⑤⑰
Trim-out/tilt-upcreep in Reverse gear (no reverse lock)	⑥⑧⑰⑱
Leaking, Up and Down (trim-in/out creep)	⑦
Slow or no trim-out/tilt-up while motor running	①③④⑤⑥⑦⑧⑪⑫⑱
No trim-out/tilt-up, motor does not run	⑬⑭⑮
Slow or no trim-in/tilt-down while motor running	①③⑥⑧⑨⑩⑫
No trim-in/tilt-down, motor does not run	⑬⑭⑮⑯
No trim/tilt Up or Down, motor runs	①③⑫
No trim/tilt Up or Down, motor runs fast	⑫
No trim/tilt Up or Down, motor does not run	⑬⑭⑮⑯
No tilt-down from full tilt-up position, motor stalls	④
Erratic operation, jumpy motion, pulsing/hissing fluid sounds	②
Continual use or loss of hydraulic fluid	⑰

① Air or low/no oil in reservoir. Fill reservoir and cycle pump
② Air in hydraulic system. Cycle pump through range 3-5 times. Fill reservoir
③ Manual release valve loose or leaking. Check seal ring, O-ring and seats. Tighten valve to 45-55 inch lbs. (5-6 Nm).
④ Thermal expansion valve sticking or damaged spring. Dirt under check ball.
⑤ Trim-out/tilt-up check valve sticking, damaged seal tip, dirt under seal, leaking O-ring, damaged spring
⑥ Pump control piston sticking, dirt under check balls or in small oil passages.
⑦ Trim/tilt cylinder internal leak. Check for damaged piston seal impact relief valves, or dirt under relief valves
⑧ Trim-in/tilt-down check valves sticking. Damaged seal tip, dirt under seal, leaking O-ring or bad spring
⑨ Manual release check ball leaking. Check for damaged or dirty ball or seat.
⑩ Trim-in relief valve leaking. Check seal tip, seat or spring. Check for sticking or dirt.
⑪ Tilt-up relief valve leaking. Check sticking valve, seal tip, seal or spring. Check valve body O-ring.
⑫ Pump coupling broken.
⑬ 10 or 50 amp circuit breaker tripped
⑭ Motor inoperative, replace motor.
⑮ Relay control box malfunction. Repllace box.
⑯ Switch circuit malfunction.
⑰ External oil leak. Check all lines, connections, cylinders and valve body.
⑱ Pump-to-manifold oil line attachment reversed.

4. If the pump motor does not operate, and the solenoids do not "click", the problem is electrical.

5. If the pump motor operates, but the stern drive does not move, the problem is most likely a hydraulic or mechanical problem. Refer to hydraulic/mechanical troubleshooting.

6. On early models, check the manual release valve— it MUST be closed and tightened to 45-55 inch lbs. (5-6 Nm).

HYDRAULIC AND MECHANICAL PROBLEMS

◆ See Figure 45

All testing procedures should be performed using a OMC Trim/Tilt In-Line Pressure Tester (# 983977) consisting of a pressure gauge, adapter, tester body, shut-off valve and hydraulic line.

To install the tester:

1. Raise the stern drive up to its full tilt-up position and block it securely with wood.

2. On early models equipped with a manual release valve, open the valve all the way for a few seconds until the pressure equalizes and then close the valve tightly. On later models without the valve, simply open the filler cap/plug for a few seconds and then close it.

3. Label the UP hydraulic line at the pump and then disconnect it; make sure you have some rags handy, safety glasses on and a suitable pan underneath the connection. Once the pressure has been released, reconnect the line.

4. The body of the tester has two ports (marked A and B). Port A is always connected to the gauge, while port B can be shut off by closing the valve on the tester. The A port must always be the one that is connected to the component or circuit thus allowing the circuit to be isolated when the vessel's line is connected to port B and the valve is closed.

Trim-Out/Tilt-Up—Complete Circuit

◆ See Figure 45

Performing this test will determine running and stall pressure for the trim-out/tilt-up circuit.

1. Disconnect the UP hydraulic hose (2) at the outlet (1) on the manifold.

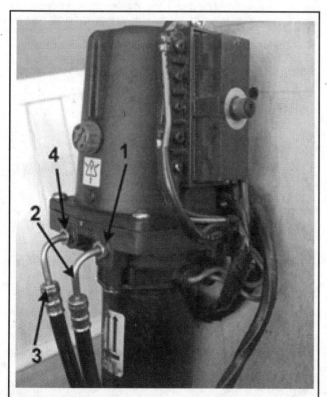

Fig. 45 Trim pump connections

2. Connect the inlet line from the tester (remember that this is connected to the A port on the tester) to the outlet (1) on the manifold.

3. Connect the UP hydraulic hose (2) to the outlet port (stamped B) on the tester and then open the shut-off valve 1-2 turns.

✱✱ WARNING

Opening the valve more than 2 turns may unseat the retaining ring and cause fluid to spray or spill.

4. Run the drive unit through a few Up and Down cycles to purge any air that may have entered the system while hooking up the tester. Check the fluid level and then move the drive unit to the full trim-out/tilt-up position and check the running and stall pressure (that point where the drive stops moving up, but the pump keeps running). Running pressure should be 225-275 psi, while the stall pressure should be 1200-1600. If within specifications, proceed to the Trim-In/Tilt-Down - Complete Circuit test; if not within specifications, move to the Trim-Out/Tilt-Up - Pump Isolation test.

Trim-Out/Tilt-Up—Pump Isolation

◆ See Figure 45

Performing this test will isolate the pump manifold trim-out/tilt-up components from the trim cylinders and will determine whether the manifold or the trim cylinders have a problem.

1. Disconnect the UP hydraulic hose (2) at the outlet (1) on the manifold.

2. Connect the inlet line from the tester (remember that this is connected to the A port on the tester) to the outlet (1) on the manifold.

3. Connect the UP hydraulic hose (2) to the outlet port (stamped B) on the tester and then open the shut-off valve 1-2 turns.

✱✱ WARNING

Opening the valve more than 2 turns may unseat the retaining ring and cause fluid to spray or spill.

4. Move the drive unit up to the full tilt position and then flick the trim switch for a moment in the DOWN position to relieve pressure. Once relieved, close the shut-off valve on the tester.

5. Run the pump in the trim-out/tilt-up direction until the drive stops moving but the pump continues to run; record the pressure at the stall.

6. Release the trim switch and check the leak-down rate as the pressure drops for at least 5 minutes. Record this and record the final pressure.

7. Stall pressure should be 1200-1600 psi. Leak-down rate should not exceed 200 psi. If the stall pressure is less than specification, or the leak-down pressure is greater than the figure, check the manifold. If the leak-down rate and stall pressure is within specifications, check each of the trim cylinders for leakage.

8. Flip the trim switch to operate the drive in the opposite direction and watch the gauge until all pressure has been relieved. Disconnect the tester and reconnect the hose.

Trim-In/Tilt-Down—Complete Circuit

◆ See Figure 45

Performing this test will determine running and stall pressure for the trim-in/tilt-down circuit.

1. Disconnect the DOWN hydraulic hose (3) at the outlet (4) on the manifold.

2. Connect the inlet line from the tester (remember that this is connected to the A port on the tester) to the outlet (4) on the manifold.

3. Connect the DOWN hydraulic hose (3) to the outlet port (stamped B) on the tester and then open the shut-off valve 1-2 turns.

✱✱ WARNING

Opening the valve more than 2 turns may unseat the retaining ring and cause fluid to spray or spill.

4. Run the drive unit through a few Up and Down cycles to purge any air that may have entered the system while hooking up the tester. Check the fluid level and then move the drive unit to the full trim-in/tilt-down position and check the running and stall pressure (that point where the drive stops moving down, but the pump keeps running). Running pressure should be 400-800 psi, while the stall pressure should be 400-800 psi. If not within specifications, move to the Trim-In/Tilt-Down - Pump Isolation test.

Trim-In/Tilt-Down—Pump Isolation

◆ See Figure 45

Performing this test will isolate the pump manifold trim-in/tilt-down components from the trim cylinders and will determine whether the manifold or the trim cylinders have a problem.

1. Disconnect the DOWN hydraulic hose (3) at the outlet (4) on the manifold.
2. Connect the inlet line from the tester (remember that this is connected to the **A** port on the tester) to the outlet (4) on the manifold.
3. Connect the DOWN hydraulic hose (3) to the outlet port (stamped **B**) on the tester and then open the shut-off valve 1-2 turns.

✳✳ WARNING

Opening the valve more than 2 turns may unseat the retaining ring and cause fluid to spray or spill.

4. Move the drive unit down to the full tilt position and then flick the trim switch for a moment in the UP position to relieve pressure. Once relieved, close the shut-off valve on the tester.
5. Run the pump in the trim-in/tilt-down direction until the drive stops moving but the pump continues to run; record the pressure at the stall.
6. Release the trim switch and check the leak-down rate as the pressure drops for at least 5 minutes. Record this and record the final pressure.
7. Stall pressure should be 400-800 psi. Leak-down rate should not exceed 200 psi. If the stall pressure is less than specification, or the leak-down pressure is greater than the figure, check the manifold. If the leak-down rate and stall pressure is within specifications, check each of the trim cylinders for leakage.
8. Flip the trim switch to operate the drive in the opposite direction and watch the gauge until all pressure has been relieved. Disconnect the tester and reconnect the hose.

Erratic Motion During Operation

Check the hydraulic fluid level and fill if necessary. Bleed the system.

Static Leak-down, Trim-In Creep in Forward

◆ See Figures 46, 47, 48 and 49

1. If equipped with a manual release valve, confirm that it is tight and not leaking. Tighten the valve to 45-55 inch lbs. (5-6 Nm). If leaking, drain the system and move the drive unit to the full Down position. Remove the valve and check the plastic seal, O-ring and seat.
2. Check the thermal expansion valve located in the end of the manual release valve (1989-96) or in the gear housing (1997-98). Look for sticking, damage to the spring or dirt under the check ball.
3. Check the trim-out/tilt-up check valve for damage. Check the seal tip, dirt under the seal, leaking O-ring or a damaged spring.
4. Check to see if the pump control piston is sticking. Check the condition of the piston and the O-ring.
5. Check for external leaks through out the system.

Trim-In/Trim-Out Creep

This condition is frequently caused by an internal leak at the trim cylinders—the internal piston seals or the impact relief valves may be dirty or damaged, causing fluid to slowly leak from one side of the piston to the other.

Trim-Out/Tilt-Out Creep In Reverse

◆ See Figure 45, 48 and 49

1. Check to see if the pump control piston is sticking. Look for dirt under the check balls or in the small oil passages in the piston itself. Check the piston and its bore for scuffing or other signs of damage. Check the ball seat for signs of wear.
2. The trim-in/tilt-down check valve may be sticking due to a damaged seal tip. Check for dirt under the seal, leaking O-ring or a damaged spring.
3. Check the manual release check ball and seat for leakage or damage (if equipped).
4. Check all components for external leaks or damage.
5. Confirm that the UP and DOWN hydraulic hoses are connected to the correct ports on the manifold.

No Tilt-Down From Full Tilt-Up Position

◆ See Figure 47

Check the O-ring and seal on the thermal expansion valve. Confirm that the valve is not sticking and that the spring is not damaged.

No Or Slow Trim-Out/Tilt-Up With Motor Running

◆ See Figures 47, 48 and 49

1. Check the hydraulic fluid level in the reservoir. Fill if necessary and bleed the system.
2. If equipped with a manual release valve, confirm that it is tight and not leaking. Tighten the valve to 45-55 inch lbs. (5-6 Nm). If leaking, drain the system and move the drive unit to the full Down position. Remove the valve and check the plastic seal, O-ring and seat.
3. Check the thermal expansion valve located in the end of the manual release valve (1989-96) or in the gear housing (1997-98). Look for sticking, damage to the spring or dirt under the check ball.
4. Check the trim-out/tilt-up check valve for damage. Check the seal tip, dirt under the seal, leaking O-ring or a damaged spring.
5. Check to see if the pump control piston is sticking. Check the condition of the piston and the O-ring. Poor piston performance may also be affected by an internal leak in the trim cylinders—the cylinder piston seals or impact relief valves may be damaged.
6. The trim-in/tilt-down check valve may be sticking due to a damaged seal tip. Check for dirt under the seal, leaking O-ring or a damaged spring.
7. Check the tilt-up relief valve for sticking, damaged spring or a dirty seal tip. Check that the manifold O-ring is not distorted.
8. Check the condition of the oil pump. Dirt can be in the pump or the coupling may have failed.
9. Check that the motor output is up to specification.

No Or Slow Trim-Out/Tilt-Down With Motor Running

◆ See Figures 48, 49 and 50

1. Check the hydraulic fluid level in the reservoir. Fill if necessary and bleed the system.
2. Check the trim-out/tilt-up check valve for damage. Check the seal tip, dirt under the seal, leaking O-ring or a damaged spring.
3. Check to see if the pump control piston is sticking. Check the condition of the piston and the O-ring. Poor piston performance may also be affected by an internal leak in the trim cylinders—the cylinder piston seals or impact relief valves may be damaged.
4. The trim-in/tilt-down check valve may be sticking due to a damaged seal tip. Check for dirt under the seal, leaking O-ring or a damaged spring.
5. The filter valve, or the check valve behind it, may be damaged or leaking.
6. Check the trim-in relief valve for sticking, damaged spring or a dirty seal tip.
7. Check the condition of the oil pump. Dirt can be in the pump or the coupling may have failed.
8. Check that the motor output is up to specification.

Fig. 46 Separating the pump from the manifold

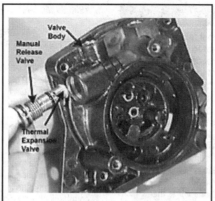

Fig. 47 On early models, the thermal expansion valve is located in the tip of the manual release valve

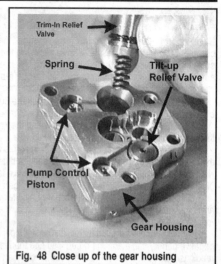

Fig. 48 Close up of the gear housing

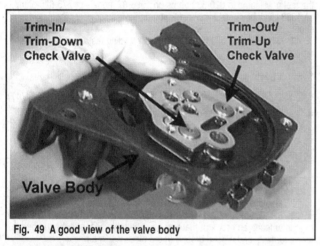

Fig. 49 A good view of the valve body

No Trim/Tilt, Motor Will Not Run

More times than not this condition is caused by an electrical problem in the switch circuit, relay/solenoid circuit, or with the motor. Please refer to the Electrical Problems section.

No Trim/Tilt, Motor Runs

MODERATE

◆ See Figure 47

1. Check the hydraulic fluid level in the reservoir. Fill if necessary and bleed the system.

2. If equipped with a manual release valve, confirm that it is tight and not leaking. Tighten the valve to 45-55 inch lbs. (5-6 Nm). If leaking, drain the system and move the drive unit to the full Down position. Remove the valve and check the plastic seal, O-ring and seat.

3. Check the thermal expansion valve located in the end of the manual release valve (1989-96) or in the gear housing (1997-98). Look for sticking, damage to the spring or dirt under the check ball.

4. Check the condition of the oil pump. Dirt can be in the pump or the coupling may have failed.

ELECTRICAL PROBLEMS

OEM *MODERATE*

■ **Please refer to the electrical schematics later in this section.**

Before performing any electrical tests, always verify that the battery has a full charge. A low battery can always give a false indication that some other problem in the system exists.

Check the circuit breakers in the main engine electrical box and in the trim/tilt relay control box have not been tripped for some reason. Push in the button and reset the breaker.

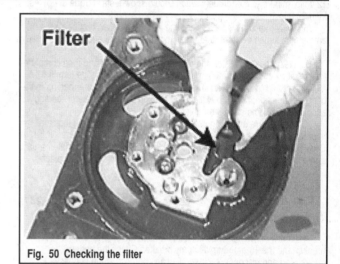

Fig. 50 Checking the filter

Trim/Tilt Switch

MODERATE

During switch operation, electrical current flows from the input terminal (red/purple wire) to the respective output terminals.

1. Check all electrical connections. Check the 50 amp trim/tilt circuit breaker and the 10 amp trim/tilt pump breaker. Check the battery.

2. Connect a test light or a multi-meter as per the manufacturer's instructions and touch the probe to the red/purple wire terminal. The tester should light or the meter should indicate battery voltage; if not, recheck the 50 amp and 10 amp circuit breakers and/or check any related electrical connections.

3. Move the probe to the green/white wire terminal and move the switch to the Down position. The tester should light or the meter should indicate battery voltage; if not, confirm that all electrical connections are OK and then replace the switch.

4. Now move the probe to the blue/white wire terminal and move the switch to the Up position. The tester should light or the meter should indicate battery voltage; if not, confirm that all electrical connections are OK and then replace the switch.

5. If all test prove to be good and the pump will still not operate, refer to the Pump Relay test later in this section.

Trim/Tilt Motor

MODERATE

◆ See Figures 51 and 52

A stall-torque test is the only reliable means of checking motor performance. Trim/tilt motors are not serviceable and must be replaced as a unit.

✷✷ WARNING

Never operate the trim/tilt motor for any longer than it takes to complete the following tests. Always allow the motor to cool sufficiently between each test. Performing the two successive tests on a hot motor will lead to inaccurate stall-torque readings.

1. Check all electrical connections. Check the 50 amp trim/tilt circuit breaker and the 10 amp trim/tilt pump breaker. Check the battery.

2. Remove the pump/motor assembly and then remove the motor from the manifold. Install it carefully in a soft-jawed vise.

3. Connect an ammeter or multi-meter in series with the **B+** terminal on the battery (make sure it's fully charged!) and the blue lead at the motor. Next connect a voltmeter (or DVOM) so the positive lead is at the motor's blue lead and the negative lead is at the motor's green lead.

4. Install an inch lb. torque wrench with a suitable adapter to the armature shaft on the pump. Have someone study the wrench while you move to the next step.

5. Now complete the circuit by running a jumper wire between the negative battery cable and the green lead and then take note of all of the readings. The ammeter should show 185 amps, the voltmeter should show 12 volts (min.) and the torque wrench should show a stall-torque of not less than 20.6 inch lbs.

6. Disconnect the 2 meters and the jumper and then reconnect the ammeter in series between the battery positive terminal and the green lead at the motor.

7. Next, make sure that the motor has cooled down and then complete the circuit by connecting the jumper wire between the negative terminal on the battery and the blue lead. This will allow you a stall-torque reading in the opposite direction; both the stall-torque reading and the ammeter reading should be the same as in the earlier step.

8. A low current reading coupled with a low torque reading indicates high resistance in the internal connections or in the brush contacts. Conversely, a high current reading and a low torque reading indicates a bad armature or field windings. Either way, the motor will have to be replaced.

Trim Sending Unit

1989-95 Models

1. Check all electrical connections. Check the 50 amp trim/tilt circuit breaker and the 10 amp trim/tilt pump breaker. Check the battery.

2. Tag and disconnect the sending unit harness at the connector.

3. Install an ohmmeter, or multi-meter, as per the manufacturer's instructions. Connect the probes to the 2 sender leads and then move the sender arm through a full Up and Down cycle while observing the meter.

4. If the meter does not read 0 to 190-210 ohms through out the cycle, the unit must be replaced.

1996-98 Models

1. Check all electrical connections. Check the 50 amp trim/tilt circuit breaker and the 10 amp trim/tilt pump breaker. Check the battery.

2. Remove the sending unit from the gimbal ring. Tag and disconnect the 3-pin sender connector inside the boat.

3. Install an ohmmeter as per the manufacturer's instructions. Connect the meter between the **A** terminal black lead in the connector and the terminal **C** black lead of the sender.

4. Rotate the unit control nut counterclockwise very slowly until the meter shows infinity. Continue rotating the nut and the meter will jump to 0 ohms, rotate it more until you achieve the highest reading, which should be 620-630 ohms on 1996-97 models or 575-625 ohms on 1998 models. Disconnect the meter.

5. Now, connect the meter between the **C** terminal black lead in the connector and the terminal **B** white lead of the sender. Slowly rotate the nut counterclockwise and confirm that the meter reading is steady at 620-630 ohms on 1996-97 models or 575-625 ohms on 1998 models. Disconnect the meter.

6. Finally, connect the meter between the **A** terminal black lead in the connector and the terminal **B** white lead of the sender. Slowly rotate the nut counterclockwise until the meter shows infinity. Continue rotating the nut and the meter will jump to 620-630 ohms on 1996-97 models or 575-625 ohms on 1998 models; rotate it more until you achieve the lowest reading, which should be 0 ohms. Disconnect the meter.

7. If the sending unit fails any of the preceding tests, it will require replacement.

Trim/Tilt System

◆ See Figures 53 and 54

■ On DP units, if the sender or electrical leads see a short or open condition, the needle on the trim gauge will show erratic movement.

1. Check all electrical connections. Check the 50 amp trim/tilt circuit breaker and the 10 amp trim/tilt pump breaker. Check the battery.

2. Remove the control relay box from the side of the pump.

3. Install a multi-meter as per the manufacturer's instructions. Connect the meter leads and confirm that you have battery voltage at the red/purple lead's terminal and a good ground at the black lead's terminal.

■ There are two red/purple leads, the above test involves the terminal labeled "red/purple1" in the accompanying illustration.

4. Connect the negative meter lead to a good engine ground. Connect the positive lead to the other red/purple lead terminal. If the meter shows battery voltage, the circuit breaker or the 10 amp fuse is good. If not, reset the breaker and check it again; if the meter is still not showing the correct voltage, replace the circuit breaker or the fuse.

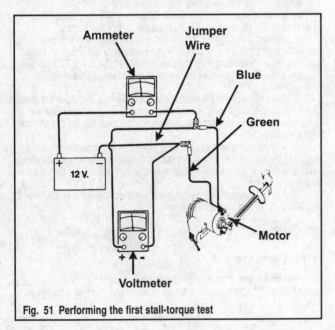

Fig. 51 Performing the first stall-torque test

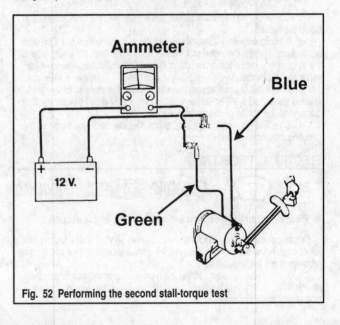

Fig. 52 Performing the second stall-torque test

Fig. 53 Removing the control box

Fig. 54 Wire colors and their related terminals

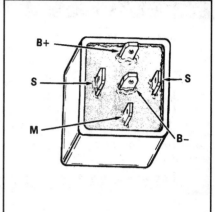

Fig. 55 Relay blade identification

■ Some later models may utilize a 10 amp inline fuse in place of the 10 amp circuit breaker.

5. Leave the negative lead connected to the engine ground and connect the positive lead to the green/white terminal. Move the trim switch to the Down position and check that the meter shows battery voltage. No voltage, or very low voltage, indicates you have a problem in the circuit between the switch and the box; check all wires and connections.

6. Now move the positive probe to the green terminal. Move the trim switch to the Down position and check that the meter shows battery voltage. No voltage, or very low voltage, indicates you have a bad trim-down relay. Replace the relay.

7. Leave the negative lead connected to the engine ground and connect the positive lead to the blue/white terminal. Move the trim switch to the Up position and check that the meter shows battery voltage. No voltage, or very low voltage, indicates you have a problem in the circuit between the switch and the box; check all wires and connections.

8. Now move the positive probe to the blue lead's terminal. Move the trim switch to the Up position and check that the meter shows battery voltage. No voltage, or very low voltage, indicates you have a bad trim-up relay. Replace the relay.

Relay Ohmmeter Test

◆ See Figure 55

Although the UP and DOWN relays may have different locations and orientations on different applications, the terminals will always be connected to the electrical system and the motor as follows:

• **S** blade (switch + or -)—blue/white, green/white or black wire
• **B+** blade (battery +)—red wire
• **B-** blade (battery -)—black wire
• **M** blade (motor)—blue or green wire

1. Mark each relay and remove them from their sockets.
2. Install an ohmmeter as per the manufacturer's instructions. Set the meter to the High scale.
3. Connect the meter leads between blades **B-** and **M** on the relay. The meter must show continuity.
4. Connect the meter leads between blades **B+** and **M** on the relay. The meter must show no continuity.
5. Now calibrate the meter to the appropriate scale and connect the leads between the two **S** blades on the relay. The meter must show 70-100 ohms.
6. Now connect a power source (12 V) to the **S** blades on the relay and then connect the meter leads across the **B+** and **M** blades on the relay. The meter must show continuity.
7. With the power source still connected to the **S** blades on the relay, connect the meter leads across the **B-** and **M** blades on the relay. The meter must show no continuity.
8. If the relay fails any of the preceding tests, it will require replacement

Service Precautions

The following nine points should always be observed when installing, testing, or servicing any part of the power trim/tilt system.

1. Coat the threads of all fittings and O-rings with Power Trim/Tilt or Steering Fluid, or an equivalent product.

2. Use only clean, new, Power Trim/Tilt and Steering fluid when filling the hydraulic fluid reservoir. If the Power Trim and Steering fluid is not available, substitute with SAE 10W-30 or 10W-40 motor oil in an emergency and then drain the system as soon as possible.

3. The stern drive must be in the full UP/OUT position to relieve hydraulic pressure when checking the fluid level or adding fluid to the reservoir. Normal fluid level should be maintained between the **MIN** and **MAX** marks on the side of the reservoir. If the quantity is low, add fluid until the level is up to the **MAX** mark on the side of the reservoir. Obviously, on models without the plastic reservoir, the level should be to the bottom of the filler hole.

4. If the pump stops during long use, allow the pump motor to cool at least five minutes before starting again. An internal thermal circuit protects the pump motor from overheating. If the pump still will not operate, the system breaker may have tripped or the fuse may have blown.

5. Keep the work area clean when servicing or disassembling parts. The smallest amount of dirt or lint can cause failure of the pump to operate.

6. The manifold, pump assembly, pressure relief valves, and thermal relief valves must be replaced as individual assemblies. There are no piece parts available for these components except for the O-rings that are sold in a kit.

7. The motor is protected from internal fluid leaks and external moisture by seals, O-rings, and a grommet on the field winding harness. If the motor fails due to hydraulic fluid, clean the motor thoroughly and replace the pump internal seal and O-ring. If the motor fails from moisture as evidenced by internal corrosion, the wire harness grommet is damaged or deteriorated, and the pump motor should be replaced. Sometimes liquid neoprene can be used on the wire harness grommet to seal minor cracks. Use liquid neoprene with care to prevent contaminating the electric motor or hydraulic valves.

8. Keep the trim cylinders attached to the forward anchor pin during servicing and repairs. Do not allow the trim cylinders to hang by their hydraulic hoses. Such practice may kink and damage the hoses. Use care when handling trim cylinders during removal/installation of the stern drive unit. Rough treatment of the hoses could result in a weakened hose, partial separation at the fittings, or bending of the metal tubing, any one of which could restrict the flow of fluid to the trim cylinders.

9. Always hold both fittings with a wrench when tightening or loosening hydraulic hoses and fittings. When installing flex hoses, be sure there are no twists or sharp bends in the flex hose.

System Testing

High pressure testing of hydraulic components requires expensive special gauges, tools and highly trained personnel to accurately interpret test results. A danger to the operator and others in the area always exists during the testing process. Therefore, it is highly recommended that the boat, with the trim/tilt system installed, be taken to a qualified repair facility having the necessary equipment to conduct the testing properly and safely.

Do not attempt to work on hydraulic hose connections without the proper tools designed for that specific purpose. The use of a common box end wrench will quite likely result in "rounding off" the flats on a hydraulic fitting. It is imperative that you have a set of "flare" or "line" wrenches that will almost guarantee the fitting will not to be damaged.

System Bleeding

■ **Please refer to the Maintenance section for procedures on filling the pump reservoir.**

The power trim/tilt system is designed to be self-purging with nothing to be done other than raising and lowering the stern drive unit several times. If, during service work, components were installed already filled with hydraulic fluid, or a line was disconnected and then reconnected, the self-purging feature of the system will adequately eliminate the small amount of unwanted air. However, if components were removed and installed "dry", the following procedures must be performed to adequately purge all air from the system.

✳✳ WARNING

Check to be sure the vent hole opening in the reservoir fill cap is free and clear. A small amount of air must be allowed to enter the reservoir tank as the fluid level drops or vacuum could develop in the tank making it difficult for the pump to draw the fluid from the reservoir.

FILLING AND BLEEDING A DRY SYSTEM

Trim Cylinders

Before installing a new or rebuilt trim cylinder:
1. Compress the trim cylinders so they are in their fully retracted (Down) position.
2. Unscrew the end caps on each cylinder (use the special spanner wrench as detailed in the Disassembly & Assembly procedure) and fill the low-volume chamber with OMC Power Trim/Tilt & Steering Fluid. Install the end caps and tighten to 25-30 ft. lbs. (34-41 Nm).
3. Install the cylinders and connect the hydraulic lines.

Hydraulic Lines & Manifold

REMOVAL & INSTALLATION

■ **It will be necessary to remove the stern drive unit in order to gain access to the inner manifold line connections.**

1. Loosen and remove the hydraulic line retaining clamps on each side of the gimbal housing.
2. Although not strictly necessary, we recommend removing the grease extension tube in order to gain access to the starboard hydraulic lines. Clamp a pair of vise grips onto the tube (not too tight please) as close to the gimbal bearing bore as possible. Insert a small prybar between the pliers and the bearing flange and then pry the tube out with the bar while rotating back and forth with the pliers.
3. Loosen each of the hydraulic line nuts at the manifold and carefully pull each line out of its recess. Remove each O-ring and throw them away. It's a good idea to plug the line ends with a golf tee or something similar to minimize fluid leakage.

4. Move inside the vessel and remove the 2 lines connected to the inner side of the manifold. Mark them for installation, plug them and secure them out of the way.
5. Back outside the boat, loosen and remove the mounting nut on the upper side of the manifold from the stud. The manifold is a tight fit in the housing with an O-ring—you may be able to pull it right out, but you'll probably need to position a small drift on the back (inner) side and force it through.

To Install:

6. Coat a new manifold O-ring with Gasket Sealing compound and slide it onto the manifold flange. Coat 2 new O-rings for the inner hydraulic lines with hydraulic fluid and slide them over the line tips.
7. Pull the 2 inside lines through the bore in the housing and then position the manifold near the bore so that the inner side is facing to Starboard. Route the lines to the manifold so they are parallel and then tighten the fittings to 14-18 ft. lbs. (19-24 Nm) on 1989-97 units, or 84-108 inch lbs. (9.5-12.2 Nm) on 1998 units. It is imperative that the lines get reinstalled on the port from which they were detached! As a check, once installed, the lower line should be attached to the **UP** inlet on the trim pump manifold.

■ **Yes, we know it is unlikely that you will be able to get a torque wrench on the line nut—do it by feel, its about as tight as a spark plug, just don't over-tighten them.**

8. Feed the lines and through the hole in the housing until the manifold seats on it's mounting stud. Install a new locknut onto the stud and tighten it to 108-132 inch lbs. (12-15 Nm).
9. Coat 4 new O-rings with hydraulic fluid and install them on each of the outer (gimbal housing side) hydraulic line tips. Attach the lower of the 2 Port lines and then the upper. The shorter of the 2 Starboard lines should be connected to the lower of the two fittings on the starboard side of the manifold and then connect the remaining (longer) line to the upper fitting. Tighten all lines to 14-18 ft. lbs. (19-24 Nm) on 1989-97 units, or 84-108 inch lbs. (9.5-12.2 Nm) on 1998 units—see the previous Note.
10. Route each pair of lines together and install the retaining clamp on each side. Tighten the bolts securely.
11. If you removed the grease extension tube, pull the grease fitting off of the outer end of the tube. Coat the last 1/4 in. of the tube on the fitting side with Gasket Sealing compound and then insert the tube through the outer wall of the housing.
12. Spray the tapered end of the tube and the hole in the housing with Loctite primer and allow it to dry completely. Once dry, coat the tapered end with Loctite and then drive the end of the tube into the bearing flange. Wipe off any excess Loctite and then install the grease fitting to the opposite end of the tube.

Trim Limiter

Some DP drives utilize a trim limiter system consisting of a limiter module, override switch and associated wiring. This system has three set points which automatically limit the trim range of the drive unit. These points are set at the factory and based on a boat with an average 13 degree transom angle. Trim is limited between +5 degrees and -2 degrees, while tilt is limited to 45 degrees.

These limits, by necessity, may require changing to accommodate different transom angles or swim platforms. Although changing the parameters is not particularly difficult, it is a serious proposition with strong consequences if reset to the wrong figures.

The boat must be out of the water and in a sling when performing this procedure—the hull must be level. You must also know the transom angle—consequently, we recommend having this procedure performed at an authorized OMC service facility.

Trim Motor And Pump

1. Mount the motor/pump assembly on the transom and connect all hydraulic hoses.
2. Remove the reservoir filler plug (or cap) and fill the unit with OMC Power Trim/Tilt & Steering Fluid until the fluid level reaches the bottom of the hole (or the **MAX** line on the side of the reservoir if you happen to have a translucent plastic reservoir).

■ You can also use GM Dexron II ATF fluid if necessary.

3. Press in the UP button on the trim switch, allow it to raise slightly and then stop it. Check the fluid level and add if necessary. Continue this "raise and stop" procedure, filling at each stop if necessary, until the drive is in the full UP/OUT position and the trim cylinders are fully extended.

4. Check the fluid level again and then install the filler plug of reservoir cap, tightening either on securely.

5. Run the drive unit through at least 5 Up/Down cycles to purge any residual air in the system, stop the drive in the full Up position and check the fluid level one last time; topping off if necessary.

Trim Sender

ADJUSTMENT

1989-92 Models

■ Before adjusting the trim sender, the sender must be installed in the gimbal ring and the drive must be installed on the boat.

1. Turn the drive over hard to port and set it to full negative (Down) trim. Turn the ignition switch to the **ON** position; if the trim gauge shows the drive to be in the Full Down position there is no need to proceed with the adjustment, if not, move to the next step.

2. Loosen the 2 trim sender mounting screws, turn the ignition switch back to the **ON** position and then slowly rotate the sender until the trim gauge registers a Full Down position. Obviously, you'll need an assistant for this step unless you've got a long neck and great eyes.

3. Tighten the sender mounting screws to 18-24 inch lbs. (2.0-2.7 Nm) and then recheck the gauge.

1993-98 Models

◆ See Figures 56 and 57

1. Turn the drive unit over hard to port and set it to the full Down Position.

2. Loosen the 2 mounting bolts and remove the sending unit from the gimbal ring.

3. Disconnect the sending unit harness connector.

4. Install an ohmmeter as per the manufacturer's instructions and connect the probes between the **A** and **C** terminals in the connector; these will both have black wires attached to them. Slowly turn the control nut on the bottom of the sending unit until the meter reads 10-12 ohms.

5. Reinsert the sending unit into the gimbal ring and tighten the bolts finger-tight. Check the gauge again and slowly rotate the installed unit until the figure is achieved once again.

6. Tighten the bolts to 18-24 inch lbs. (2.0-2.7 Nm) and then recheck the gauge. Remove the ohmmeter and connect the sender lead.

◆ See Figure 58

■ Complete removal of the sending unit will frequently damage the wiring harness, so do not remove the unit unless it is absolutely necessary.

1. Turn the drive unit over hard to port and ensure that it is in the full Down position.

2. Loosen the 2 mounting bolts and pull the sender out of the cavity in the gimbal ring.

3. Working inside the boat, disconnect the trim sender harness connector at the engine.

4. Take note of the wire positioning inside of the plug and then use a socket removal tool (# 3854350) and push the wire terminals out of the connector plug.

5. Now, back outside the vessel, locate the harness grommet where it feeds through the transom assembly—just above the water tube in the gimbal housing. Carefully pull off the retaining clip with a pair of needle nosed pliers.

6. Grasp the grommet and pull backward on it while prying it out with a small screwdriver. Once free of the housing, pull the wires all the way through the transom assembly. Cut any plastic ties securing the harness and then remove the trim assembly.

To Install:

7. Coat the harness grommet with grease and then feed the leads through the hole in the gimbal housing and press the grommet into place until it is seated. Pop the retaining clip into position over the base of the grommet.

✳✳ WARNING

Make sure that you do not route the harness leads under the extension tube or hydraulic lines.

8. Coat the connector plug with alcohol or solvent and then use the special tool and press the leads back into the connector plug until they seat—make sure they go back in the same positions as they were prior to removal. On certain 1998 models, the ribbed black wire goes into the **A** terminal, the white wire into the **B** terminal and the smooth black wire into the **C** terminal.

9. Reconnect the harness at the engine.

10. Use a new plastic tie and secure the harness to one of the hydraulic lines.

11. Install the sending unit into the gimbal ring and tighten the mounting bolts to 18-24 inch lbs. (2.0-2.7 Nm). Adjust the sending unit if necessary.

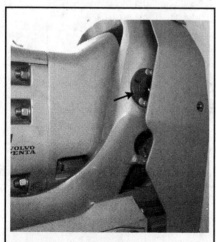

Fig. 56 Trim sender (SP shown)

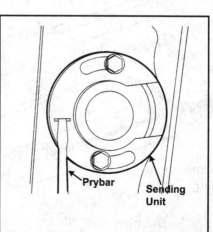

Fig. 57 Use the small slot and a pry bar to rotate the sending unit for final adjustment.

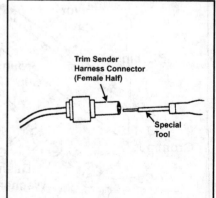

Fig. 58 Use a special tool to pop the leads out of the connector.

Trim/Tilt Cylinder

■ If troubleshooting procedures have isolated a problem to the trim/tilt cylinders, for example: a leaking oil scraper seal around the rod or a defective impact valve, it is strongly recommended that the cylinder(s) be removed and replaced with new ones, rather than attempting to disassemble the cylinders. This recommendation is based on the fact considerable difficulty may be encountered in removing the end cap from the cylinder. Even with the aid of the special tool for this purpose, the task is most difficult.

REMOVAL & INSTALLATION

◆ See Figures 59 thru 64  *MODERATE*

1. Remove the plastic cap on the aft (drive) end of each trim/tilt cylinder. Remove the elastic lock nut and flat washer; on certain 1998 units, the nut may have been replaced with an e-clip so remove the clip and flat washer from the rod.

2. Support the hydraulic cylinder and lightly tap the pivot rod out of the cylinders with a rubber mallet—tap on one side while pulling on the other side. Carefully lower the cylinders down until they rest against the spray plate.

■ If the bushings come out with the pivot rod, make sure that you keep the two grounding clips.

3. Grab the pivot ends of each cylinder and pull them out to the fully extended position. Loosen the manual release valve on models so equipped and then remove the filler plug at the reservoir to relieve any residual pressure in the system. Close the valve and install the cap. On later models without the release valve, simply loosen one of the hydraulic hoses at the pump and then remove the reservoir cap, but make sure you've got a shop towel over it and the hose connection because there may be some spray.

4. Working at the forward end of the cylinder, unscrew the bolt and disconnect the ground wire. Remove the cylinder cover retainer and lift off the cover. These models will also have a small screw and line retainer over the hydraulic line running to the aft end of the cylinder; remove them.

5. With the stern drive still in the full DOWN position, place a suitable drain pan under the hydraulic lines, tag and then disconnect the two hydraulic hoses from the end of the cylinder (cover with rag) using the correct size "Flare Nut" or "Line Wrench". These wrenches will prevent damaging the hex flats on the line or hose fittings if they are extremely tight and/or a slight amount of corrosion has built up on the fitting. Most standard open-end wrenches will flex under high torque loads, causing the wrench to slip, damaging the hex fitting on the hydraulic line.

6. Remove the O-rings on the end of each line fitting. Install plugs into the hoses and/or fittings to prevent draining the trim/tilt hydraulic system any more than necessary.

7. Pry off (or unscrew) the plastic pivot pin cap on the end of the forward pin. Remove the locknut or pull the E-Clip from the end of the pin and then slide off the flat washer and then the bushing from the pivot pin.

8. Pull the pivot pin out of the cylinder and gimbal ring, or grasp the cylinder on the inside of each pin and pull it off the anchor pins. Remove the flat washer and bushing on the inside surface of the cylinder ends or the anchor pins.

To Install:

9. Press the trim/tilt cylinder bushing into the boss on each side of the gimbal ring. Install bushings into the outer side of each cylinder. Although it is not necessary to replace the bushings unless they are worn excessively or damaged, you may want to consider replacing them anyway while you have everything apart.

10. Clean the forward pivot pin and then coat it with grease. Apply a light coat of Gasket Sealing Compound to the threaded ends of the pin and then center it in the ring.

11. Align the cylinder and then press it on to the pivot pin very carefully. You may have to give it a few taps with a rubber mallet.

12. On cylinders with lock nuts, install the washers and then the nuts (new!). Tighten the nuts until there is an equal amount of thread exposed on each side of the pin and then tighten both nuts to 32-34 ft. lbs. (43-46 Nm) on 1989-93 models, or 10-12 ft. lbs. (14-16 Nm) on all others. On models using the e-clip, install the washer and then install the e-clip into the groove on the pin. Replace the plastic caps.

13. Coat new O-rings with hydraulic fluid and then install them into the grooves on the end of each hydraulic line fitting. Remove the plugs and attach each line to its respective fitting on the cylinder. Tighten each fitting to 14-18 ft. lbs. (19-24 Nm) on 1989-93 models, or 84-108 inch lbs. (9.5-12.2 Nm) on all others, with a flare wrench.

14. Install the aft line retainer and tighten the bolt securely. Also on these models, position and install the cylinder cover.

15. Install the retainer; and then on all models, install the ground strap and tighten the bolt securely.

16. Clean the aft pivot pin thoroughly and then coat it with grease.

17. Make sure that the grounding clips are in position and then press the bushings into the end of each cylinder and into the bore boss in the drive unit. As we did with the forward end bushings, we suggest replacing the bushing as preventative maintenance, although it is perfectly acceptable to reuse the old bushing as long as they are in good shape.

18. Align the cylinder ends with the bores on each side of the drive and then insert the pivot pin into one end and press it through until it comes out the other side—you will probably have to give it a few taps with a mallet.

19. Install the flat washers and lock nuts or E-clips. On models with nuts, tighten the nuts until there is an equal amount of thread showing on each side and then tighten each nut to 32-34 ft. lbs. (43-46 Nm) on 1989-93 models, or 10-12 ft. lbs. (14-16 Nm) on all others. Install the plastic caps.

20. Fill the reservoir with Power Trim/Tilt & Steering fluid and run the drive through its full range of motion at least 5 times to purge any air in the system.

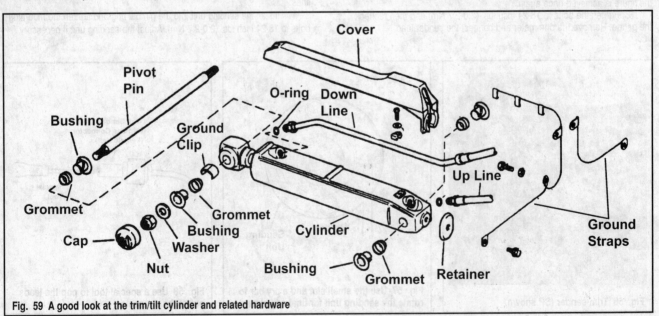

Fig. 59 A good look at the trim/tilt cylinder and related hardware

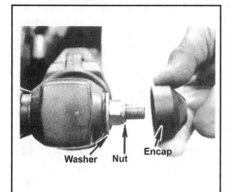

Fig. 60 Most models use a nut on the pivot pin...

Fig. 61 ...while certain later models may use an E-clip

Fig. 62 Tap one end of the pivot pin lightly...

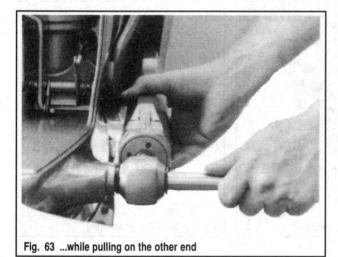

Fig. 63 ...while pulling on the other end

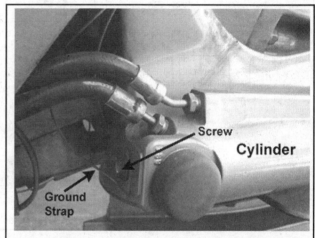

Fig. 64 A close-up of the forward end of the cylinder (later model shown)

DISASSEMBLY & ASSEMBLY

OEM   ③   DIFFICULT

◆ **See Figures 59 and 65 thru 71**

1. Remove the trim cylinder(s) as previously detailed.
2. Hold the cylinder over a drain pan with the hydraulic line ports facing down and into the pan; extend, and then retract the cylinder 2-3 times to remove all fluid from the cylinder.
3. Remove the pivot pin bushings if you intend to replace them.
4. Place the cylinder in a vise equipped with soft jaws and carefully tighten it.
5. Obtain a spanner wrench (# 912084) or an equivalent tool. Removal of the end cap is difficult using the special tool so exercise care. Insert the tangs of the spanner into the holes in the end cap. If needed, slide a long breaker bar onto the tool so the bar is in the same plane as the tool. This position will provide maximum mechanical advantage. Remove the end cap. Tap the breaker bar, if necessary, but bear in mind—if the holes in the end of the cap become damaged (elongated), the cylinder might as well be given the "deep six". Continue to unscrew the end cap until it is held by a single thread. Extend the rod, and then continue to remove the end cap and piston assembly.
6. Clean and degrease the piston rod thoroughly and then install it in a vise using a rod holder (# 983213). Carefully insert the pivot pin through the hole in the end piece and spin it off the end of the rod. If you are unable to loosen the end piece, heat the shank very carefully with a torch or heat gun and then try it again.

■ **If you find that the piston rod is slipping in the holder tool while trying to spin off the end piece, wrap a piece of paper around the rod before inserting it into the tool.**

7. Pull the end cap off of the end of the rod and then pry out the scraper with a small awl. Do the same with the 2 O-rings, one inside and one outside. Discard the scraper and the O-rings.
8. Move to the end piston and remove the 2 split rings and the O-ring; throw them all away.
9. There are no serviceable components in the assembly, so there is no service procedure. This being said though, if you suspect impact valve leakage there is a quick procedure to try before replacing everything. Install the piston rod in the holder tool again and then loosen the large bolt on the end of the piston just enough to allow the relief check balls to unseat themselves; but not too much. Clean the piston components with solvent and then blow everything dry with compressed air. Tighten the bolt securely until the balls seat themselves.

To Assemble:

10. Make an effort to keep the work area clean, because any contamination on the piston could lead to a malfunction. Inspect the interior of the cylinder for any signs of scoring or roughness. Clean all surfaces with safety solvent and blow them dry with compressed air.
11. If the piston has any signs of corrosion or damage it must be replaced with a new unit.
12. Clean the end cap threads with a wire brush until you have removed all traces of old sealant.
13. Lubricate all O-rings and interior components with Power Trim/Tilt and Steering Fluid.
14. Position a new scraper over the end cap so that the lip is facing upward and then use a small seal installer (# 326545), or an appropriately sized socket, to press the scraper into the cap.
15. Coat new O-rings with fluid and then install them into their respective grooves in side and outside the end cap.
16. Coat a new piston O-ring with fluid and roll it into place in the recess in the piston. Position new split ring retainers on each side of the O-ring so that their ends match but do not overlap. These rings are quite fragile, so be careful when installing them into the recess.

Fig. 65 Drain the cylinder into a container

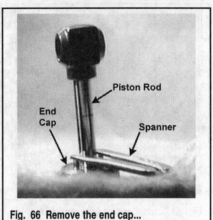
Fig. 66 Remove the end cap...

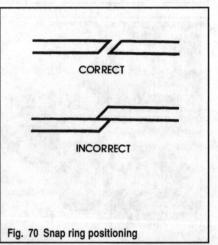
Fig. 67 ...and then pull the piston rod out of the cylinder

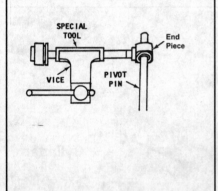
Fig. 68 Use the pivot pin to remove the end piece

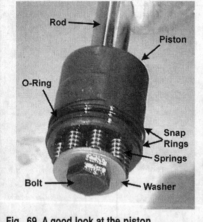
Fig. 69 A good look at the piston

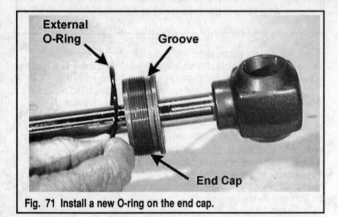
Fig. 70 Snap ring positioning

Fig. 71 Install a new O-ring on the end cap.

17. Install a seal protector tool (# 326005) over the end of the piston rod and slide on the end cap. Make sure that the rod, O-ring and cap are thoroughly lubricated with hydraulic fluid.

18. Make sure that the piston rod threads are free of all old sealant and then spray some Loctite Primer on the threads. While waiting for it to dry, secure the rod in the holding fixture again and then coat the threads with Loctite. Screw on the end piece and tighten it securely. Remove the rod and tool from the vise.

19. Re-clamp the cylinder in the vise and coat the cylinder bore and piston thoroughly with fluid. Grab the piston body so that you compress the split rings with your fingers and then feed it into the cylinder bore. Push the piston and rod all of the way into the cylinder. Once the piston is completely in, cap the hydraulic line ports and fill the cylinder with as much hydraulic fluid as possible.

20. Slide the end cap down the rod and thread it into the cylinder. Install the spanner wrench and tighten it to 25-30 ft. lbs. (34-41 Nm). Once there, continue rotating the end cap until the two holes are parallel with the pivot pin hole in the end piece.

21. Attach the cylinders, check the fluid level in the reservoir and then run the drive through its full range of motion 5 times to purge any remaining air in the system.

Trim/Tilt Pump Assembly

The pump motor is a 12-volt DC bi-directional motor. The armature of the motor rests inside the field winding and is energized through brushes contacting the commutator on the end of the armature. When the up or down switches are closed, current is directed through the motor field in one of two directions causing the motor to run in one desired direction.

The opposite end of the armature is mechanically splined to the hydraulic pump. The gear rotor action of the pump causes the fluid to flow in either direction.

The motor case is sealed with O-rings and a shaft seal to prevent the entry of water or hydraulic fluid into the motor cavity.

Before going directly to the pump as a source of trouble, check the battery to be sure it is up to a full charge. Inspect the wiring for loose connections, corrosion and damaged wires. Take a good look at the control switches and connection for evidence of a problem.

The control switches may be quickly eliminated as a source of trouble by connecting the pump directly to the battery for testing purposes.

REMOVAL & INSTALLATION

◆ See Figures 72 and 73

■ Integrated reservoir models are identifiable because they have the motor on the bottom and the reservoir on the top. There may be some models in this year span that came through with the newer style pump; motor on top and clear plastic reservoir on the bottom.

1. Using the trim switch, raise the stern drive to the full UP position and then support the bottom of the unit with a piece of wood. Remember that on early models, you can raise the drive even if the system is not working by loosening the manual release valve.

2. Disconnect the battery cables from the battery terminals.

3. Relieve the pressure in the system. If your pump is equipped with a manual release valve, loosen it a few turns; if you have a later model, carefully unscrew the reservoir cap very slowly while covered with a rag until any residual pressure has escaped.

4. Mark or tag the hydraulic hoses (right or left) before disconnecting them from the pump. The inlet and outlet ports on the manifold are also usually stamped with **UP** or **DN** markings identifying each port, but the right hose (as you face the pump) is the **UP** circuit and the left hose is for the **DOWN** circuit. Place a suitable container under the fittings and then use a line nut wrench to back the fitting out of the manifold port. Cap the hoses to prevent fluid leakage into the boat and dirt from entering the system. OMC actually sells actual line and port caps if you are interested, otherwise you can improvise with something from your shop.

5. Disconnect the green and blue wires at the terminal board on the side of the assembly. Cut the plastic tie securing the wires and the harness to the pump body and move them out of the way.

6. Remove the screw at the top of the control relay box, slide the bottom free of the lower mount and then remove the box and lay it out of the way.

7. Remove the lag bolts securing the trim/tilt pump assembly to the boat. Lift the assembly out and free of the boat.

8. Try to keep the assembly standing upright while out of the boat.

To Install:

9. Position the trim/tilt pump assembly in the boat. Install the lag bolts through the pump support bracket and into the transom. Tighten the bolts securely.

10. Insert the lower end of the control box into the notch on the pump and then secure it with the Phillips screw. Reconnect the blue and green leads and secure the harness to the bracket with a new plastic tie. Don't forget to coat the screws and terminals with liquid Neoprene.

11. Connect the hydraulic hoses to their ports in the manifold—make sure they thread into the correct port. Tighten the line nuts securely.

12. Fill the reservoir with OMC Power Trim/Tilt and Steering Fluid to the bottom of the filler hole.

13. Connect the battery cables and then cycle the stern drive Up and Down several (at least 5) times to bleed all air from the system. Check the reservoir level after each cycle and service, if needed.

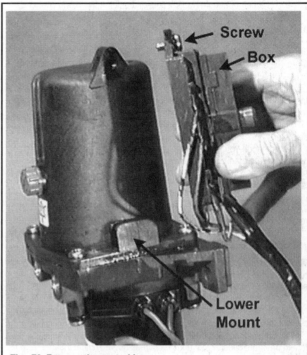

Fig. 73 Remove the control box

DISASSEMBLY & ASSEMBLY

◆ See Figurea 41 and 74 thru 93

1. Remove the pump assembly.

2. Drain all the remaining fluid in the reservoir into a suitable container. Discard the fluid in accordance with any local hazardous waste ordinances.

3. Remove the 4 reservoir-to-manifold mounting screws (Phillips) and then hold the assembly over a suitable container while you separate the two units. Remove the O-ring from the bottom of the reservoir and throw it away.

4. Turn the remaining assembly over and remove the 3 motor-to-manifold mounting screws (Phillips). Separate the motor from the manifold and then remove the O-ring and discard it.

5. Remove the round mesh filter screen from the manifold. Clean it thoroughly and set it aside for assembly.

6. Pull the drive coupling out of the manifold and check the slotted and hex ends for wear or damage—replace as necessary.

7. Loosen the 4 pump-to-manifold Allen bolts alternately and only ONE TURN at a time until you can lift off the gear housing (pump). Make sure that both pump control pistons and their check balls stay in the housing. If equipped with an oil filter, pull it out of the manifold.

✳✳ WARNING

The pump housing contains pistons with springs that exert an upward pressure on the unit; it is imperative that when loosening the mounting screws, they only be turned one complete revolution at a time, moving on to the next screw diagonally across from it each time until all spring pressure has been eliminated. Failure to follow this procedure will result in a warped housing.

■ The mating surfaces of the pump housing and manifold are precision machined and do not use a gasket; extreme care should be exercised during removal. Be very careful no to gouge, scratch or mar either of these services.

Fig. 72 Disconnect the hydraulic hoses at the pump

8. Turn the pump housing over and mark the exposed end of each pump gear with an indelible marker so that you know which end of which gear goes where. Lift the two gears out of the housing. The gears may not always be in the housing as they frequently stay in the manifold after lifting off the housing.

9. Us a pair of needle-nosed pliers and pull the two pump control pistons out of their bores in the housing. Reach in and remove the check balls.

10. Turn the housing back over and insert a small piece of stock through the hole in the tilt-up relief valve bore. Press the valve out of the bore, remove the inner spring and then pull out the O-ring from the groove on the valve. If the housing in your application has no hole in the back of the bore, carefully drill a 1/8 in. hole in the housing as indicated in the illustration. Inspect the valve, spring and check ball for wear, scoring or other damage, replacing as necessary.

11. Repeat the previous step for the trim-in relief valve (smaller diameter).

12. Although you have probably already done so, remove the pump gears from the manifold if this is where they stayed after lifting off the housing. Either way, check the excessive wear or chipped teeth. Remove the checks balls.

13. On 1989-96 models, insert a small metal pin into the manual release valve check ball hole on the valve body and force out the check ball and retainer. Pull the O-ring off of the retainer and discard it.

14. On 1989-96 models, remove the small inner snap ring in the release valve bore and unscrew the valve. Throw away all O-rings.

■ **The thermal expansion valve is located inside the tip of the manual release valve on most 1989-96 models and cannot be serviced separately.**

15. On 1997-98 models, remove the thermal expansion valve and then pull out the O-ring from the groove in the bore.

16. Squeeze some grease into the hydraulic line opening adjacent to the trim-in or tilt-out check valve until the bore is about 1/2 full. Install a 7/16 in. fine thread bolt into the opening and slowly tighten it slowly until the check valve pops out of the bore. Lift out the valve, seat and spring. If the valve does not com out all the way, remove the bolt and squeeze in some more grease; then attempt tightening the bolt again.

To Assemble:

17. Thoroughly clean all components in solvent and blow them dry with compressed air. Pay particular attention to drying all manifold passages. Keep everything clean and lint-free.

18. Inspect the pump drive coupling for excessive wear or damage. Check both gears for wear and chipped or damaged teeth. Check all valves and bores for any sign of scoring or corrosion. Check the machined mating surfaces of the pump housing and manifold for scratches, grooves or other marring.

19. Coat all components thoroughly with Power Trim/Tilt & Steering fluid before installation—do not assemble any component dry!

20. On 1989-96 models, coat new O-rings and the nylon sealing ring with hydraulic fluid and install them onto the manual release valve. Coat the valve thoroughly and thread it into the bore in the manifold, tightening it to 45-55 inch lbs. (5-6 Nm). Do not over-tighten the valve. Install the inner snap ring into the groove in the valve bore.

21. On 1989-96 models, drop the manual release valve check ball into the oil passage. Install a new O-ring on the retainer, coat them with hydraulic fluid thoroughly and then press the retainer into the bore until it is flush with the surface of the manifold. Make sure that you do not push the retainer in any further than flush or it will restrict check ball movement and affect release valve functions.

22. On 1997-98 models, install the thermal expansion valve after inserting a new O-ring into the bottom of the bore and tighten it to 45-57 inch lbs. (5-6 Nm).

23. Position a check valve spring into each bore in the manifold and insert the check valve over the spring so that the tip is facing upward. Install a new O-ring into the groove in the seat and then carefully install the seat over the valve so it is centered, pressing it in until it is fully seated. If you encounter difficulty, position an appropriately sized socket over the seat and lightly tap it into place.

24. Install a new O-ring into the groove on the trim-in relief valve seat (piston) and coat them, and the valve with hydraulic fluid. Invert the pump housing and insert the valve spring into the bore, followed by the valve itself. Position the seat over the valve and press it in until it seats itself. Repeat this procedure for the tilt-up relief valve.

25. Insert the check balls for each pump control piston into it's bore in the manifold. Position the piston over the ball and press it into the bore.

26. Coat the teeth of the two drive gears, and their bores, with grease and then insert them into their respective bores. Remember the markings that you made on each gear during disassembly—the marks must be facing away from the housing.

■ **If you removed the drive gears from the manifold instead of the housing during disassembly, then your mark would be facing away from the manifold and NOT the pump housing.**

27. Insert a check ball into each of the reservoir inlet passages.

28. Install the small oil filter into the manifold if your pump was equipped with one.

29. Position the pump housing onto the manifold mating surface and insert each of the screws. Tighten the screws to 35 inch lbs. (4 Nm), one full rotation at a time, in the sequence illustrated.

30. Insert the drive coupling into the drive gear and rotate it slowly by hand to make sure there is no binding evident. If you are unable to turn it by hand, loosen the pump housing screws a little (remember, one turn at a time) and then try it again. If the coupling now turns, you've got a problem in the housing; check all components and start over again.

31. Install the round reservoir filter screen onto the manifold so that the foam gasket is facing toward the motor.

32. Pour hydraulic fluid into the cavity surrounded by the filter screen until it reaches the top of the casting. Insert a screwdriver into the drive coupler and rotate it until you no longer see air bubbles coming out of the check ball opening. The pump is now primed.

33. Install a new O-ring onto the motor and then position it to the manifold so the mounting holes line up correctly. Now rotate the coupler until the slot lines up with the blade on the motor shaft and then install the motor. Tighten the bolts to 35-52 inch lbs. (4.0-5.9 Nm).

34. Turn the assembly over and install a new O-ring on the reservoir flange. Position the reservoir on the manifold so that the bolt holes are in alignment and then tighten the bolts to 35-52 inch lbs. (4.0-5.9 Nm).

35. Install the assembly in the boat, fill with fluid and test for leaks.

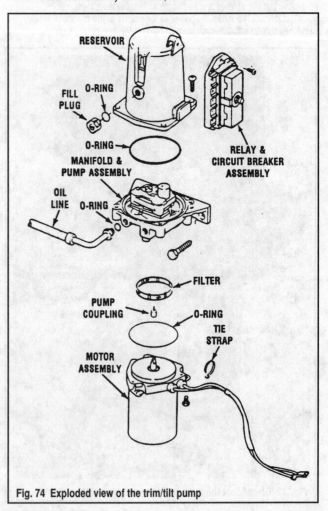

Fig. 74 Exploded view of the trim/tilt pump

Fig. 75 Drain any residual fluid into an appropriate container...

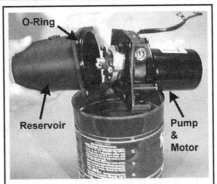

Fig. 76 ...keeping the assembly over the container when separating the reservoir

Fig. 77 Separate the motor from the manifold...

Fig. 78 ...remove the filter screen...

Fig. 79 ...and then lift out the coupling

Fig. 80 Loosen the pump housing bolts ONE TURN at a time...

Fig. 81 ...and then lift the housing from the manifold

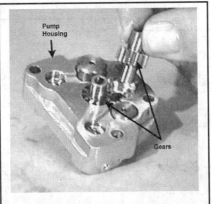

Fig. 82 Pull the drive gears out of the housing after they have been marked...

Fig. 83 ...and then remove the small oil filter from the manifold (if equipped)

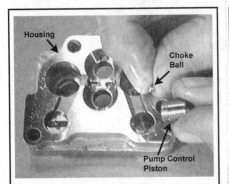

Fig. 84 Remove the pump control pistons

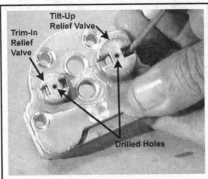

Fig. 85 Removing the relief valves

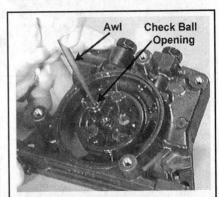

Fig. 86 Press out the manual release valve check ball and retainer (1989-96)

Fig. 87 Remove the manual release valve snap ring on 1989-96 models...

Fig. 88 ...and then unscrew the valve

Fig. 89 Thread a bolt into each bore...

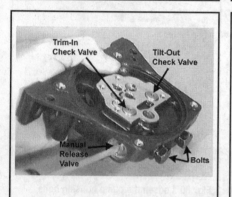

Fig. 90 ...and then turn the release valve (early models only)

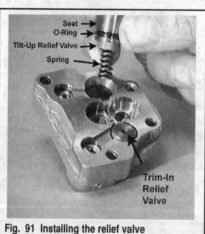

Fig. 91 Installing the relief valve

Fig. 92 Installing the manual release check valve ball retainer

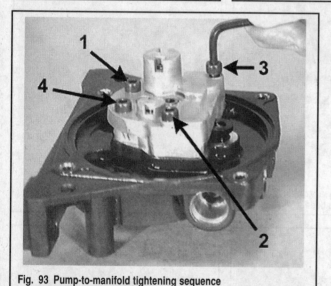

Fig. 93 Pump-to-manifold tightening sequence

PRESSURE TESTING THE MANIFOLD

1. Thread a 10/32 x 1/2 in. screw with a 5/32 in. flat washer into the pump housing mounting bolt hole nearest the trim-out-tilt-up check valve, making sure that the washer does not cover the small hole in the center of the valve seat.

2. Connect a pressure tester adapter (# 914459) to a Stevens Pressure Tester S-34. Lubricate an hydraulic hose O-ring and install it into the groove on the end of the adapter.

3. Thread the adapter end into the **UP** port on the manifold. Brush some soapy water onto the top of the manifold or submerse it in a container of Power Trim/Tilt & Steering fluid and then apply 30 psi to the manifold and check for any leaks.

4. Repeat Steps 1-3 for the trim-in/tilt-down check valve and manual release check ball retainer.

5. Replace any component showing a leak during the test.

WIRING SCHEMATICS

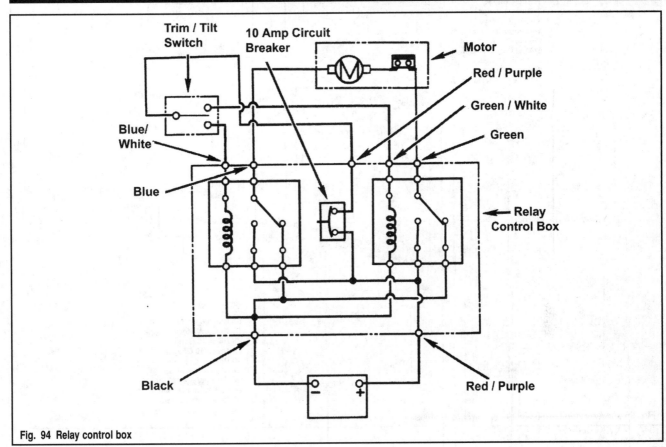

Trim / Tilt Switch

10 Amp Circuit Breaker

Motor

Red / Purple

Green / White

Green

Blue/ White

Blue

Relay Control Box

Black

Red / Purple

Fig. 94 Relay control box

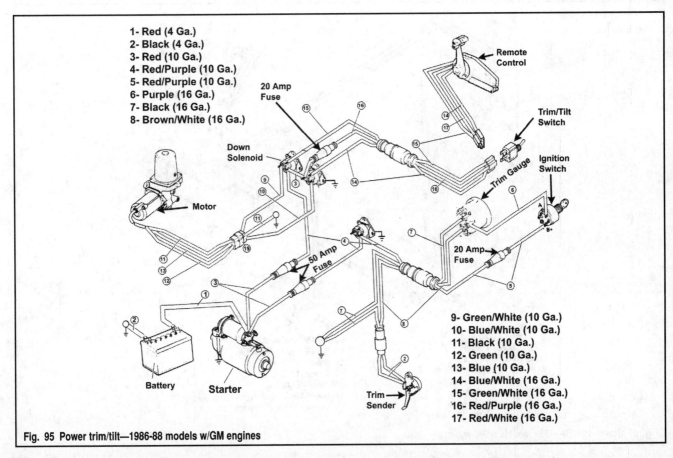

1- Red (4 Ga.)
2- Black (4 Ga.)
3- Red (10 Ga.)
4- Red/Purple (10 Ga.)
5- Red/Purple (10 Ga.)
6- Purple (16 Ga.)
7- Black (16 Ga.)
8- Brown/White (16 Ga.)

Remote Control

Trim/Tilt Switch

Ignition Switch

Trim Gauge

20 Amp Fuse

Down Solenoid

Motor

50 Amp Fuse

20 Amp Fuse

Battery

Starter

Trim Sender

9- Green/White (10 Ga.)
10- Blue/White (10 Ga.)
11- Black (10 Ga.)
12- Green (10 Ga.)
13- Blue (10 Ga.)
14- Blue/White (16 Ga.)
15- Green/White (16 Ga.)
16- Red/Purple (16 Ga.)
17- Red/White (16 Ga.)

Fig. 95 Power trim/tilt—1986-88 models w/GM engines

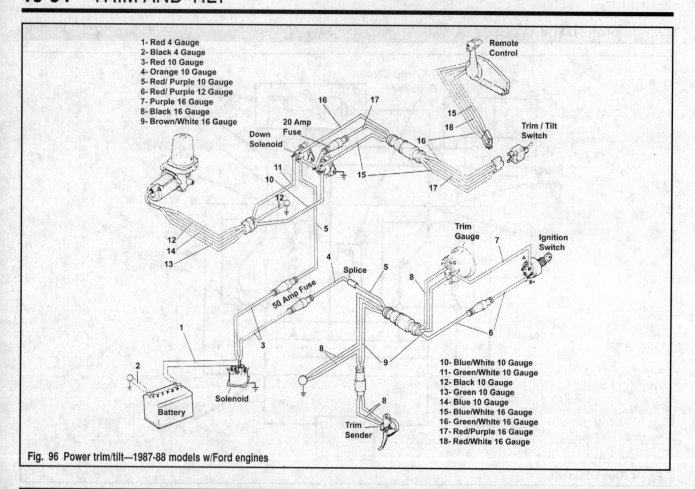

1- Red 4 Gauge
2- Black 4 Gauge
3- Red 10 Gauge
4- Orange 10 Gauge
5- Red/ Purple 10 Gauge
6- Red/ Purple 12 Gauge
7- Purple 16 Gauge
8- Black 16 Gauge
9- Brown/White 16 Gauge

10- Blue/White 10 Gauge
11- Green/White 10 Gauge
12- Black 10 Gauge
13- Green 10 Gauge
14- Blue 10 Gauge
15- Blue/White 16 Gauge
16- Green/White 16 Gauge
17- Red/Purple 16 Gauge
18- Red/White 16 Gauge

Fig. 96 Power trim/tilt—1987-88 models w/Ford engines

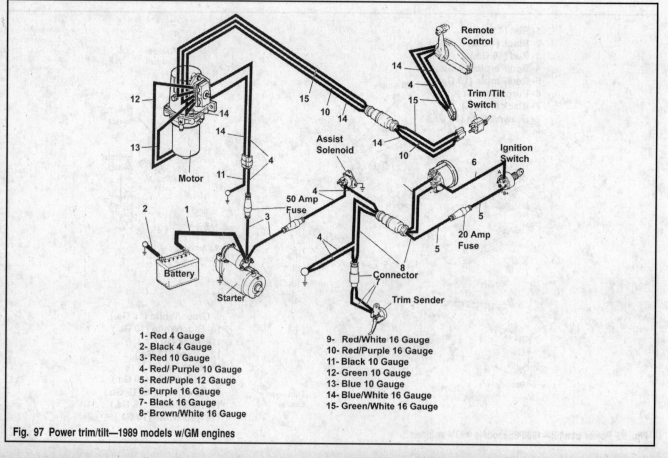

1- Red 4 Gauge
2- Black 4 Gauge
3- Red 10 Gauge
4- Red/ Purple 10 Gauge
5- Red/Puple 12 Gauge
6- Purple 16 Gauge
7- Black 16 Gauge
8- Brown/White 16 Gauge

9- Red/White 16 Gauge
10- Red/Purple 16 Gauge
11- Black 10 Gauge
12- Green 10 Gauge
13- Blue 10 Gauge
14- Blue/White 16 Gauge
15- Green/White 16 Gauge

Fig. 97 Power trim/tilt—1989 models w/GM engines

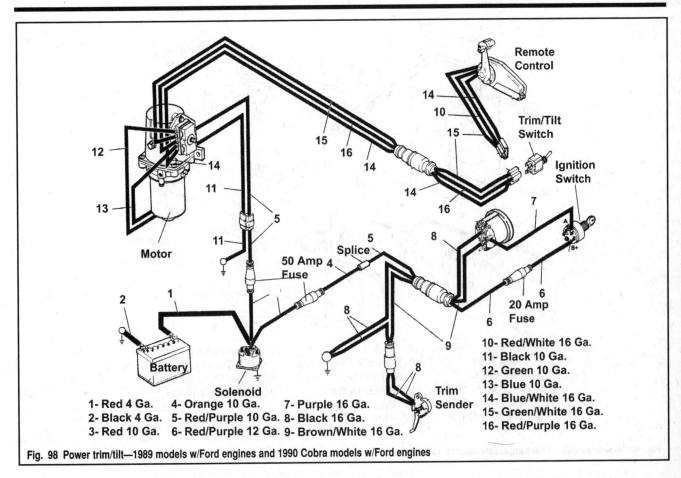

1- Red 4 Ga. **4-** Orange 10 Ga. **7-** Purple 16 Ga.
2- Black 4 Ga. **5-** Red/Purple 10 Ga. **8-** Black 16 Ga.
3- Red 10 Ga. **6-** Red/Purple 12 Ga. **9-** Brown/White 16 Ga.

10- Red/White 16 Ga.
11- Black 10 Ga.
12- Green 10 Ga.
13- Blue 10 Ga.
14- Blue/White 16 Ga.
15- Green/White 16 Ga.
16- Red/Purple 16 Ga.

Fig. 98 Power trim/tilt—1989 models w/Ford engines and 1990 Cobra models w/Ford engines

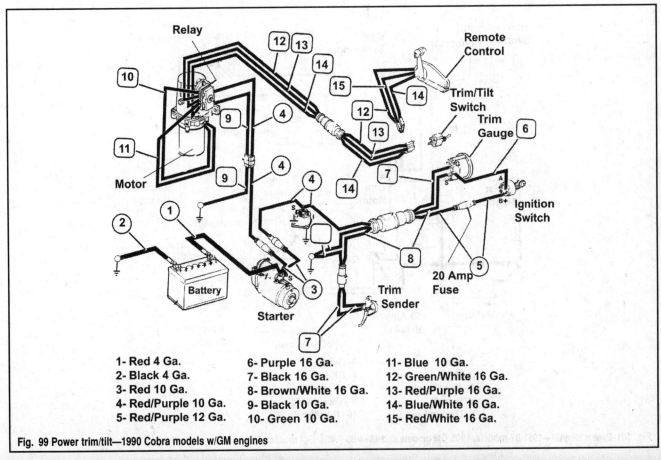

1- Red 4 Ga.   6- Purple 16 Ga.   11- Blue 10 Ga.
2- Black 4 Ga.   7- Black 16 Ga.   12- Green/White 16 Ga.
3- Red 10 Ga.   8- Brown/White 16 Ga.   13- Red/Purple 16 Ga.
4- Red/Purple 10 Ga. 9- Black 10 Ga.   14- Blue/White 16 Ga.
5- Red/Purple 12 Ga. 10- Green 10 Ga.   15- Red/White 16 Ga.

Fig. 99 Power trim/tilt—1990 Cobra models w/GM engines

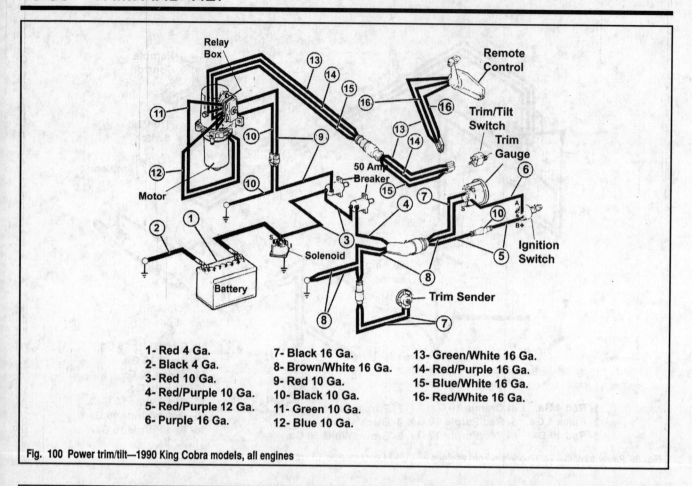

1- Red 4 Ga.
2- Black 4 Ga.
3- Red 10 Ga.
4- Red/Purple 10 Ga.
5- Red/Purple 12 Ga.
6- Purple 16 Ga.

7- Black 16 Ga.
8- Brown/White 16 Ga.
9- Red 10 Ga.
10- Black 10 Ga.
11- Green 10 Ga.
12- Blue 10 Ga.

13- Green/White 16 Ga.
14- Red/Purple 16 Ga.
15- Blue/White 16 Ga.
16- Red/White 16 Ga.

Fig. 100 Power trim/tilt—1990 King Cobra models, all engines

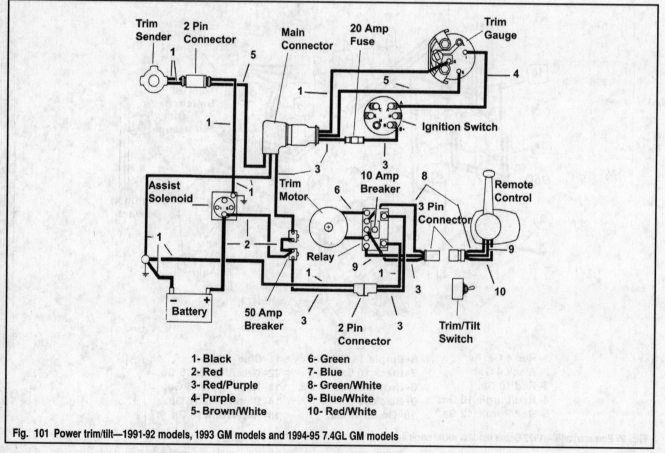

1- Black
2- Red
3- Red/Purple
4- Purple
5- Brown/White

6- Green
7- Blue
8- Green/White
9- Blue/White
10- Red/White

Fig. 101 Power trim/tilt—1991-92 models, 1993 GM models and 1994-95 7.4GL GM models

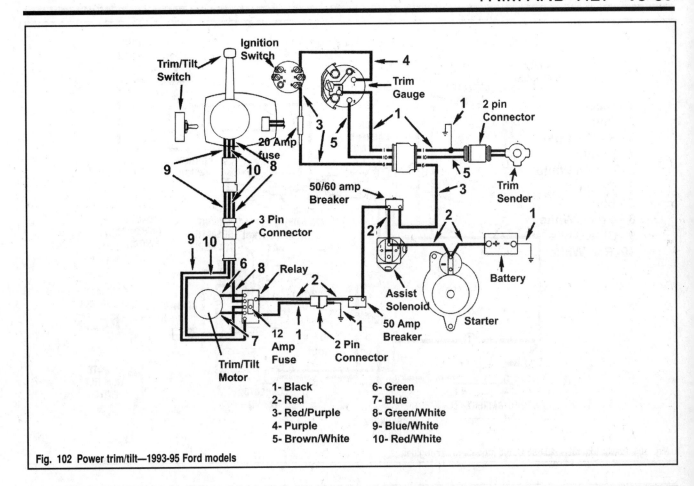

1- Black   **6- Green**
2- Red   **7- Blue**
3- Red/Purple   **8- Green/White**
4- Purple   **9- Blue/White**
5- Brown/White   **10- Red/White**

Fig. 102 Power trim/tilt—1993-95 Ford models

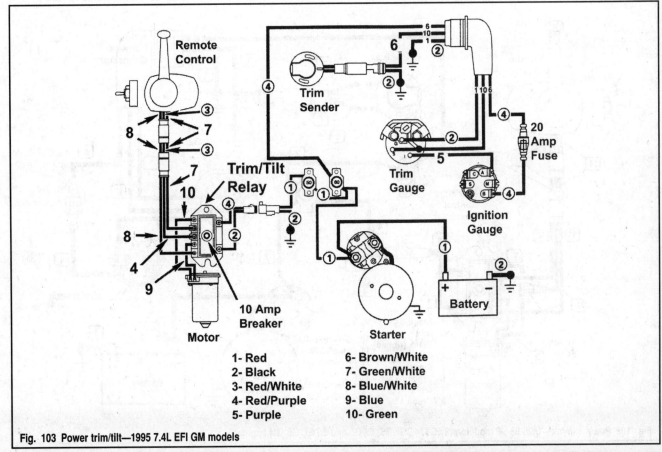

1- Red   **6- Brown/White**
2- Black   **7- Green/White**
3- Red/White   **8- Blue/White**
4- Red/Purple   **9- Blue**
5- Purple   **10- Green**

Fig. 103 Power trim/tilt—1995 7.4L EFI GM models

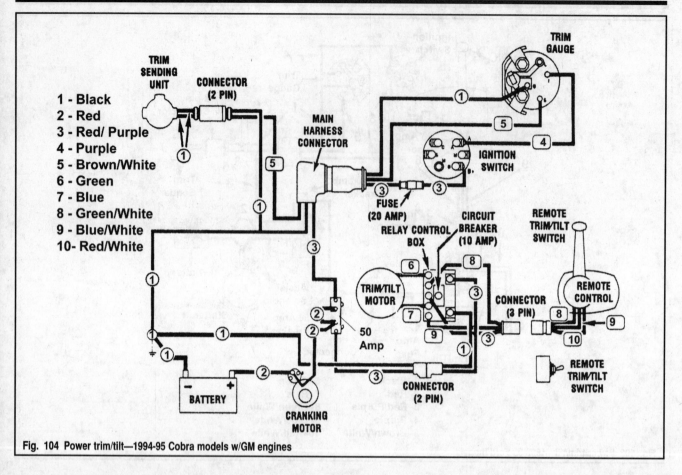

1 - Black
2 - Red
3 - Red/ Purple
4 - Purple
5 - Brown/White
6 - Green
7 - Blue
8 - Green/White
9 - Blue/White
10- Red/White

Fig. 104 Power trim/tilt—1994-95 Cobra models w/GM engines

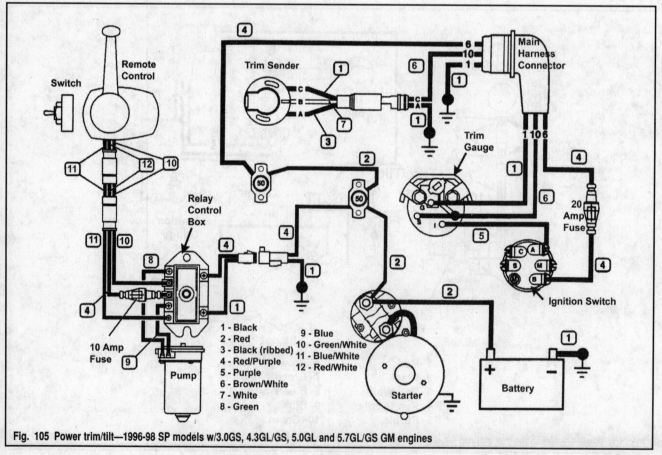

1 - Black
2 - Red
3 - Black (ribbed)
4 - Red/Purple
5 - Purple
6 - Brown/White
7 - White
8 - Green

9 - Blue
10 - Green/White
11 - Blue/White
12 - Red/White

Fig. 105 Power trim/tilt—1996-98 SP models w/3.0GS, 4.3GL/GS, 5.0GL and 5.7GL/GS GM engines

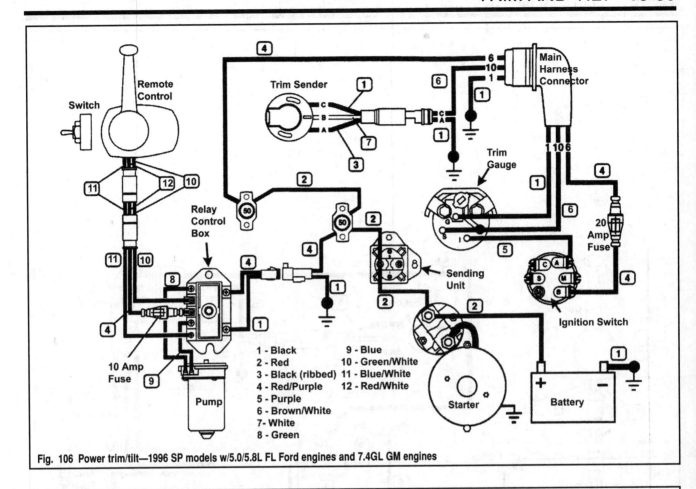

1 - Black
2 - Red
3 - Black (ribbed)
4 - Red/Purple
5 - Purple
6 - Brown/White
7 - White
8 - Green
9 - Blue
10 - Green/White
11 - Blue/White
12 - Red/White

Fig. 106 Power trim/tilt—1996 SP models w/5.0/5.8L FL Ford engines and 7.4GL GM engines

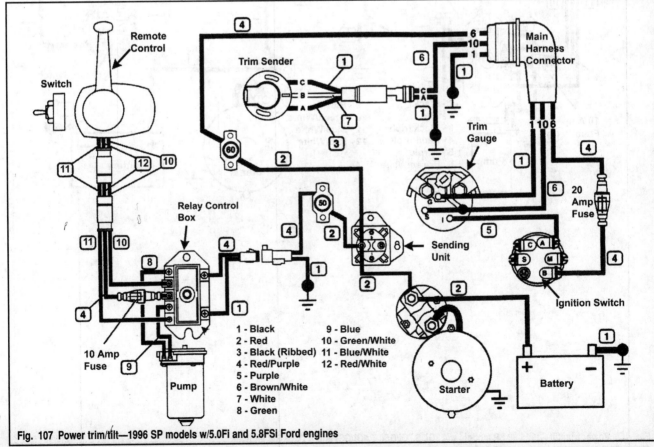

1 - Black
2 - Red
3 - Black (Ribbed)
4 - Red/Purple
5 - Purple
6 - Brown/White
7 - White
8 - Green
9 - Blue
10 - Green/White
11 - Blue/White
12 - Red/White

Fig. 107 Power trim/tilt—1996 SP models w/5.0Fi and 5.8FSi Ford engines

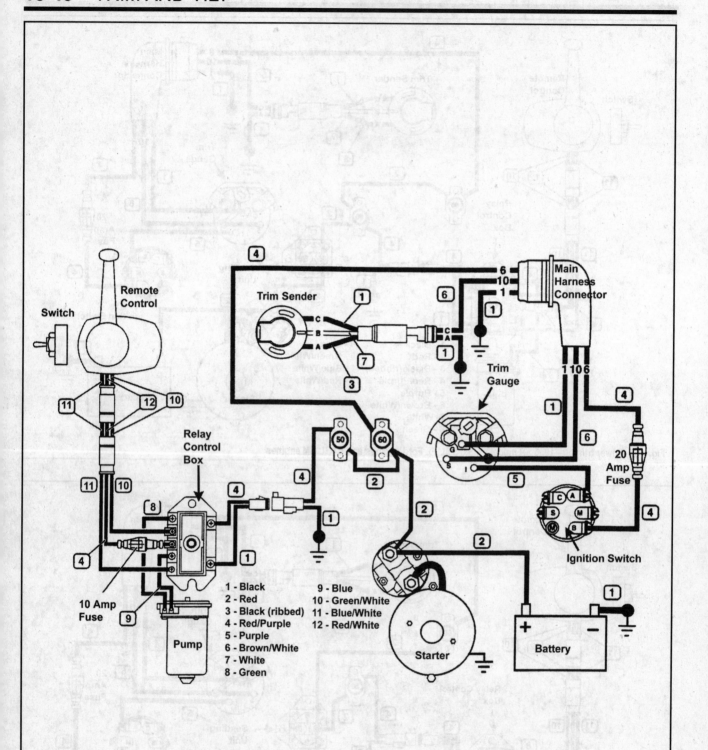

Fig. 108 Power trim/tilt—1996-98 SP models w/4.3Gi, 5.7Gi/GSi, 7.4Gi/GSi and 8.2GSi GM engines

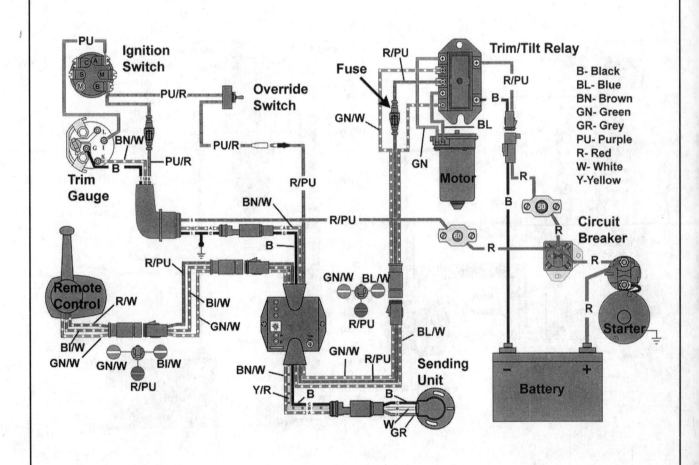

Fig. 109 Power trim/tilt—1996-98 DP models w/GM engines

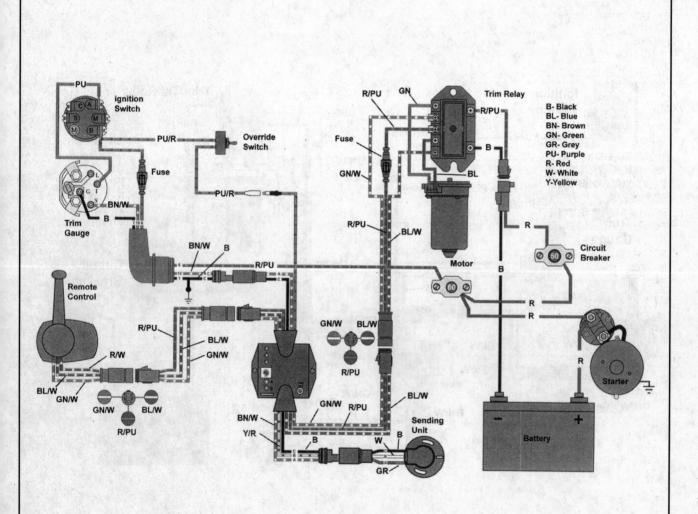

Fig. 110 Power trim/tilt—1996 DP models w/Ford engines

14

STEERING SYSTEMS

MANUAL STEERING SYSTEM

The manual steering system on your boat is very simple—a steering wheel, a steering cable and a steering tube/lever in the stern drive unit. Please refer to the Drive Systems sections for Steering Lever procedures.

The jacketed steering cable is connected on the one end to the helm (steering wheel) rack and pinion assembly. The opposite end of the outer jacket is connected to the steering tube at the transom plate. The inner cable passes through the tube and is connected to the steering arm coming from the stern drive unit. The tube and cable are connected to the transom via an anchor block, which provides the mechanical adjustment capabilities for centering of the steering system. When the steering wheel is rotated, the inner cable extends or retracts, moving the drive unit to port or starboard.

An adjustable trim tab on most models is located on the lower unit and is positioned to off-set the torsional steering (centered position) when operated at mid-range and full power settings.

Steering Cable

REMOVAL & INSTALLATION

1986 Models

◆ **See Figures 1, 2 and 3**

■ **The steering cable is serviced as an assembly only.**

1. Locate the end of the steering cable at the transom. Pull out the cotter pin securing the bottom of the clevis pin holding the cable ram to the steering lever (arm). Remove the steering arm clevis and disconnect the cable ram from the steering lever. On dual engine installations, support the tie bar and then repeat the procedure on the opposite side.
2. Remove the steering lever bushings (if equipped).
3. Turn the helm hard over to port until it reaches the full lock position and then loosen the cable anchor nut at the steering tube. Disconnect the cable.
4. Disconnect the cable at the helm and remove it.

■ **The steering tube position in the anchor block is preset at the factory—do not loosen the jam nut or attempt to change the centering position of the steering tube.**

To Install:

5. Clean the steering tube and steering lever thoroughly and inspect carefully for cracks, wear or other damage.
6. Connect the new cable to the helm as per the cable manufacturer's instructions and run it back through the boat.
7. Turn the steering wheel until the steering cable ram is extended to its full length and then coat the ram liberally with OMC Triple Guard Grease over its entire length.
8. Retract the steering cable ram completely while turning the steering wheel. Count the exact number of turns (yes, of the steering wheel) it takes until the ram is fully retracted.
9. Feed the cable ram through the steering tube and then position the steering lever at its center position of travel.
10. Now, turn the steering wheel in the other direction exactly half of the amount of turns you noted in the previous step—this will center the steering cable ram.
11. Position the steering cable end against the steering tube, loosen the lock nut and then turn the steering tube until the hole in the cable ram is lined up with the hole in the steering arm. Use your hand, not pliers or vise grips!
12. Move the steering tube jam nut so that it is next to the anchor block. Attach a 1-5/16 in. (34mm) 90 deg. crows foot wrench to a torque wrench and then tighten the jam nut to 20-25 ft. lbs. (27-34 Nm).
13. Position the steering ram in the steering lever so the holes line up and then insert the clevis pin from the top. Install a new cotter pin and bend over the arms.
14. Turn the steering wheel from lock-to-lock, extending and retracting the cable ram. The steering lever should move from stop-to-stop without coming in contact with the cut-out in the transom. Check that the steering arm has equal clearance at each stop and that it is centered when the wheel is centered.

1987-98 Models

◆ **See Figures 1, 2 and 3**

■ **The steering cable is serviced as an assembly only.**

1. Locate the end of the steering cable at the transom. Pull out the cotter pin securing the bottom of the clevis pin holding the cable ram to the steering lever (arm). Remove the steering arm clevis and disconnect the cable ram from the steering lever. On dual engine installations, support the tie bar and then repeat the procedure on the opposite side.
2. Remove the steering lever bushings.
3. Turn the helm hard over to port until it reaches the full lock position and then loosen the cable anchor nut at the steering tube. Disconnect the cable.
4. Disconnect the cable at the helm and remove it.

■ **The steering tube position in the anchor block is preset at the factory—do not loosen the jam nut or attempt to change the centering position of the steering tube.**

To Install:

5. Clean the steering tube and steering lever thoroughly and inspect carefully for cracks, wear or other damage.
6. Connect the new cable to the helm as per the cable manufacturer's instructions and run it back through the boat.
7. Turn the steering wheel until the steering cable ram is extended to its full length and then coat the ram liberally with OMC Triple Guard Grease over its entire length.
8. Feed the cable ram through the steering tube and then thread on the anchor nut (while holding the cable in the tube) until it bottoms against the end of the tube. Attach a 90 deg. crows foot wrench to a torque wrench and then tighten the anchor nut to 120 inch lbs. (14 Nm).
9. Install the steering lever bushings into the lever. It is not necessary to replace these bushings unless they are worn or damaged; we recommend replacing them anyway as preventative maintenance although it is by no means imperative.
10. Position the steering ram in the steering lever so the holes line up and then insert the clevis pin from the top. Install a new cotter pin and bend over the arms.
11. Turn the steering wheel from lock-to-lock, extending and retracting the cable ram. The steering lever should move from stop-to-stop without coming in contact with the cut-out in the transom. Check that the steering arm has equal clearance at each stop and that it is centered when the wheel is centered.

MAINTENANCE

The steering cable should be lubricated with water repellent grease (OMC Triple Guard) every 60 days during your boating season, although depending on your particular local conditions it may be necessary to shorten the interval.

Steering Tube

REMOVAL & INSTALLATION

◆ **See Figures 4 and 5**

■ **The steering tube position in the anchor block is preset at the factory—although we are providing the appropriate service procedures here, we do not recommend removing the assembly unless you are a skilled technician.**

1. Disconnect the steering cable from the steering tube as detailed in the previous section.

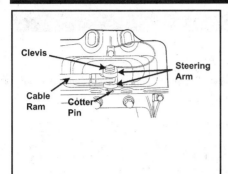

Fig. 1 Pull the cotter pin out of the steering lever clevis

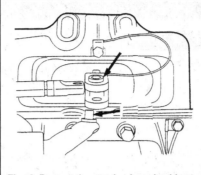

Fig. 2 Pop out the steering lever bushings

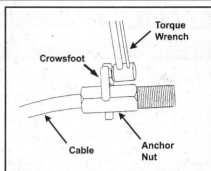

Fig. 3 Use a crowsfoot wrench at 90 deg. to the torque wrench when tightening the anchor nut

2. Locate the anchor block at the transom, loosen the jam nut and thread the steering tube out of the block. Remove the jam nut from the tube.

3. Remove the cotter pins (if equipped) and then the 2 mounting bolts and lift the anchor block out of the transom plate.

To Install:

4. Clean the anchor block and steering tube thoroughly.

5. Thread the jam nut all of the way onto the steering tube. Measure 3 1/4 in. (82.55mm) in from the threaded end of the tube. Now measure in an additional 1 1/4 in. (31.80mm) from that mark and spray Loctite primer on the 1 1/4 inch area of threads. Allow the primer to dry completely and coat the same area of threads with Loctite sealer or OMC Nut Lock.

6. Thread the anchor block onto the tube until the outer edge of the block is 3-5/32 in. (80.2mm) from the end of the tube. This should position the steering tube for accurate centering of the system.

7. Thread the jam nut up against the anchor block and then tighten it to 35-40 ft. lbs. (47-54 Nm). It is very important that the anchor block does not change position during the final tightening of the jam nut!

8. Coat the bushings LIGHTLY with grease and position the assembly into the transom bracket. Install the original mounting bolts and tighten them to 40-45 ft. lbs. (54-61 Nm). Insert new cotter pins into the holes and bend over the arms.

✳✳ WARNING

It is imperative that when greasing the bushings, you squeeze only a little grease into them—filling the bushing can cause the bolts to hydraulically lock and break the transom plate bracket.

9. Reconnect the steering cable to the steering tube.

Trim Tab

ADJUSTMENT

◆ See Figure 6

The trim tab should be pre-set at the factory and should normally require no adjustment. To confirm the setting, measure from the very end of the trim tab directly over to the port side of the anti-cavitation plate so that you measurement is taken at a point perpendicular to the centerline of the drive unit. Left-hand rotation units should measure 3 1/4 in. (83mm) (5 degrees to port of the centerline), while right-hand units should measure 2 1/4 in. (57mm) (30 degrees to port of the centerline).

If the measurement is not within specifications, loosen the trim tab bolt and re-position the tab to meet the correct dimension. Tighten the bolt to 28-32 ft. lbs. (38-43 Nm) on drives through 1994; for 1995-98 drives, tighten the bolt to 14-16 ft. lbs. (19-22 Nm).

In spite of the fact that the tab is pre-set at the factory, the boat may still steer more easily in one direction or another. If this is the case, confirm that the tab setting is correct as per factory recommendations and then perform the following test while running the boat in a straight line, with a balance load, in an area where current and wind will be as little a factor as possible.

10. While running the boat in a straight line, turn the wheel in one direction and then the other direction a few times while determining which direction requires the least amount of effort to complete the turn.

11. The trim tab will need to be moved SLIGHTLY in the direction that requires the LEAST effort. If boat is easier to turn to port, loosen the tab mounting bolt and move the tab just a bit toward the port side. If the boat turns easier to starboard, do the same, but move the tab to the starboard side. Tighten the trim tab bolt to 28-32 ft. lbs. (38-43 Nm) on drives through 1994; for 1995-98 drives, tighten the bolt to 14-16 ft. lbs. (19-22 Nm).

■ **On boats with twin engines, the trim tab on each engine must be moved the same amount and in the same direction.**

12. Recheck the steering effort after making the adjustment. Repeat the procedure until you are confident that the boat steers with the same amount of effort in each direction.

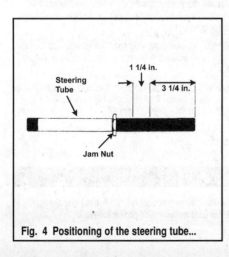

Fig. 4 Positioning of the steering tube...

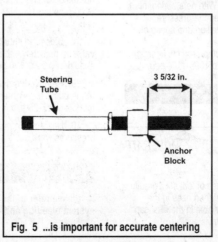

Fig. 5 ...is important for accurate centering

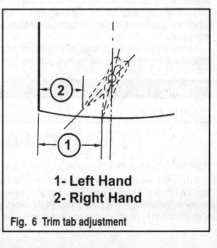

1- Left Hand
2- Right Hand

Fig. 6 Trim tab adjustment

POWER STEERING SYSTEM

Description

◆ See Figure 7

All OMC power steering systems utilize an engine-driven, vane-type hydraulic pump supplying fluid and pressure, via hoses, to a control or servo valve. The valve controls flow and pressure to a steering cylinder, the two of which make up the power steering assembly. There are three basic power steering modes: Neutral, left turn and right turn. The control valve, activated by the steering wheel via a steering cable controls all three modes.

The hydraulic pump is mounted on the front of the engine and is belt-driven by the crankshaft pulley. Some models have a separate fluid reservoir, but most utilize a reservoir contained within the pump housing. On the integrated models, the fill cap has a dipstick attached.

The control valve and steering cylinder are usually a single unit and are mounted to the transom plate at the back of the boat. This assembly is connected to the pump by means of hydraulic lines, with an oil cooler being incorporated into the low pressure return line. The assembly is also connected to the helm by means of a standard mechanical steering cable, which along with the steering ram, is connected to the steering arm on the drive unit.

When there is no steering input from the helm, the control valve is kept centered by a spring. Hydraulic fluid from the pump enters the valve at the pressure point, flows by each of the cylinder ports with almost no restriction and then through the return port, oil cooler and back to the pump reservoir.

When the helm is turned over to port, the system pulls the steering arm to starboard. The casing on the steering cable pushes the valve spool to port, fluid enters the valve and is routed to the rod-end outlet and then on to the low volume side of the cylinder. Pump pressure then moves the piston and steering arm to starboard. Fluid in the high volume side of the cylinder is forced through the valve, out the return port and back to the pump reservoir.

When the helm is turned over to starboard, the system pushes the steering arm to port. The casing on the steering cable pulls the valve spool to starboard, fluid enters the valve and is routed to the piston-end outlet and then on to the high volume side of the cylinder. Pump pressure then moves the piston and steering arm to port. Fluid in the low volume side of the cylinder is forced through the valve, out the return port and back to the pump reservoir.

Even when the power assist is not working for the steering system, the boat may still be steered, albeit not quite so easily. When turned to port at the wheel, the steering cable will still move the steering arm to starboard. This, in turn, causes the cable casing to press against the valve spool to port and against the stop at about 0.125 in. (3.18mm). The cable is now transmitting full manual effort to the steering arm. Fluid ahead of the piston is manually forced out of the cylinder through the outlet and into the valve, then through the return port and back to the pump reservoir. Additionally, fluid will be drawn out of the reservoir through the pressure inlet and back to the opposite side of the piston.

During normal operation, dual intake ports in the pump draw fluid from the reservoir. Pump output is also via ports. Vanes on the pump are held against the pump bore by pressure against the back of the vanes. Pump output should always be sufficient to perform quick boating maneuvers at idle. Pump output, flow-restricted to a maximum of 2.3 gallons/min., is always determined by resistance to flow found in the system. With no steering input, pressure should be 50-100 psi. Steering input increases the pressure, dependant upon boat attitude, trim position, throttle position and speed of maneuver.

A pressure relief valve located in the body of the pump protects against exceedingly high pressure and limits output pressure to 1000-1100 psi.

Testing

PUMP LEAKAGE

◆ See Figures 8, 9 and 10

Over-filling the pump reservoir is frequently the cause of leaks. Hydraulic fluid expands as it heats up during usage and the level then rises in the reservoir. Excess fluid is expelled through the breather hole in the filler cap where it can then be sprayed over the engine by the drive belt.

■ When using a torque wrench on line fittings, you will need the use of a 90 ° crowsfoot wrench.

1. Remove the lower hose fitting (1) with a flare wrench and install a new O-ring. Make sure you have opened the filler cap and have plenty of rags on hand. Tighten the fitting to 40 ft. lbs. (54 Nm) on 1986-89 models or 15-26 ft. lbs. (20-32 Nm) on 1990-98 models. If leakage continues, replace the hose or pump as necessary.

2. Check the torque on the lower fitting port nut (2). If the port is already tightened to 35 ft. lbs. (48 Nm) on 1986-89 models or 37-75 ft. lbs. (50-102 Nm) on all others, replace the pump; otherwise tighten the port to bring it into specifications.

3. Check the fluid level in the reservoir. If leakage continues with the fluid at the correct level and the cap on tightly, replace the cap or the pump.

4. Refer to the illustration for other likely areas of leakage. If detested in any of these areas, replace the pump.

PUMP PRESSURE

◆ See Figures 10 and 11

This test will require the use of a power steering system pressure gauge kit.

1. Ensure the power steering fluid is at the correct level in the reservoir(s).

2. Turn off the engine and connect a pressure gauge (Kent Moore #J-5176) to the high pressure hose. Disconnect the lower hose at the rear of the pump with a flare wrench. Connect the end of the hose to the pressure gauge on the shut-off valve side and then connect a spare hose between the other side of the gauge and the rear pump connection port. Open the shut-off valve.

■ The shut-off must be downstream from the gauge,

3. Remove the filler cap from the pump reservoir and check the fluid level, making sure the fluid is at the FULL line on the dipstick.

4. Start the engine and turn the wheel hard over (either way) until it is at full lock. Check the connections at the tool for any leaks.

5. Bleed the system as detailed in the Maintenance section.

6. Insert a thermometer (Kent Moore # J-5421) into the filler opening on the reservoir. Start the engine and move the steering wheel back and forth from lock-to-lock several times until the thermometer indicates a temperature of 150-190°. F (65-88 °C); this is normal operating temperature.

7. Check the fluid level and add if necessary. While the fluid is at normal operating temperature, the pressure with the valve open should be 80-125 psi on 1986-89 models or 50-100 psi on 1990-98 models. If pressure is higher, check all hoses for obstructions.

8. Close the shut-off valve momentarily and read the gauge. Open the valve immediately once you have noted the pressure. Relief valve pressure should be 1000-1100 psi.

9. Remove the gauge and spare hose. Connect the high pressure hose to the pump and tighten the fitting nut to 40 ft. lbs. (54 Nm) on 1986-89 models or 15-26 ft. lbs. (20-32 Nm) on 1990-98 models.

10. Replace the pump if the specified temperature or either of the pressures were outside of their ranges.

Maintenance Procedures

■ Please refer to the Maintenance section for fluid level checking, system bleeding and drive belt adjustment procedures.

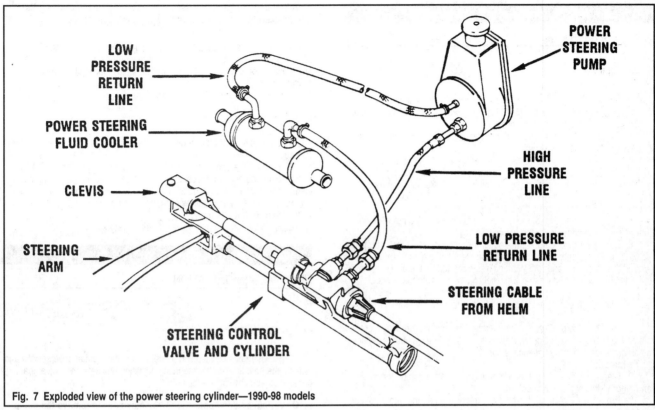

LOW
PRESSURE
RETURN
LINE

POWER STEERING
FLUID COOLER

CLEVIS

STEERING
ARM

POWER
STEERING
PUMP

HIGH
PRESSURE
LINE

LOW PRESSURE
RETURN LINE

STEERING CABLE
FROM HELM

STEERING CONTROL
VALVE AND CYLINDER

Fig. 7 Exploded view of the power steering cylinder—1990-98 models

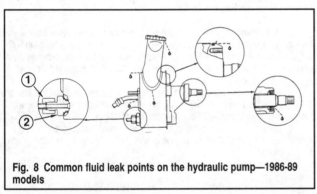

Fig. 8 Common fluid leak points on the hydraulic pump—1986-89 models

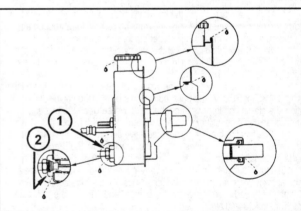

Fig. 9 Common fluid leak points on the hydraulic pump—1990-98 models

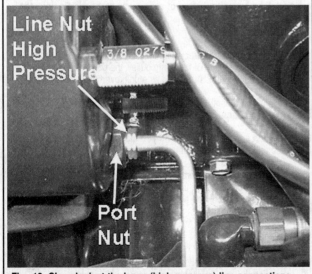

Line Nut
High
Pressure

Port
Nut

Fig. 10 Close look at the lower (high pressure) line connections (later models shown)

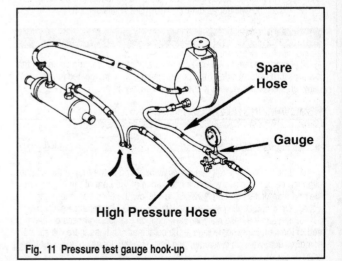

Spare
Hose

Gauge

High Pressure Hose

Fig. 11 Pressure test gauge hook-up

Actuator Valve And Cylinder Assembly

REMOVAL & INSTALLATION

1986-89 Models Only

◆ See Figure 12

1. Remove the steering cylinder as detailed elsewhere in this section.
2. Loosen and remove the guide lock nut and washer. Lift off the valve guide and steering tube as an assembly.
3. Carefully clamp the assembly in a soft-jawed vise and remove the bracket-to-valve bolt (and washer).
4. Spin two 1/2-13 nuts onto the bracket-to-valve bolt and then thread it into the bracket bushing. Tighten the inner nut against the bushing (firmly) and then cinch the other nut up against the inner nut. Remove the bushing clamp bolt and throw it away. Now turn the bracket-to-valve bolt counterclockwise and remove the bushing.
5. Aside from removal of the grease fitting nipple on the housing, no further service is possible on the actuator valve.
6. Bend a piece of cardboard around the cylinder rod so that it butts up against the inner edge of the clevis fork and then clamp a pair of Vise Grips around the cardboard and rod. Working inside the fork in the clevis, remove the cylinder lock nut and washer while holding the rod with the pliers.
7. Remove the cotter pin on the bracket and then remove the 2 cylinder trunnion pins by threading in a 2 in. long bolt—the pins have internal 10-24 threads.
8. Separate the bracket and the cylinder. Remove the remaining cotter pin on the bracket and pull out the cylinder bushing. The cylinder is serviced as an assembly, no further service is possible.

To Install:
9. Clean all components thoroughly and inspect everything for cracks and other visible signs of damage.
10. Press the cylinder rod bushing into the bracket so that the grooved end lines up with the cotter pin hole. Insert the pin and bend over the ends.
11. Slide the bracket onto the power cylinder so that the hydraulic line ports are exposed and then install the trunnion pins so the threaded end is on the outer side. Slide in a new cotter pin for the lower trunnion.
12. Slide the clevis over the end of the cylinder rod and then grasp the rod with Vise Grips as detailed in the removal procedure. Tighten the lock nut to 23-28 ft. lbs. (31-38 Nm).
13. Thread the bushing into the valve until the outer edge of the bushing is 1/16-3/32 in. BELOW the outer edge of the bore in the valve— this will allow the clamp bolt to pass through the groove between the bushing threads. Install a new clamp bolt and tighten it to 24-30 ft. lbs. (33-41 Nm).
14. Position the bracket against the valve assembly so that the slot on the valve mating surface lines up with the small rise on the bracket mating surface. Install the bracket bolt and tighten it to 50-60 ft. lbs. (68-81 Nm).
15. Install the steering tube and valve guide as an assembly. If the tube lock nut was removed, or even loosened, position the tube so that 7/8 in. (22mm) of thread is protruding on the cylinder end. Tighten the lock nut to 25-30 ft. lbs. (34-40 Nm).
16. Position the steering tube over the power cylinder and set the guide on the actuator stud. Slide a new internal toothed lock washer onto the nut and thread it onto the stud. Tighten the nut to 40-50 ft. lbs. (54-68 Nm).
17. Install the cylinder as detailed elsewhere in this section.

Power Steering Cooler

All engines covered here utilize a raw-water cooled steering cooler on the low pressure (return) side of the power steering system; located in-line between the control valve and the power steering pump.

REMOVAL& INSTALLATION

◆ See Figures 13, 14, 15, 16 and 17

1. Locate the cooler. It can generally be found at the rear of the engine, although certain engines will have it tucked in at the front of the engine on the port side. If it's not readily visible, just follow the lines from the pump.
2. Drain the cooling system as detailed in the Maintenance section.
3. Loosen the hose clamps and wiggle off the cooling hoses on each end of the unit. Have some rags and a container handy as there will be some residual water in the cooler and lines. Tie the hoses up and out of the way.

■ Although not absolutely necessary, we recommend draining the power steering system before attempting this procedure.

4. Loosen the hydraulic lines running into and out of the cooler unit. Most will simply use a hose clamp, but a few applications will use a screw-in fitting. Have some rags and a container handy as there will still be some residual fluid in the cooler and lines. Tie the hoses up and out of the way. Plug the openings to prevent contamination.
5. Loosen and remove the mounting brackets—usually just hose clamps, and lift out the cooler.

To Install:
6. Clean the cooler thoroughly and inspect it for cracks, wear or other damage.
7. Position the cooler and tighten the mounting hardware.
8. Slide the cooling hoses over the end flanges and tighten the hose clamps securely. Check the hose ends for cracks, crimping or bulges.
9. Install the hydraulic lines onto their original fitting and tighten the hose clamps or nuts securely.
10. Refill the cooling and power steering systems.

Power Steering Pump

REMOVAL & INSTALLATION

◆ See Figures 14, 15, 16, 17 and 18

1. Relieve drive belt tension as detailed in the Maintenance section and remove the belt. Loosen the mounting/adjusting brackets and swivel the pump in so you can pull the belt off.

■ On certain engine applications you may have to remove the pump before disconnecting the hydraulic lines.

2. Make sure to have a suitable container and some rags available. Loosen the pressure line fitting (lower) on the rear of the pump housing with a flare wrench and remove the line. Plug the line and tie it up so the plugged end is facing upward. Drain the fluid reservoir and then plug the port. Remove and discard the O-ring.
3. Loosen the hose clamp on the return line (upper) and pull the line off the fitting on the pump housing. Plug the line and secure it somewhere with the plugged end facing up. Plug the port on the pump.
4. Loosen the pump-to-brace bolt (if you haven't already) and disconnect the pump from the engine brace. If necessary, remove the brace from the engine.
5. Remove the two pump bracket mounting bolts and lift out the pump assembly.

■ On dual engine installations there may be a remote reservoir connected by hoses to the top of each pump reservoir. Remove the hose if necessary.

6. Remove the pump pulley with the appropriate puller (# 982270) and then remove the mounting bracket if necessary.
7. With the exception of the pump pulley, the power steering pump is a non-serviceable item and is replaced as a unit; overhaul is not possible.

To Install:
8. Install the mounting bracket and pulley if removed (use installer tool #982271 if necessary).

■ On certain engines applications you may have to connect the hydraulic lines prior to installing the pump.

9. Install the pump and bracket assembly onto the brace and tighten the bolts just tight enough to hold the pump in position.
10. If removed, attach the pump brace to the engine and tighten it to 30 ft. lbs. (41 Nm).
11. Connect the pump to the engine brace (if equipped) and tighten the bolt securely.
12. Connect the pressure line to the lower fitting and tighten it to 40 ft. lbs. (54 Nm) on 1986-89 models and 15-26 ft. lbs. (20-35 Nm) on all others; be sure to use a new O-ring coated with power steering fluid. Slide the return line over the pump fitting and tighten the hose clamp to 12-17 inch lbs. (1.4-2.0 Nm).

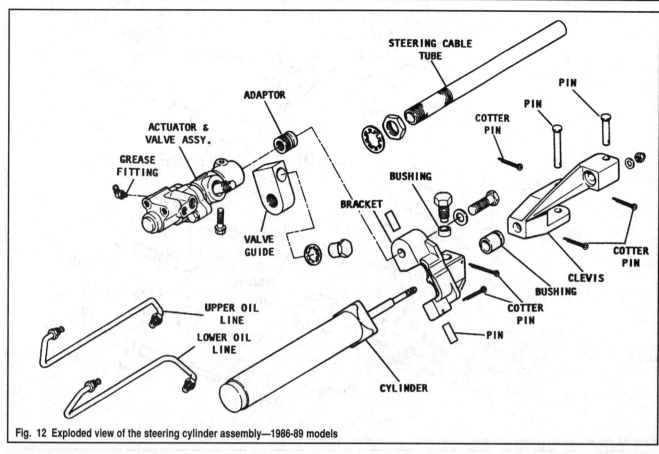

Fig. 12 Exploded view of the steering cylinder assembly—1986-89 models

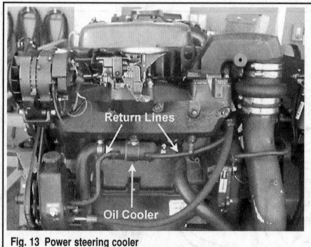

Fig. 13 Power steering cooler

13. Install and adjust the drive belt.
14. Fill the reservoir with power steering fluid and bleed the system.

Steering Cable

REMOVAL & INSTALLATION

◆ **See Figures 19, 20 and 21**

■ The steering cable is serviced as an assembly only.

1. Turn the steering wheel hard over to port until it reaches full lock.
2. Locate the end of the steering cable at the control valve/cylinder on the transom. Pull out the cotter pin securing the bottom of the clevis pin holding the cable ram to the guide tube. Remove the clevis pin.

■ There are two clevis pins in the guide clevis—the larger one in the center cut-out is for the steering arm, while the smaller one on the port side end is for the cable ram.

3. Secure the steering tube with a wrench (22mm) and then loosen the cable anchor nut at the control valve. Disconnect the cable and pull it out of the valve.
4. Disconnect the cable at the helm and remove it.
To Install:
5. Clean the control valve assembly thoroughly and inspect carefully for cracks, wear or other damage.
6. Connect the new cable to the helm as per the cable manufacturer's instructions and run it back through the boat.
7. Turn the steering wheel until the steering cable ram is extended to its full length and then coat the ram liberally with OMC Water Repellent Grease over its entire length.
8. Now retract the ram into the cable casing, feed the end of the ram through the valve while holding the anchor nut back so that the cable seats into the end of the valve.
9. Press the cable tightly against the valve and then thread on the anchor nut until it bottoms against the end of the tube.
10. Hold the cable guide tube at the flat near the valve with a 22mm wrench. Attach a 90 deg. crowsfoot wrench to a torque wrench and then tighten the anchor nut to 120 inch lbs. (14 Nm).

■ On 1986-89 models hold the jam nut, not the tube itself.

11. Position the steering ram eyelet so it lines up with the hole in the assembly and then insert the clevis pin from the top. Install a new cotter pin and bend over the arms.
12. Turn the steering wheel from lock-to-lock, extending and retracting the cable ram. The steering lever should move from stop-to-stop without coming in contact with the cut-out in the transom. Check that the steering arm has equal clearance at each stop and that it is centered when the wheel is centered.
13. Center the steering wheel and confirm that the steering lever is centered in the drive unit.
14. Check the fluid level and bleed the system.

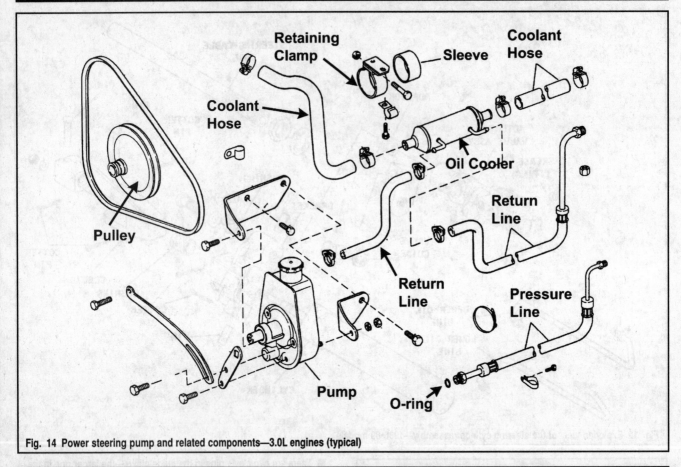

Fig. 14 Power steering pump and related components—3.0L engines (typical)

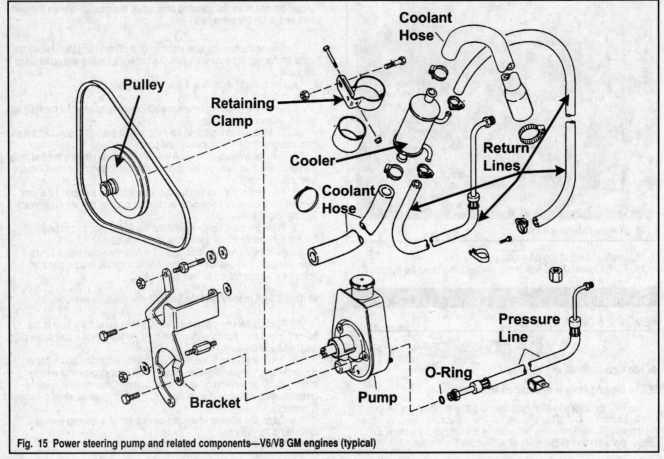

Fig. 15 Power steering pump and related components—V6/V8 GM engines (typical)

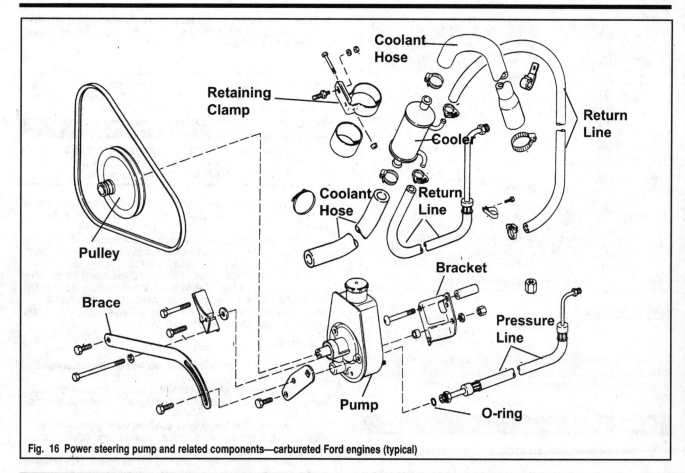

Fig. 16 Power steering pump and related components—carbureted Ford engines (typical)

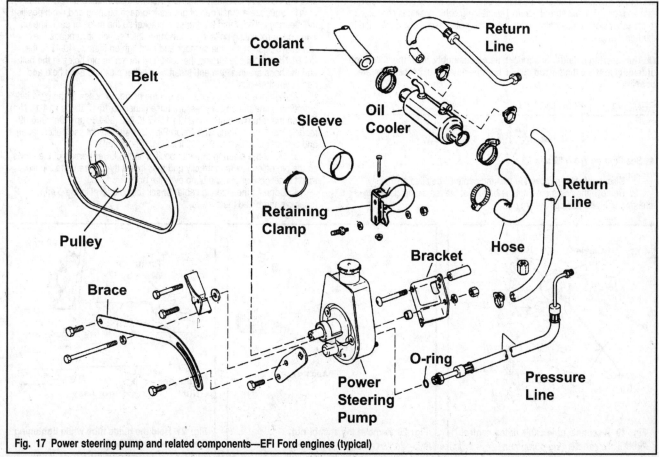

Fig. 17 Power steering pump and related components—EFI Ford engines (typical)

Fig. 18 A good shot of a typical power steering pump

MAINTENANCE

The steering cable should be lubricated with water repellent grease (OMC Triple Guard) every 60 days during the boating season, although depending on your particular local conditions it may be necessary to shorten the interval.

Steering Cylinder

The steering cylinder assembly consists of the steering power cylinder and the control valve (actuator valve on early models. A piston rod clevis at the end of the guide tube provides connection points for the steering cable ram and the drive unit steering arm. The steering cable casing is connected to the other side of the control valve, while the entire assembly is mounted in the transom plate.

■ **The steering cylinder is serviced as an assembly. With the exception of removing the cylinder rod clevis, no other overhaul procedures are possible.**

REMOVAL & INSTALLATION

◆ **See Figures 7, 12, 22 and 23**

1. Disconnect and remove the steering cable as detailed previously.
2. Remove the cotter pin from the bottom of the remaining clevis pin (steering arm pin) and pull the pin out.

■ There are two clevis pins in the guide clevis—the larger one in the center cut-out is for the steering arm, while the smaller one on the port side end is for the cable ram.

3. Position a small pan under the hydraulic hose connections and then loosen the line nuts with a flare wrench; it's a good idea to cover the fitting with a rag while loosening it to protect against any fluid spill or spray. Plug the hoses and valve ports immediately. Move the hoses out of the way.

✱✱ CAUTION

Do not move the cylinder rod until you have drained it completely. Aside from the possible resulting fluid spray being potentially dangerous, you'll create quite a mess in the engine bay.

4. Pull out the cotter pins running through the mounting bracket bosses on the transom plate and then remove the 2 mounting bolts. Lift off the assembly.
5. If the steering arm bushings require replacement, or if you wish to replace them anyway since you've got the unit out, carefully drill two 1/16 in. (1.5mm) holes in each and pull them out with a small bent-tip awl.
6. If the clevis needs service on 1990-98 models, extend the rod fully and wrap it with heavy paper. Clamp the rod in a soft-jawed vise being very careful not to scratch or damage the rod's surface. OMC offers special clamping blocks for this (# 983213) if you are so inclined. Loosen the 19mm locknut in the end of the rod and remove the clevis.

To Install:

7. Clean all components thoroughly in solvent and dry with compressed air.
8. Clamp the guide tube rod carefully in a vise as discussed previously. Position the clevis on the end of the rod and tighten locknut to 23-38 ft. lbs. (31-38 Nm).
9. Coat new bushings very lightly with Triple Guard grease and then press them into their bores until they seat themselves; you may have to tap them lightly with a small mallet.
10. Thread the two mounting bolts back into the transom plate bracket until the ends are just flush with the inside surface of the bracket.
11. Squeeze a small dab of grease into each bushing and them position the steering cylinder into the transom bracket so the holes of each line up. Thread the mounting bolts in by hand until they sufficiently engage their respective bushings in the cylinder and then tighten them to 40-45 ft. lbs. (54-61 Nm). Continue turning the bolts until the cotter pin holes in the bolt and the boss are in alignment. Insert new cotter pins into each hole and bend over the ends.
12. Remove the plugs and connect the hydraulic hoses to the cylinder. Tighten the line nut on the pressure hose (inner) to 10-12 ft. lbs. (14-16 Nm). Tighten the return line nut (outer) to 15-17 ft. lbs. (20-23 Nm). On 1986-89 models, coat the threads with Teflon Pipe Sealant (do not use Teflon tape!) and tighten the fittings to 95-105 inch lbs. (11-12 Nm).
13. Move the steering arm into position in the lower portion of the clevis so the pin holes line up correctly and then install the pin from the top side. Insert a new cotter pin and bend over the ends.
14. Connect the steering cable. Check the fluid level in the pump reservoir and bleed the system.

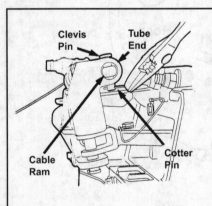

Fig. 19 Disconnect the cable at the control valve

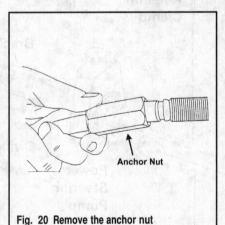

Fig. 20 Remove the anchor nut

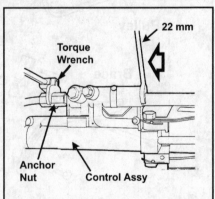

Fig. 21 Hold the guide tube while tightening the anchor nut (1990-98)

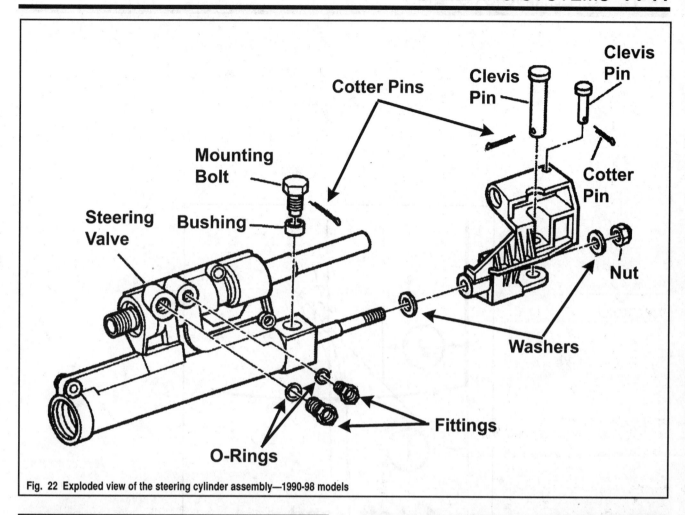

Fig. 22 Exploded view of the steering cylinder assembly—1990-98 models

Fig. 23 Hydraulic hose connections

Trim Tab

ADJUSTMENT

◆ See Figure 24

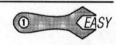

■ **Models utilizing power steering should not have been equipped with a trim tab, but if your boat IS equipped with one, confirm that the tab has been set correctly at the factory as follows.**

The trim tab should be pre-set at the factory and should normally require no adjustment. To confirm the setting, measure from the very end of the trim tab directly over to the port side of the anti-cavitation plate so that you measurement is taken at a point perpendicular to the centerline of the drive unit. Left-hand rotation units should measure 3 1/4 in. (83mm) (5 degrees to port of the centerline), while right-hand units should measure 2 1/4 in. (57mm) (30 degrees to port of the centerline).

If the measurement is not within specifications, loosen the trim tab bolt and re-position the tab to meet the correct dimension. Tighten the bolt to 28-32 ft. lbs. (38-43 Nm) on drives through 1994; for 1994-98 drives, tighten the bolt to 14-16 ft. lbs. (19-22 Nm).

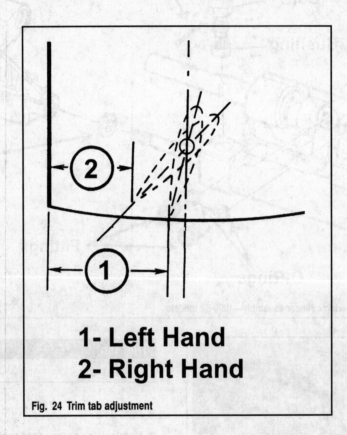

1- Left Hand
2- Right Hand

Fig. 24 Trim tab adjustment

15

REMOTE CONTROLS

REMOTE CONTROLS

Control Handle

REMOVAL & INSTALLATION

1986-89 Models

◆ See Figures 1 and 2

1. Disconnect the battery cables.

2. Move the control handle to the Neutral position and then disconnect the trim switch leads at the cable connector under the instrument panel.

3. Pry out the control handle cover. Remove the 3/4 in. handle screw and lift off the handle.

To install:

4. Install the control handle (still in the Neutral position). Use a new washer and screw, and tighten to 60-80 inch lbs. (7-9 Nm).

■ **Do not use another patchlock screw than the one supplied - the special head size is made just for this application.**

5. Snap the handle cover back into position, connect the trim leads to the cable connector and connect the battery cables. Test the system.

1990-98 Models

1. Disconnect the battery cables.

2. Move the control handle to the Neutral position and then disconnect the trim switch leads at the cable connector under the instrument panel.

3. Pull off the rubber boot on the warm-up knob at the base of the handle. Remove the retaining screw and pull off the knob.

4. Disconnect the throttle cable at the carburetor and then move the handle to the full Reverse position.

5. Insert an Allen wrench into the small opening on the bottom of the handle, loosen the set screw and pull the handle off of the shaft.

To install:

6. Position the handle onto the control box and tighten the Allen screw securely.

7. Install the warm-up knob, tighten the screw securely and then push on the rubber boot.

8. Connect the battery cables and check for proper operation.

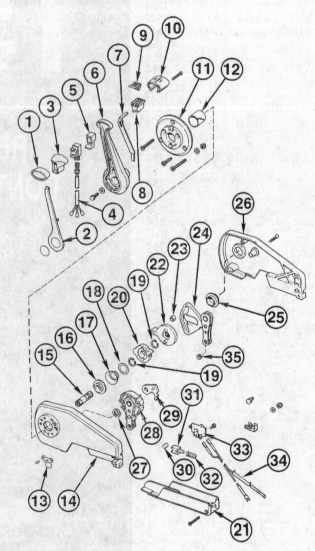

1. End Cap Knob (Non-Trim Models)
2. Handle Cover
3. Trim Switch Knob
4. Trim Switch Assembly
5. Trim Switch Grommet
6. Control Handle
7. Neutral Lock Plate
8. Lockout Button
9. Neutral Lock Spring
10. Lockout Button Knob
11. Mounting Flange-Concealed Side Mount Only
12. Handle Bushing
13. Trim Lead Retainer
14. Remote Control Housing - Port
15. Remote Control Shaft
16. Bushing
17. Drive Gear Return Spring
18. Washer
19. Drive Gear Retaining Ring
20. Shift Drive Gear
21. Control Housing Tray
22. Throttle Hub
23. Throttle Cam Roller
24. Throttle Cam and Arms Assembly
25. Throttle Hub Bushing
26. Remote Control Housing - Starboard
27. Shift Arm Washer
28. Shift Gear and Arms Assembly
29. Neutral Switch Cam
30. Detent Roller
31. Detent Roller Shoe
32. Detent Spring
33. Neutral Start Switch
34. Neutral Start Switch Lead Assembly
35. Bushing

Fig. 1 Exploded view of the remote control

Trim Switch

REMOVAL & INSTALLATION

1986-89 Models

◆ See Figures 1, 2 and 3

1. Disconnect the battery cables.
2. Move the control handle to the Neutral position and then disconnect the trim switch leads at the cable connector under the instrument panel.
3. Pry out the control handle cover. Remove the 3/4 in. handle screw and lift off the handle.
4. Remove the screws holding the trim switch knob to the lock-out knob. Separate the knobs from the handle.
5. Remove the trim lead retainer from the remote control and then press the collar out of the retainer.
6. Slide the retainer down and off of the leads and then pull the leads out through the handle and knob.
7. Discard the switch,

To install:

8. Insert the switch leads through the switch grommet and then position it on the switch.
9. Press the switch and grommet into the switch knob until they seat.
10. Position the switch knob on the control handle and then route the leads down the handle and through the hole.
11. Position the neutral lock spring into the lock-our button so that it is at a 45 degree angle. Insert eh button into the knob.
12. Now install the neutral lock plate under the center spring tab. Insert the round boss of the lock-out button through the slot in the neutral lock plate.
13. Slide the lock plate through the guide in the rear of the control handle and then secure the switch knob to the lock-out button knob with the two screws.
14. Press the trim lead collar into the retainer until it snaps into place.

■ **If the retainer is installed backwards on the leads, the handle will not fit on the remote control box.**

15. Install the retainer into the remote control and tighten the screws to 12-15 inch lbs. (1.4-1.6 Nm).
16. Install the control handle (still in the Neutral position). Use a new washer and screw, and tighten to 60-80 inch lbs. (7-9 Nm).

■ **Do not use another patchlock screw than the one supplied - the special head size is made just for this application.**

17. Snap the handle cover back into position, connect the trim leads to the cable connector and connect the battery cables. Test the system.

1990-98 Models

1. Disconnect the battery cables.
2. Move the control handle to the Neutral. Cut the plastic tie securing the switch leads to the control housing. Disconnect the trim switch leads at the cable connector under the instrument panel.
3. Pull off the rubber boot on the warm-up knob at the base of the handle. Remove the retaining screw and pull off the knob.
4. Disconnect the throttle cable at the carburetor and then move the handle to the full Reverse position.
5. Insert an Allen wrench into the small opening on the bottom of the handle, loosen the set screw and pull the handle off of the shaft.
6. Turn the handle over and remove the 2 retaining screws. Separate the cover from the handle.
7. Remove the neutral rod lock-out spring and handle from inside the control handle. Be careful, the spring is going to fly out under considerable force.
8. Remove the trim switch and leads from the handle grip.

To install:

9. Slide a new switch into the handle grip so that the leads, which are off center, are toward the top of the grip.
10. Looking at the inside of the handle, you'll notice casting ridges that create two channels running down the handle just inside of the edges. As you look at the handle in a vertical position, route the blue wire down the right channel and the red/green wire down the left side. Root them both around the hub in the bottom of the handle one full turn.
11. Reinstall the lock-out lever and spring and then route the switch leads through the handle opening.
12. Route the switch leads through the gunnel.
13. Position the handle cover on the handle and tighten the 2 screws securely.
14. Position the handle onto the control box and tighten the Allen screw securely.
15. Install the warm-up knob, tighten the screw securely and then push on the rubber boot.
16. Loop the switch leads loosely around the control unit and secure the loop with a plastic tie at the strain relief hole (next to the neutral start switch).
17. Connect the battery cables and check for proper operation.

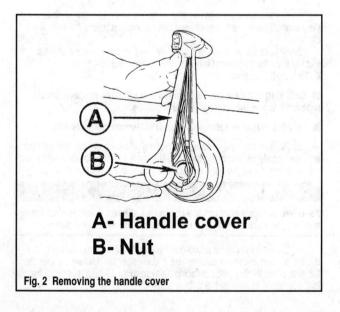

A- Handle cover
B- Nut

Fig. 2 Removing the handle cover

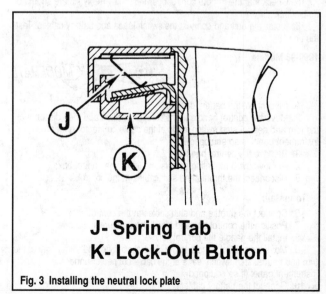

J- Spring Tab
K- Lock-Out Button

Fig. 3 Installing the neutral lock plate

Remote Control Box

REMOVAL & INSTALLATION

1986-89 Models

◆ **See Figure 1**

1. Disconnect the battery cables.
2. Move the control handle to the Neutral position and then disconnect the trim and neutral start switch leads at the cable connector under the instrument panel (if so equipped).
3. Remove the mounting screws and lift off the control box and handle.
4. Remove the control housing tray and then disconnect the shift and throttle cables.
5. Remove the 3 screws retaining the control housing. On side mount models, peel off the decorative trim strip.
6. Slowly pull the two housings apart.
7. Disconnect the neutral start switch leads and then remove the mounting screw and lift out the switch.

To assemble and install:

8. Install the neutral start switch and tighten the screw to 25-35 inch lbs. (2.8-4 Nm). Install the switch leads and route them back through the two "V" castings in the side of the housing. Next route them between the two round bosses at the back of the housing.
9. If removed, or it fell out, install the throttle hub bushing into position in the starboard side housing.
10. Install and position the throttle cam assembly as it came out.
11. Lube any and all friction point thoroughly with OMC Moly Lube. Swing the throttle cam assembly all the way toward the front of the housing and then connect the 2 housings. Tighten the screws to 25-35 inch lbs. (2.8-4 Nm).
12. Lubricate shift and throttle cable ends with OMC Triple Guard grease.
13. Insert the shift cable trunnion into the port side trunnion pocket on the control. Push the cable end forward between the shift arm halves and then shift the control into Reverse. Insert the cable pin into the shift arm and cable end. Make sure that the cable pin head is toward the center of the box and that the pin snaps into place in the plastic bushing.
   Insert the throttle cable trunnion into the starboard side trunnion pocket. Pull the control handle back to full throttle in Reverse and then insert the cable pin through the throttle control arm and the cable end. The head of the pin must be toward the center of the box and snap into the plastic bushing.
14. Replace the cover tray so that the center wed of the tray is between the shift and throttle arms. Install the mounting screws and tighten them securely.

■ **The control tray is what holds the cable trunnions in their pockets and also holds the cable pins in place.**

15. Check that the remote control operates freely freely - there should be no binding.
16. Install the unit and connect the switch leads and battery cables. Test the unit.

1990-98 Models

1. Disconnect the battery cables.
2. Move the control handle to the Neutral position and then disconnect the trim and neutral start switch leads at the cable connector under the instrument panel (if so equipped).
3. Remove the control handle.
4. Remove the 3 mounting screws and lift off the control box.
5. Disconnect the throttle and shift cables at the control.

To install:

6. Connect the throttle and shift cables at the control.
7. Position the control box and install the 3 mounting screws
8. Install the handle the control handle.
9. Move the control handle to the Neutral position and then connect the trim and neutral start switch leads at the cable connector under the instrument panel (if so equipped).
10. Connect the battery cables.

Throttle Cable

ORDERING CABLES

◆ **See Figure 4**

■ **Although OMC only began to offer the following suggestion with their 1996 and later models, we believe that as a general rule, the cable length estimating formula is a good guide for all models regardless of the year.**

■ **On 1996 and later models, the shift cable should "pull" for Forward gear on all standard rotation and dual prop applications. The shift cable should "push" for Forward gear on all counter rotation propeller applications. The throttle cable should always "pull" to close. All OMC remote controls allow for either direction of cable operation - just make sure the cable is installed properly for the individual application.**

If the remote control cables have never been installed, aren't currently installed or are being replaced, the following suggestions should help you order the correct length cables.

Measure the proposed route of the cable, remembering that all bends must have a radius **greater** than 6 in. (15cm):
• Shift Cable: Add **A** and **B**. Add an additional 20 in. (50cm) and then round up to the next available cable length.
• Throttle Cable: Add **A**, **B** and **C**. Add an additional 4 in. (10cm) and then round up to the next available cable length. Remember that **C** is the distance from the inside of the transom to the center of the throttle arm pin.

REMOVAL & INSTALLATION

1986-87 Models

1. At the engine, loosen the anchor screw and rotate the retainer out of the anchor. This is the barrel shaped component just before the end of the cable.
2. Now, at the very end of the cable, remove the cotter pin and flat washer from the carburetor arm and then disconnect the cable.

To install:

3. Position the cable end on the carburetor arm, slide in the washer and then install a new cotter pin.
4. Rotate the throttle cable trunnion (knurled knob) so it lines up with the anchor. Install the trunnion into the anchor and then secure the tab with the screw and elastic stop nut.

1988-89 Models

1. At the engine, loosen the anchor screw and rotate the retainer out of the anchor. This is the barrel shaped component just before the end of the cable.
2. Now, at the very end of the cable, remove the cotter pin and flat washer from the carburetor arm and then disconnect the cable.

To install & adjust:

■ **Both shift cables must be installed and adjusted properly before proceeding; see the appropriate procedures.**

■ **This procedure is appropriate for OMC remote controls only.**

3. Position the control handle in Neutral. Have an assistant wiggle the propeller slowly while moving the control handle into the Forward detent and then carefully move the handle HALFWAY back to the Neutral position.

✳✳ WARNING

Failure to properly position the control handle will result in shift linkage binding and hard shifting when coming out of the Neutral detent.

4. Connect the casing guide (the tube on the end of the cable) to the throttle pin and install the washer and a new cotter pin. Make sure that the flat side of the casing guide is facing away from the throttle lever and that the idle stop screw is against the throttle stop.

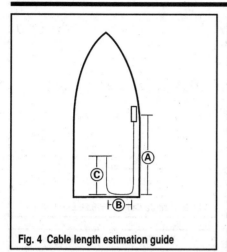

Fig. 4 Cable length estimation guide

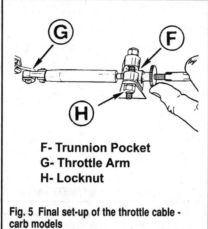

F- Trunnion Pocket
G- Throttle Arm
H- Locknut

Fig. 5 Final set-up of the throttle cable - carb models

F- Trunnion
G- Throttle Arm
H- Locknut
I- Cover Nut

Fig. 6 Final set-up of the throttle cable - EFI models

5. Grab the cable trunnion nut (small knurled knob) and pull on it firmly to remove any backlash from the cable and control. This is important because if you fail to remove any backlash, the engine will fail to return to idle when the control handle is moved to Neutral.

6. Hold the throttle closed and turn the trunnion nut until the trunnion will fit into the anchor block. Flip the cable retainer up and over the trunnion and then tighten the screw.

7. Although this completes the adjustment procedure, under normal operating conditions you may have to further adjust the idle stop screw to achieve the proper idle speed. If this is necessary, you will need to further adjust the trunnion nut to get the engine to consistently return to idle when the throttle is closed and the control handle is moved to Neutral.

1990-93 Models

◆ See Figures 5 and 6

1. Move the control handle into the Neutral position and check that the propeller moves freely.

2. At the engine end of the cable, remove the cotter pin and washer from the carburetor arm. On EFI models, remove the nut securing the throttle arm and the cable; there is no cotter pin.

3. Loosen the anchor retainer screw and rotate the retainer away from the cable trunnion. Pull the cable out of the anchor bracket and off of the carburetor.

4. Remove the control box and disconnect the cables.

To install:

5. Lubricate the control unit end of the cable with OMC Triple Guard grease.

6. Install the forward cable trunnion into the anchor and then install the anchor on the trunnion.

7. Insert the anchor into the forward spacer plate mounting hole. Install the screw and locknut, tightening securely.

■ The forward mounting hole is for OMC cables only. The rear mounting holes is used exclusively for SAE cables.

8. Pull the cable forward so it aligns with the hole in the throttle arm. Install the cable pin and a new cotter pin.

■ Both shift cables must be installed and adjusted properly before installing and adjusting the throttle cable.

■ This procedure is appropriate for OMC remote controls only.

9. Position the control handle in Neutral. Have an assistant wiggle the propeller slowly while moving the control handle into the Forward detent and then carefully move the handle HALFWAY back to the Neutral position.

✳✳ WARNING

Failure to properly position the control handle will result in shift linkage binding and hard shifting when coming out of the Neutral detent.

10. Confirm that the throttle lever is against the idle stop (carb models only). Coat the throttle lever pin with OMC Triple Guard.

11. Connect the casing guide (the tube on the end of the cable) to the throttle pin and install the washer and a new cotter pin on carb models, or install the locknut on EFI models. Make sure that the flat side of the casing guide is facing away from the throttle lever and that the idle stop screw is against the throttle stop (1990-92 carb models).

12. Grab the cable trunnion nut (small knurled knob) and pull on it firmly to remove any backlash/end play from the cable and control. This is important because if you fail to remove any backlash, the engine will fail to return to idle when the control handle is moved to Neutral.

13. Hold the throttle closed and turn the trunnion nut until the trunnion will fit into the anchor block. Flip the cable retainer up and over the trunnion and then tighten the screw securely.

14. Although this completes the adjustment procedure, under normal operating conditions you may have to further adjust the idle stop screw to achieve the proper idle speed. If this is necessary, you will need to further adjust the trunnion nut to get the engine to consistently return to idle when the throttle is closed and the control handle is moved to Neutral.

1994-98 Cobra & 1995-98 King Cobra Models

◆ See Figures 7 thru 12

1. Move the control handle into the Neutral position and check that the propeller moves freely.

2. At the engine end of the cable, remove the cotter pin and washer from the throttle arm. On EFI models, remove the nut securing the throttle arm and the cable; there is no cotter pin.

3. Loosen the anchor retainer nut and rotate the retainer away from the cable trunnion. Pull the cable out of the anchor bracket and off of the throttle arm.

4. Remove the control box and disconnect the cables. Remove the cotter pin and cable connector securing the cable to the throttle lever. Remove the screw and nut securing the anchor bracket to the housing and then remove the bracket.

To install:
OMC Cables

5. Lubricate the control unit end of the cable with OMC Triple Guard grease.

6. Install the cable trunnion into the anchor and then install the anchor on the trunnion.

7. Insert the anchor into the forward spacer plate mounting hole. Install the screw and locknut, tightening securely.

■ The forward mounting hole is for OMC cables only. The rear mounting holes is used exclusively for SAE cables.

8. Pull the cable forward so it aligns with the hole in the throttle arm. Install the cable pin and a new cotter pin.

■ Always connect the cable on the same side of the throttle lever as the cable anchor on the anchor plate. This will position the control cable parallel to the anchor plate assembly.

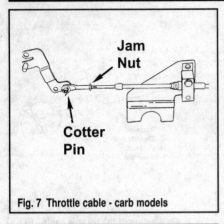

Fig. 7 Throttle cable - carb models

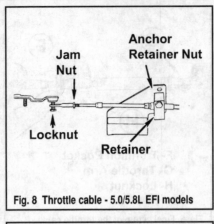

Fig. 8 Throttle cable - 5.0/5.8L EFI models

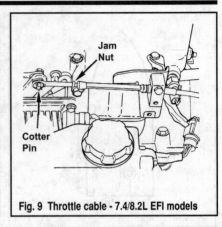

Fig. 9 Throttle cable - 7.4/8.2L EFI models

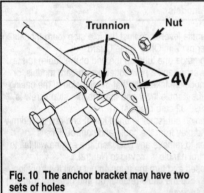

Fig. 10 The anchor bracket may have two sets of holes

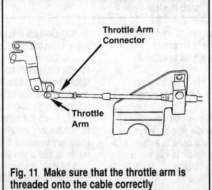

Fig. 11 Make sure that the throttle arm is threaded onto the cable correctly

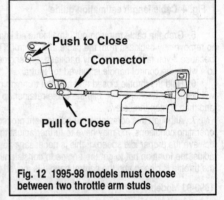

Fig. 12 1995-98 models must choose between two throttle arm studs

SAE Cables to OMC Concealed Side-Mount Remote

9. Use the cable trunnion and connector supplied with the appropriate OMC Adapter Kit to adapt SAE cables to OMC controls.

10. Install the cable trunnion and connector onto the SAE cable.

11. Install the plastic trunnion adapter and the metal trunnion adapter onto the cable trunnion. In order to retain the trunnion, the plastic adapter must go over the open end of the trunnion.

12. Attach the cable to the rear mounting position in the control with the screw and nut. Remember, the forward position is for the OMC cables

13. Thread the cable connector onto the throttle cable core wire until one full thread is showing and then lock it in place with the nut.

14. Attach the connector to the throttle arm so the cotter pin is facing the cam and anchor plate assembly.

And now for the other end of the cable

■ **Determine whether you have a "push to close" or "pull to close" remote control system. The throttle cable attachment point at the throttle arm is determined by the direction of cable movement.**

■ **The remote control shift cable must be installed.**

15. Position the control handle in Neutral. Have an assistant wiggle the propeller slowly while moving the control handle into the Forward detent and then carefully move the handle HALFWAY back to the Neutral position.

✱✱ WARNING

Failure to properly position the control handle will result in shift linkage binding and hard shifting when coming out of the Neutral detent.

16. Make sure that the throttle lever is against the idle stop.

17. Align the two internal bosses on the trunnion with the groove in the throttle cable and press the trunnion onto the cable until it seats itself.

■ **The anchor bracket on 1994 carburetted models may have two sets of holes for mounting the anchor block. The top set (should be marked) is for models with a 4 bbl carburetor, while the bottom set (unmarked) is for models with a 2 bbl carburetor. This also holds true for the 1995-96 4.3L engines, which should use the bottom set of holes. And, 1997-98 4.3L engines, where the top set is for the GS and the bottom set is for the GL.**

18. Install the open end of the trunnion in the anchor block. Thread in the screw and position the assembly in the correct set of holes. The nut must be against the anchor bracket and tightened securely.

19. Install the throttle arm connector onto the cable. Make sure that the throttle arm connector has at least 9 full turns onto the cable (or a full 1/4 in. of thread engagement). Pull the connector forward to remove all end play from the cable and then turn the connector in until the holes align with the throttle arm. On 1995-98 models, remember there is a "push to close" and "pull to close" stud on the throttle arm.

■ **If the connector hole cannot be adjusted to align with the throttle arm, check that the cable has been installed properly in the control box.**

20. On models with a carburetor, Install the connector onto the throttle arm with the washer and a new cotter pin. Tighten the jam nut up against the connector.

21. On EFI models, install the connector onto the throttle arm stud (aft on 1995-98) and tighten the locknut securely. Tighten the jam nut up against the connector.

1994 King Cobra Models

② MODERATE

1. Move the control handle into the Neutral position and check that the propeller moves freely.

2. At the engine end of the cable, remove the cotter pin and washer from the throttle arm. On EFI models, remove the nut securing the throttle arm and the cable; there is no cotter pin.

3. Loosen the anchor retainer nut and rotate the retainer away from the cable trunnion. Pull the cable out of the anchor bracket and off of the throttle arm.

4. Remove the control box and disconnect the cables. Remove the cotter pin and cable connector securing the cable to the throttle lever. Remove the screw and nut securing the anchor bracket to the housing and then remove the bracket.

To install:
OMC Cables
5. Lubricate the control unit end of the cable with OMC Triple Guard grease.
6. Install the cable trunnion into the anchor and then install the anchor on the trunnion.
7. Insert the anchor into the forward spacer plate mounting hole. Install the screw and locknut, tightening securely.

■ **The forward mounting hole is for OMC cables only. The rear mounting holes is used exclusively for SAE cables.**

8. Pull the cable forward so it aligns with the hole in the throttle arm. Install the cable pin and a new cotter pin.

■ **Always connect the cable on the same side of the throttle lever as the cable anchor on the anchor plate. This will position the control cable parallel to the anchor plate assembly.**

SAE Cables to OMC Concealed Side-Mount Remote
9. Use the cable trunnion and connector supplied with the appropriate OMC Adapter Kit to adapt SAE cables to OMC controls.
10. Install the cable trunnion and connector onto the SAE cable.
11. Install the plastic trunnion adapter and the metal trunnion adapter onto the cable trunnion. In order to retain the trunnion, the plastic adapter must go over the open end of the trunnion.
12. Attach the cable to the rear mounting position in the control with the screw and nut.
13. Thread the cable connector onto the throttle cable core wire until one full thread is showing and then lock it in place with the nut.
14. Attach the connector to the throttle arm so the cotter pin is facing the cam and anchor plate assembly.
And now for the other end of the cable
15. Position the control handle in Neutral. Have an assistant wiggle the propeller slowly while moving the control handle into the Forward detent and then carefully move the handle HALFWAY back to the Neutral position.

✹✹ WARNING

Failure to properly position the control handle will result in shift linkage binding and hard shifting when coming out of the Neutral detent.

16. Ensure that the throttle lever is against the idle stop.
17. Coat the throttle lever pin with OMC Triple Guard grease and then connect the casing guide to the throttle arm. Install the washer and a new cotter pin on models with a carburetor; or the lock nut on EFI models.
18. Grab the throttle cable trunnion and pull on it firmly to remove any backlash from the cable and control. If not done, the engine may not return to idle when moved to Neutral.
19. Hold the throttle closed while you turn the trunnion nut until the trunnion will fit into position in the anchor bracket. Move the bracket retainer over the trunnion and tighten the nut securely.
20. Although this completes the adjustment procedure, under normal operating conditions you may have to further adjust the idle stop screw to achieve the proper idle speed. If this is necessary, you will need to further adjust the trunnion nut to get the engine to consistently return to idle when the throttle is closed and the control handle is moved to Neutral.

Shift Cables

ORDERING CABLES

◆ See Figure 4

■ Although OMC only began to offer the following suggestion with their 1996 and later models, we believe that as a general rule, the cable length estimating formula is a good guide for all models regardless of the year.

■ **On 1996 and later models, the shift cable should "pull" for Forward gear on all standard rotation and dual prop applications. The shift cable should "push" for Forward gear on all counter rotation propeller applications. The throttle cable should always "pull" to close. All OMC remote controls allow for either direction of cable operation - just make sure the cable is installed properly for the individual application.**

If the remote control cables have never been installed, aren't currently installed or are being replaced, the following suggestions should help you order the correct length cables.
Measure the proposed route of the cable, remembering that all bends must have a radius **greater** than 6 in. (15cm):
• Shift Cable: Add **A** and **B**. Add an additional 20 in. (50cm) and then round up to the next available cable length.
• Throttle Cable: Add **A**, **B** and **C**. Add an additional 4 in. (10cm) and then round up to the next available cable length. Remember that **C** is the distance from the inside of the transom to the center of the throttle arm pin.

REMOVAL, INSTALLATION & ADJUSTMENT

Remote Control Cable - 1986-87 Models

1. Move the remote control handle into the Neutral position.
2. Disconnect the cable at the remote control as detailed in the Remote Control section.
3. Trace the cable back to the shift bracket on the engine and remove the cotter pin and washer from the end of the cable. Disconnect the cable from the shift lever.

■ **Remember, the shift cable running between the remote control and the shift bracket is the top cable; the lower cable runs from the bracket to the drive.**

4. Now loosen the anchor block screw (a short ways down from the cable end) and retainer. Remove the shift cable.

To install:

■ **If you are installing a new cable, the accompanying Adjustment procedures MUST be performed.**

5. Lubricate the control unit end of the cable with OMC Triple Guard grease.
6. Install the forward cable trunnion into the anchor and then install the anchor on the trunnion.
7. Insert the anchor into the forward spacer plate mounting hole. Install the screw and locknut, tightening securely.

■ **The forward mounting hole is for OMC cables only. The rear mounting holes is used exclusively for SAE cables.**

8. Pull the cable forward so it's end aligns with the inner hole in the shift arm. Install the cable pin and a new cotter pin.
9. Now, at the other end, loosen the remote control shift cable (this is the upper cable) anchor block retainer - this is the square box the cable runs through before attaching to the shift lever.
10. Move the shift lever (where the end of the cable is attached on the engine) to the vertical position.
11. Move the remote control handle to Neutral and then make sure that the end of the cable is attached to the top pin on the shift lever. Make sure the flat washer and cotter pin are installed.
12. Rotate the trunnion (knurled knob) until it seats into the anchor block and then tighten the screw and nut securely.

■ **If the shift cable of any other shift system component has been replaced, the drive unit must be removed to confirm that the following conditions are met:**

• The remote control handle must be in the Neutral position
• The shift lever on the cable bracket must be in the vertical position with a measurement of 6 1/2 in. (165mm) between the center of the shift lever pin and the center of the anchor.
• The centerline of the bellcrank lever in the drive unit should be at 90 deg. angle to the machined surface of the transom mount.
If all of the above conditions have been met, then adjustment of the remote control shift cable is necessary. Otherwise, proceed as follows:
13. With the remote control handle in the Neutral position, disconnect the cable (coming from the control) from the anchor and top of the shift lever on the shift bracket at the back of the engine. Remove the cotter pin holding the end of the cable to the shift lever and pull the cable and washer off of the lever. Loosen the anchor block screw and rotate the retainer until you can remove the cable.

14. Now, you must make sure that the shift lever on the bracket is in the vertical position. Measure the distance from the centerline of the shift lever pin for the lower cable to the center of the anchor - it must be 6 1/2 in (165mm).

15. If not, loosen the anchor and slide the lower cable until you achieve this measurement and then tighten the anchor (see the following procedure for the lower cable).

16. Once the lower cable is adjusted, and the shift lever is in the vertical position, simply reinstall the remote control shift cable (upper) to the top of the shift lever (don't forget the washer and pin). Now turn the trunnion on the cable until it aligns with the anchor block and secure it to the block with the retainer.

17. Position a 90 deg. angle (carpenters square) against the machined surface of the drive's transom mount (pivot housing).

18. Confirm that the bellcrank lever is at exactly 90 deg. perpendicular to the machined surface. The remote control should be in Neutral and the shift bracket shift lever must be vertical as in the previous step.

19. If the bellcrank lever is at 90 deg., you're through. If the bellcrank lever is not at 90 deg., move to the Transom Mount Shift Cable Adjustment procedures.

Remote Control Cable - 1988-89 Cobra

1. Move the remote control handle into the Neutral position.

2. Disconnect the cable at the remote control as detailed in the Remote Control section.

3. Trace the cable back to the shift bracket on the engine and remove the cotter pin and washer from the end of the cable. Disconnect the cable from the shift lever.

■ **Remember, the shift cable running between the remote control and the shift bracket is the top cable; the lower cable runs from the bracket to the drive.**

4. Now loosen the anchor block screw (a short ways down from the cable end) and retainer. Remove the shift cable.

To install:

5. Lubricate the control unit end of the cable with OMC Triple Guard grease.

6. Install the forward cable trunnion into the port side trunnion pocket.

7. Push the cable end forward between the shift arm halves and then shift the control handle into Reverse.

8. Insert the cable pin into the shift arm and the cable end. The head of the cable pin must be toward the center of the control box and the cable pin must snap into the plastic bushing.

9. Now, at the other end:

If the cable was removed, but not replaced

10. Loosen the remote control shift cable (this is the upper cable) anchor block retainer - this is the square box the cable runs through before attaching to the shift lever.

11. Move the shift lever (where the end of the cable is attached on the engine) to the vertical position so that the transom cable is in Neutral.

12. Move the remote control handle to Neutral and then make sure that the end of the cable is attached to the top pin on the shift lever. Make sure the flat washer and cotter pin are installed.

13. Rotate the trunnion (knurled knob) until it seats into the anchor block and then tighten the screw and nut securely.

If a new cable or control box has been installed
- The drive unit must be installed
- The throttle cable must be disconnected from the carburetor
- The transom mount shift cable must be disconnected from the shift lever bracket
- The remote control shift cable must be disconnected from the shift lever bracket

Route the transom shift cable behind the steering assembly on all engines but the 2.3L where it needs to be in the front of the assembly.

■ **If this causes interference on your particular application, you can route them in front of the assembly.**

14. Shift the control handle into Forward and move it to the full throttle position.

15. Apply OMC Triple Guard grease to the upper and lower shift pins and the actuator anchor.

16. Connect the end of the cable to the upper shift lever pin and install the washer and a new cotter pin.

17. Pop the cable trunnion into the anchor block. Tighten the retaining nut securely and then loosen it 1/2 turn.

18. Slide the trunnion back and forth slowly until the center of the fixed cam on the shift lever is aligned with the overstroke switch button.

19. Tighten the retainer nut securely.

Remote Control Cable - 1990-91 Cobra

1. Move the remote control handle into the Neutral position.

2. Disconnect the cable at the remote control as detailed in the Remote Control section.

3. Trace the cable back to the shift bracket on the engine and remove the cotter pin and washer from the end of the cable. Disconnect the cable from the shift lever.

■ **Remember, the shift cable running between the remote control and the shift bracket is the top cable; the lower cable runs from the bracket to the drive.**

4. Now loosen the anchor block screw (a short ways down from the cable end) and retainer. Remove the shift cable.

To install:

5. Lubricate the control unit end of the cable with OMC Triple Guard grease.

6. Install the forward cable trunnion into the anchor and then install the anchor on the trunnion.

7. Insert the anchor into the forward spacer plate mounting hole. Install the screw and locknut, tightening securely.

■ **The forward mounting hole is for OMC cables only. The rear mounting holes is used exclusively for SAE cables.**

8. Pull the cable forward so it's end aligns with the inner hole in the shift arm. Install the cable pin and a new cotter pin.

Whenever the cable has been removed or replaced on 1991 models, it must be adjusted as follows. If non-OMC remote controls are used, the cable stroke must be at least 2 7/8 in. (73mm) and no more than 3 3/16 in. (81mm).

■ **Prior to installing or adjusting the shift system, the following conditions must be met:**

- The drive unit must be installed
- The throttle cable must be disconnected from the carburetor
- The transom mount shift cable must be disconnected from the shift lever bracket
- The remote control shift cable must be disconnected from the shift lever bracket

Route the transom shift cable behind the steering assembly on all engines but the 2.3L where it needs to be in the front of the assembly.

■ **If this causes interference on your particular application, you can route them in front of the assembly.**

9. Shift the control handle into Forward and move it to the full throttle position.

10. Apply OMC Triple Guard grease to the upper and lower shift pins and the actuator anchor. On 1991 models, loosen the locknut to gain access to the upper shift pin.

11. Connect the end of the cable to the upper shift lever pin and install the washer and a new cotter pin.

12. On 1991 and later models, tighten the lever locknut securely.

13. Pop the cable trunnion into the anchor block. Tighten the retaining nut securely and then loosen it 1/2 turn.

14. Slide the trunnion back and forth slowly until the center of the fixed cam on the shift lever is aligned with the overstroke switch button.

15. Tighten the retainer nut securely.

Remote Control Cable - 1992-93 Cobra

 MODERATE

◆ **See Figures 13 and 14**

1. Move the remote control handle into the Neutral position.
2. Disconnect the cable at the remote control as detailed in the Remote Control section.
3. Trace the cable back to the shift bracket on the engine and remove the cotter pin and washer from the end of the cable. Disconnect the cable from the shift lever.

■ **Remember, the shift cable running between the remote control and the shift bracket is the top cable; the lower cable runs from the bracket to the drive.**

4. Now loosen the anchor block screw (a short ways down from the cable end) and retainer. Remove the shift cable.

To install:

5. Lubricate the control unit end of the cable with OMC Triple Guard grease.
6. Install the shift cable trunnion into the anchor and then install the anchor on the trunnion.
7. Insert the anchor into the forward spacer plate mounting hole. Install the screw and locknut, tightening securely.

■ **The forward mounting hole is for OMC cables only. The rear mounting holes is used exclusively for SAE cables.**

8. Pull the cable forward so it's end aligns with the inner hole in the shift arm. Install the cable pin and a new cotter pin.
9. Move the control handle to the full forward, wide open throttle position. Pull on the casing guide at the engine end of the cable to remove any excess play and then mark the cable where it enters the casing guide.
10. Return the control handle to the Neutral detent position. Measure the distance between the mark made in the last step and the end of the casing guide. This is the shift cable stroke for Forward. Record the measurement and then mark the cable where it goes into the guide once again.
11. Now move the control handle to the full reverse wide open throttle position. Push in on the cable guide to remove any excess end play and then measure the distance between the mark made in the previous step and the end of the cable guide. Mark this point as well and then record the measurement; this is the shift stroke for Reverse.
12. The distance between Forward (1st mark) and Neutral (2nd mark) must be 1 1/4-1 1/2 in. (31.8-38mm).
13. The distance between Neutral (2nd mark) and Reverse (3rd mark) must be 1 1/4-1 1/2 in. (31.8-38mm).
14. If your measurements are not within these specifications, make sure you check your attachment at the control box before moving on.
15. Make sure that the transom cable is routed into the groove in the

exhaust pipe and behind the steering assembly. If routing the cable behind the assembly causes interference, it's OK to move it to the front of the steering assembly.

16. Have an assistant rotate the propeller while you pull out on the transom cable until you get full forward gear engagement (the prop should stop moving). Mark the shift cable where it comes out of the casing guide for a future reference.
17. Connect the transom cable as detailed in the appropriate procedure.
18. Manually move the engine shift lever on the bracket into the neutral detent position - this should be almost vertical and the propeller should rotate freely in either direction. Position the control handle at the helm in the neutral detent position.

■ **If not already done, make sure that the throttle cable is disconnected from the carburetor or throttle arm and anchor block on EFI models.**

19. Connect the cable to the upper (carb) or lower (EFI) shift lever pin and secure it with the washer and a new cotter pin.
20. Loosen the bolt on the trunnion anchor and pivot the cover out of the way.
21. Make sure that the control handle, shift lever and drive unit are all still in the neutral detent position and then pull aft on the cable to remove any excess end play.

✳✳ CAUTION

Do not pull so hard as to cause any of the components to shift out of the neutral detent position.

22. Now rotate the knurled knob on the trunnion until it slides into the anchor block pocket. Once there, hold the trunnion in position while backing it off 6 full turns from the cable casing guide. Once done, pivot the cover into place and tighten the retaining bolt securely.

Remote Control Cable - 1994-96 Cobra & 1996 King Cobra

 MODERATE

1. Remove the control box and disconnect the cables. Remove the cotter pin and cable connector securing the cable to the throttle lever. Remove the screw and nut securing the anchor bracket to the housing and then remove the bracket.
2. Trace the cable back to the shift bracket on the engine and remove the cotter pin and washer from the end of the cable. Disconnect the cable guide from the shift lever pin.

■ **Take note of the positioning of the cable pin in the slot for ease of installation.**

A- Remote Shift Cable
B- Cotter Pin
C- Anchor Nut
D- Transom Cable

Fig. 13 Setting up the remote shift cable - carb

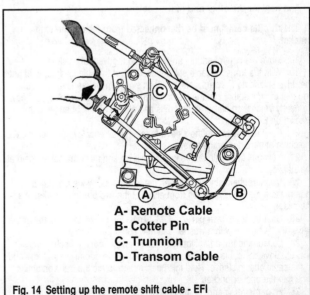

A- Remote Cable
B- Cotter Pin
C- Trunnion
D- Transom Cable

Fig. 14 Setting up the remote shift cable - EFI

3. Now loosen the anchor block screw (a short ways down from the cable end) and pivot the retainer away from the trunnion. Remove the shift cable.

To install:

OMC Cables

4. Lubricate the control unit end of the cable with OMC Triple Guard grease.

5. Install the cable trunnion into the anchor and then install the anchor on the trunnion.

6. Insert the anchor into the forward spacer plate mounting hole. Install the screw and locknut, tightening securely.

■ **The forward mounting hole is for OMC cables only. The rear mounting holes is used exclusively for SAE cables.**

7. Pull the cable forward so it aligns with the hole in the throttle arm. Install the cable pin and a new cotter pin. Install the pin with the head against the cable and then insert a new cotter pin.

■ **Always connect the cable on the same side of the throttle lever as the cable anchor on the anchor plate. This will position the control cable parallel to the anchor plate assembly.**

SAE Cables to OMC Concealed Side-Mount Remote

8. Use the cable trunnion and connector supplied with the appropriate OMC Adapter Kit to adapt SAE cables to OMC controls.

9. Install the cable trunnion and connector onto the SAE cable.

10. Install the plastic trunnion adapter and the metal trunnion adapter onto the cable trunnion. In order to retain the trunnion, the plastic adapter must go over the open end of the trunnion.

11. Attach the cable to the rear mounting position in the control with the screw and nut.

12. Mark the cable core at the housing in its fully retracted position.

13. Pull out on the cable core so it is in its fully extended position. Measure from the mark made in the previous step to the end of the cable casing. Divide the measurement in half to find the midpoint of the travel.

14. Take the last measurement (the one after you divided in half) and make a mark on the core that distance from your original mark - this is the midpoint of travel. Move the core in until the mark is even with the edge.

15. Thread the cable connector on until one thread is exposed and then adjust it to line it up with the inner hole on the shift arm. Do not adjust it in or out more than one full turn.

16. Lock the connector to the cable with the nut, being careful not to disturb the midpoint adjustment.

17. Attach the cable to the inner hole in the shift arm and secure it with a new cotter pin.

■ **Always connect the cable on the same side of the throttle lever as the cable anchor on the anchor plate. This will position the control cable parallel to the anchor plate assembly.**

And now for the other end of the cable

■ **The throttle cable must be disconnected from the engine shift bracket.**

18. Move the control handle at the helm to the Neutral position.

19. With the transom cable still attached, pivot the shift lever forward until the drive shifts into Forward gear and then slowly move the lever back until you feel the drive shift into the Neutral detent position. The lever should now be nearly vertical on carburetor models and almost at a right angle to the bracket on EFI models.

20. Push the cable into the sleeve to remove all endplay and then mark the cable at the edge of the sleeve.

21. Now pull the cable out as far as it will go and mark the cable again at the sleeve.

22. Measure the distance between the two marks, divide it in half and then make a mark this distance from either of the two previous marks; this is the midpoint.

23. Coat the shift lever pin with OMC Triple Guard grease and attach the cable end to the pin with a washer and new cotter pin.

24. Loosen the nut and pivot the retainer away from the anchor bracket pocket. Move the cable until the midpoint mark you made aligns with the end of the cable casing/sleeve. Hold this position and rotate the trunnion until it slips into the anchor pocket.

25. Pivot the retainer over the pocket and tighten the screw securely.

Remote Control Cable - 1997-98 Cobra/King Cobra

 ② MODERATE

1. Remove the control box and disconnect the cables. Remove the cotter pin and cable connector securing the cable to the throttle lever. Remove the screw and nut securing the anchor bracket to the housing and then remove the bracket.

2. Trace the cable back to the shift bracket on the engine and remove the cotter pin and washer from the end of the cable. Disconnect the cable guide from the shift lever pin.

■ **Take note of the positioning of the cable pin in the slot for ease of installation.**

✱✱ WARNING

Be very careful not to loosen or remove the shift lever pin locknut.

3. Now loosen the anchor block retainer screw (upper on standard rotation props and lower on counter rotation props) and pivot the retainer away from the trunnion. Remove the shift cable.

To install:

OMC Cables

4. Lubricate the control unit end of the cable with OMC Triple Guard grease.

5. Install the cable trunnion into the anchor and then install the anchor on the trunnion.

6. Insert the anchor into the forward spacer plate mounting hole. Install the screw and locknut, tightening securely.

■ **The forward mounting hole is for OMC cables only. The rear mounting holes is used exclusively for SAE cables.**

7. Pull the cable forward so it aligns with the hole in the throttle arm. Install the cable pin and a new cotter pin. Install the pin with the head against the cable and then insert a new cotter pin.

■ **Always connect the cable on the same side of the throttle lever as the cable anchor on the anchor plate. This will position the control cable parallel to the anchor plate assembly.**

SAE Cables to OMC Concealed Side-Mount Remote

8. Use the cable trunnion and connector supplied with the appropriate OMC Adapter Kit to adapt SAE cables to OMC controls.

9. Install the cable trunnion and connector onto the SAE cable.

10. Install the plastic trunnion adapter and the metal trunnion adapter onto the cable trunnion. In order to retain the trunnion, the plastic adapter must fo over the open end of the trunnion.

11. Attach the cable to the rear mounting position in the control with the screw and nut.

12. Mark the cable core at the housing in its fully retracted position.

13. Pull out on the cable core so it is in its fully extended position. Measure from the mark made in the previous step to the end of the cable casing. Divide the measurement in half to find the midpoint of the travel.

14. Take the last measurement (the one after you divided in half) and make a mark on the core that distance from your original mark - this is the midpoint of travel. Move the core in until the mark is even with the edge.

15. Thread the cable connector on until one thread is exposed and then adjust it to line it up with the inner hole on the shift arm. Do not adjust it in or out more than one full turn.

16. Lock the connector to the cable with the nut, being careful not to disturb the midpoint adjustment.

17. Attach the cable to the inner hole in the shift arm and secure it with a new cotter pin.

■ **Always connect the cable on the same side of the throttle lever as the cable anchor on the anchor plate. This will position the control cable parallel to the anchor plate assembly.**

And now for the other end of the cable

■ **The throttle cable must be disconnected from the engine shift bracket.**

18. Now, at the other end, move the control handle to the full forward, wide open throttle position. Pull on the casing guide at the engine end of the

cable to remove any excess play and then mark the cable where it enters the casing guide.

19. Return the control handle to the Neutral detent position. Measure the distance between the mark made in the last step and the end of the casing guide. This is the shift cable stroke for Forward. Record the measurement and then mark the cable where it goes into the guide once again.

20. Now move the control handle to the full reverse wide open throttle position. Push in on the cable guide to remove any excess end play and then measure the distance between the mark made in the previous step and the end of the cable guide. Mark this point as well and then record the measurement; this is the cable stroke for Reverse.

21. The distance between Forward (1st mark) and Neutral (2nd mark) must be 1 1/4-1 3/8 in. (31.8-35mm).

22. The distance between Neutral (2nd mark) and Reverse (3rd mark) must be 1 1/4-1 3/8 in. (31.8-35mm).

23. If your measurements are not within these specifications, make sure you check your attachment at the control box before moving on.

24. The cable attachment pin on standard rotation (right hand) engines must be in the center of the lower slot on the engine shift bracket. On counter rotation (left hand) engines, it must be in the center of the upper slot.

25. Move the control handle at the helm to the Neutral position.

26. With the transom cable still attached, pivot the shift lever forward until the drive shifts into Forward gear and then slowly move the lever back until you feel the drive shift into the Neutral detent position. The lever should now be nearly vertical on carburetor models and almost at a right angle to the bracket on EFI models.

27. Push the cable into the sleeve to remove all endplay and then mark the cable at the edge of the sleeve.

28. Now pull the cable out as far as it will go and mark the cable again at the sleeve.

29. Measure the distance between the two marks, divide it in half and then make a mark this distance from either of the two previous marks; this is the midpoint.

30. Install the trunnion cover over the trunnion and secure it with the screw.

31. Slide the shift cable in or out until the end of the casing is lined up with the midpoint mark made previously.

32. Holding the cable in position, loosen the jam nut and turn the connector in or out until it lines up with the shift lever pin. Tighten the jam nut making sure that the midpoint lark is still in position.

33. Install the cable connector onto the pin with the washer and a new cotter pin.

Remote Control Cable - 1990-91 King Cobra

1. Move the remote control handle into the Neutral position.

2. Disconnect the cable at the remote control as detailed in the Remote Control section.

3. Trace the cable back to the shift bracket on the engine and remove the cotter pin and washer from the end of the cable. Disconnect the cable from the shift lever.

■ **Remember, the shift cable running between the remote control and the shift bracket is the top cable; the lower cable runs from the bracket to the drive.**

✳✳ WARNING

Be very careful not to loosen or remove the shift lever pin locknut.

4. Now loosen the anchor block screw (a short ways down from the cable end) and retainer. Remove the shift cable.

To install:

5. Lubricate the control unit end of the cable with OMC Triple Guard grease.

6. Install the shift cable trunnion into the anchor and then install the anchor on the trunnion.

7. Insert the anchor into the forward spacer plate mounting hole. Install the screw and locknut, tightening securely.

■ **The forward mounting hole is for OMC cables only. The rear mounting holes is used exclusively for SAE cables.**

8. Pull the cable forward so it's end aligns with the inner hole in the shift arm. Install the cable pin and a new cotter pin.

9. Now, at the other end, coat the shift lever pin with OMC Triple Guard grease.

10. Connect the cable guide to the shift lever at the original pin position and install the washer and a new cotter pin.

■ **If the original pin position has been changed, the control cable will require readjustment, please refer to Determining Shift Lever Pin Position.**

11. Loosen the anchor bracket retainer and pivot it away from the anchor pocket. Position the control cable trunnion in the pocket, pivot the retainer back into position and tighten the nut securely.

■ **If a new cable has been installed, please make sure you follow the Adjustment procedures.**

Whenever the cable has been removed or replaced, it must be adjusted as follows. If non-OMC remote controls are used, the cable stroke must be at least 2 7/8 in. (73mm) and no more than 3 3/16 in. (81mm).

You must also determine if your propeller is a right hand rotation or a left hand (counter). Generally speaking:

• all single engines are right hand rotation

• the starboard engines in dual engine applications are right hand rotation

• the port engines in dual engine applications are left hand (counter)

The cable attachment pin must be located in the center of the lower slot on the bottom of the shift lever for standard rotation models and in the center of the upper slot for counter rotation models.

12. Remove the cable from the shift attachment pin and anchor block and then move the control handle at the helm into the Neutral position.

13. With the transom cable still attached, you must determine the drives neutral position. Pivot the shift lever (on the bracket) forward to shift the drive into forward gear. Now slowly move the lever back until you feel the drive shift into the neutral detent position - the lever should now be in the vertical position.

14. With the control handle still in Neutral, push the cable into the cable guide to remove any end play; mark the cable where it goes into the guide.

15. Pull the cable out and mark it at the edge of the guide as well. Measure the distance between the two marks, divide it in half and then mark this measurement on the cable - obviously, between the two previous marks.

16. Coat the shift lever pin with OMC Triple Guard grease and install the cable end. Install the washer and a new cotter pin.

17. Loosen the retainer nut and pivot the retainer away from the bracket pocket. Push or pull on the cable until the center mark you made is lined up with the end of the cable guide.

18. Hold the cable in this position and then rotate the trunnion (knurled knob) until it just slips into the anchor pocket. Install the retainer and tighten the nut securely.

Remote Control Cable - 1992-93 King Cobra

◆ **See Figure 13**

1. Move the remote control handle into the Neutral position.

2. Disconnect the cable at the remote control as detailed in the Remote Control section.

3. Trace the cable back to the shift bracket on the engine and remove the cotter pin and washer from the end of the cable. Disconnect the cable from the shift lever.

■ **Remember, the shift cable running between the remote control and the shift bracket is the top cable; the lower cable runs from the bracket to the drive.**

✳✳ WARNING

Be very careful not to loosen or remove the shift lever pin locknut.

4. Now loosen the anchor block screw (a short ways down from the cable end) and retainer. Remove the shift cable.

To install:

5. Lubricate the control unit end of the cable with OMC Triple Guard grease.

6. Install the shift cable trunnion into the anchor and then install the anchor on the trunnion.

7. Insert the anchor into the forward spacer plate mounting hole. Install the screw and locknut, tightening securely.

■ **The forward mounting hole is for OMC cables only. The rear mounting holes is used exclusively for SAE cables.**

8. Pull the cable forward so it's end aligns with the inner hole in the shift arm. Install the cable pin and a new cotter pin.

9. Now, at the other end, move the control handle to the full forward, wide open throttle position. Pull on the casing guide at the engine end of the cable to remove any excess play and then mark the cable where it enters the casing guide.

10. Return the control handle to the Neutral detent position. Measure the distance between the mark made in the last step and the end of the casing guide. This is the shift cable stroke for Forward. Record the measurement and then mark the cable where it goes into the guide once again.

11. Now move the control handle to the full reverse wide open throttle position. Push in on the cable guide to remove any excess end play and then measure the distance between the mark made in the previous step and the end of the cable guide. Mark this point as well and then record the measurement; this is the cable stroke for Reverse.

12. The distance between Forward (1st mark) and Neutral (2nd mark) must be 1 1/4-1 1/2 in. (31.8-38mm).

13. The distance between Neutral (2nd mark) and Reverse (3rd mark) must be 1 1/4-1 1/2 in. (31.8-38mm).

14. If your measurements are not within these specifications, make sure you check your attachment at the control box before moving on.

15. The cable attachment pin on standard rotation (right hand) engines must be in the center of the lower slot on the engine shift bracket. On counter rotation (left hand) engines, it must be in the center of the upper slot.

16. With the remote control cable disconnected at the shift bracket, move the control handle into the Neutral position.

17. With the transom cable still attached to the shift lever, pivot the lever forward to shift the drive into forward gear. Slowly move the lever backward until you feel the drive shift into the neutral detent position. The lever should now be in a vertical position.

18. Now push in on the engine end of the remote control cable to remove excess end play and then mark the cable.

19. Next, pull the cable out to remove any endplay in the opposite direction and make another mark on the cable.

20. Measure the distance between the marks, divide in half and make a mark on the cable equal to this measurement, between the two existing marks.

21. Coat the shift lever pin with OMC Triple Guard grease and install the cable end. Install the washer and a new cotter pin.

22. Loosen the retainer nut and pivot the retainer away from the bracket pocket. Push or pull on the cable until the center mark you made is lined up with the end of the cable guide.

23. Hold the cable in this position and then rotate the trunnion (knurled knob) until it just slips into the anchor pocket. Install the retainer and tighten the nut securely.

Remote Control Cable - 1994 King Cobra

 ②◁MODERATE

1. Remove the control box and disconnect the cables. Remove the cotter pin and cable connector securing the cable to the throttle lever. Remove the screw and nut securing the anchor bracket to the housing and then remove the bracket.

2. Trace the cable back to the shift bracket on the engine and remove the cotter pin and washer from the end of the cable. Disconnect the cable guide from the shift lever pin.

■ **Take note of the positioning of the cable pin in the slot for ease of installation.**

✲✲ WARNING

Be very careful not to loosen or remove the shift lever pin locknut.

3. Now loosen the anchor block retainer screw (upper on standard rotation props and lower on counter rotation props) and pivot the retainer away from the trunnion. Remove the shift cable.

To install:
OMC Cables

4. Lubricate the control unit end of the cable with OMC Triple Guard grease.

5. Install the cable trunnion into the anchor and then install the anchor on the trunnion.

6. Insert the anchor into the forward spacer plate mounting hole. Install the screw and locknut, tightening securely.

■ **The forward mounting hole is for OMC cables only. The rear mounting holes is used exclusively for SAE cables.**

7. Pull the cable forward so it aligns with the hole in the throttle arm. Install the cable pin and a new cotter pin. Install the pin with the head against the cable and then insert a new cotter pin.

■ **Always connect the cable on the same side of the throttle lever as the cable anchor on the anchor plate. This will position the control cable parallel to the anchor plate assembly.**

SAE Cables to OMC Concealed Side-Mount Remote

8. Use the cable trunnion and connector supplied with the appropriate OMC Adapter Kit to adapt SAE cables to OMC controls.

9. Install the cable trunnion and connector onto the SAE cable.

10. Install the plastic trunnion adapter and the metal trunnion adapter onto the cable trunnion. In order to retain the trunnion, the plastic adapter must fo over the open end of the trunnion.

11. Attach the cable to the rear mounting position in the control with the screw and nut.

12. Mark the cable core at the housing in its fully retracted position.

13. Pull out on the cable core so it is in its fully extended position. Measure from the mark made in the previous step to the end of the cable casing. Divide the measurement in half to find the midpoint of the travel.

14. Take the last measurement (the one after you divided in half) and make a mark on the core that distance from your original mark - this is the midpoint of travel. Move the core in until the mark is even with the edge.

15. Thread the cable connector on until one thread is exposed and then adjust it to line it up with the inner hole on the shift arm. Do not adjust it in or out more than one full turn.

16. Lock the connector to the cable with the nut, being careful not to disturb the midpoint adjustment.

17. Attach the cable to the inner hole in the shift arm and secure it with a new cotter pin.

■ **Always connect the cable on the same side of the throttle lever as the cable anchor on the anchor plate. This will position the control cable parallel to the anchor plate assembly.**

And now for the other end of the cable

■ **The throttle cable must be disconnected from the engine shift bracket.**

18. Now, at the other end, move the control handle to the full forward, wide open throttle position. Pull on the casing guide at the engine end of the cable to remove any excess play and then mark the cable where it enters the casing guide.

19. Return the control handle to the Neutral detent position. Measure the distance between the mark made in the last step and the end of the casing guide. This is the shift cable stroke for Forward. Record the measurement and then mark the cable where it goes into the guide once again.

20. Now move the control handle to the full reverse wide open throttle position. Push in on the cable guide to remove any excess end play and then measure the distance between the mark made in the previous step and the end of the cable guide. Mark this point as well and then record the measurement; this is the cable stroke for Reverse.

21. The distance between Forward (1st mark) and Neutral (2nd mark) must be 1 1/4-1 1/2 in. (31.8-38mm).

22. The distance between Neutral (2nd mark) and Reverse (3rd mark) must be 1 1/4-1 1/2 in. (31.8-38mm).

23. If your measurements are not within these specifications, make sure you check your attachment at the control box before moving on.

24. The cable attachment pin on standard rotation (right hand) engines must be in the center of the lower slot on the engine shift bracket. On counter rotation (left hand) engines, it must be in the center of the upper slot.

25. Move the control handle at the helm to the Neutral position.

26. With the transom cable still attached, pivot the shift lever forward until the drive shifts into Forward gear and then slowly move the lever back until

you feel the drive shift into the Neutral detent position. The lever should now be nearly vertical on carburetor models and almost at a right angle to the bracket on EFI models.

27. Push the cable into the sleeve to remove all endplay and then mark the cable at the edge of the sleeve.

28. Now pull the cable out as far as it will go and mark the cable again at the sleeve.

29. Measure the distance between the two marks, divide it in half and then make a mark this distance from either of the two previous marks; this is the midpoint.

30. Coat the shift lever pin with OMC Triple Guard grease and attach the cable end to the pin with a washer and new cotter pin.

31. Loosen the nut and pivot the retainer away from the anchor bracket pocket. Move the cable until the midpoint mark you made aligns with the end of the cable casing/sleeve. Hold this position and rotate the trunnion until it slips into the anchor pocket.

32. Pivot the retainer over the pocket and tighten the screw securely.

Remote Control Cable - 1995 King Cobra

 MODERATE

1. Remove the control box and disconnect the cables. Remove the cotter pin and cable connector securing the cable to the throttle lever. Remove the screw and nut securing the anchor bracket to the housing and then remove the bracket.

2. Trace the cable back to the shift bracket on the engine and remove the cotter pin and washer from the end of the cable. Disconnect the cable guide from the shift lever pin.

■ Take note of the positioning of the cable pin in the slot for ease of installation.

✳✳ WARNING

Be very careful not to loosen or remove the shift lever pin locknut.

3. Now loosen the anchor block retainer screw (upper on standard rotation props and lower on counter rotation props) and pivot the retainer away from the trunnion. Remove the shift cable.

To install:
OMC Cables

4. Lubricate the control unit end of the cable with OMC Triple Guard grease.

5. Install the cable trunnion into the anchor and then install the anchor on the trunnion.

6. Insert the anchor into the forward spacer plate mounting hole. Install the screw and locknut, tightening securely.

■ The forward mounting hole is for OMC cables only. The rear mounting holes is used exclusively for SAE cables.

7. Pull the cable forward so it aligns with the hole in the throttle arm. Install the cable pin and a new cotter pin. Install the pin with the head against the cable and then insert a new cotter pin.

■ Always connect the cable on the same side of the throttle lever as the cable anchor on the anchor plate. This will position the control cable parallel to the anchor plate assembly.

SAE Cables to OMC Concealed Side-Mount Remote

8. Use the cable trunnion and connector supplied with the appropriate OMC Adapter Kit to adapt SAE cables to OMC controls.

9. Install the cable trunnion and connector onto the SAE cable.

10. Install the plastic trunnion adapter and the metal trunnion adapter onto the cable trunnion. In order to retain the trunnion, the plastic adapter must fo over the open end of the trunnion.

11. Attach the cable to the rear mounting position in the control with the screw and nut.

12. Mark the cable core at the housing in its fully retracted position.

13. Pull out on the cable core so it is in its fully extended position. Measure from the mark made in the previous step to the end of the cable casing. Divide the measurement in half to find the midpoint of the travel.

14. Take the last measurement (the one after you divided in half) and make a mark on the core that distance from your original mark - this is the midpoint of travel. Move the core in until the mark is even with the edge.

15. Thread the cable connector on until one thread is exposed and then adjust it to line it up with the inner hole on the shift arm. Do not adjust it in or out more than one full turn.

16. Lock the connector to the cable with the nut, being careful not to disturb the midpoint adjustment.

17. Attach the cable to the inner hole in the shift arm and secure it with a new cotter pin.

■ Always connect the cable on the same side of the throttle lever as the cable anchor on the anchor plate. This will position the control cable parallel to the anchor plate assembly.

And now for the other end of the cable

■ The throttle cable must be disconnected from the engine shift bracket.

18. Now, at the other end, move the control handle to the full forward, wide open throttle position. Pull on the casing guide at the engine end of the cable to remove any excess play and then mark the cable where it enters the casing guide.

19. Return the control handle to the Neutral detent position. Measure the distance between the mark made in the last step and the end of the casing guide. This is the shift cable stroke for Forward. Record the measurement and then mark the cable where it goes into the guide once again.

20. Now move the control handle to the full reverse wide open throttle position. Push in on the cable guide to remove any excess end play and then measure the distance between the mark made in the previous step and the end of the cable guide. Mark this point as well and then record the measurement; this is the cable stroke for Reverse.

21. The distance between Forward (1st mark) and Neutral (2nd mark) must be 1 1/4-1 1/2 in. (31.8-38mm).

22. The distance between Neutral (2nd mark) and Reverse (3rd mark) must be 1 1/4-1 1/2 in. (31.8-38mm).

23. If your measurements are not within these specifications, make sure you check your attachment at the control box before moving on.

24. The cable attachment pin on standard rotation (right hand) engines must be in the center of the lower slot on the engine shift bracket. On counter rotation (left hand) engines, it must be in the center of the upper slot.

25. Move the control handle at the helm to the Neutral position.

26. With the transom cable still attached, pivot the shift lever forward until the drive shifts into Forward gear and then slowly move the lever back until you feel the drive shift into the Neutral detent position. The lever should now be nearly vertical on carburetor models and almost at a right angle to the bracket on EFI models.

27. Push the cable into the sleeve to remove all endplay and then mark the cable at the edge of the sleeve.

28. Now pull the cable out as far as it will go and mark the cable again at the sleeve.

29. Measure the distance between the two marks, divide it in half and then make a mark this distance from either of the two previous marks; this is the midpoint.

30. Install the trunnion cover over the trunnion and secure it with the screw.

31. Slide the shift cable in or out until the end of the casing is lined up with the midpoint mark made previously.

32. Holding the cable in position, loosen the jam nut and turn the connector in or out until it lines up with the shift lever pin. Tighten the jam nut making sure that the midpoint lark is still in position.

33. Install the cable connector onto the pin with the washer and a new cotter pin.

Transom Cable - 1986-88 Models

 DIFFICULT

1. Remove the cotter pin and washer securing the casing guide (end of cable) to the lower shift lever pin.

2. Remove the cotter pin securing the cable trunnion to the load sensing lever and remove the cable from the shift lever bracket.

■ If only the core wire is to be replaced without the cable casing itself, please refer to the Core Wire procedures in this section.

3. Cut the 2 plastic ties that secure the boot to the cable sleeve on the inside of the boat and slide off the boot.

4. Pull the split ring grommet from the boot and slide it off the end of the cable.

5. From the pivot housing side again now, loosen and remove the bolt at the end of the shift lever/cable guide.

6. Follow the Pivot Housing removal procedures and pull out the housing just enough so you can reach in and loosen the cable retaining nut. Remove the housing.

7. Pull the cable through the gimbal housing until you can slide the plastic corrugated casing off of the cable.

8. Back inside the boat again, pull the cable through the sleeve and remove it.

To install:

If the cable was removed, but not replaced

9. Route the cable between the exhaust pipe and flywheel housing, above the starter and behind the steering assembly.

■ **If this causes interference on your particular application, you can route them in front of the assembly.**

■ **Clean any paint off of the inside of the anchor pocket. Check the cable trunnion for any out-of-round condition as the trunnion must be free to pivot in the anchor pocket with no restrictions.**

10. Coat the trunnion with OMC Triple Guard grease.

11. Attach the end of the cable to the pin on the lower shift lever, install the washer and install a new cotter pin.

12. Install the cable trunnion into the anchor pocket and secure it with a new cotter pin. Make sure that the trunnion pivots freely on the pocket.

If the cable has been replaced
◆ See Figure 15

13. Slide the rubber boot onto the engine end of the new cable. Attach a new grommet to the cable and then press it into the boot.

14. Feed the cable into the sleeve from the inside of the boat. Pull it through on the gimbal housing end until there is enough cable exposed to install the plastic corrugated tubing so that its end is flush with the back of the hex nut and then push it back through into the boat a bit.

15. Coat a new O-ring with gasket sealing compound onto the drive end of the cable and up against the fitting hex nut. Wipe any excess sealing compound off of the cable end and then coat the entire length with OMC Triple Guard grease.

16. Position the pivot housing so you can insert the end of the cable through the rear of the housing and into the cable retainer/guide. Thread the fitting into the housing and tighten to 35-37 ft. lbs. (47-50 Nm) with a crowfoot adapter on the torque wrench.

■ **Make sure the crowfoot adapter is attached to the wrench at a 90 degree angle to ensure an accurate meeting.**

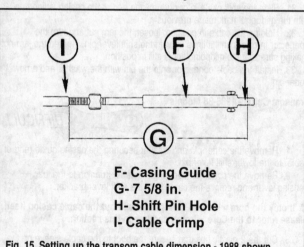

F- Casing Guide
G- 7 5/8 in.
H- Shift Pin Hole
I- Cable Crimp

Fig. 15 Setting up the transom cable dimension - 1988 shown

17. Position the pivot housing and install it as detailed in the appropriate section.

18. Now check the neutral position. On 1986-87 models, working at the engine end of the cable, move the casing guide in or out so that you achieve a measurement of 6 1/2 in. (16.5 cm) between the center of the trunnion and the center of the attachment hole. Once set, be careful not to disturb it!

19. Set the bellcrank at a 90 degree angle to the machined face of the pivot housing. Push or pull the cable retainer/guide until the axis of the bellcrank arm is properly positioned. Use a carpenter's square to check the alignment.

20. Being careful not to move the cable guide, screw the retainer onto the cable end until it just comes in contact with the guide surface. Align the retainer and guide screw holes and thread in the bolt.

21. In order to fully engage the shift cable guide and prevent binding, shift the remote control handle into full forward gear. Hold the cable guide with a 9/16 in. wrench and tighten the bolt to 10-12 ft. lbs. (14-16 nm).

22. On 1988-89 models, position the small swivel retainer at the end of its threads - unscrew it until it is on the last threads without falling out of the cable retainer/guide.

23. Move the cable guide in or out until the lower edge of the guide tab is just flush with the pivot housing gasket surface.

24. At the engine end of the cable, slide the casing guide in or out until you achieve a measurement of 7 5/8 in. (19.4 cm) +/- 1/32 in. (0.07 cm) between the center of the shift pin hole at the end of the casing guide and the edge of the cable crimp.

25. Make sure that the guide tab is still flush with the pivot housing surface and then screw in the retainer until it just contacts the edge of the shift cable guide. Install, but do not tighten, the retainer screw.

26. To prevent binding the cable while tightening the retainer screw, press the cable guide all of the way forward (toward the back of the drive) into the pivot housing. Hold the cable retainer with an open end wrench and tighten the screw to 10-12 ft. lbs. (14-16 Nm).

27. On all models now, press the boot onto the end of the cable sleeve and install new plastic ties to secure it to the sleeve and grommet.

28. For final hook-up at the engine shift bracket, go back and follow the steps detailed above under the "Cable was removed, but not replaced" procedure.

Transom Cable - 1989 Models

1. Remove the cotter pin and washer securing the casing guide (end of cable) to the lower shift lever pin.

2. Remove the cotter pin securing the cable trunnion to the load sensing lever and remove the cable from the shift lever bracket.

To install:
If the cable was removed, but not replaced

3. Route the cable between the exhaust pipe and flywheel housing, above the starter and behind the steering assembly. On 2.3L engines, route the cable in front of the steering assembly.

■ **If this causes interference on your particular application, you can route them in front of the assembly.**

■ **Clean any paint off of the inside of the anchor pocket. Check the cable trunnion for any out-of-round condition as the trunnion must be free to pivot in the anchor pocket with no restrictions.**

4. Coat the trunnion with OMC Triple Guard grease.

5. Attach the end of the cable to the pin on the lower shift lever, install the washer and install a new cotter pin.

6. Install the cable trunnion into the anchor pocket and secure it with a new cotter pin. Make sure that the trunnion pivots freely on the pocket.

If the cable has been replaced
◆ See Figure 16

7. Remove the drive unit and position the small swivel retainer at the end of its threads - unscrew it until it is on the last threads without falling out of the cable guide.

8. Position the bellcrank alignment plate (#914017) on the two lowest pivot housing studs. Slide the shift cable in or out until the bellcrank pin is captured by the slot in the alignment plate.

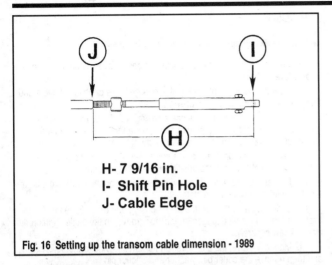

H- 7 9/16 in.
I- Shift Pin Hole
J- Cable Edge

Fig. 16 Setting up the transom cable dimension - 1989

9. Install three 1/8 in. (3.18mm) thick washers with a 7/16 in. (11.11mm) I.D. (#312972) on each stud. Follow them with a hex nut and tighten the nuts until there is no movement in the plate.

10. At the engine end of the cable, slide the casing guide in or out until you achieve a measurement of 7 -9/16 in. (19.2 cm) +/- 1/32 in. (0.07 cm) between the center of the shift pin hole at the end of the casing guide and the edge of the brass cable fitting.

11. Without disturbing the cable setting, screw in the retainer until it just contacts the edge of the shift cable guide. Install, but do not tighten, the retainer screw.

12. Remove the special alignment plate.

13. To prevent binding the cable while tightening the retainer screw, press the cable guide all of the way forward (toward the back of the drive) into the pivot housing. Hold the cable retainer with an open end wrench (9/16 in.) and tighten the screw to 10-12 ft. lbs. (14-16 Nm).

14. Recheck the casing guide dimension, repeating the previous 8 steps if it is no longer in spec.

15. Now go back and follow the steps detailed above under the "cable was removed, but not replaced" procedure.

Transom Cable - 1990-93 Cobra

◆ See Figure 17

1. Remove the cotter pin and washer securing the casing guide (end of cable) to the shift lever pin.

2. Remove the cotter pin securing the cable trunnion to the load sensing lever and remove the cable from the shift lever bracket.

3. Loosen both set screws and pull the casing guide off of the cable.

4. Loosen the locknut and unscrew the casing guide jacket from the cable.

5. Working inside the boat, cut the 2 plastic ties securing the boot to the end of the cable sleeve. Pull the boot off of the sleeve.

6. Remove the split ring grommet from the end of the boot and take it off of the cable. Slide the boot off the end of the cable.

7. Remove the drive unit.

8. Remove the collared retainer nut on the inner shift cable wire. Remove the screw and retainer from the end of the drive shift lever.

9. Disengage the U-joint boot at the pivot housing lip and push it inside.

10. Unscrew the water hose nipple retaining nut and push the nipple through the housing as far as possible.

11. Turn the gimbal ring and remove the ground strap.

12. Remove the pivot pins and then pull the pivot housing out as far as you can until you have access to unscrew the cable retaining nut.

13. Slide the plastic corrugated casing off of the cable and then remove the cable from the housing.

To install:

14. Pull the inner cable in flush with the shift cable casing. Cover the end of the casing with electrical tape.

15. Wrap the plastic tubing around the cable. The end of the tubing should just contact the back of the hex fitting. Lubricate the tubing with soapy water and then push the cable casing through the sleeve and connector.

16. Coat a new shift cable O-ring with gasket sealing compound, slide it onto the cable end (drive side) and then position it against the fitting hex.

17. Wipe any excess sealer off of the cable end and then coat the entire length with OMC Triple Guard grease.

18. Insert the cable end through the rear of the housing and into the guide. Screw in the fitting and use a crowsfoot adapter to tighten it to 35-37 ft. lbs. (47-50 Nm).

19. Push the rubber boot onto the engine end of the cable. Attach the small grommet and push it into the boot.

20. Coat the end of the sleeve with Scotch Grip Rubber Adhesive 1300 and pull the boot onto the sleeve. Attach 2 new plastic ties where they were originally installed.

21. Thread the casing guide jacket into the cable until it seats. If properly installed, it will measure no more than 4-15/16 in. (12.5cm) from the forward end to the cable tip.

22. Tighten the locknut to 35-50 inch lbs. (4-5.6 Nm).

23. Lubricate the cable and casing with OMC Triple Guard grease. Slide the casing onto the jacket until the inner cable seats against the sight hole. Turn one set screw against the core wire and tighten it to 18-20 inch lbs. (2-2.3 Nm), Now turn the other set screw in against the wire and tighten it to the same torque.

Now at the other end, if the cable was removed, but not replaced

24. Route the cable between the exhaust pipe and flywheel housing, below the starter and behind the steering assembly.

■ **If this causes interference on your particular application, you can route them in front of the assembly.**

■ **Clean any paint off of the inside of the anchor pocket. Check the cable trunnion for any out-of-round condition as the trunnion must be free to pivot in the anchor pocket with no restrictions.**

25. Coat the trunnion with OMC Triple Guard grease.

26. Attach the end of the cable to the pin on the lower shift lever, install the washer and install a new cotter pin.

27. Install the cable trunnion into the anchor pocket and secure it with a new cotter pin. Make sure that the trunnion pivots freely on the pocket.

28. On 1993 carb models, the lower cable pin must be centered in the shift lever slot. Coat the upper and lower cable pins, the upper trunnion and the lower actuator pocket with OMC Triple Guard grease. Set the trunnion-to-nut dimension at 11/16 in. (17.5mm) +/- 1/32 in. (0.78mm) and make sure this does not change when attaching the cable to the shift lever. Attach the cable with a washer and new cotter pin and then install the trunnion into the actuator pocket and secure with a new cotter pin also.

29. On 1993 EFI models, the upper cable pin must be centered in the shift lever slot. Coat the upper and lower cable pins, the upper trunnion and the lower actuator pocket with OMC Triple Guard grease. Set the trunnion-to-nut dimension at 11/16 in. (17.5mm) +/- 1/32 in. (0.78mm) and make sure this does not change when attaching the cable to the shift lever. Attach the cable with a washer and new cotter pin and then install the trunnion into the actuator pocket and secure with a new cotter pin also.

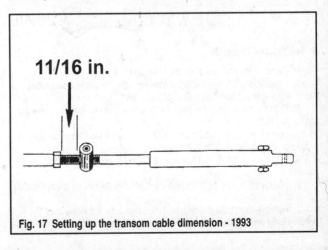

11/16 in.

Fig. 17 Setting up the transom cable dimension - 1993

If the cable has been replaced
◆ See Figure 18

30. Remove the drive unit and position the small swivel retainer (cable anchor) at the end of its threads - unscrew it until it is on the last threads without falling out of the cable guide.

31. Position the bellcrank alignment plate (#914017) on the two lowest pivot housing studs. Slide the shift cable in or out until the bellcrank pin is captured by the slot in the alignment plate.

32. Install three 1/8 in. (3.18mm) thick washers with a 7/16 in. (11.11mm) I.D. (#312972) on each stud. Follow them with a hex nut and tighten the nuts until there is no movement in the plate.

33. At the engine end of the cable, slide the casing guide in or out until you achieve a measurement of 7 -9/16 in. (19.2 cm) +/- 1/32 in. (0.07 cm) between the center of the shift pin hole at the end of the casing guide and the edge of the brass cable fitting.

34. Without disturbing the cable setting, screw in the retainer until it just contacts the edge of the shift cable guide. Install, but do not tighten, the retainer screw.

35. Remove the special alignment plate.

36. To prevent binding the cable while tightening the retainer screw, press the cable guide all of the way forward (toward the back of the drive) into the pivot housing. Hold the cable retainer with an open end wrench (9/16 in.) and tighten the screw to 10-12 ft. lbs. (14-16 Nm).

37. Recheck the casing guide dimension, repeating the previous 8 steps if it is no longer in spec.

38. Now go back and follow the steps detailed above under the "cable was removed, but not replaced" procedure.

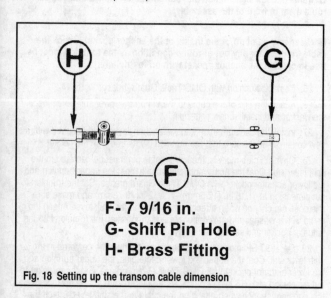

F- 7 9/16 in.
G- Shift Pin Hole
H- Brass Fitting

Fig. 18 Setting up the transom cable dimension

Transom Cable - 1994-96 Cobra & 1996-98 King Cobra

◆ See Figures 19 and 20

1. Remove the rear cover on the drive unit.

2. Remove the flat washer and cotter pin from the anchor pin (cube). Remove the jam nut from the shift cable core wire.

3. Remove the anchor pin from the bellcrank slot and then unscrew the anchor pin from the core wire.

4. Loosen the screw and slide the retainer out of the shift cable anchor groove on the starboard side of the drive.

5. Pull the shift cable into the boat through the drive and transom bracket.

6. Remove the cotter pin and washer securing the casing guide (end of cable) to the shift lever pin.

7. Remove the cotter pin securing the cable trunnion to the load sensing lever and remove the cable from the shift lever bracket.

To install:

■ SAE or OMC Snap-In/SAE cables must be used.

8. Make sure that the throttle cable has been disconnected from the throttle lever.

9. Note the positions of the small and large seals on the shift cable and then remove the jam nut and both seals.

10. Lightly coat the end of the cable casing with OMC Triple Guard grease. Slide the cable through the cable tube in the transom (from the inside) until it appears on the other side of the transom.

11. Turn the drive unit hard over to port. Reinstall the large and small seals onto the cable end in the positions you noted previously.

12. Loosen the cable clamp screw in the recess on the starboard side of the drive and then pull out the clamp.

13. Slide the cable through the pivot housing and drive. Slide the clamp back in all the way so it engages the anchor groove on the cable. Tighten the clamp screw securely.

14. Manually swivel the shift lever on the drive to extend the bellcrank and then remove the cotter pin, washer and anchor pin.

15. Thread the anchor pin (cube) halfway onto the end of the control cable.

16. Swivel the shift lever back into the neutral detent position and then ensure that the control handle at the helm is also still in the neutral detent position. Turn the anchor pin in or out until it aligns with the center of the bellcrank slot.

17. Install the washer and a new cotter pin. Install the jam nut and tighten it securely against the anchor pin.

18. To be safe, check which side of the shift lever the bellcrank link is connected to; standard rotation propellers use the port side and counter rotation models utilize the starboard side.

19. Install the drive cover.

Transom Cable - 1990-95 King Cobra

1. Shift the control handle into the Neutral position. Make sure that the propeller will rotate freely.

2. Loosen the anchor bracket retainer nut and then pivot the retainer away from the cable trunnion.

3. Remove the cotter pin from the shift lever pocket and then pull out the retaining pin to free the cable guide. Lift the cable out of the bracket.

4. Remove the stern drive unit.

5. Remove the inlet water hose barb and remove the ground strap from the port side of the pivot housing.

6. Lift up on the U-joint bellows and push it inward.

7. Lift up on the pivot housing, loosen the shift cable locknut and then pull the cable out of the bracket and pivot housing.

To install:

8. Apply OMC Adhesive M to the groove around the shift cable passage in the drive and install a new seal.

9. Coat a new O-ring with gasket sealing compound and install it on the drive end of the cable.

10. Slide the cable through the hole in the pivot housing and then thread on the nut.

11. Position the cable in the housing so that both guides are facing down and the connector pin hole is facing up. Tighten the nut to 35-37 ft. lbs. (47.5-50 Nm). Use a crowfoot on the torque wrench to ensure proper torque.

12. Slide the end of the cable under the grease tube and through the hole in the gimbal housing.

13. Install the drive unit as detailed in the appropriate section.

14. Coat the shift cable retaining pin with OMC Triple Guard grease. Position the cable guide in the lower inner shift lever pocket and then install the pin. Insert a new cotter pin so that the cotter pin head is positioned over the end of the retaining pin.

15. Place the cable trunnion in the anchor pocket, move the retainer into position and tighten the nut securely.

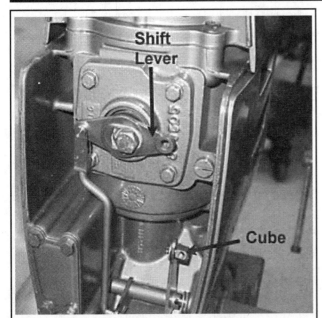

Fig. 19 A good look at the shift mechanism in the drive

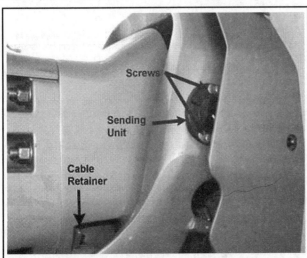

Fig. 20 The cable is held in place by a clamp at the drive

DETERMINING SHIFT LEVER PIN POSITION

1990 King Cobra

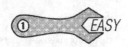

The engine shift lever (on the shift bracket) has a slot at the top of the lever and another at the bottom. The upper slot is used for counter rotation (left hand) drives and the lower slot is used for standard rotation (right hand) units. Each of the slots has multiple variations for control cable hook-up and adjustment.

1. Have an assistant observe the engine shift bracket while you move the control handle at the helm from Neutral to Forward and then to Reverse. Do this a number of times while determining if the carburetor throttle lever is moving before forward or reverse engagement occurs.

2. If it does, move the shift lever pin toward the + (plus) end of the slot until the throttle lever is beginning to move at the same time forward or reverse gears are being engaged.

3. Now move the control handle out of Reverse and/or Forward while having an assistant observe the shift bracket. Shift lever disengagement should occur just BEFORE the control handle enters the neutral detent position. If not, move the shift lever pin incrementally toward the - (minus) side of the slot until achieved.

SHIFT SYSTEM TESTS

1992-93 Cobra - Carb Models

◆ See Figures 20 and 21

■ To ensure full gear engagement through all of the following tests, have an assistant rotate the propeller during all procedures requiring you to move the control handle to full forward or reverse positions.

Interrupter Switch Roller Position

This test verifies that there will be no engine ignition interruption while at full throttle.

1. Shift from Neutral into Forward gear (while rotating the prop). The shift interrupter roller should return to the center of the V-notch on the load lever when at full throttle.

2. Shift from Neutral into Reverse gear (while rotating the prop). The shift interrupter roller should return to the center of the V-notch on the load lever when at full throttle.

3. If the roller fails to return to the V-notch while at full throttle in either gear, note which gear this happens in and what position the roller is in when it happens.

4. If the roller does not return to the V-notch in either of the tests, move on to the Interrupter Roller Adjustment.

5. If the roller returns to the V-notch in both tests, move to Check Full Forward Gear Engagement.

Interrupter Roller Adjustment - High in One Gear Only

6. If the interrupter roller is high on one side of the V-notch in a particular gear, leave the control handle at full throttle and then adjust the cable.

7. Loosen the trunnion cover screw and readjust the trunnion by turning the knurled knob until the roller falls into the center of the V-notch. Tighten the retaining screw.

8. Shift the handle into Neutral and check that the propeller rotates freely and the roller is in the center of the V-notch.

9. Repeat the previous test.

Interrupter Roller Adjustment - High in Both Gears

10. If the interrupter roller is equally high on one side of the V-notch in both gears, leave the control handle at full throttle and then adjust the cable.

■ This will work for both OMC or SAE cables.

11. Loosen the lower transom bracket cable locknut and move the shift cable pin up toward the - (minus) side of the slot until the shift roller falls into the center of the V-notch. Tighten the pin locknut

12. Shift the handle into Neutral and check that the propeller rotates freely and the roller is in the center of the V-notch.

13. Repeat the Interrupter Switch Roller Position test.

Check Full Forward Gear Engagement

14. Have someone rotate the propeller while moving the control handle into the full forward wide open throttle position. Look for the reference mark made on the cable in an earlier test. If the mark is visible, then the adjustments made so far are correct and complete.

15. If the mark is not visible, loosen the lower transom cable pin locknut and move the cable pin downward towards the + (plus) side of the slot until the mark is just visible at the end of the casing guide.

16. Retighten the locknut and then repeat the Interrupter Switch Roller Position test.

17. All adjustments are correct and complete when the roller remains in the V-notch during and after full forward and reverse gear engagements; and the reference mark is visible on the transom cable in full forward gear.

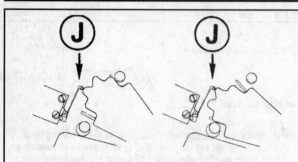

J- Interrupter Roller

Fig. 20 If the roller is on a high side of the V-notch. . .

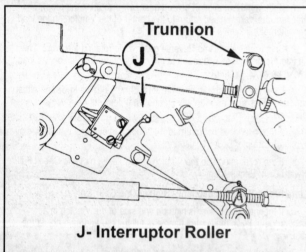

J- Interruptor Roller

Fig. 21 . . .adjust the trunnion until it falls into the center of the notch

1992-93 Cobra - EFI Models

② MODERATE

◆ See Figures 22 and 23

■ To ensure full gear engagement through all of the following tests, have an assistant rotate the propeller during all procedures requiring you to move the control handle to full forward or reverse positions.

Interrupter Switch Roller and Neutral Drive Switch Roller Position

This test verifies that there will be no engine ignition interruption while at full throttle.

1. Shift from Neutral into Forward gear (while rotating the prop). The shift interrupter roller should return to the center of the V-notch on the load lever when at full throttle and the neutral drive switch roller should be off the shift lever cam.

2. Shift from Neutral into Reverse gear (while rotating the prop). The shift interrupter roller should return to the center of the V-notch on the load lever when at full throttle and the neutral drive switch roller should be off the shift lever cam.

3. Shift into Neutral. The shift interrupter roller should now return to the center of the V-notch and the neutral drive switch roller must be on the shift lever cam (the roller does not have to be centered).

4. If the interrupter switch roller was centered in the V-notch, and the neutral drive switch roller was correctly positioned in all three checks, continue on to Check Full Forward Gear Engagement.

5. If the interrupter switch roller was centered in the V-notch, and the neutral drive switch roller was not correctly positioned in all three checks, continue on to Correct Neutral Drive Switch Roller Position.

Interrupter Roller Adjustment - High in One Gear Only

6. If the interrupter roller is high on one side of the V-notch in a particular gear, leave the control handle at full throttle and then adjust the cable.

7. Loosen the trunnion cover screw and readjust the trunnion by turning the knurled knob until the roller falls into the center of the V-notch. Tighten the retaining screw.

8. Shift the handle into Neutral and check that the propeller rotates freely and the roller is in the center of the V-notch.

9. Repeat the previous test.

Interrupter Roller Adjustment - High in Both Gears

10. If the interrupter roller is equally high on one side of the V-notch in both gears, leave the control handle at full throttle and then adjust the cable.

■ This will work for both OMC or SAE cables.

11. Loosen the transom bracket cable locknut and move the shift cable pin toward the - (minus) side of the slot until the shift roller falls into the center of the V-notch. Tighten the pin locknut

12. Shift the handle into Neutral and check that the propeller rotates freely and the roller is in the center of the V-notch.

13. Repeat the Interrupter Switch Roller and Neutral Drive Switch Roller Position test.

Check Full Forward Gear Engagement

14. Have someone rotate the propeller while moving the control handle into the full forward wide open throttle position. Look for the reference mark made on the cable in an earlier test. If the mark is visible, then the adjustments made so far are correct and complete.

15. If the mark is not visible, loosen the transom cable pin locknut and move the cable pin toward the + (plus) side of the slot until the mark is just visible at the end of the casing guide.

16. Retighten the locknut and then repeat the Interrupter Switch Roller and Neutral Drive Switch Roller Position test.

17. All adjustments are correct and complete when the roller remains in the V-notch during and after full forward and reverse gear engagements; and the reference mark is visible on the transom cable in full forward gear.

Correct Position Of Neutral Drive Switch Roller

Perform this procedure **only** if the neutral drive switch roller was not correctly positioned during the previous tests.

18. Move the control handle into the Neutral position and make sure that the propeller rotates freely in either direction. Check that the switch roller is sitting directly on the cam. If not, both shift cable trunnions will require adjustment.

19. Remove the cotter pin from the transom bracket pocket and lift out the trunnion. Remove the set screw (throw it away). Loosen the nut on the remote control trunnion anchor cover and lift out the trunnion.

20. Rotate the transom bracket trunnion while counting the number of turns it takes to position the switch roller on the cam.

■ The roller does not have to be exactly centered, just on it.

21. Now rotate the remote control trunnion the same number of turns, but in the opposite direction of how you turned the transom trunnion. If you turned the transom trunnion 4 turns **toward** the casing guide, turn the remote trunnion 4 turns **away** from the casing guide; or visa versa.

22. Install both trunnion back in there pockets.

■ After the roller and cam have been realigned, verify the adjustment. When in Neutral, the roller must be on the cam. When in Forward or Reverse, the roller must be off the cam and the interrupter roller must be in the center of the V-notch. If the roller position is correct in all three positions, repeat the Check Full Forward Gear Engagement test.

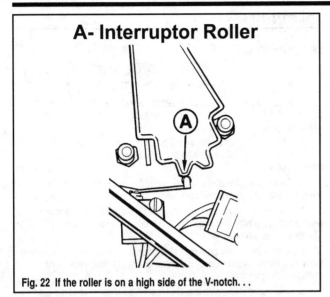

A- Interruptor Roller

Fig. 22 If the roller is on a high side of the V-notch. . .

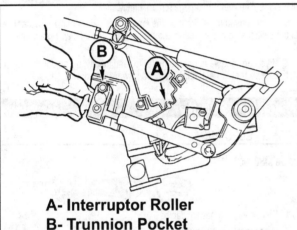

A- Interruptor Roller
B- Trunnion Pocket

Fig. 23 . . .adjust the trunnion until it falls into the center of the notch

Core Wire

REPLACEMENT

1986-89 Models

DIFFICULT

1. Disconnect the transom shift cable at the engine shift bracket and lift the cable out of the bracket.
2. Unscrew of the core wire anchors on the casing guide and pull the guide off of the cable. Clip the bent end of the core wire off with wire cutters.

■ **Never pull the core wire through the shift cable with the bent ends still attached or you will damage the inner casing.**

3. Remove the drive unit.
4. Remove the screw from the end of the transom bracket shift lever and then screw the retainer off of the cable core wire.
5. Pull the core wire out of the shift cable and discard it.

To install:

6. Insert a new core wire into the shift lever and cable. Push the wire all the way into the cable casing until the threaded ferrule seats itself in the cable guide on the drive end. There is no need to lubricate the cable.
7. Insert a cylindrical cable clamp into the casing guide - make sure the anchor screws are backed out far enough so they don't interfere with the core wire when it passes through.

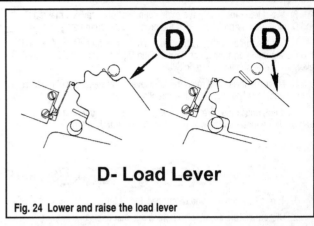

D- Load Lever

Fig. 24 Lower and raise the load lever

8. Align the cable clamp entry hole and slide the casing guide over the end of the core wire. Push the core wire all the way through the clamp.

■ **The core wire end must be visible in the inspection hole to ensure adequate anchor screw engagement.**

9. Rotate the casing guide until the shift pin attachment hole is in line with the cable trunnion. Turn both anchor screws in until they contact the core wire and then tighten both screws to 18-20 inch lbs. (2.2-3 Nm).

■ **It is imperative that both screws are touching the core wire BEFORE they are torques down.**

10. Adjust the shift cable.

Electronic Shift Assist (Interrupter/Neutral Drive) Switch

TESTING

1992-93 Cobra

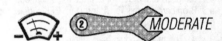 *MODERATE*

◆ See Figure 24

1. Connect an ohmmeter and set it to the High scale. Make sure the control handle is in the Neutral position.
2. Reach behind the shift cable bracket, remove the wire retainer and disconnect the receptacle plug for the switch. On EFI models, the plug is at the shift bracket at the front of the engine.
3. Insert test probes into the 2 male connector plug cavities marked **A** and **B** on carb models, or **A** and **D** on EFI models. Attach the probes to an ohmmeter.
4. The meter should read infinity.
5. Lower and raise the load lever by hand. The meter must read **0** (zero) when the lever is held in either direction.
6. If the meter does not read **0**, check the wire and probe connections and then try the test again. If it still does not read **0**, replace the overstroke/interrupter switch assembly.
7. To test the Neutral Drive switch individually on EFI models, repeat the above procedures using the **A** and **C** cavities instead.

Interrupter Switch

TESTING

1986-89 Models

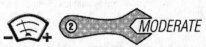 *MODERATE*

1. This is the lower switch on the bracket. Turn off the ignition and disconnect the battery.
2. Unclip the wire retainer and disconnect the switch plug behind the shift cable bracket.

3. Insert test probes into the male connector plug cavities marked **C** and **D** (blue leads) and then attach the probes to an ommeter.

4. Set the meter to the High scale. With the control system in neutral and the actuating arm roller positioned in the V-notch of the load sensing lever, the meter should read infinity.

5. Manually lower and raise the load lever - the meter should read zero when the lever is held fully in either direction.

6. If not, check the wire and probe connections at the plug and then try it again. If the meter still fails to read zero, replace the switch.

1990-92 Models

1. Turn off the ignition and disconnect the battery.

2. Disconnect the switch plug behind the shift cable bracket.

3. Insert test probes into the male connector plug cavities marked and then attach the probes to an ommeter.

4. Set the meter to the High scale. With the control system in Neutral the meter should read infinity.

5. Manually lower and raise the load sensing lever - the meter should read zero when the lever is held fully in either direction.

6. If not, check the wire and probe connections at the plug and then try it again. If the meter still fails to read zero, replace the switch.

1993-95 King Cobra

1. Connect an ohmmeter and set it to the High scale. Make sure the control handle is in the Neutral position.

2. Reach behind the shift cable bracket, remove the wire retainer and disconnect the receptacle plug for the switch. On EFI models, the plug is at the shift bracket at the front of the engine.

3. Insert test probes into the 2 male connector plug cavities marked **A** and **B** on carb models, or **A** and **D** on EFI models. Attach the probes to an ohmmeter.

4. The meter should read infinity.

5. Lower and raise the load lever by hand. The meter must read **0** (zero) when the lever is held in either direction.

6. If the meter does not read **0**, check the wire and probe connections and then try the test again. If it still does not read **0**, replace the overstroke/interrupter switch assembly.

Overstroke Switch

TESTING

1986-90 Models

1. This is the upper switch on the bracket. Turn off the ignition and disconnect the battery.

2. Unclip the wire retainer and disconnect the switch plug behind the shift cable bracket.

3. Insert test probes into the connector plug cavities marked **A** and **B** (black leads) and then attach the probes to an ommeter.

4. When the control system is in Neutral, the meter must read zero.

5. Insert a small screwdriver and depress the overstroke switch stud - the meter should read infinity.

6. If the meter does not read infinity, check the wire and probe connections at the plug and then try it again. If the meter still fails to read infinity, replace the switch.

ADJUSTMENT

1986-87 Models Only

The overstroke switch must have tested correctly before performing this adjustment. Both shift cables must be attached to the shift lever bracket while adjusting the actuating cams.

1. This is the upper switch on the bracket by the shift lever. Insert test probes into the **A** and **B** cavities on the male connector plug (these are the ones with black leads).

2. Connect an ommeter and then move the shifter to Neutral position - the meter should indicate zero continuity.

3. Now move the remote control handle into the full forward position while rotating the propeller (when it won't rotate you know it's engaged fully). The meter must show infinity.

4. If not, loosen the upper cam locknut on the shift lever and reposition the cam until the switch stud is depressed and the meter shows infinity. Retighten the cam nut.

5. Now move the remote lever to full reverse while rotating the prop until it stops. The meter must again read infinity.

6. If not, loosen the locknut on the lower cam and reposition the cam until the switch stud is depressed and the meter reads infinity. Tighten the nut.

7. If you are unable to get the meter to read infinity, try adjusting the switch itself so that the actuating tip faces forward and down.

Neutral Start Switch

TESTING - ELECTRICAL

1990-96 Models

1. Disconnect the battery cables.

2. Disconnect the Yellow/Red lead at the ignition switch and the Yellow/Red lead at the instrument wiring harness.

3. Set an ohmmeter to the High scale and connect the meter to the two leads. The meter should read zero (0).

4. Now move the control handle to the Forward position; the meter should read infinity. Move it to the Reverse position, the meter should read infinity again.

5. If the meter does not show infinity in either of the two tests, check all connections and retest. If it still fails to read infinity, replace the switch

TESTING - MECHANICAL

1990-96 Models

1. With the boat in the water and the control handle in the Neutral position, turn the ignition key to the Start position. The engine should start.

2. Pull the control handle out and into the Forward warm-up position; the engine should start.

3. Move the handle back to the Reverse warm-up position, the engine should start. Move the handle back to the Neutral position.

4. Now move the handle to the Forward gear position (past the warm-up position) and make sure that the engine will not start. Do the same with Reverse gear; the engine should not start.

5. If the engine starts in either of the two positions in the previous step, replace the switch.

Troubleshooting

 MODERATE

◆ See Figures 25, 26 and 27

A. Hard Shifting

Hard shifting coming out of gear. Check "B"

Hard Shifting from Neutral Position
1. Disconnect throttle cable.
2. Run engine and shift remote controls.

YES

Hard Shifting
1. Disconnect remote control shift cable.
2. Run engine and manually shift. Shifting effort should not require more than 16 pounds @800 rpm. Use top pin of engine shift lever for test.

YES

Shifting Effort in Excess of 16 Pounds
Cause: shifting lever or vertical drive
Remove transom shift cable from shift lever. Run engine and manually shift. Shifting effort of cable should not require more than 14 pounds @800 rpm.

YES

Shifting Effort in Excess of 14 Pounds
Cause: vertical drive or cable
Loosen six outdrive retaining nuts one full turn. Manually shift unit while manually turning propeller shaft. Shifting effort should not exceed 14 pounds.

YES

Shifting Effort in Excess of 14 Pounds
Cause: transom shift cable or lower gearcase
Remove vertical drive. Check operation of transom shift cable. Cable should operate smoothly and freely.

YES

Cable not Smooth and Free
Cause: cable guide transom bellcrank or transom shift cable.
1. Loosen lock screw of cable guide and check for free operation of cable.
2. If not smooth, check for binding in transom bellcrank or transom shift cable.
3. If smooth and free, reinstall shift cable to lower shift lever pin and secure trunnion into lower anchor block. Readjust shift cable, being sure cable guide is pushed forward into pivot housing while tightening lock screw.

No Hard Shifting
Cause: throttle cable
1. Throttle cable trunnion out of adjustment
2. Carburetor idle stop screw out of adjustment
3. Carburetor idle screws out of adjustment
4. Wrong throttle cable length
5. Throttle cable routing
6. Installation of throttle cable in remote control

NO

Unit Shifts Within Required 16 Pound Range
Cause: remote control shift cable
1. Remote shift cable trunnion anchor out of adjustment
2. Incorrect remote shift cable length
3. Incorrect remote shift cable routing
4. Incorrect installation of remote shift cable into control box
5. Remote control binding

NO

Unit Shifts Within Required 14 Pound Range
Cause: shifting lever
Inspect shifting components – they must be secure, lubricated and have a smooth movement.

NO

Unit Shifts Within Required 14 Pound Range
Cause: shift guide pin binding against shift rod in vertical drive.
Remove vertical drive and position guide pin per specifications of service manual.

NO

Cable Operation is Smooth and Free
Cause: lower gearcase
1. Check shift rod height per specifications in service bulletin No. 4022 or service manual.
2. If shift rod height is within specifications, position shift rod in neutral position. Manually turn propeller shaft and pull shift lever "up" into forward. Shifting will require between 22 to 25 pounds. Reposition shift rod in neutral position. Manually turn propeller shaft and push shift lever "down" into reverse. Shifting will require between 14 to 17 pounds. If either reading is over the required specification, the lower gearcase shifting is at fault.

Fig. 25 Hard shifting

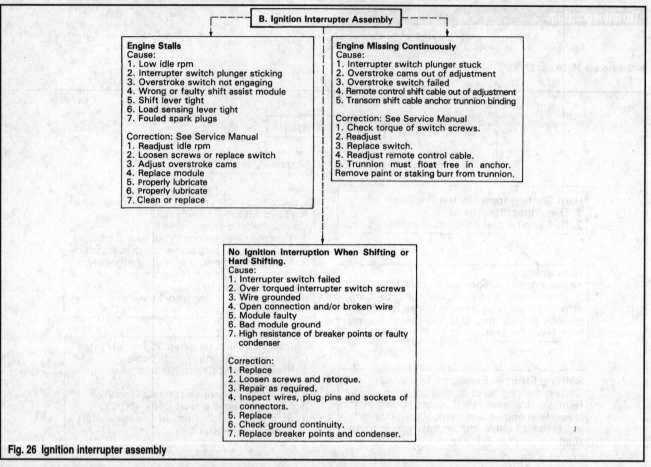

B. Ignition Interrupter Assembly

Engine Stalls
Cause:
1. Low idle rpm
2. Interrupter switch plunger sticking
3. Overstroke switch not engaging
4. Wrong or faulty shift assist module
5. Shift lever tight
6. Load sensing lever tight
7. Fouled spark plugs

Correction: See Service Manual
1. Readjust idle rpm
2. Loosen screws or replace switch
3. Adjust overstroke cams
4. Replace module
5. Properly lubricate
6. Properly lubricate
7. Clean or replace

Engine Missing Continuously
Cause:
1. Interrupter switch plunger stuck
2. Overstroke cams out of adjustment
3. Overstroke switch failed
4. Remote control shift cable out of adjustment
5. Transom shift cable anchor trunnion binding

Correction: See Service Manual
1. Check torque of switch screws.
2. Readjust
3. Replace switch.
4. Readjust remote control cable.
5. Trunnion must float free in anchor.
Remove paint or staking burr from trunnion.

No Ignition Interruption When Shifting or Hard Shifting.
Cause:
1. Interrupter switch failed
2. Over torqued interrupter switch screws
3. Wire grounded
4. Open connection and/or broken wire
5. Module faulty
6. Bad module ground
7. High resistance of breaker points or faulty condenser

Correction:
1. Replace
2. Loosen screws and retorque.
3. Repair as required.
4. Inspect wires, plug pins and sockets of connectors.
5. Replace
6. Check ground continuity.
7. Replace breaker points and condenser.

Fig. 26 Ignition interrupter assembly

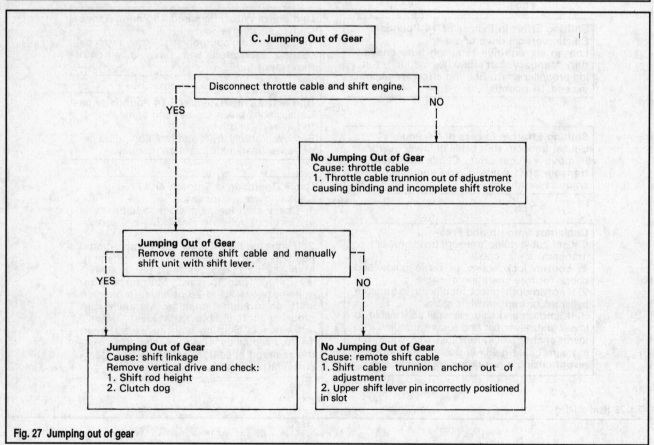

C. Jumping Out of Gear

Disconnect throttle cable and shift engine.

YES

NO

No Jumping Out of Gear
Cause: throttle cable
1. Throttle cable trunnion out of adjustment causing binding and incomplete shift stroke

Jumping Out of Gear
Remove remote shift cable and manually shift unit with shift lever.

YES

NO

Jumping Out of Gear
Cause: shift linkage
Remove vertical drive and check:
1. Shift rod height
2. Clutch dog

No Jumping Out of Gear
Cause: remote shift cable
1. Shift cable trunnion anchor out of adjustment
2. Upper shift lever pin incorrectly positioned in slot

Fig. 27 Jumping out of gear

SELOC PUBLISHING'S FULL-LINE MASTER LIST

ISBN		PART NO.	TITLE/DESCRIPTION	YEARS
			OUTBOARDS	
089330	018-7	1000	Chrysler Outboards, All Engines	1962-84
089330	055-1	1100	Force Outboards, All Engines	1984-99
089330	048-9	1200	Honda Outboards, All Engines	1978-01
089330	007-1	1300	Johnson/Evinrude Outboards, 1-2 Cyl	1956-70
089330	008-X	1302	Johnson/Evinrude Outboards, 1-2 Cyl	1971-89
089330	026-8	1304	Johnson/Evinrude Outboards, 1-2 Cyl	1990-95
089330	009-8	1306	Johnson/Evinrude Outboards, 3-4 Cyl	1958-72
089330	010-1	1308	Johnson/Evinrude Outboards, 3, 4 & 6 Cyl	1973-91
089330	063-2	1311	Johnson/Evinrude Outboards - All V Engines	1992-01
089330	052-7	1312	Johnson/Evinrude Outboards, All In-line engines/2 & 4 Stroke	1996-01
089330	015-2	1400	Mariner Outboards, 1-2 Cyl	1977-89
089330	016-0	1402	Mariner Outboards, 3, 4 & 6 Cyl	1977-89
089330	012-8	1404	Mercury Outboards, 1-2 Cyl	1965-91
089330	013-6	1406	Mercury Outboards, 3-4 Cyl	1965-89
089330	014-4	1408	Mercury Outboards, 6 Cyl	1965-89
089330	051-9	1416	Mercury/Mariner Outboards, All Engines	1990-00
089330	050-0	1600	Suzuki Outboards, All Engines	1988-99
089330	064-0	1701	Yamaha Outboards, All Engines	1984-96
089330	065-9	1703	Yamaha Outboards, All Engines	1997-03
			STERN DRIVES	
089330	029-2	3000	Marine Jet Drive	1961-96
089330	005-5	3200	Mercruiser Stern Drives	1964-91
089330	053-5	3206	Mercruiser Stern Drives	1992-01
089330	004-7	3400	OMC Stern Drives	1964-86
089330	056-X	3404	OMC Stern Drives	1986-98
089330	011-X	3600	Volvo/Penta Stern Drives	1968-91
089330	038-1	3602	Volvo/Penta Stern Drives	1992-93
089330	057-8	3606	Volvo/Penta Stern Drives	1992-03
			INBOARDS	
089330	049-7	7400	Yanmar Inboards	1975-98
			PERSONAL WATERCRAFT	
089330	032-2	9200	Kawasaki	1973-91
089330	042-X	9202	Kawasaki	1992-97
089330	045-4	9400	Polaris	1992-97
089330	033-0	9000	Sea-Doo/Bombardier	1988-91
089330	043-8	9002	Sea-Doo/Bombardier	1992-97
089330	034-9	9600	Yamaha	1987-91
08933	0044-6	9602	Yamaha	1992-97
			Seloc-On-Line (Internet Access)	
089330	075-6	5000	One Mfg/Model - Subscription per user 3 years	1990-01
			PROFESSIONAL TECHNICIANS MANUALS	
089330	060-8	4500	Labor Guide - Johnson and Evinrude	1980-00
089330	061-6	4550	Labor Guide - Yamaha	1980-00
089330	062-4	4600	Labor Guide - Mercury	1980-00
			BRIGGS AND STRATTON	
096376	610-4	610	Briggs & Stratton 4 Stroke Horizontal Crankshaft	1950+
096376	611-2	611	Briggs & Stratton 4 Stroke Vertical Crankshaft	1950+
096376	612-0	612	Briggs & Stratton 4 Stroke Overhead Crankshaft	1950+

ENGINE FINDER

The following listing contains all engines covered in this manual

Model/L	Engine/Cylinder	Coverage Years
2.3L	140 CID, 4 cyl	1987-90
2.5L	153 CID, 4 cyl	1986
3.0L	181 CID, 4 cyl	1986-98
3.0 HO	181 CID, 4 cyl	1990-93
4.3L	262 CID, V6	1986-98
4.3 HO	262 CID, V6	1990-93
5.0L	302 CID, V8 (Ford)	1989-96
5.0 HO	302 CID, V8 (Ford)	1990-93
5.0L	305 CID, V8 (GM)	1986-88, 1998
5.7L	350 CID, V8 (GM)	1986-92, 1994 & 1996-98
5.7 LE	350 CID, V8 (GM)	1990-92
5.8L	351 CID, V8 (Ford)	1989-96
5.8 HO	351 CID, V8 (Ford)	1995
5.8 LE	351 CID, V8 (Ford)	1991-93
7.4L	454 CID, V8 (GM)	1988, 1990-98
7.5L	460 CID, V8 (Ford)	1987-90
8.2L	502 CID, V8 (GM)	1992-93, 1998
262	262 CID, V6	1989
350	350 CID, V8 (GM)	1989-91
351	351 CID, V8 (Ford)	1992-94
454	454 CID, V8 (GM)	1988, 1990-93
454 HO	454 CID, V8 (GM)	1991-93
460	460 CID, V8 (Ford)	1988-90
502	502 CID, V8 (GM)	1992-93

SelocOnLine

SelocOnLine is a maintenance and repair database accessed via the Internet. Always up-to-date, skill level and special tool icons, quick access buttons to wiring diagrams, specification charts, maintenance charts, and a parts database. Contact your local marine dealer or see our demo at www.seloconline.com

SelocOnLine

Step 1
Select your manufacturer
Year / Model
Engine

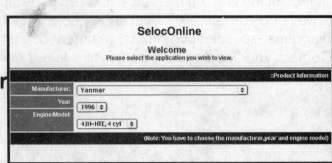

Step 2
Select an Engine System

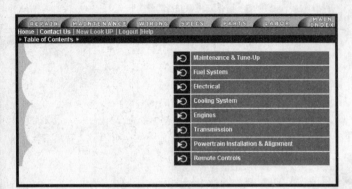

Step 3
Select the repair

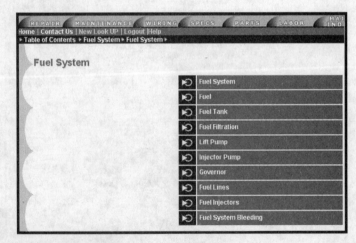

► Table of Contents ► Fuel System ► Fuel System ► Lift Pump ► Removal & Installation ►

JH Series Engines

Fig. 12 On JH engines, the fuel lift pump mounts to the injector pump ► see image

1. Turn the fuel tank petcock to the **OFF** position.
2. Place a small receptacle under the lift pump to contain any excess fuel.
3. Remove the fuel line fittings attached to the lift pump. Be careful not to lose the copper washers.

> Be sure to mark the fuel lines before they are removed from the pump. If the lines are reversed, the lift pump will not supply fuel to the injector pump.

4. Remove the two bolts that attach the lift pump to the engine block.
5. Draw the lift pump from the engine block.

To install:

6. Carefully scrape any gasket residue from the mating surfaces of the lift pump and injector pump.
7. Lightly grease the end of the lift pump piston where it rides on the cam.
8. Using a new gasket, install the lift pump to the injector pump.
9. Install the fuel lines to the lift pump in the proper direction.
10. Once the lines are attached, turn the fuel petcock to the **ON** position.
11. Loosen the air bleed screw on the secondary (between the lift pump and injector pump) fuel filter.
12. Operate the priming lever on the lift pump until fuel runs out of the air bleed screw opening, then tighten the screw.

Step 4
Print the repair procedure